Cytokines

From Basic Mechanisms of Cellular Control
to New Therapeutics

A subject collection from *Cold Spring Harbor Perspectives in Biology*

OTHER SUBJECT COLLECTIONS FROM *COLD SPRING HARBOR PERSPECTIVES IN BIOLOGY*

Circadian Rhythms

Immune Memory and Vaccines: Great Debates

Cell–Cell Junctions, Second Edition

Prion Biology

The Biology of the TGF-β Family

Synthetic Biology: Tools for Engineering Biological Systems

Cell Polarity

Cilia

Microbial Evolution

Learning and Memory

DNA Recombination

Neurogenesis

Size Control in Biology: From Organelles to Organisms

Mitosis

Glia

Innate Immunity and Inflammation

The Genetics and Biology of Sexual Conflict

The Origin and Evolution of Eukaryotes

Endocytosis

SUBJECT COLLECTIONS FROM *COLD SPRING HARBOR PERSPECTIVES IN MEDICINE*

Bone: A Regulator of Physiology

Multiple Sclerosis

Cancer Evolution

The Biology of Exercise

Prion Diseases

Tissue Engineering and Regenerative Medicine

Chromatin Deregulation in Cancer

Malaria: Biology in the Era of Eradication

Antibiotics and Antibiotic Resistance

The p53 Protein: From Cell Regulation to Cancer

Aging: The Longevity Dividend

Epilepsy: The Biology of a Spectrum Disorder

Molecular Approaches to Reproductive and Newborn Medicine

The Hepatitis B and Delta Viruses

Intellectual Property in Molecular Medicine

Retinal Disorders: Genetic Approaches to Diagnosis and Treatment

The Biology of Heart Disease

Cytokines

From Basic Mechanisms of Cellular Control to New Therapeutics

A subject collection from *Cold Spring Harbor Perspectives in Biology*

EDITED BY

Warren J. Leonard

Robert D. Schreiber

COLD SPRING HARBOR LABORATORY PRESS
Cold Spring Harbor, New York • www.cshlpress.org

Cytokines: From Basic Mechanisms of Cellular Control to New Therapeutics
A subject collection from *Cold Spring Harbor Perspectives in Biology*
Articles online at www.cshperspectives.org

Executive Editor	Richard Sever
Managing Editor	Maria Smit
Senior Project Manager	Barbara Acosta
Permissions Administrator	Carol Brown
Production Editor	Diane Schubach
Production Manager/Cover Designer	Denise Weiss
Publisher	John Inglis

Front cover artwork: Cytokines and cytokine receptor signal transduction between cells (endocrine, paracrine, and autocrine secretion). (Image reprinted with express permission from Designua via Shutterstock.)

Library of Congress Cataloging-in-Publication Data

Names: Leonard, Warren J., editor.
Title: Cytokines : from basic mechanisms of cellular control to new therapeutics / edited by
    Warren J. Leonard, National Heart, Lung, and Blood Institute, National Institutes of Health
    and Robert D. Schreiber, Washington University School of Medicine.
Description: Cold Spring Harbor, New York : Cold Spring Harbor Laboratory Press, [2018] |
    Series: Perspectives CSHL | "A subject collection from Cold Spring Harbor Perspectives
    in Biology." | Includes bibliographical references and index.
Identifiers: LCCN 2017031859 (print) | LCCN 2017039893 (ebook) | ISBN 9781621822660
    (ePublication) | ISBN 9781621821250 (hardcover)
Subjects: LCSH: Cytokines. | BISAC: SCIENCE / Life Sciences / Biology / Developmental Biology.
    | SCIENCE / Life Sciences / Biology / Molecular Biology. | MEDICAL / Immunology.
Classification: LCC QR185.8.C95 (ebook) | LCC QR185.8.C95 C986 2017 (print) | DDC 616.07/
    9--dc23
LC record available at https://lccn.loc.gov/2017031859

Contents

Contents

Preface

THE TERM "CYTOKINES" WAS COINED BY Stanley Cohen in 1974 to refer to secreted molecules, primarily of immunological importance, that both are produced by and execute actions on immune cells. Typically, cytokines act over limited distances, either directly on the cytokine-producing cell by an autocrine mechanism or instead on a proximal cell in a paracrine fashion. This attribute distinguishes cytokines from growth factors that act more distally, but only partially, as some cytokines indeed act over large distances. At times, the distinction between cytokines and other types of factors is more one of semantics. Cytokines typically act on multiple types of target cells, and accordingly can have broad, often pleiotropic actions. Although many cytokines promote proliferation and survival, some potently drive differentiation and others can be immunosuppressive or promote cell death.

The cytokine field has truly exploded over the years, beginning with observations of relatively poorly defined activities that were transformed by recombinant DNA technology into defined molecules that have since been rigorously studied in vitro and in vivo by biochemical and cell biological approaches, sophisticated imaging, structural biology, mouse models, and next generation sequencing approaches. Moreover, the development of cytokine mutants and specific antibodies have allowed additional new insights.

This volume includes a range of chapters covering a large number of cytokines that collectively span multiple families, including type I four-α-helical bundle cytokines, type II cytokines (which include the interferons and IL-10 family cytokines), as well as other structurally distinct molecules, such as members of the IL-1 and IL-17 families of cytokines, that exhibit other modes of signaling. Some chapters focus primarily on basic mechanisms of signaling or gene regulation, whereas others focus more on the biological actions and/or clinical utility.

The book is necessarily not all-inclusive, but hopefully this compendium of chapters highlights many critical aspects of and the tremendous ongoing excitement within the broader field. One must appreciate the incredible importance of cytokines related to the potency of their biological actions, the mechanisms by which they signal and control gene regulation, and how their actions are now being harnessed and manipulated—either augmented or inhibited—with clinical benefit. In fact, the manipulation of cytokines or their actions represent the basis of therapy for many diseases, including cancer, immunodeficiency, allergy, and autoimmune diseases, and it is noteworthy that the first successful human gene therapy was for X-linked severe combined immunodeficiency, a disease of defective cytokine signaling. The interested reader will presumably go well beyond the limits of this one volume, as the field is rapidly advancing, with surging numbers of publications.

I particularly wish to thank my co-editor, Bob Schreiber, who co-formulated the vision of this book. All of the authors who contributed are extremely busy, so I thank them for their enthusiasm, substantial time, and hard work in writing and revising state-of-the-art chapters in their respective fields. I also wish to thank Harvey Lodish for suggesting that Richard Sever contact me about such a book, Richard Sever for discussing the general vision, and Barbara Acosta and her colleagues at Cold Spring Harbor Press for their wonderful help, patience related to delays, and their ability to "nudge" the right amount to get things done without offending anyone. Hopefully many, including graduate and medical students as well as more senior scientists, will greatly appreciate the product that this book represents and be inspired to pursue new ideas and additional innovative research in the area of cytokines.

WARREN J. LEONARD

The Common Cytokine Receptor γ Chain Family of Cytokines

Jian-Xin Lin and Warren J. Leonard

Laboratory of Molecular Immunology and the Immunology Center, National Heart, Lung, and Blood Institute, National Institutes of Health, Bethesda, Maryland 20892-1674

Correspondence: linjx@nhlbi.nih.gov; leonardw@nhlbi.nih.gov

Interleukin (IL)-2, IL-4, IL-7, IL-9, IL-15, and IL-21 form a family of cytokines based on their sharing the common cytokine receptor γ chain (γc), which was originally discovered as the third receptor component of the IL-2 receptor, IL-2Rγ. The *IL2RG* gene is located on the X chromosome and is mutated in humans with X-linked severe combined immunodeficiency (XSCID). The breadth of the defects in XSCID could not be explained solely by defects in IL-2 signaling, and it is now clear that γc is a shared receptor component of the six cytokines noted above, making XSCID a disease of defective cytokine signaling. Janus kinase (JAK)3 associates with γc, and *JAK3*-deficient SCID phenocopies XSCID, findings that served to stimulate the development of JAK3 inhibitors as immunosuppressants. γc family cytokines collectively control broad aspects of lymphocyte development, growth, differentiation, and survival, and these cytokines are clinically important, related to allergic and autoimmune diseases and cancer as well as immunodeficiency. In this review, we discuss the actions of these cytokines, their critical biological roles and signaling pathways, focusing mainly on JAK/STAT (signal transducers and activators of transcription) signaling, and how this information is now being used in clinical therapeutic efforts.

Interleukin (IL)-2 is the prototype member of the γ chain (γc) family of cytokines. Initially identified as an activity present in the conditioned medium from normal human lymphocytes cultured with phytohemagglutinin (PHA) that could support the long-term in vitro culture of normal human T cells, IL-2 was initially known as T-cell growth factor (TCGF) (Morgan et al. 1976), but then subsequently renamed as IL-2 (Mizel and Farrar 1979). The cloning of complementary DNAs (cDNAs) encoding human (Taniguchi et al. 1983) and mouse (Kashima et al. 1985) IL-2 and the production of recombinant IL-2 allowed investigators to dis-

cover additional actions of IL-2 on T, B, and natural killer (NK) cells. Subsequent cloning of the human IL-2 receptor α chain (Leonard et al. 1984; Nikaido et al. 1984; Cosman et al. 1984) revealed that it had a short cytoplasmic tail with only 13 amino acids, making it unlikely to transduce IL-2 signals. This led to the search for additional IL-2 receptor components and the identification of IL-2Rβ (Sharon et al. 1986; Tsudo et al. 1986; Dukovich et al. 1987; Teshigawara et al. 1987) and subsequently IL-2Rγ (Takeshita et al. 1990; Saito et al. 1991) with eventual cloning of cDNAs encoding IL-2Rβ (Hatakeyama et al. 1989) and IL-2Rγ (Takeshita et al. 1992).

IL2RG was localized to the X chromosome at Xq13, the disease locus for X-linked severe combined immunodeficiency ([XSCID], also known as SCIDX1), which then led to the discovery that *IL2RG* mutations indeed cause XSCID (Noguchi et al. 1993b), a disease characterized by the absence of T and NK cells with nonfunctional B cells (Fischer et al. 2005). This finding was immediately important as it allowed earlier and more precise diagnosis of XSCID and also paved the way for successful gene therapy (Leonard 2001; Hacein-Bey-Abina et al. 2002). However, there were other major scientific implications of the XSCID discovery as well. Given that T- and NK-cell development was normal in *IL2*-deficient patients (Pahwa et al. 1989; Weinberg and Parkman 1990) and *Il2* knockout (KO) mice (Schorle et al. 1991), it was hypothesized that IL-2Rγ was a shared receptor component for other cytokines as well (Noguchi et al. 1993b), leading to the eventual demonstration that IL-2Rγ is indeed a shared receptor component for IL-2, IL-4, IL-7, IL-9, IL-15, and IL-21 (Rochman et al. 2009). Thus, it was renamed as the common cytokine receptor γc (Noguchi et al. 1993a; Russell et al. 1993), and cytokines using γc are now known as γc family cytokines. The inactivation of signaling by six cytokines in XSCID underscores that it is indeed a disease of defective cytokine signaling (Leonard 1996).

γc family cytokines all share similar three-dimensional structural features and are four α-helix-bundle type I cytokines (Bazan 1990). Although all of these cytokines were initially discovered based on specific actions for either the development or function of T, B, and NK cells (except for IL-21, which was identified based on its binding to an orphan receptor, as will be discussed below), we now know that each cytokine is pleiotropic with broad roles in the development of immune cells or related to immune responses, including some actions beyond the immune systems.

In this review, we discuss the molecular and cellular biology of this family of cytokines, their signaling pathways, actions, and the interplay among them during the development of immune cells and immune responses. We will also discuss the emerging promising approaches for rationally modulating the actions of these cytokines for treating patients with immunodeficiency, autoimmune disorders, infectious diseases, allergic conditions, and malignancies. Needless to say, the number of studies performed and wealth of information on γc family cytokines is enormous, with a huge number of publications in the field (see Fig. 1 for the number of publications just in the period from 2010 to 2017). We have necessarily been selective in our discussion, trying to highlight important early studies as well as some of the exciting progress in this field, and apologize in advance for being unable to cite large numbers of superb studies on these cytokines. However, many other articles in this collection also cover aspects of γc family cytokines, and the reader is directed to those as well.

γc FAMILY CYTOKINES—AN OVERVIEW

γc family cytokines collectively mediate biological actions on a range of immune cells (Fig. 2). CD4[+] T cells are the main producers of IL-2 in response to T-cell receptor (TCR) stimulation, whereas CD8[+] T cells, NK cells, and NK T (NKT) cells can also produce IL-2 but at much lower levels (Liao et al. 2013). Although IL-2 was initially discovered as a T-cell growth factor

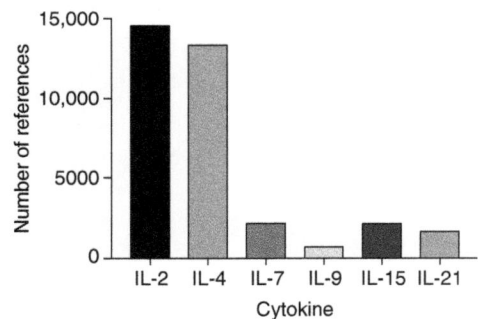

Figure 1. PubMed search results of γ chain (γc) family cytokines between 2010 and 2017. The search was performed using EndNote X7.7.1 with the key words IL-2 or interleukin 2, IL-4 or interleukin 4, IL-7 or interleukin 7, IL-9 or interleukin 9, IL-15 or interleukin 15, and IL-21 or interleukin 21, respectively, under Abstract, 2010–2017 under Year, and English under Language.

Cite this article as *Cold Spring Harb Perspect Biol* doi: 10.1101/cshperspect.a028449

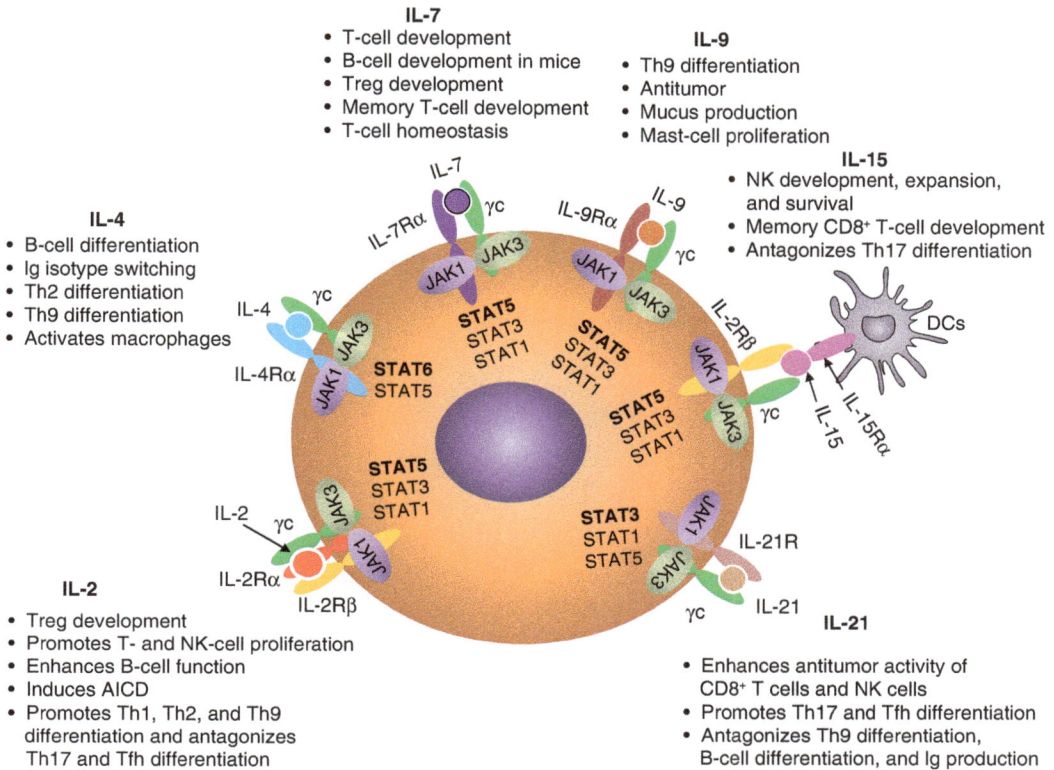

Figure 2. Schematic of γ chain (γc) family cytokines and their receptors. Shown are how Janus kinase (JAK)1 and JAK3 associate with each receptor, the signal transducers and activators of transcription (STAT) proteins activated by each cytokine, and the major actions of these cytokines on the development and function immune cells. The STAT proteins predominantly activated by each cytokine are in bold. IL, Interleukin; Th, T helper; Tfh, T follicular helper; Treg, regulatory T; NK, natural killer; Ig, immunoglobulin; DCs, dendritic cells; AICD, activation-induced cell death.

(Morgan et al. 1976), it can also promote the growth and differentiation of B cells that are stimulated by anti-immunoglobulin (Ig)M or CD40 ligand (Armitage et al. 1995) and promote NK-cell proliferation and enhance NK-cell cytotoxicity (Siegel et al. 1987). In addition to its potent proliferative activity for T cells in vitro, IL-2 can induce activation-induced cell death (AICD) (Lenardo 1991) of CD4$^+$ T cells, which is important for the maintenance of peripheral self-tolerance. In addition to its actions as a T-cell growth factor, an essential role of IL-2 in vivo is to promote the development and maintenance of regulatory T (Treg) cells whose suppressive activity is vital to control pathologic inflammatory responses (Malek et al. 2002). In addition,

IL-2 promotes the differentiation of T helper (Th)1 (Liao et al. 2011), Th2 (Zhu et al. 2003; Cote-Sierra et al. 2004; Liao et al. 2008), and Th9 (Liao et al. 2014) cells, whereas it suppresses the differentiation of Th17 (Laurence et al. 2007; Liao et al. 2011) and T follicular helper (Tfh) (Ballesteros-Tato et al. 2012; Johnston et al. 2012; Oestreich et al. 2012) cells. As is discussed below, IL-2 has been extensively used as an anticancer agent and for modulating the immune responses.

IL-4 was first identified as a B-cell differentiation factor(s) produced by T cells, which induces Ig isotype switch (Howard et al. 1982; Isakson et al. 1982). IL-4 is a signature cytokine for Th2 responses that are essential for the control of extracellular parasites infection and contribute to allergic reactions (Paul 2015). IL-4 also induces the generation of M2 (M-IL-4) macrophages, which are essential for macrophage-mediated control of infection with the protozoan *Trypanosoma cruzi* (Wirth et al. 1989). In addition, IL-4 promotes the generation of IL-9[+]IL-10[+] T cells in the presence of transforming growth factor (TGF)-β, which are subsequently designated as Th9 cells and suppress TGF-β-induced Foxp3[+] Treg-cell generation (Dardalhon et al. 2008). With its key roles in mediating allergic responses, modulating IL-4 activity is now of considerable therapeutic interest (see below).

In contrast to the production of IL-2 and IL-4 by T cells, IL-7 was discovered as a stromal-cell-derived factor that supported the growth of pre-B cells (Namen et al. 1988; Goodwin et al. 1989). IL-7 signaling also plays an essential, nonredundant role in the development of T cells in humans (Puel et al. 1998; Giliani et al. 2005) and of both B and T cells in mice (Peschon et al. 1994; von Freeden-Jeffry et al. 1995). Unlike other γc family cytokines, the expression level of IL-7 is relatively stable. IL-7 by itself does not promote proliferation of naïve T cells but is essential to ensure sustained expression of antiapoptotic proteins BCL2 and MCL1, and thereby long-term in vivo survival of T cells (Rathmell et al. 2001; Opferman et al. 2003; Sprent and Surh 2011). To maintain long-term memory-T-cell survival, IL-7-induced expression of the glycerol channel aquaporin 9 in antigen-specific memory CD8[+] T cells is essential for IL-7-directed glycerol uptake, and this is required for triglyceride synthesis and lipid storage to maintain the longevity of memory CD8[+] T cells after viral clearance (Cui et al. 2015). Although IL-7 is not required for the development of conventional NK cells (He and Malek 1996; Puel et al. 1998), it plays a critical role in the homeostasis of thymic NK cells (Vosshenrich et al. 2006). Interestingly, either osteoblast ablation or deletion of *Il7* in osteoblasts of adult mice results in significantly lower numbers of common lymphoid progenitors (CLPs) without affecting hematopoietic stem-cell numbers, and administration of IL-7 can restore normal numbers of CLPs (Terashima et al. 2016).

IL-9 was discovered as a T-cell-derived growth factor for certain Th-cell clones in the absence of either antigen or antigen-presenting cells (APCs) (Uyttenhove et al. 1988; Schmitt et al. 1989) and then shown to also be a growth factor for bone marrow mast cells (Hultner et al. 1990). Although IL-9 was initially considered to be a Th2 cytokine (Gessner et al. 1993), it is now recognized as the signature cytokine for Th9 cells, whose differentiation is induced when naïve CD4[+] T cells are cultured with IL-2, IL-4, and TGF-β (Schmitt et al. 1994) or when Th2 cells are cultured with TGF-β (Dardalhon et al. 2008; Veldhoen et al. 2008). IL-9 can enhance IL-4-induced IgE and IgG production (Dugas et al. 1993; Petit-Frere et al. 1993), supports innate lymphoid-cell (ILC) survival, and induces cytokine production by these cells (Wilhelm et al. 2011; Turner et al. 2013). IL-9 can also be produced by Th17 cells and can promote the expansion of these cells (Elyaman et al. 2009; Nowak et al. 2009). Expression of IL-9 in NKT cells can be enhanced by IL-2 but not IL-15 (Lauwerys et al. 2000). As discussed below, IL-9 also has anticancer activity.

IL-15 was codiscovered as a T-cell growth factor activity present in the supernatants from a simian kidney epithelial line CV-1/EBNA (Grabstein et al. 1994) and HTLV-1-transformed HUT-102 leukemia cells (Bamford et al. 1994; Burton et al. 1994). Although many different cell types can express IL-15 messenger RNA (mRNA), IL-15 protein is mainly produced by dendritic cells (DCs) and monocytes in response to Toll-like receptor (TLR) activation and binds to receptors on these cells (Waldmann 2006). IL-15 plays critical roles in the development and/or maintenance of memory CD8[+] T cells and preferentially can expand central memory phenotype T cells in vivo, with $Il15^{-/-}$ or $Il15ra^{-/-}$ mice having profound loss of memory phenotype CD8[+] T cells, intestinal intraepithelial lymphocytes, NKT cells, and NK cells (Lodolce et al. 1998; Kennedy et al. 2000). Although IL-2 and IL-15 both share IL-2Rβ and γc and activate the same Janus kinase (JAK)1/JAK3/signal transducers and activators of transcription (STAT)5 pathway, IL-15 can inhibit IL-2-induced AICD, and the addition of blocking

Cite this article as *Cold Spring Harb Perspect Biol* doi: 10.1101/cshperspect.a028449

antibodies to IL-15 restores IL-2-induced AICD of these CD4$^+$ T cells (Marks-Konczalik et al. 2000). As discussed below, IL-15 is under active evaluation as an anticancer agent.

IL-21 was identified by expression cloning (Parrish-Novak et al. 2000) as the ligand for a novel "orphan" type I cytokine receptor, originally also denoted as "novel interleukin receptor" (NILR) in addition to the IL-21R (Ozaki et al. 2000; Parrish-Novak et al. 2000). Interestingly, the genes encoding IL-2 and IL-21 are adjacent on human chromosome 4q27 and mouse chromosome 3. IL-21 has significant homology with IL-2, IL-4, and IL-15 (Parrish-Novak et al. 2000), is expressed by Tfh, NKT, Th1, Th2, and Th17 cells, is the key γc family cytokine that contributes to Tfh-cell differentiation in vivo, and it can also promote Th17 differentiation (reviewed in Spolski and Leonard 2014). Although IL-21 alone does not show proliferative activity for either B or T cells, it can potentiate the proliferation of human B cells stimulated by anti-CD40 (Parrish-Novak et al. 2000). IL-21 plays complex roles in the differentiation and function of B cells. For example, IL-21 can promote B-cell proliferation in the presence of anti-CD40 or anti-IgM, whereas it can induce death of resting B cells or B cells stimulated with lipopolysaccharide (LPS) or CpG (Mehta et al. 2003; Jin et al. 2004; Ozaki et al. 2004). IL-21 transgenic mice or wild-type (WT) mice overexpressing IL-21 by hydrodynamic-based gene delivery of IL-21 plasmid DNA show increased numbers of immature B cells, memory B cells, and plasma cells, with elevated serum IgG and IgM levels (Ozaki et al. 2004). IL-21, together with IL-4, plays a vital role in B-cell differentiation and Ig production (Ozaki et al. 2002). In addition to its direct role in the inhibition of Treg differentiation by suppressing Foxp3 expression (Nurieva et al. 2007), IL-21 can diminish Treg homeostasis through inhibiting IL-2 production by T cells (Attridge et al. 2012). IL-21 can also increase the numbers of CD56$^+$ CD16^{high} human NK cells generated from CD34$^+$ human hematopoietic progenitor cells cultured with IL-15 and Flt3 ligand (FLt3L) (Parrish-Novak et al. 2000). IL-21 by itself does not significantly promote the growth of naïve or memory CD8$^+$ T cells, but it can greatly synergize with IL-15 and to a lesser extent with IL-7, but not with IL-2, to enhance the proliferation of these cells (Zeng et al. 2005). These cells develop normally in mice lacking IL-21R, indicating that this cytokine is not required for their development but rather contributes to their expansion. As discussed below, IL-21 is a potent inducer of IL-10 and can drive the production of regulatory B cells and, moreover, IL-21 promotes autoimmune disease and has anticancer activity.

RECEPTORS USED BY γc FAMILY CYTOKINES

Most of the receptors for γc family cytokines (IL-2Rβ, IL-4Rα, IL-7Rα, IL-9Rα, and IL-21R, and γc) are type I cytokine receptors and share an approximately 200-amino-acid-long cytokine-binding homology region (CHR) consisting of two fibronectin type III (FNIII) domains connected by a linker (Bazan 1990). These proteins have four conserved cysteine residues at the N-terminal domain, which can form interstrand disulfide bonds, and a WSXWS (tryptophan-serine-any amino acid-tryptophan-serine) motif near the C terminus (Bazan 1990; Wang et al. 2009). IL-2Rα and IL-15Rα do not contain the CHR module and instead have "Sushi" domains that mediate ligand binding but with very different affinities (Rickert et al. 2005; Lorenzen et al. 2006). Interestingly, in addition to sharing γc, IL-2 and IL-15 receptors additionally share IL-2Rβ (Bamford et al. 1994; Giri et al. 1994).

IL-2 receptors are expressed mainly by lymphoid cells. In the absence of stimulation, IL-2Rα is mainly expressed on Treg cells and not expressed on naïve T cells, but it is potently induced on T-cell activation (Leonard et al. 1985; Malek and Castro 2010; Liao et al. 2013). IL-2Rβ and γc are constitutively expressed on some resting T cells, especially NK cells, CD8$^+$ T cells, B cells, macrophages, monocytes, and DCs. Like IL-2Rα, IL-2Rβ can be further induced on stimulation of these cells, either by antigen or by IL-2. Three IL-2 receptor chains, IL-2Rα, IL-2Rβ, and IL-2Rγ, form three different types of IL-2 receptors, binding IL-2 with low affinity (IL-2Rα only, $K_d \approx 10$ nM), intermediate affinity (IL-2Rβ + IL-2Rγ, $K_d \approx 1$ nM), and high affinity (IL-2Rα

+ IL-2Rβ + IL-2Rγ, $K_d \approx 10$ pM) (Malek and Castro 2010; Liao et al. 2013). Although IL-2Rα and IL-2Rβ together can form pseudo-high-affinity IL-2 receptors ($K_d \approx 100$ pM), they cannot transduce IL-2 signals because of the lack of γc (Arima et al. 1992). The intermediate-affinity and high-affinity IL-2 receptors are the functional receptors (Malek and Castro 2010; Liao et al. 2013). Although IL-2 mainly signals via IL-2 receptor chains coexpressed on the same cell (*cis* signaling), antigen-specific CD25$^+$ mature DCs that lack IL-2Rβ can *trans*-present IL-2 to CD25$^-$ T cells, which can be blocked by declizumab, a humanized antihuman CD25 antibody (Wuest et al. 2011).

Unlike IL-2 receptor chains, IL-4Rα is expressed on both lymphohematopoietic and non-lymphohematopoietic cells, with ∼300 IL-4 binding sites on resting lymphocytes and 10-fold more upon activation (Paul 2015). IL-4 receptors are also present on macrophages and mast cells (Ohara and Paul 1987). There are two types of IL-4Rs: type I IL-4 receptors on lymphohematopoietic cells are composed of IL-4Rα and γc (Kondo et al. 1993; Russell et al. 1993), whereas in nonhematopoietic cells, type II IL-4 receptors comprise IL-4Rα plus IL-13Rα1, which is also the functional receptor for IL-13 (Aman et al. 1996). In reconstitution experiments using COS-7 cells, IL-4 binds to IL-4Rα with high-affinity ($K_d \approx 266$ pM) and in the presence of γc, the binding affinity is further increased ($K_d \approx 79$ pM) (Russell et al. 1993). The direct interaction of the IL-4Rα chain with γc is very weak (K_d in the μM range) (LaPorte et al. 2008).

The IL-7 receptor consists of IL-7Ra and γc (Noguchi et al. 1993a; Kondo et al. 1994), and IL-7Ra) is also a component of the receptor for thymic stromal lymphopoietin (TSLP), in that context cooperating with the direct TSLP-binding protein, TSLPR (Pandey et al. 2000; Park et al. 2000). IL-7Rα is expressed on both hematopoietic cells and cells of nonlymphoid origin (Jiang et al. 2005). The expression of IL-7Rα is dynamically regulated during lymphocyte development and in response to TCR or cytokine stimulation, and IL-7Rα regulation is the major mechanism for regulating IL-7 responses (Mazzucchelli and Durum 2007). IL-7Rα can be in-duced by FLT3 ligand, glucocorticoids, type I interferons (IFNs), and tumor necrosis factor (TNF), whereas it is suppressed by two γc family cytokines, IL-2 and IL-7, as well as by IL-6 (Xue et al. 2002; Mazzucchelli and Durum 2007). Both high-affinity ($K_d \approx 65$ pM) and low-affinity ($K_d \approx 100$ nM) IL-7 receptors can be detected on peripheral blood lymphocytes, and a reconstitution experiment in COS-7 cells showed that IL-7 binds to IL-7Rα with an intermediate affinity ($K_d \approx 250$ pM) that is enhanced to high affinity when γc is present (Noguchi et al. 1993a).

IL-9 receptors contain IL-9Rα chain and γc (Russell et al. 1994; Kimura et al. 1995). A single class of IL-9 receptor with high affinity ($K_d \approx 100$ pM) can be detected on T cells, mast cells, and macrophages (Druez et al. 1990). Interestingly, IL-9Rα is also detected in nonhematopoietic cells, including airway and intestinal epithelial cells and smooth muscle cells (Goswami and Kaplan 2011; Kaplan et al. 2015). The human *IL9R* gene is located on the X chromosome, and at least four *IL9R* pseudogenes are localized at the pseudoautosomal region of X and Y chromosomes (Kermouni et al. 1995; Vermeesch et al. 1997). A study of the single-nucleotide polymorphisms (SNPs) shows that a specific haplotype of IL-9Rα gene is protective against wheezing in boys but not in girls and provides weak protection against sensitization to inhalant and/or food allergens (Melen et al. 2004).

IL-15 receptors are composed of IL-15Rα, IL-2Rβ, and γc (Bamford et al. 1994; Giri et al. 1994; Grabstein et al. 1994). IL-15Rα is expressed on a wide range of cells, including immune cells (T cells, B cells, macrophages, and stromal-cell lines) and nonimmune cells (keratinocytes and skeletal muscle cells) (Grabstein et al. 1994). IL-15 binds to IL-15Rα with high affinity ($K_d \approx 25$ pM), much higher than the affinity of IL-2 for IL-2Rα, whereas IL-15 binds IL-2Rβ and γc with a $K_d \approx 1$ nM, similar to the affinity of IL-2 to IL-2Rβ and γc. Like IL-2Rα, IL-15Rα does not transduce IL-15 signals, but IL-15Rα-expressing APCs, such as DCs and monocytes, can bind IL-15 and *trans*-present the cytokine to lymphocytes that express IL-2Rβ and γc; this *trans* signaling is the dominant mode for IL-15 action (Dubois et al. 2002).

Cite this article as *Cold Spring Harb Perspect Biol* doi: 10.1101/cshperspect.a028449

IL-21 receptors consist of IL-21Rα chain and γc (Asao et al. 2001). IL-21Rα mRNA is expressed on lymphohematopoietic cells and is potently induced on stimulation of human peripheral blood mononuclear cells with PHA (Ozaki et al. 2000) and TCR (Wu et al. 2005). At the amino acid level, IL-21Rα is most similar to IL-2Rβ (29% identity, 46% similarity), and its cytoplasmic domain is more similar to that of IL-9Rα, whereas its overall domain organization (a single cytokine-binding domain followed by a transmembrane domain and a relatively long cytoplasmic domain) is most similar to IL-4Rα (Parrish-Novak et al. 2002). IL-4Rα and IL-21R are encoded by the adjacent genes (Ozaki et al. 2000).

SIGNALING PATHWAYS USED BY γc FAMILY CYTOKINES

Unlike most growth factor receptors, type I cytokine receptors do not have intrinsic protein kinase activity in their cytoplasmic domains. Thus, the association of nonreceptor JAK1 with the cytokine-specific receptor chains (IL-2Rβ, IL-4Rα, IL-7Rα, IL-9Rα, and IL-21R but not IL-2Rα and IL-15Rα) and JAK3 with γc is essential for transducing the signals induced by γc family cytokines (Fig. 2) (Leonard and O'Shea 1998). Because JAK3 interacts with and is "downstream" from γc (Boussiotis et al. 1994; Russell et al. 1994; Kawahara et al. 1995), it was hypothesized (Russell et al. 1994) and then shown that mutations in *JAK3* cause a form of T⁻B⁺NK⁻ SCID that phenocopies XSCID (Macchi et al. 1995; Russell et al. 1995). Each γc cytokine induces the juxtaposition of the cytoplasmic domain of its cytokine-specific receptor chain with γc to trigger the activation of JAK1 and JAK3 (Nakamura et al. 1994; Nelson et al. 1994), which then phosphorylates the key tyrosine residues in the cytoplasmic domain of each unique receptor chain (IL-2Rβ, IL-4Rα, IL-7Rα, IL-9Rα, and IL-21Rα), providing the phosphotyrosine docking site(s) for STAT proteins via their SH2 domains; the STAT proteins can then be phosphorylated by JAK kinases (Leonard and O'Shea 1998). Tyrosine-phosphorylated STAT proteins dimerize via bivalent interactions between the C-terminal phosphotyrosine on each monomeric STAT protein and the SH2 domain on the other monomeric STAT protein (Fig. 3). STAT dimers then translocate to the nucleus to bind to IFN-γ-activated site (GAS) motifs, activating the transcription of their target genes (Darnell et al. 1994; Leonard and O'Shea 1998). The docking site(s) for STAT proteins on the cytoplasmic domains of each cytokine receptor determine which STAT protein will be activated by a given cytokine (Lin et al. 1995; Demoulin et al. 1996). For example, IL-2, IL-7, IL-9, and IL-15 predominantly activate STAT5A and STAT5B and to a lesser extent STAT3 and STAT1 (Lin et al. 1995; Demoulin et al. 1999), IL-4 mainly activates STAT6 and to a lesser extent STAT5 (Hou et al. 1994; Quelle et al. 1995; Rolling et al. 1996), and IL-21 activates STAT3 and to a lesser extent STAT1 and STAT5 (Fig. 2) (Zeng et al. 2007; Wan et al. 2013; Wan et al. 2015).

An early study of the IFN-γ enhancer revealed cooperative binding of STAT proteins and the recognition of STAT tetramers. STAT tetramers form based on interactions via their highly conserved N-terminal regions ("N-domains") in addition to the SH2-phosphotyrosine interactions required for dimerization, thus allowing the dimerization of dimers. Tetramers can bind to less-well-conserved STAT-binding motifs (Vinkemeier et al. 1996; Xu et al. 1996; Soldaini et al. 2000), and the cooperative binding of STAT proteins in this context is caused by N-domain-mediated tetramerization (Vinkemeier et al. 1996; Xu et al. 1996). The crystal structure of the STAT4 N-domain identified the key residues involved in the N-domain-mediated interactions (Vinkemeier et al. 1998; Chen et al. 2003). The nonredundant in vivo function of STAT dimers and tetramers was shown in STAT5 tetramer-deficient mice (Lin et al. 2012). Whereas a complete deletion of both *Stat5a* and *Stat5b* results in fetal lethality, at least in part caused by defective erythropoietin-based STAT5 activation and red-cell formation (Socolovsky et al. 1999; Yao et al. 2006), STAT5 tetramer-deficient mice are viable, indicating STAT5 tetramers are dispensable for survival of mice (Lin et al.

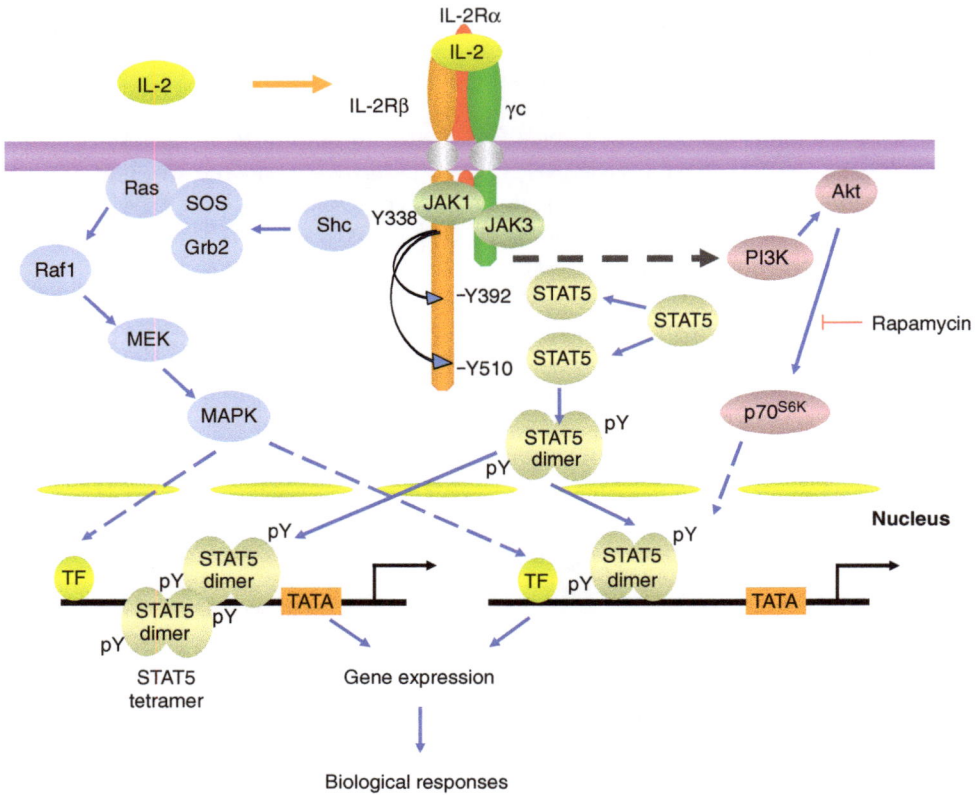

Figure 3. The interleukin (IL)-2 signaling pathway. Activation of Janus kinase (JAK)/signal transducers and activators of transcription (STAT), mitogen-activated protein kinase (MAPK), and phosphatidylinositol 3-kinase (PI3K) pathways by γ chain (γc) family cytokines. As indicated in the text, IL-7 was reported to not activate the MAPK pathway. TF, Transcription factor.

2012). Moreover, STAT5 tetramers are not required for the development of B cells, CD4[+], CD8[+], and CD4[+]Foxp3[+] T cells, but are required for normal numbers of peripheral CD8[+] T-cell and NK-cell numbers, in vitro proliferation of CD8[+] T cells stimulated by IL-2 or IL-15, and homeostasis of CD4[+] and especially CD8[+] T cells in lymphopenic hosts (Lin et al. 2012). In addition, STAT5 tetramer-deficient virus antigen-specific CD8[+] T cells show decreased expansion in response to lymphocytic choriomeningitis virus (LCMV) or adenovirus 5. Consistent with the potent induction of IL-2Rα mRNA by IL-2, ~13 major STAT5-binding sites were identified in the *Il2ra* gene by ChIP-Seq analysis, but only a few of them bind to STAT5 tetramers. Interestingly, IL-2-induced

Il2ra mRNA is markedly decreased in STAT5-tetramer-deficient T cells, indicating that STAT5 tetramers are essential for normal *Il2ra* transcription (Lin et al. 2012). Despite the normal numbers of CD4[+]Foxp3[+] T cells in STAT5 tetramer-deficient mice, IL-2Rα expression in these cells is greatly diminished and STAT5-tetramer-deficient Treg cells show decreased suppressive activity in an adoptive transfer colitis model (Lin et al. 2012). Interestingly, STAT1-tetramer-deficient mice have also been generated and show normal type I IFN responses and antiviral activity, but abolished IFN-γ signaling and antibacterial immunity (Begitt et al. 2014). Thus, there are key roles for STAT1 and STAT5 dimers versus tetramers in mediating cytokine signaling, indicating that targeting the formation of STAT tetramers could be a means of modulating cytokine actions.

JAK/STAT signaling is critical for a broad range of cellular functions, including proliferation, survival, and differentiation, but γc family cytokines, except IL-7, have been shown to activate mitogen-activated protein kinase (MAPK) to promote cell growth, and all γc family cytokines activate phosphatidylinositol 3-kinase (PI3K) to support cell survival (see Fig. 3 for IL-2 activation of these signaling pathways). IL-4 (Wang et al. 1993) and IL-9 (Yin et al. 1995) additionally activate insulin receptor substrates (IRSs) in a manner that is dependent on JAK1 (Yin et al. 1995; Nelms et al. 1999).

To further characterize IL-2 signaling, mass-spectrometry-based quantitative phosphoproteomics were used to identify IL-2 signaling networks in preactivated mouse CD8$^+$ T cells that were cultured in IL-12 to keep the cells viable (Ross et al. 2016). An IL-2-JAK-dependent network appeared to account for most the phosphoproteins identified, including those crucial for the cellular fitness and functions, such as transcription factors, regulators of chromatin structure, mRNA translation machinery, GTPases, vesicle trafficking, and the actin and microtubule cytoskeleton. About 10% of the phosphoproteins identified in these cells were JAK-independent (tofacitinib resistant), comprising those involved in the generation of phosphatidylinositol (3,4,5)-trisphosphate (PIP3) and the AKT pathway.

Although tofacitinib was developed as a JAK3 inhibitor (Changelian et al. 2003), it also inhibits JAK1 and JAK2 as well as JAK3. A recent study using a more specific JAK3 inhibitor, JAK3i (3000-fold more selective for JAK3 than for JAK1, JAK2, and TYK2), unexpectedly revealed biphasic catalytic activity of JAK3 in STAT5 activation in CD4$^+$ T cells stimulated by IL-2 (Smith et al. 2016). Interestingly, the second wave is required for the expression of cyclins and cell-cycle progression to the S phase, and more sensitive than the first wave of STAT5 activation to JAK3i (Smith et al. 2016). Additional work is needed to clarify the role of the first wave in IL-2 action and whether the second wave is directly or indirectly activated by IL-2. Recently, an additional JAK3-specific inhibitor has been reported (Telliez et al. 2016).

GENE EXPRESSION MEDIATED BY γc FAMILY CYTOKINES

STAT-mediated activation by γ_c family cytokines is vital for the development and function of immune cells. As discussed below, STAT6 activated by IL-4 and STAT3 activated by IL-21 play essential roles in B-cell differentiation and Ig switching (Linehan et al. 1998; Diehl et al. 2012); STAT5 plays nonredundant roles in T- and B-cell development (IL-7) (Yao et al. 2006), in NK-cell development, expansion, and survival (IL-15) (Yao et al. 2006), and in Treg development (IL-2) (Burchill et al. 2007). STAT5 is also critically important for T-cell homeostasis (IL-7) in peripheral and for T-cell expansion in response to antigen stimulation (IL-2 and IL-15). Furthermore, STAT activation by this family of cytokines is also critical for CD4$^+$ Th-cell differentiation. For example, STAT3 activated by IL-21 is vital for Tfh differentiation (Nurieva et al. 2008), STAT5 activated by IL-2 is required for normal Th1, Th2, and Th9 differentiation (Zhu et al. 2003; Liao et al. 2008, 2011, 2014), and STAT6 activated by IL-4 is essential for Th2 and Th9 differentiation (Kuperman et al. 1998; Goswami et al. 2012).

In addition to promoting CD4$^+$ T-cell differentiation, IL-21-activated STAT3 and IL-2-activated STAT5 show opposing actions on Th9, Th17, and Tfh differentiation (Laurence et al. 2007; Yang et al. 2011; Johnston et al. 2012; Liao et al. 2014). These opposing actions of IL-2 and IL-21 during Th17 differentiation can be achieved either by IL-2-mediated inhibition of *Il6ra* and *Il6st* expression or enhanced T-bet (Liao et al. 2011), which should augment binding of RUNX1, thereby diminishing association of RUNX1 with RORγt (Lazarevic et al. 2011), or by competition between STAT3 versus STAT5 for binding to GAS (IFN-γ-activated sequence, TTCN$_3$GAA) motifs on *Il17a* and *Il17f* (Yang et al. 2011). Because the GAS motifs recognized by STAT proteins are fairly conserved, except for STAT6 using the motif TTCN$_4$GAA (with N$_4$ instead of N$_3$) (Leonard and O'Shea 1998), binding of STAT3 versus STAT5 is likely influenced by the local concentration of each STAT protein. For Th9 differentiation, BCL6 expression suppressed by IL-2 and enhanced by IL-21 are at least, in part, attributed to their opposing actions (Liao et al. 2014). For Tfh differentiation, induction of BLIMP1 via IL-2-activated STAT5 and maintaining BCL6 levels via IL-21-activated STAT3 are responsible for the opposing actions of these cytokines (Johnston et al. 2012; Oestreich et al. 2012). Thus, STAT proteins activated by γc family cytokines are essential to restrain the inflammation triggered by immune responses and to protect the host from harm by controlling the balance between Treg and Th cells.

In addition to the activation of different STAT proteins by different γc family cytokines, different STAT proteins activated by the same cytokine can also result in distinct biological actions. For example, IL-21 activates both STAT3 and STAT1 in mouse CD4$^+$ T cells, with STAT1 promoting expression of the *Ifng* and *Tbx21* genes and STAT3 inhibiting their expression. Correspondingly, expression of *IFNG* and *TBX21* genes are higher in patients with loss-of-function mutation in STAT3 or with gain-of-function mutation in STAT1 (Wan et al. 2015). Moreover, STAT proteins are involved in the assembly of complexes of transcription factors at enhancers, as first shown for IL-21-induced expression of the *Prdm1* gene, which requires STAT3 and IRF4 to act in concert and cooperatively binding during Tfh differentiation (Kwon et al. 2009). Such assembly is essential to initiate and maintain the transcription program during Th differentiation (Ciofani et al. 2012; Vahedi et al. 2012). Moreover, during Th17 differentiation, IL-21 signals via AP1-IRF composite elements (AICEs) (Li et al. 2012) and binding of large complexes of factors (e.g., STAT3, Jun family proteins, basic leucine zipper ATF-like transcription factor (BATF), RORγt, c-Maf, p300, and CTCF) to the regulatory regions of key Th17 genes is required for the initiation and maintenance of the Th17 transcription program (Ciofani et al. 2012).

γc FAMILY CYTOKINES AND T-CELL DEVELOPMENT

The essential roles of γc family cytokines in the development, homeostasis, and survival of T cells are shown by the findings in humans with gene mutations in either the human *IL2RG* gene encoding γc or the *JAK3* gene, as well as in corresponding KO mice. IL-7 is the most critical cytokine for the development of T cells in both humans and mice, with normal T-cell development in *Il2* (Schorle et al. 1991), *Il4* (Kuhn et al. 1991), *Il9* (Townsend et al. 2000), *Il15* (Kennedy et al. 2000), and *Il21* (Nurieva et al. 2007) KO mice.

IL-7Rα is mainly expressed on CLPs, pre–T cells, and thymic and peripheral CD4$^+$ and CD8$^+$ single-positive T cells. The key roles for IL-7 in T-cell development are to maintain the expression of the antiapoptotic protein, BCL2, for double-negative (DN) thymocyte survival, and of *Rag1* for the rearrangement of T-cell antigen receptors in thymic CD4$^-$CD8$^-$ DN T cells (Mazzucchelli and Durum 2007). The absence of γδ T cells in *Il7ra$^{-/-}$* mice and the observation that they cannot be restored by expression of a *Bcl2* transgene show the indispensable role of IL-7 in RAG-mediated rearrangement of TCR-γ (Maki et al. 1996; Mazzucchelli and Durum 2007). Regulation of the accessibility of the *Tcrg* locus to recruit STAT5 proteins activated by IL-7 and other cofactors is attributed to the essential role of IL-7 in T-cell development (Ye et al. 2001). In addition, there is a profound decrease in αβ T-cell numbers in *Il7$^{-/-}$* or *Il7ra$^{-/-}$* mice (Peschon et al. 1994; von Freeden-Jeffry et al. 1995), but normal T-cell numbers could be restored by a *Bcl2* transgene or partially restored by the deletion of the gene encoding pro-apoptotic BAX in *Il7ra$^{-/-}$* mice (Akashi et al. 1997; Khaled et al. 2002). The absence of γδ T cells but presence of some αβ T cells results from the lack of initiation of recombination of the *Tcrg* locus, with normal recombination of Vβ2 but greatly reduced recombination of Vβ14 (Schlissel et al. 2000).

Interestingly, the defect in T-cell development appears somewhat less severe in *Il7$^{-/-}$* mice than in *Il7ra$^{-/-}$* mice (Peschon et al.

1994; von Freeden-Jeffry et al. 1995), and TSLP also uses IL-7Rα as a receptor component, even though it is not a γc family cytokine. Despite normal development of T cells in $Crlf2^{-/-}$ (TSLPR KO) mice, TSLP contributes to T-cell development to some degree as $Il2rg^{-/y}Crlf2^{-/-}$ mice display a greater defect in T-cell development than observed in $Il2rg^{-/y}$ mice (Al-Shami et al. 2004).

The levels of expression of cytokine receptors is well-correlated with mature T-cell homeostasis. For example, naïve T-cell homeostasis is predominantly dependent on survival signals provided by IL-7, with high expression of IL-7Rα but very low expression of IL-2Rα and IL-2Rβ on these cells. TCR activation of naïve T cells results in decreased expression of IL-7Rα but potently induces IL-2Rα and IL-2Rβ, and IL-2 and IL-15 can also support memory T-cell homeostasis (Surh and Sprent 2008).

γc FAMILY CYTOKINES AND Treg-CELL DEVELOPMENT AND BIOLOGY

Approximately 10% of peripheral mature CD4$^+$ T cells express IL-2Rα (CD25), and these Treg cells show potent suppressive activity as shown by the ability of adoptively transferred CD4$^+$CD25$^+$ T cells to prevent CD4$^+$CD25$^-$ T-cell-induced autoimmunity in athymic nude mice (Sakaguchi et al. 1995). Naturally occurring Treg cells constitutively express the forkhead family transcription factor FOXP3. Treg cells play essential roles in the maintenance of a balanced immunity under various physiological and pathophysiological conditions (Ohkura et al. 2013; Li and Rudensky 2016). For example, they are essential for suppressing autoreactive T cells that have escaped negative selection in the thymus, controlling potentially harmful excessive immune responses, and suppressing the antitumor immunity that helps tumors to escape immune surveillance. The gene encoding FOXP3 is localized on the X chromosome, and FOXP3 is vital for the development and function of Treg cells. Mutation of the *FOXP3* gene in humans causes severe autoimmune disease (Bennett et al. 2001), and a truncation mutation of the mouse *Foxp3* gene caused by a frameshift

attributed to an insertion of two nucleotides in exon VIII (*Scurfy* mouse) results in fatal lymphoproliferative disorders (Brunkow et al. 2001). FOXP3 is required for the development of these cells in the thymus, as shown by the complete absence of CD4$^+$CD25$^+$ Treg cells in *Foxp3*-deficient mice (Fontenot et al. 2003). STAT5 proteins are required for optimal expression of FOXP3 in vitro, with binding of STAT5 to three GAS motifs in the *Foxp3* gene from CD4$^+$CD25$^+$ but not from CD4$^+$CD25$^-$ splenic T cells (Yao et al. 2007).

Finding that *Il2*, *Il2ra*, and *Il2rb* KO mice do not show defective T-cell development but rather have lymphoproliferative autoimmune disorders caused by dysregulated expansion of effector phenotype CD4$^+$ T cells (Sadlack et al. 1993; Suzuki et al. 1995; Willerford et al. 1995) was unexpected, given that IL-2 can potently promote the in vitro proliferation of T cells that express IL-2 receptors. However, this can be explained by markedly decreased Treg-cell numbers in these mice. Importantly, the severe autoimmunity in $Il2rb^{-/-}$ mice can be corrected either by thymic-specific expression of IL-2Rβ or by adoptive transfer of normal Treg cells into $Il2rb^{-/-}$ newborn mice (Malek et al. 2000, 2002; Malek and Castro 2010), demonstrating the important roles of IL-2 in the development and/or homeostasis of Treg cells in vivo. Subsequent studies showed that IL-2 signaling is critical for the expansion and survival of Treg cells (Bayer et al. 2005, 2007; Fontenot et al. 2005; Setoguchi et al. 2005) but is not required for their suppressive function, given that the residual Treg cells from $Il2^{-/-}$ or $Il2ra^{-/-}$ mice show similar suppressive activity in vitro as cells from WT mice (Fontenot et al. 2005). Together with TGF-β, IL-2 also plays important roles in the generation of inducible Treg (iTreg) cells from naïve CD4$^+$ T cells, especially in peripheral mucosal tissues like lungs and gut (Davidson et al. 2007; Zheng et al. 2007).

Foxp3$^+$ Treg cells are essentially absent in thymus and spleen of $Il2rg^{-/-}$, $Jak3^{-/-}$, or $Stat5^{-/-}$ mice (Fontenot et al. 2005; Mayack and Berg 2006; Yao et al. 2007), indicating the vital roles of γc family cytokines in the development and homeostasis of Treg cells in vivo. However, although CD4$^+$Foxp3$^+$ Treg cells are markedly decreased in $Il2^{-/-}$, $Il2ra^{-/-}$, or $Il2rb^{-/-}$ mice, some Treg cells are still present, suggesting the involvement of other γc family cytokine(s) in maintaining normal Treg numbers. Indeed, there is a moderate decrease of both thymic and splenic CD4⁺Foxp3⁺ Treg cells in $Il7ra^{-/-}$ mice (Burchill et al. 2007), and the severity of the Treg defects in $Il2rb^{-/-}Il7ra^{-/-}$ double KO mice is similar to that observed in $Il2rg^{-/Y}$ mice, whereas $Il2rb^{-/-}Il4ra^{-/-}$ double KO mice had more Treg cells than $Il2rg^{-/Y}$ mice (Bayer et al. 2008; Vang et al. 2008), indicating that IL-7R-dependent signaling also contributes to normal Treg numbers. In addition, both IL-7 and IL-15, but not TSLP, can induce the differentiation of Treg cells from CD4⁺CD25⁺Foxp3⁻ thymic Treg progenitor cells in vitro, although less potently than IL-2 (Vang et al. 2008). IL-9 was also reported to enhance in vitro Treg function, as a neutralizing antibody to IL-9 blocked such a function (Elyaman et al. 2009). Consistent with this finding, $Il9ra^{-/-}$ mice show decreased in vivo Treg function in a mouse experimental autoimmune encephalomyelitis (EAE) model, and injection of anti-IL-9-blocking antibodies in WT mice also have diminished Treg function (Elyaman et al. 2009).

THE ROLES OF γc FAMILY CYTOKINES IN CD4⁺ T HELPER–CELL DIFFERENTIATION

The adaptive immune system plays major roles in eliminating pathogens and protecting the host from harm. Key components of this system include peripheral naïve CD4⁺ T cells, which in response to antigen stimulation can differentiate into at least five types of Th cells, including Th1, Th2, Th9, Th17, and Tfh cells (Fig. 4). Which cells form is determined based on the type and strength of the antigen encountered, the presence of cytokines produced by innate immune cells, key transcriptional profiles, and the signature cytokines produced by these cells. Although IL-4 and IL-9 are the only γc family cytokines to serve as signature cytokines (for Th2 and Th9, respectively), other γc family cytokines play indispensable roles in these processes as well, either to facilitate or antagonize a given differentiation process (Fig. 4).

Th1 responses are critical for controlling intracellular pathogens and serve major roles in developing inflammatory disorders. STAT4-dependent IL-12-induced T-bet expression is a key step for the differentiation of naïve CD4⁺ T cells into Th1 cells, ensuring expression of IFN-γ, and suppressing the expression of IL-4 and IL-17. Among γc family cytokines, IL-2 plays critical roles in promoting Th1 differentiation. IFN-γ production is markedly impaired in $Jak3^{-/-}$ or $Stat5^{fl/fl}$CD2-Cre⁺ CD4⁺ T cells, despite intact IFN-γ-STAT1 and IL-12-STAT4 signaling pathways in these cells. Moreover, antibodies to IL-2, IL-2Rα, and IL-2Rβ also greatly diminish IFN-γ expression under Th1 differentiation conditions in WT CD4⁺ T cells (Shi et al. 2008). IL-7 can partially restore the decreased IFN-γ expression when IL-2 signaling is blocked, but the effect is only partial, correlating with less potent STAT5 activation by IL-7 than by IL-2. IL-2 induces the expression of IL-12Rβ2 as well as IFN-γ and T-bet, at least in part via STAT5 binding to the regulatory regions of these genes (Liao et al. 2011). The importance of IL-2-induced IL-12Rβ2 expression in Th1 differentiation is further shown by the ability of retroviral transduction of IL-12Rβ2 into $Il2^{-/-}$ CD4⁺ T cells to restore impaired Th1 differentiation in these cells. Interestingly, the anti-inflammatory cytokine IL-27 inhibits Th1 differentiation, limiting the production of IL-2 during Th1 differentiation (Villarino et al. 2006).

Th2 differentiation is critical for eliminating extracellular parasites and involves the production of the signature Th2 cytokines IL-4, IL-5, and IL-13. Together, these cytokines play important roles in Ig isotype switch, IgG1 and IgE production, and mediate allergic inflammatory responses, including asthma and atopic dermatitis. Both the sustained expression of GATA3, at least in part via STAT6 (activated by IL-4) and STAT5 proteins (activated by IL-2 or potentially other γc family cytokines), are essential to initiate and maintain Th2 differentiation (Zhu et al. 2003, 2004; Pai et al. 2004). Early findings that the neutralization of IL-2 inhibited in vivo IL-4 production and that overexpression of constitutively activated STAT5A restored defective IL-4 production in the ab-

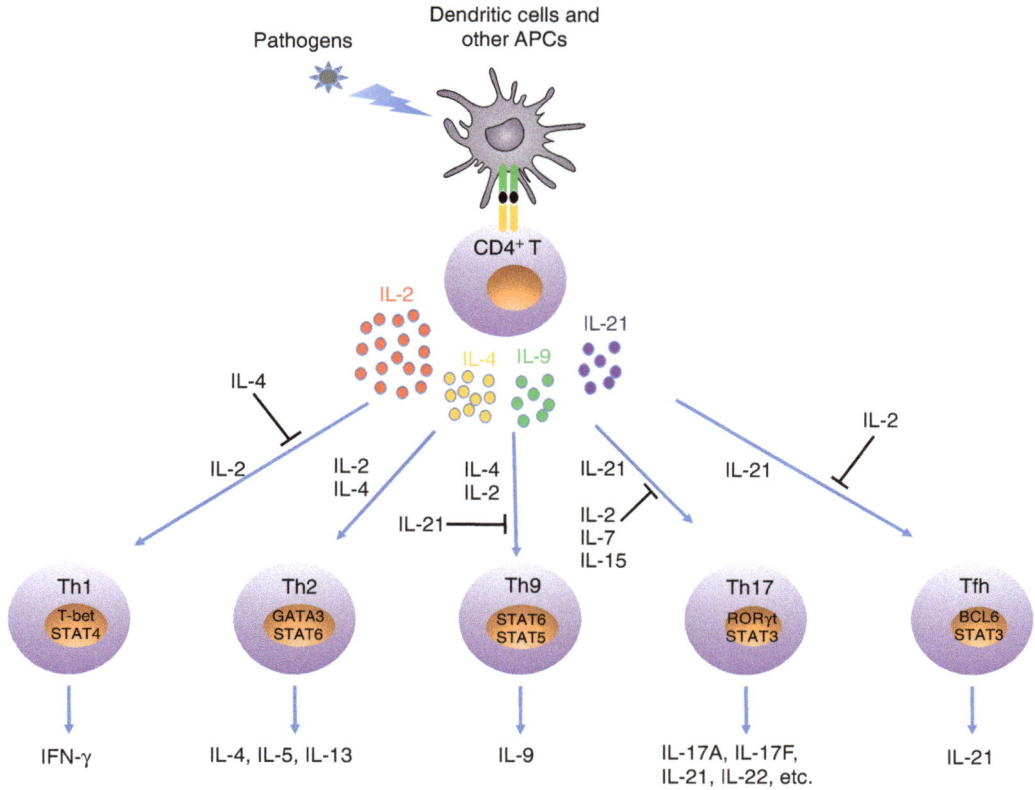

Figure 4. Schematic of roles of γ chain (γc) family cytokines in T helper (Th) differentiation. On antigen stimulation, CD4$^+$ T cells can produce interleukin (IL)-2, IL-4, IL-9, and IL-21. Under different differentiation conditions, each γc family cytokine can either promote or suppress a given differentiation process. APCs, antigen-presenting cells; Tfh, T follicular helper; STAT, signal transducers and activators of transcription; IFN, interferon.

sence of IL-2 signaling demonstrate the indispensable role of IL-2 for normal Th2 differentiation (Cote-Sierra et al. 2004). These results were extended by the findings that early during the Th2 differentiation, IL-2 potently induces the expression of IL-4Rα chain in both WT and $Il4^{-/-}$ CD4$^+$ T cells activated by TCR stimulation, and that this is critical for the cells to respond to IL-4 for sustained GATA3 expression. The induction of IL-4Rα by IL-2 is attributed to the direct binding of STAT5 proteins to the GAS motifs in the first intron of the $Il4ra$ gene, and, under Th2 differentiation conditions, retroviral transduction of $Il4ra$ cDNA into $Il2^{-/-}$ CD4$^+$ T cells could fully restore the defective IL-4 expression in these cells (Liao et al. 2008). These findings together show essential

roles for both IL-4 and IL-2 for normal Th2 differentiation.

Th9 cells express the signature cytokine IL-9, which is involved in inflammatory disorders, including allergic inflammation, autoimmunity, and antitumor immunity. A number of transcription factors are required for Th9 differentiation, including SMAD family proteins, E-26-specific (ETS) family proteins, IRF4, and the BATF (Kaplan et al. 2015). IL-4 and IL-2 are the key γc family cytokines required for Th9 differentiation, as evidenced by the greatly diminished IL-9 expression in $Stat6^{-/-}$ (Goswami et al. 2012) or $Il2^{-/-}$ CD4$^+$ T cells (Liao et al. 2014). Although IL-7, IL-9, and IL-15 can also induce IL-9 expression in $Il2^{-/-}$ CD4$^+$ T cells in the presence of IL-4 and TGF-β, they are much less potent than IL-2 (Liao et al. 2014). Both IL-4-STAT6 and IL-2-STAT5 pathways promote Th9 differentiation through the regulation of different transcription factors. IL-4-STAT6 induces IRF4 to increase IL-9 expression while suppressing the expression of FOXP3 and T-bet to inhibit Treg and Th1 differentiation, respectively (Goswami et al. 2012), whereas IL-2-STAT5 suppresses BCL6 to antagonize the potent inhibitory action of IL-21 on Th9 differentiation, indicating that IL-2-STAT5 and IL-21-STAT3 signaling pathways differentially regulate Th9 differentiation via their opposing effect on BCL6 expression (Liao et al. 2014). In addition, IL-2-STAT5 can promote IL-9 production by directly regulating IRF4 and IL-9 expression (Gomez-Rodriguez et al. 2016). Furthermore, the progressive loss of IL-9 production during in vitro Th9 differentiation is observed in WT but not in $Stat3^{-/-}$ cells, suggesting that the diminished IL-9 results from IL-10-induced activation of STAT3 (Ulrich et al. 2017).

In a mouse melanoma model, IL-9 and Th9 cells show potent antitumor activity (Lu et al. 2012; Purwar et al. 2012). IL-1β can also potentiate Th9 differentiation via activation of STAT1, which in turn induces the expression of IRF1 and thereby IRF1 binding to regulatory regions in both the *Il9* and *Il21* genes, with augmented expression of both IL-9 and IL-21. The indispensable role of IL-21 in the antitumor activity of Th9 cells is shown by the ability of an anti-IL-21 antibody to abolish the antitumor activity of Th9 cells induced by IL-1β (Vegran et al. 2014).

Both Th2 and Th9 cells are involved in the clearance of parasite infections, but IL-9 is expressed in vivo earlier during *Nippostrongylus brasiliensis* infection than are IL-4, IL-5, and IL-13 (Licona-Limon et al. 2013). In fact, IL-9 expression is required for IL-5 and IL-13 expression, with impaired expression of these cytokines in *Il9*$^{-/-}$ mice. Moreover, adoptive transfer of Th9 cells, but not Th2 cells, improves worm expulsion in *Rag2*$^{-/-}$ mice and the administration of neutralizing IL-9 antibody to these mice abolishes the ability of adoptively transferred Th9 cells to clear worm infection, leading to the conclusion that Th9 cells more efficiently clear worm

infection than Th2 cells (Licona-Limon et al. 2013).

Th17 cells express transcription factor RORγt and the signature cytokines IL-17A and IL-17F, as well as other cytokines, to eliminate extracellular pathogens and fungal infections. Early during Th17 differentiation, the cooperative binding of BATF and IRF4 to their target genes plays crucial roles for chromatin accessibility and the subsequent binding of STAT3 activated by IL-6 and IL-21 and other transcription factors to initiate the Th17 transcriptional program (Ciofani et al. 2012; Li et al. 2012). Depending on how Th17 cells are generated, they can be either pathogenic and mediate autoimmune disorders or nonpathogenic and protect the host from harm caused by inflammation. For example, nonpathogenic Th17 cells are generated in the presence of IL-6 and TGF-β (McGeachy et al. 2007), whereas pathogenic Th17 cells are generated when proinflammatory IL-23 is additionally present, thereby inhibiting Th17 cells from producing IL-10 and up-regulating granulocyte macrophage colony-stimulating factor (GM-CSF) (Codarri et al. 2011; El-Behi et al. 2011). In contrast to the positive effects of IL-2 on Th1, Th2, and Th9 differentiation described above, IL-2 suppresses Th17 differentiation as shown by the observation that $Il2^{-/-}$ CD4$^+$ T cells or blocking IL-2 signaling in WT CD4$^+$ T cells results in increased Th17 differentiation (Laurence et al. 2007). Similarly, IL-15 can also inhibit Th17 differentiation as evidenced by the lower IL-17A expression observed when IL-15 is added and increased IL-17A seen when anti-IL-15 antibody is added to neutralize IL-15 produced by APC or as is observed in $Il15^{-/-}$ or $Il15r\alpha^{-/-}$ CD4$^+$ T cells (Pandiyan et al. 2012). The inhibitory effects of IL-2 and IL-15 on Th17 differentiation can potentially be attributed to the competition of the shared binding sites in *Il17* by STAT3 activated by IL-6 and IL-21 versus STAT5 activated by IL-2 and IL-15 (Laurence et al. 2007; Yang et al. 2011; Pandiyan et al. 2012) as well as to the suppressive effect of IL-2 on the expression of IL-6Rα and the IL-6 signal-transducing molecule, gp130, which is shown by the finding that retroviral transduction of gp130 can partially reverse the inhibitory effect of IL-2 on Th17 differentiation (Liao et al. 2011). In contrast to its inhibitory actions on Th17 differentiation, IL-2 can expand Th17 cells isolated from peripheral blood of healthy human donors or patients with scleritis and increase IL-17 expression in these cells (Amadi-Obi et al. 2007). It was also reported that IL-2, IL-7, and IL-15 can each enhance the expression of IL-17A, IL-17F, IL-22, and IL-26 in CCR6$^+$, but not in CCR6$^-$, human CD4$^+$CD45RO$^+$ CD25$^-$ memory T cells, and this was attributed to the ability of these cytokines to activate PI3K, as inhibition of PI3K signaling selectively abolished the expression of IL-17 signature cytokines induced by these γc family cytokines (Wan et al. 2011). In addition, memory phenotype Th17 cells isolated from human peripheral blood constitutively express low levels of IL-2, and an IL-2-neutralizing antibody can induce apoptosis of these cells, indicating a critical role for IL-2 in Th17-cell survival as well (Yu et al. 2011). Although IL-6 and TGF-β can initiate Th17 differentiation, IL-21 produced by TCR stimulation also contribute to Th17 differentiation, with impaired in vitro differentiation of Th17 cells in $Il21^{-/-}$ CD4$^+$ T cells; IL-21 can also partially compensate for the absence of IL-6 signaling during Th17 differentiation (Korn et al. 2007; Nurieva et al. 2007; Zhou et al. 2007).

Tfh cells are localized in germinal centers and are characterized by coexpression of transcription factor BCL6, inhibitory receptor PD1, IL-21, the chemokine receptor CXCR5, and costimulatory protein ICOS, but these cells do not express BLIMP1 (Crotty 2014; Qi 2016). Tfh cells play crucial roles in T-cell-dependent humoral immunity by directing B-cell differentiation into plasma cells, the selection of affinity-matured antibody-producing B cells, Ig isotype switch, and production of Ig (Crotty 2014; Qi 2016). BCL6 is the key regulator that promotes Tfh differentiation, whereas BLIMP1 antagonizes BCL6 and thus suppresses Tfh differentiation (Johnston et al. 2009). ICOS ligand, IL-6, IL-21, and STAT3 are all required for the development of Tfh cells; indeed, the lack of any of these proteins in mice impairs the development of CD4$^+$CXCR5$^+$ T cells after immunization

(Nurieva et al. 2008). Interestingly, IL-21 uniquely up-regulates the expression of both BCL6 and BLIMP1, which result in different outcomes: germinal center B-cell maintenance mediated by BCL6 (Johnston et al. 2009; Yu et al. 2009) versus plasma-cell differentiation mediated by BLIMP1 (Shaffer et al. 2002). IL-21-induced BLIMP1 expression requires the cooperative binding of STAT3 and IRF4 in the regulatory region of $Prdm1$ (gene encoding for BLIMP1) (Kwon et al. 2009). In contrast, IL-2 potently suppresses Tfh-cell differentiation by activating STAT5 and thereby inducing BLIMP1, which down-regulates BCL6 (Johnston et al. 2009, 2012; Ballesteros-Tato et al. 2012; Oestreich et al. 2012). In addition, quenching IL-2 from activated CD4$^+$ T cells by soluble IL-2Rα produced by Tfh-cell-priming ICOSL^{hi}CD25$^+$ DCs favors Tfh-cell differentiation at the follicle outer T-cell zone (Li et al. 2016). Interestingly, IL-2-activated AKT and mTORc1 kinases play a critical role in determining Th1 versus Tfh differentiation. The lower levels of IL-2 expression in Tfh cells are correlated with lower proliferation, glycolysis, and mitochondrial respiration than is observed in Th1 cells during acute LCMV infection (Ray et al. 2015). Down-regulation of BCL6 in Tfh cells late after immunization correlates with decreased proliferation and increased IL-7 responses, in part via increased IL-7Rα expression in these cells (Kitano et al. 2011); however, unlike IL-2, IL-7 does not increase BLIMP1 expression (McDonald et al. 2016).

Besides their opposing actions on Tfh cells, IL-2 and IL-21 regulate Tfh-cell function through differential actions on follicular regulatory T (Tfr) cells, which suppress Tfh-cell-mediated Ig production by B cells (Sage and Sharpe 2015). In addition to other signature molecules expressed by Tfh cells, Tfr cells additionally express Foxp3$^+$ and high levels of ICOS and PD1 (Chung et al. 2011; Linterman et al. 2011). A transcriptomic analysis of Tfh cells in response to Tfr cells revealed that transcription factors essential for Tfh cells, including $Bcl6$, $Ascl2$, and $Tcf1$, are not affected by Tfr-cell suppression, whereas $Prdm1$ and especially $Il4$ and $Il21$ mRNA levels in Tfh cells are markedly reduced

(Sage and Sharpe 2016). In addition, Tfr cells alter Myc signals and mechanistic target of rapamycin (mTOR) pathway in B cells. Importantly, IL-21 or IL-6, but not IL-4, can rescue Tfh- and B-cell functions suppressed by Tfr cells (Sage et al. 2016). In human patients with loss-of-function mutations in *IL21RA* (Kotlarz et al. 2013; Stepensky et al. 2015), there are increased numbers of IL-2-dependent total Treg and Tfr cells in peripheral blood (Jandl et al. 2017). Correspondingly, in the mouse, IL-21 suppresses the expansion of Tfr cells by down-regulating IL-2Rα expression, which is mediated by BCL6 (Jandl et al. 2017). Thus, IL-2 and IL-21 act either directly on Tfh cells or on Tfr cells to maintain B-cell functions via their opposing actions.

ROLES OF IL-2, IL-7, IL-15, AND IL-21 IN THE DEVELOPMENT AND DIFFERENTIATION OF CD8+ T CELLS

IL-2, IL-7, IL-15, and IL-21 play important roles in the homeostasis, expansion, survival, and function of CD8+ T cells. IL-2 and IL-15 potently promote CD8+ T-cell proliferation and induce effector molecules and inflammatory cytokines (Waldmann 2006; Liao et al. 2013), and IL-7 is essential for the survival and homeostasis of naïve CD8+ T cells (Schluns et al. 2000); IL-15 can also promote naïve CD8+ T-cell homeostasis albeit less potently (Kennedy et al. 2000; Berard et al. 2003). IL-7 also regulates the homeostasis of memory CD8+ T cells after viral infection (Kaech et al. 2003), whereas IL-2 can suppress the homeostasis and function of memory CD8+ T cells by Treg cells (Ku et al. 2000; Murakami et al. 2002; Waldmann 2006). When combined with IL-15 or to a lesser extent with IL-7 but not with IL-2, IL-21 can synergistically expand CD8+ T cells in vitro, and the combination of IL-21 and IL-15 potently increases antigen-specific CD8+ T-cell numbers in vivo and inhibits tumor progression in a B16 melanoma model (Zeng et al. 2005). IL-2 and IL-15 enhance expression of transcription factor eomesodermin (*Eomes*) by antigen-primed CD8+ T cells and their development into effector CD8+ T cells, whereas IL-21 induces *Tcf7* and *Lef1* while suppressing Eomes and effector CD8+

T-cell development (Hinrichs et al. 2008). Moreover, antigen-primed CD8+ T cells cultured with IL-21 but not with IL-2 or IL-15 show enhanced antitumor activity when adoptively transferred into B16 melanoma-bearing mice (Hinrichs et al. 2008).

Cytotoxic CD8+ T cells can directly kill virus-infected cells via perforin and granzymes. During LCMV infection, a sustained presence of IL-2 up-regulates Eomes and perforin but down-regulates expression of BCL6 and IL-7Rα, resulting in the expansion of effector cells while suppressing the development of memory cells (Pipkin et al. 2010). Dynamic expression of IL-2Rα in antigen-specific CD8+ T cells appears to correlate with the fate of CD8+ T-cell differentiation in response to LCMV infection. IL-2Rα^low CD8+ T cells expressing high levels of IL-7Rα and CD62L show enhanced survival and are thought to become long-lived functional memory cells, whereas IL-2Rα^high CD8+ T cells expand more rapidly but are more susceptible to apoptosis (Kalia et al. 2010). In chronic viral infections, including human immunodeficiency virus (HIV) in humans and LCMV in mice, CD8+ T cells become exhausted (Zajac et al. 1998), and this is characterized by their greatly diminished secretion of IL-2 and decreased proliferation but increased expression of the inhibitory receptor PD1 (Barber et al. 2006). Besides PD1, transcription factor BATF is coexpressed, potently suppressing T-cell proliferation and cytokine secretion, whereas BATF small interfering RNA (siRNA) can rescue the defective secretion of IL-2 in T cells from HIV patients (Quigley et al. 2010). Furthermore, in mice chronically infected with LCMV, low-dose IL-2 can enhance CD8+ T-cell responses, decrease PD1 expression, and increase IL-7Rα and CD44 expression on virus-specific CD8+ T cells. Moreover, a combination of IL-2 with anti-PD-L1 greatly enhances virus-specific CD8+ T-cell responses (West et al. 2013).

ROLES OF IL-4, IL-7, AND IL-21 IN THE DEVELOPMENT AND BIOLOGY OF B CELLS

B-cell numbers in *Il2rg*$^{-/-}$ and *Jak3*$^{-/-}$ mice are markedly diminished, and the cells that develop are nonfunctional because of the lack of T-cell help, given the defective T-cell development in these animals (Cao et al. 1995; Nosaka et al. 1995). IL-7 is the key γc family cytokine for mouse B-cell development, with a severe block at the early stages of B-cell development in mice deficient in either IL-7 or IL-7Rα, including in CLPs and during the transition from pro-B cells to pre-B cells, with absent mature follicular B cells in both $Il7$- or $Il7r$-deficient mice (Peschon et al. 1994; von Freeden-Jeffry et al. 1995). Interestingly, B-cell development is more defective in $Il7ra^{-/-}$ mice than in $Il7^{-/-}$ mice, suggesting that TSLP, which also uses IL-7Rα as a receptor component, might also be involved. However, unlike the observation indicating a role for TSLP in T-cell development, B-cell development is similar in $Il7/Tslpr$ double KO mice and $Il7$ KO mice, indicating that TSLP may not be involved in B-cell development (Jensen et al. 2008). Unlike its ability to restore αβ T-cell development in $Il7r^{-/-}$ mice, expression of the $Bcl2$ transgene could not rescue B-cell development defect in the absence of an IL-7 signal (Maraskovsky et al. 1998). However, the $Bcl2$ transgene could compensate for the markedly decreased expression of antiapoptotic protein MCL1 and restore pro-B-cell development in the absence of STAT5 (Malin et al. 2010). IL-7 produced by bone marrow endothelial cells is crucial in B-cell lymphogenesis, as the conditional deletion of $Il7$ in mouse mesenchymal progenitor cells results in a decrease of both pro- and pre-B-cell numbers in bone marrow (Cordeiro Gomes et al. 2016). In contrast to its importance for B-cell development in the mouse, IL-7 signaling is not essential for human B-cell development, as patients with mutations in $IL7RA$ have normal B-cell numbers, although these B cells are not functional because of severe defects in T cells (Puel et al. 1998; Giliani et al. 2005). Thus, either IL-7 does not contribute to human B-cell development or there is a redundant pathway.

IL-4 signaling is not involved in B-cell development but rather is crucial for the differentiation of mature B cells, Ig isotype switching, and IgE production, as $Il4^{-/-}$ and $Il4ra^{-/-}$ mice show normal development of T and B cells but diminished IgG1 production, greatly diminished IgE production, and normal levels of IgM and other Ig isotypes after infection with $N.\ brasiliensis$ (Kuhn et al. 1991; Noben-Trauth et al. 1997). Similarly, the development of T and B cells are normal in $Il21r^{-/-}$ mice but IgG1 levels are greatly diminished, indicating the involvement of IL-21 in Ig switch as well. However, IL-21 is also required for IgG3 production by CD40-activated naïve human B cells (Pene et al. 2004; Avery et al. 2008). Interestingly, IL-4 favors the production of IgG1, whereas IL-21 favors the IgG3 production by CD40-activated naïve human B cells (Avery et al. 2008). However, unlike the mice lacking IL-4 signaling, IgE levels are actually elevated in $Il21r^{-/-}$ mice after immunization (Ozaki et al. 2002), possibly caused by an inhibitory effect of IL-21 on IL-4-induced germ line Cε transcription (Suto et al. 2002). Possible cooperative actions between IL-4 and IL-21 for Ig production are further provided by the findings of a pan-hypogammaglobulinemia and poorly organized germinal centers in mice lacking both IL-4 and IL-21 signaling (Ozaki et al. 2002). These observations indicate that both IL-4 and IL-21 contribute to the differentiation of B cells into plasma cells, Ig isotype switch, and Ig production. In addition to actions of IL-21 on T cells to promote B-cell function, a study using a mouse multiple sclerosis (MS) model showed that IL-21 and CD40 are required for the expansion and maturation of regulatory B cells into IL-10-producing effector "B10" cells to inhibit autoimmune disorders by suppressing effector T (Teff)-cell function (Yoshizaki et al. 2012).

As discussed above, IL-4 and IL-21 are the most important γc family cytokines for promoting Tfh development and function, which are essential for B-cell differentiation and Ig isotype switching and production (Ozaki et al. 2002) and they play nonredundant roles in these processes (Ozaki et al. 2004). However, the kinetics of when each of these two key γc family cytokines is expressed during Tfh differentiation was unclear. By using mice expressing IL-21 and IL-4 reporters, it was shown that Tfh cells in germinal centers go through different transcriptional and functional stages to provide different signals for B-cell differentiation and function in germinal centers. For example, cells secreting only IL-21 are most efficient at promoting somatic hypermutation and affinity maturation in B cells, whereas cells expressing only IL-4 are more efficient at promoting plasma-cell differentiation and IgG1 class switch, and cells expressing both IL-4 and IL-21 can serve both functions (Weinstein et al. 2016). During LCMV infection, IL-21 signaling is dispensable for Tfh differentiation but is required for the generation of long-lived plasma cells and sustained antibody production (Rasheed et al. 2013).

Defective T cells in *IL2RG*- or *JAK3*-deficient patients can be reconstituted after hematopoietic cell transplantation, but B cells are still not functional in a significant portion of patients who require Ig replacement therapy. Interestingly, the naïve B cells from these patients respond well to IL-4 plus CD40L, but respond poorly to IL-21 plus CD40L, indicating that IL-21 is the major γc family cytokine to initiate humoral immunity in humans (Recher et al. 2011). Consistent with this, *IL7R*-deficient SCID patients, where IL-21 signaling is intact, show more efficient reconstitution of normal B-cell function after bone marrow transplantation (Buckley 2011).

ROLES OF γc FAMILY CYTOKINES FOR NK-CELL DEVELOPMENT AND CYTOLYTIC ACTIVITY

Unlike T and B cells, NK cells can rapidly secrete inflammatory cytokines, chemokines, and, importantly, proteinases on encountering virus-infected cells or tumor cells to eliminate these cells (Vivier et al. 2011; Cerwenka and Lanier 2016). The essential roles of γc family cytokines in the maturation, expansion, and survival of NK cells are shown by the profoundly decreased numbers of immature and mature NK cells in $Il2rg^{-/Y}$ mice. However, NK progenitor numbers in mutant mice are similar to those in WT mice, indicating that γc family cytokines are dispensable for NK-cell commitment in bone marrow. IL-15 is the principal γc family cytokine for the generation and maintenance of normal numbers of immature and mature

NK cells, as $Il15^{-/-}$, $Il15ra^{-/-}$, and $Il2rb^{-/-}$ mice each show profoundly decreased NK-cell numbers (Vosshenrich et al. 2005). IL-7 is dispensable for conventional NK-cell development, as mice deficient in either *Il7* or *Il7ra* show normal bone marrow and splenic NK numbers (He and Malek 1996; Moore et al. 1996), but IL-7 plays an important role in thymic NK-cell homeostasis, as there is a higher frequency but lower number of thymic NK cells in $Il7^{-/-}$ mice (Moore et al. 1996) and nearly a complete absence of IL-7Rα⁺ NK cells in $Rag2^{-/-}Il7^{-/-}$ mice (Vosshenrich et al. 2006).

Although IL-2, IL-4, and IL-21 are dispensable for the generation, differentiation, expansion, and survival of NK cells in vivo, they can nevertheless either promote NK-cell proliferation and survival or enhance NK-cell function. In mice infected with *Leishmania major*, IL-2 and IL-12, but not IL-4 produced by antigen-specific CD4⁺ T cells, are required for early IFN-γ production by NK cells, which depends on the presence of CD40/CD40L (Bihl et al. 2010). In response to virus rechallenge after vaccination, IL-2 secreted by human memory T cells can potently induce a sustained cytokine secretion and degranulation by NK cells, indicating the important roles of IL-2-induced NK-cell function in the initial control of the viral infection after vaccination (Horowitz et al. 2010). Wiskott–Aldrich syndrome protein (WASp) is an actin regulator, and NK cells from patients with loss-of-function mutations in WASp cannot form lytic synapses because of decreased degranulation and expression of IFN-γ (Huang et al. 2005; Orange et al. 2011). IL-2-treated WASp KO NK cells can rescue the defective NK function to eliminate major histocompatibility complex (MHC) class I negative hematopoietic tumor cells (Kritikou et al. 2016). Overexpression of IL-4 in mice results in altered expression of cell-surface markers and function of conventional NK cells via a cooperation with macrophage-produced IL-15, with lower expression of CD11b and IL-18Rα but increased expression of IFN-γ, IL-10, and GM-CSF. These IL-4-stimulated NK cells show higher cytotoxicity than conventional NK cells, and NK cells with a phenotype similar

Cite this article as *Cold Spring Harb Perspect Biol* doi: 10.1101/cshperspect.a028449

to the IL-4-stimulated NK cells are found in mice following infection with *N. brasiliensis* (Kiniwa et al. 2016), suggesting the involvement of NK cells in Th2 responses during the clearance of this parasite.

IL-21 can promote differentiation and expansion of bone marrow human NK progenitors in the presence of Flt3L and IL-15, and it can also enhance NK cytotoxicity, albeit less potently than IL-2 or IL-15, and IL-21 has a cooperative effect when combined with IL-2 or IL-15 (Parrish-Novak et al. 2000). NK function in human patients with loss-of-function mutations in IL-21R is variable, ranging from normal to profoundly impaired NK-cell cytotoxicity (Kotlarz et al. 2013); the reason(s) for this variation in NK function remains unclear. Interestingly, IL-21-dependent expansion of the IFN-γ-producing memory-like NK cells during bacille Calmett–Guérin (BCG) vaccination is important for protective immunity against *Mycobacterium tuberculosis* (Venkatasubramanian et al. 2017).

ROLE OF IL-4 IN MACROPHAGE BIOLOGY

Activation of macrophages by cytokines and TLRs is important for host defense and immunity. IL-4-activated macrophages play a role in type 2 immunity in allergic inflammation, helminth infection, and wound healing, and these cells are referred to as M2 or M(IL-4) macrophages (Van Dyken and Locksley 2013; Wynn and Vannella 2016; Eming et al. 2017). M(IL-4) macrophages also play a critical role in wound repair following helminth infection (Chen et al. 2012) and are involved in the repair of infarcts in the adult mouse heart (Shiraishi et al. 2016). Tumor-associated macrophages (TAMs) share the immune tolerance and immunosuppressive phenotype with M2 macrophages and are associated with tumor progression (Wynn et al. 2013). Local injection of IL-21 into tumors can suppress the expression of genes associated with M2 macrophages in TAMs, including *Vegf* and *Tgfb*. In contrast, *Ccl2* expression is associated with tumor-inhibiting M1 macrophages, which then act through CD8[+] T cells to enhance antitumor therapy (Xu et al. 2015).

ROLES OF IL-15 AND IL-21 IN DENDRITIC CELL FUNCTION

DCs play vital roles in both innate and adaptive immunities and can be divided in two major types based on the cell-surface markers and their functions, including conventional DCs (cDCs) and plasmacytoid DCs (pDCs) (Pulendran 2015; Durai and Murphy 2016). cDCs, which can be subdivided into cDC1 and cDC2 cells and express high levels of MHC class II molecules, are professional APCs, whereas pDCs rapidly secrete high levels of type I IFNs on encountering pathogens mediated by activation of TLRs. γc family cytokines critically regulate the development and function of DCs. For example, IL-15 can promote the differentiation of monocytes into Langerhans cells in the presence of GM-CSF (Mohamadzadeh et al. 2001), a subset of immature DCs, and can support the maturation of monocytes into DCs (Saikh et al. 2001). IL-15 also promotes DC activation and maturation, whereas IL-21 inhibits these processes and blocks the maturation and activation of LPS-induced DCs (Brandt et al. 2003). In addition, IL-21 induces apoptosis of splenic cDCs but not GM-CSF-induced DCs, and this is dependent on its activation of STAT3 and increased expression of proapoptotic protein BIM. GM-CSF via STAT5 activation can antagonize the apoptotic effect of IL-21 on cDCs (Wan et al. 2013). Although IL-21 does not affect expression of type I IFNs, IL-6, or TNF-α nor the maturation of pDCs, it can potently induce granzyme B (GZMB) expression in human pDCs via a STAT3-dependent pathway. Increased GZMB production induced by IL-21 in pDC is, at least in part, responsible for the inhibition of CD4[+] T-cell proliferation mediated by TLR-activated pDCs (Karrich et al. 2013).

γc FAMILY CYTOKINES AND AUTOIMMUNE DISEASES: KEY ROLES FOR IL-2 AND IL-21

As discussed above, γc family cytokines are essential for the development of T and NK cells and the function of B cells and other immune cells and are involved in every aspect of the immune response. Balanced actions of these cytokines are crucial for protecting the host from harm, not only by pathogens but also by inflammation caused by an immune response. Accumulating evidence has shown that effector CD4$^+$ T cells play key roles in mediating autoimmune pathology, whereas Tregs play important roles in controlling the immune response (Grant et al. 2015; Suarez-Fueyo et al. 2017). Dysregulated signals by γc family cytokines are significantly associated with the progression and outcome of the human autoimmune disorders (Spolski and Leonard 2014; Tangye 2015; Suarez-Fueyo et al. 2017); thus, modulating γc family signals is an important means to manage autoimmunity.

Systemic lupus erythematosus (SLE) is a systemic autoimmune disease of unknown cause, characterized by massive production of autoantibodies and proinflammatory cytokines and can affect every organ. A study of 1318 SLE patients and 1318 matched controls found that two of three SNPs in the introns of *IL21* were significantly associated with SLE (Sawalha et al. 2008). Another study in two large cohorts found that one of the 17 SNPs in the *IL21RA* gene was significantly associated with SLE (Webb et al. 2009). There is also a significant association of the decreased expression of IL-21Rα in peripheral B cells with nephritis and high-titer anti-double-stranded DNA antibodies in SLE patients. B cells from some SLE patients fail to proliferate in response to IL-21 plus anti-CD40 antibody stimulation (Mitoma et al. 2005). Increased IL-21 mRNA levels were also detected in skin biopsies of SLE patients (Caruso et al. 2009). In addition to this association in human SLE, IL-21 was shown to play a vital role in the development of SLE in mouse models. In the BXSB-*Yaa* mouse model of SLE, both IL-10 and IL-21 levels are elevated and the increased IL-21 is not produced by Th17 cells but instead is from ICOS$^+$CD4$^+$ T cells, and increased IL-10 production by T cells requires IL-21 (Pot et al. 2009; Spolski et al. 2009). Importantly, *Il21r*$^{-/-}$ BXSB-*Yaa* mice have decreased antinuclear antibodies, do not develop histological and immunological characteristics of SLE, and survive for more than 250 days (Bubier et al. 2009), and blocking IL-21 signaling in SLE-prone MRL-

Fas^{lpr} mice with an IL-21R-Fc fusion protein can reduce disease progression (Herber et al. 2007). IL-21 also plays a pathological role in mouse experimental autoimmune uveitis (EAU), which shares the pathological features with human uveitis and is a Th17-cell-related disease, with attenuated EAU in *Il21r*$^{-/-}$ mice as compared with WT mice (Wang et al. 2011).

Rheumatoid arthritis (RA) is a chronic autoimmune inflammatory disorder characterized by autoantibody-induced joint inflammation and systemic inflammation. A number of proinflammatory cytokines are associated with the development and disease activity of RA, although the exact etiology of RA is still not understood. IL-21 is associated with pathogenesis of RA. For example, expression of IL-21Rα was detected in synovial fibroblasts and synovial macrophages of RA patients (Jungel et al. 2004), and IL-21 expression in synovial fluid and peripheral blood in RA patients correlates with the presence of Th17 cells (Niu et al. 2010). In addition, increased plasma IL-21 level correlates with the disease activity of RA (Rasmussen et al. 2010). Interestingly, treatment with tocilizumab, a blocking antibody to the IL-6 receptor, selectively reduced IL-21 and IgG4 anti-CCP autoantibody levels, with an improved RA disease activity (Carbone et al. 2013). Based on mass cytometry, multidimensional cytometry, transcriptomics, and functional assays, a marked increase was noted in PD1^{hi}CXCR5$^-$CD4$^+$ T cells in the synovium of RA patients. These cells express IL-21, CXCL13, ICOS, and MAF and induce plasma-cell differentiation to promote B-cell responses and Ig production only within pathologically inflamed nonlymphoid tissues (Rao et al. 2017). Autoantibodies can be detected years before the onset of RA, but titers do not well correlate with disease activity. A strong correlation was noted between the expression levels of IL-17 and increased Th17-cell numbers and the systemic disease activity in both the onset and the progression of RA (Leipe et al. 2010). In a mouse RA model, IL-23 can suppress the expression of *St6gal1*, which encodes for a rate-limiting enzyme to control the IgG glycosylation in antibody-producing cells during plasma-cell development. The suppressive effect of *St6gal1* by IL-23 is mediated by IL-21 and IL-22, but not by IL-17A, IL-17F, or GM-CSF (Pfeifle et al. 2017).

Although the etiology for MS is not fully understood, infiltration of the central nervous system (CNS) by autoreactive T cells is important for the initiation of chronic inflammation and neurodegeneration causing damage to myelin and axons. Among the factors associated with MS, Th1- and Th17-mediated inflammation in both peripheral tissues and CNS play important roles in the pathogenesis of MS. Among proinflammatory cytokines and chemokines aberrantly produced during inflammation, IL-2 and IL-21 are the only γc family cytokines. Interestingly, genome-wide association studies (GWAS) reveal that genetic variations in *IL2RA, IL7, IL7R, STAT3, STAT4,* and *TYK2* are risk factors that confer susceptibility in MS patients (International Multiple Sclerosis Genetics et al. 2011, 2013). Furthermore, in mouse EAE, a mouse model of MS, mice with *Stat3*-deficient CD4$^+$ T cells are resistant to EAE owing to defective Th17 differentiation (Liu et al. 2008), and, unexpectedly, mice with *Stat5* deleted in CD4$^+$ T cells also show diminished development of EAE because of their impaired expression of GM-CSF in CD4$^+$ T cells activated by IL-7 (Sheng et al. 2014). To dampen IL-2's potent proliferative activity on IL-2Rα^{high} Teff cells during inflammation mediated by T cells, a humanized neutralizing IgG1 monoclonal antibody to IL-2Rα (daclizumab) was developed (Queen et al. 1989). Daclizumab is well tolerated and results in a significant improvement in treating relapsing-remitting or secondary progressive forms of MS (Bielekova et al. 2004), which account for a majority of MS patients. Daclizumab treatment causes a gradual decrease in circulating T-cell numbers and significantly expands CD56^{bright} NK-cell numbers, which correlates with a positive response to daclizumab therapy (Bielekova et al. 2006; Wynn et al. 2010). In 2016, daclizumab (Zinbryta; previously known as daclizumab high-yield process) was approved by the European Medicines Agency and the U.S. Food and Drug Administration (FDA) for the treatment of relapsing forms of MS in adults (Shirley 2017a).

Inflammatory bowel diseases (IBDs) are characterized by chronic inflammation in the intestine and are mediated by increased proinflammatory cytokine secretion. Environmental, genetic, and commensal microbial factors are involved in the development of IBD. Crohn's disease (CD) and ulcerative colitis (UC) are two major forms of IBD and a meta-analysis identified 163 IBD susceptibility loci, including signaling molecules for cytokines *STAT1, STAT3, STAT4,* and *JAK2* and γc family cytokines and cytokine receptors *IL2, IL21, IL2RA,* and *IL15RA* (Jostins et al. 2012). In addition to genetic and environmental factors, T cells play important roles in the outcome of IBD. As discussed above, mice with defective IL-2 signaling spontaneously develop severe colitis (Sadlack et al. 1993; Suzuki et al. 1995; Willerford et al. 1995). A homozygous mutation of *IL21* (c.T147C, p.Leu49Pro) was identified in early-onset IBD patients with decreased circulating B-cell numbers, including naïve and memory B cells and increased transitional B-cell numbers. Although the mutant IL-21 was predicted to have decreased protein stability, in vitro–produced IL-21^{Leu49Pro} cannot activate STAT3, promote normal B-cell proliferation, nor mediate B-cell activation in vitro, suggesting that the mutant IL-21 may not be able to stably bind to the IL-21 receptor. Indeed, the B cells from these patients show normal B-cell activation and Ig class switch in response to either WT IL-21 plus CD40L or IL-4 plus CD40L stimulation, indicating that the defective B-cell function is not caused by intrinsic B-cell defects (Salzer et al. 2014).

Type I diabetes (T1D) is a chronic autoimmune disease characterized by the loss of insulin-producing pancreatic β cells, and both genetic susceptibility and environmental factors contribute to the onset of T1D. Inflammatory processes driven by self-reactive T cells in response to infectious agents and commensal organisms can either be pathogenic or protective (Herold et al. 2013). In human GWAS, among γc family cytokine and cytokine receptor gene loci, *IL2RA, IL2, IL21,* and *IL7RA* are risk loci for T1D patients (Todd et al. 2007; Concannon et al. 2009). The essential role of IL-2 in Treg cells was further shown by the findings that the ratios of Treg and Teff cells progressively decrease in inflamed pancreatic islets but not in pancreatic lymph nodes. A low dose of IL-2 promotes survival of Treg cells and protects nonobese diabetic (NOD) mice from T1D onset (Tang et al. 2008; Grinberg-Bleyer et al. 2010). In addition, injection of NOD mice with anti-IL-7Rα antibody alone can reverse established T1D (Lee et al. 2012). Treg cells from either healthy individuals or from T1D patients can respond to IL-2 at a 10-fold lower concentration than that required for memory T cells (Yu et al. 2015). Low-dose IL-2 mainly increases Treg-cell numbers and suppresses IFN-γ production by pancreas-infiltrating T cells, whereas anti-IL-7Rα antibody reduces Th1 and Tc1-cell numbers and increases PD1 expression in Teff cells. In addition, circulating anti-IL-2 autoantibodies are also detected in NOD mice and patients with T1D, and their titer is positively correlated with age and disease onset (Perol et al. 2016). The *Il2/Il21* loci are within the diabetes-associated *Idd3* locus, and increased levels of IL-21 were detected in NOD mice, suggesting that IL-21 is a pathogenic cytokine in this model of T1D. Indeed, NOD mice crossed with *Il21r*$^{-/-}$ mice show low lymphocyte infiltration into the pancreas and reduced Th17 cells, indicating critical roles of IL-21 in the development of T1D in NOD mice (Spolski et al. 2008; McGuire et al. 2009). Incubation of bone marrow from NOD mice with CpG can induce a B-cell population that resembles innate pro-B cells (Montandon et al. 2013). Adoptive transfer of these CpG-induced pro-B cells can protect NOD mice from disease onset, which is attributed to the ability of these CpG-pro-B to suppress the proliferation of Teff cells and induce their apoptosis. Furthermore, the expression of IL-21 by Teff cells cocultured with CpG-induced pro-B cells was markedly reduced, but the expression of IL-2 and IL-10 were increased (Montandon et al. 2013).

MODULATION OF γc FAMILY CYTOKINE SIGNALING AND THERAPEUTIC RAMIFICATIONS

Given their indispensable roles in the development and function of T, B, and NK cells and broad actions on the immune response, rationally modulating the actions of γc family cytokines has become an important approach for managing patients with a range of diseases, including immunodeficiency, allergy, infection, autoimmune disorders, transplant rejection, and cancer. For example, giving cytokines can enhance their actions, whereas blocking antibodies against cytokines or cytokine receptors, or small molecules that either inhibit cytokine production or kinase activity like JAK inhibitors, can dampen their actions to protect the host from harm mediated by overreactive immune responses. Indeed, γc family cytokines have been effective in completed and/or ongoing clinical trials for cancer immunotherapy (Rosenberg 2014; Spolski and Leonard 2014; Waldmann 2015; Pulliam et al. 2016; Tran et al. 2017). Moreover, blocking IL-4 signaling is effective for treating adult patients with moderate-to-severe atopic dermatitis (Wenzel et al. 2013; Beck et al. 2014) and adult patients with uncontrolled persistent asthma (Wenzel et al. 2016).

IL-2's potent ability to expand Teff cells and enhance NK-cell cytotoxicity led to the use of high-dose recombinant IL-2 (proleukin) to treat metastatic melanoma and renal-cell carcinoma. Although 15% of the patients with melanoma or renal-cell carcinoma responded well to the treatment and some of them had a long-term tumor-free period, the use of high-dose IL-2 therapy was limited because of its severe side effects, including vascular leak syndrome (VLS) and increased serum creatinine and bilirubin levels (Rosenberg 2014). However, adoptive cell transfer of autologous tumor-infiltrating lymphocytes (TILs) expanded in vitro by IL-2 also results in significant improvement in treating patients with metastatic melanoma (Rosenberg 2014; Tran et al. 2017).

The discovery of IL-2's essential role in Treg function led to observations that low-dose IL-2 preferentially expands Treg cells in vivo, thereby suppressing graft-versus-host disease (GVHD) (Koreth et al. 2011; Asano et al. 2017) and autoimmune disorders, like T1D (Hartemann et al. 2013; Yu et al. 2015) and SLE (He et al. 2016). Importantly, when combined with IL-2, two monoclonal antibodies against mouse IL-2

Cite this article as *Cold Spring Harb Perspect Biol* doi: 10.1101/cshperspect.a028449

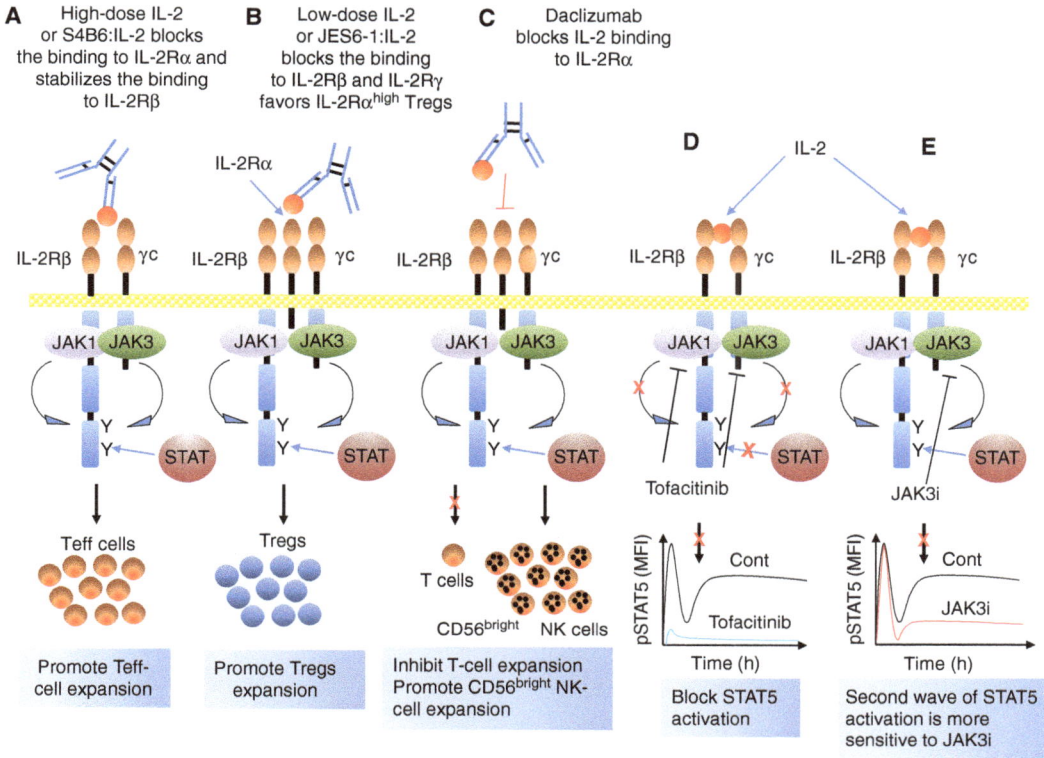

Figure 5. Modulation of interleukin (IL)-2 signaling by antibodies or Janus kinase (JAK) inhibitors. (*A*) High-dose IL-2 or IL-2 complexed with anti-IL-2 antibody S4B6 preferentially promotes the expansion of effector T cells to enhance the immune response, whereas (*B*) low-dose IL-2 or IL-2 complexed with anti-IL-2 antibody JES6-1 preferentially expands Treg cells to suppress the immune response. (*C*) The humanized anti-IL-2Rα antibody, daclizumab, can block IL-2 binding to IL-2Rα to favor the expansion of CD56^{bright} natural killer (NK) cells while diminishing T-cell expansion. (*D*) The JAK inhibitor, tofacitinib, can potently inhibit signal transducers and activators of transcription (STAT) activation by γ chain (γc) family cytokines because of its inhibitory effect on JAK3 and JAK1, whereas (*E*) JAK3i more efficiently inhibits JAK3 and blocks the second wave of STAT5 activation and proliferation of CD4^{+} T cells stimulated with IL-2.

can selectively mimic either high-dose or low-dose IL-2 treatment. The IL-2–JES6-1 complexes, like low-dose IL-2, can expand Treg and the IL-2–S4B6 complexes, similar to high-dose IL-2, can expand Teff cells (Fig. 5) (Boyman et al. 2006). The reason for the differential effects of these two anti-IL-2 antibodies has been elucidated by structural and biochemical studies (Spangler et al. 2015). JES6-1–IL-2 complexes sterically block the interaction of IL-2 with IL-2Rβ and γc and lower the affinity of IL-2–IL-2Rα interaction to favor IL-2Rα^{high} cells, including Treg cells. In contrast, S4B6–IL-2 complexes sterically block the interaction

of IL-2 with IL-2Rα, resulting in increased affinity and stability of IL-2–IL-2Rβ interaction to favor the expansion of IL-2Rα^{low} T cells (Spangler et al. 2015).

To extend these mouse studies into human cancer immunotherapy clinical trials, a monoclonal antihuman IL-2 antibody, NARA1, has been used. NARA1 binds IL-2Rα with high affinity, therefore blocking the interaction of IL-2 with IL-2Rα (Arenas-Ramirez et al. 2016). In vivo administration of IL-2–NARA1 complexes into mice preferentially expands CD8^{+} T, CD44^{hi}CD8^{+} T, and NK cells over CD25^{+} CD4^{+} T cells, resulting in superior antimelanoma CD8$^+$ T-cell responses than those achieved with high-dose IL-2. In another study, adoptive transfer of tumor-reactive CD8$^+$ T cells treated with IL-2–anti-IL-2 complexes, but not IL-15-soluble IL-15Rα complexes, showed sustained antitumor immunity (Su et al. 2015); this is attributed to a vigorous and sustained expansion of IL-2–anti-IL-2 complex-treated cells in vivo versus a poor and transient expansion of IL-15-soluble IL-15Rα-treated cells. The sustained expression of high-level IL-2Rα on these tumor-reactive CD8$^+$ T cells is caused by IL-2Rα's ability to sustain IL-2 signaling by recycling IL-2 to the cell surface. Consistent with this finding and the critical role of IL-15 in maintaining memory CD8$^+$ T cells in vivo, membrane-bound chimeric IL-15 on CD19-specific chimeric antigen receptor (CAR)$^+$ T cells can maintain long-lived

CD45RO$^-$CCR7$^+$CD95$^+$ T cells in vivo; these cells most resemble T-memory stem cells and show potent antitumor activity for established CD19$^+$ leukemias (Hurton et al. 2016). STAT5 activated by IL-15 plays a role in maintaining these cells in vivo even after tumor clearance.

The detailed analyses of the IL-2–IL-2 receptor complexes also led to the creation of IL-2 molecules that can differentially act on different T-cell populations by changing the binding affinity of IL-2 with one of the IL-2 receptor chains (Fig. 6). For example, "super-2" (also known as H9) is more potent than WT IL-2 based on its augmented affinity for IL-2Rβ, where it no longer requires the presence of IL-2Rα to mediate STAT5 activation (Levin et al. 2012). H9 more potently expands cytotoxic T cells than Treg cells and shows less-severe VLS

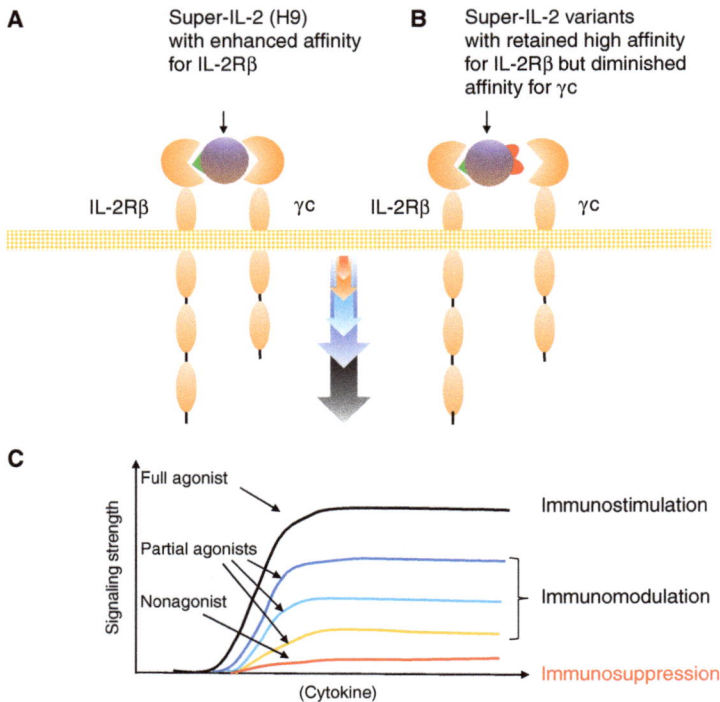

Figure 6. An interleukin (IL)-2 superkine and IL-2 partial agonists. (A) An IL-2 superkine (super-IL-2, H9) (green triangle) shows enhanced affinity for IL-2Rβ and more potent activity. (B) By retaining the high-affinity binding for IL-2Rβ of H9 (in green) but by changing additional amino acids (in red) on H9 to reduce the affinity for γc, a number of partial agonists with different potencies for activating IL-2 signaling pathways were created as well as a nonagonist (H9-RETR) that could potently inhibit the actions of both IL-2 and IL-15. (C) Schematic of full, partial, and nonagonists of IL-2. (Based on Fig. 1A in Mitra et al. 2015.)

as assessed by pulmonary edema in mice than WT IL-2. A number of H9 variants were also generated, which retain high-affinity binding for IL-2Rβ but have altered binding affinity for γc, thereby outcompeting endogenous IL-2 and IL-15 and showing distinct activities on different T cells (Fig. 6) (Mitra et al. 2015). As such, these molecules are partial agonists, and indeed represent the first partial agonists for a type I cytokine. Interestingly, H9-RET, which contains three amino acid substitutions at the IL-2–γc interface, can promote the proliferation of pre-activated but not freshly isolated CD8+ T cells; whereas H9-RETR, with four amino acid substitutions at the interface, only minimally activates STAT5 and does not promote T-cell proliferation. However, H9-RETR can potently antagonize IL-2 or IL-15 activity—if anything—more potently than blocking antibodies to IL-2Rα or IL-2Rβ and similar to the combination of such agents, so that it inhibits IL-2-induced NK cytolytic activity, prolongs survival in a mouse GVHD model, and inhibits spontaneous proliferation of malignant cells from patients with the chronic/smoldering form adult T-cell leukemia (Mitra et al. 2015).

Engineered IL-4 superkines have also been generated. A type I IL-4 receptor–selected IL-4 superkine shows a much higher binding affinity for IL-2Rγ chain and is 3- to 10-fold more potent than WT IL-4 to activate STAT6 phosphorylation and cytokine production, whereas a variant selected with high affinity to IL-13Rα1 more potently induces differentiation of monocyte-derived DCs (Junttila et al. 2012). Dupilumab (developed by Regeneron Pharmaceuticals and Sanofi) is a fully humanized IgG4 monoclonal antibody to IL-4Rα that inhibits signaling by both IL-4 and IL-13 and has been approved by the FDA for treating adult patients with moderate-to-severe atopic dermatitis (Chang and Nadeau 2017; Shirley 2017b). Collectively, these results show that using monoclonal antibodies, engineered cytokines, and potentially small molecules are ways to fine-tune cytokine signal strength and selectively promote certain functions of immune cells. Such molecules represent next-generation approaches to treat infectious, allergic, and autoimmune disorders, and cancer.

Given the key roles of JAKs in mediating cytokine signaling, including by γc family cytokines, a number of JAK inhibitors have been developed to suppress cytokine-mediated inflammation. Because of their potent immunosuppressive activity, these JAK inhibitors have been tested in clinical trials for autoimmune disorders, including psoriasis, diabetic nephropathy, atopic dermatitis, myelofibrosis, juvenile idiopathic arthritis (JIA), RA, SLE, and IBD (Winthrop 2017). Among them, tofacitinib preferentially inhibits JAK1 and JAK3 and to a lesser extent JAK2 (Changelian et al. 2003; Meyer et al. 2010) and was approved by the FDA for treating RA and is being tested in a phase III trial for psoriasis and UC and in a phase I trial for JIA (Winthrop 2017). In a large cohort phase III trial, tofacitinib showed promising efficacy as an induction and maintenance therapy for patients with moderate-to-severe UC (Sandborn et al. 2017).

Thus, a wide variety of approaches to more efficiently modulate γc family cytokine signals have been and are being developed, which hopefully will further enhance the treatment of patients with various immune disorders and cancers. For example, high-dose IL-2 and super IL-2 (H9) can potently promote Teff-cell expansion and function to treat malignancies; low-dose IL-2 and H9 agonist variants preferentially expand Treg cells to better treat autoimmune disorders like T1D; the humanized anti-IL-2Rα antibody, daclizumab, can block IL-2 binding to IL-2Rα to inhibit T-cell expansion while serving to expand CD56bright NK cells because of enhanced availability of IL-2; and the H9-RETR variant can block IL-2 and IL-15 binding to IL-2Rβ, thus suppressing the growth of chronic/smoldering acute T-cell leukemia ex vivo. In mice, anti-IL-2 antibody S4B6 complexed with IL-2 mimics the action of high-dose IL-2, whereas anti-IL-2 antibody JES6-1 complexed with IL-2 mimics the action of low-dose IL-2. Moreover, in addition to tofacitinib, which can inhibit JAK1 and JAK2 in addition to JAK3, more specific JAK3 inhibitors have been developed but have not yet been evaluated clinically. Moreover, it is possible that specific inhibitors for STAT proteins may also be useful immunosuppressants.

CONCLUSIONS

In this review, we have discussed the roles of γc family cytokines, including their biological actions, mechanisms of signaling and gene regulation, and the therapeutic applications and potential of administering or blocking the actions of these cytokines. The study of this system has had profound impact, from elucidating the basis of human inherited immunodeficiency, spawning new medications (e.g., IL-2 and JAK inhibitors), and providing the basis for the first human gene therapy (for XSCID), as well as providing tremendous insight into the molecular basis of gene regulation and cell differentiation. Future mechanistic studies will be invaluable for greater understanding of the basic science of these cytokines as well as evolving new therapeutic approaches to a range of diseases. Furthermore, the approaches used for modulation of IL-2 signals and the lessons learned may be broadly applicable to other γc family cytokines as well as other cytokines and growth factors, with implications for scientific investigation and potentially for therapeutic advances as well.

REFERENCES

Akashi K, Kondo M, von Freeden-Jeffry U, Murray R, Weissman IL. 1997. Bcl-2 rescues T lymphopoiesis in interleukin-7 receptor-deficient mice. *Cell* **89:** 1033–1041.

Al-Shami A, Spolski R, Kelly J, Fry T, Schwartzberg PL, Pandey A, Mackall CL, Leonard WJ. 2004. A role for thymic stromal lymphopoietin in CD4+ T cell development. *J Exp Med* **200:** 159–168.

Amadi-Obi A, Yu CR, Liu X, Mahdi RM, Clarke GL, Nussenblatt RB, Gery I, Lee YS, Egwuagu CE. 2007. T_H17 cells contribute to uveitis and scleritis and are expanded by IL-2 and inhibited by IL-27/STAT1. *Nat Med* **13:** 711–718.

Aman MJ, Tayebi N, Obiri NI, Puri RK, Modi WS, Leonard WJ. 1996. cDNA cloning and characterization of the human interleukin 13 receptor α chain. *J Biol Chem* **271:** 29265–29270.

Arenas-Ramirez N, Zou C, Popp S, Zingg D, Brannetti B, Wirth E, Calzascia T, Kovarik J, Sommer L, Zenke G, et al. 2016. Improved cancer immunotherapy by a CD25-mimobody conferring selectivity to human interleukin-2. *Sci Transl Med* **8:** 367ra166.

Arima N, Kamio M, Imada K, Hori T, Hattori T, Tsudo M, Okuma M, Uchiyama T. 1992. Pseudo-high affinity interleukin 2 (IL-2) receptor lacks the third component that is essential for functional IL-2 binding and signaling. *J Exp Med* **176:** 1265–1272.

Armitage RJ, Macduff BM, Eisenman J, Paxton R, Grabstein KH. 1995. IL-15 has stimulatory activity for the induction of B cell proliferation and differentiation. *J Immunol* **154:** 483–490.

Asano T, Meguri Y, Yoshioka T, Kishi Y, Iwamoto M, Nakamura M, Sando Y, Yagita H, Koreth J, Kim HT, et al. 2017. PD-1 modulates regulatory T-cell homeostasis during low-dose interleukin-2 therapy. *Blood* **129:** 2186–2197.

Asao H, Okuyama C, Kumaki S, Ishii N, Tsuchiya S, Foster D, Sugamura K. 2001. Cutting edge: The common γ-chain is an indispensable subunit of the IL-21 receptor complex. *J Immunol* **167:** 1–5.

Attridge K, Wang CJ, Wardzinski L, Kenefeck R, Chamberlain JL, Manzotti C, Kopf M, Walker LS. 2012. IL-21 inhibits T cell IL-2 production and impairs Treg homeostasis. *Blood* **119:** 4656–4664.

Avery DT, Bryant VL, Ma CS, de Waal Malefyt R, Tangye SG. 2008. IL-21-induced isotype switching to IgG and IgA by human naïve B cells is differentially regulated by IL-4. *J Immunol* **181:** 1767–1779.

Ballesteros-Tato A, Leon B, Graf BA, Moquin A, Adams PS, Lund FE, Randall TD. 2012. Interleukin-2 inhibits germinal center formation by limiting T follicular helper cell differentiation. *Immunity* **36:** 847–856.

Bamford RN, Grant AJ, Burton JD, Peters C, Kurys G, Goldman CK, Brennan J, Roessler E, Waldmann TA. 1994. The interleukin (IL) 2 receptor β chain is shared by IL-2 and a cytokine, provisionally designated IL-T, that stimulates T-cell proliferation and the induction of lymphokine-activated killer cells. *Proc Natl Acad Sci* **91:** 4940–4944.

Barber DL, Wherry EJ, Masopust D, Zhu B, Allison JP, Sharpe AH, Freeman GJ, Ahmed R. 2006. Restoring function in exhausted CD8 T cells during chronic viral infection. *Nature* **439:** 682–687.

Bayer AL, Yu A, Adeegbe D, Malek TR. 2005. Essential role for interleukin-2 for CD4+CD25+ T regulatory cell development during the neonatal period. *J Exp Med* **201:** 769–777.

Bayer AL, Yu A, Malek TR. 2007. Function of the IL-2R for thymic and peripheral CD4+CD25+ Foxp3+ T regulatory cells. *J Immunol* **178:** 4062–4071.

Bayer AL, Lee JY, de la Barrera A, Surh CD, Malek TR. 2008. A function for IL-7R for CD4+CD25+Foxp3+ T regulatory cells. *J Immunol* **181:** 225–234.

Bazan JF. 1990. Structural design and molecular evolution of a cytokine receptor superfamily. *Proc Natl Acad Sci* **87:** 6934–6938.

Beck LA, Thaci D, Hamilton JD, Graham NM, Bieber T, Rocklin R, Ming JE, Ren H, Kao R, Simpson E, et al. 2014. Dupilumab treatment in adults with moderate-to-severe atopic dermatitis. *N Engl J Med* **371:** 130–139.

Begitt A, Droescher M, Meyer T, Schmid CD, Baker M, Antunes F, Knobeloch KP, Owen MR, Naumann R, Decker T, et al. 2014. STAT1-cooperative DNA binding distinguishes type 1 from type 2 interferon signaling. *Nat Immunol* **15:** 168–176.

Bennett CL, Christie J, Ramsdell F, Brunkow ME, Ferguson PJ, Whitesell L, Kelly TE, Saulsbury FT, Chance PF, Ochs HD. 2001. The immune dysregulation, polyendocrinopathy, enteropathy, X-linked syndrome (IPEX) is caused by mutations of *FOXP3*. *Nat Genet* **27:** 20–21.

Cite this article as *Cold Spring Harb Perspect Biol* doi: 10.1101/cshperspect.a028449

Berard M, Brandt K, Bulfone-Paus S, Tough DF. 2003. IL-15 promotes the survival of naïve and memory phenotype CD8[+] T cells. *J Immunol* **170**: 5018–5026.

Bielekova B, Richert N, Howard T, Blevins G, Markovic-Plese S, McCartin J, Frank JA, Wurfel J, Ohayon J, Waldmann TA, et al. 2004. Humanized anti-CD25 (daclizumab) inhibits disease activity in multiple sclerosis patients failing to respond to interferon β. *Proc Natl Acad Sci* **101**: 8705–8708.

Bielekova B, Catalfamo M, Reichert-Scrivner S, Packer A, Cerna M, Waldmann TA, McFarland H, Henkart PA, Martin R. 2006. Regulatory CD56[bright] natural killer cells mediate immunomodulatory effects of IL-2Rα-targeted therapy (daclizumab) in multiple sclerosis. *Proc Natl Acad Sci* **103**: 5941–5946.

Bihl F, Pecheur J, Breart B, Poupon G, Cazareth J, Julia V, Glaichenhaus N, Braud VM. 2010. Primed antigen-specific CD4[+] T cells are required for NK cell activation in vivo upon *Leishmania major* infection. *J Immunol* **185**: 2174–2181.

Boussiotis VA, Barber DL, Nakarai T, Freeman GJ, Gribben JG, Bernstein GM, D'Andrea AD, Ritz J, Nadler LM. 1994. Prevention of T cell anergy by signaling through the γc chain of the IL-2 receptor. *Science* **266**: 1039–1042.

Boyman O, Kovar M, Rubinstein MP, Surh CD, Sprent J. 2006. Selective stimulation of T cell subsets with antibody-cytokine immune complexes. *Science* **311**: 1924–1927.

Brandt K, Bulfone-Paus S, Foster DC, Ruckert R. 2003. Interleukin-21 inhibits dendritic cell activation and maturation. *Blood* **102**: 4090–4098.

Brunkow ME, Jeffery EW, Hjerrild KA, Paeper B, Clark LB, Yasayko SA, Wilkinson JE, Galas D, Ziegler SF, Ramsdell F. 2001. Disruption of a new forkhead/winged-helix protein, scurfin, results in the fatal lymphoproliferative disorder of the scurfy mouse. *Nat Genet* **27**: 68–73.

Bubier JA, Sproule TJ, Foreman O, Spolski R, Shaffer DJ, Morse HC 3rd, Leonard WJ, Roopenian DC. 2009. A critical role for IL-21 receptor signaling in the pathogenesis of systemic lupus erythematosus in BXSB-*Yaa* mice. *Proc Natl Acad Sci* **106**: 1518–1523.

Buckley RH. 2011. Transplantation of hematopoietic stem cells in human severe combined immunodeficiency: Longterm outcomes. *Immunol Res* **49**: 25–43.

Burchill MA, Yang J, Vogtenhuber C, Blazar BR, Farrar MA. 2007. IL-2 receptor β-dependent STAT5 activation is required for the development of Foxp3[+] regulatory T cells. *J Immunol* **178**: 280–290.

Burton JD, Bamford RN, Peters C, Grant AJ, Kurys G, Goldman CK, Brennan J, Roessler E, Waldmann TA. 1994. A lymphokine, provisionally designated interleukin T and produced by a human adult T-cell leukemia line, stimulates T-cell proliferation and the induction of lymphokine-activated killer cells. *Proc Natl Acad Sci* **91**: 4935–4939.

Cao X, Shores EW, Hu-Li J, Anver MR, Kelsall BL, Russell SM, Drago J, Noguchi M, Grinberg A, Bloom ET, et al. 1995. Defective lymphoid development in mice lacking expression of the common cytokine receptor γ chain. *Immunity* **2**: 223–238.

Carbone G, Wilson A, Diehl SA, Bunn J, Cooper SM, Rincon M. 2013. Interleukin-6 receptor blockade selectively reduces IL-21 production by CD4 T cells and IgG4 autoantibodies in rheumatoid arthritis. *Int J Biol Sci* **9**: 279–288.

Caruso R, Botti E, Sarra M, Esposito M, Stolfi C, Diluvio L, Giustizieri ML, Pacciani V, Mazzotta A, Campione E, et al. 2009. Involvement of interleukin-21 in the epidermal hyperplasia of psoriasis. *Nat Med* **15**: 1013–1015.

Cerwenka A, Lanier LL. 2016. Natural killer cell memory in infection, inflammation and cancer. *Nat Rev Immunol* **16**: 112–123.

Chang HY, Nadeau KC. 2017. IL-4Rα inhibitor for atopic disease. *Cell* **170**: 222.

Changelian PS, Flanagan ME, Ball DJ, Kent CR, Magnuson KS, Martin WH, Rizzuti BJ, Sawyer PS, Perry BD, Brissette WH, et al. 2003. Prevention of organ allograft rejection by a specific Janus kinase 3 inhibitor. *Science* **302**: 875–878.

Chen X, Bhandari R, Vinkemeier U, Van Den Akker F, Darnell JE Jr, Kuriyan J. 2003. A reinterpretation of the dimerization interface of the N-terminal domains of STATs. *Protein Sci* **12**: 361–365.

Chen F, Liu Z, Wu W, Rozo C, Bowdridge S, Millman A, Van Rooijen N, Urban JF Jr, Wynn TA, Gause WC. 2012. An essential role for TH2-type responses in limiting acute tissue damage during experimental helminth infection. *Nat Med* **18**: 260–266.

Chung Y, Tanaka S, Chu F, Nurieva RI, Martinez GJ, Rawal S, Wang YH, Lim H, Reynolds JM, Zhou XH, et al. 2011. Follicular regulatory T cells expressing Foxp3 and Bcl-6 suppress germinal center reactions. *Nat Med* **17**: 983–988.

Ciofani M, Madar A, Galan C, Sellars M, Mace K, Pauli F, Agarwal A, Huang W, Parkhurst CN, Muratet M, et al. 2012. A validated regulatory network for Th17 cell specification. *Cell* **151**: 289–303.

Codarri L, Gyulveszi G, Tosevski V, Hesske L, Fontana A, Magnenat L, Suter T, Becher B. 2011. RORγt drives production of the cytokine GM-CSF in helper T cells, which is essential for the effector phase of autoimmune neuroinflammation. *Nat Immunol* **12**: 560–567.

Concannon P, Rich SS, Nepom GT. 2009. Genetics of type 1A diabetes. *N Engl J Med* **360**: 1646–1654.

Cordeiro Gomes A, Hara T, Lim VY, Herndler-Brandstetter D, Nevius E, Sugiyama T, Tani-Ichi S, Schlenner S, Richie E, Rodewald HR, et al. 2016. Hematopoietic stem cell niches produce lineage-instructive signals to control multipotent progenitor differentiation. *Immunity* **45**: 1219–1231.

Cosman D, Cerretti DP, Larsen A, Park L, March C, Dower S, Gillis S, Urdal D. 1984. Cloning, sequence and expression of human interleukin-2 receptor. *Nature* **312**: 768–771.

Cote-Sierra J, Foucras G, Guo L, Chiodetti L, Young HA, Hu-Li J, Zhu J, Paul WE. 2004. Interleukin 2 plays a central role in Th2 differentiation. *Proc Natl Acad Sci* **101**: 3880–3885.

Crotty S. 2014. T follicular helper cell differentiation, function, and roles in disease. *Immunity* **41**: 529–542.

Cui G, Staron MM, Gray SM, Ho PC, Amezquita RA, Wu J, Kaech SM. 2015. IL-7-induced glycerol transport and TAG synthesis promotes memory CD8[+] T cell longevity. *Cell* **161**: 750–761.

Dardalhon V, Awasthi A, Kwon H, Galileos G, Gao W, Sobel RA, Mitsdoerffer M, Strom TB, Elyaman W, Ho IC, et al. 2008. IL-4 inhibits TGF-β-induced Foxp3⁺ T cells and, together with TGF-β, generates IL-9⁺ IL-10⁺ Foxp3⁻ effector T cells. *Nat Immunol* **9:** 1347–1355.

Darnell JE Jr, Kerr IM, Stark GR. 1994. Jak-STAT pathways and transcriptional activation in response to IFNs and other extracellular signaling proteins. *Science* **264:** 1415–1421.

Davidson TS, DiPaolo RJ, Andersson J, Shevach EM. 2007. Cutting Edge: IL-2 is essential for TGF-β-mediated induction of Foxp3⁺ T regulatory cells. *J Immunol* **178:** 4022–4026.

Demoulin JB, Uyttenhove C, Van Roost E, DeLestre B, Donckers D, Van Snick J, Renauld JC. 1996. A single tyrosine of the interleukin-9 (IL-9) receptor is required for STAT activation, antiapoptotic activity, and growth regulation by IL-9. *Mol Cell Biol* **16:** 4710–4716.

Demoulin JB, Van Roost E, Stevens M, Groner B, Renauld JC. 1999. Distinct roles for STAT1, STAT3, and STAT5 in differentiation gene induction and apoptosis inhibition by interleukin-9. *J Biol Chem* **274:** 25855–25861.

Diehl SA, Schmidlin H, Nagasawa M, Blom B, Spits H. 2012. IL-6 triggers IL-21 production by human CD4⁺ T cells to drive STAT3-dependent plasma cell differentiation in B cells. *Immunol Cell Biol* **90:** 802–811.

Druez C, Coulie P, Uyttenhove C, Van Snick J. 1990. Functional and biochemical characterization of mouse P40/IL-9 receptors. *J Immunol* **145:** 2494–2499.

Dubois S, Mariner J, Waldmann TA, Tagaya Y. 2002. IL-15Rα recycles and presents IL-15 in *trans* to neighboring cells. *Immunity* **17:** 537–547.

Dugas B, Renauld JC, Pene J, Bonnefoy JY, Peti-Frere C, Braquet P, Bousquet J, Van Snick J, Mencia-Huerta JM. 1993. Interleukin-9 potentiates the interleukin-4-induced immunoglobulin (IgG, IgM and IgE) production by normal human B lymphocytes. *Eur J Immunol* **23:** 1687–1692.

Dukovich M, Wano Y, Le thi Bich T, Katz P, Cullen BR, Kehrl JH, Greene WC. 1987. A second human interleukin-2 binding protein that may be a component of high-affinity interleukin-2 receptors. *Nature* **327:** 518–522.

Durai V, Murphy KM. 2016. Functions of murine dendritic cells. *Immunity* **45:** 719–736.

El-Behi M, Ciric B, Dai H, Yan Y, Cullimore M, Safavi F, Zhang GX, Dittel BN, Rostami A. 2011. The encephalitogenicity of Tₕ17 cells is dependent on IL-1- and IL-23-induced production of the cytokine GM-CSF. *Nat Immunol* **12:** 568–575.

Elyaman W, Bradshaw EM, Uyttenhove C, Dardalhon V, Awasthi A, Imitola J, Bettelli E, Oukka M, van Snick J, Renauld JC, et al. 2009. IL-9 induces differentiation of Tₕ17 cells and enhances function of FoxP3⁺ natural regulatory T cells. *Proc Natl Acad Sci* **106:** 12885–12890.

Eming SA, Wynn TA, Martin P. 2017. Inflammation and metabolism in tissue repair and regeneration. *Science* **356:** 1026–1030.

Fischer A, Le Deist F, Hacein-Bey-Abina S, Andre-Schmutz I, Basile Gde S, de Villartay JP, Cavazzana-Calvo M. 2005. Severe combined immunodeficiency. A model disease for molecular immunology and therapy. *Immunol Rev* **203:** 98–109.

Fontenot JD, Gavin MA, Rudensky AY. 2003. Foxp3 programs the development and function of CD4⁺CD25⁺ regulatory T cells. *Nat Immunol* **4:** 330–336.

Fontenot JD, Rasmussen JP, Gavin MA, Rudensky AY. 2005. A function for interleukin 2 in Foxp3-expressing regulatory T cells. *Nat Immunol* **6:** 1142–1151.

Gessner A, Blum H, Rollinghoff M. 1993. Differential regulation of IL-9-expression after infection with Leishmania major in susceptible and resistant mice. *Immunobiology* **189:** 419–435.

Giliani S, Mori L, de Saint Basile G, Le Deist F, Rodriguez-Perez C, Forino C, Mazzolari E, Dupuis S, Elhasid R, Kessel A, et al. 2005. Interleukin-7 receptor α (IL-7Rα) deficiency: Cellular and molecular bases. Analysis of clinical, immunological, and molecular features in 16 novel patients. *Immunol Rev* **203:** 110–126.

Giri JG, Ahdieh M, Eisenman J, Shanebeck K, Grabstein K, Kumaki S, Namen A, Park LS, Cosman D, Anderson D. 1994. Utilization of the β and γ chains of the IL-2 receptor by the novel cytokine IL-15. *EMBO J* **13:** 2822–2830.

Gomez-Rodriguez J, Meylan F, Handon R, Hayes ET, Anderson SM, Kirby MR, Siegel RM, Schwartzberg PL. 2016. Itk is required for Th9 differentiation via TCR-mediated induction of IL-2 and IRF4. *Nat Commun* **7:** 10857.

Goodwin RG, Lupton S, Schmierer A, Hjerrild KJ, Jerzy R, Clevenger W, Gillis S, Cosman D, Namen AE. 1989. Human interleukin 7: Molecular cloning and growth factor activity on human and murine B-lineage cells. *Proc Natl Acad Sci* **86:** 302–306.

Goswami R, Kaplan MH. 2011. A brief history of IL-9. *J Immunol* **186:** 3283–3288.

Goswami R, Jabeen R, Yagi R, Pham D, Zhu J, Goenka S, Kaplan MH. 2012. STAT6-dependent regulation of Th9 development. *J Immunol* **188:** 968–975.

Grabstein KH, Eisenman J, Shanebeck K, Rauch C, Srinivasan S, Fung V, Beers C, Richardson J, Schoenborn MA, Ahdieh M, et al. 1994. Cloning of a T cell growth factor that interacts with the β chain of the interleukin-2 receptor. *Science* **264:** 965–968.

Grant CR, Liberal R, Mieli-Vergani G, Vergani D, Longhi MS. 2015. Regulatory T-cells in autoimmune diseases: Challenges, controversies and—yet—unanswered questions. *Autoimmun Rev* **14:** 105–116.

Grinberg-Bleyer Y, Baeyens A, You S, Elhage R, Fourcade G, Gregoire S, Cagnard N, Carpentier W, Tang Q, Bluestone J, et al. 2010. IL-2 reverses established type 1 diabetes in NOD mice by a local effect on pancreatic regulatory T cells. *J Exp Med* **207:** 1871–1878.

Hacein-Bey-Abina S, Le Deist F, Carlier F, Bouneaud C, Hue C, De Villartay JP, Thrasher AJ, Wulffraat N, Sorensen R, Dupuis-Girod S, et al. 2002. Sustained correction of X-linked severe combined immunodeficiency by ex vivo gene therapy. *N Engl J Med* **18:** 1185–1193.

Hartemann A, Bensimon G, Payan CA, Jacqueminet S, Bourron O, Nicolas N, Fonfrede M, Rosenzwajg M, Bernard C, Klatzmann D. 2013. Low-dose interleukin 2 in patients with type 1 diabetes: A phase 1/2 randomised, double-blind, placebo-controlled trial. *Lancet Diabetes Endocrinol* **1:** 295–305.

Hatakeyama M, Tsudo M, Minamoto S, Kono T, Doi T, Miyata T, Miyasaka M, Taniguchi T. 1989. Interleukin-2 receptor β chain gene: Generation of three receptor forms

Cite this article as *Cold Spring Harb Perspect Biol* doi: 10.1101/cshperspect.a028449

by cloned human α and β chain cDNA's. *Science* **244:** 551–556.

He YW, Malek TR. 1996. Interleukin-7 receptor α is essential for the development of γδ⁺ T cells, but not natural killer cells. *J Exp Med* **184:** 289–293.

He J, Zhang X, Wei Y, Sun X, Chen Y, Deng J, Jin Y, Gan Y, Hu X, Jia R, et al. 2016. Low-dose interleukin-2 treatment selectively modulates CD4⁺ T cell subsets in patients with systemic lupus erythematosus. *Nat Med* **22:** 991–993.

Herber D, Brown TP, Liang S, Young DA, Collins M, Dunussi-Joannopoulos K. 2007. IL-21 has a pathogenic role in a lupus-prone mouse model and its blockade with IL-21R.Fc reduces disease progression. *J Immunol* **178:** 3822–3830.

Herold KC, Vignali DA, Cooke A, Bluestone JA. 2013. Type 1 diabetes: Translating mechanistic observations into effective clinical outcomes. *Nat Rev Immunol* **13:** 243–256.

Hinrichs CS, Spolski R, Paulos CM, Gattinoni L, Kerstann KW, Palmer DC, Klebanoff CA, Rosenberg SA, Leonard WJ, Restifo NP. 2008. IL-2 and IL-21 confer opposing differentiation programs to CD8⁺ T cells for adoptive immunotherapy. *Blood* **111:** 5326–5333.

Horowitz A, Behrens RH, Okell L, Fooks AR, Riley EM. 2010. NK cells as effectors of acquired immune responses: Effector CD4⁺ T cell-dependent activation of NK cells following vaccination. *J Immunol* **185:** 2808–2818.

Hou J, Schindler U, Henzel WJ, Ho TC, Brasseur M, McKnight SL. 1994. An interleukin-4-induced transcription factor: IL-4 Stat. *Science* **265:** 1701–1706.

Howard M, Farrar J, Hilfiker M, Johnson B, Takatsu K, Hamaoka T, Paul WE. 1982. Identification of a T cell-derived B cell growth factor distinct from interleukin 2. *J Exp Med* **155:** 914–923.

Huang W, Ochs HD, Dupont B, Vyas YM. 2005. The Wiskott–Aldrich syndrome protein regulates nuclear translocation of NFAT2 and NF-κB (RelA) independently of its role in filamentous actin polymerization and actin cytoskeletal rearrangement. *J Immunol* **174:** 2602–2611.

Hultner L, Druez C, Moeller J, Uyttenhove C, Schmitt E, Rude E, Dormer P, Van Snick J. 1990. Mast cell growth-enhancing activity (MEA) is structurally related and functionally identical to the novel mouse T cell growth factor P40/TCGFIII (interleukin 9). *Eur J Immunol* **20:** 1413–1416.

Hurton LV, Singh H, Najjar AM, Switzer KC, Mi T, Maiti S, Olivares S, Rabinovich B, Huls H, Forget MA, et al. 2016. Tethered IL-15 augments antitumor activity and promotes a stem-cell memory subset in tumor-specific T cells. *Proc Natl Acad Sci* **113:** E7788–E7797.

International Multiple Sclerosis Genetics Consortium; Wellcome Trust Case Control Consortium; Sawcer S, Hellenthal G, Pirinen M, Spencer CC, Patsopoulos NA, Moutsianas L, Dilthey A, Su Z, et al. 2011. Genetic risk and a primary role for cell-mediated immune mechanisms in multiple sclerosis. *Nature* **476:** 214–219.

International Multiple Sclerosis Genetics Consortium; Beecham AH, Patsopoulos NA, Xifara DK, Davis MF, Kemppinen A, Cotsapas C, Shah TS, Spencer C, Booth D, et al. 2013. Analysis of immune-related loci identifies 48 new susceptibility variants for multiple sclerosis. *Nat Genet* **45:** 1353–1360.

Isakson PC, Pure E, Vitetta ES, Krammer PH. 1982. T cell-derived B cell differentiation factor(s). Effect on the isotype switch of murine B cells. *J Exp Med* **155:** 734–748.

Jandl C, Liu SM, Canete PF, Warren J, Hughes WE, Vogelzang A, Webster K, Craig ME, Uzel G, Dent A, et al. 2017. IL-21 restricts T follicular regulatory T cell proliferation through Bcl-6 mediated inhibition of responsiveness to IL-2. *Nat Commun* **8:** 14647.

Jensen CT, Kharazi S, Boiers C, Cheng M, Lubking A, Sitnicka E, Jacobsen SE. 2008. FLT3 ligand and not TSLP is the key regulator of IL-7-independent B-1 and B-2 B lymphopoiesis. *Blood* **112:** 2297–2304.

Jiang Q, Li WQ, Aiello FB, Mazzucchelli R, Asefa B, Khaled AR, Durum SK. 2005. Cell biology of IL-7, a key lymphotrophin. *Cytokine Growth Factor Rev* **16:** 513–533.

Jin H, Carrio R, Yu A, Malek TR. 2004. Distinct activation signals determine whether IL-21 induces B cell costimulation, growth arrest, or Bim-dependent apoptosis. *J Immunol* **173:** 657–665.

Johnston RJ, Poholek AC, DiToro D, Yusuf I, Eto D, Barnett B, Dent AL, Craft J, Crotty S. 2009. Bcl6 and Blimp-1 are reciprocal and antagonistic regulators of T follicular helper cell differentiation. *Science* **325:** 1006–1010.

Johnston RJ, Choi YS, Diamond JA, Yang JA, Crotty S. 2012. STAT5 is a potent negative regulator of T_FH cell differentiation. *J Exp Med* **209:** 243–250.

Jostins L, Ripke S, Weersma RK, Duerr RH, McGovern DP, Hui KY, Lee JC, Schumm LP, Sharma Y, Anderson CA, et al. 2012. Host–microbe interactions have shaped the genetic architecture of inflammatory bowel disease. *Nature* **491:** 119–124.

Jungel A, Distler JH, Kurowska-Stolarska M, Seemayer CA, Seibl R, Forster A, Michel BA, Gay RE, Emmrich F, Gay S, et al. 2004. Expression of interleukin-21 receptor, but not interleukin-21, in synovial fibroblasts and synovial macrophages of patients with rheumatoid arthritis. *Arthritis Rheum* **50:** 1468–1476.

Junttila IS, Creusot RJ, Moraga I, Bates DL, Wong MT, Alonso MN, Suhoski MM, Lupardus P, Meier-Schellersheim M, Engleman EG, et al. 2012. Redirecting cell-type specific cytokine responses with engineered interleukin-4 superkines. *Nat Chem Biol* **8:** 990–998.

Kaech SM, Tan JT, Wherry EJ, Konieczny BT, Surh CD, Ahmed R. 2003. Selective expression of the interleukin 7 receptor identifies effector CD8 T cells that give rise to long-lived memory cells. *Nat Immunol* **4:** 1191–1198.

Kalia V, Sarkar S, Subramaniam S, Haining WN, Smith KA, Ahmed R. 2010. Prolonged interleukin-2Rα expression on virus-specific CD8⁺ T cells favors terminal-effector differentiation in vivo. *Immunity* **32:** 91–103.

Kaplan MH, Hufford MM, Olson MR. 2015. The development and in vivo function of T helper 9 cells. *Nat Rev Immunol* **15:** 295–307.

Karrich JJ, Jachimowski LC, Nagasawa M, Kamp A, Balzarolo M, Wolkers MC, Uittenbogaart CH, Marieke van Ham S, Blom B. 2013. IL-21-stimulated human plasmacytoid dendritic cells secrete granzyme B, which impairs their capacity to induce T-cell proliferation. *Blood* **121:** 3103–3111.

Kashima N, Nishi-Takaoka C, Fujita T, Taki S, Yamada G, Hamuro J, Taniguchi T. 1985. Unique structure of murine interleukin-2 as deduced from cloned cDNAs. *Nature* **313:** 402–404.

Kawahara A, Minami Y, Miyazaki T, Ihle JN, Taniguchi T. 1995. Critical role of the interleukin 2 (IL-2) receptor γ-chain-associated Jak3 in the IL-2-induced *c-fos* and *c-myc*, but not *bcl-2*, gene induction. *Proc Natl Acad Sci* **92:** 8724–8728.

Kennedy MK, Glaccum M, Brown SN, Butz EA, Viney JL, Embers M, Matsuki N, Charrier K, Sedger L, Willis CR, et al. 2000. Reversible defects in natural killer and memory CD8 T cell lineages in interleukin 15-deficient mice. *J Exp Med* **191:** 771–780.

Kermouni A, Van Roost E, Arden KC, Vermeesch JR, Weiss S, Godelaine D, Flint J, Lurquin C, Szikora JP, Higgs DR, et al. 1995. The IL-9 receptor gene (*IL9R*): Genomic structure, chromosomal localization in the pseudoautosomal region of the long arm of the sex chromosomes, and identification of IL9R pseudogenes at 9qter, 10pter, 16pter, and 18pter. *Genomics* **29:** 371–382.

Khaled AR, Li WQ, Huang J, Fry TJ, Khaled AS, Mackall CL, Muegge K, Young HA, Durum SK. 2002. Bax deficiency partially corrects interleukin-7 receptor α deficiency. *Immunity* **17:** 561–573.

Kimura Y, Takeshita T, Kondo M, Ishii N, Nakamura M, Van Snick J, Sugamura K. 1995. Sharing of the IL-2 receptor γ chain with the functional IL-9 receptor complex. *Int Immunol* **7:** 115–120.

Kiniwa T, Enomoto Y, Terazawa N, Omi A, Miyata N, Ishiwata K, Miyajima A. 2016. NK cells activated by Interleukin-4 in cooperation with Interleukin-15 exhibit distinctive characteristics. *Proc Natl Acad Sci* **113:** 10139–10144.

Kitano M, Moriyama S, Ando Y, Hikida M, Mori Y, Kurosaki T, Okada T. 2011. Bcl6 protein expression shapes pregerminal center B cell dynamics and follicular helper T cell heterogeneity. *Immunity* **34:** 961–972.

Kondo M, Takeshita T, Ishii N, Nakamura M, Watanabe S, Arai K, Sugamura K. 1993. Sharing of the interleukin-2 (IL-2) receptor γ chain between receptors for IL-2 and IL-4. *Science* **262:** 1874–1877.

Kondo M, Takeshita T, Higuchi M, Nakamura M, Sudo T, Nishikawa S, Sugamura K. 1994. Functional participation of the IL-2 receptor γ chain in IL-7 receptor complexes. *Science* **263:** 1453–1454.

Koreth J, Matsuoka K, Kim HT, McDonough SM, Bindra B, Alyea EP 3rd, Armand P, Cutler C, Ho VT, Treister NS, et al. 2011. Interleukin-2 and regulatory T cells in graft-versus-host disease. *N Engl J Med* **365:** 2055–2066.

Korn T, Bettelli E, Gao W, Awasthi A, Jager A, Strom TB, Oukka M, Kuchroo VK. 2007. IL-21 initiates an alternative pathway to induce proinflammatory T$_H$17 cells. *Nature* **448:** 484–487.

Kotlarz D, Zietara N, Uzel G, Weidemann T, Braun CJ, Diestelhorst J, Krawitz PM, Robinson PN, Hecht J, Puchalka J, et al. 2013. Loss-of-function mutations in the IL-21 receptor gene cause a primary immunodeficiency syndrome. *J Exp Med* **210:** 433–443.

Kritikou JS, Dahlberg CI, Baptista MA, Wagner AK, Banerjee PP, Gwalani LA, Poli C, Panda SK, Karre K, Kaech SM, et al. 2016. IL-2 in the tumor microenvironment is necessary for Wiskott–Aldrich syndrome protein deficient NK cells to respond to tumors in vivo. *Sci Rep* **6:** 30636.

Ku CC, Murakami M, Sakamoto A, Kappler J, Marrack P. 2000. Control of homeostasis of CD8$^+$ memory T cells by opposing cytokines. *Science* **288:** 675–678.

Kuhn R, Rajewsky K, Muller W. 1991. Generation and analysis of interleukin-4 deficient mice. *Science* **254:** 707–710.

Kuperman D, Schofield B, Wills-Karp M, Grusby MJ. 1998. Signal transducer and activator of transcription factor 6 (Stat6)-deficient mice are protected from antigen-induced airway hyperresponsiveness and mucus production. *J Exp Med* **187:** 939–948.

Kwon H, Thierry-Mieg D, Thierry-Mieg J, Kim HP, Oh J, Tunyaplin C, Carotta S, Donovan CE, Goldman ML, Tailor P, et al. 2009. Analysis of interleukin-21-induced Prdm1 gene regulation reveals functional cooperation of STAT3 and IRF4 transcription factors. *Immunity* **31:** 941–952.

LaPorte SL, Juo ZS, Vaclavikova J, Colf LA, Qi X, Heller NM, Keegan AD, Garcia KC. 2008. Molecular and structural basis of cytokine receptor pleiotropy in the interleukin-4/13 system. *Cell* **132:** 259–272.

Laurence A, Tato CM, Davidson TS, Kanno Y, Chen Z, Yao Z, Blank RB, Meylan F, Siegel R, Hennighausen L, et al. 2007. Interleukin-2 signaling via STAT5 constrains T helper 17 cell generation. *Immunity* **26:** 371–381.

Lauwerys BR, Garot N, Renauld JC, Houssiau FA. 2000. Cytokine production and killer activity of NK/T-NK cells derived with IL-2, IL-15, or the combination of IL-12 and IL-18. *J Immunol* **165:** 1847–1853.

Lazarevic V, Chen X, Shim JH, Hwang ES, Jang E, Bolm AN, Oukka M, Kuchroo VK, Glimcher LH. 2011. T-bet represses T$_H$17 differentiation by preventing Runx1-mediated activation of the gene encoding RORγt. *Nat Immunol* **12:** 96–104.

Lee LF, Logronio K, Tu GH, Zhai W, Ni I, Mei L, Dilley J, Yu J, Rajpal A, Brown C, et al. 2012. Anti-IL-7 receptor-α reverses established type 1 diabetes in nonobese diabetic mice by modulating effector T-cell function. *Proc Natl Acad Sci* **109:** 12674–12679.

Leipe J, Grunke M, Dechant C, Reindl C, Kerzendorf U, Schulze-Koops H, Skapenko A. 2010. Role of Th17 cells in human autoimmune arthritis. *Arthritis Rheum* **62:** 2876–2885.

Lenardo MJ. 1991. Interleukin-2 programs mouse αβ T lymphocytes for apoptosis. *Nature* **353:** 858–861.

Leonard WJ. 1996. The molecular basis of X-linked severe combined immunodeficiency: Defective cytokine receptor signaling. *Annu Rev Med* **47:** 229–239.

Leonard WJ. 2001. Cytokines and immunodeficiency diseases. *Nat Rev Immunol* **1:** 200–208.

Leonard WJ, O'Shea JJ. 1998. Jaks and STATs: Biological implications. *Annu Rev Immunol* **16:** 293–322.

Leonard WJ, Depper JM, Crabtree GR, Rudikoff S, Pumphrey J, Robb RJ, Kronke M, Svetlik PB, Peffer NJ, Waldmann TA, et al. 1984. Molecular cloning and expression of cDNAs for the human interleukin-2 receptor. *Nature* **311:** 626–631.

Leonard WJ, Kronke M, Peffer NJ, Depper JM, Greene WC. 1985. Interleukin 2 receptor gene expression in normal human T lymphocytes. *Proc Natl Acad Sci* **82:** 6281–6285.

Levin AM, Bates DL, Ring AM, Krieg C, Lin JT, Su L, Moraga I, Raeber ME, Bowman GR, Novick P, et al. 2012. Exploiting a natural conformational switch to engineer an interleukin-2 "superkine." *Nature* **484**: 529–533.

Li MO, Rudensky AY. 2016. T cell receptor signalling in the control of regulatory T cell differentiation and function. *Nat Rev Immunol* **16**: 220–233.

Li P, Spolski R, Liao W, Wang L, Murphy TL, Murphy KM, Leonard WJ. 2012. BATF-JUN is critical for IRF4-mediated transcription in T cells. *Nature* **490**: 543–546.

Li J, Lu E, Yi T, Cyster JG. 2016. EBI2 augments Tfh cell fate by promoting interaction with IL-2-quenching dendritic cells. *Nature* **533**: 110–114.

Liao W, Schones DE, Oh J, Cui Y, Cui K, Roh TY, Zhao K, Leonard WJ. 2008. Priming for T helper type 2 differentiation by interleukin 2-mediated induction of interleukin 4 receptor α-chain expression. *Nat Immunol* **9**: 1288–1296.

Liao W, Lin JX, Wang L, Li P, Leonard WJ. 2011. Modulation of cytokine receptors by IL-2 broadly regulates differentiation into helper T cell lineages. *Nat Immunol* **12**: 551–559.

Liao W, Lin JX, Leonard WJ. 2013. Interleukin-2 at the crossroads of effector responses, tolerance, and immunotherapy. *Immunity* **38**: 13–25.

Liao W, Spolski R, Li P, Du N, West EE, Ren M, Mitra S, Leonard WJ. 2014. Opposing actions of IL-2 and IL-21 on Th9 differentiation correlate with their differential regulation of BCL6 expression. *Proc Natl Acad Sci* **111**: 3508–3513.

Licona-Limon P, Henao-Mejia J, Temann AU, Gagliani N, Licona-Limon I, Ishigame H, Hao L, Herbert DR, Flavell RA. 2013. Th9 cells drive host immunity against gastrointestinal worm infection. *Immunity* **39**: 744–757.

Lin JX, Migone TS, Tsang M, Friedmann M, Weatherbee JA, Zhou L, Yamauchi A, Bloom ET, Mietz J, John S, et al. 1995. The role of shared receptor motifs and common Stat proteins in the generation of cytokine pleiotropy and redundancy by IL-2, IL-4, IL-7, IL-13, and IL-15. *Immunity* **2**: 331–339.

Lin JX, Li P, Liu D, Jin HT, He J, Ata Ur Rasheed M, Rochman Y, Wang L, Cui K, Liu C, et al. 2012. Critical role of STAT5 transcription factor tetramerization for cytokine responses and normal immune function. *Immunity* **36**: 586–599.

Linehan LA, Warren WD, Thompson PA, Grusby MJ, Berton MT. 1998. STAT6 is required for IL-4-induced germline Ig gene transcription and switch recombination. *J Immunol* **161**: 302–310.

Linterman MA, Pierson W, Lee SK, Kallies A, Kawamoto S, Rayner TF, Srivastava M, Divekar DP, Beaton L, Hogan JJ, et al. 2011. Foxp3+ follicular regulatory T cells control the germinal center response. *Nat Med* **17**: 975–982.

Liu X, Lee YS, Yu CR, Egwuagu CE. 2008. Loss of STAT3 in CD4+ T cells prevents development of experimental autoimmune diseases. *J Immunol* **180**: 6070–6076.

Lodolce JP, Boone DL, Chai S, Swain RE, Dassopoulos T, Trettin S, Ma A. 1998. IL-15 receptor maintains lymphoid homeostasis by supporting lymphocyte homing and proliferation. *Immunity* **9**: 669–676.

Lorenzen I, Dingley AJ, Jacques Y, Grotzinger J. 2006. The structure of the interleukin-15α receptor and its implications for ligand binding. *J Biol Chem* **281**: 6642–6647.

Lu Y, Hong S, Li H, Park J, Hong B, Wang L, Zheng Y, Liu Z, Xu J, He J, et al. 2012. Th9 cells promote antitumor immune responses in vivo. *J Clin Invest* **122**: 4160–4171.

Macchi P, Villa A, Giliani S, Sacco MG, Frattini A, Porta F, Ugazio AG, Johnston JA, Candotti F, O'Shea JJ, et al. 1995. Mutations of Jak-3 gene in patients with autosomal severe combined immune deficiency (SCID). *Nature* **377**: 65–68.

Maki K, Sunaga S, Komagata Y, Kodaira Y, Mabuchi A, Karasuyama H, Yokomuro K, Miyazaki JI, Ikuta K. 1996. Interleukin 7 receptor-deficient mice lack γδ T cells. *Proc Natl Acad Sci* **93**: 7172–7177.

Malek TR, Castro I. 2010. Interleukin-2 receptor signaling: At the interface between tolerance and immunity. *Immunity* **33**: 153–165.

Malek TR, Porter BO, Codias EK, Scibelli P, Yu A. 2000. Normal lymphoid homeostasis and lack of lethal autoimmunity in mice containing mature T cells with severely impaired IL-2 receptors. *J Immunol* **164**: 2905–2914.

Malek TR, Yu A, Vincek V, Scibelli P, Kong L. 2002. CD4 regulatory T cells prevent lethal autoimmunity in IL-2Rβ-deficient mice. Implications for the nonredundant function of IL-2. *Immunity* **17**: 167–178.

Malin S, McManus S, Cobaleda C, Novatchkova M, Delogu A, Bouillet P, Strasser A, Busslinger M. 2010. Role of STAT5 in controlling cell survival and immunoglobulin gene recombination during pro-B cell development. *Nat Immunol* **11**: 171–179.

Maraskovsky E, Peschon JJ, McKenna H, Teepe M, Strasser A. 1998. Overexpression of Bcl-2 does not rescue impaired B lymphopoiesis in IL-7 receptor-deficient mice but can enhance survival of mature B cells. *Int Immunol* **10**: 1367–1375.

Marks-Konczalik J, Dubois S, Losi JM, Sabzevari H, Yamada N, Feigenbaum L, Waldmann TA, Tagaya Y. 2000. IL-2-induced activation-induced cell death is inhibited in IL-15 transgenic mice. *Proc Natl Acad Sci* **97**: 11445–11450.

Mayack SR, Berg LJ. 2006. Cutting edge: An alternative pathway of CD4+ T cell differentiation is induced following activation in the absence of γ-chain-dependent cytokine signals. *J Immunol* **176**: 2059–2063.

Mazzucchelli R, Durum SK. 2007. Interleukin-7 receptor expression: Intelligent design. *Nat Rev Immunol* **7**: 144–154.

McDonald PW, Read KA, Baker CE, Anderson AE, Powell MD, Ballesteros-Tato A, Oestreich KJ. 2016. IL-7 signalling represses Bcl-6 and the T_FH gene program. *Nat Commun* **7**: 10285.

McGeachy MJ, Bak-Jensen KS, Chen Y, Tato CM, Blumenschein W, McClanahan T, Cua DJ. 2007. TGF-β and IL-6 drive the production of IL-17 and IL-10 by T cells and restrain T_H-17 cell-mediated pathology. *Nat Immunol* **8**: 1390–1397.

McGuire HM, Vogelzang A, Hill N, Flodstrom-Tullberg M, Sprent J, King C. 2009. Loss of parity between IL-2 and IL-21 in the NOD Idd3 locus. *Proc Natl Acad Sci* **106**: 19438–19443.

Mehta DS, Wurster AL, Whitters MJ, Young DA, Collins M, Grusby MJ. 2003. IL-21 induces the apoptosis of resting and activated primary B cells. *J Immunol* **170**: 4111–4118.

Melen E, Gullsten H, Zucchelli M, Lindstedt A, Nyberg F, Wickman M, Pershagen G, Kere J. 2004. Sex specific protective effects of interleukin-9 receptor haplotypes on childhood wheezing and sensitisation. *J Med Genet* **41**: e123.

Meyer DM, Jesson MI, Li X, Elrick MM, Funckes-Shippy CL, Warner JD, Gross CJ, Dowty ME, Ramaiah SK, Hirsch JL, et al. 2010. Anti-inflammatory activity and neutrophil reductions mediated by the JAK1/JAK3 inhibitor, CP-690,550, in rat adjuvant-induced arthritis. *J Inflamm (Lond)* **7**: 41.

Mitoma H, Horiuchi T, Kimoto Y, Tsukamoto H, Uchino A, Tamimoto Y, Miyagi Y, Harada M. 2005. Decreased expression of interleukin-21 receptor on peripheral B lymphocytes in systemic lupus erythematosus. *Int J Mol Med* **16**: 609–615.

Mitra S, Ring AM, Amarnath S, Spangler JB, Li P, Ju W, Fischer S, Oh J, Spolski R, Weiskopf K, et al. 2015. Interleukin-2 activity can be fine tuned with engineered receptor signaling clamps. *Immunity* **42**: 826–838.

Mizel SB, Farrar JJ. 1979. Revised nomenclature for antigen-nonspecific T-cell proliferation and helper factors. *Cell Immunol* **48**: 433–436.

Mohamadzadeh M, Berard F, Essert G, Chalouni C, Pulendran B, Davoust J, Bridges S, Palucka AK, Banchereau J. 2001. Interleukin 15 skews monocyte differentiation into dendritic cells with features of Langerhans cells. *J Exp Med* **194**: 1013–1020.

Montandon R, Korniotis S, Layseca-Espinosa E, Gras C, Megret J, Ezine S, Dy M, Zavala F. 2013. Innate pro-B-cell progenitors protect against type 1 diabetes by regulating autoimmune effector T cells. *Proc Natl Acad Sci* **110**: E2199–E2208.

Moore TA, von Freeden-Jeffry U, Murray R, Zlotnik A. 1996. Inhibition of $\gamma\delta$ T cell development and early thymocyte maturation in IL-7$^{-/-}$ mice. *J Immunol* **157**: 2366–2373.

Morgan DA, Ruscetti FW, Gallo R. 1976. Selective in vitro growth of T lymphocytes from normal human bone marrows. *Science* **193**: 1007–1008.

Murakami M, Sakamoto A, Bender J, Kappler J, Marrack P. 2002. CD25$^+$CD4$^+$ T cells contribute to the control of memory CD8$^+$ T cells. *Proc Natl Acad Sci* **99**: 8832–8837.

Nakamura Y, Russell SM, Mess SA, Friedmann M, Erdos M, Francois C, Jacques Y, Adelstein S, Leonard WJ. 1994. Heterodimerization of the IL-2 receptor β- and γ-chain cytoplasmic domains is required for signalling. *Nature* **369**: 330–333.

Namen AE, Lupton S, Hjerrild K, Wignall J, Mochizuki DY, Schmierer A, Mosley B, March CJ, Urdal D, Gillis S. 1988. Stimulation of B-cell progenitors by cloned murine interleukin-7. *Nature* **333**: 571–573.

Nelms K, Keegan AD, Zamorano J, Ryan JJ, Paul WE. 1999. The IL-4 receptor: Signaling mechanisms and biologic functions. *Annu Rev Immunol* **17**: 701–738.

Nelson BH, Lord JD, Greenberg PD. 1994. Cytoplasmic domains of the interleukin-2 receptor β and γ chains mediate the signal for T-cell proliferation. *Nature* **369**: 333–336.

Nikaido T, Shimizu A, Ishida N, Sabe H, Teshigawara K, Maeda M, Uchiyama T, Yodoi J, Honjo T. 1984. Molecular cloning of cDNA encoding human interleukin-2 receptor. *Nature* **311**: 631–635.

Niu X, He D, Zhang X, Yue T, Li N, Zhang JZ, Dong C, Chen G. 2010. IL-21 regulates Th17 cells in rheumatoid arthritis. *Hum Immunol* **71**: 334–341.

Noben-Trauth N, Shultz LD, Brombacher F, Urban JF Jr, Gu H, Paul WE. 1997. An interleukin 4 (IL-4)-independent pathway for CD4$^+$ T cell IL-4 production is revealed in IL-4 receptor-deficient mice. *Proc Natl Acad Sci* **94**: 10838–10843.

Noguchi M, Nakamura Y, Russell SM, Ziegler SF, Tsang M, Cao X, Leonard WJ. 1993a. Interleukin-2 receptor γ chain: A functional component of the interleukin-7 receptor. *Science* **262**: 1877–1880.

Noguchi M, Yi H, Rosenblatt HM, Filipovich AH, Adelstein S, Modi WS, McBride OW, Leonard WJ. 1993b. Interleukin-2 receptor γ chain mutation results in X-linked severe combined immunodeficiency in humans. *Cell* **73**: 147–157.

Nosaka T, van Deursen JM, Tripp RA, Thierfelder WE, Witthuhn BA, McMickle AP, Doherty PC, Grosveld GC, Ihle JN. 1995. Defective lymphoid development in mice lacking Jak3. *Science* **270**: 800–802.

Nowak EC, Weaver CT, Turner H, Begum-Haque S, Becher B, Schreiner B, Coyle AJ, Kasper LH, Noelle RJ. 2009. IL-9 as a mediator of Th17-driven inflammatory disease. *J Exp Med* **206**: 1653–1660.

Nurieva R, Yang XO, Martinez G, Zhang Y, Panopoulos AD, Ma L, Schluns K, Tian Q, Watowich SS, Jetten AM, et al. 2007. Essential autocrine regulation by IL-21 in the generation of inflammatory T cells. *Nature* **448**: 480–483.

Nurieva RI, Chung Y, Hwang D, Yang XO, Kang HS, Ma L, Wang YH, Watowich SS, Jetten AM, Tian Q, et al. 2008. Generation of T follicular helper cells is mediated by interleukin-21 but independent of T helper 1, 2, or 17 cell lineages. *Immunity* **29**: 138–149.

Oestreich KJ, Mohn SE, Weinmann AS. 2012. Molecular mechanisms that control the expression and activity of Bcl-6 in T_H1 cells to regulate flexibility with a T_{FH}-like gene profile. *Nat Immunol* **13**: 405–411.

Ohara J, Paul WE. 1987. Receptors for B-cell stimulatory factor-1 expressed on cells of haematopoietic lineage. *Nature* **325**: 537–540.

Ohkura N, Kitagawa Y, Sakaguchi S. 2013. Development and maintenance of regulatory T cells. *Immunity* **38**: 414–423.

Opferman JT, Letai A, Beard C, Sorcinelli MD, Ong CC, Korsmeyer SJ. 2003. Development and maintenance of B and T lymphocytes requires antiapoptotic MCL-1. *Nature* **426**: 671–676.

Orange JS, Roy-Ghanta S, Mace EM, Maru S, Rak GD, Sanborn KB, Fasth A, Saltzman R, Paisley A, Monaco-Shawver L, et al. 2011. IL-2 induces a WAVE2-dependent pathway for actin reorganization that enables WASp-independent human NK cell function. *J Clin Invest* **121**: 1535–1548.

Ozaki K, Kikly K, Michalovich D, Young PR, Leonard WJ. 2000. Cloning of a type I cytokine receptor most related to the IL-2 receptor β chain. *Proc Natl Acad Sci* **97**: 11439–11444.

Ozaki K, Spolski R, Feng CG, Qi CF, Cheng J, Sher A, Morse HC III, Liu C, Schwartzberg PL, Leonard WJ. 2002. A critical role for IL-21 in regulating immunoglobulin production. *Science* **298**: 1630–1634.

Cite this article as *Cold Spring Harb Perspect Biol* doi: 10.1101/cshperspect.a028449

Ozaki K, Spolski R, Ettinger R, Kim HP, Wang G, Qi CF, Hwu P, Shaffer DJ, Akilesh S, Roopenian DC, et al. 2004. Regulation of B cell differentiation and plasma cell generation by IL-21, a novel inducer of Blimp-1 and Bcl-6. *J Immunol* **173:** 5361–5371.

Pahwa R, Chatila T, Pahwa S, Paradise C, Day NK, Geha R, Schwartz SA, Slade H, Oyaizu N, Good RA. 1989. Recombinant interleukin 2 therapy in severe combined immunodeficiency disease. *Proc Natl Acad Sci* **86:** 5069–5073.

Pai SY, Truitt ML, Ho IC. 2004. GATA-3 deficiency abrogates the development and maintenance of T helper type 2 cells. *Proc Natl Acad Sci* **101:** 1993–1998.

Pandey A, Ozaki K, Baumann H, Levin SD, Puel A, Farr AG, Ziegler SF, Leonard WJ, Lodish HF. 2000. Cloning of a receptor subunit required for signaling by thymic stromal lymphopoietin. *Nat Immunol* **1:** 59–64.

Pandiyan P, Yang XP, Saravanamuthu SS, Zheng L, Ishihara S, O'Shea JJ, Lenardo MJ. 2012. The role of IL-15 in activating STAT5 and fine-tuning IL-17A production in CD4 T lymphocytes. *J Immunol* **189:** 4237–4246.

Park LS, Martin U, Garka K, Gliniak B, Di Santo JP, Muller W, Largaespada DA, Copeland NG, Jenkins NA, Farr AG, et al. 2000. Cloning of the murine thymic stromal lymphopoietin (TSLP) receptor: Formation of a functional heteromeric complex requires interleukin 7 receptor. *J Exp Med* **192:** 659–670.

Parrish-Novak J, Dillon SR, Nelson A, Hammond A, Sprecher C, Gross JA, Johnston J, Madden K, Xu W, West J, et al. 2000. Interleukin 21 and its receptor are involved in NK cell expansion and regulation of lymphocyte function. *Nature* **408:** 57–63.

Parrish-Novak J, Foster DC, Holly RD, Clegg CH. 2002. Interleukin-21 and the IL-21 receptor: Novel effectors of NK and T cell responses. *J Leukoc Biol* **72:** 856–863.

Paul WE. 2015. History of interleukin-4. *Cytokine* **75:** 3–7.

Pene J, Gauchat JF, Lecart S, Drouet E, Guglielmi P, Boulay V, Delwail A, Foster D, Lecron JC, Yssel H. 2004. Cutting edge: IL-21 is a switch factor for the production of IgG1 and IgG3 by human B cells. *J Immunol* **172:** 5154–5157.

Perol L, Lindner JM, Caudana P, Nunez NG, Baeyens A, Valle A, Sedlik C, Loirat D, Boyer O, Creange A, et al. 2016. Loss of immune tolerance to IL-2 in type 1 diabetes. *Nat Commun* **7:** 13027.

Peschon JJ, Morrissey PJ, Grabstein KH, Ramsdell FJ, Maraskovsky E, Gliniak BC, Park LS, Ziegler SF, Williams DE, Ware CB, et al. 1994. Early lymphocyte expansion is severely impaired in interleukin 7 receptor-deficient mice. *J Exp Med* **180:** 1955–1960.

Petit-Frere C, Dugas B, Braquet P, Mencia-Huerta JM. 1993. Interleukin-9 potentiates the interleukin-4-induced IgE and IgG1 release from murine B lymphocytes. *Immunology* **79:** 146–151.

Pfeifle R, Rothe T, Ipseiz N, Scherer HU, Culemann S, Harre U, Ackermann JA, Seefried M, Kleyer A, Uderhardt S, et al. 2017. Regulation of autoantibody activity by the IL-23-T$_H$17 axis determines the onset of autoimmune disease. *Nat Immunol* **18:** 104–113.

Pipkin ME, Sacks JA, Cruz-Guilloty F, Lichtenheld MG, Bevan MJ, Rao A. 2010. Interleukin-2 and inflammation induce distinct transcriptional programs that promote the differentiation of effector cytolytic T cells. *Immunity* **32:** 79–90.

Pot C, Jin H, Awasthi A, Liu SM, Lai CY, Madan R, Sharpe AH, Karp CL, Miaw SC, Ho IC, et al. 2009. Cutting edge: IL-27 induces the transcription factor c-Maf, cytokine IL-21, and the costimulatory receptor ICOS that coordinately act together to promote differentiation of IL-10-producing Tr1 cells. *J Immunol* **183:** 797–801.

Puel A, Ziegler SF, Buckley RH, Leonard WJ. 1998. Defective *IL7R* expression in T$^-$B$^+$NK$^+$ severe combined immunodeficiency. *Nat Genet* **20:** 394–397.

Pulendran B. 2015. The varieties of immunological experience: Of pathogens, stress, and dendritic cells. *Annu Rev Immunol* **33:** 563–606.

Pulliam SR, Uzhachenko RV, Adunyah SE, Shanker A. 2016. Common γ chain cytokines in combinatorial immune strategies against cancer. *Immunol Lett* **169:** 61–72.

Purwar R, Schlapbach C, Xiao S, Kang HS, Elyaman W, Jiang X, Jetten AM, Khoury SJ, Fuhlbrigge RC, Kuchroo VK, et al. 2012. Robust tumor immunity to melanoma mediated by interleukin-9-producing T cells. *Nat Med* **18:** 1248–1253.

Qi H. 2016. T follicular helper cells in space-time. *Nat Rev Immunol* **16:** 612–625.

Queen C, Schneider WP, Selick HE, Payne PW, Landolfi NF, Duncan JF, Avdalovic NM, Levitt M, Junghans RP, Waldmann TA. 1989. A humanized antibody that binds to the interleukin 2 receptor. *Proc Natl Acad Sci* **86:** 10029–10033.

Quelle FW, Shimoda K, Thierfelder W, Fischer C, Kim A, Ruben SM, Cleveland JL, Pierce JH, Keegan AD, Nelms K, et al. 1995. Cloning of murine Stat6 and human Stat6, Stat proteins that are tyrosine phosphorylated in responses to IL-4 and IL-3 but are not required for mitogenesis. *Mol Cell Biol* **15:** 3336–3343.

Quigley M, Pereyra F, Nilsson B, Porichis F, Fonseca C, Eichbaum Q, Julg B, Jesneck JL, Brosnahan K, Imam S, et al. 2010. Transcriptional analysis of HIV-specific CD8$^+$ T cells shows that PD-1 inhibits T cell function by upregulating BATF. *Nat Med* **16:** 1147–1151.

Rao DA, Gurish MF, Marshall JL, Slowikowski K, Fonseka CY, Liu Y, Donlin LT, Henderson LA, Wei K, Mizoguchi F, et al. 2017. Pathologically expanded peripheral T helper cell subset drives B cells in rheumatoid arthritis. *Nature* **542:** 110–114.

Rasheed MA, Latner DR, Aubert RD, Gourley T, Spolski R, Davis CW, Langley WA, Ha SJ, Ye L, Sarkar S, et al. 2013. Interleukin-21 is a critical cytokine for the generation of virus-specific long-lived plasma cells. *J Virol* **87:** 7737–7746.

Rasmussen TK, Andersen T, Hvid M, Hetland ML, Horslev-Petersen K, Stengaard-Pedersen K, Holm CK, Deleuran B. 2010. Increased interleukin 21 (IL-21) and IL-23 are associated with increased disease activity and with radiographic status in patients with early rheumatoid arthritis. *J Rheumatol* **37:** 2014–2020.

Rathmell JC, Farkash EA, Gao W, Thompson CB. 2001. IL-7 enhances the survival and maintains the size of naïve T cells. *J Immunol* **167:** 6869–6876.

Ray JP, Staron MM, Shyer JA, Ho PC, Marshall HD, Gray SM, Laidlaw BJ, Araki K, Ahmed R, Kaech SM, et al. 2015. The interleukin-2-mTORc1 kinase axis defines the signaling, differentiation, and metabolism of T helper 1 and follicular B helper T cells. *Immunity* **43:** 690–702.

Recher M, Berglund LJ, Avery DT, Cowan MJ, Gennery AR, Smart J, Peake J, Wong M, Pai SY, Baxi S, et al. 2011. IL-21 is the primary common γ chain-binding cytokine required for human B-cell differentiation in vivo. *Blood* **118**: 6824–6835.

Rickert M, Wang X, Boulanger MJ, Goriatcheva N, Garcia KC. 2005. The structure of interleukin-2 complexed with its α receptor. *Science* **308**: 1477–1480.

Rochman Y, Spolski R, Leonard WJ. 2009. New insights into the regulation of T cells by γc family cytokines. *Nat Rev Immunol* **9**: 480–490.

Rolling C, Treton D, Pellegrini S, Galanaud P, Richard Y. 1996. IL4 and IL13 receptors share the γc chain and activate STAT6, STAT3 and STAT5 proteins in normal human B cells. *FEBS Lett* **393**: 53–56.

Rosenberg SA. 2014. IL-2: The first effective immunotherapy for human cancer. *J Immunol* **192**: 5451–5458.

Ross SH, Rollings C, Anderson KE, Hawkins PT, Stephens LR, Cantrell DA. 2016. Phosphoproteomic analyses of interleukin 2 signaling reveal integrated JAK kinase-dependent and -independent networks in CD8+ T cells. *Immunity* **45**: 685–700.

Russell SM, Keegan AD, Harada N, Nakamura Y, Noguchi M, Leland P, Friedmann MC, Miyajima A, Puri RK, Paul WE, et al. 1993. Interleukin-2 receptor γ chain: A functional component of the interleukin-4 receptor. *Science* **262**: 1880–1883.

Russell SM, Johnston JA, Noguchi M, Kawamura M, Bacon CM, Friedmann M, Berg M, McVicar DW, Witthuhn BA, Silvennoinen O, et al. 1994. Interaction of IL-2R β and γc chains with Jak1 and Jak3: Implications for XSCID and XCID. *Science* **266**: 1042–1045.

Russell SM, Tayebi N, Nakajima H, Riedy MC, Roberts JL, Aman MJ, Migone TS, Noguchi M, Markert ML, Buckley RH, et al. 1995. Mutation of Jak3 in a patient with SCID: Essential role of Jak3 in lymphoid development. *Science* **270**: 797–800.

Sadlack B, Merz H, Schorle H, Schimpl A, Feller AC, Horak I. 1993. Ulcerative colitis-like disease in mice with a disrupted interleukin-2 gene. *Cell* **75**: 253–261.

Sage PT, Sharpe AH. 2015. T follicular regulatory cells in the regulation of B cell responses. *Trends Immunol* **36**: 410–418.

Sage PT, Sharpe AH. 2016. T follicular regulatory cells. *Immunol Rev* **271**: 246–259.

Sage PT, Ron-Harel N, Juneja VR, Sen DR, Maleri S, Sungnak W, Kuchroo VK, Haining WN, Chevrier N, Haigis M, et al. 2016. Suppression by TFR cells leads to durable and selective inhibition of B cell effector function. *Nat Immunol* **17**: 1436–1446.

Saikh KU, Khan AS, Kissner T, Ulrich RG. 2001. IL-15-induced conversion of monocytes to mature dendritic cells. *Clin Exp Immunol* **126**: 447–455.

Saito Y, Tada H, Sabe H, Honjo T. 1991. Biochemical evidence for a third chain of the interleukin-2 receptor. *J Biol Chem* **266**: 22186–22191.

Sakaguchi S, Sakaguchi N, Asano M, Itoh M, Toda M. 1995. Immunologic self-tolerance maintained by activated T cells expressing IL-2 receptor α-chains (CD25). Breakdown of a single mechanism of self-tolerance causes various autoimmune diseases. *J Immunol* **155**: 1151–1164.

Salzer E, Kansu A, Sic H, Majek P, Ikinciogullari A, Dogu FE, Prengemann NK, Santos-Valente E, Pickl WF, Bilic I, et al. 2014. Early-onset inflammatory bowel disease and common variable immunodeficiency-like disease caused by IL-21 deficiency. *J Allergy Clin Immunol* **133**: 1651–1659 e1612.

Sandborn WJ, Su C, Sands BE, D'Haens GR, Vermeire S, Schreiber S, Danese S, Feagan BG, Reinisch W, Niezychowski W, et al. 2017. Tofacitinib as induction and maintenance therapy for ulcerative colitis. *N Engl J Med* **376**: 1723–1736.

Sawalha AH, Kaufman KM, Kelly JA, Adler AJ, Aberle T, Kilpatrick J, Wakeland EK, Li QZ, Wandstrat AE, Karp DR, et al. 2008. Genetic association of interleukin-21 polymorphisms with systemic lupus erythematosus. *Ann Rheum Dis* **67**: 458–461.

Schlissel MS, Durum SD, Muegge K. 2000. The interleukin 7 receptor is required for T cell receptor γ locus accessibility to the V(D)J recombinase. *J Exp Med* **191**: 1045–1050.

Schluns KS, Kieper WC, Jameson SC, Lefrancois L. 2000. Interleukin-7 mediates the homeostasis of naïve and memory CD8 T cells in vivo. *Nat Immunol* **1**: 426–432.

Schmitt E, Van Brandwijk R, Van Snick J, Siebold B, Rude E. 1989. TCGF III/P40 is produced by naïve murine CD4+ T cells but is not a general T cell growth factor. *Eur J Immunol* **19**: 2167–2170.

Schmitt E, Germann T, Goedert S, Hoehn P, Huels C, Koelsch S, Kuhn R, Muller W, Palm N, Rude E. 1994. IL-9 production of naïve CD4+ T cells depends on IL-2, is synergistically enhanced by a combination of TGF-β and IL-4, and is inhibited by IFN-γ. *J Immunol* **153**: 3989–3996.

Schorle H, Holtschke T, Hunig T, Schimpl A, Horak I. 1991. Development and function of T cells in mice rendered interleukin-2 deficient by gene targeting. *Nature* **352**: 621–624.

Setoguchi R, Hori S, Takahashi T, Sakaguchi S. 2005. Homeostatic maintenance of natural Foxp3+ CD25+ CD4+ regulatory T cells by interleukin (IL)-2 and induction of autoimmune disease by IL-2 neutralization. *J Exp Med* **201**: 723–735.

Shaffer AL, Lin KI, Kuo TC, Yu X, Hurt EM, Rosenwald A, Giltnane JM, Yang L, Zhao H, Calame K, et al. 2002. Blimp-1 orchestrates plasma cell differentiation by extinguishing the mature B cell gene expression program. *Immunity* **17**: 51–62.

Sharon M, Klausner RD, Cullen BR, Chizzonite R, Leonard WJ. 1986. Novel interleukin-2 receptor subunit detected by cross-linking under high-affinity conditions. *Science* **234**: 859–863.

Sheng W, Yang F, Zhou Y, Yang H, Low PY, Kemeny DM, Tan P, Moh A, Kaplan MH, Zhang Y, et al. 2014. STAT5 programs a distinct subset of GM-CSF-producing T helper cells that is essential for autoimmune neuroinflammation. *Cell Res* **24**: 1387–1402.

Shi M, Lin TH, Appell KC, Berg LJ. 2008. Janus-kinase-3-dependent signals induce chromatin remodeling at the Ifng locus during T helper 1 cell differentiation. *Immunity* **28**: 763–773.

Shiraishi M, Shintani Y, Shintani Y, Ishida H, Saba R, Yamaguchi A, Adachi H, Yashiro K, Suzuki K. 2016. Alternatively activated macrophages determine repair of the infarcted adult murine heart. *J Clin Invest* **126**: 2151–2166.

Shirley M. 2017a. Daclizumab: A review in relapsing multiple sclerosis. *Drugs* **77**: 447–458.

Shirley M. 2017b. Dupilumab: First global approval. *Drugs* **77**: 1115–1121.

Siegel JP, Sharon M, Smith PL, Leonard WJ. 1987. The IL-2 receptor β chain (p70): Role in mediating signals for LAK, NK, and proliferative activities. *Science* **238**: 75–78.

Smith GA, Uchida K, Weiss A, Taunton J. 2016. Essential biphasic role for JAK3 catalytic activity in IL-2 receptor signaling. *Nat Chem Biol* **12**: 373–379.

Socolovsky M, Fallon AE, Wang S, Brugnara C, Lodish HF. 1999. Fetal anemia and apoptosis of red cell progenitors in Stat5a$^{-/-}$5b$^{-/-}$ mice: A direct role for Stat5 in Bcl-X$_L$ induction. *Cell* **98**: 181–191.

Soldaini E, John S, Moro S, Bollenbacher J, Schindler U, Leonard WJ. 2000. DNA binding site selection of dimeric and tetrameric Stat5 proteins reveals a large repertoire of divergent tetrameric Stat5a binding sites. *Mol Cell Biol* **20**: 389–401.

Spangler JB, Tomala J, Luca VC, Jude KM, Dong S, Ring AM, Votavova P, Pepper M, Kovar M, Garcia KC. 2015. Antibodies to interleukin-2 elicit selective T cell subset potentiation through distinct conformational mechanisms. *Immunity* **42**: 815–825.

Spolski R, Leonard WJ. 2014. Interleukin-21: A double-edged sword with therapeutic potential. *Nat Rev Drug Discov* **13**: 379–395.

Spolski R, Kashyap M, Robinson C, Yu Z, Leonard WJ. 2008. IL-21 signaling is critical for the development of type I diabetes in the NOD mouse. *Proc Natl Acad Sci* **105**: 14028–14033.

Spolski R, Kim HP, Zhu W, Levy DE, Leonard WJ. 2009. IL-21 mediates suppressive effects via its induction of IL-10. *J Immunol* **182**: 2859–2867.

Sprent J, Surh CD. 2011. Normal T cell homeostasis: The conversion of naïve cells into memory-phenotype cells. *Nat Immunol* **12**: 478–484.

Stepensky P, Keller B, Abuzaitoun O, Shaag A, Yaacov B, Unger S, Seidl M, Rizzi M, Weintraub M, Elpeleg O, et al. 2015. Extending the clinical and immunological phenotype of human interleukin-21 receptor deficiency. *Haematologica* **100**: e72–e76.

Su EW, Moore CJ, Suriano S, Johnson CB, Songalia N, Patterson A, Neitzke DJ, Andrijauskaite K, Garrett-Mayer E, Mehrotra S, et al. 2015. IL-2Rα mediates temporal regulation of IL-2 signaling and enhances immunotherapy. *Sci Transl Med* **7**: 311ra170.

Suarez-Fueyo A, Bradley SJ, Klatzmann D, Tsokos GC. 2017. T cells and autoimmune kidney disease. *Nat Rev Nephrol* **13**: 329–343.

Surh CD, Sprent J. 2008. Homeostasis of naïve and memory T cells. *Immunity* **29**: 848–862.

Suto A, Nakajima H, Hirose K, Suzuki K, Kagami S, Seto Y, Hoshimoto A, Saito Y, Foster DC, Iwamoto I. 2002. Interleukin 21 prevents antigen-induced IgE production by inhibiting germ line Cε transcription of IL-4-stimulated B cells. *Blood* **100**: 4565–4573.

Suzuki H, Kundig TM, Furlonger C, Wakeham A, Timms E, Matsuyama T, Schmits R, Simard JJ, Ohashi PS, Griesser

H, et al. 1995. Deregulated T cell activation and autoimmunity in mice lacking interleukin-2 receptor β. *Science* **268**: 1472–1476.

Takeshita T, Asao H, Suzuki J, Sugamura K. 1990. An associated molecule, p64, with high-affinity interleukin 2 receptor. *Int Immunol* **2**: 477–480.

Takeshita T, Asao H, Ohtani K, Ishii N, Kumaki S, Tanaka N, Munakata H, Nakamura M, Sugamura K. 1992. Cloning of the γ chain of the human IL-2 receptor. *Science* **257**: 379–382.

Tang Q, Adams JY, Penaranda C, Melli K, Piaggio E, Sgouroudis E, Piccirillo CA, Salomon BL, Bluestone JA. 2008. Central role of defective interleukin-2 production in the triggering of islet autoimmune destruction. *Immunity* **28**: 687–697.

Tangye SG. 2015. Advances in IL-21 biology—Enhancing our understanding of human disease. *Curr Opin Immunol* **34**: 107–115.

Taniguchi T, Matsui H, Fujita T, Takaoka C, Kashima N, Yoshimoto R, Hamuro J. 1983. Structure and expression of a cloned cDNA for human interleukin-2. *Nature* **302**: 305–310.

Telliez JB, Dowty ME, Wang L, Jussif J, Lin T, Li F, Moy E, Balbo P, Li W, Zhao Y, et al. 2016. Discovery of a JAK3-selective inhibitor: Functional differentiation of JAK3-selective inhibition over pan-JAK or JAK1-selective inhibition. *ACS Chem Biol* **11**: 3442–3451.

Terashima A, Okamoto K, Nakashima T, Akira S, Ikuta K, Takayanagi H. 2016. Sepsis-induced osteoblast ablation causes immunodeficiency. *Immunity* **44**: 1434–1443.

Teshigawara K, Wang HM, Kato K, Smith KA. 1987. Interleukin 2 high-affinity receptor expression requires two distinct binding proteins. *J Exp Med* **165**: 223–238.

Todd JA, Walker NM, Cooper JD, Smyth DJ, Downes K, Plagnol V, Bailey R, Nejentsev S, Field SF, Payne F, et al. 2007. Robust associations of four new chromosome regions from genome-wide analyses of type 1 diabetes. *Nat Genet* **39**: 857–864.

Townsend JM, Fallon GP, Matthews JD, Smith P, Jolin EH, McKenzie NA. 2000. IL-9-deficient mice establish fundamental roles for IL-9 in pulmonary mastocytosis and goblet cell hyperplasia but not T cell development. *Immunity* **13**: 573–583.

Tran E, Robbins PF, Rosenberg SA. 2017. "Final common pathway" of human cancer immunotherapy: Targeting random somatic mutations. *Nat Immunol* **18**: 255–262.

Tsudo M, Kozak RW, Goldman CK, Waldmann TA. 1986. Demonstration of a non-Tac peptide that binds interleukin 2: A potential participant in a multichain interleukin 2 receptor complex. *Proc Natl Acad Sci* **83**: 9694–9698.

Turner JE, Morrison PJ, Wilhelm C, Wilson M, Ahlfors H, Renauld JC, Panzer U, Helmby H, Stockinger B. 2013. IL-9-mediated survival of type 2 innate lymphoid cells promotes damage control in helminth-induced lung inflammation. *J Exp Med* **210**: 2951–2965.

Ulrich BJ, Verdan FF, McKenzie AN, Kaplan MH, Olson MR. 2017. STAT3 activation impairs the stability of Th9 cells. *J Immunol* **198**: 2302–2309.

Uyttenhove C, Simpson RJ, Van Snick J. 1988. Functional and structural characterization of P40, a mouse glycoprotein with T-cell growth factor activity. *Proc Natl Acad Sci* **85:** 6934–6938.

Vahedi G, Takahashi H, Nakayamada S, Sun HW, Sartorelli V, Kanno Y, O'Shea JJ. 2012. STATs shape the active enhancer landscape of T cell populations. *Cell* **151:** 981–993.

Van Dyken SJ, Locksley RM. 2013. Interleukin-4- and inter-leukin-13-mediated alternatively activated macrophages: Roles in homeostasis and disease. *Annu Rev Immunol* **31:** 317–343.

Vang KB, Yang J, Mahmud SA, Burchill MA, Vegoe AL, Farrar MA. 2008. IL-2, -7, and -15, but not thymic stromal lymphopoietin, redundantly govern CD4+Foxp3+ regula-tory T cell development. *J Immunol* **181:** 3285–3290.

Vegran F, Berger H, Boidot R, Mignot G, Bruchard M, Dos-set M, Chalmin F, Rebe C, Derangere V, Ryffel B, et al. 2014. The transcription factor IRF1 dictates the IL-21-dependent anticancer functions of T_H9 cells. *Nat Immu-nol* **15:** 758–766.

Veldhoen M, Uyttenhove C, van Snick J, Helmby H, West-endorf A, Buer J, Martin B, Wilhelm C, Stockinger B. 2008. Transforming growth factor-β "reprograms" the differentiation of T helper 2 cells and promotes an inter-leukin 9-producing subset. *Nat Immunol* **9:** 1341–1346.

Venkatasubramanian S, Cheekatla S, Paidipally P, Tripathi D, Welch E, Tvinnereim AR, Nurieva R, Vankayalapati R. 2017. IL-21-dependent expansion of memory-like NK cells enhances protective immune responses against *My-cobacterium tuberculosis*. *Mucosal Immunol* **10:** 1031–1042.

Vermeesch JR, Petit P, Kermouni A, Renauld JC, Van Den Berghe H, Marynen P. 1997. The IL-9 receptor gene, lo-cated in the Xq/Yq pseudoautosomal region, has an au-tosomal origin, escapes X inactivation and is expressed from the Y. *Hum Mol Genet* **6:** 1–8.

Villarino AV, Stumhofer JS, Saris CJ, Kastelein RA, de Sau-vage FJ, Hunter CA. 2006. IL-27 limits IL-2 production during Th1 differentiation. *J Immunol* **176:** 237–247.

Vinkemeier U, Cohen SL, Moarefi I, Chait BT, Kuriyan J, Darnell JE Jr. 1996. DNA binding of in vitro activated Stat1 α, Stat1 β and truncated Stat1: Interaction between NH2-terminal domains stabilizes binding of two dimers to tandem DNA sites. *EMBO J* **15:** 5616–5626.

Vinkemeier U, Moarefi I, Darnell JE Jr, Kuriyan J. 1998. Structure of the amino-terminal protein interaction do-main of STAT-4. *Science* **279:** 1048–1052.

Vivier E, Raulet DH, Moretta A, Caligiuri MA, Zitvogel L, Lanier LL, Yokoyama WM, Ugolini S. 2011. Innate or adaptive immunity? The example of natural killer cells. *Science* **331:** 44–49.

von Freeden-Jeffry U, Vieira P, Lucian LA, McNeil T, Bur-dach SE, Murray R. 1995. Lymphopenia in interleukin (IL)-7 gene-deleted mice identifies IL-7 as a nonredun-dant cytokine. *J Exp Med* **181:** 1519–1526.

Vosshenrich CA, Ranson T, Samson SI, Corcuff E, Colucci F, Rosmaraki EE, Di Santo JP. 2005. Roles for common cy-tokine receptor γ-chain-dependent cytokines in the gen-eration, differentiation, and maturation of NK cell pre-cursors and peripheral NK cells in vivo. *J Immunol* **174:** 1213–1221.

Vosshenrich CA, Garcia-Ojeda ME, Samson-Villeger SI, Pasqualetto V, Enault L, Richard-Le Goff O, Corcuff E,

Guy-Grand D, Rocha B, Cumano A, et al. 2006. A thymic pathway of mouse natural killer cell development charac-terized by expression of GATA-3 and CD127. *Nat Immu-nol* **7:** 1217–1224.

Waldmann TA. 2006. The biology of interleukin-2 and in-terleukin-15: Implications for cancer therapy and vaccine design. *Nat Rev Immunol* **6:** 595–601.

Waldmann TA. 2015. The shared and contrasting roles of IL2 and IL15 in the life and death of normal and neoplas-tic lymphocytes: Implications for cancer therapy. *Cancer Immunol Res* **3:** 219–227.

Wan Q, Kozhaya L, ElHed A, Ramesh R, Carlson TJ, Djuretic IM, Sundrud MS, Unutmaz D. 2011. Cytokine signals through PI-3 kinase pathway modulate Th17 cytokine production by CCR6+ human memory T cells. *J Exp Med* **208:** 1875–1887.

Wan CK, Oh J, Li P, West EE, Wong EA, Andraski AB, Spolski R, Yu ZX, He J, Kelsall BL, et al. 2013. The cyto-kines IL-21 and GM-CSF have opposing regulatory roles in the apoptosis of conventional dendritic cells. *Immunity* **38:** 514–527.

Wan CK, Andraski AB, Spolski R, Li P, Kazemian M, Oh J, Samsel L, Swanson PA II, McGavern DB, Sampaio EP, et al. 2015. Opposing roles of STAT1 and STAT3 in IL-21 function in CD4+ T cells. *Proc Natl Acad Sci* **112:** 9394–9399.

Wang LM, Keegan AD, Li W, Lienhard GE, Pacini S, Gut-kind JS, Myers MG Jr, Sun XJ, White MF, Aaronson SA, et al. 1993. Common elements in interleukin 4 and insulin signaling pathways in factor-dependent hematopoietic cells. *Proc Natl Acad Sci* **90:** 4032–4036.

Wang X, Lupardus P, Laporte SL, Garcia KC. 2009. Struc-tural biology of shared cytokine receptors. *Annu Rev Im-munol* **27:** 29–60.

Wang L, Yu CR, Kim HP, Liao W, Telford WG, Egwuagu CE, Leonard WJ. 2011. Key role for IL-21 in experimental autoimmune uveitis. *Proc Natl Acad Sci* **108:** 9542–9547.

Webb R, Merrill JT, Kelly JA, Sestak A, Kaufman KM, Lan-gefeld CD, Ziegler J, Kimberly RP, Edberg JC, Ramsey-Goldman R, et al. 2009. A polymorphism within IL21R confers risk for systemic lupus erythematosus. *Arthritis Rheum* **60:** 2402–2407.

Weinberg K, Parkman R. 1990. Severe combined immuno-deficiency due to a specific defect in the production of interleukin-2. *N Engl J Med* **322:** 1718–1723.

Weinstein JS, Herman EI, Lainez B, Licona-Limon P, Es-plugues E, Flavell R, Craft J. 2016. T_{FH} cells progressively differentiate to regulate the germinal center response. *Nat Immunol* **17:** 1197–1205.

Wenzel S, Ford L, Pearlman D, Spector S, Sher L, Skobie-randa F, Wang L, Kirkesseli S, Rocklin R, Bock B, et al. 2013. Dupilumab in persistent asthma with elevated eo-sinophil levels. *N Engl J Med* **368:** 2455–2466.

Wenzel S, Castro M, Corren J, Maspero J, Wang L, Zhang B, Pirozzi G, Sutherland ER, Evans RR, Joish VN, et al. 2016. Dupilumab efficacy and safety in adults with uncontrolled persistent asthma despite use of medium-to-high-dose inhaled corticosteroids plus a long-acting β2 agonist: A randomised double-blind placebo-controlled pivotal phase 2b dose-ranging trial. *Lancet* **388:** 31–44.

West EE, Jin HT, Rasheed AU, Penaloza-Macmaster P, Ha SJ, Tan WG, Youngblood B, Freeman GJ, Smith KA, Ahmed R. 2013. PD-L1 blockade synergizes with IL-2 therapy in reinvigorating exhausted T cells. *J Clin Invest* **123**: 2604–2615.

Wilhelm C, Hirota K, Stieglitz B, Van Snick J, Tolaini M, Lahl K, Sparwasser T, Helmby H, Stockinger B. 2011. An IL-9 fate reporter demonstrates the induction of an innate IL-9 response in lung inflammation. *Nat Immunol* **12**: 1071–1077.

Willerford DM, Chen J, Ferry JA, Davidson L, Ma A, Alt FW. 1995. Interleukin-2 receptor α chain regulates the size and content of the peripheral lymphoid compartment. *Immunity* **3**: 521–530.

Winthrop KL. 2017. The emerging safety profile of JAK inhibitors in rheumatic disease. *Nat Rev Rheumatol* **13**: 234–243.

Wirth JJ, Kierszenbaum F, Zlotnik A. 1989. Effects of IL-4 on macrophage functions: Increased uptake and killing of a protozoan parasite (*Trypanosoma cruzi*). *Immunology* **66**: 296–301.

Wu Z, Kim HP, Xue HH, Liu H, Zhao K, Leonard WJ. 2005. Interleukin-21 receptor gene induction in human T cells is mediated by T-cell receptor-induced Sp1 activity. *Mol Cell Biol* **25**: 9741–9752.

Wuest SC, Edwan JH, Martin JF, Han S, Perry JS, Cartagena CM, Matsuura E, Maric D, Waldmann TA, Bielekova B. 2011. A role for interleukin-2 *trans*-presentation in dendritic cell-mediated T cell activation in humans, as revealed by daclizumab therapy. *Nat Med* **17**: 604–609.

Wynn TA, Vannella KM. 2016. Macrophages in tissue repair, regeneration, and fibrosis. *Immunity* **44**: 450–462.

Wynn D, Kaufman M, Montalban X, Vollmer T, Simon J, Elkins J, O'Neill G, Neyer L, Sheridan J, Wang C, et al. 2010. Daclizumab in active relapsing multiple sclerosis (CHOICE study): A phase 2, randomised, double-blind, placebo-controlled, add-on trial with interferon β. *Lancet Neurol* **9**: 381–390.

Wynn TA, Chawla A, Pollard JW. 2013. Macrophage biology in development, homeostasis and disease. *Nature* **496**: 445–455.

Xu X, Sun YL, Hoey T. 1996. Cooperative DNA binding and sequence-selective recognition conferred by the STAT amino-terminal domain. *Science* **273**: 794–797.

Xu M, Liu M, Du X, Li S, Li H, Li X, Li Y, Wang Y, Qin Z, Fu YX, et al. 2015. Intratumoral delivery of IL-21 overcomes anti-Her2/Neu resistance through shifting tumor-associated macrophages from M2 to M1 phenotype. *J Immunol* **194**: 4997–5006.

Xue HH, Kovanen PE, Pise-Masison CA, Berg M, Radovich MF, Brady JN, Leonard WJ. 2002. IL-2 negatively regulates IL-7 receptor α chain expression in activated T lymphocytes. *Proc Natl Acad Sci* **99**: 13759–13764.

Yang XP, Ghoreschi K, Steward-Tharp SM, Rodriguez-Canales J, Zhu J, Grainger JR, Hirahara K, Sun HW, Wei L, Vahedi G, et al. 2011. Opposing regulation of the locus encoding IL-17 through direct, reciprocal actions of STAT3 and STAT5. *Nat Immunol* **12**: 247–254.

Yao Z, Cui Y, Watford WT, Bream JH, Yamaoka K, Hissong BD, Li D, Durum SK, Jiang Q, Bhandoola A, et al. 2006. Stat5a/b are essential for normal lymphoid development and differentiation. *Proc Natl Acad Sci* **103**: 1000–1005.

Yao Z, Kanno Y, Kerenyi M, Stephens G, Durant L, Watford WT, Laurence A, Robinson GW, Shevach EM, Moriggl R, et al. 2007. Nonredundant roles for Stat5a/b in directly regulating Foxp3. *Blood* **109**: 4368–4375.

Ye SK, Agata Y, Lee HC, Kurooka H, Kitamura T, Shimizu A, Honjo T, Ikuta K. 2001. The IL-7 receptor controls the accessibility of the TCRγ locus by Stat5 and histone acetylation. *Immunity* **15**: 813–823.

Yin T, Keller SR, Quelle FW, Witthuhn BA, Tsang ML, Lienhard GE, Ihle JN, Yang YC. 1995. Interleukin-9 induces tyrosine phosphorylation of insulin receptor substrate-1 via JAK tyrosine kinases. *J Biol Chem* **270**: 20497–20502.

Yoshizaki A, Miyagaki T, DiLillo DJ, Matsushita T, Horikawa M, Kountikov EI, Spolski R, Poe JC, Leonard WJ, Tedder TF. 2012. Regulatory B cells control T-cell autoimmunity through IL-21-dependent cognate interactions. *Nature* **491**: 264–268.

Yu D, Rao S, Tsai LM, Lee SK, He Y, Sutcliffe EL, Srivastava M, Linterman M, Zheng L, Simpson N, et al. 2009. The transcriptional repressor Bcl-6 directs T follicular helper cell lineage commitment. *Immunity* **31**: 457–468.

Yu CR, Oh HM, Golestaneh N, Amadi-Obi A, Lee YS, Eseonu A, Mahdi RM, Egwuagu CE. 2011. Persistence of IL-2 expressing Th17 cells in healthy humans and experimental autoimmune uveitis. *Eur J Immunol* **41**: 3495–3505.

Yu A, Snowhite I, Vendrame F, Rosenzwajg M, Klatzmann D, Pugliese A, Malek TR. 2015. Selective IL-2 responsiveness of regulatory T cells through multiple intrinsic mechanisms supports the use of low-dose IL-2 therapy in type 1 diabetes. *Diabetes* **64**: 2172–2183.

Zajac AJ, Blattman JN, Murali-Krishna K, Sourdive DJ, Suresh M, Altman JD, Ahmed R. 1998. Viral immune evasion due to persistence of activated T cells without effector function. *J Exp Med* **188**: 2205–2213.

Zeng R, Spolski R, Finkelstein SE, Oh S, Kovanen PE, Hinrichs CS, Pise-Masison CA, Radonovich MF, Brady JN, Restifo NP, et al. 2005. Synergy of IL-21 and IL-15 in regulating CD8[+] T cell expansion and function. *J Exp Med* **201**: 139–148.

Zeng R, Spolski R, Casas E, Zhu W, Levy DE, Leonard WJ. 2007. The molecular basis of IL-21-mediated proliferation. *Blood* **109**: 4135–4142.

Zheng SG, Wang J, Wang P, Gray JD, Horwitz DA. 2007. IL-2 is essential for TGF-β to convert naïve CD4[+]CD25[−] cells to CD25[+]Foxp3[+] regulatory T cells and for expansion of these cells. *J Immunol* **178**: 2018–2027.

Zhou L, Ivanov II, Spolski R, Min R, Shenderov K, Egawa T, Levy DE, Leonard WJ, Littman DR. 2007. IL-6 programs T$_H$-17 cell differentiation by promoting sequential engagement of the IL-21 and IL-23 pathways. *Nat Immunol* **8**: 967–974.

Zhu J, Cote-Sierra J, Guo L, Paul WE. 2003. Stat5 activation plays a critical role in Th2 differentiation. *Immunity* **19**: 739–748.

Zhu J, Min B, Hu-Li J, Watson CJ, Grinberg A, Wang Q, Killeen N, Urban JF Jr, Guo L, Paul WE. 2004. Conditional deletion of *Gata3* shows its essential function in T$_H$1-T$_H$2 responses. *Nat Immunol* **5**: 1157–1165.

Role of the β Common (βc) Family of Cytokines in Health and Disease

Timothy R. Hercus,[1] Winnie L.T. Kan,[1] Sophie E. Broughton,[2] Denis Tvorogov,[1]
Hayley S. Ramshaw,[1] Jarrod J. Sandow,[3] Tracy L. Nero,[2] Urmi Dhagat,[2] Emma J. Thompson,[1]
Karen S. Cheung Tung Shing,[2,4] Duncan R. McKenzie,[1] Nicholas J. Wilson,[5]
Catherine M. Owczarek,[5] Gino Vairo,[5] Andrew D. Nash,[5] Vinay Tergaonkar,[1,6]
Timothy Hughes,[1,7] Paul G. Ekert,[8] Michael S. Samuel,[1,9] Claudine S. Bonder,[1]
Michele A. Grimbaldeston,[1,10] Michael W. Parker,[2,4] and Angel F. Lopez[1]

[1]The Centre for Cancer Biology, SA Pathology and the University of South Australia, Adelaide, South Australia 5000, Australia

[2]ACRF Rational Drug Discovery Centre, St. Vincent's Institute of Medical Research, Fitzroy, Victoria 3065, Australia

[3]Division of Systems Biology and Personalised Medicine, The Walter and Eliza Hall Institute; and Department of Medical Biology, The University of Melbourne, Parkville, Victoria 3052, Australia

[4]Department of Biochemistry and Molecular Biology, Bio21 Molecular Science and Biotechnology Institute, The University of Melbourne, Parkville, Victoria 3010, Australia

[5]CSL Limited, Parkville, Victoria 3010, Australia

[6]Institute of Molecular and Cell Biology, Agency for Science, Technology and Research, Singapore 138673, Singapore

[7]South Australian Health and Medical Research Institute and University of Adelaide, North Terrace, Adelaide, South Australia 5000, Australia

[8]Murdoch Childrens Research Institute, Royal Children's Hospital, Parkville, Victoria 3052, Australia

[9]School of Medicine, University of Adelaide, Adelaide, South Australia 5000, Australia

[10]OMNI-Biomarker Development, Genentech Inc., South San Francisco, California 94080

Correspondence: angel.lopez@sa.gov.au

The β common ([βc]/CD131) family of cytokines comprises granulocyte macrophage colony-stimulating factor (GM-CSF), interleukin (IL)-3, and IL-5, all of which use βc as their key signaling receptor subunit. This is a prototypic signaling subunit-sharing cytokine family that has unveiled many biological paradigms and structural principles applicable to the IL-2, IL-4, and IL-6 receptor families, all of which also share one or more signaling subunits. Originally identified for their functions in the hematopoietic system, the βc cytokines are now known to be truly pleiotropic, impacting on multiple cell types, organs, and biological systems, and thereby controlling the balance between health and disease. This review will focus on the emerging biological roles for the βc cytokines, our progress toward understanding the mechanisms of receptor assembly and signaling, and the application of this knowledge to develop exciting new therapeutic approaches against human disease.

The β common (βc) cytokines are produced by many cell types in the body and act in the immediate microenvironment or at a distance. They are recognized by specific cell-surface receptors present on hemopoietic and nonhemopoietic cell types (Fig. 1). On binding to their specific receptors, granulocyte macrophage colony-stimulating factor (GM-CSF), interleukin (IL)-3, and IL-5 signal through heterodimeric receptor complexes comprising βc as the major signaling subunit and cytokine-specific α subunits (Broughton et al. 2012). Cytokine binding

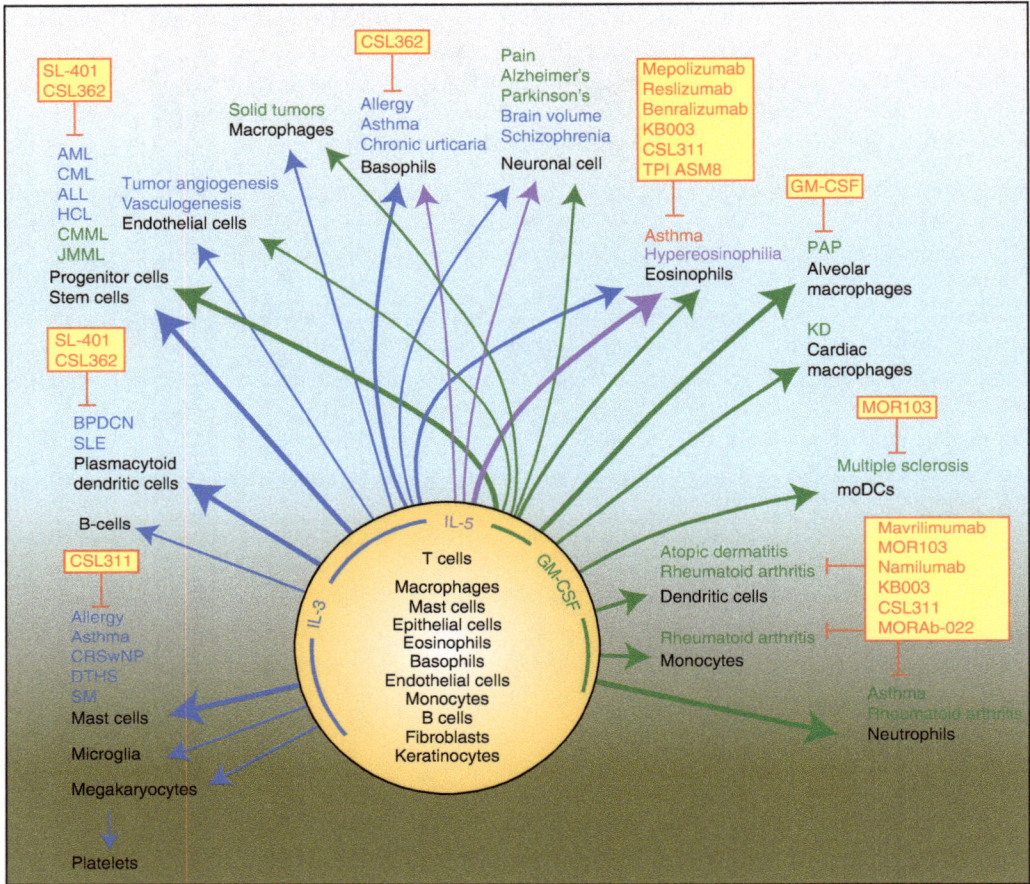

Figure 1. Diverse functions in health and disease for the β common (βc) cytokine family. The βc cytokines are predominantly expressed by activated T cells, but many other cell types have been reported to express selected βc cytokines. Among the βc cytokines, interleukin (IL)-3 and granulocyte macrophage colony-stimulating factor (GM-CSF) have actions on the widest range of cell types, whereas IL-5 is largely restricted to actions on eosinophils. Cytokines are coded IL-3 (blue), IL-5 (purple), GM-CSF (green), and the line thickness indicates the relative biological significance. All βc cytokines have been reported to contribute to diseases in humans, either directly or indirectly, and this includes asthma, AML (acute myeloid leukemia), AMML (acute myelo-monocytic leukemia), ALL (acute lymphoblastic leukemia), BPDCN (blastic plasmacytoid dendritic cell neoplasm), CML (chronic myeloid leukemia), CMML (chronic myelomonocytic leukemia), CRSwNP (chronic rhinosinusitis with nasal polyps), DTHS (delayed-type hypersensitivity), HCL (hairy cell leukemia), KD (Kawasaki disease), PAP (pulmonary alveolar proteinosis), RA (rheumatoid arthritis), SLE (systemic lupus erythematosus), and SM (systemic mastocytosis). Agents that block the function of βc cytokines or target their receptors have been generated and are undergoing clinical development in a range of disease settings (red boxes). See Table 1 for details of the listed agents. moDCs, Monocyte-derived dendritic cells.

Table 1. Therapeutic agents targeting the βc family of cytokines or their receptors

Target	Disease	Type	Therapy name	Mode of action	Publications	Clinical trial reference
GM-CSF	RA	MAb	GSK3196165/MOR103	Blocks GM-CSF	Behrens et al. 2015	NCT01023256 (c)
GM-CSF	MS RA	MAb	Namilumab	Blocks GM-CSF	Constantinescu et al. 2015 -	NCT01517282 (c) NCT01317797 (c) NCT02379091 (o) NCT02393378 (p) NCT02129777 (c)
GM-CSF	PP RA	MAb	Lenzilumab/KB003	Blocks GM-CSF	- -	NCT00995449 (t)
GM-CSF	Asthma CMML RA	MAb	MORAb-022	Blocks GM-CSF	Molfino et al. 2016 -	NCT01603277 (c) NCT02546284 (p) NCT01357759 (c)
GMRα	RA	MAb	Mavrilimumab/CAM-3001	Blocks GM-CSF	- Burmester et al. 2011 Burmester et al. 2013	NCT00771420 (c) NCT01050998 (c) NCT01715896 (c) NCT01706926 (c)
IL3Rα + CD3	AML	DART	MGD006/S680880	Targets cell, ADCC	Al-Hussaini et al. 2016	NCT02152956 (p)
IL3Rα	AML, MDS	MAb	KHK2823	ADCC	Akiyama et al. 2015	NCT02181699 (p)
IL3Rα	AML	MAb	CSL362/JNJ-56022473	Blocks IL-3 and ADCC	Busfield et al. 2014	NCT01633852 (c) NCT02472145 (p) -
IL3Rα	SLE AML BPDCN AML, BPDCN AML SM, CMML, PED, MS	Immunotoxin	SL-401/DT388IL3	Cytokine targets cell that is killed by toxin	Oon et al. 2016a Feuring-Buske et al. 2002; Frankel et al. 2008; Tettamanti et al. 2013 Frankel et al. 2014; Angelot-Delettre et al. 2015	NCT00397579 (c) NCT02113982 (p) NCT02270463 (p) NCT02268253 (p)

Continued

Table 1. *Continued*

Target	Disease	Type	Therapy name	Mode of action	Publications	Clinical trial reference
IL3Rα	AML	CAR		T cell kills AML	Mardiros et al. 2013	NCT02159495 (p)
IL-5	Asthma	MAb	Mepolizumab/Nucala/SB-240563	Blocks IL-5	Keating 2015; Patterson et al. 2015; Pavord et al. 2012	NCT01000506 (c)
						NCT02654145 (p)
						NCT02594332 (p)
	HES				Rothenberg et al. 2008	NCT00086658 (c)
						NCT02836496 (n)
	NP				-	NCT01362244 (c)
						NCT02105961 (p)
						NCT02105948 (p)
	COPD				-	NCT00274703 (c)
	EO				Straumann et al. 2010	NCT02452190 (p)
						NCT02559791 (p)
IL-5	Asthma	MAb	Reslizumab/Cinqair/SCH-55700	Blocks IL-5	Patterson et al. 2015; Pelaia et al. 2016	NCT02821416 (n)
IL5Rα	Asthma	MAb	Benralizumab	Blocks IL-5 and ADCC	Patterson et al. 2015; Tan et al. 2016	NCT02075255 (o)
					FitzGerald et al. 2016	NCT01914757 (c)
					Bleecker et al. 2016	NCT01928771 (c)
						NCT02417961 (c)
						NCT02258542 (o)
	COPD				Castro et al. 2014	NCT01227278 (c)
						NCT02155660 (p)
βc	Asthma	MAb	CSL311	Blocks GM-CSF, IL-3, and IL-5	Panousis et al. 2016	NCT01759849*
βc + CCR3	Asthma	Oligo	TPI ASM8	Blocks βc expression	Gauvreau et al. 2011	NCT01158898 (c)
					Imaoka et al. 2011	NCT00050797 (c)

AML, Acute myeloid leukemia; BPDCN, blastic plasmacytoid dendritic cell neoplasm; CMML, chronic myelomonocytic leukemia; COPD, chronic obstructive pulmonary disease; EO, eosinophilic oesophagitis; HES, hypereosinophilic syndrome; MDS, myelodysplastic syndrome; MF, myelofibrosis; MS, multiple sclerosis; NP, nasal polyposis; PED, advanced symptomatic hypereosinoophic disorder; PP, plaque psoriasis; RA, rheumatoid arthritis; SLE, systemic lupus erythematosus; SM, systemic mastocytosis. CAR, chimeric antigen receptor expressed on transduced T cells; DART, dual-affinity retargeting proteins; Immunotoxin, cytokine conjugated to diphtheria toxin; MAb, monoclonal antibody; Oligo, antisense oligonucleotide. ADCC, antibody-dependent cell cytotoxicity.

Clinical Trials identifier and status: c, completed; n, not started; o, ongoing, not recruiting; p, in progress; t, terminated; *, preclinical.

Cite this article as *Cold Spring Harb Perspect Biol* doi: 10.1101/cshperspect.a028514

triggers assembly of the cytokine–receptor complex in a distinct stoichiometry that initiates multiple signaling pathways. These include emerging roles for signaling cross talk with other receptor systems and novel kinases as well as the more familiar Janus kinase (JAK) 2/signal transducer and activator of transcription (STAT) 5, mitogen-activated protein kinase (MAPK), and phosphoinositide 3-kinase (PI3K) pathways (Martinez-Moczygemba and Huston 2003; Hercus et al. 2013). Signaling by the βc cytokines gives rise to diverse biological responses, including cell survival, proliferation, differentiation, migration, and mature leukocyte effector functions. A putative fourth member of this family, KK34, has recently been described in some species (Yamaguchi et al. 2016), but its ortholog in humans and mice is a pseudogene.

BIOLOGY OF THE βc CYTOKINES

Among the plethora of cell types that secrete βc cytokines, there is a clear distinction between GM-CSF, which is produced by multiple cell types, including macrophages, fibroblasts, and epithelial cells, and IL-3 and IL-5, which are more T-cell restricted (Metcalf 2008; Broughton et al. 2012; Hercus et al. 2013). The spectrum of GM-CSF-producing cells immediately suggests a role for this cytokine in various innate and adaptive immune responses, and in inflammatory diseases (Fig. 1) (Broughton et al. 2012; Bhattacharya et al. 2015; Wicks and Roberts 2016). This concept has been supported by studies in rheumatoid arthritis (RA) (Campbell et al. 1998; Cook et al. 2001; Hamilton 2008), multiple sclerosis (MS) (McQualter et al. 2001; Codarri et al. 2011; El-Behi et al. 2011; Croxford et al. 2015a), asthma (Yamashita et al. 2002; Asquith et al. 2008; Saha et al. 2009), and chronic obstructive pulmonary disease (COPD) (Saha et al. 2009).

In addition to their role in inflammation, the βc cytokines are increasingly being recognized as contributing to several forms of leukemia. In human myeloid leukemia, GM-CSF acts as a growth and survival factor (Hercus et al. 2013) and promotes the survival of chronic myeloid leukemia (CML) even in the presence of

BCR-ABL kinase inhibitors (Hiwase et al. 2010). Blast cells from patients with acute myeloid leukemia (AML) show enhanced GM-CSF receptor expression and proliferation (Riccioni et al. 2009; Mathew et al. 2013), whereas juvenile myelomonocytic leukemia (JMML) (Bernard et al. 2002; Bunda et al. 2013) and chronic myelomonocytic leukemia (CMML) (Ramshaw et al. 2002; Padron et al. 2013) display a characteristic GM-CSF hypersensitivity. This suggests a pathogenic role for GM-CSF and highlights a focus for targeted therapies in leukemias (Fig. 1).

Whereas GM-CSF produced during inflammation promotes hemopoietic cell production and immunity, its role in steady-state hematopoiesis does not appear to be critical, with the notable exception of normal alveolar macrophage function. GM-CSF signaling blockade by autoantibodies (Kitamura et al. 1999; Inoue et al. 2008; Wang et al. 2013) or disruption of GM-CSF signaling by mutation of the genes encoding GM-CSF or its receptor (Suzuki et al. 2008, 2010, 2011; Griese et al. 2011; Tanaka et al. 2011) is associated with pulmonary alveolar proteinosis (PAP), revealing a nonredundant role for GM-CSF in lung function. Thus, emerging therapies that antagonize GM-CSF must factor in the need to maintain lung homeostasis.

An important role for GM-CSF in at least some forms of antiviral immunity is indicated by the existence of a viral protein that antagonizes GM-CSF and IL-2 function. The orf virus infects many species, including humans and sheep, in which it causes a contagious ecthyma, also known as orf (Felix et al. 2016). Intriguingly, the orf virus encodes a protein, GM-CSF/IL-2-inhibition factor (GIF), that is able to inhibit ovine GM-CSF and ovine IL-2 function but does not appear to bind the human or murine orthologs or the other βc cytokines (Deane et al. 2000). GIF has recently been shown to bind ovine GM-CSF and ovine IL-2 through mutually exclusive binding sites and is likely to act as a competitive decoy receptor (Felix et al. 2016). The species-restricted function of GIF has been suggested to be an adaptation by the orf virus to its principal host, the sheep (Deane et al. 2000), but further studies are desirable to determine the role of GIF in human orf virus infections.

The role of IL-3 as a growth and survival factor for multiple lineages of normal and malignant hemopoietic cells is well recognized (Hercus et al. 2013). Unlike GM-CSF and IL-5, IL-3 also regulates hemopoietic stem cells and megakaryocytes, desirable properties for bone marrow reconstitution, although its stimulation of mast cell and basophil production and activity has restricted its clinical application (Fig. 1) (Eder et al. 1997). Of particular significance is the overexpression of the IL-3 receptor α subunit (IL3Rα/CD123) in several hematological malignancies. Originally described in leukemic stem-like cells of patients with AML (Jordan et al. 2000; Testa et al. 2002; Jin et al. 2009; Vergez et al. 2011), this overexpression is also observed in progenitor cells of CML (Nievergall et al. 2014), blastic plasmacytoid dendritic cell neoplasm (BPDCN) (Frankel et al. 2014; Angelot-Delettre et al. 2015), neoplastic mast cells (Pardanani et al. 2015), and systemic mastocytosis (Fig. 1) (Pardanani et al. 2016). IL3Rα overexpression is particularly important in AML where IL3Rα expression levels strongly correlate with reduced patient survival (Jordan et al. 2000; Testa et al. 2002; Vergez et al. 2011), and this is being exploited therapeutically through targeting of IL3Rα by antibodies (Jin et al. 2009; Busfield et al. 2014; Al-Hussaini et al. 2016), chimeric antigen receptor (CAR) T cells (Tettamanti et al. 2013), and IL-3-toxin conjugates (Frankel et al. 2008), which specifically recognize IL3Rα. Of these approaches, the IL3Rα-specific antibody CSL362, which blocks IL-3-mediated signaling (Sun et al. 1996; Jin et al. 2009) through a dual mechanism (Broughton et al. 2014) and enhances natural killer (NK) cell–mediated antibody-dependent cell-mediated cytotoxicity (ADCC) (Busfield et al. 2014), is the most advanced and is currently in phase II clinical trials for the treatment of AML.

IL-3 has also been reported to contribute to the development of lupus nephritis in a mouse model of the human autoimmune disease systemic lupus erythematosus (SLE), a process that could be blocked by anti-IL-3 antibodies (Renner et al. 2015). Interestingly, along with leuke-mic stem-like cells and basophils, human plasmacytoid dendritic cells (pDCs) also show high expression of IL3Rα. These cells play a central role in autoimmunity through their capacity to produce type I interferons (Liu 2005; Llanos et al. 2013) and their high expression of IL3Rα may present an opportunity to control their function. This notion is supported by recent experiments in which blocking IL3Rα or depleting pDCs showed a marked reduction in expression of both type I and III interferons and IL-6 (Oon et al. 2016a) and suggests a new approach to manage autoimmune diseases such as SLE (Oon et al. 2016b).

In contrast to IL-3 and GM-CSF, the actions of IL-5 are restricted to eosinophils and basophils, with IL-5 being directly linked to the hypereosinophilic syndrome (Fig. 1) (Broughton et al. 2012; Tan et al. 2016). Interestingly, although IL-5 is eosinophil-specific, all three βc cytokines stimulate eosinophil production and function, both under steady-state (Nishinakamura et al. 1996) and inflammatory conditions such as asthma (Panousis et al. 2016). Thus, anti-IL-5 therapies that target eosinophil-mediated inflammation may be successful in cases in which IL-5 is the predominant cytokine involved (Broughton et al. 2012; Tan et al. 2016), but ultimately blocking all three cytokines by targeting βc may yield the best results.

While mutations of the βc cytokine receptors are only infrequently observed, a recent publication reported a germline mutation in the βc gene (*CSF2RB*) from a patient with T-cell acute lymphoblastic leukemia (T-ALL) that is able to stimulate factor-independent cell proliferation in vitro (Watanabe-Smith et al. 2016). This is consistent with previous in vitro studies that identified the potential for mutations in βc to give rise to factor-independent cell growth (D'Andrea et al. 1994; Jenkins et al. 1998) and represents the first demonstration in vivo that such mutations are associated with leukemia.

A ROLE FOR βc CYTOKINES IN SEPSIS AND INFLAMMATORY DISEASE

Sepsis is a serious and often fatal complication of infection that has been attributed to uncon-

Cite this article as *Cold Spring Harb Perspect Biol* doi: 10.1101/cshperspect.a028514

trolled inflammatory responses. Clinical trials to control sepsis have been largely unsuccessful, probably owing to the heterogeneous nature of the septic syndrome. As the mechanisms underpinning sepsis are beginning to be dissected, it appears that βc cytokines play distinct roles in this setting. On the one hand, GM-CSF can be viewed as a beneficial factor owing to its antimicrobial immune function, as sepsis-associated immunosuppression is becoming recognized as a critical feature in disease progression (Hotchkiss and Sherwood 2015). In fact, clinical trials with GM-CSF have shown some reversal of sepsis-associated immunosuppression (Meisel et al. 2009; Leentjens et al. 2012). This protective role is consistent with the identification of GM-CSF-expressing innate response activator (IRA) B cells that protect mice from death in a cecal ligation and puncture (CLP) model of sepsis (Rauch et al. 2012) and that produce immunoglobulin M (IgM) through an autocrine GM-CSF loop to protect against pulmonary bacterial infections (Weber et al. 2014). On the other hand, IL-3 produced by B cells appears to play a pathogenic role in sepsis. A recent study reported that IL-3 amplifies acute inflammation in a mouse model of sepsis and could be blocked by anti-IL-3 agents (Weber et al. 2015). These findings may suggest differences in inflammatory responses mediated by GM-CSF and IL-3, fundamental differences between chronic human disease and the acute animal models (Hotchkiss and Sherwood 2015), or may simply be a reflection of sepsis representing a complex set of related diseases rather than a single condition.

Experimental autoimmune encephalomyelitis (EAE) is a rodent model of MS, an inflammatory disease of the central nervous system (CNS). EAE is driven by autoreactive CD4$^+$ T helper (Th) cells and it has become clear that Th cell production of GM-CSF in response to IL-23 and IL-1β or IL-7 (McQualter et al. 2001; Codarri et al. 2011; El-Behi et al. 2011; Sheng et al. 2014) is a critical factor in the recruitment and activation of myeloid cells and ultimately CNS demyelination (Croxford et al. 2015b). Recent studies have reported elevated numbers of Th cells expressing GM-CSF in patients with MS

(Rasouli et al. 2015) and, in some patients, this has been linked to MS-associated polymorphic variants of the IL-2 receptor α gene (*IL2RA*) (Hartmann et al. 2014). Expression of GM-CSF by a subset of human memory B cells has also been linked to MS (Li et al. 2015). GM-CSF production by human Th cells is regulated differently from their murine counterparts, as IL-2 and IL-23 promote and repress its expression, respectively (Noster et al. 2014). The essential role of GM-CSF in EAE pathogenesis is because of its activation of CCR2$^+$Ly6C^{hi} monocytes and subsequent programming of inflammatory function in their progeny (Croxford et al. 2015a). The presence of these activated or inflammatory monocyte-derived dendritic cells (moDCs) in the CNS has been directly linked to demyelination in EAE (Yamasaki et al. 2014). GM-CSF stimulation of moDCs induces expression of multiple genes linked to EAE, including IL-1β, which expands the GM-CSF-expressing Th cell pool and thereby amplifies the inflammatory response (Croxford et al. 2015b; Paré et al. 2016). These studies highlight a potential role for pathogenic GM-CSF activity in MS, and prompted a recently completed phase Ib clinical trial in MS patients using a human anti-GM-CSF antibody (Constantinescu et al. 2015).

Recent reports have indicated an unexpected role of GM-CSF in cardiomyopathies. GM-CSF has recently been identified as a primary initiator of cardiac inflammation in a model of Kawasaki disease (KD), which is the leading cause of pediatric heart disease in developed countries (Stock et al. 2016). Using a CAWS (*Candida albicans* water-soluble fraction) model of KD, Stock et al. reported that expression of GM-CSF by cardiac fibroblasts was required for the expression of multiple proinflammatory chemokines by cardiac macrophages that then promote neutrophil and monocyte recruitment into the heart where they mediate cardiac pathology. Although using different cellular sources, the indirect role of GM-CSF in the KD model is reminiscent of the indirect role of GM-CSF in EAE (Croxford et al. 2015a) and suggests the therapeutic potential of GM-CSF antagonists in patients with KD.

THE βc CYTOKINES IN CANCER

In recent years, there has been increasing interest in the role of the βc cytokines outside the hematopoietic compartment. GM-CSF activity is reported to be relevant in a number of solid tumor settings and through multiple mechanisms. Although the balance between antitumor and protumor effects is complex, the activity of GM-CSF appears to be predominantly protumorigenic based on its expression by certain tumors and their recruitment of tumor suppressor cells. Expression of GM-CSF is associated with protumorigenic outcomes in squamous cell carcinomas (Obermueller et al. 2004; Gutschalk et al. 2006), breast cancer (Su et al. 2014; Vilalta et al. 2014), colon cancer (Rigo et al. 2010), and glioma (Revoltella et al. 2012). GM-CSF may act directly on tumors in an autocrine manner (Obermueller et al. 2004; Gutschalk et al. 2006; Revoltella et al. 2012; Vilalta et al. 2014) or indirectly through GM-CSF-mediated responses from tumor-associated macrophages (TAMs) (Rigo et al. 2010; Su et al. 2014). Several studies have also found that tumor-derived GM-CSF recruits, expands, and activates myeloid-derived suppressor cells (MDSCs), a heterogeneous group of immature myeloid cells that suppress other immune cell types, such as T cells, to contribute to the metastasis of glioma, pancreatic, and liver cancers (Bayne et al. 2012; Pylayeva-Gupta et al. 2012; Kohanbash et al. 2013; Thorn et al. 2016). Furthermore, the recent recognition that neutrophils can be manipulated by the tumor microenvironment to facilitate tumor growth (Coffelt et al. 2016) and the associated expression of GM-CSF in some tumors (Garcia-Mendoza et al. 2016) suggests an unanticipated role for GM-CSF on tumor-associated neutrophils. In contrast, GM-CSF has been reported to contribute to antitumor responses in colon cancer (Urdinguio et al. 2013), and, through expression in oncolytic viral therapies (Andtbacka et al. 2016; Kuryk et al. 2016), is being used to stimulate host antitumor responses. It would be important to clarify the contribution of GM-CSF activity in different solid tumor settings to harness its protective effects or to inhibit its function.

In certain solid tumors, IL-3 may also play a pathogenic role by acting on both the vasculature and on the tumor cells directly. IL-3 receptors are expressed on endothelial cells (Korpelainen et al. 1993), endothelial progenitor cells, mediating their rapid expansion (Moldenhauer et al. 2015), and on tumor-derived endothelial cells thus contributing to angiogenesis and tumor growth (Dentelli et al. 2011). Furthermore, the concomitant expression of IL-3 receptors in certain breast cancers (C Bonder and A Lopez, unpubl.) and the activating effects of IL-3 on these cells suggest that IL-3 and its receptor may constitute previously unappreciated targets to simultaneously inhibit the growth and activity of some breast cancer cells directly and also the angiogenic microenvironment.

βc CYTOKINES IN THE NERVOUS SYSTEM

A fascinating aspect of βc cytokine biology has emerged from reports that describe a role for the βc cytokines in the nervous system. Whereas the molecular basis for many of these actions remains to be firmly established, the data suggest an important role for βc cytokines in neural development, diseases such as MS and schizophrenia, as well as a direct role in nociception. Expression of GM-CSF, IL-3, and IL-5 receptors in brain tissue (Lins and Borojevic 2001; Schabitz et al. 2008; Luo et al. 2012) support a role for these cytokines in the CNS. GM-CSF promotes the in vitro regeneration of retinal ganglion cells (Legacy et al. 2013; Hanea et al. 2016), has been shown to have neuroprotective effects in a transient focal cerebral ischemia model in rats (Kong et al. 2009), restores motor function in a mouse model of stroke (Shanmugalingam et al. 2016; Theoret et al. 2016), and provides a neuroprotective response in mouse models of Alzheimer's disease (Boyd et al. 2010) and Parkinson's disease (Kim et al. 2009; Kosloski et al. 2013). GM-CSF may also provide a neuroprotective effect by triggering the entry of microglia, tissue-resident macrophages of the CNS, into the brain across the

Cite this article as *Cold Spring Harb Perspect Biol* doi: 10.1101/cshperspect.a028514

blood–brain barrier (Boyd et al. 2010; Shang et al. 2016), suggesting a beneficial role in patients with Alzheimer's disease or Parkinson's disease (Boyd et al. 2010; Heinzelman and Priebe 2015).

IL-3 and the IL-3 receptor subunits, IL3Rα and βc, are also expressed in mouse embryonic brain (Luo et al. 2012), and IL-3 has been shown to stimulate the proliferation of microglial cells (Frei et al. 1986) as well as neuronal progenitor cells (Luo et al. 2012). Intriguingly, a number of recent studies have now linked the expression and function of IL-3 with CNS development and schizophrenia. Genome-wide association studies (GWAS) have identified single-nucleotide polymorphisms (SNPs) upstream of or within the *IL3* gene sequence that are associated with variations in human brain volume (Luo et al. 2012; Li et al. 2016b) or are linked to schizophrenia (Chen et al. 2007; Edwards et al. 2008). Other studies have identified variants in the genes for βc (*CSF2RB*) and IL3Rα (*IL3RA*) that are associated with schizophrenia (Lencz et al. 2007; Chen et al. 2008) and depression (Chen et al. 2011). Consistent with these observations, elevated levels of serum IL-3 are associated with more severe symptoms among patients with chronic schizophrenia (Xiu et al. 2015; Fu et al. 2016). It remains to be seen whether these observations are causally linked.

In addition to their actions on the CNS, the βc cytokines also play a role in the peripheral nervous system, in particular in regulating pain. Models of inflammatory pain and arthritis in GM-CSF-deficient *Csf2*$^{-/-}$ mice have shown that pain is dependent on GM-CSF activity and that GM-CSF-mediated pain but not arthritis is cyclooxygenase-dependent (Cook et al. 2013). Pain associated with bone cancer was mediated by GM-CSF in a mouse sarcoma model and specific knockdown of GM-CSF receptor α subunit (GMRα/CD116) expression in the sensory nerves was able to significantly reduce tumor-induced pain, suggesting that GM-CSF may act directly on peripheral nerves (Schweizerhof et al. 2009) and potentially open up new indications for anti-GM-CSF therapies.

INTERACTIONS OF βc CYTOKINE RECEPTORS WITH OTHER MEMBRANE RECEPTORS

Although the βc cytokines are the classical activators of the βc receptor family, a surprising number of membrane proteins have been reported to interact and signal with βc (Broughton et al. 2012). These include the receptor for stem-cell factor (c-Kit), β1 integrin, the FcRγ subunit of the high-affinity immunoglobulin E (IgE) receptor, vascular endothelial growth factor (VEGF) receptor 2 (VEGFR2), and the receptor for erythropoietin (EPOR). As a result of these interactions there is emerging evidence that signaling through the βc subunit can be regulated by noncanonical cytokines. Transforming growth factor β1 has been shown to regulate cellular functions controlled by IL-3 (Broide et al. 1989) and the cell-surface density of GM-CSF receptor (Jacobsen et al. 1993). Furthermore, VEGFR-2 has been shown to interact with βc, leading to enhanced p38 activation (Saulle et al. 2009) and tumor angiogenesis (Dentelli et al. 2005). However, the mechanisms underlying the cross talk between these receptors remain to be elucidated. Gene expression data obtained from publicly available datasets suggest that, in some epithelial and hematological cancers, the ratio of βc and α subunit gene expression differs significantly from that observed in corresponding normal tissues (Fig. 2). Whereas it is well known that IL3Rα is overexpressed relative to βc in AML (Jordan et al. 2000; Testa et al. 2002; Jin et al. 2009; Vergez et al. 2011), the genes for GMRα and the IL-5 receptor α subunit (IL5Rα/CD125) are also differentially expressed relative to the βc subunit in certain malignant tissues (Fig. 2). These observations led us to hypothesize that noncanonical interactions between βc and other receptors may be a feature of malignancy, with specific aberrant interactions driving the phenotype in different cancers.

The best characterized of the noncanonical βc interactions is the innate repair receptor (IRR), a complex of EPOR and βc (Brines et al. 2004) that is expressed in many tissues under stress and triggers a tissue protection and repair response (Collino et al. 2015). Tissue-protective activation of the IRR in vivo is thought to preferentially use hyposialated EPO (hsEPO). EPO variants that activate the IRR, but not the classical EPOR homodimer (Leist et al. 2004) have also been developed. A short peptide fragment of EPO (pHBSP/ARA-290) (Brines et al. 2008) selectively activates IRR and is undergoing clinical development to reduce symptoms of small fiber neuropathy in patients with sarcoidosis (Heij et al. 2012; Dahan et al. 2013) and as a therapy in type 2 diabetes (Brines et al. 2014). Although the interaction of βc with other cell-surface receptors may not be required for classical or established signaling outcomes from these receptors (Scott et al. 2000), it is important to keep in mind that noncanonical heteromeric receptors, such as the IRR complex, may exist. Novel proteomic-based approaches coupled to appropriate biological characterizations will be needed to fully appreciate the importance of βc signaling beyond hematopoiesis.

Although all known activities of the βc cytokines use α/βc heteromeric receptor complexes, a recent report describes the direct sensitization of persister cells of *Pseudomonas aeruginosa* strains to a range of antibiotics by incubation with low doses of human GM-CSF (Choudhary et al. 2015). Although the mechanism that drives this surprising response is unknown, this novel action for GM-CSF could represent a new activity for the βc cytokine family and perhaps other cytokine families.

βc CYTOKINE RECEPTOR ASSEMBLY AND INITIATION OF SIGNALING

A major advance in understanding how the βc family of receptors assemble and signal has been the determination of the three-dimensional (3D) structure of the receptor's extracellular components as apo, binary, and ternary complexes. Crystal structures of the GM-CSF and IL-5 cytokines (PDB IDs: 1CSG, 2GMF, 1HUL [Milburn et al. 1993; Rozwarski et al. 1996]), the βc subunit (PDB IDs: 1GH7, 2GYS [Carr et al. 2001, 2006]), the IL-5:IL5Rα binary complex (PDB IDs: 3QT2, 3VA2 [Patino et al. 2011; Ku-

sano et al. 2012]), and the GM-CSF binary and ternary receptor complexes (PDB IDs: 4RS1, 4NKQ, respectively [Fig. 3A] [Hansen et al. 2008; Broughton et al. 2016]) have been solved. Apart from unbound βc, the only published components of the IL-3 receptor signaling complex are a nuclear magnetic resonance (NMR) structure of IL-3 (PDB ID: 1JLI [Feng et al. 1996]) and IL3Rα bound to a therapeutic antibody (Broughton et al. 2014). IL-3 and GM-CSF adopt a typical cytokine four-helical bundle motif (helices A–D), whereas IL-5 forms an intertwined dimer of two four-helix bundles, with the fourth helix of each monomer being swapped (Milburn et al. 1993). The three α subunits share a similar architecture consisting of three fibronectin III (FNIII) domains (NTD, D2, and D3) that adopt a "wrench-like" conformation relative to the bound cytokine (Fig. 3A). The shared βc subunit has four linked FNIII domains (D1–D4), which, in solution and the solid state, form an arched intertwined homodimer (Fig. 3A).

Receptor activation is initiated when the cytokine binds to its specific α subunit (site 1, Fig. 3A). Residues from all three FNIII domains of the α subunit participate in the interaction with the cytokine to form a binary complex. This is a low-affinity interaction that differs markedly between the three cytokines with K_D values of ∼100 nM for IL-3, 1-10 nM for GM-CSF, and 1 nM for IL-5 (Broughton et al. 2012). In both the GM-CSF and IL-5 binary complexes, the α subunit amino-terminal domain (NTD) interacts with its specific cytokine via extensive inter-β-strand main-chain hydrogen bonds (Patino et al. 2011; Kusano et al. 2012; Broughton et al. 2016). Residues within GMRα D2 and D3 make significant polar interactions with GM-CSF (Fig. 3B), whereas D2 and D3 of IL5Rα make very few polar interactions with IL-5 and instead interact with the cytokine via numerous van der Waals interactions. Mutagenesis studies have shown that a small patch of residues in the IL5Rα NTD appears to drive the formation of the binary complex, whereas residues in D2 and D3 are critical in the GM-CSF binary complex formation. In contrast to GMRα and IL5Rα, residues in the IL3Rα

Cite this article as *Cold Spring Harb Perspect Biol* doi: 10.1101/cshperspect.a028514

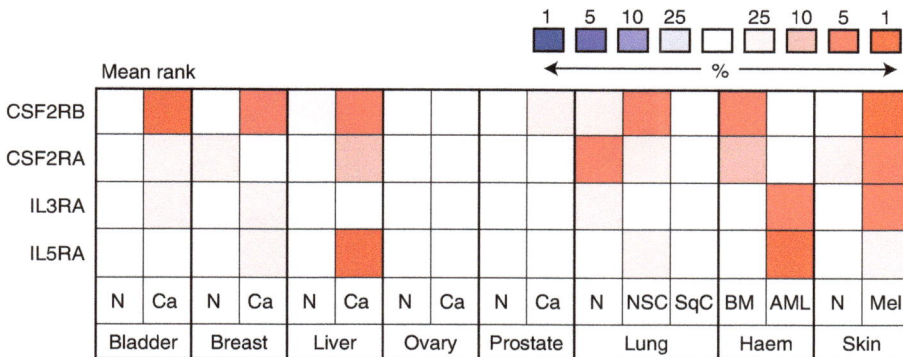

Figure 2. Relative gene expression for the β common (βc) and α subunits in normal and malignant tissue. Mean rank gene expression data extracted from Oncomine of genes for βc (*CSF2RB*), granulocyte macrophage colony-stimulating factor (GM-CSF) receptor α (GMRα) (*CSF2RA*), IL3Rα (*IL3RA*), and IL5Rα (*IL5RA*) in normal versus malignant tissue. Key indicates percentage mean rank enrichment. Datasets used: Roth normal, Roth normal 2, Su multicancer, Wouters leukemia, Xu melanoma, and Ge normal. N, Normal; Ca, cancer; NSC, non-small-cell lung cancer; SqC, squamous cell lung cancer; BM, bone marrow; AML, acute myeloid leukemia; Mel, melanoma; haem, hematopoietic system.

NTD and D3 have been shown to be important in the formation of the IL-3 binary complex (Broughton et al. 2014). The binary complexes then interact with the βc subunit to form a ternary complex, as observed in the GM-CSF ternary structure (sites 2 and 3, Fig. 3A) (Hansen et al. 2008; Broughton et al. 2016); this is a high-affinity interaction with K_D values of 100 pM for GM-CSF, 250 pM for IL-5, and 140 pM for IL-3 (Broughton et al. 2012). The symmetrical nature of the βc homodimer allows two binary complexes to bind to one βc subunit and to form either a hexameric complex (IL-3 and GM-CSF) or an octameric complex (IL-5). Crystal packing in the GM-CSF receptor structure revealed a second hexamer stacked against the first to create a dodecamer (sites 4 and 5, Fig. 3A). The physiological relevance of this dodecamer in JAK/STAT signaling has been shown by (1) functional studies of site 4 mutants, and (2) site-4-specific antibodies, in which both abolished JAK/STAT signaling (Hansen et al. 2008; Broxmeyer et al. 2012). Although the IL-3 and IL-5 ternary complexes are yet to be solved, it is highly likely that a similar higher-order receptor assembly occurs during IL-3 (dodecamer) and IL-5 (hexadecamer) signaling.

During the assembly of the ternary complex, the structure shows that GM-CSF appears to actively transition toward the βc subunit, becoming locked into place by numerous interactions in the site 2 interface (Fig. 3C) while maintaining critical site 1 interactions with GMRα (Fig. 3B) (Broughton et al. 2016). The recently solved structures of the GM-CSF: GMRα binary and the improved GM-CSF: GMRα:βc ternary complexes (Broughton et al. 2016) have revealed key residues involved in the transition from the binary to ternary complex. These include GMRα R283, which acts as a pivot point for GM-CSF (Fig. 3B), and βc Y421, which forms critical interactions with GM-CSF E21 (Fig. 3C). E21 is structurally conserved across the βc family of cytokines and mutation of the equivalent residues in IL-3 (E22) and IL-5 (E13) has been shown to abrogate binding and signaling (Barry et al. 1994; Tavernier et al. 1995). The ternary complex structure showed a shift in the amino-terminal region of GM-CSF helix A and reorientation of the βc D4 BC loop to prevent steric clashes between residues in βc and GM-CSF (Broughton et al. 2016) during the binary to ternary transition. These movements accommodate βc D4 on complexation, and avoid steric hindrance with residues in the membrane proximal domain of βc and facilitate βc contact with E21 in GM-CSF (site 2, Fig. 3A,C). In addition, significant remodeling

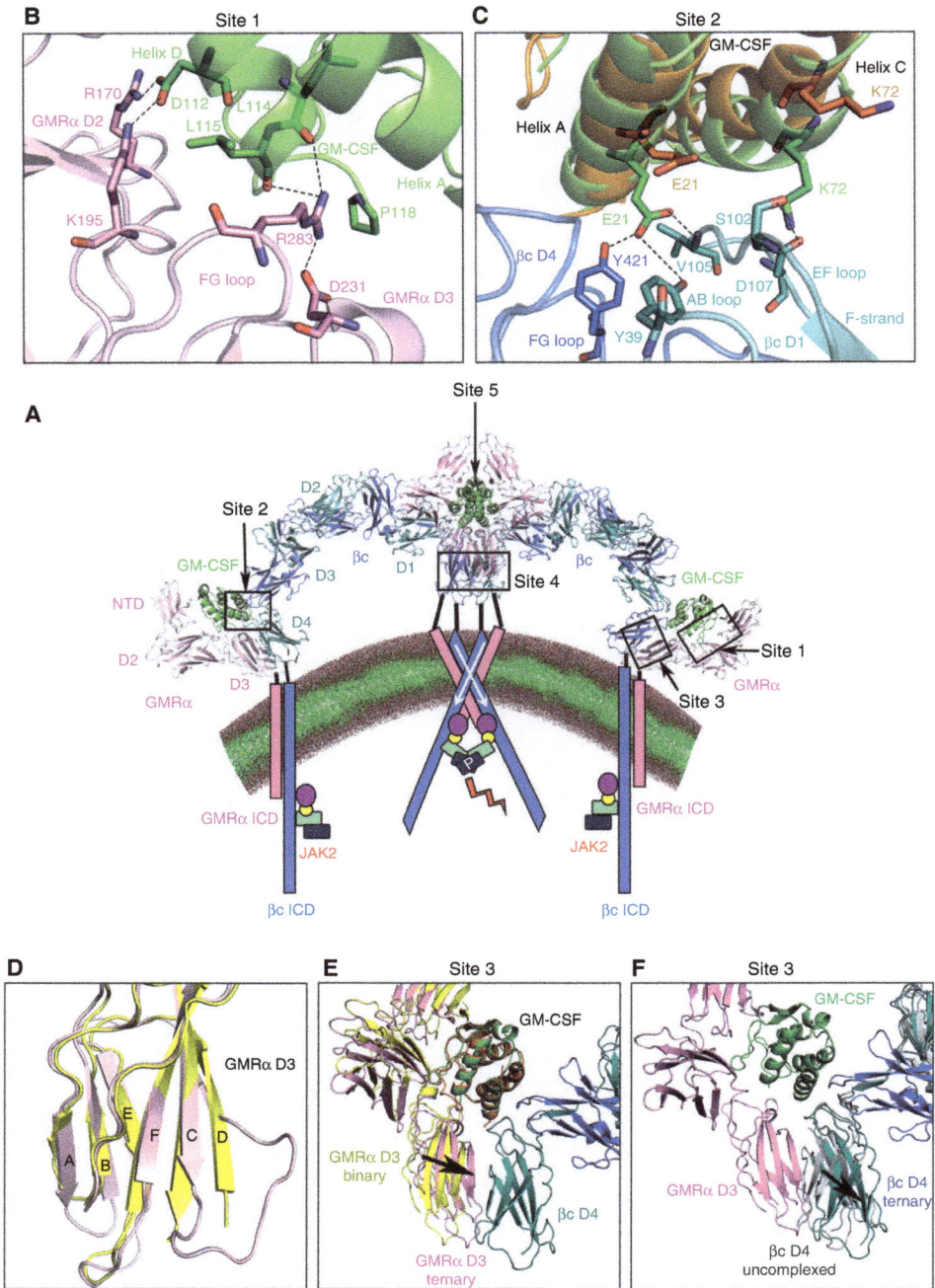

Figure 3. β Common (βc) cytokine receptor complex assembly and initiation of signaling. (*A*) The α and βc receptor subunits for this cytokine family each consist of an extracellular domain connected, via a short juxtamembrane region and helical transmembrane domain, to an intracellular tail. The granulocyte macrophage colony-stimulating factor (GM-CSF) dodecamer is shown bound to the cell membrane as a cartoon. GM-CSF is colored green, GM-CSF receptor α (GMRα) pink, and the βc dimer blue and teal. (*Legend continues on following page.*)

 Cite this article as *Cold Spring Harb Perspect Biol* doi: 10.1101/cshperspect.a028514

within the membrane proximal region of GMRα D3 was observed during the binary to ternary transition (Fig. 3D). These different structures illustrate the dynamic nature of receptor assembly and how conformational changes in the receptor complexes initiate signaling (Broughton et al. 2016).

To understand how movements of the βc-family heterodimeric receptor extracellular domains might be transmitted through to the intracellular domains, we can examine the simpler homodimeric human growth hormone receptor (hGHR), the archetypal cytokine receptor system. Waters and coworkers recently showed that adjacent juxtamembrane and transmembrane domains interact to keep hGHR as a dimer (Brooks et al. 2014). In this preformed homodimer, the two hGHR intracellular domains are held in a parallel inactive state, with the pseudokinase domain of one JAK2 molecule inhibiting the kinase domain of the JAK2 molecule bound to the adjacent hGHR intracellular domain. Binding of human growth hormone (hGH) to the hGHR extracellular domains results in the rotation of the juxtamembrane regions and transmembrane helices from the inactive parallel state to a left-handed crossover configuration, splaying apart of the carboxy-terminal end of the transmembrane helices and separation of the intracellular domains. The movement of the hGHR intracellular domains uncouples the JAK2 pseudokinase domain from the adjacent JAK2 kinase domain, thereby allowing the two adjacent kinase domains to interact and trigger cross-activation of JAK2.

The receptor activation mechanism for the βc cytokine family is more complicated than for the hGH by virtue of their heterodimeric nature; however, the basic signaling principles appear similar. For the GM-CSF receptor the major signaling configuration appears to be a dodecamer or higher-order complex (Hansen et al. 2008; Broughton et al. 2016). The distance between the βc juxtamembrane domains in the center of the GM-CSF dodecameric complex (i.e., immediately below site 4, Fig. 3A) is similar to the distance between the juxtamembrane domains of the hGHR homodimer (∼10 Å). The movement of the membrane proximal domains of GMRα (D3) and the βc subunit (D4) observed on cytokine binding and affinity conversion (Fig. 3A,E,F) suggests that the transmembrane helices of the GM-CSF receptor

---

Figure 3. (*Continued*) The juxtamembrane domains are shown as black rods, whereas the transmembrane and intracellular domains are indicated as rectangles and colored as above. Janus kinase 2 (JAK2) (FERM, purple circle; SH2, yellow circle; JH2, light green rectangle; JH1, dark blue rectangle) is shown bound via the FERM domain to the intracellular domain of βc. The direction of signaling through the transmembrane and intracellular regions of βc is indicated by white arrows, and the productive interaction of the JH1–JH2 domains of the adjacent JAK2 molecules resulting in signaling is indicated by a white "P" and jagged red arrow. The sites of interaction within the extracellular dodecamer complex are indicated. (*B*) Interaction of GMRα D2 and D3 with GM-CSF (part of site 1) in the ternary complex. R283 in the FG loop of GMRα D3 acts as a pivot point for GM-CSF movement during ternary complex formation. This region of site 1 is almost identical in both the binary and ternary complexes. Some residues key to the formation of the binary complex are shown in the panel. Polar interactions are depicted by black dashed lines, colors as in *A*. (*C*) View of site 2 illustrating the reorientation of GM-CSF helices A and C on ternary complex formation. The position of GM-CSF in the binary complex is shown in orange. Ternary complex colored as in *A*. (*D*) Remodeling of the GMRα D3 domain D strand to a two-turn coiled loop on ternary complex formation. GMRα D3 in the binary complex is colored yellow and pink in the ternary complex. (*E*) Movement of the GMRα D3 domain on recruitment of βc (site 3) to form the ternary complex. The binary and ternary complexes have been aligned via GM-CSF. In the binary complex, GMRα is colored yellow and GM-CSF is orange. Components of the ternary complex are colored as in *A*. Direction of domain movement is indicated by the arrow. (*F*) Movement of the βc D4 domain on formation of the ternary complex (site 3). The uncomplexed βc dimer was aligned with the βc dimer in the GM-CSF ternary complex. Ternary complex colored as in *A*. The uncomplexed βc dimer is colored gray. Direction of domain movement is indicated by the arrow. A full description of the structural changes that occur during the binary to ternary transition can be found in Broughton et al. (2016).

---

might also cross over into an active state as the cytokine "rolls" from its binary to ternary complex orientation. At the center of the GM-CSF receptor dodecamer, the rotation of the βc D4 domains is likely to translate to a splaying out of the central transmembrane domains, effectively positioning the JAK2 molecules associated with the βc intracellular domains to initiate JAK/ STAT signaling (Fig. 3A).

RECEPTOR PROXIMAL ACTIVATION OF SIGNALING

The βc subunit is the main signaling subunit of the GM-CSF, IL-3, and IL-5 receptors and mutational analysis of βc has provided insight into the mechanisms of βc activation and the location of distinct functional domains (Hercus et al. 2013). However, it is becoming increasingly clear that while the α subunits provide ligand specificity they also regulate some signaling outcomes and this may explain some of the signaling diversity arising from the βc receptor family.

Formation of the multimeric receptor complex, consisting of ligand, a ligand-specific α subunit, and βc, triggers the activation of kinases associated with these receptors and leads initially to activation of the JAK2/STAT5 and PI3 kinase pathways (Fig. 4). JAK2 is thought to be the tyrosine kinase responsible for βc subunit phosphorylation and data from JAK2-deficient mouse models support this view because JAK2 is essential for IL-3-dependent biological responses (Silvennoinen et al. 1993). However, it is now clear that other tyrosine kinases associate with the receptor complex and contribute to βc signal transduction. For example, whereas fetal liver myeloid progenitor cells from JAK2-deficient mice fail to respond to IL-3 in a colony-formation assay (Parganas et al. 1998), IL-3 induced colony formation from the cells isolated from JAK1-deficient mice, albeit with reduced number and size compared with those from wild-type mice. This suggests that JAK1 is contributing to the full effect of IL-3 receptor signaling, but is not required for signaling. This may be specific for the IL-3 receptor because deletion of JAK1 had no observable effect on

GM-CSF- and IL-5-dependent colony formation (Rodig et al. 1998).

The tyrosine kinase Lyn interacts with the GM-CSF receptor complex, having been shown to bind βc (Adachi et al. 1999; Dahl et al. 2000) and coimmunoprecipitate with GMRα (Perugini et al. 2010). The functional role of Lyn in βc receptor signaling and whether it is required for specific signaling outcomes is not yet known. Some indications of a functional link between Lyn and the βc receptor come from in vivo studies of Lyn-deficient mice in which bone marrow–derived macrophages showed increased sensitivity to GM-CSF stimulation and B cells showed increased IL-3-dependent cell survival and STAT5 phosphorylation (Infantino et al. 2014). These data support a model in which Lyn acts as a negative regulator of GM-CSF and IL-3 receptor signaling (Scapini et al. 2009). There are also data from the use of Src-family kinase inhibitors suggesting that Lyn kinase recruitment to the GM-CSF receptor contributes to GM-CSF-dependent survival signaling; a caveat, however, is that the specificity of such inhibitors is not known (Perugini et al. 2010).

IL-3/IL-5/GM-CSF stimulation also activates the PI3K signaling pathway. In other receptor systems, PI3K binds to phosphotyrosine residues of an activated receptor through its SH2 domain. However, in βc cytokine signaling, most studies suggest that PI3K interacts with a proline-rich motif within the intracellular region of the α subunit, probably through the SH3 domain of PI3K (Dhar-Mascareno et al. 2005; Perugini et al. 2010). Recruitment of PI3K to the receptor complex results in phosphorylation of Ser585 of βc and binding of the chaperone molecule 14-3-3, giving rise to survival signals that specifically repress apoptosis (Guthridge et al. 2000, 2004). The βc Ser585 residue is part of a bidentate motif that includes an Shc-binding site at the βc Tyr577 residue (Guthridge et al. 2006). Phosphorylation of these two residues is mutually exclusive, and they act as a binary switch to activate distinct signaling pathways associated with either survival alone or both proliferation and survival (Guthridge et al. 2006). One experimental

Cite this article as *Cold Spring Harb Perspect Biol* doi: 10.1101/cshperspect.a028514

method that modulates this binary switch is to vary the ligand concentration. At low concentrations of GM-CSF (<10 pM), βc Ser585 is selectively phosphorylated and this results in a cell survival signal through recruitment of 14-3-3 and activation of the PI3K pathway. At higher GM-CSF concentrations, βc Tyr577 is phosphorylated, resulting in recruitment of Shc and activation of cell survival and cell-proliferation pathways (Guthridge et al. 2004). The ability to maintain a survival-only signal at low concentrations of GM-CSF may be critical for steady-state maintenance of cell viability, with an emergency response initiated by high GM-CSF levels in response to stress, injury, or infection. Recent reports have shown that physiological concentrations of cytokine in the picomolar range were sufficient to activate PI3K protein kinase activity, leading to βc Ser585 phosphorylation and hemopoietic cell survival but did not activate PI3K lipid kinase signaling or promote proliferation (Thomas et al. 2013). The recently determined GM-CSF receptor complex structure (Broughton et al. 2016) can allow one to speculate whether altering the concentration of ligand can influence the probability of the formation of intermediate and multimeric receptor complexes.

Despite numerous examples of βc phosphorylation regulating receptor activity, there is no convincing evidence to suggest that the α subunits undergo similar posttranslational modifications. However, it is also clear that the short intracellular domain of the α subunits is absolutely required for signaling (Sakamaki et al. 1992; Weiss et al. 1993; Takaki et al. 1994; Cornelis et al. 1995; Barry et al. 1997) and contribute to cytokine-specific ligand stimulation (Evans et al. 2002). These data suggest that the intracellular domains of the α subunits are important for the assembly of an active receptor complex or provide an initiating role in activation through interactions with kinases or signaling molecules required to initiate downstream signaling events (Polotskaya et al. 1993; Ebner et al. 2003; Perugini et al. 2010; Liontos et al. 2011). Identifying the proteins recruited to the α subunits, and how these fit into the active conformation of the receptor complexes, will provide a more complete understanding of how signaling through βc may have specific outcomes dependent on the α subunit and the ligand engaged.

DEVELOPMENT OF βc CYTOKINE THERAPEUTIC MOLECULES: WHY, WHERE, AND HOW

The βc cytokines offer direct therapeutic utility by their ability to stimulate the production and function of macrophages and neutrophils. Conversely, they can also contribute to pathology in a number of cancer and inflammation settings thereby prompting the development of various targeted therapies. Because GM-CSF enhances host responses to certain cancers by recruitment and activation of antigen-presenting cells (APCs) such as dendritic cells (Mach et al. 2000), talimogene laherparepvec (T-VEC) has been developed as a vaccine engineered from the herpes simplex virus to kill cancer cells and to express GM-CSF to stimulate the host anti-immune response (Andtbacka et al. 2016). T-VEC is the first oncolytic virus to gain approval for use in the treatment of advanced melanoma in the United States and Europe (Liu et al. 2003; Andtbacka et al. 2015; Hoeller et al. 2016; Kaufman et al. 2016). Recent clinical trials have shown that T-VEC, in combination with ipilimumab, had increased efficacy compared with either agent alone (Puzanov et al. 2016). Another oncolytic virus that expresses GM-CSF, ONCOS-102, reduced tumor growth in a murine mesothelioma xenograft model, whereas 40% of patients in phase I clinical trials responded to the ONCOS-102 treatment (Kuryk et al. 2016; Ranki et al. 2016). In contrast, IL-3 is not used therapeutically owing to undesirable side effects (Eder et al. 1997), although a recent report from our group suggests a role for IL-3 in the repair of cardiac tissue in patients following acute myocardial infarction (AMI). We showed that IL-3 has potent growth factor activity on human CD133[+] cells, which have proangiogenic properties while maintaining low immunogenic potential (Moldenhauer et al. 2015).

Monoclonal antibodies (MAbs) are versatile therapeutic molecules increasingly used clini-

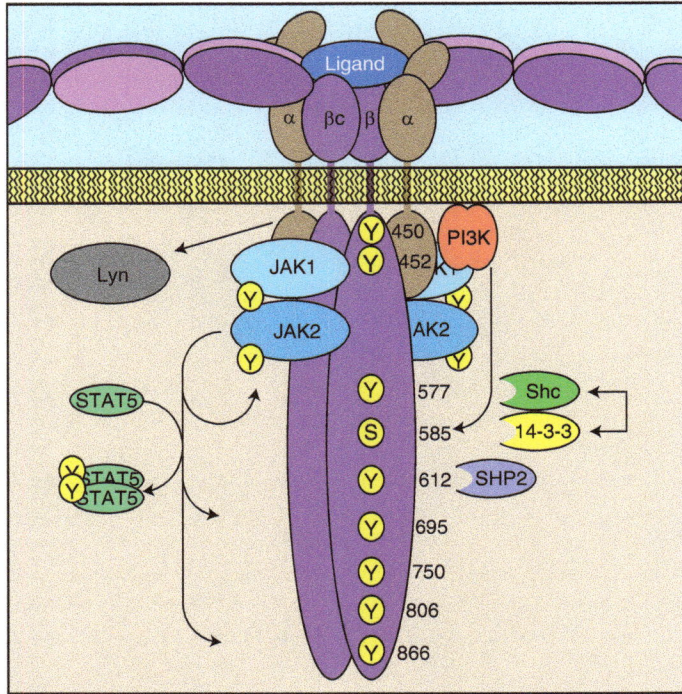

Figure 4. Intracellular signaling pathways arising from β common (βc) receptor activation. Schematic of key protein interactions following activation of the βc cytokine receptor family. Low concentrations of cytokine promote the activation of phosphoinositide 3-kinase (PI3K) and subsequent phosphorylation of βc on Ser585 (yellow circle) resulting in the recruitment of 14-3-3 and promoting cell survival. High concentrations of cytokine result in Janus kinase (JAK) 1 and JAK2 transphosphorylation and subsequent phosphorylation of βc at eight intracellular tyrosine residues (yellow circles), activating signal transducers and activators of transcription 5 (STAT5) signaling. Cytokine binding activates the tyrosine kinase, Lyn, which interacts with either βc or the α subunit. Phosphorylation of βc on Tyr577 is mutually exclusive with Ser585 phosphorylation and leads to the recruitment of Shc and the loss of 14-3-3 binding, while promoting cell survival and proliferation. SHP2 has been shown to interact with Tyr612 to promote activation of the mitogen-activated protein kinase (MAPK) signaling pathway.

cally for their specificity and well-characterized in vivo properties that have allowed U.S. Food and Drug Administration (FDA) fast tracking. Several MAbs are being developed against βc cytokines and their receptors (Table 1). The GM-CSF blocking antibody GSK3196165/ MOR103 has shown preliminary evidence for efficacy in patients with RA (Behrens et al. 2015; Shiomi and Usui 2015) and was well tolerated in a phase Ib clinical trial in patients with MS (Constantinescu et al. 2015). A number of other GM-CSF blocking antibodies have been developed, including lenzilumab/KB003 (Molfino et al. 2016), MORAb-022, and namilumab,

and are currently progressing through clinical trials in patients with RA, plaque psoriasis, CMML, or asthma (Table 1). The anti-GMRα antibody, mavrilimumab/CAM-3001, has produced clinically significant responses in a phase IIa study in patients with RA with generally mild-to-moderate adverse events (Burmester et al. 2011, 2013). Importantly, there was no evidence of lung toxicity in patients treated with mavrilimumab, which might arise from GM-CSF inhibition, possibly because of limited exposure of the lung to systemically administered antibody (Campbell et al. 2016) or the possibility that effective GM-CSF blockade in patients with PAP requires a polyclonal GM-CSF autoantibody response (Piccoli et al. 2015).

Several clinical trials are underway to develop antibody therapies that target IL3Rα for patients with leukemia or myelodysplastic syndromes (MDSs). CSL362, an anti-IL3Rα MAb that blocks IL-3 binding and signaling (Sun et al. 1996; Jin et al. 2009) through a dual mechanism (Broughton et al. 2014), is optimized for antibody-dependent cell-mediated cytotoxicity (ADCC) (Busfield et al. 2014). CSL362/JNJ-56022473 has recently progressed to phase II clinical trials in patients with AML in combination with decitabine (ClinicalTrials.gov identifier: NCT02472145). CSL362 may also be effective at targeting leukemic stem cells in patients with CML as these cells are thought to be insensitive to tyrosine kinase inhibitors (TKIs) (Corbin et al. 2011). We recently observed that IL3Rα is overexpressed in $CD34^+/CD38^-$ cells from patients with CML and that CSL362-mediated killing of these cells by ADCC was enhanced by treatment with the TKI nilotinib (Nievergall et al. 2014). This MAb is being progressed in parallel as a potential therapy for SLE given its ability to deplete pDCs and, therefore, down-regulate type I and III interferon pathways (Oon et al. 2016a). The anti-IL3Rα MAb, KHK2823 (Akiyama et al. 2015), is currently in phase I clinical trials in patients with AML or MDS (ClinicalTrials.gov identifier: NCT02181699) as is the bispecific antibody MGD006 (Al-Hussaini et al. 2016) that targets IL3Rα and CD3 (ClinicalTrials.gov identifier: NCT02152956). The development of an anti-CD123 antibody–drug conjugate that directs a chemotherapy agent, camptothecin, to the target cell (Li et al. 2016a) represents an alternative use for anti-IL3Rα MAbs. In related approaches, engineered T cells expressing CARs have been developed to target IL3Rα expressed on AML blast cells and leukemic stem cells (Testa et al. 2002). These studies indicate that CD123-directed CAR T cells (CART123) are effective at inducing cell killing (Tettamanti et al. 2013) and are able to eradicate primary AML cells engrafted in immunodeficient mice (Mardiros et al. 2013; Gill et al. 2014). Although normal hemopoietic stem cells are not reported to express IL3Rα, CART123 impact on normal human myelopoiesis in one study (Gill et al. 2014) suggests some caution is needed with this approach.

An alternate to antibody therapy for hematological malignancies that overexpress IL3Rα is the fusion protein SL-401/DT388IL3, which fuses the catalytic and translocation domains of diphtheria toxin (DT) to IL-3 (Feuring-Buske et al. 2002). Initial phase I clinical trials with DT388IL3 in patients with chemorefractory AML or myelodysplasia showed that the treatment was well tolerated and some favorable clinical responses were observed (Frankel et al. 2008). Phase II clinical trials of SL-401 are continuing in patients with BPDCN (Frankel et al. 2014; Angelot-Delettre et al. 2015) and AML while improved fusion proteins are being developed (Liu et al. 2004; Testa et al. 2005; Hogge et al. 2006; Frankel et al. 2008; Frolova et al. 2014).

The prominent role played by IL-5 and eosinophils in asthma has prompted development and extensive clinical testing of a number of antibodies to IL-5 itself and to the IL-5 receptor (Table 1; and in Keating 2015; Patterson et al. 2015; Nixon et al. 2016). Initial clinical studies with these antibodies in patients with mild-to-moderate asthma were disappointing until a responder subset of asthma patients were identified with elevated blood eosinophil counts ($\geq$ 300 cells/µl). This patient population showed successful depletion of the key pharmacodynamic biomarkers, eosinophils (in blood, bone marrow, and sputum), and basophils (in blood), which confirmed target engagement and provided an understanding of mechanism of action (Haldar et al. 2009; Nair et al. 2009). In more recent multicenter phase III trials, an anti-IL5Rα MAb, benralizumab, showed efficacy in reducing annual exacerbation rates and safety for uncontrolled severe asthma patients with elevated eosinophils, further indicating that this patient population will likely receive the greatest benefit from benralizumab treatment (Bleecker et al. 2016; FitzGerald et al. 2016). Mepolizumab/Nucala/SB-240563, an anti-IL-5 MAb, was the first candidate therapeutic targeting the IL-5 pathway to enter clinical studies and, since the early 2000s, there have been multiple trials in a range of asthmatic populations. Although in earlier studies mepolizumab showed evidence of target engagement, as evidenced by a reduction in circulating eosinophils, there was no improvement in FEV_1 or other clinical asthma measures (Leckie et al. 2000; Menzies-Gow et al. 2007). Similar to benralizumab, the efficacy of mepolizumab is now being explored for treatment of severe asthma with an eosinophilic phenotype and it is also currently in clinical trials for a number of other conditions (Table 1), including hypereosinophilic syndrome (HES) (Rothenberg et al. 2008) and COPD.

The failure of IL-5 targeting therapies to treat endotypes of asthma other than eosinophilic asthma reflects the heterogeneous nature of asthma, the role of multiple cell types including neutrophils, as well as redundancy in βc cytokine signaling. Targeting the βc subunit allows simultaneous targeting of all three βc cytokines, which may be a more effective therapeutic approach. In one approach, TPI ASM8 contains two antisense oligonucleotides targeting the expression of βc and the chemokine receptor CCR3. Preliminary clinical studies in patients with mild asthma indicated that TPI ASM8 was safe and reduced eosinophil accumulation after allergen challenge (Imaoka et al. 2011). Alternatively, MAbs against the common cytokine-binding site in βc, site 2, are being developed. Initially, MAb BION-1 (Sun et al. 1999) showed proof-of-principle for blocking site 2 in βc. This has been followed by MAb CSL311 (Panousis et al. 2016), which also blocks the function of all three βc cytokines. CSL311 is a fully human MAb that binds βc at the cytokine-binding surface (site 2, Fig. 3A) with high affinity and is a potent antagonist of GM-CSF, IL-3, and IL-5 function, inhibiting the survival of cells isolated from inflammatory airway disease tissue. Although the pharmacodynamic and pharmacokinetic properties of CSL311 and TPI ASM8 are likely to be quite different, the indicated safety profile of TPI ASM8 is encouraging and suggests that blocking βc function with CSL311 might yield a safe and efficacious intervention and be of broad utility in a range of asthma endotypes owing to its ability to simultaneously block activity of all three βc family cytokines.

ACKNOWLEDGMENTS

This work is supported by grants from the National Health and Medical Research Council of Australia (NHMRC), the Australian Cancer Research Foundation, the Leukemia Foundation of Australia, and CSL Limited, as well as funding from Victorian State Government Operational Infrastructure Support and Australian Government NHMRC Independent Research Institute Infrastructure Support Scheme.

REFERENCES

Adachi T, Pazdrak K, Stafford S, Alam R. 1999. The mapping of the Lyn kinase binding site of the common β subunit of IL-3/granulocyte-macrophage colony-stimulating factor/IL-5 receptor. *J Immunol* **162:** 1496–1501.

Akiyama T, Takayanagi SI, Maekawa Y, Miyawaki K, Jinnouchi F, Jiromaru T, Sugio T, Daitoku S, Kusumoto H, Shimabe M, et al. 2015. First preclinical report of the efficacy and PD results of KHK2823, a non-fucosylated fully human monoclonal antibody against IL-3Rα. *Blood* **126:** 1349.

Al-Hussaini M, Rettig MP, Ritchey JK, Karpova D, Uy GL, Eissenberg LG, Gao F, Eades WC, Bonvini E, Chichili GR, et al. 2016. Targeting CD123 in acute myeloid leukemia using a T-cell-directed dual-affinity retargeting platform. *Blood* **127:** 122–131.

Andtbacka RH, Kaufman HL, Collichio F, Amatruda T, Senzer N, Chesney J, Delman KA, Spitler LE, Puzanov I, Agarwala SS, et al. 2015. Talimogene Laherparepvec improves durable response rate in patients with advanced melanoma. *J Clin Oncol* **33:** 2780–2788.

Andtbacka RH, Agarwala SS, Ollila DW, Hallmeyer S, Milhem M, Amatruda T, Nemunaitis JJ, Harrington KJ, Chen L, Shilkrut M, et al. 2016. Cutaneous head and neck melanoma in OPTiM, a randomized phase 3 trial of talimogene laherparepvec versus granulocyte-macrophage colony-stimulating factor for the treatment of unresected stage IIIB/IIIC/IV melanoma. *Head Neck* **38:** 1752–1758.

Angelot-Delettre F, Roggy A, Frankel AE, Lamarthee B, Seilles E, Biichle S, Royer B, Deconinck E, Rowinsky EK, Brooks C, et al. 2015. In vivo and in vitro sensitivity of blastic plasmacytoid dendritic cell neoplasm to SL-401, an interleukin-3 receptor targeted biologic agent. *Haematologica* **100:** 223–230.

Asquith KL, Ramshaw HS, Hansbro PM, Beagley KW, Lopez AF, Foster PS. 2008. The IL-3/IL-5/GM-CSF common receptor plays a pivotal role in the regulation of Th2 immunity and allergic airway inflammation. *J Immunol* **180:** 1199–1206.

Barry SC, Bagley CJ, Phillips J, Dottore M, Cambareri B, Moretti P, D'Andrea R, Goodall GJ, Shannon MF, Vadas MA, et al. 1994. Two contiguous residues in human interleukin-3, Asp21 and Glu22, selectively interact with the α- and β-chains of its receptor and participate in function. *J Biol Chem* **269**: 8488–8492.

Barry SC, Korpelainen E, Sun Q, Stomski FC, Moretti PA, Wakao H, D'Andrea RJ, Vadas MA, Lopez AF, Goodall GJ. 1997. Roles of the N and C terminal domains of the interleukin-3 receptor α chain in receptor function. *Blood* **89**: 842–852.

Bayne LJ, Beatty GL, Jhala N, Clark CE, Rhim AD, Stanger BZ, Vonderheide RH. 2012. Tumor-derived granulocyte-macrophage colony-stimulating factor regulates myeloid inflammation and T cell immunity in pancreatic cancer. *Cancer Cell* **21**: 822–835.

Behrens F, Tak PP, Ostergaard M, Stoilov R, Wiland P, Huizinga TW, Berenfus VY, Vladeva S, Rech J, Rubbert-Roth A, et al. 2015. MOR103, a human monoclonal antibody to granulocyte-macrophage colony-stimulating factor, in the treatment of patients with moderate rheumatoid arthritis: Results of a phase Ib/IIa randomised, double-blind, placebo-controlled, dose-escalation trial. *Ann Rheum Dis* **74**: 1058–1064.

Bernard F, Thomas C, Emile JF, Hercus T, Cassinat B, Chomienne C, Donadieu J. 2002. Transient hematologic and clinical effect of E21R in a child with end-stage juvenile myelomonocytic leukemia. *Blood* **99**: 2615–2616.

Bhattacharya P, Budnick I, Singh M, Thiruppathi M, Alharshawi K, Elshabrawy H, Holterman MJ, Prabhakar BS. 2015. Dual role of GM-CSF as a pro-inflammatory and a regulatory cytokine: Implications for immune therapy. *J Interferon Cytokine Res* **35**: 585–599.

Bleecker ER, FitzGerald JM, Chanez P, Papi A, Weinstein SF, Barker P, Sproule S, Gilmartin G, Aurivillius M, Werkstrom V, et al. 2016. Efficacy and safety of benralizumab for patients with severe asthma uncontrolled with high-dosage inhaled corticosteroids and long-acting β2-agonists (SIROCCO): A randomised, multicentre, placebo-controlled phase 3 trial. *Lancet* **388**: 2115–2127.

Boyd TD, Bennett SP, Mori T, Governatori N, Runfeldt M, Norden M, Padmanabhan J, Neame P, Wefes I, Sanchez-Ramos J, et al. 2010. GM-CSF upregulated in rheumatoid arthritis reverses cognitive impairment and amyloidosis in Alzheimer mice. *J Alzheimers Dis* **21**: 507–518.

Brines M, Grasso G, Fiordaliso F, Sfacteria A, Ghezzi P, Fratelli M, Latini R, Xie QW, Smart J, Su-Rick CJ, et al. 2004. Erythropoietin mediates tissue protection through an erythropoietin and common β-subunit heteroreceptor. *Proc Natl Acad Sci* **101**: 14907–14912.

Brines M, Patel NS, Villa P, Brines C, Mennini T, De Paola M, Erbayraktar Z, Erbayraktar S, Sepodes B, Thiemermann C, et al. 2008. Nonerythropoietic, tissue-protective peptides derived from the tertiary structure of erythropoietin. *Proc Natl Acad Sci* **105**: 10925–10930.

Brines M, Dunne AN, van Velzen M, Proto PL, Ostenson CG, Kirk RI, Petropoulos IN, Javed S, Malik RA, Cerami A, et al. 2014. ARA 290, a nonerythropoietic peptide engineered from erythropoietin, improves metabolic control and neuropathic symptoms in patients with type 2 diabetes. *Mol Med* **20**: 658–666.

Broide DH, Wasserman SI, Alvaro-Gracia J, Zvaifler NJ, Firestein GS. 1989. Transforming growth factor-β1 selectively inhibits IL-3-dependent mast cell proliferation without affecting mast cell function or differentiation. *J Immunol* **143**: 1591–1597.

Brooks AJ, Dai W, O'Mara ML, Abankwa D, Chhabra Y, Pelekanos RA, Gardon O, Tunny KA, Blucher KM, Morton CJ, et al. 2014. Mechanism of activation of protein kinase JAK2 by the growth hormone receptor. *Science* **344**: 1249783.

Broughton SE, Dhagat U, Hercus TR, Nero TL, Grimbaldeston MA, Bonder CS, Lopez AF, Parker MW. 2012. The GM-CSF/IL-3/IL-5 cytokine receptor family: From ligand recognition to initiation of signaling. *Immunol Rev* **250**: 277–302.

Broughton SE, Hercus TR, Hardy MP, McClure BJ, Nero TL, Dottore M, Huynh H, Braley H, Barry EF, Kan WL, et al. 2014. Dual mechanism of interleukin-3 receptor blockade by an anti-cancer antibody. *Cell Rep* **8**: 410–419.

Broughton SE, Hercus TR, Nero TL, Dottore M, McClure BJ, Dhagat U, Taing H, Gorman MA, King-Scott J, Lopez AF, et al. 2016. Conformational changes in the GM-CSF receptor suggest a molecular mechanism for affinity conversion and receptor signaling. *Structure* **24**: 1271–1281.

Broxmeyer HE, Hoggatt J, O'Leary HA, Mantel C, Chitteti BR, Cooper S, Messina-Graham S, Hangoc G, Farag S, Rohrabaugh SL, et al. 2012. Dipeptidylpeptidase 4 negatively regulates colony-stimulating factor activity and stress hematopoiesis. *Nat Med* **18**: 1786–1796.

Bunda S, Kang MW, Sybingco SS, Weng J, Favre H, Shin DH, Irwin MS, Loh ML, Ohh M. 2013. Inhibition of SRC corrects GM-CSF hypersensitivity that underlies juvenile myelomonocytic leukemia. *Cancer Res* **73**: 2540–2550.

Burmester GR, Feist E, Sleeman MA, Wang B, White B, Magrini F. 2011. Mavrilimumab, a human monoclonal antibody targeting GM-CSF receptor-α, in subjects with rheumatoid arthritis: A randomised, double-blind, placebo-controlled, phase I, first-in-human study. *Ann Rheum Dis* **70**: 1542–1549.

Burmester GR, Weinblatt ME, McInnes IB, Porter D, Barbarash O, Vatutin M, Szombati I, Esfandiari E, Sleeman MA, Kane CD, et al. 2013. Efficacy and safety of mavrilimumab in subjects with rheumatoid arthritis. *Ann Rheum Dis* **72**: 1445–1452.

Busfield SJ, Biondo M, Wong M, Ramshaw HS, Lee EM, Ghosh S, Braley H, Panousis C, Roberts AW, He SZ, et al. 2014. Targeting of acute myeloid leukemia in vitro and in vivo with an anti-CD123 mAb engineered for optimal ADCC. *Leukemia* **28**: 2213–2221.

Campbell IK, Rich MJ, Bischof RJ, Dunn AR, Grail D, Hamilton JA. 1998. Protection from collagen-induced arthritis in granulocyte-macrophage colony-stimulating factor-deficient mice. *J Immunol* **161**: 3639–3644.

Campbell J, Nys J, Eghobamien L, Cohen ES, Robinson MJ, Sleeman MA. 2016. Pulmonary pharmacodynamics of an anti-GM-CSFRα antibody enables therapeutic dosing that limits exposure in the lung. *mAbs* **8**: 1398–1406.

Carr PD, Gustin SE, Church AP, Murphy JM, Ford SC, Mann DA, Woltring DM, Walker I, Ollis DL, Young IG. 2001. Structure of the complete extracellular domain of the common β subunit of the human GM-CSF, IL-3, and

IL-5 receptors reveals a novel dimer configuration. *Cell* **104:** 291–300.

Carr PD, Conlan F, Ford S, Ollis DL, Young IG. 2006. An improved resolution structure of the human β common receptor involved in IL-3, IL-5 and GM-CSF signalling which gives better definition of the high-affinity binding epitope. *Acta Crystallogr Sect F Struct Biol Cryst Commun* **62:** 509–513.

Castro M, Wenzel SE, Bleecker ER, Pizzichini E, Kuna P, Busse WW, Gossage DL, Ward CK, Wu Y, Wang B, et al. 2014. Benralizumab, an anti-interleukin 5 receptor α monoclonal antibody, versus placebo for uncontrolled eosinophilic asthma: A phase 2b randomised dose-ranging study. *Lancet Respir Med* **2:** 879–890.

Chen X, Wang X, Hossain S, O'Neill FA, Walsh D, van den Oord E, Fanous A, Kendler KS. 2007. Interleukin 3 and schizophrenia: The impact of sex and family history. *Mol Psychiatry* **12:** 273–282.

Chen Q, Wang X, O'Neill FA, Walsh D, Fanous A, Kendler KS, Chen X. 2008. Association study of *CSF2RB* with schizophrenia in Irish family and case—Control samples. *Mol Psychiatry* **13:** 930–938.

Chen P, Huang K, Zhou G, Zeng Z, Wang T, Li B, Wang Y, He L, Feng G, Shi Y. 2011. Common SNPs in *CSF2RB* are associated with major depression and schizophrenia in the Chinese Han population. *World J Biol Psychiatry* **12:** 233–238.

Choudhary GS, Yao X, Wang J, Peng B, Bader RA, Ren D. 2015. Human granulocyte macrophage colony-stimulating factor enhances antibiotic susceptibility of pseudomonas aeruginosa persister cells. *Sci Rep* **5:** 17315.

Codarri L, Gyulveszi G, Tosevski V, Hesske L, Fontana A, Magnenat L, Suter T, Becher B. 2011. RORγt drives production of the cytokine GM-CSF in helper T cells, which is essential for the effector phase of autoimmune neuroinflammation. *Nat Immunol* **12:** 560–567.

Coffelt SB, Wellenstein MD, de Visser KE. 2016. Neutrophils in cancer: Neutral no more. *Nat Rev Cancer* **16:** 431–446.

Collino M, Thiemermann C, Cerami A, Brines M. 2015. Flipping the molecular switch for innate protection and repair of tissues: Long-lasting effects of a non-erythropoietic small peptide engineered from erythropoietin. *Pharmacol Ther* **151:** 32–40.

Constantinescu CS, Asher A, Fryze W, Kozubski W, Wagner F, Aram J, Tanasescu R, Korolkiewicz RP, Dirnberger-Hertweck M, Steidl S, et al. 2015. Randomized phase 1b trial of MOR103, a human antibody to GM-CSF, in multiple sclerosis. *Neurol Neuroimmunol Neuroinflamm* **2:** e117.

Cook AD, Braine EL, Campbell IK, Rich MJ, Hamilton JA. 2001. Blockade of collagen-induced arthritis post-onset by antibody to granulocyte-macrophage colony-stimulating factor (GM-CSF): Requirement for GM-CSF in the effector phase of disease. *Arthritis Res* **3:** 293–298.

Cook AD, Pobjoy J, Sarros S, Steidl S, Durr M, Lacey DC, Hamilton JA. 2013. Granulocyte-macrophage colony-stimulating factor is a key mediator in inflammatory and arthritic pain. *Ann Rheum Dis* **72:** 265–270.

Corbin AS, Agarwal A, Loriaux M, Cortes J, Deininger MW, Druker BJ. 2011. Human chronic myeloid leukemia stem cells are insensitive to imatinib despite inhibition of BCR-ABL activity. *J Clin Invest* **121:** 396–409.

Cornelis S, Fache I, Van der Heyden J, Guisez Y, Tavernier J, Devos R, Fiers W, Plaetinck G. 1995. Characterization of critical residues in the cytoplasmic domain of the human interleukin-5 receptor α chain required for growth signal transduction. *Eur J Immunol* **25:** 1857–1864.

Croxford AL, Lanzinger M, Hartmann FJ, Schreiner B, Mair F, Pelczar P, Clausen BE, Jung S, Greter M, Becher B. 2015a. The cytokine GM-CSF drives the inflammatory signature of CCR2+ monocytes and licenses autoimmunity. *Immunity* **43:** 502–514.

Croxford AL, Spath S, Becher B. 2015b. GM-CSF in neuroinflammation: Licensing myeloid cells for tissue damage. *Trend Immunol* **36:** 651–662.

Dahan A, Dunne A, Swartjes M, Proto PL, Heij L, Vogels O, van Velzen M, Sarton E, Niesters M, Tannemaat MR, et al. 2013. ARA 290 improves symptoms in patients with sarcoidosis-associated small nerve fiber loss and increases corneal nerve fiber density. *Mol Med* **19:** 334–345.

Dahl ME, Arai KI, Watanabe S. 2000. Association of Lyn tyrosine kinase to the GM-CSF and IL-3 receptor common βc subunit and role of Src tyrosine kinases in DNA synthesis and anti-apoptosis. *Genes Cells* **5:** 143–153.

D'Andrea R, Rayner J, Moretti P, Lopez A, Goodall GJ, Gonda TJ, Vadas MA. 1994. A mutation of the common receptor subunit for interleukin-3 (IL-3), granulocyte-macrophage colony-stimulating factor, and IL-5 that leads to ligand independence and tumorigenicity. *Blood* **83:** 2802–2808.

Deane D, McInnes CJ, Percival A, Wood A, Thomson J, Lear A, Gilray J, Fleming S, Mercer A, Haig D. 2000. Orf virus encodes a novel secreted protein inhibitor of granulocyte-macrophage colony-stimulating factor and interleukin-2. *J Virol* **74:** 1313–1320.

Dentelli P, Rosso A, Garbarino G, Calvi C, Lombard E, Di Stefano P, Defilippi P, Pegoraro L, Brizzi MF. 2005. The interaction between KDR and interleukin-3 receptor (IL-3R) β common modulates tumor neovascularization. *Oncogene* **24:** 6394–6405.

Dentelli P, Rosso A, Olgasi C, Camussi G, Brizzi MF. 2011. IL-3 is a novel target to interfere with tumor vasculature. *Oncogene* **30:** 4930–4940.

Dhar-Mascareno M, Pedraza A, Golde DW. 2005. PI3-kinase activation by GM-CSF in endothelium is upstream of Jak/Stat pathway: Role of αGMR. *Biochem Biophys Res Commun* **337:** 551–556.

Ebner K, Bandion A, Binder BR, de Martin R, Schmid JA. 2003. GMCSF activates NF-κB via direct interaction of the GMCSF receptor with IκB kinase β. *Blood* **102:** 192–199.

Eder M, Geissler G, Ganser A. 1997. IL-3 in the clinic. *Stem Cells* **15:** 327–333.

Edwards TL, Wang X, Chen Q, Wormly B, Riley B, O'Neill FA, Walsh D, Ritchie MD, Kendler KS, Chen X. 2008. Interaction between interleukin 3 and dystrobrevin-binding protein 1 in schizophrenia. *Schizophr Res* **106:** 208–217.

El-Behi M, Ciric B, Dai H, Yan Y, Cullimore M, Safavi F, Zhang GX, Dittel BN, Rostami A. 2011. The encephalitogenicity of T$_H$17 cells is dependent on IL-1- and IL-23-induced production of the cytokine GM-CSF. *Nat Immunol* **12:** 568–575.

Evans CA, Ariffin S, Pierce A, Whetton AD. 2002. Identification of primary structural features that define the differential actions of IL-3 and GM-CSF receptors. *Blood* **100:** 3164–3174.

Felix J, Kandiah E, De Munck S, Bloch Y, van Zundert GC, Pauwels K, Dansercoer A, Novanska K, Read RJ, Bonvin AM, et al. 2016. Structural basis of GM-CSF and IL-2 sequestration by the viral decoy receptor GIF. *Nat Commun* **7:** 13228.

Feng Y, Klein BK, McWherter CA. 1996. Three-dimensional solution structure and backbone dynamics of a variant of human interleukin-3. *J Mol Biol* **259:** 524–541.

Feuring-Buske M, Frankel AE, Alexander RL, Gerhard B, Hogge DE. 2002. A diphtheria toxin–interleukin 3 fusion protein is cytotoxic to primitive acute myeloid leukemia progenitors but spares normal progenitors. *Cancer Res* **62:** 1730–1736.

FitzGerald JM, Bleecker ER, Nair P, Korn S, Ohta K, Lommatzsch M, Ferguson GT, Busse WW, Barker P, Sproule S, et al. 2016. Benralizumab, an anti-interleukin-5 receptor α monoclonal antibody, as add-on treatment for patients with severe, uncontrolled, eosinophilic asthma (CALIMA): A randomised, double-blind, placebo-controlled phase 3 trial. *Lancet* **16:** 31322–31328.

Frankel A, Liu JS, Rizzieri D, Hogge D. 2008. Phase I clinical study of diphtheria toxin–interleukin 3 fusion protein in patients with acute myeloid leukemia and myelodysplasia. *Leuk Lymphoma* **49:** 543–553.

Frankel AE, Woo JH, Ahn C, Pemmaraju N, Medeiros BC, Carraway HE, Frankfurt O, Forman SJ, Yang XA, Konopleva M, et al. 2014. Activity of SL-401, a targeted therapy directed to interleukin-3 receptor, in blastic plasmacytoid dendritic cell neoplasm patients. *Blood* **124:** 385–392.

Frei K, Bodmer S, Schwerdel C, Fontana A. 1986. Astrocyte-derived interleukin 3 as a growth factor for microglia cells and peritoneal macrophages. *J Immunol* **137:** 3521–3527.

Frolova O, Benito J, Brooks C, Wang RY, Korchin B, Rowinsky EK, Cortes J, Kantarjian H, Andreeff M, Frankel AE, et al. 2014. SL-401 and SL-501, targeted therapeutics directed at the interleukin-3 receptor, inhibit the growth of leukaemic cells and stem cells in advanced phase chronic myeloid leukaemia. *Br J Haematol* **166:** 862–874.

Fu YY, Zhang T, Xiu MH, Tang W, Han M, Yun LT, Chen DC, Chen S, Tan SP, Soares JC, et al. 2016. Altered serum levels of interleukin-3 in first-episode drug-naive and chronic medicated schizophrenia. *Schizophr Res* **176:** 196–200.

Garcia-Mendoza MG, Inman DR, Ponik SM, Jeffery JJ, Sheerar DS, Van Doorn RR, Keely PJ. 2016. Neutrophils drive accelerated tumor progression in the collagen-dense mammary tumor microenvironment. *Breast Cancer Res* **18:** 49.

Gauvreau GM, Pageau R, Seguin R, Carballo D, Gauthier J, D'Anjou H, Campbell H, Watson R, Mistry M, Parry-Billings M, et al. 2011. Dose-response effects of TPI ASM8 in asthmatics after allergen. *Allergy* **66:** 1242–1248.

Gill S, Tasian SK, Ruella M, Shestova O, Li Y, Porter DL, Carroll M, Danet-Desnoyers G, Scholler J, Grupp SA, et al. 2014. Preclinical targeting of human acute myeloid leukemia and myeloablation using chimeric antigen receptor-modified T cells. *Blood* **123:** 2343–2354.

Griese M, Ripper J, Sibbersen A, Lohse P, Lohse P, Brasch F, Schams A, Pamir A, Schaub B, Muensterer OJ, et al. 2011. Long-term follow-up and treatment of congenital alveolar proteinosis. *BMC Pediatr* **11:** 72.

Guthridge MA, Stomski FC, Barry EF, Winnall W, Woodcock JM, McClure BJ, Dottore M, Berndt MC, Lopez AF. 2000. Site-specific serine phosphorylation of the IL-3 receptor is required for hemopoietic cell survival. *Mol Cell* **6:** 99–108.

Guthridge MA, Barry EF, Felquer FA, McClure BJ, Stomski FC, Ramshaw H, Lopez AF. 2004. The phosphoserine-585-dependent pathway of the GM-CSF/IL-3/IL-5 receptors mediates hematopoietic cell survival through activation of NF-κB and induction of bcl-2. *Blood* **103:** 820–827.

Guthridge MA, Powell JA, Barry EF, Stomski FC, McClure BJ, Ramshaw H, Felquer FA, Dottore M, Thomas DT, To B, et al. 2006. Growth factor pleiotropy is controlled by a receptor Tyr/Ser motif that acts as a binary switch. *EMBO J* **25:** 479–489.

Gutschalk CM, Herold-Mende CC, Fusenig NE, Mueller MM. 2006. Granulocyte colony-stimulating factor and granulocyte-macrophage colony-stimulating factor promote malignant growth of cells from head and neck squamous cell carcinomas in vivo. *Cancer Res* **66:** 8026–8036.

Haldar P, Brightling CE, Hargadon B, Gupta S, Monteiro W, Sousa A, Marshall RP, Bradding P, Green RH, Wardlaw AJ, et al. 2009. Mepolizumab and exacerbations of refractory eosinophilic asthma. *N Engl J Med* **360:** 973–984.

Hamilton JA. 2008. Colony-stimulating factors in inflammation and autoimmunity. *Nat Rev Immunol* **8:** 533–544.

Hanea ST, Shanmugalingam U, Fournier AE, Smith PD. 2016. Preparation of embryonic retinal explants to study CNS neurite growth. *Exp Eye Res* **146:** 304–312.

Hansen G, Hercus TR, McClure BJ, Stomski FC, Dottore M, Powell J, Ramshaw H, Woodcock JM, Xu Y, Guthridge M, et al. 2008. The structure of the GM-CSF receptor complex reveals a distinct mode of cytokine receptor activation. *Cell* **134:** 496–507.

Hartmann FJ, Khademi M, Aram J, Ammann S, Kockum I, Constantinescu C, Gran B, Piehl F, Olsson T, Codarri L, et al. 2014. Multiple sclerosis-associated IL2RA polymorphism controls GM-CSF production in human TH cells. *Nat Commun* **5:** 5056.

Heij L, Niesters M, Swartjes M, Hoitsma E, Drent M, Dunne A, Grutters JC, Vogels O, Brines M, Cerami A, et al. 2012. Safety and efficacy of ARA 290 in sarcoidosis patients with symptoms of small fiber neuropathy: A randomized, double-blind pilot study. *Mol Med* **18:** 1430–1436.

Heinzelman P, Priebe MC. 2015. Engineering superactive granulocyte macrophage colony-stimulating factor transferrin fusion proteins as orally delivered candidate agents for treating neurodegenerative disease. *Biotechnol Prog* **31:** 668–677.

Hercus TR, Dhagat U, Kan WL, Broughton SE, Nero TL, Perugini M, Sandow JJ, D'Andrea RJ, Ekert PG, Hughes T, et al. 2013. Signalling by the βc family of cytokines. *Cytokine Growth Factor Rev* **24:** 189–201.

Hiwase DK, White DL, Powell JA, Saunders VA, Zrim SA, Frede AK, Guthridge MA, Lopez AF, D'Andrea RJ, To LB, et al. 2010. Blocking cytokine signaling along with intense Bcr-Abl kinase inhibition induces apoptosis in primary CML progenitors. *Leukemia* 24: 771–778.

Hoeller C, Michielin O, Ascierto PA, Szabo Z, Blank CU. 2016. Systematic review of the use of granulocyte-macrophage colony-stimulating factor in patients with advanced melanoma. *Cancer Immunol Immunother* 65: 1015–1034.

Hogge DE, Yalcintepe L, Wong SH, Gerhard B, Frankel AE. 2006. Variant diphtheria toxin–interleukin-3 fusion proteins with increased receptor affinity have enhanced cytotoxicity against acute myeloid leukemia progenitors. *Clin Cancer Res* 12: 1284–1291.

Hotchkiss RS, Sherwood ER. 2015. Immunology. Getting sepsis therapy right. *Science* 347: 1201–1202.

Imaoka H, Campbell H, Babirad I, Watson RM, Mistry M, Sehmi R, Gauvreau GM. 2011. TPI ASM8 reduces eosinophil progenitors in sputum after allergen challenge. *Clin Exp Allergy* 41: 1740–1746.

Infantino S, Jones SA, Walker JA, Maxwell MJ, Light A, O'Donnell K, Tsantikos E, Peperzak V, Phesse T, Ernst M, et al. 2014. The tyrosine kinase Lyn limits the cytokine responsiveness of plasma cells to restrict their accumulation in mice. *Sci Signal* 7: ra77.

Inoue Y, Trapnell BC, Tazawa R, Arai T, Takada T, Hizawa N, Kasahara Y, Tatsumi K, Hojo M, Ichiwata T, et al. 2008. Characteristics of a large cohort of patients with autoimmune pulmonary alveolar proteinosis in Japan. *Am J Respir Crit Care Med* 177: 752–762.

Jacobsen SE, Ruscetti FW, Roberts AB, Keller JR. 1993. TGF-β is a bidirectional modulator of cytokine receptor expression on murine bone marrow cells. Differential effects of TGF-β1 and TGF-β3. *J Immunol* 151: 4534–4544.

Jenkins BJ, Blake TJ, Gonda TJ. 1998. Saturation mutagenesis of the β subunit of the human granulocyte-macrophage colony-stimulating factor receptor shows clustering of constitutive mutations, activation of ERK MAP kinase and STAT pathways, and differential β subunit tyrosine phosphorylation. *Blood* 92: 1989–2002.

Jin L, Lee EM, Ramshaw HS, Busfield SJ, Peoppl AG, Wilkinson L, Guthridge MA, Thomas D, Barry EF, Boyd A, et al. 2009. Monoclonal antibody-mediated targeting of CD123, IL-3 receptor α chain, eliminates human acute myeloid leukemic stem cells. *Cell Stem Cell* 5: 31–42.

Jordan CT, Upchurch D, Szilvassy SJ, Guzman ML, Howard DS, Pettigrew AL, Meyerrose T, Rossi R, Grimes B, Rizzieri DA, et al. 2000. The interleukin-3 receptor α chain is a unique marker for human acute myelogenous leukemia stem cells. *Leukemia* 14: 1777–1784.

Kaufman HL, Amatruda T, Reid T, Gonzalez R, Glaspy J, Whitman E, Harrington K, Nemunaitis J, Zloza A, Wolf M, et al. 2016. Systemic versus local responses in melanoma patients treated with talimogene laherparepvec from a multi-institutional phase II study. *J Immunother Cancer* 4: 12.

Keating GM. 2015. Mepolizumab: First global approval. *Drugs* 75: 2163–2169.

Kim NK, Choi BH, Huang X, Snyder BJ, Bukhari S, Kong TH, Park H, Park HC, Park SR, Ha Y. 2009. Granulocyte-macrophage colony-stimulating factor promotes survival of dopaminergic neurons in the 1-methyl-4-phenyl-1,2,3,6-tetrahydropyridine-induced murine Parkinson's disease model. *Eur J Neurosci* 29: 891–900.

Kitamura T, Tanaka N, Watanabe J, Uchida, Kanegasaki S, Yamada Y, Nakata K. 1999. Idiopathic pulmonary alveolar proteinosis as an autoimmune disease with neutralizing antibody against granulocyte/macrophage colony-stimulating factor. *J Exp Med* 190: 875–880.

Kohanbash G, McKaveney K, Sakaki M, Ueda R, Mintz AH, Amankulor N, Fujita M, Ohlfest JR, Okada H. 2013. GM-CSF promotes the immunosuppressive activity of glioma-infiltrating myeloid cells through interleukin-4 receptor-α. *Cancer Res* 73: 6413–6423.

Kong T, Choi JK, Park H, Choi BH, Snyder BJ, Bukhari S, Kim NK, Huang X, Park SR, Park HC, et al. 2009. Reduction in programmed cell death and improvement in functional outcome of transient focal cerebral ischemia after administration of granulocyte-macrophage colony-stimulating factor in rats. Laboratory investigation. *J Neurosurg* 111: 155–163.

Korpelainen EI, Gamble JR, Smith WB, Goodall GJ, Qiyu S, Woodcock JM, Dottore M, Vadas MA, Lopez AF. 1993. The receptor for interleukin 3 is selectively induced in human endothelial cells by tumor necrosis factor α and potentiates interleukin 8 secretion and neutrophil transmigration. *Proc Natl Acad Sci* 90: 11137–11141.

Kosloski LM, Kosmacek EA, Olson KE, Mosley RL, Gendelman HE. 2013. GM-CSF induces neuroprotective and anti-inflammatory responses in 1-methyl-4-phenyl-1, 2,3,6-tetrahydropyridine intoxicated mice. *J Neuroimmunol* 265: 1–10.

Kuryk L, Haavisto E, Garofalo M, Capasso C, Hirvinen M, Pesonen S, Ranki T, Vassilev L, Cerullo V. 2016. Synergistic anti-tumor efficacy of immunogenic adenovirus ONCOS-102 (Ad5/3-D24-GM-CSF) and standard of care chemotherapy in preclinical mesothelioma model. *Int J Cancer* 139: 1883–1893.

Kusano S, Kukimoto-Niino M, Hino N, Ohsawa N, Ikutani M, Takaki S, Sakamoto K, Hara-Yokoyama M, Shirouzu M, Takatsu K, et al. 2012. Structural basis of interleukin-5 dimer recognition by its α receptor. *Protein Sci* 21: 850–864.

Leckie MJ, ten Brinke A, Khan J, Diamant Z, O'Connor BJ, Walls CM, Mathur AK, Cowley HC, Chung KF, Djukanovic R, et al. 2000. Effects of an interleukin-5 blocking monoclonal antibody on eosinophils, airway hyper-responsiveness, and the late asthmatic response. *Lancet* 356: 2144–2148.

Leentjens J, Kox M, Koch RM, Preijers F, Joosten LA, van der Hoeven JG, Netea MG, Pickkers P. 2012. Reversal of immunoparalysis in humans in vivo: A double-blind, placebo-controlled, randomized pilot study. *Am J Respir Crit Care Med* 186: 838–845.

Legacy J, Hanea S, Theoret J, Smith PD. 2013. Granulocyte macrophage colony-stimulating factor promotes regeneration of retinal ganglion cells in vitro through a mammalian target of rapamycin-dependent mechanism. *J Neurosci Res* 91: 771–779.

Leist M, Ghezzi P, Grasso G, Bianchi R, Villa P, Fratelli M, Savino C, Bianchi M, Nielsen J, Gerwien J, et al. 2004. Derivatives of erythropoietin that are tissue protective but not erythropoietic. *Science* 305: 239–242.

Cite this article as *Cold Spring Harb Perspect Biol* doi: 10.1101/cshperspect.a028514

Lencz T, Morgan TV, Athanasiou M, Dain B, Reed CR, Kane JM, Kucherlapati R, Malhotra AK. 2007. Converging evidence for a pseudoautosomal cytokine receptor gene locus in schizophrenia. *Mol Psychiatry* **12**: 572–580.

Li R, Rezk A, Miyazaki Y, Hilgenberg E, Touil H, Shen P, Moore CS, Michel L, Althekair F, Rajasekharan S, et al. 2015. Proinflammatory GM-CSF-producing B cells in multiple sclerosis and B cell depletion therapy. *Sci Transl Med* **7**: 310ra166.

Li B, Zhao W, Zhang X, Wang J, Luo X, Baker SD, Jordan CT, Dong Y. 2016a. Design, synthesis and evaluation of anti-CD123 antibody drug conjugates. *Bioorg Med Chem* **24**: 5855–5860.

Li M, Huang L, Li K, Huo Y, Chen C, Wang J, Liu J, Luo Z, Chen C, Dong Q, et al. 2016b. Adaptive evolution of interleukin-3 (IL3), a gene associated with brain volume variation in general human populations. *Hum Genet* **135**: 377–392.

Lins C, Borojevic R. 2001. Interleukin-5 receptor α chain expression and splicing during brain development in mice. *Growth Factors* **19**: 145–152.

Liontos LM, Dissanayake D, Ohashi PS, Weiss A, Dragone LL, McGlade CJ. 2011. The Src-like adaptor protein regulates GM-CSFR signaling and monocytic dendritic cell maturation. *J Immunol* **186**: 1923–1933.

Liu YJ. 2005. IPC: Professional type 1 interferon-producing cells and plasmacytoid dendritic cell precursors. *Annu Rev Immunol* **23**: 275–306.

Liu BL, Robinson M, Han ZQ, Branston RH, English C, Reay P, McGrath Y, Thomas SK, Thornton M, Bullock P, et al. 2003. ICP34.5 deleted herpes simplex virus with enhanced oncolytic, immune stimulating, and anti-tumour properties. *Gene Ther* **10**: 292–303.

Liu TF, Urieto JO, Moore JE, Miller MS, Lowe AC, Thorburn A, Frankel AE. 2004. Diphtheria toxin fused to variant interleukin-3 provides enhanced binding to the interleukin-3 receptor and more potent leukemia cell cytotoxicity. *Exp Hematol* **32**: 277–281.

Llanos C, Mackern-Oberti JP, Vega F, Jacobelli SH, Kalergis AM. 2013. Tolerogenic dendritic cells as a therapy for treating lupus. *Clin Immunol* **148**: 237–245.

Luo XJ, Li M, Huang L, Nho K, Deng M, Chen Q, Weinberger DR, Vasquez AA, Rijpkema M, Mattay VS, et al. 2012. The interleukin 3 gene (IL3) contributes to human brain volume variation by regulating proliferation and survival of neural progenitors. *PLoS ONE* **7**: e50375.

Mach N, Gillessen S, Wilson SB, Sheehan C, Mihm M, Dranoff G. 2000. Differences in dendritic cells stimulated in vivo by tumors engineered to secrete granulocyte-macrophage colony-stimulating factor or Flt3-ligand. *Cancer Res* **60**: 3239–3246.

Mardiros A, Dos Santos C, McDonald T, Brown CE, Wang X, Budde LE, Hoffman L, Aguilar B, Chang WC, Bretzlaff W, et al. 2013. T cells expressing CD123-specific chimeric antigen receptors exhibit specific cytolytic effector functions and antitumor effects against human acute myeloid leukemia. *Blood* **122**: 3138–3148.

Martinez-Moczygemba M, Huston DP. 2003. Biology of common β receptor-signaling cytokines: IL-3, IL-5, and GM-CSF. *J Allergy Clin Immunol* **112**: 653–665.

Mathew M, Zaineb KC, Verma RS. 2013. GM-CSF-DFF40: A novel humanized immunotoxin induces apoptosis in acute myeloid leukemia cells. *Apoptosis* **18**: 882–895.

McQualter JL, Darwiche R, Ewing C, Onuki M, Kay TW, Hamilton JA, Reid HH, Bernard CC. 2001. Granulocyte macrophage colony-stimulating factor: A new putative therapeutic target in multiple sclerosis. *J Exp Med* **194**: 873–882.

Meisel C, Schefold JC, Pschowski R, Baumann T, Hetzger K, Gregor J, Weber-Carstens S, Hasper D, Keh D, Zuckermann H, et al. 2009. Granulocyte-macrophage colony-stimulating factor to reverse sepsis-associated immunosuppression: A double-blind, randomized, placebo-controlled multicenter trial. *Am J Respir Crit Care Med* **180**: 640–648.

Menzies-Gow AN, Flood-Page PT, Robinson DS, Kay AB. 2007. Effect of inhaled interleukin-5 on eosinophil progenitors in the bronchi and bone marrow of asthmatic and non-asthmatic volunteers. *Clin Exp Allergy* **37**: 1023–1032.

Metcalf D. 2008. Hematopoietic cytokines. *Blood* **111**: 485–491.

Milburn MV, Hassel AM, Lambert MH, Jordan SR, Proudfoot AEI, Graber P, Wells TNC. 1993. A novel dimer configuration revealed by the crystal structure at 2.4 Å resolution of human interleukin-5. *Nature* **363**: 172–176.

Moldenhauer LM, Cockshell MP, Frost L, Parham KA, Tvorogov D, Tan LY, Ebert LM, Tooley K, Worthley S, Lopez AF, et al. 2015. Interleukin-3 greatly expands non-adherent endothelial forming cells with pro-angiogenic properties. *Stem Cell Res* **14**: 380–395.

Molfino NA, Kuna P, Leff JA, Oh CK, Singh D, Chernow M, Sutton B, Yarranton G. 2016. Phase 2, randomised placebo-controlled trial to evaluate the efficacy and safety of an anti-GM-CSF antibody (KB003) in patients with inadequately controlled asthma. *BMJ Open* **6**: e007709.

Nair P, Pizzichini MM, Kjarsgaard M, Inman MD, Efthimiadis A, Pizzichini E, Hargreave FE, O'Byrne PM. 2009. Mepolizumab for prednisone-dependent asthma with sputum eosinophilia. *N Engl J Med* **360**: 985–993.

Nievergall E, Ramshaw HS, Yong AS, Biondo M, Busfield SJ, Vairo G, Lopez AF, Hughes TP, White DL, Hiwase DK. 2014. Monoclonal antibody targeting of IL-3 receptor α with CSL362 effectively depletes CML progenitor and stem cells. *Blood* **123**: 1218–1228.

Nishinakamura R, Miyajima A, Mee PJ, Tybulewicz VLJ, Murray R. 1996. Hematopoiesis in mice lacking the entire granulocyte-macrophage colony-stimulating factor/interleukin-3/interleukin-5 functions. *Blood* **88**: 2458–2464.

Nixon J, Newbold P, Mustelin T, Anderson GP, Kolbeck R. 2016. Monoclonal antibody therapy for the treatment of asthma and chronic obstructive pulmonary disease with eosinophilic inflammation. *Pharmacol Ther* **169**: 57–77.

Noster R, Riedel R, Mashreghi MF, Radbruch H, Harms L, Haftmann C, Chang HD, Radbruch A, Zielinski CE. 2014. IL-17 and GM-CSF expression are antagonistically regulated by human T helper cells. *Sci Transl Med* **6**: 241ra280.

Obermueller E, Vosseler S, Fusenig NE, Mueller MM. 2004. Cooperative autocrine and paracrine functions of granulocyte colony-stimulating factor and granulocyte-macrophage colony-stimulating factor in the progression of skin carcinoma cells. *Cancer Res* **64**: 7801–7812.

Oon S, Huynh H, Tai TY, Ng M, Monaghan K, Biondo M, Vairo G, Maraskovsky E, Nash AD, Wicks IP, et al. 2016a. A cytotoxic anti-IL-3Rα antibody targets key cells and cytokines implicated in systemic lupus erythematosus. *JCI Insight* **1**: e86131.

Oon S, Wilson NJ, Wicks I. 2016b. Targeted therapeutics in SLE: Emerging strategies to modulate the interferon pathway. *Clin Transl Immunol* **5**: e79.

Padron E, Painter JS, Kunigal S, Mailloux AW, McGraw K, McDaniel JM, Kim E, Bebbington C, Baer M, Yarranton G, et al. 2013. GM-CSF-dependent pSTAT5 sensitivity is a feature with therapeutic potential in chronic myelomonocytic leukemia. *Blood* **121**: 5068–5077.

Panousis C, Dhagat U, Edwards KM, Rayzman V, Hardy MP, Braley H, Gauvreau GM, Hercus TR, Smith S, Sehmi R, et al. 2016. CSL311, a novel, potent, therapeutic monoclonal antibody for the treatment of diseases mediated by the common β chain of the IL-3, GM-CSF and IL-5 receptors. *mAbs* **8**: 436–453.

Pardanani A, Lasho T, Chen D, Kimlinger TK, Finke C, Zblewski D, Patnaik MM, Reichard KK, Rowinsky E, Hanson CA, et al. 2015. Aberrant expression of CD123 (interleukin-3 receptor-α) on neoplastic mast cells. *Leukemia* **29**: 1605–1608.

Pardanani A, Reichard KK, Zblewski D, Abdelrahman RA, Wassie EA, Morice WG II, Brooks C, Grogg KL, Hanson CA, Tefferi A, et al. 2016. CD123 immunostaining patterns in systemic mastocytosis: Differential expression in disease subgroups and potential prognostic value. *Leukemia* **30**: 914–918.

Paré A, Mailhot B, Lévesque SA, Lacroix S. 2016. Involvement of the IL-1 system in experimental autoimmune encephalomyelitis and multiple sclerosis: Breaking the vicious cycle between IL-1β and GM-CSF. *Brain Behav Immun* **62**: 1–8.

Parganas E, Wang D, Stravopodis D, Topham DJ, Marine JC, Tegland S, Vanin EF, Bodner S, Colamonici OR, van Deursen JM, et al. 1998. Jak2 is essential for signaling through a variety of cytokine receptors. *Cell* **93**: 385–395.

Patino E, Kotzsch A, Saremba S, Nickel J, Schmitz W, Sebald W, Mueller TD. 2011. Structure analysis of the IL-5 ligand-receptor complex reveals a wrench-like architecture for IL-5Rα. *Structure* **19**: 1864–1875.

Patterson MF, Borish L, Kennedy JL. 2015. The past, present, and future of monoclonal antibodies to IL-5 and eosinophilic asthma: A review. *J Asthma Allergy* **8**: 125–134.

Pavord ID, Korn S, Howarth P, Bleecker ER, Buhl R, Keene ON, Ortega H, Chanez P. 2012. Mepolizumab for severe eosinophilic asthma (DREAM): A multicentre, double-blind, placebo-controlled trial. *Lancet* **380**: 651–659.

Pelaia G, Vatrella A, Busceti MT, Gallelli L, Preiano M, Lombardo N, Terracciano R, Maselli R. 2016. Role of biologics in severe eosinophilic asthma—Focus on reslizumab. *Ther Clin Risk Manag* **12**: 1075–1082.

Perugini M, Brown AL, Salerno DG, Booker GW, Stojkoski C, Hercus TR, Lopez AF, Hibbs ML, Gonda TJ, D'Andrea RJ. 2010. Alternative modes of GM-CSF receptor activation revealed using activated mutants of the common β-subunit. *Blood* **115**: 3346–3353.

Piccoli L, Campo I, Fregni CS, Rodriguez BM, Minola A, Sallusto F, Luisetti M, Corti D, Lanzavecchia A. 2015. Neutralization and clearance of GM-CSF by autoantibodies in pulmonary alveolar proteinosis. *Nat Commun* **6**: 7375.

Polotskaya A, Zhao Y, Lilly ML, Kraft AS. 1993. A critical role for the cytoplasmic domain of the granulocyte macrophage colony-stimulating factor α receptor in mediating cell growth. *Cell Growth Differ* **4**: 523–531.

Puzanov I, Milhem MM, Minor D, Hamid O, Li A, Chen L, Chastain M, Gorski KS, Anderson A, Chou J, et al. 2016. Talimogene laherparepvec in combination with ipilimumab in previously untreated, unresectable stage IIIB-IV melanoma. *J Clin Oncol* **34**: 2619–2626.

Pylayeva-Gupta Y, Lee KE, Hajdu CH, Miller G, Bar-Sagi D. 2012. Oncogenic Kras-induced GM-CSF production promotes the development of pancreatic neoplasia. *Cancer Cell* **21**: 836–847.

Ramshaw HS, Bardy PG, Lee M, Lopez AF. 2002. Chronic myelomonocytic leukaemia requires granulocyte-macrophage colony-stimulating factor for growth in vitro and in vivo. *Exp Hematol* **30**: 1124–1131.

Ranki T, Pesonen S, Hemminki A, Partanen K, Kairemo K, Alanko T, Lundin J, Linder N, Turkki R, Ristimaki A, et al. 2016. Phase I study with ONCOS-102 for the treatment of solid tumors—An evaluation of clinical response and exploratory analyses of immune markers. *J Immunother Cancer* **4**: 17.

Rasouli J, Ciric B, Imitola J, Gonnella P, Hwang D, Mahajan K, Mari ER, Safavi F, Leist TP, Zhang GX, et al. 2015. Expression of GM-CSF in T cells is increased in multiple sclerosis and suppressed by IFN-β therapy. *J Immunol* **194**: 5085–5093.

Rauch PJ, Chudnovskiy A, Robbins CS, Weber GF, Etzrodt M, Hilgendorf I, Tiglao E, Figueiredo JL, Iwamoto Y, Theurl I, et al. 2012. Innate response activator B cells protect against microbial sepsis. *Science* **335**: 597–601.

Renner K, Hermann FJ, Schmidbauer K, Talke Y, Rodriguez Gomez M, Schiechl G, Schlossmann J, Bruhl H, Anders HJ, Mack M. 2015. IL-3 contributes to development of lupus nephritis in MRL/lpr mice. *Kidney Int* **88**: 1088–1098.

Revoltella RP, Menicagli M, Campani D. 2012. Granulocyte-macrophage colony-stimulating factor as an autocrine survival-growth factor in human gliomas. *Cytokine* **57**: 347–359.

Riccioni R, Diverio D, Riti V, Buffolino S, Mariani G, Boe A, Cedrone M, Ottone T, Foa R, Testa U. 2009. Interleukin (IL)-3/granulocyte macrophage-colony stimulating factor/IL-5 receptor α and β chains are preferentially expressed in acute myeloid leukaemias with mutated FMS-related tyrosine kinase 3 receptor. *Br J Haematol* **144**: 376–387.

Rigo A, Gottardi M, Zamo A, Mauri P, Bonifacio M, Krampera M, Damiani E, Pizzolo G, Vinante F. 2010. Macrophages may promote cancer growth via a GM-CSF/HB-EGF paracrine loop that is enhanced by CXCL12. *Mol Cancer* **9**: 273.

Rodig SJ, Meraz MA, White JM, Lampe PA, Riley JK, Arthur CD, King KL, Sheehan KC, Yin L, Pennica D, et al. 1998. Disruption of the *Jak1* gene demonstrates obligatory and nonredundant roles of the Jaks in cytokine-induced biologic responses. *Cell* **93**: 373–383.

Rothenberg ME, Klion AD, Roufosse FE, Kahn JE, Weller PF, Simon HU, Schwartz LB, Rosenwasser LJ, Ring J, Griffin EF, et al. 2008. Treatment of patients with the hypereosinophilic syndrome with mepolizumab. *N Engl J Med* **358**: 1215–1228.

Rozwarski DA, Diederichs K, Hecht R, Boone T, Karplus PA. 1996. Refined crystal structure and mutagenesis of human granulocyte-macrophage colony-stimulating factor. *Proteins* **26**: 304–313.

Saha S, Doe C, Mistry V, Siddiqui S, Parker D, Sleeman M, Cohen ES, Brightling CE. 2009. Granulocyte-macrophage colony-stimulating factor expression in induced sputum and bronchial mucosa in asthma and COPD. *Thorax* **64**: 671–676.

Sakamaki K, Miyajima I, Kitamura T, Miyajima A. 1992. Critical cytoplasmic domains of the common b subunit of the human GM-CSF, IL-3 and IL-5 receptors for growth signal transduction and tyrosine phosphorylation. *EMBO J* **11**: 3541–3549.

Saulle E, Riccioni R, Coppola S, Parolini I, Diverio D, Riti V, Mariani G, Laufer S, Sargiacomo M, Testa U. 2009. Co-localization of the VEGF-R2 and the common IL-3/GM-CSF receptor β chain to lipid rafts leads to enhanced p38 activation. *Br J Haematol* **145**: 399–411.

Scapini P, Pereira S, Zhang H, Lowell CA. 2009. Multiple roles of Lyn kinase in myeloid cell signaling and function. *Immunol Rev* **228**: 23–40.

Schabitz WR, Kruger C, Pitzer C, Weber D, Laage R, Gassler N, Aronowski J, Mier W, Kirsch F, Dittgen T, et al. 2008. A neuroprotective function for the hematopoietic protein granulocyte-macrophage colony stimulating factor (GM-CSF). *J Cereb Blood Flow Metab* **28**: 29–43.

Schweizerhof M, Stosser S, Kurejova M, Njoo C, Gangadharan V, Agarwal N, Schmelz M, Bali KK, Michalski CW, Brugger S, et al. 2009. Hematopoietic colony-stimulating factors mediate tumor-nerve interactions and bone cancer pain. *Nat Med* **15**: 802–807.

Scott CL, Robb L, Papaevangeliou B, Mansfield R, Nicola NA, Begley CG. 2000. Reassessment of interactions between hematopoietic receptors using common β-chain and interleukin-3-specific receptor β-chain-null cells: No evidence of functional interactions with receptors for erythropoietin, granulocyte colony-stimulating factor, or stem cell factor. *Blood* **96**: 1588–1590.

Shang S, Yang YM, Zhang H, Tian L, Jiang JS, Dong YB, Zhang K, Li B, Zhao WD, Fang WG, et al. 2016. Intracerebral GM-CSF contributes to transendothelial monocyte migration in APP/PS1 Alzheimer's disease mice. *J Cereb Blood Flow Metab* **36**: 1978–1991.

Shanmugalingam U, Jadavji NM, Smith PD. 2016. Role of granulocyte macrophage colony stimulating factor in regeneration of the central nervous system. *Neural Regen Res* **11**: 902–903.

Sheng W, Yang F, Zhou Y, Yang H, Low PY, Kemeny DM, Tan P, Moh A, Kaplan MH, Zhang Y, et al. 2014. STAT5 programs a distinct subset of GM-CSF-producing T helper cells that is essential for autoimmune neuroinflammation. *Cell Res* **24**: 1387–1402.

Shiomi A, Usui T. 2015. Pivotal roles of GM-CSF in autoimmunity and inflammation. *Mediators Inflamm* **2015**: 568543.

Silvennoinen O, Witthuhn BA, Quelle FW, Cleveland JL, Yi T, Ihle JN. 1993. Structure of the murine Jak2 protein-tyrosine kinase and its role in interleukin 3 signal transduction. *Proc Natl Acad Sci* **90**: 8429–8433.

Stock AT, Hansen JA, Sleeman MA, McKenzie BS, Wicks IP. 2016. GM-CSF primes cardiac inflammation in a mouse model of Kawasaki disease. *J Exp Med* **213**: 1983–1998.

Straumann A, Conus S, Grzonka P, Kita H, Kephart G, Bussmann C, Beglinger C, Smith DA, Patel J, Byrne M, et al. 2010. Anti-interleukin-5 antibody treatment (mepolizumab) in active eosinophilic oesophagitis: A randomised, placebo-controlled, double-blind trial. *Gut* **59**: 21–30.

Su S, Liu Q, Chen J, Chen J, Chen F, He C, Huang D, Wu W, Lin L, Huang W, et al. 2014. A positive feedback loop between mesenchymal-like cancer cells and macrophages is essential to breast cancer metastasis. *Cancer Cell* **25**: 605–620.

Sun Q, Woodcock JM, Rapoport A, Stomski FC, Korpelainen EI, Bagley CJ, Goodall GJ, Smith WB, Gamble JR, Vadas MA, et al. 1996. Monoclonal antibody 7G3 recognizes the *N*-terminal domain of the human interleukin-3 (IL-3) receptor α-chain and functions as a specific IL-3 receptor antagonist. *Blood* **87**: 83–92.

Sun Q, Jones K, McClure B, Cambareri B, Zacharakis B, Iversen PO, Stomski F, Woodcock JM, Bagley CJ, D'Andrea R, et al. 1999. Simultaneous antagonism of interleukin-5, granulocyte-macrophage colony-stimulating factor, and interleukin-3 stimulation of human eosinophils by targetting the common cytokine binding site of their receptors. *Blood* **94**: 1943–1951.

Suzuki T, Sakagami T, Rubin BK, Nogee LM, Wood RE, Zimmerman SL, Smolarek T, Dishop MK, Wert SE, Whitsett JA, et al. 2008. Familial pulmonary alveolar proteinosis caused by mutations in *CSF2RA*. *J Exp Med* **205**: 2703–2710.

Suzuki T, Sakagami T, Young LR, Carey BC, Wood RE, Luisetti M, Wert SE, Rubin BK, Kevill K, Chalk C, et al. 2010. Hereditary pulmonary alveolar proteinosis: Pathogenesis, presentation, diagnosis, and therapy. *Am J Respir Crit Care Med* **182**: 1292–1304.

Suzuki T, Maranda B, Sakagami T, Catellier P, Couture CY, Carey BC, Chalk C, Trapnell BC. 2011. Hereditary pulmonary alveolar proteinosis caused by recessive *CSF2RB* mutations. *Eur Respir J* **37**: 201–204.

Takaki S, Kanazawa H, Shiiba M, Takatsu K. 1994. A critical cytoplasmic domain of the interleukin-5 (IL-5) receptor α chain and its function in IL-5-mediated growth signal transduction. *Mol Cell Biol* **14**: 7404–7413.

Tan LD, Bratt JM, Godor D, Louie S, Kenyon NJ. 2016. Benralizumab: A unique IL-5 inhibitor for severe asthma. *J Asthma Allergy* **9**: 71–81.

Tanaka T, Motoi N, Tsuchihashi Y, Tazawa R, Kaneko C, Nei T, Yamamoto T, Hayashi T, Tagawa T, Nagayasu T, et al. 2011. Adult-onset hereditary pulmonary alveolar proteinosis caused by a single-base deletion in *CSF2RB*. *J Med Genet* **48**: 205–209.

Tavernier J, Cornelis S, Devos R, Guisez Y, Plaetinck G, Van der Heyden J. 1995. Structure/function analysis of human interleukin 5 and its receptor. *Agents Actions Suppl* **46:** 23–34.

Testa U, Riccioni R, Militi S, Coccia E, Stellacci E, Samoggia P, Latagliata R, Mariani G, Rossini A, Battistini A, et al. 2002. Elevated expression of IL-3Rα in acute myelogenous leukemia is associated with enhanced blast proliferation, increased cellularity, and poor prognosis. *Blood* **100:** 2980–2988.

Testa U, Riccioni R, Biffoni M, Diverio D, Lo-Coco F, Foa R, Peschle C, Frankel AE. 2005. Diphtheria toxin fused to variant human interleukin-3 induces cytotoxicity of blasts from patients with acute myeloid leukemia according to the level of interleukin-3 receptor expression. *Blood* **106:** 2527–2529.

Tettamanti S, Marin V, Pizzitola I, Magnani CF, Giordano Attianese GM, Cribioli E, Maltese F, Galimberti S, Lopez AF, Biondi A, et al. 2013. Targeting of acute myeloid leukaemia by cytokine-induced killer cells redirected with a novel CD123-specific chimeric antigen receptor. *Br J Haematol* **161:** 389–401.

Theoret JK, Jadavji NM, Zhang M, Smith PD. 2016. Granulocyte macrophage colony-stimulating factor treatment results in recovery of motor function after white matter damage in mice. *Eur J Neurosci* **43:** 17–24.

Thomas D, Powell JA, Green BD, Barry EF, Ma Y, Woodcock J, Fitter S, Zannettino AC, Pitson SM, Hughes TP, et al. 2013. Protein kinase activity of phosphoinositide 3-kinase regulates cytokine-dependent cell survival. *PLoS Biol* **11:** e1001515.

Thorn M, Guha P, Cunetta M, Espat NJ, Miller G, Junghans RP, Katz SC. 2016. Tumor-associated GM-CSF overexpression induces immunoinhibitory molecules via STAT3 in myeloid-suppressor cells infiltrating liver metastases. *Cancer Gene Ther* **23:** 188–198.

Urdinguio RG, Fernandez AF, Moncada-Pazos A, Huidobro C, Rodriguez RM, Ferrero C, Martinez-Camblor P, Obaya AJ, Bernal T, Parra-Blanco A, et al. 2013. Immune-dependent and independent antitumor activity of GM-CSF aberrantly expressed by mouse and human colorectal tumors. *Cancer Res* **73:** 395–405.

Vergez F, Green AS, Tamburini J, Sarry JE, Gaillard B, Cornillet-Lefebvre P, Pannetier M, Neyret A, Chapuis N, Ifrah N, et al. 2011. High levels of CD34⁺CD38^{low/-}CD123⁺ blasts are predictive of an adverse outcome in acute myeloid leukemia: A Groupe Ouest-Est des Leucemies Aigues et Maladies du Sang (GOELAMS) study. *Haematologica* **96:** 1792–1798.

Vilalta M, Rafat M, Giaccia AJ, Graves EE. 2014. Recruitment of circulating breast cancer cells is stimulated by radiotherapy. *Cell Rep* **8:** 402–409.

Wang Y, Thomson CA, Allan LL, Jackson LM, Olson M, Hercus TR, Nero TL, Turner A, Parker MW, Lopez AL, et al. 2013. Characterization of pathogenic human monoclonal autoantibodies against GM-CSF. *Proc Natl Acad Sci* **110:** 7832–7837.

Watanabe-Smith K, Tognon C, Tyner JW, Meijerink JP, Druker BJ, Agarwal A. 2016. Discovery and functional characterization of a germline, *CSF2RB*-activating mutation in leukemia. *Leukemia* **30:** 1950–1953.

Weber GF, Chousterman BG, Hilgendorf I, Robbins CS, Theurl I, Gerhardt LM, Iwamoto Y, Quach TD, Ali M, Chen JW, et al. 2014. Pleural innate response activator B cells protect against pneumonia via a GM-CSF-IgM axis. *J Exp Med* **211:** 1243–1256.

Weber GF, Chousterman BG, He S, Fenn AM, Nairz M, Anzai A, Brenner T, Uhle F, Iwamoto Y, Robbins CS, et al. 2015. Interleukin-3 amplifies acute inflammation and is a potential therapeutic target in sepsis. *Science* **347:** 1260–1265.

Weiss M, Yokoyama C, Shikama Y, Naugle C, Druker B, Sieff CA. 1993. Human granulocyte-macrophage colony-stimulating factor receptor signal transduction requires the proximal cytoplasmic domains of the α and β subunits. *Blood* **82:** 3298–3306.

Wicks IP, Roberts AW. 2016. Targeting GM-CSF in inflammatory diseases. *Nat Rev Rheumatol* **12:** 37–48.

Xiu MH, Lin CG, Tian L, Tan YL, Chen J, Chen S, Tan SP, Wang ZR, Yang FD, Chen da C, et al. 2015. Increased IL-3 serum levels in chronic patients with schizophrenia: Associated with psychopathology. *Psychiatry Res* **229:** 225–229.

Yamaguchi T, Schares S, Fischer U, Dijkstra JM. 2016. Identification of a fourth ancient member of the IL-3/IL-5/GM-CSF cytokine family, KK34, in many mammals. *Dev Comp Immunol* **65:** 268–279.

Yamasaki R, Lu H, Butovsky O, Ohno N, Rietsch AM, Cialic R, Wu PM, Doykan CE, Lin J, Cotleur AC, et al. 2014. Differential roles of microglia and monocytes in the inflamed central nervous system. *J Exp Med* **211:** 1533–1549.

Yamashita N, Tashimo H, Ishida H, Kaneko F, Nakano J, Kato H, Hirai K, Horiuchi T, Ohta K. 2002. Attenuation of airway hyperresponsiveness in a murine asthma model by neutralization of granulocyte-macrophage colony-stimulating factor (GM-CSF). *Cell Immunol* **219:** 92–97.

Cite this article as *Cold Spring Harb Perspect Biol* doi: 10.1101/cshperspect.a028514

Interleukin-6 Family Cytokines

Stefan Rose-John

Institute of Biochemistry, Kiel University, Olshausenstrasse 40, Kiel, Germany

Correspondence: rosejohn@biochem.uni-kiel.de

The interleukin (IL)-6 family cytokines is a group of cytokines consisting of IL-6, IL-11, ciliary neurotrophic factor (CNTF), leukemia inhibitory factor (LIF), oncostatin M (OSM), cardiotrophin 1 (CT-1), cardiotrophin-like cytokine (CLC), and IL-27. They are grouped into one family because the receptor complex of each cytokine contains two (IL-6 and IL-11) or one molecule (all others cytokines) of the signaling receptor subunit gp130. IL-6 family cytokines have overlapping but also distinct biologic activities and are involved among others in the regulation of the hepatic acute phase reaction, in B-cell stimulation, in the regulation of the balance between regulatory and effector T cells, in metabolic regulation, and in many neural functions. Blockade of IL-6 family cytokines has been shown to be beneficial in autoimmune diseases, but bacterial infections and metabolic side effects have been observed. Recent advances in cytokine blockade might help to minimize such side effects during therapeutic blockade.

Cytokines are small (15–20 kDa) and short-lived proteins important in autocrine, paracrine, and endocrine signaling. Cytokines coordinate the development and the activity of the immune system (Gandhi et al. 2016). Many cytokines belong to the four α-helical class of mediators, which share a common up-up–down-down topology of the four helices. Furthermore, cytokines are grouped into families according to the structure and the specificity and composition of their receptor complexes. Cytokines bind to multimeric receptor complexes in which often one subunit is also found in the receptor complexes for other cytokines (Spangler et al. 2015). This is a reasonable classification because common receptor subunits imply similarity or even identity in intracellular signal transduction.

The class of four-helical cytokines consists of more than 35 interleukins and many mediators with trivial names such as growth hormone, prolactin, leptin, erythropoietin, thrombopoietin, leukemia inhibitory factor (LIF), and oncostatin M (OSM). Moreover, all interferons and many colony-stimulating factors (CSFs) belong to this class of cytokines, which altogether contains far more than 60 members (Spangler et al. 2015).

Interleukin (IL)-6 family cytokines are defined as cytokines that use the common signaling receptor subunit glycoprotein 130 kDa (gp130). Presently, eight cytokines fulfill this criterion although, as will be explained below, the group of IL-6 family cytokines is still expanding and the definition of gp130-containing complexes needs to be revised (Rose-John et al. 2015).

IL-6 family cytokines have been implicated in many functions, including B-cell stimulation and induction of the hepatic acute phase proteins. Moreover, metabolic functions and neurotrophic functions have been ascribed to this group of cytokines. Lately, an IL-6 receptor (IL-6R)-neutralizing monoclonal antibody (tocilizumab) has been approved in more than 100 countries for the treatment of autoimmune diseases (Tanaka et al. 2014), and blockade of IL-6 activity was observed to be at least as efficient as the blockade of tumor necrosis factor α in patients with rheumatoid arthritis (Gabay et al. 2013).

This review covers the identification and complementary DNA (cDNA) cloning of IL-6 family cytokines, the identification of their cognate receptors, and the recognition that these cytokines form a large family of mediators, which are involved in the coordination of the immune system but also in many other physiologic functions as well.

CLONING OF IL-6, IL-6R, AND gp130: THE IL-6R COMPLEX

When the cDNA coding for B-cell stimulatory factor 2 (BSF-2) was cloned in 1986 by the Kishimoto group, the comparison of its protein sequence with the protein sequences of IL-1, IL-2, IL-4, interferon γ (IFN-γ), granulocyte macrophage colony-stimulating factor (GM-CSF), and granulocyte colony-stimulating factor (G-CSF) yielded only some faint similarities with G-CSF (Hirano et al. 1986). It became, however, immediately clear from the cDNA sequence that BSF-2 was identical to several other proteins with completely different biologic activities (Table 1), indicating that the activity of the newly cloned BSF-2 was not restricted to the immune system. In December 1988, a group of 19 leading scientists in the field, including Dr. Kishimoto, agreed to refer to the protein as IL-6 (The New York Academy of Sciences Conference 1988).

No structural or functional information on IL-6 or the IL-6R complex was available until, in 1988, the Kishimoto group cloned the

Table 1. Synonyms of interleukin-6 as evidenced on complementary DNA (cDNA) cloning of B-cell stimulatory factor 2 (BSF-2)

B-cell stimulatory factor-2 (Hirano et al. 1986)
Hepatocyte-stimulating factor (Gauldie et al. 1987)
Hybridoma-plasmacytoma growth factor (Brakenhoff et al. 1987)
Interferon β₂ (Zilberstein et al. 1986)
26 kDa protein (Haegeman et al. 1986)

cDNA coding for the human IL-6R (Yamasaki et al. 1988). The receptor belonged to the immunoglobulin (Ig) superfamily but the cytoplasmic portion lacked a domain with discernible enzymatic activity, thus failing to provide any clue regarding the mechanism of signal transduction induced by IL-6. Figure 1 shows how IL-6 and IL-6R were seen during these early days of the cytokine field when there was no information on the structure of these molecules.

Because cytokines show very little homology at the protein level, it was not clear at the time that cytokines share common features and actually belong to a common protein family. At this time, a seminal article was published by Bazan (1990), who hypothesized that cytokines such as growth hormone, prolactin, IL-6, G-CSF, and erythropoietin belong to the protein family of four-helical cytokines. Bazan stated that lymphokines, interleukins, CSFs, growth hormones, and interferons, which he collectively designated as cytokines, display no (or at best fragmentary) similarities in amino acid sequence. All these proteins, however, were predicted or had been shown to be rich in α-helices. Based on the then already-published X-ray structure of G-CSF (Abdel-Meguid et al. 1987) and the high predicted helical content of growth hormone, prolactin, IL-6, G-CSF, and erythropoietin, Bazan suggested a four-helical tertiary fold for cytokines, an idea that he also backed by the conserved gene exon/intron boundaries flanking the helices and by conserved disulfide bridge patterns (Bazan 1990). Based on his model, Bazan even detected that the protein topology in the then-published X-ray structure of IL-2 was incorrect and needed correction (Bazan 1992). Now we know that all cytokines of the IL-6 family, but also essentially all other cytokines, share the four-helices with an up-up–down-down topology (Fig. 1). It was recognized that, instead of showing clear sequence homology, cytokines are characterized by a high structural homology (Spangler et al. 2015). Such structural homology extends to cytokine receptors, which all belong to the Ig superfamily and which contain a cytokine-binding module consisting of tandem fibronectin III domains where the amino-terminal domain contains a set of

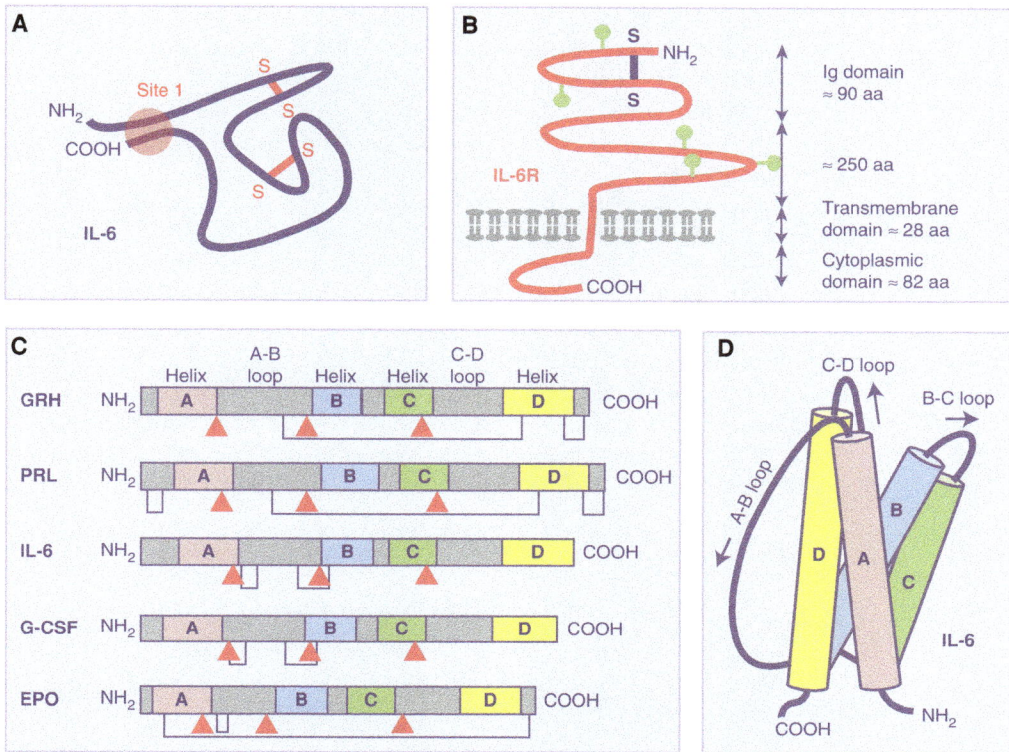

Figure 1. Recognition of interleukin-6 (IL-6) as a four-helical cytokine with up-up–down-down topology. (*A*) IL-6 was functionally characterized by sets of monoclonal antibodies and it was stated that the NH$_2$ terminus and the COOH terminus must be in close proximity in the biologic active IL-6 protein (Brakenhoff et al. 1990). Hypothesized disulfide bridges are indicated in red. (*B*) When the complementary DNA (cDNA) coding for the IL-6 receptor (IL-6R) was cloned, no details about structure or function were available and only a schematic model of the IL-6R as a member of the immunoglobulin (Ig) superfamily could be envisioned (Hirano et al. 1989). A single hypothesized disulfide bridge is shown in blue. Postulated *N*-glycosylation sites are shown in green. (*C*) It was recognized by Bazan (1990) that, although growth hormone (GRH), prolactin (PRL), IL-6, granulocyte colony-stimulating factor (G-CSF), and erythropoietin (EPO) (Bazan 1990) display no (or at best fragmentary) similarities in amino acid sequence, they share exon–intron boundaries at predicted secondary structure elements (red arrowheads) and predicted disulfide bridges. Based on the known three-dimensional structure of G-CSF (Abdel-Meguid et al. 1987), Bazan postulated that GRH, PRL, IL-6, G-CSF, and EPO formed a structurally related family of cytokines. (*D*) Model of the IL-6 as a four-helical cytokine with up-up–down-down topology (Bazan 1990).

four conserved cysteines and the membrane proximal domain contains a tryptophan-serine-X-tryptophan-serine motif (Spangler et al. 2015).

In 1990, the Kishimoto group cloned the cDNA coding for a 130 kDa glycoprotein (gp130) (Hibi et al. 1990), which acted as a signal-transducing coreceptor for IL-6. The gp130 protein is expressed on all cells of the human body (Hibi et al. 1990; Oberg et al. 2006). It turned out that IL-6 signaling required both IL-6R and gp130. Only when IL-6 bound to the IL-6R, this complex bound to gp130, induced dimerization of gp130 and induced intracellular signaling (Fig. 2) (Taga et al. 1989; Hibi et al. 1990). Interestingly, the cytoplasmic portion of IL-6R was not needed for IL-6 signaling and, curiously, a truncated IL-6R lacking cytoplasmic and transmembrane domain in the presence of IL-6 was still able to bind to and activate gp130 (Taga et al. 1989; Hibi et al. 1990). Remarkably, it was shown that neither IL-6 nor IL-6R alone have a measurable affinity for gp130. Only the complex of IL-6 and IL-6R can bind to and activate gp130 (Taga et al. 1989; Hibi et al. 1990). This characteristic of the IL-6 system has important consequences for the therapeutic inhibition of IL-6 responses (see below).

Subsequently, the X-ray structure of the complex of IL-6, IL-6R, and gp130 was solved by the Garcia group and revealed a hexameric assembly of two molecules of IL-6, IL-6R, and gp130, which was proposed to serve as a blueprint of the other IL-6 family cytokines (Boulanger et al. 2003). The entire complex of IL-6,

IL-6R, gp130, and Janus kinase 1 (JAK1) has been reconstituted by electron microscopy imaging (Lupardus et al. 2011). The structural understanding of cytokine binding and activation of receptor complexes has recently made it possible to generate modified cytokines with altered (agonistic and antagonistic) properties (Spangler et al. 2015).

During the 1990s, intracellular signaling induced by activated gp130 was recognized to be mainly mediated by JAKs constitutively associated with the cytoplasmic portion of gp130, which led to the recruitment and activation of signal transducers and activators of transcription (STATs), which on phosphorylation by JAKs dimerize and translocate into the nucleus

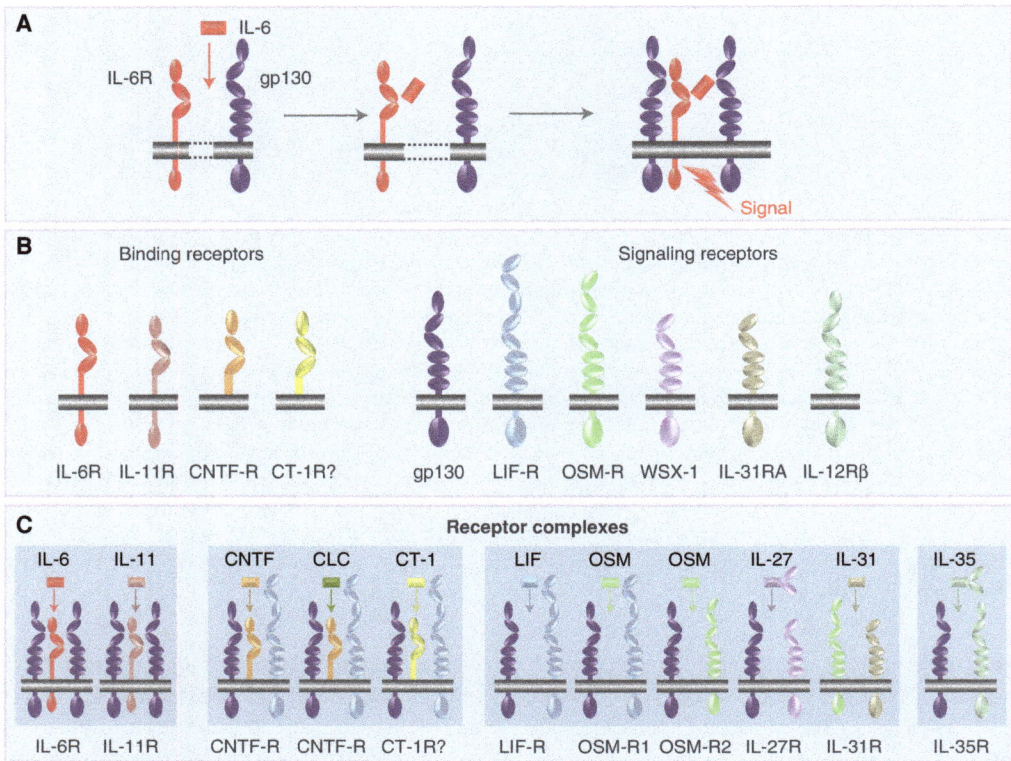

Figure 2. Molecular set-up of the interleukin-6 receptor (IL-6R) complex and comparison with other receptor complexes of IL-6 family cytokines. (A) IL-6 (red) binds to the membrane-bound IL-6R (red) and the complex of IL-6 and IL-6R associate with gp130 (lilac), which dimerizes and initiates signaling. (B) The IL-6 family cytokines act via four different ligand-binding receptors (left) and six different signaling receptors (right). (C) Assembly of the receptor complexes of the IL-6 family cytokines. CNTF, Ciliary neurotrophic factor; CLC, cardiotrophin-like cytokine; CT-1, cardiotrophin 1; LIF, leukemia inhibitory factor; OSM, oncostatin M.

to act as transcription factors (Levy and Darnell 2002). Thereby, gp130 signaling showed striking similarities with interferon signaling via JAKs and STATs (Lutticken et al. 1994). In the case of gp130, the main Janus kinase is JAK1 and the major signal transducer and activator of transcription is STAT3 (Schaper and Rose-John 2015). Interferon and cytokine signaling via JAKs and STATs is discussed in Stark et al. (2017).

gp130 IS A PLEIOTROPIC SIGNALING RECEPTOR: EMERGENCE OF A CYTOKINE FAMILY

Shortly after the cloning of the IL-6 cDNA, several additional cytokine cDNAs were molecularly cloned, including IL-11 (Paul et al. 1990), LIF (Gearing et al. 1987), ciliary neurotrophic factor (CNTF) (Lin et al. 1989; Stockli et al. 1989), OSM (Malik et al. 1989), and cardiotrophin 1 (CT-1) (Pennica et al. 1995). All of these cytokines showed similar but also distinct biologic activities, which could be explained by the fact that the receptor complexes of each of these cytokines contained gp130 (Gearing et al. 1992). All cytokines that use gp130 as a receptor subunit were referred to as IL-6 family cytokines or gp130 cytokines (Jones and Rose-John 2002; Jones et al. 2011). As detailed in Figure 2, the second signaling receptor in the case of CNTF, LIF, CT-1, and cardiotrophin-like cytokine (CLC) is the LIF receptor (LIF-R), a protein structurally related to gp130 (Gearing et al. 1991). OSM can alternatively bind to a heterodimer of gp130 and LIF-R or of gp130 and OSM receptor (OSM-R) (Mosley et al. 1996; Hermanns 2015). The dimeric cytokine IL-27 binds to a heterodimer of gp130 and WSX-1 (Aparicio-Siegmund and Garbers 2015). The signal transduction of all of these receptor complexes is similar, although subtle differences exist (Schaper and Rose-John 2015).

Cytokine specificity is brought about by the unique cell-surface expression of the receptor subunits. No natural cytokine can activate gp130 in the absence of other receptor subunits. Gp130 is the only receptor subunit that is expressed on all cells of the body, whereas all other receptor subunits show a more restricted expression pattern. It follows that the expression of these second receptor subunits determines whether a given cell will be able to respond to a given cytokine (Jones and Rose-John 2002; Jones et al. 2011).

Interestingly, the IL-6 family cytokines IL-6, IL-11, CNTF, CLC, and possibly CT-1 (Pennica et al. 1996) bind to specific cytokine-binding receptors, which are not signaling competent but rather present their ligand to the homodimeric or heterodimeric gp130-containing receptor complex. In contrast, the IL-6 family cytokines LIF, OSM, IL-27, and IL-31 directly interact with the two signaling receptor subunits without the help of a ligand-binding receptor subunit (Fig. 2). As will be explained below, the presence of nonsignaling receptor subunits enables a type of signaling that is not possible for cytokines without such specific receptors.

IL-6 TRANS-SIGNALING VIA SOLUBLE RECEPTORS: A NEW PARADIGM NOT FOR IL-6 AND OTHER CYTOKINES AS WELL

In the early 1990s, it was found that the IL-6R was efficiently cleaved from the cell surface by a then-unknown protease (Mullberg et al. 1992, 1993). With the help of pulse-chase experiments, it was shown that IL-6R cleavage was complete after 24 h but could be drastically stimulated on stimulation of protein kinase C with the phorbol ester, PMA. Around the same time, a human messenger RNA (mRNA) species coding for an IL-6R devoid of a transmembrane domain was isolated (Lust et al. 1992). This mRNA originated from alternative splicing of the gene encoding IL-6R. In the mouse, no alternative splicing of the IL-6R mRNA was detected (Schumacher et al. 2015). Interestingly, it was shown that the soluble IL-6R (sIL-6R) cleaved from one cell, in the presence of IL-6, could stimulate cells, which did not express IL-6R and therefore were completely unresponsive to IL-6 (Mackiewicz et al. 1992). This process was called "trans-signaling" and it was speculated that this type of signaling via a soluble receptor contributed significantly to the biology of IL-6 because it dramatically enlarged the spectrum of target cells for IL-6 (Fig. 3A) (Rose-John and Heinrich 1994).

The new paradigm of IL-6 trans-signaling was initially met with some skepticism in the scientific community because soluble receptors were assumed to act as antagonists by competing for their cognate ligands with membrane-bound receptors. This made it necessary to generate molecular tools to convincingly demonstrate the existence of trans-signaling not only in vitro but also in vivo.

The first tool we designed was "hyper-IL-6" (Fig. 3A), a protein in which IL-6 and sIL-6R are covalently connected by a flexible peptide linker

(Fischer et al. 1997). Comparing the response of many cells to IL-6 and hyper-IL-6 allowed us to show that many cells, including hematopoietic stem cells (Fischer et al. 1997; Audet et al. 2001), embryonic stem cells (Viswanathan et al. 2002; Humphrey et al. 2004), and smooth muscle cells (Klouche et al. 1999), required sIL-6R for their response to IL-6. Moreover, we could show that efficient liver regeneration required IL-6 in combination with sIL-6R (Galun et al. 2000; Peters et al. 2000). Although the use of hyper-IL-6 showed the potential of IL-6 trans-signaling, it did not prove that IL-6 trans-signaling actually occurred in vivo.

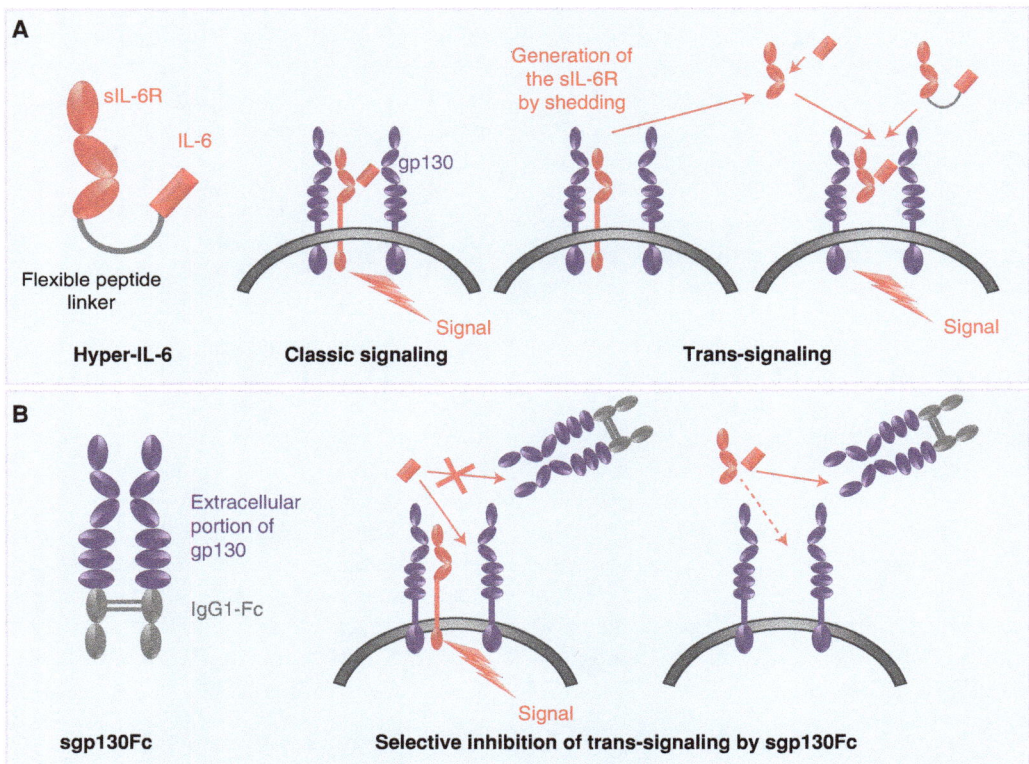

Figure 3. Molecular tools for the analysis of interleukin-6 (IL-6) trans-signaling. (A) IL-6 cannot only bind to the membrane-bound IL-6 receptor (IL-6R) (classic signaling) but also to a soluble IL-6R (sIL-6R), which can be generated by limited proteolysis via ADAM proteases. Cells, which express gp130 but no IL-6R, are not responsive to IL-6 but can be stimulated by the complex of IL-6 and sIL-6R (trans-signaling). IL-6 trans-signaling can be mimicked by the designer cytokine hyper-IL-6 (*left*) in which IL-6 and sIL-6R are covalently connected by a flexible peptide linker. (B) In the sgp130Fc protein, the extracellular portion of gp130 is fused to the constant portion of a human immunoglobulin G (IgG)1 antibody leading to dimerization by disulfide bridges. Since gp130 has no measurable affinity for IL-6 but binds the IL-6/sIL-6R complex with high affinity, sgp130Fc selectively blocks IL-6 trans-signaling without affecting IL-6 classic signaling via the membrane-bound IL-6R.

A second protein we generated consisted of the entire extracellular portion of gp130, which were dimerized using the Fc portion of human immunoglobulin G (IgG)1 antibodies, resulting in the protein sgp130Fc (Fig. 3B). Because IL-6 shows no measurable affinity to gp130, sgp130Fc did not affect IL-6 signaling via the membrane-bound IL-6R (classic signaling), but it efficiently blocked IL-6 trans-signaling (Jostock et al. 2001). Thus, sgp130Fc is a molecular tool to distinguish between IL-6 classic- and trans-signaling (Fig. 3B). In many animal models of human inflammatory or inflammation-associated cancer diseases, we side-by-side tested the consequences of global IL-6 blockade with the help of neutralizing antibodies versus specific IL-6 trans-signaling blockade with sgp130Fc. In addition, many models were performed in $Il6^{-/-}$ mice and in mice expressing the sgp130Fc as a transgene (Rabe et al. 2008). In all cases, the blockade of IL-6 trans-signaling was sufficient to block/inhibit the inflammatory state (Table 2).

Global blockade of IL-6 in patients leads to an increased susceptibility to bacterial infections (Selmi et al. 2015) as well as to an increase in serum cholesterol, serum triglycerides, and weight gain (Febbraio et al. 2010). Strikingly, selective blockade of IL-6 trans-signaling did not compromise host defense to mycobacteria (Sodenkamp et al. 2012), although infection was lethal in $Il6^{-/-}$ mice (Ladel et al. 1997). *Listeria* infection of mice was aggravated on global blockade of IL-6, whereas selective blockade of IL-6 trans-signaling by sgp130Fc did not affect the defense of the body against this infection (Hoge et al. 2013). In mice fed a high-fat diet, selective blockade of IL-6 trans-signaling prevented the inflammatory infiltration of macrophages into adipose tissue (Kraakman et al. 2015) without leading to increased insulin resistance or decreased glucose tolerance, which had been observed in $IL-6^{-/-}$ mice (Matthews et al. 2010; Wunderlich et al. 2010). These data indicated that IL-6 signaling via the membrane-bound IL-6R but not via the sIL-6R are responsible for the metabolic side effects seen in patients after global IL-6 blockade with tocilizumab (Febbraio et al. 2010). Furthermore, comparison of global blockade of IL-6 with selective blockade of IL-6 trans-signaling by sgp130Fc showed that IL-6 via the membrane-bound IL-6R shows regenerative activity in the intestine (Grivennikov et al. 2009), in the kidney (Luig et al. 2015), and in the pancreas (Zhang et al. 2013).

These accumulated data indicated that the proinflammatory activities of the cytokine IL-6 seem to depend mainly on trans-signaling via the sIL-6R, whereas the anti-inflammatory activities of the cytokine are mediated via the membrane-bound IL-6 (Fig. 4) (Jones et al. 2011; Calabrese and Rose-John 2014; Schmidt-Arras and Rose-John 2016). Furthermore, the sgp130Fc protein has been further developed and has passed clinical phase I trials with no adverse effects. Phase II clinical trials with inflammatory bowel disease patients began in 2016.

Table 2. Efficacy of sgp130Fc-mediated blockade of interleukin-6 (IL-6) trans-signaling in preclinical models of inflammation and inflammation-associated cancer

Intestinal inflammation (Atreya et al. 2000; Mitsuyama et al. 2006)	Pancreatic cancer (Lesina et al. 2011)	Ovarian hyperstimulation (Wei et al. 2013)
Rheumatoid arthritis (Nowell et al. 2003, 2009; Richards et al. 2006)	Acute inflammation (Chalaris et al. 2007; Rabe et al. 2008)	Lupus erythematosus (Tsantikos et al. 2013)
Asthma (Doganci et al. 2005)	Sepsis (Barkhausen et al. 2011; Greenhill et al. 2011)	Nephrotoxic nephritis (Luig et al. 2015; Braun et al. 2016)
Colon cancer (Grivennikov et al. 2009; Matsumoto et al. 2010)	Arterosclerosis (Schuett et al. 2012)	Lung cancer (Brooks et al. 2016)
Ovarian cancer (Greenhill et al. 2011; Lo et al. 2011)	Pancreatitis-lung failure (Zhang et al. 2013)	Lung emphysema (Ruwanpura et al. 2016)

THE EXPANDING IL-6 FAMILY OF CYTOKINES

As shown in Fig. 2, there are seven four-helical cytokines of the IL-6 family (IL-6, IL-11, LIF, OSM, CNTF, CT-1, and CLC), which bind to receptor complexes containing gp130. Additionally, IL-27 is a heterodimeric cytokine consisting of the four-helical protein p28 and a soluble cytokine receptor named EBI3, which is an Epstein–Barr virus-induced gene (Aparicio-Siegmund and Garbers 2015). EBI3 is also part of the heterodimeric cytokine IL-35, in which it forms a complex with p35, the four-helical subunit of the heterodimeric cytokine IL-12. IL-35 interacts with different receptor complexes, which can be composed of gp130 and the β subunit of the IL-12R, a homodimer of gp130, a homodimer of the β subunit of the IL-12R or of the IL-27-specific receptor subunit WSX-1 complexed with the β subunit of the IL-

12R (Egwuagu et al. 2015). The biology of the emerging family of IL-12 type cytokines is discussed in Yan et al. (2017). A cytokine related to the IL-6 cytokine family is the four-helical cytokine IL-31 (Fig. 2), which binds to a receptor complex formed by the OSM-R and the IL-31RA (Dillon et al. 2004). Because the receptor complex for the cytokine does not contain gp130, IL-31 is not formally part of the IL-6 cytokine family, but it is highly IL-6-family-related because it interacts with a gp130 family receptor (Hermanns 2015).

PLASTICITY OF THE IL-6 FAMILY OF CYTOKINES

The IL-6R was believed to be specific for the cytokine IL-6 (Yamasaki et al. 1988). Therefore, it was surprising that the IL-6 family cytokine CNTF also bound to the IL-6R and induced sig-

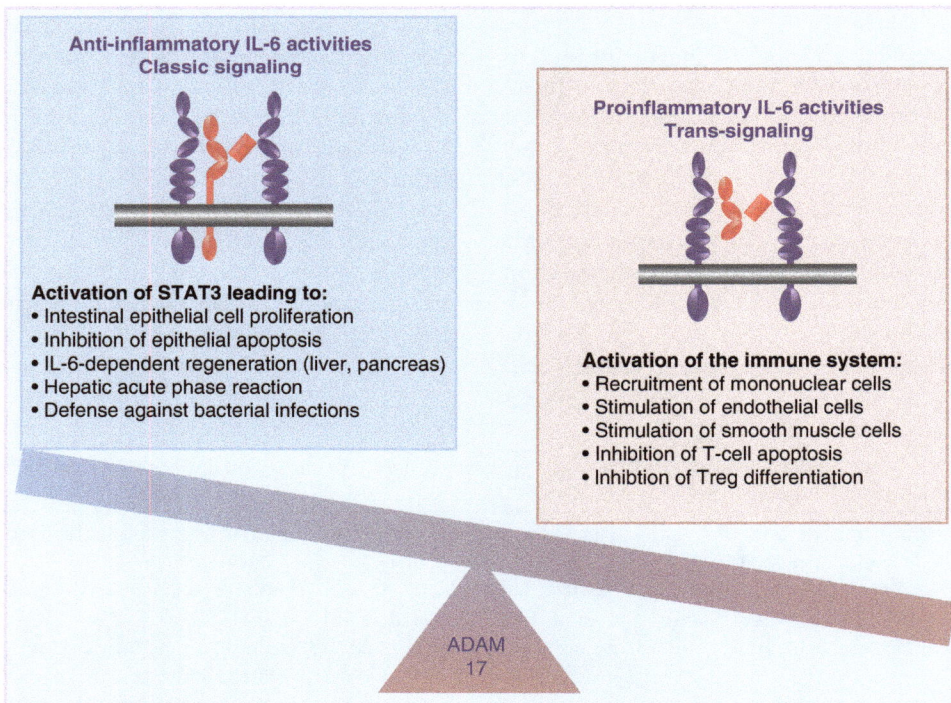

Figure 4. Pro- and anti-inflammatory activities of interleukin-6 (IL-6). IL-6 classic signaling (*left*) via signal transducer and activator of transcription 3 (STAT3) phosphorylation leads to protective and regenerative activities, whereas IL-6 trans-signaling (*right*) leads to an activation of the immune system resulting in proinflammatory IL-6 activities. Treg, Regulatory T cell.

naling. In this case, CNTF bound to the IL-6R but signaled via a receptor complex consisting of gp130 and LIF-R (Schuster et al. 2003). This was, however, not the first reported case of receptor promiscuity in the IL-6 cytokine family. It was known that the cytokine CNTF and CLC and possibly also CT-1 acted via the CNTF receptor (CNTF-R) (Garbers et al. 2012). Recently, it was reported that p28, the four-helical subunit of the heterodimeric cytokine IL-27, could act via the membrane-bound or sIL-6R in the absence of EBI. Interestingly, when acting via the IL-6R, p28 formed a signaling complex of a homodimeric gp130, whereas, when acting via EBI3, p28 signaled via a heterodimeric receptor complex consisting of gp130 and WSX-1 (Garbers et al. 2013). This might have important functional consequences because a homodimeric complex of gp130 preferentially activates STAT3, whereas the heterodimeric complex of gp130 and WSX-1 rather activates STAT1, leading to a different physiologic outcome (Pflanz et al. 2004).

VIRAL IL-6 FROM HUMAN HERPES VIRUS 8

In the late 1990s, a gene of the Kaposi sarcoma virus, also called human herpes virus 8 (HHV8) was discovered, which coded for a protein that shared 25% sequence identity with human IL-6 (Parravicini et al. 1997). It was shown that viral IL-6 (vIL-6) activated gp130 even in the absence of the IL-6R, leading to activation of the JAK/STAT signaling pathway (Molden et al. 1997). Furthermore, it was shown that vIL-6 directly bound to and stimulated gp130, leading to the persistent proliferation of cells, which were dependent on the IL-6/sIL-6R complex, indicating that vIL-6 mimicked IL-6 trans-signaling. Consequently, the biologic activity of vIL-6 could be blocked by sgp130Fc (Mullberg et al. 2000). The direct binding of vIL-6 to gp130 without the assistance of the IL-6R was shown by the crystal structure of the complex of gp130 and vIL-6 (Chow et al. 2001). A comparison with the structure of the complex of IL-6, IL-6R, and gp130 (Boulanger et al. 2003) shows that vIL-6 directly binds to gp130 in a similar molecular set-up as the IL-6/sIL-6R complex binds to gp130.

HHV-8 is associated with B-lymphoproliferative disorders, such as multicentric Castleman disease. Transgenic mice expressing serum levels of vIL-6 comparable to HHV8-infected patients spontaneously developed key features of human plasma cell-type multicentric Castleman disease (Suthaus et al. 2012). Interestingly, this disease pattern only developed in the presence of endogenous IL-6, whereas vIL-6 transgenic mice on an IL-6$^{-/-}$ genetic background showed no phenotype. These results indicated that vIL-6 induced Castleman disease only in the presence of endogenous IL-6 and it explained why in HHV8-infected patients with Castleman disease the IL-6R neutralizing antibody tocilizumab showed a therapeutic benefit although vIL-6 is not neutralized by this antibody (Nishimoto et al. 2000). These results demonstrate that HHV8 mimicked the IL-6 trans-signaling paradigm and that systemic induction of IL-6 trans-signaling might have pathophysiologic consequences (Suthaus et al. 2011).

THE IL-6 BUFFER IN THE BLOOD

The levels of IL-6 in the blood of healthy individuals are in the range of 1–5 pg/mL. IL-6 levels increase several thousand-fold during inflammatory states and can even reach levels of several µg/mL under lethal septic conditions. Levels of sIL-6R were found to be in the range of 40–75 ng/mL and sgp130 levels are approximately 250–400 ng/mL (Rose-John 2015). IL-6 secreted by cells will bind in the blood to the sIL-6R with an affinity of 1 nM and the complex of IL-6 and sIL-6R will bind to sgp130 with an affinity of 10 pM, leading to neutralization of IL-6 activity. Thus, sIL-6R and sgp130 in the blood can form a buffer for IL-6. We hypothesized that this is the mechanism by which the body is protected from overstimulation by IL-6 trans-signaling, because all cells of the body express gp130 and could be stimulated by the IL-6/sIL-6R complex (Schaper and Rose-John 2015).

A single nucleotide polymorphism (SNP) has been identified in the human IL-6R gene, which alters Asp358 into Ala358. Asp358 is directly adjacent to the proposed proteolytic cleavage site of the IL-6R (Mullberg et al.

1994). Consequently, shedding of the Ala358 carrying IL-6R protein by the protease ADAM17 is more efficient and levels of sIL-6R are significantly higher in homozygous individuals (Garbers et al. 2014). Remarkably, individuals with the Ala358 variant IL-6R are less susceptible to cardiovascular and arthritic diseases (Ferreira et al. 2013). This might be explained by the higher capacity of the sIL-6R/sgp130 buffer in the blood of these individuals, caused by the higher sIL-6R levels. Moreover, these data demonstrate the relevance of sIL-6R levels for the pathophysiology of inflammatory diseases.

TRANS-SIGNALING AS A SPECIFICITY OVERRIDING EMERGENCY REACTION

All cells in the body express the gp130 receptor subunit but not a single natural cytokine of the IL-6 family (with the exception of vIL-6) stimulates gp130 without the help of an additional receptor subunit (Fig. 2). The complex of IL-6 and sIL-6R, however, binds to and stimulates membrane-bound gp130. This might be the reason for the existence of the IL-6 buffer in the blood (see above). Signal transduction via the gp130 homodimer and the heterodimer of gp130 and LIF-R or OSM-R are not identical but very similar and are mainly driven by the activation of the JAK/STAT pathway (Hermanns 2015). Stimulation of membrane-bound gp130 by the IL-6/sIL-6R complex (in the absence of IL-11R, CNTF-R, LIF-R, or OSM-R) can therefore also be seen as an overriding reaction, which could fulfill functions of all other IL-6 family cytokines (Fig. 5). ADAM17 activity, which leads to the generation of the sIL-6R, is low under normal conditions but is strongly increased in inflammatory states and in cancer (Scheller et al. 2011). Consequently, levels of sIL-6R are increased under inflammatory conditions (Rabe et al. 2008; Nowell et al. 2009; Braun et al. 2016). IL-27 may represent an exception to this because signaling via WSX-1 is mainly characterized by STAT1 activation, whereas gp130 signaling is typically dominated

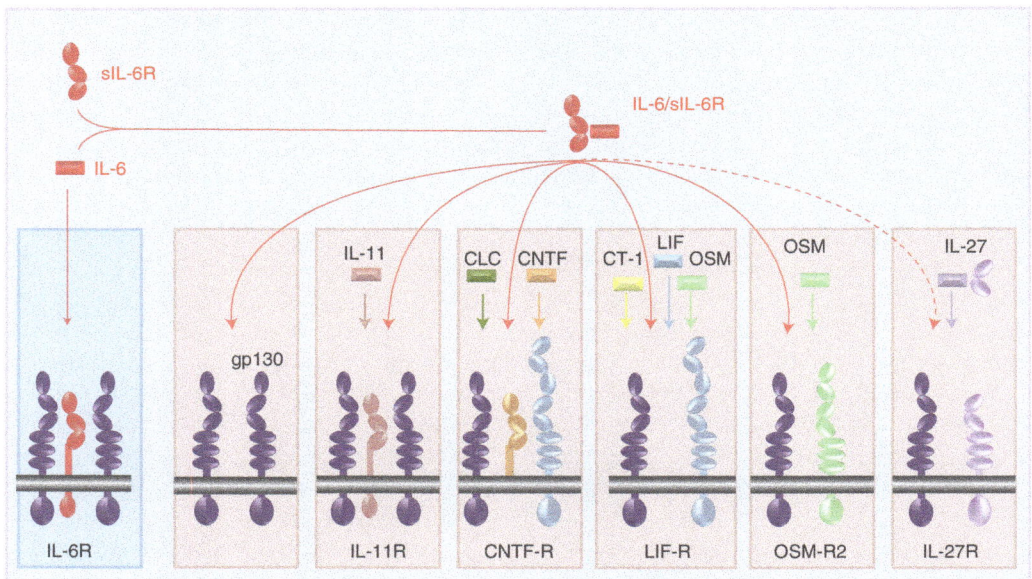

Figure 5. General gp130 activation by the interleukin-6 (IL-6)/soluble IL-6 receptor (sIL-6R) complex. Because all depicted receptor complexes of IL-6 family cytokines contain at least one molecule of gp130, the IL-6/sIL-6R complex can substitute for the activity of all depicted IL-6 family cytokines leading to signal transducer and activator of transcription 3 (STAT3) activation. CLC, Cardiotrophin-like cytokine; CNTF, ciliary neurotrophic factor; LIF, leukemia inhibitory factor; OSM, oncostatin M.

Cite this article as *Cold Spring Harb Perspect Biol* doi: 10.1101/cshperspect.a028415

by STAT3 signaling. In this respect, it is interesting that the four-helical subunit p28 is involved in both gp130/WSX-1 and gp130/gp130 signaling (see above).

TRANS-SIGNALING FOR OTHER CYTOKINES?

An interesting question is whether other IL-6 family cytokines also use trans-signaling. As is evident from Fig. 2, only cytokines that bind to a ligand-binding receptor (e.g., IL-6, IL-11, CNTF, and CLC) can theoretically perform trans-signaling, provided the respective receptor also exists in a soluble form. For IL-11, it was recently shown that the metalloprotease ADAM10 but not ADAM17 cleaves membrane-bound IL-11R, thereby generating a soluble IL-11R (sIL-11R). IL-11 bound to the sIL-11R could stimulate cells, which only expressed gp130 but no IL-11R. Furthermore, it was shown that IL-11 trans-signaling could be blocked by the sgp130Fc protein. Finally, evidence for the existence of sIL-11R in healthy individuals was provided (Lokau et al. 2016).

The CNTF-R is a glycosylphosphatidylinositol-anchored membrane protein and can be released from the cell membrane by phosphatidylinositol-specific phospholipase C (Davis et al. 1993). A soluble form of the CNTF-R (sCNTF-R) was released by muscle tissue in response to denervation and the combination of CNTF and sCNTF-R (but not CNTF or sCNTF-R alone) stimulated cells expressing gp130 and LIF-R but no CNTF-R leading to autophosphorylation of the LIF-R and transcription of the known CNTF target gene *tis11*.

These data clearly indicate that IL-11 and CNTF can stimulate cells via trans-signaling. This was further corroborated by experiments with recombinant fusion proteins of IL-11 and sIL-11R (Pflanz et al. 1999) or CNTF and sCNTF-R (Marz et al. 2002). Both fusion proteins, designed after the fusion protein hyper-IL-6 (Fischer et al. 1997), were fully active on cells expressing gp130 (in the case of hyper-IL-11) or gp130 and LIF-R (in the case of hyper-CNTF) but not IL-11R of CNTF-R. Stimulation of the cells with fusion proteins led to

STAT3 activation and to the proliferation of murine pre-B cells (Pflanz et al. 1999; Marz et al. 2002).

In addition, other cytokines such as IL-2 and IL-15 act via ligand-binding receptor subunits (IL-2Rα, IL-15Rα) and the signaling receptor subunits IL-2Rβ and IL-2Rγ (Waldmann 2006). An IL-15/sIL-15R fusion protein has been shown to strongly stimulate target cells expressing the IL-2Rβ and IL-2Rγ receptor subunits (Desbois et al. 2016), and a soluble IL-15Rα protein has been shown to be generated by a cellular metalloprotease (Mortier et al. 2004). A soluble form of the IL-2Rα (sIL-2Rα) is produced by proteolysis and the levels of sIL-2Rα are elevated in the blood of cancer patients. Furthermore, the sIL-2Rα enhanced the biologic activity of IL-2 on several target cells (Yang et al. 2011). These data indicate that IL-2 and IL-15 can act via trans-signaling, although little is known yet about the functional importance of this pathway. Theoretically, all cytokines or growth factors could also act via trans-signaling provided they bind to a nonsignaling receptor subunit.

MUTATIONS IN IL-6 FAMILY CYTOKINES AND THEIR RECEPTORS

The physiologic functions of cytokines can often be deduced from natural mutations or from genetically engineered gene-deficient animals. Although IL-6 family cytokines often can compensate for the loss of one or more cytokines, there are phenotypes in animals with gene deficiency or with mutations in cytokine or cytokine receptor genes, which can apparently not be compensated for by other cytokines of the same family.

IL-6-deficient mice are phenotypically normal but they show an impaired hepatic acute phase response and were significantly more susceptible to bacterial infection (Kopf et al. 1994). Moreover, $Il6^{-/-}$ mice had difficulties regenerating the liver on hepatectomy (Cressman et al. 1996).

Although $Il11r^{-/-}$ mice develop normally and show undisturbed hematopoiesis (Nandurkar et al. 1997), mutations in the human *IL11R* gene led to a severe loss of function of the receptor and caused a craniosynostosis syndrome (Keupp et al. 2013). Female $Il11r^{-/-}$ mice are infertile because of defective decidualization (Robb et al. 1998).

Although animals with a loss of a functional CNTF gene show only a very mild phenotype (Masu et al. 1993), mice with a deficiency in CNTF-R are unable to initiate feeding and die shortly after birth. Moreover, these mice show a dramatic loss of motor neurons (DeChiara et al. 1995). These data indicated at the time that an additional ligand for the CNTF-R should exist. This additional ligand was later identified as CLC (Elson et al. 2000). As described above, CLC acts via the CNTF-R (Fig. 2). Mutations in CLC, which result in loss of CNTF-R binding, caused cold-induced sweating syndrome. These patients sweat in the cold but are unable to sweat in hot weather (Rousseau et al. 2006).

Animals with a gene disruption in the LIF gene have been generated. These animals were phenotypically normal but they had a defect in implantation of the developing embryo. Female $Lif^{-/-}$ mice are fertile but they could not implant and failed to develop their blastocysts. Nevertheless, the blastocysts are viable and can be transferred to wild-type recipients in which they normally develop to term (Stewart et al. 1992).

Mice with a gp130-sensitizing mutation have been generated by replacing the membrane proximal cytoplasmic tyrosine at position 757 of gp130 with the amino acid phenylalanine (Ernst and Jenkins 2004). This membrane proximal cytoplasmic tyrosine of gp130 is needed for negative feedback regulation of gp130 signaling by SOCS3. Consequently, the gp130-sensitized mice show a twofold higher STAT3 activation on a given gp130 stimulus as compared with wild-type mice (Nowell et al. 2009). These gp130-sensitized mice spontaneously developed gastric tumors (Judd et al. 2004) and lung emphysema (Brooks et al. 2016). Moreover, these mice were more sensitive than wild-type mice in a lipopolysaccharide (LPS)-induced sepsis model (Greenhill et al. 2011). Naturally occurring somatic mutations in human gp130, which led to ligand-independent constitutive gp130 activation, have been described in 60% of inflammatory hepatocellular adenomas (Rebouissou et al. 2009), demonstrating the importance of gp130 signaling in liver pathophysiology.

CONCLUDING REMARKS

IL-6 family cytokines are four-helical proteins and they have been grouped into a common family because of their common use of the gp130 receptor subunit. As outlined above, this strict definition does, however, not fully apply to family members such as IL-31 and IL-35. Signal transduction of IL-6 family cytokines is very similar and is dominated by STAT3 activation. The only exception is IL-27, which predominantly signals via activated STAT1.

Levels of IL-6 are very low under normal conditions, but these levels can raise many thousand-fold in inflammatory states. Autoimmune diseases such as rheumatoid arthritis are characterized by elevated IL-6 levels, and neutralization of IL-6 activity by the IL-6R-specific monoclonal antibody tocilizumab as a treatment of autoimmune disease has been approved in more than 100 countries. Interestingly, neutralization of IL-6 activity is at least as efficient in rheumatoid arthritis patients as neutralization of tumor necrosis factor α.

IL-6 (and possibly also other members of the IL-6 family cytokines) apparently has pro- and anti-inflammatory activities. Proinflammatory activities of IL-6 are mediated by IL-6 trans-signaling via the sIL-6R, whereas protective and anti-inflammatory activities of IL-6 are mainly executed via the membrane-bound IL-6R (classic signaling). Because IL-6 trans-signaling can be blocked by the sgp130Fc protein without affecting classic signaling, future IL-6-based therapies might use a specific blockade of IL-6 trans-signaling rather than a global blockade of all IL-6 activities.

ACKNOWLEDGMENTS

The work described in this review is supported by grants from the Deutsche Forschungsgemeinschaft, Bonn, Germany (SFB 841, project

C1; SFB 877, project A1 and the Cluster of Excellence "Inflammation at Interfaces"). S.R.-J. is an inventor of patents owned by CONARIS Research Institute, which develops the sgp130Fc protein together with Ferring Pharmaceuticals and he has stock ownership in CONARIS.

REFERENCES

Reference is also in this collection.

Abdel-Meguid SS, Shieh HS, Smith WW, Dayringer HE, Violand BN, Bentle LA. 1987. Three-dimensional structure of a genetically engineered variant of porcine growth hormone. *Proc Natl Acad Sci* **84:** 6434–6437.

Aparicio-Siegmund S, Garbers C. 2015. The biology of interleukin-27 reveals unique pro- and anti-inflammatory functions in immunity. *Cytokine Growth Factor Rev* **26:** 579–586.

Atreya R, Mudter J, Finotto S, Mullberg J, Jostock T, Wirtz S, Schutz M, Bartsch B, Holtmann M, Becker C, et al. 2000. Blockade of interleukin 6 trans signaling suppresses T-cell resistance against apoptosis in chronic intestinal inflammation: Evidence in Crohn's disease and experimental colitis in vivo. *Nat Med* **6:** 583–588.

Audet J, Miller CL, Rose-John S, Piret JM, Eaves CJ. 2001. Distinct role of gp130 activation in promoting self-renewal divisions by mitogenically stimulated murine hematopoietic stem cells. *Proc Natl Acad Sci* **98:** 1757–1762.

Barkhausen T, Tschernig T, Rosenstiel P, van Griensven M, Vonberg RP, Dorsch M, Mueller-Heine A, Chalaris A, Scheller J, Rose-John S, et al. 2011. Selective blockade of interleukin-6 trans-signaling improves survival in a murine polymicrobial sepsis model. *Crit Care Med* **39:** 1407–1413.

Bazan JF. 1990. Haemopoietic receptors and helical cytokines. *Immunol Today* **11:** 350–354.

Bazan JF. 1992. Unraveling the structure of IL-2. *Science* **257:** 410–413.

Boulanger MJ, Chow DC, Brevnova EE, Garcia KC. 2003. Hexameric structure and assembly of the interleukin-6/IL-6α-receptor/gp130 complex. *Science* **300:** 2101–2104.

Brakenhoff JP, de Groot ER, Evers RF, Pannekoek H, Aarden LA. 1987. Molecular cloning and expression of hybridoma growth factor in *Escherichia coli*. *J Immunol* **139:** 4116–4121.

Brakenhoff JP, Hart M, De Groot ER, Di Padova F, Aarden LA. 1990. Structure–function analysis of human IL-6. Epitope mapping of neutralizing monoclonal antibodies with amino- and carboxyl-terminal deletion mutants. *J Immunol* **145:** 561–568.

Braun GS, Nagayama Y, Maruta Y, Heymann F, van Roeyen CR, Klinkhammer BM, Boor P, Villa L, Salant DJ, Raffetseder U, et al. 2016. IL-6 trans-signaling drives murine crescentic GN. *J Am Soc Nephrol* **27:** 132–142.

Brooks GD, McLeod L, Alhayyani S, Miller A, Russell PA, Ferlin W, Rose-John S, Ruwanpura S, Jenkins BJ. 2016.

IL6 trans-signaling promotes KRAS-driven lung carcinogenesis. *Cancer Res* **76:** 866–876.

Calabrese LH, Rose-John S. 2014. IL-6 biology: Implications for clinical targeting in rheumatic disease. *Nat Rev Rheumatol* **10:** 720–727.

Chalaris A, Rabe B, Paliga K, Lange H, Laskay T, Fielding CA, Jones SA, Rose-John S, Scheller J. 2007. Apoptosis is a natural stimulus of IL6R shedding and contributes to the proinflammatory trans-signaling function of neutrophils. *Blood* **110:** 1748–1755.

Chow D, He X, Snow AL, Rose-John S, Garcia KC. 2001. Structure of an extracellular gp130 cytokine receptor signaling complex. *Science* **291:** 2150–2155.

Cressman DE, Greenbaum LE, DeAngelis RA, Ciliberto G, Furth EE, Poli V, Taub R. 1996. Liver failure and defective hepatocyte regeneration in interleukin-6-deficient mice. *Science* **274:** 1379–1383.

Davis S, Aldrich TH, Ip NY, Stahl N, Scherer S, Farruggella T, DiStefano PS, Curtis R, Panayotatos N, Gascan H, et al. 1993. Released form of CNTF receptor α component as a soluble mediator of CNTF responses. *Science* **259:** 1736–1739.

DeChiara TM, Vejsada R, Poueymirou WT, Acheson A, Suri C, Conover JC, Friedman B, McClain J, Pan L, Stahl N, et al. 1995. Mice lacking the CNTF receptor, unlike mice lacking CNTF, exhibit profound motor neuron deficits at birth. *Cell* **83:** 313–322.

Desbois M, Le Vu P, Coutzac C, Marcheteau E, Beal C, Terme M, Gey A, Morisseau S, Teppaz G, Boselli L, et al. 2016. IL-15 trans-signaling with the superagonist RLI promotes effector/memory CD8+ T cell responses and enhances antitumor activity of PD-1 antagonists. *J Immunol* **197:** 168–178.

Dillon SR, Sprecher C, Hammond A, Bilsborough J, Rosenfeld-Franklin M, Presnell SR, Haugen HS, Maurer M, Harder B, Johnston J, et al. 2004. Interleukin 31, a cytokine produced by activated T cells, induces dermatitis in mice. *Nat Immunol* **5:** 752–760.

Doganci A, Eigenbrod T, Krug N, De Sanctis GT, Hausding M, Erpenbeck VJ, Haddad el B, Lehr HA, Schmitt E, Bopp T, et al. 2005. The IL-6R α chain controls lung CD4+CD25+ Treg development and function during allergic airway inflammation in vivo. *J Clin Invest* **115:** 313–325.

Egwuagu CE, Yu CR, Sun L, Wang R. 2015. Interleukin 35: Critical regulator of immunity and lymphocyte-mediated diseases. *Cytokine Growth Factor Rev* **26:** 587–593.

Elson GC, Lelievre E, Guillet C, Chevalier S, Plun-Favreau H, Froger J, Suard I, de Coignac AB, Delneste Y, Bonnefoy JY, et al. 2000. CLF associates with CLC to form a functional heteromeric ligand for the CNTF receptor complex. *Nat Neurosci* **3:** 867–872.

Ernst M, Jenkins BJ. 2004. Acquiring signalling specificity from the cytokine receptor gp130. *Trends Genet* **20:** 23–32.

Febbraio MA, Rose-John S, Pedersen BK. 2010. Is interleukin-6 receptor blockade the Holy Grail for inflammatory diseases? *Clin Pharmacol Ther* **87:** 396–398.

Ferreira RC, Freitag DF, Cutler AJ, Howson JM, Rainbow DB, Smyth DJ, Kaptoge S, Clarke P, Boreham C, Coulson RM, et al. 2013. Functional IL6R 358Ala allele impairs classical IL-6 receptor signaling and influences risk of diverse inflammatory diseases. *PLoS Genet* **9:** e1003444.

Fischer M, Goldschmitt J, Peschel C, Brakenhoff JP, Kallen KJ, Wollmer A, Grotzinger J, Rose-John S. 1997. A bioactive designer cytokine for human hematopoietic progenitor cell expansion. *Nat Biotechnol* **15:** 142–145.

Gabay C, Emery P, van Vollenhoven R, Dikranian A, Alten R, Pavelka K, Klearman M, Musselman D, Agarwal S, Green J, et al. 2013. Tocilizumab monotherapy versus adalimumab monotherapy for treatment of rheumatoid arthritis (ADACTA): A randomised, double-blind, controlled phase 4 trial. *Lancet* **381:** 1541–1550.

Galun E, Zeira E, Pappo O, Peters M, Rose-John S. 2000. Liver regeneration induced by a designer human IL-6/sIL-6R fusion protein reverses severe hepatocellular injury. *FASEB J* **14:** 1979–1987.

Gandhi NA, Bennett BL, Graham NM, Pirozzi G, Stahl N, Yancopoulos GD. 2016. Targeting key proximal drivers of type 2 inflammation in disease. *Nat Rev Drug Discov* **15:** 35–50.

Garbers C, Hermanns HM, Schaper F, Muller-Newen G, Grotzinger J, Rose-John S, Scheller J. 2012. Plasticity and cross-talk of interleukin 6-type cytokines. *Cytokine Growth Factor Rev* **23:** 85–97.

Garbers C, Spudy B, Aparicio-Siegmund S, Waetzig GH, Sommer J, Holscher C, Rose-John S, Grotzinger J, Lorenzen I, Scheller J. 2013. An interleukin-6 receptor-dependent molecular switch mediates signal transduction of the IL-27 cytokine subunit p28 (IL-30) via a gp130 protein receptor homodimer. *J Biol Chem* **288:** 4346–4354.

Garbers C, Monhasery N, Aparicio-Siegmund S, Lokau J, Baran P, Nowell MA, Jones SA, Rose-John S, Scheller J. 2014. The interleukin-6 receptor Asp358Ala single nucleotide polymorphism rs2228145 confers increased proteolytic conversion rates by ADAM proteases. *Biochim Biophys Acta* **1842:** 1485–1494.

Gauldie J, Richards C, Harnish D, Lansdorp P, Baumann H. 1987. Interferon β2/B-cell stimulatory factor type 2 shares identity with monocyte-derived hepatocyte-stimulating factor and regulates the major acute phase protein response in liver cells. *Proc Natl Acad Sci* **84:** 7251–7255.

Gearing DP, Gough NM, King JA, Hilton DJ, Nicola NA, Simpson RJ, Nice EC, Kelso A, Metcalf D. 1987. Molecular cloning and expression of cDNA encoding a murine myeloid leukaemia inhibitory factor (LIF). *EMBO J* **6:** 3995–4002.

Gearing DP, Thut CJ, VandeBos T, Gimpel SD, Delaney PB, King J, Price V, Cosman D, Beckmann MP. 1991. Leukemia inhibitory factor receptor is structurally related to the IL-6 signal transducer, gp130. *EMBO J* **10:** 2839–2848.

Gearing DP, Comeau MR, Friend DJ, Gimpel SD, Thut CJ, McGourty J, Brasher KK, King JA, Gillis S, Mosley B, et al. 1992. The IL-6 signal transducer, gp130: An oncostatin M receptor and affinity converter for the LIF receptor. *Science* **255:** 1434–1437.

Greenhill CJ, Rose-John S, Lissilaa R, Ferlin W, Ernst M, Hertzog PJ, Mansell A, Jenkins BJ. 2011. IL-6 trans-signaling modulates TLR4-dependent inflammatory responses via STAT3. *J Immunol* **186:** 1199–1208.

Grivennikov S, Karin E, Terzic J, Mucida D, Yu GY, Vallabhapurapu S, Scheller J, Rose-John S, Cheroutre H, Eckmann L, et al. 2009. IL-6 and Stat3 are required for survival of intestinal epithelial cells and development of colitis-associated cancer. *Cancer Cell* **15:** 103–113.

Haegeman G, Content J, Volckaert G, Derynck R, Tavernier J, Fiers W. 1986. Structural analysis of the sequence coding for an inducible 26-kDa protein in human fibroblasts. *Eur J Biochem* **159:** 625–632.

Hermanns HM. 2015. Oncostatin M and interleukin-31: Cytokines, receptors, signal transduction and physiology. *Cytokine Growth Factor Rev* **26:** 545–558.

Hibi M, Murakami M, Saito M, Hirano T, Taga T, Kishimoto T. 1990. Molecular cloning and expression of an IL-6 signal transducer, gp130. *Cell* **63:** 1149–1157.

Hirano T, Yasukawa K, Harada H, Taga T, Watanabe Y, Matsuda T, Kashiwamura S, Nakajima K, Koyama K, Iwamatsu A, et al. 1986. Complementary DNA for a novel human interleukin (BSF-2) that induces B lymphocytes to produce immunoglobulin. *Nature* **324:** 73–76.

Hirano T, Taga T, Yamasaki K, Matsuda T, Yasukawa K, Hirata Y, Yawata H, Tanabe O, Akira S, Kishimoto T. 1989. Molecular cloning of the cDNAs for interleukin-6/B cell stimulatory factor 2 and its receptor. *Ann NY Acad Sci* **557:** 167–178; discussion 178–180.

Hoge J, Yan I, Janner N, Schumacher V, Chalaris A, Steinmetz OM, Engel DR, Scheller J, Rose-John S, Mittrucker HW. 2013. IL-6 controls the innate immune response against *Listeria* monocytogenes via classical IL-6 signaling. *J Immunol* **190:** 703–711.

Humphrey RK, Beattie GM, Lopez AD, Bucay N, King CC, Firpo MT, Rose-John S, Hayek A. 2004. Maintenance of pluripotency in human embryonic stem cells is STAT3 independent. *Stem Cells* **22:** 522–530.

Jones SA, Rose-John S. 2002. The role of soluble receptors in cytokine biology: The agonistic properties of the sIL-6R/IL-6 complex. *Biochim Biophys Acta* **1592:** 251–263.

Jones SA, Scheller J, Rose-John S. 2011. Therapeutic strategies for the clinical blockade of IL-6/gp130 signaling. *J Clin Invest* **121:** 3375–3383.

Jostock T, Mullberg J, Ozbek S, Atreya R, Blinn G, Voltz N, Fischer M, Neurath MF, Rose-John S. 2001. Soluble gp130 is the natural inhibitor of soluble interleukin-6 receptor transsignaling responses. *Eur J Biochem* **268:** 160–167.

Judd LM, Alderman BM, Howlett M, Shulkes A, Dow C, Moverley J, Grail D, Jenkins BJ, Ernst M, Giraud AS. 2004. Gastric cancer development in mice lacking the SHP2 binding site on the IL-6 family co-receptor gp130. *Gastroenterology* **126:** 196–207.

Keupp K, Li Y, Vargel I, Hoischen A, Richardson R, Neveling K, Alanay Y, Uz E, Elcioglu N, Rachwalski M, et al. 2013. Mutations in the interleukin receptor IL11RA cause autosomal recessive Crouzon-like craniosynostosis. *Mol Genet Genomic Med* **1:** 223–237.

Klouche M, Bhakdi S, Hemmes M, Rose-John S. 1999. Novel path to activation of vascular smooth muscle cells: Upregulation of gp130 creates an autocrine activation loop by IL-6 and its soluble receptor. *J Immunol* **163:** 4583–4589.

Kopf M, Baumann H, Freer G, Freudenberg M, Lamers M, Kishimoto T, Zinkernagel R, Bluethmann H, Kohler G.

1994. Impaired immune and acute-phase responses in interleukin-6-deficient mice. *Nature* **368**: 339–342.

Kraakman MJ, Kammoun HL, Allen TL, Deswaerte V, Henstridge DC, Estevez E, Matthews VB, Neill B, White DA, Murphy AJ, et al. 2015. Blocking IL-6 trans-signaling prevents high-fat diet-induced adipose tissue macrophage recruitment but does not improve insulin resistance. *Cell Metab* **21**: 403–416.

Ladel CH, Blum C, Dreher A, Reifenberg K, Kopf M, Kaufmann SH. 1997. Lethal tuberculosis in interleukin-6-deficient mutant mice. *Infect Immun* **65**: 4843–4849.

Lesina M, Kurkowski MU, Ludes K, Rose-John S, Treiber M, Kloppel G, Yoshimura A, Reindl W, Sipos B, Akira S, et al. 2011. Stat3/Socs3 activation by IL-6 transsignaling promotes progression of pancreatic intraepithelial neoplasia and development of pancreatic cancer. *Cancer Cell* **19**: 456–469.

Levy DE, Darnell JEJr. 2002. STATs: Transcriptional control and biological impact. *Nat Rev Mol Cell Biol* **3**: 651–662.

Lin LF, Mismer D, Lile JD, Armes LG, Butler ET III, Vannice JL, Collins F. 1989. Purification, cloning, and expression of ciliary neurotrophic factor (CNTF). *Science* **246**: 1023–1025.

Lo CW, Chen MW, Hsiao M, Wang S, Chen CA, Hsiao SM, Chang JS, Lai TC, Rose-John S, Kuo ML, et al. 2011. IL-6 trans-signaling in formation and progression of malignant ascites in ovarian cancer. *Cancer Res* **71**: 424–434.

Lokau J, Nitz R, Agthe M, Monhasery N, Aparicio-Siegmund S, Schumacher N, Wolf J, Moller-Hackbarth K, Waetzig GH, Grotzinger J, et al. 2016. Proteolytic cleavage governs interleukin-11 trans-signaling. *Cell Rep* **14**: 1761–1773.

Luig M, Kluger MA, Goerke B, Meyer M, Nosko A, Yan I, Scheller J, Mittrucker HW, Rose-John S, Stahl RA, et al. 2015. Inflammation-induced IL-6 functions as a natural brake on macrophages and limits GN. *J Am Soc Nephrol* **26**: 1597–1607.

Lupardus PJ, Skiniotis G, Rice AJ, Thomas C, Fischer S, Walz T, Garcia KC. 2011. Structural snapshots of full-length Jak1, a transmembrane gp130/IL-6/IL-6Rα cytokine receptor complex, and the receptor-Jak1 holocomplex. *Structure* **19**: 45–55.

Lust JA, Donovan KA, Kline MP, Greipp PR, Kyle RA, Maihle NJ. 1992. Isolation of an mRNA encoding a soluble form of the human interleukin-6 receptor. *Cytokine* **4**: 96–100.

Lutticken C, Wegenka UM, Yuan J, Buschmann J, Schindler C, Ziemiecki A, Harpur AG, Wilks AF, Yasukawa K, Taga T, et al. 1994. Association of transcription factor APRF and protein kinase Jak1 with the interleukin-6 signal transducer gp130. *Science* **263**: 89–92.

Mackiewicz A, Schooltink H, Heinrich PC, Rose-John S. 1992. Complex of soluble human IL-6-receptor/IL-6 up-regulates expression of acute-phase proteins. *J Immunol* **149**: 2021–2027.

Malik N, Kallestad JC, Gunderson NL, Austin SD, Neubauer MG, Ochs V, Marquardt H, Zarling JM, Shoyab M, Wei CM, et al. 1989. Molecular cloning, sequence analysis, and functional expression of a novel growth regulator, oncostatin M. *Mol Cell Biol* **9**: 2847–2853.

Marz P, Ozbek S, Fischer M, Voltz N, Otten U, Rose-John S. 2002. Differential response of neuronal cells to a fusion

protein of ciliary neurotrophic factor/soluble CNTF-receptor and leukemia inhibitory factor. *Eur J Biochem* **269**: 3023–3031.

Masu Y, Wolf E, Holtmann B, Sendtner M, Brem G, Thoenen H. 1993. Disruption of the CNTF gene results in motor neuron degeneration. *Nature* **365**: 27–32.

Matsumoto S, Hara T, Mitsuyama K, Yamamoto M, Tsuruta O, Sata M, Scheller J, Rose-John S, Kado S, Takada T. 2010. Essential roles of IL-6 trans-signaling in colonic epithelial cells, induced by the IL-6/soluble-IL-6 receptor derived from lamina propria macrophages, on the development of colitis-associated premalignant cancer in a murine model. *J Immunol* **184**: 1543–1551.

Matthews VB, Allen TL, Risis S, Chan MH, Henstridge DC, Watson N, Zaffino LA, Babb JR, Boon J, Meikle PJ, et al. 2010. Interleukin-6-deficient mice develop hepatic inflammation and systemic insulin resistance. *Diabetologia* **53**: 2431–2441.

Mitsuyama K, Matsumoto S, Rose-John S, Suzuki A, Hara T, Tomiyasu N, Handa K, Tsuruta O, Funabashi H, Scheller J, et al. 2006. STAT3 activation via interleukin 6 trans-signalling contributes to ileitis in SAMP1/Yit mice. *Gut* **55**: 1263–1269.

Molden J, Chang Y, You Y, Moore PS, Goldsmith MA. 1997. A Kaposi's sarcoma-associated herpesvirus-encoded cytokine homolog (vIL-6) activates signaling through the shared gp130 receptor subunit. *J Biol Chem* **272**: 19625–19631.

Mortier E, Bernard J, Plet A, Jacques Y. 2004. Natural, proteolytic release of a soluble form of human IL-15 receptor α-chain that behaves as a specific, high affinity IL-15 antagonist. *J Immunol* **173**: 1681–1688.

Mosley B, De Imus C, Friend D, Boiani N, Thoma B, Park LS, Cosman D. 1996. Dual oncostatin M (OSM) receptors. Cloning and characterization of an alternative signaling subunit conferring OSM-specific receptor activation. *J Biol Chem* **271**: 32635–32643.

Mullberg J, Schooltink H, Stoyan T, Heinrich PC, Rose-John S. 1992. Protein kinase C activity is rate limiting for shedding of the interleukin-6 receptor. *Biochem Biophys Res Commun* **189**: 794–800.

Mullberg J, Schooltink H, Stoyan T, Gunther M, Graeve L, Buse G, Mackiewicz A, Heinrich PC, Rose-John S. 1993. The soluble interleukin-6 receptor is generated by shedding. *Eur J Immunol* **23**: 473–480.

Mullberg J, Oberthur W, Lottspeich F, Mehl E, Dittrich E, Graeve L, Heinrich PC, Rose-John S. 1994. The soluble human IL-6 receptor. Mutational characterization of the proteolytic cleavage site. *J Immunol* **152**: 4958–4968.

Mullberg J, Geib T, Jostock T, Hoischen SH, Vollmer P, Voltz N, Heinz D, Galle PR, Klouche M, Rose-John S. 2000. IL-6 receptor independent stimulation of human gp130 by viral IL-6. *J Immunol* **164**: 4672–4677.

Nandurkar HH, Robb L, Tarlinton D, Barnett L, Kontgen F, Begley CG. 1997. Adult mice with targeted mutation of the interleukin-11 receptor (IL11Ra) display normal hematopoiesis. *Blood* **90**: 2148–2159.

Nishimoto N, Sasai M, Shima Y, Nakagawa M, Matsumoto T, Shirai T, Kishimoto T, Yoshizaki K. 2000. Improvement in Castleman's disease by humanized anti-interleukin-6 receptor antibody therapy. *Blood* **95**: 56–61.

Nowell MA, Richards PJ, Horiuchi S, Yamamoto N, Rose-John S, Topley N, Williams AS, Jones SA. 2003. Soluble IL-6 receptor governs IL-6 activity in experimental arthritis: Blockade of arthritis severity by soluble glycoprotein 130. *J Immunol* **171:** 3202–3209.

Nowell MA, Williams AS, Carty SA, Scheller J, Hayes AJ, Jones GW, Richards PJ, Slinn S, Ernst M, Jenkins BJ, et al. 2009. Therapeutic targeting of IL-6 trans signaling counteracts STAT3 control of experimental inflammatory arthritis. *J Immunol* **182:** 613–622.

Oberg HH, Wesch D, Grussel S, Rose-John S, Kabelitz D. 2006. Differential expression of CD126 and CD130 mediates different STAT-3 phosphorylation in CD4$^+$ CD25$-$ and CD25^{high} regulatory T cells. *Int Immunol* **18:** 555–563.

Parravicini C, Corbellino M, Paulli M, Magrini U, Lazzarino M, Moore PS, Chang Y. 1997. Expression of a virus-derived cytokine, KSHV vIL-6, in HIV-seronegative Castleman's disease. *Am J Pathol* **151:** 1517–1522.

Paul SR, Bennett F, Calvetti JA, Kelleher K, Wood CR, O'Hara RMJr, Leary AC, Sibley B, Clark SC, Williams DA, et al. 1990. Molecular cloning of a cDNA encoding interleukin 11, a stromal cell-derived lymphopoietic and hematopoietic cytokine. *Proc Natl Acad Sci* **87:** 7512–7516.

Pennica D, King KL, Shaw KJ, Luis E, Rullamas J, Luoh SM, Darbonne WC, Knutzon DS, Yen R, Chien KR, et al. 1995. Expression cloning of cardiotrophin 1, a cytokine that induces cardiac myocyte hypertrophy. *Proc Natl Acad Sci* **92:** 1142–1146.

Pennica D, Arce V, Swanson TA, Vejsada R, Pollock RA, Armanini M, Dudley K, Phillips HS, Rosenthal A, Kato AC, et al. 1996. Cardiotrophin-1, a cytokine present in embryonic muscle, supports long-term survival of spinal motoneurons. *Neuron* **17:** 63–74.

Peters M, Blinn G, Jostock T, Schirmacher P, Meyer zum Buschenfelde KH, Galle PR, Rose-John S. 2000. Combined interleukin 6 and soluble interleukin 6 receptor accelerates murine liver regeneration. *Gastroenterology* **119:** 1663–1671.

Pflanz S, Tacken I, Grotzinger J, Jacques Y, Minvielle S, Dahmen H, Heinrich PC, Muller-Newen G. 1999. A fusion protein of interleukin-11 and soluble interleukin-11 receptor acts as a superagonist on cells expressing gp130. *FEBS Lett* **450:** 117–122.

Pflanz S, Hibbert L, Mattson J, Rosales R, Vaisberg E, Bazan JF, Phillips JH, McClanahan TK, de Waal Malefyt R, Kastelein RA. 2004. WSX-1 and glycoprotein 130 constitute a signal-transducing receptor for IL-27. *J Immunol* **172:** 2225–2231.

Rabe B, Chalaris A, May U, Waetzig GH, Seegert D, Williams AS, Jones SA, Rose-John S, Scheller J. 2008. Transgenic blockade of interleukin 6 transsignaling abrogates inflammation. *Blood* **111:** 1021–1028.

Rebouissou S, Amessou M, Couchy G, Poussin K, Imbeaud S, Pilati C, Izard T, Balabaud C, Bioulac-Sage P, Zucman-Rossi J. 2009. Frequent in-frame somatic deletions activate gp130 in inflammatory hepatocellular tumours. *Nature* **457:** 200–204.

Richards PJ, Nowell MA, Horiuchi S, McLoughlin RM, Fielding CA, Grau S, Yamamoto N, Ehrmann M, Rose-John S, Williams AS, et al. 2006. Functional characterization of a soluble gp130 isoform and its therapeutic capacity in an experimental model of inflammatory arthritis. *Arthritis Rheum* **54:** 1662–1672.

Robb L, Li R, Hartley L, Nandurkar HH, Koentgen F, Begley CG. 1998. Infertility in female mice lacking the receptor for interleukin 11 is due to a defective uterine response to implantation. *Nat Med* **4:** 303–308.

Rose-John S. 2015. The soluble interleukin-6 receptor and related proteins. *Best Pract Res Clin Endocrinol Metab* **29:** 787–797.

Rose-John S, Heinrich PC. 1994. Soluble receptors for cytokines and growth factors: Generation and biological function. *Biochem J* **300:** 281–290.

Rose-John S, Scheller J, Schaper F. 2015. "Family reunion"— A structured view on the composition of the receptor complexes of interleukin-6-type and interleukin-12-type cytokines. *Cytokine Growth Factor Rev* **26:** 471–474.

Rousseau F, Gauchat JF, McLeod JG, Chevalier S, Guillet C, Guilhot F, Cognet I, Froger J, Hahn AF, Knappskog PM, et al. 2006. Inactivation of cardiotrophin-like cytokine, a second ligand for ciliary neurotrophic factor receptor, leads to cold-induced sweating syndrome in a patient. *Proc Natl Acad Sci* **103:** 10068–10073.

Ruwanpura SM, McLeod L, Dousha LF, Seow HJ, Alhayyani S, Tate MD, Deswaerte V, Brooks GD, Bozinovski S, MacDonald M, et al. 2016. Therapeutic targeting of the IL-6 trans-signaling/mechanistic target of rapamycin complex 1 axis in pulmonary emphysema. *Am J Respir Crit Care Med* doi: 10.1164/rccm.201512-2368OC.

Schaper F, Rose-John S. 2015. Interleukin-6: Biology, signaling and strategies of blockade. *Cytokine Growth Factor Rev* **26:** 475–487.

Scheller J, Chalaris A, Garbers C, Rose-John S. 2011. ADAM17: A molecular switch to control inflammation and tissue regeneration. *Trends Immunol* **32:** 380–387.

Schmidt-Arras D, Rose-John S. 2016. IL-6 pathway in the liver: From physiopathology to therapy. *J Hepatol* **64:** 1403–1415.

Schuett H, Oestreich R, Waetzig GH, Annema W, Luchtefeld M, Hillmer A, Bavendiek U, von Felden J, Divchev D, Kempf T, et al. 2012. Transsignaling of interleukin-6 crucially contributes to atherosclerosis in mice. *Arterioscler Thromb Vasc Biol* **32:** 281–290.

Schumacher N, Meyer D, Mauermann A, von der Heyde J, Wolf J, Schwarz J, Knittler K, Murphy G, Michalek M, Garbers C, et al. 2015. Shedding of endogenous interleukin-6 receptor (IL-6R) is governed by a disintegrin and metalloproteinase (ADAM) proteases while a full-length IL-6R isoform localizes to circulating microvesicles. *J Biol Chem* **290:** 26059–26071.

Schuster B, Kovaleva M, Sun Y, Regenhard P, Matthews V, Grotzinger J, Rose-John S, Kallen KJ. 2003. Signaling of human ciliary neurotrophic factor (CNTF) revisited. The interleukin-6 receptor can serve as an α-receptor for CTNF. *J Biol Chem* **278:** 9528–9535.

Selmi C, Ceribelli A, Naguwa SM, Cantarini L, Shoenfeld Y. 2015. Safety issues and concerns of new immunomodulators in rheumatology. *Expert Opin Drug Saf* **14:** 389–399.

Sodenkamp J, Waetzig GH, Scheller J, Seegert D, Grotzinger J, Rose-John S, Ehlers S, Holscher C. 2012. Therapeutic targeting of interleukin-6 trans-signaling does not affect the outcome of experimental tuberculosis. *Immunobiology* 217: 996–1004.

Spangler JB, Moraga I, Mendoza JL, Garcia KC. 2015. Insights into cytokine-receptor interactions from cytokine engineering. *Annu Rev Immunol* 33: 139–167.

* Stark GR, Cheon HJ, Wang Y. 2017. Responses to cytokines and interferons that depend upon JAKs and STATs. *Cold Spring Harb Perspect Biol* doi: 10.1101/cshperspect. a028555.

Stewart CL, Kaspar P, Brunet LJ, Bhatt H, Gadi I, Kontgen F, Abbondanzo SJ. 1992. Blastocyst implantation depends on maternal expression of leukaemia inhibitory factor. *Nature* 359: 76–79.

Stockli KA, Lottspeich F, Sendtner M, Masiakowski P, Carroll P, Gotz R, Lindholm D, Thoenen H. 1989. Molecular cloning, expression and regional distribution of rat ciliary neurotrophic factor. *Nature* 342: 920–923.

Suthaus J, Adam N, Grotzinger J, Scheller J, Rose-John S. 2011. Viral Interleukin-6: Structure, pathophysiology and strategies of neutralization. *Eur J Cell Biol* 90: 495–504.

Suthaus J, Stuhlmann-Laeisz C, Tompkins VS, Rosean TR, Klapper W, Tosato G, Janz S, Scheller J, Rose-John S. 2012. HHV-8-encoded viral IL-6 collaborates with mouse IL-6 in the development of multicentric Castleman disease in mice. *Blood* 119: 5173–5181.

Taga T, Hibi M, Hirata Y, Yamasaki K, Yasukawa K, Matsuda T, Hirano T, Kishimoto T. 1989. Interleukin-6 triggers the association of its receptor with a possible signal transducer, gp130. *Cell* 58: 573–581.

Tanaka T, Narazaki M, Ogata A, Kishimoto T. 2014. A new era for the treatment of inflammatory autoimmune diseases by interleukin-6 blockade strategy. *Semin Immunol* 26: 88–96.

The New York Academy of Sciences Conference. 1988. *Regulation of the acute phase and immune responses: Interleukin-6* (ed. Sehgal PB, Grieninger G, Tosato G). The New York Academy of Sciences, New York.

Tsantikos E, Maxwell MJ, Putoczki T, Ernst M, Rose-John S, Tarlinton DM, Hibbs ML. 2013. Interleukin-6 trans-signaling exacerbates inflammation and renal pathology in lupus-prone mice. *Arthritis Rheum* 65: 2691–2702.

Viswanathan S, Benatar T, Rose-John S, Lauffenburger DA, Zandstra PW. 2002. Ligand/receptor signaling threshold (LIST) model accounts for gp130-mediated embryonic stem cell self-renewal responses to LIF and HIL-6. *Stem Cells* 20: 119–138.

Waldmann TA. 2006. The biology of interleukin-2 and interleukin-15: Implications for cancer therapy and vaccine design. *Nat Rev Immunol* 6: 595–601.

Wei LH, Chou CH, Chen MW, Rose-John S, Kuo ML, Chen SU, Yang YS. 2013. The role of IL-6 trans-signaling in vascular leakage: Implications for ovarian hyperstimulation syndrome in a murine model. *J Clin Endocrinol Metab* 98: E472–E484.

Wunderlich FT, Ströhle P, Könner AC, Gruber S, Tovar S, Brönneke HS, Juntti-Berggren L, Li LS, van Rooijen N, Libert C, et al. 2010. Interleukin-6 signaling in liver-parenchymal cells suppresses hepatic inflammation and improves systemic insulin action. *Cell Metab* 12: 237–249.

Yamasaki K, Taga T, Hirata Y, Yawata H, Kawanishi Y, Seed B, Taniguchi T, Hirano T, Kishimoto T. 1988. Cloning and expression of the human interleukin-6 (BSF-2/IFN β2) receptor. *Science* 241: 825–828.

* Yan J, Smyth MJ, Teng MWL. 2017. IL-12 and IL-23 and their conflicting roles in cancer. *Cold Spring Harb Perspect Biol* doi: 10.1101/cshperspect.a028530.

Yang ZZ, Grote DM, Ziesmer SC, Manske MK, Witzig TE, Novak AJ, Ansell SM. 2011. Soluble IL-2Rα facilitates IL-2-mediated immune responses and predicts reduced survival in follicular B-cell non-Hodgkin lymphoma. *Blood* 118: 2809–2820.

Zhang H, Neuhofer P, Song L, Rabe B, Lesina M, Kurkowski MU, Treiber M, Wartmann T, Regner S, Thorlacius H, et al. 2013. IL-6 trans-signaling promotes pancreatitis-associated lung injury and lethality. *J Clin Invest* 123: 1019–1031.

Zilberstein A, Ruggieri R, Korn JH, Revel M. 1986. Structure and expression of cDNA and genes for human interferon-β-2, a distinct species inducible by growth-stimulatory cytokines. *EMBO J* 5: 2529–2537.

Interleukin (IL-6) Immunotherapy

Toshio Tanaka,[1,2] Masashi Narazaki,[2,3] and Tadamitsu Kishimoto[4]

[1]Department of Clinical Application of Biologics, Osaka University Graduate School of Medicine, Osaka University, Osaka 565-0871, Japan

[2]Department of Immunopathology, World Premier International Immunology Frontier Research Center, Osaka University, Osaka 565-0871, Japan

[3]Deparment of Respiratory Medicine, Allergy and Rheumatic Diseases, Osaka University Graduate School of Medicine, Osaka University, Osaka 565-0871, Japan

[4]Laboratory of Immune Regulation, World Premier International Immunology Frontier Research Center, Osaka University, Osaka 565-0871, Japan

Correspondence: kishimoto@ifrec.osaka-u.ac.jp

Interleukin 6 (IL-6) is a prototypical cytokine for maintaining homeostasis. When homeostasis is disrupted by infections or tissue injuries, IL-6 is produced immediately and contributes to host defense against such emergent stress through activation of acute-phase and immune responses. However, dysregulated excessive and persistent synthesis of IL-6 has a pathological effect on, respectively, acute systemic inflammatory response syndrome and chronic immune-mediated diseases. The IL-6 inhibitor, tocilizumab, a humanized anti-IL-6 receptor antibody, is currently being used for the treatment of rheumatoid arthritis, juvenile idiopathic arthritis, and Castleman disease. Lines of recent evidence strongly suggest IL-6 blockade can provide broader therapeutic strategy for various diseases included in acute systemic and chronic inflammatory diseases.

Therapeutic strategy by either targeting immune-mediated molecules or their supplementation has been successful in various clinical applications and produced outstanding results. Interleukin 6 (IL-6) blockade therapy has also been successful for the treatment of several chronic immune-mediated diseases and is expected to be widely used for various diseases. The focus of this review is on current IL-6 blockade immunotherapy and future perspectives.

BIOLOGICAL FUNCTIONS OF IL-6

The human IL-6 gene was first cloned as B-cell stimulatory factor 2 (BSF-2), which induces B cells to produce immunoglobulin (Ig) (Kishimoto 1985; Hirano et al. 1986). Subsequently, IL-6 was shown to be a prototypical cytokine with a pleiotropic effect on inflammation, immune response, and hematopoiesis (Kishimoto 1989; Akira et al. 1993). When IL-6 acts on hepatocytes, it induces a broad range of acute phase proteins such as C-reactive protein (CRP), complement C3, serum amyloid A (SAA), fibrinogen, thrombopoietin, hepcidin, haptoglobin, and α1-antichymotrypsin, thus making IL-6 a vital mediator of acute phase response (Fig. 1) (Heinrich et al. 1990). However, if high-level concentrations of SAA in the serum persist for long periods, it can lead to a

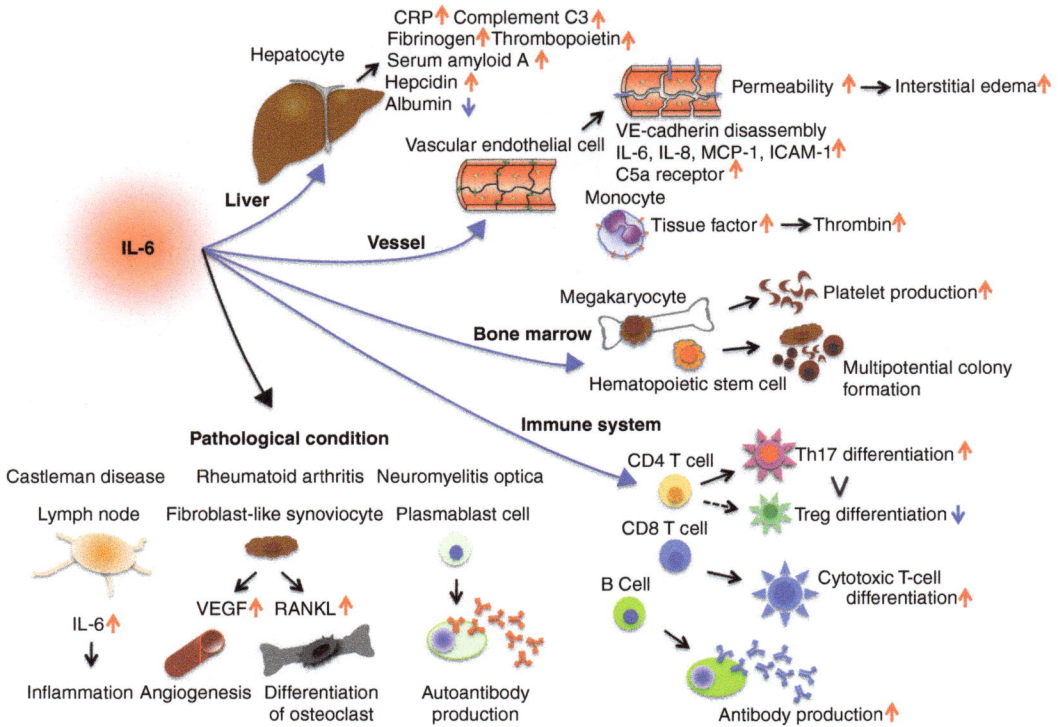

Figure 1. Pleiotropic effect of interleukin 6 (IL-6) (a cytokine featuring pleiotropic activity). It induces synthesis of acute phase proteins in liver such as C-reactive protein (CRP), complement C3, fibrinogen, thrombopoietin, serum amyloid A, and hepcidin, whereas it inhibits production of albumin. IL-6 also plays an important role in acquired immune responses by stimulating antibody production and inducing the differentiation of naïve CD4[+] T cells into effector T cells. IL-6 activates vascular endothelial cells to produce IL-6, IL-8, monocyte chemoattractant protein-1 (MCP-1), intercellular adhesion molecule (ICAM)-1, and C5a receptors, and induces vascular endothelial cadherin disassembly. In addition, IL-6 can promote differentiation or proliferation of several nonimmune cells. Because of its pleiotropic activity, dysregulated persistent or excessive production of IL-6 leads to the onset or development of various diseases. Excessive production of IL-6 is pathologically involved in the swollen lymph nodes of Castleman disease, whereas excessive IL-6 in synovial fluid stimulates fibroblast-like synoviocytes to produce vascular endothelial growth factor (VEGF) and receptor activator of nuclear factor of κB (NF-κB) ligand (RANKL), which enhance angiogenesis and osteoporosis in patients with rheumatoid arthritis. IL-6 supports the survival of plasmablasts, which produce anti-aquaporin 4 antibodies in patients with neuromyelitis optica. Treg, Regulatory T cell.

serious complication: amyloid A amyloidosis (Gillmore et al. 2001). Moreover, IL-6-induced hepcidin production blocks the action of iron transporter ferroportin 1 on gut and thus reduces serum iron levels (Nemeth et al. 2004), resulting in hypoferremia and anemia associated with chronic inflammation. In hematopoiesis, IL-6 promotes hematopoietic stem cell differentiation as well as megakaryocyte maturation, which leads to the release of platelets (Ishibashi et al. 1989).

In addition to promoting differentiation of activated B cells into Ig-producing cells, in an acquired immune response, IL-6 regulates the direction of specific differentiation of naïve CD4[+] T cells. IL-6 in combination with transforming growth factor β (TGF-β) is indispensable for T helper (Th)17 differentiation, whereas IL-6 inhibits TGF-β-induced regulatory T-cell (Treg) differentiation (Kimura and Kishimoto 2010). This effect causes up-regulation of the Th17/Treg balance, which is pathologically involved in the development of various autoimmune and chronic inflammatory diseases. It has also been shown that IL-6 can promote T follicular helper cell differentiation as well as production of IL-21 (Ma et al. 2012), which also regulates Ig synthesis.

Furthermore, IL-6 induces a variety of biological activities. When IL-6 is produced in bone marrow stromal cells, it stimulates fibroblast-like synoviocytes to produce the receptor activator of the nuclear factor κB (NF-κB) ligand (RANKL) (Hashizume et al. 2008), which is essential for the differentiation and activation of osteoclasts (Kotake et al. 1996). IL-6 can also induce excess production of vascular endothelial growth factor (VEGF), which leads to both enhanced angiogenesis and increased vascular permeability (Cohen et al. 1996).

Not only does IL-6 promote production of fibrinogen and platelet release but it also activates the coagulation system. IL-6 induces tissue factor (TF) on the cell surface of monocytes (Fig. 1) (Neumann et al. 1997), which promotes coagulation by initiating the extrinsic coagulation pathway, and in turn leads to thrombin production. These activities of IL-6 lead to a hypercoagulable state and thrombosis. IL-6 has also been shown to activate vascular endothelial (VE) cells. VE-cadherin is an important molecule, which promotes endothelial adhesion of adjacent cells by means of homophilic binding, while its disassembly by IL-6 leads to vascular leakage (Kruttgen and Rose-John 2011). Moreover, IL-6 increases production of VEGF, which induces the phosphorylation and internalization of VE-cadherin, and thus has a potent vascular permeability effect on endothelial cells (Esser et al. 1998; Desai et al. 2002). Vascular permeability by IL-6 itself or via induction of VEGF then leads to interstitial edema and elevates tissue pressure resulting in tissue damage. The complex IL-6/sIL-6R (soluble interleukin 6 receptor) can activate VE cells to produce IL-6, IL-8, and MCP-1 as well as augment intercellular adhesion molecule (ICAM)-1 expression, which results in leukocyte recruitment (Romano et al. 1997). IL-6 up-regulates the complement 5a receptor (C5aR) on endothelial cells, thus increasing their responsiveness to C5a, which functions as an anaphylatoxin, causes smooth muscle contraction and histamine release from mast cells, and results in further enhancement of vascular permeability (Laudes et al. 2002; Riedemann et al. 2003). Furthermore, IL-6 has been shown to weaken papillary muscle contraction, leading to myocardial dysfunction (Finkel et al. 1992; Pathan et al. 2004). These multiple effects of IL-6 in coagulation cascade, vascular permeability, and myocardial dysfunction thus play a pathological role in tissue hypoxia, hypotension, disseminated intravascular coagulation (DIC), and multiple organ dysfunction, all characteristic features of systemic inflammatory response syndrome (SIRS).

IL-6-MEDIATED SIGNALING PATHWAY

The function of IL-6 is initiated by its binding to an IL-6 receptor (IL-6R), and exists in two forms, 80-kDa transmembrane and 50–55-kDa sIL-6R (Fig. 2) (Yamasaki et al. 1988; Narazaki et al. 1993). Transmembrane IL-6R is expressed in limited cells such as leukocytes and hepatocytes, whereas sIL-6R is in human serum. Once IL-6 binds to transmembrane or sIL-6R, the complex in turn induces homodimerization of gp130 triggering a downstream signal cascade (Fig. 2) (Hibi et al. 1990; Kishimoto et al. 1992; Murakami et al. 1993). The functional receptor is a hexamer with two molecules of IL-6 and two of IL-6R as well as two molecules of gp130 (Boulanger et al. 2003). Tocilizumab, the IL-6 inhibitor developed first, is a humanized anti-IL-6R monoclonal antibody, which binds to transmembrane and sIL-6R and inhibits IL-6 binding to both receptors. The pleiotropic effect of IL-6 can be explained by the broad expression of gp130 on various cells (Taga and Kishimoto 1997). Homodimerization of gp130 allows the Janus kinases (JAKs) such as JAK1, JAK2, and tyrosine kinase 2 (TYK2) to come close to each other and to lead to phosphorylation in the cytoplasmic tyrosine residues of gp130. The Src homology 2 (SH2)-domain-containing molecules, signal transducers and activators of transcription (STAT)3, STAT1, and SH2-domain-containing protein-tyrosine

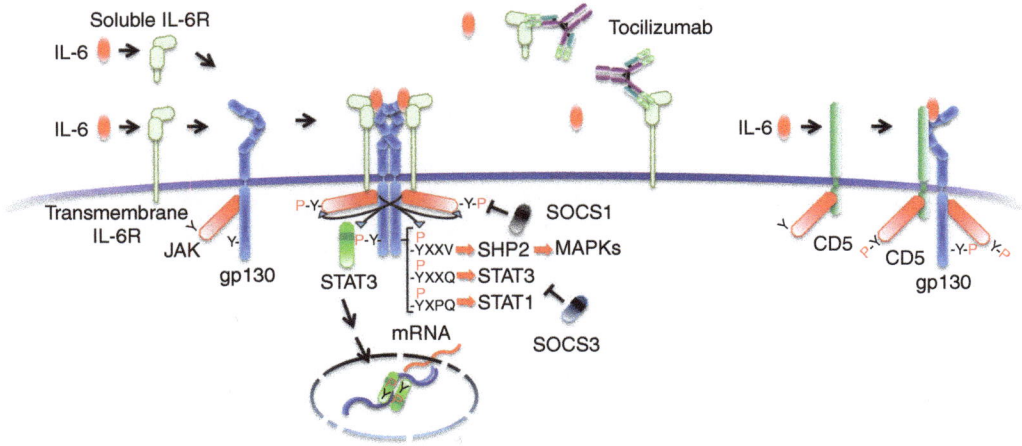

Figure 2. Interleukin 6 (IL-6) receptor-mediated signaling pathway and an IL-6 inhibitor, the humanized anti-IL-6 receptor antibody tocilizumab. IL-6 binds to soluble and transmembrane IL-6R and the resultant complex induces homodimerization of gp130, leading to activation of Janus kinase (JAK)1, JAK2, and tyrosine kinase 2 (TYK2). JAKs, in turn, phosphorylate cytoplasmic tyrosine-based motifs of gp130, followed by attracting Src homology 2 (SH2)-containing molecules SH2 domain-containing protein-tyrosine phosphatase 2 (SHP2) to YXXV, signal transducers and activators of transcription (STAT)3 to YXXQ, or STAT1 to YXPQ (Y, tyrosine; V, valine; Q, glutamine; P, proline; X, any amino acid), which results in a downstream signal. Of the negative feedback molecules in cytoplasm, negative feedback molecules, suppressor of cytokine signaling (SOCS)1 binds to and inhibits JAK, whereas SOCS3 binds to gp130 and inhibits STAT3 activation. A humanized anti-IL-6 receptor (IL-6R) antibody, tocilizumab, blocks the IL-6-mediated signaling pathway by inhibiting IL-6 binding to both the soluble and transmembrane form of IL-6Rs. IL-6 can also bind to CD5 with low affinity, which leads to STAT3 activation. MAPKs, Mitogen-activated protein kinases.

phosphatase 2 (SHP2) are attracted to the tyrosine-phosphorylated motifs of gp130, YXXV for SHP2, YXXQ for STAT3, and YXPQ for STAT1 (Y, tyrosine; V, valine; Q, glutamine; P, proline; X, any amino acid). STAT3 and STAT1 are then phosphorylated by the JAKs and translocate to the nucleus, which generates the transcriptional output (Fig. 2). SHP2 activates the mitogen-activated protein (MAP) kinase pathway (Kishimoto et al. 1994). This is followed by the induction of the various IL-6-responsive genes, which include acute phase proteins, by STAT3 activation. STAT3 also induces the suppressor of cytokine signaling (SOCS)1 and SOCS3, which bind to tyrosine-phosphorylated JAK (Naka et al. 1997) and tyrosine-phosphorylated gp130 (Schmitz et al. 2000), respectively, to stop IL-6 signaling by means of a negative feedback loop.

It was found recently that IL-6 can bind to another surface molecule, CD5, and also acti-

vate STAT3 via gp130 and JAK2 (Zhang et al. 2016). STAT3 activation up-regulates CD5 expression, forming a feedforward loop that may play a critical role in the promotion of cancer development by CD5[+] B cells. However, the binding affinity between IL-6 and sIL-6R is around 1 nM, whereas IL-6 binds to CD5 at 100 nM. This means that interaction of IL-6 with CD5 is doubtful under physiological conditions, but can occur under pathological conditions when the IL-6 level is highly elevated (Masuda and Kishimoto 2016).

REGULATORY MECHANISM OF IL-6 SYNTHESIS

When emergent stress such as infections or tissue injuries occur, IL-6 is produced immediately by innate immune cells such as macrophages and monocytes and plays a major role in removal of infectious agents and in tissue repair by

Cite this article as *Cold Spring Harb Perspect Biol* doi: 10.1101/cshperspect.a028456

activating immune and acute-phase responses. When such stress has been eliminated from the host and homeostasis is fully recovered, IL-6 synthesis is terminated. IL-6 production is therefore tightly regulated in response to environmental stress and its regulation is controlled through transcriptional and posttranscriptional mechanisms (Tanaka and Kishimoto 2014; Tanaka et al. 2014a). However, if such regulation is disrupted, excessive or persistent production of IL-6 is induced and causes the development of various diseases. Not only immune-mediated cells but also mesenchymal cells, VE cells, fibroblasts, and many other cells have been found to produce IL-6 under physiological and pathological conditions (Akira et al. 1993).

It has been shown that a number of transcription factors regulate IL-6 gene transcription. The functional *cis*-regulatory elements in the 5' flanking region of the human IL-6 gene have been identified as binding sites for NF-κB, specificity protein 1, nuclear factor IL-6 ([NF-IL-6], also known as C/EBP-β), activator protein 1, and interferon regulatory factor 1 (Liebermann and Baltimore 1990; Akira and Kishimoto 1992). Interestingly, certain viral products can enhance the DNA-binding activity of NF-κB and NF-IL-6, which results in increased IL-6 messenger RNA (mRNA) transcription. For example, interaction with NF-κB of the transactivator protein (TAX), derived from the human T lymphotropic virus 1 enhances IL-6 production (Ballard et al. 1988), whereas the DNA-binding activity of both NF-κB and NF-IL-6 is enhanced by the transactivator of the transcription protein (TAT), of human immunodeficiency virus 1 (HIV1) (Scala et al. 1994). Moreover, some microRNAs (miRNAs) directly or indirectly regulate transcription activity. For example, interaction of miRNA-155 with the 3'-untranslated region (UTR) of NF-IL-6 suppressing NF-IL-6 expression (He et al. 2009).

IL-6 expression is also controlled by posttranscriptional regulation (Chen and Shyu 1995; Anderson 2008). Initiation of mRNA translation is controlled by the 5' UTR, and the stability of mRNA by the 3' UTR, which in turn is regulated by modulation of AU-rich elements (AREs) located in the 3'-UTR region. A number of RNA-binding proteins and miRNAs bind to the 3' UTRs and regulate the stability of IL-6 mRNA. The nuclease regulatory RNase-1 (Regnase-1) plays a role in destabilizing IL-6 mRNA, and it was found that the relevant knockout mice spontaneously developed autoimmune diseases accompanied by splenomegaly and lymphadenopathy (Matsushita et al. 2009). Another RNA-binding protein, Roquin, recognizes target mRNAs overlapping with Regnase-1 (Mino et al. 2015). However, their localizations and states of the target mRNAs are spatiotemporally different. Regnase-1 degrades transcriptionally active mRNA in cytoplasm, endoplasmic reticulum, and ribosome, whereas Roquin degrades transcriptionally inactive mRNA in stress granules and processing bodies. On the other hand, we identified a unique RNA-binding protein, AT-rich interactive domain-containing protein 5a (Arid5a), which binds to the 3' UTR of IL-6 mRNA, resulting in selective stabilization of IL-6 but not of tumor necrosis factor α (TNF-α) or IL-12 mRNA (Masuda et al. 2013). Arid5a expression is enhanced in macrophages in response to lipopolysaccharide (LPS), IL-1β, and IL-6, and also induced under Th17-polarizing conditions in T cells. *Arid5a* gene deficiency inhibits elevation of IL-6 levels in LPS-injected mice and development in experimental autoimmune encephalomyelitis. Arid5a counteracts the degrading effect of Regnase-1 on IL-6 mRNA (Fig. 3), indicating that there is a balance between Arid5a and Regnase-1, playing an important role in IL-6 mRNA stability. Moreover, we also found that Arid5a is an important regulator for differentiation of naïve CD4[+] T cells into Th17 cells through selective stabilization of STAT3 mRNA (Masuda et al. 2016). A similar observation was made in IL-6 mRNA metabolism, in which Arid5a can counteract Regnase-1-mediated STAT3 degradation. These findings strongly suggest that the formation of a positive loop, consisting of IL-6, STAT3, and Arid5a, plays an important role in the exaggerated expression of IL-6 and accelerated IL-6R-mediated signaling, indicating that it conceivably plays a pathological role in various IL-6-mediated diseases.

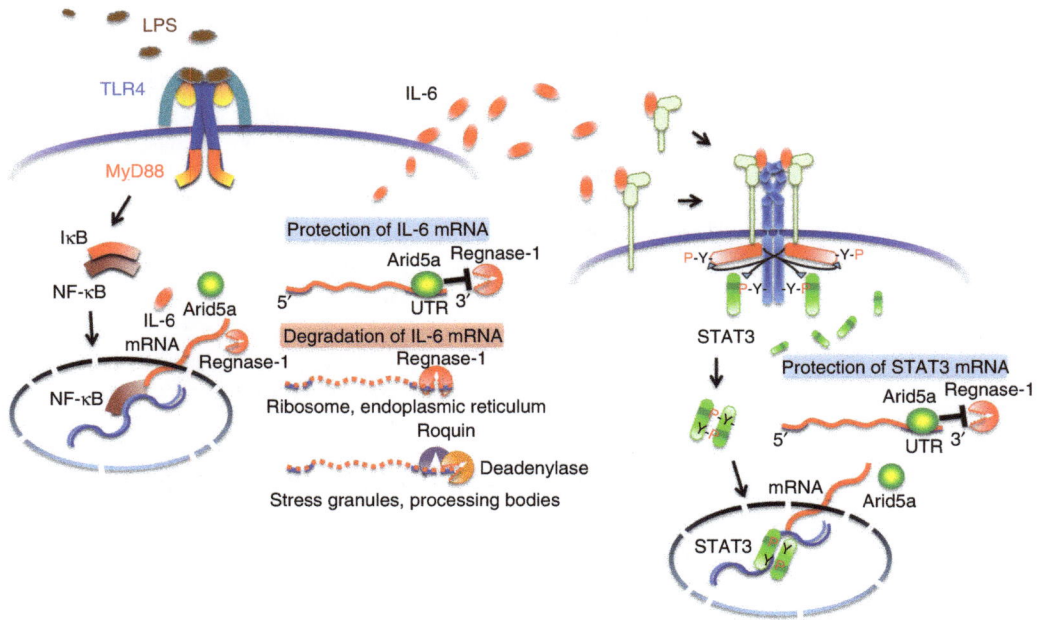

Figure 3. Arid5a stabilizes interleukin 6 (IL-6) and Stat3 messenger RNAs (mRNAs). Toll-like receptor (TLR)4 recognizes lipopolysaccharide (LPS) and induces IL-6 mRNA via activation of the signaling pathway of nuclear factor of κB (NF-κB). Regnase-1 promotes IL-6 mRNA degradation in ribosomes and the endoplasmic reticulum, whereas Arid5a inhibits the destabilizing effect of Regnase-1. Roquin, which requires deadenylase for its function, degrades inactive mRNA in stress granules and processing bodies. In naïve $CD4^+$ T cells, IL-6 activates signal transducers and activators of transcription (STAT)3 and induces transcription of IL-6-target genes, including Arid5a, which protects STAT3 mRNA from the degrading effect of Regnase-1. The regulation of IL-6 and STAT3 mRNAs by Arid5a is important for the IL-6 production and strength of the IL-6R-mediated signaling. MyD88, Myeloid differentiation primary response 88; IκB, inhibitor of NF-κB; UTR, untranslated region.

IL-6 BLOCKADE IMMUNOTHERAPY

The association of IL-6 overexpression and disease development was first found in a patient with cardiac myxoma, who presented with fever, polyarthritis with positivity for antinuclear antibody, elevated CRP level, and hypergammaglobulinemia. The myxoma tissue was highly stained with anti-IL-6 antibody, which suggested that continuous production of IL-6 from myxoma tissue might have contributed to chronic inflammation and autoimmunity (Hirano et al. 1987). Follow-up studies have shown that dysregulation of IL-6 production occurs in the synovial cells of rheumatoid arthritis (RA) (Hirano et al. 1988), myeloma cells (Kawano et al. 1988), germinal center B cells in swollen lymph nodes of Castleman disease (Yoshizaki

et al. 1989), and peripheral blood cells or tissues involved in various other autoimmune and chronic inflammatory diseases and even malignant cells in cancers (Akira et al. 1993). Serum concentrations of IL-6 are <4 pg/mL under healthy conditions, but IL-6 levels can increase in various ways, depending on the disease and its severity, usually up to several tens or even hundreds of pg/mL in chronic diseases, whereas in a cytokine storm the level can increase dramatically to >1000 pg/mL and, in severe cases, reach a level measured in μg/mL.

In addition, the pathological role of IL-6 in disease development has been confirmed in a number of animal disease models, in which it has been shown that IL-6 blockade by means of gene knockout or administration of neutralizing anti-IL-6 or anti-IL-6R antibody is preventive or leads to therapeutic suppression of disease development (Kishimoto 2005; Tanaka et al. 2012). These findings led to the notion that IL-6 targeting might constitute a novel treatment strategy for various diseases, leading to the development of tocilizumab, a humanized anti-IL-6R monoclonal antibody of the IgG1 class (Sato et al. 1993) that blocks IL-6-mediated signal transduction by inhibiting IL-6 binding to transmembrane and sIL-6R (Fig. 2).

DISEASES FOR WHICH TREATMENT WITH TOCILIZUMAB HAS BEEN APPROVED

Currently, tocilizumab is approved for the treatment of Castleman disease, RA, and juvenile idiopathic arthritis (JIA).

Castleman Disease

Following the first in-human clinical study in which tocilizumab was administered to a patient with multiple myeloma, the clinical trial of tocilizumab was performed with seven patients with Castleman disease, a chronic lymphoproliferative disorder, in which IL-6 is pathologically produced in germinal center B cells in the involved lymph node(s). These patients presented with severe inflammatory symptoms and laboratory findings such as high fever, anemia, increased levels of acute-phase proteins, hypoalbuminemia, and hypergammaglobulinemia. Administration of tocilizumab promptly ameliorated clinical symptoms, normalized serum CRP levels, and improved anemia, serum albumin concentration, and hypergammaglobulinemia (Nishimoto et al. 2000). The outstanding efficacy of tocilizumab was confirmed in another clinical trial with an enrollment of 28 patients with Castleman disease (Nishimoto et al. 2005), and these results led to the approval of tocilizumab in Japan in 2005. Moreover, a chimeric monoclonal antibody to IL-6, siltuximab, was approved by the U.S. Food and Drug Administration (FDA) in 2014 for the treatment of patients with multicentric Castleman disease who are human immunodeficiency virus negative and human herpesvirus-8 negative (Deisseroth et al. 2015).

Rheumatoid Arthritis

The first randomized controlled trial of tocilizumab for RA was performed with 45 patients who received a single intravenous dose of either 0.1, 1, 5, or 10 mg/kg of tocilizumab or placebo (Choy et al. 2002). At week 2, a significant difference was observed between the group treated with 5 mg/kg of tocilizumab and the placebo group. Five patients (56%) in the tocilizumab cohort and none in the placebo cohort attained 20% improvement based on the American College of Rheumatology Criteria (ACR20), or 20% improvement in tender or swollen joints as well as 20% improvement of three of the other five criteria. At 12 weeks, a multicenter, double-blind, and placebo-controlled phase II trial was then performed in Japan (Nishimoto et al. 2004). In this trial, 164 patients with RA were randomized for intravenous administration of either 8 or 4 mg/kg of tocilizumab or placebo every 4 weeks. By week 12, an ACR20% response had been attained for 78%, 57%, and 11% of the patients injected with 8 mg/kg, 4 mg/kg of tocilizumab, and placebo, respectively. Seven subsequent independent phase III clinical trials performed worldwide as well as in Japan also verified the outstanding efficacy of tocilizumab for the suppression of disease activity and joint destruction progression with satisfactory safety, so that this biologic is currently approved for RA treatment in more than 130 countries (Tanaka et al. 2013, 2014b). Guidelines published by the European League Against Rheumatism (EULAR) and ACR recommends tocilizumab as one of eight first-line biologics to be used for RA patients with an inadequate response to the standard disease-modifying antirheumatic drug (DMARD) methotrexate (MTX) (Smolen et al. 2014; Singh et al. 2016). First-line biologics include five TNF inhibitors, a T-cell activation blocker (CTLA4-Ig), and a B-cell depletory agent (rituximab), as well as tocilizumab. However, tocilizumab is the only biologic that has proved to be more efficacious as monotherapy than MTX or other DMARDs (Tanaka et al. 2013). TNF inhibitors require the concomitant use of MTX to achieve their maximal effects, whereas tocilizumab monotherapy is as effective as the combination therapy of tocilizumab plus MTX for the suppression of disease activity. Moreover, a direct head-to-head comparison of tocilizumab and adalimumab, a fully human anti-TNFα antibody, showed that tocilizumab as monotherapy was superior to adalimumab, as assessed with several indices of disease activity (Gabay et al. 2013). Thus, the most powerful antirheumatic biological currently available is tocilizumab.

Juvenile Idiopathic Arthritis

The third disease for which tocilizumab is on-label is systemic JIA (sJIA), which is a subtype of chronic childhood arthritis leading to joint destruction and functional disability and is accompanied by systemic inflammation. A clinical trial, consisting of a 6-week open-label lead-in phase and a 12-week double-blind phase, was performed with an enrollment of 56 children with sJIA (Yokota et al. 2008). By week 6, tocilizumab treatment (8 mg/kg, every 2 weeks, intravenously) had resulted in ACR Pediatric 30%, 50%, and 70% responses for 91%, 86%, and 68% of the patients, respectively. Forty-three patients continued to the double-blind phase and 16 of the 20 patients (80%) in the tocilizumab group could maintain an ACR Pediatric 30% response, while only four of the 23 patients (17%) in the placebo group could maintain the same response. Moreover, a global phase III trial, with an enrollment of 112 children with active sJIA, has also shown that tocilizumab is highly efficacious for the suppression of disease activity of sJIA (De Benedetti et al. 2012).

BROAD APPLICATION OF IL-6 BLOCKADE IMMUNOTHERAPY FOR VARIOUS CHRONIC IMMUNE-MEDIATED DISEASES

In addition, a variety of case studies, series, and pilot studies of off-label use with tocilizumab have produced favorable results, suggesting tocilizumab may become established as a novel drug for the treatment of various chronic intractable immune-mediated diseases (Fig. 4) (Tanaka et al. 2012, 2014b). These include au-

toimmune diseases, chronic inflammatory diseases, and other diseases such as atherosclerosis, diabetes mellitus, orthopedic diseases, cancers, and psychological diseases, and various clinical trials of tocilizumab are in progress. Specifically, there are strong indications based on accumulated evidence that tocilizumab appears to be highly promising for the treatment of systemic sclerosis (SSc), neuromyelitis optica (NMO), large-vessel vasculitis, and polymyalgia rheumatica (PMR).

Systemic Sclerosis

SSc is a connective tissue disease, characterized by skin and tissue fibrosis, vasculopathy, and immune abnormalities. Various studies analyzed the pathogenic mechanisms of SSc, but no effective treatment has been established. However, because IL-6 plays a role in formation of these characteristics and IL-6 elevation was found in serum as well as involved tissues of patients, IL-6 may well be a potential target molecule for SSc.

We were the first to report the beneficial effect of tocilizumab for two patients with SSc (Shima et al. 2010), and the results of the subsequent phase II/III, multicenter, randomized, double-blind, and placebo-controlled study (NCT01532869) have recently been reported (Khanna et al. 2016). In this trial, 87 patients were enrolled: 43 were assigned to receive weekly 162 mg of subcutaneous tocilizumab and 44 to receive placebo. The primary end point was a significant difference in the mean change from baseline in the modified Rodnan skin score (MRSS), an index of evaluating skin thickness. The least squares mean change in the score at 24 weeks was -3.92 for the tocilizumab group and -1.22 for the placebo group (difference -2.70, 95% CI: -5.85 to 0.45; $p = 0.09$), so that the primary end point was not attained. However, fewer patients administered with tocilizumab than placebo showed a decline in the percentage of predicted forced vital capacity at 48 weeks ($p = 0.0373$), indicating that worsening in respiratory disturbance is significantly suppressed by tocilizumab. These results led to the granting by the FDA to tocilizumab of a Breakthrough

Tocilizumab

IL-6

IL-6 receptor

Approved diseases
1. Rheumatoid arthritis (in more than 130 countries)
2. Juvenile idiopathic arthritis (in Japan, U.S., EU, and India)
3. Castleman disease (in Japan and India)

The most promising candidate diseases
Chronic diseases
1. Systemic sclerosis (phase II/III completed → phase III)
2. Neuromyelitis optica (phase II completed → phase III by SA237)
3. Large vessel vasculitis
   Giant cell arteritis (phase II completed → phase III)
   Takayasu arteritis (phase II completed → phase III)
4. Polymyalgia rheumatica (phase II completed → phase III)
Acute diseases
1. Cytokine release syndrome (pilot studies)

Candidate diseases
Autoimmune diseases
- Polymyositis/dermatomyositis
- Systemic lupus erythematosus
- Relapsing polychondritis
- Autoimmune hemolytic anemia
- Acquired hemophilia A
- Graves' orbitopathy
- Sjogren's syndrome

Inflammatory diseases
- Adult-onset Still's disease
- Amyloid A amyloidosis
- RS$_3$PE
- Behcet's disease
- Noninfectious uveitis, uveitic macular edema

- Crohn's disease
- Graft-versus-host disease
- Autoinflammatory syndrome
- Pulmonary arterial hypertension
- IgG4-related disease

Acute diseases
- Systemic inflammatory response syndrome, including severe sepsis, septic shock, tissue injury (burn, crush syndrome, heat stroke), acute pancreatitis
- Hemophagocytic lymphohistiocytosis
- Macrophage activation syndrome
- Myocardial Infarction

Other diseases
- Atherosclerosis
- Diabetes mellitus (type I and II)
- Sciatica
- Osteoarthritis
- Fibrous dysplasia of bone
- Amyotrophic lateral sclerosis
- Schizophrenia, depression
- Multiple myeloma
- B-CLL AML
- Pancreatic cancer
- Cancer-related cachexia
- HIV infection

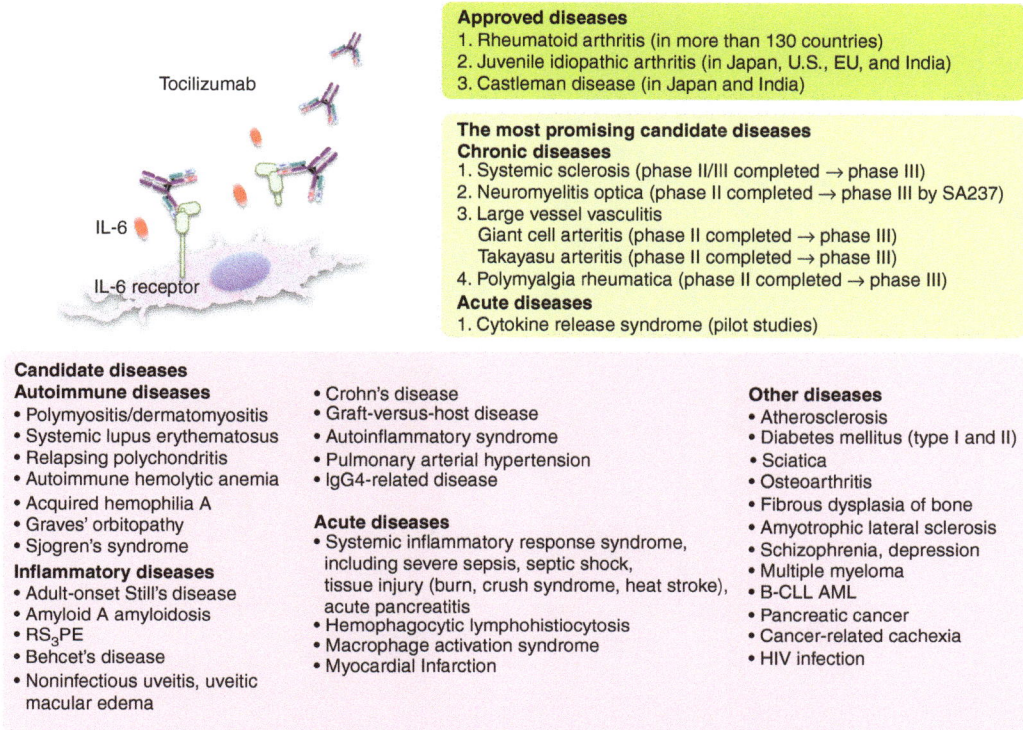

Figure 4. Interleukin 6 (IL-6) blockade immunotherapy for various diseases. Currently, tocilizumab is approved for the treatment of rheumatoid arthritis, systemic and polyarticular juvenile idiopathic arthritis, and Castleman diseases. It is expected that IL-6 blockade immunotherapy will constitute a novel therapeutic strategy for a wide range of diseases. In particular, the most promising candidate diseases are systemic sclerosis, neuromyelitis, large vessel vasculitis, polymyalgia rheumatica, and cytokine release syndrome (CRS), and ongoing clinical trials are in progress. RS3PE, Remitting seronegative symmetrical synovitis with pitting edema; B-CLL, B-cell chronic lymphocytic leukemia; AML, acute myelogenous leukemia; HIV, human immunodeficiency virus.

Therapy Designation for SSc and a subsequent clinical trial (NCT02453256) was started recently. In this phase II/III study, analyses of gene expression using microarray technology for skin biopsy specimens showed that, by 24 weeks, 16 genes had been definitely down-regulated by tocilizumab and 12 of these genes belonged to the M2 macrophage cluster. M2 macrophages are known to play a pathological role in fibrosis through the release of inflammatory and fibrotic factors. In addition, a recent study reported that IL-6 primes macrophages for IL-4-dependent M2 polarization by inducing IL-4 receptor expression (Mauer et al. 2014), so that the effect of tocilizumab on SSc may be mediated by modulating M2 macrophage activity. This trial did not evaluate the efficacy of tocilizumab for pulmonary arterial hypertension (PAH), which is a severe, refractory disease caused by an increase of blood pressure in the pulmonary artery, often complicated with SSc. In a hypoxia-induced PAH model, however, IL-6 blockade resulted in a striking amelioration in PAH and prevented IL-21-mediated polarization of primary alveolar macrophages to M2 macrophages (Hashimoto-Kataoka et al. 2015). Because this finding suggests that IL-6 blockade could also become a novel therapeutic strategy for PAH, a phase II clinical trial (NCT02676947) has been initiated.

Neuromyelitis Optica

NMO is a chronic inflammatory demyelinating disease of the central nervous system primarily affecting the spinal cord and optic nerves. Autoantibodies against aquaporin-4 (AQP-4), an astrocyte water channel protein, play a major part in the disease development. Anti-AQP-4 antibodies are produced by the plasmablast population showing a $CD19^{intermediate}CD29^{high}CD38^{high}CD180^{negative}$ phenotype, which is increased in the peripheral blood of NMO patients (Chihara et al. 2011). An in vitro study showed that IL-6 enhanced the survival of the plasmablast population, but the addition of tocilizumab into the culture diminished its survival and inhibited the production of anti-AQP-4 antibody, thus suggesting that tocilizumab is promising for the treatment of NMO. Indeed, prominent beneficial effects of tocilizumab in the suppression of relapse rate and neuropathic pain have been reported for refractory patients with NMO (Araki et al. 2014). This has led to clinical trials (NCT02028884 and NCT02073279, now in progress) for treatment of NMO with a new humanized antibody against IL-6R (SA237) with a longer half-life, which was generated from tocilizumab using an antibody structural optimization technique (Igawa et al. 2010).

Giant Cell Arteritis and Polymyalgia Rheumatica

Giant cell arteritis (GCA) is a chronic inflammatory disease of large and medium-sized arteries that affects persons over 50 years of age. Although its etiology remains unknown, IL-6 has a central role in the pathogenesis of GCA. The serum concentrations of IL-6 are elevated at the onset and during clinical relapse, while it has been reported that tissue-infiltrating cells produce major quantities of IL-6 as well as Th1-type cytokine interferon γ (IFN-γ) in patients with GCA. Subsequent to various reports regarding off-label use of tocilizumab for intractable GCA patients, a phase II, randomized, double-blind placebo-controlled trial (NCT01450137) was performed in Switzerland (Villiger et al. 2016).

Thirty patients with new-onset or relapsing GCA were assigned (2:1) to receive tocilizumab (8 mg/kg, every 4 weeks) or placebo intravenously until week 52. Both groups received oral prednisolone, starting at 1 mg/kg per day and tapered down to 0. Seventeen (85%) of patients achieved relapse-free survival patterns in the tocilizumab group and two (20%) in the placebo group (risk difference: 65%, 95% CI: 36–94; $p = 0.0010$). The mean survival time difference to stop glucocorticoids was 12 weeks in favor of tocilizumab (95% CI: 7–17; $p < 0.0001$), leading to a cumulative prednisolone dose of 43 mg/kg in the tocilizumab group versus 110 mg/kg in the placebo group ($p = 0.0005$) after 52 weeks. Serious adverse events affected seven patients (35%) in the tocilizumab group and five (50%) in the placebo group. These results show the marked efficacy of tocilizumab for the induction and maintenance of remission in patients with GCA.

PMR is a chronic inflammatory disease affecting the elderly. It is characterized by aching and morning stiffness in the shoulders, neck, and pelvic girdles. Onset of this disorder often occurs in association with GCA. IL-6 has been recognized as the most sensitive marker of disease activity and course. In a prospective open-label study (NCT01713842), 20 glucocorticoid-free patients with symptom onset within the previous 12 months and a PMR activity score (PMR-AS) of >10 received three infusions of 8 mg/kg tocilizumab every 4 weeks, followed by prednisolone from weeks 12 to 24 (0.15 mg/kg if PMR-AS <10 and 0.30 mg/kg otherwise) (Devauchelle-Pensec et al. 2016). Baseline median PMR-AS was 36.6, but by week 12 the score had become <10 in all patients, so that all patients subsequently received the low dose of prednisolone. Median PMR-AS was 4.5 by week 12 and 0.95 by week 24. In another open-label clinical trial (NCT01396317), 10 patients with newly diagnosed PMR and prior treatment with less than 1 month of glucocorticoids were treated with 8 mg/kg tocilizumab every 4 weeks intravenously plus a standardized rapid steroid taper (Lally et al. 2016). The primary end point was the percentage of subjects in relapse-free remission without glucocorticoids at 6 months.

One patient withdrew after 2 months, but the other nine patients attained the primary end point and remained in remission without relapse throughout the entire 15-month study. These findings indicate that tocilizumab is effective for recent-onset PMR.

APPLICATION OF IL-6 BLOCKADE IMMUNOTHERAPY FOR ACUTE INFLAMMATORY DISEASES

In addition to these potential indications for treatment with tocilizumab of a variety of chronic diseases, it could potentially serve as rescue therapy for acute life-threatening conditions such as cytokine storm and myocardial infarction.

Cytokine Storm

A cytokine storm is a potentially fatal immune reaction with highly elevated levels of various cytokines caused by extremely activated immune cells such as T cells, macrophages, and/or histiocytes in response to infections, tissue injuries, and autoimmune reaction.

Cytokine release syndrome (CRS) entails severe or fatal acute complications and is induced by nonphysiologic T-cell activation after T-cell-engaging therapies using chimeric antigen receptor–modified T cells (CAR-T) or a CD19/CD3-bi-specific antibody (blinatumomab) (Maude et al. 2014a). IL-6, IL-8, IL-10, and MCP-1, as well as the effector cytokine, IFN-γ, have been shown to be markedly elevated in patients with CRS. However, one administration of tocilizumab to the first patient with CRS as a complication of CAR-T therapy dramatically and unexpectedly resolved her serious condition (Grupp et al. 2013). Moreover, CRS induced by blinatumomab treatment of another patient, who presented with hemophagocytic lymphohistiocytosis with multisystem organ failure, was rapidly alleviated after one injection of tocilizumab (Teachey et al. 2013). Before tocilizumab therapy, this patient's serum levels of IFN-γ, IL-6, IL-8, IL-10, and MCP-1 were highly elevated, but tocilizumab injection resulted in a reduction in all elevated cytokine levels. In a subsequent study, 30 patients with acute lymphoblastic leukemia treated with CAR-T therapy had mild-to-severe CRS, with severe CRS in 27% of the patients, who required intensive care because of respiratory failure, coagulopathy, and hypotension (Maude et al. 2014b). In the patients with severe CRS, serum IL-6 levels were >1000 pg/mL, but again tocilizumab could produce a rapid and profound improvement in their severe clinical manifestations in only 1 to 3 days. These findings point to the possibility that IL-6 blockade may constitute a novel therapeutic strategy for emergent fatal complications mediated by "cytokine storm" (Tanaka et al. 2016).

A cytokine storm is induced by numerous conditions, which include, in addition to CRS, SIRS, macrophage-activation syndrome, and hemophagocytic lymphohistiocytosis. SIRS is composed of a variety of diseases: infection-induced SIRS (sepsis) and noninfectious SIRS such as trauma, burns, acute pancreatitis, heat stroke, crush syndrome, blast injury, graft-versus-host disease, ischemia, and hemorrhage. Surgery-related complications, adrenal insufficiency, pulmonary embolism, and aortic aneurysm can also lead to SIRS, and a biological drug against CD28 was also reported to induce SIRS (Suntharalingam et al. 2006). Although the mortality rate for SIRS is very high, an effective immunomodulation therapy has not yet been established. In fact, more than 100 clinical trials of biologics, including TNF and IL-1 inhibitors and other agents for the treatment of sepsis, have been conducted, yet no effective drug has been found.

IL-6 is an attractive target molecule for SIRS. As mentioned previously, the complex IL-6/sIL-6R is implicated in the induction of thrombosis, vascular leakage, and myocardial dysfunction, leading to multiple organ dysfunction and DIC (Kruttgen and Rose-John 2011). Moreover, numerous studies have shown IL-6 to be an excellent biomarker of severity and prognostic indicator of outcome for patients with SIRS. In patients with severe multiple organ dysfunction, serum levels of IL-6 can increase to >1000 pg/mL. Initially, CRS-related cytokines (IL-6, IL-8, IL-10, IFN-γ, and MCP-1) as well as other cytokines, including TNF-α and IL-1, are highly expressed in patients with SIRS and sepsis in particular. However, the serum levels of TNF-α and IL-1 return to normal levels within the first few hours and this is one reason why the clinical trials of both TNF and IL-1 inhibitors failed to show effectiveness for sepsis. However, levels of circulating IL-6 as well as of IL-8, IL-10, and MCP-1 persist much longer in patients with sepsis and other SIRS (Kang et al., in prep.). These findings indicate that IL-6 blockade features a broader therapeutic window than other procedures and thus may constitute a novel therapeutic strategy for SIRS. Because the K_d of IL-6 binding to sIL-6R is $\sim$1 nM and the resultant complex binds to gp130 with a K_d of 10 pM, such highly elevated IL-6 can activate various cells, especially VE cells, through the sIL-6R-mediated *trans*-signaling mechanism, resulting in the further secretion of IL-6, IL-8, and MCP-1. Thus, as seen in CRS complicated with T-cell engaging therapy, tocilizumab therapy may possibly lead to suppression of all these cytokines in SIRS, irrespective of underlying diseases, by suppressing activation of VE cells. If this is the case, IL-6 blockade will become a novel therapeutic approach for a wide variety of acute severe systemic inflammatory diseases presenting with "cytokine storm," although additional clinical studies are required to verify this possibility.

Myocardial Infarction

IL-6 contributes to atherosclerotic plaque development and is involved in ischemia-reperfusion myocardial injury. In a two-center, double-blind, placebo-controlled trial, 117 patients with non-ST-elevation myocardial infarction were randomized at a median of 2 days after symptom onset to receive 240 mg of tocilizumab or placebo intravenously before coronary angiography (NCT01491074) (Kleveland et al. 2016). The area under the curve (AUC) for CRP and troponin T (a marker of myocardial damage) at seven time points between days 1 and 3 was measured. Median AUC for CRP was 2.1 times higher for the placebo group than for the tocilizumab group (4.2 vs. 2.0 mg/L/h; p

< 0.001), whereas median AUC for troponin T was also higher for the placebo group compared with the tocilizumab group (234 vs. 159 ng/L/h; $p = 0.007$). These findings indicate that tocilizumab can protect against inflammation as well as myocardial damage induced by myocardial infarction, leading to a clinical trial of short-term application of tocilizumab following myocardial infarction (NCT 02419937).

CONCLUDING REMARKS

In parallel with the discovery of IL-6 and subsequent elucidation of the IL-6 signaling system, the pathological involvement of IL-6 in various diseases was also ascertained. This was followed by the development of the IL-6 inhibitor tocilizumab, which is currently used for the treatment of Castleman disease, RA, and systemic and polyarticular JIA. It is anticipated that, during the next decade, the IL-6 inhibitor will be widely used for the treatment of various as yet intractable diseases, including cytokine storm, and that its use will overcome the refractoriness of such diseases. To achieve this goal, however, further clinical evaluations will be essential.

Nevertheless, one conundrum remains to be solved, namely, why is IL-6 excessively or persistently expressed in various diseases? The IL-6 synthesis is controlled by transcription and posttranscriptional regulations. As described elsewhere, Arid5a stabilizes IL-6 as well as STAT3 mRNA by competing with degrading functions of Regnase-1 and forms a positive loop between IL-6 and STAT3. Accurate and detailed analyses of such RNA-binding proteins and of other regulators, which affect IL-6 synthesis, will certainly contribute to solving this mystery, while clarification of the mechanism(s) involved will facilitate the identification of more specific target molecules and investigation into the pathogenesis of specific diseases.

ACKNOWLEDGMENTS

T.K. holds a patent for tocilizumab and has received royalties for Actemra. T.T. has received a grant and payment for lectures, including services for speakers' bureaus from Chugai Pharmaceutical Co., Ltd. The Department of Clinical Application of Biologics of Osaka University Graduate School of Medicine is an endowment department, supported with an unrestricted grant from Chugai Pharmaceutical Co., Ltd. M.N. has received payment for lectures, including services for speakers' bureaus from Chugai Pharmaceutical Co., Ltd.

REFERENCES

Akira S, Kishimoto T. 1992. IL-6 and NF-IL6 in acute-phase response and viral infection. *Immunol Rev* **127**: 25–50.

Akira S, Taga T, Kishimoto T. 1993. Interleukin-6 in biology and medicine. *Adv Immunol* **54**: 1–78.

Anderson P. 2008. Post-transcriptional control of cytokine production. *Nature Immunol* **9**: 353–359.

Araki M, Matsuoka T, Miyamoto K, Kusunoki S, Okamoto T, Murata M, Miyake S, Aranami T, Yamamura T. 2014. Efficacy of the anti-IL-6 receptor antibody tocilizumab in neuromyelitis optica: A pilot study. *Neurology* **82**: 1302–1306.

Ballard DW, Bohnlein E, Lowenthal JW, Wano Y, Franza BR, Greene WC. 1988. HTLV-I tax induces cellular proteins that activate the κB element in the IL-2 receptor α gene. *Science* **241**: 1652–1655.

Boulanger MJ, Chow DC, Brevnova EE, Garcia KC. 2003. Hexameric structure and assembly of the interleukin-6/IL-6α-receptor/gp130 complex. *Science* **300**: 2101–2104.

Chen CY, Shyu AB. 1995. AU-rich elements: Characterization and importance in mRNA degradation. *Trends Biochem Sci* **20**: 465–470.

Chihara N, Aranami T, Sato W, Miyazaki Y, Miyake S, Okamoto T, Ogawa M, Toda T, Yamamura T. 2011. Interleukin 6 signaling promotes anti-aquaporin 4 autoantibody production from plasmablasts in neuromyelitis optica. *Proc Natl Acad Sci* **108**: 3701–3706.

Choy EH, Isenberg DA, Garrood T, Farrow S, Ioannou Y, Bird H, Cheung N, Williams B, Hazleman B, Price R, et al. 2002. Therapeutic benefit of blocking interleukin-6 activity with an anti-interleukin-6 receptor monoclonal antibody in rheumatoid arthritis: A randomized, double-blind, placebo-controlled, dose-escalation trial. *Arthritis Rheum* **46**: 3143–3150.

Cohen T, Nahari D, Cerem LW, Neufeld G, Levi BZ. 1996. Interleukin 6 induces the expression of vascular endothelial growth factor. *J Biol Chem* **271**: 736–741.

De Benedetti F, Brunner HI, Ruperto N, Kenwright A, Wright S, Calvo I, Cuttica R, Ravelli A, Schneider R, Woo P, et al. 2012. Randomized trial of tocilizumab in systemic juvenile idiopathic arthritis. *N Engl J Med* **367**: 2385–2395.

Deisseroth A, Ko CW, Nie L, Zirkelbach JF, Zhao L, Bullock J, Mehrotra N, Del Valle P, Saber H, Sheth C, et al. 2015. FDA approval: Siltuximab for the treatment of patients with multicentric Castleman diseases. *Clin Cancer Res* **21**: 950–954.

Desai TR, Leeper NJ, Hynes KL, Gewertz BL. 2002. Interleukin-6 causes endothelial barrier dysfunction via the protein kinase C pathway. *J Surg Res* **104**: 118–123.

Devauchelle-Pensec V, Berthelot JM, Cornec D, Renaudineau Y, Marhadour T, Jousse-Joulin S, Querellou S, Garrigues F, De Bandt M, Gouillou M, et al. 2016. Efficacy of first-line tocilizumab therapy in early polymyalgia rheumatica: A prospective longitudinal study. *Ann Rheum Dis* **75**: 1506–1510.

Esser S, Lampugnani MG, Corada M, Dejana E, Risau W. 1998. Vascular endothelial growth factor induces VE-cadherin tyrosine phosphorylation in endothelial cells. *J Cell Sci* **111**: 1853–1865.

Finkel MS, Oddis CV, Jacob TD, Watkins SC, Hattler BG, Simmons RL. 1992. Negative inotropic effects of cytokines on heart mediated by nitric oxide. *Science* **257**: 387–389.

Gabay C, Emery P, van Vollenhoven R, Dikranian A, Alten R, Pavelka K, Klearman M, Musselman D, Agrawal S, Green J, et al. 2013. Tocilizumab monotherapy versus adalimumab monotherapy for treatment of rheumatoid arthritis (ADACTA): A randomised, double-blind, controlled phase 4 trial. *Lancet* **381**: 1541–1550.

Gillmore JD, Lovat LB, Persey MR, Pepys MB, Hawkins PN. 2001. Amyloid load and clinical outcome in AA amyloidosis in relation to circulating concentration of serum amyloid A protein. *Lancet* **358**: 24–29.

Grupp SA, Kalos M, Barrett D, Aplenc R, Porter DL, Rheingold SR, Teachey DT, Chew A, Hauck B, Wright JF, et al. 2013. Chimeric antigen receptor-modified T cells for acute lymphoid leukemia. *N Engl J Med* **368**: 1509–1518.

Hashimoto-Kataoka T, Hosen N, Sonobe T, Arita Y, Yasui T, Masaki T, Minami M, Inagaki T, Miyagawa S, Sawa Y, et al. 2015. Interleukin-6/interleukin-21 signaling axis is critical in the pathogenesis of pulmonary arterial hypertension. *Proc Natl Acad Sci* **112**: 2677–2686.

Hashizume M, Hayakawa N, Mihara M. 2008. IL-6 trans-signalling directly induces RANKL on fibroblast-like synovial cells and is involved in RANKL induction by TNF-α and IL-17. *Rheumatology (Oxford)* **47**: 1635–1640.

He M, Xu Z, Ding T, Kuang DM, Zheng L. 2009. microRNA-155 regulates inflammatory cytokine production in tumor-associated macrophages via targeting C/EBPβ. *Cell Mol Immunol* **6**: 343–352.

Heinrich PC, Castell JV, Andus T. 1990. Interleukin-6 and the acute phase response. *Biochem J* **265**: 621–636.

Hibi M, Murakami M, Saito M, Hirano T, Taga T, Kishimoto T. 1990. Molecular cloning and expression of an IL-6 signal transducer, gp130. *Cell* **63**: 1149–1157.

Hirano T, Yasukawa K, Harada H, Taga T, Watanabe Y, Matsuda T, Kashiwamura S, Nakajima K, Koyama K, Iwamatsu A, et al. 1986. Complementary DNA for a novel human interleukin (BSF-2) that induces B lymphocytes to produce immunoglobulin. *Nature* **324**: 73–76.

Hirano T, Taga T, Yasukawa K, Nakajima K, Nakano N, Takatsuki F, Shimizu M, Murashima A, Tsunasawa S, Sakiyama F, et al. 1987. Human B-cell differentiation factor defined by an anti-peptide antibody and its possible role in autoantibody production. *Proc Natl Acad Sci* **84**: 228–231.

Hirano T, Matsuda T, Turner M, Miyasaka N, Buchan G, Tang B, Sato K, Shimizu M, Maini R, Feldmann M, et al. 1988. Excessive production of interleukin 6/B cell stimulatory factor-2 in rheumatoid arthritis. *Eur J Immunol* **18:** 1797–1801.

Igawa T, Ishii S, Tachibana T, Maeda A, Higuchi Y, Shimaoka S, Moriyama C, Watanabe T, Takubo R, Doi Y, et al. 2010. Antibody recycling by engineered pH-dependent antigen binding improves the duration of antigen neutralization. *Nat Biotechnol* **28:** 1203–1207.

Ishibashi T, Kimura H, Shikama Y, Uchida T, Kariyone S, Hirano T, Kishimoto T, Takatsuki F, Akiyama Y. 1989. Interleukin-6 is a potent thrombopoietic factor in vivo in mice. *Blood* **74:** 1241–1244.

Kawano M, Hirano T, Matsuda T, Taga T, Horii Y, Iwato K, Asaoku H, Tang B, Tanabe O, Tanaka H, et al. 1988. Autocrine generation and requirement of BSF-2/IL-6 for human multiple myelomas. *Nature* **332:** 83–85.

Khanna D, Denton CP, Jahreis A, van Laar JM, Frech TM, Anderson ME, Baron M, Chung L, Fierlbeck G, Lakshminarayanan S, et al. 2016. Safety and efficacy of subcutaneous tocilizumab in adults with systemic sclerosis (faSScinate): A phase 2, randomised controlled trial. *Lancet* **387:** 2630–2640.

Kimura A, Kishimoto T. 2010. IL-6: Regulator of Treg/Th17 balance. *Eur J Immunol* **40:** 1830–1835.

Kishimoto T. 1985. Factors affecting B-cell growth and differentiation. *Annu Rev Immunol* **3:** 133–157.

Kishimoto T. 1989. The biology of interleukin-6. *Blood* **74:** 1–10.

Kishimoto T. 2005. Interleukin-6: From basic science to medicine—40 years in immunology. *Annu Rev Immunol* **23:** 1–21.

Kishimoto T, Akira S, Taga T. 1992. Interleukin-6 and its receptor: A paradigm for cytokines. *Science* **258:** 593–597.

Kishimoto T, Taga T, Akira S. 1994. Cytokine signal transduction. *Cell* **76:** 253–262.

Kleveland O, Kunszt G, Bratlie M, Ueland T, Brock K, Holte E, Michelsen AE, Bendz B, Amundsen BH, Espevik T, et al. 2016. Effect of a single dose of the interleukin-6 receptor antagonist tocilizumab on inflammation and troponin T release in patients with non-ST-elevation myocardial infarction: A double-blind, randomized, placebo-controlled phase 2 trial. *Eur Heart J* **37:** 2406–2413.

Kotake S, Sato K, Kim KJ, Takahashi N, Udagawa N, Nakamura I, Yamaguchi A, Kishimoto T, Suda T, Kashiwazaki S. 1996. Interleukin-6 and soluble interleukin-6 receptors in the synovial fluids from rheumatoid arthritis patients are responsible for osteoclast-like cell formation. *J Bone Miner Res* **11:** 88–95.

Kruttgen A, Rose-John S. 2011. Interleukin-6 in sepsis and capillary leakage syndrome. *J Interferon Cytokine Res* **32:** 60–65.

Lally L, Forbess L, Hatzis C, Spiera R. 2016. A prospective open label phase IIa trial of tocilizumab in the treatment of polymyalgia rheumatica. *Arthritis Rheumatol* **68:** 2550–2554.

Laudes IJ, Chu JC, Huber-Lang M, Guo RF, Riedemann NC, Sarma JV, Mahdi F, Murphy HS, Speyer C, Lu KT, et al. 2002. Expression and function of C5a receptor in mouse microvascular endothelial cells. *J Immunol* **169:** 5962–5970.

Libermann TA, Baltimore D. 1990. Activation of interleukin-6 gene expression through the NF-κB transcription factor. *Mol Cell Biol* **10:** 2327–2334.

Ma CS, Deenick EK, Batten M, Tangye SG. 2012. The origins, function, and regulation of T follicular helper cells. *J Exp Med* **209:** 1241–1253.

Masuda K, Kishimoto T. 2016. CD5: A new partner for IL-6. *Immunity* **44:** 720–722.

Masuda K, Ripley B, Nishimura R, Mino T, Takeuchi O, Shioi G, Kiyonari H, Kishimoto T. 2013. Arid5a controls IL-6 mRNA stability, which contributes to elevation of IL-6 level in vivo. *Proc Natl Acad Sci* **110:** 9409–9414.

Masuda K, Ripley B, Nyati KK, Dubey PK, Zaman MM, Hanieh H, Higa M, Yamashita K, Standley DM, Mashima T, et al. 2016. Arid5a regulates naïve CD4$^+$ T cell fate through selective stabilization of Stat3 mRNA. *J Exp Med* **213:** 605–619.

Matsushita K, Takeuchi O, Standley DM, Kumagai Y, Kawagoe T, Miyake T, Satoh T, Kato H, Tsujimura T, Nakamura H, et al. 2009. Zc3h12a is an RNase essential for controlling immune responses by regulating mRNA decay. *Nature* **458:** 1185–1190.

Maude SL, Barrett D, Teachey DT, Grupp SA. 2014a. Managing cytokine release syndrome associated with novel T cell-engaging therapies. *Cancer J* **20:** 119–122.

Maude SL, Frey N, Shaw PA, Aplenc R, Barrett DM, Bunin NJ, Chew A, Gonzalez VE, Zheng Z, Lacey SF, et al. 2014b. Chimeric antigen receptor T cells for sustained remissions in leukemia. *N Engl J Med* **371:** 1507–1517.

Mauer J, Chaurasia B, Goldau J, Vogt MC, Ruud J, Nguyen KD, Theurich S, Hausen AC, Schmitz J, Bronneke HS, et al. 2014. Signaling by IL-6 promotes alternative activation of macrophages to limit endotoxemia and obesity-associated resistance to insulin. *Nat Immunol* **15:** 423–430.

Mino T, Murakawa Y, Fukao A, Vandenbon A, Wessels HH, Ori D, Uehata T, Tartey S, Akira S, Suzuki Y, et al. 2015. Regnase-1 and roquin regulate a common element in inflammatory mRNAs by spatiotemporally distinct mechanisms. *Cell* **161:** 1058–1073.

Murakami M, Hibi M, Nakagawa N, Nakagawa T, Yasukawa K, Yamanishi K, Taga T, Kishimoto T. 1993. IL-6-induced homodimerization of gp130 and associated activation of a tyrosine kinase. *Science* **260:** 1808–1810.

Naka T, Narazaki M, Hirata M, Matsumoto T, Minamoto S, Aono A, Nishimoto N, Kajita T, Taga T, Yoshizaki K, et al. 1997. Structure and function of a new STAT-induced STAT inhibitor. *Nature* **387:** 924–929.

Narazaki M, Yasukawa K, Saito T, Ohsugi Y, Fukui H, Koishihara Y, Yancopoulos GD, Taga T, Kishimoto T. 1993. Soluble forms of the interleukin-6 signal-transducing receptor component gp130 in human serum possessing a potential to inhibit signals through membrane-anchored gp130. *Blood* **82:** 1120–1126.

Nemeth E, Rivera S, Gabayan V, Keller C, Taudorf S, Pedersen BK, Ganz T. 2004. IL-6 mediates hypoferremia of inflammation by inducing the synthesis of the iron regulatory hormone hepcidin. *J Clin Invest* **113:** 1271–1276.

Neumann FJ, Ott I, Marx N, Luther T, Kenngott S, Gawaz M, Kotzsch M, Schomig A. 1997. Effect of human recombinant interleukin-6 and interleukin-8 on monocyte procoagulant activity. *Arterioscler Thromb Vasc Biol* **17:** 3399–3405.

Nishimoto N, Sasai M, Shima Y, Nakagawa M, Matsumoto T, Shirai T, Kishimoto T, Yoshizaki K. 2000. Improvement in Castleman's disease by humanized anti-interleukin-6 receptor antibody therapy. *Blood* **95:** 56–61.

Nishimoto N, Yoshizaki K, Miyasaka N, Yamamoto K, Kawai S, Takeuchi T, Hashimoto J, Azuma J, Kishimoto T. 2004. Treatment of rheumatoid arthritis with humanized anti-interleukin-6 receptor antibody: A multicenter, double-blind, placebo-controlled trial. *Arthritis Rheum* **50:** 1761–1769.

Nishimoto N, Kanakura Y, Aozasa K, Johkoh T, Nakamura M, Nakano S, Nakano N, Ikeda Y, Sasaki T, Nishioka K, et al. 2005. Humanized anti-interleukin-6 receptor antibody treatment of multicentric Castleman disease. *Blood* **106:** 2627–2632.

Pathan N, Hemingway CA, Alizadeh AA, Stephens AC, Boldrick JC, Oragui EE, McCabe C, Welch SB, Whitney A, O'Gara P, et al. 2004. Role of interleukin 6 in myocardial dysfunction of meningococcal septic shock. *Lancet* **363:** 203–209.

Riedemann NC, Neff TA, Guo RF, Bernacki KD, Laudes IJ, Sarma JV, Lambris JD, Ward PA. 2003. Protective effects of IL-6 blockade in sepsis are linked to reduced C5a receptor expression. *J Immunol* **170:** 503–507.

Romano M, Sironi M, Toniatti C, Polentarutti N, Fruscella P, Ghezzi P, Faggioni R, Luini W, van Hinsbergh V, Sozzani S, et al. 1997. Role of IL-6 and its soluble receptor in induction of chemokines and leukocyte recruitment. *Immunity* **6:** 315–325.

Sato K, Tsuchiya M, Saldanha J, Koishihara Y, Ohsugi Y, Kishimoto T, Bendig MM. 1993. Reshaping a human antibody to inhibit the interleukin 6-dependent tumor cell growth. *Cancer Res* **53:** 851–856.

Scala G, Ruocco MR, Ambrosino C, Mallardo M, Giordano V, Baldassarre F, Dragonetti E, Quinto I, Venuta S. 1994. The expression of the interleukin 6 gene is induced by the human immunodeficiency virus 1 TAT protein. *J Exp Med* **179:** 961–971.

Schmitz J, Weissenbach M, Haan S, Heinrich PC, Schaper F. 2000. SOCS3 exerts its inhibitory function on interleukin-6 signal transduction through the SHP2 recruitment sites of gp130. *J Biol Chem* **275:** 12848–12856.

Shima Y, Kuwahara Y, Murota H, Kitaba S, Kawai M, Hirano T, Arimitsu J, Narazaki M, Hagihara K, Ogata A, et al. 2010. The skin of patients with systemic sclerosis softened during the treatment with anti-IL-6 receptor antibody tocilizumab. *Rheumatology (Oxford)* **49:** 2408–2412.

Singh JA, Saag KG, Bridges SL Jr, Akl EA, Bannuru RR, Sullivan MC, Vaysbrot E, McNaughton C, Osani M, Shmerling RH, et al. 2016. 2015 American College of Rheumatology guideline for the treatment of rheumatoid arthritis. *Arthritis Rheumatol* **68:** 1–26.

Smolen JS, Landewe R, Breedveld FC, Buch M, Burmester G, Dougados M, Emery P, Gaujoux-Viala C, Gosses L, Nam J, et al. 2014. EULAR recommendations for the manage-

ment of rheumatoid arthritis with synthetic and biological disease-modifying antirheumatic drugs. 2013 update. *Ann Rheum Dis* **73:** 492–509.

Suntharalingam G, Perry MR, Ward S, Brett SJ, Castello-Cortes A, Brunner MD, Panoskaltsis N. 2006. Cytokine storm in a phase 1 trial of the anti-CD28 monoclonal antibody TGN1412. *N Engl J Med* **355:** 1018–1028.

Taga T, Kishimoto T. 1997. Gp130 and the interleukin-6 family of cytokines. *Annu Rev Immunol* **15:** 797–819.

Tanaka T, Kishimoto T. 2014. The biology and medical implications of interleukin-6. *Cancer Immunol Res* **2:** 288–294.

Tanaka T, Narazaki M, Kishimoto T. 2012. Therapeutic targeting of the interleukin-6 receptor. *Annu Rev Pharmacol Toxicol* **52:** 199–219.

Tanaka T, Ogata A, Narazaki M. 2013. Tocilizumab: An updated review of its use in the treatment of rheumatoid arthritis and its application for other immune-mediated diseases. *Clin Med Insights Ther* **5:** 33–52.

Tanaka T, Narazaki M, Kishimoto T. 2014a. IL-6 in inflammation, immunity, and disease. *Cold Spring Harb Perspect Biol* **6:** a016295.

Tanaka T, Narazaki M, Ogata A, Kishimoto T. 2014b. A new era for the treatment of inflammatory autoimmune diseases by interleukin-6 blockade strategy. *Semin Immunol* **26:** 88–96.

Tanaka T, Narazaki M, Kishimoto T. 2016. Immunotherapeutic implications of IL-6 blockade for cytokine storm. *Immunotherapy* **8:** 959–970.

Teachey DT, Rheingold SR, Maude SL, Zugmaier G, Barrett DM, Seif AE, Nichols KE, Suppa EK, Kalos M, Berg RA, et al. 2013. Cytokine release syndrome after blinatumomab treatment related to abnormal macrophage activation and ameliorated with cytokine-directed therapy. *Blood* **121:** 5154–5157.

Villiger PM, Adler S, Kuchen S, Wermelinger F, Dan D, Fiege V, Butikofer L, Seitz M, Reichenbach S. 2016. Tocilizumab for induction and maintenance of remission in giant cell arteritis: A phase 2, randomised, double-blind, placebo-controlled trial. *Lancet* **387:** 1921–1927.

Yamasaki K, Taga T, Hirata Y, Yawata H, Kawanishi Y, Seed B, Taniguchi T, Hirano T, Kishimoto T. 1988. Cloning and expression of the human interleukin-6 (BSF-2/IFNβ2) receptor. *Science* **241:** 825–828.

Yokota S, Imagawa T, Mori M, Miyamae T, Aihara Y, Takei S, Iwata N, Umebayashi H, Murata T, Miyoshi M, et al. 2008. Efficacy and safety of tocilizumab in patients with systemic-onset juvenile idiopathic arthritis: A randomized, double-blind, placebo-controlled, withdrawal phase III trial. *Lancet* **371:** 998–1006.

Yoshizaki K, Matsuda T, Nishimoto N, Kuritani T, Taeho L, Aozasa K, Nakahata T, Kawai H, Tagoh H, Komori T, et al. 1989. Pathogenic significance of interleukin-6 (IL-6/BSF-2) in Castleman's disease. *Blood* **74:** 1360–1367.

Zhang C, Xin H, Zhang W, Yazaki PJ, Zhang Z, Le K, Li W, Lee H, Kwak L, Forman S, et al. 2016. CD5 binds to interleukin-6 and induces a feed-forward loop with the transcription factor STAT3 in B cells to promote cancer. *Immunity* **44:** 913–923.

The Tumor Necrosis Factor Family: Family Conventions and Private Idiosyncrasies

David Wallach

Department of Biomolecular Sciences, The Weizmann Institute of Science, 76100 Rehovot, Israel

Correspondence: d.wallach@weizmann.ac.il

The tumor necrosis factor (TNF) cytokine family and the TNF/nerve growth factor (NGF) family of their cognate receptors together control numerous immune functions, as well as tissue-homeostatic and embryonic-development processes. These diverse functions are dictated by both shared and distinct features of family members, and by interactions of some members with nonfamily ligands and coreceptors. The spectra of their activities are further expanded by the occurrence of the ligands and receptors in both membrane-anchored and soluble forms, by "re-anchoring" of soluble forms to extracellular matrix components, and by signaling initiation via intracellular domains (IDs) of both receptors and ligands. Much has been learned about shared features of the receptors as well as of the ligands; however, we still have only limited knowledge of the mechanistic basis for their functional heterogeneity and for the differences between their functions and those of similarly acting cytokines of other families.

The study of protein families and their individual members contribute cooperatively to the assembly of knowledge, providing insights into the features shared by family members as well as their distinctive features. The benefit of such a cooperative endeavor is lavishly demonstrated by the huge advances in understanding the mechanisms of action of the tumor necrosis factor (TNF) ligand and TNF/nerve growth factor receptor (NGFR) families. The founding members of these families—the cytokine TNF and the low-affinity NGFR—were isolated and cloned three decades ago (Old 1985; Johnson et al. 1986; Radeke et al. 1987). In this review, I will present a brief overview of the knowledge of common structural, mechanistic, and functional features of the TNF ligand and TNF/NGFR families. I will also refer to interactions that are known for only a few family members, but whose occurrence raises the possibility that other family members participate in similar associations.

THE WIDE RANGE OF FUNCTIONS OF THE TNF FAMILY

There are 18 known human genes for the TNF ligand family and 29 for the TNF/NGFR family (see genenames.org/genefamilies/TNFSF and genenames.org/genefamilies/TNFRSF; see Table 1 for their major known functions). Practically, all cells in the body express receptors, and many also express some of the ligands of these families. Each receptor–ligand pair controls a wide range of cellular activities. Assessing the impact of functional arrest of specific members of the TNF and TNF/NGF families has revealed some key physiological roles served ex-

Table 1. Principal known functions of receptors of the tumor necrosis factor (TNF)/nerve growth factor receptor (NGFR) family and ligands of the TNF family

Receptor			Ligand			Main known functions of the receptor
Widely used name	Formal name	Other names	Widely used name	Formal name	Other names	
TNFR1	TNFRSF1A	p55, CD120a	TNF / LT	Tumor necrosis factor / Lymphotoxin α		Orchestration of inflammation; cell killing
TNFR2	TNFRSF1A	p75, CD120b	TNF / LT	Tumor necrosis factor / Lymphotoxin		Costimulation of T lymphocytes (information on various other functions is sporadic)
Fn14	TNFRSF12A	TweakR	TWEAK	TNFSF12	APO3L	Control of tissue regeneration in response to injury; inflammation
DR3	TNFRSF12	TRAMP, WSL-1, APO-3	TL1	TNFSF15	TL1A, VEGI	Costimulation of T lymphocytes; inflammation
FAS	FAS	APO-1, CD95	Fas ligand	FASL, Fas ligand	CD178	Cell killing; cell growth; inflammation
DcR3	TNFRSF6B	TR6	Fas ligand, LIGHT, TL1			Secreted decoy receptor that blocks the function of the three indicated ligands; seems also capable of triggering signaling through its binding to HSPG
TRAIL-R1	TNFRSF10A	Apo2, DR4	TRAIL	TNFSF10	Apo-2L, CD253	Cell killing
TRAIL-R2	TNFRSF10B	DR5	TRAIL	TNFSF10		Cell killing
DcR1	TNFRSF10C	TRAILR3	TRAIL	TNFSF10		GPI-linked decoy receptor restraining the function of TRAIL
DcR2	TNFRSF10D	TRAILR4	TRAIL	TNFSF10		Transmembrane truncated decoy receptor restraining the function of TRAIL
LTβR	LTβR (lymphotoxin β receptor)		LTα1β2 (lymphotoxin β [LTβ] in complex with lymphotoxin α), LIGHT	LTβ / TNFSF14	CD258	Development and function of lymphoid organs; amplification of antiviral response and various other innate and adaptive immune functions

Cite this article as *Cold Spring Harb Perspect Biol* doi: 10.1101/cshperspect.a028431

HVEM	TNFRSF14	ATAR, CD270	LIGHT; LT (weak binding)	TNFSF14		Costimulation in T and B lymphocytes and in NK cells; triggering of immunosuppression through BTLA, and immunosuppression, inflammation, and NK activation through CD160
OX40	TNFRSF4	CD134	OX40 ligand (OX40L)	TNFSF4	CD252	Costimulation of T lymphocytes
CD27	TNFRSF7		CD70	CD70	CD27 ligand	Costimulation of T lymphocytes
CD30	TNFRSF8		CD30 ligand (CD30L)	TNFSF8		Costimulation of T lymphocytes
4-1BB	TNFRSF9	CD137	4-1BB ligand (4-1BBL)	TNFSF9		Costimulation of T lymphocytes
GITR	TNFRSF18	AITR, CD357	GITR ligand (GITRL)	TNFSF18	AITR	Costimulation of T lymphocytes
RELT	RELT		?			Unknown
CD40	TNFRSF5		CD40 ligand (CD40L)	CD40LG	CD154	Wide range of effects contributing to initiation and progression of adaptive immunity
TACI	TNFRSF13B	CD267	BAFF; APRIL	TNFSF20; TNFSF13	BLYS, CD257, CD256	B-lymphocyte growth, survival, and differentiation; costimulation of T lymphocytes
BAFFR	TNFRSF13C	CD268	BAFF	TNFSF20		B-lymphocyte survival, growth, and differentiation
BCMA	TNFRSF17	CD269	BAFF; APRIL	TNFSF13		B-lymphocyte survival, growth, and differentiation
RANK	TNFRSF11A		RANK ligand (RANKL)	TNFSF11	RANKL, TRANCE, OPGL, CD254	Osteoclast differentiation and activation; mammary gland development; development of lymph nodes; various immune functions
OPG	TNFRSF11B	OCIF, TR1	RANK ligand TRAIL	TNFSF11		Secretory decoy receptor that blocks the functions of RANKL and TRAIL; seems also capable of triggering signaling through its binding to HSPG
EDAR	EDAR (ectodysplasin A receptor)	EDA1R	EDA	EDA (ectodysplasin A)		Embryonic development of ectodermal appendages

Continued

Table 1. *Continued*

Receptor			Ligand			Main known functions of the receptor
Widely used name	Formal name	Other names	Widely used name	Formal name	Other names	
XEDAR	EDA2R (ectodysplasin A2 receptor)		EDA-A2 (a shorter splice variant of the EDA gene)			Unknown
NGFR	Nerve growth factor receptor	p75NTR, CD271	Neurotrophins; myelin-associated inhibitory factors; fragments of β-amyloid precursor protein; prion peptide			Neuronal growth and death; arrest of axonal degeneration; pain sensation
TROY	TNFRSF19	TAJ	Myelin-associated inhibitory factors			Repression of axonal regeneration
DR6	TNFRSF21	CD358	?			Probably contributes to amyloid-β-induced axonal pruning

TWEAK, TNF-related weak inducer of apoptosis; HSPG, heparan sulfate proteoglycan; TRAIL, TNF-related apoptosis-inducing ligand; GPI, glycosylphosphatidylinositol; HVEM, herpesvirus entry mediator; NK, natural killer; BTLA, B- and T-lymphocyte attenuator; GITR, glucocorticoid-induced TNFR-related protein; AITR, activation-inducible TNFR family receptor; RELT, receptor expressed in lymphoid tissue; TACI, transmembrane activator and calcium-modulating cyclophilin ligand interactor; BAFF, B-cell-activating factor of the tumor necrosis factor family; APRIL, a proliferation-inducing ligand; BCMA, B-cell maturation antigen; RANK, receptor activator of nuclear factor κB; OPG, osteoprotegerin; EDA, ectodermal dysplasia.

Cite this article as *Cold Spring Harb Perspect Biol* doi: 10.1101/cshperspect.a028431

clusively by them. Listed below and briefly discussed are the functions of family members that have attracted the greatest attention, either because they appear to be unique to the family or because of known pathological consequences of their deficiency or hyperactivity.

Control of Cell Survival

Stimulation of Cell Death

The ability of certain ligands of the TNF family to trigger death of cells independently of protein synthesis was the family's first cellular effect to be described (Granger and Kolb 1968; Ruddle and Waksman 1968), and to this day it remains the only function not known to be shared with other cytokines. Several TNF family receptors (TNFR1, DR3, FAS, TNF-related apoptosis-inducing ligand [TRAIL]-R1, TRAIL-R2) trigger rapid cell death. Others signal for cell death less effectively, either through induction of TNF (Grell et al. 1999; Burkly 2014) or by poorly defined cell-autonomous mechanisms (Force et al. 2000; Georgopoulos et al. 2006; Elmetwali et al. 2010).

The cytotoxic functions of the TNF family contribute to immune-mediated cell killing. They also seem to contribute to the control of expansion and to the duration of activities of immune-cell populations and to the shaping of leukocyte repertoires (Falschlehner et al. 2009; Strasser et al. 2009).

Providing Survival Signaling

One way in which the TNF family members facilitate maintenance and amplification of immune responses is by providing the relevant cells with survival signals. Best documented are the crucial roles of several members of the family in maintaining the survival of B and T lymphocytes (Croft 2014; Figgett et al. 2014). TNF family members are also capable of inducing resistance of cells to the cytotoxic activities that they themselves activate (e.g., Wallach 1984; Hahn et al. 1985; Blomberg et al. 2008; Chen et al. 2010; Jeon et al. 2015).

Orchestration of Inflammation

The best-documented pivotal role of a ligand of the TNF family in pathological disorders is the contribution of TNF to chronic inflammatory diseases. This has been demonstrated by the therapeutic effects of TNF-blocking agents observed in millions of patients, as well as in experimental animal models for both chronic and acute inflammatory diseases (Tracey et al. 1987; Apostolaki et al. 2010; Sfikakis 2010). TNF contributes to the initiation, progression, and termination of inflammation, while displaying antagonistic effects: for example, induction of cell death but also cell growth and resistance to cell death, and obstruction and destruction of capillaries but also stimulation of angiogenesis. TNF is also an important player in systemic manifestations of inflammation, for example, in activating the acute-phase response in the liver (Wallach and Kovalenko 2016).

Also contributing to inflammation, although in more restricted ways, are several TNF family ligands, including TKA1, TNF-related weak inducer of apoptosis (TWEAK), TRAIL, CD40L, LIGHT, and receptor activator of nuclear factor (NF)-κB ligand (RANKL), which are known primarily for other functions.

Tissue Modeling

Tissue Remodeling in Response to Injury

Like TNF, TWEAK exerts a wide range of antagonistic effects on cell functions; it contributes to both inflammation and its arrest, as well as to both destruction and regeneration of tissues. It thus serves important roles in coordinating tissue remodeling in response to injury (Burkly 2014).

Bone Homeostasis

Calcified bone matter undergoes constant turnover as a result of the antagonistic effects of osteoblasts, which construct bones, and osteoclasts, which resorb them. Local inflammation in the bone facilitates bone destruction. RANKL serves a crucial role in maintaining the constitutive activity of the osteoclasts. Its soluble receptor, osteoprotegerin (OPG), blocks RANKL function by competing for the binding of RANK (Boyce and Xing 2007; Walsh and Choi 2014). CD40L potentiates this inhibition via several effects, including induction of OPG in B lymphocytes (Li et al. 2007). In contrast, TNF enhances bone dissolution. It does this by triggering the egress of osteoclast precursors from the bone marrow as well as by enhancing RANKL generation and triggering osteoclast differentiation and activation synergistically with this ligand as well as with the non-TNF family cytokine interleukin (IL)-1 (Li et al. 2004).

Control of the Development of Ectodermal Tissues

Ectodysplasin (EDA) and its receptor EDAR are the only known ligand–receptor pair within the TNF families that seems to make no contribution to immune regulation. They signal for embryonic development of ectodermal appendages such as the hair, teeth, and sweat glands. This role is evolutionarily conserved in other metazoan phyla in which EDA and its receptor control the development of ectodermal appendages such as feathers, scales, and fins (Lefebvre and Mikkola 2014).

EDA-A2, a slightly shorter splice variant encoded by the *Eda* gene, binds to a distinct receptor of the TNF/NGF family, XEDAR. Although expressed in the developing hair follicle, this receptor does not seem to be required for hair growth. Limited evidence suggests that it might serve to control skeletal muscle homeostasis (Lefebvre and Mikkola 2014). TROY, an orphan receptor of the TNF/NGF family, is also coexpressed with EDAR in hair follicles and in embryonic skin, but its function at these sites is not known (Kojima et al. 2000).

Control of Adaptive Immunity

Elicitation of an adaptive immune response necessitates fine-tuning of the development of lymphocytes and of their interaction with antigen-presenting cells as well as regulation of the development and function of specific organs in which these processes occur. In all of these func-

tions, members of the TNF family play pivotal roles.

Control of the Generation and Maintenance of Lymphoid Organs

Embryonic development of the secondary lymphoid organs crucially depends on signaling by lymphotoxin β receptor (LTβR) and RANK. Both LTβR-triggered and TNF-triggered signaling are required for maintenance of the microarchitecture of the lymphoid organs and their appropriate function in the adult (Ware 2005; McCarthy et al. 2006). Both are also required for the neogenesis of lymphoid assemblages at sites of chronic inflammation (Drayton et al. 2006). A similar "morphogenic" role is served by TNF in dictating the generation of granuloma (Kindler et al. 1989).

Control of the Development and Function of T Lymphocytes

Signaling by several TNF/NGF family receptors, including TNFR1, LTβR, RANK, herpesvirus entry mediator (HVEM), OX40, and CD40, controls the migration, maturation, and activation of dendritic cells (Ware 2005; Summers de-Luca and Gommerman 2012; Walsh and Choi 2014). Signaling by receptors of the TNF/NGF family contributes to the selection of lymphocytes in the thymus. TNFR1, TNFR2, DR3, HVEM, OX40, CD27, CD30, 4-1BB, B-cell-activating factor of the tumor necrosis factor family (BAFF), and transmembrane activator and calcium-modulating cyclophilin ligand interactor (TACI) provide costimulatory signals in antigen-stimulated T lymphocytes. Some of those receptors also initiate stimulatory signals in T lymphocytes whose antigen receptors have not been activated. These stimulatory signals enhance lymphocyte survival, growth, and effector functions. The various costimulatory family members apparently serve distinct roles at different phases of T-lymphocyte response, and the relative contribution of their costimulatory effects to defense varies depending on the nature of the particular pathogenic challenge (Croft 2014; Mbanwi and Watts 2014).

Control of the Development and Function of B Lymphocytes

Activation of the TNF/NGF family receptors BAFFR, B-cell maturation antigen (BCMA), TACI, and CD40 provide B lymphocytes with survival and growth signals at distinct stages of their development. CD40 signaling is crucially required also for antibody isotope switching and for the generation of memory B lymphocytes. Various other TNF/NGF family members also affect B-cell biology (Bishop and Hostager 2003; Dillon et al. 2006; Elgueta et al. 2009; Figgett et al. 2014).

Functions of the TNF Family in the Brain

At times of injury or autoimmune response within the brain, the affected cells display the same modes of TNF family–induced regulation as those observed in such situations elsewhere in the body (Akassoglou et al. 1999; Shohami et al. 1999). Some of these cytokine effects result in tissue damage. However, TNF also provides protective and survival signals in nervous tissue, in part through TNFR2 (Bruce et al. 1996; Fontaine et al. 2002). Effects of TNF on brain functions contribute to several behavioral responses to disease, including enhanced slow-wave sleep (Shoham et al. 1987), fever (Dinarello et al. 1986), anorexia (Cerami et al. 1985; Plata-Salaman et al. 1988), increased pain perception (Hess et al. 2011), and others (Dantzer 2001). Fever induction by TNF, as well as by several other inflammatory cytokines, is triggered indirectly through induction of RANK-mediated signaling in astrocytes (Hanada et al. 2009).

Emerging knowledge indicates that TNF family members also contribute to brain functions unrelated to immune defense. TNF produced by glial cells stabilizes neuronal circuits by dictating homeostatic synaptic scaling (Stellwagen and Malenka 2006). Some evidence indicates that FAS signals for neurogenesis in the adult brain (Corsini et al. 2009). The receptor DR3 is expressed in motor neurons and its tonic signaling seems to be necessary for their survival (Richard et al. 2015).

The function of NGFR, a TNF/NGF family receptor for which no ligand of the TNF family is known, serves cooperatively with several co-receptors to control the growth and survival of neurons in response to simulation by several ligands and also to control pain sensation (Hempstead 2002; Chao 2003; Ibanez and Simi 2012). Limited evidence suggests that TROY and DR6, additional members of the TNF/NGFR family not known to bind any ligand of the TNF family, also serve to signal for neuronal death (Shao et al. 2005; Nikolaev et al. 2009; Olsen et al. 2014).

TRIGGERING OF SIGNALING: INDUCED JUXTAPOSITION

Members of the TNF ligand and TNF/NGFR families share conserved extracellular motifs by which they bind each other. With the exception of lymphotoxin (LT), a secretory protein, all ligands are produced as type II transmembrane (TM) proteins in which the receptor-binding motif, whose structure consists of two packed sheets of eight antiparallel β strands, is located at the carboxyl terminus. Most ligands also occur in soluble forms, generated by proteolytic processing of the TM forms to yield soluble ligand-binding molecules. Both in their membrane-bound and in their soluble forms, the ligand molecules associate constitutively in trimers.

The receptors are produced as type I TM proteins, whose amino-terminal ligand-binding motif consists of a variable number of two conserved modules that together form a 40-amino-acid structure containing several cysteines, the "cysteine-rich domain" (CRD).

Distinct parts of receptors' extracellular domains (EDs) serve opposing roles in controlling signaling. The amino-terminal CRD serves as a "pre-ligand assembly domain" (PLAD), which safeguards against ligand-independent signaling (Chan 2007). In the absence of ligands, the PLADs in the EDs of three or (more likely) two (Naismith et al. 1995) receptor molecules associate in a way that keeps their intracellular domains (IDs) apart. In contrast, ligand binding, which occurs downstream from the PLAD, imposes juxtaposition of the receptor IDs. In receptors whose IDs contain a death domain (DD) (see below), such juxtaposition is fostered by the propensity of the DD to self-associate (Boldin et al. 1995). Proline-containing motifs found in the transmembrane domain (TD) in FAS, and apparently also in the other receptors of the family, also tend to self-associate (forming homotrimers), further strengthening the ligand-binding effect (Fu et al. 2016).

Juxtaposition of the IDs of the receptors exposes their binding surfaces to signaling proteins. Their binding to the receptors and consequent juxtaposition triggers their enzymatic activity and/or their association with downstream signaling proteins (Fig. 1A).

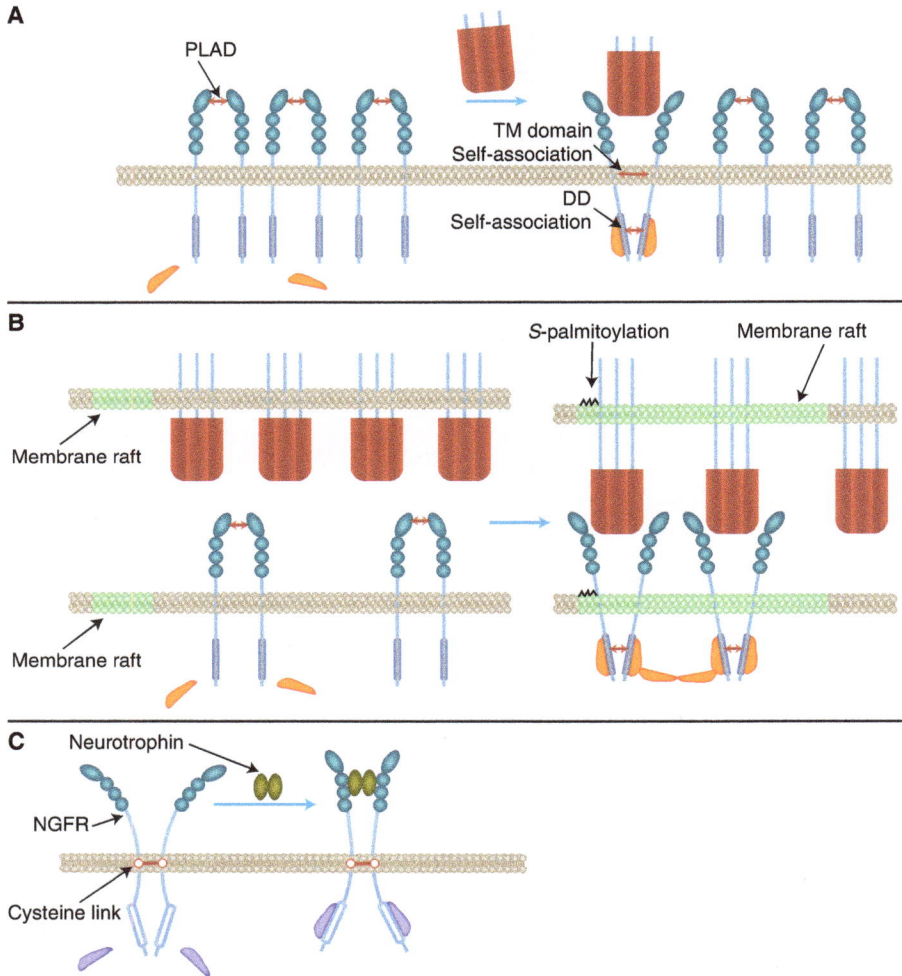

Figure 1. Signaling triggering: proposed mechanisms. (A) Triggering by intracellular domain (ID) juxtaposition imposed by tumor necrosis factor (TNF) family ligands, and its restriction by the pre-ligand assembly domain (PLAD). (B) Triggering potentiation by anchorage of ligands to membranes. (C) Triggering by ID distancing: A fraction of the nerve growth factor receptor (NGFR) molecules occurs as dimers of receptor molecules that are covalently linked through a conserved transmembrane (TM) cysteine residue. In these dimers, the intracellular death domains (DDs) seem to be constitutively associated. Neurotrophin-induced triggering of these dimers to signal for death was suggested to occur by distancing of the IDs, imposed rather like a snail tongue, with the TM cysteine link serving as a fulcrum (Vilar et al. 2009). (See discussion of these findings in the last paragraph of this review.)

Cite this article as *Cold Spring Harb Perspect Biol* doi: 10.1101/cshperspect.a028431

The self-association of four of the receptors (CD40, NGFR, CD27, and the soluble receptor OPG) is facilitated by constitutive cysteine links (found in the ID of CD40, in the TD of NGFR, at the carboxyl terminus of OPG, and apparently in the ED of CD27) (Van Lier et al. 1987; Schneeweis et al. 2005; Vilar et al. 2009; Nadiri et al. 2015). Constitutive dimerization of the soluble receptor OPG is also dictated by two DD motifs found at its carboxyl terminus (Schneeweis et al. 2005).

Structural studies of glucocorticoid-induced TNFR-related protein (GITR) suggest that the trimeric ligands of the TNF family may tend to undergo further clustering through noncovalent homotypic self-association (Zhou et al. 2008). However, formation of a soluble supercluster (of 20 trimers) has been documented only in the case of one TNF family member, BAFF (Liu et al. 2002; Cachero et al. 2006). The trimers of EDA are exceptional in that they are constitutively dimerized by homotypic self-association of an amino-terminal collagen-like domain unique to this ligand (Swee et al. 2009).

SIGNALING FACILITATION BY MEMBRANE ANCHORAGE

The ability of membrane-anchored forms of the ligands to induce clustering of the receptors that they bind is greater than that of the soluble forms. This ability is enhanced by positioning of both the membrane-bound ligands and their receptors within membrane rafts (Legler et al. 2003; Muppidi et al. 2004; So and Croft 2013) and by *S*-palmitoylation of the ligands and their receptors (Chakrabandhu et al. 2007; Feig et al. 2007; Rossin et al. 2009; Poggi et al. 2013). Formation of larger aggregates stabilizes binding of the ligands to the receptors. It can also dictate higher-order organization of signaling proteins (Fig. 1B).

The various signaling functions activated by the TNF family differ in the extents of their dependence on such higher-order signaling-protein organization. Accordingly, membrane and soluble forms of the ligands have differential abilities to trigger different cellular responses. For example, although TNFR1 can be activated both by soluble and by membrane-anchored TNF, TNFR2 is activated only by the latter. FAS can be triggered to signal for death by a dimer of FASL trimers, but not by a single trimer. Fn14, when massively aggregated, activates both the canonical and the alternative NF-κB transcription factor pathways, but only the alternative one when mildly aggregated (Grell et al. 1995; Schneider et al. 1998; Muhlenbeck et al. 2000; Bishop and Hostager 2003; Holler et al. 2003; Stone et al. 2006; O'Reilly et al. 2009; Wyzgol et al. 2009; Burkly 2014).

BIDIRECTIONAL SIGNALING: MULTIPLE FORMS AND FUNCTIONS OF SOLUBLE AND MEMBRANE-ANCHORED LIGANDS AND RECEPTORS

Similar to the IDs of TNF/NGF family receptors, the IDs of membrane-anchored TNF family ligands are found to recruit and activate signaling proteins on receptor–ligand interaction. Thus, they trigger "reverse signaling" within the ligand-producing cells (Stuber et al. 1995; Arens et al. 2004; Eissner et al. 2004; Grohmann et al. 2007; Kang et al. 2007; Sun et al. 2007; Juhasz et al. 2013). Various other similarities in action of the TNF and TNF/NGF families further blur the distinction between their identities as ligands and as receptors. As with the ligands, many of the receptors can be proteolytically cleaved, yielding ligand-binding "soluble receptors." Two of these receptors, OPG and DcR3, occur only in soluble forms, while two others, DcR1 and DcR2, despite their anchorage to the membrane, are devoid of the molecular structures within the IDs that are required for signaling.

The various forms of soluble and membrane-anchored ligands and receptors, and the functions served by the transitions in these forms, are shown in Figure 2.

"RE-ANCHORING" OF SOLUBLE FORMS OF LIGANDS AND RECEPTORS VIA THEIR BINDING TO EXTRACELLULAR MATRIX COMPONENTS

Several ligands of the TNF family and the two TNF-family receptors that occur only in soluble form (OPG and DCR3) contain sites

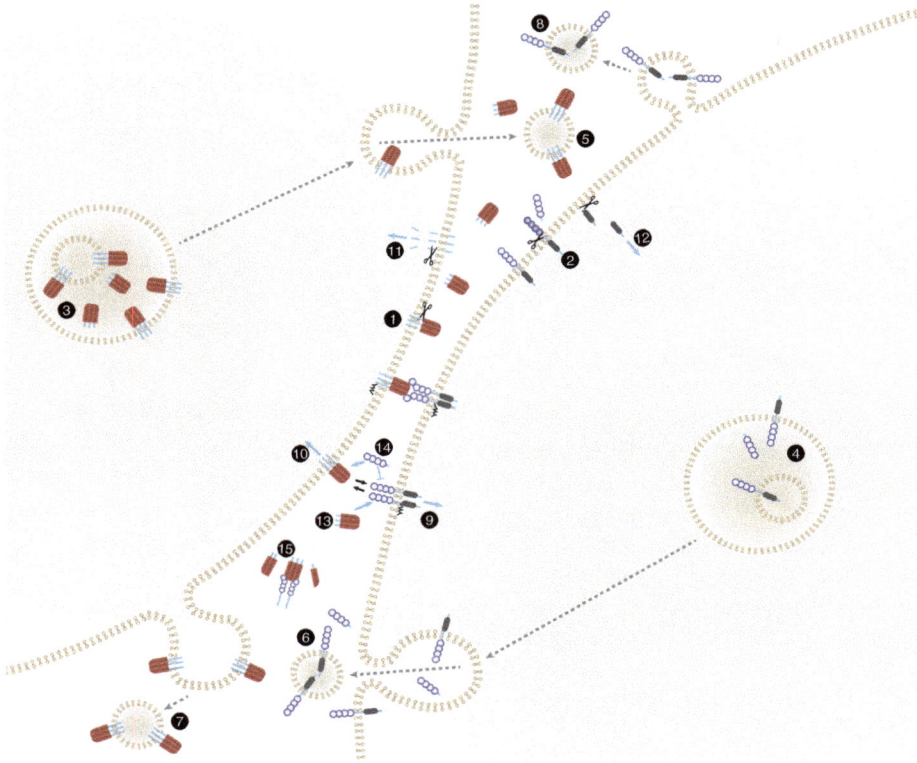

Figure 2. Membrane-anchored and soluble forms of ligands and receptors of the tumor necrosis factor (TNF) families, and their functional roles. The figure shows an intriguing symmetry of the spectra of soluble and membrane-anchored forms attained by TNF family ligands and TNF/nerve growth factor (NGF) family receptors, and of the functions of these forms. Both ligands (*1*) and receptors (*2*) are proteolytically cleaved to yield soluble forms (Kriegler et al. 1988; Engelmann et al. 1990; Nophar et al. 1990; Black et al. 1997; Moss et al. 1997). The cleavage occurs constitutively or inducibly, either on the cell surface as illustrated (Black et al. 1997; Moss et al. 1997; Becker-Pauly and Rose-John 2013) or within the cell (Lopez-Fraga et al. 2001). (*3,4*) Some of the ligands (Gordon and Galli 1990; Bossi and Griffiths 1999; Koguchi et al. 2007) and receptors (Wang et al. 2003) accumulate within intracellular vesicles from which they are secreted in response to specific stimuli, thus supplementing either the cell-surface-expressed or the soluble pools. Some are released while anchored to membranes that might correspond either to exosomes that have accumulated in intracellular multivesicular bodies (*5,6*) or to microvesicles exfoliating from the cell surface (*7,8*) (Albanese et al. 1998; Martinez-Lorenzo et al. 1999; Islam et al. 2007). Binding of membrane-anchored ligands of the TNF family to their receptors, besides triggering receptor signaling (*9*), also triggers "reverse signaling" by the ligand molecules (*10*) (Stuber et al. 1995; Arens et al. 2004; Eissner et al. 2004; Grohmann et al. 2007; Kang et al. 2007; Sun et al. 2007; Juhasz et al. 2013). One mechanism contributing to this reverse signaling is intramembrane proteolytic cleavage, yielding ligand intracellular domain (ID) fragments that apparently mediate signaling following their translocation to the nucleus (Domonkos et al. 2001; Fluhrer et al. 2006; Friedmann et al. 2006; Kirkin et al. 2007) (*11*). Intramembrane cleavage that apparently contributes to signaling has also been reported for two receptors of the TNF/NGF family, nerve growth factor receptor (NGFR) and TNFR1 (Kanning et al. 2003; Kenchappa et al. 2006; Chhibber-Goel et al. 2016) (*12*). Accumulation of the soluble forms of both receptors and ligands interferes with the binding of membrane-bound forms to the ligands and the receptors, respectively, and can thus block signaling (*13,14*). To the extent that the soluble forms are incapable of triggering signaling, they may also interfere with signaling activation by their membrane-bound forms. However, the association of soluble receptors with soluble ligands also stabilizes the trimeric structures of the soluble ligands, so that they function not as mere inhibitors but rather as buffering agents. While decreasing the intensity of signaling activation, they also extend its duration (*15*) (Aderka et al. 1992; Eliaz et al. 1996).

Cite this article as *Cold Spring Harb Perspect Biol* doi: 10.1101/cshperspect.a028431

that bind extracellular matrix components such as heparan sulfate proteoglycans (HSPGs) and fibronectin. Their associations with these extracellular compounds have several functional consequences. (The numbering that follows corresponds to that shown in Fig. 3.) (1) Binding of TNF (Alon et al. 1994) and of FASL (Aoki et al. 2001; Zanin-Zhorov et al. 2003) to extracellular matrix proteins such as fibronectin, and of APRIL to HSPGs (Ingold et al. 2005; Kimberley et al. 2009), facilitate signaling activation by these ligands, apparently by imposing an oligomeric state on them. Such binding can also impose local restrictions. The anchorage of APRIL to HSPGs in mucosal lymphoid tissues not only augments its function but also restricts this ligand to target cells at this site (Huard et al. 2008), while the binding of EDA to proteoglycan

interferes with EDA function by preventing its access to EDAR. (2) Binding of OPG to HSPGs blocks the function of OPG by also causing its uptake into the cell and its subsequent degradation (Standal et al. 2002). (3) Binding of OPG, APRIL, or DcR3 to HSPG appears to also trigger signaling by cell-surface molecules such as syndecans, which are linked to HSPG moieties, or by cell-surface molecules with which these moieties associate noncovalently (Couchman 2003; Mosheimer et al. 2005; Chang et al. 2006; Dillon et al. 2006; You et al. 2008). (4) Conversely, binding of cell-associated HSPG to TACI triggers TACI signaling, apparently independently of any ligand of the TNF family (Bischof et al. 2006). Because both APRIL and its receptor TACI bind HSPG, they probably form tripartite complexes in which the three pro-

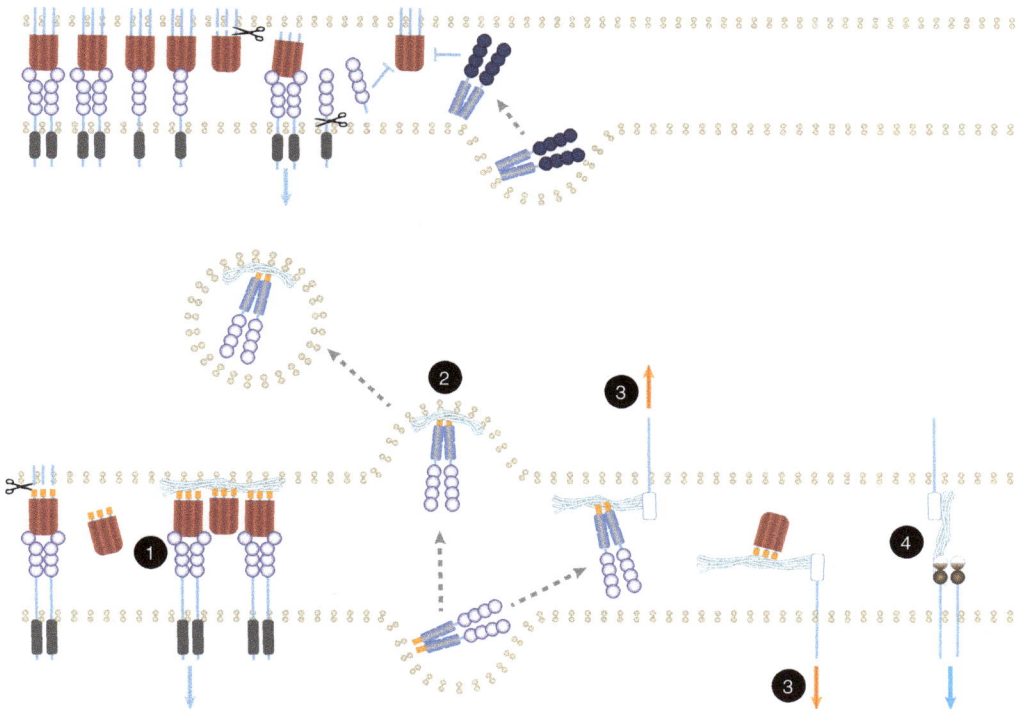

Figure 3. "Re-anchoring" of soluble forms of ligands and receptors by their binding to extracellular matrix components. Various reported effects of extracellular matrix components on the functions of soluble ligands and receptors, shown in the *lower* part of the figure, are compared with the functions of soluble ligands and receptors in the absence of such components, as shown in the *upper* part. See text for details and for numbering of the illustrated effects. Undulating lines correspond to extracellular matrix components. Orange forms correspond to structural motifs that dictate binding to such components.

teins modulate one another's function (Moreaux et al. 2009).

MECHANISMS OF SIGNALING ACTIVATION BY THE TNF FAMILY

The cellular responses initiated by the TNF family reflect a gamut of molecular changes, including altered expression of many proteins, and extensive changes in their patterns of phosphorylation, ubiquitination, association, localization, and rates of turnover, as well as alterations in the expression of various lipid metabolites. Because neither the IDs of TNF/NGF family receptors nor those of TNF family ligands possess enzymatic activity, they need to bind to signaling proteins that do express such activities, which are stimulated as a consequence of ligand–receptor association.

Protein-Binding Motifs in the Receptors and Ligands

The receptors of the TNF family can be subclassified into two groups with distinct binding motifs in their IDs. The IDs of TNFR1, DR3, FAS, TRAIL-R1, TRAIL-R2, NGFR, and EDAR each contain a DD motif of about 80 amino acids that bind adapter proteins containing the same motif. All other receptors contain short (five to eight amino acid residues) motifs of binding to adapter proteins of the TNF receptor–associated factors (TRAFs) family. NGFR is the only receptor that contains both an (atypical) DD and a TRAF-binding motif. The TRAF-binding receptors are incapable of triggering signaling pathways activated by the DD adapter proteins. In contrast, because two of the DD-containing adapter proteins—TNF receptor-associated death domain (TRADD) and EDAR-associated death domain (EDARADD)—also contain a TRAF-binding motif, the DD-containing receptors are capable of triggering at least part of the signaling pathways activated by the TRAF-binding receptors.

Binding of both DD-containing adapter proteins and the TRAFs to their receptors dictates the formation of large complexes containing several receptor molecules and multiple adapter molecules. As to the associations of the receptors with TRAFs (associations whose structural basis has so far been explored only for TRAF2 and TRAF6), assembly of the signaling complexes is fostered by constitutive trimeric oligomerization of the carboxyl-terminal leucine zipper TRAF motifs and by induced dimeric association of the TRAFs' amino-terminal regions (McWhirter et al. 1999; Park et al. 1999; Yin et al. 2009a). In the initiation of signaling through DD association, an event so far explored only for FAS association with the adapter protein FADD/MORT1, oligomerization is fostered by homotypic associations of the DDs of both FAS and FADD/MORT1 as well as by their heterotypic associations (Scott et al. 2009; Wang et al. 2010; Kersse et al. 2011; Li et al. 2013).

These associations between receptors and their adapter proteins initiate signaling by exposing, within the adapter proteins, binding surfaces for signaling proteins. With the exception of TRAF1, and—as was recently indicated, probably also with the exception of TRAF2 (Yin et al. 2009b)—members of the TRAF adapter protein family possess ubiquitin ligase activity, which is triggered by their induced associations with receptors.

Some adapter proteins initiate the above associations and activities while still bound to receptors expressed on the cell surface. Others do so only after uptake of the receptors into the cells (Schutze et al. 2008; Ganeff et al. 2011). Some of the signaling complexes that they generate remain associated with the receptors. Others dissociate from the receptors and generate signaling complexes in the cytoplasm (Sessler et al. 2013).

Sporadic evidence indicates that various other evolutionarily conserved amino acid residues within the IDs of receptors and of ligands serve—either without modification or after undergoing phosphorylation or ubiquitination—as binding sites for additional cytoplasmic proteins. In so doing, they contribute to the initiation of additional signaling pathways and to the control of trafficking of the receptors and ligands (e.g., Pocsik et al. 1995; Adam Klages et al. 1996; Watts et al. 1999; Eissner et al. 2004; Kimura et al. 2004; Sun et al. 2007; Juhasz et al. 2013; Ma et al. 2013; Fritsch et al. 2014; Chakrabandhu et al. 2016).

In addition to their ligand–receptor interaction motifs, the EDs of both the ligands and the receptors contain, in closer proximity to their TDs, specific sequences that determine their mode of shedding. Several ligands, including EDA, BAFF, and APRIL, contain consensus sequences for furin cleavage and are constitutively shed by intracellularly located furin. Other ligands as well as several receptors, including TNF and its receptors, are shed inducibly after they reach the cell membrane, mainly via the effects of metalloproteinases such as TACE/ADAM17. Little is known of the sequence determinants that dictate this selective shedding (Brakebusch et al. 1994; Thomas 2002; Hayashida et al. 2010).

Signaling Pathways

Studies of the mechanisms of signaling for apoptotic and for necrotic cell death as well as for activation of NF-κB transcription factors by TNF family members provided the foundation for exploring the regulation of these functions by external inducers. These activities of the TNF family are the most extensively studied to date. However, several others are known, including signaling for activation of the extracellular signal-regulated kinase (ERK), c-Jun amino terminal kinase (JNK), and p38 mitogen-activated protein (MAP) kinase cascades, other serine/threonine kinases, the phosphoinositide 3-kinase (PI3K)/Akt pathway, superoxide generation, soluble Src-family tyrosine kinases, the neutral and acidic sphingomyelinases, and phospholipase C. Space constraints preclude further attention in this review to these signaling mechanisms, which have been extensively reviewed elsewhere (Eissner et al. 2004; Hayden and Ghosh 2012; Juhasz et al. 2013; Li et al. 2013; Sessler et al. 2013; So and Croft 2013; Sabio and Davis 2014; Wallach 2016b).

FOREIGN ENCOUNTERS: NONCANONICAL CORECEPTORS AND LIGANDS

The information presented above delineates a shared set of mechanistic principles by which the different members of the TNF ligand and receptor families associate. However, several members of the families can also participate in unique interactions with coreceptors and ligands, including some that do not belong to the TNF ligand and TNF/NGFR families. Examples of these noncanonical associations, some extensively documented and others for which the evidence is limited and requires confirmation, are presented in Figure 4. Although such interactions have been noticed for only a few members of the TNF ligand and receptor families up to now, they might be found to occur with other members as well.

HOW IS REGULATION BY THE TNF FAMILY REGULATED?

As in the case of other cytokines, signaling by members of the TNF family is triggered when ligand and receptor molecules are brought close enough to allow their binding to each other. This approximation was found to be dictated at multiple mechanistic levels. In most cases, it occurs by induced up-regulation of the ligand molecules, allowing them to bind to receptor molecules that are constitutively expressed. Such is the case with TNF, whose induction mostly occurs in a transient manner, although TNFR1 to which it binds is widely expressed constitutively. However, the inverse type of modulation is also observed. TWEAK, for example, is a ligand that is constitutively expressed in macrophages, whereas the expression of its receptor Fn14 is induced in injured tissues. Up-regulation of the expression of both ligands and receptors is induced by specific signaling pathways that are activated in response to insults or developmental cues. The best-documented mode of up-regulation is by enhanced transcription. However, it also occurs on other mechanistic levels, including splicing, RNA transport, and altered messenger RNA (mRNA) stability and mRNA translation rates. Also contributing to this regulation are associations of proteins, microRNAs and long noncoding RNAs with transcripts (Wallach 2016a; Wallach and Kovalenko 2016). Once synthesized, the ligands and receptors can be subjected to further regulation by mechanisms controlling their translocation to the cell surface (Bossi and Grif-

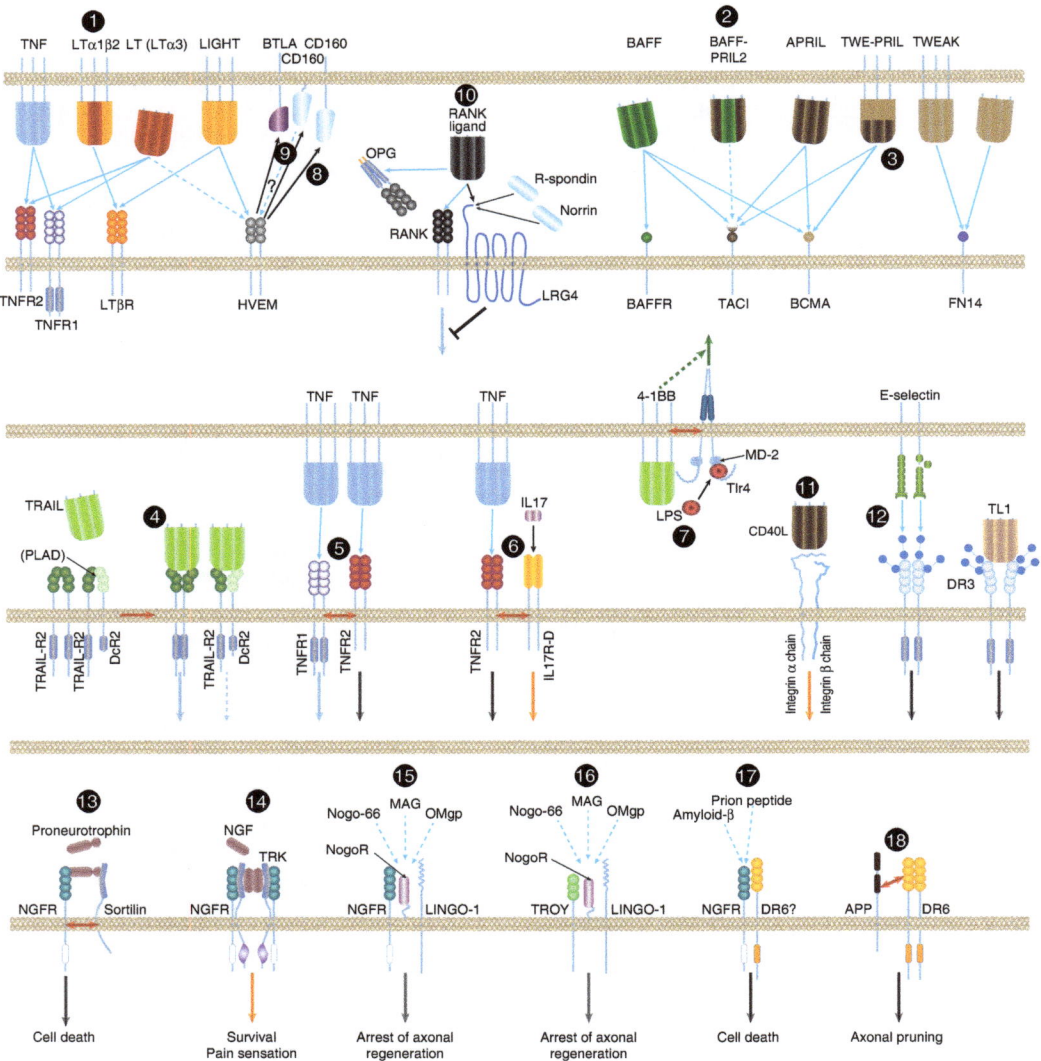

Figure 4. Noncanonical associations of the tumor necrosis factor (TNF) ligand and the TNF/ nerve growth factor receptor (NGFR) families. Most of the interactions of ligands and receptors of the TNF ligand and TNF/NGFR families are known to occur between homotrimeric molecules of a particular ligand and molecules of a particular receptor. A few exceptions, however, are known: (1) Lymphotoxin β (LTβ) functions only within heterotrimers that it forms with lymphotoxin α (LTα). Whereas homotrimers of LT bind to the TNF receptors, the LTα1β2 heterotrimer binds to the LTβ receptor (LTβR) (Ware 2005). (2) B-cell-activating factor of the tumor necrosis factor family (BAFF) and a proliferation-inducing ligand (APRIL) also form heterotrimers (so far discerned only in patients with autoimmune diseases), and these heterotrimers apparently activate only transmembrane activator and calcium-modulating cyclophilin ligand interactor (TACI) (Roschke et al. 2002; Dillon et al. 2006; Schuepbach-Mallepell et al. 2015). (3) Although APRIL is constitutively shed and its trimers therefore occur only in soluble form, its gene also yields a fused joint splice variant with the neighboring *TWEAK* gene. The protein encoded by this transcript—TWE-PRIL—is anchored to the cell membrane (Pradet-Balade et al. 2002). (4) Association of molecules of TRAIL-R2 and the membrane-anchored truncated receptor DcR2 through their preligand assembly domain (PLAD) dictate association of TRAIL with the two receptors in mixed complexes wherein TRAIL-R2 signaling is suppressed by DcR2 (Clancy et al. 2005). (*Legend continues on following page.*)

fiths 1999) and, in the case of their soluble forms, by the proteolytic activity through which these forms are generated. For some family members, this proteolytic activity is exerted constitutively (Lopez-Fraga et al. 2001), whereas for others it is activated by specific signals (Black et al. 1997; Moss et al. 1997; Becker-Pauly and Rose-John 2013). Approximation of receptors to mem-

brane-bound ligands can also be dictated merely by induced juxtapositioning of the cells that express the receptors and ligands.

Some of the ligands accumulate within intracellular reservoirs. Signaling initiation by these ligands is thus dictated by their exocytotic secretion (Gordon and Galli 1990; Bossi and Griffiths 1999; Koguchi et al. 2007).

Figure 4. (*Continued*) (*5*) Molecules of the two receptors for TNF (TNFR1 and TNFR2) reportedly also associate on their binding to ligand. However, this association does not occur through binding of the two receptors to the same ligand molecule. In fact, the two receptor species are incapable of binding simultaneously to the same TNF molecule; molecules of the two receptors rather associate only after binding independently to distinct TNF molecules (Pinckard et al. 1997). (*6*) Molecules of TNFR2 reportedly also associate with molecules of IL-17 receptor D, a member of an unrelated cytokine-receptor family. This association, which occurs on triggering by the ligands of the two receptors, imposes assembly of aggregates of the two receptors and functional cooperation between them (Yang et al. 2015). (*7*) Another example of functional interaction with a structurally unrelated receptor is the association of the ligand for 4-1BB with a coexpressed TLR4–MD2 complex. This apparently occurs through the TMs of 4-1BBL and Tlr4 and the consequent potentiation of lipopolysaccharide (LPS)-induced Tlr4 signaling in a way that depends on the function of the intracellular domain (ID) of 4-1BBL but independently of the association of 4-1BBL with 4-1BB (Kang et al. 2007; Ma et al. 2013). (*8*) A different kind of noncanonical association is observed in the function of herpesvirus entry mediator (HVEM), a receptor of the TNF family. Besides its association with two TNF family ligands (with LIGHT, and [weakly] with LT homotrimers) it also binds to B- and T-lymphocyte attenuator (BTLA) and CD160, two cell-surface proteins of the immunoglobulin superfamily, thereby triggering inhibitory signaling by those two proteins (Shui et al. 2011). (*9*) Whether this association also triggers signaling by the ID of HVEM is not known. (*10*) RANKL, besides binding to RANK and to the soluble receptor osteoprotegerin (OPG), also binds to the seven-transmembrane (TM), leucine-rich repeat containing G protein-coupled receptor 4 (LGR4), which also serves as receptor for R-spondins and for Norrin. It thus triggers signaling antagonistic to that initiated by RANK (Luo et al. 2016). (*11*) Binding of soluble CD40L to integrin αIIbβ3 and to integrin α5β1 triggers signaling by those two membrane–protein complexes (Andre et al. 2002; Leveille et al. 2007). (*12*) Binding of the cell-adhesion lectin, E-selectin, to DR3-linked sialic-acid-linked sugar chains triggers signaling by DR3 (Porquet et al. 2011). Juxtaposition and activation of the two TRAIL receptors, DR4 and DR4, by TRAIL also depends, for a reason not yet clear, on glycosylation of these receptors (Wagner et al. 2007). The most elaborate known set of noncanonical associations is observed in the function of the NGFR, a receptor of the TNF/NGF family for which no ligand of the TNF family is known. NGFR contributes to signaling for different effects in response to different inducers through association with a series of different coreceptors. (*13*) It contributes to signaling for death in response to proneurotrophins, to which it binds in association with a sortilin family receptor. (*14*) When associating with a Trk tyrosine kinase receptor, apparently through the TMs and IDs of both receptors (Esposito et al. 2001), NGFR contributes to high-affinity binding of NGF, and signals for cell survival and for pain sensation. (*15*) NGFR is also found to form, through extracellular domain (ED) associations, a ternary complex with the glycosylphosphatidylinositol (GPI)-linked Nogo-66 receptor (NogoR) and the TM receptor LINGO-1. In response to myelin-associated inhibitory factors (a 66-amino-acid fragment of the oligodendrocyte-derived growth inhibitory protein Nogo, the oligodendrocyte myelin glycoprotein [OMgp] or the myelin-associated glycoprotein [MAG]), this complex signals for arrest of axonal regeneration following injury (Hempstead 2002; Chao 2003; Ibanez and Simi 2012). (*16*) In that complex, NGFR can be replaced by the orphan TNF family receptor TROY (Shao et al. 2005). (*17*) Finally, direct binding of amyloid-β to NGFR (Hempstead 2002; Chao 2003; Ibanez and Simi 2012), and probably also to a complex of NGFR and another orphan receptor of the TNF/NGF family, DR6 (Hu et al. 2013), reportedly triggers signaling for neuronal death. (*18*) DR6 was found to bind to a carboxyl-terminal region in the ED of the amyloid precursor protein (APP). APP and DR6 cooperate in the induction of axonal pruning. The mechanism underlying this cooperation is not clear, nor is it known whether the cooperation occurs between proteins expressed in the same cell, as illustrated in the figure, or in distinct cells (Olsen et al. 2014). BCMA, B-cell maturation antigen.

Increased expression of the receptors or ligands, either artificially (Boldin et al. 1995) or in response to natural stimuli (Lu et al. 2014), is in some situations sufficient to allow the IDs of these molecules to encounter one another and hence to trigger signaling independently of ligand–receptor association.

CONCLUDING THEOLOGICAL REMARKS

Members of the TNF cytokine family and of the TNF/NGFR family are used by almost all metazoan phyla. Some of their activities have been well preserved over more than 500 million years of evolution (Igaki and Miura 2014; Quistad et al. 2014). Their major proximal signaling enzymes originated even earlier (Uren et al. 2000; Zapata et al. 2007; Yuan et al. 2009; Zmasek and Godzik 2013; Sakamaki et al. 2014). Nature, it would seem, had good reason to preserve this cytokine family so well and for so long. At face value, however, the known activities of this family do not seem to be sufficiently unique to warrant such preservation. Other than in the case of the cytotoxic activity of a few family members, the various individual cellular effects of the TNF family and the signaling mechanisms that account for them do not seem to differ radically from those of various other cytokines. The pattern of cellular effects of IL-1 in inflammation, for example, as well as the signaling mechanisms that it activates, greatly resemble those of TNF. Why would nature preserve two cytokines that seem to serve the same set of functions, and why would it choose to use them differentially in different situations?

Also puzzling is the remarkable similarity among the ranges of signaling activities triggered by the various family members. How can the heterogeneous and distinct biological activity patterns of the different family members be explained in terms of a set of mechanisms so limited and so invariant? These enigmas suggest that the mechanisms of action of the TNF families possess a greater degree of sophistication than has so far met the eye.

An example of the insight gained in attempting to deepen our perception of this area was the discovery that, although all TNF family members signal for activation of NF-κB, some are able to do so via a distinct route, the so-called "alternative pathway," which yields molecular targets and functional consequences that differ from those of the more widely used "canonical pathways" (Hayden and Ghosh 2012). The alternative pathway, and the protein kinase NF-κB-inducing kinase (NIK) that initiates it, appear to have evolved relatively late in the history of the TNF family, contemporaneously with emergence of the vertebrates and of their adaptive immunity that this pathway regulates. Other signaling pathways shared by different TNF family members may likewise be found to have evolved into several different forms that are affected differentially by these different members and serve distinct functions.

The various "noncanonical" interactions of the TNF and TNF/NGF families (Fig. 4) doubtless also endow individual members of the family with unique mechanisms of action. An example of a unique mechanism that might depend on such noncanonical interactions is presented in Figure 1C. As opposed to the usual mode of signaling initiation by the TNF/NGF family (by induced juxtaposition), NGFR—which participates in several such noncanonical associations—was suggested rather to trigger signaling by imposing separation of the receptor IDs. This mechanism is dictated by covalent linkage of a conserved cysteine in the TM of NGFR. Conserved cysteines also occur in the TDs of several other TNF/NGF family members. It was therefore suggested that portions of these other receptors also occur as covalently linked dimers and serve similarly to mediate unique functions, which likewise may depend on some noncanonical associations (Vilar et al. 2009).

We have come a long way in clarifying some general mechanisms of action of the various families of cytokines, including the TNF family. It is now time to focus on clarifying the mechanisms that endow each family, and each family member, with uniqueness. Progress in this regard will undoubtedly increase our ability to design selective therapeutic modulation of specific functions of these fascinating molecules.

Cite this article as *Cold Spring Harb Perspect Biol* doi: 10.1101/cshperspect.a028431

REFERENCES

Adam Klages S, Adam D, Wiegmann K, Struve S, Kolanus W, Schneider Mergener J, Kronke M. 1996. FAN, a novel WD-repeat protein, couples the p55 TNF-receptor to neutral sphingomyelinase. *Cell* **86:** 937–947.

Aderka D, Engelmann H, Maor Y, Brakebusch C, Wallach D. 1992. Stabilization of the bioactivity of tumor necrosis factor by its soluble receptors. *J Exp Med* **175:** 323–329.

Akassoglou K, Bauer J, Kassiotis G, Lassmann H, Kollias G, Probert L. 1999. Transgenic models of TNF induced demyelination. *Adv Exp Med Biol* **468:** 245–259.

Albanese J, Meterissian S, Kontogiannea M, Dubreuil C, Hand A, Sorba S, Dainiak N. 1998. Biologically active Fas antigen and its cognate ligand are expressed on plasma membrane-derived extracellular vesicles. *Blood* **91:** 3862–3874.

Alon R, Cahalon L, Hershkoviz R, Elbaz D, Reizis B, Wallach D, Akiyama SK, Yamada KM, Lider O. 1994. TNF-α binds to the N-terminal domain of fibronectin and augments the β1-integrin-mediated adhesion of CD4+ T lymphocytes to the glycoprotein. *J Immunol* **152:** 1304–1313.

Andre P, Prasad KS, Denis CV, He M, Papalia JM, Hynes RO, Phillips DR, Wagner DD. 2002. CD40L stabilizes arterial thrombi by a β3 integrin-dependent mechanism. *Nat Med* **8:** 247–252.

Aoki K, Kurooka M, Chen JJ, Petryniak J, Nabel EG, Nabel GJ. 2001. Extracellular matrix interacts with soluble CD95L: Retention and enhancement of cytotoxicity. *Nat Immunol* **2:** 333–337.

Apostolaki M, Armaka M, Victoratos P, Kollias G. 2010. Cellular mechanisms of TNF function in models of inflammation and autoimmunity. *Curr Dir Autoimmun* **11:** 1–26.

Arens R, Nolte MA, Tesselaar K, Heemskerk B, Reedquist KA, van Lier RA, van Oers MH. 2004. Signaling through CD70 regulates B cell activation and IgG production. *J Immunol* **173:** 3901–3908.

Becker-Pauly C, Rose-John S. 2013. TNFα cleavage beyond TACE/ADAM17: Matrix metalloproteinase 13 is a potential therapeutic target in sepsis and colitis. *EMBO Mol Med* **5:** 902–904.

Bischof D, Elsawa SF, Mantchev G, Yoon J, Michels GE, Nilson A, Sutor SL, Platt JL, Ansell SM, von Bulow G, et al. 2006. Selective activation of TACI by syndecan-2. *Blood* **107:** 3235–3242.

Bishop GA, Hostager BS. 2003. The CD40–CD154 interaction in B cell–T cell liaisons. *Cytokine Growth Factor Rev* **14:** 297–309.

Black RA, Rauch CT, Kozlosky CJ, Peschon JJ, Slack JL, Wolfson MF, Castner BJ, Stocking KL, Reddy P, Srinivasan S, et al. 1997. A metalloproteinase disintegrin that releases tumour-necrosis factor-α from cells. *Nature* **385:** 729–733.

Blomberg J, Ruuth K, Santos D, Lundgren E. 2008. Acquired resistance to Fas/CD95 ligation in U937 cells is associated with multiple molecular mechanisms. *Anticancer Res* **28:** 593–599.

Boldin MP, Mett IL, Varfolomeev EE, Chumakov I, Shemer AY, Camonis JH, Wallach D. 1995. Self-association of the "death domains" of the p55 tumor necrosis factor (TNF)

receptor and Fas/APO1 prompts signaling for TNF and Fas/APO1 effects. *J Biol Chem* **270:** 387–391.

Bossi G, Griffiths GM. 1999. Degranulation plays an essential part in regulating cell surface expression of Fas ligand in T cells and natural killer cells. *Nat Med* **5:** 90–96.

Boyce BF, Xing L. 2007. Biology of RANK, RANKL, and osteoprotegerin. *Arthritis Res Ther* **9:** S1.

Brakebusch C, Varfolomeev EE, Batkin M, Wallach D. 1994. Structural requirements for inducible shedding of the p55 tumor necrosis factor receptor. *J Biol Chem* **269:** 32488–32496.

Bruce AJ, Boling W, Kindy MS, Peschon J, Kraemer PJ, Carpenter MK, Holtsberg FW, Mattson MP. 1996. Altered neuronal and microglial responses to excitotoxic and ischemic brain injury in mice lacking TNF receptors. *Nat Med* **2:** 788–794.

Burkly LC. 2014. TWEAK/Fn14 axis: The current paradigm of tissue injury-inducible function in the midst of complexities. *Semin Immunol* **26:** 229–236.

Cachero TG, Schwartz IM, Qian F, Day ES, Bossen C, Ingold K, Tardivel A, Krushinskie D, Eldredge J, Silvian L, et al. 2006. Formation of virus-like clusters is an intrinsic property of the tumor necrosis factor family member BAFF (B-cell-activating factor). *Biochemistry* **45:** 2006–2013.

Cerami A, Ikeda Y, Le Trang N, Hotez PJ, Beutler B. 1985. Weight loss associated with an endotoxin-induced mediator from peritoneal macrophages: The role of cachectin (tumor necrosis factor). *Immunol Lett* **11:** 173–177.

Chakrabandhu K, Herincs Z, Huault S, Dost B, Peng L, Conchonaud F, Marguet D, He HT, Hueber AO. 2007. Palmitoylation is required for efficient Fas cell death signaling. *EMBO J* **26:** 209–220.

Chakrabandhu K, Huault S, Durivault J, Lang K, Ta Ngoc L, Bole A, Doma E, Derijard B, Gerard JP, Pierres M, et al. 2016. An evolution-guided analysis reveals a multi-signaling regulation of Fas by tyrosine phosphorylation and its implication in human cancers. *PLoS Biol* **14:** e1002401.

Chan FK. 2007. Three is better than one: Pre-ligand receptor assembly in the regulation of TNF receptor signaling. *Cytokine* **37:** 101–107.

Chang YC, Chan YH, Jackson DG, Hsieh SL. 2006. The glycosaminoglycan-binding domain of decoy receptor 3 is essential for induction of monocyte adhesion. *J Immunol* **176:** 173–180.

Chao MV. 2003. Neurotrophins and their receptors: A convergence point for many signalling pathways. *Nat Rev Neurosci* **4:** 299–309.

Chen L, Park SM, Tumanov AV, Hau A, Sawada K, Feig C, Turner JR, Fu YX, Romero IL, Lengyel E, et al. 2010. CD95 promotes tumour growth. *Nature* **465:** 492–496.

Chhibber-Goel J, Coleman-Vaughan C, Agrawal V, Sawhney N, Hickey E, Powell JC, McCarthy JV. 2016. γ-Secretase activity is required for regulated intramembrane proteolysis of tumor necrosis factor (TNF) receptor 1 and TNF-mediated pro-apoptotic signaling. *J Biol Chem* **291:** 5971–5985.

Clancy L, Mruk K, Archer K, Woelfel M, Mongkolsapaya J, Screaton G, Lenardo MJ, Chan FK. 2005. Preligand assembly domain-mediated ligand-independent association between TRAIL receptor 4 (TR4) and TR2 regulates

TRAIL-induced apoptosis. *Proc Natl Acad Sci* **102:** 18099–18104.

Corsini NS, Sancho-Martinez I, Laudenklos S, Glagow D, Kumar S, Letellier E, Koch P, Teodorczyk M, Kleber S, Klussmann S, et al. 2009. The death receptor CD95 activates adult neural stem cells for working memory formation and brain repair. *Cell Stem Cell* **5:** 178–190.

Couchman JR. 2003. Syndecans: Proteoglycan regulators of cell-surface microdomains? *Nat Rev Mol Cell Biol* **4:** 926–937.

Croft M. 2014. The TNF family in T cell differentiation and function—Unanswered questions and future directions. *Semin Immunol* **26:** 183–190.

Dantzer R. 2001. Cytokine-induced sickness behavior: Mechanisms and implications. *Ann NY Acad Sci* **933:** 222–234.

Dillon SR, Gross JA, Ansell SM, Novak AJ. 2006. An APRIL to remember: Novel TNF ligands as therapeutic targets. *Nat Rev Drug Discov* **5:** 235–246.

Dinarello CA, Cannon JG, Wolff SM, Bernheim HA, Beutler B, Cerami A, Figari IS, Palladino MA Jr, O'Connor JV. 1986. Tumor necrosis factor (cachectin) is an endogenous pyrogen and induces production of interleukin 1. *J Exp Med* **163:** 1433–1450.

Domonkos A, Udvardy A, Laszlo L, Nagy T, Duda E. 2001. Receptor-like properties of the 26 kDa transmembrane form of TNF. *Eur Cytokine Netw* **12:** 411–419.

Drayton DL, Liao S, Mounzer RH, Ruddle NH. 2006. Lymphoid organ development: From ontogeny to neogenesis. *Nat Immunol* **7:** 344–353.

Eissner G, Kolch W, Scheurich P. 2004. Ligands working as receptors: Reverse signaling by members of the TNF superfamily enhance the plasticity of the immune system. *Cytokine Growth Factor Rev* **15:** 353–366.

Elgueta R, Benson MJ, de Vries VC, Wasiuk A, Guo Y, Noelle RJ. 2009. Molecular mechanism and function of CD40/CD40L engagement in the immune system. *Immunol Rev* **229:** 152–172.

Eliaz R, Wallach D, Kost J. 1996. Long-term protection against tumor necrosis factor (TNF) effects by controlled delivery of the soluble p55 tumor necrosis factor receptor. *Cytokine* **8:** 482–487.

Elmetwali T, Young LS, Palmer DH. 2010. CD40 ligand-induced carcinoma cell death: A balance between activation of TNFR-associated factor (TRAF) 3-dependent death signals and suppression of TRAF6-dependent survival signals. *J Immunol* **184:** 1111–1120.

Engelmann H, Novick D, Wallach D. 1990. Two tumor necrosis factor-binding proteins purified from human urine. Evidence for immunological cross-reactivity with cell surface tumor necrosis factor receptors. *J Biol Chem* **265:** 1531–1536.

Esposito D, Patel P, Stephens RM, Perez P, Chao MV, Kaplan DR, Hempstead BL. 2001. The cytoplasmic and transmembrane domains of the p75 and Trk A receptors regulate high affinity binding to nerve growth factor. *J Biol Chem* **276:** 32687–32695.

Falschlehner C, Schaefer U, Walczak H. 2009. Following TRAIL's path in the immune system. *Immunology* **127:** 145–154.

Feig C, Tchikov V, Schutze S, Peter ME. 2007. Palmitoylation of CD95 facilitates formation of SDS-stable receptor aggregates that initiate apoptosis signaling. *EMBO J* **26:** 221–231.

Figgett WA, Vincent FB, Saulep-Easton D, Mackay F. 2014. Roles of ligands from the TNF superfamily in B cell development, function, and regulation. *Semin Immunol* **26:** 191–202.

Fluhrer R, Grammer G, Israel L, Condron MM, Haffner C, Friedmann E, Bohland C, Imhof A, Martoglio B, Teplow DB, et al. 2006. A γ-secretase-like intramembrane cleavage of TNFα by the GxGD aspartyl protease SPPL2b. *Nat Cell Biol* **8:** 894–896.

Fontaine V, Mohand-Said S, Hanoteau N, Fuchs C, Pfizenmaier K, Eisel U. 2002. Neurodegenerative and neuroprotective effects of tumor necrosis factor (TNF) in retinal ischemia: Opposite roles of TNF receptor 1 and TNF receptor 2. *J Neurosci* **22:** RC216.

Force WR, Glass AA, Benedict CA, Cheung TC, Lama J, Ware CF. 2000. Discrete signaling regions in the lymphotoxin-β receptor for tumor necrosis factor receptor-associated factor binding, subcellular localization, and activation of cell death and NF-κB pathways. *J Biol Chem* **275:** 11121–11129.

Friedmann E, Hauben E, Maylandt K, Schleeger S, Vreugde S, Lichtenthaler SF, Kuhn PH, Stauffer D, Rovelli G, Martoglio B. 2006. SPPL2a and SPPL2b promote intramembrane proteolysis of TNFα in activated dendritic cells to trigger IL-12 production. *Nat Cell Biol* **8:** 843–848.

Fritsch J, Stephan M, Tchikov V, Winoto-Morbach S, Gubkina S, Kabelitz D, Schutze S. 2014. Cell fate decisions regulated by K63 ubiquitination of tumor necrosis factor receptor 1. *Mol Cell Biol* **34:** 3214–3228.

Fu Q, Fu TM, Cruz AC, Sengupta P, Thomas SK, Wang S, Siegel RM, Wu H, Chou JJ. 2016. Structural basis and functional role of intramembrane trimerization of the Fas/CD95 death receptor. *Mol Cell* **61:** 602–613.

Ganeff C, Remouchamps C, Boutaffala L, Benezech C, Galopin G, Vandepaer S, Bouillenne F, Ormenese S, Chariot A, Schneider P, et al. 2011. Induction of the alternative NF-κB pathway by lymphotoxin αβ (LTαβ) relies on internalization of LTβ receptor. *Mol Cell Biol* **31:** 4319–4334.

Georgopoulos NT, Steele LP, Thomson MJ, Selby PJ, Southgate J, Trejdosiewicz LK. 2006. A novel mechanism of CD40-induced apoptosis of carcinoma cells involving TRAF3 and JNK/AP-1 activation. *Cell Death Differ* **13:** 1789–1801.

Gordon JR, Galli SJ. 1990. Mast cells as a source of both preformed and immunologically inducible TNF-α/cachectin. *Nature* **346:** 274–276.

Granger GA, Kolb WP. 1968. Lymphocyte in vitro cytotoxicity: Mechanisms of immune and non-immune small lymphocyte mediated target L cell destruction. *J Immunol* **101:** 111–120.

Grell M, Douni E, Wajant H, Löhden M, Clauss M, Baxeiner B, Georgopoulos S, Lesslauer W, Kollias G, Pfizenmaier K, et al. 1995. The transmembrane form of tumor necrosis factor is the prime activating ligand of the 80 kDa tumor necrosis factor receptor. *Cell* **83:** 793–802.

Grell M, Zimmermann G, Gottfried E, Chen CM, Grunwald U, Huang DC, Wu Lee YH, Durkop H, Engelmann H,

Scheurich P, et al. 1999. Induction of cell death by tumour necrosis factor (TNF) receptor 2, CD40 and CD30: A role for TNF-R1 activation by endogenous membrane-anchored TNF. *EMBO J* **18:** 3034–3043.

Grohmann U, Volpi C, Fallarino F, Bozza S, Bianchi R, Vacca C, Orabona C, Belladonna ML, Ayroldi E, Nocentini G, et al. 2007. Reverse signaling through GITR ligand enables dexamethasone to activate IDO in allergy. *Nat Med* **13:** 579–586.

Hahn T, Toker L, Budilovsky S, Aderka D, Eshhar Z, Wallach D. 1985. Use of monoclonal antibodies to a human cytotoxin for its isolation and for examining the self-induction of resistance to this protein. *Proc Natl Acad Sci* **82:** 3814–3818.

Hanada R, Leibbrandt A, Hanada T, Kitaoka S, Furuyashiki T, Fujihara H, Trichereau J, Paolino M, Qadri F, Plehm R, et al. 2009. Central control of fever and female body temperature by RANKL/RANK. *Nature* **462:** 505–509.

Hayashida K, Bartlett AH, Chen Y, Park PW. 2010. Molecular and cellular mechanisms of ectodomain shedding. *Anat Rec (Hoboken)* **293:** 925–937.

Hayden MS, Ghosh S. 2012. NF-κB, the first quarter-century: Remarkable progress and outstanding questions. *Genes Dev* **26:** 203–234.

Hempstead BL. 2002. The many faces of p75NTR. *Curr Opin Neurobiol* **12:** 260–267.

Hess A, Axmann R, Rech J, Finzel S, Heindl C, Kreitz S, Sergeeva M, Saake M, Garcia M, Kollias G, et al. 2011. Blockade of TNF-α rapidly inhibits pain responses in the central nervous system. *Proc Natl Acad Sci* **108:** 3731–3736.

Holler N, Tardivel A, Kovacsovics-Bankowski M, Hertig S, Gaide O, Martinon F, Tinel A, Deperthes D, Calderara S, Schulthess T, et al. 2003. Two adjacent trimeric Fas ligands are required for Fas signaling and formation of a death-inducing signaling complex. *Mol Cell Biol* **23:** 1428–1440.

Hu Y, Lee X, Shao Z, Apicco D, Huang G, Gong BJ, Pepinsky RB, Mi S. 2013. A DR6/p75(NTR) complex is responsible for β-amyloid-induced cortical neuron death. *Cell Death Dis* **4:** e579.

Huard B, McKee T, Bosshard C, Durual S, Matthes T, Myit S, Donze O, Frossard C, Chizzolini C, Favre C, et al. 2008. APRIL secreted by neutrophils binds to heparan sulfate proteoglycans to create plasma cell niches in human mucosa. *J Clin Invest* **118:** 2887–2895.

Ibanez CF, Simi A. 2012. p75 neurotrophin receptor signaling in nervous system injury and degeneration: Paradox and opportunity. *Trends Neurosci* **35:** 431–440.

Igaki T, Miura M. 2014. The *Drosophila* TNF ortholog Eiger: Emerging physiological roles and evolution of the TNF system. *Semin Immunol* **26:** 267–274.

Ingold K, Zumsteg A, Tardivel A, Huard B, Steiner QG, Cachero TG, Qiang F, Gorelik L, Kalled SL, Acha-Orbea H, et al. 2005. Identification of proteoglycans as the APRIL-specific binding partners. *J Exp Med* **201:** 1375–1383.

Islam A, Shen X, Hiroi T, Moss J, Vaughan M, Levine SJ. 2007. The brefeldin A-inhibited guanine nucleotide-exchange protein, BIG2, regulates the constitutive release of TNFR1 exosome-like vesicles. *J Biol Chem* **282:** 9591–9599.

Jeon YJ, Middleton J, Kim T, Lagana A, Piovan C, Secchiero P, Nuovo GJ, Cui R, Joshi P, Romano G, et al. 2015. A set of NF-κB-regulated microRNAs induces acquired TRAIL resistance in lung cancer. *Proc Natl Acad Sci* **112:** E3355–E3364.

Johnson D, Lanahan A, Buck C R, Sehgal A, Morgan C, Mercer E, Bothwell M, Chao M. 1986. Expression and structure of the human NGF receptor. *Cell* **47:** 545–554.

Juhasz K, Buzas K, Duda E. 2013. Importance of reverse signaling of the TNF superfamily in immune regulation. *Exp Rev Clin Immunol* **9:** 335–348.

Kang YJ, Kim SO, Shimada S, Otsuka M, Seit-Nebi A, Kwon BS, Watts TH, Han J. 2007. Cell surface 4-1BBL mediates sequential signaling pathways "downstream" of TLR and is required for sustained TNF production in macrophages. *Nat Immunol* **8:** 601–609.

Kanning KC, Hudson M, Amieux PS, Wiley JC, Bothwell M, Schecterson LC. 2003. Proteolytic processing of the p75 neurotrophin receptor and two homologs generates C-terminal fragments with signaling capability. *J Neurosci* **23:** 5425–5436.

Kenchappa RS, Zampieri N, Chao MV, Barker PA, Teng HK, Hempstead BL, Carter BD. 2006. Ligand-dependent cleavage of the P75 neurotrophin receptor is necessary for NRIF nuclear translocation and apoptosis in sympathetic neurons. *Neuron* **50:** 219–232.

Kersse K, Verspurten J, Vanden Berghe T, Vandenabeele P. 2011. The death-fold superfamily of homotypic interaction motifs. *Trends Biochem Sci* **36:** 541–552.

Kimberley FC, van Bostelen L, Cameron K, Hardenberg G, Marquart JA, Hahne M, Medema JP. 2009. The proteoglycan (heparan sulfate proteoglycan) binding domain of APRIL serves as a platform for ligand multimerization and cross-linking. *FASEB J* **23:** 1584–1595.

Kimura A, Naka T, Nagata S, Kawase I, Kishimoto T. 2004. SOCS-1 suppresses TNF-α-induced apoptosis through the regulation of Jak activation. *Int Immunol* **16:** 991–999.

Kindler V, Sappino AP, Grau GE, Piguet PF, Vassalli P. 1989. The inducing role of tumor necrosis factor in the development of bactericidal granulomas during BCG infection. *Cell* **56:** 731–740.

Kirkin V, Cahuzac N, Guardiola-Serrano F, Huault S, Luckerath K, Friedmann E, Novac N, Wels WS, Martoglio B, Hueber AO, et al. 2007. The Fas ligand intracellular domain is released by ADAM10 and SPPL2a cleavage in T-cells. *Cell Death Differ* **14:** 1678–1687.

Koguchi Y, Thauland TJ, Slifka MK, Parker DC. 2007. Preformed CD40 ligand exists in secretory lysosomes in effector and memory CD4⁺ T cells and is quickly expressed on the cell surface in an antigen-specific manner. *Blood* **110:** 2520–2527.

Kojima T, Morikawa Y, Copeland NG, Gilbert DJ, Jenkins NA, Senba E, Kitamura T. 2000. TROY, a newly identified member of the tumor necrosis factor receptor superfamily, exhibits a homology with Edar and is expressed in embryonic skin and hair follicles. *J Biol Chem* **275:** 20742–20747.

Kriegler M, Perez C, DeFay K, Albert I, Lu SD. 1988. A novel form of TNF/cachectin is a cell surface cytotoxic transmembrane protein: Ramifications for the complex physiology of TNF. *Cell* **53:** 45–53.

Lefebvre S, Mikkola ML. 2014. Ectodysplasin research—Where to next? *Semin Immunol* **26**: 220–228.

Legler DF, Micheau O, Doucey MA, Tschopp J, Bron C. 2003. Recruitment of TNF receptor 1 to lipid rafts is essential for TNFα-mediated NF-κB activation. *Immunity* **18**: 655–664.

Leveille C, Bouillon M, Guo W, Bolduc J, Sharif-Askari E, El-Fakhry Y, Reyes-Moreno C, Lapointe R, Merhi Y, Wilkins JA, et al. 2007. CD40 ligand binds to α5β1 integrin and triggers cell signaling. *J Biol Chem* **282**: 5143–5151.

Li P, Schwarz EM, O'Keefe RJ, Ma L, Looney RJ, Ritchlin CT, Boyce BF, Xing L. 2004. Systemic tumor necrosis factor α mediates an increase in peripheral CD11b^high osteoclast precursors in tumor necrosis factor α-transgenic mice. *Arthritis Rheum* **50**: 265–276.

Li Y, Toraldo G, Li A, Yang X, Zhang H, Qian WP, Weitzmann MN. 2007. B cells and T cells are critical for the preservation of bone homeostasis and attainment of peak bone mass in vivo. *Blood* **109**: 3839–3848.

Li J, Yin Q, Wu H. 2013. Structural basis of signal transduction in the TNF receptor superfamily. *Adv Immunol* **119**: 135–153.

Liu Y, Xu L, Opalka N, Kappler J, Shu HB, Zhang G. 2002. Crystal structure of sTALL-1 reveals a virus-like assembly of TNF family ligands. *Cell* **108**: 383–394.

Lopez-Fraga M, Fernandez R, Albar JP, Hahne M. 2001. Biologically active APRIL is secreted following intracellular processing in the Golgi apparatus by furin convertase. *EMBO Rep* **2**: 945–951.

Lu M, Lawrence DA, Marsters S, Acosta-Alvear D, Kimmig P, Mendez AS, Paton AW, Paton JC, Walter P, Ashkenazi A. 2014. Opposing unfolded-protein-response signals converge on death receptor 5 to control apoptosis. *Science* **345**: 98–101.

Luo J, Yang Z, Ma Y, Yue Z, Lin H, Qu G, Huang J, Dai W, Li C, Zheng C, et al. 2016. LGR4 is a receptor for RANKL and negatively regulates osteoclast differentiation and bone resorption. *Nat Med* **22**: 539–546.

Ma J, Bang BR, Lu J, Eun SY, Otsuka M, Croft M, Tobias P, Han J, Takeuchi O, Akira S, et al. 2013. The TNF family member 4-1BBL sustains inflammation by interacting with TLR signaling components during late-phase activation. *Sci Signal* **6**: ra87.

Martinez-Lorenzo MJ, Anel A, Gamen S, Monle NI, Lasierra P, Larrad L, Pineiro A, Alava MA, Naval J. 1999. Activated human T cells release bioactive Fas ligand and APO2 ligand in microvesicles. *J Immunol* **163**: 1274–1281.

Mbanwi AN, Watts TH. 2014. Costimulatory TNFR family members in control of viral infection: Outstanding questions. *Semin Immunol* **26**: 210–219.

McCarthy DD, Summers-Deluca L, Vu F, Chiu S, Gao Y, Gommerman JL. 2006. The lymphotoxin pathway: Beyond lymph node development. *Immunol Res* **35**: 41–54.

McWhirter SM, Pullen SS, Holton JM, Crute JJ, Kehry MR, Alber T. 1999. Crystallographic analysis of CD40 recognition and signaling by human TRAF2. *Proc Natl Acad Sci* **96**: 8408–8413.

Moreaux J, Sprynski AC, Dillon SR, Mahtouk K, Jourdan M, Ythier A, Moine P, Robert N, Jourdan E, Rossi JF, et al. 2009. APRIL and TACI interact with syndecan-1 on the surface of multiple myeloma cells to form an essential survival loop. *Eur J Haematol* **83**: 119–129.

Mosheimer BA, Kaneider NC, Feistritzer C, Djanani AM, Sturn DH, Patsch JR, Wiedermann CJ. 2005. Syndecan-1 is involved in osteoprotegerin-induced chemotaxis in human peripheral blood monocytes. *J Clin Endocrinol Metab* **90**: 2964–2971.

Moss ML, Jin SL, Milla ME, Bickett DM, Burkhart W, Carter HL, Chen WJ, Clay WC, Didsbury JR, Hassler D, et al. 1997. Cloning of a disintegrin metalloproteinase that processes precursor tumour-necrosis factor-α. *Nature* **385**: 733–736.

Muhlenbeck F, Schneider P, Bodmer JL, Schwenzer R, Hauser A, Schubert G, Scheurich P, Moosmayer D, Tschopp J, Wajant H. 2000. The tumor necrosis factor-related apoptosis-inducing ligand receptors TRAIL-R1 and TRAIL-R2 have distinct cross-linking requirements for initiation of apoptosis and are non-redundant in JNK activation. *J Biol Chem* **275**: 32208–32213.

Muppidi JR, Tschopp J, Siegel RM. 2004. Life and death decisions: Secondary complexes and lipid rafts in TNF receptor family signal transduction. *Immunity* **21**: 461–465.

Nadiri A, Jundi M, El Akoum S, Hassan GS, Yacoub D, Mourad W. 2015. Involvement of the cytoplasmic cysteine-238 of CD40 in its up-regulation of CD23 expression and its enhancement of TLR4-triggered responses. *Int Immunol* **27**: 555–565.

Naismith JH, Devine TQ, Brandhuber BJ, Sprang SR. 1995. Crystallographic evidence for dimerization of unliganded tumor necrosis factor receptor. *J Biol Chem* **270**: 13303–13307.

Nikolaev A, McLaughlin T, O'Leary DD, Tessier-Lavigne M. 2009. APP binds DR6 to trigger axon pruning and neuron death via distinct caspases. *Nature* **457**: 981–989.

Nophar Y, Kemper O, Brakebusch C, Engelmann H, Zwang R, Aderka D, Holtmann H, Wallach D. 1990. Soluble forms of tumor necrosis factor receptors (TNF-Rs). The cDNA for the type I TNF-R cloned using amino acid data of its soluble form, encodes for both the cell surface and a soluble form of the receptor. *EMBO J* **9**: 3269–3278.

Old LJ. 1985. Tumor necrosis factor (TNF). *Science* **230**: 630–632.

Olsen O, Kallop DY, McLaughlin T, Huntwork-Rodriguez S, Wu Z, Duggan CD, Simon DJ, Lu Y, Easley-Neal C, Takeda K, et al. 2014. Genetic analysis reveals that amyloid precursor protein and death receptor 6 function in the same pathway to control axonal pruning independent of β-secretase. *J Neurosci* **34**: 6438–6447.

O'Reilly LA, Tai L, Lee L, Kruse EA, Grabow S, Fairlie WD, Haynes NM, Tarlinton DM, Zhang JG, Belz GT, et al. 2009. Membrane-bound Fas ligand only is essential for Fas-induced apoptosis. *Nature* **461**: 659–663.

Park YC, Burkitt V, Villa AR, Tong L, Wu H. 1999. Structural basis for self-association and receptor recognition of human TRAF2. *Nature* **398**: 533–538.

Pinckard JK, Sheehan KC, Schreiber RD. 1997. Ligand-induced formation of p55 and p75 tumor necrosis factor receptor heterocomplexes on intact cells. *J Biol Chem* **272**: 10784–10789.

Plata-Salaman CR, Oomura Y, Kai Y. 1988. Tumor necrosis factor and interleukin-1β: Suppression of food intake by direct action in the central nervous system. *Brain Res* **448:** 106–114.

Pocsik E, Duda E, Wallach D. 1995. Phosphorylation of the 26 kDa TNF precursor in monocytic cells and in transfected HeLa cells. *J Inflamm* **45:** 152–160.

Poggi M, Kara I, Brunel JM, Landrier JF, Govers R, Bonardo B, Fluhrer R, Haass C, Alessi MC, Peiretti F. 2013. Palmitoylation of TNFα is involved in the regulation of TNF receptor 1 signalling. *Biochim Biophys Acta* **1833:** 602–612.

Porquet N, Poirier A, Houle F, Pin AL, Gout S, Tremblay PL, Paquet ER, Klinck R, Auger FA, Huot J. 2011. Survival advantages conferred to colon cancer cells by E-selectin-induced activation of the PI3K-NF-κB survival axis downstream of death receptor-3. *BMC Cancer* **11:** 285.

Pradet-Balade B, Medema JP, Lopez-Fraga M, Lozano JC, Kolfschoten GM, Picard A, Martinez AC, Garcia-Sanz JA, Hahne M. 2002. An endogenous hybrid mRNA encodes TWE-PRIL, a functional cell surface TWEAK-APRIL fusion protein. *EMBO J* **21:** 5711–5720.

Quistad SD, Stotland A, Barott KL, Smurthwaite CA, Hilton BJ, Grasis JA, Wolkowicz R, Rohwer FL. 2014. Evolution of TNF-induced apoptosis reveals 550 My of functional conservation. *Proc Natl Acad Sci* **111:** 9567–9572.

Radeke MJ, Misko TP, Hsu C, Herzenberg LA, Shooter EM. 1987. Gene transfer and molecular cloning of the rat nerve growth factor receptor. *Nature* **325:** 593–597.

Richard AC, Ferdinand JR, Meylan F, Hayes ET, Gabay O, Siegel RM. 2015. The TNF-family cytokine TL1A: From lymphocyte costimulator to disease co-conspirator. *J Leukoc Biol* **98:** 333–345.

Roschke V, Sosnovtseva S, Ward CD, Hong JS, Smith R, Albert V, Stohl W, Baker KP, Ullrich S, Nardelli B, et al. 2002. BLyS and APRIL form biologically active heterotrimers that are expressed in patients with systemic immune-based rheumatic diseases. *J Immunol* **169:** 4314–4321.

Rossin A, Derouet M, Abdel-Sater F, Hueber AO. 2009. Palmitoylation of the TRAIL receptor DR4 confers an efficient TRAIL-induced cell death signalling. *Biochem J* **419:** 185–192.

Ruddle NH, Waksman BH. 1968. Cytotoxicity mediated by soluble antigen and lymphocytes in delayed hypersensitivity. III: Analysis of mechanisms. *J Exp Med* **128:** 1267–1279.

Sabio G, Davis RJ. 2014. TNF and MAP kinase signalling pathways. *Semin Immunol* **26:** 237–245.

Sakamaki K, Shimizu K, Iwata H, Imai K, Satou Y, Funayama N, Nozaki M, Yajima M, Nishimura O, Higuchi M, et al. 2014. The apoptotic initiator caspase-8: Its functional ubiquity and genetic diversity during animal evolution. *Mol Biol Evol* **31:** 3282–3301.

Schneeweis LA, Willard D, Milla ME. 2005. Functional dissection of osteoprotegerin and its interaction with receptor activator of NF-κB ligand. *J Biol Chem* **280:** 41155–41164.

Schneider P, Holler N, Bodmer JL, Hahne M, Frei K, Fontana A, Tschopp J. 1998. Conversion of membrane-bound Fas (CD95) ligand to its soluble form is associated with downregulation of its proapoptotic activity and loss of liver toxicity. *J Exp Med* **187:** 1205–1213.

Schuepbach-Mallepell S, Das D, Willen L, Vigolo M, Tardivel A, Lebon L, Kowalczyk-Quintas C, Nys J, Smulski C, Zheng TS, et al. 2015. Stoichiometry of heteromeric BAFF and APRIL cytokines dictates their receptor binding and signaling properties. *J Biol Chem* **290:** 16330–16342.

Schutze S, Tchikov V, Schneider-Brachert W. 2008. Regulation of TNFR1 and CD95 signalling by receptor compartmentalization. *Nat Rev Mol Cell Biol* **9:** 655–662.

Scott FL, Stec B, Pop C, Dobaczewska MK, Lee JJ, Monosov E, Robinson H, Salvesen GS, Schwarzenbacher R, Riedl SJ. 2009. The Fas-FADD death domain complex structure unravels signalling by receptor clustering. *Nature* **457:** 1019–1022.

Sessler T, Healy S, Samali A, Szegezdi E. 2013. Structural determinants of DISC function: New insights into death receptor-mediated apoptosis signalling. *Pharmacol Therapeut* **140:** 186–199.

Sfikakis PP. 2010. The first decade of biologic TNF antagonists in clinical practice: Lessons learned, unresolved issues and future directions. *Curr Dir Autoimmun* **11:** 180–210.

Shao Z, Browning JL, Lee X, Scott ML, Shulga-Morskaya S, Allaire N, Thill G, Levesque M, Sah D, McCoy JM, et al. 2005. TAJ/TROY, an orphan TNF receptor family member, binds Nogo-66 receptor 1 and regulates axonal regeneration. *Neuron* **45:** 353–359.

Shoham S, Davenne D, Cady AB, Dinarello CA, Krueger JM. 1987. Recombinant tumor necrosis factor and interleukin 1 enhance slow-wave sleep. *Am J Physiol* **253:** R142–R149.

Shohami E, Ginis I, Hallenbeck JM. 1999. Dual role of tumor necrosis factor α in brain injury. *Cytokine Growth Factor Rev* **10:** 119–130.

Shui JW, Steinberg MW, Kronenberg M. 2011. Regulation of inflammation, autoimmunity, and infection immunity by HVEM-BTLA signaling. *J Leukoc Biol* **89:** 517–523.

So T, Croft M. 2013. Regulation of PI-3-kinase and Akt signaling in T lymphocytes and other cells by TNFR family molecules. *Front Immunol* **4:** 139.

Standal T, Seidel C, Hjertner O, Plesner T, Sanderson RD, Waage A, Borset M, Sundan A. 2002. Osteoprotegerin is bound, internalized, and degraded by multiple myeloma cells. *Blood* **100:** 3002–3007.

Stellwagen D, Malenka RC. 2006. Synaptic scaling mediated by glial TNF-α. *Nature* **440:** 1054–1059.

Stone GW, Barzee S, Snarsky V, Kee K, Spina CA, Yu XF, Kornbluth RS. 2006. Multimeric soluble CD40 ligand and GITR ligand as adjuvants for human immunodeficiency virus DNA vaccines. *J Virol* **80:** 1762–1772.

Strasser A, Jost PJ, Nagata S. 2009. The many roles of FAS receptor signaling in the immune system. *Immunity* **30:** 180–192.

Stuber E, Neurath M, Calderhead D, Fell HP, Strober W. 1995. Cross-linking of OX40 ligand, a member of the TNF/NGF cytokine family, induces proliferation and differentiation in murine splenic B cells. *Immunity* **2:** 507–521.

Summers deLuca L, Gommerman JL. 2012. Fine-tuning of dendritic cell biology by the TNF superfamily. *Nat Rev Immunol* **12:** 339–351.

Sun M, Lee S, Karray S, Levi-Strauss M, Ames KT, Fink PJ. 2007. Cutting edge: Two distinct motifs within the Fas ligand tail regulate Fas ligand-mediated costimulation. *J Immunol* 179: 5639–5643.

Swee LK, Ingold-Salamin K, Tardivel A, Willen L, Gaide O, Favre M, Demotz S, Mikkola M, Schneider P. 2009. Biological activity of ectodysplasin A is conditioned by its collagen and heparan sulfate proteoglycan-binding domains. *J Biol Chem* 284: 27567–27576.

Thomas G. 2002. Furin at the cutting edge: From protein traffic to embryogenesis and disease. *Nat Rev Mol Cell Biol* 3: 753–766.

Tracey KJ, Fong Y, Hesse DG, Manogue KR, Lee AT, Kuo GC, Lowry SF, Cerami A. 1987. Anti-cachectin/TNF monoclonal antibodies prevent septic shock during lethal bacteraemia. *Nature* 330: 662–664.

Uren AG, O'Rourke K, Aravind LA, Pisabarro MT, Seshagiri S, Koonin EV, Dixit VM. 2000. Identification of paracaspases and metacaspases: Two ancient families of caspase-like proteins, one of which plays a key role in MALT lymphoma. *Mol Cell* 6: 961–967.

Van Lier RAW, Borst J, Vroom TM, Klein H, Van Mourik P, Zeijlemaker WP, Melief CJ. 1987. Tissue distribution and biochemical and functional properties of Tp55 (CD27), a novel T cell differentiation antigen. *J Immunol* 139: 1589–1596.

Vilar M, Charalampopoulos I, Kenchappa RS, Simi A, Karaca E, Reversi A, Choi S, Bothwell M, Mingarro I, Friedman WJ, et al. 2009. Activation of the p75 neurotrophin receptor through conformational rearrangement of disulphide-linked receptor dimers. *Neuron* 62: 72–83.

Wagner KW, Punnoose EA, Januario T, Lawrence DA, Pitti RM, Lancaster K, Lee D, von Goetz M, Yee SF, Totpal K, et al. 2007. Death-receptor O-glycosylation controls tumor-cell sensitivity to the proapoptotic ligand Apo2L/TRAIL. *Nat Med* 13: 1070–1077.

Wallach D. 1984. Preparations of lymphotoxin induce resistance to their own cytotoxic effect. *J Immunol* 132: 2464–2469.

Wallach D. 2016a. The cybernetics of TNF: Old views and newer ones. *Semin Cell Dev Biol* 50: 105–114.

Wallach D. 2016b. TNFα. In *Encyclopedia of inflammatory diseases* (ed. Parnham M). Springer, Basel.

Wallach D, Kovalenko A. 2016. Roles of TNF and other members of the TNF family in the regulation of innate immunity. In *Encyclopedia of immunobiology* (ed. Ratcliffe MJH), pp. 454–465. Academic, Oxford.

Walsh MC, Choi Y. 2014. Biology of the RANKL-RANK-OPG system in immunity, bone, and beyond. *Front Immunol* 5: 511.

Wang J, Al-Lamki RS, Zhang H, Kirkiles-Smith N, Gaeta ML, Thiru S, Pober JS, Bradley JR. 2003. Histamine antagonizes tumor necrosis factor (TNF) signaling by stimulating TNF receptor shedding from the cell surface and Golgi storage pool. *J Biol Chem* 278: 21751–21760.

Wang L, Yang JK, Kabaleeswaran V, Rice AJ, Cruz AC, Park AY, Yin Q, Damko E, Jang SB, Raunser S, et al. 2010. The Fas-FADD death domain complex structure reveals the basis of DISC assembly and disease mutations. *Nat Struct Mol Biol* 17: 1324–1329.

Ware CF. 2005. Network communications: Lymphotoxins, LIGHT, and TNF. *Annu Rev Immunol* 23: 787–819.

Watts AD, Hunt NH, Wanigasekara Y, Bloomfield G, Wallach D, Roufogalis BD, Chaudhri G. 1999. A casein kinase I motif present in the cytoplasmic domain of members of the tumour necrosis factor ligand family is implicated in "reverse signalling." *EMBO J* 18: 2119–2126.

Wyzgol A, Muller N, Fick A, Munkel S, Grigoleit GU, Pfizenmaier K, Wajant H. 2009. Trimer stabilization, oligomerization, and antibody-mediated cell surface immobilization improve the activity of soluble trimers of CD27L, CD40L, 41BBL, and glucocorticoid-induced TNF receptor ligand. *J Immunol* 183: 1851–1861.

Yang S, Wang Y, Mei K, Zhang S, Sun X, Ren F, Liu S, Yang Z, Wang X, Qin Z, et al. 2015. Tumor necrosis factor receptor 2 (TNFR2)·interleukin-17 receptor D (IL-17RD) heteromerization reveals a novel mechanism for NF-κB activation. *J Biol Chem* 290: 861–871.

Yin Q, Lin SC, Lamothe B, Lu M, Lo YC, Hura G, Zheng L, Rich RL, Campos AD, Myszka DG, et al. 2009a. E2 interaction and dimerization in the crystal structure of TRAF6. *Nat Struct Mol Biol* 16: 658–666.

Yin Q, Lamothe B, Lamothe B, Wu H. 2009b. Structural basis for the lack of E2 interaction in the ring of TRAF2. *Biochemistry* 48: 10558–10567.

You RI, Chang YC, Chen PM, Wang WS, Hsu TL, Yang CY, Lee CT, Hsieh SL. 2008. Apoptosis of dendritic cells induced by decoy receptor 3 (DcR3). *Blood* 111: 1480–1488.

Yuan S, Liu T, Huang S, Wu T, Huang L, Liu H, Tao X, Yang M, Wu K, Yu Y, et al. 2009. Genomic and functional uniqueness of the TNF receptor-associated factor gene family in amphioxus, the basal chordate. *J Immunol* 183: 4560–4568.

Zanin-Zhorov A, Hershkoviz R, Hecht I, Cahalon L, Lider O. 2003. Fibronectin-associated Fas ligand rapidly induces opposing and time-dependent effects on the activation and apoptosis of T cells. *J Immunol* 171: 5882–5889.

Zapata JM, Martinez-Garcia V, Lefebvre S. 2007. Phylogeny of the TRAF/MATH domain. *Adv Exp Med Biol* 597: 1–24.

Zhou Z, Song X, Berezov A, Zhang G, Li Y, Zhang H, Murali R, Li B, Greene MI. 2008. Human glucocorticoid-induced TNF receptor ligand regulates its signaling activity through multiple oligomerization states. *Proc Natl Acad Sci* 105: 5465–5470.

Zmasek CM, Godzik A. 2013. Evolution of the animal apoptosis network. *Cold Spring Harb Perspect Biol* 5: a008649.

Cite this article as *Cold Spring Harb Perspect Biol* doi: 10.1101/cshperspect.a028431

Interleukin 17 Family Cytokines: Signaling Mechanisms, Biological Activities, and Therapeutic Implications

Leticia Monin[1] and Sarah L. Gaffen

Division of Rheumatology and Clinical Immunology, University of Pittsburgh, Pittsburgh, Pennsylvania 15261

Correspondence: sarah.gaffen@pitt.edu

The cytokines of the interleukin 17 (IL-17) family play a central role in the control of infections, especially extracellular fungi. Conversely, if unrestrained, these inflammatory cytokines contribute to the pathology of numerous autoimmune and chronic inflammatory conditions. Recent advances have led to the approval of IL-17A-blocking biologics for the treatment of moderate to severe plaque psoriasis, but much remains to be understood about the biological functions, regulation, and signaling pathways downstream of these factors. In this review, we outline the current knowledge of signal transduction and known physiological activities of IL-17 family cytokines. We will highlight in particular the current understanding of these cytokines in the context of skin manifestations of disease.

Interleukin (IL)-17A, the founding and most studied member of the IL-17 family, was cloned in 1993 and initially named cytotoxic T lymphocyte-associated antigen 8 (CTLA-8). Its sequence and predicted structure were markedly different from other known cytokines, but interestingly was homologous to an open reading frame (ORF) in the T-cell tropic *Herpesvirus saimiri* virus (Rouvier et al. 1993). A decade later, IL-17A took central stage with the discovery of Th17 cells as a T helper (Th) subset distinct from Th1 and Th2 cells (Langrish et al. 2005; Park et al. 2005). Five additional family members have been described, designated IL-17B, C, D, E, and F. Of these, IL-17F shares the greatest degree of conservation to IL-17A (55%) and is commonly produced by the same cell types. IL-17F was the first member of this family for which a crystallographic structure was elucidated. Interestingly, structural analysis revealed the formation of a cysteine-knot fold, similar to that adopted by neurotrophins such as nerve growth factor (NGF) (Hymowitz et al. 2001). IL-17E, also known as IL-25, displays the lowest degree of sequence conservation (16%) (Huang et al. 2015). In turn, other family members derive from different cellular sources and are associated with varying functions. IL-17A, IL-17F, IL-17C, and IL-17E function in host defense against pathogens and play various but not fully understood roles in mediating inflammation in autoimmune, allergic, and chronic inflammatory conditions. Given the central role of IL-17A in autoimmunity, much effort has focused on the development of neutralizing antibodies for therapeutic use. Indeed, IL-17A-blocking antibodies secukinumab and ixekizumab recently received U.S. Food and Drug Administration (FDA) approval for the treatment of psoriasis, ankylosing sponylitis (AS), and

---

[1]Present address: The Francis Crick Institute, London NW1 1AT, United Kingdom.

Cite this article as *Cold Spring Harb Perspect Biol* doi: 10.1101/cshperspect.a028522

psoriatic arthritis (PsA) (Langley et al. 2014; Gordon et al. 2016). Nonetheless, many aspects of IL-17A function, and especially of other cytokines in this family, remain poorly defined.

All known IL-17 family cytokines signal via a receptor family that is distinct from other known cytokine receptors (Yao et al. 1995). The IL-17R family contains five members, IL-17RA-E, all of which are single-pass transmembrane receptors with conserved structural features (Aggarwal and Gurney 2002). Specifically, all family members encode two extracellular fibronectin II–like domains and an intracellular SEFIR domain; the name alludes to the presence of this domain in SEF/IL-17RD and other IL-17 receptor proteins. The SEFIR is structurally related to the TIR domain found in the TLR/IL-1R family and is crucial for triggering downstream signaling events (see also the section "IL-17 Cytokine Signaling and Regulation") (Novatchkova et al. 2003). The prevailing paradigm for most IL-17 cytokines is that signaling occurs through heterodimeric receptors composed of a common IL-17RA chain and a second chain that determines ligand or signaling specificity. The second receptor chains are as follows: IL-17RC for IL-17A and IL-17F (Toy et al. 2006), IL-17RB for IL-17E (Rickel et al. 2008), and IL-17RE for IL-17C (Fig. 1) (Ramirez-Carrozzi et al. 2011). IL-17B is also reported to bind IL-17RB, albeit less strongly than IL-17E (Shi et al. 2000). In addition, the requirement for IL-17RA in IL-17B signaling is still under debate, and the receptor for IL-17D remains undefined. Here, we review the current understanding of cellular sources of the IL-17 family of cytokines, signal transduction mechanisms that govern their function, and the cutaneous biological processes in which these cytokines participate.

CELLULAR SOURCES OF IL-17 FAMILY CYTOKINES

IL-17A and IL-17F

More than 30 years ago, the paradigm of Th differentiation postulated that two discrete Th populations, Th1 and Th2 cells, acquired the ability to produce canonical cytokines, and were thus "tuned" to control biologically dissimilar pathogens (Mosmann et al. 1986). Although a useful model, there were numerous discrepancies that called this view into question (Steinman 2007). Indeed, in 2005, a third Th cell subset was described that produced IL-17A, IL-17F, as well as IL-21, IL-22, and granulocyte macrophage colony-stimulating factor (GM-CSF) (Park et al. 2005; Liang et al. 2006; Korn et al. 2007; Nurieva et al. 2007), and hence came to be known as "Th17." Like other Th subsets, naïve CD4$^+$ T cells become committed to the Th17 lineage via cytokine cues received during antigen presentation in secondary lymphoid organs. For Th17 cells, this is a combination of IL-1b, IL-6, transforming growth factor β (TGF-β), and IL-21 for initial commitment (Bettelli et al. 2006; Mangan et al. 2006; Veldhoen et al. 2006; Zhou et al. 2007) and IL-23 for full acquisition of their pathogenic capacity (Cua et al. 2003; Awasthi et al. 2009; McGeachy et al. 2009). Like Th1 and Th2 cells, Th17 cells express a lineage-determining "master" transcription factor, Rorγt, which directs the production of their hallmark cytokines (Ivanov et al. 2006).

More recently, it has become clear that additional populations of cells are also important sources of IL-17A and IL-17F. These include CD8$^+$ cytotoxic T (Tc) cells (He et al. 2006; Huber et al. 2013) and innate tissue-resident cells that are rapidly activated on injury or pathogenic insult. Among these innate subsets are γδ T cells (including Vg4$^+$ and Vg6$^+$ cells [Cua and Tato 2010]) innate lymphoid cells ([ILCs], specifically the ILC3 subset) (Villanova et al. 2014), "natural" CD4$^+$ Th17 cells (Marks et al. 2009), and natural killer T (NKT) cells (Kronenberg 2005). All IL-17-producing cells share a common dependence on IL-23 and on the transcription factor Rorγt, and express the chemokine receptor CCR6 (Cua and Tato 2010). In addition, given their positioning at barrier sites and their fast responsiveness, these innate-like cells constitute important early sources of IL-17 during infection and tissue damage. Recent reports have also proposed the expression of IL-17A by myeloid cells, including macrophages, neutrophils, and mast cells (Hoshino et al. 2008; Cua and Tato 2010; Li

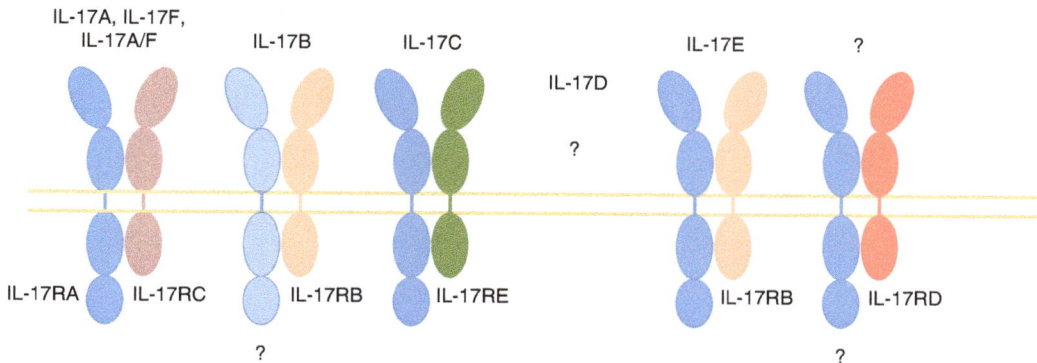

Figure 1. Interleukin 17 (IL-17) family cytokines and their receptors. Most IL-17 family cytokines signal via a heterodimeric receptor composed of IL-17RA and a second chain that varies depending on ligand, as indicated. Despite advances in the characterization of receptor–ligand interactions, several questions remain. Namely, a role for IL-17RA in IL-17B signaling has not been fully shown. In addition, the receptor for IL-17D, as well as the ligand for IL-17RD, remain unknown.

et al. 2010). However, these findings remain controversial, especially given the low levels of IL-17 detected in these cells and their propensity for phagocytosis, which might internalize IL-17 found in the environment. Indeed, a recent article showed that mast cells can take up IL-17A from the extracellular environment via receptor-mediated endocytosis and subsequently release it to promote inflammation (Noordenbos et al. 2016). Similarly, neutrophils and mast cells have been proposed to release IL-17 via extracellular traps (Lin et al. 2011).

IL-17E (IL-25)

IL-17E, also known as IL-25, was discovered through a bioinformatics search for proteins homologous to IL-17A (Lee et al. 2001). At the protein level, IL-17E bears 16%−20% sequence similarity to IL-17A, B, and C. IL-17E derives from both hematopoietic and nonhematopoietic cells (Lee et al. 2001). In mice, IL-17E is expressed by innate immune cells such as mast cells and alveolar macrophages in response to allergic stimuli (Morita et al. 2015). This also seems to be true in humans, as blood eosinophils and basophils from normal and allergic subjects expressed IL-17E messenger RNA (mRNA), which was further boosted following IL-5 treatment (Wang et al. 2007). In addition,

tissue stromal cells can express IL-17E. Human lung epithelial cells and murine primary type II alveolar epithelial cells express IL-17E following challenge with *Aspergillus oryzae*, ragweed allergens, and allergen proteases (Angkasekwinai et al. 2007; Kouzaki et al. 2013). Concordantly, IL-17E was detected at higher levels via immunohistochemistry (IHC) in the bronchial mucosa of asthmatics (Corrigan et al. 2011). The triggers for IL-17E production in many of these cells remains an active area of investigation.

IL-17E is a pleiotropic cytokine, acting on stromal, innate immune, and adaptive immune cells to orchestrate Th2-type inflammation. Consistent with the association of dysregulated Th2 responses with the development of allergy, IL-17E production is linked to the severity of chronic allergic conditions (Cheng et al. 2014). Thus, IL-17E-induced inflammation can be distinguished from IL-17A- and IL-17F-induced inflammation through the nature of the immune infiltrate, which mostly consists of eosinophils for the former and neutrophils for the latter (Morita et al. 2015). However, IL-17E expression can be advantageous in some situations, as IL-17E can inhibit Th17 development through the induction of IL-13 by dendritic cells (DCs) and by inhibiting macrophage-derived IL-23 production (Kleinschek et al. 2007). In addition, IL-17E delivery ameliorates autoimmune diabetes in animal models (Emamaullee et al. 2009; Saadoun et al. 2011). IL-17E, therefore, seems to be an atypical IL-17 family member, both in terms of low sequence homology and different biological actions.

IL-17C

IL-17C was identified during the search for IL-17A-related cytokines (Li et al. 2000). IL-17C is mainly expressed by epithelial cells following stimulation with TLR2 and TLR5 ligands or with the proinflammatory cytokines IL-1β and tumor necrosis factor α (TNF-α) (Ramirez-Carrozzi et al. 2011). Its expression has been reported to be induced in CD4$^+$ T cells, dendritic cells, and macrophages in inflamed tissues (Li et al. 2000; Hwang and Kim 2005). IL-17C has been suggested to act via a heterodimeric receptor composed of IL-17RA and IL-17RE, mediating a seemingly overlapping function to that of IL-17A and IL-17F (Ramirez-Carrozzi et al. 2011). Indeed, intranasal delivery of IL-17C-expressing adenovirus triggers neutrophilia and drives the expression of a set of proinflammatory molecules that overlaps considerably with IL-17A-dependent target genes (Hurst et al. 2002). Its role in mediating inflammation in several inflammatory and infection settings is just beginning to be unraveled.

IL-17B and IL-17D

IL-17B and D were also found through a search for IL-17A homologs (Li et al. 2000; Starnes et al. 2002). IL-17B is expressed at the transcriptional level in many cell types, including chondrocytes, neurons, intestinal epithelial cells, and breast cancer cells. Similar to IL-17E, IL-17B can bind to IL-17RB, albeit with a lower affinity (Chang and Dong 2011). However, its function in the context of these cells is still enigmatic. IL-17D mRNA is detected in various tissues, including brain, heart, lung, pancreas, skeletal muscle, and adipose tissue (Starnes et al. 2002). In the immune system, expression seems to be restricted to naïve CD4$^+$ T cells and B cells. IL-17D most closely resembles IL-17B, with which it shares 27% homology. Its carboxy-terminal motif is absent in other IL-17 family members (Starnes

et al. 2002). To date, its receptor remains unknown.

IL-17 CYTOKINE SIGNALING AND REGULATION

Most IL-17 family members characterized to date mediate signaling through heterodimeric receptors composed of IL-17RA and a subunit that confers ligand or signaling specificity. IL-17RA is widely expressed among cells of both hematopoietic and nonhematopoietic compartments (Yao et al. 1995; Ishigame et al. 2009). Other IL-17R family receptors generally exhibit expression more restricted to specific cell types, which helps explain the target cell specificity of different ligands. This situation is analogous to signaling by IL-6 or βc family cytokines, which use the common gp130 subunit or the common β subunit for signaling (Ozaki and Leonard 2002; Hercus et al. 2013). The existence of conserved mechanisms of receptor binding in the IL-17 family is reinforced by crystallographic analyses of IL-17RA in complex with IL-17A and IL-17F. These analyses revealed the acquisition of a similar conformation by the receptor on cytokine binding, and the requirement for the same amino acid residues for receptor–ligand interactions (Liu et al. 2013). Stoichiometry of the receptor complex seems to be dimeric. The lack of further receptor chains may be explained by the induction of conformational changes in the receptors on cytokine binding, which disfavor binding to a second homotypic receptor chain (Liu et al. 2013).

Signaling pathways downstream of IL-17 cytokine family members are beginning to be unraveled, with IL-17A-targeted signaling mechanisms having been most thoroughly studied. In this section, we will focus on current knowledge regarding the molecular actions downstream of IL-17A, and point out commonalities, divergences, and gaps in our understanding of IL-17 family cytokines.

IL-17A, IL-17F, and IL-17A/F

IL-17A and IL-17F signal through the IL-17RA/RC heterodimer, evidenced by a complete loss of responsiveness in $Il17ra^{-/-}$ and $Il17rc^{-/-}$ mice or cell lines derived from them (Gaffen 2009). Importantly, this receptor can bind to three different covalent cytokine dimers: IL-17A homodimers, IL-17F homodimers, or IL-17A/F heterodimers, albeit with varying affinities (Wright et al. 2007). IL-17RA has a 100-fold weaker affinity for IL-17F and an intermediate affinity for the IL-17A/F heterodimer and bears weaker affinity for IL-17B, C, D, and E. Conversely, IL-17RC has a higher affinity for IL-17F than for IL-17A (Kuestner et al. 2007). Overall, IL-17A signaling induces stronger responses than IL-17F (10–30 times more potent, as assessed by downstream gene induction), which may explain its dominant role in driving autoimmunity (Zrioual et al. 2009). Receptor expression patterns also differ between the two chains, with IL-17RA being expressed more highly in the immune compartment, and IL-17RC expression being largely restricted to nonimmune cells (Kuestner et al. 2007; Ishigame et al. 2009). Whether varying expression patterns coupled with the different affinity of each receptor chain for IL-17A or IL-17F underlies their diverging biological functions remains an open question.

Detailed sequence analysis of IL-17R family members revealed the presence of a conserved intracellular subdomain with homology to Toll-IL-1R (TIR) domains, which are essential for signaling downstream of the IL-1 receptor and Toll-like receptors (TLRs). These motifs share sequence homology with boxes 1 and 2 of the TIR domain, but lack box 3. Interestingly, this motif was discovered in "similar expression to fibroblast (SEF) growth factor" proteins (an IL-17RD ortholog) from zebrafish and chicken and hence became known as the SEFIR domain (Novatchkova et al. 2003). On cytokine ligation, the IL-17 receptor complex is thought to undergo a conformational change enabling the establishment of homotypic interactions between the SEFIR domains of the receptor and the signaling adaptor Act1 (Qian et al. 2007). Act1, also known as CIKS (connection to IκB kinase and stress-activated protein kinases), is an adapter required for all known downstream IL-17A signaling pathways. The canonical pathway relies on the E3 ligase activity of Act1, which mediates Lys63-linked ubiquitylation of TRAF6 (Schwandner et al. 2000). This event leads to activation of the canonical nuclear factor κB (NF-κB) and mitogen-activated protein kinase (MAPK) pathways, which include extracellular signal-regulated kinase (ERK), p38, and c-Jun amino-terminal kinase (JNK), as well as the CCAAT-enhancer-binding proteins (C/EBPs) pathway (Yao et al. 1995; Ruddy et al. 2004). Together, these transcription factors drive transcriptional activation of IL-17A target genes, which play key roles in inflammation.

In contrast, a second, noncanonical pathway is elicited by IL-17A, which leads to the stabilization of mRNA transcripts, particularly those encoding for intrinsically unstable targets such as cytokines and chemokines. This mRNA stabilization pathway is dependent on IκB kinase (IKKi) and TBK1-mediated phosphorylation of Act1 (Bulek et al. 2011; Qu et al. 2012). TRAF2 and TRAF5 are thereby recruited to the receptor complex, which results in the recruitment of molecules that control mRNA turnover (Sun et al. 2011). In particular, TRAF2 and TRAF5 can sequester the RNA-destabilizing factor ASF/SF2 and recruit the mRNA-stabilizing factor HuR, enhancing the half-life of various mRNAs (Sun et al. 2011; Herjan et al. 2013). In addition, Act1 is reported to interact with Hsp90 to activate IL-17 activity (Wang et al. 2013). A psoriasis-associated genetic variant in Act1 carrying the D10N mutation abrogates this interaction (Ellinghaus et al. 2010; Genetic Analysis of Psoriasis Consortium & the Wellcome Trust Case Control Consortium 2 et al. 2010; Hüffmeier et al. 2010). Together, IL-17-mediated events at both the transcriptional and posttranscriptional levels enhance production of genes that underlie its functions, including cytokines and chemokines, antimicrobial peptides (AMPs), acute phase proteins, and other inflammatory effectors (Onishi and Gaffen 2010). IL-17RA/RC signaling is summarized in Figure 2.

IL-17E (IL-25)

IL-17E signals through a heterodimer of IL-17RA and IL-17RB (Rickel et al. 2008), as

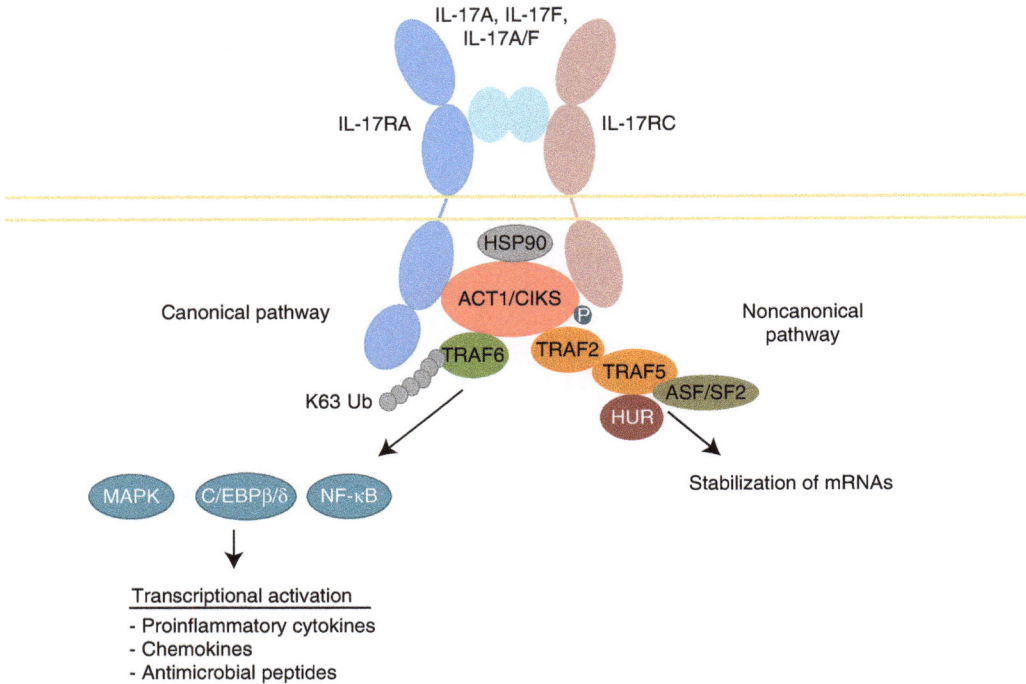

Figure 2. Interleukin (IL)-17RA/RC signaling pathways. IL-17A/IL-17F/IL-17A/F binding to the receptor complex enables homotypic interactions between the SEF/IL-17R (SEFIR) domains in the receptor and in the adapter Act1/CIKS. The canonical IL-17 signaling pathway initiates signaling through Act1-induced K63-linked ubiquitylation of TRAF6, thereby activating the mitogen-activated protein kinase (MAPK), CCAAT-enhancer-binding protein β (C/EBPβ), and nuclear factor κB (NF-κB) pathways. This triggers transcriptional activation of downstream target genes, including proinflammatory cytokines, chemokines, and antimicrobial peptides. In turn, noncanonical signaling relies on Act1 phosphorylation at amino acid 311. This recruits TRAF2 and TRAF5, which sequesters the messenger RNA (mRNA)-destabilizing factor ASF/SF2 and recruits the mRNA-stabilizing factor HuR. Together, these two pathways mediate the proinflammatory functions of IL-17A, IL-17F, and IL-17A/F.

summarized in Figure 3. Unlike the relatively stromal-restricted activity of IL-17A and IL-17F, IL-17E acts mainly on immune cells, including Th2, Th9, and NKT cells. IL-17E induces the production of classical type 2 cytokines, such as IL-4, IL-5, IL-9, and IL-13, in a Gata3-, c-MAF-, and JunB-dependent fashion (Wang et al. 2007). IL-17RB is also expressed on monocytes, certain populations of type 2 innate lymphocytes such as nuocytes, non-T/non-B cells, multipotent progenitor type 2 cells, and innate type 2 helper cells (Dolgachev et al. 2009; Moro et al. 2010; Neill et al. 2010; Price et al. 2010). In addition, stromal cells such as intestinal and pulmonary epithelial cells also respond to IL-17E. Similar to IL-17A/F signaling, IL-

17RB interacts with Act1 via homotypic SEFIR interactions (Claudio et al. 2009; Swaidani et al. 2009). Act1 recruits TRAF6, enabling NF-κB activation (Maezawa et al. 2006). However, the pathways diverge in that IL-17RB can recruit TRAF4 via Act1, leading to the further recruitment of the E3 ligase SMURF2 (Zepp et al. 2015). This leads to the ubiquitylation and subsequent degradation of the IL-17RB inhibitor DAZAP2, consequently reinforcing IL-17E-mediated signaling (Zepp et al. 2015). Further, IL-17E is reported to activate STAT5 in an Act1-independent manner, which further potentiates a Th2 response (Wu et al. 2015b). The precise stoichiometry of the receptor required for signaling via IL-17E is currently unclear, as there

Cite this article as *Cold Spring Harb Perspect Biol* doi: 10.1101/cshperspect.a028522

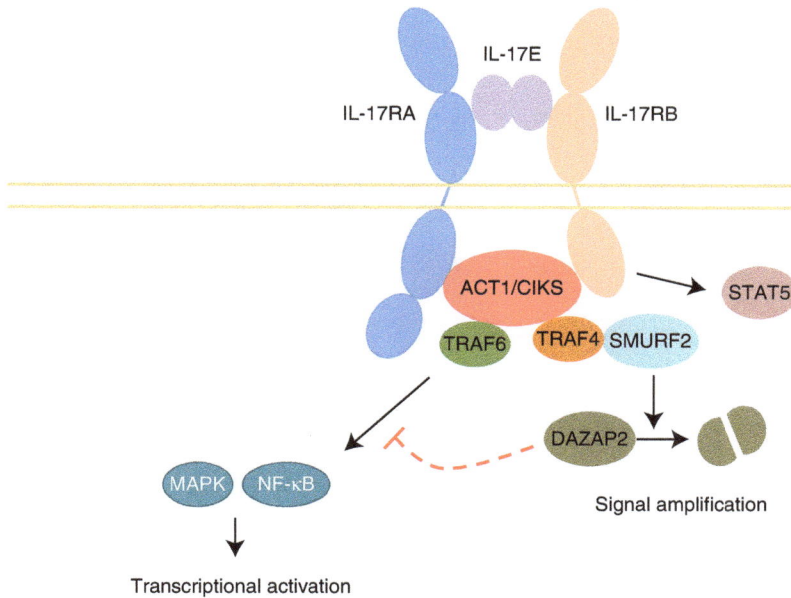

Figure 3. Interleukin (IL)-17RA/RB signaling. On IL-17E binding to its receptor, homotypic interactions between the SEF/IL-17R (SEFIR) domains in the receptor and in the adapter Act1/CIKS are established. This leads to the recruitment of TRAF6, activating the mitogen-activated protein kinase (MAPK) and nuclear factor κB (NF-κB) signaling pathways. In turn, Act1 can recruit TRAF4, which activates the E3 ligase SMURF2. This leads to the ubiquitylation and subsequent degradation of the inhibitor DAZAP2, amplifying IL-17E-mediated signaling. In addition, IL-17RB can elicit STAT5 activation in an Act1-independent manner.

are reports of IL-17E being unable to bind IL-17RA in vitro (Hymowitz et al. 2001). Whether the nature of the receptor varies depending on the cell type is another area of inquiry.

IL-17B, C, and D

Our current understanding of signaling downstream of IL-17 family members other than IL-17A, F, and E remains very limited. IL-17B has been shown to induce proinflammatory cytokine secretion by the THP-1 acute monocytic leukemic cell line, and to enhance inflammation, survival, and metastasis in breast and pancreatic cancer (Huang et al. 2014; Wu et al. 2015a). IL-17RB engagement in these cells recruited the Act1-TRAF6-TAK1 complex to the receptor (Wu et al. 2015a). Interestingly, IL-17B and IL-17E seem to present antagonic activities, despite reportedly binding to the same receptor (Reynolds et al. 2015). IL-17C has been reported to signal via a heterodimeric IL-17RA/RE

complex (Ramirez-Carrozzi et al. 2011). Expression of IL-17RE is restricted to epithelial cells, specialized epithelial cells like keratinocytes and Th17 cells (reviewed in Song et al. 2016). In line with other proinflammatory cytokines in the family, IL-17C signaling activates the NF-κB and MAPK pathways (Song et al. 2011). Similar to other cytokines in the family, IL-17C signaling also seems to be dependent on Act1, and Song et al. have recently reported that IL-17RE associates with Act1 (Song et al. 2016). As noted above, the action of IL-17D, its receptor, and the signaling mechanisms it elicits remain obscure.

Regulation of IL-17 Family Cytokines

Given its central role in inflammation, numerous mechanisms have evolved to restrict the IL-17A signaling pathway, presumably to curtail bystander inflammation. For example, TRAF3 and TRAF4 interfere with early events in IL-17A signaling by competing with Act1 or TRAF6 for IL-17RA binding (Zhu et al. 2010; Zepp et al. 2012). The deubiquitinase A20 is induced downstream of IL-17A and dampens the activation of NF-κB and MAPK pathways by removal of K63-linked ubiquitin chains on TRAF6 (Garg et al. 2013). Thus, A20 serves as a feedback regulator of the IL-17 pathway, analogous to its effect for TNF-α and IL-1 signaling as well (He and Ting 2002; Duong et al. 2015; Luo et al. 2015). Similarly, the deubiquitinase USP25 acts on TRAF5 and TRAF6, suppressing IL-17A signaling (Johnston et al. 2013). GSK-3β-mediated phosphorylation of the transcription factor C/EBPβ inhibits IL-17 target gene expression (Shen et al. 2009). Genome-wide association study (GWAS) analysis of psoriasis has revealed genetic associations with known regulators of immune signaling, including *TNFAIP3* (A20), *TNIP1* (ABIN-1, NAF1), and *NFKBIA* (IκBα) (Harden et al. 2015). Importantly, ABIN-1 was recently shown to regulate IL-17A signaling in keratinocytes. Correspondingly, *Tnip1*-deficient mice develop cutaneous inflammation with psoriasiform characteristics, linking findings in this mouse model to the enhanced susceptibility to psoriasis of individuals with *TNIP1* single-nucleotide polymorphisms (SNPs) (Ippagunta et al. 2016). Finally, the endoribonuclease MCPIP1 (also known as Regnase-1, encoded by *ZC3H12A*) limits IL-17 signaling through the degradation of IL-17-driven genes, including *Il6*, *Nfkbiz*, and *Il17ra* and *Il17rc* (Sonder et al. 2011; Garg et al. 2015; Somma et al. 2015). To date, 11 nonsynonymous SNPs have been described for the *ZC3H12A* gene, but so far none are associated with human disease (Cifuentes et al. 2010).

With an emerging role in inflammatory diseases (Johnston et al. 2013), IL-17C may similarly be subject to robust control mechanisms, although little is currently known about this issue. We recently showed that the action of MCPIP1 can also curb IL-17C-mediated inflammation in murine keratinocytes both in vitro and in vivo, thereby limiting skin inflammation in the imiquimod-driven psoriasis model (Monin et al. 2017). Given the high degree of conservation across signaling pathways in the IL-17 family of cytokines, it is tempting to speculate that other members of the IL-17 will share common regulatory mechanisms.

IL-17 FAMILY CYTOKINES IN HOST PROTECTION AND INFLAMMATION

IL-17A, F, and A/F in Infection

IL-17A and IL-17F evolved to protect from infection, and it is now clear that they orchestrate protective responses against infections at mucosal and epithelial surfaces, including the intestine, skin, lung, and oral cavity. Their central role in mediating protective immunity relies on the induction of molecules that stimulate epithelial barrier function. Signaling downstream of IL-17RA/RC elicits the expression of AMPs, including β-defensins, S100 proteins, and lipocalin-2 ([Lcn2], also known as 24p3 or NGAL) (Yang et al. 1999). Lcn2 competes with bacterial siderophores for acquisition of free iron and thus limits bacterial growth (Yang et al. 2002). In addition, IL-17A and IL-17F induce a proinflammatory milieu with enhanced cytokine and chemokine, and matrix metalloproteinase (MMP) production. These factors mediate the activation and recruitment of immune cells to the site of infection, promoting a potent immune response to the invading pathogen. One of the hallmarks of IL-17A-driven inflammation is neutrophil accumulation. Indeed, induction of granulocyte colony-stimulating factor (G-CSF) regulates neutrophil production, whereas chemokines such as CXCL1, CXCL5, and CCL2 stimulate neutrophil chemotaxis (Shen et al. 2005). In addition, IL-17 induces CCL20, which recruits CCR6-expressing cells such as Th17 and ILC3s (Acosta-Rodriguez et al. 2007). In this manner, IL-17A and to a lesser extent IL-17F regulate the coordinated action of stromal, innate, and adaptive immune cells.

The central role of IL-17A and IL-17F in protective immunity against infections is highlighted by the increased susceptibility of IL-17A or IL-17F-deficient mice to pathogens. For example, IL-17RA−/− mice are unable to control lung infection with *Klebsiella pneumoniae*

(Ye et al. 2001). In addition, IL-17A stimulates macrophage-derived IL-12, which is required to promote protective Th1 responses against pulmonary infection with *Francisella tularensis* live vaccine strain (Lin et al. 2009). Furthermore, IL-17 levels are elevated during acute lung infection with *Pseudomonas aeruginosa*, which contributes to neutrophil recruitment and bacterial containment (Liu et al. 2011; Dubin et al. 2012). Similarly, a deficiency in IL-17 or the IL-17-promoting cytokine IL-23 renders mice more susceptible to *Citrobacter rodentium* intestinal infection (Mangan et al. 2006), as well as to a number of other bacterial pathogens (reviewed in Curtis and Way 2009; Manni et al. 2014).

Candida albicans is a commensal fungal organism in about 70% of healthy individuals, residing in the skin, mouth, gastrointestinal tract, and vagina without causing disease. However, following loss of immune control mechanisms, *C. albicans* can become an opportunistic pathogen. Chronic mucocutaneous candidiasis (CMC) can ensue in individuals with primary and acquired immunodeficiencies, leading to oropharyngeal candidiasis ([OPC] or thrush), or to cutaneous lesions. Importantly, defects in the IL-17/IL-23 axis render the host exquisitely susceptible to CMC, highlighting the importance of this pathway in controlling *C. albicans* infections. Genetic variants in IL-12Rβ1 and STAT3, which compromise IL-23 signaling, have also been associated with diminished Th17 responses in humans and accordingly to CMC (Milner and Holland 2013).

IL-17A is required for control of *C. albicans* OPC infection in mice, although IL-17F and IL-17AF may also contribute to protection (Whibley et al. 2016). Notably, a mutation in the *IL17F* gene was recently reported in a family with CMC (Puel et al. 2011). This point mutation at position 65 in the polypeptide chain leads to the production of a dominant-negative variant, which abrogates IL-17F and IL-17A/IL-17F heterodimer signaling. Likewise, mutations in IL-17RA, IL-17RC, and Act1 lead to CMC in humans. Mutations in the *AIRE* gene (autoimmune regulator) lead to the development of the multiorgan autoimmune disease APECED (autoimmune polyendocrinopathy candidiasis ectodermal dystrophy). One of the main manifestations of APECED is an enhanced susceptibility to CMC (Milner and Holland 2013). Interestingly, compromised negative selection in the thymus because of AIRE deficiency leads to the development of neutralizing autoantibodies against IL-17A, IL-17F, and IL-22 (Puel et al. 2010). Importantly, the dependence of the host on IL-17 for containment of *Candida* infections is dependent on colonization route and tissue. For instance, vulvovaginal candidiasis is associated with alterations in other host factors, such as pH and microbial flora composition. In turn, control of systemic candidiasis seems to be more reliant on Th1 and natural killer (NK) cell responses (Conti and Gaffen 2015).

Staphylococcus aureus dermatitis has been reported in patients with *ACT1* or *IL17RA* null variants (Puel et al. 2011; Boisson et al. 2013). In line with a role for IL-17 in *S. aureus* control, *Il17ra*-deficient mice exhibit an increased susceptibility to cutaneous *S. aureus* infection (Cho et al. 2010; Chan et al. 2015). Given the emergence in recent years of methicillin-resistant *S. aureus* (MRSA) strains, harnessing the IL-17 axis in vaccination strategies may be of prophylactic promise.

IL-17A, F, and A/F in Chronic Skin Inflammatory and Autoimmune Diseases

Up-regulation of inflammatory and tissue-remodeling molecules can lead to tissue damage if IL-17 activity is left uncontrolled. Indeed, IL-17A and related cytokines are up-regulated in numerous autoimmune conditions, including psoriasis, rheumatoid arthritis (RA), multiple sclerosis, scleroderma, and lupus, among others. Similarly, GWAS studies have associated SNPs in genes of the IL-17 pathway with autoimmunity. A number of reviews have recently addressed the role of IL-17 in other autoimmune conditions (Iwakura et al. 2011; Song et al. 2016). We will focus here on the role of IL-17A and IL-17F in driving cutaneous inflammation (Table 1).

Psoriasis is a chronic inflammatory skin condition characterized by epidermal hyperplasia, affecting 2%–3% of the world's population.

Table 1. Cytokines and receptors driving cutaneous inflammation

Cytokine	Receptor	Infection	Skin inflammatory phenotype(s)
IL-17A and IL-17F	IL-17RA/RC	*Candida albicans, Staphylococcus aureus*	Psoriasis, atopic dermatitis, skin cancer
IL-17B	IL-17RB/?	Undefined	Undefined
IL-17C	IL-17RA/RE	*S. aureus*	Psoriasis
IL-17D	Undefined	Undefined	Undefined
IL-17E (IL-25)	IL-17RA/RB	Undefined	Psoriasis, atopic dermatitis

IL, Interleukin.

One of the hallmarks of disease is neutrophilic infiltration and formation of neutrophil microabscesses (Nestle et al. 2009). Elevated IL-17A and Th17-related cytokines such as IL-22 and IL-23 are found in human psoriasis skin lesions (Wilson et al. 2007; Johansen et al. 2009; Johnston et al. 2013). In addition, IL-17A can directly act on human keratinocytes stimulated to upregulate AMPs and neutrophil-attracting chemokines (Liang et al. 2006; Nograles et al. 2008). Consistently, GWAS studies have identified psoriasis-associated variants in genes participating in Th17 differentiation and IL-17A signaling, such as *IL23R* and *TRAF3IP2* (encoding Act1) (Cargill et al. 2007; Ellinghaus et al. 2010; Hüffmeier et al. 2010; Sonder et al. 2012; Tsoi et al. 2012). Mouse preclinical models of psoriasis have confirmed a role for IL-17 family cytokines in mediating disease. In the imiquimod-driven dermatitis model (driven by a TLR7 agonist), IL-17RA-deficient mice show dramatically diminished skin involvement (van der Fits et al. 2009). Interestingly, IL-17 signaling plays a dual role during imiquimod-driven psoriasis, its role varying by cell type. Mice deficient in IL-17 signaling in keratinocytes present dampened keratinocyte proliferation and neutrophilic microabscess formation. In turn, Act1 deficiency in skin fibroblasts limits the recruitment of IL-17-producing cells, thereby controlling the amplification of skin inflammation (Ha et al. 2014). Interestingly, intradermal injection of IL-22 or IL-23 into mouse ear also elicits the development of psoriasis-like disease, indicating that other cytokines in the IL-23/IL-17 axis can initiate disease (Song et al. 2016). The importance of IL-17A-mediated inflammation in psoriasis has been more recently highlighted by the

clinical success of biologic drugs, including IL-17A-blocking antibodies secukinumab and ixekizumab and the IL-17RA-targeting antibody brodalumab (Leonardi et al. 2012; Langley et al. 2014; Baeten et al. 2015; Durham et al. 2015; Sanford and McKeage 2015).

Atopic dermatitis affects 10%–20% of children and 1%–3% of adults in the Western world (Schultz Larsen 1996), and is characterized by chronic skin inflammation because of exacerbated responses to environmental antigens. The IL-17 axis has been reported to participate in allergic skin reactions, including atopic dermatitis and contact dermatitis. Serum levels of IL-17A and F are increased in children with atopic dermatitis and positively correlated with disease severity (Leonardi et al. 2015). Expression of IL-17A at the mRNA level is increased in the skin of patients with nickel allergy (Albanesi et al. 1999). In addition, increased Th17 cell infiltration was detected in a mouse model of contact dermatitis, and IL-17A-deficient mice displayed reduced pathology (Nakae et al. 2002). Whether the enhancement of IL-17 responses is a driver of pathology or reflects the immune efforts to limit colonization of skin lesions by bacteria remains an open question.

Strikingly, IL-17 signaling has been associated with the promotion of skin cancer development during chemical carcinogenesis in mouse models. Indeed, IL-17 or IL-17RA-deficient mice show considerably diminished incidence of DMBA/TPA-induced skin tumors (Wang et al. 2010; He et al. 2012). This protumorigenic effect of IL-17 is thought to occur via the promotion of epithelial proliferation and the antiapoptotic effect of STAT-3, which may be downstream of IL-17-induced genes such as

IL-6. Importantly, IL-17RA blocking in mice with established tumors blocked further tumor progression (He et al. 2012). Thus, IL-17A blockade may be useful for controlling at least some cancers.

IL-17B

IL-17B is expressed by neutrophils in the synovial tissue of RA patients (Kouri et al. 2014). Treatment of human fibroblasts with IL-17B synergized with TNF-α to induce G-CSF and IL-6 (Kouri et al. 2014). Intriguingly, IL-17B is expressed in limb buds during mouse embryonic development, suggesting a role in chondrogenesis and osteogenesis that may be dysregulated in autoimmune processes affecting the joints (You et al. 2005). IL-17B, like IL-17E (IL-25), has been shown to bind to IL-17RB. IL-17B can oppose IL-25-driven inflammation, and has been shown to play an antagonistic role. In a dextran sulphate sodium (DSS)-driven colitis mouse model, IL-25 administration exacerbated colonic damage (McHenga et al. 2008). In contrast, in a second report, IL-25-deficient mice exhibited reduced weight loss, inflammation, and tissue damage (Reynolds et al. 2015). These discrepant findings may result from differences in microbiome composition between the two studies. Interestingly, IL-17B-deficient mice developed increased susceptibility to DSS colitis, with enhanced weight loss, proinflammatory cytokine production, and colonic tissue destruction (Reynolds et al. 2015). Similarly, IL-17B and IL-25 play opposing roles in the context of *Citrobacter rodentium* infection and ovalbumin (OVA)-induced lung inflammation (Reynolds et al. 2015). In vitro, cotreatment of primary colonic epithelial cells with IL-17B diminished IL-25-, but not IL-17A-driven IL-6 production. IL-17B remains among the most obscure of all IL-17 family members, with a role in skin immunity and pathology yet to be ascribed. The described antagonism to IL-25 function is interesting in this context, particularly given the association between IL-25 expression and skin atopy. The potential role of IL-17B in skin immunity and pathology should therefore be explored.

IL-17C

As noted above, IL-17C induces a similar pattern of gene expression to IL-17A, which poses the question of functional redundancy. However, despite the overlap in target gene induction, IL-17C-deficient mice do not exhibit a compromised ability to control oral, dermal, or disseminated candidiasis, in contrast to IL-17RA-deficient mice (Conti et al. 2015). In addition, the IL-17RA-dependent gene signature associated with immunity against *C. albicans* was unchanged in IL-17C-deficient mice. Concordantly, IL-17RE deficiency did not lead to enhanced susceptibility to candidiasis. Interestingly, IL-17C can be induced in keratinocytes infected with *S. aureus* via a NOD2-dependent mechanism. Using this in vitro system, suppression of IL-17C expression rendered keratinocytes slightly more permissive to *S. aureus* survival (Roth et al. 2014). Thus, IL-17C may contribute to the control of infections, potentially through the activation of common mechanisms with IL-17A and F.

A common feature for IL-17 family cytokines or indeed all inflammatory stimuli is their propensity to promote protective immunity while simultaneously exacerbating tissue damage. IL-17C is the most highly expressed IL-17 family member in psoriatic lesions (Wilson et al. 2007; Johansen et al. 2009; Johnston et al. 2013), and drives the expression of AMPs, proinflammatory cytokines, and neutrophil-attracting chemokines in keratinocytes (Liang et al. 2006; Nograles et al. 2008). Consistent with a role in mediating cutaneous pathology, intradermal delivery of recombinant IL-17C into mouse ears led to epidermal thickening and neutrophil recruitment, whereas IL-17C-deficient mice developed less skin inflammation on imiquimod treatment (Ramirez-Carrozzi et al. 2011). In line with these findings, keratinocyte-specific IL-17C transgenic mice develop spontaneous psoriasiform dermatitis with epidermal hyperplasia, increased leukocytosis, and overexpression of proinflammatory cytokines (Johnston et al. 2013). Consistent with manifestations in human psoriasis, these mice display an enhanced proclivity toward thrombotic arterial occlusion, indicating the potential systemic effects of a skin inflammatory process (Golden et al. 2015). Therefore, IL-17C is clearly a driver of psoriasis and could be a safer target for blockade because its role in immunity to infection, at least in mice, appears to be less central (Conti et al. 2015).

IL-17D

As mentioned, the orphan cytokine IL-17D is poorly understood, with reports showing that this cytokine can induce IL-6, IL-8, and GM-CSF expression in endothelial cells (Starnes et al. 2002) and IL-6 and IL-8 in chicken fibroblasts (Hong et al. 2008). Interestingly, the stress-sensing protein NRF2 induces IL-17D expression in cancer cells. IL-17D-deficient mice displayed increased tumor growth when compared to wild-type mice (Saddawi-Konefka et al. 2016). Recent studies have also linked IL-17D to the recruitment of NK cells into the tumor microenvironment and subsequent activation (O'Sullivan et al. 2014; Saddawi-Konefka et al. 2014). Indeed, IL-17D plays a dual role in promoting human NK cell cytotoxicity and inducing NK-recruiting MCP-1 by tumor endothelial cells, thus placing this cytokine in a central role in tumor surveillance (O'Sullivan et al. 2014). The potential involvement of IL-17D in cutaneous surveillance mechanisms remains an open question.

IL-17E

IL-17E is an interesting example of an IL-17-family cytokine that possesses a divergent function to its founding member IL-17A. Like IL-17A, IL-17E can activate NF-κB and induce the production of IL-8. In addition, transgenic mice overexpressing IL-17E develop common features of IL-17A-driven inflammation, including neutrophilia and elevated circulating G-CSF (Pan et al. 2001). However, IL-17E functions mainly to stimulate Th2 responses, promoting Th2 cytokine secretion, class switch recombination to IgE, IgG1, and IgA, and the recruitment and activation of eosinophils in both mice and humans. Indeed, IL-17E transgenic mice

present with eosinophilia, increased IgE and IgG1, and elevated serum IL-5 and IL-13 (Pan et al. 2001). Given its role in promoting Th2-mediated immunity, IL-17E plays a central role in protection against helminth infection (Fallon et al. 2006; Owyang et al. 2006). In turn, *IL17E* mRNA expression is enhanced in the lungs of asthmatic patients (Wang et al. 2007), and IL-17E delivery promotes Th2 cytokine and IgE production as well as eosinophil infiltration in a mouse model of asthma (Fort et al. 2001).

IL-17E expression has been reported in patients with several skin conditions. In particular, an SNP in the *IL17E* gene is positively correlated with severe forms of psoriasis in a Spanish cohort of patients (Batalla et al. 2015). However, the effect of this polymorphism on IL-17E expression and/or function remains unknown. Atopic dermatitis often presents in association with mutations in the gene encoding filaggrin (Palmer et al. 2006; Weidinger et al. 2006; Barker et al. 2007). IL-25 was overexpressed in the epidermis of atopic dermatitis patients and in corresponding mouse models (Hvid et al. 2011; Aktar et al. 2015). In cultured keratinocytes, IL-25 treatment inhibited the expression of filaggrin, which may account for the loss of skin barrier function associated with atopic dermatitis (Hvid et al. 2011). In addition, IL-25 can mediate the recruitment of "type 2" cytokine-producing ILC2s in atopic dermatitis (Salimi et al. 2013). Therefore, IL-25 plays a dual role in promoting atopic dermatitis via stimulation of type 2 responses and through its direct action on keratinocytes.

CONCLUDING REMARKS

The IL-17 family of cytokines plays a central part in the induction of inflammation to limit numerous pathogenic insults. Here, we have reviewed the prominent role of IL-17A in orchestrating protective responses against cutaneous bacterial and fungal infections and the emerging roles of other IL-17 family members in boosting immunity. Given the recent development of novel therapies to block IL-17A and IL-17RA signals in chronic inflammatory diseases, the potential long-term consequences of such treatments vis-à-vis the exacerbation of fungal and extracellular bacterial infections should be examined. In that light, dissecting the commonalities and divergences in signaling pathways that drive protective versus tissue disruptive functions could provide alternative therapeutic strategies for at-risk populations. Given the pleiotropic roles of IL-17 family members, an in-depth analysis of individual cytokines' roles during infection and inflammation could provide insight into the advantages of the therapeutic alternatives that are currently under study, including IL-17A- versus IL-17RA-blocking strategies.

ACKNOWLEDGMENTS

S.L.G. is supported by the National Institutes of Health (Grants AI107825, AR062546, and DE022550). S.L.G. received research grants from Janssen and Novartis, and serves on the Scientific Advisory Board of Lycera Corporation. There are no other conflicts of interest.

REFERENCES

Acosta-Rodriguez EV, Rivino L, Geginat J, Jarrossay D, Gattorno M, Lanzavecchia A, Sallusto F, Napolitani G. 2007. Surface phenotype and antigenic specificity of human interleukin 17-producing T helper memory cells. *Nat Immunol* 8: 639–646.

Aggarwal S, Gurney AL. 2002. IL-17: Prototype member of an emerging cytokine family. *J Leukoc Biol* 71: 1–8.

Aktar MK, Kido-Nakahara M, Furue M, Nakahara T. 2015. Mutual upregulation of endothelin-1 and IL-25 in atopic dermatitis. *Allergy* 70: 846–854.

Albanesi C, Cavani A, Girolomoni G. 1999. IL-17 is produced by nickel-specific T lymphocytes and regulates ICAM-1 expression and chemokine production in human keratinocytes: Synergistic or antagonist effects with IFN-γ and TNF-α. *J Immunol* 162: 494–502.

Angkasekwinai P, Park H, Wang YH, Wang YH, Chang SH, Corry DB, Liu YJ, Zhu Z, Dong C. 2007. Interleukin 25 promotes the initiation of proallergic type 2 responses. *J Exp Med* 204: 1509–1517.

Awasthi A, Riol-Blanco L, Jager A, Korn T, Pot C, Galileos G, Bettelli E, Kuchroo VK, Oukka M. 2009. Cutting edge: IL-23 receptor gfp reporter mice reveal distinct populations of IL-17-producing cells. *J Immunol* 182: 5904–5908.

Baeten D, Sieper J, Braun J, Baraliakos X, Dougados M, Emery P, Deodhar A, Porter B, Martin R, Andersson M, et al. 2015. Secukinumab, an Interleukin-17A inhibitor, in ankylosing spondylitis. *N Engl J Med* 373: 2534–2548.

Barker JN, Palmer CN, Zhao Y, Liao H, Hull PR, Lee SP, Allen MH, Meggitt SJ, Reynolds NJ, Trembath RC, et al. 2007. Null mutations in the filaggrin gene (FLG) determine major susceptibility to early-onset atopic dermatitis that persists into adulthood. *J Invest Dermatol* 127: 564–567.

Batalla A, González-Lara L, González-Fernández D, Gómez J, Aranguren TF, Queiro R, Santos-Juanes J, López-Larrea C, Coto-Segura P. 2015. Association between single nucleotide polymorphisms IL17RA rs4819554 and IL17E rs79877597 and psoriasis in a Spanish cohort. *J Dermatol Sci* 80: 111–115.

Bettelli E, Carrier Y, Gao W, Korn T, Strom TB, Oukka M, Weiner HL, Kuchroo VK. 2006. Reciprocal developmental pathways for the generation of pathogenic effector TH17 and regulatory T cells. *Nature* 441: 235–238.

Boisson B, Wang C, Pedergnana V, Wu L, Cypowyj S, Rybojad M, Belkadi A, Picard C, Abel L, Fieschi C, et al. 2013. An ACT1 mutation selectively abolishes interleukin-17 responses in humans with chronic mucocutaneous candidiasis. *Immunity* 39: 676–686.

Bulek K, Liu C, Swaidani S, Wang L, Page RC, Gulen MF, Herjan T, Abbadi A, Qian W, Sun D, et al. 2011. The inducible kinase IKKi is required for IL-17-dependent signaling associated with neutrophilia and pulmonary inflammation. *Nat Immunol* 12: 844–852.

Cargill M, Schrodi SJ, Chang M, Garcia VE, Brandon R, Callis KP, Matsunami N, Ardlie KG, Civello D, Catanese JJ, et al. 2007. A large-scale genetic association study confirms IL12B and leads to the identification of IL23R as psoriasis-risk genes. *Am J Hum Genet* 80: 273–290.

Chan LC, Chaili S, Filler SG, Barr K, Wang H, Kupferwasser D, Edwards JEJr, Xiong YQ, Ibrahim AS, Miller LS, et al. 2015. Nonredundant roles of interleukin-17A (IL-17A) and IL-22 in murine host defense against cutaneous and hematogenous infection due to methicillin-resistant *Staphylococcus aureus*. *Infect Immun* 83: 4427–4437.

Chang SH, Dong C. 2011. Signaling of interleukin-17 family cytokines in immunity and inflammation. *Cell Signal* 23: 1069–1075.

Cheng D, Xue Z, Yi L, Shi H, Zhang K, Huo X, Bonser LR, Zhao J, Xu Y, Erle DJ, et al. 2014. Epithelial interleukin-25 is a key mediator in Th2-high, corticosteroid-responsive asthma. *Am J Respir Crit Care Med* 190: 639–648.

Cho JS, Pietras EM, Garcia NC, Ramos RI, Farzam DM, Monroe HR, Magorien JE, Blauvelt A, Kolls JK, Cheung AL, et al. 2010. IL-17 is essential for host defense against cutaneous *Staphylococcus aureus* infection in mice. *J Clin Invest* 120: 1762–1773.

Cifuentes RA, Cruz-Tapias P, Rojas-Villarraga A, Anaya JM. 2010. ZC3H12A (MCPIP1): Molecular characteristics and clinical implications. *Clin Chim Acta* 411: 1862–1868.

Claudio E, Sønder SU, Saret S, Carvalho G, Ramalingam TR, Wynn TA, Chariot A, Garcia-Perganeda A, Leonardi A, Paun A, et al. 2009. The adaptor protein CIKS/Act1 is essential for IL-25-mediated allergic airway inflammation. *J Immunol* 182: 1617–1630.

Conti HR, Gaffen SL. 2015. IL-17-mediated immunity to the opportunistic fungal pathogen *Candida albicans*. *J Immunol* 195: 780–788.

Conti HR, Whibley N, Coleman BM, Garg AV, Jaycox JR, Gaffen SL. 2015. Signaling through IL-17C/IL-17RE is dispensable for immunity to systemic, oral and cutaneous candidiasis. *PLoS ONE* **10:** e0122807.

Corrigan CJ, Wang W, Meng Q, Fang C, Eid G, Caballero MR, Lv Z, An Y, Wang YH, Liu YJ, et al. 2011. Allergen-induced expression of IL-25 and IL-25 receptor in atopic asthmatic airways and late-phase cutaneous responses. *J Allergy Clin Immunol* **128:** 116–124.

Cua DJ, Tato CM. 2010. Innate IL-17-producing cells: The sentinels of the immune system. *Nat Rev Immunol* **10:** 479–489.

Cua DJ, Sherlock J, Chen Y, Murphy CA, Joyce B, Seymour B, Lucian L, To W, Kwan S, Churakova T, et al. 2003. Interleukin-23 rather than interleukin-12 is the critical cytokine for autoimmune inflammation of the brain. *Nature* **421:** 744–748.

Curtis MM, Way SS. 2009. Interleukin-17 in host defense against bacterial, mycobacterial and fungal pathogens. *Immunology* **126:** 177–185.

Dolgachev V, Petersen BC, Budelsky AL, Berlin AA, Lukacs NW. 2009. Pulmonary IL-17E (IL-25) production and IL-17RB⁺ myeloid cell-derived Th2 cytokine production are dependent upon stem cell factor-induced responses during chronic allergic pulmonary disease. *J Immunol* **183:** 5705–5715.

Dubin PJ, Martz A, Eisenstatt JR, Fox MD, Logar A, Kolls JK. 2012. Interleukin-23-mediated inflammation in *Pseudomonas aeruginosa* pulmonary infection. *Infect Immun* **80:** 398–409.

Duong BH, Onizawa M, Oses-Prieto JA, Advincula R, Burlingame A, Malynn BA, Ma A. 2015. A20 restricts ubiquitination of pro-interleukin-1β protein complexes and suppresses NLRP3 inflammasome activity. *Immunity* **42:** 55–67.

Durham LE, Kirkham BW, Taams LS. 2015. Contribution of the IL-17 pathway to psoriasis and psoriatic arthritis. *Curr Rheumatol Rep* **17:** 55.

Ellinghaus E, Ellinghaus D, Stuart PE, Nair RP, Debrus S, Raelson JV, Belouchi M, Fournier H, Reinhard C, Ding J, et al. 2010. Genome-wide association study identifies a psoriasis susceptibility locus at TRAF3IP2. *Nat Genet* **42:** 991–995.

Emamaullee JA, Davis J, Merani S, Toso C, Elliott JF, Thiesen A, Shapiro AM. 2009. Inhibition of Th17 cells regulates autoimmune diabetes in NOD mice. *Diabetes* **58:** 1302–1311.

Fallon PG, Ballantyne SJ, Mangan NE, Barlow JL, Dasvarma A, Hewett DR, McIlgorm A, Jolin HE, McKenzie AN. 2006. Identification of an interleukin (IL)-25-dependent cell population that provides IL-4, IL-5, and IL-13 at the onset of helminth expulsion. *J Exp Med* **203:** 1105–1116.

Fort MM, Cheung J, Yen D, Li J, Zurawski SM, Lo S, Menon S, Clifford T, Hunte B, Lesley R, et al. 2001. IL-25 induces IL-4, IL-5, and IL-13 and Th2-associated pathologies in vivo. *Immunity* **15:** 985–995.

Gaffen SL. 2009. Structure and signalling in the IL-17 receptor family. *Nat Rev Immunol* **9:** 556–567.

Garg AV, Ahmed M, Vallejo AN, Ma A, Gaffen SL. 2013. The deubiquitinase A20 mediates feedback inhibition of interleukin-17 receptor signaling. *Sci Signal* **6:** ra44.

Garg AV, Amatya N, Chen K, Cruz JA, Grover P, Whibley N, Conti HR, Hernandez Mir G, Sirakova T, Childs EC, et al. 2015. MCPIP1 endoribonuclease activity negatively regulates interleukin-17-mediated signaling and inflammation. *Immunity* **43:** 475–487.

Genetic Analysis of Psoriasis Consortium & the Wellcome Trust Case Control Consortium 2; Strange A, Capon F, Spencer CC, Knight J, Weale ME, Allen MH, Barton A, Band G, Bellenguez C, et al. 2010. A genome-wide association study identifies new psoriasis susceptibility loci and an interaction between HLA-C and ERAP1. *Nat Genet* **42:** 985–990.

Golden JB, Wang Y, Fritz Y, Diaconu D, Zhang X, Debanne SM, Simon DI, McCormick TS, Ward NL. 2015. Chronic, not acute, skin-specific inflammation promotes thrombosis in psoriasis murine models. *J Transl Med* **13:** 382.

Gordon KB, Blauvelt A, Papp KA, Langley RG, Luger T, Ohtsuki M, Reich K, Amato D, Ball SG, Braun DK, et al. 2016. Phase 3 trials of ixekizumab in moderate-to-severe plaque psoriasis. *N Engl J Med* **375:** 345–356.

Ha HL, Wang H, Pisitkun P, Kim JC, Tassi I, Tang W, Morasso MI, Udey MC, Siebenlist U. 2014. IL-17 drives psoriatic inflammation via distinct, target cell-specific mechanisms. *Proc Natl Acad Sci* **111:** E3422–E3431.

Harden JL, Krueger JG, Bowcock AM. 2015. The immunogenetics of psoriasis: A comprehensive review. *J Autoimmun* **64:** 66–73.

He KL, Ting AT. 2002. A20 inhibits tumor necrosis factor (TNF) α-induced apoptosis by disrupting recruitment of TRADD and RIP to the TNF receptor 1 complex in Jurkat T cells. *Mol Cell Biol* **22:** 6034–6045.

He D, Wu L, Kim HK, Li H, Elmets CA, Xu H. 2006. CD8⁺ IL-17-producing T cells are important in effector functions for the elicitation of contact hypersensitivity responses. *J Immunol* **177:** 6852–6858.

He D, Li H, Yusuf N, Elmets CA, Athar M, Katiyar SK, Xu H. 2012. IL-17 mediated inflammation promotes tumor growth and progression in the skin. *PLoS ONE* **7:** e32126.

Hercus TR, Dhagat U, Kan WL, Broughton SE, Nero TL, Perugini M, Sandow JJ, D'Andrea RJ, Ekert PG, Hughes T, et al. 2013. Signalling by the βc family of cytokines. *Cytokine Growth Factor Rev* **24:** 189–201.

Herjan T, Yao P, Qian W, Li X, Liu C, Bulek K, Sun D, Yang WP, Zhu J, He A, et al. 2013. HuR is required for IL-17-induced Act1-mediated CXCL1 and CXCL5 mRNA stabilization. *J Immunol* **191:** 640–649.

Hong YH, Lillehoj HS, Park DW, Lee SH, Han JY, Shin JH, Park MS, Kim JK. 2008. Cloning and functional characterization of chicken interleukin-17D. *Vet Immunol Immunopathol* **126:** 1–8.

Hoshino A, Nagao T, Nagi-Miura N, Ohno N, Yasuhara M, Yamamoto K, Nakayama T, Suzuki K. 2008. MPO-ANCA induces IL-17 production by activated neutrophils in vitro via classical complement pathway-dependent manner. *J Autoimmun* **31:** 79–89.

Huang CK, Yang CY, Jeng YM, Chen CL, Wu HH, Chang YC, Ma C, Kuo WH, Chang KJ, Shew JY, et al. 2014. Autocrine/paracrine mechanism of interleukin-17B receptor promotes breast tumorigenesis through NF-κB-mediated antiapoptotic pathway. *Oncogene* **33:** 2968–2977.

Huang XD, Zhang H, He MX. 2015. Comparative and evolutionary analysis of the interleukin 17 gene family in invertebrates. *PLos ONE* **10:** e0132802.

Huber M, Heink S, Pagenstecher A, Reinhard K, Ritter J, Visekruna A, Guralnik A, Bollig N, Jeltsch K, Heinemann C, et al. 2013. IL-17A secretion by CD8$^+$ T cells supports Th17-mediated autoimmune encephalomyelitis. *J Clin Invest* **123:** 247–260.

Hüffmeier U, Uebe S, Ekici AB, Bowes J, Giardina E, Korendowych E, Juneblad K, Apel M, McManus R, Ho P, et al. 2010. Common variants at TRAF3IP2 are associated with susceptibility to psoriatic arthritis and psoriasis. *Nat Genet* **42:** 996–999.

Hurst SD, Muchamuel T, Gorman DM, Gilbert JM, Clifford T, Kwan S, Menon S, Seymour B, Jackson C, Kung TT, et al. 2002. New IL-17 family members promote Th1 or Th2 responses in the lung: In vivo function of the novel cytokine IL-25. *J Immunol* **169:** 443–453.

Hvid M, Vestergaard C, Kemp K, Christensen GB, Deleuran B, Deleuran M. 2011. IL-25 in atopic dermatitis: A possible link between inflammation and skin barrier dysfunction? *J Invest Dermatol* **131:** 150–157.

Hwang SY, Kim HY. 2005. Expression of IL-17 homologs and their receptors in the synovial cells of rheumatoid arthritis patients. *Mol Cell* **19:** 180–184.

Hymowitz SG, Filvaroff EH, Yin JP, Lee J, Cai L, Risser P, Maruoka M, Mao W, Foster J, Kelley RF, et al. 2001. IL-17s adopt a cystine knot fold: Structure and activity of a novel cytokine, IL-17F, and implications for receptor binding. *EMBO J* **20:** 5332–5341.

Ippagunta SK, Gangwar R, Finkelstein D, Vogel P, Pelletier S, Gingras S, Redecke V, Häcker H. 2016. Keratinocytes contribute intrinsically to psoriasis upon loss of Tnip1 function. *Proc Natl Acad Sci* **113:** E6162–E6171.

Ishigame H, Kakuta S, Nagai T, Kadoki M, Nambu A, Komiyama Y, Fujikado N, Tanahashi Y, Akitsu A, Kotaki H, et al. 2009. Differential roles of interleukin-17A and -17F in host defense against mucoepithelial bacterial infection and allergic responses. *Immunity* **30:** 108–119.

Ivanov II, McKenzie BS, Zhou L, Tadokoro CE, Lepelley A, Lafaille JJ, Cua DJ, Littman DR. 2006. The orphan nuclear receptor RORγt directs the differentiation program of proinflammatory IL-17$^+$ T helper cells. *Cell* **126:** 1121–1133.

Iwakura Y, Ishigame H, Saijo S, Nakae S. 2011. Functional specialization of interleukin-17 family members. *Immunity* **34:** 149–162.

Johansen C, Usher PA, Kjellerup RB, Lundsgaard D, Iversen L, Kragballe K. 2009. Characterization of the interleukin-17 isoforms and receptors in lesional psoriatic skin. *Br J Dermatol* **160:** 319–324.

Johnston A, Fritz Y, Dawes SM, Diaconu D, Al-Attar PM, Guzman AM, Chen CS, Fu W, Gudjonsson JE, McCormick TS, et al. 2013. Keratinocyte overexpression of IL-17C promotes psoriasiform skin inflammation. *J Immunol* **190:** 2252–2262.

Kleinschek MA, Owyang AM, Joyce-Shaikh B, Langrish CL, Chen Y, Gorman DM, Blumenschein WM, McClanahan T, Brombacher F, Hurst SD, et al. 2007. IL-25 regulates Th17 function in autoimmune inflammation. *J Exp Med* **204:** 161–170.

Korn T, Oukka M, Kuchroo V, Bettelli E. 2007. Th17 cells: Effector T cells with inflammatory properties. *Semin Immunol* **19:** 362–371.

Kouri VP, Olkkonen J, Ainola M, Li TF, Björkman L, Konttinen YT, Mandelin J. 2014. Neutrophils produce interleukin-17B in rheumatoid synovial tissue. *Rheumatology* **53:** 39–47.

Kouzaki H, Tojima I, Kita H, Shimizu T. 2013. Transcription of interleukin-25 and extracellular release of the protein is regulated by allergen proteases in airway epithelial cells. *Am J Respir Cell Mol Biol* **49:** 741–750.

Kronenberg M. 2005. Toward an understanding of NKT cell biology: Progress and paradoxes. *Annu Rev Immunol* **23:** 877–900.

Kuestner RE, Taft DW, Haran A, Brandt CS, Brender T, Lum K, Harder B, Okada S, Ostrander CD, Kreindler JL, et al. 2007. Identification of the IL-17 receptor related molecule IL-17RC as the receptor for IL-17F. *J Immunol* **179:** 5462–5473.

Langley RG, Elewski BE, Lebwohl M, Reich K, Griffiths CE, Papp K, Puig L, Nakagawa H, Spelman L, Sigurgeirsson B, et al. 2014. Secukinumab in plaque psoriasis—Results of two phase 3 trials. *N Engl J Med* **371:** 326–338.

Langrish CL, Chen Y, Blumenschein WM, Mattson J, Basham B, Sedgwick JD, McClanahan T, Kastelein RA, Cua DJ. 2005. IL-23 drives a pathogenic T cell population that induces autoimmune inflammation. *J Exp Med* **201:** 233–240.

Lee J, Ho WH, Maruoka M, Corpuz RT, Baldwin DT, Foster JS, Goddard AD, Yansura DG, Vandlen RL, Wood WI, et al. 2001. IL-17E, a novel proinflammatory ligand for the IL-17 receptor homolog IL-17Rh1. *J Biol Chem* **276:** 1660–1664.

Leonardi C, Matheson R, Zachariae C, Cameron G, Li L, Edson-Heredia E, Braun D, Banerjee S. 2012. Anti-interleukin-17 monoclonal antibody ixekizumab in chronic plaque psoriasis. *N Engl J Med* **366:** 1190–1199.

Leonardi S, Cuppari C, Manti S, Filippelli M, Parisi GF, Borgia F, Briuglia S, Cannavò P, Salpietro A, Arrigo T, et al. 2015. Serum interleukin 17, interleukin 23, and interleukin 10 values in children with atopic eczema/dermatitis syndrome (AEDS): Association with clinical severity and phenotype. *Allergy Asthma Proc* **36:** 74–81.

Li H, Chen J, Huang A, Stinson J, Heldens S, Foster J, Dowd P, Gurney AL, Wood WI. 2000. Cloning and characterization of IL-17B and IL-17C, two new members of the IL-17 cytokine family. *Proc Natl Acad Sci* **97:** 773–778.

Li L, Huang L, Vergis AL, Ye H, Bajwa A, Narayan V, Strieter RM, Rosin DL, Okusa MD. 2010. IL-17 produced by neutrophils regulates IFN-γ-mediated neutrophil migration in mouse kidney ischemia-reperfusion injury. *J Clin Invest* **120:** 331–342.

Liang SC, Tan XY, Luxenberg DP, Karim R, Dunussi-Joannopoulos K, Collins M, Fouser LA. 2006. Interleukin (IL)-22 and IL-17 are coexpressed by Th17 cells and cooperatively enhance expression of antimicrobial peptides. *J Exp Med* **203:** 2271–2279.

Lin Y, Ritchea S, Logar A, Slight S, Messmer M, Rangel-Moreno J, Guglani L, Alcorn JF, Strawbridge H, Park SM, et al. 2009. Interleukin-17 is required for T helper 1 cell immunity and host resistance to the intracellular pathogen *Francisella tularensis*. *Immunity* **31:** 799–810.

Lin AM, Rubin CJ, Khandpur R, Wang JY, Riblett M, Yala-varthi S, Villanueva EC, Shah P, Kaplan MJ, Bruce AT. 2011. Mast cells and neutrophils release IL-17 through extracellular trap formation in psoriasis. *J Immunol* **187:** 490–500.

Liu J, Feng Y, Yang K, Li Q, Ye L, Han L, Wan H. 2011. Early production of IL-17 protects against acute pulmonary *Pseudomonas aeruginosa* infection in mice. *FEMS Immunol Med Microbiol* **61:** 179–188.

Liu S, Song X, Chrunyk BA, Shanker S, Hoth LR, Marr ES, Griffor MC. 2013. Crystal structures of interleukin 17A and its complex with IL-17 receptor A. *Nat Commun* **4:** 1888.

Luo H, Liu Y, Li Q, Liao L, Sun R, Liu X, Jiang M, Hu J. 2015. A20 regulates IL-1-induced tolerant production of CXC chemokines in human mesangial cells via inhibition of MAPK signaling. *Sci Rep* **5:** 18007.

Maezawa Y, Nakajima H, Suzuki K, Tamachi T, Ikeda K, Inoue J, Saito Y, Iwamoto I. 2006. Involvement of TNF receptor-associated factor 6 in IL-25 receptor signaling. *J Immunol* **176:** 1013–1018.

Mangan PR, Harrington LE, O'Quinn DB, Helms WS, Bullard DC, Elson CO, Hatton RD, Wahl SM, Schoeb TR, Weaver CT. 2006. Transforming growth factor-β induces development of the T$_H$17 lineage. *Nature* **441:** 231–234.

Manni ML, Robinson KM, Alcorn JF. 2014. A tale of two cytokines: IL-17 and IL-22 in asthma and infection. *Exp Rev Respir Med* **8:** 25–42.

Marks BR, Nowyhed HN, Choi JY, Poholek AC, Odegard JM, Flavell RA, Craft J. 2009. Thymic self-reactivity selects natural interleukin 17-producing T cells that can regulate peripheral inflammation. *Nat Immunol* **10:** 1125–1132.

McGeachy MJ, Chen Y, Tato CM, Laurence A, Joyce-Shaikh B, Blumenschein WM, McClanahan T, O'Shea JJ, Cua DJ. 2009. The interleukin 23 receptor is essential for the terminal differentiation of interleukin 17-producing effector T helper cells in vivo. *Nat Immunol* **10:** 314–324.

McHenga SS, Wang D, Li C, Shan F, Lu C. 2008. Inhibitory effect of recombinant IL-25 on the development of dextran sulfate sodium-induced experimental colitis in mice. *Cell Mol Immunol* **5:** 425–431.

Milner JD, Holland SM. 2013. The cup runneth over: Lessons from the ever-expanding pool of primary immunodeficiency diseases. *Nat Rev Immunol* **13:** 635–648.

Monin L, Gudjonsson JE, Childs EE, Amatya N, Xing X, Verma AH, Coleman BM, Garg AV, Killeen M, Mathers A, et al. 2017. MCPIP1/regnase-1 restricts IL-17A- and IL-17C-dependent skin inflammation. *J Immunol* **198:** 767–775.

Morita H, Arae K, Unno H, Toyama S, Motomura K, Matsuda A, Suto H, Okumura K, Sudo K, Takahashi T, et al. 2015. IL-25 and IL-33 Contribute to development of eosinophilic airway inflammation in epicutaneously antigen-sensitized mice. *PLos ONE* **10:** e0134226.

Moro K, Yamada T, Tanabe M, Takeuchi T, Ikawa T, Kawamoto H, Furusawa J, Ohtani M, Fujii H, Koyasu S. 2010. Innate production of T$_H$2 cytokines by adipose tissue-associated c-Kit$^+$Sca-1$^+$ lymphoid cells. *Nature* **463:** 540–544.

Mosmann TR, Cherwinski H, Bond MW, Giedlin MA, Coffman RL. 1986. Two types of murine helper T cell clone. I: Definition according to profiles of lymphokine activities and secreted proteins. *J Immunol* **136:** 2348–2357.

Nakae S, Komiyama Y, Nambu A, Sudo K, Iwase M, Homma I, Sekikawa K, Asano M, Iwakura Y. 2002. Antigen-specific T cell sensitization is impaired in IL-17-deficient mice, causing suppression of allergic cellular and humoral responses. *Immunity* **17:** 375–387.

Neill DR, Wong SH, Bellosi A, Flynn RJ, Daly M, Langford TK, Bucks C, Kane CM, Fallon PG, Pannell R, et al. 2010. Nuocytes represent a new innate effector leukocyte that mediates type-2 immunity. *Nature* **464:** 1367–1370.

Nestle FO, Kaplan DH, Barker J. 2009. Psoriasis. *N Engl J Med* **361:** 496–509.

Nograles KE, Zaba LC, Guttman-Yassky E, Fuentes-Duculan J, Suarez-Farinas M, Cardinale I, Fuentes-Duculan J, Suárez-Fariñas M, Cardinale I, Khatcherian A, et al. 2008. Th17 cytokines interleukin (IL)-17 and IL-22 modulate distinct inflammatory and keratinocyte-response pathways. *Br J Dermatol* **159:** 1092–1102.

Noordenbos T, Blijdorp I, Chen S, Stap J, Mul E, Cañete JD, Lubberts E, Yeremenko N, Baeten D. 2016. Human mast cells capture, store, and release bioactive, exogenous IL-17A. *J Leukoc Biol* **100:** 453–462.

Novatchkova M, Leibbrandt A, Werzowa J, Neubuser A, Eisenhaber F. 2003. The STIR-domain superfamily in signal transduction, development and immunity. *Trends Biochem Sci* **28:** 226–229.

Nurieva R, Yang XO, Martinez G, Zhang Y, Panopoulos AD, Ma L, Schluns K, Tian Q, Watowich SS, Jetten AM, et al. 2007. Essential autocrine regulation by IL-21 in the generation of inflammatory T cells. *Nature* **448:** 480–483.

Onishi RM, Gaffen SL. 2010. Interleukin-17 and its target genes: Mechanisms of interleukin-17 function in disease. *Immunology* **129:** 311–321.

O'Sullivan T, Saddawi-Konefka R, Gross E, Tran M, Mayfield SP, Ikeda H, Bui JD. 2014. Interleukin-17D mediates tumor rejection through recruitment of natural killer cells. *Cell Rep* **7:** 989–998.

Owyang AM, Zaph C, Wilson EH, Guild KJ, McClanahan T, Miller HR, Cua DJ, Goldschmidt M, Hunter CA, Kastelein RA, et al. 2006. Interleukin 25 regulates type 2 cytokine-dependent immunity and limits chronic inflammation in the gastrointestinal tract. *J Exp Med* **203:** 843–849.

Ozaki K, Leonard WJ. 2002. Cytokine and cytokine receptor pleiotropy and redundancy. *J Biol Chem* **277:** 29355–29358.

Palmer CN, Irvine AD, Terron-Kwiatkowski A, Zhao Y, Liao H, Lee SP, Goudie DR, Sandilands A, Campbell LE, Smith FJ, et al. 2006. Common loss-of-function variants of the epidermal barrier protein filaggrin are a major predisposing factor for atopic dermatitis. *Nat Genet* **38:** 441–446.

Pan G, French D, Mao W, Maruoka M, Risser P, Lee J, Foster J, Aggarwal S, Nicholes K, Guillet S, et al. 2001. Forced expression of murine IL-17E induces growth retardation, jaundice, a Th2-biased response, and multiorgan inflammation in mice. *J Immunol* **167:** 6559–6567.

Park H, Li Z, Yang XO, Chang SH, Nurieva R, Wang YH, Wang Y, Hood L, Zhu Z, Tian Q, et al. 2005. A distinct lineage of CD4 T cells regulates tissue inflammation by producing interleukin 17. *Nat Immunol* **6:** 1133–1141.

Cite this article as *Cold Spring Harb Perspect Biol* doi: 10.1101/cshperspect.a028522

Price AE, Liang HE, Sullivan BM, Reinhardt RL, Eisley CJ, Erle DJ, Locksley RM. 2010. Systemically dispersed innate IL-13-expressing cells in type 2 immunity. *Proc Natl Acad Sci* **107:** 11489–11494.

Puel A, Döffinger R, Natividad A, Chrabieh M, Barcenas-Morales G, Picard C, Cobat A, Ouachée-Chardin M, Toulon A, Bustamante J, et al. 2010. Autoantibodies against IL-17A, IL-17F, and IL-22 in patients with chronic mucocutaneous candidiasis and autoimmune polyendocrine syndrome type I. *J Exp Med* **207:** 291–297.

Puel A, Cypowyj S, Bustamante J, Wright JF, Liu L, Lim HK, Migaud M, Israel L, Chrabieh M, Audry M, et al. 2011. Chronic mucocutaneous candidiasis in humans with inborn errors of interleukin-17 immunity. *Science* **332:** 65–68.

Qian Y, Liu C, Hartupee J, Altuntas CZ, Gulen MF, Jane-Wit D, Xiao J, Lu Y, Giltiay N, Liu J, et al. 2007. The adaptor Act1 is required for interleukin 17-dependent signaling associated with autoimmune and inflammatory disease. *Nat Immunol* **8:** 247–256.

Qu F, Gao H, Zhu S, Shi P, Zhang Y, Liu Y, Jallal B, Yao Y, Shi Y, Qian Y. 2012. TRAF6-dependent Act1 phosphorylation by the IκB kinase-related kinases suppresses interleukin-17-induced NF-κB activation. *Mol Cell Biol* **32:** 3925–3937.

Ramirez-Carrozzi V, Sambandam A, Luis E, Lin Z, Jeet S, Lesch J, Hackney J, Kim J, Zhou M, Lai J, et al. 2011. IL-17C regulates the innate immune function of epithelial cells in an autocrine manner. *Nat Immunol* **12:** 1159–1166.

Reynolds JM, Lee YH, Shi Y, Wang X, Angkasekwinai P, Nallaparaju KC, Flaherty S, Chang SH, Watarai H, Dong C. 2015. Interleukin-17B antagonizes interleukin-25-mediated mucosal inflammation. *Immunity* **42:** 692–703.

Rickel EA, Siegel LA, Yoon BR, Rottman JB, Kugler DG, Swart DA, Anders PM, Tocker JE, Comeau MR, Budelsky AL. 2008. Identification of functional roles for both IL-17RB and IL-17RA in mediating IL-25-induced activities. *J Immunol* **181:** 4299–4310.

Roth SA, Simanski M, Rademacher F, Schroder L, Harder J. 2014. The pattern recognition receptor NOD2 mediates *Staphylococcus aureus*-induced IL-17C expression in keratinocytes. *J Invest Dermatol* **134:** 374–380.

Rouvier E, Luciani MF, Mattei MG, Denizot F, Golstein P. 1993. CTLA-8, cloned from an activated T cell, bearing AU-rich messenger RNA instability sequences, and homologous to a herpesvirus saimiri gene. *J Immunol* **150:** 5445–5456.

Ruddy MJ, Wong GC, Liu XK, Yamamoto H, Kasayama S, Kirkwood KL, Gaffen SL. 2004. Functional cooperation between interleukin-17 and tumor necrosis factor-α is mediated by CCAAT/enhancer-binding protein family members. *J Biol Chem* **279:** 2559–2567.

Saadoun D, Terrier B, Cacoub P. 2011. Interleukin-25: Key regulator of inflammatory and autoimmune diseases. *Curr Pharm Des* **17:** 3781–3785.

Saddawi-Konefka R, O'Sullivan T, Gross ET, Washington AJr, Bui JD. 2014. Tumor-expressed IL-17D recruits NK cells to reject tumors. *Oncoimmunology* **3:** e954853.

Saddawi-Konefka R, Seelige R, Gross ET, Levy E, Searles SC, Washington AJr, Santosa EK, Liu B, O'Sullivan TE, Har-

ismendy O, et al. 2016. Nrf2 induces IL-17D to mediate tumor and virus surveillance. *Cell Rep* **16:** 2348–2358.

Salimi M, Barlow JL, Saunders SP, Xue L, Gutowska-Owsiak D, Wang X, Huang LC, Johnson D, Scanlon ST, McKenzie AN, et al. 2013. A role for IL-25 and IL-33-driven type-2 innate lymphoid cells in atopic dermatitis. *J Exp Med* **210:** 2939–2950.

Sanford M, McKeage K. 2015. Secukinumab: First global approval. *Drugs* **75:** 329–338.

Schultz Larsen F. 1996. Atopic dermatitis: An increasing problem. *Pediatr Allergy Immunol* **7:** 51–53.

Schwandner R, Yamaguchi K, Cao Z. 2000. Requirement of tumor necrosis factor receptor-associated factor (TRAF)6 in interleukin 17 signal transduction. *J Exp Med* **191:** 1233–1240.

Shen F, Ruddy MJ, Plamondon P, Gaffen SL. 2005. Cytokines link osteoblasts and inflammation: Microarray analysis of interleukin-17- and TNF-α-induced genes in bone cells. *J Leukoc Biol* **77:** 388–399.

Shen F, Li N, Gade P, Kalvakolanu DV, Weibley T, Doble B, Woodgett JR, Wood TD, Gaffen SL. 2009. IL-17 receptor signaling inhibits C/EBPβ by sequential phosphorylation of the regulatory 2 domain. *Sci Signal* **2:** ra8.

Shi Y, Ullrich SJ, Zhang J, Connolly K, Grzegorzewski KJ, Barber MC, Wang W, Wathen K, Hodge V, Fisher CL, et al. 2000. A novel cytokine receptor-ligand pair. Identification, molecular characterization, and in vivo immunomodulatory activity. *J Biol Chem* **275:** 19167–19176.

Somma D, Mastrovito P, Grieco M, Lavorgna A, Pignalosa A, Formisano L, Salzano AM, Scaloni A, Pacifico F, Siebenlist U, et al. 2015. CIKS/DDX3X interaction controls the stability of the Zc3h12a mRNA induced by IL-17. *J Immunol* **194:** 3286–3294.

Sonder SU, Saret S, Tang W, Sturdevant DE, Porcella SF, Siebenlist U. 2011. IL-17-induced NF-κB activation via CIKS/Act1: Physiologic significance and signaling mechanisms. *J Biol Chem* **286:** 12881–12890.

Sonder SU, Paun A, Ha HL, Johnson PF, Siebenlist U. 2012. CIKS/Act1-mediated signaling by IL-17 cytokines in context: Implications for how a CIKS gene variant may predispose to psoriasis. *J Immunol* **188:** 5906–5914.

Song X, Zhu S, Shi P, Liu Y, Shi Y, Levin SD, Qian Y. 2011. IL-17RE is the functional receptor for IL-17C and mediates mucosal immunity to infection with intestinal pathogens. *Nat Immunol* **12:** 1151–1158.

Song X, He X, Li X, Qian Y. 2016. The roles and functional mechanisms of interleukin-17 family cytokines in mucosal immunity. *Cell Mol Immunol* **13:** 418–431.

Starnes T, Broxmeyer HE, Robertson MJ, Hromas R. 2002. Cutting edge: IL-17D, a novel member of the IL-17 family, stimulates cytokine production and inhibits hemopoiesis. *J Immunol* **169:** 642–646.

Steinman L. 2007. A brief history of T$_H$17, the first major revision in the T$_H$1/T$_H$2 hypothesis of T cell-mediated tissue damage. *Nat Med* **13:** 139–145.

Sun D, Novotny M, Bulek K, Liu C, Li X, Hamilton T. 2011. Treatment with IL-17 prolongs the half-life of chemokine CXCL1 mRNA via the adaptor TRAF5 and the splicing-regulatory factor SF2 (ASF). *Nat Immunol* **12:** 853–860.

Swaidani S, Bulek K, Kang Z, Liu C, Lu Y, Yin W, Aronica M, Li X. 2009. The critical role of epithelial-derived Act1 in

IL-17- and IL-25-mediated pulmonary inflammation. *J Immunol* **182:** 1631–1640.

Toy D, Kugler D, Wolfson M, Vanden Bos T, Gurgel J, Derry J, Tocker J, Peschon J. 2006. Cutting edge: Interleukin 17 signals through a heteromeric receptor complex. *J Immunol* **177:** 36–39.

Tsoi LC, Spain SL, Knight J, Ellinghaus E, Stuart PE, Capon F, Ding J, Li Y, Tejasvi T, Gudjonsson JE, et al. 2012. Identification of 15 new psoriasis susceptibility loci highlights the role of innate immunity. *Nat Genet* **44:** 1341–1348.

van der Fits L, Mourits S, Voerman JS, Kant M, Boon L, Laman JD, Cornelissen F, Mus AM, Florencia E, Prens EP, et al. 2009. Imiquimod-induced psoriasis-like skin inflammation in mice is mediated via the IL-23/IL-17 axis. *J Immunol* **182:** 5836–5845.

Veldhoen M, Hocking RJ, Atkins CJ, Locksley RM, Stockinger B. 2006. TGFβ in the context of an inflammatory cytokine milieu supports de novo differentiation of IL-17-producing T cells. *Immunity* **24:** 179–189.

Villanova F, Flutter B, Tosi I, Grys K, Sreeneebus H, Perera GK, Chapman A, Smith CH, Di Meglio P, Nestle FO. 2014. Characterization of innate lymphoid cells in human skin and blood demonstrates increase of NKp44+ ILC3 in psoriasis. *J Invest Dermatol* **134:** 984–991.

Wang YH, Angkasekwinai P, Lu N, Voo KS, Arima K, Hanabuchi S, Hippe A, Corrigan CJ, Dong C, Homey B, et al. 2007. IL-25 augments type 2 immune responses by enhancing the expansion and functions of TSLP-DC-activated Th2 memory cells. *J Exp Med* **204:** 1837–1847.

Wang L, Yi T, Zhang W, Pardoll DM, Yu H. 2010. IL-17 enhances tumor development in carcinogen-induced skin cancer. *Cancer Res* **70:** 10112–10120.

Wang C, Wu L, Bulek K, Martin BN, Zepp JA, Kang Z, Liu C, Herjan T, Misra S, Carman JA, et al. 2013. The psoriasis-associated D10N variant of the adaptor Act1 with impaired regulation by the molecular chaperone hsp90. *Nat Immunol* **14:** 72–81.

Weidinger S, Illig T, Baurecht H, Irvine AD, Rodriguez E, Diaz-Lacava A, Klopp N, Wagenpfeil S, Zhao Y, Liao H, et al. 2006. Loss-of-function variations within the filaggrin gene predispose for atopic dermatitis with allergic sensitizations. *J Allergy Clin Immunol* **118:** 214–219.

Whibley N, Tritto E, Traggiai E, Kolbinger F, Moulin P, Brees D, Coleman BM, Mamo AJ, Garg AV, Jaycox JR, et al. 2016. Antibody blockade of IL-17 family cytokines in immunity to acute murine oral mucosal candidiasis. *J Leukoc Biol* **99:** 1153–1164.

Wilson NJ, Boniface K, Chan JR, McKenzie BS, Blumenschein WM, Mattson JD, Basham B, Smith K, Chen T, Morel F, et al. 2007. Development, cytokine profile and function of human interleukin 17-producing helper T cells. *Nat Immunol* **8:** 950–957.

Wright JF, Guo Y, Quazi A, Luxenberg DP, Bennett F, Ross JF, Qiu Y, Whitters MJ, Tomkinson KN, Dunussi-Joanno-

poulos K, et al. 2007. Identification of an interleukin 17F/17A heterodimer in activated human CD4+ T cells. *J Biol Chem* **282:** 13447–13455.

Wu HH, Hwang-Verslues WW, Lee WH, Huang CK, Wei PC, Chen CL, Shew JY, Lee EY, Jeng YM, Tien YW, et al. 2015a. Targeting IL-17B-IL-17RB signaling with an anti-IL-17RB antibody blocks pancreatic cancer metastasis by silencing multiple chemokines. *J Exp Med* **212:** 333–349.

Wu L, Zepp JA, Qian W, Martin BN, Ouyang W, Yin W, Bunting KD, Aronica M, Erzurum S, Li X. 2015b. A novel IL-25 signaling pathway through STAT5. *J Immunol* **194:** 4528–4534.

Yang D, Chertov O, Bykovskaia SN, Chen Q, Buffo MJ, Shogan J, Anderson M, Schröder JM, Wang JM, Howard OM, et al. 1999. β-Defensins: Linking innate and adaptive immunity through dendritic and T cell CCR6. *Science* **286:** 525–528.

Yang J, Goetz D, Li JY, Wang W, Mori K, Setlik D, Du T, Erdjument-Bromage H, Tempst P, Strong R, et al. 2002. An iron delivery pathway mediated by a lipocalin. *Mol Cell* **10:** 1045–1056.

Yao Z, Fanslow WC, Seldin MF, Rousseau AM, Painter SL, Comeau MR, Cohen JI, Spriggs MK. 1995. Herpesvirus Saimiri encodes a new cytokine, IL-17, which binds to a novel cytokine receptor. *Immunity* **3:** 811–821.

Ye P, Garvey PB, Zhang P, Nelson S, Bagby G, Summer WR, Schwarzenberger P, Shellito JE, Kolls JK. 2001. Interleukin-17 and lung host defense against Klebsiella pneumoniae infection. *Am J Respir Cell Mol Biol* **25:** 335–340.

You Z, DuRaine G, Tien JY, Lee C, Moseley TA, Reddi AH. 2005. Expression of interleukin-17B in mouse embryonic limb buds and regulation by BMP-7 and bFGF. *Biochem Biophys Res Commun* **326:** 624–631.

Zepp JA, Liu C, Qian W, Wu L, Gulen MF, Kang Z, Li X. 2012. Cutting edge: TNF receptor-associated factor 4 restricts IL-17-mediated pathology and signaling processes. *J Immunol* **189:** 33–37.

Zepp JA, Wu L, Qian W, Ouyang W, Aronica M, Erzurum S, Li X. 2015. TRAF4-SMURF2-mediated DAZAP2 degradation is critical for IL-25 signaling and allergic airway inflammation. *J Immunol* **194:** 2826–2837.

Zhou L, Ivanov II, Spolski R, Min R, Shenderov K, Egawa T, Levy DE, Leonard WJ, Littman DR. 2007. IL-6 programs T$_H$-17 cell differentiation by promoting sequential engagement of the IL-21 and IL-23 pathways. *Nat Immunol* **8:** 967–974.

Zhu S, Pan W, Shi P, Gao H, Zhao F, Song X, Liu Y, Zhao L, Li X, Shi Y, et al. 2010. Modulation of experimental autoimmune encephalomyelitis through TRAF3-mediated suppression of interleukin 17 receptor signaling. *J Exp Med* **207:** 2647–2662.

Zrioual S, Ecochard R, Tournadre A, Lenief V, Cazalis MA, Miossec P. 2009. Genome-wide comparison between IL-17A- and IL-17F-induced effects in human rheumatoid arthritis synoviocytes. *J Immunol* **182:** 3112–3120.

Interleukin (IL)-12 and IL-23 and Their Conflicting Roles in Cancer

Juming Yan,[1,2] Mark J. Smyth,[2,3] and Michele W.L. Teng[1,2]

[1]Cancer Immunoregulation and Immunotherapy Laboratory, QIMR Berghofer Medical Research Institute, Herston 4006, Queensland, Australia

[2]School of Medicine, University of Queensland, Herston 4006, Queensland, Australia

[3]Immunology in Cancer and Infection Laboratory, QIMR Berghofer Medical Research Institute, Herston 4006, Queensland, Australia

Correspondence: michele.teng@qimrberghofer.edu.au

The balance of proinflammatory cytokines interleukin (IL)-12 and IL-23 plays a key role in shaping the development of antitumor or protumor immunity. In this review, we discuss the role IL-12 and IL-23 plays in tumor biology from preclinical and clinical data. In particular, we discuss the mechanism by which IL-23 promotes tumor growth and metastases and how the IL-12/IL-23 axis of inflammation can be targeted for cancer therapy.

The recognized interleukin (IL)-12 cytokine family currently consists of IL-12, IL-23, IL-27, and IL-35 and these cytokines play important roles in the development of appropriate immune responses in various disease conditions (Vignali and Kuchroo 2012). They act as a link between the innate and adaptive immune system through mediating the appropriate differentiation of naïve CD4[+] T cells into various T helper (Th) subsets and regulating the functions of different effector cell types. IL-12, IL-23, and IL-27 are secreted by activated antigen-presenting cells (APCs), such as dendritic cells (DCs) and macrophages (Vignali and Kuchroo 2012), whereas IL-35 is generally thought to be produced by regulatory T (Treg) and B cells, although they were recently detected in human tolerogenic DCs (Pylayeva-Gupta 2016). A unique feature of these cytokines is their heterodimeric subunit composition whereby the α-subunit (p19, p28, p35) and β-subunit (p40, Ebi3) are differentially shared to generate IL-12 (p40-p35), IL-23 (p40-p19), IL-27 (Ebi3-p28), and IL-35 (p40-p35) (Fig. 1A). Given their ability to share α- and β-subunits, it has been predicted that combinations such as Ebi3-p19 and p28-p40 could exist and serve physiological function (Fig. 1B) (Wang et al. 2012; Flores et al. 2015; Ramnath et al. 2015). Similarly, the subunit-sharing feature of the IL-12 cytokine family also extends to its receptor chain usage (Fig. 1B). Although they share structural similarities and downstream signaling components, members of the IL-12 cytokine family mediate distinct biological functions. IL-12 and IL-23 are proinflammatory cytokines and are required for the development of Th1 and Th17 cells (Vignali and Kuchroo 2012). In contrast, IL-27 has im-

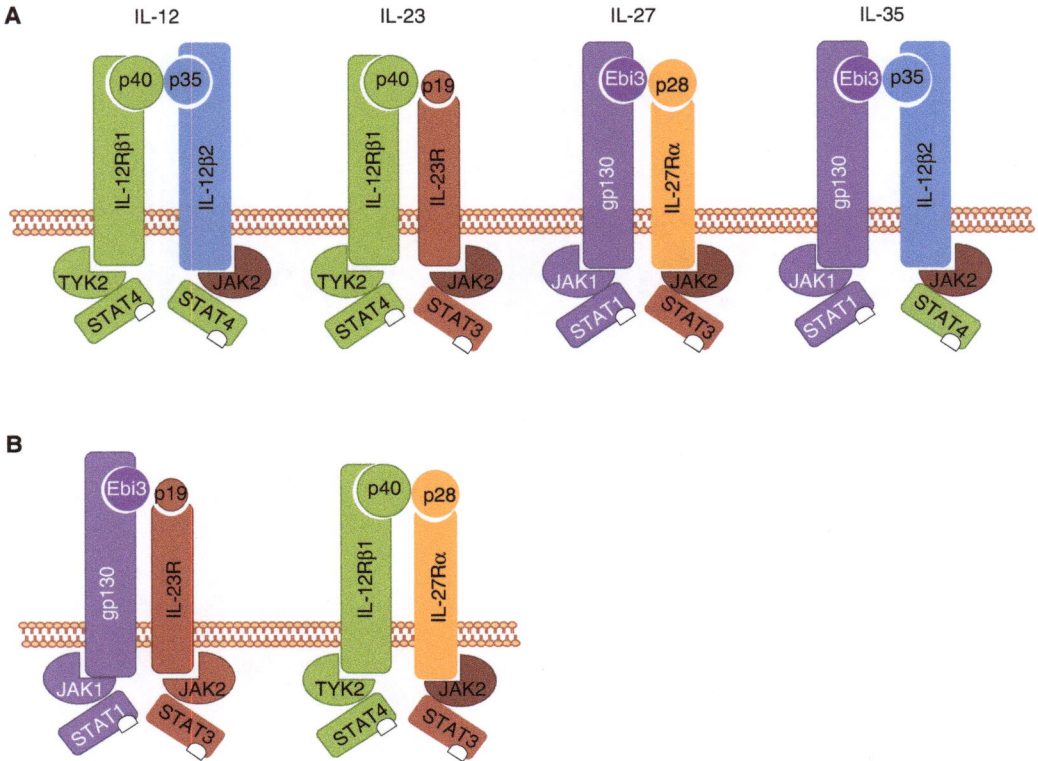

Figure 1. Schematic representation of the interleukin (IL)-12 cytokine family and their receptors, and associated Janus kinase and signal transducers and activators of transcription (JAK-STAT) signaling partners. (*A*) Current members of the IL-12 family with defined physiological function. (*B*) Potential new members of the IL-12 cytokine family that can be generated following subunit pairing. Ebi3, Epstein–Barr virus-induced gene 3 protein (also known as IL-27β); IL-12R, IL-12 receptor; IL-23R, IL-23 receptor; TYK2, tyrosine kinase 2.

munosuppressive function and can curtail different classes of inflammation through its ability to directly modify CD4[+] and CD8[+] T-cell effector functions, to induce IL-10, and to promote specialized Treg cell responses (Yoshida and Hunter 2015). Likewise, IL-35 has an immunosuppressive function as shown by its ability to induce the development of IL-35-producing induced Tregs and the suppression of T-cell effector function and proliferative capacity in a variety of in vitro and in vivo systems (Vignali and Kuchroo 2012; Pylayeva-Gupta 2016).

In cancer, inflammation has been shown to play a critical role in tumor initiation, growth, and metastasis (Grivennikov et al. 2010; Elinav et al. 2013). It is now appreciated that the balance between the proinflammatory cytokine IL-12

and IL-23 in tumors can shape the development of antitumor or protumor immunity. Given that the importance of IL-12 in promoting antitumor immunity is well recognized and recently reviewed (Tugues et al. 2015), this review will particularly focus on discussing the role of IL-23 in tumor biology and its mechanism of action in promoting tumor growth and metastases. Finally, we discuss how IL-12 and IL-23 are cross-regulated and how the IL-12/IL-23 axis of inflammation can be targeted for cancer therapy.

ROLE OF IL-12 AND IL-23 IN TUMOR BIOLOGY

APCs, such as DCs and macrophages, are thought to be the predominant source of IL-

12 and IL-23 (Hunter 2005). A major role of IL-12 is to promote differentiation of Th1 cells and to induce type II interferon (IFN)-γ production. The requirement for host IL-12 and its downstream cytokine in activating antitumor immunity is well recognized and has been previously reviewed extensively (Colombo and Trinchieri 2002; Dunn et al. 2006; Tugues et al. 2015). A number of studies have clearly illustrated the importance of endogenous IL-12 and IFN-γ in preventing cancer initiation, growth, and metastasis. In contrast, IL-23 plays an important role in promoting the proliferation and effector function of Th17 cells, which are characterized by expression of the IL-17 family cytokines (Aggarwal et al. 2003; Langrish et al. 2004; Langrish et al. 2005). Mice that were deficient for IL-12/23p40 or IFN-γ and challenged with methylcholanthrene-A (MCA), a chemical carcinogen, displayed increased rate and frequency of tumor growth compared to wild-type (WT) controls (Kaplan et al. 1998; Smyth et al. 2000). In contrast, mice deficient in IL-23p19 were strongly protected from developing MCA-induced fibrosarcomas (Teng et al. 2010). In another study, using a mouse model of dimethylbenz[a]anthracene (DMBA)-initiated and 12-*O*-tetradecanoylphorbol-13-acetate (TPA)-promoted two-stage skin carcinogenesis, IL-12/23p40-deficient and IL-23-p19-deficient mice displayed significantly decreased numbers of carcinogen-induced papillomas compared to WT mice, whereas the opposite was observed in IL-12p35-deficient mice (Langowski et al. 2006). Similarly, mice lacking IL-12p35 and IFN-γ were more susceptible to mortality caused by *N*-methyl-*N*-nitrosourea (MNU)-induced lymphoma compared to IL-12/23p40-deficient mice (Liu et al. 2004). Although naïve IL-12/23p40-deficient mice did not display any increase in tumor development compared to WT mice when monitored over their normal life span (Street et al. 2002), a proportion of aged mice deficient for IL-12Rβ2 and, hence, nonresponsive to IL-12, developed plasmacytoma and lung epithelial tumors (Airoldi et al. 2005). In addition to directly activating T-cell and natural killer (NK) cell effector function, IL-12 and IFN-γ can modulate the tumor mi-

croenvironment to be more conducive to antitumor immunity by inhibiting angiogenesis and expanding intratumoral Tregs (Cao et al. 2009; Tugues et al. 2015). Nevertheless, IL-12 may have IFN-γ-independent tumor-suppressive properties. A study using IL-12-producing B16 melanomas, in which innate lymphoid tissue-inducer cells but not T and NK cells induced tumor suppression (Eisenring et al. 2010), showed the pleiotropic ability of IL-12 to activate multiple arms of antitumor immunity.

The data that support the tumor-promoting effect of host IL-23 is also strong. A seminal paper by Langowski et al. (2006) provided the first demonstration that mice deficient in IL-23p19 were resistant to DMBA/TPA-induced skin papillomas and this resistance correlated with a significant increase in CD8+ T cells infiltrating the skin and a reduction in IL-17A, matrix metallopeptidase 9 (MMP9), CD31, granulocytes (Gr-1+), and macrophages (CD11b+, F4/80+). A study by Teng et al. (2010) further confirmed the resistance phenotype of these mice to DMBA/TPA-induced skin papillomas and also highlighted their resistance to MCA-induced fibrosarcomas. Interestingly, IL-17A did not promote the formation of MCA-induced fibrosarcomas, showing that IL-23p19 had tumor-promoting properties independent of IL-17A (Teng et al. 2010). In contrast to the MCA model, mice deficient for IL-12/IL-23p40 were still protected from tumor development following DMBA/TPA treatment, suggesting in this model that the loss of IL-23 was more important than the loss of IL-12 (Teng et al. 2010). In this model, *Il17a*-deficient mice displayed a protective phenotype albeit weaker than *Il23a*-deficient mice. Furthermore, this study also showed that, not only did IL-23 suppress the antitumor function of T cells as first uncovered by Langowski et al. 2006, it also suppressed the antimetastatic function of NK cells (Teng et al. 2010). In two other de novo mouse models of colon carcinogenesis (CPC-APC, Apc$^{Min/+}$), IL-23 and IL-17A both had tumor-promoting effects as loss of IL-23/IL-23R/IL-17R or IL-23R/IL-17A blockade resulted in reduced tumor load (Wu et al. 2009; Grivennikov et al. 2012). Interestingly, one study reported that expression of IL-23 alone in mice was sufficient to induce rapid (3–4 wk) de novo development of intestinal adenomas with 100% incidence. Tumorigenesis was mediated by type 3 innate lymphoid cells (ILC3) but independent of exogenous carcinogens, *Helicobacter* colonization, or preexisting tumor-suppressor gene mutations (Chan et al. 2014). Similarly, a role for IL-23 has also been shown for the T-cell-mediated tumor dormancy phase (equilibrium) of cancer immunoediting (Schreiber et al. 2011), best characterized in the MCA-induced fibrosarcoma model (Koebel et al. 2007). In mice bearing dormant tumors induced by MCA, anti-IL-23 resulted in the elimination of the residual tumor cells, whereas anti-IL-12p40 (IL-12 and IL-23) led to their outgrowth (Teng et al. 2012), thus showing that loss of IL-23 was not sufficient to compensate for loss of IL-12 in this setting. Overall, these studies suggest the hierarchy of dominance between IL-12 in tumor suppression and IL-23 in tumor promotion in de novo models of inflammation-induced carcinogenesis varies, and this can be because of the type and location of the tumor and the immune cell infiltrates.

On the contrary, there have been a number of studies suggesting that IL-23 can have tumor suppressing effects when it is overexpressed in different tumor cell lines and implanted into mice (Lo et al. 2003; Chiyo et al. 2004; Hu et al. 2006, 2009; Oniki et al. 2006; Shan et al. 2006; Yuan et al. 2006; Kaiga et al. 2007; Reay et al. 2012; Ngiow et al. 2013). A recent study also showed that injection of IL-23 in combination with a transforming growth factor β (TGF-β) receptor inhibitor could suppress the progression of premalignant lesions to cancer in a 4-nitroquinoline-1-oxide (4NQO) carcinogen-induced mouse model of oral cancer due to maintenance of Th17 cells and preventing their shift to a Treg phenotype (Young et al. 2016). In an acute UV-induced immunosuppression model, IL-23 was shown to be important in reducing UV-induced DNA damage and inhibiting UV-induced Tregs (Majewski et al. 2010). Similarly, a later study reported that mice deficient for IL-23p19 and chronically exposed to UVB were found to have a higher likelihood of developing tumors, particularly nonepithelial sarcomas, compared to the WT controls (Jantschitsch et al. 2012). One explanation could be that Th17 cells or IL-17A may have a role in tumor suppression similar to the DMBA/TPA model, although the role of Th17/IL-17A was not examined in this study. Although these data appear contradictory at first, the caveat of most of these experiments is that the expression of IL-23 is in a nonphysiological manner.

CLINICAL RELEVANCE OF IL-23 EXPRESSION IN HUMAN CANCERS

In agreement with the role of endogenous IL-23 in promoting mouse tumor growth, IL-23 was found to be overexpressed in many human cancers (Table 1). Similarly, two functional genetic variants of the IL-23R (IL-23R rs1884444 T>G and rs6682925 T>C) have been found to contribute to susceptibility to solid cancer and blood malignancies. However, the functional consequences of these variants in modulating IL-23R signaling are not clear (Chu et al. 2012; Qian et al. 2013a; Xu et al. 2013). In contrast, there has only been one clinical study in ovarian cancer in which a higher level of intratumoral IL-23p19 transcript correlated with improved patient overall survival. However, higher levels of intratumoral IL-12p35 transcripts were also measured in this study (Wolf et al. 2010). Overall, these studies provide compelling evidence that supports the involvement of IL-23 in the pathogenesis of different cancers, particularly those that are of inflammation-induced origin (Elinav et al. 2013).

CELLULAR SOURCES OF IL-23 IN TUMORS

In mice and humans, the main sources of IL-23 are thought to be produced by myeloid cells in response to exogenous or endogenous signals, such as damage-associated molecular patterns (DAMPs), pathogen-associated molecular patterns (PAMPs), or tumor-secreted factors such as prostaglandin E2 (PGE$_2$) (Qian et al. 2013b; von Scheidt et al. 2014; Chang et al. 2015; Teng et al. 2015; Kvedaraite et al. 2016). However, it remains unclear which myeloid subsets in tu-

Table 1. Overexpression of interleukin (IL)-23 in different human cancers

Cancer	Correlation	IL-23 levels in patients	IL-23 levels in control	Sample size	Method of detection	References
NSCLC	Increased levels associated with increased risk of NSCLC	0.27 (0.15–0.73) pg/mL	0.17 (0.10–0.28) pg/mL	218	CBA	Liao et al. 2015
NSCLC	N/A	Higher compared to controls		53	qPCR	Baird et al. 2013
Lung (NSCLC and SCLC)	N/A	491.27 ± 1263.38 pg/mL (serum)	240.51 ± 233.18 pg/mL (serum)	46	ELISA	Cam et al. 2016
CRC	N/A	Higher compared to normal tissue		7	qPCR	Grivennikov et al. 2012
CRC	N/A	~5000 ng/mL from homogenized tumors	~2000 ng/mL normal tissues	13	qPCR ELISA	Lan et al. 2011
CRC	Significantly higher at all tumor stages (I–IV) compared with healthy donors	31.5 ± 10.5 pg/mL (serum)	8.48 ± 13.3 pg/mL (serum)	48	ELISA	Stanilov et al. 2010
CRC	Elevated expression with concomitant VEGF overexpression positively correlated with histological grade 2	189.46 pg/mL (serum)	34.77 pg/mL (serum)	40	ELISA	Ljujic et al. 2010
HCC	Positively correlated with metastasis	Higher compared to normal tissue		81 49	qPCR and IHC	Li et al. 2012
Ovarian cancer	Higher levels positively correlated with OS	Higher compared to normal tissue		112	qPCR	Wolf et al. 2010
Breast cancer	Higher levels positively correlated with OS	14.522 ± 11.39 pg/mL (serum)	6.34 ± 4.6 pg/mL (serum)	50	ELISA	Gangemi et al. 2012
Pancreatic cancer	Higher levels negatively correlated with OS	266.5 ± 98.1 pg/mL (serum)	95.1 ± 37.2 pg/mL (serum)	20	ELISA	He et al. 2011
MM	N/A	~300 pg/mL (bone marrow)	~50 pg/mL (bone marrow)	5	ELISA	Prabhala et al. 2010
Bladder cancer	N/A	Higher IL-23 producing CD123+ pDCs compared to healthy volunteers		20	IHC	Wang et al. 2016
Melanoma	N/A	Higher than normal tissue, benign melanocytic nevi, and Spitz nevi		35	IHC	Ganzetti et al. 2015
Advanced gastric cancer	N/A	293 ± 132 pg/mL (serum)	102 ± 55 pg/mL (serum)	36	qPCR ELISA	Zhang et al. 2008

NSCLC, Non-small-cell lung carcinoma; SCLC, small-cell lung carcinoma; CRC, colorectal carcinoma; HCC, hepatocellular carcinoma; MM, multiple myeloma; OS, overall survival; pDC, plasmacytoid dendritic cells; CBA, cytometric bead array; qPCR, real-time polymerase chain reaction; ELISA, enzyme-linked immunosorbent assay; IHC, immunohistochemistry; VEGF, vascular endothelial growth factor.

mors secrete IL-23, and questions remain as to whether tumors themselves can also secrete IL-23. In mice, most studies have quantified IL-23p19 messenger RNA (mRNA) expression, given the lack of current reagents that can robustly detect IL-23 protein levels by ELISA or intracellular staining. However, a caveat is that increased expression of p19 mRNA may not correlate with the translation of p19 protein and/or formation of secreted bioactive IL-23, which requires coexpression and disulfide bond formation between the IL-23p19 and IL-12p40 subunits in the same cell (Waibler et al. 2007; Brentano et al. 2009; Floss et al. 2015). IL-23 has been measured in the supernatant from tumor, infiltrating CD11c$^+$DCs and CD11b$^+$ CD11c$^-$ macrophages sorted from mice bearing subcutaneous B16 melanoma tumors (Kortylewski et al. 2009). In CPC-APC/IL-23$^{-/-}$ mice harboring the green fluorescent protein (GFP) gene in the IL-23p19 locus, GFP was detected in both CD11b$^+$ or CD11b$^-$ cells derived from the mesenteric lymph node or tumor, although this was not quantitated and control staining was not shown (Grivennikov et al. 2012). In B16F10 and RM-1 tumor-bearing lungs, expression of IL-23, as measured by flow cytometry, was found to be restricted to MHC-II$^+$CD11b$^+$CD11c$^+$ cells rather than MHC-II$^+$CD11c$^+$CD11b$^-$ cells, following lipopolysaccharide (LPS) restimulation (von Scheidt et al. 2014). Overall, this indicates that mouse IL-23 production can be restricted to specific immune cell subsets and depends on the context and environment in which it is measured.

To date, there have been no convincing studies showing that mouse tumor cells themselves produce IL-23. IL-23 (as measured by ELISA) was not detected in in vitro culture of various experimental mouse tumor cell lines such as EG7, B16F10, and RM-1 (von Scheidt et al. 2014). Similarly, in another study, very little IL-23 was detected in in vitro cultures of B16 and C4 melanoma tumor cell lines (Kortylewski et al. 2009). However, whether these cell lines can produce IL-23 following exposure to various metabolites or cytokine stimulation or following in vivo inoculation remains to be examined. Similarly, only a few studies have reported that human tumor cell lines themselves could secrete IL-23 (Table 2). However, these studies generally used real-time polymerase chain reaction (qPCR) and/or Western blotting to measure the presence of IL-23p19; whether bioactive IL-23 is secreted remains to be validated.

EXPRESSION PATTERN OF IL-12R AND IL-23R

The IL-12 and IL-23 receptors are made up of IL-12Rβ1 and IL-12Rβ2 or IL-12Rβ1 and IL-

Table 2. Interleukin (IL)-23-producing human tumor cell lines

Human tumor cell lines	IL-23 detection method	References

Human lung carcinoma cell lines
 A549 (adenocarcinoma)
 SK-MES-1 (squamous-cell carcinoma)
 H1299, H460, and H647 (large-cell carcinoma)
 BEAS2B (SV40 transformed normal bronchoepithelia)
   HBEC3, HBEC4, and HBEC5 (normal bronchial epithelial
   cell lines immortalized in the absence of viral oncoproteins)
	RT-PCR	Baird et al. 2013

Human oral squamous cell carcinoma HSC-2, HSC-3, HSC-4, and Ca9-22	qPCR	Fukuda et al. 2010a,b
Human hepatocellular carcinoma HepG2, PLC8024, QGY7703, H2P, H2M, Huh7, and MHCC-97L	WB	Li et al. 2012
Human pancreatic carcinoma cell line PANC-1	qPCR, cytokine array	Chang et al. 2015

RT-PCR, Reverse transcription polymerase chain reaction; qPCR, real-time polymerase chain reaction; WB, Western blot.

 Cite this article as *Cold Spring Harb Perspect Biol* doi: 10.1101/cshperspect.a028530

23R, respectively (Fig. 1A). Signaling by IL-12 stimulates nonreceptor Janus kinase 2 (JAK2) and tyrosine kinase 2 (TYK2) activity resulting in the phosphorylation of signal transducers and activators of transcription (STAT) family members, particularly STAT4 (Zundler and Neurath 2015), whereas STAT3 is preferentially activated following IL-23 stimulation (Floss et al. 2015). In mice and humans, a range of innate and adaptive immune cells have been shown to express IL-12 and IL-23 receptors constitutively as measured by qPCR, flow cytometry, or using an IL-23R GFP KI reporter mouse (Tables 3 and 4), although they appear to be expressed at varying levels, which can be further up-regulated following activation (Ivanov et al. 2006; Ghoreschi et al. 2010; Gaffen et al. 2014). In addition, it was suggested that the IL-12 and IL-23 receptors may not be expressed on the same cell population (Chognard et al. 2014). Therefore, it will be interesting to assess the IL-23R-expressing cells present in tumors and whether their composition differs in different tumor microenvironments. Although mouse tumor cell lines generally have not been reported to express IL-23R, a number of studies have reported that some human tumor cells lines expressed IL-23R and could respond to IL-23 (Table 5). Interestingly, varying levels of IL-23R (as measured by flow cytometry) were also detected on primary tumors, such as pediatric B-ALL cell samples compared to their normal counterparts (Cocco et al. 2010) in diffuse large B-cell lymphoma (DLBCL) and follicular lymphoma (FL) (Cocco et al. 2012) and in a proportion of colorectal carcinoma (CRC) (Suzuki et al. 2012).

MECHANISM OF IL-23 IN PROMOTING TUMOR GROWTH AND METASTASES

The antitumor and antimetastatic activities of IL-12 are thought to be mediated by STAT4 activation of IFN-γ (Colombo and Trinchieri 2002; Trinchieri 2003; Zundler and Neurath 2015). In contrast, the mechanism of action of IL-23 is not fully elucidated. Although IL-23 has been linked with Th17 cells and is crucial for their function and cytokine production (such as IL-17A) in vivo (Muranski and Restifo

2013), IL-23 can have tumor-promoting function independent of IL-17A (as discussed above). Interestingly, a recent study reported that mice lacking IL-17A had reduced lung metastases (Kulig et al. 2016), although mice lacking IL-23 were not examined in the same assay. Nevertheless, it is likely that IL-17A will be involved in the protumor activity of IL-23 because, in some models, anti-IL-17A alone or in combination with anti-IL-23R could reduce bacterial-induced colon carcinogenesis (Wu et al. 2009). Similarly, in a mouse model of Apc-driven colorectal cancer model, IL-23 signaling was shown to promote tumor growth and progression and development of an intratumoral IL-17 response (Grivennikov et al. 2012). However, Th17 cells and IL-17A have also been reported to have tumor-suppressing function in some mouse models of cancer and in certain human cancers (Zou and Restifo 2010; Wilke et al. 2011b). It would appear that the requirement of IL-17 for the tumor-promoting activity of IL-23 may be tumor-dependent. In one study, it was reported that tumor-infiltrating Tregs expressed IL-23R and that blocking IL-23R signaling could reduce Treg numbers and their capacity to secrete IL-10 in a number of experimental mouse tumor models (Kortylewski et al. 2009). However, expression of IL-23R on Tregs has not been reported elsewhere. Nevertheless, this study also showed that STAT3 enhanced the expression of IL-23 in macrophages but inhibited IL-12 in DCs in the tumor microenvironment (Kortylewski et al. 2009). Thus, one mechanism by which IL-23 promotes tumorigenesis may be through driving protumor inflammation to suppress antitumor effector cells.

In addition to IL-17A and IL-10, other Th17 cytokines that have been reported to be regulated by IL-23, including other IL-17 isoforms and IL-22 (Eyerich et al. 2010; Cornelissen et al. 2011). Similar to IL-17A, IL-22 has been shown in different mouse models of inflammation/carcinogen-induced cancer to mediate both tumor-promoting and -suppressing functions (Blake and Teng 2014). Importantly, it is clear that the function of Th17 cells in tumor immunity cannot be linked just to the function of the IL-17A given that other cells of the innate and

Table 3. IL-23R expression on immune cells in mice

Immune cells	Organ	Mouse	Methods	References
γδ T (γδ$^+$)	Spleen, LP	Naïve IL-23R GFP B6	qPCR, FC	Chognard et al. 2014
γδ T (γδ$^+$)	Lung	Naïve IL-23R GFP B6	FC	Paget et al. 2015
γδ T (γδ$^+$) in vitro cultured	Spleen	Naïve B6	qPCR, FC	Sutton et al. 2009
γδ T (γδ$^+$)	LN, LP	Naïve IL-23R GFP B6	FC	Awasthi et al. 2009
γδ T (CD3$^+$ TCRδ$^+$)	Spleen, LN	EAE B6 mice	qPCR, FC	Raverdeau et al. 2016
Memory T cells (DX5$^-$ TCRβ$^+$ CD62L$^-$)	Spleen	Naïve B6	qPCR	Rachitskaya et al. 2008
CD4$^+$ T cells (CD3$^+$ CD4$^+$)	LP	Naïve IL-23R GFP B6	FC	Chan et al. 2014
CD4$^+$ T (CD4$^+$ TCRβ$^+$)	Spleen, LN	Naïve IL-23R GFP B6	FC	Awasthi et al. 2009; Chognard et al. 2014
Th17 cells (CD4$^+$) cultured in in vitro Th17 conditions	Spleen	Naïve B6	qPCR	Ciric et al. 2009; El-Behi et al. 2011
LTi cells (CD3$^-$ CD4$^+$)	LP	Naïve IL-23R GFP B6	FC	Awasthi et al. 2009
Treg (CD4$^+$ FOXP3$^+$)	Tumor, DLN	B16, MC38 and MB49 tumor-bearing B6	FC	Kortylewski et al. 2009
Tc17 (CD8$^+$) Cultured in in vitro Tc17 conditions	Spleen	Naïve B6	qPCR	Ciric et al. 2009
B cells (CD19$^+$ B220$^+$)	Spleen, LP	Naïve IL-23R GFP B6	FC	Chognard et al. 2014
NKT (DX5$^+$ TCRβ$^+$)	Spleen	Naïve B6	qPCR	Rachitskaya et al. 2008
NKT (α-GC/CD1d tetramer$^+$ TCRβ$^+$ NK1.1$^-$) stimulated with anti-CD3 and anti-CD28	Thymus	Naïve B6	qPCR	Coquet et al. 2008
DC (CD11c$^+$)	Spleen, LN	Naïve B6, naïve IL-23R GFP B6	qPCR, FC	Awasthi et al. 2009, El-Behi et al. 2011
Macrophage (CD11b$^+$)	LN, LP	Naïve IL-23R GFP B6	FC	Awasthi et al. 2009
Inflammatory macrophage (CD4$^-$ CD11b$^+$ CD45$^+$)	CNS	EAE B6	qPCR	Cua et al. 2003
Neutrophil thioglycollate induced	Peritoneal cavity	Naïve B6	qPCR	Chen et al. 2016
Neutrophil (Ly6G$^+$)	BM	Naïve B6	qPCR, FC	Taylor et al. 2014
Neutrophils (stimulated)	Colon	DSS-treated B6	qPCR	Zindl et al. 2013
LTβR$^+$ cells	LP	DSS-treated mice	qPCR	Macho-Fernandez et al. 2015
ILC3 (NKp46$^+$ CD127$^+$ CD117$^+$ CD49b$^-$)	Spleen	Naïve IL-23R GFP B6	FC	Chognard et al. 2014
ILC3 (Thy1$^+$ NKp46$^+$ CD3$^-$)	LP	Naïve IL-23R GFP mice	FC	Chan et al. 2014
LTi-like cells (CD3$^-$ CD127$^+$ CD117$^+$)	Spleen	Naïve IL-23R GFP mice	FC	Chognard et al. 2014
LTi (Thy1$^+$ cKIT$^+$ NKp46$^-$ CD3$^-$)	LP	Naïve IL-23R GFP mice	FC	Chan et al. 2014
Thymocytes (Thy1.2$^+$)	Thymus	Naïve B6	qPCR, FC	Li et al. 2014
Thymic epithelial cells	Thymus	Naïve B6	qPCR	Li et al. 2014

FC, Flow cytometry; DLN, draining lymph node; LN, lymph node; LP, lamina propria; EAE, experimental autoimmune encephalomyelitis; B6, C57BL/6 mice; Treg, T regulatory cell; LTi, lymphoid tissue inducer cell; CNS, central nervous system; DSS, dextran sodium sulfate; LTβR, lymphotoxin β receptor; ILC3, type 3 innate lymphoid cells; BM, bone marrow; NKT, natural killer T cell; DC, dendritic cell; GFP, green fluorescent protein; IL, interleukin; qPCR, real-time polymerase chain reaction.

Cite this article as *Cold Spring Harb Perspect Biol* doi: 10.1101/cshperspect.a028530

Table 4. Interleukin (IL)-23R expression on human immune cells

Immune cells	Organ cells were isolated from	Method of detection	References
$\gamma\delta$ T cells (CD3$^+$ TCRδ^+)	PBMC	RT-PCR	Chognard et al. 2014
$\gamma\delta$ T cells (CD3$^+$ Vγ9$^+$)	PBMC, cord blood	FC	Moens et al. 2011
CD4$^+$ T cells (CD3$^+$ CD4$^+$)	PBMC	RT-PCR	Chognard et al. 2014
CD4$^+$ memory T cells (CD4$^+$ CD45RO$^+$)	PBMC	FC	Sarin et al. 2011
Memory T cells (DX5$^-$ TCR$^+$ CD62L$^-$)	PBMC	qPCR	Rachitskaya et al. 2008
Tc17 cells (CD8$^+$)	PBMC	FC	Sarin et al. 2011
CD8$^+$ T cells (CD3$^+$ CD8$^+$)	PBMC	RT-PCR	Chognard et al. 2014
ILC3 (Lin$^-$CD127$^+$ c-Kit$^+$NKp44$^+$)	Tonsil	qPCR	Bernink et al. 2013
Neutrophil	PBMC	IF	Taylor et al. 2014
NKT cells (DX5$^+$ TCRβ^+)	PBMC	qPCR	Rachitskaya et al. 2008
MAIT cells (CD3$^+$ CD161^{high}Vα7.2$^+$CD4$^-$ TCR$\gamma\delta^-$)	Liver blood	Nanostring	Tang et al. 2013

PBMC, Peripheral blood mononuclear cells; MAIT, mucosal-associated invariant T cells; ILC3, type 3 innate lymphoid cells; FC, flow cytometry; IF, immunofluorescence; RT-PCR, reverse transcription polymerase chain reaction; qPCR, real-time polymerase chain reaction; NKT, natural killer T cell.

adaptive immune system, such as $\gamma\delta$ T cells, NK-T cells, and ILCs, also produce substantial quantities of this and other cytokines such as IL-22 (Wilke et al. 2011a,b; Sabat et al. 2013). Other cytokines, including IL-6 and granulocyte macrophage colony-stimulating factor (GM-CSF), can also be induced downstream from IL-23/IL-23R signaling and have also been implicated in the pathogenesis of several autoinflammatory and autoimmune diseases (Yen et al. 2006; Lindroos et al. 2011; Wu et al. 2016). Whether these cytokines contribute to the protumor effect of IL-23 remains to be investigated. IL-23 is also thought to negatively regulate the functions and infiltration of CD8$^+$ T cells into tumor tissue. IL-23p19-deficiency or anti-IL-23p19 antibody treatment has been shown to increase CD8$^+$ T-cell infiltration into DMBA/TPA-treated skin (Langowski et al. 2006), as measured by immunohistochemistry (IHC). Indeed, cytotoxic markers such as FasL, perforin, and granzymes were shown to be upregulated in carcinogen-treated skin of IL-23p19-deficient mice (Langowski et al. 2006). Importantly, within tumors, CD8$^+$ T cells have been shown to be required for tumor suppression mediated by ablation of IL-23/IL-23R signaling (Langowski et al. 2006; Teng et al. 2011; von Scheidt et al. 2014). However, so far, there is no clear evidence that IL-23R is expressed by tumor-infiltrating CD8$^+$ T cells, and it is most

likely that IL-23 suppresses the antitumor activity of CD8$^+$ T cells indirectly. In addition to their effects on immune cells, IL-23R expression on human tumor cell lines has been reported. Coculture with IL-23 generally promoted their proliferation except in the case of B-cell malignancies in which a high dose of IL-23 was used (Table 5). Interestingly, it has also been shown that different concentrations of IL-23 can exert opposite effects on the capability of IL-23R expressing human lung cancer cells to proliferate (Li et al. 2013).

A key function of IL-23 appears to be its ability to promote tumor metastases through up-regulation of proangiogenic factors. Evidence suggests that IL-23 overexpression can induce metastasis of hepatocellular carcinoma (HCC), CRC, melanoma, and esophageal and thyroid cancer (Li et al. 2012; Suzuki et al. 2012; Zhang et al. 2014; Chen et al. 2015; Ganzetti et al. 2015; Klein et al. 2015; Mei et al. 2015). IL-23 was reported to directly up-regulate expression of MMP9 and vascular endothelial growth factor (VEGF)-C in human esophageal cancer cell lines to facilitate epithelial–mesenchymal transition and migratory ability in vitro (Chen et al. 2015). In the same study, patients with esophageal cancer ($n = 23$) who had lymphatic and distant metastasis compared to those who did not had significantly higher IL-23p19 expression (although the source of the anti-IL-

Table 5. Interleukin (IL)-23R expressing human tumor cell lines

Human tumor cell lines expressing IL-23R	IL-23R detection method	Effect of coculture with IL-23	References
Lung adenocarcinoma: A549, SPCA-1	IHC, IF WB	Low concentration promoted proliferation, whereas high concentration suppressed proliferation	Li et al. 2013 Baird et al. 2013
Oral squamous cell carcinoma: HSC-2, HSC-3, HSC-4, Ca9-22	WB, RT-PCR	Promoted proliferation of HSC-3	Fukuda et al. 2010b
Pro-B-ALL cell line: RS4;11 Pre-B-ALL cell line: Nalm-6, Nalm-697	FC	High concentration suppressed proliferation and promoted apoptosis	Cocco et al. 2010
B-cell lymphoma: SU-DHL-4, DOHH-2, OCI-LY8 cell	FC	Suppressed proliferation of SU-DHL-4	Cocco et al. 2012
Colon carcinoma: SW480	RT-PCR, IHC	Not performed	Lan et al. 2011
Colon carcinoma: SW480, HCT116, and HT29	ND	Promoted proliferation of HCT116, HT29, but not SW480	Shapiro et al. 2016
Colon carcinoma: MIP101, DLD-1, and KM12c	WB	Promoted proliferation and invasive capability of DLD-1 only	Suzuki et al. 2012
Hepatocellular carcinoma: PLC8024 and QGY-7703	ND	Promoted invasion and migration and production of MMP9	Li et al. 2012

WB, Western blot; IHC, immunohistochemistry; RT-PCR, reverse transcription polymerase chain reaction; MMP9, matrix metallopeptidase 9; FC, flow cytometry; IF; immunofluorescence; ND, not detected; B-ALL, B-acute lymphoblastic leukemia.

23 antibody used for IHC was not listed [Chen et al. 2015]). Similarly, another study also showed that IL-23 could promote the migration and invasive ability of HCC cell lines through up-regulation of MMP9 expression via activation of NF-κB/p65 (Li et al. 2012). Higher IL-23p19 levels, as measured by mRNA expression and IHC, were also detected in primary HCC tumor tissues with metastasis compared with paired nontumor tissue (Li et al. 2012). IL-23 was also reported to correlate with IL-17A and MMP9 expression in these clinical samples. Similar findings were also reported for IL-23 in facilitating the migration and invasive ability of human thyroid cancer cell lines to migrate and this was mediated via an miR-25/*SOCS4* signaling pathway (Mei et al. 2015). Additionally, higher IL-23p19 mRNA expression was present in thyroid cancer patients who had lymphatic and distant metastasis. It was also reported that IL-23 promoted the metastasis of CRC with impaired *SOCS3* expression via a STAT5 pathway (Zhang et al. 2014). In melanoma, astrocytes (glial cells) were reported to fa-

cilitate human melanoma brain metastasis via secretion of IL-23 in an orthotopic brain melanoma metastases mouse model (Klein et al. 2015). Using IHC, Ganzetti et al. (2015) found that the intensity and percentage of IL-17 and IL-23 was significantly higher in malignant melanomas than in benign melanocytic or Spitz nevi. Overall, it appears that the direct effects of IL-23 are most likely mediated through induction of cell-cycle pathways (cyclin-dependent kinases and cyclin D) and oncogenic and associated genes that regulate growth-factor-induced signaling (AKT, NF-κB, AP-1) (Kortylewski et al. 2009; Chan et al. 2014).

CROSS-REGULATION OF IL-12 AND IL-23

In tumors, the ratio of IL-12 and IL-23 produced by DCs and macrophages will be determined by the balance of endogenous Toll-like receptor (TLR) agonists, danger signals, and/or tumor-derived mediators in the tumor microenvironment, leading to activation of their respective downstream pathways, which ultimately dictate tumor growth outcome (Gerosa et al. 2008; Ngiow et al. 2013). The mechanism by which IL-12 and IL-23 regulate each other is not well elucidated and may be quite complex. Some studies have reported that IFN-γ can negatively regulate IL-23 production by inhibiting mouse *Il23a* gene-promoter activity (Sheikh et al. 2010, 2011). Conversely, another study showed that IL-23 antagonized IL-12-induced secretion of IFN-γ (Sieve et al. 2010). Adding to the complexity, a recent study in mice suggested signaling through IL-23R may potentially promote IL-12 production, whereas signaling through IL-12Rβ2 suppresses IL-1β and IL-23 (Chognard et al. 2014). Furthermore, a dichotomous pattern of expression for IL-12 and IL-23 receptors in both mouse and humans was reported in this study, suggesting that immune cells involved in antitumor responses may be quite distinct from those that are tumor promoting. Finally, IL-12 and IL-23 can also be regulated both genetically and epigenetically. It was reported that conventional DCs (cDCs) and plasmacytoid DCs (pDCs) from *Grm4*$^{-/-}$ mice produced higher amounts of IL-6 and IL-23, but less IL-12 and IL-27 compared to their WT counterparts in response to LPS or CpG-ODN, respectively (Fallarino et al. 2010). In addition, phosphatase 2A has been shown to negatively regulate IL-23 but not IL-12 in LPS stimulated DC by suppressing *IL-23p19* gene expression (Chang et al. 2010). In contrast, Trabid, a deubiquitinase, was reported to mediate epigenetic regulation of both *Il12* and *Il23* gene expression (Balhara et al. 2016), while deubiquitinases ADAM10 and ADAM17 have been reported to mediate IL-23R ectodomain shedding (Franke et al. 2016). In mice and humans, differential splicing of the *IL-23R* gene has also been reported to generate antagonistic soluble IL-23R variants (Zhang et al. 2006; Kan et al. 2008; Mancini et al. 2008; Floss et al. 2015). These soluble IL-23R may potentially reduce the cellular responsiveness of IL-23R expressing cells toward IL-23 and can also bind to IL-23 and act as competitive antagonist of IL-23 signaling (Franke et al. 2016).

Recently, it was reported that IL-4 had opposing effects on the production of either IL-12 or IL-23 in which it promoted the IL-12-producing capacity of DCs while abrogating IL-23 production (Guenova et al. 2015). It is also possible that the IL-12 and IL-23 pathways can be regulated by their shared IL-12p40, p35, p19 subunits, and corresponding receptors. Free IL-12p40 exists in either homodimeric IL-12p80 or monomeric IL-12p40 forms in mice and human, and can act as natural antagonists of IL-12 and IL-23 by competing for binding to IL-12Rβ1 (Mattner et al. 1993; Gillessen et al. 1995; Ling et al. 1995; Trinchieri 2003; Shimozato et al. 2006). Similarly, in mice, it has been reported that IL-12p80 suppressed splenic Tregs via induction of nitric oxide from APCs, and this suppression was dependent on IL-12Rβ1 rather than IL-12Rβ2 in vitro (Brahmachari and Pahan 2009). Given that IL-12p35 is also used as a shared subunit by the inhibitory cytokine IL-35 (IL-12p35/Ebi3 [IL-27p28]) produced by Tregs and B cells (Collison et al. 2007; Shen et al. 2014), one must also consider how the balance between IL-12 and IL-35 impacts on tumorigenesis (Banchereau et al. 2012). Similarly, a recent study showing that IL-23p19 can interact with Ebi3 (Ramnath et al. 2015) raises new questions regarding its physiological function and how it may regulate the other IL-12 family cytokines in which common subunits are shared.

Interestingly, a number of papers have also reported that immune checkpoint receptors such as TIM-3 and PD-1 expressed on monocytes/macrophages can regulate the balance of IL-12/IL-23 in viral infections such as hepatitis C (Zhang et al. 2011a,b; Wang et al. 2013). Similarly, a recent study reported that agonistic anti-BTLA inhibited DC-induced Th17- and Th1-cell responses because of decreased production of the Th17- and Th1-related cytokines IL-1β, IL-6, IL-23, and IL-12p70 and reduced CD40 expression in DCs (Ye et al. 2016). Given that checkpoint receptors are a major pathway of tumor-induced immune suppression (Pardoll 2012), this represents another mechanism by which the IL-12/IL-23 balance can be affected. Overall, the subunit sharing between the IL-12 family of cytokines makes it difficult to definitively delineate the function of each cytokine. Generating antibodies that can specifically neutralize the specific cytokine rather than the shared subunits might be a better approach for dissecting out the endogenous role these family members play in promoting or suppressing tumorigenesis.

TARGETING THE IL-12/IL-23 AXIS OF INFLAMMATION FOR CANCER THERAPY

It is now recognized that the immune system is actively suppressed in the tumor microenvironment and that lymphoid, myeloid, and granulocytic cells contribute to this suppression. In the past 5 years, therapies targeting T-cell immune checkpoint receptors have achieved remarkable success in the clinic, particularly in combination, and will increasingly become incorporated into standard-of-care for many cancer types (Smyth et al. 2016). Nevertheless, not all cancer types respond to checkpoint blockade even when targeted in combination. Given that human tumors are extremely heterogeneous with respect to their proportions of immune cells, blocking immunosuppressive pathways mediated by both T and myeloid cells may be required in certain cancer types to release full endogenous antitumor immunity.

A number of preclinical mouse studies have shown that changing the balance of IL-12 and IL-23 can promote tumor progression or suppression and thus targeting this axis may be beneficial particularly in combination immunotherapies. In one study, intratumoral IL-12 application in combination with systemic checkpoint blockade of CTLA-4 resulted in eradication of mice with orthotopic gliomas (Vom Berg et al. 2013). Similarly, another study showed a combination of agonistic anti-CD40 monoclonal antibodies (mAbs) to drive IL-12 production and anti-IL-23 mAbs to counter the tumor-promoting effects of IL-23 had greater antitumor activity than either agent alone (von Scheidt et al. 2014). The efficacy of this combination may potentially be effective in patients whose cancer display rich myeloid infiltrates and up-regulated IL-23 (e.g., sarcomas), where neutralization of IL-23 in

the tumor microenvironment may allow the agonistic activity of anti-CD40 to be fully maximized. More recently, agonistic CD40 mAb-driven IL-12 was shown to reverse resistance to anti-PD-1 therapy in T-cell-rich tumors (Ngiow et al. 2016). Similarly, in a preclinical murine model of bladder cancer, Tasquinimod, a small molecule that binds S100A9 increased tumor mRNA expression levels of *Il12b*, *Tbet*, and *Ifng* and synergized with anti-PD-L1 to suppress tumor growth (Nakhle et al. 2016).

Prophylactic neutralization of IL-23 has been shown to significantly suppress experimental lung metastases of B16F10 melanoma, RM-1 prostate carcinoma, and 3LL lung carcinoma, which are controlled by host NK cells rather than CD8$^+$ T cells (Teng et al. 2010, 2011). In mice bearing established lung metastases, IL-2 immunotherapy was enhanced in mice deficient for IL-23p19, suggesting neutralizing anti-IL-23 mAbs may synergize with immunotherapies that activate NK cells such as IL-2 (Teng et al. 2010). Neutralization of IL-23 also synergized with anti-ERBB2 mAb in suppressing subcutaneous growth of established Her-2/neu-positive breast tumors in mice (Teng et al. 2010). In addition, targeting of activating receptors such as CD137 or CD226 (DNAM-1) (Kohrt et al. 2012; Blake et al. 2016) or inhibitory receptors such as CD96 and TIGIT may also synergize with IL-23 neutralization to suppress metastases (Blake et al. 2016). Alternatively, anti-IL-23p19 mAbs can be combined with chemotherapy, such as gemcitabine, that has been shown to induce IL-23p19 and IL-23R mRNA expression in non-small-cell lung carcinoma (NSCLC) cell lines (Baird et al. 2013).

CONCLUSIONS

In this review, we have discussed how the balance of proinflammatory cytokines IL-12 and IL-23 plays a key role in shaping the development of antitumor or protumor immunity, respectively. Although the antitumor efficacy of IL-12 is generally mediated via downstream activation of IFN-γ, IL-23 can have a tumor-promoting function independent of IL-17A, such as directly up-regulating proangiogenic factors to facilitate the epithelial–mesenchymal transition and migratory ability of tumor cells. Preclinically, strategies to alter the ratio of IL-12 and IL-23 in the tumor microenvironment have been shown to synergize in combination with other anticancer therapies. Clinically, IL-23 is overexpressed in a number of cancer types, which could potentially be targeted, particularly in those that display rich myeloid infiltrates. Although there are obvious caveats in the interpretation of mouse studies on the role of IL-12 and IL-23 in tumor biology, clinical studies have reported the favorable safety profile of anti-IL-12/23p40 mAbs (Ustekinumab). However, there was an increased risk in developing malignancies in psoriasis patients treated when high doses of anti-IL-12/23p40 mAbs were administered (Young and Czarnecki 2012). In contrast, major adverse cardiac events were observed in psoriasis patients treated with another anti-IL-12/23p40 mAb (Briakinumab) (Teng et al. 2015). To date, a number of IL-23-specific antagonists have showed efficacy and safety in the treatment of psoriasis patients in late-stage clinical trials (Gordon et al. 2015; Teng et al. 2015). Should larger and longer-term trials show that neutralization of IL-23 impacts minimally on the risk of malignancies and infection development, anti-IL-23 could potentially be repositioned for use in immuno-oncology in combination with other immunotherapies.

ACKNOWLEDGMENTS

M.W.L.T. is supported by National Health and Medical Research Council of Australia (NHMRC) and Cancer Council of Queensland (CCQ) project grants. M.J.S. is supported by an NHMRC Project Grant, an NHMRC Senior Principal Research Fellowship, and the CCQ. J.Y. is supported by a University of Queensland International Scholarship and Queensland Institute of Medical Research (QIMR) Berghofer International Scholarship. M.J.S. has scientific research agreements with Bristol-Myers Squibb, Corvus Pharmaceuticals, and Aduro Biotech. The other authors declare that they have no conflicting financial interests.

REFERENCES

Aggarwal S, Ghilardi N, Xie MH, de Sauvage FJ, Gurney AL. 2003. Interleukin-23 promotes a distinct CD4 T cell activation state characterized by the production of interleukin-17. *J Biol Chem* **278:** 1910–1914.

Airoldi I, Di Carlo E, Cocco C, Sorrentino C, Fais F, Cilli M, D'Antuono T, Colombo MP, Pistoia V. 2005. Lack of Il12rb2 signaling predisposes to spontaneous autoimmunity and malignancy. *Blood* **106:** 3846–3853.

Awasthi A, Riol-Blanco L, Jager A, Korn T, Pot C, Galileos G, Bettelli E, Kuchroo VK, Oukka M. 2009. Cutting edge: IL-23 receptor gfp reporter mice reveal distinct populations of IL-17-producing cells. *J Immunol* **182:** 5904–5908.

Baird AM, Leonard J, Naicker KM, Kilmartin L, O'Byrne KJ, Gray SG. 2013. IL-23 is pro-proliferative, epigenetically regulated and modulated by chemotherapy in non-small cell lung cancer. *Lung Cancer* **79:** 83–90.

Balhara J, Shan L, Zhang J, Muhuri A, Halayko AJ, Almiski MS, Doeing D, McConville J, Matzuk MM, Gounni AS. 2016. Pentraxin 3 deletion aggravates allergic inflammation through a TH17-dominant phenotype and enhanced CD4 T-cell survival. *J Allergy Clin Immunol* **139:** 950–963.

Banchereau J, Pascual V, O'Garra A. 2012. From IL-2 to IL-37: The expanding spectrum of anti-inflammatory cytokines. *Nat Immunol* **13:** 925–931.

Bernink JH, Peters CP, Munneke M, te Velde AA, Meijer SL, Weijer K, Hreggvidsdottir HS, Heinsbroek SE, Legrand N, Buskens CJ, et al. 2013. Human type 1 innate lymphoid cells accumulate in inflamed mucosal tissues. *Nat Immunol* **14:** 221–229.

Blake SJ, Teng MW. 2014. Role of IL-17 and IL-22 in autoimmunity and cancer. *Actas Dermosifiliogr* **105:** 41–50.

Blake SJ, Stannard K, Liu J, Allen S, Yong MC, Mittal D, Aguilera AR, Miles JJ, Lutzky VP, de Andrade LF, et al. 2016. Suppression of metastases using a new lymphocyte checkpoint target for cancer immunotherapy. *Cancer Discov* **6:** 446–459.

Brahmachari S, Pahan K. 2009. Suppression of regulatory T cells by IL-12p40 homodimer via nitric oxide. *J Immunol* **183:** 2045–2058.

Brentano F, Ospelt C, Stanczyk J, Gay RE, Gay S, Kyburz D. 2009. Abundant expression of the interleukin (IL)23 subunit p19, but low levels of bioactive IL23 in the rheumatoid synovium: Differential expression and Toll-like receptor-(TLR) dependent regulation of the IL23 subunits, p19 and p40, in rheumatoid arthritis. *Ann Rheum Dis* **68:** 143–150.

Cam C, Karagoz B, Muftuoglu T, Bigi O, Emirzeoglu L, Celik S, Ozgun A, Tuncel T, Top C. 2016. The inflammatory cytokine interleukin-23 is elevated in lung cancer, particularly small cell type. *Contemp Oncol (Pozn)* **20:** 215–219.

Cao X, Leonard K, Collins LI, Cai SF, Mayer JC, Payton JE, Walter MJ, Piwnica-Worms D, Schreiber RD, Ley TJ. 2009. Interleukin 12 stimulates IFN-γ-mediated inhibition of tumor-induced regulatory T-cell proliferation and enhances tumor clearance. *Cancer Res* **69:** 8700–8709.

Chan IH, Jain R, Tessmer MS, Gorman D, Mangadu R, Sathe M, Vives F, Moon C, Penaflor E, Turner S, et al. 2014. Interleukin-23 is sufficient to induce rapid de novo gut tumorigenesis, independent of carcinogens, through activation of innate lymphoid cells. *Mucosal Immunol* **7:** 842–856.

Chang J, Voorhees TJ, Liu Y, Zhao Y, Chang CH. 2010. Interleukin-23 production in dendritic cells is negatively regulated by protein phosphatase 2A. *Proc Natl Acad Sci* **107:** 8340–8345.

Chang HH, Young SH, Sinnett-Smith J, Chou CE, Moro A, Hertzer KM, Hines OJ, Rozengurt E, Eibl G. 2015. Prostaglandin E2 activates the mTORC1 pathway through an EP4/cAMP/PKA- and EP1/Ca^{2+}-mediated mechanism in the human pancreatic carcinoma cell line PANC-1. *Am J Physiol Cell Physiol* **309:** C639–C649.

Chen D, Li W, Liu S, Su Y, Han G, Xu C, Liu H, Zheng T, Zhou Y, Mao C. 2015. Interleukin-23 promotes the epithelial-mesenchymal transition of oesophageal carcinoma cells via the Wnt/β-catenin pathway. *Sci Rep* **5:** 8604.

Chen F, Cao A, Yao S, Evans-Marin HL, Liu H, Wu W, Carlsen ED, Dann SM, Soong L, Sun J, et al. 2016. mTOR mediates IL-23 induction of neutrophil IL-17 and IL-22 production. *J Immunol* **196:** 4390–4399.

Chiyo M, Shimozato O, Iizasa T, Fujisawa T, Tagawa M. 2004. Antitumor effects produced by transduction of dendritic cells-derived heterodimeric cytokine genes in murine colon carcinoma cells. *Anticancer Res* **24:** 3763–3767.

Chognard G, Bellemare L, Pelletier AN, Dominguez-Punaro MC, Beauchamp C, Guyon AJ, Charron G, Morin N, Sivanesan D, Kuchroo V, et al. 2014. The dichotomous pattern of IL-12R and IL-23R expression elucidates the role of IL-12 and IL-23 in inflammation. *PLoS ONE* **9:** e89092.

Chu H, Cao W, Chen W, Pan S, Xiao Y, Liu Y, Gu H, Guo W, Xu L, Hu Z, et al. 2012. Potentially functional polymorphisms in IL-23 receptor and risk of esophageal cancer in a Chinese population. *Int J Cancer* **130:** 1093–1097.

Ciric B, El-behi M, Cabrera R, Zhang GX, Rostami A. 2009. IL-23 drives pathogenic IL-17-producing $CD8^+$ T cells. *J Immunol* **182:** 5296–5305.

Cocco C, Canale S, Frasson C, Di Carlo E, Ognio E, Ribatti D, Prigione I, Basso G, Airoldi I. 2010. Interleukin-23 acts as antitumor agent on childhood B-acute lymphoblastic leukemia cells. *Blood* **116:** 3887–3898.

Cocco C, Di Carlo E, Zupo S, Canale S, Zorzoli A, Ribatti D, Morandi F, Ognio E, Airoldi I. 2012. Complementary IL-23 and IL-27 anti-tumor activities cause strong inhibition of human follicular and diffuse large B-cell lymphoma growth in vivo. *Leukemia* **26:** 1365–1374.

Collison LW, Workman CJ, Kuo TT, Boyd K, Wang Y, Vignali KM, Cross R, Sehy D, Blumberg RS, Vignali DA. 2007. The inhibitory cytokine IL-35 contributes to regulatory T-cell function. *Nature* **450:** 566–569.

Colombo MP, Trinchieri G. 2002. Interleukin-12 in antitumor immunity and immunotherapy. *Cytokine Growth Factor Rev* **13:** 155–168.

Coquet JM, Chakravarti S, Kyparissoudis K, McNab FW, Pitt LA, McKenzie BS, Berzins SP, Smyth MJ, Godfrey DI. 2008. Diverse cytokine production by NKT cell subsets and identification of an IL-17-producing $CD4^-NK1.1^-$ NKT cell population. *Proc Natl Acad Sci* **105:** 11287–11292.

Cornelissen F, Aparicio Domingo P, Reijmers RM, Cupedo T. 2011. Activation and effector functions of human

RORC$^+$ innate lymphoid cells. *Curr Opin Immunol* **23:** 361–367.

Cua DJ, Sherlock J, Chen Y, Murphy CA, Joyce B, Seymour B, Lucian L, To W, Kwan S, Churakova T, et al. 2003. Interleukin-23 rather than interleukin-12 is the critical cytokine for autoimmune inflammation of the brain. *Nature* **421:** 744–748.

Dunn GP, Koebel CM, Schreiber RD. 2006. Interferons, immunity and cancer immunoediting. *Nat Rev Immunol* **6:** 836–848.

Eisenring M, vom Berg J, Kristiansen G, Saller E, Becher B. 2010. IL-12 initiates tumor rejection via lymphoid tissue-inducer cells bearing the natural cytotoxicity receptor NKp46. *Nat Immunol* **11:** 1030–1038.

El-Behi M, Ciric B, Dai H, Yan Y, Cullimore M, Safavi F, Zhang GX, Dittel BN, Rostami A. 2011. The encephalitogenicity of T_H17 cells is dependent on IL-1- and IL-23-induced production of the cytokine GM-CSF. *Nat Immunol* **12:** 568–575.

Elinav E, Nowarski R, Thaiss CA, Hu B, Jin C, Flavell RA. 2013. Inflammation-induced cancer: Crosstalk between tumours, immune cells and microorganisms. *Nat Rev Cancer* **13:** 759–771.

Eyerich S, Eyerich K, Cavani A, Schmidt-Weber C. 2010. IL-17 and IL-22: Siblings, not twins. *Trends Immunol* **31:** 354–361.

Fallarino F, Volpi C, Fazio F, Notartomaso S, Vacca C, Busceti C, Bicciato S, Battaglia G, Bruno V, Puccetti P, et al. 2010. Metabotropic glutamate receptor-4 modulates adaptive immunity and restrains neuroinflammation. *Nat Med* **16:** 897–902.

Flores RR, Kim E, Zhou L, Yang C, Zhao J, Gambotto A, Robbins PD. 2015. IL-Y, a synthetic member of the IL-12 cytokine family, suppresses the development of type 1 diabetes in NOD mice. *Eur J Immunol* **45:** 3114–3125.

Floss DM, Schroder J, Franke M, Scheller J. 2015. Insights into IL-23 biology: From structure to function. *Cytokine Growth Factor Rev* **26:** 569–578.

Franke M, Schröder J, Monhasery N, Ackfeld T, Hummel TM, Rabe B, Garbers C, Becker-Pauly C, Floss DM, Scheller J. 2016. Human and murine interleukin 23 receptors are novel substrates for a disintegrin and metalloproteases ADAM10 and ADAM17. *J Biol Chem* **291:** 10551–10561.

Fukuda M, Ehara M, Suzuki S, Ohmori Y, Sakashita H. 2010a. IL-23 promotes growth and proliferation in human squamous cell carcinoma of the oral cavity. *Int J Oncol* **36:** 1355–1365.

Fukuda M, Ehara M, Suzuki S, Sakashita H. 2010b. Expression of interleukin-23 and its receptors in human squamous cell carcinoma of the oral cavity. *Mol Med Rep* **3:** 89–93.

Gaffen SL, Jain R, Garg AV, Cua DJ. 2014. The IL-23-IL-17 immune axis: From mechanisms to therapeutic testing. *Nat Rev Immunol* **14:** 585–600.

Gangemi S, Minciullo P, Adamo B, Franchina T, Ricciardi GR, Ferraro M, Briguglio R, Toscano G, Saitta S, Adamo V. 2012. Clinical significance of circulating interleukin-23 as a prognostic factor in breast cancer patients. *J Cell Biochem* **113:** 2122–2125.

Cite this article as *Cold Spring Harb Perspect Biol* doi: 10.1101/cshperspect.a028530

Ganzetti G, Rubini C, Campanati A, Zizzi A, Molinelli E, Rosa L, Simonacci F, Offidani A. 2015. IL-17, IL-23, and p73 expression in cutaneous melanoma: A pilot study. *Melanoma Res* **25**: 232–238.

Gerosa F, Baldani-Guerra B, Lyakh LA, Batoni G, Esin S, Winkler-Pickett RT, Consolaro MR, De Marchi M, Giachino D, Robbiano A, et al. 2008. Differential regulation of interleukin 12 and interleukin 23 production in human dendritic cells. *J Exp Med* **205**: 1447–1461.

Ghoreschi K, Laurence A, Yang XP, Tato CM, McGeachy MJ, Konkel JE, Ramos HL, Wei L, Davidson TS, Bouladoux N, et al. 2010. Generation of pathogenic T_H17 cells in the absence of TGF-β signalling. *Nature* **467**: 967–971.

Gillessen S, Carvajal D, Ling P, Podlaski FJ, Stremlo DL, Familletti PC, Gubler U, Presky DH, Stern AS, Gately MK. 1995. Mouse interleukin-12 (IL-12) p40 homodimer: A potent IL-12 antagonist. *Eur J Immunol* **25**: 200–206.

Gordon KB, Duffin KC, Bissonnette R, Prinz JC, Wasfi Y, Li S, Shen YK, Szapary P, Randazzo B, Reich K. 2015. A phase 2 trial of guselkumab versus adalimumab for plaque psoriasis. *N Engl J Med* **373**: 136–144.

Grivennikov SI, Greten FR, Karin M. 2010. Immunity, inflammation, and cancer. *Cell* **140**: 883–899.

Grivennikov SI, Wang K, Mucida D, Stewart CA, Schnabl B, Jauch D, Taniguchi K, Yu GY, Osterreicher CH, Hung KE, et al. 2012. Adenoma-linked barrier defects and microbial products drive IL-23/IL-17-mediated tumour growth. *Nature* **491**: 254–258.

Guenova E, Skabytska Y, Hoetzenecker W, Weindl G, Sauer K, Tham M, Kim KW, Park JH, Seo JH, Ignatova D, et al. 2015. IL-4 abrogates T_H17 cell-mediated inflammation by selective silencing of IL-23 in antigen-presenting cells. *Proc Natl Acad Sci* **112**: 2163–2168.

He S, Fei M, Wu Y, Zheng D, Wan D, Wang L, Li D. 2011. Distribution and clinical significance of Th17 cells in the tumor microenvironment and peripheral blood of pancreatic cancer patients. *Int J Mol Sci* **12**: 7424–7437.

Hu J, Yuan X, Belladonna ML, Ong JM, Wachsmann-Hogiu S, Farkas DL, Black KL, Yu JS. 2006. Induction of potent antitumor immunity by intratumoral injection of interleukin 23-transduced dendritic cells. *Cancer Res* **66**: 8887–8896.

Hu P, Hu HD, Chen M, Peng ML, Tang L, Tang KF, Matsui M, Belladonna ML, Yoshimoto T, Zhang DZ, et al. 2009. Expression of interleukins-23 and 27 leads to successful gene therapy of hepatocellular carcinoma. *Mol Immunol* **46**: 1654–1662.

Hunter CA. 2005. New IL-12-family members: IL-23 and IL-27, cytokines with divergent functions. *Nat Rev Immunol* **5**: 521–531.

Ivanov II, McKenzie BS, Zhou L, Tadokoro CE, Lepelley A, Lafaille JJ, Cua DJ, Littman DR. 2006. The orphan nuclear receptor RORγt directs the differentiation program of proinflammatory IL-17+ T helper cells. *Cell* **126**: 1121–1133.

Jantschitsch C, Weichenthal M, Proksch E, Schwarz T, Schwarz A. 2012. IL-12 and IL-23 affect photocarcinogenesis differently. *J Invest Dermatol* **132**: 1479–1486.

Kaiga T, Sato M, Kaneda H, Iwakura Y, Takayama T, Tahara H. 2007. Systemic administration of IL-23 induces potent antitumor immunity primarily mediated through Th1-type response in association with the endogenously expressed IL-12. *J Immunol* **178**: 7571–7580.

Kan SH, Mancini G, Gallagher G. 2008. Identification and characterization of multiple splice forms of the human interleukin-23 receptor α chain in mitogen-activated leukocytes. *Genes Immun* **9**: 631–639.

Kaplan DH, Shankaran V, Dighe AS, Stockert E, Aguet M, Old LJ, Schreiber RD. 1998. Demonstration of an interferon γ-dependent tumor surveillance system in immunocompetent mice. *Proc Natl Acad Sci* **95**: 7556–7561.

Klein A, Schwartz H, Sagi-Assif O, Meshel T, Izraely S, Ben Menachem S, Bengaiev R, Ben-Shmuel A, Nahmias C, Couraud PO, et al. 2015. Astrocytes facilitate melanoma brain metastasis via secretion of IL-23. *J Pathol* **236**: 116–127.

Koebel CM, Vermi W, Swann JB, Zerafa N, Rodig SJ, Old LJ, Smyth MJ, Schreiber RD. 2007. Adaptive immunity maintains occult cancer in an equilibrium state. *Nature* **450**: 903–907.

Kohrt HE, Houot R, Weiskopf K, Goldstein MJ, Scheeren F, Czerwinski D, Colevas AD, Weng WK, Clarke MF, Carlson RW, et al. 2012. Stimulation of natural killer cells with a CD137-specific antibody enhances trastuzumab efficacy in xenotransplant models of breast cancer. *J Clin Invest* **122**: 1066–1075.

Kortylewski M, Xin H, Kujawski M, Lee H, Liu Y, Harris T, Drake C, Pardoll D, Yu H. 2009. Regulation of the IL-23 and IL-12 balance by Stat3 signaling in the tumor microenvironment. *Cancer Cell* **15**: 114–123.

Kulig P, Burkhard S, Mikita-Geoffroy J, Croxford AL, Hovelmeyer N, Gyulveszi G, Gorzelanny C, Waisman A, Borsig L, Becher B. 2016. IL17A-mediated endothelial breach promotes metastasis formation. *Cancer Immunol Res* **4**: 26–32.

Kvedaraite E, Lourda M, Idestrom M, Chen P, Olsson-Akefeldt S, Forkel M, Gavhed D, Lindforss U, Mjosberg J, Henter JI, et al. 2016. Tissue-infiltrating neutrophils represent the main source of IL-23 in the colon of patients with IBD. *Gut* **65**: 1632–1641.

Lan F, Zhang L, Wu J, Zhang J, Zhang S, Li K, Qi Y, Lin P. 2011. IL-23/IL-23R: Potential mediator of intestinal tumor progression from adenomatous polyps to colorectal carcinoma. *Int J Colorectal Dis* **26**: 1511–1518.

Langowski JL, Zhang X, Wu L, Mattson JD, Chen T, Smith K, Basham B, McClanahan T, Kastelein RA, Oft M. 2006. IL-23 promotes tumour incidence and growth. *Nature* **442**: 461–465.

Langrish CL, McKenzie BS, Wilson NJ, de Waal Malefyt R, Kastelein RA, Cua DJ. 2004. IL-12 and IL-23: Master regulators of innate and adaptive immunity. *Immunol Rev* **202**: 96–105.

Langrish CL, Chen Y, Blumenschein WM, Mattson J, Basham B, Sedgwick JD, McClanahan T, Kastelein RA, Cua DJ. 2005. IL-23 drives a pathogenic T cell population that induces autoimmune inflammation. *J Exp Med* **201**: 233–240.

Li J, Lau G, Chen L, Yuan YF, Huang J, Luk JM, Xie D, Guan XY. 2012. Interleukin 23 promotes hepatocellular carcinoma metastasis via NF-κB induced matrix metalloproteinase 9 expression. *PLoS ONE* **7**: e46264.

Li J, Zhang L, Zhang J, Wei Y, Li K, Huang L, Zhang S, Gao B, Wang X, Lin P. 2013. Interleukin 23 regulates proliferation of lung cancer cells in a concentration-dependent way in association with the interleukin-23 receptor. *Carcinogenesis* **34:** 658–666.

Li H, Hsu HC, Wu Q, Yang P, Li J, Luo B, Oukka M, Steele CH 3rd, Cua DJ, Grizzle WE, et al. 2014. IL-23 promotes TCR-mediated negative selection of thymocytes through the upregulation of IL-23 receptor and RORγt. *Nat Commun* **5:** 4259.

Liao C, Yu ZB, Meng G, Wang L, Liu QY, Chen LT, Feng SS, Tu HB, Li YF, Bai L. 2015. Association between Th17-related cytokines and risk of non–small cell lung cancer among patients with or without chronic obstructive pulmonary disease. *Cancer* **121:** 3122–3129.

Lindroos J, Svensson L, Norsgaard H, Lovato P, Moller K, Hagedorn PH, Olsen GM, Labuda T. 2011. IL-23-mediated epidermal hyperplasia is dependent on IL-6. *J Invest Dermatol* **131:** 1110–1118.

Ling P, Gately MK, Gubler U, Stern AS, Lin P, Hollfelder K, Su C, Pan YC, Hakimi J. 1995. Human IL-12 p40 homodimer binds to the IL-12 receptor but does not mediate biologic activity. *J Immunol* **154:** 116–127.

Liu J, Xiang Z, Ma X. 2004. Role of IFN regulatory factor-1 and IL-12 in immunological resistance to pathogenesis of *N*-methyl-*N*-nitrosourea-induced T lymphoma. *J Immunol* **173:** 1184–1193.

Ljujic B, Radosavljevic G, Jovanovic I, Pavlovic S, Zdravkovic N, Milovanovic M, Acimovic L, Knezevic M, Bankovic D, Zdravkovic D, et al. 2010. Elevated serum level of IL-23 correlates with expression of VEGF in human colorectal carcinoma. *Arch Med Res* **41:** 182–189.

Lo CH, Lee SC, Wu PY, Pan WY, Su J, Cheng CW, Roffler SR, Chiang BL, Lee CN, Wu CW, et al. 2003. Antitumor and antimetastatic activity of IL-23. *J Immunol* **171:** 600–607.

Macho-Fernandez E, Koroleva EP, Spencer CM, Tighe M, Torrado E, Cooper AM, Fu YX, Tumanov AV. 2015. Lymphotoxin β receptor signaling limits mucosal damage through driving IL-23 production by epithelial cells. *Mucosal Immunol* **8:** 403–413.

Majewski S, Jantschitsch C, Maeda A, Schwarz T, Schwarz A. 2010. IL-23 antagonizes UVR-induced immunosuppression through two mechanisms: Reduction of UVR-induced DNA damage and inhibition of UVR-induced regulatory T cells. *J Invest Dermatol* **130:** 554–562.

Mancini G, Kan SH, Gallagher G. 2008. A novel insertion variant of the human IL-23 receptor-α chain transcript. *Genes Immun* **9:** 566–569.

Mattner F, Fischer S, Guckes S, Jin S, Kaulen H, Schmitt E, Rude I, Germann T. 1993. The interleukin-12 subunit p40 specifically inhibits effects of the interleukin-12 heterodimer. *Eur J Immunol* **23:** 2202–2208.

Mei Z, Chen S, Chen C, Xiao B, Li F, Wang Y, Tao Z. 2015. Interleukin-23 facilitates thyroid cancer cell migration and invasion by inhibiting *SOCS4* expression via microRNA-25. *PLoS ONE* **10:** e0139456.

Moens E, Brouwer M, Dimova T, Goldman M, Willems F, Vermijlen D. 2011. IL-23R and TCR signaling drives the generation of neonatal Vγ9Vδ2 T cells expressing high levels of cytotoxic mediators and producing IFN-γ and IL-17. *J Leukoc Biol* **89:** 743–752.

Muranski P, Restifo NP. 2013. Essentials of Th17 cell commitment and plasticity. *Blood* **121:** 2402–2414.

Nakhle J, Pierron V, Bauchet AL, Plas P, Thiongane A, Meyer-Losic F, Schmidlin F. 2016. Tasquinimod modulates tumor-infiltrating myeloid cells and improves the antitumor immune response to PD-L1 blockade in bladder cancer. *Oncoimmunology* **5:** e1145333.

Ngiow SF, Teng MWL, Smyth MJ. 2013. A balance of interleukin-12 and -23 in cancer. *Trends Immunol* **34:** 548–555.

Ngiow SF, Young A, Blake SJ, Hill GR, Yagita H, Teng MW, Korman AJ, Smyth MJ. 2016. Agonistic CD40 mAb-driven IL-12 reverses resistance to anti-PD1 in a T cell-rich tumor. *Cancer Res* **76:** 6266–6277.

Oniki S, Nagai H, Horikawa T, Furukawa J, Belladonna ML, Yoshimoto T, Hara I, Nishigori C. 2006. Interleukin-23 and interleukin-27 exert quite different antitumor and vaccine effects on poorly immunogenic melanoma. *Cancer Res* **66:** 6395–6404.

Paget C, Chow MT, Gherardin NA, Beavis PA, Uldrich AP, Duret H, Hassane M, Souza-Fonseca-Guimaraes F, Mogilenko DA, Staumont-Salle D, et al. 2015. CD3^{bright} signals on γδ T cells identify IL-17A-producing Vγ6Vδ1$^+$ T cells. *Immunol Cell Biol* **93:** 198–212.

Pardoll DM. 2012. The blockade of immune checkpoints in cancer immunotherapy. *Nat Rev Cancer* **12:** 252–264.

Prabhala RH, Pelluru D, Fulciniti M, Prabhala HK, Nanjappa P, Song W, Pai C, Amin S, Tai YT, Richardson PG, et al. 2010. Elevated IL-17 produced by T$_H$17 cells promotes myeloma cell growth and inhibits immune function in multiple myeloma. *Blood* **115:** 5385–5392.

Pylayeva-Gupta Y. 2016. Molecular pathways: Interleukin-35 in autoimmunity and cancer. *Clin Cancer Res* **22:** 4973–4978.

Qian X, Cao S, Yang G, Pan Y, Yin C, Chen X, Zhu Y, Zhuang Y, Shen Y, Hu Z. 2013a. Potentially functional polymorphism and risk of acute myeloid leukemia in a Chinese population. *PLoS ONE* **8:** e55473.

Qian X, Gu L, Ning H, Zhang Y, Hsueh EC, Fu M, Hu X, Wei L, Hoft DF, Liu J. 2013b. Increased Th17 cells in the tumor microenvironment is mediated by IL-23 via tumor-secreted PGE$_2$. *J Immunol* **190:** 5894–5902.

Rachitskaya AV, Hansen AM, Horai R, Li Z, Villasmil R, Luger D, Nussenblatt RB, Caspi RR. 2008. Cutting edge: NKT cells constitutively express IL-23 receptor and RORγt and rapidly produce IL-17 upon receptor ligation in an IL-6-independent fashion. *J Immunol* **180:** 5167–5171.

Ramnath D, Tunny K, Hohenhaus DM, Pitts CM, Bergot AS, Hogarth PM, Hamilton JA, Kapetanovic R, Sturm RA, Scholz GM, et al. 2015. TLR3 drives IRF6-dependent IL-23p19 expression and p19/EBI3 heterodimer formation in keratinocytes. *Immunol Cell Biol* **93:** 771–779.

Raverdeau M, Breen CJ, Misiak A, Mills KH. 2016. Retinoic acid suppresses IL-17 production and pathogenic activity of γδ T cells in CNS autoimmunity. *Immunol Cell Biol* **94:** 763–773.

Reay J, Gambotto A, Robbins PD. 2012. The antitumor effects of adenoviral-mediated, intratumoral delivery of interleukin 23 require endogenous IL-12. *Cancer Gene Ther* **19:** 135–143.

Sabat R, Ouyang W, Wolk K. 2013. Therapeutic opportunities of the IL-22-IL-22R1 system. *Nat Rev Drug Discov* **13:** 21–38.

Cite this article as *Cold Spring Harb Perspect Biol* doi: 10.1101/cshperspect.a028530

Sarin R, Wu X, Abraham C. 2011. Inflammatory disease protective R381Q IL23 receptor polymorphism results in decreased primary CD4$^+$ and CD8$^+$ human T-cell functional responses. *Proc Natl Acad Sci* **108:** 9560–9565.

Schreiber RD, Old LJ, Smyth MJ. 2011. Cancer immuno-editing: Integrating immunity's roles in cancer suppression and promotion. *Science* **331:** 1565–1570.

Shan BE, Hao JS, Li QX, Tagawa M. 2006. Antitumor activity and immune enhancement of murine interleukin-23 expressed in murine colon carcinoma cells. *Cell Mol Immunol* **3:** 47–52.

Shapiro M, Nandi B, Pai C, Samur MK, Pelluru D, Fulciniti M, Prabhala RH, Munshi NC, Gold JS. 2016. Deficiency of IL-17A, but not the prototypical Th17 transcription factor RORγt, decreases murine spontaneous intestinal tumorigenesis. *Cancer Immunol Immunother* **65:** 13–24.

Sheikh SZ, Matsuoka K, Kobayashi T, Li F, Rubinas T, Plevy SE. 2010. Cutting edge: IFN-γ is a negative regulator of IL-23 in murine macrophages and experimental colitis. *J Immunol* **184:** 4069–4073.

Sheikh SZ, Kobayashi T, Matsuoka K, Onyiah JC, Plevy SE. 2011. Characterization of an interferon-stimulated response element (ISRE) in the Il23a promoter. *J Biol Chem* **286:** 1174–1180.

Shen P, Roch T, Lampropoulou V, O'Connor RA, Stervbo U, Hilgenberg E, Ries S, Dang VD, Jaimes Y, Daridon C, et al. 2014. IL-35-producing B cells are critical regulators of immunity during autoimmune and infectious diseases. *Nature* **507:** 366–370.

Shimozato O, Ugai S, Chiyo M, Takenobu H, Nagakawa H, Wada A, Kawamura K, Yamamoto H, Tagawa M. 2006. The secreted form of the p40 subunit of interleukin (IL)-12 inhibits IL-23 functions and abrogates IL-23-mediated antitumour effects. *Immunology* **117:** 22–28.

Sieve AN, Meeks KD, Lee S, Berg RE. 2010. A novel immunoregulatory function for IL-23: Inhibition of IL-12-dependent IFN-γ production. *Eur J Immunol* **40:** 2236–2247.

Smyth MJ, Thia KY, Street SE, Cretney E, Trapani JA, Taniguchi M, Kawano T, Pelikan SB, Crowe NY, Godfrey DI. 2000. Differential tumor surveillance by natural killer (NK) and NKT cells. *J Exp Med* **191:** 661–668.

Smyth MJ, Ngiow SF, Ribas A, Teng MW. 2016. Combination cancer immunotherapies tailored to the tumour microenvironment. *Nat Rev Clin Oncol* **13:** 143–158.

Stanilov N, Miteva L, Deliysky T, Jovchev J, Stanilova S. 2010. Advanced colorectal cancer is associated with enhanced IL-23 and IL-10 serum levels. *Lab Med* **41:** 159–163.

Street SE, Trapani JA, MacGregor D, Smyth MJ. 2002. Suppression of lymphoma and epithelial malignancies effected by interferon γ. *J Exp Med* **196:** 129–134.

Sutton CE, Lalor SJ, Sweeney CM, Brereton CF, Lavelle EC, Mills KH. 2009. Interleukin-1 and IL-23 induce innate IL-17 production from γδ T cells, amplifying Th17 responses and autoimmunity. *Immunity* **31:** 331–341.

Suzuki H, Ogawa H, Miura K, Haneda S, Watanabe K, Ohnuma S, Sasaki H, Sase T, Kimura S, Kajiwara T, et al. 2012. IL-23 directly enhances the proliferative and invasive activities of colorectal carcinoma. *Oncol Lett* **4:** 199–204.

Tang XZ, Jo J, Tan AT, Sandalova E, Chia A, Tan KC, Lee KH, Gehring AJ, De Libero G, Bertoletti A. 2013. IL-7 licenses activation of human liver intrasinusoidal mucosal-associated invariant T cells. *J Immunol* **190:** 3142–3152.

Taylor PR, Roy S, Leal SM Jr, Sun Y, Howell SJ, Cobb BA, Li X, Pearlman E. 2014. Activation of neutrophils by autocrine IL-17A-IL-17RC interactions during fungal infection is regulated by IL-6, IL-23, RORγt and dectin-2. *Nat Immunol* **15:** 143–151.

Teng MW, Andrews DM, McLaughlin N, von Scheidt B, Ngiow SF, Moller A, Hill GR, Iwakura Y, Oft M, Smyth MJ. 2010. IL-23 suppresses innate immune response independently of IL-17A during carcinogenesis and metastasis. *Proc Natl Acad Sci* **107:** 8328–8333.

Teng MW, von Scheidt B, Duret H, Towne JE, Smyth MJ. 2011. Anti-IL-23 monoclonal antibody synergizes in combination with targeted therapies or IL-2 to suppress tumor growth and metastases. *Cancer Res* **71:** 2077–2086.

Teng MW, Vesely MD, Duret H, McLaughlin N, Towne JE, Schreiber RD, Smyth MJ. 2012. Opposing roles for IL-23 and IL-12 in maintaining occult cancer in an equilibrium state. *Cancer Res* **72:** 3987–3996.

Teng MW, Bowman EP, McElwee JJ, Smyth MJ, Casanova JL, Cooper AM, Cua DJ. 2015. IL-12 and IL-23 cytokines: From discovery to targeted therapies for immune-mediated inflammatory diseases. *Nat Med* **21:** 719–729.

Trinchieri G. 2003. Interleukin-12 and the regulation of innate resistance and adaptive immunity. *Nat Rev Immunol* **3:** 133–146.

Tugues S, Burkhard SH, Ohs I, Vrohlings M, Nussbaum K, Vom Berg J, Kulig P, Becher B. 2015. New insights into IL-12-mediated tumor suppression. *Cell Death Differ* **22:** 237–246.

Vignali DA, Kuchroo VK. 2012. IL-12 family cytokines: Immunological playmakers. *Nat Immunol* **13:** 722–728.

Vom Berg J, Vrohlings M, Haller S, Haimovici A, Kulig P, Sledzinska A, Weller M, Becher B. 2013. Intratumoral IL-12 combined with CTLA-4 blockade elicits T cell-mediated glioma rejection. *J Exp Med* **210:** 2803–2811.

von Scheidt B, Leung PS, Yong MC, Zhang Y, Towne JE, Smyth MJ, Teng MW. 2014. Combined anti-CD40 and anti-IL-23 monoclonal antibody therapy effectively suppresses tumor growth and metastases. *Cancer Res* **74:** 2412–2421.

Waibler Z, Kalinke U, Will J, Juan MH, Pfeilschifter JM, Radeke HH. 2007. TLR-ligand stimulated interleukin-23 subunit expression and assembly is regulated differentially in murine plasmacytoid and myeloid dendritic cells. *Mol Immunol* **44:** 1483–1489.

Wang RX, Yu CR, Mahdi RM, Egwuagu CE. 2012. Novel IL27p28/IL12p40 cytokine suppressed experimental autoimmune uveitis by inhibiting autoreactive Th1/Th17 cells and promoting expansion of regulatory T cells. *J Biol Chem* **287:** 36012–36021.

Wang JM, Shi L, Ma CJ, Ji XJ, Ying RS, Wu XY, Wang KS, Li G, Moorman JP, Yao ZQ. 2013. Differential regulation of interleukin-12 (IL-12)/IL-23 by Tim-3 drives T$_H$17 cell development during hepatitis C virus infection. *J Virol* **87:** 4372–4383.

Wang P, Yang B, Zhou B, Zhang J, Li S, Jiang J, Sun Z, Jin F. 2016. Distribution and expression profiles of dendritic cell subpopulations in human bladder cancer. *Int J Clin Exp Pathol* 9: 7180–7187.

Wilke CM, Bishop K, Fox D, Zou W. 2011a. Deciphering the role of Th17 cells in human disease. *Trends Immunol* 32: 603–611.

Wilke CM, Kryczek I, Wei S, Zhao E, Wu K, Wang G, Zou W. 2011b. Th17 cells in cancer: Help or hindrance? *Carcinogenesis* 32: 643–649.

Wolf AM, Rumpold H, Reimer D, Marth C, Zeimet AG, Wolf D. 2010. High IL-12 p35 and IL-23 p19 mRNA expression is associated with superior outcome in ovarian cancer. *Gynecol Oncol* 118: 244–250.

Wu S, Rhee KJ, Albesiano E, Rabizadeh S, Wu X, Yen HR, Huso DL, Brancati FL, Wick E, McAllister F, et al. 2009. A human colonic commensal promotes colon tumorigenesis via activation of T helper type 17 T cell responses. *Nat Med* 15: 1016–1022.

Wu L, Diny NL, Ong S, Barin JG, Hou X, Rose NR, Talor MV, Čiháková D. 2016. Pathogenic IL-23 signaling is required to initiate GM-CSF-driven autoimmune myocarditis in mice. *Eur J Immunol* 46: 582–592.

Xu Y, Liu Y, Pan S, Liu L, Liu J, Zhai X, Shen H, Hu Z. 2013. IL-23R polymorphisms, HBV infection, and risk of hepatocellular carcinoma in a high-risk Chinese population. *J Gastroenterol* 48: 125–131.

Ye Z, Deng B, Wang C, Zhang D, Kijlstra A, Yang P. 2016. Decreased B and T lymphocyte attenuator in Behcet's disease may trigger abnormal Th17 and Th1 immune responses. *Sci Rep* 6: 20401.

Yen D, Cheung J, Scheerens H, Poulet F, McClanahan T, McKenzie B, Kleinschek MA, Owyang A, Mattson J, Blumenschein W, et al. 2006. IL-23 is essential for T cell-mediated colitis and promotes inflammation via IL-17 and IL-6. *J Clin Invest* 116: 1310–1316.

Yoshida H, Hunter CA. 2015. The immunobiology of interleukin-27. *Annu Rev Immunol* 33: 417–443.

Young L, Czarnecki D. 2012. The rapid onset of multiple squamous cell carcinomas in two patients commenced on ustekinumab as treatment of psoriasis. *Australas J Dermatol* 53: 57–60.

Young MRI, Levingston CA, Johnson SD. 2016. Treatment to sustain a Th17-type phenotype to prevent skewing toward Treg and to limit premalignant lesion progression to cancer. *Int J Cancer* 138: 2487–2498.

Yuan X, Hu J, Belladonna ML, Black KL, Yu JS. 2006. Interleukin-23-expressing bone marrow-derived neural stem-like cells exhibit antitumor activity against intracranial glioma. *Cancer Res* 66: 2630–2638.

Zhang XY, Zhang HJ, Zhang Y, Fu YJ, He J, Zhu LP, Wang SH, Liu L. 2006. Identification and expression analysis of alternatively spliced isoforms of human interleukin-23 receptor gene in normal lymphoid cells and selected tumor cells. *Immunogenetics* 57: 934–943.

Zhang B, Rong G, Wei H, Zhang M, Bi J, Ma L, Xue X, Wei G, Liu X, Fang G. 2008. The prevalence of Th17 cells in patients with gastric cancer. *Biochem Biophys Res Commun* 374: 533–537.

Zhang Y, Ma CJ, Ni L, Zhang CL, Wu XY, Kumaraguru U, Li CF, Moorman JP, Yao ZQ. 2011a. Cross-talk between programmed death-1 and suppressor of cytokine signaling-1 in inhibition of IL-12 production by monocytes/macrophages in hepatitis C virus infection. *J Immunol* 186: 3093–3103.

Zhang Y, Ma CJ, Wang JM, Ji XJ, Wu XY, Jia ZS, Moorman JP, Yao ZQ. 2011b. Tim-3 negatively regulates IL-12 expression by monocytes in HCV infection. *PLoS ONE* 6: e19664.

Zhang L, Li J, Li L, Zhang J, Wang X, Yang C, Li Y, Lan F, Lin P. 2014. IL-23 selectively promotes the metastasis of colorectal carcinoma cells with impaired Socs3 expression via the STAT5 pathway. *Carcinogenesis* 35: 1330–1340.

Zindl CL, Lai JF, Lee YK, Maynard CL, Harbour SN, Ouyang W, Chaplin DD, Weaver CT. 2013. IL-22-producing neutrophils contribute to antimicrobial defense and restitution of colonic epithelial integrity during colitis. *Proc Natl Acad Sci* 110: 12768–12773.

Zou W, Restifo NP. 2010. T_H17 cells in tumour immunity and immunotherapy. *Nat Rev Immunol* 10: 248–256.

Zundler S, Neurath MF. 2015. Interleukin-12: Functional activities and implications for disease. *Cytokine Growth Factor Rev* 26: 559–568.

Cite this article as *Cold Spring Harb Perspect Biol* doi: 10.1101/cshperspect.a028530

Targeting IL-10 Family Cytokines for the Treatment of Human Diseases

Xiaoting Wang,[1] Kit Wong,[2] Wenjun Ouyang,[3] and Sascha Rutz[4]

[1]Department of Comparative Biology and Safety Sciences, Amgen, South San Francisco, California 94080

[2]Department of Biomarker Development, Genentech, South San Francisco, California 94080

[3]Department of Inflammation and Oncology, Amgen, South San Francisco, California 94080

[4]Department of Cancer Immunology, Genentech, South San Francisco, California 94080

Correspondence: wouyang@amgen.com; saschar@gene.com

Members of the interleukin (IL)-10 family of cytokines play important roles in regulating immune responses during host defense but also in autoimmune disorders, inflammatory diseases, and cancer. Although IL-10 itself primarily acts on leukocytes and has potent immunosuppressive functions, other family members preferentially target nonimmune compartments, such as tissue epithelial cells, where they elicit innate defense mechanisms to control viral, bacterial, and fungal infections, protect tissue integrity, and promote tissue repair and regeneration. As cytokines are prime drug targets, IL-10 family cytokines provide great opportunities for the treatment of autoimmune diseases, tissue damage, and cancer. Yet no therapy in this space has been approved to date. Here, we summarize the diverse biology of the IL-10 family as it relates to human disease and review past and current strategies and challenges to target IL-10 family cytokines for clinical use.

Interleukin (IL)-10, a cytokine with pleiotropic immunosuppressive functions, is also the founding member of the IL-10 family of cytokines (Fig. 1). In addition to IL-10 itself, this group of cytokines encompasses IL-19, IL-20, IL-22, IL-24, and IL-26, which are collectively referred to as the IL-20 subfamily, as well as the more distantly related members IL-28A, IL-28B, and IL-29, also known as the interferon (IFN)-λ family or type III IFNs (Pestka et al. 2004; Ouyang et al. 2011; Rutz et al. 2014).

IL-10 was initially described as a secreted cytokine synthesis inhibitory factor (CSIF) produced by T helper (Th)2 T-cell clones with the ability to inhibit the production of Th1 cytokines (Fiorentino et al. 1989). It has since been found that IL-10 is expressed by a wide variety of cell types of both the innate and the adaptive arms of the immune system, including macrophages, monocytes, dendritic cells (DCs), mast cells, eosinophils, neutrophils, natural killer (NK) cells, CD4$^+$ and CD8$^+$ T cells, and B cells (Moore et al. 2001; Ouyang et al. 2011). IL-20 subfamily cytokines are also produced mainly by immune cells, such as myeloid cells and lymphocytes (Rutz et al. 2014). Myeloid cells are the primary source for IL-19, IL-20, and IL-24 (Wolk et al. 2002). Epithelial cells, the main target cells of IL-20 family cytokines, also produce IL-19, IL-20, and IL-24 in response to cytokines

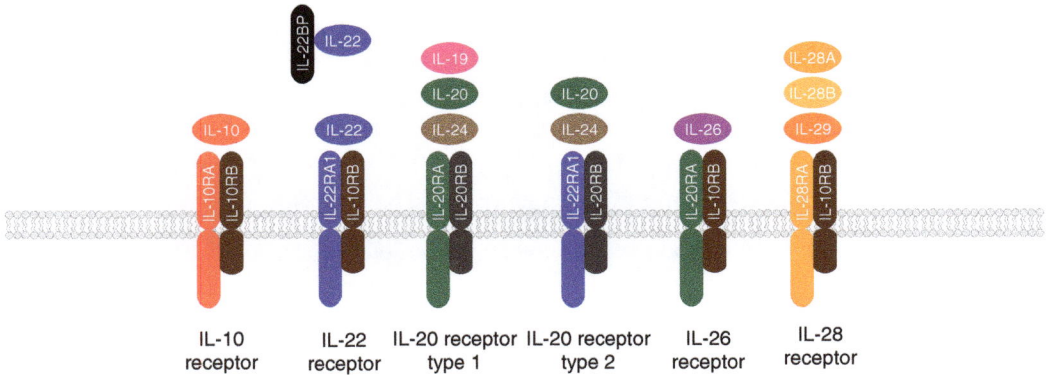

Figure 1. Interleukin (IL)-10 family cytokines and their receptors.

secreted by immune cells (Hunt et al. 2006; Sa et al. 2007; Wolk et al. 2009b). T cells are a primary source for IL-22, IL-24, and IL-26. Additionally, IL-22 is produced by various other lymphoid populations, including γδ-T cells, NK cells, and innate lymphoid cells (ILCs) (Rutz et al. 2013, 2014). Finally, both leukocytes and epithelial cells are major sources of IFN-λ family cytokines (Fig. 1) (Kotenko et al. 2003; Sheppard et al. 2003; Uzé and Monneron 2007).

Although the biological functions of the other IL-10 family cytokines are quite distinct from IL-10 itself, all family members share significant structural homology, having evolved through gene duplication. Most IL-10 family cytokines form homodimers, whereas some family members, such as IL-22, exist in a monomeric form. IL-10 family cytokines signal through heterodimeric receptors, composed of class II receptor α- and β-subunits (Pestka et al. 2004). The prototypical class II receptor structure consists of tandem β–sheet-rich immunoglobulin (Ig)-like domains with fibronectin type III connectivity. The α-receptor subunits show higher affinity for the cytokine ligand than the β-subunits. Interestingly, the receptor-binding mode is virtually identical for monomeric or dimeric IL-10 family cytokines (Pestka et al. 2004).

The shared usage of common receptor subunits is a defining feature of the IL-10 cytokine family (Fig. 1). All members bind either the IL-10RB or IL-20RB β-receptor subunits in combination with varying α-subunits. IL-10 uniquely uses IL-10RA, whereas IL-19, IL-20, and IL-24 use IL-20RA as receptor α-subunits. IL-22 binds IL-22RA1, which can also be bound by IL-20 and IL-24, collectively defining the IL-20 subfamily. IL-28A, IL-28B, and IL-29, on the other hand, use a unique IL-28RA subunit. In addition to these membrane-bound receptors, a soluble form of the IL-22 receptor (IL-22BP or IL-22RA2) with homology to the extracellular domain of IL-22RA1, binds IL-22 with high affinity and blocks its activity (Ouyang et al. 2011; Rutz et al. 2014).

IL-10 family receptors signal through Janus tyrosine kinases (JAKs) and signal transducers and activators of transcription (STATs). Receptor α-subunits are constitutively associated with Jak1, whereas Jak2 or Tyk2 are bound to the β-subunits. Ligand binding initiates recruitment and phosphorylation of STATs, which in turn form homo- and heterodimers that translocate into the nucleus to induce transcription. IL-10 and the IL-20 subfamily cytokines signal primarily through STAT3 and STAT1, whereas IL-28A, IL-28B, and IL-29 activate the ISGF3 complex (Pestka et al. 2004; Ouyang et al. 2011; Rutz et al. 2014).

Distinct receptor expression patterns drive the diverse biology of IL-10 family cytokines (Fig. 1). The IL-10RB β-subunit is ubiquitously expressed throughout the body, whereas expression of the IL-20RB β-subunit is more restricted. IL-10RA is mainly expressed on leukocytes. In contrast, IL-20 subfamily receptors show a rather restricted expression pattern. In particular the α-receptor subunits, IL-20RA and IL-22RA1, are preferentially expressed on epithelial cells and fibroblasts, but absent from hematopoietic cells (Aggarwal et al. 2001; Blumberg et al. 2001; Wolk et al. 2002). IL-22RA1 is highly expressed in the skin, pancreas, kidney, lung, intestine, and liver. IL-20RA is expressed in the skin, lung, ovary, testes, and placenta, and at lower levels in the intestine and liver (Rutz et al. 2013, 2014).

TARGETING IL-10 FOR THE TREATMENT OF HUMAN AUTOIMMUNE DISEASES

As a major immune regulatory cytokine, IL-10 can be produced by many leukocyte subsets and is under the control of various signal transduction pathways and transcriptional networks (Saraiva and O'Garra 2010; Rutz and Ouyang 2011, 2016). According to the expression of its receptor, IL-10 acts on many cells of the immune system (Fig. 2), where it has profound anti-inflammatory functions (Moore et al. 2001; Rutz and Ouyang 2011). IL-10 mainly targets antigen-presenting cells (APCs), such as monocytes and macrophages, and inhibits their release of proinflammatory cytokines, such as tumor necrosis factor α (TNF-α), IL-1β, IL-6, IL-8, granulocyte colony-stimulating factor (G-CSF), and granulocyte macrophage colony-stimulating factor (GM-CSF), as well as chemokines, including MCP1, IL-8, and IP-10 (de Waal-Malefyt et al. 1991a; Fiorentino et al. 1991; Moore et al. 2001). IL-10 also interferes with antigen presentation by reducing the expression of major histocompatibility complex (MHC)-II and costimulatory and adhesion molecules (de Waal-Malefyt et al. 1991b; Willems et al. 1994; Creery et al. 1996). Furthermore, IL-10 suppresses cytokines, such as IL-12 and IL-23, which are required for CD4$^+$ T-cell differentiation (D'Andrea et al. 1993; Schuetze et al. 2005). Similarly, IL-10 attenuates the production of inflammatory mediators, including cytokines and chemokines from neutrophils (Cassatella et al. 1993; Kasama et al. 1994). In addition, IL-10 can act directly on T cells to inhibit both their proliferation and cytokine production, and to induce nonresponsiveness or anergy (Groux et al. 1996). However,

IL-10 also has stimulatory effects on CD8$^+$ T cells, and augments their proliferation and cytotoxic activity (Groux et al. 1998; Mumm et al. 2011). It enhances the survival of human B cells and promotes B-cell proliferation (Levy and Brouet 1994; Itoh and Hirohata 1995) and contributes to the differentiation of B cells and their production and isotype switch of antibodies (Rousset et al. 1992).

The Role of IL-10 in Inflammatory Bowel Disease

Given its multiple anti-inflammatory functions, it is not surprising that IL-10 exerts essential regulatory roles in many human autoimmune diseases. Inflammatory bowel disease (IBD) comprises ulcerative colitis (UC) and Crohn's disease (CD), both of which show uncontrolled inflammation in the intestinal tract but differ in pathophysiology. Mice deficient in either IL-10 or the IL-10 receptor α or β chains develop spontaneous colitis (Kühn et al. 1993), which is dependent on the presence of the intestinal microbiota and involves the up-regulation of IL-23 (Sellon et al. 1998; Yen et al. 2006). Exogenously provided recombinant IL-10 can delay and attenuate colitis development in these IL-10-deficient mice. In addition, IL-10 administration showed therapeutic value in several other colitis models (Hagenbaugh et al. 1997; Steidler et al. 2000).

IBD has a strong genetic association with the IL-10 pathway. Genome-wide association studies (GWAS) linked the *IL10* locus with the susceptibility to both UC and CD (Franke et al. 2008, 2010). Furthermore, individuals carrying missense mutations in the *IL10*, *IL10RA*, or *IL10RB* genes develop very early-onset UC or neonatal CD (Glocker et al. 2009; Engelhardt et al. 2013; Shim et al. 2013). Restoring IL-10RA or IL-10 expression by hematopoietic stem-cell transplantation rapidly alleviates clinical symptoms (Engelhardt et al. 2013).

As mentioned earlier, IL-10 represses several key proinflammatory cytokines, including TNF-α and IL-23, which are clinically validated to participate in the pathogenesis of IBD. TNF-blocking agents, including anti-TNF antibodies,

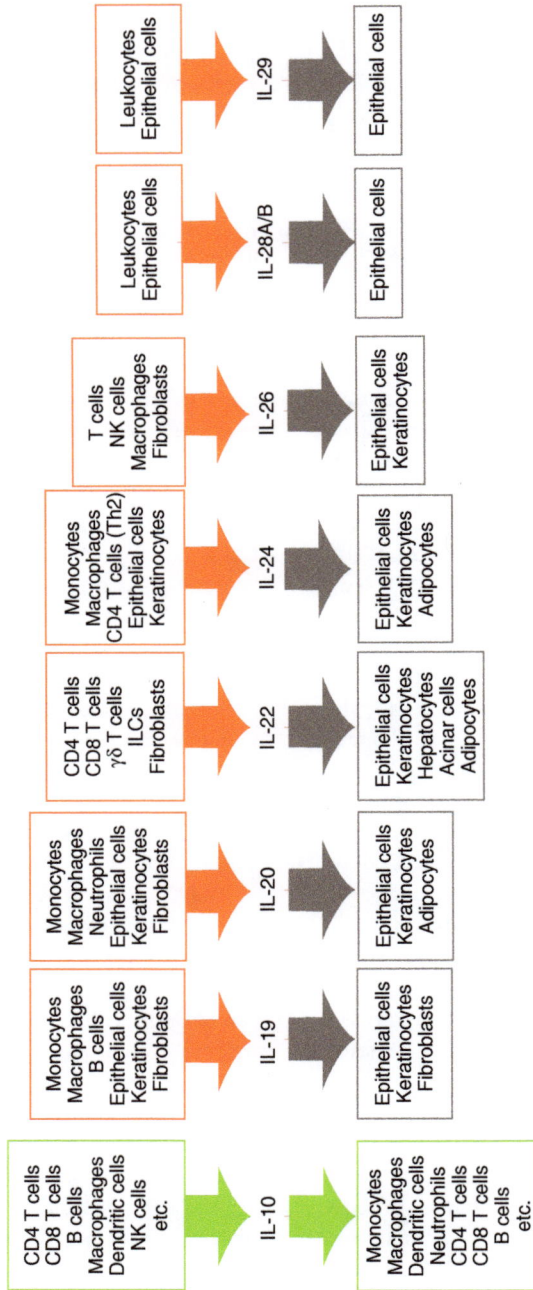

Figure 2. Cellular sources and target cells of interleukin (IL)-10 family cytokines. NK, Natural killer; ILCs, innate lymphoid cells.

Cite this article as *Cold Spring Harb Perspect Biol* doi: 10.1101/cshperspect.a028548

such as infliximab, adalimumab, and certolizumab, have shown good clinical efficacy in the treatment of both CD and UC, and are currently the standard of care for patients with moderate-to-severe disease (Hanauer et al. 2002, 2006; Järnerot et al. 2005; Sandborn et al. 2007). Recently, an anti-IL-12/IL-23 antibody, ustekinumab, also showed promising efficacy in CD (Sandborn et al. 2008; Feagan et al. 2016). Collectively, these findings provide a strong scientific rationale for targeting IL-10 for the therapy of IBD.

Clinical Experience in Treating Inflammatory Bowel Disease with Recombinant IL-10

Recombinant IL-10 has been tested in the clinic for the treatment of IBD and other inflammatory diseases (Table 1). The first recombinant human (rhu)IL-10 (Tenovil) tested in clinical trials was produced in a genetically engineered *Escherichia coli* strain. This rhuIL-10 was tolerated in multiple dose toxicity studies in mice and monkey (Rosenblum et al. 2002). It was also tolerated up to 100 μg/kg in a single intravenous dose in healthy volunteers (Chernoff et al. 1995; Huhn et al. 1996). The pharmacokinetics of rhuIL-10 are similar to many other cytokines, with serum half-life ranging from 2.3 to 3.7 h (Huhn et al. 1996). A single dose of either intravenously or

subcutaneously administered rhuIL-10 induced a transient increase in neutrophils and monocytes and a reduction in lymphocytes, especially at higher doses (Chernoff et al. 1995; Huhn et al. 1996, 1997). A modest decrease in circulating platelets was also observed (Huhn et al. 1997). Supporting the immunosuppressive functions of IL-10, the production of TNF-α and IL-1β was significantly reduced from lipopolysaccharide (LPS)-stimulated peripheral blood cells isolated from rhuIL-10-treated volunteers (Huhn et al. 1997). In addition, although rhuIL-10 increased the expression of FcγRI on monocytes and neutrophils, the expression of activation markers IL-2Rα and HLA-DR on T cells was inhibited (Dejaco et al. 2000).

Because rhuIL-10 was relatively well tolerated, it was tested in multiple clinical trials in IBD patients (Table 1). In the first reported trial, 46 patients with steroid refractory CD were treated with one of five doses of rhuIL-10 (0.5, 1, 5, 10, or 25 μg/kg) or placebo administered intravenously once daily for 7 days. A mild reduction in the Crohn's Disease Activity Index (CDAI) and an increase in remissions within a 3-week follow-up period were observed in the rhuIL-10-treated groups (van Deventer et al. 1997). A clinical response was observed in a trial in patients with mild-to-moderate CD, who

Table 1. Clinical trials targeting interleukin (IL)-10

Intervention	Indication	Clinical stage	Sponsor
Tenovil (rhuIL-10)	Crohn's disease	Phase I/II N/A	Schering-Plough Research Institute
Dekavil (F8-IL-10)	Rheumatoid arthritis	Phase II NCT02076659 Phase II NCT02270632	Philogen/Pfizer
Tenovil TM (IL-10)	Acute pancreatitis	Phase II NCT00040131 (terminated)	Merck Sharp & Dohme
IL-10	Psoriasis	Phase II NCT00001797	National Cancer Institute
Prevascar (rhuIL-10)	Cicatrix, wound healing	Phase II NCT00984646	Renovo
AG011 (engineered *Lactococcus lactis* secreting human IL-10)	Ulcerative colitis	Phase I/II NCT00729872	ActoGeniX N.V.
AM0010: PEGylated human IL-10	Solid tumors/pancreatic cancer	Phase I NCT02009449 Phase III NCT02923921	ARMO BioSciences
BT063 (antibody to neutralize IL-10)	Systemic lupus erythematosus	Phase II NCT02554019	Biotest

Data source: clinicaltrials.gov.
rhu, Recombinant human.

were treated with subcutaneous rhuIL-10 (1, 5, 10, or 20 µg/kg) or placebo once daily for 28 days (Fedorak et al. 2000). Interestingly, in this study, only patients treated with 5 µg/kg, but not higher doses, displayed clinical improvement. However, in a larger multicenter, double-blinded, placebo-controlled prospective study no difference in clinical remission, defined as a reduction in CDAI by more than 150 points, was observed between rhuIL-10 and placebo-treated groups (Schreiber et al. 2000). In this study, patients were subcutaneously dosed with rhuIL-10 (1, 4, 8, or 20 µg/kg) or placebo daily for 28 days, and were followed for an additional 4 weeks after treatment. A significant clinical improvement (reduction of CDAI by more than 100 points) was observed only in patients who received 8 µg/kg rhuIL-10, but not in patients who were treated with the higher dose of 20 µg/kg. In addition, patients with high disease activity responded better (Schreiber et al. 2000). Clinical response was accompanied by a decrease in inflammatory signals as measured by nuclear factor (NF)-κB activation in ileocolic biopsies. In both trials, only the intermediate, but not higher doses of rhuIL-10, induced clinical responses, suggesting a more complex biology of IL-10 in IBD. Finally, rhuIL-10 was also tested in the prevention of relapse after patients underwent a curative colon resection. In this study, both 4 and 8 µg/kg/d did not show significant benefit in comparison to the control-treated group (Colombel et al. 2001).

Although rhuIL-10 was generally tolerated in these studies, the major side effects included a decrease in hemoglobin and thrombocyte counts, which led to significantly more withdrawals in some of the rhuIL-10-treated groups (Buruiana et al. 2010). Adverse events, such as anemia, were observed in a dose-dependent manner. Patients receiving higher doses of rhuIL-10 (>4 µg/Kg) presented progressively and significantly decreased hemoglobin values and platelet counts (Tilg et al. 2002). These symptoms were reversible after discontinuation of therapy in all studies. The mechanism through which IL-10 induces anemia is not clear. The anemia was associated with an altered iron metabolism, as evidenced by increased serum ferritin and soluble transferrin receptor levels (Fedorak et al. 2000; Schreiber et al. 2000; Colombel et al. 2001). An earlier study investigating the effects of the anti-inflammatory cytokines IL-4 and IL-13 had shown that ferritin translation was enhanced by these cytokines in IFN-γ-treated activated macrophages (Weiss et al. 1997). IL-10 may act through a similar mechanism. A study investigating IL-10-induced thrombocytopenia in healthy adult volunteers suggested that IL-10 might affect platelet production indirectly through the inhibition of cytokines produced from monocytes and macrophages (Sosman et al. 2000).

As a general conclusion, systemic administration of rhuIL-10 did not result in significant clinical benefits compared with placebo groups in CD patients (Buruiana et al. 2010). The lack of efficacy may be because of an inability of IL-10 alone to sufficiently suppress all inflammatory mediators or caused by patient-to-patient variabilities in responding to exogenous IL-10. However, an important caveat in these studies is that the local IL-10 concentration in the intestine after systemic dosing may be too low to induce meaningful biological responses.

Therapeutic Potential for IL-10 in Other Autoimmune Diseases

TNF-α production is pathogenic in psoriasis and rheumatoid arthritis (RA), as drugs blocking the TNF-α pathway provide a clear therapeutic benefit in both diseases (Sfikakis 2010). Psoriasis is an inflammatory skin disease characterized by infiltration of epidermis and dermis with leukocytes that produce various inflammatory cytokines, such as TNF-α, IL-17, and IL-22, which stimulate keratinocyte proliferation and promote epidermal hyperplasia (Nestle et al. 2009). Antibodies blocking IL-12/IL-23, TNF-α, or IL-17 have shown good efficacy for the treatment of psoriasis (Kofoed et al. 2015). IL-10 might also provide therapeutic benefit in this disease, given its role in repressing proinflammatory cytokines. Indeed, in multiple clinical studies (Table 1), administration of recombinant IL-10 provided mild-to-moderate benefits for patients with psoriasis (Asadullah et al. 1998). However, further development of IL-10

Cite this article as *Cold Spring Harb Perspect Biol* doi: 10.1101/cshperspect.a028548

as a therapy for psoriasis was hindered by its poor in vivo pharmacokinetics and far superior efficacy of other cytokine-blocking therapies.

IL-10 has a dual role in RA, a disease characterized by synovitis associated with bone and cartilage loss, production of rheumatoid factor and autoantibodies, and systemic inflammation. Although IL-10 represses pathogenic cytokines, such as IL-6 and TNF-α, it also stimulates B cells and drives autoantibody production. Recombinant IL-10 has proven to be relatively safe in clinical trials in human (Chernoff et al. 1995). A new modified IL-10 that can be targeted to the site of inflammation and has better in vivo pharmacokinetics did alleviate disease severity in preclinical RA models (Trachsel et al. 2007). Currently it is being tested in RA patients (Table 1) (Galeazzi et al. 2014).

A pathogenic function for IL-10 has been proposed for type 1 diabetes (T1D) and multiple sclerosis (MS) (Asadullah et al. 2003; Saxena et al. 2015). However, given rather inconsistent results in preclinical models (Wogensen et al. 1994; Cannella et al. 1996; Nagelkerken et al. 1997; Zheng et al. 1997), targeting IL-10 in MS or T1D has not been tested in the clinic.

Systemic lupus erythematosus (SLE) is another autoimmune disease in which IL-10 might have a pathogenic role, as it is a growth factor for human B cells, promotes antibody production, class switching, and plasma cell differentiation (Rousset et al. 1992, 1995). Elevated IL-10 levels have been reported in SLE patients (Koenig et al. 2012). In a small clinical trial, six SLE patients were treated with an IL-10 blocking antibody for 21 days and followed over 6 months (Llorente et al. 2000). Although an improvement of symptoms was observed that allowed a reduction in steroid use, the lack of a control arm demands caution in interpreting the data. A placebo-controlled phase II clinical trial evaluating a neutralized anti-IL-10 antibody in SLE is currently ongoing (Table 1). In all autoimmune disorders, however, the potential risk of developing IBD when blocking the IL-10 pathway needs further evaluation.

In conclusion, although there is strong scientific rationale supporting the use of recombinant IL-10 as a therapy for various autoimmune and inflammatory diseases, the limited clinical experience thus far has not revealed clear clinical benefit.

Strategies for Improved Local Exposure and Pharmacokinetic Properties of IL-10

There are a number of possible explanations for the lack of efficacy of recombinant IL-10 therapy in patients. First, IL-10 is a pleotropic cytokine with both immune-repressive and immune-stimulatory properties. For example, whereas IL-10 represses IL-12 production from macrophages and limits Th1 differentiation and IFN-γ production induced by IL-12, it also promotes proliferation and activation of $CD8^+$ T cells and their IFN-γ production. Indeed, increased IFN-γ production has been observed in healthy volunteers and IBD patients following rhuIL-10 administration (Lauw et al. 2000; Tilg et al. 2002). By lowering the serum concentration of the administered IL-10 or by reducing its activity through protein engineering, it might be possible to overcome this problem because myeloid cells, the main therapeutic targets, have higher IL-10 receptor expression and are thus more sensitive to IL-10 (Moore et al. 2001). However, to enable this strategy, the poor pharmacokinetic properties of rhuIL-10 have to be improved to achieve sufficient coverage at lower doses. Several approaches, such as using PEGylated IL-10 or IL-10-Fc fusion proteins, are being pursued (Schwager et al. 2009; Naing et al. 2016). Another concern is that systemic delivery of rhuIL-10 will not result in sufficiently high local concentrations in the inflamed tissues. Several local delivery strategies, such as ectopic expression of IL-10 by commensal bacteria or IL-10/antibody fusion proteins, have been devised and are being tested (Table 1).

Engineered Bacteria for Intestinal Delivery of IL-10

To increase the concentration of IL-10 in the gut while minimizing systemic exposure, local delivery of IL-10 by engineered bacterial strains has been considered (Steidler et al. 2000, 2003; Miyoshi et al. 2004; Innocentin et al. 2009; del

Carmen et al. 2011). In this strategy, the *Il10* gene is inserted into the genome of *Lactococcus lactis*, a widely used dairy microbe, and intragastrically administered to mice. The local delivery of mouse IL-10 substantially reduces histology scores in both dextran sulfate sodium (DSS)-induced and spontaneous colitis in IL-10-deficient mice, at a significantly lower therapeutic dose (Steidler et al. 2000). This strategy provides several potential benefits. First, because oral delivery of IL-10 is not a viable option as cytokines are extremely sensitive to the acidic environment of the stomach, the engineered bacteria are able to produce recombinant IL-10-locally in the intestine (Steidler et al. 2000). To further minimize degradation of IL-10 in the gastrointestinal track, an *L. lactis* strain expressing fibronectin-binding protein A from *Staphylococcus aureus* was engineered to carry an IL-10-expression vector, which can efficiently deliver recombinant DNA to human epithelial cells, thereby triggering IL-10 expression in situ (Innocentin et al. 2009). IL-10 produced locally is effective in ameliorating gut inflammation in both DSS-induced and trinitrobenzene sulfonic acid (TNBS)-mediated IBD mouse models (del Carmen et al. 2014; Zurita-Turk et al. 2014). Second, this strategy minimizes systemic exposure and toxicity. For example, *L. lactis* is a nonpathogenic and noninvasive Gram$^+$ bacterium that survives in the digestive tract of animals or humans (Drouault et al. 1999). To control potential transgene escape into the environment, a biologically contained *L. lactis* has been developed in which the thymidylate synthase gene (*thyA*) has been replaced by the human *IL10* gene. This strain is unable to grow in the absence of thymidine or thymine (Steidler et al. 2003). Lastly, the expression of IL-10 in bacteria can be controlled through further engineering. For example, a xylose-inducible expression system has been used to limit IL-10 production by *L. lactis* to the mucosal epithelium. Anti-inflammatory effects of milk fermented with this bacteria were observed in an acute TNBS-induced IBD mouse model (Miyoshi et al. 2004; del Carmen et al. 2011). Another inducible system uses stress-inducible controlled expression (SICE) based on a stress-inducible promoter (pGroESL) that allows the

production of murine IL-10 locally at mucosal surfaces (Benbouziane et al. 2013). Here, IL-10 is produced locally without the need for extrinsic inducers. IL-10 levels achieved through this approach result in significant protective effects as measured by gut permeability, immune responses, and gut function in a mouse IBD model (Martin et al. 2014).

In a first human trial with biologically contained *L. lactis* expressing human IL-10, patients with moderate-to-severe CD were dosed twice daily for 7 days (Table 1). The results showed that the approach was both safe and controllable. The genetically modified *L. lactis* could only be recovered from feces with the addition of thymidine. However, the study was conducted in a small patient population, and clinical outcomes revealed no statistically significant differences between those individuals who received the IL-10-expressing strain and those patients in the control group (Steidler et al. 2003). For this strategy to succeed, several parameters need to be optimized further. A sufficient amount of IL-10 has to be produced at the site of inflammation over a prolonged period of time with minimal variations across the patient population. IL-10 needs to remain intact and functional under the harsh conditions in the gastrointestinal tract, so that biologically meaningful amounts of IL-10 can cross the intestinal barrier and act on inflammatory leukocytes. The development of sensitive methods to assess the pharmacokinetics and pharmacodynamics of bacterial IL-10, both locally and systemically, will be key to the success of this approach in patients.

Antibody-Conjugated IL-10 for Improved Targeted Delivery and Pharmacokinetics

To improve the pharmacokinetic properties of IL-10 and to deliver it in a more targeted way to the site of inflammation, antibody-mediated delivery strategies have been conceived. Four antibodies, L19 (binding the extra-domain B of fibronectin), F8 (binding the extra-domain A of fibronectin), G11 (binding the extra-domain C of tenascin-C), and F16 (binding the extra-domain A1 of tenascin-C), which specifically recognize angiogenic but not normal tissue, were considered as fusion partners for IL-10 (Trachsel et al. 2007; Schwager et al. 2009). L19 selectively accumulates at sites of inflammation in a collagen-induced arthritis (CIA) mouse model (Trachsel et al. 2007). As a proof-of-concept, IL-10 as well as the proinflammatory cytokines IL-2 and TNF-α were fused to the carboxyl terminus of a single-chain Fv fragment of L19. As expected, the L19-IL-10 fusion protein was found to be enriched at the site of angiogenesis. L19-IL-2 and L19-TNF-α exacerbate disease in the CIA model, whereas L19-IL-10 significantly reduces disease severity (Trachsel et al. 2007). More importantly, the repression of inflammation by L19-IL-10 is more profound than the effect mediated by a systemically distributed fusion protein of a control antibody single chain fused to IL-10 (Trachsel et al. 2007). In a comparative study, the four antibodies were used to stain synovial tissues of RA patients (Schwager et al. 2009). F8 showed the strongest staining. The corresponding IL-10 immunocytokine, F8-IL-10 (Dekavil), also showed efficacy in the CIA model (Doll et al. 2013). The combination of a murine F8-mIL-10 and a TNF-blocking muTNFR-Fc fusion protein displayed superior efficacy compared with either agent alone (Doll et al. 2013). Interestingly, F8-IL-10 also showed efficacy in endometriosis in a syngeneic mouse model and in a chronic cardiac allograft model (Schwager et al. 2011; Franz et al. 2015).

F8-IL-10 was combined with methotrexate in a preclinical toxicity study in cynomolgus monkeys. In this study, 180 μg/kg (equivalent to 60 μg/kg IL-10) of F8-IL-10 was administered subcutaneously three times a week for 8 weeks (Schwager et al. 2009). No major toxicity findings, other than transient regenerative anemia, have been reported. In pharmacokinetic measurements, about 20 ng/ml of F8-IL-10 was detected in the serum 3 h after subcutaneous infection with undetectable levels after 24 h. In a phase Ib study, 24 RA patients were dosed subcutaneously with 6 μg/kg to 300 μg/kg F8-IL-10 in combination with methotrexate every week for 8 weeks. No major side effects have been reported, except for one case of progressive anemia in the 160-μg/kg dose group (Galeazzi

et al. 2014). Among the 23 patients evaluated in this study, 15 patients achieved American College of Rheumatology (ACR)20 improvement, seven patients achieved ACR50, and three achieved ACR70 responses. Impressively, two patients experienced long-lasting remission (ACR70 maintained for more than 1 year after last dose) (Galeazzi et al. 2014). Placebo-controlled phase II clinical trials in RA patients are currently ongoing (Table 1).

TARGETING IL-10 AND IL-24 FOR THE TREATMENT OF CANCER

A Potential Role for IL-10 in Cancer

The recent success of therapies aimed at modulating the immune system to elicit antitumor immunity has completely changed the landscape of cancer therapies (Sharma and Allison 2015). Anti-PD1/PD-L1 therapies are now being considered as first-line treatment for major cancer types, such as non-small-cell lung cancer (NSCLC), at least in subsets of patients (Topalian et al. 2012, 2015; Borghaei et al. 2015; Motzer et al. 2015; Robert et al. 2015). However, many cancer patients have yet to benefit from novel cancer immunotherapies. To further unleash the immune system and promote antitumor immunity, targeting additional checkpoints, the indoleamine 2,3-dioxygenase (IDO) pathway and other strategies are being considered (Chen and Mellman 2017; Sharma et al. 2017). IL-10 has been detected in the tumor microenvironment of many cancer types, and has been correlated with poor prognosis (Nemunaitis et al. 2001; O'Garra et al. 2008; Mannino et al. 2015). Based on its strong immunosuppressive functions, especially in inhibiting IL-12 production and Th1 differentiation, IL-10 has been considered as a target for cancer immunotherapy. However, evidence for both tumor-promoting and tumor-repressing functions of IL-10 have been presented (Groux et al. 1998). On the one hand, IL-10 represses cytotoxic T-cell activation by down-regulating MHC expression on cancer cells and on professional APCs, thereby preventing the recognition of cancer cells by antigen-specific T cells (Adris et al. 1999; Steinbrink et al. 1999). IL-10 also inhibits IL-12 production from APCs, a cytokine that strongly promotes Th1 differentiation and cytotoxicity. On the other hand, high doses of IL-10 enhance the proliferation of CD8$^+$ T cells and their cytotoxic activity (Groux et al. 1998; Mumm et al. 2011). Certain inflammatory cytokines and conditions may promote tissue damage and oncogenesis. For example, UC is associated with an increased risk of colon cancer (Sturlan et al. 2001; Huang et al. 2006). IL-10 may repress these inflammatory conditions, thus preventing subsequent oncogenesis. It will therefore be necessary to evaluate a potential therapeutic intervention by either inhibiting or promoting the IL-10 pathway on a case-by-case basis in specific cancer types and patient subpopulations.

The role of IL-10 in various cancer models has been examined since it was first discovered (Asadullah et al. 2003; Vicari and Trinchieri 2004). In an IL-2 promoter-driven IL-10 transgenic mouse model, elevated expression of IL-10 results in failure to control an immunogenic tumor, an effect that can be reversed by administering an anti-IL-10 antibody (Hagenbaugh et al. 1997). Similarly, IL-10-deficient mice are resistant to UV-induced skin carcinogenesis, a finding that is correlated with a potent Th1 response in these mice (Loser et al. 2007). In contrast, in a chemically induced skin carcinoma model, IL-10 deficiency results in significantly increased tumor burden and accelerated mortality (Mumm et al. 2011). In tumor-bearing mouse mammary tumor virus–polyoma middle T oncogene (MMTV-PyMT) mice, antibody blockade of IL-10RA combined with paclitacel chemotherapy provides significant therapeutic benefit (Ruffell et al. 2014). On the other hand, the administration of PEGylated murine IL-10 in both a Her2 transgenic mouse breast cancer model and in a squamous cell carcinoma model can suppress tumor growth, supporting the rationale for IL-10 therapy in these cancers (Mumm et al. 2011). IL-10 induces antigen-specific CD8$^+$ T-cell responses and increases IFN-γ production from these cells. Furthermore, the antitumor activity of PEGylated IL-10 is dependent on IL-10 receptor expression on CD8$^+$ T cells, further supporting the notion of these cells being direct IL-10 targets in this setting (Emmerich et al. 2012).

Clinical Experience with IL-10 Therapies in Cancer

Based on the potential antitumor effects of IL-10, a PEGylated human IL-10 (PEG-rhuIL-10) was engineered for clinical use (Naing et al. 2016). Whereas PEG-rhuIL-10 inhibited IFN-γ production in peripheral blood mononuclear cells (PBMCs) in vitro, it stimulated IFN-γ secretion from CD8$^+$ T cells. PEG-rhuIL-10 also promoted perforin and granzyme B production from CD8$^+$ T cells, supporting its role in enhancing cytotoxicity (Naing et al. 2016). This molecule has recently been tested in clinical trials in cancer patients (Table 1). The PEG-rhuIL-10 has a prolonged half-life and exposure compared with rhuIL-10. In this study, patients were treated with one of six different doses (1, 2.5, 5, 10, 20, and 40 µg/kg) of PEG-rhuIL-10. In addition to increased IFN-γ levels in the serum, elevated IL-18 and reduced TGF-β levels were observed in treated patients, supporting the immune stimulatory function of this molecule in vivo. Interestingly, some partial clinical responses have been observed in this small cohort of patients (Naing et al. 2016). A phase III clinical trial investigating PEG-rhuIL-10 in combination with FOLFOX in metastatic pancreatic cancer patients is currently ongoing (Table 1).

Antitumor Activity Elicited by Ectopic Expression of IL-24 in Cancer Cells

IL-24 has been ascribed unique antitumor activity when intrinsically expressed in tumor cells. Although the primary cellular sources of IL-24 include T cells and myeloid cells (Fig. 2), IL-24 was first discovered with elevated expression levels in terminally differentiated human melanoma cells treated with recombinant IFN-β and the protein kinase C (PKC) activator mezerein (Jiang et al. 1995). Enhanced IL-24 expression causes an irreversible growth arrest and suppression of tumorigenic properties. Similar results were obtained in a variety of human cancer cells, including lung, colorectal, breast, and prostate cancer (Emdad et al. 2009; Whitaker et al. 2012). Interestingly, IL-24 is also readily induced in some normal epithelial and fibroblast cells with insignificant effects on growth and survival (Jiang et al. 1996; Su et al. 1998). IL-24 was categorized as an IL-10 cytokine family member based on its conserved genomic structure, chromosome localization, and cytokine-like properties (Caudell et al. 2002; Sauane et al. 2003). It shares the same receptor complex with IL-20 (Rutz et al. 2014). However, no direct antitumor effects have been reported for any other family member. Only ectopic expression of IL-24 in tumor cells, but not its addition to the culture media, shows antitumor activity in cultured cancer cells (Whitaker et al. 2012). Several potential mechanisms for these surprising findings have been discussed, including endoplasmic reticulum stress, the unfolded protein response, apoptosis, autophagy, and the production of reactive oxygen species, seemingly independent of IL-24 receptor signaling (Whitaker et al. 2012). However, whether the antitumor activities of IL-

24 are linked to its properties as a cytokine is still a matter of debate (Kreis et al. 2007, 2008).

The therapeutic potential of IL-24 for cancer treatment has been evaluated in clinical trials (Table 2). Efficacy was observed in a phase I/II clinical trial in patients with multiple advanced cancers following intratumoral injection of INGN 241, an adenovirus expressing IL-24 (Cunningham et al. 2005; Tong et al. 2005). Up to 80% of tumor cells at the injection sites showed positive terminal deoxynucleotidyl transferase dUTP nick end labeling (TUNEL) staining, and 67% of tumors showed reduced Ki-67 staining postinjection. A transiently elevated level of $CD8^+$ T cells and cytokines IL-6, IL-10, and TNF-α were noticed in peripheral blood, consistent with increased tumor apoptosis. The intratumoral injection of INGN 241 was well tolerated with some mild side effects, including pain and erythema at the injection sites. Only one of the patients suffered from grade 3 serious adverse events (SAEs) and discontinued the trial. It is unclear, however, whether the observed activity in this study was mediated solely

Table 2. Clinical trials targeting interleukin (IL)-20, IL-22, and IL-24

Intervention	Indication	Clinical stage	Sponsor
NNC109-0012: anti-IL-20	Rheumatoid arthritis	Phase I NCT00818064	Novo Nordisk A/S
		Phase I NCT01038674	
		Phase II	
		NCT01282255	
		NCT01636817 (terminated)	
		NCT01636843 (terminated)	
	Psoriasis	Phase I	
		NCT01261767 (terminated)	
F-652:	Alcoholic hepatitis	Phase I/II NCT02655510	Mayo Clinic Generon
IL-22 IgG2-Fc	Acute GVHD	Phase I/II NCT02406651	(Shanghai)
ILV-094: anti-IL22	HV	Phase I NCT00434746	Pfizer
	HV	Phase I NCT00447681	
	Psoriasis	Phase I NCT00563524	
	Atopic dermatitis	Phase II NCT01941537	Rockefeller University
	Rheumatoid arthritis	Phase II NCT00883896	Pfizer
ILV-095: anti-IL22	HV	Phase I NCT00822835	Pfizer
	HV	Phase I NCT00822484	
	Psoriasis	Phase I NCT01010542	
		(terminated)	
INGN 241 (Ad-mda-7)	Melanoma	Phase II NCT00116363	Introgen Therapeutics

Data source: clinicaltrials.gov.
GVHD, Graft-versus-host disease.

by IL-24, or by the immune response elicited by the virus, or both.

THE IL-20 SUBFAMILY OF CYTOKINES IN HUMAN DISEASE

IL-20 subfamily cytokines mainly facilitate the communication between leukocytes and epithelial cells (Fig. 2), where they function to enhance innate defense mechanisms, wound healing, and tissue repair at epithelial surfaces (Rutz et al. 2014). Although it is evident how these functions can be beneficial to the host during infections, a clear role in host defense has only been established for IL-22. IL-22 is indispensable during infections with extracellular pathogens such as *Citrobacter rodentium*, *Klebsiella pneumonia*, or yeast at mucosal surfaces (Ouyang et al. 2011; Rutz et al. 2013; Eidenschenk et al. 2014). Recent studies also reveal important functions of IL-22 in shaping intestinal homeostasis and commensal communities (Sonnenberg et al. 2011). However, IL-20 subfamily cytokines have been widely studied for their roles in inflammation and autoimmunity, most prominently in the context of skin inflammation, with well-defined pathogenic roles in psoriasis, atopic dermatitis (AD) (Sa et al. 2007; Ouyang 2010; Sabat et al. 2013), and RA. Exciting new approaches are aimed at harnessing the tissue-protective and wound-healing functions of IL-22 for the treatment of IBD and diabetic foot ulcer (DFU).

Role of IL-20 Subfamily Cytokines in Psoriasis

As briefly discussed above, psoriasis is a chronic inflammatory disease of the skin characterized by abnormal keratinocyte differentiation and proliferation, leukocyte infiltration into the dermis and epidermis, and increased dilation and growth of blood vessels. The cross talk between keratinocytes and leukocytes is essential during the pathogenesis of psoriasis. Various leukocytes, including T cells, neutrophils, DCs, and macrophages, infiltrate into the skin and contribute to inflammation. IL-19, IL-20, IL-22, and IL-24 are expressed in psoriatic but not in healthy skin (Rømer et al. 2003; Wolk et al. 2004; Otkjaer et al. 2005). More importantly, the receptors for IL-20 subfamily cytokines are highly expressed on keratinocytes (Sa et al. 2007). T cells appear to be the major source of IL-22 in psoriatic skin lesions. T cells isolated from skin lesions produce much higher levels of IL-22 than T cells in circulation (Boniface et al. 2007). In addition, T-cell clones generated from psoriatic tissue are largely CCR6[+] IL-22[+], presumably Th17 cells or Th22 cells (Pène et al. 2008; Kagami et al. 2010). Myeloid cells and keratinocytes themselves are potential sources of IL-19, IL-20, and IL-24 (Fig. 1) (Sa et al. 2007; Tohyama et al. 2009; Wolk et al. 2009b).

The role for IL-20 subfamily cytokines in psoriatic skin inflammation has been extensively studied in preclinical models in mice. Transgenic mice that ectopically express IL-20, IL-22, or IL-24 develop skin lesions and histological features similar to those seen in human psoriasis (Blumberg et al. 2001; Wolk et al. 2009a; He and Liang 2010). Direct injection of IL-23 induces ear thickening, acanthosis, and dermal infiltrates, similar to some features in psoriatic skin (Kopp et al. 2003; Chan et al. 2006; Zheng et al. 2007), which are dependent on IL-22 and other IL-20 family cytokines. In addition, injection of IL-22 into the skin of normal mice induces S100 and β-defensin, as well as keratinocyte hyperplasia (Ma et al. 2008), whereas neutralization of IL-22 or IL-23 prevents psoriasis-like symptoms induced by the transfer of CD4[+]CD45RB[hi] CD25[−] cells into severe combined immunodeficiency (SCID) mice (Ma et al. 2008). Furthermore, the treatment of immunodeficient mice engrafted with human psoriatic skin with blocking antibodies to IL-20 or IL-22 resolves the psoriasis condition (Stenderup et al. 2009; Perera et al. 2014).

IL-20 subfamily cytokines induce several factors that either promote or sustain psoriatic lesions. All IL-20 subfamily members, except IL-26, induce the expression of various antimicrobial peptides, including S100 family genes and β-defensin family genes (Boniface et al. 2005; Liang et al. 2006; Wolk et al. 2006; Sa et al. 2007). Furthermore, the IL-20 subfamily cytokines, especially IL-22, regulate genes in keratinocytes and fibroblasts involved in tissue remodeling, including proteases and extracellular matrix proteins, such as MMP1, MMP3, kallikrienes, marapsin, and platelet-derived growth factor (PDGF) (Boniface et al. 2005; Wolk and Sabat 2006; Sa et al. 2007; Li et al. 2009). IL-20 subfamily cytokines promote the re-epithelialization process by acting on epidermal keratinocytes, partly through the induction of keratinocyte growth factor (KGF) and epidermal growth factor (EGF). IL-20 subfamily cytokines also activate proinflammatory responses through the induction of chemokines and cytokines from keratinocytes, including CXCL1, CXCL5, and CXCL7 (Boniface et al. 2005; Sa et al. 2007). The effects of IL-20 subfamily cytokines can be further synergized by other proinflammatory cytokines, such as IL-17A, TNF-α, or IFN-γ (Liang et al. 2006; Tohyama et al. 2009; Guilloteau et al. 2010), suggesting that IL-20 subfamily cytokines in conjunction with other proinflammatory cytokines orchestrate the pathogenesis of psoriasis.

Clinical Experience with Targeting IL-20 Subfamily Cytokines in Psoriasis

Despite the convincing preclinical data suggesting a disease-promoting role for IL-20 subfamily cytokines in psoriasis, neutralizing antibodies targeting either IL-20 or IL-22 did not advance in clinical trials (Table 2). A randomized, placebo-controlled, phase I/IIa dose-escalation trial was conducted to evaluate a fully human IgG4 anti-IL-20 antibody in patients with moderate-to-severe stable chronic plaque psoriasis. However, the trial's expansion phase was terminated early owing to an apparent lack of Psoriasis Area and Severity Index (PASI) improvement (Gottlieb et al. 2015). Similarly, early-stage clinical trials evaluating two anti-IL22 antibodies (fezakinumab, ILV-095) were ended, presumably because of lack of efficacy. Although the failure of these trials may be simply a result of redundant functions of IL-20 and IL-22, the clinical data suggest that IL-20 subfamily cytokines might not be the key engine driving the inflammatory cascade in psoriasis.

Targeting IL-22 for the Treatment of Atopic Dermatitis

Atopic dermatitis (AD) is the most common inflammatory skin disease, affecting up to 25% of children and up to 3% of the adult population (Williams 2005; Montes-Torres et al. 2015). AD is characterized by itchy, red, and flaky lesions that often occur on bending sides of the limbs. The lesions are infiltrated with immune cells in the dermis and epidermis. Acanthosis, fibrosis, and collagen deposition are observed during the chronic phase. IL-22 has been shown to contribute to the epidermal barrier dysfunction in AD, and also seems to be responsible for the characteristic epidermal hyperplasia (Nograles et al. 2009; Gittler et al. 2012). IL-22 is highly expressed in the affected skin of patients (Wolk et al. 2004). In contrast to psoriasis, the expression of IL-22 in AD is mainly derived from Th22 cells and from IL-22-producing CD8[+] cells (Nograles et al. 2009). The role of IL-22 in the development and maintenance of AD needs further exploration. However, a phase II clinical trial with an anti-IL-22 antibody (ILV-094) in patients with moderate-to-severe AD is currently ongoing (Table 2). Data from this study is not yet available.

Role of IL-20 Subfamily Cytokines in Rheumatoid Arthritis

IL-20 subfamily cytokines have been studied extensively for their role in RA. All family members are up-regulated in synovial fluid of RA patients (Ikeuchi et al. 2005; Hsu et al. 2006; Kragstrup et al. 2008; Sakurai et al. 2008; Alanärä et al. 2010; Corvaisier et al. 2012). The corresponding receptors are expressed in synovial tissues (Ikeuchi et al. 2005; Kragstrup et al. 2008; Sakurai et al. 2008; Corvaisier et al. 2012). Increased frequencies of IL-22- and IL-26-producing Th17 cells and Th22 cells are found in peripheral blood and joints of RA patients (Pène et al. 2008; Shen et al. 2009; Leipe et al. 2011; Zhang et al. 2012). IL-19, IL-20, and IL-22 are thought to be pathogenic in RA owing to their ability to enhance the proliferation of synovial cells and to induce proinflammatory cytokines and chemokines, including IL-6, IL-8, and CCL2 (Ikeuchi et al. 2005; Sakurai et al. 2008). Additionally, IL-20 promotes neutrophil chemotaxis (Hsu et al. 2006).

Preclinical data further support a pathogenic role of IL-20 subfamily cytokines in arthritis. Blockade of IL-19 or the administration of a soluble IL-20RA attenuates disease in collagen-induced arthritis in rats (Hsu et al. 2006, 2012).

Based on these data, a neutralizing anti-IL-20 antibody has been evaluated in a phase II clinical study in patients with active RA (Table 2). Sixty-seven patients with RA received either anti-IL-20 (3 mg/kg per week, subcutaneously) or a placebo over 12 weeks with a 13-week follow-up. A significant proportion of patients with seropositive RA receiving anti-IL-20, compared with those receiving placebo, achieved treatment responses according to the American College of Rheumatology 20% (ACR20) (59% vs. 21%), ACR50 (48% vs. 14%), and ACR70 (35% vs. 0%) levels of improvement (Šenolt et al. 2015).

An elevated plasma concentration of IL-22 is associated with disease severity and progression of erosive RA in patients (Leipe et al. 2011; Zhang et al. 2011; da Rocha et al. 2012), suggesting that IL-22 blockade might be equally efficacious in the treatment of RA. However, preclinical arthritis models paint a more complicated picture. Blockade of IL-22 with neutralizing antibodies before disease onset increases the incidence and severity of collagen-induced arthritis in mice, whereas treatment after disease onset is beneficial (Justa et al. 2014). In a different model, anti-IL-22 reduces inflammation and bone erosion, but does not affect overall arthritis severity (Marijnissen et al. 2011).

Nonetheless, a neutralizing anti-IL-22 antibody (ILV-094) has been evaluated in a phase II clinical trial for the treatment of RA (Table 2), but the results of the study have not been published to date.

Therapeutic Potential for IL-22 in Inflammatory Bowel Disease

Genetic studies in IBD not only revealed alterations in pathways regulating inflammation, such as the IL-10 pathway, but also showed a strong link between defects in epithelial innate defense and integrity and disease onset (Xavier and Podolsky 2007; Kaser et al. 2010). The IL-22 receptor, but not the IL-20 receptor, is highly expressed in the gastrointestinal tract (Zheng et al. 2008; Wang et al. 2014), suggesting a predominant role for IL-22 in regulating intestinal epithelial cells during inflammation and host defense (Ouyang 2010). Although more pronounced in CD, UC patients also show increased numbers of IL-22-producing cells in affected tissues (Andoh et al. 2005; Yu et al. 2013). Sources of IL-22 in the intestine comprise Th cells and innate cells, such as Th17, Th22, ILCs, and NK22 cells (Andoh et al. 2005; Geremia et al. 2011). More importantly, changes in IL-22 expression occur during acute phases of disease. In UC patients, a significant reduction in IL-22$^+$ CD4$^+$ T cells has been observed in actively inflamed tissues compared with normal tissues from UC patients or from healthy controls (Leung et al. 2013). The reduction in IL-22

Figure 3. Therapeutic potential of interleukin (IL)-22-Fc fusion protein in inflammatory bowel disease and diabetic foot ulcer patients. (*A*) In inflammatory bowel disease, severe epithelium damage and uncontrolled microbiota result in chronic inflammation in the gut. Systemic administration of IL-22-Fc is thought to promote epithelial cell regeneration, stimulate the production of antimicrobial peptides to control the microbiota, enhance the mucin production to restore mucus layer, and boost chemokine secretion to recruit immune cells for host defense. As a consequence, such therapy may reduce gut inflammation and restore epithelial integrity. (*B*) In diabetic foot ulcer, the cutaneous wound-healing process is significantly impaired, and the wounded skin is often associated with bacterial infections. IL-22-Fc is thought to increase re-epithelialization through directly stimulating kerantinocytes or indirectly promoting epidermal growth factor (EGF) and keratinocyte growth factor (KGF) production, thereby augmenting antimicrobial peptide production, angiogenesis, chemokine secretion to recruit immune cells, and tissue remodeling, with accelerated wound repair. IEL, Intraepithelial lymphocytes; IL, interleukin; TNF, tumor necrosis factor; DCs, dendritic cells; MMPs, matrix metalloproteinases.

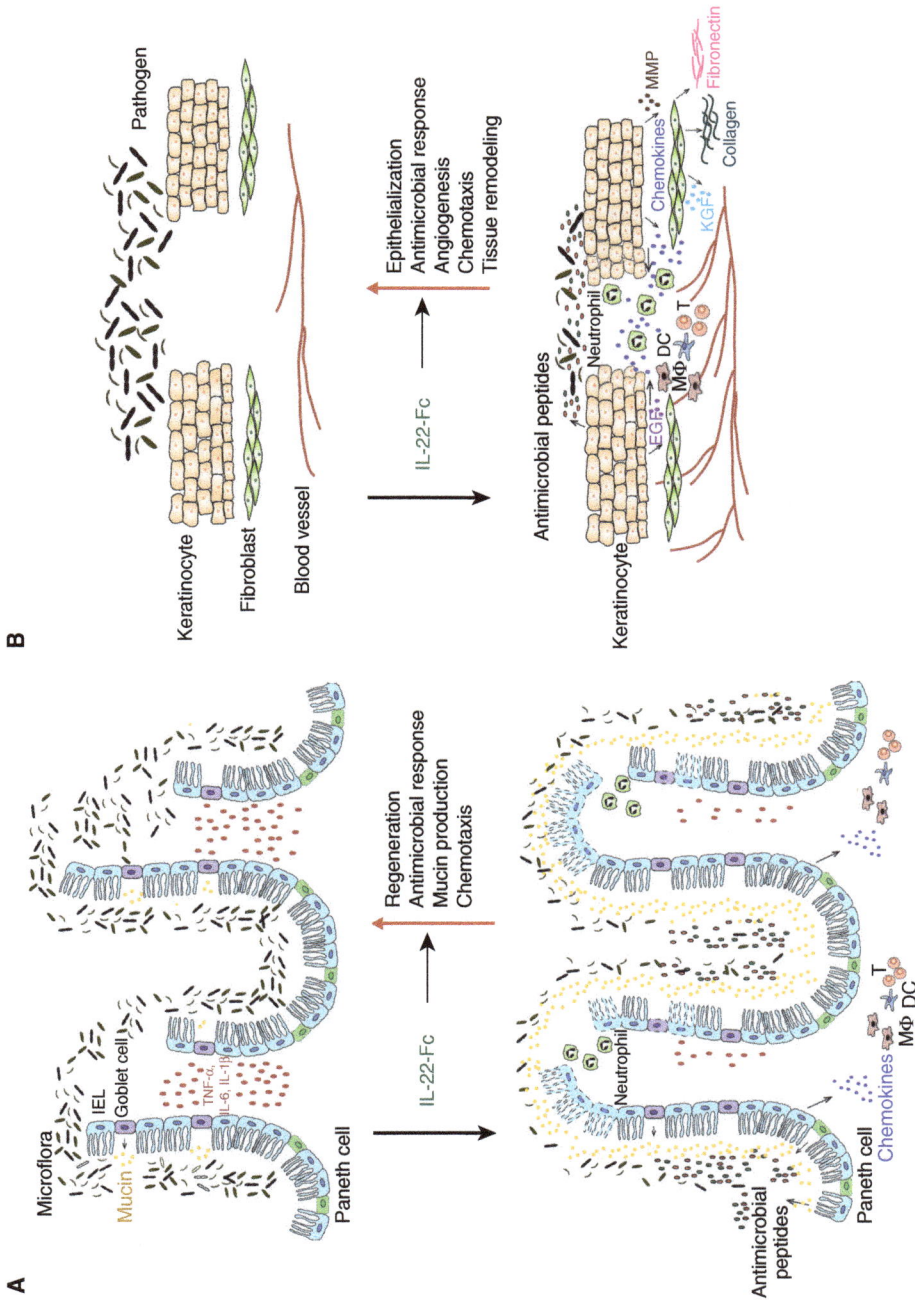

Figure 3. (*See legend on facing page.*)

production might result from an increased expression of TGF-β, as TGF-β represses IL-22 production from T cells (Zheng et al. 2008; Rutz et al. 2011). Furthermore, expression of IL-22BP, acting as a natural IL-22 antagonist, has been detected in mucosal samples from IBD patients (Pelczar et al. 2016). Blockade of the TNF-α pathway can reduce IL-22BP production, suggesting that the enhanced IL-22 activity might be partially attributed to the efficacy of anti-TNF-α therapies in IBD. The beneficial effects of IL-22 in IBD were further supported by a case study of a UC patient (Broadhurst et al. 2010). The reported patient infected himself with the nematode *Trichuris trichiura* to treat his symptoms. The disease went into remission accompanied by an accumulation of IL-22-producing CD4[+] T cells in the intestinal mucosa. Goblet cell hyperplasia and increased mucus production were also observed (Broadhurst et al. 2010).

IL-22 exerts its protective effects at mucosal surfaces (Fig. 3A) by increasing the expression of mucus-associated molecules and restoring goblet cells (Sugimoto et al. 2008). It contributes to intestinal immunity by promoting the formation of secondary lymphoid structures, such as colonic patches, in the gut (Ota et al. 2011). Furthermore, IL-22 induces the production of various antimicrobial peptides and enhances innate intestinal defense functions (Zheng et al. 2008; Zenewicz et al. 2013). Finally, IL-22 fosters the repair of epithelial barriers by promoting epithelial cell proliferation and the expansion of intestinal stem cells (Pickert et al. 2009; Huber et al. 2012; Lindemans et al. 2015).

Studies in mice have shown a beneficial role for IL-22 in various colitis models. IL-22-deficient mice (Zenewicz et al. 2008) or mice receiving neutralizing anti-IL-22 antibodies (Sugimoto et al. 2008) show increased disease severity, as evidenced by extensive epithelial destruction and inflammation in the colon, increased weight loss, and delayed recovery in a DSS-induced acute colitis model. In a T-cell transfer model of colitis, the transfer of IL-22-deficient naïve CD4[+] T cells cause more severe colitis than the transfer of wild-type CD4[+] T cells (Zenewicz et al. 2008). In a *C. rodentium*–

dependent colitis model, in which wild-type mice normally recover from the infection, IL-22-deficient mice do not (Zheng et al. 2008; Ota et al. 2011). Conversely, an increase in IL-22 levels either by administration of recombinant cytokine or hydrodynamic tail vein injection of IL-22-encoding plasmids ameliorates disease in various colitis models (Sugimoto et al. 2008; Zenewicz et al. 2008; Ota et al. 2011).

However, several potential caveats must be considered carefully before moving IL-22 forward as a treatment for IBD. Despite its tissue-protective role, IL-22 is a proinflammatory cytokine that can enhance inflammation, especially in synergy with other cytokines (Rutz et al. 2014). Indeed, a pathogenic role of IL-22 has been described in certain colitis models (Kamanaka et al. 2011). Furthermore, IL-22 targets many other organs such as skin, liver, and lung, and the downstream effects on these organs following systemic administration of IL-22 need to be considered (Park et al. 2011). Lastly, IL-22 activates STAT3 and promotes proliferation of many epithelial cell types. STAT3 is well known for its oncogenic activity in many epithelial tumors (Yu et al. 2014). Many cell types expressing IL-22R have been associated with neoplastic proliferation and IL-22-expressing cells have been detected in carcinomas of the colon (Jiang et al. 2013), stomach (Zhuang et al. 2012), liver (Jiang et al. 2013), and lung (Kobold et al. 2013). Increased colon tumorigenesis is observed when IL-22 is elevated in mice with colitis undergoing carcinogen treatment (Kirchberger et al. 2013) or in mice bearing a genetic alteration (Apc[Min/+]) that causes colorectal cancer (Huber et al. 2012). However, studies in transgenic mouse models suggest that long-term ectopic expression of IL-22 itself is insufficient to cause cancer (Park et al. 2011; Wang et al. 2011). Given that chronic inflammation is a risk factor for colon cancer, the beneficial effect of IL-22 in reducing tissue damage and inflammation might actually decrease the risk of early oncogenesis (Huber et al. 2012).

In conclusion, strong genetic and mechanistic data support a novel strategy for treating IBD by enhancing barrier function and intestinal in-

Cite this article as *Cold Spring Harb Perspect Biol* doi: 10.1101/cshperspect.a028548

nate defense, which is an approach that is differentiated from current therapies that mainly rely on systemic anti-inflammatory agents, such as TNF-α blockers (Ouyang 2010). To modulate the activity of IL-22 and to prolong its half-life, we generated murine IL-22-Fc fusion proteins (Ota et al. 2011; Wang et al. 2014). These molecules have a longer half-life in vivo, are able to induce downstream effector functions similar to IL-22, and protect intestinal integrity in various models.

Tissue Protective and Regenerative Effects of IL-22 in Pancreatitis and Hepatitis

In addition to skin and intestinal epithelial cells, the IL-22 receptor is also expressed in liver, pancreas, lung, kidney, and many other cell types of epithelial origin (Ouyang et al. 2011; Rutz et al. 2014). Recent advances in IL-22 biology revealed profound effects of IL-22 in tissue protection and regeneration in many organs including liver, pancreas, thymus, and skin (Radaeva et al. 2004; Zenewicz et al. 2008; Dudakov et al. 2012; Feng et al. 2012b; Kolumam et al. 2017). Acinar cells in the pancreas have the highest expression level of IL-22 receptor in the body (Xie et al. 2000; Aggarwal et al. 2001; Huan et al. 2016). Pancreatitis, an inflammatory condition, can occur in acute or chronic forms (Braganza et al. 2011; Banks et al. 2013). Excessive alcohol consumption, gallstones, and autoimmune diseases are some of its leading causes. The current treatment for pancreatitis is supportive and no curative therapy is available. A protective role of IL-22 in murine pancreatitis models has been described, suggesting a therapeutic potential for IL-22-based therapies (Feng et al. 2012b; Huan et al. 2016). Elevated IL-22 levels protect mice from cerulein-induced acute pancreatitis, as evidenced by a reduction in serum digestive enzymes, apoptosis, and infiltration with inflammatory cells (Feng et al. 2012b). Increased IL-22 levels inhibit autophagosome formation through STAT3-dependent regulation of Bcl-2 and Bcl-XL (Feng et al. 2012b; Huan et al. 2016). Furthermore, IL-22 up-regulates the expression of Reg3 (Hill et al. 2013; Huan et al. 2016), which protects acinar cells

against injury and inflammation. IL-22 has not yet been tested clinically for the treatment of pancreatitis.

Hepatitis is commonly associated with viral infections or excessive alcohol consumption. Patients with chronic hepatitis B virus (HBV) or hepatitis C virus (HCV) infection show increased numbers of IL-22-producing cells and corresponding elevated levels of IL-22 (Dambacher et al. 2008; Park et al. 2011). Administration of recombinant IL-22 protein (Radaeva et al. 2004), delivery of an IL-22-expression construct via hydrodynamic tail vein injection (Pan et al. 2004), or transgenic expression of IL-22 under a liver-specific albumin promoter (Park et al. 2011) protect mice from liver damage in hepatitis models, such as concanavalin A–induced T-cell-mediated hepatitis. In contrast, inhibition of IL-22 by neutralizing antibodies (Radaeva et al. 2004) or genetic ablation of IL-22 (Zenewicz et al. 2007) worsens liver damage in these models. In acute and chronic mouse models of alcohol-induced liver damage, treatment with IL-22 ameliorates alcoholic fatty liver, liver injury, and hepatic oxidative stress (Ki et al. 2010; Xing et al. 2011). The protective effects of IL-22 in the liver are thought to be mediated through STAT3 and the induction of antiapoptotic, mitogenic, and antioxidant pathways in hepatocytes and liver stem cells (Pan et al. 2004; Radaeva et al. 2004; Feng et al. 2012a). Based on these preclinical data, an IL-22-IgG2 fusion protein is currently being tested in the clinic for the treatment of alcoholic hepatitis (Table 2).

A Protective Role for IL-22 in Graft-Versus-Host Disease

Allogeneic hematopoietic stem-cell transplantation holds great promise for the treatment of malignant and nonmalignant hematologic diseases. The major obstacle that hinders successful transplantation is graft-versus-host disease (GVHD) (Ferrara et al. 2009), which is associated with severe morbidity and mortality (Wingard et al. 2011). Both innate and adaptive immune cells are thought to be involved in the pathophysiology of GVHD (Ferrara et al. 2009; Konya and Mjösberg 2015). Interestingly, an increased number of donor-derived ILCs in the blood correlates with reduced severity of GVHD in patients (Munneke et al. 2014). A recent study in mice reported a protective role of IL-22 in GVHD during transplantation of intestinal stem cells (Hanash et al. 2012). Intestinal stem cells and downstream progenitor cells express the IL-22 receptor (Hanash et al. 2012), which can promote epithelial regeneration through STAT3-dependent mechanisms in response to IL-22 secreted from various sources, including ILC3s (Hanash et al. 2012). Furthermore, administration of IL-22 promotes the recovery of intestinal stem cells and intestinal regeneration, and reduces mortality from GVHD in transplanted animals (Hanash et al. 2012). IL-22 deficiency in the recipient, on the other hand, increases tissue damage and mortality as the result of acute GVHD (Hanash et al. 2012). A clinic trial is currently ongoing to evaluate the therapeutic potential of IL-22 in GVHD (Table 2).

Potential Therapeutic Applications for IL-20 Subfamily Cytokines to Enhance Wound Healing in Diabetic Foot Ulcer

A regulated topical wound-healing response restores tissue homeostasis and immune defense mechanisms in the skin (Singer and Clark 1999; Li et al. 2007). The healing process can be divided into several overlapping phases: hemostasis, inflammation, angiogenesis, re-epithelialization, and remodeling or maturation. These phases involve many types of tissue cells, immune cells, and soluble factors, such as cytokines and growth factors. DFU is a chronic wound with impeded healing. As one of the major complications of type II diabetes, diabetic ulcers lead to a significant number of amputations, and increase the morbidity and mortality of diabetes patients (Jeffcoate and Harding 2003; Boulton et al. 2005). The impaired wound healing is in part because of deficiencies in leukocyte recruitment, macrophage function, angiogenesis, extracellular matrix deposition, epidermal barrier function, and fibroblast activities under diabetic conditions (Falanga 2005; Brem and Tomic-Canic 2007). Frequently, wound

site infections in DFU patients are difficult to control and further hinder the wound-healing process.

Several IL-20 subfamily cytokines participate in the different stages of the wound-healing response (Fig. 3B) by regulating re-epithelialization, immune cell infiltration, production of growth factors, and tissue remodeling (Rutz et al. 2014). Mice ectopically expressing IL-20 subfamily cytokines develop skin pathogenesis that resembles human psoriasis, which can be viewed as an excessive wound-healing response (Blumberg et al. 2001; Wolk et al. 2009a; He and Liang 2010). All IL-20 subfamily cytokines are induced during normal wound healing in mouse models, and elevated IL-24 levels have been identified in biopsies of human wounded skin (Bosanquet et al. 2012; Kolumam et al. 2017). The receptors for IL-20 subfamily cytokines are expressed primarily on keratinocytes in the skin (Sa et al. 2007). Studies conducted in vitro and in vivo have shown several downstream pathways induced by IL-20 subfamily cytokines to contribute to different stages of wound healing (Wolk and Sabat 2006; Sa et al. 2007; Rutz et al. 2014). All IL-20 subfamily cytokines promote re-epithelialization by stimulating the proliferation of epidermal keratinocytes and by inducing EGF and KGF production from keratinocytes (Sa et al. 2007). They also trigger chemokine production from keratinocytes and thereby enhance inflammatory infiltration of leukocytes, especially macrophages. In addition, all IL-20 subfamily cytokines induce vascular endothelial growth factor (VEGF), an important growth factor that facilitates angiogenesis and neovascularization (Sa et al. 2007; Kolumam et al. 2017). Importantly, they induce the expression of various antimicrobial peptides from keratinocytes to dampen infections that are commonly associated with wounded skin. Finally, IL-20 subfamily cytokines induce the production of many proteases and extracellular matrix proteins from keratinocytes and fibroblasts to mediate scar formation and tissue remodeling. IL-22 is the most potent, followed by IL-24 and IL-20, with IL-19 being the least effective family member with regard to its wound-healing activity. Yet, various IL-20 subfamily members appear to have redundant functions (McGee et al. 2013; Kolumam et al. 2017).

In preclinical models, topical treatment of wounded skin with recombinant IL-19, IL-20, or Fc-fusion proteins of IL-20, IL-22, and IL-24 accelerates the cutaneous healing process in mice (Sun et al. 2013; Kolumam et al. 2017). The therapeutic potential of IL-22, IL-20, and IL-24 has also been studied in diabetic wound-healing models. In streptozotocin-induced type I diabetic mice, topical treatment with recombinant IL-22 accelerates wound closure compared with control-treated diabetic wounds (Avitabile et al. 2015). Re-epithelialization of wounds treated with IL-22 is significantly improved. Another study examined the therapeutic effect of IL-22R-binding cytokines in diabetic wound repair in db/db mice (Kolumam et al. 2017), a commonly used mouse model for type II diabetes, which shows much slower wound closure compared with other diabetes models (Michaels et al. 2007). Both systemic and topical application of Fc fusion versions of IL-20, IL-22, and IL-24 significantly accelerates wound closure in db/db mice, independent of their metabolic functions (Wang et al. 2014; Kolumam et al. 2017). Gene-expression analysis revealed that IL-22-Fc treatment specifically promotes the activity of critical pathways for re-epithelialization, tissue remodeling, host defense, and fatty acid/lipid metabolism compared with PDGF or VEGF treatment. Many genes that are up-regulated by IL-22-Fc been reported to participate in cutaneous wound healing, such as *Cxcr2*, *Grhl3*, and *Klk8* (Kolumam et al. 2017).

THE INTERFERON λ SUBFAMILY IL-28A, IL-28B, IL-29

The genes encoding IL-28A (IFN-λ2), IL-28B (IFN-λ3), and IL-29 (IFN-λ1) were identified based on their homology with genes of the IFN/IL-10 superfamily of cytokines (Kotenko et al. 2003; Sheppard et al. 2003; Fox et al. 2009). Structurally, these three cytokines are more closely related to IL-10 family cytokines, in particular to IL-22 (Gad et al. 2009). They signal through the IL-10R2 chain, also used by other IL-10 family cytokines, but they use a unique IL-28R as their receptor α chain (Kotenko et al. 2003; Sheppard et al. 2003). Despite differences in receptor usage, downstream signaling shows substantial overlap with IFN-α/β (Ouyang et al. 2011; Durbin et al. 2013). Accordingly, IFN-λ and type I IFNs trigger similar biological effects and antiviral responses (Onoguchi et al. 2007; Kotenko et al. 2003). However, substantial differences exist with regard to the magnitude and kinetics of IFN-sensitive gene (ISG) induction. Whereas IFN-α strongly induces ISGs with an early peak followed by rapid down-regulation, IFN-λ triggers a weak but sustained ISG response (Kotenko et al. 2003; Meager et al. 2005; Marcello et al. 2006). More importantly, in contrast to a broad range of cells that respond to type I IFNs, only a few cell types, in particular epithelial cells (Fig. 1), respond to type III IFNs (Sommereyns et al. 2008; Witte et al. 2009). This has led to the idea that IFN-λ and IL-20 subfamily members serve parallel functions in protecting epithelial tissue against viral and bacterial infections, respectively (Gad et al. 2009).

IFN-λ for the Treatment of Viral Infections

Generally speaking, IFN-λs induce relatively weak antiviral responses. However, GWAS identified a role for polymorphisms near the *IFNL3* gene in the control of HCV infection, as well as other viruses with epithelial cell tropism. IFN-λs are in fact the dominant IFN subclass produced in the liver of humans, chimpanzees, and in primary human hepatocyte cultures infected with HCV (Park et al. 2012; Thomas et al. 2012). A genetic polymorphism linked to the gene for IL-28B has been associated with clinical response to IFN-α and ribavirin therapy in patients with chronic HCV infection (Ge et al. 2009; Suppiah et al. 2009; Tanaka et al. 2009; Thomas et al. 2009), suggesting cooperation between the IFN-λ and type I IFN responses during host defense. A critical function for IFN-λ has been reported for a variety of preclinical models of pulmonary infections in mice, including influenza A, influenza B, severe acute respiratory syndrome (SARS), coronavirus, and H1N1 influenza virus (Mordstein et al. 2008, 2010). Furthermore, IFN-λs have the ability to inhibit the replication of multiple other viruses, including human immunodeficiency virus (HIV) (Hou et al. 2009), herpes simplex virus type 2 (HSV2) (Ank et al. 2006), cytomegalovirus (CMV) (Brand et al. 2005), and encephalomyocarditis virus (EMCV) (Hou et al. 2009).

IFN-λ is an attractive therapeutic target for the treatment of viral infections. The more restricted expression of IFN-λ receptors predominantly on epithelial cells suggests that IFN-λ, although sharing the same therapeutic advantages, might avoid many of the systemic side effects of IFN-α/β. Furthermore, there is genetic evidence that certain *IFNL3/IFNL4* polymorphisms are linked to the clearance of HCV infection and potentially other types of infections (Miller et al. 2009). Type I and II IFNs are currently approved for the treatment of HCV and HBV (Gibbert et al. 2013; Lin and Young 2014).

A PEGylated IFN-λ1 has been evaluated in clinical trials for the treatment of HCV (Table 3). A phase II trial showed that IFN-λ1 was as effective as PEGylated IFN-α with significantly fewer extrahepatic adverse events, such as neutropenia, thrombocytopenia, flu-like symptoms, autoimmune thyroid disease, and pulmonary arterial hypertension (Muir et al. 2014; Fredlund et al. 2015). Several phase III clinical trials are currently ongoing (Friborg et al. 2013).

However, the availability of highly efficacious direct-acting antiviral agents (DAAs) makes it highly unlikely for IFN-λ to play a major role for HCV therapy in the future. Although efficacy will have to be shown in the clinic, IFN-λ

could potentially hold promise for the treatment of select infections with epithelial tropism, such as intestinal infection caused by rotaviruses or noroviruses, and respiratory infections caused by viruses, such as influenza virus or coronaviruses.

Role for IFN-λ in Cancer

IFN-λ is induced in the tumor microenvironment, either through viral infections or other mechanisms, and shows antitumor activity (Steen and Gamero 2010; Burkart et al. 2013; Cannella et al. 2014). Depending on receptor expression, IFN-λ induces apoptosis in various murine and human cancer cells. Responsive tumors include neuroendocrine tumors, colorectal/intestinal carcinoma, hepatocellular carcinoma, esophageal carcinoma, lung adenocarcinomas, Burkitt lymphoma, and melanoma (Sato et al. 2006; Zitzmann et al. 2006; Guenterberg et al. 2010; Li et al. 2010; Steen and Gamero 2010). In a mouse model of breast cancer, for example, IFN-λ expression on mammary epithelial cells inversely correlates with tumor growth (Burkart et al. 2013). IFN-λs induce apoptosis in colorectal cancer cells more potently than IFN-α/β or IFN-γ (Li et al. 2008), and show comparable potency in a mouse hepatoma model (Abushahba et al. 2010). Interestingly, the combination of IFN-λ and IFN-α showed more substantial antitumor activity than either IFN alone (Lasfar et al. 2016). Like other IFNs, IFN-λs possess immune stimulatory activities and enhance antitumor immunity through various mechanisms (Numasaki et al. 2007; Li et al. 2010; Steen

Table 3. Clinical trials targeting interleukin (IL)-29

Intervention	Indication	Clinical stage	Sponsor
PEG-rIL-29	Hepatitis C	Phase I NCT00565539	ZymoGenetics
		Phase II NCT01001754	
PEG-rIL-29	Hepatitis C	Phase III NCT01718158	Bristol-Myers Squibb
		Phase III NCT01598090	
		Phase III NCT01754974	
		Phase III NCT01866930	
		Phase III NCT01616524	
PEG-rIL-29	Hepatitis D	Phase II NCT02765802	Eiger BioPharmaceuticals

Data source: clinicaltrials.gov.

Cite this article as *Cold Spring Harb Perspect Biol* doi: 10.1101/cshperspect.a028548

and Gamero 2010). Accordingly, antitumor activity has also been observed in models in which the tumor cells themselves are not responsive to IFN-λ (Sato et al. 2006). To date, IFN-λ has not been studied clinically as a cancer therapy.

CONCLUDING REMARKS

Over the past three decades, a number of cytokines have been targeted successfully for the treatment of a wide spectrum of human disorders (O'Shea et al. 2014). In fact, since the approval of recombinant IFN-α2b for the treatment of chronic HBV infection, therapies targeting members of most cytokine families, including IL-1, IL-2 γc, IL-6, IL-12, IL-17, and TNF-α, have been developed for clinical use in autoimmune disorders, infectious diseases, and cancer. However, despite their known essential immune regulatory functions in host defense and tissue protection, the IL-10 family of cytokines is currently an exception with no approved therapies for any indication. As we have discussed, part of the reason is that we are still in the early stages of understanding the pleiotropic functions of many cytokines in this family. The degree of functional redundancy within the IL-20 subfamily further complicates targeting these cytokines. However, given the recent advances in our understanding of the diverse biology of IL-10 family cytokines, combined with technical progress in protein engineering, and innovative targeting and delivery strategies, we are increasingly able to modulate these cytokine pathways in a cell-type- and disease-specific manner. Novel therapies targeting IL-10 family cytokines that will improve the lives of patients may therefore be within reach.

REFERENCES

Abushahba W, Balan M, Castaneda I, Yuan Y, Reuhl K, Raveche E, la Torre de A, Lasfar A, Kotenko SV. 2010. Antitumor activity of type I and type III interferons in BNL hepatoma model. *Cancer Immunol Immunother* **59**: 1059–1071.

Adris S, Klein S, Jasnis M, Chuluyan E, Ledda M, Bravo A, Carbone C, Chernajovsky Y, Podhajcer O. 1999. IL-10 expression by CT26 colon carcinoma cells inhibits their malignant phenotype and induces a T-cell-mediated tu-mor rejection in the context of a systemic Th2 response. *Gene Ther* **6**: 1705–1712.

Aggarwal S, Xie MH, Maruoka M, Foster J, Gurney AL. 2001. Acinar cells of the pancreas are a target of interleukin-22. *J Interferon Cytokine Res* **21**: 1047–1053.

Alanärä T, Karstila K, Moilanen T, Silvennoinen O, Isomäki P. 2010. Expression of IL-10 family cytokines in rheumatoid arthritis: Elevated levels of IL-19 in the joints. *Scand J Rheumatol* **39**: 118–126.

Andoh A, Zhang Z, Inatomi O, Fujino S, Deguchi Y, Araki Y, Tsujikawa T, Kitoh K, Kim-Mitsuyama S, Takayanagi A, et al. 2005. Interleukin-22, a member of the IL-10 subfamily, induces inflammatory responses in colonic subepithelial myofibroblasts. *Gastroenterology* **129**: 969–984.

Ank N, West H, Bartholdy C, Eriksson K, Thomsen AR, Paludan SR. 2006. Lambda interferon (IFN-λ), a type III IFN, is induced by viruses and IFNs and displays potent antiviral activity against select virus infections in vivo. *J Virol* **80**: 4501–4509.

Asadullah K, Sterry W, Stephanek K, Jasulaitis D, Leupold M, Audring H, Volk H-D, Döcke WD. 1998. IL-10 is a key cytokine in psoriasis. Proof of principle by IL-10 therapy: A new therapeutic approach. *J Clin Invest* **101**: 783–794.

Asadullah K, Sterry W, Volk H-D. 2003. Interleukin-10 therapy—Review of a new approach. *Pharmacol Rev* **55**: 241–269.

Avitabile S, Odorisio T, Madonna S, Eyerich S, Guerra L, Eyerich K, Zambruno G, Cavani A, Cianfarani F. 2015. Interleukin-22 promotes wound repair in diabetes by improving keratinocyte pro-healing functions. *J Invest Dermatol* **135**: 2862–2870.

Banks PA, Bollen TL, Dervenis C, Gooszen HG, Johnson CD, Sarr MG, Tsiotos GG, Vege SS; Acute Pancreatitis Classification Working Group. 2013. Classification of acute pancreatitis—2012: Revision of the Atlanta classification and definitions by international consensus. *Gut* **62**: 102–111.

Benbouziane B, Ribelles P, Aubry C, Martin R, Kharrat P, Riazi A, Langella P, Bermúdez-Humarán LG. 2013. Development of a stress-inducible controlled expression (SICE) system in *Lactococcus lactis* for the production and delivery of therapeutic molecules at mucosal surfaces. *J Biotechnol* **168**: 120–129.

Blumberg H, Conklin D, Xu WF, Grossmann A, Brender T, Carollo S, Eagan M, Foster D, Haldeman BA, Hammond A, et al. 2001. Interleukin 20: Discovery, receptor identification, and role in epidermal function. *Cell* **104**: 9–19.

Boniface K, Bernard FX, Garcia M, Gurney AL, Lecron JC, Morel F. 2005. IL-22 inhibits epidermal differentiation and induces proinflammatory gene expression and migration of human keratinocytes. *J Immunol* **174**: 3695–3702.

Boniface K, Guignouard E, Pedretti N, Garcia M, Delwail A, Bernard FX, Nau F, Guillet G, Dagregorio G, Yssel H, et al. 2007. A role for T cell-derived interleukin 22 in psoriatic skin inflammation. *Clin Exp Immunol* **150**: 407–415.

Borghaei H, Paz-Ares L, Horn L, Spigel DR, Steins M, Ready NE, Chow LQ, Vokes EE, Felip E, Holgado E, et al. 2015. Nivolumab versus docetaxel in advanced nonsquamous non-small-cell lung cancer. *N Engl J Med* **373**: 1627–1639.

Bosanquet DC, Harding KG, Ruge F, Sanders AJ, Jiang WG. 2012. Expression of IL-24 and IL-24 receptors in human wound tissues and the biological implications of IL-24 on keratinocytes. *Wound Repair Regen* **20**: 896–903.

Boulton AJM, Vileikyte L, Ragnarson-Tennvall G, Apelqvist J. 2005. The global burden of diabetic foot disease. *Lancet* **366**: 1719–1724.

Braganza JM, Lee SH, McCloy RF, McMahon MJ. 2011. Chronic pancreatitis. *Lancet* **377**: 1184–1197.

Brand S, Beigel F, Olszak T, Zitzmann K, Eichhorst ST, Otte JM, Diebold J, Diepolder H, Adler B, Auernhammer CJ, et al. 2005. IL-28A and IL-29 mediate antiproliferative and antiviral signals in intestinal epithelial cells and murine CMV infection increases colonic IL-28A expression. *Am J Physiol Gastrointest Liver Physiol* **289**: G960–G968.

Brem H, Tomic-Canic M. 2007. Cellular and molecular basis of wound healing in diabetes. *J Clin Invest* **117**: 1219–1222.

Broadhurst MJ, Leung JM, Kashyap V, McCune JM, Mahadevan U, McKerrow JH, Loke P. 2010. IL-22+ CD4+ T cells are associated with therapeutic *Trichuris trichiura* infection in an ulcerative colitis patient. *Sci Transl Med* **2**: 60ra88.

Burkart C, Arimoto KI, Tang T, Cong X, Xiao N, Liu YC, Kotenko SV, Ellies LG, Zhang DE. 2013. Usp18 deficient mammary epithelial cells create an antitumour environment driven by hypersensitivity to IFN-λ and elevated secretion of Cxcl10. *EMBO Mol Med* **5**: 1035–1050.

Buruiana FE, Solà I, Alonso-Coello P. 2010. Recombinant human interleukin 10 for induction of remission in Crohn's disease. *Cochrane Database Syst Rev* **119**: CD005109.

Cannella B, Gao YL, Brosnan C, Raine CS. 1996. IL-10 fails to abrogate experimental autoimmune encephalomyelitis. *J Neurosci Res* **45**: 735–746.

Cannella F, Scagnolari C, Selvaggi C, Stentella P, Recine N, Antonelli G, Pierangeli A. 2014. Interferon λ1 expression in cervical cells differs between low-risk and high-risk human papillomavirus-positive women. *Med Microbiol Immunol* **203**: 177–184.

Cassatella MA, Meda L, Bonora S, Ceska M, Constantin G. 1993. Interleukin 10 (IL-10) inhibits the release of proinflammatory cytokines from human polymorphonuclear leukocytes. Evidence for an autocrine role of tumor necrosis factor and IL-1 β in mediating the production of IL-8 triggered by lipopolysaccharide. *J Exp Med* **178**: 2207–2211.

Caudell EG, Mumm JB, Poindexter N, Ekmekcioglu S, Mhashilkar AM, Yang XH, Retter MW, Hill P, Chada S, Grimm EA. 2002. The protein product of the tumor suppressor gene, melanoma differentiation-associated gene 7, exhibits immunostimulatory activity and is designated IL-24. *J Immunol* **168**: 6041–6046.

Chan JR, Blumenschein W, Murphy E, Diveu C, Wiekowski M, Abbondanzo S, Lucian L, Geissler R, Brodie S, Kimball AB, et al. 2006. IL-23 stimulates epidermal hyperplasia via TNF and IL-20R2-dependent mechanisms with implications for psoriasis pathogenesis. *J Exp Med* **203**: 2577–2587.

Chen DS, Mellman I. 2017. Elements of cancer immunity and the cancer-immune set point. *Nature* **541**: 321–330.

Chernoff AE, Granowitz EV, Shapiro L, Vannier E, Lonnemann G, Angel JB, Kennedy JS, Rabson AR, Wolff SM, Dinarello CA. 1995. A randomized, controlled trial of IL-

10 in humans. Inhibition of inflammatory cytokine production and immune responses. *J Immunol* **154**: 5492–5499.

Colombel JF, Rutgeerts P, Malchow H, Jacyna M, Nielsen OH, Rask-Madsen J, Van Deventer S, Ferguson A, Desreumaux P, Forbes A, et al. 2001. Interleukin 10 (Tenovil) in the prevention of postoperative recurrence of Crohn's disease. *Gut* **49**: 42–46.

Corvaisier M, Delneste Y, Jeanvoine H, Preisser L, Blanchard S, Garo E, Hoppe E, Barré B, Audran M, Bouvard B, et al. 2012. IL-26 is overexpressed in rheumatoid arthritis and induces proinflammatory cytokine production and Th17 cell generation. *PLoS Biol* **10**: e1001395.

Creery WD, Diaz-Mitoma F, Filion L, Kumar A. 1996. Differential modulation of B7-1 and B7-2 isoform expression on human monocytes by cytokines which influence the development of T helper cell phenotype. *Eur J Immunol* **26**: 1273–1277.

Cunningham CC, Chada S, Merritt JA, Tong A, Senzer N, Zhang Y, Mhashilkar A, Parker K, Vukelja S, Richards D, et al. 2005. Clinical and local biological effects of an intratumoral injection of *mda-7* (IL24; INGN 241) in patients with advanced carcinoma: A phase I study. *Mol Ther* **11**: 149–159.

Dambacher J, Beigel F, Zitzmann K, Heeg MHJ, Göke B, Diepolder HM, Auernhammer CJ, Brand S. 2008. The role of interleukin-22 in hepatitis C virus infection. *Cytokine* **41**: 209–216.

D'Andrea A, Aste-Amezaga M, Valiante NM, Ma X, Kubin M, Trinchieri G. 1993. Interleukin 10 (IL-10) inhibits human lymphocyte interferon λ-production by suppressing natural killer cell stimulatory factor/IL-12 synthesis in accessory cells. *J Exp Med* **178**: 1041–1048.

da Rocha LF, Duarte ALBP, Dantas AT, Mariz HA, Pitta IDR, Galdino SL, Pitta MGDR. 2012. Increased serum interleukin 22 in patients with rheumatoid arthritis and correlation with disease activity. *J Rheumatol* **39**: 1320–1325.

Dejaco C, Reinisch W, Lichtenberger C, Waldhoer T, Kuhn I, Tilg H, Gasche C. 2000. In vivo effects of recombinant human interleukin-10 on lymphocyte phenotypes and leukocyte activation markers in inflammatory bowel disease. *J Investig Med* **48**: 449–456.

del Carmen S, de Moreno de LeBlanc A, Perdigon G, Bastos Pereira V, Miyoshi A, Azevedo V, LeBlanc JG. 2011. Evaluation of the anti-inflammatory effect of milk fermented by a strain of IL-10-producing *Lactococcus lactis* using a murine model of Crohn's disease. *J Mol Microbiol Biotechnol* **21**: 138–146.

del Carmen S, Martín Rosique R, Saraiva T, Zurita-Turk M, Miyoshi A, Azevedo V, de Moreno de LeBlanc A, Langella P, Bermúdez-Humarán LG, LeBlanc JG. 2014. Protective effects of lactococci strains delivering either IL-10 protein or cDNA in a TNBS-induced chronic colitis model. *J Clin Gastroenterol* **48** (Suppl 1): S12–S17.

de Waal-Malefyt R, Abrams J, Bennett B, Figdor CG, de Vries JE. 1991a. Interleukin 10(IL-10) inhibits cytokine synthesis by human monocytes: An autoregulatory role of IL-10 produced by monocytes. *J Exp Med* **174**: 1209–1220.

de Waal-Malefyt R, Haanen J, Spits H, Roncarolo MG, Velde Te A, Figdor C, Johnson K, Kastelein R, Yssel H, de Vries

 Cite this article as *Cold Spring Harb Perspect Biol* doi: 10.1101/cshperspect.a028548

JE. 1991b. Interleukin 10 (IL-10) and viral IL-10 strongly reduce antigen-specific human T cell proliferation by diminishing the antigen-presenting capacity of monocytes via downregulation of class II major histocompatibility complex expression. *J Exp Med* **174:** 915–924.

Doll F, Schwager K, Hemmerle T, Neri D. 2013. Murine analogues of etanercept and of F8-IL10 inhibit the progression of collagen-induced arthritis in the mouse. *Arthritis Res Ther* **15:** R138.

Drouault S, Corthier G, Ehrlich SD, Renault P. 1999. Survival, physiology, and lysis of *Lactococcus lactis* in the digestive tract. *Appl Environ Microbiol* **65:** 4881–4886.

Dudakov JA, Hanash AM, Jenq RR, Young LF, Ghosh A, Singer NV, West ML, Smith OM, Holland AM, Tsai JJ, et al. 2012. Interleukin-22 drives endogenous thymic regeneration in mice. *Science* **336:** 91–95.

Durbin RK, Kotenko SV, Durbin JE. 2013. Interferon induction and function at the mucosal surface. *Immunol Rev* **255:** 25–39.

Eidenschenk C, Rutz S, Liesenfeld O, Ouyang W. 2014. Role of IL-22 in microbial host defense. *Curr Top Microbiol Immunol* **380:** 213–236.

Emdad L, Lebedeva IV, Su ZZ, Gupta P, Sauane M, Dash R, Grant S, Dent P, Curiel DT, Sarkar D, et al. 2009. Historical perspective and recent insights into our understanding of the molecular and biochemical basis of the antitumor properties of mda-7/IL-24. *Cancer Biol Ther* **8:** 391–400.

Emmerich J, Mumm JB, Chan IH, LaFace D, Truong H, McClanahan T, Gorman DM, Oft M. 2012. IL-10 directly activates and expands tumor-resident CD8$^+$ T cells without de novo infiltration from secondary lymphoid organs. *Cancer Res* **72:** 3570–3581.

Engelhardt KR, Shah N, Faizura-Yeop I, Kocacik Uygun DF, Frede N, Muise AM, Shteyer E, Filiz S, Chee R, Elawad M, et al. 2013. Clinical outcome in IL-10- and IL-10 receptor-deficient patients with or without hematopoietic stem cell transplantation. *J Allergy Clin Immunol* **131:** 825–830.

Falanga V. 2005. Wound healing and its impairment in the diabetic foot. *Lancet* **366:** 1736–1743.

Feagan BG, Sandborn WJ, Gasink C, Jacobstein D, Lang Y, Friedman JR, Blank MA, Johanns J, Gao LL, Miao Y, et al. 2016. Ustekinumab as induction and maintenance therapy for Crohn's disease. *N Engl J Med* **375:** 1946–1960.

Fedorak RN, Gangl A, Elson CO, Rutgeerts P, Schreiber S, Wild G, Hanauer SB, Kilian A, Cohard M, LeBeaut A, et al. 2000. Recombinant human interleukin 10 in the treatment of patients with mild to moderately active Crohn's disease. The Interleukin 10 Inflammatory Bowel Disease Cooperative Study Group. *Gastroenterology* **119:** 1473–1482.

Feng D, Kong X, Weng H, Park O, Wang H, Dooley S, Gershwin ME, Gao B. 2012a. Interleukin-22 promotes proliferation of liver stem/progenitor cells in mice and patients with chronic hepatitis B virus infection. *Gastroenterology* **143:** 188–198.e7.

Feng D, Park O, Radaeva S, Wang H, Yin S, Kong X, Zheng M, Zakhari S, Kolls JK, Gao B. 2012b. Interleukin-22 ameliorates cerulein-induced pancreatitis in mice by inhibiting the autophagic pathway. *Int J Biol Sci* **8:** 249–257.

Ferrara JLM, Levine JE, Reddy P, Holler E. 2009. Graft-versus-host disease. *Lancet* **373:** 1550–1561.

Fiorentino DF, Bond MW, Mosmann TR. 1989. Two types of mouse T helper cell. IV: Th2 clones secrete a factor that inhibits cytokine production by Th1 clones. *J Exp Med* **170:** 2081–2095.

Fiorentino DF, Zlotnik A, Mosmann TR, Howard M, O'Garra A. 1991. IL-10 inhibits cytokine production by activated macrophages. *J Immunol* **147:** 3815–3822.

Fox BA, Sheppard PO, O'Hara PJ. 2009. The role of genomic data in the discovery, annotation and evolutionary interpretation of the interferon-λ family. *PloS ONE* **4:** e4933.

Franke A, Balschun T, Karlsen TH, Sventoraityte J, Nikolaus S, Mayr G, Domingues FS, Albrecht M, Nothnagel M, Ellinghaus D, et al. 2008. Sequence variants in IL10, ARPC2 and multiple other loci contribute to ulcerative colitis susceptibility. *Nat Genet* **40:** 1319–1323.

Franke A, McGovern DPB, Barrett JC, Wang K, Radford-Smith GL, Ahmad T, Lees CW, Balschun T, Lee J, Roberts R, et al. 2010. Genome-wide meta-analysis increases to 71 the number of confirmed Crohn's disease susceptibility loci. *Nat Genet* **42:** 1118–1125.

Franz M, Doll F, Grün K, Richter P, Köse N, Ziffels B, Schubert H, Figulla HR, Jung C, Gummert J, et al. 2015. Targeted delivery of interleukin-10 to chronic cardiac allograft rejection using a human antibody specific to the extra domain A of fibronectin. *Int J Cardiol* **195:** 311–322.

Fredlund P, Hillson J, Gray T, Shemanski L, Dimitrova D, Srinivasan S. 2015. Peginterferon Lambda-1a is associated with a low incidence of autoimmune thyroid disease in chronic hepatitis C. *J Interferon Cytokine Res* **35:** 841–843.

Friborg J, Levine S, Chen C, Sheaffer AK, Chaniewski S, Voss S, Lemm JA, McPhee F. 2013. Combinations of λ interferon with direct-acting antiviral agents are highly efficient in suppressing hepatitis C virus replication. *Antimicrob Agents Chemother* **57:** 1312–1322.

Gad HH, Dellgren C, Hamming OJ, Vends S, Paludan SR, Hartmann R. 2009. Interferon-λ is functionally an interferon but structurally related to the interleukin-10 family. *J Biol Chem* **284:** 20869–20875.

Galeazzi M, Bazzichi L, Sebastiani GD, Neri D, Garcia E, Ravenni N, Giovannoni L, Wilton J, Bardelli M, Baldi C, et al. 2014. A phase IB clinical trial with Dekavil (F8-IL10), an immunoregulatory "armed antibody" for the treatment of rheumatoid arthritis, used in combination wiIh methotrexate. *Isr Med Assoc J* **16:** 666.

Ge D, Fellay J, Thompson AJ, Simon JS, Shianna KV, Urban TJ, Heinzen EL, Qiu P, Bertelsen AH, Muir AJ, et al. 2009. Genetic variation in *IL28B* predicts hepatitis C treatment-induced viral clearance. *Nature* **461:** 399–401.

Geremia A, Arancibia-Cárcamo CV, Fleming MPP, Rust N, Singh B, Mortensen NJ, Travis SPL, Powrie F. 2011. IL-23-responsive innate lymphoid cells are increased in inflammatory bowel disease. *J Exp Med* **208:** 1127–1133.

Gibbert K, Schlaak JF, Yang D, Dittmer U. 2013. IFN-α subtypes: Distinct biological activities in anti-viral therapy. *Br J Pharmacol* **168:** 1048–1058.

Gittler JK, Shemer A, Suárez-Fariñas M, Fuentes-Duculan J, Gulewicz KJ, Wang CQF, Mitsui H, Cardinale I, de Guzman Strong C, Krueger JG, et al. 2012. Progressive activation of T$_H$2/T$_H$22 cytokines and selective epidermal proteins characterizes acute and chronic atopic dermatitis. *J Allergy Clin Immunol* **130:** 1344–1354.

Glocker E-O, Kotlarz D, Boztug K, Gertz EM, Schäffer AA, Noyan F, Perro M, Diestelhorst J, Allroth A, Murugan D, et al. 2009. Inflammatory bowel disease and mutations affecting the interleukin-10 receptor. *N Engl J Med* **361:** 2033–2045.

Gottlieb AB, Krueger JG, Sandberg Lundblad M, Göthberg M, Skolnick BE. 2015. First-in-human, phase 1, randomized, dose-escalation trial with recombinant anti-IL-20 monoclonal antibody in patients with psoriasis. *PloS ONE* **10:** e0134703.

Groux H, Bigler M, de Vries JE, Roncarolo MG. 1996. Interleukin-10 induces a long-term antigen-specific anergic state in human CD4+ T cells. *J Exp Med* **184:** 19–29.

Groux H, Bigler M, de Vries JE, Roncarolo MG. 1998. Inhibitory and stimulatory effects of IL-10 on human CD8+ T cells. *J Immunol* **160:** 3188–3193.

Guenterberg KD, Grignol VP, Raig ET, Zimmerer JM, Chan AN, Blaskovits FM, Young GS, Nuovo GJ, Mundy BL, Lesinski GB, et al. 2010. Interleukin-29 binds to melanoma cells inducing Jak-STAT signal transduction and apoptosis. *Mol Cancer Ther* **9:** 510–520.

Guilloteau K, Paris I, Pedretti N, Boniface K, Juchaux F, Huguier V, Guillet G, Bernard F-X, Lecron J-C, Morel F. 2010. Skin inflammation induced by the synergistic action of IL-17A, IL-22, oncostatin M, IL-1α, and TNF-α recapitulates some features of psoriasis. *J Immunol* **184:** 5263–5270.

Hagenbaugh A, Sharma S, Dubinett SM, Wei SH, Aranda R, Cheroutre H, Fowell DJ, Binder S, Tsao B, Locksley RM, et al. 1997. Altered immune responses in interleukin 10 transgenic mice. *J Exp Med* **185:** 2101–2110.

Hanash AM, Dudakov JA, Hua G, O'Connor MH, Young LF, Singer NV, West ML, Jenq RR, Holland AM, Kappel LW, et al. 2012. Interleukin-22 protects intestinal stem cells from immune-mediated tissue damage and regulates sensitivity to graft versus host disease. *Immunity* **37:** 339–350.

Hanauer SB, Feagan BG, Lichtenstein GR, Mayer LF, Schreiber S, Colombel JF, Rachmilewitz D, Wolf DC, Olson A, Bao W, et al. 2002. Maintenance infliximab for Crohn's disease: The ACCENT I randomised trial. *Lancet* **359:** 1541–1549.

Hanauer SB, Sandborn WJ, Rutgeerts P, Fedorak RN, Lukas M, MacIntosh D, Panaccione R, Wolf D, Pollack P. 2006. Human anti-tumor necrosis factor monoclonal antibody (adalimumab) in Crohn's disease: The CLASSIC-I trial. *Gastroenterology* **130:** 323–33; quiz 591.

He M, Liang P. 2010. IL-24 transgenic mice: In vivo evidence of overlapping functions for IL-20, IL-22, and IL-24 in the epidermis. *J Immunol* **184:** 1793–1798.

Hill T, Krougly O, Nikoopour E, Bellemore S, Lee-Chan E, Fouser LA, Hill DJ, Singh B. 2013. The involvement of interleukin-22 in the expression of pancreatic β cell regenerative *Reg* genes. *Cell Regen (Lond)* **2:** 2.

Hou W, Wang X, Ye L, Zhou L, Yang Z-Q, Riedel E, Ho WZ. 2009. Lambda interferon inhibits human immunodeficiency virus type 1 infection of macrophages. *J Virol* **83:** 3834–3842.

Hsu YH, Li HH, Hsieh MY, Liu MF, Huang KY, Chin LS, Chen PC, Cheng HH, Chang MS. 2006. Function of interleukin-20 as a proinflammatory molecule in rheumatoid and experimental arthritis. *Arthritis Rheum* **54:** 2722–2733.

Hsu YH, Hsieh PP, Chang MS. 2012. Interleukin-19 blockade attenuates collagen-induced arthritis in rats. *Rheumatology (Oxford)* **51:** 434–442.

Huan C, Kim D, Ou P, Alfonso A, Stanek A. 2016. Mechanisms of interleukin-22's beneficial effects in acute pancreatitis. *World J Gastrointest Pathophysiol* **7:** 108–116.

Huang EH, Park JC, Appelman H, Weinberg AD, Banerjee M, Logsdon CD, Schmidt AM. 2006. Induction of inflammatory bowel disease accelerates adenoma formation in Min +/- mice. *Surgery* **139:** 782–788.

Huber S, Gagliani N, Zenewicz LA, Huber FJ, Bosurgi L, Hu B, Hedl M, Zhang W, O'Connor W, Murphy AJ, et al. 2012. IL-22BP is regulated by the inflammasome and modulates tumorigenesis in the intestine. *Nature* **491:** 259–263.

Huhn RD, Radwanski E, O'Connell SM, Sturgill MG, Clarke L, Cody RP, Affrime MB, Cutler DL. 1996. Pharmacokinetics and immunomodulatory properties of intravenously administered recombinant human interleukin-10 in healthy volunteers. *Blood* **87:** 699–705.

Huhn RD, Radwanski E, Gallo J, Affrime MB, Sabo R, Gonyo G, Monge A, Cutler DL. 1997. Pharmacodynamics of subcutaneous recombinant human interleukin-10 in healthy volunteers. *Clin Pharmacol Ther* **62:** 171–180.

Hunt DWC, Boivin WA, Fairley LA, Jovanovic MM, King DE, Salmon RA, Utting OB. 2006. Ultraviolet B light stimulates interleukin-20 expression by human epithelial keratinocytes. *Photochem Photobiol* **82:** 1292–1300.

Ikeuchi H, Kuroiwa T, Hiramatsu N, Kaneko Y, Hiromura K, Ueki K, Nojima Y. 2005. Expression of interleukin-22 in rheumatoid arthritis: Potential role as a proinflammatory cytokine. *Arthritis Rheum* **52:** 1037–1046.

Innocentin S, Guimarães V, Miyoshi A, Azevedo V, Langella P, Chatel JM, Lefèvre F. 2009. *Lactococcus lactis* expressing either *Staphylococcus aureus* fibronectin-binding protein A or *Listeria monocytogenes* internalin A can efficiently internalize and deliver DNA in human epithelial cells. *Appl Environ Microbiol* **75:** 4870–4878.

Itoh K, Hirohata S. 1995. The role of IL-10 in human B cell activation, proliferation, and differentiation. *J Immunol* **154:** 4341–4350.

Järnerot G, Hertervig E, Friis-Liby I, Blomquist L, Karlén P, Grännö C, Vilien M, Ström M, Danielsson A, Verbaan H, et al. 2005. Infliximab as rescue therapy in severe to moderately severe ulcerative colitis: A randomized, placebo-controlled study. *Gastroenterology* **128:** 1805–1811.

Jeffcoate WJ, Harding KG. 2003. Diabetic foot ulcers. *Lancet* **361:** 1545–1551.

Jiang H, Lin JJ, Su ZZ, Goldstein NI, Fisher PB. 1995. Subtraction hybridization identifies a novel melanoma differentiation associated gene, *mda-7*, modulated during human melanoma differentiation, growth and progression. *Oncogene* **11:** 2477–2486.

Jiang H, Su ZZ, Lin JJ, Goldstein NI, Young CS, Fisher PB. 1996. The melanoma differentiation associated gene *mda-7* suppresses cancer cell growth. *Proc Natl Acad Sci* **93:** 9160–9165.

Jiang R, Wang H, Deng L, Hou J, Shi R, Yao M, Gao Y, Yao A, Wang X, Yu L, et al. 2013. IL-22 is related to development of human colon cancer by activation of STAT3. *BMC Cancer* **13:** 59.

Justa S, Zhou X, Sarkar S. 2014. Endogenous IL-22 plays a dual role in arthritis: Regulation of established arthritis via IFN-γ responses. *PloS ONE* **9**: e93279.

Kagami S, Rizzo HL, Lee JJ, Koguchi Y, Blauvelt A. 2010. Circulating Th17, Th22, and Th1 cells are increased in psoriasis. *J Invest Dermatol* **130**: 1373–1383.

Kamanaka M, Huber S, Zenewicz LA, Gagliani N, Rathinam C, O'Connor W, Wan YY, Nakae S, Iwakura Y, Hao L, et al. 2011. Memory/effector (CD45RB^lo) CD4 T cells are controlled directly by IL-10 and cause IL-22-dependent intestinal pathology. *J Exp Med* **208**: 1027–1040.

Kasama T, Strieter RM, Lukacs NW, Burdick MD, Kunkel SL. 1994. Regulation of neutrophil-derived chemokine expression by IL-10. *J Immunol* **152**: 3559–3569.

Kaser A, Zeissig S, Blumberg RS. 2010. Inflammatory bowel disease. *Annu Rev Immunol* **28**: 573–621.

Ki SH, Park O, Zheng M, Morales-Ibanez O, Kolls JK, Bataller R, Gao B. 2010. Interleukin-22 treatment ameliorates alcoholic liver injury in a murine model of chronic-binge ethanol feeding: Role of signal transducer and activator of transcription 3. *Hepatology* **52**: 1291–1300.

Kirchberger S, Royston DJ, Boulard O, Thornton E, Franchini F, Szabady RL, Harrison O, Powrie F. 2013. Innate lymphoid cells sustain colon cancer through production of interleukin-22 in a mouse model. *J Exp Med* **210**: 917–931.

Kobold S, Völk S, Clauditz T, Küpper NJ, Minner S, Tufman A, Düwell P, Lindner M, Koch I, Heidegger S, et al. 2013. Interleukin-22 is frequently expressed in small- and large-cell lung cancer and promotes growth in chemotherapy-resistant cancer cells. *J Thorac Oncol* **8**: 1032–1042.

Koenig KF, Groeschl I, Pesickova SS, Tesar V, Eisenberger U, Trendelenburg M. 2012. Serum cytokine profile in patients with active lupus nephritis. *Cytokine* **60**: 410–416.

Kofoed K, Skov L, Zachariae C. 2015. New drugs and treatment targets in psoriasis. *Acta Derm Venereol* **95**: 133–139.

Kolumam G, Wu X, Lee WP, Hackney JA, Zavala-Solorio J, Gandham V, Danilenko DM, Arora P, Wang X, Ouyang W. 2017. IL-22R ligands IL-20, IL-22, and IL-24 promote wound healing in diabetic db/db mice. *PloS ONE* **12**: e0170639.

Konya V, Mjösberg J. 2015. Innate lymphoid cells in graft-versus-host disease. *Am J Transplant* **15**: 2795–2801.

Kopp T, Lenz P, Bello-Fernandez C, Kastelein RA, Kupper TS, Stingl G. 2003. IL-23 production by cosecretion of endogenous p19 and transgenic p40 in keratin 14/p40 transgenic mice: Evidence for enhanced cutaneous immunity. *J Immunol* **170**: 5438–5444.

Kotenko SV, Gallagher G, Baurin VV, Lewis-Antes A, Shen M, Shah NK, Langer JA, Sheikh F, Dickensheets H, Donnelly RP. 2003. IFN-λs mediate antiviral protection through a distinct class II cytokine receptor complex. *Nat Immunol* **4**: 69–77.

Kragstrup TW, Otkjaer K, Holm C, Jørgensen A, Hokland M, Iversen L, Deleuran B. 2008. The expression of IL-20 and IL-24 and their shared receptors are increased in rheumatoid arthritis and spondyloarthropathy. *Cytokine* **41**: 16–23.

Kreis S, Philippidou D, Margue C, Rolvering C, Haan C, Dumoutier L, Renauld JC, Behrmann I. 2007. Recombi-

nant interleukin-24 lacks apoptosis-inducing properties in melanoma cells. *PloS ONE* **2**: e1300.

Kreis S, Philippidou D, Margue C, Behrmann I. 2008. IL-24: A classic cytokine and/or a potential cure for cancer? *J Cell Mol Med* **12**: 2505–2510.

Kühn R, Löhler J, Rennick D, Rajewsky K, Muller W. 1993. Interleukin-10-deficient mice develop chronic enterocolitis. *Cell* **75**: 263–274.

Lasfar A, de laTorre A, Abushahba W, Cohen-Solal KA, Castaneda I, Yuan Y, Reuhl K, Zloza A, Raveche E, Laskin DL, et al. 2016. Concerted action of IFN-α and IFN-λ induces local NK cell immunity and halts cancer growth. *Oncotarget* **7**: 49259–49267.

Lauw FN, Pajkrt D, Hack CE, Kurimoto M, van Deventer SJ, van der Poll T. 2000. Proinflammatory effects of IL-10 during human endotoxemia. *J Immunol* **165**: 2783–2789.

Leipe J, Schramm MA, Grunke M, Baeuerle M, Dechant C, Nigg AP, Witt MN, Vielhauer V, Reindl CS, Schulze-Koops H, et al. 2011. Interleukin 22 serum levels are associated with radiographic progression in rheumatoid arthritis. *Ann Rheum Dis* **70**: 1453–1457.

Leung JM, Davenport M, Wolff MJ, Wiens KE, Abidi WM, Poles MA, Cho I, Ullman T, Mayer L, Loke P. 2013. IL-22-producing CD4^+ cells are depleted in actively inflamed colitis tissue. *Mucosal Immunol* **7**: 124–133.

Levy Y, Brouet JC. 1994. Interleukin-10 prevents spontaneous death of germinal center B cells by induction of the bcl-2 protein. *J Clin Invest* **93**: 424–428.

Li J, Chen J, Kirsner R. 2007. Pathophysiology of acute wound healing. *Clin Dermatol* **25**: 9–18.

Li W, Lewis-Antes A, Huang J, Balan M, Kotenko SV. 2008. Regulation of apoptosis by type III interferons. *Cell Prolif* **41**: 960–979.

Li W, Danilenko DM, Bunting S, Ganesan R, Sa S, Ferrando R, Wu TD, Kolumam GA, Ouyang W, Kirchhofer D. 2009. The serine protease marapsin is expressed in stratified squamous epithelia and is up-regulated in the hyperproliferative epidermis of psoriasis and regenerating wounds. *J Biol Chem* **284**: 218–228.

Li Q, Kawamura K, Ma G, Iwata F, Numasaki M, Suzuki N, Shimada H, Tagawa M. 2010. Interferon-λ induces G$_1$ phase arrest or apoptosis in oesophageal carcinoma cells and produces anti-tumour effects in combination with anti-cancer agents. *Eur J Cancer* **46**: 180–190.

Liang SC, Tan XY, Luxenberg DP, Karim R, Dunussi-Joannopoulos K, Collins M, Fouser LA. 2006. Interleukin (IL)-22 and IL-17 are coexpressed by Th17 cells and cooperatively enhance expression of antimicrobial peptides. *J Exp Med* **203**: 2271–2279.

Lin FC, Young HA. 2014. Interferons: Success in anti-viral immunotherapy. *Cytokine Growth Factor Rev* **25**: 369–376.

Lindemans CA, Calafiore M, Mertelsmann AM, O'Connor MH, Dudakov JA, Jenq RR, Velardi E, Young LF, Smith OM, Lawrence G, et al. 2015. Interleukin-22 promotes intestinal-stem-cell-mediated epithelial regeneration. *Nature* **528**: 560–564.

Llorente L, Richaud-Patin Y, García-Padilla C, Claret E, Jakez-Ocampo J, Cardiel MH, Alcocer-Varela J, Grangeot-Keros L, Alarcón-Segovia D, Wijdenes J, et al. 2000. Clinical and biologic effects of anti-interleukin-10 monoclonal antibody administration in systemic lupus erythematosus. *Arthritis Rheum* **43:** 1790–1800.

Loser K, Apelt J, Voskort M, Mohaupt M, Balkow S, Schwarz T, Grabbe S, Beissert S. 2007. IL-10 controls ultraviolet-induced carcinogenesis in mice. *J Immunol* **179:** 365–371.

Ma H-L, Liang S, Li J, Napierata L, Brown T, Benoit S, Senices M, Gill D, Dunussi-Joannopoulos K, Collins M, et al. 2008. IL-22 is required for Th17 cell-mediated pathology in a mouse model of psoriasis-like skin inflammation. *J Clin Invest* **118:** 597–607.

Mannino MH, Zhu Z, Xiao H, Bai Q, Wakefield MR, Fang Y. 2015. The paradoxical role of IL-10 in immunity and cancer. *Cancer Lett* **367:** 103–107.

Marcello T, Grakoui A, Barba-Spaeth G, Machlin ES, Kotenko SV, MacDonald MR, Rice CM. 2006. Interferons α and λ inhibit hepatitis C virus replication with distinct signal transduction and gene regulation kinetics. *Gastroenterology* **131:** 1887–1898.

Marijnissen RJ, Koenders MI, Smeets RL, Stappers MHT, Nickerson-Nutter C, Joosten LAB, Boots AMH, van den Berg WB. 2011. Increased expression of interleukin-22 by synovial Th17 cells during late stages of murine experimental arthritis is controlled by interleukin-1 and enhances bone degradation. *Arthritis Rheum* **63:** 2939–2948.

Martin R, Chain F, Miquel S, Natividad JM, Sokol H, Verdu EF, Langella P, Bermúdez-Humarán LG. 2014. Effects in the use of a genetically engineered strain of *Lactococcus lactis* delivering in situ IL-10 as a therapy to treat low-grade colon inflammation. *Hum Vaccin Immunother* **10:** 1611–1621.

McGee HM, Schmidt BA, Booth CJ, Yancopoulos GD, Valenzuela DM, Murphy AJ, Stevens S, Flavell RA, Horsley V. 2013. IL-22 promotes fibroblast-mediated wound repair in the skin. *J Invest Dermatol* **133:** 1321–1329.

Meager A, Visvalingam K, Dilger P, Bryan D, Wadhwa M. 2005. Biological activity of interleukins-28 and -29: Comparison with type I interferons. *Cytokine* **31:** 109–118.

Michaels J, Churgin SS, Blechman KM, Greives MR, Aarabi S, Galiano RD, Gurtner GC. 2007. db/db mice exhibit severe wound-healing impairments compared with other murine diabetic strains in a silicone-splinted excisional wound model. *Wound Repair Regen* **15:** 665–670.

Miller DM, Klucher KM, Freeman JA, Hausman DF, Fontana D, Williams DE. 2009. Interferon λ as a potential new therapeutic for hepatitis C. *Ann NY Acad Sci* **1182:** 80–87.

Miyoshi A, Jamet E, Commissaire J, Renault P, Langella P, Azevedo V. 2004. A xylose-inducible expression system for *Lactococcus lactis*. *FEMS Microbiol Lett* **239:** 205–212.

Montes-Torres A, Llamas-Velasco M, Pérez-Plaza A, Solano-López G, Sánchez-Pérez J. 2015. Biological treatments in atopic dermatitis. *J Clin Med* **4:** 593–613.

Moore K, de Waal M, Coffman R, O'Garra A. 2001. Interleukin-10 and the interleukin-10 receptor. *Annu Rev Immunol* **19:** 683–765.

Mordstein M, Kochs G, Dumoutier L, Renauld JC, Paludan SR, Klucher K, Staeheli P. 2008. Interferon-λ contributes to innate immunity of mice against influenza A virus but not against hepatotropic viruses. *PloS Pathog* **4:** e1000151.

Mordstein M, Neugebauer E, Ditt V, Jessen B, Rieger T, Falcone V, Sorgeloos F, Ehl S, Mayer D, Kochs G, et al.

2010. λ interferon renders epithelial cells of the respiratory and gastrointestinal tracts resistant to viral infections. *J Virol* **84:** 5670–5677.

Motzer RJ, Escudier B, McDermott DF, George S, Hammers HJ, Srinivas S, Tykodi SS, Sosman JA, Procopio G, Plimack ER, et al. 2015. Nivolumab versus everolimus in advanced renal-cell carcinoma. *N Engl J Med* **373:** 1803–1813.

Muir AJ, Arora S, Everson G, Flisiak R, George J, Ghalib R, Gordon SC, Gray T, Greenbloom S, Hassanein T, et al. 2014. A randomized phase 2b study of peginterferon λ-1a for the treatment of chronic HCV infection. *J Hepatol* **61:** 1238–1246.

Mumm JB, Emmerich J, Zhang X, Chan I, Wu L, Mauze S, Blaisdell S, Basham B, Dai J, Grein J, et al. 2011. IL-10 elicits IFNγ-dependent tumor immune surveillance. *Cancer Cell* **20:** 781–796.

Munneke JM, Björklund AT, Mjösberg JM, Garming-Legert K, Bernink JH, Blom B, Huisman C, van Oers MHJ, Spits H, Malmberg K-J, et al. 2014. Activated innate lymphoid cells are associated with a reduced susceptibility to graft-versus-host disease. *Blood* **124:** 812–821.

Nagelkerken L, Blauw B, Tielemans M. 1997. IL-4 abrogates the inhibitory effect of IL-10 on the development of experimental allergic encephalomyelitis in SJL mice. *Int Immunol* **9:** 1243–1251.

Naing A, Papadopoulos KP, Autio KA, Ott PA, Patel MR, Wong DJ, Falchook GS, Pant S, Whiteside M, Rasco DR, et al. 2016. Safety, antitumor activity, and immune activation of pegylated recombinant human interleukin-10 (AM0010) in patients with advanced solid tumors. *J Clin Oncol* **34:** 3562–3569.

Nemunaitis J, Fong T, Shabe P, Martineau D, Ando D. 2001. Comparison of serum interleukin-10 (IL-10) levels between normal volunteers and patients with advanced melanoma. *Cancer Invest* **19:** 239–247.

Nestle FO, Kaplan DH, Barker J. 2009. Psoriasis. *N Engl J Med* **361:** 496–509.

Nograles KE, Zaba LC, Shemer A, Fuentes-Duculan J, Cardinale I, Kikuchi T, Ramon M, Bergman R, Krueger JG, Guttman-Yassky E. 2009. IL-22-producing "T22" T cells account for upregulated IL-22 in atopic dermatitis despite reduced IL-17-producing TH17 T cells. *J Allergy Clin Immunol* **123:** 1244–1252.e2.

Numasaki M, Tagawa M, Iwata F, Suzuki T, Nakamura A, Okada M, Iwakura Y, Aiba S, Yamaya M. 2007. IL-28 elicits antitumor responses against murine fibrosarcoma. *J Immunol* **178:** 5086–5098.

O'Garra A, Barrat FJ, Castro AG, Vicari A, Hawrylowicz C. 2008. Strategies for use of IL-10 or its antagonists in human disease. *Immunol Rev* **223:** 114–131.

Onoguchi K, Yoneyama M, Takemura A, Akira S, Taniguchi T, Namiki H, Fujita T. 2007. Viral infections activate types I and III interferon genes through a common mechanism. *J Biol Chem* **282:** 7576–7581.

O'Shea JJ, Kanno Y, Chan AC. 2014. In search of magic bullets: The golden age of immunotherapeutics. *Cell* **157:** 227–240.

Ota N, Wong K, Valdez PA, Zheng Y, Crellin NK, Diehl L, Ouyang W. 2011. IL-22 bridges the lymphotoxin pathway with the maintenance of colonic lymphoid structures

Cite this article as *Cold Spring Harb Perspect Biol* doi: 10.1101/cshperspect.a028548

during infection with *Citrobacter rodentium*. *Nat Immunol* **12**: 941–948.

Otkjaer K, Kragballe K, Funding AT, Clausen JT, Noerby PL, Steiniche T, Iversen L. 2005. The dynamics of gene expression of interleukin-19 and interleukin-20 and their receptors in psoriasis. *Br J Dermatol* **153**: 911–918.

Ouyang W. 2010. Distinct roles of IL-22 in human psoriasis and inflammatory bowel disease. *Cytokine Growth Factor Rev* **21**: 435–441.

Ouyang W, Rutz S, Crellin NK, Valdez PA, Hymowitz SG. 2011. Regulation and functions of the IL-10 family of cytokines in inflammation and disease. *Annu Rev Immunol* **29**: 71–109.

Pan H, Hong F, Radaeva S, Gao B. 2004. Hydrodynamic gene delivery of interleukin-22 protects the mouse liver from concanavalin A-, carbon tetrachloride-, and Fas ligand-induced injury via activation of STAT3. *Cell Mol Immunol* **1**: 43–49.

Park O, Wang H, Weng H, Feigenbaum L, Li H, Yin S, Ki SH, Yoo SH, Dooley S, Wang FS, et al. 2011. In vivo consequences of liver-specific interleukin-22 expression: Implications for human liver disease progression. *Hepatology* **54**: 252–261.

Park H, Serti E, Eke O, Muchmore B, Prokunina-Olsson L, Capone S, Folgori A, Rehermann B. 2012. IL-29 is the dominant type III interferon produced by hepatocytes during acute hepatitis C virus infection. *Hepatology* **56**: 2060–2070.

Pelczar P, Witkowski M, Perez LG, Kempski J, Hammel AG, Brockmann L, Kleinschmidt D, Wende S, Haueis C, Bedke T, et al. 2016. A pathogenic role for T cell-derived IL-22BP in inflammatory bowel disease. *Science* **354**: 358–362.

Pène J, Chevalier S, Preisser L, Vénéreau E, Guilleux MH, Ghannam S, Molès JP, Danger Y, Ravon E, Lesaux S, et al. 2008. Chronically inflamed human tissues are infiltrated by highly differentiated Th17 lymphocytes. *J Immunol* **180**: 7423–7430.

Perera GK, Ainali C, Semenova E, Hundhausen C, Barinaga G, Kassen D, Williams AE, Mirza MM, Balazs M, Wang X, et al. 2014. Integrative biology approach identifies cytokine targeting strategies for psoriasis. *Sci Transl Med* **6**: 223ra22.

Pestka S, Krause CD, Sarkar D, Walter MR, Shi Y, Fisher PB. 2004. Interleukin-10 and related cytokines and receptors. *Annu Rev Immunol* **22**: 929–979.

Pickert G, Neufert C, Leppkes M, Zheng Y, Wittkopf N, Warntjen M, Lehr HA, Hirth S, Weigmann B, Wirtz S, et al. 2009. STAT3 links IL-22 signaling in intestinal epithelial cells to mucosal wound healing. *J Exp Med* **206**: 1465–1472.

Radaeva S, Sun R, Pan HN, Hong F, Gao B. 2004. Interleukin 22 (IL-22) plays a protective role in T cell-mediated murine hepatitis: IL-22 is a survival factor for hepatocytes via STAT3 activation. *Hepatology* **39**: 1332–1342.

Robert C, Schachter J, Long GV, Arance A, Grob JJ, Mortier L, Daud A, Carlino MS, McNeil C, Lotem M, et al. 2015. Pembrolizumab versus ipilimumab in advanced melanoma. *N Engl J Med* **372**: 2521–2532.

Rømer J, Hasselager E, Nørby PL, Steiniche T, Thorn Clausen J, Kragballe K. 2003. Epidermal overexpression of interleukin-19 and -20 mRNA in psoriatic skin disap-

pears after short-term treatment with cyclosporine a or calcipotriol. *J Invest Dermatol* **121**: 1306–1311.

Rosenblum IY, Johnson RC, Schmahai TJ. 2002. Preclinical safety evaluation of recombinant human interleukin-10. *Regul Toxicol Pharmacol* **35**: 56–71.

Rousset F, Garcia E, Defrance T, Péronne C, Vezzio N, Hsu DH, Kastelein R, Moore KW, Banchereau J. 1992. Interleukin 10 is a potent growth and differentiation factor for activated human B lymphocytes. *Proc Natl Acad Sci* **89**: 1890–1893.

Rousset F, Peyrol S, Garcia E, Vezzio N, Andujar M, Grimaud JA, Banchereau J. 1995. Long-term cultured CD40-activated B lymphocytes differentiate into plasma cells in response to IL-10 but not IL-4. *Int Immunol* **7**: 1243–1253.

Ruffell B, Chang-Strachan D, Chan V, Rosenbusch A, Ho CMT, Pryer N, Daniel D, Hwang ES, Rugo HS, Coussens LM. 2014. Macrophage IL-10 blocks CD8$^+$ T cell-dependent responses to chemotherapy by suppressing IL-12 expression in intratumoral dendritic cells. *Cancer Cell* **26**: 623–637.

Rutz S, Ouyang W. 2011. Regulation of interleukin-10 and interleukin-22 expression in T helper cells. *Curr Opin Immunol* **23**: 605–612.

Rutz S, Ouyang W. 2016. Regulation of interleukin-10 expression. *Adv Exp Med Biol* **941**: 89–116.

Rutz S, Noubade R, Eidenschenk C, Ota N, Zeng W, Zheng Y, Hackney J, Ding J, Singh H, Ouyang W. 2011. Transcription factor c-Maf mediates the TGF-β-dependent suppression of IL-22 production in T_H17 cells. *Nat Immunol* **12**: 1238–1245.

Rutz S, Eidenschenk C, Ouyang W. 2013. IL-22, not simply a Th17 cytokine. *Immunol Rev* **252**: 116–132.

Rutz S, Wang X, Ouyang W. 2014. The IL-20 subfamily of cytokines—From host defence to tissue homeostasis. *Nat Rev Immunol* **14**: 783–795.

Sa SM, Valdez PA, Wu J, Jung K, Zhong F, Hall L, Kasman I, Winer J, Modrusan Z, Danilenko DM, et al. 2007. The effects of IL-20 subfamily cytokines on reconstituted human epidermis suggest potential roles in cutaneous innate defense and pathogenic adaptive immunity in psoriasis. *J Immunol* **178**: 2229–2240.

Sabat R, Ouyang W, Wolk K. 2013. Therapeutic opportunities of the IL-22-IL-22R1 system. *Nat Rev Drug Discov* **13**: 21–38.

Sakurai N, Kuroiwa T, Ikeuchi H, Hiramatsu N, Maeshima A, Kaneko Y, Hiromura K, Nojima Y. 2008. Expression of IL-19 and its receptors in RA: Potential role for synovial hyperplasia formation. *Rheumatology (Oxford)* **47**: 815–820.

Sandborn WJ, Feagan BG, Stoinov S, Honiball PJ, Rutgeerts P, Mason D, Bloomfield R, Schreiber S, PRECISE 1 Study Investigators. 2007. Certolizumab pegol for the treatment of Crohn's disease. *N Engl J Med* **357**: 228–238.

Sandborn WJ, Feagan BG, Fedorak RN, Scherl E, Fleisher MR, Katz S, Johanns J, Blank M, Rutgeerts P, Ustekinumab Crohn's Disease Study. 2008. A randomized trial of Ustekinumab, a human interleukin-12/23 monoclonal antibody, in patients with moderate-to-severe Crohn's disease. *Gastroenterology* **135**: 1130–1141.

Saraiva M, O'Garra A. 2010. The regulation of IL-10 production by immune cells. *Nat Rev Immunol* **10:** 170–181.

Sato A, Ohtsuki M, Hata M, Kobayashi E, Murakami T. 2006. Antitumor activity of IFN-λ in murine tumor models. *J Immunol* **176:** 7686–7694.

Sauane M, Gopalkrishnan RV, Sarkar D, Su ZZ, Lebedeva IV, Dent P, Pestka S, Fisher PB. 2003. MDA-7/IL-24: Novel cancer growth suppressing and apoptosis inducing cytokine. *Cytokine Growth Factor Rev* **14:** 35–51.

Saxena A, Khosraviani S, Noel S, Mohan D, Donner T, Hamad ARA. 2015. Interleukin-10 paradox: A potent immunoregulatory cytokine that has been difficult to harness for immunotherapy. *Cytokine* **74:** 27–34.

Schreiber S, Fedorak RN, Nielsen OH, Wild G, Williams CN, Nikolaus S, Jacyna M, Lashner BA, Gangl A, Rutgeerts P, et al. 2000. Safety and efficacy of recombinant human interleukin 10 in chronic active Crohn's disease. Crohn's Disease IL-10 Cooperative Study Group. *Gastroenterology* **119:** 1461–1472.

Schuetze N, Schoeneberger S, Mueller U, Freudenberg MA, Alber G, Straubinger RK. 2005. IL-12 family members: Differential kinetics of their TLR4-mediated induction by *Salmonella enteritidis* and the impact of IL-10 in bone marrow-derived macrophages. *Int Immunol* **17:** 649–659.

Schwager K, Kaspar M, Bootz F, Marcolongo R, Paresce E, Neri D, Trachsel E. 2009. Preclinical characterization of DEKAVIL (F8-IL10), a novel clinical-stage immunocytokine which inhibits the progression of collagen-induced arthritis. *Arthritis Res Ther* **11:** R142.

Schwager K, Bootz F, Imesch P, Kaspar M, Trachsel E, Neri D. 2011. The antibody-mediated targeted delivery of interleukin-10 inhibits endometriosis in a syngeneic mouse model. *Hum Reprod* **26:** 2344–2352.

Sellon RK, Tonkonogy S, Schultz M, Dieleman LA, Grenther W, Balish E, Rennick DM, Sartor RB. 1998. Resident enteric bacteria are necessary for development of spontaneous colitis and immune system activation in interleukin-10-deficient mice. *Infect Immun* **66:** 5224–5231.

Šenolt L, Leszczynski P, Dokoupilová E, Göthberg M, Valencia X, Hansen BB, Cañete JD. 2015. Efficacy and safety of anti-interleukin-20 monoclonal antibody in patients with rheumatoid arthritis: A randomized phase IIa trial. *Arthritis Rheumatol* **67:** 1438–1448.

Sfikakis PP. 2010. The first decade of biologic TNF antagonists in clinical practice: Lessons learned, unresolved issues and future directions. *Curr Dir Autoimmun* **11:** 180–210.

Sharma P, Allison JP. 2015. The future of immune checkpoint therapy. *Science* **348:** 56–61.

Sharma P, Hu-Lieskovan S, Wargo JA, Ribas A. 2017. Primary, adaptive, and acquired resistance to cancer immunotherapy. *Cell* **168:** 707–723.

Shen H, Goodall JC, Hill Gaston JS. 2009. Frequency and phenotype of peripheral blood Th17 cells in ankylosing spondylitis and rheumatoid arthritis. *Arthritis Rheum* **60:** 1647–1656.

Sheppard P, Kindsvogel W, Xu W, Henderson K, Schlutsmeyer S, Whitmore TE, Kuestner R, Garrigues U, Birks C, Roraback J, et al. 2003. IL-28, IL-29 and their class II cytokine receptor IL-28R. *Nat Immunol* **4:** 63–68.

Shim JO, Hwang S, Yang HR, Moon JS, Chang JY, Ko JS, Park SS, Kang GH, Kim WS, Seo JK. 2013. Interleukin-10 receptor mutations in children with neonatal-onset Crohn's disease and intractable ulcerating enterocolitis. *Eur J Gastroenterol Hepatol* **25:** 1235–1240.

Singer AJ, Clark RA. 1999. Cutaneous wound healing. *N Engl J Med* **341:** 738–746.

Sommereyns C, Paul S, Staeheli P, Michiels T. 2008. IFN-lambda (IFN-λ) is expressed in a tissue-dependent fashion and primarily acts on epithelial cells in vivo. *PloS Pathog* **4:** e1000017.

Sonnenberg GF, Fouser LA, Artis D. 2011. Border patrol: Regulation of immunity, inflammation and tissue homeostasis at barrier surfaces by IL-22. *Nat Immunol* **12:** 383–390.

Sosman JA, Verma A, Moss S, Sorokin P, Blend M, Bradlow B, Chachlani N, Cutler D, Sabo R, Nelson M, et al. 2000. Interleukin 10-induced thrombocytopenia in normal healthy adult volunteers: Evidence for decreased platelet production. *Br J Haematol* **111:** 104–111.

Steen HC, Gamero AM. 2010. Interferon-λ as a potential therapeutic agent in cancer treatment. *J Interferon Cytokine Res* **30:** 597–602.

Steidler L, Hans W, Schotte L, Neirynck S, Obermeier F, Falk W, Fiers W, Remaut E. 2000. Treatment of murine colitis by *Lactococcus lactis* secreting interleukin-10. *Science* **289:** 1352–1355.

Steidler L, Neirynck S, Huyghebaert N, Snoeck V, Vermeire A, Goddeeris B, Cox E, Remon JP, Remaut E. 2003. Biological containment of genetically modified *Lactococcus lactis* for intestinal delivery of human interleukin 10. *Nat Biotechnol* **21:** 785–789.

Steinbrink K, Jonuleit H, Müller G, Schuler G, Knop J, Enk AH. 1999. Interleukin-10-treated human dendritic cells induce a melanoma-antigen-specific anergy in CD8[+] T cells resulting in a failure to lyse tumor cells. *Blood* **93:** 1634–1642.

Stenderup K, Rosada C, Worsaae A, Dagnaes-Hansen F, Steiniche T, Hasselager E, Iversen LF, Zahn S, Wöldike H, Holmberg HL, et al. 2009. Interleukin-20 plays a critical role in maintenance and development of psoriasis in the human xenograft transplantation model. *Br J Dermatol* **160:** 284–296.

Sturlan S, Oberhuber G, Beinhauer BG, Tichy B, Kappel S, Wang J, Rogy MA. 2001. Interleukin-10-deficient mice and inflammatory bowel disease associated cancer development. *Carcinogenesis* **22:** 665–671.

Su ZZ, Madireddi MT, Lin JJ, Young CS, Kitada S, Reed JC, Goldstein NI, Fisher PB. 1998. The cancer growth suppressor gene *mda-7* selectively induces apoptosis in human breast cancer cells and inhibits tumor growth in nude mice. *Proc Natl Acad Sci* **95:** 14400–14405.

Sugimoto K, Ogawa A, Mizoguchi E, Shimomura Y, Andoh A, Bhan AK, Blumberg RS, Xavier RJ, Mizoguchi A. 2008. IL-22 ameliorates intestinal inflammation in a mouse model of ulcerative colitis. *J Clin Invest* **118:** 534–544.

Sun DP, Yeh CH, So E, Wang LY, Wei TS, Chang MS, Hsing CH. 2013. Interleukin (IL)-19 promoted skin wound healing by increasing fibroblast keratinocyte growth factor expression. *Cytokine* **62:** 360–368.

Suppiah V, Moldovan M, Ahlenstiel G, Berg T, Weltman M, Abate ML, Bassendine M, Spengler U, Dore GJ, Powell E, et al. 2009. *IL28B* is associated with response to chronic hepatitis C interferon-α and ribavirin therapy. *Nat Genet* **41**: 1100–1104.

Tanaka Y, Nishida N, Sugiyama M, Kurosaki M, Matsuura K, Sakamoto N, Nakagawa M, Korenaga M, Hino K, Hige S, et al. 2009. Genome-wide association of *IL28B* with response to pegylated interferon-α and ribavirin therapy for chronic hepatitis C. *Nat Genet* **41**: 1105–1109.

Thomas DL, Thio CL, Martin MP, Qi Y, Ge D, O'Huigin C, Kidd J, Kidd K, Khakoo SI, Alexander G, et al. 2009. Genetic variation in *IL28B* and spontaneous clearance of hepatitis C virus. *Nature* **461**: 798–801.

Thomas E, Gonzalez VD, Li Q, Modi AA, Chen W, Noureddin M, Rotman Y, Liang TJ. 2012. HCV infection induces a unique hepatic innate immune response associated with robust production of type III interferons. *Gastroenterology* **142**: 978–988.

Tilg H, Van Montfrans C, van den Ende A, Kaser A, van Deventer SJH, Schreiber S, Gregor M, Ludwiczek O, Rutgeerts P, Gasche C, et al. 2002. Treatment of Crohn's disease with recombinant human interleukin 10 induces the proinflammatory cytokine interferon γ. *Gut* **50**: 191–195.

Tohyama M, Hanakawa Y, Shirakata Y, Dai X, Yang L, Hirakawa S, Tokumaru S, Okazaki H, Sayama K, Hashimoto K. 2009. IL-17 and IL-22 mediate IL-20 subfamily cytokine production in cultured keratinocytes via increased IL-22 receptor expression. *Eur J Immunol* **39**: 2779–2788.

Tong AW, Nemunaitis J, Su D, Zhang Y, Cunningham C, Senzer N, Netto G, Rich D, Mhashilkar A, Parker K, et al. 2005. Intratumoral injection of INGN 241, a nonreplicating adenovector expressing the melanoma-differentiation associated gene-7 (*mda-7/IL24*): Biologic outcome in advanced cancer patients. *Mol Ther* **11**: 160–172.

Topalian SL, Hodi FS, Brahmer JR, Gettinger SN, Smith DC, McDermott DF, Powderly JD, Carvajal RD, Sosman JA, Atkins MB, et al. 2012. Safety, activity, and immune correlates of anti-PD-1 antibody in cancer. *N Engl J Med* **366**: 2443–2454.

Topalian SL, Drake CG, Pardoll DM. 2015. Immune checkpoint blockade: A common denominator approach to cancer therapy. *Cancer Cell* **27**: 450–461.

Trachsel E, Bootz F, Silacci M, Kaspar M, Kosmehl H, Neri D. 2007. Antibody-mediated delivery of IL-10 inhibits the progression of established collagen-induced arthritis. *Arthritis Res Ther* **9**: R9.

Uzé G, Monneron D. 2007. IL-28 and IL-29: Newcomers to the interferon family. *Biochimie* **89**: 729–734.

van Deventer SJ, Elson CO, Fedorak RN; Crohn's Disease Study Group. 1997. Multiple doses of intravenous interleukin 10 in steroid-refractory Crohn's disease. *Gastroenterology* **113**: 383–389.

Vicari AP, Trinchieri G. 2004. Interleukin-10 in viral diseases and cancer: Exiting the labyrinth? *Immunol Rev* **202**: 223–236.

Wang Z, Yang L, Jiang Y, Ling ZQ, Li Z, Cheng Y, Huang H, Wang L, Pan Y, Wang Z, et al. 2011. High fat diet induces formation of spontaneous liposarcoma in mouse adipose tissue with overexpression of interleukin 22. *PloS ONE* **6**: e23737.

Wang X, Ota N, Manzanillo P, Kates L, Zavala-Solorio J, Eidenschenk C, Zhang J, Lesch J, Lee WP, Ross J, et al.

2014. Interleukin-22 alleviates metabolic disorders and restores mucosal immunity in diabetes. *Nature* **514**: 237–241.

Weiss G, Bogdan C, Hentze MW. 1997. Pathways for the regulation of macrophage iron metabolism by the anti-inflammatory cytokines IL-4 and IL-13. *J Immunol* **158**: 420–425.

Whitaker EL, Filippov VA, Duerksen-Hughes PJ. 2012. Interleukin 24: Mechanisms and therapeutic potential of an anti-cancer gene. *Cytokine Growth Factor Rev* **23**: 323–331.

Willems F, Marchant A, Delville JP, Gerard C, Delvaux A, Velu T, de Boer M, Goldman M. 1994. Interleukin-10 inhibits B7 and intercellular adhesion molecule-1 expression on human monocytes. *Eur J Immunol* **24**: 1007–1009.

Williams HC. 2005. Clinical practice. Atopic dermatitis. *N Engl J Med* **352**: 2314–2324.

Wingard JR, Majhail NS, Brazauskas R, Wang Z, Sobocinski KA, Jacobsohn D, Sorror ML, Horowitz MM, Bolwell B, Rizzo JD, et al. 2011. Long-term survival and late deaths after allogeneic hematopoietic cell transplantation. *J Clin Oncol* **29**: 2230–2239.

Witte W, Gruetz G, Volk H-D, Looman AC, Asadullah K, Sterry W, Sabat R, Wolk K. 2009. Despite IFN-λ receptor expression, blood immune cells, but not keratinocytes or melanocytes, have an impaired response to type III interferons: Implications for therapeutic applications of these cytokines. *Genes Immun* **10**: 702–714.

Wogensen L, Lee MS, Sarvetnick N. 1994. Production of interleukin 10 by islet cells accelerates immune-mediated destruction of β cells in nonobese diabetic mice. *J Exp Med* **179**: 1379–1384.

Wolk K, Sabat R. 2006. Interleukin-22: A novel T- and NK-cell derived cytokine that regulates the biology of tissue cells. *Cytokine Growth Factor Rev* **17**: 367–380.

Wolk K, Kunz S, Asadullah K, Sabat R. 2002. Cutting edge: Immune cells as sources and targets of the IL-10 family members? *J Immunol* **168**: 5397–5402.

Wolk K, Kunz S, Witte E, Friedrich M, Asadullah K, Sabat R. 2004. IL-22 increases the innate immunity of tissues. *Immunity* **21**: 241–254.

Wolk K, Witte E, Wallace E, Döcke WD, Kunz S, Asadullah K, Volk HD, Sterry W, Sabat R. 2006. IL-22 regulates the expression of genes responsible for antimicrobial defense, cellular differentiation, and mobility in keratinocytes: A potential role in psoriasis. *Eur J Immunol* **36**: 1309–1323.

Wolk K, Haugen HS, Xu W, Witte E, Waggie K, Anderson M, Baur Vom E, Witte K, Warszawska K, Philipp S, et al. 2009a. IL-22 and IL-20 are key mediators of the epidermal alterations in psoriasis while IL-17 and IFN-λ are not. *J Mol Med* **87**: 523–536.

Wolk K, Witte E, Warszawska K, Schulze-Tanzil G, Witte K, Philipp S, Kunz S, Döcke WD, Asadullah K, Volk HD, et al. 2009b. The Th17 cytokine IL-22 induces IL-20 production in keratinocytes: A novel immunological cascade with potential relevance in psoriasis. *Eur J Immunol* **39**: 3570–3581.

Xavier RJ, Podolsky DK. 2007. Unravelling the pathogenesis of inflammatory bowel disease. *Nature* **448**: 427–434.

Xie MH, Aggarwal S, Ho WH, Foster J, Zhang Z, Stinson J, Wood WI, Goddard AD, Gurney AL. 2000. Interleukin (IL)-22, a novel human cytokine that signals through the interferon receptor-related proteins CRF2–4 and IL-22R. *J Biol Chem* **275**: 31335–31339.

Xing WW, Zou MJ, Liu S, Xu T, Wang JX, Xu DG. 2011. Interleukin-22 protects against acute alcohol-induced hepatotoxicity in mice. *Biosci Biotechnol Biochem* **75**: 1290–1294.

Yen D, Cheung J, Scheerens H, Poulet F, McClanahan T, McKenzie B, Kleinschek MA, Owyang A, Mattson J, Blumenschein W, et al. 2006. IL-23 is essential for T cell-mediated colitis and promotes inflammation via IL-17 and IL-6. *J Clin Invest* **116**: 1310–1316.

Yu LZ, Wang HY, Yang SP, Yuan ZP, Xu FY, Sun C, Shi RH. 2013. Expression of interleukin-22/STAT3 signaling pathway in ulcerative colitis and related carcinogenesis. *World J Gastroenterol* **19**: 2638–2649.

Yu H, Lee H, Herrmann A, Buettner R, Jove R. 2014. Revisiting STAT3 signalling in cancer: New and unexpected biological functions. *Nat Rev Cancer* **14**: 736–746.

Zenewicz LA, Yancopoulos GD, Valenzuela DM, Murphy AJ, Karow M, Flavell RA. 2007. Interleukin-22 but not interleukin-17 provides protection to hepatocytes during acute liver inflammation. *Immunity* **27**: 647–659.

Zenewicz LA, Yancopoulos GD, Valenzuela DM, Murphy AJ, Stevens S, Flavell RA. 2008. Innate and adaptive interleukin-22 protects mice from inflammatory bowel disease. *Immunity* **29**: 947–957.

Zenewicz LA, Yin X, Wang G, Elinav E, Hao L, Zhao L, Flavell RA. 2013. IL-22 deficiency alters colonic microbiota to be transmissible and colitogenic. *J Immunol* **190**: 5306–5312.

Zhang L, Li JM, Liu XG, Ma DX, Hu NW, Li YG, Li W, Hu Y, Yu S, Qu X, et al. 2011. Elevated Th22 cells correlated with Th17 cells in patients with rheumatoid arthritis. *J Clin Immunol* **31**: 606–614.

Zhang L, Li Y-G, Li Y-H, Qi L, Liu X-G, Yuan C-Z, Hu N-W, Ma D-X, Li Z-F, Yang Q, et al. 2012. Increased frequencies of Th22 cells as well as Th17 cells in the peripheral blood of patients with ankylosing spondylitis and rheumatoid arthritis. *PloS ONE* **7**: e31000.

Zheng XX, Steele AW, Hancock WW, Stevens AC, Nickerson PW, Roy-Chaudhury P, Tian Y, Strom TB. 1997. A noncytolytic IL-10/Fc fusion protein prevents diabetes, blocks autoimmunity, and promotes suppressor phenomena in NOD mice. *J Immunol* **158**: 4507–4513.

Zheng Y, Danilenko DM, Valdez P, Kasman I, Eastham-Anderson J, Wu J, Ouyang W. 2007. Interleukin-22, a T$_H$17 cytokine, mediates IL-23-induced dermal inflammation and acanthosis. *Nature* **445**: 648–651.

Zheng Y, Valdez PA, Danilenko DM, Hu Y, Sa SM, Gong Q, Abbas AR, Modrusan Z, Ghilardi N, de Sauvage FJ, et al. 2008. Interleukin-22 mediates early host defense against attaching and effacing bacterial pathogens. *Nat Med* **14**: 282–289.

Zhuang Y, Peng LS, Zhao YL, Shi Y, Mao XH, Guo G, Chen W, Liu XF, Zhang JY, Liu T, et al. 2012. Increased intratumoral IL-22-producing CD4$^+$ T cells and Th22 cells correlate with gastric cancer progression and predict poor patient survival. *Cancer Immunol Immunother* **61**: 1965–1975.

Zitzmann K, Brand S, Baehs S, Göke B, Meinecke J, Spöttl G, Meyer H, Auernhammer CJ. 2006. Novel interferon-λs induce antiproliferative effects in neuroendocrine tumor cells. *Biochem Biophys Res Commun* **344**: 1334–1341.

Zurita-Turk M, del Carmen S, Santos ACG, Pereira VB, Cara DC, Leclercq SY, de LeBlanc AD, Azevedo V, Chatel JM, LeBlanc JG, et al. 2014. *Lactococcus lactis* carrying the pValac DNA expression vector coding for IL-10 reduces inflammation in a murine model of experimental colitis. *BMC Biotechnol* **14**: 73.

Responses to Cytokines and Interferons that Depend upon JAKs and STATs

George R. Stark, HyeonJoo Cheon, and Yuxin Wang

Department of Cancer Biology, Lerner Research Institute of the Cleveland Clinic, Cleveland, Ohio 44195

Correspondence: starkg@ccf.org

Many cytokines and all interferons activate members of a small family of kinases (the Janus kinases [JAKs]) and a slightly larger family of transcription factors (the signal transducers and activators of transcription [STATs]), which are essential components of pathways that induce the expression of specific sets of genes in susceptible cells. JAK-STAT pathways are required for many innate and acquired immune responses, and the activities of these pathways must be finely regulated to avoid major immune dysfunctions. Regulation is achieved through mechanisms that include the activation or induction of potent negative regulatory proteins, posttranslational modification of the STATs, and other modulatory effects that are cell-type specific. Mutations of JAKs and STATs can result in gains or losses of function and can predispose affected individuals to autoimmune disease, susceptibility to a variety of infections, or cancer. Here we review recent developments in the biochemistry, genetics, and biology of JAKs and STATs.

Because the basic biochemistry of Janus kinase–signal transducers and activators of transcription (JAK-STAT) signaling pathways has been frequently and extensively reviewed (see, for example, Stark and Darnell 2012; Cai et al. 2015; O'Shea et al. 2015; Villarino et al. 2015), we present here only a brief summary. After a cytokine or interferon binds to its specific receptor, the receptor forms homodimers, heterodimers, or trimers, depending on the cytokine, thus activating the tightly bound JAKs to cross-phosphorylate each other. The activated JAKs then phosphorylate specific tyrosine residues in the cytoplasmic domains of the receptors, providing binding sites for the STATs through their highly conserved SH2 domains. The receptor-bound STATs are phosphorylated, each on a highly conserved tyrosine residue, af- ter which they leave the receptor as homo- or heterodimers whose association is strengthened by SH2-phosphotyrosine interactions. The phosphorylated STAT dimers are then transported to the nucleus, where they bind to and activate specific promoters. The basic outline of JAK-STAT signaling (Fig. 1) shows that interferon (IFN)-γ (type II IFN) and all of the cytokines primarily drive the formation of specific STAT homodimers, which then bind to DNA directly. However, in some cases, heterodimers involving STAT1 and STAT3 or STAT5A and STAT5B can also form. In contrast, IFN-β and the subtypes of IFN-α (collectively type I IFNs) and the subtypes of IFN-λ (collectively type III IFNs) drive the formation of STAT1-STAT2 heterodimers, which then associate with the DNA-binding interferon regulatory protein 9 (IRF9) to form in-

General cytokine response

Type I and III IFN response

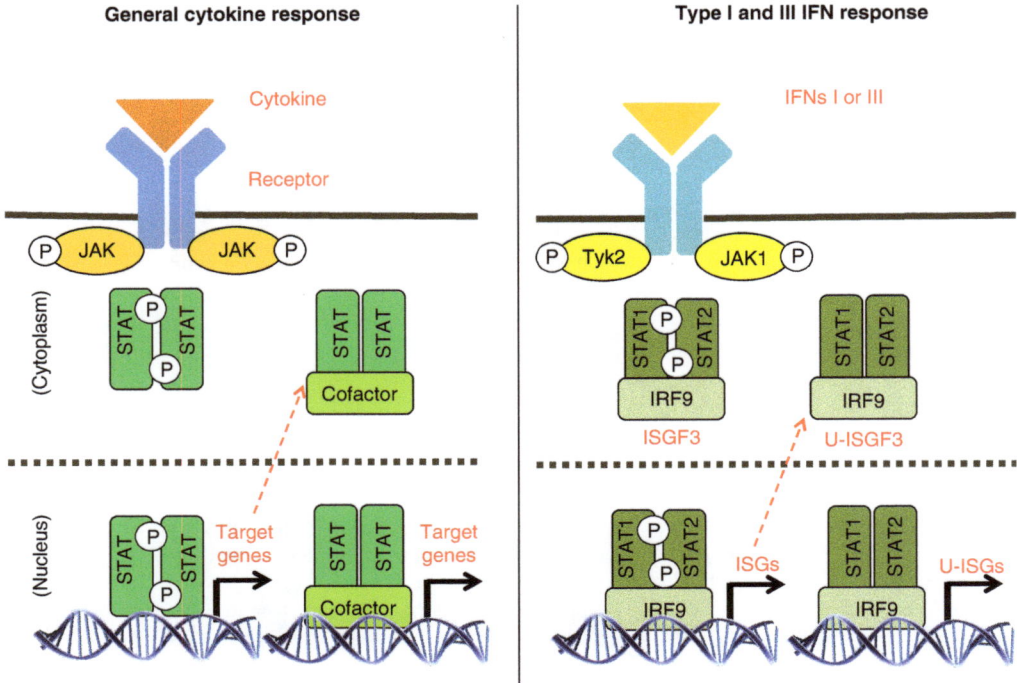

Figure 1. Outline of Janus kinase−signal transducers and activators of transcription (JAK-STAT) signaling shows that interferon (IFN)-γ (type II IFN) and all of the cytokines primarily drive the formation of specific STAT homodimers, which then bind to DNA directly. However, in some cases, heterodimers involving STAT1 and STAT3 or STAT5A and STAT5B can also form. In contrast, IFN-β and the subtypes of IFN-α (collectively type I IFNs), and the subtypes of IFN-λ (collectively type III IFNs) drive the formation of STAT1-STAT2 heterodimers, which then associate with the DNA-binding interferon regulatory protein 9 (IRF9) to form interferon-stimulated gene factor 3 (ISGF3). Also shown are STAT transcription factors that form and function without tyrosine phosphorylation (unphosphorylated STATs [U-STATs]).

terferon-stimulated gene factor 3 (ISGF3). The DNA sequences to which STAT homodimers bind (γ-activated sequences [GAS] elements) are related to, but distinct from, the sequences to which IRF9 binds (interferon-stimulated regulatory elements [ISREs]). The consensus GAS sequences are 5′TTCCNGGAA3′ for STAT1, 3, 4, and 5 and 5′TTCCNNGGAA for STAT6. Note that these motifs are palindromic, consistent with the need for each STAT subunit of the dimer to contact DNA. The consensus ISRE sequence is RRTTTCNNTTTCY (Decker and Kovarik 1999). Also shown in Figure 1 are STAT transcription factors that form and function without tyrosine phosphorylation (unphosphorylated STATs [U-STATs]). The genes encoding STAT3, STAT1, STAT2, and IRF9 are themselves tran-

scriptional targets of the primary signals generated by phosphorylated STAT3 and ISGF3, respectively, leading to secondary increases in the concentrations of these U-STATs in response to the initial signal, driving STAT-STAT association and promoter binding even in the absence of tyrosine phosphorylation. This aspect is discussed further below.

As summarized in Table 1, many cytokines and all IFNs use JAK tyrosine kinases and STAT transcription factors to connect their cell-surface receptors to the activation of their specific gene targets. The assignments of specific JAKs and STATs to specific cytokines and IFNs in the table should be understood to represent the major pathways only. In reality, the situation is much more complex, and a single ligand−

Table 1. Summary of cytokines and IFNs that use JAK tyrosine kinases and STAT transcription factors to connect their cell-surface receptors to the activation of specific gene targets

STATs	JAKs	Major cytokines
STAT1	JAK1, JAK2, TYK2	Type I, II, and III IFNs
STAT2	JAK1, TYK2	Type I and III IFNs
STAT3	JAK1, JAK2, JAK3, TYK2	IL-6 family cytokines, IL-10, IL-27, IL-21
STAT4	JAK2, TYK2	IL-12, IL-23
STAT5A/B	JAK1, JAK2, JAK3	IL-2, IL-7, IL-9, IL-15, EPO, TPO, GM-CSF, GH, PRL
STAT6	JAK1, JAK2, JAK3, TYK2	IL-4, IL-13

STATs, Signal transducers and activators of transcription; JAKs, Janus kinases; IFN, interferon; IL, interleukin; EPO, erythropoietin; TPO, thrombopoietin; GM-CSF, granulocyte macrophage colony-stimulating factor; GH, growth hormone; PRL, prolactin.

receptor pair may activate more than one STAT. The ratio of activated STATs can depend on their relative intracellular concentrations (specific examples for interleukin (IL)-6, IL-21, and IL-27 are cited below), on the specific cell type, on whether or not the cell has received prior signals ("priming"), and probably on other variables as well. As one example of cell-type-specific responses, van Boxel-Dezaire et al. (2010) found that different primary human leukocyte subsets respond quite differently to IFN-β. In B cells and CD4$^+$ T cells, IFN-β activates STAT3 and STAT5 primarily, with biological effects opposite from those driven by the "canonical" activation of STAT1 and STAT2 in the other leukocyte subtypes that were studied.

FUNCTIONALLY IMPORTANT CHEMICAL MODIFICATIONS OF THE STATs

The STATs are substrates for phosphorylation, methylation, and other posttranslational modifications that facilitate both positive and negative fine-tuning of the transcriptional responses.

Carboxy-Terminal Serine Phosphorylations

All the STATs except STAT2 share a functionally important serine phosphorylation site within a P(M)SP motif located near their carboxyl termini. Carboxy-terminal serine phosphorylation is stimulated by many different cytokines and growth factors, and is mediated by many different kinases, including extracellular signal-regulated kinase (ERK), p38, c-Jun amino-

terminal kinase (JNK), mechanistic target of rapamycin (mTOR), nemo-like kinase (NLK), calcium/calmodulin-dependent protein kinase II (CaMKII), IκB kinase ε (IKKε), and protein kinase C δ (PKC-δ) (Schindler et al. 2007). Phosphorylation increases the transactivation potential of these proteins. The most intensively investigated phosphorylated serine residues are S727 in both STAT1 and STAT3. Phosphorylation of S727 of STAT1 and STAT3 is necessary for full activation of transcription in response to IFNs and IL-6 family cytokines, but the phosphorylation of S727 in either STAT1 or STAT3 is not associated with increased tyrosine phosphorylation (Wen et al.1995). The serine to alanine mutation of S727 in STAT1 or STAT3 leads to reduction of the cytokine-induced transcription of specific genes, the extent of which is likely to vary in different cellular contexts. In addition to altering the transcriptional activation of STATs, serine phosphorylation of S727 in STAT1 and STAT3 has been correlated with enhanced DNA-binding ability (Eilers et al. 1995; Zhang et al. 1995; Ng and Cantrell 1997). Visconti et al. (2000) showed that mutating S721 of STAT4 decreased its transcriptional activity in IL-12-treated cells. Phosphorylation of both S725 and S779 of STAT5A and of S730 of STAT5B negatively regulate transactivation in response to stimulation of mammary glands with prolactin (Yamashita et al. 1998; Benitah et al. 2003). Wang et al. (2004) showed that IL-4 and IL-13 promote STAT6 phosphorylation on S756 in human T cells. However, the contribution of this phosphorylation to function is not yet clear.

Additional Serine and Threonine Phosphorylations

Recent advances in mass spectrometry have revolutionized the analysis of protein phosphorylation by allowing rapid identification of the sites modified with precision and sensitivity. Several additional serine and threonine modifications sites on STATs have been found.

STAT1

Phosphorylation of S708 by TRIM6-activated IKKε regulates STAT1 homodimerization but not ISGF3 formation in response to IFN (Tenoever et al. 2007; Rajsbaum et al. 2014). Ultimately, the phosphorylation of S708 facilitates the induction of a subset of ISGs whose protein products are essential for the antiviral response in vitro and in vivo. Phosphorylation of S744/S747 has also been reported, but the binding of ISGF3 to ISREs was unaffected by a carboxy-terminal deletion of STAT1 that removed both of these residues (Tenoever et al. 2007).

STAT2

The first serine phosphorylation of STAT2 (S287) was revealed by the work of Steen et al. (2013). Phosphorylation-defective mutants of S287 of STAT2 enhanced the ability of ISGF3 to bind to DNA, revealing that this phosphorylation is a negative regulatory event. We have shown that the phosphorylation of T387 regulates the ability of ISGF3 to bind to DNA (Wang et al. 2017). This phosphorylation negatively regulates the expression of most genes induced by type I IFN, inhibiting the ability of IFN to protect cells against virus infection and to inhibit cell growth. In most untreated cell types, the great majority of STAT2 is phosphorylated on T387 constitutively. T387 lies in a cyclin-dependent kinase (CDK) consensus sequence, and CDK inhibitors decrease T387 phosphorylation markedly. Using CDK inhibitors to reverse the constitutive inhibitory phosphorylation of T387 of STAT2 might enhance the efficacy of type I IFNs.

STAT3

Waitkus et al. (2014) have provided evidence that GSK3 α/β directly phosphorylates STAT3, simultaneously on T714 and S727, and that these modifications are required for STAT3-dependent gene expression in response to simultaneous activation of epidermal growth factor receptor (EGFR) and protease-activated receptor 1 (PAR-1) in endothelial cells. Levels of both T714 and S727 phosphorylation of STAT3 are significantly elevated in renal tumor tissues, suggesting that the GSK3-activated STAT3 signaling may be important in this disease.

STAT5A

S127/S128 phosphorylation of STAT5A is required for ERBB4-induced Y694 phosphorylation and has a substantial impact on ERBB4-dependent regulation of STAT5A activity (Clark et al. 2005). The expression of a STAT5 mutant in which the S725 and S779 phosphorylation sites were altered prohibited transformation and induced apoptosis in bone marrow cells (Pircher et al. 1999; Xue et al. 2002; Friedbichler et al. 2010; Berger et al. 2014). S779 is phosphorylated by p21-activated kinase (PAK) in human myeloid malignancies.

STAT5B

STAT5B constitutively phosphorylated on S193 has been found in hematopoietic cancers (Mitra et al. 2012). This phosphorylation is dependent on the mTOR signaling pathway and positively regulates STAT5B DNA binding and transcriptional activity.

STAT6

Phosphorylation of STAT6 S707 is triggered by the virus infection–responsive protein STING, which is located in the endoplasmic reticulum. Homodimers of STAT6 phosphorylated on Y641 and S407 then activate specific target genes in the nucleus that mediate immune cell homing (Chen et al. 2011). S707 is phosphorylated by JNK, which can be activated in response to cel-

Table 2. Summary of lysine and arginine modifications of STAT1 and STAT3

STATs	Modifications	Sites
STAT1	Methylation	R31me1 (Zhu et al. 2002)
		R31me2 (Mowen et al. 2001)
	Acetylation	K410/413ac (Kramer et al. 2006; Antunes et al. 2011; Kotla and Rao, 2015), controversial
	Sumoylation	K703sm (Ungureanu et al. 2003; Ungureanu et al. 2005; Gronholm et al. 2012)
STAT3	Methylation	R31me1 (Iwasaki et al. 2010)
		K49me2 (Dasgupta et al. 2015b)
		K140 (Yang et al. 2010)
	Acetylation	K49/87ac (Ray et al. 2005; Hou et al. 2008; Nie et al. 2009)
		K679ac (Nie et al. 2009)
		K685ac (Yuan et al. 2005; Lee et al. 2009, 2016; Dasgupta et al. 2014; Kang et al. 2015)
		K707ac (Nie et al. 2009)
	Sumoylation	K451sm (Zhou et al. 2016b)

Listed posttranslational modifications (PTMs) can be found at www.phosphosite.org.
STATs, Signal transducers and activators of transcription.

lular stress or IL-1β. Phosphorylation of S707 is a negative regulatory event that decreases the DNA-binding ability of STAT6 following its activation by IL-4 (Shirakawa et al. 2011).

Lysine and Arginine Modifications

The lysine residues of proteins can be modified by acetylation or by the addition of one, two, or three methyl groups, and the arginine residues can be modified by methylation. Furthermore, these reactions are reversible, providing rich opportunities to modify function. Many enzymes carry out the reversible acetylation and methylation of histones, providing the chemical basis of the modification of chromatin structure and function known as the histone code (Allis and Jenuwein 2016). In many cases, the specifically modified lysine or arginine residues provide docking sites for the binding of accessory proteins that modulate function. As summarized in Table 2, several different lysine and arginine residues of STAT1 and STAT3 are methylated or acetylated, but we are not aware of reports of these modifications for any other STAT. In every case, the enzymes responsible for STAT1 or STAT3 modification were previously known to modify histones. One of the earliest papers reports the dimethylation of R31 of STAT1, which facilitates IFN-dependent gene expression by

inhibiting the association of STAT1 with the negative regulator PIAS1 (Mowen et al. 2001). It is very interesting that the effects of gain-of-function mutations of STAT1 that cause disseminated yeast infections in patients can be ameliorated by reducing the level of PIAS1 or by facilitating STAT1 arginine methylation (Sampaio et al. 2013).

We have described the reversible dimethylation of K140 of STAT3, which regulates STAT3-dependent gene expression negatively (Yang et al. 2010). In this case, the reaction is catalyzed by the histone lysine methyltransferase SET9 and occurs only after STAT3 has been bound to a promoter. The docking site for SET9 is provided by the phosphorylated S727 residue of STAT3, because the S727A mutant of STAT3 fails to recruit SET9 to the promoter. We reviewed several additional examples of the lysine methylation of promoter-bound transcription factors (Stark et al. 2011). Our working hypothesis, which needs to be tested further, is that the promoter-bound factor provides a docking site for a histone-modifying enzyme that then catalyzes functionally important modifications, not only of the transcription factor but potentially also of local histones and the transcriptional machinery itself. Another important modification of STAT3 is the dimethylation of K49, carried out by the lysine methyltransferase EZH2 (Dasgupta et al. 2015b). Failure to carry out this reaction inhibits the ability of STAT3 to activate the expression of a substantial fraction of its target genes by an as-yet-unknown mechanism.

SUMOylation, Glycosylation, and Additional Tyrosine Phosphorylation

Glycosylation (Gewinner et al. 2004) and tyrosine phosphorylation of STATs at additional sites, including STAT2, human, Y631 (Scarzello et al. 2007), STAT5A, mouse, Y682/683 (Schaller-Schonitz et al. 2014), and STAT5B, rat, Y679 (Kabotyanski and Rosen 2003), have been reported, but the functional importance of these modifications is not yet well established. On the other hand, SUMOylation of STAT1 on K703 is a modification of great functional importance (references in Table 2). This modification, catalyzed by PIAS1 (Rogers et al. 2003; Ungureanu et al. 2003), leads to inhibition of STAT1 function (Rogers et al. 2003; Ungureanu et al. 2005), and modulation of the response to IFN-γ is facilitated by the SUMOylation of STAT1 (Begitt et al. 2011; Maarifi et al. 2015).

NEGATIVE REGULATION

Failure to regulate cytokine-stimulated responses leads to catastrophic hyperinflammatory responses, and therefore elaborate mechanisms exist to achieve the necessary negative regulation. A major mechanism involves the suppressor of cytokine signaling (SOCS) proteins (reviewed by Kazi et al. 2014). These potent negative regulators are typically induced in response to acute exposure to specific cytokines, and they typically function by inhibiting STAT activation at the receptors (see Babon et al. 2014 for an example of the SOCS3 and the IL-6 family of cytokines). Defective SOCS3 function has been reported to contribute to many diseases, including allergy, autoimmune diseases such as rheumatoid arthritis, vascular inflammatory diseases, insulin resistance, and cancer, as reviewed by Yin et al. (2015). Many additional mechanisms contribute to adequate regulation of the induction of and responses to IFNs, including the PIAS proteins, which inhibit the function of activated STATs in the nucleus and also catalyze the inhibitory SUMOylation of STAT1, and protein tyrosine phosphatases, which inactivate the STATs (Hertzog and Williams 2013; Porritt and Hertzog 2015). Although the STAT3-induced expression of SOCS3 is important for dampening acute responses to IL-6 and other gp130-linked cytokines, tumor cells use an IL-6-induced association between the IL-6 receptor:gp130 complex and the EGFR to nullify the inhibitory effect of SOCS3 and thus to sustain STAT3 activation constitutively (Wang et al. 2013). It seems likely that specific mechanisms to prevent negative regulation of STAT activation will be used whenever sustained STAT activation is necessary in normal physiology.

UNPHOSPHORYLATED STATs

Many studies have shown that U-STATs, which lack phosphorylation of their highly conserved tyrosine residues, are located in nuclei, bind to promoters, and activate gene expression (Chatterjee-Kishore et al. 2000; Yang et al. 2005, 2007; Cui et al. 2007; Cheon and Stark 2009; Cheon et al. 2013; Park et al. 2015). Whether the phosphorylation of other residues affects the function of U-STATs is not clear, but K685 of U-STAT3 is acetylated, and mutation of this residue results in loss of expression of many U-STAT3-induced genes (Dasgupta et al. 2014). U-STATs 1, 2, 3, 5, and 6 regulate gene expression through their interactions with partner cofactors. The expression of U-STATs 1, 2, and 3 is increased in response to cytokines that induce their tyrosine phosphorylation, but how the expression of the other U-STATs is regulated is not yet known. The roles of U-STATs in regulating gene expression were originally controversial, but it is now generally accepted that U-STATs are critical transcription factors that are involved in many biological events in both normal and pathological situations.

U-STATs as Positive Regulators of Gene Expression

The levels of the U-STAT1, U-STAT2, and U-STAT3 proteins are induced in response to signals that lead to the tyrosine phosphorylation of each STAT, because each phosphorylated STAT binds to the promoter of its own gene to induce expression (Yang et al. 2005; Cheon et al. 2013). The U-STAT proteins then accumulate, translocate into nuclei, bind to target gene promoters together with their cofactors, and induce the expression of these genes. U-STATs contribute to the steady-state constitutive expression of specific genes, while tyrosine-phosphorylated STATs induce rapid and transient responses in response to cytokine stimulation. Many target genes of U-STATs 1, 2, 3, and 6 encode proteins that promote cell survival and resistance to cell death, suggesting that the U-STAT system helps to sustain the survival of cells in stressful environments.

U-STAT1 induces its target genes as U-ISGF3, a tripartite complex with U-STAT2 and IRF9, and the expression of all three U-ISGF3 components is increased in response to either type I or type III IFN (Cheon et al. 2013; Sung et al. 2015). U-STAT1 does not activate gene expression as a homodimer, because high levels of U-STAT1 do not induce target gene expression if U-STAT2 and IRF9 are not present in sufficient quantities (Cheon and Stark 2009; Cheon et al. 2013). U-ISGF3 induces a subset of ISGs that are also induced in the initial response to IFNs, resulting in their prolonged expression. The phenotypes of cells that express high levels of only the U-ISGF3-induced genes are different from those of cells treated with high levels of IFNs, which express all of the ISGs in response to phosphorylated ISGF3. Cancer cells that express high levels of U-ISGF3 are more resistant to DNA damage, while IFNs inhibit cancer cell proliferation and increase their apoptosis (Borden et al. 2007). Similarly, hepatocytes expressing high levels of U-ISGF3 are more resistant to IFN-α therapy, although U-ISGF3 itself suppresses viral replication by inducing antiviral genes (Sung et al. 2015). Phosphorylated STAT3 mediates the induction of U-STAT3 in response to IL-6 and other cytokines that activate gp130-linked receptors (Yang et al. 2005). However, in strong contrast to the situation for U-STAT1, U-STAT3 induces the expression of a set of genes that is completely different from the set induced by phosphorylated STAT3 (Yang et al. 2005). Some of the genes specifically induced by U-STAT3 are important oncogenes, including *SRC*, *MET*, and *MRAS*. The mechanism by which U-STAT3 induces expression of target genes is only partially understood. However, for the expression of a subset of the induced genes, including *RANTES*, *IL6*, and *IL8*, U-STAT3 forms a complex with nuclear factor κB (NF-κB), which then binds to κB elements in the promoters of a small fraction of NF-κB target genes (Yang et al. 2007). Cui et al. (2007) show that U-STAT6 constitutively activates the expression of the *COX2* gene. Similarly to the other U-STATs, U-STAT6 also forms a complex with a cofactor protein, p300, facilitating binding to the *COX2* promoter.

U-STATs as Negative Regulators of Gene Expression

In contrast to their ability to regulate gene expression positively, some U-STATs repress gene expression instead. ChIP-seq data showing genome-wide distribution reveals that U-STAT5 and phosphorylated STAT5 bind to different *cis*-acting elements in the genome, and differently regulate gene expression (Park et al. 2015). In this study, phosphorylated STAT5 was shown to bind to GAS elements in canonical STAT5-induced promoters in response to thrombopoietin (TPO) in mouse hematopoietic stem cells. In the absence of TPO, however, a large portion of U-STAT5 occupies binding sites for early growth response (EGR), an activator that promotes the expression of megakaryocytic genes, thus repressing EGR-induced gene expression. Only a small portion of U-STAT5 binds to promoters that are occupied by phosphorylated STAT5 in response to TPO, but the role of U-STAT5 is not known in that situation. Differently from U-STAT5, U-ISGF3 and phosphorylated ISGF3 bind to similar ISRE elements in the promoters of ISGs that are induced by both transcription factors (Cheon et al. 2013). Using a ChIP-on-chip analysis, Testoni et al. (2011) showed that U-STAT2, possibly as a component of U-ISGF3, binds to more than half of the ISG promoters investigated before IFN-α treatment, but the effect on the expression of ISGs is not clear. Although U-ISGF3 itself increases the expression of some ISGs in the absence of IFN treatment, we do not yet know whether or not U-ISGF3 inhibits the expression of other ISGs in response to IFN.

Nongenomic Activity of U-STAT3 in Mitochondria

U-STAT3 plays important roles not only in the nucleus but also in mitochondria (Garama et al. 2016). The mitochondrial activity of STAT3 is not dependent on Y705 phosphorylation, but phosphorylation of S727 is necessary (Gough et al. 2009, 2014). Mitochondrial U-STAT3 regulates the activity of the electron transport chain, which is required for adenosine triphosphate (ATP) production, and the opening of the mitochondrial permeability transition pore (Gough et al. 2009; Wegrzyn et al. 2009). In RAS-transformed cells, the MEK−ERK pathway drives the phosphorylation of STAT3 on S727, and RAS-transformed cells carrying a mutation of S727 are partially resistant to inhibitors of the ERK pathway (Gough et al. 2013).

STRUCTURES AND INTRACELLULAR LOCATIONS OF STATS

Phosphorylated STATs are translocated into the nucleus after cytokine stimulation, but U-STATs shuttle constitutively between cytoplasm and nucleus (Meyer and Vinkemeier 2004; Pranada et al. 2004; Liu et al. 2005; Iyer and Reich 2007; Vogt et al. 2011). U-STATs 1, 3, and 5 form antiparallel dimers, whereas phosphorylated STAT dimers are in a parallel conformation that is stabilized by phosphotyrosine−SH2 domain interactions, allowing both DNA-binding domains to contact the GAS sequences simultaneously, resulting in strong binding (Mao et al. 2005; Neculai et al. 2005; Wenta et al. 2008; Timofeeva et al. 2012; Nkansah et al. 2013). In addition, phosphorylated homodimers of STAT1 bind to each other to form tetramers, facilitating gene expression in response to type

II (but not type I) IFNs. Consistently, the F77A mutation of murine STAT1, which disrupts dimer−dimer interactions, blunts signaling in response to type II IFNs in mice (Begitt et al. 2014). Interestingly, Droescher et al. (2011) found that all activated STATs can form paracrystalline arrays in the nuclei of cytokine-stimulated cells, with important biological consequences. However, for STAT1 only, this phenomenon can be prevented by modification of the protein by SUMO (Begitt et al. 2011; Droescher et al. 2011). STAT2 forms a stable antiparallel heterodimer with either the unphosphorylated or tyrosine-phosphorylated forms of STAT1; these heterodimers are not transported to the nucleus and have no transcriptional activity, so STAT2 is a pervasive negative regulator of STAT1-dependent functions (Ho et al. 2016). However, when IRF9 is present in sufficient quantity, antiparallel STAT1-STAT2 heterodimers will be converted to U-ISGF3 (Cheon et al. 2013) or to hemiphosphorylated ISGF3 (Morrow et al. 2011), in which the relative orientation of the two STATs becomes parallel. They are then transported to the nucleus and bind to DNA to activate transcription. Thus, the availability of free STAT1 to signal in response, for example, to IFN-γ or IL-27 (Ho et al. 2016), will depend on the relative concentrations of both STAT2 and IRF9.

CHROMATIN REMODELING BY STATs

There are a few intriguing reports that STATs function to affect chromatin structure independently of their ability to activate transcription. U-STAT92E (the only *Drosophila* STAT) stabilizes heterochromatin in association with heterochromatin protein 1 (HP1) (Shi et al. 2008). Phosphorylation of U-STAT92E causes heterochromatin instability and promotes gene expression. Human U-STAT5A binds to HP1α, repressing the expression of multiple oncogenes (Hu et al. 2013). The overexpression of STAT5A Y704F, which cannot be phosphorylated at this site, has effects on global gene expression similar to the effects of over expressing HP1α. The phosphorylation of STAT1 remodels chromatin to generate a local environment appropriate for the activation of target gene expression. When human histocompatibility (MHC) genes are activated by IFN-γ, phosphorylated STAT1, which binds to specific elements of the target gene promoter and recruits the chromatin remodeling enzyme BRG1, causes the release of entire MHC locus loops from compacted chromatin (Christova et al. 2007).

JAK AND STAT MUTATIONS

Naturally Occurring STAT Mutations

Because the STATs have essential roles in infectious disease, immunity, and cancer, many naturally occurring mutations have been discovered that affect human health, providing a rich source of structural and functional information. All seven STATs form homo- and heterodimers following phosphorylation of their tyrosine residues, and thus there is great potential for dominant effects in which only one partner is mutated. These mutations can affect many different protein–protein interactions, leading to dominant phenotypes that reflect either loss of function (LOF) or gain of function (GOF), as summarized recently for STAT3 (Chandrasekaran et al. 2016). However, STAT mutations will be manifest in human disease only if they lead to a discernable biological phenotype and, although many dominant mutations have been described for STAT1 and STAT3, far fewer have emerged for the other STATs. Because the structures of the STAT dimers are similar, it seems likely that germline dominant mutations will occur with similar frequencies for all the STATs, but with far less frequent phenotypic consequences for STATs 2, 4, 5, and 6. In their summary of all human primary immunodeficiencies, Boisson et al. (2015) point out that the known autosomal-recessive deficiencies are all caused by alleles with LOF, and that 44 out of 61 autosomal-dominant defects are caused by LOF mutations. Negative regulation of cytokine-dependent signaling is vital to prevent overstimulation of immune responses, so that, within the 17 examples of GOF-dominant mutations, those affecting STAT1 lead to infection, autoimmunity, and malignancy, whereas GOF mutations of STAT3 affect autoimmunity, allergy, and autoinflammation. Y705, S727, and K49 mutations of STAT3 have not yet been reported, presumably because they would not be consistent with survival.

Gain-of-Function Somatic Mutations in Cancer

The constitutive activation of JAK-STAT signaling pathways in cancer cells, through somatic mutation and other mechanisms, drives proliferation and resistance to stresses, and helps to overcome barriers to perpetual cell growth. Because this topic has been well reviewed recently (O'Shea et al. 2015; Pilati and Zucman-Rossi 2015; Thomas et al. 2015), we present only a few illustrative examples here. As recent examples for the STATs, activating somatic mutations of STAT5B lead to leukemia (Rajala et al. 2013) and activating mutations of STAT6 lead to follicular lymphoma (Yildiz et al. 2015). For JAK2, the activating V617F mutation causes polycythemia vera and other myeloproliferative diseases (Spivak 2010).

The Gain-of-Function STAT1 Paradox in Infectious Disease

Resolution of infection depends heavily on IFN responses, and specific STAT1 and JAK mutations that lead to increased susceptibility to infection have been extensively reviewed (Casanova et al. 2012; Boisson et al. 2015). Because STAT1 is required for all known responses to all three IFN subtypes, dominant STAT1 mutations lead to increased susceptibility to a wide range of infectious agents. For example, some dominant LOF mutations of STAT1 underlie chronic infections, such as candidiasis (van de Veerdonk et al. 2011) and disseminated mycobacterial disease (Sampaio et al. 2012). It is surprising that germline GOF mutations that lead to constitutive STAT1 activity predispose affected individuals to diseases resulting from infection with a variety of mycobacteria and fungi (Sampaio et al. 2013; Uzel et al. 2013; Kumar et al. 2014). As summarized by Zerbe et al. (2016),

"STAT1 LOF mutations are associated with viral, mycobacterial ... and bacterial infections, while GOF mutations are associated with mucocutaneous and invasive fungal infections and viral infections." STAT1 GOF mutations also cause failure to control persistent JC virus infections in the central nervous system, leading to progressive multifocal leukoencelphalopathy in humans (Zerbe et al. 2016). Furthermore, persistent activation of IFN signaling, which depends on STAT1 activation, facilitates persistent lymphocytic choriomeningitis virus (LCMV) infection in mice (Teijaro et al. 2013; Oldstone 2015). Why are we betrayed by STAT1-dependent systems whose primary roles should be to protect us from infection? As summarized by Michael Oldstone (2015) for the example of persistent LCMV infection in mice, production of type I IFN leads to the expression of IL-10 and PD-1/PD-L1, which in turn cause loss of antiviral T-cell function, so that inhibiting the IFN response helps to cure the persistent infection! The general problem is that the up-regulation of strong inhibitory responses to acute or persistent infections must be balanced by the need to limit these responses, to avoid killing uninfected cells, and to avoid autoimmunity, resulting in a complex system that is poised on a veritable knife-edge and thus susceptible to misregulation.

There are a few examples of human germline mutations in STATs other than STAT1 or STAT3 that affect infection or inflammation. Patients with an abnormally low level of STAT2 expression are susceptible to virus infections (Shahni et al. 2015) but, very surprisingly, the complete loss of STAT2 expression does not seem to eliminate completely the host defense against many viruses, but does sensitize affected individuals to measles (Hambleton et al. 2013). For STAT4, some polymorphisms influence the risk of developing juvenile arthritis (Fan et al. 2015).

STATs AND IFNs

Mechanisms of Misregulation of IFN Signaling

The complexity of how negative regulators fine-tune IFN responses and the consequences of failure to regulate these responses effectively have been well reviewed recently (Hertzog and Williams 2013; Porritt and Hertzog 2015). In hepatitis C virus (HCV) infections, the virus persists even though many IFN-induced proteins are expressed at a high level in the livers of chronically infected patients. However, the levels of tyrosine-phosphorylated STAT1 and STAT2 are low, and the patients respond poorly to exogenous IFN (Shin et al. 2016). The underlying mechanism is complex, revealing some important general principles (Sung et al. 2015). The levels of U-ISGF3 are high in these livers because of chronic exposure to IFN-λ and, as a result, the downstream gene ISG15 is constitutively activated. Increased concentrations of the ISG15 protein stabilize USP18, a negative regulator of the response to type I IFNs (Zhang et al. 2015). Negative regulators play an essential role in modulating IFN responses that, if not well controlled, are extremely deleterious. The expected phenotype of a cell with a low level of IFN-dependent signaling, as a result of either chronic exposure to a low level of IFN or to a GOF mutation of STAT1, is failure to respond effectively to the high level of IFN that would be produced in response to an infection. As examples, we note GOF STAT1 mutations that lead to increased constitutive STAT1 tyrosine phosphorylation and STAT1-dependent gene expression, but decreased ability of the affected cells to respond to restimulation by IFN (Sampaio et al. 2013; Uzel et al. 2013). We anticipate that increased expression of prominent negative regulators, such as the SOCS and PIAS proteins in response to GOF STAT mutations, will help to explain why cells bearing these mutations fail to respond well to a high level of the relevant cytokine, especially IFN. We also suspect that U-ISGF3, which activates some promoters (Cheon et al. 2013), also binds to other promoters without activating them, thus competing with phosphorylated ISGF3 to inhibit the IFN-induced responses of these promoters.

It is fascinating to observe that not all mammals use the IFN system in the same way. Zhou et al. (2016a) found that at least one species of bats expresses type I IFNs constitutively in the absence of exogenous stimulation, leading

Cite this article as *Cold Spring Harb Perspect Biol* doi: 10.1101/cshperspect.a028555

to the constitutive expression of U-ISGF3 and providing resistance to viruses that are pathogenic in other mammals. Some bats are especially sensitive to fungal infections (Verant et al. 2014), and it seems possible that desensitization to increased production of IFN on infection may contribute to this situation. It is also relevant that constitutive IFN signaling at very low levels in normal individuals modulates profound biological effects (Gough et al. 2012), including the regulation of basal STAT1 expression, as shown by the failure of Y701F STAT1 to sustain STAT1 gene expression in transgenic mice (Majoros et al. 2016). Another possibility for the deleterious effects of GOF mutations in STAT1 and STAT3 is the competition of these two proteins for activation by specific receptors, including the receptors for IL-21 (Wan et al. 2015), IL-6 (Costa-Pereira et al. 2002), and IFN-γ (Qing and Stark 2004). In the case of IL-21, STAT1 and STAT3 have opposing roles in regulating the function of CD4$^+$ T cells (Wan et al. 2015). The relative concentrations of STAT1 and STAT3 are determined not only by endogenous factors but also by the actions of cytokines, because activated STAT3 drives the expression of the STAT3 gene (Yang et al. 2007) and ISGF3 drives the expression of not only the STAT1 gene but also the STAT2 and IRF9 genes (Cheon et al. 2013). Therefore, GOF mutations in STAT1 or STAT3 are likely to alter the steady-state levels of these two proteins and thus affect the biological responses to cytokines such as IL-21.

Good and Bad IFNs in Cancer

Cancers are constitutively exposed to IFNs that are produced by immune cells in the tumor microenvironment, especially macrophages and dendritic cells. In addition, cancer cells make type I IFNs themselves in response to endogenous or induced DNA damage and in response to the enhanced expression of double-stranded RNAs that are encoded by endogenous retrovirus-like DNA sequences, following reduction in the extent of the DNA methylation that normally suppresses their expression (reviewed by Cheon et al. 2014 and Borden 2017). Acute exposure to endogenous or exogenous (therapeutic) IFN often leads to the arrest or death of cancer cells (Borden 2017). In particular, the IFN that is induced in response to DNA damage facilitates the arrest or killing of cancer cells (Widau et al. 2014; Yu et al. 2015). On the other hand, chronic exposure to a low level of IFN leads to the expression of genes comprising the IFN-related DNA damage-resistance signature (IRDS) (Weichselbaum et al. 2008), which is virtually identical to the pattern of gene expression observed in response to U-ISGF3 (see above). The IRDS phenotype in cancer is characterized by resistance to DNA damage, very likely because of the action of one or more IFN-induced protein. Cancer cells have to survive the toxic effects of endogenous IFNs that arise from constitutive DNA damage and the formation of endogenous double-stranded RNA (Leonova et al. 2013) and exogenous IFNs produced by immune cells in the tumor microenvironment. They do this by desensitizing the full response to IFNs, which otherwise would induce the expression of many cytotoxic or cytostatic proteins while retaining the partial response to IFN that is driven by U-ISGF3, which induces the expression of proteins that provide protection against DNA damage. How the cancer cells manage to achieve such a selective response to IFNs remains to be elucidated. Important factors are likely to be the amounts and types of IFN that are present and the modulatory effects of the many other signaling pathways that are activated by other cytokines in the tumor microenvironment.

SUMMARY AND CONCLUSIONS

Many complex mechanisms are required for appropriate control of how cells respond to cytokines and IFNs, including multiple post-translational modifications of the STATs and inhibition by many constitutive and induced negative regulators. Furthermore, all the responses are time-dependent, with kinetics that are regulated in many ways as well, including changing levels of STAT expression and modulation of the effects of the negative regulators. How a cell responds to a specific cytokine or IFN is determined not only by the cell type but also by the experience of that cell, in which all the signals that the cell is receiving from the environment are integrated into a specific pattern of behavior. Defects in control become evident in patients with rare germline defects in the STATs and in the abnormal responses of cancer cells to extracellular signals. We can anticipate many more years of important new discoveries as the many layers of this amazing system are exposed to our view by ongoing research.

REFERENCES

Allis CD, Jenuwein T. 2016. The molecular hallmarks of epigenetic control. *Nat Rev Genet* **17:** 487–500.

Antunes F, Marg A, Vinkemeier U. 2011. STAT1 signaling is not regulated by a phosphorylation-acetylation switch. *Mol Cell Biol* **31:** 3029–3037.

Babon JJ, Varghese LN, Nicola NA. 2014. Inhibition of IL-6 family cytokines by SOCS3. *Semin Immunol* **26:** 13–19.

Begitt A, Droescher M, Knobeloch KP, Vinkemeier U. 2011. SUMO conjugation of STAT1 protects cells from hyper-responsiveness to IFNγ. *Blood* **118:** 1002–1007.

Begitt A, Droescher M, Meyer T, Schmid CD, Baker M, Antunes F, Knobeloch KP, Owen MR, Naumann R, Decker T, et al. 2014. STAT1-cooperative DNA binding distinguishes type 1 from type 2 interferon signaling. *Nat Immunol* **15:** 168–176.

Benitah SA, Valeron PF, Rui H, Lacal JC. 2003. STAT5a activation mediates the epithelial to mesenchymal transition induced by oncogenic RhoA. *Mol Biol Cell* **14:** 40–53.

Berger A, Hoelbl-Kovacic A, Bourgeais J, Hoefling L, Warsch W, Grundschober E, Uras IZ, Menzl I, Putz EM, Hoermann G, et al. 2014. PAK-dependent STAT5 serine phosphorylation is required for BCR-ABL-induced leukemogenesis. *Leukemia* **28:** 629–641.

Boisson B, Quartier P, Casanova JL. 2015. Immunological loss-of-function due to genetic gain-of-function in humans: Autosomal dominance of the third kind. *Curr Opin Immunol* **32:** 90–105.

Borden EC. 2017. Interferon signaling in cancer therapy: New opportunities. *Nat Rev Drug Discov* (in press).

Borden EC, Sen GC, Uze G, Silverman RH, Ransohoff RM, Foster GR, Stark GR. 2007. Interferons at age 50: Past, current and future impact on biomedicine. *Nat Rev Drug Discov* **6:** 975–990.

Cai B, Cai JP, Luo YL, Chen C, Zhang S. 2015. The specific roles of JAK/STAT signaling pathway in sepsis. *Inflammation* **38:** 1599–1608.

Casanova JL, Holland SM, Notarangelo LD. 2012. Inborn errors of human JAKs and STATs. *Immunity* **36:** 515–528.

Chandrasekaran P, Zimmerman O, Paulson M, Sampaio EP, Freeman AF, Sowerwine KJ, Hurt D, Alcantara-Montiel JC, Hsu AP, Holland SM. 2016. Distinct mutations at the same positions of STAT3 cause either loss or gain of function. *J Allergy Clin Immunol* **138:** 1222–1224.

Chatterjee-Kishore M, Wright KL, Ting JPY, Stark GR. 2000. How Stat1 mediates constitutive gene expression: A complex of unphosphorylated Stat1 and IRF1 supports transcription of the LMP2 gene. *EMBO J* **19:** 4111–4122.

Chen H, Sun H, You F, Sun W, Zhou X, Chen L, Yang J, Wang Y, Tang H, Guan Y, et al. 2011. Activation of STAT6 by STING is critical for antiviral innate immunity. *Cell* **147:** 436–446.

Cheon H, Stark GR. 2009. Unphosphorylated STAT1 prolongs the expression of interferon-induced immune regulatory genes. *Proc Natl Acad Sci* **106:** 9373–9378.

Cheon H, Holvey-Bates EG, Schoggins JW, Forster S, Hertzog P, Imanaka N, Rice CM, Jackson MW, Junk DJ, Stark GR. 2013. IFNβ-dependent increases in STAT1, STAT2, and IRF9 mediate resistance to viruses and DNA damage. *EMBO J* **32:** 2751–2763.

Cheon H, Borden EC, Stark GR. 2014. Interferons and their stimulated genes in the tumor microenvironment. *Semin Oncol* **41:** 156–173.

Christova R, Jones T, Wu PJ, Bolzer A, Costa-Pereira AP, Watling D, Kerr IM, Sheer D. 2007. P-STAT1 mediates higher-order chromatin remodelling of the human MHC in response to IFN. *J Cell Sci* **120:** 3262–3270.

Clark DE, Williams CC, Duplessis TT, Moring KL, Notwick AR, Long W, Lane WS, Beuvink I, Hynes NE, Jones FE. 2005. ERBB4/HER4 potentiates STAT5A transcriptional activity by regulating novel STAT5A serine phosphorylation events. *J Biol Chem* **280:** 24175–24180.

Costa-Pereira AP, Tininini S, Strobl B, Alonzi T, Schlaak JF, Is'harc H, Gesualdo I, Newman SJ, Kerr IM, Poli V. 2002. Mutational switch of an IL-6 response to an interferon-γ-like response. *Proc Natl Acad Sci* **99:** 8043–8047.

Cui X, Zhang L, Luo J, Rajasekaran A, Hazra S, Cacalano N, Dubinett SM. 2007. Unphosphorylated STAT6 contributes to constitutive cyclooxygenase-2 expression in human non-small cell lung cancer. *Oncogene* **26:** 4253–4260.

Dasgupta M, Unal H, Willard B, Yang J, Karnik SS, Stark GR. 2014. Critical role for lysine 685 in gene expression mediated by transcription factor unphosphorylated STAT3. *J Biol Chem* **289:** 30763–30771.

Dasgupta A, Chen KH, Tian L, Archer SL. 2015a. Gone fission: An asymptomatic STAT2 mutation elongates mitochondria and causes human disease following viral infection. *Brain* **138:** 2802–2806.

Dasgupta M, Dermawan JK, Willard B, Stark GR. 2015b. STAT3-driven transcription depends upon the dimethylation of K49 by EZH2. *Proc Natl Acad Sci* **112:** 3985–3990.

Decker T. 2016. Emancipation from transcriptional latency: Unphosphorylated STAT5 as guardian of hematopoietic differentiation. *EMBO J* **35:** 555–557.

Decker T, Kovarik P. 1999. Transcription factor activity of STAT proteins: Structural requirements and regulation by phosphorylation and interacting proteins. *Cell Mol Life Sci* **55:** 1535–1546.

Droescher M, Begitt A, Marg A, Zacharias M, Vinkemeier U. 2011. Cytokine-induced paracrystals prolong the activity of signal transducers and activators of transcription (STAT) and provide a model for the regulation of protein solubility by small ubiquitin-like modifier (SUMO). *J Biol Chem* **286:** 18731–18746.

Eilers A, Georgellis D, Klose B, Schindler C, Ziemiecki A, Harpur AG, Wilks AF, Decker T. 1995. Differentiation-regulated serine phosphorylation of STAT1 promotes GAF activation in macrophages. *Mol Cell Biol* **15**: 3579–3586.

Fan ZD, Wang FF, Huang H, Huang N, Ma HH, Guo YH, Zhang YY, Qian XQ, Yu HG. 2015. STAT4 rs7574865 G/T and PTPN22 rs2488457 G/C polymorphisms influence the risk of developing juvenile idiopathic arthritis in Han Chinese patients. *PLoS ONE* **10**: e0117389.

Friedbichler K, Kerenyi MA, Kovacic B, Li G, Hoelbl A, Yahiaoui S, Sexl V, Mullner EW, Fajmann S, Cerny-Reiterer S, et al. 2010. Stat5a serine 725 and 779 phosphorylation is a prerequisite for hematopoietic transformation. *Blood* **116**: 1548–1558.

Garama DJ, White CL, Balic JJ, Gough DJ. 2016. Mitochondrial STAT3: Powering up a potent factor. *Cytokine* **87**: 20–25.

Gewinner C, Hart G, Zachara N, Cole R, Beisenherz-Huss C, Groner B. 2004. The coactivator of transcription CREB-binding protein interacts preferentially with the glycosylated form of Stat5. *J Biol Chem* **279**: 3563–3572.

Gough DJ, Corlett A, Schlessinger K, Wegrzyn J, Larner AC, Levy DE. 2009. Mitochondrial STAT3 supports Ras-dependent oncogenic transformation. *Science* **324**: 1713–1716.

Gough DJ, Messina NL, Clarke CJ, Johnstone RW, Levy DE. 2012. Constitutive type I interferon modulates homeostatic balance through tonic signaling. *Immunity* **36**: 166–174.

Gough DJ, Koetz L, Levy DE. 2013. The MEK–ERK pathway is necessary for serine phosphorylation of mitochondrial STAT3 and Ras-mediated transformation. *PLoS ONE* **8**: e83395.

Gough DJ, Marie IJ, Lobry C, Aifantis I, Levy DE. 2014. STAT3 supports experimental K-RasG12D-induced murine myeloproliferative neoplasms dependent on serine phosphorylation. *Blood* **124**: 2252–2261.

Gronholm J, Vanhatupa S, Ungureanu D, Valiaho J, Laitinen T, Valjakka J, Silvennoinen O. 2012. Structure-function analysis indicates that sumoylation modulates DNA-binding activity of STAT1. *BMC Biochem* **13**: 20.

Hambleton S, Goodbourn S, Young DF, Dickinson P, Mohamad SM, Valappil M, McGovern N, Cant AJ, Hackett SJ, Ghazal P, et al. 2013. STAT2 deficiency and susceptibility to viral illness in humans. *Proc Natl Acad Sci* **110**: 3053–3058.

Hertzog PJ, Williams BR. 2013. Fine tuning type I interferon responses. *Cytokine Growth Factor Rev* **24**: 217–225.

Ho J, Pelzel C, Begitt A, Mee M, Elsheikha HM, Scott DJ, Vinkemeier U. 2016. STAT2 is a pervasive cytokine regulator due to its inhibition of STAT1 in multiple signaling pathways. *PLoS Biol* **14**: e2000117.

Hou T, Ray S, Lee C, Brasier AR. 2008. The STAT3 NH2-terminal domain stabilizes enhanceosome assembly by interacting with the p300 bromodomain. *J Biol Chem* **283**: 30725–30734.

Hu X, Dutta P, Tsurumi A, Li J, Wang J, Land H, Li WX. 2013. Unphosphorylated STAT5A stabilizes heterochromatin and suppresses tumor growth. *Proc Natl Acad Sci* **110**: 10213–10218.

Iwasaki H, Kovacic JC, Olive M, Beers JK, Yoshimoto T, Crook MF, Tonelli LH, Nabel EG. 2010. Disruption of protein arginine N-methyltransferase 2 regulates leptin signaling and produces leanness in vivo through loss of STAT3 methylation. *Circ Res* **107**: 992–1001.

Iyer J, Reich NC. 2007. Constitutive nuclear import of latent and activated STAT5a by its coiled coil domain. *FASEB J* **22**: 391–400.

Kabotyanski EB, Rosen JM. 2003. Signal transduction pathways regulated by prolactin and Src result in different conformations of activated Stat5b. *J Biol Chem* **278**: 17218–17227.

Kang HJ, Yi YW, Hou SJ, Kim HJ, Kong Y, Bae I, Brown ML. 2015. Disruption of STAT3-DNMT1 interaction by SH-I-14 induces re-expression of tumor suppressor genes and inhibits growth of triple-negative breast tumor. *Oncotarget* doi: 10.18632/oncotarget.4054.

Kazi JU, Kabir NN, Flores-Morales A, Ronnstrand L. 2014. SOCS proteins in regulation of receptor tyrosine kinase signaling. *Cell Mol Life Sci* **71**: 3297–3310.

Kotla S, Rao GN. 2015. Reactive oxygen species (ROS) mediate p300–dependent STAT1 protein interaction with peroxisome proliferator-activated receptor (PPAR)-γ in CD36 protein expression and foam cell formation. *J Biol Chem* **290**: 30306–30320.

Kramer OH, Baus D, Knauer SK, Stein S, Jager E, Stauber RH, Grez M, Pfitzner E, Heinzel T. 2006. Acetylation of Stat1 modulates NF-κB activity. *Genes Dev* **20**: 473–485.

Kumar N, Hanks ME, Chandrasekaran P, Davis BC, Hsu AP, Van Wagoner NJ, Merlin JS, Spalding C, La Hoz RM, Holland SM, et al. 2014. Gain-of-function signal transducer and activator of transcription 1 (STAT1) mutation-related primary immunodeficiency is associated with disseminated mucormycosis. *J Allergy Clin Immunol* **134**: 236–239.

Lee JL, Wang MJ, Chen JY. 2009. Acetylation and activation of STAT3 mediated by nuclear translocation of CD44. *J Cell Biol* **185**: 949–957.

Lee SC, Min HY, Jung HJ, Park KH, Hyun SY, Cho J, Woo JK, Kwon SJ, Lee HJ, Johnson FM, et al. 2016. Essential role of insulin-like growth factor 2 in resistance to histone deacetylase inhibitors. *Oncogene* **35**: 5515–5526.

Leonova KI, Brodsky L, Lipchick B, Pal M, Novototskaya L, Chenchik AA, Sen GC, Komarova EA, Gudkov AV. 2013. p53 cooperates with DNA methylation and a suicidal interferon response to maintain epigenetic silencing of repeats and noncoding RNAs. *Proc Natl Acad Sci* **110**: E89–E98.

Levine RL, Gilliland DG. 2008. Myeloproliferative disorders. *Blood* **112**: 2190–2198.

Liu L, McBride KM, Reich NC. 2005. STAT3 nuclear import is independent of tyrosine phosphorylation and mediated by importin-3. *Proc Natl Acad Sci* **102**: 8150–8155.

Maarifi G, Maroui MA, Dutrieux J, Dianoux L, Nisole S, Chelbi-Alix MK. 2015. Small ubiquitin-like modifier alters IFN response. *J Immunol* **195**: 2312–2324.

Majoros A, Platanitis E, Szappanos D, Cheon H, Vogl C, Shukla P, Stark GR, Sexl V, Schreiber R, Schindler C, et al. 2016. Response to interferons and antibacterial innate immunity in the absence of tyrosine-phosphorylated STAT1. *EMBO Rep* **17**: 367–382.

Mao X, Ren Z, Parker GN, Sondermann H, Pastorello MA, Wang W, McMurray JS, Demeler B, Darnell JE, Chen X. 2005. Structural bases of unphosphorylated STAT1 association and receptor binding. *Mol Cell* **17:** 761–771.

Meyer T, Vinkemeier U. 2004. Nucleocytoplasmic shuttling of STAT transcription factors. *Eur J Biochem* **271:** 4606–4612.

Meyer T, Begitt A, Lödige I, van Rossum M, Vinkemeier U. 2002. Constitutive and IFN-γ-induced nuclear import of STAT1 proceed through independent pathways. *EMBO J* **21:** 344–354.

Mitra A, Ross JA, Rodriguez G, Nagy ZS, Wilson HL, Kirken RA. 2012. Signal transducer and activator of transcription 5b (Stat5b) serine 193 is a novel cytokine-induced phospho-regulatory site that is constitutively activated in primary hematopoietic malignancies. *J Biol Chem* **287:** 16596–16608.

Morrow AN, Schmeisser H, Tsuno T, Zoon KC. 2011. A novel role for IFN-stimulated gene factor 3II in IFN-γ signaling and induction of antiviral activity in human cells. *J Immunol* **186:** 1685–1693.

Mowen KA, Tang J, Zhu W, Schurter BT, Shuai K, Herschman HR, David M. 2001. Arginine methylation of STAT1 modulates IFNα/β-induced transcription. *Cell* **104:** 731–741.

Neculai D, Neculai AM, Verrier S, Straub K, Klumpp K, Pfitzner E, Becker S. 2005. Structure of the unphosphorylated STAT5a dimer. *J Biol Chem* **280:** 40782–40787.

Ng J, Cantrell D. 1997. STAT3 is a serine kinase target in T lymphocytes. Interleukin 2 and T cell antigen receptor signals converge upon serine 727. *J Biol Chem* **272:** 24542–24549.

Nie Y, Erion DM, Yuan Z, Dietrich M, Shulman GI, Horvath TL, Gao Q. 2009. STAT3 inhibition of gluconeogenesis is downregulated by SirT1. *Nat Cell Biol* **11:** 492–500.

Nkansah E, Shah R, Collie GW, Parkinson GN, Palmer J, Rahman KM, Bui TT, Drake AF, Husby J, Neidle S, et al. 2013. Observation of unphosphorylated STAT3 core protein binding to target dsDNA by PEMSA and X-ray crystallography. *FEBS Lett* **587:** 833–839.

Oldstone MB. 2015. A Jekyll and Hyde profile: Type 1 interferon signaling plays a prominent role in the initiation and maintenance of a persistent virus infection. *J Infect Dis* **212:** S31–S36.

O'Shea JJ, Schwartz DM, Villarino AV, Gadina M, McInnes IB, Laurence A. 2015. The JAK-STAT pathway: Impact on human disease and therapeutic intervention. *Annu Rev Med* **66:** 311–328.

Park HJ, Li J, Hannah R, Biddie S, Leal-Cervantes AI, Kirschner K, Flores Santa Cruz D, Sexl V, Gottgens B, Green AR. 2015. Cytokine-induced megakaryocytic differentiation is regulated by genome-wide loss of a uSTAT transcriptional program. *EMBO J* **35:** 580–594.

Pilati C, Zucman-Rossi J. 2015. Mutations leading to constitutive active gp130/JAK1/STAT3 pathway. *Cytokine Growth Factor Rev* **26:** 499–506.

Pircher TJ, Petersen H, Gustafsson JA, Haldosen LA. 1999. Extracellular signal-regulated kinase (ERK) interacts with signal transducer and activator of transcription (STAT) 5a. *Mol Endocrinol* **13:** 555–565.

Porritt RA, Hertzog PJ. 2015. Dynamic control of type I IFN signalling by an integrated network of negative regulators. *Trends Immunol* **36:** 150–160.

Pranada AL, Metz S, Herrmann A, Heinrich PC, Muller-Newen G. 2003. Real time analysis of STAT3 nucleocytoplasmic shuttling. *J Biol Chem* **279:** 15114–15123.

Qing Y, Stark GR. 2004. Alternative activation of STAT1 and STAT3 in response to interferon-γ. *J Biol Chem* **279:** 41679–41685.

Rajala HL, Eldfors S, Kuusanmaki H, van Adrichem AJ, Olson T, Lagstrom S, Andersson EI, Jerez A, Clemente MJ, Yan Y, et al. 2013. Discovery of somatic STAT5b mutations in large granular lymphocytic leukemia. *Blood* **121:** 4541–4550.

Rajsbaum R, Versteeg GA, Schmid S, Maestre AM, Belicha-Villanueva A, Martinez-Romero C, Patel JR, Morrison J, Pisanelli G, Miorin L, et al. 2014. Unanchored K48-linked polyubiquitin synthesized by the E3-ubiquitin ligase TRIM6 stimulates the interferon-IKKε kinase-mediated antiviral response. *Immunity* **40:** 880–895.

Ray S, Boldogh I, Brasier AR. 2005. STAT3 NH2-terminal acetylation is activated by the hepatic acute-phase response and required for IL-6 induction of angiotensinogen. *Gastroenterology* **129:** 1616–1632.

Rogers RS, Horvath CM, Matunis MJ. 2003. SUMO modification of STAT1 and its role in PIAS-mediated inhibition of gene activation. *J Biol Chem* **278:** 30091–30097.

Sampaio EP, Bax HI, Hsu AP, Kristosturyan E, Pechacek J, Chandrasekaran P, Paulson ML, Dias DL, Spalding C, Uzel G, et al. 2012. A novel STAT1 mutation associated with disseminated mycobacterial disease. *J Clin Immunol* **32:** 681–689.

Sampaio EP, Hsu AP, Pechacek J, Bax HI, Dias DL, Paulson ML, Chandrasekaran P, Rosen LB, Carvalho DS, Ding L, et al. 2013. Signal transducer and activator of transcription 1 (STAT1) gain-of-function mutations and disseminated coccidioidomycosis and histoplasmosis. *J Allergy Clin Immunol* **131:** 1624–1634.

Scarzello AJ, Romero-Weaver AL, Maher SG, Veenstra TD, Zhou M, Qin A, Donnelly RP, Sheikh F, Gamero AM. 2007. A mutation in the SH2 domain of STAT2 prolongs tyrosine phosphorylation of STAT1 and promotes type I IFN-induced apoptosis. *Mol Biol Cell* **18:** 2455–2462.

Schaller-Schonitz M, Barzan D, Williamson AJ, Griffiths JR, Dallmann I, Battmer K, Ganser A, Whetton AD, Scherr M, Eder M. 2014. BCR-ABL affects STAT5A and STAT5B differentially. *PLoS ONE* **9:** e97243.

Schindler C, Levy DE, Decker T. 2007. JAK-STAT signaling: From interferons to cytokines. *J Biol Chem* **282:** 20059–20063.

Shahni R, Cale CM, Anderson G, Osellame LD, Hambleton S, Jacques TS, Wedatilake Y, Taanman JW, Chan E, Qasim WV, et al. 2015. Signal transducer and activator of transcription 2 deficiency is a novel disorder of mitochondrial fission. *Brain* **138:** 2834–2846.

Shi S, Larson K, Guo D, Lim SJ, Dutta P, Yan SJ, Li WX. 2008. *Drosophila* STAT is required for directly maintaining HP1 localization and heterochromatin stability. *Nat Cell Biol* **10:** 489–496.

Shin EC, Sung PS, Park SH. 2016. Immune responses and immunopathology in acute and chronic viral hepatitis. *Nat Rev Immunol* **16:** 509–523.

Shirakawa T, Kawazoe Y, Tsujikawa T, Jung D, Sato S, Uesugi M. 2011. Deactivation of STAT6 through serine 707 phosphorylation by JNK. *J Biol Chem* **286:** 4003–4010.

Spivak JL. 2010. Narrative review: Thrombocytosis, polycythemia vera, and JAK2 mutations: The phenotypic mimicry of chronic myeloproliferation. *Ann Intern Med* **152:** 300–306.

Stark GR, Darnell JE Jr. 2012. The JAK-STAT pathway at twenty. *Immunity* **36:** 503–514.

Stark GR, Wang Y, Lu T. 2011. Lysine methylation of promoter-bound transcription factors and relevance to cancer. *Cell Res* **21:** 375–380.

Steen HC, Nogusa S, Thapa RJ, Basagoudanavar SH, Gill AL, Merali S, Barrero CA, Balachandran S, Gamero AM. 2013. Identification of STAT2 serine 287 as a novel regulatory phosphorylation site in type I interferon-induced cellular responses. *J Biol Chem* **288:** 747–758.

Sung PS, Cheon H, Cho CH, Hong SH, Park DY, Seo HI, Park SH, Yoon SK, Stark GR, Shin EC. 2015. Roles of unphosphorylated ISGF3 in HCV infection and interferon responsiveness. *Proc Natl Acad Sci* **112:** 10443–10448.

Teijaro JR, Ng C, Lee AM, Sullivan BM, Sheehan KC, Welch M, Schreiber RD, de la Torre JC, Oldstone MB. 2013. Persistent LCMV infection is controlled by blockade of type I interferon signaling. *Science* **340:** 207–211.

Tenoever BR, Ng SL, Chua MA, McWhirter SM, Garcia-Sastre A, Maniatis T. 2007. Multiple functions of the IKK-related kinase IKKε in interferon-mediated antiviral immunity. *Science* **315:** 1274–1278.

Testoni B, Vollenkle C, Guerrieri F, Gerbal-Chaloin S, Blandino G, Levrero M. 2011. Chromatin dynamics of gene activation and repression in response to interferon (IFN) reveal new roles for phosphorylated and unphosphorylated forms of the transcription factor STAT2. *J Biol Chem* **286:** 20217–20227.

Thomas SJ, Snowden JA, Zeidler MP, Danson SJ. 2015. The role of JAK/STAT signalling in the pathogenesis, prognosis and treatment of solid tumours. *Br J Cancer* **113:** 365–371.

Timofeeva OA, Chasovskikh S, Lonskaya I, Tarasova NI, Khavrutskii L, Tarasov SG, Zhang X, Korostyshevskiy VR, Cheema A, Zhang L, et al. 2012. Mechanisms of unphosphorylated STAT3 transcription factor binding to DNA. *J Biol Chem* **287:** 14192–14200.

Ungureanu D, Vanhatupa S, Kotaja N, Yang J, Aittomaki S, Janne OA, Palvimo JJ, Silvennoinen O. 2003. PIAS proteins promote SUMO-1 conjugation to STAT1. *Blood* **102:** 3311–3313.

Ungureanu D, Vanhatupa S, Gronholm J, Palvimo JJ, Silvennoinen O. 2005. SUMO-1 conjugation selectively modulates STAT1-mediated gene responses. *Blood* **106:** 224–226.

Uzel G, Sampaio EP, Lawrence MG, Hsu AP, Hackett M, Dorsey MJ, Noel RJ, Verbsky JW, Freeman AF, Janssen E, et al. 2013. Dominant gain-of-function STAT1 mutations in FOXP3 wild-type immune dysregulation-polyendocrinopathy-enteropathy-X-linked-like syndrome. *J Allergy Clin Immunol* **131:** 1611–1623.

van Boxel-Dezaire AH, Zula JA, Xu Y, Ransohoff RM, Jacobberger JW, Stark GR. 2010. Major differences in the responses of primary human leukocyte subsets to IFN-β. *J immunol* **185:** 5888–5899.

van de Veerdonk FL, Plantinga TS, Hoischen A, Smeekens SP, Joosten LA, Gilissen C, Arts P, Rosentul DC, Carmichael AJ, Smits-van der Graaf CA, et al. 2011. STAT1 mutations in autosomal dominant chronic mucocutaneous candidiasis. *N Engl J Med* **365:** 54–61.

Verant ML, Meteyer CU, Speakman JR, Cryan PM, Lorch JM, Blehert DS. 2014. White-nose syndrome initiates a cascade of physiologic disturbances in the hibernating bat host. *BMC Physiol* **14:** 10.

Villarino AV, Kanno Y, Ferdinand JR, O'Shea JJ. 2015. Mechanisms of Jak/STAT signaling in immunity and disease. *J Immunol* **194:** 21–27.

Visconti R, Gadina M, Chiariello M, Chen EH, Stancato LF, Gutkind JS, O'Shea JJ. 2000. Importance of the MKK6/p38 pathway for interleukin-12-induced STAT4 serine phosphorylation and transcriptional activity. *Blood* **96:** 1844–1852.

Vogt M, Domoszlai T, Kleshchanok D, Lehmann S, Schmitt A, Poli V, Richtering W, Muller-Newen G. 2011. The role of the N-terminal domain in dimerization and nucleocytoplasmic shuttling of latent STAT3. *J Cell Sci* **124:** 900–909.

Waitkus MS, Chandrasekharan UM, Willard B, Tee TL, Hsieh JK, Przybycin CG, Rini BI, Dicorleto PE. 2014. Signal integration and gene induction by a functionally distinct STAT3 phosphoform. *Mol Cell Biol* **34:** 1800–1811.

Wan CK, Andraski AB, Spolski R, Li P, Kazemian M, Oh J, Samsel L, Swanson PA2nd, McGavern DB, Sampaio EP, et al. 2015. Opposing roles of STAT1 and STAT3 in IL-21 function in CD4⁺ T cells. *Proc Natl Acad Sci* **112:** 9394–9399.

Wang Y, Malabarba MG, Nagy ZS, Kirken RA. 2004. Interleukin 4 regulates phosphorylation of serine 756 in the transactivation domain of Stat6. Roles for multiple phosphorylation sites and Stat6 function. *J Biol Chem* **279:** 25196–25203.

Wang Y, van Boxel-Dezaire AH, Cheon H, Yang J, Stark GR. 2013. STAT3 activation in response to IL-6 is prolonged by the binding of IL-6 receptor to EGF receptor. *Proc Natl Acad Sci* **110:** 16975–16980.

Wang Y, Nan J, Willard B, Wang X, Yang J, Stark GR. 2017. Negative regulation of type I IFN signaling by phosphorylation of STAT2 on T387. *EMBO J* **36:** 202–212.

Wegrzyn J, Potla R, Chwae YJ, Sepuri NBV, Zhang Q, Koeck T, Derecka M, Szczepanek K, Szelag M, Gornicka A, et al. 2009. Function of mitochondrial Stat3 in cellular respiration. *Science* **323:** 793–797.

Weichselbaum RR, Ishwaran H, Yoon T, Nuyten DS, Baker SW, Khodarev N, Su AW, Shaikh AY, Roach P, Kreike B, et al. 2008. An interferon-related gene signature for DNA damage resistance is a predictive marker for chemotherapy and radiation for breast cancer. *Proc Natl Acad Sci* **105:** 18490–18495.

Wen Z, Zhong Z, Darnell JE Jr. 1995. Maximal activation of transcription by Stat1 and Stat3 requires both tyrosine and serine phosphorylation. *Cell* **82:** 241–250.

Wenta N, Strauss H, Meyer S, Vinkemeier U. 2008. Tyrosine phosphorylation regulates the partitioning of STAT1 between different dimer conformations. *Proc Natl Acad Sci* **105:** 9238–9243.

Widau RC, Parekh AD, Ranck MC, Golden DW, Kumar KA, Sood RF, Pitroda SP, Liao Z, Huang X, Darga TE, et al. 2014. RIG-I-like receptor LGP2 protects tumor cells from ionizing radiation. *Proc Natl Acad Sci* **111:** E484–E491.

Xue HH, Fink DW Jr, Zhang X, Qin J, Turck CW, Leonard WJ. 2002. Serine phosphorylation of Stat5 proteins in lymphocytes stimulated with IL-2. *Int Immunol* **14:** 1263–1271.

Yamashita H, Xu J, Erwin RA, Farrar WL, Kirken RA, Rui H. 1998. Differential control of the phosphorylation state of proline-juxtaposed serine residues Ser725 of Stat5a and Ser730 of Stat5b in prolactin-sensitive cells. *J Biol Chem* **273:** 30218–30224.

Yang J, Chatterjee-Kishore M, Staugaitis SM, Nguyen H, Schlessinger K, Levy DE, Stark GR. 2005. Novel roles of unphosphorylated STAT3 in oncogenesis and transcriptional regulation. *Cancer Res* **65:** 939–947.

Yang J, Liao X, Agarwal MK, Barnes L, Auron PE, Stark GR. 2007. Unphosphorylated STAT3 accumulates in response to IL-6 and activates transcription by binding to NFκB. *Genes Dev* **21:** 1396–1408.

Yang J, Huang J, Dasgupta M, Sears N, Miyagi M, Wang B, Chance MR, Chen X, Du Y, Wang Y, et al. 2010. Reversible methylation of promoter-bound STAT3 by histone-modifying enzymes. *Proc Natl Acad Sci* **107:** 21499–21504.

Yildiz M, Li H, Bernard D, Amin NA, Ouillette P, Jones S, Saiya-Cork K, Parkin B, Jacobi K, Shedden K, et al. 2015. Activating STAT6 mutations in follicular lymphoma. *Blood* **125:** 668–679.

Yin Y, Liu W, Dai Y. 2015. SOCS3 and its role in associated diseases. *Hum Immunol* **76:** 775–780.

Yu Q, Katlinskaya YV, Carbone CJ, Zhao B, Katlinski KV, Zheng H, Guha M, Li N, Chen Q, Yang T, et al. 2015. DNA-damage-induced type I interferon promotes senescence and inhibits stem cell function. *Cell Rep* **11:** 785–797.

Yuan ZL, Guan YJ, Chatterjee D, Chin YE. 2005. Stat3 dimerization regulated by reversible acetylation of a single lysine residue. *Science* **307:** 269–273.

Zerbe CS, Marciano BE, Katial RK, Santos CB, Adamo N, Hsu AP, Hanks ME, Darnell DN, Quezado MM, Frein C, et al. 2016. Progressive multifocal leukoencephalopathy in primary immune deficiencies: Stat1 gain of function and review of the literature. *Clin Infect Dis* **62:** 986–994.

Zhang X, Blenis J, Li HC, Schindler C, Chen-Kiang S. 1995. Requirement of serine phosphorylation for formation of STAT-promoter complexes. *Science* **267:** 1990–1994.

Zhang X, Bogunovic D, Payelle-Brogard B, Francois-Newton V, Speer SD, Yuan C, Volpi S, Li Z, Sanal O, Mansouri D, et al. 2015. Human intracellular ISG15 prevents interferon-α/β over-amplification and auto-inflammation. *Nature* **517:** 89–93.

Zhou P, Tachedjian M, Wynne JW, Boyd V, Cui J, Smith I, Cowled C, Ng JH, Mok L, Michalski WP, et al. 2016a. Contraction of the type I IFN locus and unusual constitutive expression of IFN-α in bats. *Proc Natl Acad Sci* **113:** 2696–2701.

Zhou Z, Wang M, Li J, Xiao M, Chin YE, Cheng J, Yeh ET, Yang J, Yi J. 2016b. SUMOylation and SENP3 regulate STAT3 activation in head and neck cancer. *Oncogene* **35:** 5826–5838.

Zhu W, Mustelin T, David M. 2002. Arginine methylation of STAT1 regulates its dephosphorylation by T cell protein tyrosine phosphatase. *J Biol Chem* **277:** 35787–35790.

The Interferon (IFN) Class of Cytokines and the IFN Regulatory Factor (IRF) Transcription Factor Family

Hideo Negishi,[1] Tadatsugu Taniguchi,[1,2] and Hideyuki Yanai[1,2]

[1]Department of Molecular Immunology, Institute of Industrial Science, The University of Tokyo, Komaba 4-6-1, Meguro-ku, Tokyo 153-8505, Japan

[2]Max Planck-The University of Tokyo Center for Integrative Inflammology, Komaba 4-6-1, Meguro-ku, Tokyo 153-8505, Japan

Correspondence: tada@m.u-tokyo.ac.jp

Interferons (IFNs) are a broad class of cytokines elicited on challenge to the host defense and are essential for mobilizing immune responses to pathogens. Divided into three classes, type I, type II, and type III, all IFNs share in common the ability to evoke antiviral activities initiated by the interaction with their cognate receptors. The nine-member IFN regulatory factor (IRF) family, first discovered in the context of transcriptional regulation of type I IFN genes following viral infection, are pivotal for the regulation of the IFN responses. In this review, we briefly describe cardinal features of the three types of IFNs and then focus on the role of the IRF family members in the regulation of each IFN system.

All three classes of interferons (IFNs) are so named because of the shared property to interfere with viral replication in the host (Weissmann and Weber 1986; Taniguchi 1988; Stark et al. 1998; Levy and Garcia-Sastre 2001; Samuel 2001; Katze et al. 2002; Pestka et al. 2004; Platanias 2005; Vilcek 2006). It is perhaps not an exaggeration to say that the discovery of IFNs provided the biggest impetus for the study of all cytokine research. Type I IFNs were initially discovered as soluble factors that mediated viral interference. This is defined as the resistance to virus infection induced by a prior viral infection (Isaacs and Lindenmann 1957; Nagano and Kojima 1958). It was not until after two decades of study that type I IFN genes were cloned, se-

quenced, and formally recognized to comprise a related gene family (Knight 1975; Taniguchi et al. 1979; Maeda et al. 1980; Rubinstein et al. 1981; Weissmann and Weber 1986; Taniguchi 1988; Stark et al. 1998; Levy and Garcia-Sastre 2001; Samuel 2001; Katze et al. 2002; Pestka et al. 2004; Platanias 2005; Vilcek 2006). Type I IFNs are principally expressed by innate immune cells (Hervas-Stubbs et al. 2011). Type II IFN, represented by a single gene product, IFN-γ, was recognized for its antiviral activity induced by activated immune cells, typically by natural killer (NK) and T cells (Wheelock 1965; Gray et al. 1982). Type III IFNs (also called IFN-λ and initially also called interleukin [IL]-28 and IL-29) are restricted in their tissue distribution (e.g., are not highly expressed in hematopoietic cells), and act predominantly at epithelial surfaces (Sheppard et al. 2003; Iversen and Paludan 2010).

In this review, we first describe briefly the cardinal features of the three types of IFNs in the context of the signal transduction pathways they trigger and biological activities elicited and then focus on the regulation of these IFN responses by the IFN regulatory factor (IRF) transcription factor family. Of particular note, the discovery of so-called innate immune receptors beginning with Toll-like receptor 4 (Medzhitov et al. 1997) has provided the stimulus and context for understanding the regulatory mechanisms of IFNs by the IRF family transcription factors. Because of space limitations, we are unable to touch on various biological activities manifested by IFNs but instead point the reader to the many excellent reviews on the topic (Pestka et al. 2004; Vilcek 2006; Tamura et al. 2008; Trinchieri 2010).

THREE TYPES OF IFNs: A BRIEF OVERVIEW

The molecular characterization of IFNs began when type I IFN genes were recognized to comprise a gene family (Taniguchi et al. 1980; Pestka et al. 2004; Vilcek 2006). The two best-characterized and most broadly expressed genes of this subtype are IFN-α, encoded by more than a dozen genes, and IFN-β (single gene family) (Decker et al. 2005). Other type I IFN subtypes are known, namely, IFN-ω, and IFN-τ, but remain poorly characterized because of limited tissue expression, overlapping function with IFN-α and IFN-β, and species-to-species differences (Pestka et al. 2004; Capobianchi et al. 2015).

All type I IFN classes bind to a heterodimeric receptor composed of two chains, IFNAR1 and IFNAR2, which signal through recruitment of Janus-activated kinases (JAKs) TYK2 and JAK1, respectively (Fig. 1) (Taniguchi and Takaoka 2001; Pestka et al. 2004; Decker et al. 2005; Ivashkiv and Donlin 2014). Activation of these JAKs causes tyrosine phosphorylation of signal transducers and activators of transcription (STAT)1 and STAT2, which in turn leads to the formation of a trimeric complex, called ISGF3 (IFN-stimulated gene [ISG] factor 3) that consists of STAT1, STAT2, and IRF9 (Taniguchi and Takaoka 2001; Pestka et al. 2004; Decker et al. 2005; Ivashkiv and Donlin 2014). ISGF3 is a transcriptional activator that on translocation to the nucleus binds IFN-stimulated response elements (ISREs) in gene promoters of IFN-inducible genes (ISGs) to initiate gene transcription. On the other hand, IRF2, localized in the nucleus, functions as a transcriptional attenuator of ISGF3-mediated transcriptional activation; hence, absence of IRF2 results in excessive type I IFN signaling (Hida et al. 2000). IFNAR activation also results in the activation of STAT1 homodimers that bind and activate γ-activated sequence (GAS) (IFN-γ activated sequence) motifs, leading to the induction of gene transcription (Decker et al. 2005). Although less well characterized, type I IFN signaling may also induce signaling of mitogen-activated protein kinase (MAPK)/c-Jun amino-terminal kinase (JNK) pathways (Platanias 2005).

Type II IFN or IFN-γ is structurally unrelated to the other two classes of IFN genes and is best known as a critical cytokine secreted during activated NK- and T-cell responses (Wheelock 1965; Gray et al. 1982). Further, IFN-γ binds a different cell-surface receptor consisting of IFNGR1 and IFNGR2 subunits, which in turn associate with JAK1 and JAK2, respectively (Fig. 1). The activation of these kinases results in STAT1 homodimerization and nuclear translocation that targets GAS DNA sequences. It is worth noting that ISGF3 is also activated by the IFN-γ signaling, albeit weakly (Takaoka et al. 2000). It was also shown that effective IFN-γ signaling is contingent on a weak type I IFN signaling caused by a low constitutive production of type I IFNs (Takaoka et al. 2000).

Type III IFN or IFN-λs are the least characterized IFN family. Structurally related to type I IFNs and to the IL-10 family (Pestka et al. 2004; Uze and Monneron 2007; Lazear et al. 2015), this family includes IFN-λ1 (IL-28A), IFN-λ2 (IL-28B), and IFN-λ3 (IL-29). IFN-λs bind to a heterodimeric cytokine receptor composed of an IL-28R-binding chain and IL-10R2 that is shared with the IL-10 family of cytokines (Fig. 1) (Uze and Monneron 2007). Similar to

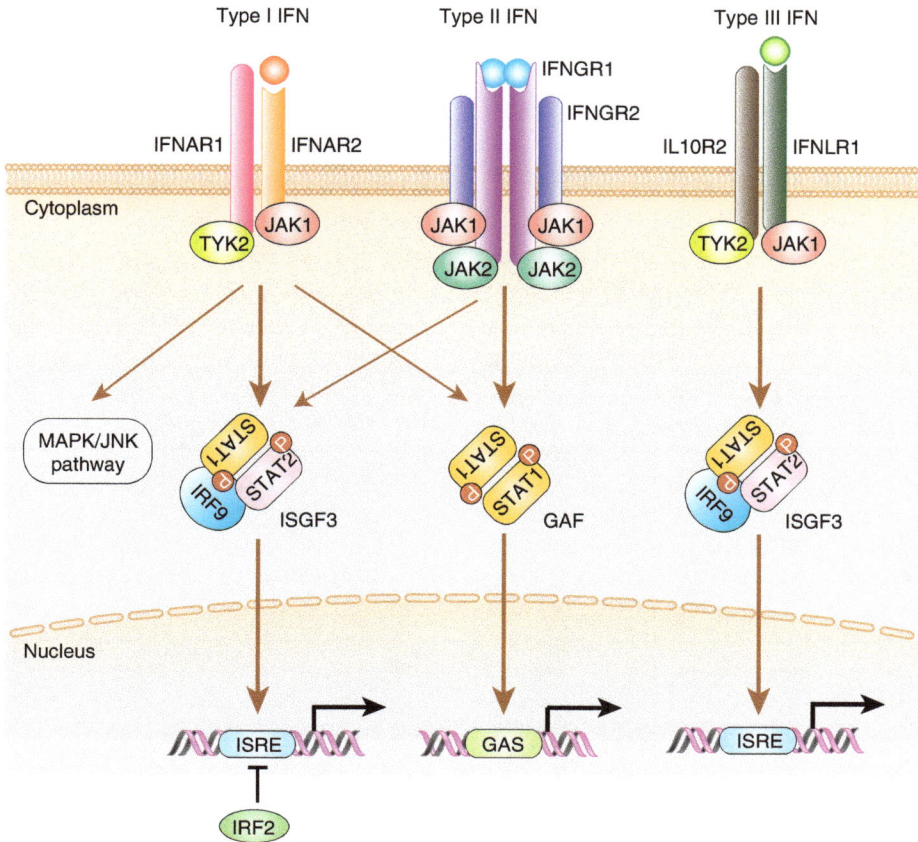

Figure 1. Signal transduction by type I, type II, and type III interferon (IFN) receptors. A schematic view of the signal transduction pathways for the three types of IFN is shown. Type I IFN binds IFNAR2, leading to the subsequent recruitment of IFNAR1. IFN-β also forms a stable complex with IFNAR1 in an IFNAR2-indepe-nent manner, whereas IFN-α does not (de Weerd et al. 2013). The binding of type I IFN induces formation of a receptor complex between IFNAR-1 and IFNAR-2, leading to activation of the receptor-associated TYK2 and JAK1 kinases. This is followed by the tyrosine phosphorylation of signal transducers and activators of transcription (STAT)1 and STAT2 and, on recruitment of IFN regulatory factor (IRF)9, forms the heterotri-meric IFN-stimulated gene factor 3 (ISGF3) transcription factor complex. In addition, a STAT1 homodimer, termed IFN-γ-activated factor (GAF), is also formed. These transcriptional–activator complexes translocate into the nucleus and activate the IFN-stimulated regulatory elements (ISREs) or γ-activated sequences (GASs) promoter elements, for ISGF3, or GAF, respectively. IRF2 functions as a transcriptional attenuator of the ISGF3-mediated transcriptional activation. Type I IFN signaling may also induce signaling of mitogen-acti-vated protein kinase (MAPK)/c-Jun amino-terminal kinase (JNK) pathways. Type II IFN binds as a homo-dimer and induces dimerization of IFNGR1 subunits and recruitment of IFNGR2 subunits. This association causes the phosphorylation of JAK1 and JAK2 kinases, leading to phosphorylation of STAT1. Phosphorylated STAT1 forms the GAF complex. IFN-γ signaling also activates ISGF3, albeit weakly. Type III IFN receptor signaling cascade causes activation of JAK1 and TYK2, which causes the recruitment of STAT1 and STAT2 to form the ISGF3 transcription factor complex that binds to ISRE elements in gene promoters to induce transcription of IFN-inducible genes (ISGs).

type I IFNs, the type III IFN receptor signaling cascade causes activation of JAK1 and TYK2 to cross-tyrosine phosphorylate the receptor complex, which causes the recruitment of STAT1 and STAT2 to form the ISGF3 transcription factor complex that activates ISGs (Uze and Monneron 2007; Lazear et al. 2015).

THE IRF FAMILY TRANSCRIPTION FACTORS

The IRF family of transcription factors is best known for its involvement in the regulation of gene expression that underlies IFN responses. Because there are available numerous review articles that describe in detail the function of each (Paun and Pitha 2007; Tamura et al. 2008; Yanai et al. 2012; Mancino and Natoli 2016), only the cardinal features of the IRF family are described in brief below. There are nine IRF proteins in mammals: IRF1, IRF2, IRF3, IRF4/PIP/LSIRF/ ICSAT, IRF5, IRF6, IRF7, IRF8/ICSBP, and IRF9/ISGF3γ (Paun and Pitha 2007; Tamura et al. 2008; Yanai et al. 2012; Mancino and Natoli 2016). In addition, IRF10 was found in birds and teleost fish but it is absent humans and mice (Suzuki et al. 2011).

All IRF proteins have a conserved amino-terminal DNA-binding domain (DBD) with a wing-type helix–loop–helix structure and a motif containing five regularly spaced tryptophan residues, which resemble the Myb DBD (Veals et al. 1992; Honda et al. 2006). The DBD recognizes a consensus DNA sequence element (A/ GNGAAANNGAAACT) (Tanaka et al. 1993), termed ISRE, which is found in the gene promoters of type I IFN, type III IFN, and ISG genes. The carboxy-terminal region of IRFs contains a conserved IRF-association domain (IAD)1 or IAD2, which mediates homodimeric and heterodimeric intramolecular interactions with other IRF members, transcription factors, and/or cofactors (Takahasi et al. 2003).

REGULATION OF THE TYPE I IFN SYSTEM BY IRFs

Clearly, the IRF family members are best studied for their intimate critical involvement in type I IFN gene induction. Below, we summarize the current status of the IRF-mediated type I IFN gene induction by the activation of distinct types of innate pattern-recognition receptors, as well as IRF-mediated IFN responses.

Regulation of Type I IFNs by IRFs via Toll-Like Receptor (TLR) Signaling

TLRs were the first discovered pattern recognition receptors (PRRs) that recognize a variety of pathogen-associated molecular patterns (PAMPs) derived from pathogens to trigger the gene induction of type I IFNs and proinflammatory cytokines, resulting in the activation of immune responses (Medzhitov 2001; Kawai and Akira 2006). Thirteen TLR proteins (12 in mice and 10 in human) have been identified to date.

TLRs are localized to cell membranes on the cell surface or within endosomes. For instance, TLR3 is mainly localized to endosomes and recognizes double-stranded (ds)RNA derived from viruses or self-RNA derived from dead cells (Alexopoulou et al. 2001; Wang et al. 2004; Rudd et al. 2006), whereas TLR4 is located on the cell surface and recognizes a variety of PAMPs from bacteria or damage-associated molecular patterns (DAMPs) derived from dead cells (Kawai and Akira 2006; Andersson and Tracey 2011). Upon binding to cognate ligand, TLRs recruit adaptor proteins such as myeloid differentiation primary-response protein 88 (MyD88) and/or Toll/interleukin-1 receptor (TIR) domain-containing adaptor, including IFN-β (TRIF) to activate IRF proteins and other transcription factors. TLR3 and TLR4 induce type I IFN gene expression via signaling by TRIF-TANK (TNF receptor-associated factor [TRAF]-associated nuclear factor [NF]-κB activator)-binding kinase 1 (TBK1)-IRF3 (Fig. 2), albeit weakly as compared to retinoic acid-inducible gene I (RIG-I)-like receptors (RLRs) or stimulator of IFN genes (STING)-mediated induction (Negishi et al. 2012). Upon the recognition of their ligands, TRIF is phosphorylated and recruits downstream signaling molecules such as TRAF3, NAP1, and TBK1, resulting in the phosphorylation, dimerization, and nuclear translocation of IRF3 (Hemmi et al. 2004; McWhirter et al. 2004; Mori et al. 2004; Perry et al. 2004; Oganesyan

Cite this article as *Cold Spring Harb Perspect Biol* doi: 10.1101/cshperspect.a028423

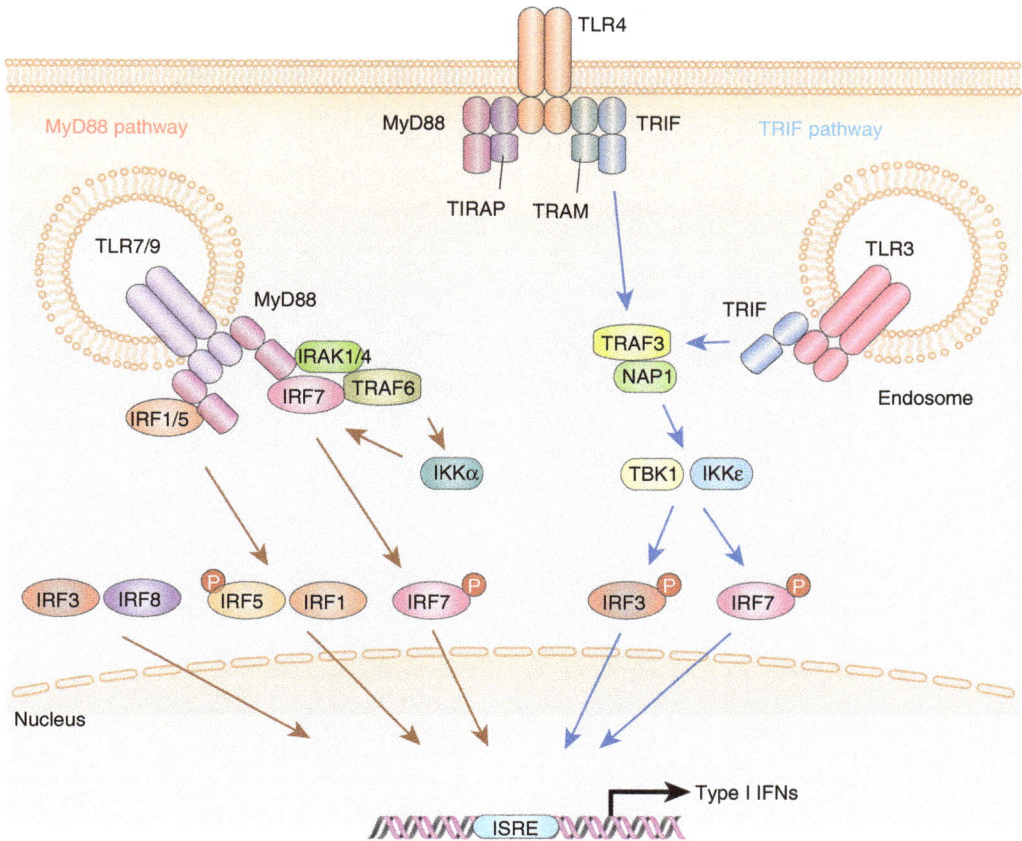

Figure 2. Schematic view of Toll-like receptor (TLR)-mediated type I interferon (IFN) gene induction by IFN regulatory factors (IRFs). The presence of RNA or double-stranded (ds)DNA in the cytosol triggers host responses via specific cytoplasmic pattern recognition receptors (PRRs). The binding of uncapped 5′-triphosphate RNA or dsRNA to the helicase domain of retinoic acid-inducible gene I (RIG-I)/melanoma differentiation-associated gene 5 (MDA5) induces the interaction between the caspase activation and recruitment domain (CARD) of RIG-I/MDA5 and the CARD-like domain of the adaptor mitochondrial antiviral signaling protein (MAVS), which is located on the mitochondrial membrane. This receptor–adaptor interaction results in the activation of tumor necrosis factor (TNF) receptor-associated factor (TRAF)-associated nuclear factor [NF]-κB activator (TANK)-binding kinase 1 (TBK1) and inhibitor of NF-κB kinase (IKK)ε. Activated TBK1 induces the phosphorylation of the specific serine residues of IRF3 and IRF7. These IRFs then translocate into the nucleus and activate the type I IFN genes. NF-κB is also activated and involved in type I IFN gene induction. In some cases, IRF5 or IRF8 participate in this IFN gene-induction pathway. dsDNA such as B-DNA is recognized by cGAS, IFI16, DDX41, and DAI. The stimulator of IFN genes (STING) adaptor protein on the endoplasmic reticulum membrane signals downstream of these DNA receptors. STING provides a scaffold for recruitment of TBK1, which phosphorylates IRF3 leading to the activation of type I IFN gene expression.

et al. 2006). Downstream of TLR4 signaling, IRF3 is critical for *Ifnb* gene induction, as expression was abolished in IRF3 knockout (*Irf3*$^{-/-}$) cells, but is essentially normal in *Irf7*$^{-/-}$ cells (Sakaguchi et al. 2003; Honda et al. 2005). On the other hand, both IRF3 and IRF7 are involved

in TLR3-dependent type I IFN gene induction (Negishi et al. 2012).

Plasmacytoid dendritic cells (pDCs) are a subset of myeloid cells that express high amounts of TLR7 and TLR9 and secrete very high levels of type I IFNs (Colonna et al. 2004). TLR7 and

TLR9 recognize single-stranded RNA (ssRNA) and nonmethylated CpG oligonucleotide DNA, respectively, to induce strong type I IFN responses (Colonna et al. 2004; Kawai and Akira 2006). In contrast to TLR3 and TLR4, TLR7- and TLR9-mediated type I IFN gene induction is entirely dependent on the MyD88 adaptor protein. Further, these receptors recruit IRF7, but not IRF3, downstream of MyD88 (Fig. 2) (Honda et al. 2004; Kawai et al. 2004) as type I IFN gene induction was completely abrogated in $Irf7^{-/-}$ pDCs but normal in $Irf3^{-/-}$ pDCs, following virus infection or treatment with synthetic ligands for TLR7 and TLR9 (Honda et al. 2005). However, in the case of a model of experimental *Listeria monocytogenes* infection in pDCs, evidence suggested that IRF3 was involved downstream of TLR9-mediated type I IFN gene induction, although the underlying molecular mechanism remains unclear (Stockinger et al. 2009). Also downstream of TLR7 and TLR9 signaling, MyD88 recruits interleukin 1 receptor associated kinase (IRAK)1 and IRAK4 protein kinases, which are critical signal transducers required for the induction of type I IFN (Honda et al. 2004; Kawai and Akira 2006). The activation of IRAK kinases activates additional protein kinase cascades, including inhibitor of NF-κB kinase (IKK)α, which is an essential kinase for the activation of IRF7 (Hoshino et al. 2006). MyD88 also recruits the adaptor protein TRAF6 (Kawai and Akira 2006). MyD88/TRAF6-dependent K63-linked ubiquitination of IRF7 is also required for TLR-mediated type I IFN gene induction in pDCs (Kawai et al. 2004).

Like IRF7, IRF5 is also involved in TLR-dependent induction of type I IFNs via recruitment of MyD88, although the strength of the type I IFN induction is relatively weak as compared to IRF7 (Fig. 2) (Takaoka et al. 2005; Yasuda et al. 2013). In contrast, IRF5's main role following MyD88-dependent signaling is the induction of proinflammatory cytokine genes such as IL-12β, IL-6, and TNF-α (Takaoka et al. 2005). Several regulatory mechanisms have been identified for modulating the TLR-dependent activation of IRF5. IKKβ phosphorylates and activates IRF5 in response to TLR stimulation (Balkhi et al. 2010; Bergstrøm et al. 2015).

TRAF6-mediated K63-linked ubiquitination is also required for IRF5 activation (Balkhi et al. 2008). In contrast, IKKα-induced IRF5 phosphorylation inhibits IRF5 ubiquitination and attenuates its transcriptional activity (Balkhi et al. 2010). In addition, IRF4 negatively regulates IRF5 activation by competing with IRF5 for its interaction with the central region of MyD88 (Negishi et al. 2005). It has also been reported that Lyn, an Src family protein tyrosine kinase, interacts with IRF5 and suppresses its transcriptional activity by inhibiting phosphorylation and ubiquitination of IRF5 (Ban et al. 2016).

In addition to IRF7 and IRF5, IRF1 also interacts with MyD88 (Negishi et al. 2006). IRF1 is not involved in type I IFN gene induction in pDCs but is critical for CpG-B oligonucleotide-mediated type I IFN gene induction in conventional DCs (cDCs) (Fig. 2) (Negishi et al. 2006). IRF1 is also required for expression of proinflammatory cytokine genes in macrophages that are augmented by type II IFN. In other words, IRF1 is strongly induced by a combination of IFN-γ and TLR signaling, which leads to the strong induction of IRF1-dependent genes such as *Il12a* and *iNos* (Negishi et al. 2006). Interestingly, IRF8 does not interact with MyD88 (Negishi et al. 2005), but is involved in the induction of type I IFN genes in TLR-stimulated pDCs (Tailor et al. 2007). Here, IRF8 binds to TRAF6 and is also involved in the activation of NF-κB for the induction of proinflammatory cytokines (Tsujimura et al. 2004; Zhao et al. 2006).

Type I IFN Gene Induction by IRFs on Innate Recognition of Cytosolic RNA

Two RNA helicase enzymes, RIG-I and melanoma differentiation-associated gene 5 (MDA5), are essential cytosolic receptors for detection of RNA, in particular uncapped 5′-triphosphate RNA and dsRNA, including poly(rI:rC) (Yoneyama et al. 2005; Kato et al. 2006). Both helicases contain a carboxy terminal DExD/H box RNA helicase domain that is responsible for detection of viral RNA, as well as two amino-terminal caspase activation and recruitment domains (CARDs) that activate downstream

Cite this article as *Cold Spring Harb Perspect Biol* doi: 10.1101/cshperspect.a028423

signaling pathways (Yoneyama et al. 2005). The adaptor molecule that links the sensing of viral RNA by RIG-I or MDA5 to downstream signaling is mitochondrial antiviral signaling protein (MAVS, also known as VISA, IPS-1, or Cardif) (Seth et al. 2005; Tamura et al. 2008). MAVS contains an amino-terminal CARD domain that mediates CARD–CARD interactions with

CARD domains of RIG-I and MDA5 to transmit downstream signaling. MAVS relays signals from RIG-I and MDA5 to TANK-binding kinase 1 (TBK1) and IKKε that are known to phosphorylate IRF3 and IRF7 (Fig. 3) (Kumar et al. 2006; Sun et al. 2006).

IRF3 and IRF7, two IRFs with the greatest sequence homology with one another, are essen-

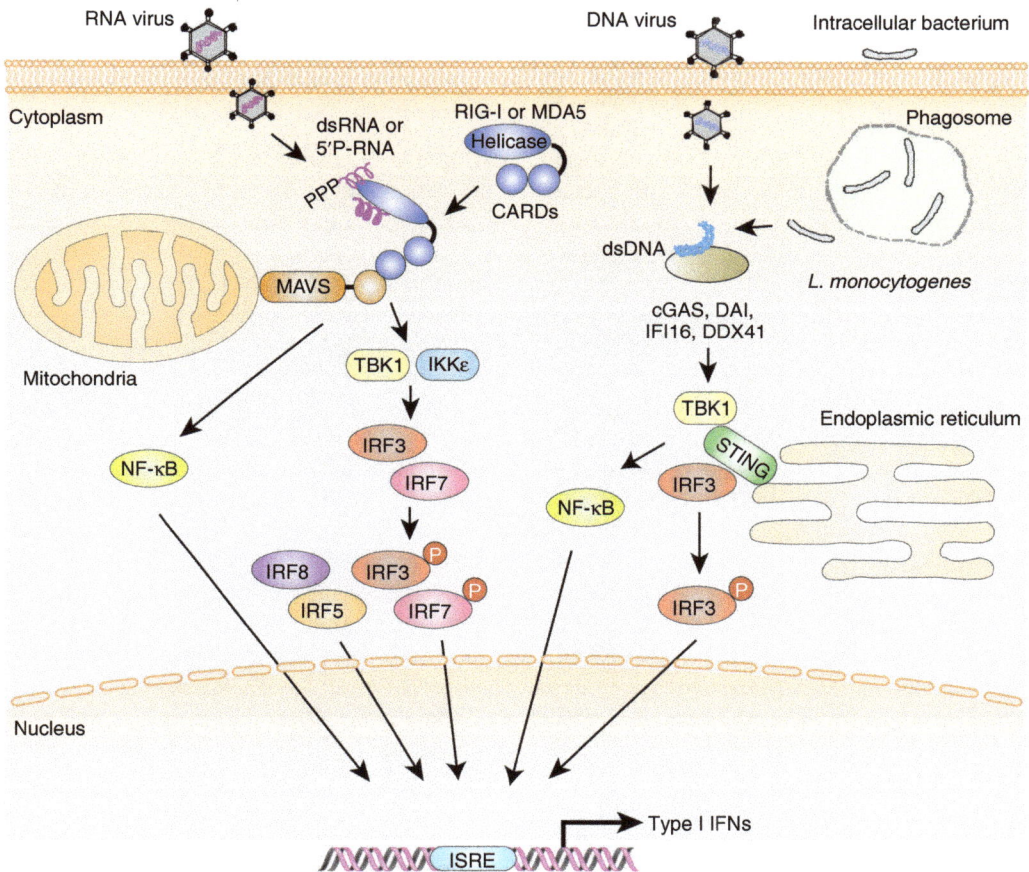

Figure 3. Schematic illustration of type I interferon (IFN) gene induction by IFN regulatory factors (IRFs) on innate recognition of cytosolic nucleic acids. Type I IFN genes, particularly IFN-β, are induced via TIR domain-containing adaptor, including IFN-β (TRIF) downstream of TLR3 and TLR4 signaling pathways, and via myeloid differentiation primary-response protein 88 (MyD88) downstream of TLR7 and TLR9 signaling pathways. Further, the TRIF pathway signals exclusively to IRF3 to induce IFN-β gene expression. IRF7 may also be involved for TLR3-, but not for retinoic acid-inducible gene I (RIG-I)-like receptor (RLR)4-mediated type I IFN gene induction. In the MyD88 pathway, IRF1, IRF5, and IRF7 form a complex with MyD88. Among MyD88-bound IRFs, IRF7 is critical for the robust induction of type I IFN in plasmacytoid dendritic cells (pDCs). IRF5 is partially involved in type I IFN gene induction in pDCs, whereas IRF1 is important for expression of type I IFNs in conventional DCs (cDCs). In contrast, IRF8 is involved in the expression of type I IFN genes in pDCs, but does not bind to MyD88. IRF3 also does not bind to MyD88, but is involved in *Listeria monocytogenes*–mediated type I IFN gene induction in pDCs.

Cite this article as *Cold Spring Harb Perspect Biol* doi: 10.1101/cshperspect.a028423

tial for the RIG-I/MDA5-mediated type I IFN gene-induction pathway. Initially, IRF3 and IRF7 reside in latent forms in the cytosol of uninfected cells. Upon virus infection, RIG-I- or MDA5-activated TBK1 phosphorylates IRF3 at Ser396, 398, 402, 404, and 405 in site 2 of the carboxy-terminal regulatory region, which alleviates autoinhibition and permits IRF3 nuclear translocation and interaction with the coactivator CBP (Servant et al. 2003; Mori et al. 2004; Honda et al. 2006; Tamura et al. 2008). Then CBP facilitates phosphorylation of Ser385 or Ser386 at site 1 within the regulatory region, permitting IRF dimerization (Mori et al. 2004; Panne et al. 2007). A similar mechanism involving IRF7 is presumed to occur. As a result, a holocomplex containing dimerized IRF3 and IRF7, either as a homodimer or heterodimer, and coactivators such as CBP or p300 is formed in the nucleus. This holocomplex binds to target ISRE DNA sequences within the promoters of type I IFN genes (Lin et al. 1998; Sato et al. 1998; Yoneyama et al. 1998; Honda et al. 2006). Additionally, IRF5 is involved in the RIG-I/MAVS signaling pathway (Yanai et al. 2007; Paun et al. 2008; Lazear et al. 2013). IRF5 translocates from the cytoplasm to the nucleus on infection by vesicular stomatitis virus (VSV) or Newcastle disease virus (NDV) (Barnes et al. 2002). Challenged with these RNA viruses, $Irf5^{-/-}$ mice show a reduction in the serum levels of type I IFN (Yanai et al. 2007). Moreover, $Irf5^{-/-}$ mice are highly vulnerable to VSV infection. Because $Irf5^{-/-}$ macrophages are defective in the production of type I IFNs by VSV, whereas $Irf5^{-/-}$ mouse embryonic fibroblasts (MEFs) are not (possibly because of a higher expression of IRF5 in hematopoietic cells), there is a cell-type-specific requirement for IRF5 (Yanai et al. 2007; Paun et al. 2008). In addition, $Irf5^{-/-}$ mice show reduced levels of proinflammatory cytokines, such as IL-6 and TNF-α on virus infection (Yanai et al. 2007; Paun et al. 2008). The precise mechanism by which IRF5 is activated by RIG-I and the nature of IRF5's contribution to the transcriptional regulation of type I IFN and proinflammatory genes remains to be clarified. Nevertheless, like IRF3 and IRF7, IRF5 can also be phosphorylated by TBK1 or IKKβ to exert its

transcriptional activities (Lin et al. 2005; Paun et al. 2008; Balkhi et al. 2010; Chang Foreman et al. 2012; Bergstrøm et al. 2015).

IRF8 is also required for type I IFN induction in virus-stimulated DCs (Tailor et al. 2007). It appears that IRF8 is involved in the transcriptional regulation of type I IFN genes; IRF8 binds to promoters of IFN-α/β genes and is required for the second amplifying phase of IFN transcription.

Type I IFN Gene Induction by IRFs on Innate Recognition of Cytosolic DNA

In addition to the cytosolic RNA-sensing mechanisms, attention has been focused on characterizing cytosolic DNA-sensing systems because they can also evoke protective and pathological immune responses. Cytoplasmic recognition of bacterial genomic DNA from *L. monocytogenes* results in IFN-β induction through the TBK1-IRF3 pathway (Stetson and Medzhitov 2006). The transfection of cells with synthetic dsDNA, such as poly(dA-dT)·poly(dT-dA) (termed B-DNA hereafter), results in the induction of type I IFN in the absence of all TLR signaling (Ishii et al. 2006). These observations indicate the presence of a cytosolic DNA sensor(s) that can independently initiate innate immune responses, including the induction of type I IFN genes. B-DNA stimulation results in the activation of IRF3 and NF-κB signaling pathways (Fig. 3) (Ishii et al. 2006; Tamura et al. 2008). A required role for IRF3 is seen in the observation that the B-DNA induction of IFN-β was abolished in $Irf3^{-/-}$ MEFs but normal in $Irf7^{-/-}$ or $Irf5^{-/-}$ MEFs. The induction of IFN-α, however, requires both IRF3 and IRF7, because both $Irf3^{-/-}$ and $Irf7^{-/-}$ MEFs showed impairment in its induction (Takaoka et al. 2007).

A candidate DNA sensor called DNA-dependent activator of IRFs (DAI), also known as DLM-1- or Z-DNA-binding protein 1 (ZBP1), has been identified and characterized (Takaoka et al. 2007). More recent studies indicate the presence of additional DNA sensors. These include cyclic CMP-AMP synthase (cGAS), IFN-inducible protein 16 (IFI16)/IFN-activated gene 204 (IFI204), and (Asp-Glu-Ala-Asp) (DEAD)

Cite this article as *Cold Spring Harb Perspect Biol* doi: 10.1101/cshperspect.a028423

box polypeptide 41 (DDX41) (Fig. 3) (Ishikawa et al. 2009; Unterholzner et al. 2010; Zhang et al. 2011; Sun et al. 2013). In addition, an adaptor protein termed STING/mediator of IRF3 activation (MITA) has been reported (Zhong et al. 2008; Ishikawa et al. 2009). The induction of dsDNA-mediated type I IFN gene expression is markedly impaired in STING/MITA-deficient cells (Zhong et al. 2008; Ishikawa et al. 2009). STING/MITA provides a scaffold to TBK1, which phosphorylates IRF3 leading to its activation and downstream activation of type I IFN gene expression.

TYPE II IFN AND IRFs IN IMMUNOLOGICAL AND INFLAMMATORY RESPONSES

IFN-γ is a multifunctional 34-kDa homodimer cytokine that is the only member of the type II class of IFNs (Gray and Goeddel 1982; Naylor et al. 1983; Farrar and Schreiber 1993). Initially, it was thought that CD4[+] T helper cell type 1 (Th1) lymphocytes, CD8[+] cytotoxic lymphocytes, and NK cells exclusively produce IFN-γ (Wheelock 1965; Gray et al. 1982; Farrar and Schreiber 1993; Schoenborn and Wilson 2007). However, recent reports provide evidence that other cells, such as B cells, NK T cells, dendritic cells, and macrophages also secrete IFN-γ. The IFN-γ gene contains binding sites for several transcription factors in its promoter region (Glimcher et al. 2004; Schoenborn and Wilson 2007). They include AP-1, CREB/ATF, NFAT, NF-κB, T-bet, Eomes, STATs, but not IRFs. Stimulation of cells via the T-cell receptor or with cytokines such as type I IFN, IL-12, or IL-18 induces IFN-γ gene expression through the activation of these transcription factors (Schoenborn and Wilson 2007). IFN-γ performs its biological functions through an IFN-γ receptor (IFNGR1/2)-mediated signal transduction pathway. The IFN-γ receptor activation causes phosphorylation of STAT1 by the receptor-associated JAK1 and JAK2 (Farrar and Schreiber 1993; Schoenborn and Wilson 2007). Phosphorylated STAT1 forms homodimers, translocates into the nucleus, and activates several IFN-γ response genes (IRGs) by binding to GAS motifs within their promoters (Fig. 1).

IFN-γ stimulation induces several IRGs, including IRF1. IRF1 is critical for the IFN-γ enhancement of a TLR-dependent gene-induction program (Negishi et al. 2006). This is underscored by the observation that Irf1[−/−] cDCs and macrophages stimulated with IFN-γ in combination with CpG are impaired in their induction of genes encoding IFN-β, iNOS, and IL-12p35 (Negishi et al. 2006). Although IFN-γ strongly induces IRF1 transcription, it is insufficient to fully activate IRF1. Rather, TLR9 engagement causes a MyD88-dependent "IRF1 licensing" event to occur in which IRF1 is posttranslationally modified to migrate into the nucleus more efficiently than non-MyD88-associated IRF1 (Negishi et al. 2006). Clarification of the modifications of IRF1 requires further investigation. Interestingly, IFN-γ promotes Th1 and attenuates IL-4-driven Th2 responses via the induction of IRF1 and IRF2, respectively (Elser et al. 2002). In addition, IRF1 is important for apoptosis that is activated or enhanced by IFN-γ (Kano et al. 1999; Tamura et al. 2008). The target gene(s) of IRF1 responsible for apoptotic responses have not been firmly identified, but may include genes encoding caspase-1, caspase-7, caspase-8, and/or TNF-related apoptosis-inducing ligand (TRAIL) (Tamura et al. 2008).

IRF8 is another member of the IRF family that is induced by IFN-γ (Kantakamalakul et al. 1999; Ozato et al. 2007; Tamura et al. 2008). IRF8 directly contributes to the induction of numerous genes in IFN-γ-treated cells (Ozato et al. 2007; Tamura et al. 2008). Moreover, numerous genes are regulated by multiprotein complexes containing IFN-γ-induced IRF8 and other transcription factors, especially PU.1, STAT1, and, in some cases, IRF1; these genes include those encoding IL-12p40, IL-12p35, gp91phox, p67phox, TLR4, TLR9, iNOS, Fcγ receptor I (FcγRI), IRF8 itself, IL-18, CCL5/RANTES, and Nramp1 (Ozato et al. 2007; Tamura et al. 2008). In addition, IRF8 has also been reported to manifest antitumor activity. IFN-γ-induced IRF8 sensitizes human colon carcinoma cells to Fas-mediated apoptosis (Liu and Abrams 2003), and IRF8 represses the PTPN13 gene that encodes a ubiquitously expressed protein-tyrosine phosphatase, namely, Fas-associated phosphatase 1 (Huang et al. 2008).

TYPE III IFN AND IRF FAMILY

As with type I IFNs, type III IFNs are induced on the recognition of a variety of PAMPs derived from viruses (Kotenko et al. 2003; Sheppard et al. 2003; Ank et al. 2006, 2008; Durbin et al. 2013; Lazear et al. 2015; Wack et al. 2015). Although nucleic acid–sensing PRRs such as TLR3, 7/8, 9, and cGAS are known to be involved in the type III IFNs gene induction, the molecular mechanism(s) remains unclear (Odendall et al. 2014; Lazear et al. 2015). On the other hand, the molecular mechanism underlying RLR-mediated induction of type III IFN genes is well studied and involves similar signaling molecules as is the case with type I IFNs (Odendall et al. 2014; Lazear et al. 2015). However, in the control of gene transcription, there are some differences between type I and III IFNs. For the induction of the *Ifnb* gene, IRFs along with AP-1 and NF-κB bind to the promoter region and cooperate to activate gene transcription, whereas IRF3, IRF7, and NF-κB, but not AP-1, are necessary for the activation of type III IFN gene transcription in spite of the existence of AP-1-binding site within the promoter regions (Osterlund et al. 2005; Onoguchi et al. 2007; Iversen and Paludan 2010; Durbin et al. 2013). Although NF-κB is involved, but not essential, in the transcription of type I IFN genes, type III IFN gene transcription is more dependent on NF-κB (Iversen and Paludan 2010). In addition, IRF1 is involved in the gene induction of type III IFNs but not type I IFNs (Odendall et al. 2014). IRF1 is activated downstream of MAVS and required for expression of type III IFN only, whereas IRF3 and IRF7 are activated in both mitochondrial and peroxisomal MAVS pathways and are involved in both type I IFN and type III IFN gene induction (Siegel et al. 2011; Odendall et al. 2014). Recently, type III IFN-specific regulators targeting IRF family proteins have been reported. For instance, Med23 interacts with IRF7 and these factors synergistically augment gene expression of type III IFNs but not type I IFNs (Griffiths et al.

2013). Further, BLIMP-1 acts as a negative regulator by binding to the IRF1-binding site on the promoter of type III IFN genes (Odendall et al. 2014). IRF1-mediated type III IFN gene is also inhibited by virus-induced activation of epidermal growth factor receptor (EGFR) signaling (Ueki et al. 2013).

CONCLUDING REMARKS

Remarkable achievements have been made in the past four decades, with the demonstration of the existence of multiple IFN types and the underlying mechanisms as to how IFN genes are regulated and show multiple biological functions. In this review, we describe how three types of IFNs are regulated by the IRF family transcription factors. Clearly, the versatile function of these IFNs is of great interest, particularly in terms of the regulation of immune responses and oncogenesis (Decker et al. 2005; Tamura et al. 2008; Ivashkiv and Donlin 2014; Zitvogel et al. 2015). Of particular note, although IFN systems are basically beneficial to the host, aberrant operation of IFNs, particularly type I IFNs, is associated with autoimmunity, indicating the pathogenic aspects of the IFN system when left uncontrolled (Vilcek 2006; Trinchieri 2010).

Another interesting issue is the evolutional aspects of the IFN systems and IRFs. It is well known that the IFN systems are found only in jawed vertebrates, whereas other cytokines such as IL-17 and IL-8 are found in jawless vertebrates (Guo et al. 2009). In the oldest literature for the type I IFN family, it was concluded that IFN-α and -β genes diverged about the time of the start for vertebrate evolution, which is the time when adaptive immunity involving T and B lymphocytes became operational (Taniguchi et al. 1980; Pancer and Cooper 2006). On the other hand, it has been shown that the IRF family members, along with the innate receptors such as TLRs, are found in animals other than vertebrates (Roach et al. 2005; Nehyba et al. 2009). It has been reported that the origin of the IRF family coincides with the appearance of multicellularity in animals (Nehyba et al. 2009; Yuan et al. 2015), and IRF genes are

 Cite this article as *Cold Spring Harb Perspect Biol* doi: 10.1101/cshperspect.a028423

present in all principal metazoan groups. Interestingly, these genes are not found in the two groups that include most of metazoan species (i.e., roundworms and insects), suggesting that they disappeared during the evolution of these animals (Nehyba et al. 2009). As such, one may envisage that the IRF family members evolved first and then these factors became operational on the acquisition of the IFN system for its regulation.

In this regard, it is also interesting that the IRF family coevolved with the NF-κB family (Takaoka et al. 2008; Nehyba et al. 2009). It is well known that these families are critical for the metazoan organism for its host defense; therefore, these families might have coevolved as the consequence of changing selection pressure in the given environments. Of note, these family members cooperate in the gene-induction programs, often by direct interactions (Honda et al. 2006; Tamura et al. 2008; Iwanaszko and Kimmel 2015). In addition, these two families also function in reproduction and development (McDonald et al. 2006; Ozato et al. 2007). As such, these families could have been acquired during the evolution of multicellular organisms so as to cooperate to ensure effective cellar responses, in particular for host defense against external threats. On the other hand, dysregulation of these families can critically contribute to the development of oncogenesis (Karin 2006; Takaoka et al. 2008; Tamura et al. 2008). Clearly, the IRF members are less characterized compared to the NF-κB members and further future studies are desired to reveal more detailed mechanisms on the regulation of IRF functions in the regulation of IFN responses as well as other biological processes.

ACKNOWLEDGMENTS

This work is supported in part by a Grant-In-Aid for Scientific Research (S) 15638461 from the Ministry of Education, Culture, Sports, Science (MEXT) and the Japan Agency for Medical Research and Development (A-MED) 15656877. The Department of Molecular Immunology is supported by BONAC Corporation and Kyowa Hakko Kirin Co., Ltd.

REFERENCES

Alexopoulou L, Holt AC, Medzhitov R, Flavell RA. 2001. Recognition of double-stranded RNA and activation of NF-κB by Toll-like receptor 3. *Nature* **413**: 732–738.

Andersson U, Tracey KJ. 2011. HMGB1 is a therapeutic target for sterile inflammation and infection. *Annu Rev Immunol* **29**: 139–162.

Ank N, West H, Bartholdy C, Eriksson K, Thomsen AR, Paludan SR. 2006. Lambda interferon (IFN-λ), a type III IFN, is induced by viruses and IFNs and displays potent antiviral activity against select virus infections in vivo. *J Virol* **80**: 4501–4509.

Ank N, Iversen MB, Bartholdy C, Staeheli P, Hartmann R, Jensen UB, Dagnaes-Hansen F, Thomsen AR, Chen Z, Haugen H, et al. 2008. An important role for type III interferon (IFN-λ/IL-28) in TLR-induced antiviral activity. *J Immunol* **180**: 2474–2485.

Balkhi MY, Fitzgerald KA, Pitha PM. 2008. Functional regulation of MyD88-activated interferon regulatory factor 5 by K63-linked polyubiquitination. *Mol Cell Biol* **28**: 7296–7308.

Balkhi MY, Fitzgerald KA, Pitha PM. 2010. IKKα negatively regulates IRF-5 function in a MyD88-TRAF6 pathway. *Cell Signal* **22**: 117–127.

Ban T, Sato GR, Nishiyama A, Akiyama A, Takasuna M, Umehara M, Suzuki S, Ichino M, Matsunaga S, Kimura A, et al. 2016. Lyn kinase suppresses the transcriptional activity of IRF5 in the TLR-MyD88 pathway to restrain the development of autoimmunity. *Immunity* **45**: 319–332.

Barnes BJ, Kellum MJ, Field AE, Pitha PM. 2002. Multiple regulatory domains of IRF-5 control activation, cellular localization, and induction of chemokines that mediate recruitment of T lymphocytes. *Mol Cell Biol* **22**: 5721–5740.

Bergstrøm B, Aune MH, Awuh JA, Kojen JF, Blix KJ, Ryan L, Flo TH, Mollnes TE, Espevik T, Stenvik J. 2015. TLR8 senses *Staphylococcus aureus* RNA in human primary monocytes and macrophages and induces IFN-β production via a TAK1-IKK β-IRF5 signaling pathway. *J Immunol* **195**: 1100–1111.

Capobianchi MR, Uleri E, Caglioti C, Dolei A. 2015. Type I IFN family members: Similarity, differences and interaction. *Cytokine Growth Factor Rev* **26**: 103–111.

Chang Foreman HC, Van Scoy S, Cheng TF, Reich NC. 2012. Activation of interferon regulatory factor 5 by site specific phosphorylation. *PLoS ONE* **7**: e33098.

Colonna M, Trinchieri G, Liu YJ. 2004. Plasmacytoid dendritic cells in immunity. *Nat Immunol* **5**: 1219–1226.

Decker T, Müller M, Stockinger S. 2005. The yin and yang of type I interferon activity in bacterial infection. *Nat Rev Immunol* **5**: 675–687.

de Weerd NA, Vivian JP, Nguyen TK, Mangan NE, Gould JA, Braniff SJ, Zaker-Tabrizi L, Fung KY, Forster SC, Beddoe T, et al. 2013. Structural basis of a unique interferon-β signaling axis mediated via the receptor IFNAR1. *Nat Immunol* **14**: 901–907.

Durbin RK, Kotenko SV, Durbin JE. 2013. Interferon induction and function at the mucosal surface. *Immunol Rev* **255**: 25–39.

Elser B, Lohoff M, Kock S, Giaisi M, Kirchhoff S, Krammer PH, Li-Weber M. 2002. IFN-γ represses IL-4 expression via IRF-1 and IRF-2. *Immunity* 17: 703–712.

Farrar MA, Schreiber RD. 1993. The molecular cell biology of interferon-γ and its receptor. *Annu Rev Immunol* 11: 571–611.

Glimcher LH, Townsend MJ, Sullivan BM, Lord GM. 2004. Recent developments in the transcriptional regulation of cytolytic effector cells. *Nat Rev Immunol* 4: 900–911.

Gray PW, Goeddel DV. 1982. Structure of the human immune interferon gene. *Nature* 298: 859–863.

Gray PW, Leung DW, Pennica D, Yelverton E, Najarian R, Simonsen CC, Derynck R, Sherwood PJ, Wallace DM, Berger SL, et al. 1982. Expression of human immune interferon cDNA in *E. coli* and monkey cells. *Nature* 295: 503–508.

Griffiths SJ, Koegl M, Boutell C, Zenner HL, Crump CM, Pica F, Gonzalez O, Friedel CC, Barry G, Martin K, et al. 2013. A systematic analysis of host factors reveals a Med23-interferon-λ regulatory axis against herpes simplex virus type 1 replication. *PLoS Pathog* 9: e1003514.

Guo P, Hirano M, Herrin BR, Li J, Yu C, Sadlonova A, Cooper MD. 2009. Dual nature of the adaptive immune system in lampreys. *Nature* 459: 796–801.

Hemmi H, Takeuchi O, Sato S, Yamamoto M, Kaisho T, Sanjo H, Kawai T, Hoshino K, Takeda K, Akira S. 2004. The roles of two IκB kinase-related kinases in lipopolysaccharide and double stranded RNA signaling and viral infection. *J Exp Med* 199: 1641–1650.

Hervas-Stubbs S, Perez-Gracia JL, Rouzaut A, Sanmamed MF, Le Bon A, Melero I. 2011. Direct effects of type I interferons on cells of the immune system. *Clin Cancer Res* 17: 2619–2627.

Hida S, Ogasawara K, Sato K, Abe M, Takayanagi H, Yokochi T, Sato T, Hirose S, Shirai T, Taki S, et al. 2000. CD8+ T cell-mediated skin disease in mice lacking IRF-2, the transcriptional attenuator of interferon-α/β signaling. *Immunity* 13: 643–655.

Honda K, Yanai H, Mizutani T, Negishi H, Shimada N, Suzuki N, Ohba Y, Takaoka A, Yeh WC, Taniguchi T. 2004. Role of a transductional–transcriptional processor complex involving MyD88 and IRF-7 in Toll-like receptor signaling. *Proc Natl Acad Sci* 101: 15416–15421.

Honda K, Yanai H, Negishi H, Asagiri M, Sato M, Mizutani T, Shimada N, Ohba Y, Takaoka A, Yoshida N, et al. 2005. IRF-7 is the master regulator of type-I interferon-dependent immune responses. *Nature* 434: 772–777.

Honda K, Takaoka A, Taniguchi T. 2006. Type I interferon gene induction by the interferon regulatory factor family of transcription factors. *Immunity* 25: 349–360.

Hoshino K, Sugiyama T, Matsumoto M, Tanaka T, Saito M, Hemmi H, Ohara O, Akira S, Kaisho T. 2006. IκB kinase-α is critical for interferon-α production induced by Toll-like receptors 7 and 9. *Nature* 440: 949–953.

Huang W, Zhu C, Wang H, Horvath E, Eklund EA. 2008. The interferon consensus sequence-binding protein (ICSBP/IRF8) represses *PTPN13* gene transcription in differentiating myeloid cells. *J Biol Chem* 283: 7921–7935.

Isaacs A, Lindenmann J. 1957. Virus interference. I: The interferon. *Proc R Soc Lond B Biol Sci* 147: 258–267.

Ishii KJ, Coban C, Kato H, Takahashi K, Torii Y, Takeshita F, Ludwig H, Sutter G, Suzuki K, Hemmi H, et al. 2006. A Toll-like receptor-independent antiviral response induced by double-stranded B-form DNA. *Nat Immunol* 7: 40–48.

Ishikawa H, Ma Z, Barber GN. 2009. STING regulates intracellular DNA-mediated, type I interferon-dependent innate immunity. *Nature* 461: 788–792.

Ivashkiv LB, Donlin LT. 2014. Regulation of type I interferon responses. *Nat Rev Immunol* 14: 36–49.

Iversen MB, Paludan SR. 2010. Mechanisms of type III interferon expression. *J Interferon Cytokine Res* 30: 573–578.

Iwanaszko M, Kimmel M. 2015. NF-κB and IRF pathways: Cross-regulation on target genes promoter level. *BMC Genomics* 16: 307.

Kano A, Haruyama T, Akaike T, Watanabe Y. 1999. IRF-1 is an essential mediator in IFN-γ-induced cell cycle arrest and apoptosis of primary cultured hepatocytes. *Biochem Biophys Res Commun* 257: 672–677.

Kantakamalakul W, Politis AD, Marecki S, Sullivan T, Ozato K, Fenton MJ, Vogel SN. 1999. Regulation of IFN consensus sequence binding protein expression in murine macrophages. *J Immunol* 162: 7417–7425.

Karin M. 2006. Nuclear factor-κB in cancer development and progression. *Nature* 441: 431–436.

Kato H, Takeuchi O, Sato S, Yoneyama M, Yamamoto M, Matsui K, Uematsu S, Jung A, Kawai T, Ishii KJ, et al. 2006. Differential roles of MDA5 and RIG-I helicases in the recognition of RNA viruses. *Nature* 441: 101–105.

Katze MG, He Y, Gale M Jr. 2002. Viruses and interferon: A fight for supremacy. *Nat Rev Immunol* 2: 675–687.

Kawai T, Akira S. 2006. TLR signaling. *Cell Death Differ* 13: 816–825.

Kawai T, Sato S, Ishii KJ, Coban C, Hemmi H, Yamamoto M, Terai K, Matsuda M, Inoue J, Uematsu S, et al. 2004. Interferon-α induction through Toll-like receptors involves a direct interaction of IRF7 with MyD88 and TRAF6. *Nat Immunol* 5: 1061–1068.

Knight E Jr. 1975. Heterogeneity of purified mouse interferons. *J Biol Chem* 250: 4139–4144.

Kotenko SV, Gallagher G, Baurin VV, Lewis-Antes A, Shen M, Shah NK, Langer JA, Sheikh F, Dickensheets H, Donnelly RP. 2003. IFN-λs mediate antiviral protection through a distinct class II cytokine receptor complex. *Nat Immunol* 4: 69–77.

Kumar H, Kawai T, Kato H, Sato S, Takahashi K, Coban C, Yamamoto M, Uematsu S, Ishii KJ, Takeuchi O, et al. 2006. Essential role of IPS-1 in innate immune responses against RNA viruses. *J Exp Med* 203: 1795–1803.

Lazear HM, Lancaster A, Wilkins C, Suthar MS, Huang A, Vick SC, Clepper L, Thackray L, Brassil MM, Virgin HW, et al. 2013. IRF-3, IRF-5, and IRF-7 coordinately regulate the type I IFN response in myeloid dendritic cells downstream of MAVS signaling. *PLoS Pathog* 9: e1003118.

Lazear HM, Nice TJ, Diamond MS. 2015. Interferon-λ: Immune functions at barrier surfaces and beyond. *Immunity* 43: 15–28.

Levy DE, Garcia-Sastre A. 2001. The virus battles: IFN induction of the antiviral state and mechanisms of viral evasion. *Cytokine Growth Factor Rev* 12: 143–156.

Lin R, Heylbroeck C, Pitha PM, Hiscott J. 1998. Virus-dependent phosphorylation of the IRF-3 transcription factor regulates nuclear translocation, transactivation potential, and proteasome-mediated degradation. *Mol Cell Biol* **18:** 2986–2996.

Lin R, Yang L, Arguello M, Penafuerte C, Hiscott J. 2005. A CRM1-dependent nuclear export pathway is involved in the regulation of IRF-5 subcellular localization. *J Biol Chem* **280:** 3088–3095.

Liu K, Abrams SI. 2003. Coordinate regulation of IFN consensus sequence-binding protein and caspase-1 in the sensitization of human colon carcinoma cells to Fas-mediated apoptosis by IFN-γ. *J Immunol* **170:** 6329–6337.

Maeda S, McCandliss R, Gross M, Sloma A, Familletti PC, Tabor JM, Evinger M, Levy WP, Pestka S. 1980. Construction and identification of bacterial plasmids containing nucleotide sequence for human leukocyte interferon. *Proc Natl Acad Sci* **77:** 7010–7013.

Mancino A, Natoli G. 2016. Specificity and function of IRF family transcription factors: Insights from genomics. *J Interferon Cytokine Res* **36:** 462–469.

McDonald DR, Janssen R, Geha R. 2006. Lessons learned from molecular defects in nuclear factor κB dependent signaling. *Microbes Infect* **8:** 1151–1156.

McWhirter SM, Fitzgerald KA, Rosains J, Rowe DC, Golenbock DT, Maniatis T. 2004. IFN-regulatory factor 3-dependent gene expression is defective in *Tbk1*-deficient mouse embryonic fibroblasts. *Proc Natl Acad Sci* **101:** 233–238.

Medzhitov R. 2001. Toll-like receptors and innate immunity. *Nat Rev Immunol* **1:** 135–145.

Medzhitov R, Preston-Hurlburt P, Janeway CA Jr. 1997. A human homologue of the *Drosophila* Toll protein signals activation of adaptive immunity. *Nature* **388:** 394–397.

Mori M, Yoneyama M, Ito T, Takahashi K, Inagaki F, Fujita T. 2004. Identification of Ser-386 of interferon regulatory factor 3 as critical target for inducible phosphorylation that determines activation. *J Biol Chem* **279:** 9698–9702.

Nagano Y, Kojima Y. 1958. Inhibition of vaccinia infection by a liquid factor in tissues infected by homologous virus. *C R Seances Soc Biol Fil* **152:** 1627–1629.

Naylor SL, Sakaguchi AY, Shows TB, Law ML, Goeddel DV, Gray PW. 1983. Human immune interferon gene is located on chromosome 12. *J Exp Med* **157:** 1020–1027.

Negishi H, Ohba Y, Yanai H, Takaoka A, Honma K, Yui K, Matsuyama T, Taniguchi T, Honda K. 2005. Negative regulation of Toll-like-receptor signaling by IRF-4. *Proc Natl Acad Sci* **102:** 15989–15994.

Negishi H, Fujita Y, Yanai H, Sakaguchi S, Ouyang X, Shinohara M, Takayanagi H, Ohba Y, Taniguchi T, Honda K. 2006. Evidence for licensing of IFN-γ-induced IFN regulatory factor 1 transcription factor by MyD88 in Toll-like receptor-dependent gene induction program. *Proc Natl Acad Sci* **103:** 15136–15141.

Negishi H, Yanai H, Nakajima A, Koshiba R, Atarashi K, Matsuda A, Matsuki K, Miki S, Doi T, Aderem A, et al. 2012. Cross-interference of RLR and TLR signaling pathways modulates antibacterial T cell responses. *Nat Immunol* **13:** 659–666.

Nehyba J, Hrdlicková R, Bose HR. 2009. Dynamic evolution of immune system regulators: The history of the interferon regulatory factor family. *Mol Biol Evol* **26:** 2539–2550.

Odendall C, Dixit E, Stavru F, Bierne H, Franz KM, Durbin AF, Boulant S, Gehrke L, Cossart P, Kagan JC. 2014. Diverse intracellular pathogens activate type III interferon expression from peroxisomes. *Nat Immunol* **15:** 717–726.

Oganesyan G, Saha SK, Guo B, He JQ, Shahangian A, Zarnegar B, Perry A, Cheng G. 2006. Critical role of TRAF3 in the Toll-like receptor-dependent and -independent antiviral response. *Nature* **439:** 208–211.

Onoguchi K, Yoneyama M, Takemura A, Akira S, Taniguchi T, Namiki H, Fujita T. 2007. Viral infections activate types I and III interferon genes through a common mechanism. *J Biol Chem* **282:** 7576–7581.

Osterlund P, Veckman V, Sirén J, Klucher KM, Hiscott J, Matikainen S, Julkunen I. 2005. Gene expression and antiviral activity of α/b interferons and interleukin-29 in virus-infected human myeloid dendritic cells. *J Virol* **79:** 9608–9617.

Ozato K, Tailor P, Kubota T. 2007. The interferon regulatory factor family in host defense: Mechanism of action. *J Biol Chem* **282:** 20065–20069.

Pancer Z, Cooper MD. 2006. The evolution of adaptive immunity. *Annu Rev Immunol* **24:** 497–518.

Panne D, McWhirter SM, Maniatis T, Harrison SC. 2007. Interferon regulatory factor 3 is regulated by a dual phosphorylation-dependent switch. *J Biol Chem* **282:** 22816–22822.

Paun A, Pitha PM. 2007. The IRF family, revisited. *Biochimie* **89:** 744–753.

Paun A, Reinert JT, Jiang Z, Medin C, Balkhi MY, Fitzgerald KA, Pitha PM. 2008. Functional characterization of murine interferon regulatory factor 5 (IRF-5) and its role in the innate antiviral response. *J Biol Chem* **283:** 14295–14308.

Perry AK, Chow EK, Goodnough JB, Yeh WC, Cheng G. 2004. Differential requirement for TANK-binding kinase-1 in type I interferon responses to Toll-like receptor activation and viral infection. *J Exp Med* **199:** 1651–1658.

Pestka S, Krause CD, Walter MR. 2004. Interferons, interferon-like cytokines, and their receptors. *Immunol Rev* **202:** 8–32.

Platanias LC. 2005. Mechanisms of type-I- and type-II-interferon-mediated signalling. *Nat Rev Immunol* **5:** 375–386.

Roach JC, Glusman G, Rowen L, Kaur A, Purcell MK, Smith KD, Hood LE, Aderem A. 2005. The evolution of vertebrate Toll-like receptors. *Proc Natl Acad Sci* **102:** 9577–9582.

Rubinstein M, Levy WP, Moschera JA, Lai CY, Hershberg RD, Bartlett RT, Pestka S. 1981. Human leukocyte interferon: Isolation and characterization of several molecular forms. *Arch Biochem Biophys* **210:** 307–318.

Rudd BD, Smit JJ, Flavell RA, Alexopoulou L, Schaller MA, Gruber A, Berlin AA, Lukacs NW. 2006. Deletion of TLR3 alters the pulmonary immune environment and mucus production during respiratory syncytial virus infection. *J Immunol* **176:** 1937–1942.

Sakaguchi S, Negishi H, Asagiri M, Nakajima C, Mizutani T, Takaoka A, Honda K, Taniguchi T. 2003. Essential role of

IRF-3 in lipopolysaccharide-induced interferon-β gene expression and endotoxin shock. *Biochem Biophys Res Commun* **306:** 860–866.

Samuel CE. 2001. Antiviral actions of interferons. *Clin Microbiol Rev* **14:** 778–809.

Sato M, Tanaka N, Hata N, Oda E, Taniguchi T. 1998. Involvement of the IRF family transcription factor IRF-3 in virus-induced activation of the IFN-β gene. *FEBS Lett* **425:** 112–116.

Schoenborn JR, Wilson CB. 2007. Regulation of interferon-γ during innate and adaptive immune responses. *Adv Immunol* **96:** 41–101.

Servant MJ, Grandvaux N, tenOever BR, Duguay D, Lin R, Hiscott J. 2003. Identification of the minimal phosphoacceptor site required for in vivo activation of interferon regulatory factor 3 in response to virus and double-stranded RNA. *J Biol Chem* **278:** 9441–9447.

Seth RB, Sun L, Ea C, Chen ZJ. 2005. Identification and characterization of MAVS, a mitochondrial antiviral signaling protein that activates NF-κB and IRF 3. *Cell* **122:** 669–682.

Sheppard P, Kindsvogel W, Xu W, Henderson K, Schlutsmeyer S, Whitmore TE, Kuestner R, Garrigues U, Birks C, Roraback J, et al. 2003. IL-28, IL-29 and their class II cytokine receptor IL-28R. *Nat Immunol* **4:** 63–68.

Siegel R, Eskdale J, Gallagher G. 2011. Regulation of IFN-λ1 promoter activity (IFN-λ1/IL-29) in human airway epithelial cells. *J Immunol* **187:** 5636–5644.

Stark GR, Kerr IM, Williams BR, Silverman RH, Schreiber RD. 1998. How cells respond to interferons. *Annu Rev Biochem* **67:** 227–264.

Stetson DB, Medzhitov R. 2006. Recognition of cytosolic DNA activates an IRF3-dependent innate immune response. *Immunity* **24:** 93–103.

Stockinger S, Kastner R, Kernbauer E, Pilz A, Westermayer S, Reutterer B, Soulat D, Stengl G, Vogl C, Frenz T, et al. 2009. Characterization of the interferon-producing cell in mice infected with *Listeria monocytogenes*. *PLoS Pathog* **5:** e1000355.

Sun Q, Sun L, Liu HH, Chen X, Seth RB, Forman J, Chen ZJ. 2006. The specific and essential role of MAVS in antiviral innate immune responses. *Immunity* **24:** 633–642.

Sun L, Wu J, Du F, Chen X, Chen ZJ. 2013. Cyclic GMP-AMP synthase is a cytosolic DNA sensor that activates the type I interferon pathway. *Science* **339:** 786–791.

Suzuki Y, Ysuike M, Kondo H, Aoki T, Hirono I. 2011. Molecular cloning and expression analysis of interferon regulatory factor 10 (IRF10) in Japanese flounder, *Paralichthys olivaceus*. *Fish Shellfish Immunol* **30:** 67–76.

Tailor P, Tamura T, Kong HJ, Kubota T, Kubota M, Borhi P, Gabriele L, Ozato K. 2007. The feedback phase of type I interferon induction in dendritic cells requires interferon regulatory factor 8. *Immunity* **27:** 228–239.

Takahasi K, Suzuki NN, Horiuchi M, Mori M, Suhara W, Okabe Y, Fukuhara Y, Terasawa H, Akira S, Fujita T, et al. 2003. X-ray crystal structure of IRF-3 and its functional implications. *Nat Struct Biol* **10:** 922–927.

Takaoka A, Mitani Y, Suemori H, Sato M, Yokochi T, Noguchi S, Tanaka N, Taniguchi T. 2000. Cross talk between interferon-γ and -α/β signaling components in caveolar membrane domains. *Science* **288:** 2357–2360.

Takaoka A, Yanai H, Kondo S, Duncan G, Negishi H, Mizutani T, Kano S, Honda K, Ohba Y, Mak TW, et al. 2005. Integral role of IRF-5 in the gene induction programme activated by Toll-like receptors. *Nature* **434:** 243–249.

Takaoka A, Wang Z, Choi MK, Yanai H, Negishi H, Ban T, Lu Y, Miyagishi M, Kodama T, Honda K, et al. 2007. DAI (DLM-1/ZBP1) is a cytosolic DNA sensor and an activator of innate immune response. *Nature* **448:** 501–505.

Takaoka A, Tamura T, Taniguchi T. 2008. Interferon regulatory factor family of transcription factors and regulation of oncogenesis. *Cancer Sci* **99:** 467–478.

Tamura T, Yanai H, Savitsky D, Taniguchi T. 2008. The IRF family transcription factors in immunity and oncogenesis. *Annu Rev Immunol* **26:** 535–584.

Tanaka N, Kawakami T, Taniguchi T. 1993. Recognition DNA sequences of interferon regulatory factor 1 (IRF-1) and IRF-2, regulators of cell growth and the interferon system. *Mol Cell Biol* **13:** 4531–4538.

Taniguchi T. 1988. Regulation of cytokine gene expression. *Annu Rev Immunol* **6:** 439–464.

Taniguchi T, Takaoka A. 2001. A weak signal for strong responses: Interferon-α/b revisited. *Nat Rev Mol Cell Biol* **2:** 378–386.

Taniguchi T, Sakai M, Fuji-Kuriyama Y, Muramatsu M, Kobayashi S, Sudo T. 1979. Construction and identification of a bacterial plasmid containing the human fibroblast interferon gene sequence. *Proc Jpn Acad* **55B:** 464–469.

Taniguchi T, Mantei N, Schwarzstein M, Nagata S, Muramatsu M, Weissmann C. 1980. Human leukocyte and fibroblast interferons are structurally related. *Nature* **285:** 547–549.

Trinchieri G. 2010. Type I interferon: Friend or foe? *J Exp Med* **207:** 2053–2063.

Tsujimura H, Tamura T, Kong HJ, Nishiyama A, Ishii KJ, Klinman DM, Ozato K. 2004. Toll-like receptor 9 signaling activates NF-κB through IFN regulatory factor-8/IFN consensus sequence binding protein in dendritic cells. *J Immunol* **172:** 6820–6827.

Ueki IF, Min-Oo G, Kalinowski A, Ballon-Landa E, Lanier LL, Nadel JA, Koff JL. 2013. Respiratory virus-induced EGFR activation suppresses IRF1-dependent interferon λ and antiviral defense in airway epithelium. *J Exp Med* **210:** 1929–1936.

Unterholzner L, Keating SE, Baran M, Horan KA, Jensen SB, Sharma S, Sirois CM, Jin T, Latz E, Xiao TS, et al. 2010. IFI16 is an innate immune sensor for intracellular DNA. *Nat Immunol* **11:** 997–1004.

Uze G, Monneron D. 2007. IL-28 and IL-29: Newcomers to the interferon family. *Biochimie* **89:** 729–734.

Veals SA, Schindler C, Leonard D, Fu XY, Aebersold R, Darnell JE Jr, Levy DE. 1992. Subunit of an α-interferon-responsive transcription factor is related to interferon regulatory factor and Myb families of DNA-binding proteins. *Mol Cell Biol* **12:** 3315–3324.

Vilcek J. 2006. Fifty years of interferon research: Aiming at a moving target. *Immunity* **25:** 343–348.

Wack A, Terczyńska-Dyla E, Hartmann R. 2015. Guarding the frontiers: The biology of type III interferons. *Nat Immunol* **16:** 802–809.

Wang T, Town T, Alexopoulou L, Anderson JF, Fikrig E, Flavell RA. 2004. Toll-like receptor 3 mediates West

Cite this article as *Cold Spring Harb Perspect Biol* doi: 10.1101/cshperspect.a028423

Nile virus entry into the brain causing lethal encephalitis. *Nat Med* **10**: 1366–1373.

Weissmann C, Weber H. 1986. The interferon genes. *Prog Nucleic Acid Res Mol Biol* **33**: 251–300.

Wheelock EF. 1965. Interferon-like virus-inhibitor induced in human leukocytes by phytohemagglutinin. *Science* **149**: 310–311.

Yanai H, Chen HM, Inuzuka T, Kondo S, Mak TW, Takaoka A, Honda K, Taniguchi T. 2007. Role of IFN regulatory factor 5 transcription factor in antiviral immunity and tumor suppression. *Proc Natl Acad Sci* **104**: 3402–3407.

Yanai H, Negishi H, Taniguchi T. 2012. The IRF family of transcription factors: Inception, impact and implications in oncogenesis. *Oncoimmunology* **1**: 1376–1386.

Yasuda K, Nündel K, Watkins AA, Dhawan T, Bonegio RG, Ubellacker JM, Marshak-Rothstein A, Rifkin IR. 2013. Phenotype and function of B cells and dendritic cells from interferon regulatory factor 5-deficient mice with and without a mutation in DOCK2. *Int Immunol* **25**: 295–306.

Yoneyama M, Suhara W, Fukuhara Y, Fukuda M, Nishida E, Fujita T. 1998. Direct triggering of the type I interferon system by virus infection: Activation of a transcription factor complex containing IRF-3 and CBP/p300. *EMBO J* **17**: 1087–1095.

Yoneyama M, Kikuchi M, Matsumoto K, Imaizumi T, Miyagishi M, Taira K, Foy E, Loo YM, Gale M Jr, Akira S, et al. 2005. Shared and unique functions of the DExD/H-box helicases RIG-I, MDA5, and LGP2 in antiviral innate immunity. *J Immunol* **175**: 2851–2858.

Yuan S, Zheng T, Li P, Yang R, Ruan J, Huang S, Wu Z, Xu A. 2015. Characterization of amphioxus IFN regulatory factor family reveals an archaic signaling framework for innate immune response. *J Immunol* **195**: 5657–5666.

Zhang Z, Yuan B, Bao M, Lu N, Kim T, Liu YJ. 2011. The helicase DDX41 senses intracellular DNA mediated by the adaptor STING in dendritic cells. *Nat Immunol* **12**: 959–965.

Zhao J, Kong HJ, Li H, Huang B, Yang M, Zhu C, Bogunovic M, Zheng F, Mayer L, Ozato K, et al. 2006. IRF-8/interferon (IFN) consensus sequence-binding protein is involved in Toll-like receptor (TLR) signaling and contributes to the cross-talk between TLR and IFN-γ signaling pathways. *J Biol Chem* **281**: 10073–10080.

Zhong B, Yang Y, Li S, Wang YY, Li Y, Diao F, Lei C, He X, Zhang L, Tien P, et al. 2008. The adaptor protein MITA links virus-sensing receptors to IRF3 transcription factor activation. *Immunity* **29**: 538–550.

Zitvogel L, Galluzzi L, Kepp O, Smyth MJ, Kroemer G. 2015. Type I interferons in anticancer immunity. *Nat Rev Immunol* **15**: 405–414.

Interferon γ and Its Important Roles in Promoting and Inhibiting Spontaneous and Therapeutic Cancer Immunity

Elise Alspach, Danielle M. Lussier, and Robert D. Schreiber

Department of Pathology and Immunology, Washington University School of Medicine, St. Louis, Missouri 63110

Correspondence: rdschreiber@wustl.edu

Originally identified in studies of cellular resistance to viral infection, interferon (IFN)-γ is now known to represent a distinct member of the IFN family and plays critical roles not only in orchestrating both innate and adaptive immune responses against viruses, bacteria, and tumors, but also in promoting pathologic inflammatory processes. IFN-γ production is largely restricted to T lymphocytes and natural killer (NK) cells and can ultimately lead to the generation of a polarized immune response composed of T helper (Th)1 CD4$^+$ T cells and CD8$^+$ cytolytic T cells. In contrast, the temporally distinct elaboration of IFN-γ in progressively growing tumors also promotes a state of adaptive resistance caused by the up-regulation of inhibitory molecules, such as programmed-death ligand 1 (PD-L1) on tumor cell targets, and additional host cells within the tumor microenvironment. This review focuses on the diverse positive and negative roles of IFN-γ in immune cell activation and differentiation leading to protective immune responses, as well as the paradoxical effects of IFN-γ within the tumor microenvironment that determine the ultimate fate of that tumor in a cancer-bearing individual.

THE STRUCTURE AND BIOSYNTHESIS OF INTERFERON γ (IFN-γ)

The IFN-γ Gene

The mouse and human IFN-γ proteins are encoded by single 6-kb genes comprised of four exons and three introns located on chromosomes 10 and 12, respectively (Trent et al. 1982; Naylor et al. 1983, 1984). The 2.3 kb of DNA upstream, 1 kb of DNA downstream, and portions within of the human IFN-γ gene have all been characterized as having regulatory elements that control expression of IFN-γ (Young et al. 1989). When expressed in transgenic mice, the 5′ promoter elements were shown to have cell-type-specific regulatory capabilities in driving expression of an IFN-γ reporter construct in CD4$^+$ versus CD8$^+$ T cells. Binding sites for many transcription factors have been located in these regulatory elements, including Fos, Jun, GATA3, NFAT, nuclear factor (NF)-κB, and T-bet (Penix et al. 1993, 1996; Sica et al. 1997; Szabo et al. 2000). Despite the structural similarities of the mouse and human IFN-γ genes, they share only 60% homology at the genetic sequence level.

The IFN-γ Protein

The IFN-γ polypeptides from both mice and humans are approximately 160 amino acids long, but share only 40% sequence homology, a result that explains the strict species specificity displayed by the mouse and human proteins for their respective target cells (Gray and Goeddel 1982, 1983; Farrar and Schreiber 1993). Biologically active IFN-γ is a noncovalently linked homodimer consisting of two identical 17 kDa polypeptides. The individual polypeptides associate in a helical, antiparallel fashion to form a compact and globular molecule displaying a twofold axis of symmetry (Ealick et al. 1991; Farrar and Schreiber 1993). The mature molecule is glycosylated such that each mature subunit displays a molecular weight of 25 kDa yielding a 50 kDa homodimer (Ealick et al. 1991). IFN-γ glycosylation does not directly contribute to the functional activity of the mature protein but plays a role in protecting it from proteolytic degradation. The finding that the IFN-γ homodimer does not contain any covalent attachments helped explain the experimental findings that IFN-γ is particularly sensitive to extremes of both temperature and pH. Both the amino and carboxyl termini of the IFN-γ polypeptide are required for binding to the IFN-γ receptor (IFNGR) and induction of biological responses. Proteolytic cleavage studies of IFN-γ revealed that amino acid residues 1–10 on the mature amino terminus are critically important for biologic activity of the molecule (Zavodny et al. 1988; Hogrefe et al. 1989). Likewise, enzymatic removal of the carboxyl-terminal 14 amino acids (residues 129–143) results in significant activity reduction (Arakawa et al. 1986; Leinikki et al. 1987). The results of these truncation experiments were further bolstered by results from experiments using blocking antibodies directed against the amino and carboxyl termini of the IFN-γ polypeptide; antibody-mediated blocking of either the amino or carboxyl termini resulted in inhibition of IFN-γ binding to IFNGR-expressing cells and IFN-γ signaling (Johnson et al. 1982; Russell et al. 1986; Jarpe and Johnson 1990). The observations that IFN-γ is a homodimer and that both the amino and carboxyl termini of the molecule are required for biological activity led to the proposal that each IFN-γ molecule interacts with two IFNGRs, which was not fully validated until the structure of the IFNGR was elucidated.

Cellular Production of IFN-γ

IFN-γ has been shown to have profound effects on both innate and adaptive immunity that facilitate host protection. However, aberrant production of IFN-γ has been associated with disease pathology, including chronic autoimmune diseases such as inflammatory bowel disease and diabetes. Therefore, the initiation, timing, and amount of IFN-γ produced must be closely regulated (Schoenborn and Wilson 2007).

Natural killer (NK) cells and natural killer T (NKT) cells are the major IFN-γ-producing cells of the innate response. Mature NK cells contain epigenetic marks that open the IFNG locus leading to their constitutive expression of IFN-γ transcripts, which allows for rapid production and secretion of IFN-γ during infection or cancer (Stetson et al. 2003; Mah and Cooper 2016). NK cells can be stimulated to secrete IFN-γ through both receptor-mediated pathways and cytokine-mediated pathways. Receptor-mediated NK cell activation is the result of stronger signaling through the NK cell–activating receptors than through the NK cell inhibitory receptors (Wang and Yokoyama 1998; Smith et al. 2001). NK cell–activating receptor signaling is mediated through immunoreceptor tyrosine-based activating motifs (ITAMs), which phosphorylate members of the Src family of protein tyrosine kinases. Src activation eventually leads to the activation of mitogen-activated protein kinases (MAPKs) such as extracellular signal-regulated kinase (ERK) and p38, which are critical for receptor-mediated IFN-γ release and, in turn, activate transcription factors like Fos and Jun (Schoenborn and Wilson 2007).

Cytokine-mediated NK cell activation is regulated by interleukin (IL)-12, which was first shown in experiments using *Listeria monocytogenes*–infected severe combined immunodeficiency (SCID) mice (Tripp et al. 1993). During early infection, macrophages secrete IL-12 that, after interacting with its receptor on NK cells, results in activation of signal transducers and activators of transcription (STAT)4, and eventual activation of NF-κB. NK cell–derived IFN-γ promotes activation of macrophages leading to their increased secretion of IL-12 and thereby establishes a positive feedback loop that results in a cytokine milieu within the tissue or tumor microenvironment that favors T helper (Th)1-type immune responses (Hoshino et al. 1999; Afkarian et al. 2002; Tassi and Colonna 2005; Tassi et al. 2005). Conversely, transforming growth factor (TGF)-β and IL-10 can inhibit the production of IFN-γ by NK cells via SMAD3/4 phosphorylation, leading to inhibition of transcription from the *Ifng* promoter (Li et al. 2006). Indeed, when SCID splenocytes were infected with *L. monocytogenes* in the presence of blocking antibodies to IL-10 IFN-γ secretion was enhanced, although antibodies that neutralized IL-12 led to a complete ablation of IFN-γ production (Tripp et al. 1993).

NKT cells are activated via their T-cell receptor (TCR), which recognizes CD1d-presented lipid antigens. NKT cells are uniquely capable of producing both Th1- and Th2-type cytokines, although the mechanism by which they do this is still unknown. Although it is thought that IL-12 and IL-18 can enhance IFN-γ production by NKT cells, IL-4 receptor- and IL-12 receptor-deficient mice have fully functional NKT cells (Matsuda et al. 2006; Nagarajan and Kronenberg 2007).

Subsets of CD4$^+$ T cells and CD8$^+$ T cells are the producers of IFN-γ in the adaptive immune response. Although all CD8$^+$ T cells are programmed to produce IFN-γ on activation, only Th1-differentiated CD4$^+$ T cells produce IFN-γ efficiently. Therefore, the cytokine milieu present at the time of initial infection can determine the fate of CD4$^+$ T cells and controls the amount of IFN-γ-producing CD4$^+$ T cells. Th1 CD4$^+$ and CD8$^+$ T cells begin to rapidly produce and secrete IFN-γ after activation, proliferation, and differentiation. IFN-γ secretion by T cells is transient, reaching its peak at approximately 24 h after cell stimulation and then returning to baseline levels (Farrar and Schreiber 1993).

IFN-γ secretion from both CD4$^+$ and CD8$^+$ T cells can be mediated by both receptor- and cytokine-dependent mechanisms, and the cellular pathways that mediate these two mechanisms are similar to the pathways already described for NK cell activation. Following the interaction between the TCR and its cognate antigen, a variety of Src family tyrosine kinases are activated that eventually result in activation of MAPKs such as p38 and ERK, which in turn leads to the activation of transcription factors Jun and Fos that transcriptionally up-regulate IFN-γ (Kane et al. 2000). IFN-γ production by previously differentiated Th1 CD4$^+$ and cytolytic CD8$^+$ T cells can also occur via cytokines such as IL-12 and IL-18 (Okamura et al. 1995; Yang et al. 1999). Interestingly, the treatment of Th1-differentiated CD4$^+$ T cells in vitro with the combination of IL-12 and IL-18 was capable of eliciting IFN-γ secretion even in the absence of TCR activation by antigen or antibodies against CD3, establishing that cytokine-mediated activation of T cells can occur in a TCR-independent manner (Yang et al. 1999). Th1 CD4$^+$ T cells not only produce IFN-γ but are made to do so in part by IFN-γ itself. The positive feedback loop established by IFN-γ signaling during differentiation and activation of CD4$^+$ T cells is discussed elsewhere (Negishi et al. 2017; Wallach 2017).

THE STRUCTURE AND FUNCTION OF THE IFN-γ RECEPTOR

The IFN-γ receptor (IFNGR) is a member of the class II cytokine receptor family and is expressed at moderate levels on nearly every cell type, except mature erythrocytes. The receptor is composed of two subunits termed IFNGR1 and IFNGR2, which show high degrees of species specificity in their capacity to interact with one another as well as their cognate ligand. Experiments using mouse–human cell fusions indicated that mouse cell reactivity against human IFN-γ was not achieved until the genes for both receptor subunits were present in the mouse cell at the same time (Jung et al. 1987; Hibino et al. 1991; Bach et al. 1997). The ~25 kb and ~17 kb genes for IFNGR1 and IFNGR2, respectively, are both composed of seven exons. Although the gene encoding IFNGR1 is constitutively expressed, IFNGR2 expression can be regulated in certain cell populations based on their state of differentiation (Pfizenmaier et al. 1988; Bach et al. 1995, 1997; Pernis et al. 1995).

The IFNGR1 protein is approximately 90 kDa, is responsible for the majority of ligand binding, and also plays critical roles in signal transduction (Fountoulakis et al. 1991; Farrar and Schreiber 1993). In the human IFNGR1 protein, amino acids 1–17 comprise the signal sequence, 18–245 comprise the extracellular domain, 246–266 comprise the transmembrane domain, and 267–489 create the intracellular domain. The IFNGR2 protein is approximately 62 kDa, enhances the affinity of ligand binding to receptor-expressing cells, and is also required for signal transduction (Marsters et al. 1995). The mature human IFNGR2 protein is composed of an extracellular domain consisting of amino acids 28–247, a transmembrane domain made up of amino acids 248–268 and an intracellular domain produced from amino acids 269–337. Both subunits of the IFNGR are translated in the endoplasmic reticulum and are extensively glycosylated before expression on the plasma membrane (Hershey and Schreiber 1989). Immunoprecipitation experiments of untreated or IFN-γ-treated cells indicate that IFNGR1 and IFNGR2 are not tightly associated before ligand binding, but gain signaling competency on interaction with IFN-γ (Bach et al. 1996).

Neither IFNGR subunit possesses intrinsic tyrosine phosphorylation activity, yet IFN-γ signaling occurs as a consequence of tyrosine kinase activation. Detailed deletion mutagenesis experiments aimed at elucidating the functionally important regions of IFNGR1 revealed that the vast majority of the IFNGR1 extracellular domain is required for appropriate IFN-γ binding (Fountoulakis et al. 1991). Alanine-scanning and deletion mutagenesis analyses performed on the IFNGR1 intracellular domain revealed that it contains two functionally important sequences. The first is comprised of the membrane proximal amino acids $_{266}LPKS_{269}$, with the proline at position 267 playing the most important

role in receptor function (Farrar et al. 1991; Kaplan et al. 1996). This sequence represents the constitutive point of attachment on IFNGR1 for Janus kinase (JAK)1 (Igarashi et al. 1994; Kaplan et al. 1996). The second functionally critical IFNGR1 domain is a $_{440}YDKPH_{444}$ motif that resides in a membrane distal position closer to the carboxyl terminus of the IFNGR1 polypeptide. Within this sequence, Y440, D441, and H444 are the functionally important amino acids (Farrar et al. 1991, 1992). The tyrosine at position 440 is phosphorylated on ligation of the intact receptor and thereby nucleates the induced site of attachment on IFNGR1 for the SH2 domain on the latent, cytosolic transcription factor STAT1 (Greenlund et al. 1995). The IFNGR2 intracellular domain contains two closely spaced sequences located in the membrane-proximal portion of the polypeptide that are required for IFN-γ-induced biologic responses. Experiments involving both alanine-scanning mutagenesis and truncation of the IFNGR2 intracellular domain defined these two regions as $_{263}PPSIP_{267}$ and $_{270}IEEYL_{274}$ (Kotenko et al. 1995; Bach et al. 1996). These sequences form the constitutive site of attachment of the JAK2 Janus kinase on IFNGR2 (Sakatsume et al. 1995). As opposed to the functional domains within the IFNGR1 intracellular domain, no single amino acid in either of the IFNGR2 intracellular regions was observed as having a dominant functional role.

The observation that IFN-γ is a homodimeric molecule suggested that it would be able to bind to two IFNGRs. This hypothesis was supported by stoichiometry experiments showing that the IFNGR–IFN-γ complex was formed at a 2:1 ratio, indicating that two receptors are bound to every one IFN-γ homodimeric molecule (Fountoulakis et al. 1992; Greenlund et al. 1993). Additionally, expression of dominant-negative signaling mutants of IFNGR1 completely inhibits IFN-γ signaling (Dighe et al. 1993). These results suggest that functional IFN-γ signaling requires multiple heterodimeric IFNGRs working together. In 1995, the structure of IFNGR1 bound to IFN-γ was solved by X-ray crystallography. Researchers observed that, just as had been hypothesized by the previously described experiments, two IFNGR1 molecules were bound to each homodimeric molecule of IFN-γ (Walter et al. 1995). Therefore, the structure of the complete, biologically active IFNGR–IFN-γ complex consists of two IFNGR1 molecules and two IFNGR2 molecules, with one of each receptor subunit bound to each end of the IFN-γ homodimer (Fig. 1). The importance of IFNGR dimerization on ligand binding and the functionally critical regions of both the IFNGR1 and IFNGR2 intracellular domains became apparent on the elucidation of the signaling pathway activated by IFN-γ.

THE IFN-γ SIGNALING PATHWAY

The signaling pathways activated downstream from the IFNGR rely almost entirely on two families of proteins: protein tyrosine kinases that are members of the JAK family and one member of the STAT family of transcription factors (Darnell et al. 1994). This signaling pathway, known as the JAK-STAT signaling pathway, was largely elucidated by studying signaling through the distinct receptors for IFN-γ and type I IFNs. Additionally, amino acid sequence comparison of the IFNGR intracellular domain and the intracellular domain of the IL-6 receptor revealed similar motifs known to be necessary for JAK binding (Murakami et al. 1991; Tanner et al. 1995). Researchers used coimmunoprecipitation experiments to identify that JAK1 binds IFNGR1, and alanine-scanning mutagenesis identified its binding site as the functionally important $_{266}$LPKS$_{269}$ motif previously described (Igarashi et al. 1994; Kaplan et al. 1996). Similar experiments were performed that showed that JAK2 binds the IFNGR2 intracellular domain at $_{263}$PPSIPLQIEEYL$_{274}$, a region encompassing the functionally important motifs of IFNGR2 (Kotenko et al. 1995; Sakatsume et al. 1995; Bach et al. 1996).

Before encountering IFN-γ, JAK1 and JAK2 are constitutively bound to IFNGR1 and IFNGR2, respectively, but are not enzymatically active. On ligation, the IFNGR1 and IFNGR2 subunits form strongly associated receptor molecules (Bach et al. 1996). This reorientation of IFNGR1 and IFNGR2 bring their associated JAK proteins

into close enough proximity to one another, allowing for their auto- and transphosphorylation, leading to the activation of both JAK1 and JAK2 (Bach et al. 1997). The functional role of the activated JAKs was revealed in part by the observation that mutagenesis of a single tyrosine, Y440, in the intracellular domain of IFNGR1 completely inhibited IFN-γ signaling. Furthermore, mutagenesis of Y440 also resulted in loss of IFN-γ-mediated STAT1 activation (Greenlund et al. 1994; Sakatsume et al. 1995). Finally, STAT1 could be immunoprecipitated with short peptide sequences mimicking the region around phosphorylated Y440 (Greenlund et al. 1994, 1995) but not with nonphosphorylated versions of the peptide. Together, these observations provided the experimental evidence supporting the concept that ligand-dependent JAK1 and JAK2 activation following IFNGR1 and IFNGR2 reorientation leads to phosphorylation of Y440, thereby providing a docking site for STAT1 on the IFNGR (Fig. 1).

STAT1 binds to the IFNGR through its SH2 domain. Indeed, antibodies that bind to the SH2 domain of STAT1 prevent the transcription factor from interacting with phosphopeptides that mimic phosphorylated IFNGR1 (Greenlund et al. 1995). On binding to IFNGR1, STAT1 is phosphorylated by JAK1 and JAK2 on tyrosine 701 (Bach et al. 1997). Further phosphorylation of STAT1 occurs on serine 727 by serine kinases, and is required for complete activation of STAT1 (Wen et al. 1995; Nguyen et al. 2001). Because of the dimeric nature of the ligand-activated IFNGR, two STAT1 molecules are activated for each molecule of IFN-γ bound (Fig. 1). The two STAT1 molecules form a homodimer through reciprocal interactions between the SH2 domain of one STAT1 protein and the phosphorylated tyrosine at position 701 of the second STAT1 and thereby dissociate from their receptor-binding site (Ramana et al. 2002). Oligomerized STAT1 translocates to the nucleus where it binds to specific DNA sequences called "γ-activated sites" (GASs) and performs its role as a transcriptional activator. GAS elements have the consensus sequence TTNCNNNAA, which was identified after sequence analysis of promoter regions from

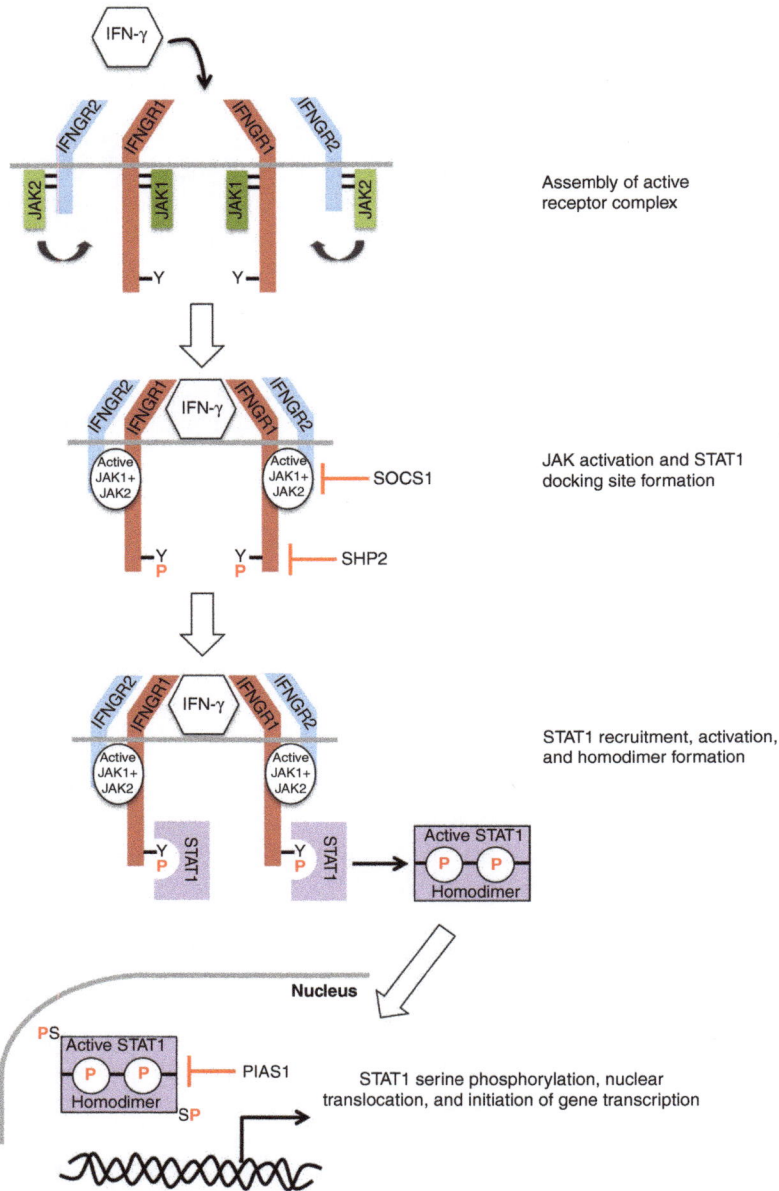

Figure 1. Assembly of the interferon (IFN)-γ receptor (IFNGR) and activation of Janus kinase (JAK)-signal transducers and activators of transcription (STAT) signaling. Binding of IFN-γ to the IFNGR complex results in tight association of IFNGR1 and IFNGR2 and a reorientation of the intracellular domains of these subunits, thereby bringing their associated JAK1 and JAK2 proteins into close proximity to facilitate auto- and trans-phosphorylation and enzymatic activation. The activated JAK proteins then phosphorylate the intracellular domain of IFNGR1 on tyrosine 440 to create a binding site for STAT1. On JAK-mediated phosphorylation, STAT1 dissociates from its IFNGR1 tether, forms a homodimer via reciprocal interactions between the SH2 domain of one STAT1 molecule and the phosphorylated tyrosine-701 of the other, and translocates to the nucleus where it binds to γ-activated site (GAS) elements and promotes gene transcription. The multiple levels of JAK-STAT pathway negative regulation are indicated. SOCS, Suppressors of cytokine signaling; PIAS, protein inhibitor of activated STATs. (From Bach et al. 1997; adapted, with permission, from the authors.)

genes activated rapidly on treatment of cells with IFN-γ (Darnell et al. 1994; Schindler and Darnell 1995). STAT1 can associate with several other proteins to enhance its transcriptional activity, including the histone acetyl transferase CBP/p300 and breast cancer susceptibility gene 1 (BRCA1) (Zhang et al. 1996; Ouchi et al. 2000). The genes that are targeted by the IFN-γ signaling pathway vary based on the cell type being targeted.

There are several molecular mechanisms that control signaling through the JAK-STAT pathway (Fig. 1). The most straightforward inhibitory pathway was first identified after the observation that phosphorylation of the IFNGR, JAK, and STAT are transient. A family of SH2-domain containing phosphatases called SHP proteins are constitutively associated with the IFNGR complex and induce removal of the many activating phosphorylated tyrosine residues responsible for mediating IFN-γ signaling (You et al. 1999). Indeed, treatment of SHP2$^{-/-}$ cells with IFN-γ results in heightened activation of JAK1 and STAT1 and ultimately results in cell death. In addition to the SHP phosphatases, there are several JAK-STAT pathway inhibitors that are up-regulated and function to negatively regulate IFN-γ signaling. The suppressors of cytokine signaling (SOCS) family of proteins is not expressed in resting cells but is rapidly up-regulated transcriptionally after exposure to IFN-γ (Endo et al. 1997; Naka et al. 1997; Starr et al. 1997). SOCS proteins mediate not only the inhibition of JAK proteins but also their ubiquitination and destruction, and SOCS1$^{-/-}$ mice have elevated serum levels of IFN-γ and die shortly after birth. The lethality of SOCS1 absence can be ameliorated by treatment with IFN-γ-neutralizing antibodies (Alexander et al. 1999; Marine et al. 1999; Naka et al. 2001). Finally, the protein inhibitor of activated STATs (PIAS) family of proteins is involved in directly interfering with STAT-mediated gene transcription. PIAS1 inhibits transcription downstream from STAT1 by binding to and preventing STAT1 from interacting with DNA (Liu et al. 1998).

Although the vast majority of IFN-γ signaling is mediated through the JAK-STAT pathway, other signaling pathways are activated downstream from the IFNGR. Examples of such pathways include the MAPK Pyk2 and ERK1/2 (Ramana et al. 2002). Microarray analysis has also identified genes that are up-regulated independently of STAT1 activation. For example, the CCAAT-enhancer-binding protein C/EBPβ is transcriptionally up-regulated on IFN-γ treatment in both wild-type and STAT1-null cells (Roy et al. 2000; Ramana et al. 2002). Likewise, macrophage inflammatory proteins MIP-1α and MIP-1β are up-regulated by IFN-γ regardless of STAT1 expression (Gil et al. 2001; Ramana et al. 2002).

THE ROLE OF IFN-γ IN IMMUNE CELL ACTIVATION

Macrophages

Macrophages serve a crucial role in both responding to pathogens and orchestrating the adaptive immune response through antigen processing and presentation. They also are major targets of IFN-γ's actions. IFN-γ was the first identified macrophage-activating factor (MAF), responsible for rendering these cells capable of increased proinflammatory cytokine synthesis, enhanced phagocytosis, enhanced antigen-presenting capacity, and enhanced nonspecific cytocidal activity toward microbial pathogens and tumors (Schreiber et al. 1985). The importance of IFN-γ's role in activating macrophages in vivo was first shown by experiments using IFN-γ-neutralizing antibodies. Mice treated with IFN-γ-neutralizing antibodies are unable to resist infection with sublethal doses of a variety of bacterial pathogens, showing the critical role that IFN-γ plays in inducing microbicidal activity in macrophages (Buchmeier and Schreiber 1985; Nacy et al. 1985; Suzuki et al. 1988).

The classically activated M1 macrophage relies on IFN-γ to activate expression of inducible nitric oxide synthase (iNOS), which catalyzes the production of nitric oxide from L-arginine. Nitric oxide produced by macrophages downstream from IFN-γ signaling, in combination with NF-κB signaling, directly confers resistance against the growth of intracellular pathogens (Liew et al. 1990a,b). Additionally, macrophages polarized to the M1 phenotype after exposure to IFN-γ produce increased levels of IL-1β, tumor necrosis factor (TNF), IL-12, IL-18, and IL-23, and up-regulate key proteins involved in Th1-antigen-specific responses and proimmune responses, including major histocompatibility complex (MHC) class II, CD68, and the co-stimulatory proteins CD80 and CD86 (Sica and Bronte 2007; Sica and Mantovani 2012). These M1 macrophages show proinflammatory functions, enhanced antigen presentation functions, induction of cytotoxic T lymphocyte (CTL) activity in CD8$^+$ T cells, and direct tumoricidal activities, all of which facilitate the immune elimination of tumors.

Antigen-Presenting Cells

Antigen-presenting cells (APCs), like dendritic cells and macrophages, are necessary for efficient antigen presentation and activation of the adaptive immune response, in particular, CTL responses against extracellular pathogens and tumors. IFN-γ signaling influences the entire process of antigen processing and presentation, starting with the generation of the peptides that become targets for the immune system. Under homeostatic conditions, the proteasome is responsible for cleaving proteins into the shorter peptides that are loaded onto the MHC class I complex. During immune responses, the composition of the proteasome changes, creating a specialized protein degradation complex called the immunoproteasome. IFN-γ signaling in APCs results in up-regulation of the components of the immunoproteasome, including LMP1, LMP7, and MECL1 (Gaczynska et al. 1994; York and Rock 1996; Pamer and Cresswell 1998). Incorporation of these IFN-γ-inducible subunits results in alteration to the proteasome cleavage pattern and changes the repertoire of peptides available for presentation by MHC class I molecules (Gaczynska et al. 1994).

Multiple other proteins involved in antigen presentation are also up-regulated downstream from IFN-γ signaling. The transporters associated with antigen processing TAP-1 and TAP-2 are responsible for moving peptides from the cytosol where they are generated into the endo-

plasmic reticulum and loaded onto MHC class I molecules. Both of the TAP proteins are up-regulated after encounters with IFN-γ (Pamer and Cresswell 1998). Furthermore, expression of both MHC class I and class II proteins are increased by IFN-γ. STAT1, activated by the IFNGR, in turn activates the transcription factor interferon regulatory factor (IRF)-1, which transcriptionally up-regulates expression of the MHC class I locus (Chang et al. 1992; Reis et al. 1992). The ability to up-regulate MHC class II is unique to IFN-γ, which does so through the up-regulation of the MHC class II transactivator CIITA and the invariant chain (Ii) (Cresswell 1994; Wolf and Ploegh 1995).

Finally, IFN-γ signaling in APCs results in the up-regulation of costimulatory molecules and cytokines involved in the production of effective T-cell responses. Indeed, lack of IFN-γ signaling on mature IFNGR$^{-/-}$ dendritic cells leads to the reduced expression of ICAM-1, CD86, IL-1β, and IL-12p70 (Pan et al. 2004).

B Cells

B-cell proliferation and antibody class switching are both regulated by IFN-γ. The relationship between IFN-γ signaling and B-cell proliferation is complex, and the effects that IFN-γ has on B-cell division are dictated by the activation state of the B cell when it encounters IFN-γ (Jurado et al. 1989; Abed et al. 1994). Before exposure to antigen, B-cell proliferation is inhibited by IFN-γ. In contrast, IFN-γ signaling can promote B-cell division during the early proliferative response after primary antigen exposure. During later stages of proliferation and during B-cell maturation, IFN-γ again works to inhibit B-cell division.

The class of antibody produced by B cells is heavily influenced by the microenvironment that the B cell occupies during maturation. For example, Th2-polarized CD4$^+$ T cells are involved in the generation of allergy, and the cytokines they produce (IL-4 and IL-5) promote B-cell class switching to immunoglobulin (Ig)E. The presence of Th1-polarized CD4$^+$ T cells and the IFN-γ they produce inhibits class switching to IgE and instead promotes class switching to

Cite this article as *Cold Spring Harb Perspect Biol* doi: 10.1101/cshperspect.a028480

IgG$_{2a}$ (Snapper and Paul 1987; Snapper et al. 1988, 1991), the class of antibody that is particularly important for mediating antibody-dependent cellular cytotoxicity, which is a mechanism that can confer resistance to tumor outgrowth. In this way, IFN-γ signaling connects the type of CD4$^+$ T-cell response present to the class of antibodies that B cells produce, and thereby facilitates the generation of a cohesive immune response.

Helper CD4$^+$ T Cells

IFN-γ plays a critical role in the generation of Th1 CD4$^+$ helper T cells. Although IL-12 is the canonical cytokine required to drive Th1 CD4$^+$ T-cell activation, IFN-γ signaling is required to achieve maximum production of IFN-γ and also stabilizes the Th1 phenotype (Bradley et al. 1996). Indeed, Th1 CD4$^+$ T cells generated in IFN-γ$^{-/-}$ mice continue to express the Th2 cytokine IL-4, indicating that Th1 polarization is not complete without IFN-γ signaling (Zhang et al. 2001; Agnello et al. 2003). IFN-γ signaling in CD4$^+$ T cells results in the activation of STAT1 and its downstream transcriptional target T-bet. The T-bet transcription factor is a master regulator of the Th1 phenotype and transcriptionally up-regulates the expression of both the IL-12 receptor and IFN-γ (Lighvani et al. 2001; Afkarian et al. 2002; Agnello et al. 2003). In this way, IFN-γ establishes a positive feedback loop for its own production by Th1 CD4$^+$ T cells and promotes the establishment of the Th1 phenotype by increasing signaling via IL-12 and the IL-12 receptor. IFN-γ signaling is also involved in actively inhibiting CD4$^+$ T-cell differentiation into other subclasses, including Th2 and Th17. In mice, Th1 CD4$^+$ T cells down-regulate the expression of certain IFNGR subunits and thus become IFN-γ-insensitive (Bach et al. 1995; Pernis et al. 1995). In contrast, Th2 and Th17 CD4$^+$ T cells maintain expression of functional IFNGRs, and IFN-γ signaling in these CD4$^+$ T-cell subsets results in an inhibition of their proliferation, leading to a net skewing of the CD4$^+$ T-cell population toward the Th1 phenotype (Oriss et al. 1997; Harrington et al. 2006).

Regulatory CD4$^+$ T Cells

The effects that IFN-γ has on regulatory CD4$^+$ T-cell (Treg) function are only now beginning to be identified. It was shown recently that Treg cell suppression of effector T-cell activation was inhibited in IFN-γ-rich environments, and that this inhibition required the expression of the IFNGR on the Treg cells. Furthermore, adoptive cell transfer of IFNGR1$^{-/-}$ Treg cells into mice bearing B16.F10 melanoma tumors resulted in enhanced tumor incidence and decreased overall survival compared with mice that received IFN-γ-sensitive Treg cells, indicating that IFN-γ signaling is capable of inhibiting the protumorigenic effects of Treg cells in the tumor microenvironment (Overacre-Delgoffe et al. 2017). The mechanisms by which IFN-γ inhibits Treg cell function remain to be elucidated.

CD8$^+$ T Cells

IFN-γ signaling directly regulates several aspects of CD8$^+$ T-cell biology. Most importantly for the biological function of CD8$^+$ T cells is the requirement for IFN-γ in the generation of cytolytic capabilities. Indeed, early experiments using recombinant protein showed that full cytolytic capability was not achieved until CD8$^+$ T cells were exposed to IFN-γ (Siegel 1988; Maraskovsky et al. 1989). IFN-γ signaling in CD8$^+$ T cells results in the up-regulation of the IL-2 receptor, the transcription factor T-bet, and granzyme. IL-2 responsiveness is critical for the generation of cytolytic CD8$^+$ T cells and granzyme is responsible for mediating cytolysis of CD8$^+$ T-cell targets. IFN-γ also regulates CD8$^+$ T-cell proliferation in either a positive or negative manner following antigen exposure (Siegel 1988; Maraskovsky et al. 1989; Haring et al. 2005).

CANCER IMMUNOEDITING: AN IFN-γ-DRIVEN PROCESS LEADING TO HOST PROTECTION AND SCULPTING OF TUMOR IMMUNOGENICITY

The process known as cancer immunoediting best exemplifies the actions of IFN-γ in the antitumor response (Fig. 2). "Cancer immunoedit-

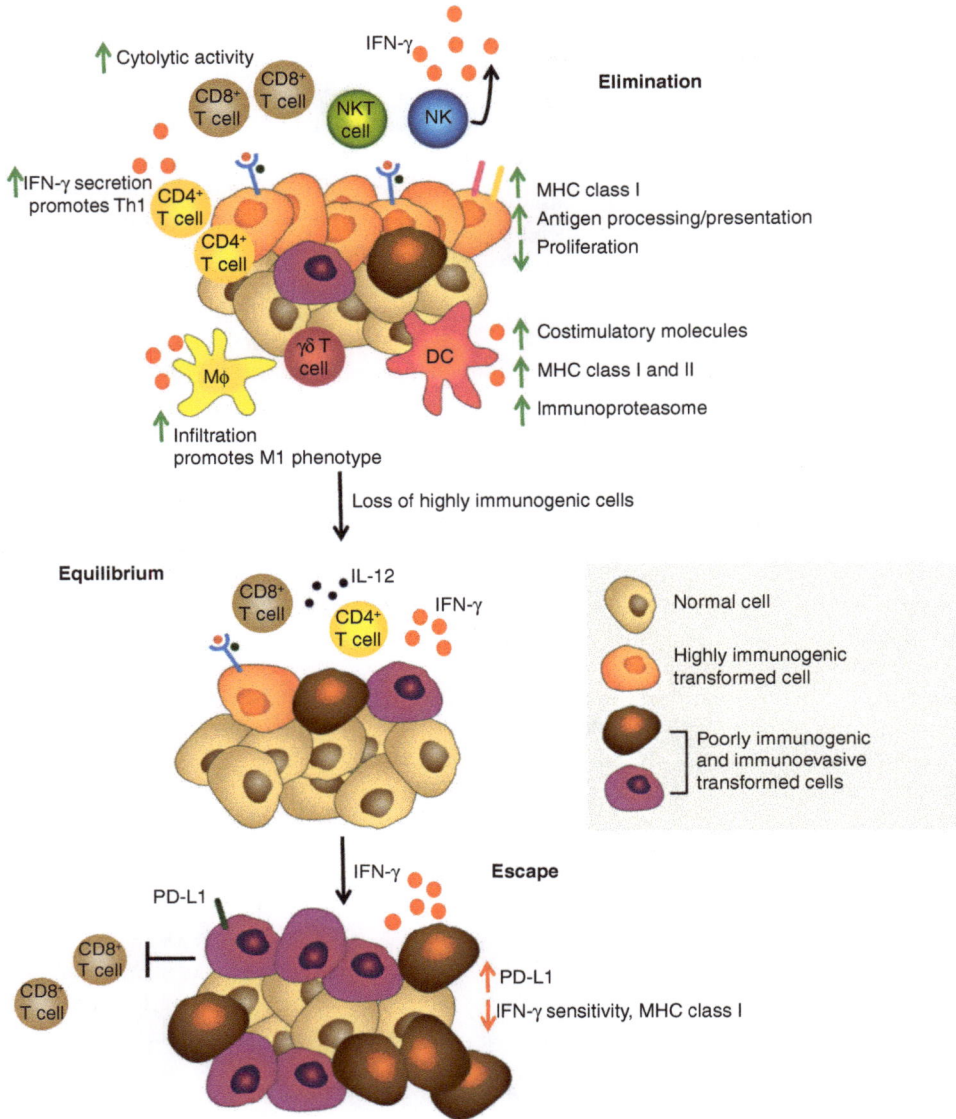

Figure 2. Interferon (IFN)-γ influences all stages of tumor immunoediting. Tumor immunoediting involves three stages: (1) elimination, in which the immune system recognizes and destroys tumor cells expressing strong neoantigens; (2) equilibrium, in which the immune system controls the outgrowth of remaining immunogenic tumor cells by manifesting in them a state of immune-mediated dormancy; and (3) escape, in which the immune system can no longer control the outgrowth of edited tumor cells resulting in the progressive outgrowth of clinically apparent tumors and the establishment of an immunosuppressive tumor microenvironment. The roles played by IFN-γ are highlighted, with green arrows indicating its immune-stimulatory roles and red arrows indicating its immune-suppressive roles. NKT, Natural killer T cells; NK, natural killer; MHC, major histocompatibility complex; DC, dendritic cell; PD-L1, programmed-death ligand 1. (From Vesely et al. 2011; adapted, with permission, from the authors.)

ing" represents an evolution of the older and more controversial concept of cancer immunosurveillance in which the adaptive immune system was thought to play a purely protective role in controlling tumor outgrowth. "Cancer immunoediting" acknowledges this protective function of adaptive immunity but now emphasizes that immune control of tumor outgrowth is a result of the collaborative actions of cells of both innate and adaptive immunity and that cancer outgrowth can be controlled either by immune-mediated destruction of cancer cells (elimination phase) or by maintaining residual cancer cells in a state of immune-mediated dormancy (equilibrium phase). Tumor cells that transit through these two stages are subjected to severe immune-mediated selection pressures that lead to the removal of highly immunogenic tumor cell clones expressing strong tumor antigens and the outgrowth of tumor cells that are less immunogenic, and therefore are more fit to survive in an immunocompetent host (escape phase). Thus, the cancer immunoediting concept stresses both the host-protective and tumor-sculpting actions of immunity and explains, at least in part, how cancer can form in an immunocompetent individual.

Elimination

The elimination phase of cancer immunoediting is critically dependent on the activity of CD4$^+$ and CD8$^+$ T cells that are reactive against tumor antigens. Early observations showed that tumors generated in immunodeficient Rag2$^{-/-}$ mice were more readily rejected, and thus more immunogenic, when retransplanted into naïve syngeneic immunocompetent mice than tumors generated in wild-type mice. Additionally, Rag2$^{-/-}$ mice developed more tumors at a faster rate than their wild-type counterparts (Shankaran et al. 2001). These results not only reconfirmed the hypothesis that the immune system was capable of restraining tumor growth, but also indicated that the immune system was responsible for shaping tumor immunogenicity and suggested that T cells played a critical role in this process.

Subsequent use of next-generation sequencing technologies to identify the expressed muta-

tions in tumors derived from Rag2$^{-/-}$ mice established that T cells do in fact generate antigen-specific responses against tumor-specific mutant neoantigens. Specifically, a foundational study in 2012 conducted on the d42m1 methylcholanthrene-A (MCA) sarcoma line showed that CD8$^+$ T-cell responses where generated against a tumor-specific MHC class I–restricted mutation in the spectrin β2 protein (Matsushita et al. 2012). Subsequent experiments in wild-type mice found that d42m1 tumor cell variants that escaped immune elimination lacked expression of this strong neoantigen. Mixing differentially labeled mutant spectrin β2-expressing tumor cells and cells lacking mutant spectrin β2 resulted in CD8$^+$ T-cell-dependent outgrowth only of cells lacking the neoantigen, indicating that CD8$^+$ T cells are responsible for eliminating tumor cells expressing strong neoantigens (Matsushita et al. 2012). Similar studies were also performed using oncogene-driven tumor models expressing the model antigens OVA and SIY, and further showed that outgrown tumors in wild-type mice lacked expression of T-cell-targetable antigens (DuPage et al. 2012). Although the elimination phase of cancer immunoediting is most well documented in mouse models, recent reports from human patients have detailed the elimination of neoantigen-expressing cells. Indeed, recurrent metastatic lesions from patients with melanoma and cervical cancer who were treated with adoptive cell transfer showed loss of neoantigen expression compared with the neoantigen repertoire observed in the primary lesions (Verdegaal et al. 2016; Stevanović et al. 2017). Together, the results from experiments in mouse models and recent observations in humans establish that immune-mediated tumor elimination is the process by which the adaptive immune system recognizes and eliminates tumor cells based on their expression of strong tumor antigens.

The first indication that endogenous IFN-γ was necessary in the elimination phase was observed from a series of experiments performed on Meth A fibrosarcomas induced after treatment with MCA, and predates the suggestion of the tumor immunoediting hypothesis (Dighe et al. 1994). Subcutaneous transplantation of

Meth A tumors into naïve syngeneic, immuno-competent mice led to tumor progression and outgrowth. However, these progressively growing tumors could be completely eliminated via immune-dependent mechanisms following treatment with lipopolysaccharide (LPS). Although it was known that LPS-induced rejection of Meth A occurred via TNF-α-dependent mechanisms, a striking observation made in 1994 showed that rejection was also blocked in the presence of neutralizing monoclonal antibodies specific for IFN-γ (Dighe et al. 1994). This was an important indication that host IFN-γ plays an important role in generating antitumor immune responses that can lead to complete tumor elimination in mouse models.

Although the above experiment showed that host-derived IFN-γ is necessary in promoting antitumor immune responses against transplantable Meth A sarcomas, it was important to determine whether host IFN-γ signaling was also necessary in driving elimination of primary tumors developing in the host. To address this question, tumor formation was compared between MCA-treated, IFN-γ-insensitive mice lacking either IFNGR1 (IFNGR1$^{-/-}$) or Stat1 (Stat1$^{-/-}$) and sex-, age-, and strain-matched wild-type mice (Kaplan et al. 1998). Importantly, a higher percentage of IFNGR1$^{-/-}$ and Stat1$^{-/-}$ mice developed tumors than wild-type mice following treatment, and the tumors from the IFN-γ-insensitive mice developed more rapidly. These experiments showed the role of IFN-γ signaling in host cells in preventing primary tumor development within mice (Kaplan et al. 1998).

Because of the demonstration that IFN-γ signaling within the host was important during tumor elimination, the requirement for IFN-γ signaling in the tumor cells was then examined. Using the transplantable Meth A tumors described above, IFN-γ sensitivity of the tumor cells was ablated by enforced expression of a truncated dominant-negative mutant form of the IFNGR1 protein (Dighe et al. 1994). Although both IFN-γ-sensitive and -insensitive tumors grew progressively in immunodeficient hosts, wild-type mice showed more rapid growth of IFN-γ-insensitive Meth A tumors compared with control IFN-γ-sensitive tumors. Even more importantly, IFN-γ-insensitive Meth A tumors no longer regressed after LPS treatment (Dighe et al. 1994). Together, this work established that the immune system could detect and respond to tumor cells and indicated that IFN-γ signaling in both the host and the tumor cells is required for the control of tumor growth (Dighe et al. 1994; Kaplan et al. 1998).

The effects of IFN-γ in the tumor environment during the elimination phase of immunoediting are predominantly antitumorigenic (Fig. 2). Innate lymphocytes (e.g., NK and NKT cells and γδ T cells) are the earliest sources of IFN-γ to enter the tumor microenvironment. In the C57BL/6 background, NK and NKT cells are necessary in generating effective antitumor immune responses, as shown in MCA experiments in mice lacking NK or NKT cells. NK-cell-deficient mice develop tumors more frequently and rapidly compared with wild-type mice (O'Sullivan et al. 2012; Guillerey and Smyth 2016). Therefore, these innate cell subsets are important in establishing the cytokine milieu that promotes effective antitumor T-cells responses.

The presence of IFN-γ in the tumor milieu during elimination is an important factor in establishing a nonpermissive microenvironment. Indeed, the extent to which tumor-associated macrophages (TAMs) are exposed to IFN-γ is partially responsible for polarizing TAMs to either the protumorigenic M2 or myeloid-derived suppressor cell (MDSC) phenotypes versus the antitumorigenic M1 phenotype (Sica and Bronte 2007; Sica and Mantovani 2012). Although it is now appreciated that macrophage polarization in vivo occurs more along a gradient than stark M1 versus M2 phenotypes, it is generally accepted that IFN-γ-exposed macrophages are more capable of controlling tumor cell outgrowth either directly or via promoting the development of helper CD4$^+$ and CTL CD8$^+$ T-cell responses through the secretion of cytokines and efficient antigen presentation via MHC class II (Lewis and Pollard 2006; Caux et al. 2016). As tumors progress, however, the alternatively activated M2 macrophage and MDSC becomes more predominant, resulting

Cite this article as *Cold Spring Harb Perspect Biol* doi: 10.1101/cshperspect.a028480

in immune suppression. Indeed, reprogramming M2 macrophages or MDSCs toward an IFN-γ-experienced M1 phenotype has been suggested as a potential tumor immunotherapeutic approach (Noy and Pollard 2014).

In addition to creating a proinflammatory microenvironment, IFN-γ signaling plays major roles in enhancing tumor immunogenicity. As discussed previously, IFN-γ has been shown to alter components of the antigen processing and presentation pathway leading to an altered epitope repertoire and up-regulation of MHC class I and class II and on the surface of APCs (Chang et al. 1992; Reis et al. 1992; Gaczynska et al. 1994). IFN-γ-mediated enhancement of MHC class I expression also occurs on tumor cells (Shankaran et al. 2001; Ikeda et al. 2002). The net result of these changes is a tumor environment containing APCs fully capable of activating the adaptive immune response and tumor cells that are targetable through MHC class I–dependent presentation of antigens (Fig. 2).

The IFN-γ-dependent establishment of a proinflammatory tumor environment and the enhanced presentation of antigens promotes the development of the adaptive antitumor response. Recent results have indicated that IFN-γ in the tumor microenvironment inhibits the activities of CD4$^+$ Treg cells (Overacre-Delgoffe et al. 2017). At the same time, the presence of IL-12 secreted by cells like M1 macrophages and the up-regulation of MHC class II promotes the development of a Th1-polarized CD4$^+$ T-cell response. In turn, IL-2 and IFN-γ secreted by Th1 CD4$^+$ T cells promote the development of completely functional cytolytic CD8$^+$ T cells that are capable of directly lysing MHC class I–expressing tumor cells. Together, these IFN-γ-mediated alterations produce a tumor microenvironment composed of cell subsets from both the innate and adaptive immune system that is decidedly antitumorigenic (Fig. 2).

Equilibrium

After elimination of highly immunogenic tumor cell clones, the immune system is capable of maintaining residual, less immunogenic tumor cell clones in a state of growth dormancy. This is the hallmark of the equilibrium phase of cancer immunoediting. The molecular mechanisms that drive immune-mediated tumor dormancy have been difficult to elucidate given the challenges of modeling the equilibrium phase in mouse models. However, several successful studies have highlighted the importance of adaptive immunity and the requirement for IFN-γ in this process.

The equilibrium phase was experimentally documented using a modification to the primary MCA-mediated tumorigenesis protocol described above, except that large groups of wild-type mice were treated with a low dose of MCA, resulting in only 20% of the treated animals developing tumors (Koebel et al. 2007). Mice that were free of clinically apparent tumors at the 200-day time point were divided into two groups. One group received control monoclonal antibody, whereas the second group was rendered immunodeficient by treatment with a cocktail of monoclonal antibodies that either depleted critical immune cell populations such as CD4$^+$ and CD8$^+$ T cells and/or blocked critical cytokines such as IFN-γ. Tumor outgrowth in the two groups was then observed for an additional 100-day period. Although tumor outgrowth was extremely rare in mice treated with control mAb, 40%–50% of the mice in groups treated with mAb that ablated or depleted adaptive immunity (e.g., mAb that neutralized IL-12 or IFN-γ or depleted CD4$^+$ or CD8$^+$ T cells) showed the rapid appearance of tumors at the original site of MCA injection (Koebel et al. 2007; Teng et al. 2012). Interestingly, mAbs that blocked innate immunity were not effective. Additional experiments showed that tumor antigen-specific T cells were capable of establishing tumor cell dormancy in the inducible RIP1-Tag2 model of pancreatic cancer. Treatment of mice with neutralizing antibodies against IFN-γ prevented the induction of tumor dormancy (Müller-Hermelink et al. 2008). Together, these results showed that maintenance of equilibrium was a sole function of adaptive immunity, and that IFN-γ signaling is required for the maintenance of tumor dormancy.

Case studies in human renal transplant patients have also revealed that immunosuppressed individuals receiving organ transplants can develop tumors that are of donor origin. In one particularly illuminating study, two transplant recipients each received a kidney isolated from the same cadaver donor (MacKie et al. 2003). Both recipients later returned to the clinic with fulminant secondary malignant melanoma. Neither recipient was found to have a primary lesion. When the clinical history of the donor was inspected, it was found that she was diagnosed with melanoma 16 years before her death and, after treatment, had been considered cured. However, the findings imply that the transplanted kidneys from the cadaver donor may have contained donor-derived tumor cells held in equilibrium by the donor's adaptive immune system (MacKie et al. 2003). On transfer into a "naïve" recipient who is immunosuppressed, the immune pressure keeping the donor tumor cells in equilibrium is removed thereby permitting the donor tumor cells to grow progressively in the recipient patient.

Escape

A tumor that is progressively growing has circumvented both the elimination and equilibrium phases of tumor immunoediting and has now reached the escape phase. Thus, the tumor sculpting actions of immunity that can occur in the elimination and equilibrium phases explain, in part, the reduced immunogenicity of tumors entering the escape phase. Researchers have characterized two general mechanisms through which these tumors now escape immune detection: adaptive immune resistance and acquired immune resistance (Pardoll 2012). Both mechanisms of escape incorporate IFN-γ signaling or its downstream targets, and the effects of IFN-γ during immune escape are generally protumorigenic (Fig. 2).

Adaptive immune resistance is based on the ability of IFN-γ signaling to up-regulate the expression of membrane-bound immune inhibitory molecules, namely, programmed-death ligand 1 (PD-L1), on the surface of tumor cells (Pardoll 2012). In addition to observing that treatment of human tumor cell lines with IFN-γ in vitro results in the up-regulation of

PD-L1, similar observations have now been made for tumor cells and myeloid cells in close proximity to infiltrating immune cells producing IFN-γ in vivo (Taube et al. 2012; Noguchi et al. 2017). PD-L1 expressed on tumor cells can bind to programmed cell death protein 1 (PD1) expressed on activated CD8[+] T cells and stimulate apoptotic T-cell death in vivo and in vitro (Dong et al. 2002). Therefore, IFN-γ-mediated up-regulation of PD-L1 on tumor cells results in the inhibition of antitumor CD8[+] T-cell cytotoxicity and enhanced tumor cell survival (Fig. 2). Indeed, blocking PD1–PD-L1 interactions with monoclonal antibodies inhibits this adaptive immune resistance and results in enhanced antitumor immunity and tumor rejection (Dong et al. 2002). These observations led the way to the development of immune checkpoint blockade therapies based on the mAb targeting of either PD1 or PD-L1. Importantly, the up-regulation of PD-L1 in response to IFN-γ occurs on both tumor cells and normal cells and does not require cellular transformation to occur (Taube et al. 2012).

In addition to the IFN-γ-mediated up-regulation of PD-L1, increased expression of several other inhibitory surface molecules in microenvironments with an activated immune response have been observed. Indeed, prolonged signaling through the TCR leads to up-regulation of CTLA-4 on T cells, which formed the basis for all current immune checkpoint blockade therapies (Leach et al. 1996). Additionally, a recent publication highlighted the compensatory action of TIM-3 up-regulation in the tumor microenvironment in mediating immune escape following the use of PD-1 blockade in a mouse model of lung adenocarcinoma (Koyama et al. 2016). Likewise, up-regulation of the recently characterized inhibitory molecule VISTA was observed in prostate cancer patients treated with ipilimumab (Gao et al. 2017). Although the up-regulation of these other immune inhibitory molecules is not directly mediated by IFN-γ signaling, their presence in the tumor microenvironment is an important mechanism of adaptive immune resistance.

As previously discussed, IFN-γ signaling in tumor cells is of critical importance to expression of tumor immunogenicity. Additionally, the IFN-γ signaling pathway up-regulates proteins involved in arresting the cell cycle (like p21 and p27), and caspase-1, a major component of the apoptotic pathway, the net result of which is decreased tumor cell viability (Chin et al. 1997; Mandal et al. 1998; Ikeda et al. 2002). Given these effects, it is not surprising that tumor cells expressing the dominant negative IFNGR1 protein grew more aggressively than tumors with intact IFNGR signaling. Furthermore, human melanoma and lung tumor cell lines also display significant reductions in or complete loss of sensitivity to IFN-γ based on a failure to up-regulate MHC class I after stimulation (Kaplan et al. 1998). These early observations generated the hypothesis that tumor cells that inactivate the IFN-γ signaling pathway will be less prone to detection by the immune system.

As opposed to adaptive immune resistance, acquired immune resistance involves the generation of additional mutations to knock out the IFN-γ signaling pathway or its downstream targets in a tumor-cell-intrinsic manner. This functional state is sometimes observed on tumor relapse or recurrence following treatment with immune therapies (Pardoll 2012; Sharma et al. 2017). The experimental evidence for acquired adaptive resistance was reported as early as 1998, when it was observed that certain IFN-γ-insensitive human lung adenocarcinoma lines contained defects in IFNGR1, JAK1, or JAK2 (Kaplan et al. 1998). Recent clinical trials using immune checkpoint blockade therapies in human patients with melanoma have highlighted additional mechanisms for acquired resistance. One group of researchers was studying the molecular mechanisms of relapse after treatment with αPD1 (Zaretsky et al. 2016). Whole-exome sequencing was used to compare the patients' relapsed tumors to their tumors before treatment. Like the 1998 study, two of the patients' tumors acquired loss-of-function mutations in JAK1 or JAK2. Another patient displayed a truncating mutation in β2 microglobulin (B2M), an accessory protein in the antigen presentation pathway that is required for surface expression of MHC class I (Zaretsky et al. 2016).

In another report, longitudinal examination of multiple recurrent melanoma metastases also revealed acquisition of mutations in B2M that resulted in loss of MHC class I expression on the surface of tumor cells (Zhao et al. 2016). These observations from clinical trials indicate that for tumors to escape an enhanced immune response generated by immune checkpoint blockade therapy they must acquire mutations that make them resistant to the biological impacts of IFN-γ signaling (Fig. 2).

CONCLUDING REMARKS

Work over the last 30 years has revealed the critical functions of immunity in preventing and shaping cancer and the critical roles that IFN-γ plays in this process. The impacts of IFN-γ on tumor–immune system interaction were originally viewed as exclusively antitumor, because this pleotropic cytokine can have antiproliferative effects on tumor cells, promote myeloid cell activation and antigen presentation, induce directed cellular migration indirectly through production of a variety of chemokines, and facilitate CD4$^+$ T-cell polarization to Th1 and maturation of CD8$^+$ T cells into cytolytic T cells. However, it is now appreciated that IFN-γ also induces paradoxical responses in tumor cells and host cells that facilitate tumor outgrowth. IFN-γ drives the up-regulation of PD-L1 on tumor cells and host myeloid cells and promotes expression of the immunosuppressive metabolite IDO in these cells, thereby helping to establish a state of adaptive resistance that leads to suppression of tumor-specific T cells and formation of an immunosuppressive tumor microenvironment. It is important to realize that these antitumor and protumor effects of IFN-γ are separated temporally and therefore we still have a lot to learn about the timing and quantitative relationships that differentially drive IFN-γ's antitumor versus protumor effects. Thus, even though we now have a much better picture of IFN-γ's importance in natural and therapeutic immune responses to cancer, we still have much work to do to learn how to modulate these effects to improve cancer immunotherapy efficacy.

ACKNOWLEDGMENTS

R.D.S receives research support from the National Cancer Institute (RO1 CA043059, RO1 CA190700, U01CA141541), the Cancer Research Institute, the WWWW Foundation, the Siteman Cancer Center/Barnes-Jewish Hospital (Cancer Frontier Fund), Bristol-Myers Squibb, Janssen, and Stand Up to Cancer. E.A. and D.M.L. are supported by a postdoctoral training grant (T32 CA00954729) from the National Cancer Institute. D.M.L. is supported by a postdoctoral training grant (Irvington Postdoctoral Fellowship) from the National Cancer Institute.

REFERENCES

Reference is also in this collection.

Abed NS, Chace JH, Fleming AL, Cowdery JS. 1994. Interferon-γ regulation of B lymphocyte differentiation: Activation of B cells is a prerequisite for IFN-γ-mediated inhibition of B cell differentiation. *Cell Immunol* **153:** 356–366.

Afkarian M, Sedy JR, Yang J, Jacobson NG, Cereb N, Yang SY, Murphy TL, Murphy KM. 2002. T-bet is a STAT1-induced regulator of IL-12R expression in naïve CD4⁺ T cells. *Nat Immunol* **3:** 549–557.

Agnello D, Lankford CSR, Bream J, Morinobu A, Gadina M, O'Shea JJ, Frucht DM. 2003. Cytokines and transcription factors that regulate T helper cell differentiation: New players and new insights. *J Clin Immunol* **23:** 147–161.

Alexander WS, Starr R, Fenner JE, Scott CL, Handman E, Sprigg NS, Corbin JE, Cornish AL, Darwiche R, Owczarek CM, et al. 1999. SOCS1 is a critical inhibitor of interferon γ signaling and prevents the potentially fatal neonatal actions of this cytokine. *Cell* **98:** 597–608.

Arakawa T, Hsu YR, Parker CG, Lai PH. 1986. Role of polycationic C-terminal portion in the structure and activity of recombinant human interferon-γ. *J Biol Chem* **261:** 8534–8539.

Bach EA, Szabo SJ, Dighe AS, Ashkenazi A, Aguet M, Murphy KM, Schreiber RD. 1995. Ligand-induced autoregulation of IFN-γ receptor β chain expression in T helper cell subsets. *Science* **270:** 1215–1218.

Bach EA, Tanner JW, Marsters S, Ashkenazi A, Aguet M, Shaw AS, Schreiber RD. 1996. Ligand-induced assembly and activation of the γ interferon receptor in intact cells. *Mol Cell Biol* **16:** 3214–3221.

Bach EA, Aguet M, Schreiber RD. 1997. The IFN γ receptor: A paradigm for cytokine receptor signaling. *Annu Rev Immunol* **15:** 563–591.

Bradley LM, Dalton DK, Croft M. 1996. A direct role for IFN-γ in regulation of Th1 cell development. *J Immunol* **157:** 1350–1358.

Buchmeier NA, Schreiber RD. 1985. Requirement of endogenous interferon-γ production for resolution of *Listeria monocytogenes* infection. *Proc Natl Acad Sci* **82:** 7404–7408.

Caux C, Ramos RN, Prendergast GC, Bendriss-Vermare N, Ménétrier-Caux C. 2016. A milestone review on how macrophages affect tumor growth. *Cancer Res* **76:** 6439–6442.

Chang CH, Hammer J, Loh JE, Fodor WL, Flavell RA. 1992. The activation of major histocompatibility complex class I genes by interferon regulatory factor-1 (IRF-1). *Immunogenetics* **35:** 378–384.

Chin YE, Kitagawa M, Kuida K, Flavell RA, Fu XY. 1997. Activation of the STAT signaling pathway can cause expression of caspase 1 and apoptosis. *Mol Cell Biol* **17:** 5328–5337.

Cresswell P. 1994. Assembly, transport, and function of MHC class II molecules. *Annu Rev Immunol* **12:** 259–293.

Darnell JE, Kerr IM, Stark GR. 1994. Jak-STAT pathways and transcriptional activation in response to IFNs and other extracellular signaling proteins. *Science* **264:** 1415–1421.

Dighe AS, Farrar MA, Schreiber RD. 1993. Inhibition of cellular responsiveness to interferon-γ (IFNγ) induced by overexpression of inactive forms of the IFNγ receptor. *J Biol Chem* **268:** 10645–10653.

Dighe AS, Richards E, Old LJ, Schreiber RD. 1994. Enhanced in vivo growth and resistance to rejection of tumor cells expressing dominant negative IFN γ receptors. *Immunity* **1:** 447–456.

Dong H, Strome SE, Salomao DR, Tamura H, Hirano F, Flies DB, Roche PC, Lu J, Zhu G, Tamada K, et al. 2002. Tumor-associated B7-H1 promotes T-cell apoptosis: A potential mechanism of immune evasion. *Nat Med* **8:** 793–800.

DuPage M, Mazumdar C, Schmidt LM, Cheung AF, Jacks T. 2012. Expression of tumour-specific antigens underlies cancer immunoediting. *Nature* **482:** 405–409.

Ealick SE, Cook WJ, Vijay-Kumar S, Carson M, Nagabhushan TL, Trotta PP, Bugg CE. 1991. Three-dimensional structure of recombinant human interferon-γ. *Science* **252:** 698–702.

Endo TA, Masuhara M, Yokouchi M, Suzuki R, Sakamoto H, Mitsui K, Matsumoto A, Tanimura S, Ohtsubo M, Misawa H, et al. 1997. A new protein containing an SH2 domain that inhibits JAK kinases. *Nature* **387:** 921–924.

Farrar MA, Schreiber RD. 1993. The molecular cell biology of interferon-γ and its receptor. *Annu Rev Immunol* **11:** 571–611.

Farrar MA, Fernandez-Luna J, Schreiber RD. 1991. Identification of two regions within the cytoplasmic domain of the human interferon-γ receptor required for function. *J Biol Chem* **266:** 19626–19635.

Farrar MA, Campbell JD, Schreiber RD. 1992. Identification of a functionally important sequence in the C terminus of the interferon-γ receptor. *Proc Natl Acad Sci* **89:** 11706–11710.

Fountoulakis M, Lahm HW, Maris A, Friedlein A, Manneberg M, Stueber D, Garotta G. 1991. A 25-kDa stretch of the extracellular domain of the human interferon γ receptor is required for full ligand binding capacity. *J Biol Chem* **266:** 14970–14977.

Fountoulakis M, Zulauf M, Lustig A, Garotta G. 1992. Stoichiometry of interaction between interferon γ and its receptor. *Eur J Biochem* **208:** 781–787.

Gaczynska M, Rock KL, Spies T, Goldberg AL. 1994. Peptidase activities of proteasomes are differentially regulated by the major histocompatibility complex-encoded genes for LMP2 and LMP7. *Proc Natl Acad Sci* **91:** 9213–9217.

Gao J, Ward JF, Pettaway CA, Shi LZ, Subudhi SK, Vence LM, Zhao H, Chen J, Chen H, Efstathiou E, et al. 2017. VISTA is an inhibitory immune checkpoint that is increased after ipilimumab therapy in patients with prostate cancer. *Nat Med* **23:** 551.

Gil MP, Bohn E, O'Guin AK, Ramana CV, Levine B, Stark GR, Virgin HW, Schreiber RD. 2001. Biologic consequences of Stat1-independent IFN signaling. *Proc Natl Acad Sci* **98:** 6680–6685.

Gray PW, Goeddel DV. 1982. Structure of the human immune interferon gene. *Nature* **298:** 859–863.

Gray PW, Goeddel DV. 1983. Cloning and expression of murine immune interferon cDNA. *Proc Natl Acad Sci* **80:** 5842–5846.

Greenlund AC, Schreiber RD, Goeddel DV, Pennica D. 1993. Interferon-γ induces receptor dimerization in solution and on cells. *J Biol Chem* **268:** 18103–18110.

Greenlund AC, Farrar MA, Viviano BL, Schreiber RD. 1994. Ligand-induced IFN γ receptor tyrosine phosphorylation couples the receptor to its signal transduction system (p91). *EMBO J* **13:** 1591–1600.

Greenlund AC, Morales MO, Viviano BL, Yan H, Krolewski J, Schreiber RD. 1995. Stat recruitment by tyrosine-phosphorylated cytokine receptors: An ordered reversible affinity-driven process. *Immunity* **2:** 677–687.

Guillerey C, Smyth MJ. 2016. NK cells and cancer immunoediting. *Curr Top Microbiol Immunol* **395:** 115–145.

Haring JS, Corbin GA, Harty JT. 2005. Dynamic regulation of IFN-γ signaling in antigen-specific CD8[+] T cells responding to infection. *J Immunol* **174:** 6791–6802.

Harrington LE, Mangan PR, Weaver CT. 2006. Expanding the effector CD4 T-cell repertoire: The Th17 lineage. *Curr Opin Immunol* **18:** 349–356.

Hershey GK, Schreiber RD. 1989. Biosynthetic analysis of the human interferon-γ receptor. Identification of N-linked glycosylation intermediates. *J Biol Chem* **264:** 11981–11988.

Hibino Y, Mariano TM, Kumar CS, Kozak CA, Pestka S. 1991. Expression and reconstitution of a biologically active mouse interferon γ receptor in hamster cells. Chromosomal location of an accessory factor. *J Biol Chem* **266:** 6948–6951.

Hogrefe HH, McPhie P, Bekisz JB, Enterline JC, Dyer D, Webb DS, Gerrard TL, Zoon KC. 1989. Amino terminus is essential to the structural integrity of recombinant human interferon-γ. *J Biol Chem* **264:** 12179–12186.

Hoshino K, Tsutsui H, Kawai T, Takeda K, Nakanishi K, Takeda Y, Akira S. 1999. Cutting edge: Generation of IL-18 receptor-deficient mice: Evidence for IL-1 receptor-related protein as an essential IL-18 binding receptor. *J Immunol* **162:** 5041–5044.

Igarashi K, Garotta G, Ozmen L, Ziemiecki A, Wilks AF, Harpur AG, Larner AC, Finbloom DS. 1994. Interferon-γ induces tyrosine phosphorylation of interferon-γ receptor and regulated association of protein tyrosine kinases, Jak1 and Jak2, with its receptor. *J Biol Chem* **269:** 14333–14336.

Ikeda H, Old LJ, Schreiber RD. 2002. The roles of IFNγ in protection against tumor development and cancer immunoediting. *Cytokine Growth Factor Rev* **13:** 95–109.

Jarpe MA, Johnson HM. 1990. Topology of receptor binding domains of mouse IFN-γ. *J Immunol* **145:** 3304–3309.

Johnson HM, Langford MP, Lakhchaura B, Chan TS, Stanton GJ. 1982. Neutralization of native human γ interferon (HuIFNγ) by antibodies to a synthetic peptide encoded by the 5′ end of HuIFNγ cDNA. *J Immunol* **129:** 2357–2359.

Jung V, Rashidbaigi A, Jones C, Tischfield JA, Shows TB, Pestka S. 1987. Human chromosomes 6 and 21 are required for sensitivity to human interferon γ. *Proc Natl Acad Sci* **84:** 4151–4155.

Jurado A, Carballido J, Griffel H, Hochkeppel HK, Wetzel GD. 1989. The immunomodulatory effects of interferon-γ on mature B-lymphocyte responses. *Experientia* **45:** 521–526.

Kane LP, Lin J, Weiss A. 2000. Signal transduction by the TCR for antigen. *Curr Opin Immunol* **12:** 242–249.

Kaplan DH, Greenlund AC, Tanner JW, Shaw AS, Schreiber RD. 1996. Identification of an interferon-γ receptor α chain sequence required for JAK-1 binding. *J Biol Chem* **271:** 9–12.

Kaplan DH, Shankaran V, Dighe AS, Stockert E, Aguet M, Old LJ, Schreiber RD. 1998. Demonstration of an interferon γ-dependent tumor surveillance system in immunocompetent mice. *Proc Natl Acad Sci* **95:** 7556–7561.

Koebel CM, Vermi W, Swann JB, Zerafa N, Rodig SJ, Old LJ, Smyth MJ, Schreiber RD. 2007. Adaptive immunity maintains occult cancer in an equilibrium state. *Nature* **450:** 903–907.

Kotenko SV, Izotova LS, Pollack BP, Mariano TM, Donnelly RJ, Muthukumaran G, Cook JR, Garotta G, Silvennoinen O, Ihle JN. 1995. Interaction between the components of the interferon γ receptor complex. *J Biol Chem* **270:** 20915–20921.

Koyama S, Akbay EA, Li YY, Herter-Sprie GS, Buczkowski KA, Richards WG, Gandhi L, Redig AJ, Rodig SJ, Asahina H, et al. 2016. Adaptive resistance to therapeutic PD-1 blockade is associated with upregulation of alternative immune checkpoints. *Nat Commun* **7:** 10501.

Leach DR, Krummel MF, Allison JP. 1996. Enhancement of antitumor immunity by CTLA-4 blockade. *Science* **271:** 1734–1736.

Leinikki PO, Calderon J, Luquette MH, Schreiber RD. 1987. Reduced receptor binding by a human interferon-γ fragment lacking 11 carboxyl-terminal amino acids. *J Immunol* **139:** 3360–3366.

Lewis CE, Pollard JW. 2006. Distinct role of macrophages in different tumor microenvironments. *Cancer Res* **66:** 605–612.

Li MO, Wan YY, Sanjabi S, Robertson AKL, Flavell RA. 2006. Transforming growth factor-β regulation of immune responses. *Annu Rev Immunol* **24:** 99–146.

Liew FY, Li Y, Millott S. 1990a. Tumor necrosis factor-α synergizes with IFN-γ in mediating killing of *Leishmania major* through the induction of nitric oxide. *J Immunol* **145:** 4306–4310.

Liew FY, Millott S, Parkinson C, Palmer RM, Moncada S. 1990b. Macrophage killing of *Leishmania* parasite in vivo is mediated by nitric oxide from L-arginine. *J Immunol* **144:** 4794–4797.

Lighvani AA, Frucht DM, Jankovic D, Yamane H, Aliberti J, Hissong BD, Nguyen BV, Gadina M, Sher A, Paul WE, et al. 2001. T-bet is rapidly induced by interferon-γ in lymphoid and myeloid cells. *Proc Natl Acad Sci* **98:** 15137–15142.

Liu B, Liao J, Rao X, Kushner SA, Chung CD, Chang DD, Shuai K. 1998. Inhibition of Stat1-mediated gene activation by PIAS1. *Proc Natl Acad Sci* **95:** 10626–10631.

MacKie RM, Reid R, Junor B. 2003. Fatal melanoma transferred in a donated kidney 16 years after melanoma surgery. *N Engl J Med* **348:** 567–568.

Mah AY, Cooper MA. 2016. Metabolic regulation of natural killer cell IFN-γ production. *Crit Rev Immunol* **36:** 131–147.

Mandal M, Bandyopadhyay D, Goepfert TM, Kumar R. 1998. Interferon-induces expression of cyclin-dependent kinase-inhibitors p21WAF1 and p27Kip1 that prevent activation of cyclin-dependent kinase by CDK-activating kinase (CAK). *Oncogene* **16:** 217–225.

Maraskovsky E, Chen WF, Shortman K. 1989. IL-2 and IFN-γ are two necessary lymphokines in the development of cytolytic T cells. *J Immunol* **143:** 1210–1214.

Marine JC, Topham DJ, McKay C, Wang D, Parganas E, Stravopodis D, Yoshimura A, Ihle JN. 1999. SOCS1 deficiency causes a lymphocyte-dependent perinatal lethality. *Cell* **98:** 609–616.

Marsters SA, Pennica D, Bach E, Schreiber RD, Ashkenazi A. 1995. Interferon γ signals via a high-affinity multisubunit receptor complex that contains two types of polypeptide chain. *Proc Natl Acad Sci* **92:** 5401–5405.

Matsuda JL, Zhang Q, Ndonye R, Richardson SK, Howell AR, Gapin L. 2006. T-bet concomitantly controls migration, survival, and effector functions during the development of Vα14i NKT cells. *Blood* **107:** 2797–2805.

Matsushita H, Vesely MD, Koboldt DC, Rickert CG, Uppaluri R, Magrini VJ, Arthur CD, White JM, Chen YS, Shea LK, et al. 2012. Cancer exome analysis reveals a T-cell-dependent mechanism of cancer immunoediting. *Nature* **482:** 400–404.

Müller-Hermelink N, Braumüller H, Pichler B, Wieder T, Mailhammer R, Schaak K, Ghoreschi K, Yazdi A, Haubner R, Sander CA, et al. 2008. TNFR1 signaling and IFN-γ signaling determine whether T cells induce tumor dormancy or promote multistage carcinogenesis. *Cancer Cell* **13:** 507–518.

Murakami M, Narazaki M, Hibi M, Yawata H, Yasukawa K, Hamaguchi M, Taga T, Kishimoto T. 1991. Critical cytoplasmic region of the interleukin 6 signal transducer gp130 is conserved in the cytokine receptor family. *Proc Natl Acad Sci* **88:** 11349–11353.

Nacy CA, Fortier AH, Meltzer MS, Buchmeier NA, Schreiber RD. 1985. Macrophage activation to kill *Leishmania major*: Activation of macrophages for intracellular destruction of amastigotes can be induced by both recombinant interferon-γ and non-interferon lymphokines. *J Immunol* **135:** 3505–3511.

Nagarajan NA, Kronenberg M. 2007. Invariant NKT cells amplify the innate immune response to lipopolysaccharide. *J Immunol* **178:** 2706–2713.

Naka T, Narazaki M, Hirata M, Matsumoto T, Minamoto S, Aono A, Nishimoto N, Kajita T, Taga T, Yoshizaki K, et al. 1997. Structure and function of a new STAT-induced STAT inhibitor. *Nature* **387:** 924–929.

Naka T, Tsutsui H, Fujimoto M, Kawazoe Y, Kohzaki H, Morita Y, Nakagawa R, Narazaki M, Adachi K, Yoshimoto T, et al. 2001. SOCS-1/SSI-1-deficient NKT cells participate in severe hepatitis through dysregulated cross-talk inhibition of IFN-γ and IL-4 signaling in vivo. *Immunity* **14:** 535–545.

Naylor SL, Sakaguchi AY, Shows TB, Law ML, Goeddel DV, Gray PW. 1983. Human immune interferon gene is located on chromosome 12. *J Exp Med* **157:** 1020–1027.

Naylor SL, Gray PW, Lalley PA. 1984. Mouse immune interferon (IFN-γ) gene is on chromosome 10. *Somat Cell Mol Genet* **10:** 531–534.

* Negishi H, Taniguchi T, Yanai H. 2017. The interferon (IFN) class of cytokines and the IFN regulatory factor (IRF) transcription factor family. *Cold Spring Harb Perspect Biol* doi: 10.1101/cshperspect.a028423.

Nguyen H, Ramana CV, Bayes J, Stark GR. 2001. Roles of phosphatidylinositol 3-kinase in interferon-γ-dependent phosphorylation of STAT1 on serine 727 and activation of gene expression. *J Biol Chem* **276:** 33361–33368.

Noguchi T, Ward JP, Gubin MM, Arthur CD, Lee SH, Hundal J, Selby MJ, Graziano RF, Mardis ER, Korman AJ, et al. 2017. Temporally distinct PD-L1 expression by tumor and host cells contributes to immune escape. *Cancer Immunol Res* **5:** 106–117.

Noy R, Pollard JW. 2014. Tumor-associated macrophages: From mechanisms to therapy. *Immunity* **41:** 49–61.

Okamura H, Tsutsi H, Komatsu T, Yutsudo M, Hakura A, Tanimoto T, Torigoe K, Okura T, Nukada Y, Hattori K. 1995. Cloning of a new cytokine that induces IFN-γ production by T cells. *Nature* **378:** 88–91.

Oriss TB, McCarthy SA, Morel BF, Campana MA, Morel PA. 1997. Crossregulation between T helper cell (Th)1 and Th2: Inhibition of Th2 proliferation by IFN-γ involves interference with IL-1. *J Immunol Baltim Md 1950* **158:** 3666–3672.

O'Sullivan T, Saddawi-Konefka R, Vermi W, Koebel CM, Arthur C, White JM, Uppaluri R, Andrews DM, Ngiow SF, Teng MWL, et al. 2012. Cancer immunoediting by the innate immune system in the absence of adaptive immunity. *J Exp Med* **209:** 1869–1882.

Ouchi T, Lee SW, Ouchi M, Aaronson SA, Horvath CM. 2000. Collaboration of signal transducer and activator of transcription 1 (STAT1) and BRCA1 in differential regulation of IFN-γ target genes. *Proc Natl Acad Sci* **97:** 5208–5213.

Overacre-Delgoffe AE, Chikina M, Dadey RE, Yano H, Brunazzi EA, Shayan G, Horne W, Moskovitz JM, Kolls JK, Sander C, et al. 2017. Interferon-γ drives Treg fragility to promote anti-tumor immunity. *Cell* **169:** 1130–1141.e11.

Pamer E, Cresswell P. 1998. Mechanisms of MHC class I–restricted antigen processing. *Annu Rev Immunol* **16:** 323–358.

Pan J, Zhang M, Wang J, Wang Q, Xia D, Sun W, Zhang L, Yu H, Liu Y, Cao X. 2004. Interferon-γ is an autocrine mediator for dendritic cell maturation. *Immunol Lett* **94:** 141–151.

Pardoll DM. 2012. The blockade of immune checkpoints in cancer immunotherapy. *Nat Rev Cancer* **12:** 252–264.

Penix L, Weaver WM, Pang Y, Young HA, Wilson CB. 1993. Two essential regulatory elements in the human interferon γ promoter confer activation specific expression in T cells. *J Exp Med* **178:** 1483–1496.

Penix LA, Sweetser MT, Weaver WM, Hoeffler JP, Kerppola TK, Wilson CB. 1996. The proximal regulatory element of the interferon-γ promoter mediates selective expression in T cells. *J Biol Chem* **271:** 31964–31972.

Pernis A, Gupta S, Gollob KJ, Garfein E, Coffman RL, Schindler C, Rothman P. 1995. Lack of interferon γ receptor β chain and the prevention of interferon γ signaling in T$_H$1 cells. *Science* **269:** 245–247.

Pfizenmaier K, Wiegmann K, Scheurich P, Krönke M, Merlin G, Aguet M, Knowles BB, Ucer U. 1988. High affinity human IFN-γ-binding capacity is encoded by a single receptor gene located in proximity to *c-ros* on human chromosome region 6q16 to 6q22. *J Immunol* **141:** 856–860.

Ramana CV, Gil MP, Schreiber RD, Stark GR. 2002. Stat1-dependent and -independent pathways in IFN-γ-dependent signaling. *Trends Immunol* **23:** 96–101.

Reis LF, Harada H, Wolchok JD, Taniguchi T, Vilcek J. 1992. Critical role of a common transcription factor, IRF-1, in the regulation of IFN-β and IFN-inducible genes. *EMBO J* **11:** 185–193.

Roy SK, Wachira SJ, Weihua X, Hu J, Kalvakolanu DV. 2000. CCAAT/enhancer-binding protein-β regulates interferon-induced transcription through a novel element. *J Biol Chem* **275:** 12626–12632.

Russell JK, Hayes MP, Carter JM, Torres BA, Dunn BM, Russell SW, Johnson HM. 1986. Epitope and functional specificity of monoclonal antibodies to mouse interferon-γ: The synthetic peptide approach. *J Immunol* **136:** 3324–3328.

Sakatsume M, Igarashi K, Winestock KD, Garotta G, Larner AC, Finbloom DS. 1995. The Jak kinases differentially associate with the α and β (accessory factor) chains of the interferon γ receptor to form a functional receptor unit capable of activating STAT transcription factors. *J Biol Chem* **270:** 17528–17534.

Schindler C, Darnell JE. 1995. Transcriptional responses to polypeptide ligands: The JAK-STAT pathway. *Annu Rev Biochem* **64:** 621–651.

Schoenborn JR, Wilson CB. 2007. Regulation of interferon-γ during innate and adaptive immune responses. *Adv Immunol* **96:** 41–101.

Schreiber RD, Hicks LJ, Celada A, Buchmeier NA, Gray PW. 1985. Monoclonal antibodies to murine γ-interferon which differentially modulate macrophage activation and antiviral activity. *J Immunol* **134:** 1609–1618.

Shankaran V, Ikeda H, Bruce AT, White JM, Swanson PE, Old LJ, Schreiber RD. 2001. IFNγ and lymphocytes prevent primary tumour development and shape tumour immunogenicity. *Nature* **410:** 1107–1111.

Sharma P, Hu-Lieskovan S, Wargo JA, Ribas A. 2017. Primary, adaptive, and acquired resistance to cancer immunotherapy. *Cell* **168:** 707–723.

Sica A, Bronte V. 2007. Altered macrophage differentiation and immune dysfunction in tumor development. *J Clin Invest* **117:** 1155–1166.

Sica A, Mantovani A. 2012. Macrophage plasticity and polarization: In vivo veritas. *J Clin Invest* **122:** 787–795.

Sica A, Dorman L, Viggiano V, Cippitelli M, Ghosh P, Rice N, Young HA. 1997. Interaction of NF-κB and NFAT with the interferon-γ promoter. *J Biol Chem* **272:** 30412–30420.

Siegel JP. 1988. Effects of interferon-γ on the activation of human T lymphocytes. *Cell Immunol* **111:** 461–472.

Smith HR, Idris AH, Yokoyama WM. 2001. Murine natural killer cell activation receptors. *Immunol Rev* **181:** 115–125.

Snapper CM, Paul WE. 1987. Interferon-γ and B cell stimulatory factor-1 reciprocally regulate Ig isotype production. *Science* **236:** 944–947.

Snapper CM, Peschel C, Paul WE. 1988. IFN-γ stimulates IgG2a secretion by murine B cells stimulated with bacterial lipopolysaccharide. *J Immunol* **140:** 2121–2127.

Snapper CM, Peçanha LM, Levine AD, Mond JJ. 1991. IgE class switching is critically dependent upon the nature of the B cell activator, in addition to the presence of IL-4. *J Immunol* **147:** 1163–1170.

Starr R, Willson TA, Viney EM, Murray LJ, Rayner JR, Jenkins BJ, Gonda TJ, Alexander WS, Metcalf D, Nicola NA, et al. 1997. A family of cytokine-inducible inhibitors of signalling. *Nature* **387:** 917–921.

Stetson DB, Mohrs M, Reinhardt RL, Baron JL, Wang Z-E, Gapin L, Kronenberg M, Locksley RM. 2003. Constitutive cytokine mRNAs mark natural killer (NK) and NK T cells poised for rapid effector function. *J Exp Med* **198:** 1069–1076.

Stevanović S, Pasetto A, Helman SR, Gartner JJ, Prickett TD, Howie B, Robins HS, Robbins PF, Klebanoff CA, Rosenberg SA, et al. 2017. Landscape of immunogenic tumor antigens in successful immunotherapy of virally induced epithelial cancer. *Science* **356:** 200–205.

Suzuki Y, Orellana MA, Schreiber RD, Remington JS. 1988. Interferon-γ: The major mediator of resistance against *Toxoplasma gondii*. *Science* **240:** 516–518.

Szabo SJ, Kim ST, Costa GL, Zhang X, Fathman CG, Glimcher LH. 2000. A novel transcription factor, T-bet, directs Th1 lineage commitment. *Cell* **100:** 655–669.

Tanner JW, Chen W, Young RL, Longmore GD, Shaw AS. 1995. The conserved box 1 motif of cytokine receptors is required for association with JAK kinases. *J Biol Chem* **270:** 6523–6530.

Tassi I, Colonna M. 2005. The cytotoxicity receptor CRACC (CS-1) recruits EAT-2 and activates the PI3K and phospholipase Cγ signaling pathways in human NK cells. *J Immunol* **175:** 7996–8002.

Tassi I, Presti R, Kim S, Yokoyama WM, Gilfillan S, Colonna M. 2005. Phospholipase C-γ 2 is a critical signaling mediator for murine NK cell activating receptors. *J Immunol* **175:** 749–754.

Taube JM, Anders RA, Young GD, Xu H, Sharma R, McMiller TL, Chen S, Klein AP, Pardoll DM, Topalian SL, et al.

2012. Colocalization of inflammatory response with B7-h1 expression in human melanocytic lesions supports an adaptive resistance mechanism of immune escape. *Sci Transl Med* **4:** 127ra37.

Teng MWL, Vesely MD, Duret H, McLaughlin N, Towne JE, Schreiber RD, Smyth MJ. 2012. Opposing roles for IL-23 and IL-12 in maintaining occult cancer in an equilibrium state. *Cancer Res* **72:** 3987–3996.

Trent JM, Olson S, Lawn RM. 1982. Chromosomal localization of human leukocyte, fibroblast, and immune interferon genes by means of in situ hybridization. *Proc Natl Acad Sci* **79:** 7809–7813.

Tripp CS, Wolf SF, Unanue ER. 1993. Interleukin 12 and tumor necrosis factor α are costimulators of interferon γ production by natural killer cells in severe combined immunodeficiency mice with listeriosis, and interleukin 10 is a physiologic antagonist. *Proc Natl Acad Sci* **90:** 3725–3729.

Verdegaal EME, de Miranda NFCC, Visser M, Harryvan T, van Buuren MM, Andersen RS, Hadrup SR, van der Minne CE, Schotte R, Spits H, et al. 2016. Neoantigen landscape dynamics during human melanoma-T cell interactions. *Nature* **536:** 91–95.

Vesely MD, Kershaw MH, Schreiber RD, Smyth MJ. 2011. Natural, innate and adaptive immunity to cancer. *Annu Rev Immunol* **29:** 235–271.

* Wallach D. 2017. The tumor necrosis factor family: Family conventions and private idiosyncrasies. *Cold Spring Harb Perspect Biol* doi: 10.1101/cshperspect.a028431.

Walter MR, Windsor WT, Nagabhushan TL, Lundell DJ, Lunn CA, Zauodny PJ, Narula SK. 1995. Crystal structure of a complex between interferon-γ and its soluble high-affinity receptor. *Nature* **376:** 230–235.

Wang LL, Yokoyama WM. 1998. Regulation of mouse NK cells by structurally divergent inhibitory receptors. *Curr Top Microbiol Immunol* **230:** 3–13.

Wen Z, Zhong Z, Darnell JE. 1995. Maximal activation of transcription by Stat1 and Stat3 requires both tyrosine and serine phosphorylation. *Cell* **82:** 241–250.

Wolf PR, Ploegh HL. 1995. How MHC class II molecules acquire peptide cargo: Biosynthesis and trafficking through the endocytic pathway. *Annu Rev Cell Dev Biol* **11:** 267–306.

Yang J, Murphy TL, Ouyang W, Murphy KM. 1999. Induction of interferon-γ production in Th1 CD4$^+$ T cells: Evidence for two distinct pathways for promoter activation. *Eur J Immunol* **29:** 548–555.

York IA, Rock KL. 1996. Antigen processing and presentation by the class I major histocompatibility complex. *Annu Rev Immunol* **14:** 369–396.

You M, Yu DH, Feng GS. 1999. Shp-2 tyrosine phosphatase functions as a negative regulator of the interferon-stimulated Jak/STAT pathway. *Mol Cell Biol* **19:** 2416–2424.

Young HA, Komschlies KL, Ciccarone V, Beckwith M, Rosenberg M, Jenkins NA, Copeland NG, Durum SK. 1989. Expression of human IFN-γ genomic DNA in transgenic mice. *J Immunol* **143:** 2389–2394.

Zaretsky JM, Garcia-Diaz A, Shin DS, Escuin-Ordinas H, Hugo W, Hu-Lieskovan S, Torrejon DY, Abril-Rodriguez G, Sandoval S, Barthly L, et al. 2016. Mutations associated with acquired resistance to PD-1 blockade in melanoma. *N Engl J Med* **375:** 819–829.

Zavodny PJ, Petro ME, Chiang TR, Narula SK, Leibowitz PJ. 1988. Alterations of the amino terminus of murine interferon-γ: Expression and biological activity. *J Interferon Res* **8:** 483–494.

Zhang JJ, Vinkemeier U, Gu W, Chakravarti D, Horvath CM, Darnell JE. 1996. Two contact regions between Stat1 and CBP/p300 in interferon γ signaling. *Proc Natl Acad Sci* **93:** 15092–15096.

Zhang Y, Apilado R, Coleman J, Ben-Sasson S, Tsang S, Hu-Li J, Paul WE, Huang H. 2001. Interferon γ stabilizes the T helper cell type 1 phenotype. *J Exp Med* **194:** 165–172.

Zhao F, Sucker A, Horn S, Heeke C, Bielefeld N, Schrörs B, Bicker A, Lindemann M, Roesch A, Gaudernack G, et al. 2016. Melanoma lesions independently acquire T-cell resistance during metastatic latency. *Cancer Res* **76:** 4347–4358.

Interleukin (IL)-33 and the IL-1 Family of Cytokines—Regulators of Inflammation and Tissue Homeostasis

Ajithkumar Vasanthakumar[1,2] and Axel Kallies[1,2,3]

[1]Department of Medical Biology, University of Melbourne, Melbourne, Victoria 3052, Australia

[2]The Walter and Eliza Hall Institute of Medical Research, Melbourne, Victoria 3052, Australia

[3]The Peter Doherty Institute for Infection and Immunity, University of Melbourne, Melbourne, Victoria 3000, Australia

Correspondence: kallies@wehi.edu.au

Cytokines play an integral role in shaping innate and adaptive immune responses. Members of the interleukin (IL)-1 family regulate a plethora of immune-cell-mediated processes, which include pathogen defense and tissue homeostasis. Notably, the IL-1 family cytokine IL-33 promotes adaptive and innate type 2 immune responses, confers viral protection and facilitates glucose metabolism and tissue repair. At the cellular level, IL-33 stimulates differentiation, maintenance, and function of various immune cell types, including regulatory T cells, effector CD4[+] and CD8[+] T cells, macrophages, and type 2 innate lymphoid cells (ILC2s). Other IL-1 family members, such as IL-1β and IL-18 promote type 1 responses, while IL-37 limits immune activation. Although IL-1 cytokines play critical roles in immunity and tissue repair, their deregulated expression is often linked to autoimmune and inflammatory diseases. Therefore, IL-1 cytokines are regulated tightly by posttranscriptional mechanisms and decoy receptors. In this review, we discuss the biology and function of IL-1 family cytokines, with a specific focus on regulation and function of IL-33 in immune and tissue homeostasis.

Interleukin (IL)-1 family cytokines are potent initiators and amplifiers of immune responses and inflammation. Because they are secreted by damaged cells and are crucial for eliciting inflammatory responses, these cytokines are also called "alarmins." Deregulated expression of IL-1 family members aggravates inflammation, autoimmunity, and allergic responses. Consequently, IL-1 blockade is used for the treatment of a variety of inflammatory diseases, including rheumatoid arthritis, psoriasis, arthrosclerosis, ischemia, and reperfusion and graft rejection (Garlanda et al. 2013; Jesus and Goldbach-Man-sky 2014). However, IL-1 family cytokines also play critical roles in facilitating tissue repair and preserving homeostasis. Thus, expression of IL-1 family members is tightly regulated at multiple levels, including transcriptional and posttranslational mechanisms as well as the expression of decoy and antagonist receptors (Garlanda et al. 2013). IL-1 family cytokines can also function as nuclear factors and influence gene expression by interacting with transcription factors or the chromatin (Martin and Martin 2016). IL-1 family members also act as RNA-splicing factors to regulate the expression of genes that are involved in diverse physiological roles, which include survival and thermogenesis (Pollock et al. 2003; Odegaard et al. 2016).

IL-33 is a prominent member of the IL-1 family, which was discovered in 2003 (Baekkevold et al. 2003) after more than two decades of search to identify a ligand for the orphan IL-1 receptor ST2 (Tominaga 1989). IL-33 was first recognized for its critical role in the induction of allergic asthma by coordinating the expansion and function of adaptive and innate immune cells (Hardman and Ogg 2016). Consistent with this notion, genome-wide association studies (GWAS) have implicated polymorphisms in IL-33 or its receptor ST2 (IL1RL1) in a range of diseases that includes asthma, cardiovascular diseases, and allergy (Hardman and Ogg 2016). More recently, however, novel roles for IL-33 were uncovered that extend beyond immunity. For example, IL-33 was identified to play a prominent role in adipose tissue homeostasis, where it facilitates the expansion of anti-inflammatory immune cells, including type 2 innate lymphoid cells (ILC2)s and regulatory T (Treg) cells, thereby suppressing obesity-induced inflammation and preserving insulin sensitivity (Brestoff et al. 2015; Lee et al. 2015; Vasanthakumar et al. 2015). Similar functions in repair and maintenance of nonlymphoid tissues were also described for other tissues such as the colon or injured muscle tissue (Burzyn et al. 2013b; Schiering et al. 2014; Kuswanto et al. 2016). IL-33 therefore plays critical and context-specific roles in lymphoid and nonlymphoid tissues by preserving tissue homeostasis or mediating inflammation. This review focuses on the multifaceted functions of IL-33; however, we will also discuss the biology of other IL-1 family cytokines.

SOURCES OF IL-33 AND OTHER IL-1 FAMILY CYTOKINES

The IL-1 family consists of 11 members with diverse and complex functions (Garlanda et al. 2013). Most members of this family are expressed constitutively as full-length isoforms in a wide range of hematopoietic and nonhematopoietic cells, but are only processed and released by cells stimulated through Toll-like receptors

(TLRs) or by damaged cells undergoing necrosis (Sims and Smith 2010). IL-33 was first identified as a nuclear factor expressed in high endothelial venules of the human Peyer's patches, tonsils, and lymph nodes (Baekkevold et al. 2003). It was later found to be produced by endothelial and epithelial cells and by some hematopoietic cell types, including dendritic cells (Schmitz et al. 2005), mast cells (Hsu et al. 2010), and macrophages (Fock et al. 2013). IL-33 protein is constitutively expressed in astrocytes and oligodendrocytes within the central nervous system (CNS); however, its expression and secretion is elevated upon CNS injury (Yasuoka et al. 2011; Gadani et al. 2015). High expression of IL-33 was also found in endothelial cells of various tissues, including the adipose tissue, liver, ovaries, vagina, skin, lung, stomach, and salivary gland (Villaret et al. 2010; Lopetuso et al. 2012; Pichery et al. 2012; Byers et al. 2013; Carlock et al. 2014). In steady state, IL-33 is found predominantly in the nucleus of these cells with no evidence of cytoplasmic localization. Inflammation induces IL-33 expression in epithelial cells, but only tissue damage is thought to result in efficient IL-33 release (Martin and Martin 2016). Similarly, expression of IL-33 is heightened in the liver during lipopolysaccharide (LPS)-induced endotoxin shock, and in the lung alveoli during papain-induced allergic airway inflammation (Pichery et al. 2012). A similar phenomenon was observed in colonic epithelium where IL-33 expression increased after inflammation and injury (Schiering et al. 2014) and in retinal Müller cells during macular degeneration (Xi et al. 2016). IL-33 can be detected in the serum of patients with inflammatory diseases or cancer and can potentially be used as a diagnostic and prognostic marker for a range of diseases, including hepatitis B infection (Wang et al. 2012), psoriasis (Mitsui et al. 2016), acute ischemic stroke (Qian et al. 2016), and non-small-cell lung cancer (Hu et al. 2013).

Similar to IL-33, other members of the IL-1 family are produced by a variety of hematopoietic and nonhematopoietic cells. Full-length IL-1α protein is expressed constitutively in healthy epithelial cells of the intestine, kidney, liver, endothelial cells, and astrocytes (Sims and Smith

2010). Sterile inflammation caused by tissue injury facilitates the release of IL-1α, aiding the recruitment of macrophages to the injured site, which may augment tissue damage (Rider et al. 2011). In contrast, during programmed apoptosis, IL-1α is retained in the nucleus to prevent tissue damage (Cohen et al. 2010). In macrophages, de novo synthesis of IL-1α occurs after stimulation with LPS (Arango Duque and Descoteaux 2014). IL-1β is produced predominantly by cells of the hematopoietic compartment, including macrophages, blood monocytes, and dendritic cells. In contrast to IL-1α, which is largely constitutively expressed, transcription of *Il1b* is induced by components of the complement system, tumor necrosis factor (TNF), TLR ligands, and IL-1β itself (Sims and Smith 2010). IL-18 is produced by two major immune cell populations, macrophages, and dendritic cells (Dinarello et al. 2013). However, stromal cells, including endothelial cells, keratinocytes, and intestinal epithelial cells constitutively express the IL-18 precursor (Dinarello et al. 2013). Similar to IL-1β and IL-33, mature IL-18 is released from tissues undergoing necrosis (Mommsen et al. 2009). IL-36 is expressed in several tissues, including skin, gut, lung, keratinocytes, and brain, and in some hematopoietic cells, including macrophages, neutrophils (Bozoyan et al. 2015), and T cells (Vigne et al. 2011). Similar to other IL-1 family members, IL-36 expression can be up-regulated by inflammatory stimuli, including LPS. In skin epithelial cells, IL-36 is upregulated during psoriasis, and in bronchial cells during viral or bacterial infection (Boutet et al. 2016). Unlike other IL-1 family members, IL-37 expression can be induced by the anti-inflammatory cytokine transforming growth factor (TGF)-β (Nold et al. 2010). However, in peripheral blood monocytic cells and dendritic cells, IL-37 expression was also elevated by stimulation through TLRs (Sharma et al. 2008; Nold et al. 2010). Expression of the little-known IL-38 cytokine is restricted to few tissue sites, including skin and the spleen (Lin et al. 2001). In summary, while IL-1 family cytokines are expressed by a wide variety of tissues, they are usually released only in response to "danger" signals.

IL-33 AND T HELPER (Th)2 RESPONSES

Extracellular pathogens such as helminths and nematodes induce a Th2-type inflammation mediated by the cytokines IL-4 and IL-5. IL-33 was originally classified as a Th2 cytokine owing to its role in amplifying type 2 immune responses and exacerbating allergic asthma. This idea was supported by the finding that expression of IL-33 receptor ST2 appeared to be restricted to the Th2 lineage (Lohning et al. 1998). Indeed, IL-33 stimulation increases IL-5 and IL-13 but not IL-4 secretion by Th2 cells (Kurowska-Stolarska et al. 2008). In line with this finding, Th2 cells expressing ST2 secrete IL-5 and are highly pathogenic (Endo et al. 2015). Consistent with a central role in Th2 responses, ST2 was indispensable for Th2 responses in specific helminth infection models (Hoshino et al. 1999). Notably, IL-33 also acts as a chemoattractant for human Th2 cells (Komai-Koma et al. 2007).

In addition to Th2 cells, ILC2s play a key role in promoting airway inflammation. They express high amounts of ST2 and are exquisitely responsive to IL-33, which induces secretion of IL-5 and IL-13 (Moro et al. 2010; Neill et al. 2010). IL-33 administration expands the lung ILC2s, leading to airway hyperresponsiveness (Moro et al. 2010; Barlow et al. 2012). RORα-deficient mice, which lack ILC2, were unable to initiate Th2 responses when treated with papain (Halim et al. 2014). Notably, IL-33-expanded ILC2s also produce the tissue repair and growth factor amphiregulin (AREG), an epidermal growth factor (EGF) receptor ligand (Monticelli et al. 2011). IL-5 and IL-13, secreted by both Th2 cells and ILC2s, contribute to the recruitment of eosinophils, which amplifies allergic airway inflammation (Yamaguchi et al. 1988; Pope et al. 2001; Nussbaum et al. 2013). In allergic asthma, eosinophils are a major source of IL-4 and mediate inflammation by releasing cytotoxic granules and lipid mediators that induce tissue damage and bronchoconstriction (Kita 2011). IL-33 contributes to this process not only by mediating the expansion of Th2 cells and ILC2s, but additionally it can promote the differentiation of eosinophils from CD117[+] progenitors in an IL-

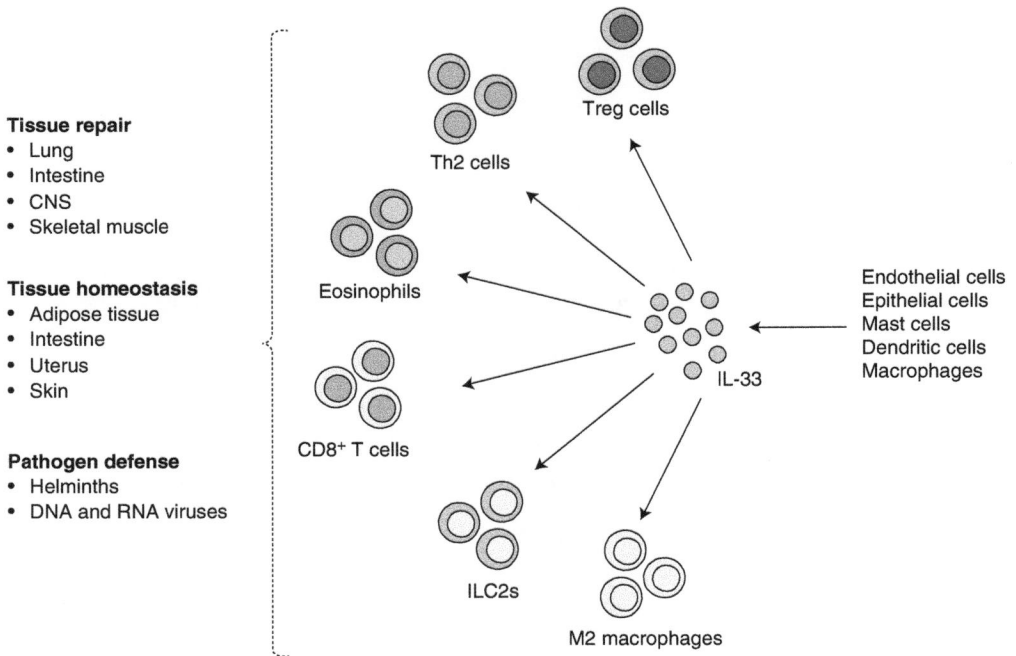

Figure 1. Illustration shows the diverse cellular sources of interleukin (IL)-33, sensing of IL-33 by adaptive and innate immune cells, and the plethora of immunological and nonimmunological functions regulated by IL-33 to preserve immune and tissue homeostasis. CNS, Central nervous system; Th, T helper; Treg, regulatory T cells; ILC2s, type 2 innate lymphoid cells.

5-dependent manner (Stolarski et al. 2010). Eosinophils in the airway express high levels of ST2 and respond to IL-33 secreted by lung epithelial cells (Kurowska-Stolarska et al. 2008). IL-33 not only promotes differentiation of eosinophils, but also improves their function by augmenting the expression of IL-13, IL-6, CCL-17, and TGF-β (Stolarski et al. 2010). IL-33 also induces the differentiation and expansion of alternatively activated M2 macrophages, which express the IL-4R, secrete CCL-24 and CCL-17 and cooperate with Th2 cells and eosinophils in airway inflammation (Kurowska-Stolarska et al. 2009). In agreement with its role in promoting and amplifying airway inflammation, mutations in *Il33* and *Il1rl1* (encoding ST2) are closely associated with increased asthma risk (Gudbjartsson et al. 2009; Moffatt et al. 2010). Taken together, IL-33 initiates and amplifies Th2 inflammation by promoting the differentiation, maintenance, and function of Th2 cells, ILC2s, eosinophils, and alveolar macrophages (Fig. 1; Table 1).

IL-33 AND Th1 RESPONSES

Intracellular pathogens, including bacteria and viruses, induce type 1 inflammation executed by CD4[+] and CD8[+] T cells, macrophages, and B cells. Recent studies have shown that IL-33 is also involved in the establishment of Th1 inflammation and cytotoxic T-cell responses. During lymphocytic choriomeningitis virus (LCMV) infection, Th1 cells were found to express ST2 in a T-bet- and signal transducers and activators of transcription (STAT)4-dependent manner (Baumann et al. 2015). ST2 expression in Th1 cells, however, was transient compared to the constitutive high levels observed in Th2 cells. ST2 contributed to the expansion of antigen-specific Th1 cells and promoted the expression of Th1 transcription factor T-bet (Baumann et al. 2015). Another study found that IL-33 induced Th1 cell differentiation by potentiating the activity of the Th1-polarizing cytokine IL-12; however, IL-33 by itself was unable to pro-

Table 1. Cytokine and receptor nomenclature of the interleukin (IL)-1 family and their functions mediated by immune cells.

Cytokine	Receptor	Coreceptor	Immune cell target	Function
IL-33	ST2	IL-1RAcP	Th2 and Th1 cells, Treg cells, macrophages, CD8$^+$ T cells, eosinophils, ILC2s, NK cells, NKT cells	Type 1 and type 2 immunity, maintenance of organismal metabolism, thermogenesis, tissue repair
IL-1α	IL-1R1 IL1-R2	IL-1RAcP	Macrophages	Sterile inflammation
IL-1β	IL-1R1 IL1-R2	IL-1RAcP	Th17 cells, Th1 cells, ILC1s, regulatory B cells	CNS inflammation, type 1 immunity, gut homeostasis
IL-18	IL-18Rα	IL-18Rβ	Th1 cells, CD8$^+$ T cells, NK cells, ILC1s	Type 1 immunity
IL-36α, β, γ	IL-1Rrp2	IL-1RAcP	CD8$^+$ T cells, NK cells, and γδ T cells	Type 1 immunity, antitumor immunity
IL-37	IL-18Rα		Dendritic cells, macrophages, Treg cells	Immune suppression
IL-38	IL-1Rrp2		T cells	Immune suppression
IL-1RA	IL-1R1		Macrophages, dendritic cells, Th17	Restrains inflammation
IL-36RA	IL-1Rrp2		Unknown	Restrains inflammation

Th, T helper; Treg cell, regulatory T cell; ILC2, type 2 innate lymphoid cell; NK, natural killer; NKT cell, natural killer T cell; CNS, central nervous system.

mote Th1 differentiation (Komai-Koma et al. 2016). IL-33 was also found to induce ST2 expression in CD8$^+$ T cells in vitro and augmented TCR-signaling-dependent interferon (IFN)-γ production (Yang et al. 2011). In line with these findings, during infection with an acute strain of the LCMV, IL-33 promoted the clonal expansion and function of effector CD8$^+$ T cells (Fig. 1) (Bonilla et al. 2012). IL-33 also expanded a population of ST2$^+$ natural killer T (NKT) cells and augmented TCR signaling and IL-12-dependent expression of IFN-γ (Bourgeois et al. 2009). Similarly, IL-33 also induced IFN-γ production in NK cells and promoted the NK cell response against murine cytomegalovirus (Nabekura et al. 2015). Overall, these data suggest that IL-33 is a potent but context-specific amplifier of Th1 inflammation; however, its precise role needs to be further evaluated.

IL-33 IN ORGANISMAL METABOLISM

Visceral adipose tissue (VAT) is an important endocrine organ that secretes soluble factors such as leptin and adiponectin to regulate organismal metabolism (Coelho et al. 2013). VAT

is also a reservoir for multiple pro- and anti-inflammatory immune cells (Cipolletta et al. 2011; Wensveen et al. 2015). Obesity leads to the expansion of proinflammatory immune cells, which induce insulin resistance and, subsequently, type 2 diabetes (Cipolletta et al. 2011; Wensveen et al. 2015). Over the last few years, it has become apparent that IL-33 plays a central role in VAT homeostasis. IL-33 facilitates the differentiation and maintenance of Foxp3$^+$ Treg cells and ILC2s in the VAT, both of which play critical roles in limiting VAT inflammation and maintaining insulin sensitivity (Table 1) (Feuerer et al. 2009; Cipolletta et al. 2012; Brestoff et al. 2015; Vasanthakumar et al. 2015). Treg cells are a subset of CD4$^+$ T cells that potently suppress autoreactive and inflammatory T cells. The majority of Treg cells develops in the thymus; however, they undergo further differentiation and diversification in the periphery, which allows them to acquire full suppressive function and the ability to enter nonlymphoid tissues, including the VAT (Burzyn et al. 2013a; Teh et al. 2015). VAT Treg cells co-opt the adipose tissue transcription factor PPAR-γ for their differentiation and maintenance (Cipolletta et al.

2012). While development and survival of Treg cells in most tissues is strictly dependent on IL-2 (Liston and Gray 2014), VAT-resident Treg cells in addition require IL-33 for their survival and maintenance (Vasanthakumar et al. 2015). VAT Treg cell numbers significantly decrease in the absence of IL-33 signaling. In vitro, IL-33 facilitated population expansion of VAT Treg cells even in the absence of IL-2 or TCR signaling (Vasanthakumar et al. 2015). IL-33 expression in the adipose tissue increased with age, which also correlates with an increase in VAT Treg cells, suggesting that IL-33 is the rate-limiting cytokine for Treg-cell expansion in the adipose tissue. In keeping with this notion, administration of IL-33 expanded Treg cells specifically in the adipose tissue (Kolodin et al. 2015; Molofsky et al. 2015; Vasanthakumar et al. 2015). Obesity leads to the decline of VAT Treg cell numbers, which coincides with exacerbation of inflammation in the VAT and development of insulin resistance (Fig. 1) (Feuerer et al. 2009; Vasanthakumar et al. 2015). When administered to obese mice, IL-33 rescued Treg cell numbers in the VAT, which coincided with suppression of inflammation and improved glucose tolerance (Han et al. 2015; Vasanthakumar et al. 2015). Besides facilitating survival and proliferation, IL-33 also enhances the expression of GATA3 and Foxp3 in VAT Treg cells, and preserves expression of PPAR-γ, suggesting a pivotal role for IL-33 in preserving VAT Treg cell function and phenotype (Vasanthakumar et al. 2015).

IL-33 is also a critical growth factor for VAT ILC2s, which promote the conversion of white adipose tissue to brown adipose tissue (BAT) (Brestoff et al. 2015; Lee et al. 2015). BAT is critically important for thermogenesis of newborns but also contributes to weight loss as it generates heat from chemical energy (calories). ILC2-derived methionine–enkephalin peptides induce the expression of brown adipocyte factor uncoupling protein-1 (UCP-1), required for thermogenesis (Brestoff et al. 2015). It has also been proposed that ILC2s were involved in mediating the IL-33-dependent VAT Treg cell expansion, as VAT Treg cells from IL-5-deficient mice, which lacked ILC2s, failed to expand upon

IL-33 treatment (Molofsky et al. 2015). However, such a mechanism has not yet been supported by additional studies. IL-33 also directly acted on adipose tissue to induce the expression of UCP-1 by an unconventional mechanism as discussed later (Odegaard et al. 2016). Thus, while it is clear that IL-33 plays a central role in maintaining adipose homeostasis, precisely how IL-33-responsive cells and IL-33-dependent mechanisms cooperate in the adipose to maintain tissue health remains to be elucidated. Of note, IL-18 also plays a protective role in obesity (Lee et al. 2016; Murphy et al. 2016), while IL-1β drives adipose tissue inflammation and insulin resistance (Lee et al. 2016).

IL-33 IN TISSUE HOMEOSTASIS AND REPAIR

As outlined earlier, IL-33 is released predominantly by inflamed or injured tissue. Remarkably however, IL-33 is also involved in tissue repair, a mechanism that is likely to have evolved as a feedback to restrain tissue damage. IL-33-dependent Treg-cell-mediated tissue repair has been demonstrated in skeletal muscle and the lung (Fig. 1). Skeletal muscle regeneration critically depends on a self-renewing population of cells called satellite cells. Upon injury, satellite cells proliferate and differentiate into new myofibers or fuse to existing ones and promote regeneration. While inflammatory CD4[+] and CD8[+] T cells impede this process (Tidball and Villalta 2010), IL-33-dependent Treg cells maintain an environment conducive for satellite cell differentiation and muscle regeneration (Burzyn et al. 2013b). Treg cells that reside and expand in the muscle after injury critically depend on IL-33 (Kuswanto et al. 2016). Similar to the injured muscle, IL-33-dependent Treg cells play an important role in restraining inflammation in the gastrointestinal tract (Schiering et al. 2014). Consistent with this finding, Treg cells deficient for ST2 could not control experimental colitis. IL-23, a cytokine known to drive intestinal inflammation and inflammatory bowel disease, inhibited the function of IL-33 (Schiering et al. 2014).

Finally, IL-33 has been shown to induce expression of AREG in Treg cells. This function was shown to play a role in Treg-cell-mediated lung tissue repair after influenza virus infection. Lung epithelial cell–derived IL-33 induced up-regulation of AREG in lung-resident Treg cells, which facilitated repair of damaged lung epithelium and improved lung function after resolution of the infection (Fig. 1) (Arpaia et al. 2015). AREG is also highly expressed in Treg cells in the adipose tissue (Vasanthakumar et al. 2015), suggesting that tissue repair is a common IL-33-induced mechanism of action for Treg cells residing in nonlymphoid tissues. Indeed, IL-33 also induces AREG expression in ILC2s to promote tissue repair in the lung and intestinal epithelia (Monticelli et al. 2011, 2015). Treatment of mice with recombinant IL-33 increased the expression of tight junction proteins claudin1 and mucin and improved intestinal integrity, thereby preserving barrier function (Monticelli et al. 2015). IL-33 also plays an important role in limiting brain tissue damage. Upon brain injury, damaged oligodendrocytes release IL-33, which acts on microglial cells and astrocytes to secrete chemokines and cytokines to attract neuroprotective M2 macrophages that facilitate neural tissue repair (Gadani et al. 2015). Similarly, ILC2s that reside in the meninges are activated by IL-33 secreted upon CNS injury and contribute to limiting CNS injury (Gadani et al. 2017). Finally, IL-33 has also been shown to induce the expansion of IL-10-producing B cells, thereby limiting inflammation in the intestine and protecting from inflammatory bowel disease (Sattler et al. 2014).

Inflammatory mediators within the tumor microenvironment can either promote an antitumor immune response or support tumor pathogenesis. In keeping with this notion, IL-33 plays a context-dependent role in tumor biology where it can be protective or deleterious. For example, IL-33 facilitates the expansion of NK cells and cytotoxic $CD8^+$ T cells to promote tumor eradication (Gao et al. 2015). On the other hand, IL-33 augments tumor growth by promoting colorectal cancer cell stemness (Fang et al. 2017) and metastasis in lung cancer (Wang et al. 2016a). The multifaceted role of IL-33 in the tumor environment was reviewed in more detail recently (Wasmer and Krebs

2016). Thus, while IL-33 maintains tissue homeostasis by augmenting population expansion and function of immune cells that facilitate tissue repair, it can also drive tumor growth (Fig. 1; Table 1).

IMMUNE REGULATION BY OTHER IL-1 FAMILY MEMBERS

In addition to IL-33, other IL-1 family cytokines play critical roles in the differentiation and function of multiple immune cells. IL-18 was originally identified as a costimulator of IFN-γ production and subsequently established to contribute to Th1 differentiation (Yoshimoto et al. 1998). IL-18 was also required for the efficient production of IFN-γ and TNF in antigen-specific $CD8^+$ T cells, suggesting a prominent role for IL-18 in viral defense (Denton et al. 2007). Similarly, IL-18 promotes IFN-γ production in NK cells and ILC1 (Table 1) (Chaix et al. 2008; Maloy and Uhlig 2013).

IL-1β promotes the differentiation of IL-17-producing Th17 cells and γδ T cells in vivo and have been implicated in autoimmune and inflammatory diseases that affect the CNS (Chung et al. 2009; Sutton et al. 2009). IL-6 induced IL-1R1 expression on $CD4^+$ T cells, thereby facilitating Th17 differentiation and induction of experimental autoimmune encephalomyelitis (Chung et al. 2009). Notably, Th17 cells themselves are equipped with NLRP3 and caspase-8, allowing for the secretion of bioactive IL-1β, creating an autocrine loop that facilitates differentiation of Th17 cells (Martin et al. 2016). IL-1β also contributes to Th1 inflammation and cytotoxic T-cell differentiation by activating dendritic cells to produce IL-12 (Wesa and Galy 2001). Interestingly, IL-1β cooperates with IL-33 in ILC2 biology. While IL-33 is critical for ILC2 cell differentiation, IL-1β is required for their activation. IL-1β also induces the expression of IL-12 receptor, which facilitates T-bet expression and transdifferentiation to ILC1 lineage cells (Ohne et al. 2016). Finally, gut microbiota–induced IL-1β is required for the differentiation of regulatory B cells, which contribute to immune homeostasis by producing the anti-inflammatory cytokine IL-10 (Rosser et al. 2014).

Several lines of evidence suggest that IL-1α and IL-1β promote tumorigenesis by inducing the expression of growth factors that facilitate tumor growth and metastasis. IL-1 induces the expression of several tumorigenic and metastatic mediators, such as matrix metalloproteinases (MMPs), vascular endothelial growth factor (VEGF), IL-8, IL-6, TNF, and TGF-β. Tumor tissue–derived IL-1β itself attracts immune cells, including macrophages and neutrophils, which further produce IL-1β to promote angiogenic response, tumor invasiveness, and metastasis (Apte et al. 2006; Perrier et al. 2009; Voronov et al. 2014). In contrast, IL-36γ, which is also highly expressed in the tumor environment, plays a protective role in cancer by polarizing CD8[+] T cells, NK cells, and γδ T cells to produce IFN-γ and establishes a conducive setting for tumor eradication (Wang et al. 2015). Th1 cells have also been shown to express IL-36R, signaling through which synergized with IL-12 in Th1 differentiation in vivo (Vigne et al. 2012).

The anti-inflammatory cytokine of the IL-1 family, IL-37, plays a role in restraining activation of antigen-specific T cells by suppressing dendritic cell activation and limiting the expression of proinflammatory cytokines IL-1β, IL-6, and IL-12 (Luo et al. 2014). IL-37 also repressed the expression of IL-6 in LPS-induced macrophages, suggesting its role in dampening innate immune responses (Li et al. 2015). Finally, IL-37 promoted the suppressive properties of Treg cells by inducing up-regulation of IL-10, TGF-β, and CTLA-4 (Wang et al. 2016b). Thus, IL-1 cytokines play diverse roles in infection autoimmunity and cancer.

REGULATION OF EXPRESSION OF ACTIVE IL-33 AND OTHER IL-1 FAMILY CYTOKINES

Production of IL-1 family members is tightly regulated at several levels. Notably, all IL-1 family members with the exception of IL-1 receptor antagonist (IL-1Ra) are translated without a signal peptide that would be required for active secretion via the endoplasmic reticulum and Golgi apparatus (Sims and Smith 2010; Afonina et al. 2015). Therefore, they need to be cleaved

by caspases and proteases to release bioactive forms (Garlanda et al. 2013). For example, IL-1β is secreted as a pro-IL-1β, which is cleaved by caspase-1 to release the bioactive IL-1β into circulation. While most immune cells express caspase-1, its activity is regulated by the inflammasome NLRP3. Neutrophil-secreted enzymes elastase and proteinase-3 also cleave pro-IL-1β efficiently to generate the bioactive forms. In contrast, the IL-1α precursor, which is released upon necrosis of source tissues, is readily available for the induction of inflammation, as it can activate signaling directly by binding to its receptor.

Similar to IL-1β, IL-18 is translated as a 24-kDa protein, which then is cleaved by NLRP3-activated caspase-1 to become 17-kDa active IL-18. Thus, NLRP1-deficient mice also have reduced amounts of active IL-18, manifesting in spontaneous obesity (Murphy et al. 2016). Pro-IL-18 can be also cleaved by neutrophil-derived proteases and proteinase-3 to generate the active form. The biological activity of IL-36 also requires posttranslational processing (Towne et al. 2011); however, precisely how pro-IL-36 is cleaved is unclear. Similarly, the mechanisms underlying IL-37 processing are unknown, although they are likely to involve caspase-1 (Li et al. 2015).

IL-33, unlike most other members of the family, is released by necrotic cells as a 29-kDa full-length protein that has bioactivity and does not require proteolytic processing (Martin and Martin 2016). Although originally believed to be processed by caspase-1, later studies revealed caspase-1 to be redundant for IL-33 activity (Talabot-Ayer et al. 2009). In contrast, proteases secreted by neutrophils, macrophages, and mast cells cleave the amino-terminal end of human IL-33, resulting in a 10-fold increase in activity (Lefrancais et al. 2012). On the other hand, caspase-3 and -7 released by apoptotic cells can effectively cleave and inactivate IL-33. Thus, the expression of the active forms of IL-1 family cytokines is controlled by multiple posttranslational mechanisms that have evolved to balance the fine line between the beneficial and detrimental functions of the members of this cytokine family.

Cite this article as *Cold Spring Harb Perspect Biol* doi: 10.1101/cshperspect.a028506

REGULATION OF IL-33 AND IL-1 FAMILY SIGNALING

Most members of the IL-1 family signal by binding to heterodimeric receptors. IL-33, IL-1α, IL-1β, and IL-36 utilize a common receptor chain called IL-1RAcP. IL-1α, IL-1β, and IL-1RA share yet another common receptor chain, IL-1R1. In contrast, ST2 is exclusively used by IL-33, while IL-1Rrp2 is specific to IL-36. IL-36RA is an IL-36R antagonist that has more than 50% similarity to IL-1RA. IL-1α and IL-1β can also signal by binding to IL-1R2. IL-18 and IL-37 bind and signal through a heterodimeric receptor composed of IL-18Rα and IL-18Rβ (Fig. 2; Table 1) (Garlanda et al. 2013). Despite the diversity in receptor recognition, all IL-1 family receptors possess a cytoplasmic Toll-IL-1R (TIR) domain and initiate a common signaling cascade, which is similar to TLRs and requires the adaptor Myd88 (O'Neill 2000). Signaling downstream of Myd88 involves IRAKs, TRAF6, and TAK1, leading to the activation of Janus kinase (JNK), extracellular signal-regulated kinase (ERK), and mitogen-activated protein kinase (MAPK) pathways. Eventually, transcription factors, including RelA/p65 (nuclear factor [NF]-κB) and c-Jun establish the transcriptional program activated by IL-1 family members (O'Neill and Greene 1998).

Proximal IL-1 family signaling is regulated tightly by decoy and antagonistic receptors (Sims and Smith 2010; Garlanda et al. 2013). For example, the IL-33 receptor is generated in two distinct forms that differ functionally: membrane-bound ST2 that allows functional IL-33 signaling and a shorter soluble form that sequesters available IL-33 (Yanagisawa et al. 1993; Garlanda et al. 2013; Martin and Martin 2016). Notably, increased soluble ST2 is a prognostic marker for several inflammatory diseases (Kim et al. 2015). Similarly, the soluble receptor isoform of IL-1R2 can bind to IL-1β extracellularly and render it unavailable for signaling. IL-1R signaling is also inhibited by the IL-1R antagonist, IL-1RA, which competes with IL-1β for IL-1R1 receptor binding (Aksentijevich et al. 2009). Furthermore, soluble IL-1R2 and IL-1RAcP can also bind to pro-IL-1β efficiently and prevent its proteolytic processing. Similar mechanisms exist for other IL-1-like cytokines, for example, for IL-36, through expression of an antagonist receptor for IL-36, and for IL-18 through the expression of IL-18-binding protein (IL-18BP), which prevents IL-18 interaction with the receptor (Fig. 2; Table 1) (Sims and Smith 2010; Garlanda et al. 2013).

In addition to the above-described mechanisms that control the availability of IL-1-like

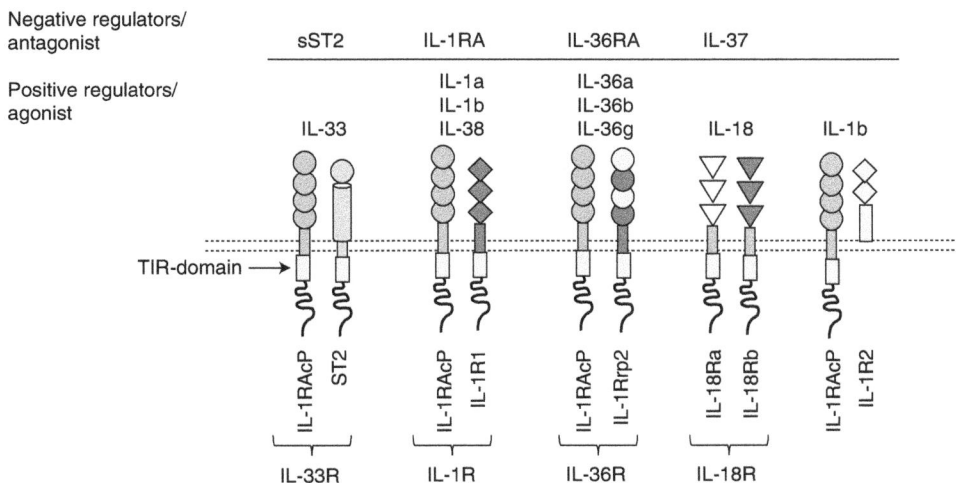

Figure 2. A schematic representation of interleukin (IL)-1 family receptors, coreceptors, ligands, and negative regulators. TIR, Toll-IL-1R.

cytokines, the expression of some receptors is tightly governed by transcriptional mechanisms. This is particularly well known for ST2, the expression of which is controlled by the Th2 lineage–determining transcription factor Gata3 (Hayakawa et al. 2005). Similarly, ST2 expression in ILC2s (Hoyler et al. 2012; Mjosberg et al. 2012) and colonic Treg cells (Schiering et al. 2014) is regulated by Gata3. In contrast, Th1 cells expressed ST2 in a T-bet- and STAT4-dependent manner (Baumann et al. 2015). ST2 expression on Treg cells also required the TCR signaling-induced transcription factor IRF4 and its binding partner BATF (Vasanthakumar et al. 2015). A recent study has shown that PPAR-γ is required for efficient ST2 expression on Th2 cells (Chen et al. 2017). It remains to be seen whether PPAR-γ contributes to regulation of ST2 expression in other cell types.

NUCLEAR FUNCTIONS OF IL-33, IL-1α, AND IL-37

Unlike other cytokines, some IL-1 family members are known to have nuclear functions. IL-33, IL-1α, and IL-37 possess a nuclear localization sequence (NLS) that facilitates nuclear entry. IL-33 also possesses an amino-terminal homeodomain like helix–turn–helix (HTH) DNA-binding domain (Carriere et al. 2007). Nuclear IL-33 can act both as a transcriptional activator and repressor in a context-specific manner. It can interact with histone H2A and H2B and regulate chromatin architecture and partner with the methyltransferase SUV39H1 to facilitate repression of genes (Carriere et al. 2007). Furthermore, IL-33 can cooperate with NF-κB transcription factors p65/RelA and p50 and sequester them to the cytoplasm, thereby, preventing the expression of inflammatory genes (Ali et al. 2011). Although not a nuclear function, a recent study has also shown that intracellular IL-33 can regulate the splicing of uncoupling protein-1 (UCP-1) gene in adipocytes, thereby regulating thermogenesis and brown fat differentiation (Odegaard et al. 2016). Similar to IL-33, IL-1α can also translocate to the nucleus and activates transcription by interacting with chromatin (Garlanda et al. 2013). The range of functions regulated by nuclear IL-1α includes proliferation, cell migration, apoptosis, and expression of cytokines, including IL-18 and TNF (Garlanda et al. 2013; van de Veerdonk and Netea 2013). Nuclear IL-1α also interacts with NF-κB transcription factors to regulate gene expression and, similar to IL-33, facilitates the splicing of prosurvival molecule Bcl-xL (Pollock et al. 2003). Finally, IL-37 was shown to translocate to the nucleus and to repress the expression of inflammatory genes in LPS-treated macrophages (Sharma et al. 2008).

CONCLUDING REMARKS

IL-1 has long been known to play a central role in inflammation. More recently, IL-33 has emerged as another crucial regulator of immunity and tissue homeostasis (Fig. 1). However, while substantial advances were made over the last decade, there are many aspects of the biology of IL-1 family cytokines that remain unexplored. Processing of the procytokine forms into biologically active forms is a key and the foremost regulatory mechanism in IL-1 biology. Yet, the precise mechanism by which IL-33 is cleaved or the exact function of long versus short forms of IL-33 in immunity and inflammation are still a matter of intense debate. It is also incompletely known how specificity of signaling is achieved by members of the IL-1 cytokine family, given that the receptor chains are shared by many of the IL-1-like cytokines. For example, it is unclear whether ST2 or IL-1R compete for IL-1RAcP or interact with different affinities. In such a scenario, ST2 or IL-1R may sequester IL-1RAcP and thereby limit signaling through the other receptor. This may be critical when designing blocking antibodies for a particular cytokine as such antibodies may affect the activity of other cytokines that share the receptor chain. Similarly, while it is clear that soluble ST2 can limit the signaling of IL-33, it is unknown how the expression of the soluble and membrane-bound forms of ST2 is regulated. Importantly, although IL-33 is localized in the nucleus, its role in transcription has not been studied fully. It is also unclear whether the IL-33 released extracellularly by necrotic or inflamed tissue can act as a transcription factor in the target cell. Given the considerable "resource sharing" between IL-1 family members, which share molecules that activate them as well as receptors and signaling components, fully understanding the biology of IL-1 family cytokines, will be essential to exploit their therapeutic potential.

ACKNOWLEDGMENTS

We thank the members of the Kallies Laboratory for discussion. This work is supported by a Senior Medical Research Fellowship from the Sylvia and Charles Viertel Foundation (to A.K.).

REFERENCES

Afonina IS, Muller C, Martin SJ, Beyaert R. 2015. Proteolytic processing of interleukin-1 family cytokines: Variations on a common theme. *Immunity* **42:** 991–1004.

Aksentijevich I, Masters SL, Ferguson PJ, Dancey P, Frenkel J, van Royen-Kerkhoff A, Laxer R, Tedgard U, Cowen EW, Pham TH, et al. 2009. An autoinflammatory disease with deficiency of the interleukin-1-receptor antagonist. *N Engl J Med* **360:** 2426–2437.

Ali S, Mohs A, Thomas M, Klare J, Ross R, Schmitz ML, Martin MU. 2011. The dual function cytokine IL-33 interacts with the transcription factor NF-κB to dampen NF-κB-stimulated gene transcription. *J Immunol* **187:** 1609–1616.

Apte RN, Dotan S, Elkabets M, White MR, Reich E, Carmi Y, Song X, Dvozkin T, Krelin Y, Voronov E. 2006. The involvement of IL-1 in tumorigenesis, tumor invasiveness, metastasis and tumor–host interactions. *Cancer Metastasis Rev* **25:** 387–408.

Arango Duque G, Descoteaux A. 2014. Macrophage cytokines: Involvement in immunity and infectious diseases. *Front Immunol* **5:** 491.

Arpaia N, Green JA, Moltedo B, Arvey A, Hemmers S, Yuan S, Treuting PM, Rudensky AY. 2015. A distinct function of regulatory T cells in tissue protection. *Cell* **162:** 1078–1089.

Baekkevold ES, Roussigne M, Yamanaka T, Johansen FE, Jahnsen FL, Amalric F, Brandtzaeg P, Erard M, Haraldsen G, Girard JP. 2003. Molecular characterization of NF-HEV, a nuclear factor preferentially expressed in human high endothelial venules. *Am J Pathol* **163:** 69–79.

Barlow JL, Bellosi A, Hardman CS, Drynan LF, Wong SH, Cruickshank JP, McKenzie AN. 2012. Innate IL-13-producing nuocytes arise during allergic lung inflammation and contribute to airways hyperreactivity. *J Allergy Clin Immunol* **129:** 191–198, e191–194.

Baumann C, Bonilla WV, Frohlich A, Helmstetter C, Peine M, Hegazy AN, Pinschewer DD, Lohning M. 2015. T-bet- and STAT4-dependent IL-33 receptor expression directly promotes antiviral Th1 cell responses. *Proc Natl Acad Sci* **112:** 4056–4061.

Bonilla WV, Frohlich A, Senn K, Kallert S, Fernandez M, Johnson S, Kreutzfeldt M, Hegazy AN, Schrick C, Fallon PG, et al. 2012. The alarmin interleukin-33 drives protective antiviral CD8[+] T cell responses. *Science* **335:** 984–989.

Bourgeois E, Van LP, Samson M, Diem S, Barra A, Roga S, Gombert JM, Schneider E, Dy M, Gourdy P, et al. 2009. The pro-Th2 cytokine IL-33 directly interacts with invariant NKT and NK cells to induce IFN-γ production. *Eur J Immunol* **39:** 1046–1055.

Boutet MA, Bart G, Penhoat M, Amiaud J, Brulin B, Charrier C, Morel F, Lecron JC, Rolli-Derkinderen M, Bourreille A, et al. 2016. Distinct expression of interleukin (IL)-36α, β and γ, their antagonist IL-36Ra and IL-38 in psoriasis, rheumatoid arthritis and Crohn's disease. *Clin Exp Immunol* **184:** 159–173.

Bozoyan L, Dumas A, Patenaude A, Vallieres L. 2015. Interleukin-36γ is expressed by neutrophils and can activate microglia, but has no role in experimental autoimmune encephalomyelitis. *J Neuroinflammation* **12:** 173.

Brestoff JR, Kim BS, Saenz SA, Stine RR, Monticelli LA, Sonnenberg GF, Thome JJ, Farber DL, Lutfy K, Seale P, et al. 2015. Group 2 innate lymphoid cells promote beiging of white adipose tissue and limit obesity. *Nature* **519:** 242–246.

Burzyn D, Benoist C, Mathis D. 2013a. Regulatory T cells in nonlymphoid tissues. *Nat Immunol* **14:** 1007–1013.

Burzyn D, Kuswanto W, Kolodin D, Shadrach JL, Cerletti M, Jang Y, Sefik E, Tan TG, Wagers AJ, Benoist C, et al. 2013b. A special population of regulatory T cells potentiates muscle repair. *Cell* **155:** 1282–1295.

Byers DE, Alexander-Brett J, Patel AC, Agapov E, Dang-Vu G, Jin X, Wu K, You Y, Alevy Y, Girard JP, et al. 2013. Long-term IL-33-producing epithelial progenitor cells in chronic obstructive lung disease. *J Clin Invest* **123:** 3967–3982.

Carlock CI, Wu J, Zhou C, Tatum K, Adams HP, Tan F, Lou Y. 2014. Unique temporal and spatial expression patterns of IL-33 in ovaries during ovulation and estrous cycle are associated with ovarian tissue homeostasis. *J Immunol* **193:** 161–169.

Carriere V, Roussel L, Ortega N, Lacorre DA, Americh L, Aguilar L, Bouche G, Girard JP. 2007. IL-33, the IL-1-like cytokine ligand for ST2 receptor, is a chromatin-associated nuclear factor in vivo. *Proc Natl Acad Sci* **104:** 282–287.

Chaix J, Tessmer MS, Hoebe K, Fuseri N, Ryffel B, Dalod M, Alexopoulou L, Beutler B, Brossay L, Vivier E, et al. 2008. Cutting edge: Priming of NK cells by IL-18. *J Immunol* **181:** 1627–1631.

Chen T, Tibbitt CA, Feng X, Stark JM, Rohrbeck L, Rausch L, Sedimbi SK, Karlsson MCI, Lambrecht BN, Karlsson Hedestam GB, et al. 2017. PPAR-γ promotes type 2 immune responses in allergy and nematode infection. *Sci Immunol* **2:** aal5196.

Chung Y, Chang SH, Martinez GJ, Yang XO, Nurieva R, Kang HS, Ma L, Watowich SS, Jetten AM, Tian Q, et al. 2009. Critical regulation of early Th17 cell differentiation by interleukin-1 signaling. *Immunity* **30:** 576–587.

Cipolletta D, Kolodin D, Benoist C, Mathis D. 2011. Tissular Tregs: A unique population of adipose-tissue-resident

Foxp3[+] CD4[+] T cells that impacts organismal metabolism. *Semin Immunol* 23: 431–437.

Cipolletta D, Feuerer M, Li A, Kamei N, Lee J, Shoelson SE, Benoist C, Mathis D. 2012. PPAR-γ is a major driver of the accumulation and phenotype of adipose tissue Treg cells. *Nature* 486: 549–553.

Coelho M, Oliveira T, Fernandes R. 2013. Biochemistry of adipose tissue: An endocrine organ. *Arch Med Sci* 9: 191–200.

Cohen I, Rider P, Carmi Y, Braiman A, Dotan S, White MR, Voronov E, Martin MU, Dinarello CA, Apte RN. 2010. Differential release of chromatin-bound IL-1α discriminates between necrotic and apoptotic cell death by the ability to induce sterile inflammation. *Proc Natl Acad Sci* 107: 2574–2579.

Denton AE, Doherty PC, Turner SJ, La Gruta NL. 2007. IL-18, but not IL-12, is required for optimal cytokine production by influenza virus-specific CD8[+] T cells. *Eur J Immunol* 37: 368–375.

Dinarello CA, Novick D, Kim S, Kaplanski G. 2013. Interleukin-18 and IL-18 binding protein. *Front Immunol* 4: 289.

Endo Y, Hirahara K, Iinuma T, Shinoda K, Tumes DJ, Asou HK, Matsugae N, Obata-Ninomiya K, Yamamoto H, Motohashi S, et al. 2015. The interleukin-33-p38 kinase axis confers memory T helper 2 cell pathogenicity in the airway. *Immunity* 42: 294–308.

Fang M, Li Y, Huang K, Qi S, Zhang J, Zgodzinski W, Majewski M, Wallner G, Gozdz S, Macek P, et al. 2017. IL33 promotes colon cancer cell stemness via JNK activation and macrophage recruitment. *Cancer Res* 77: 2735–2745.

Feuerer M, Herrero L, Cipolletta D, Naaz A, Wong J, Nayer A, Lee J, Goldfine AB, Benoist C, Shoelson S, et al. 2009. Lean, but not obese, fat is enriched for a unique population of regulatory T cells that affect metabolic parameters. *Nat Med* 15: 930–939.

Fock V, Mairhofer M, Otti GR, Hiden U, Spittler A, Zeisler H, Fiala C, Knofler M, Pollheimer J. 2013. Macrophage-derived IL-33 is a critical factor for placental growth. *J Immunol* 191: 3734–3743.

Gadani SP, Walsh JT, Smirnov I, Zheng J, Kipnis J. 2015. The glia-derived alarmin IL-33 orchestrates the immune response and promotes recovery following CNS injury. *Neuron* 85: 703–709.

Gadani SP, Smirnov I, Smith AT, Overall CC, Kipnis J. 2017. Characterization of meningeal type 2 innate lymphocytes and their response to CNS injury. *J Exp Med* 214: 285–296.

Gao X, Wang X, Yang Q, Zhao X, Wen W, Li G, Lu J, Qin W, Qi Y, Xie F, et al. 2015. Tumoral expression of IL-33 inhibits tumor growth and modifies the tumor microenvironment through CD8[+] T and NK cells. *J Immunol* 194: 438–445.

Garlanda C, Dinarello CA, Mantovani A. 2013. The interleukin-1 family: Back to the future. *Immunity* 39: 1003–1018.

Gudbjartsson DF, Bjornsdottir US, Halapi E, Helgadottir A, Sulem P, Jonsdottir GM, Thorleifsson G, Helgadottir H, Steinthorsdottir V, Stefansson H, et al. 2009. Sequence variants affecting eosinophil numbers associate with asthma and myocardial infarction. *Nat Genet* 41: 342–347.

Halim TY, Steer CA, Matha L, Gold MJ, Martinez-Gonzalez I, McNagny KM, McKenzie AN, Takei F. 2014. Group 2 innate lymphoid cells are critical for the initiation of adaptive T helper 2 cell-mediated allergic lung inflammation. *Immunity* 40: 425–435.

Han JM, Wu D, Denroche HC, Yao Y, Verchere CB, Levings MK. 2015. IL-33 reverses an obesity-induced deficit in visceral adipose tissue ST2 + T regulatory cells and ameliorates adipose tissue inflammation and insulin resistance. *J Immunol* 194: 4777–4783.

Hardman C, Ogg G. 2016. Interleukin-33, friend and foe in type-2 immune responses. *Curr Opin Immunol* 42: 16–24.

Hayakawa M, Yanagisawa K, Aoki S, Hayakawa H, Takezako N, Tominaga S. 2005. T-helper type 2 cell-specific expression of the ST2 gene is regulated by transcription factor GATA-3. *Biochim Biophys Acta* 1728: 53–64.

Hoshino K, Kashiwamura S, Kuribayashi K, Kodama T, Tsujimura T, Nakanishi K, Matsuyama T, Takeda K, Akira S. 1999. The absence of interleukin 1 receptor-related T1/ST2 does not affect T helper cell type 2 development and its effector function. *J Exp Med* 190: 1541–1548.

Hoyler T, Klose CS, Souabni A, Turqueti-Neves A, Pfeifer D, Rawlins EL, Voehringer D, Busslinger M, Diefenbach A. 2012. The transcription factor GATA-3 controls cell fate and maintenance of type 2 innate lymphoid cells. *Immunity* 37: 634–648.

Hsu CL, Neilsen CV, Bryce PJ. 2010. IL-33 is produced by mast cells and regulates IgE-dependent inflammation. *PLoS ONE* 5: e11944.

Hu LA, Fu Y, Zhang DN, Zhang J. 2013. Serum IL-33 as a diagnostic and prognostic marker in non-small cell lung cancer. *Asian Pac J Cancer Prev* 14: 2563–2566.

Jesus AA, Goldbach-Mansky R. 2014. IL-1 blockade in autoinflammatory syndromes. *Annu Rev Med* 65: 223–244.

Kim MS, Jeong TD, Han SB, Min WK, Kim JJ. 2015. Role of soluble ST2 as a prognostic marker in patients with acute heart failure and renal insufficiency. *J Korean Med Sci* 30: 569–575.

Kita H. 2011. Eosinophils: Multifaceted biological properties and roles in health and disease. *Immunol Rev* 242: 161–177.

Kolodin D, van Panhuys N, Li C, Magnuson AM, Cipolletta D, Miller CM, Wagers A, Germain RN, Benoist C, Mathis D. 2015. Antigen- and cytokine-driven accumulation of regulatory T cells in visceral adipose tissue of lean mice. *Cell Metab* 21: 543–557.

Komai-Koma M, Xu D, Li Y, McKenzie AN, McInnes IB, Liew FY. 2007. IL-33 is a chemoattractant for human Th2 cells. *Eur J Immunol* 37: 2779–2786.

Komai-Koma M, Wang E, Kurowska-Stolarska M, Li D, McSharry C, Xu D. 2016. Interleukin-33 promoting Th1 lymphocyte differentiation dependents on IL-12. *Immunobiology* 221: 412–417.

Kurowska-Stolarska M, Kewin P, Murphy G, Russo RC, Stolarski B, Garcia CC, Komai-Koma M, Pitman N, Li Y, Niedbala W, et al. 2008. IL-33 induces antigen-specific IL-5[+] T cells and promotes allergic-induced airway inflammation independent of IL-4. *J Immunol* 181: 4780–4790.

Cite this article as *Cold Spring Harb Perspect Biol* doi: 10.1101/cshperspect.a028506

Kurowska-Stolarska M, Stolarski B, Kewin P, Murphy G, Corrigan CJ, Ying S, Pitman N, Mirchandani A, Rana B, van Rooijen N, et al. 2009. IL-33 amplifies the polarization of alternatively activated macrophages that contribute to airway inflammation. *J Immunol* **183:** 6469–6477.

Kuswanto W, Burzyn D, Panduro M, Wang KK, Jang YC, Wagers AJ, Benoist C, Mathis D. 2016. Poor repair of skeletal muscle in aging mice reflects a defect in local, interleukin-33-dependent accumulation of regulatory T cells. *Immunity* **44:** 355–367.

Lee MW, Odegaard JI, Mukundan L, Qiu Y, Molofsky AB, Nussbaum JC, Yun K, Locksley RM, Chawla A. 2015. Activated type 2 innate lymphoid cells regulate beige fat biogenesis. *Cell* **160:** 74–87.

Lee MK, Yvan-Charvet L, Masters SL, Murphy AJ. 2016. The modern interleukin-1 superfamily: Divergent roles in obesity. *Semin Immunol* **28:** 441–449.

Lefrancais E, Roga S, Gautier V, Gonzalez-de-Peredo A, Monsarrat B, Girard JP, Cayrol C. 2012. IL-33 is processed into mature bioactive forms by neutrophil elastase and cathepsin G. *Proc Natl Acad Sci* **109:** 1673–1678.

Li S, Neff CP, Barber K, Hong J, Luo Y, Azam T, Palmer BE, Fujita M, Garlanda C, Mantovani A, et al. 2015. Extracellular forms of IL-37 inhibit innate inflammation in vitro and in vivo but require the IL-1 family decoy receptor IL-1R8. *Proc Natl Acad Sci* **112:** 2497–2502.

Lin H, Ho AS, Haley-Vicente D, Zhang J, Bernal-Fussell J, Pace AM, Hansen D, Schweighofer K, Mize NK, Ford JE. 2001. Cloning and characterization of IL-1HY2, a novel interleukin-1 family member. *J Biol Chem* **276:** 20597–20602.

Liston A, Gray DH. 2014. Homeostatic control of regulatory T cell diversity. *Nat Rev Immunol* **14:** 154–165.

Lohning M, Stroehmann A, Coyle AJ, Grogan JL, Lin S, Gutierrez-Ramos JC, Levinson D, Radbruch A, Kamradt T. 1998. T1/ST2 is preferentially expressed on murine Th2 cells, independent of interleukin 4, interleukin 5, and interleukin 10, and important for Th2 effector function. *Proc Natl Acad Sci* **95:** 6930–6935.

Lopetuso LR, Scaldaferri F, Pizarro TT. 2012. Emerging role of the interleukin (IL)-33/ST2 axis in gut mucosal wound healing and fibrosis. *Fibrogenesis Tissue Repair* **5:** 18.

Luo Y, Cai X, Liu S, Wang S, Nold-Petry CA, Nold MF, Bufler P, Norris D, Dinarello CA, Fujita M. 2014. Suppression of antigen-specific adaptive immunity by IL-37 via induction of tolerogenic dendritic cells. *Proc Natl Acad Sci* **111:** 15178–15183.

Maloy KJ, Uhlig HH. 2013. ILC1 populations join the border patrol. *Immunity* **38:** 630–632.

Martin NT, Martin MU. 2016. Interleukin 33 is a guardian of barriers and a local alarmin. *Nat Immunol* **17:** 122–131.

Martin BN, Wang C, Zhang CJ, Kang Z, Gulen MF, Zepp JA, Zhao J, Bian G, Do JS, Min B, et al. 2016. T cell-intrinsic ASC critically promotes T_H17-mediated experimental autoimmune encephalomyelitis. *Nat Immunol* **17:** 583–592.

Mitsui A, Tada Y, Takahashi T, Shibata S, Kamata M, Miyagaki T, Fujita H, Sugaya M, Kadono T, Sato S, et al. 2016. Serum IL-33 levels are increased in patients with psoriasis. *Clin Exp Dermatol* **41:** 183–189.

Mjosberg J, Bernink J, Golebski K, Karrich JJ, Peters CP, Blom B, te Velde AA, Fokkens WJ, van Drunen CM, Spits H. 2012. The transcription factor GATA3 is essential for the function of human type 2 innate lymphoid cells. *Immunity* **37:** 649–659.

Moffatt MF, Gut IG, Demenais F, Strachan DP, Bouzigon E, Heath S, von Mutius E, Farrall M, Lathrop M, Cookson WO, et al. 2010. A large-scale, consortium-based genomewide association study of asthma. *N Engl J Med* **363:** 1211–1221.

Molofsky AB, Van Gool F, Liang HE, Van Dyken SJ, Nussbaum JC, Lee J, Bluestone JA, Locksley RM. 2015. Interleukin-33 and interferon-γ counter-regulate group 2 innate lymphoid cell activation during immune perturbation. *Immunity* **43:** 161–174.

Mommsen P, Frink M, Pape HC, van Griensven M, Probst C, Gaulke R, Krettek C, Hildebrand F. 2009. Elevated systemic IL-18 and neopterin levels are associated with post-traumatic complications among patients with multiple injuries: A prospective cohort study. *Injury* **40:** 528–534.

Monticelli LA, Sonnenberg GF, Abt MC, Alenghat T, Ziegler CG, Doering TA, Angelosanto JM, Laidlaw BJ, Yang CY, Sathaliyawala T, et al. 2011. Innate lymphoid cells promote lung-tissue homeostasis after infection with influenza virus. *Nat Immunol* **12:** 1045–1054.

Monticelli LA, Osborne LC, Noti M, Tran SV, Zaiss DM, Artis D. 2015. IL-33 promotes an innate immune pathway of intestinal tissue protection dependent on amphiregulin-EGFR interactions. *Proc Natl Acad Sci* **112:** 10762–10767.

Moro K, Yamada T, Tanabe M, Takeuchi T, Ikawa T, Kawamoto H, Furusawa J, Ohtani M, Fujii H, Koyasu S. 2010. Innate production of T_H2 cytokines by adipose tissue-associated c-Kit$^+$Sca-1$^+$ lymphoid cells. *Nature* **463:** 540–544.

Murphy AJ, Kraakman MJ, Kammoun HL, Dragoljevic D, Lee MK, Lawlor KE, Wentworth JM, Vasanthakumar A, Gerlic M, Whitehead LW, et al. 2016. IL-18 production from the NLRP1 inflammasome prevents obesity and metabolic syndrome. *Cell Metab* **23:** 155–164.

Nabekura T, Girard JP, Lanier LL. 2015. IL-33 receptor ST2 amplifies the expansion of NK cells and enhances host defense during mouse cytomegalovirus infection. *J Immunol* **194:** 5948–5952.

Neill DR, Wong SH, Bellosi A, Flynn RJ, Daly M, Langford TK, Bucks C, Kane CM, Fallon PG, Pannell R, et al. 2010. Nuocytes represent a new innate effector leukocyte that mediates type-2 immunity. *Nature* **464:** 1367–1370.

Nold MF, Nold-Petry CA, Zepp JA, Palmer BE, Bufler P, Dinarello CA. 2010. IL-37 is a fundamental inhibitor of innate immunity. *Nat Immunol* **11:** 1014–1022.

Nussbaum JC, Van Dyken SJ, von Moltke J, Cheng LE, Mohapatra A, Molofsky AB, Thornton EE, Krummel MF, Chawla A, Liang HE, et al. 2013. Type 2 innate lymphoid cells control eosinophil homeostasis. *Nature* **502:** 245–248.

Odegaard JI, Lee MW, Sogawa Y, Bertholet AM, Locksley RM, Weinberg DE, Kirichok Y, Deo RC, Chawla A. 2016. Perinatal licensing of thermogenesis by IL-33 and ST2. *Cell* **166:** 841–854.

Ohne Y, Silver JS, Thompson-Snipes L, Collet MA, Blanck JP, Cantarel BL, Copenhaver AM, Humbles AA, Liu YJ.

2016. IL-1 is a critical regulator of group 2 innate lymphoid cell function and plasticity. *Nat Immunol* **17**: 646–655.

O'Neill L. 2000. The Toll/interleukin-1 receptor domain: A molecular switch for inflammation and host defence. *Biochem Soc Trans* **28**: 557–563.

O'Neill LA, Greene C. 1998. Signal transduction pathways activated by the IL-1 receptor family: Ancient signaling machinery in mammals, insects, and plants. *J Leukoc Biol* **63**: 650–657.

Perrier S, Caldefie-Chezet F, Vasson MP. 2009. IL-1 family in breast cancer: Potential interplay with leptin and other adipocytokines. *FEBS Lett* **583**: 259–265.

Pichery M, Mirey E, Mercier P, Lefrancais E, Dujardin A, Ortega N, Girard JP. 2012. Endogenous IL-33 is highly expressed in mouse epithelial barrier tissues, lymphoid organs, brain, embryos, and inflamed tissues: In situ analysis using a novel Il-33-LacZ gene trap reporter strain. *J Immunol* **188**: 3488–3495.

Pollock AS, Turck J, Lovett DH. 2003. The prodomain of interleukin 1α interacts with elements of the RNA processing apparatus and induces apoptosis in malignant cells. *FASEB J* **17**: 203–213.

Pope SM, Brandt EB, Mishra A, Hogan SP, Zimmermann N, Matthaei KI, Foster PS, Rothenberg ME. 2001. IL-13 induces eosinophil recruitment into the lung by an IL-5- and eotaxin-dependent mechanism. *J Allergy Clin Immunol* **108**: 594–601.

Qian L, Yuanshao L, Wensi H, Yulei Z, Xiaoli C, Brian W, Wanli Z, Zhengyi C, Jie X, Wenhui Z, et al. 2016. Serum IL-33 is a novel diagnostic and prognostic biomarker in acute ischemic stroke. *Aging Dis* **7**: 614–622.

Rider P, Carmi Y, Guttman O, Braiman A, Cohen I, Voronov E, White MR, Dinarello CA, Apte RN. 2011. IL-1α and IL-1β recruit different myeloid cells and promote different stages of sterile inflammation. *J Immunol* **187**: 4835–4843.

Rosser EC, Oleinika K, Tonon S, Doyle R, Bosma A, Carter NA, Harris KA, Jones SA, Klein N, Mauri C. 2014. Regulatory B cells are induced by gut microbiota-driven interleukin-1β and interleukin-6 production. *Nat Med* **20**: 1334–1339.

Sattler S, Ling GS, Xu D, Hussaarts L, Romaine A, Zhao H, Fossati-Jimack L, Malik T, Cook HT, Botto M, et al. 2014. IL-10-producing regulatory B cells induced by IL-33 (Breg^{IL-33}) effectively attenuate mucosal inflammatory responses in the gut. *J Autoimmun* **50**: 107–122.

Schiering C, Krausgruber T, Chomka A, Frohlich A, Adelmann K, Wohlfert EA, Pott J, Griseri T, Bollrath J, Hegazy AN, et al. 2014. The alarmin IL-33 promotes regulatory T-cell function in the intestine. *Nature* **513**: 564–568.

Schmitz J, Owyang A, Oldham E, Song Y, Murphy E, McClanahan TK, Zurawski G, Moshrefi M, Qin J, Li X, et al. 2005. IL-33, an interleukin-1-like cytokine that signals via the IL-1 receptor-related protein ST2 and induces T helper type 2-associated cytokines. *Immunity* **23**: 479–490.

Sharma S, Kulk N, Nold MF, Graf R, Kim SH, Reinhardt D, Dinarello CA, Bufler P. 2008. The IL-1 family member 7b translocates to the nucleus and down-regulates proinflammatory cytokines. *J Immunol* **180**: 5477–5482.

Sims JE, Smith DE. 2010. The IL-1 family: Regulators of immunity. *Nat Rev Immunol* **10**: 89–102.

Stolarski B, Kurowska-Stolarska M, Kewin P, Xu D, Liew FY. 2010. IL-33 exacerbates eosinophil-mediated airway inflammation. *J Immunol* **185**: 3472–3480.

Sutton CE, Lalor SJ, Sweeney CM, Brereton CF, Lavelle EC, Mills KH. 2009. Interleukin-1 and IL-23 induce innate IL-17 production from γδ T cells, amplifying Th17 responses and autoimmunity. *Immunity* **31**: 331–341.

Talabot-Ayer D, Lamacchia C, Gabay C, Palmer G. 2009. Interleukin-33 is biologically active independently of caspase-1 cleavage. *J Biol Chem* **284**: 19420–19426.

Teh PP, Vasanthakumar A, Kallies A. 2015. Development and function of effector regulatory T cells. *Prog Mol Biol Transl Sci* **136**: 155–174.

Tidball JG, Villalta SA. 2010. Regulatory interactions between muscle and the immune system during muscle regeneration. *Am J Physiol Regul Integr Comp Physiol* **298**: R1173–R1187.

Tominaga S. 1989. A putative protein of a growth specific cDNA from BALB/c-3T3 cells is highly similar to the extracellular portion of mouse interleukin 1 receptor. *FEBS Lett* **258**: 301–304.

Towne JE, Renshaw BR, Douangpanya J, Lipsky BP, Shen M, Gabel CA, Sims JE. 2011. Interleukin-36 (IL-36) ligands require processing for full agonist (IL-36α, IL-36β, and IL-36γ) or antagonist (IL-36Ra) activity. *J Biol Chem* **286**: 42594–42602.

van de Veerdonk FL, Netea MG. 2013. New insights in the immunobiology of IL-1 family members. *Front Immunol* **4**: 167.

Vasanthakumar A, Moro K, Xin A, Liao Y, Gloury R, Kawamoto S, Fagarasan S, Mielke LA, Afshar-Sterle S, Masters SL, et al. 2015. The transcriptional regulators IRF4, BATF and IL-33 orchestrate development and maintenance of adipose tissue-resident regulatory T cells. *Nat Immunol* **16**: 276–285.

Vigne S, Palmer G, Lamacchia C, Martin P, Talabot-Ayer D, Rodriguez E, Ronchi F, Sallusto F, Dinh H, Sims JE, et al. 2011. IL-36R ligands are potent regulators of dendritic and T cells. *Blood* **118**: 5813–5823.

Vigne S, Palmer G, Martin P, Lamacchia C, Strebel D, Rodriguez E, Olleros ML, Vesin D, Garcia I, Ronchi F, et al. 2012. IL-36 signaling amplifies Th1 responses by enhancing proliferation and Th1 polarization of naïve CD4+ T cells. *Blood* **120**: 3478–3487.

Villaret A, Galitzky J, Decaunes P, Esteve D, Marques MA, Sengenes C, Chiotasso P, Tchkonia T, Lafontan M, Kirkland JL, et al. 2010. Adipose tissue endothelial cells from obese human subjects: Differences among depots in angiogenic, metabolic, and inflammatory gene expression and cellular senescence. *Diabetes* **59**: 2755–2763.

Voronov E, Carmi Y, Apte RN. 2014. The role IL-1 in tumor-mediated angiogenesis. *Front Physiol* **5**: 114.

Wang S, Ding L, Liu SS, Wang C, Leng RX, Chen GM, Fan YG, Pan HF, Ye DQ. 2012. IL-33: A potential therapeutic target in autoimmune diseases. *J Investig Med* **60**: 1151–1156.

Wang X, Zhao X, Feng C, Weinstein A, Xia R, Wen W, Lv Q, Zuo S, Tang P, Yang X, et al. 2015. IL-36γ transforms the tumor microenvironment and promotes type 1 lymphocyte-mediated antitumor immune responses. *Cancer Cell* **28**: 296–306.

Wang C, Chen Z, Bu X, Han Y, Shan S, Ren T, Song W. 2016a. IL-33 signaling fuels outgrowth and metastasis of human lung cancer. *Biochem Biophys Res Commun* **479**: 461–468.

Wang DW, Dong N, Wu Y, Zhu XM, Wang CT, Yao YM. 2016b. Interleukin-37 enhances the suppressive activity of naturally occurring CD4$^+$ CD25$^+$ regulatory T cells. *Sci Rep* **6**: 38955.

Wasmer MH, Krebs P. 2016. The role of IL-33-dependent inflammation in the tumor microenvironment. *Front Immunol* **7**: 682.

Wensveen FM, Valentic S, Sestan M, Turk Wensveen T, Polic B. 2015. The "big bang" in obese fat: Events initiating obesity-induced adipose tissue inflammation. *Eur J Immunol* **45**: 2446–2456.

Wesa AK, Galy A. 2001. IL-1β induces dendritic cells to produce IL-12. *Int Immunol* **13**: 1053–1061.

Xi H, Katschke KJ Jr., Li Y, Truong T, Lee WP, Diehl L, Rangell L, Tao J, Arceo R, Eastham-Anderson J, et al. 2016. IL-33 amplifies an innate immune response in the degenerating retina. *J Exp Med* **213**: 189–207.

Yamaguchi Y, Hayashi Y, Sugama Y, Miura Y, Kasahara T, Kitamura S, Torisu M, Mita S, Tominaga A, Takatsu K. 1988. Highly purified murine interleukin 5 (IL-5) stimulates eosinophil function and prolongs in vitro survival. IL-5 as an eosinophil chemotactic factor. *J Exp Med* **167**: 1737–1742.

Yanagisawa K, Takagi T, Tsukamoto T, Tetsuka T, Tominaga S. 1993. Presence of a novel primary response gene ST2L, encoding a product highly similar to the interleukin 1 receptor type 1. *FEBS Lett* **318**: 83–87.

Yang Q, Li G, Zhu Y, Liu L, Chen E, Turnquist H, Zhang X, Finn OJ, Chen X, Lu B. 2011. IL-33 synergizes with TCR and IL-12 signaling to promote the effector function of CD8$^+$ T cells. *Eur J Immunol* **41**: 3351–3360.

Yasuoka S, Kawanokuchi J, Parajuli B, Jin S, Doi Y, Noda M, Sonobe Y, Takeuchi H, Mizuno T, Suzumura A. 2011. Production and functions of IL-33 in the central nervous system. *Brain Res* **1385**: 8–17.

Yoshimoto T, Takeda K, Tanaka T, Ohkusu K, Kashiwamura S, Okamura H, Akira S, Nakanishi K. 1998. IL-12 up-regulates IL-18 receptor expression on T cells, Th1 cells, and B cells: Synergism with IL-18 for IFN-γ production. *J Immunol* **161**: 3400–3407.

Inflammasome-Dependent Cytokines at the Crossroads of Health and Autoinflammatory Disease

Hanne Van Gorp,[1,2] **Nina Van Opdenbosch,**[1,2] **and Mohamed Lamkanfi**[1,2]

[1]Center for Inflammation Research, VIB, Zwijnaarde B-9052, Belgium

[2]Department of Internal Medicine, Ghent University, Ghent B-9000, Belgium

Correspondence: mohamed.lamkanfi@irc.vib-ugent.be

As key regulators of both innate and adaptive immunity, it is unsurprising that the activity of interleukin (IL)-1 cytokine family members is tightly controlled by decoy receptors, antagonists, and a variety of other mechanisms. Additionally, inflammasome-mediated proteolytic maturation is a prominent and distinguishing feature of two important members of this cytokine family, IL-1β and IL-18, because their full-length gene products are biologically inert. Although vital in antimicrobial host defense, deregulated inflammasome signaling is linked with a growing number of autoimmune and autoinflammatory diseases. Here, we focus on introducing the diverse inflammasome types and discussing their causal roles in periodic fever syndromes. Therapies targeting IL-1 or IL-18 show great efficacy in some of these autoinflammatory diseases, although further understanding of the molecular mechanisms leading to unregulated production of these key cytokines is required to benefit more patients.

The interleukin (IL)-1 cytokine family plays a central role in both innate and adaptive immunity because its many members exert a wide range of biological functions. The importance of these cytokines is underscored by the fact that their activity is tightly controlled through selective protein synthesis, the requirement for proteolytic processing, as well as the existence of receptor antagonists, decoys, and intracellular signaling inhibitors. Disturbance, however, of this well-oiled machinery leads to malfunctioning and ultimately may instigate or contribute to the onset of clinical or subclinical disease. In this review, we will highlight two prominent IL-1 cytokines, IL-1β and IL-18, that share the requirement for proteolytic maturation by a set of multiprotein complexes named inflammasomes. Apart from providing an overview of their roles in maintaining homeostasis, we will focus the discussion on their contributions to autoinflammatory diseases.

IL-1 CYTOKINE FAMILY

The IL-1 family consists of seven proinflammatory cytokines (IL-1α, IL-1β, IL-18, IL-33, IL-36α, IL-36β, and IL-36γ), and two less characterized family members (IL-37 and IL-38) that were suggested to act as antagonists within the IL-1 cytokine family (Table 1) (Lin et al. 2001;

Table 1. Interleukin (IL)-1 cytokine family members with their receptors, antagonists, and main functions

Cytokine	Receptor/coreceptor	Antagonist/decoy	Function
IL-1α (IL-1F1)	IL-1R1/IL-1RAcP	IL-1Ra/IL-1R2	Alarmin, Th17 response
IL-1β (IL-1F2)	IL-1R1/IL-1RAcP	IL-1Ra/IL-1R2	Inflammation, fever, Th17 response
IL-18 (IL-1F4)	IL-18Rα/IL-18Rβ	IL-18BP	Inflammation, Th1 response
IL-33 (IL-1F11)	ST2/IL-1RAcP	sST2	Inflammation, Th2 response
IL-36α (IL-1F6)	IL-36R/IL-1RAcP	IL-36Ra	Proinflammatory functions
IL-36β (IL-1F8)	IL-36R/IL-1RAcP	IL-36Ra	Proinflammatory functions
IL-36γ (IL-1F9)	IL-36R/IL-1RAcP	IL-36Ra	Proinflammatory functions
IL-37 (IL-1F7)	IL-18Rα	Unknown	IL-18 antagonist
IL-38 (IL-1F10)	IL-36R	Unknown	IL-36 antagonist

Sharma et al. 2008; Palomo et al. 2015). All IL-1 cytokine family members are composed of an amino-terminal prodomain with variable length and a carboxy-terminal cytokine domain. Apart from IL-18 and IL-33, all genes encoding IL-1 cytokines are located on syntenic regions of human and mouse chromosome 2 (Taylor et al. 2002). The gene for human IL-37 also resides in this cluster, although a murine homolog has not been identified (Boraschi et al. 2011). Unlike most cytokines, the full-length gene products of several IL-1 cytokines (IL-1β, IL-18, IL-36α, IL-36β, and IL-36γ) are biologically inert, and proteolytic processing by a select number of proteases (e.g., caspase-1, caspase-8, proteinase-3, elastase, calpain, cathepsin G, and granzyme B) greatly enhances their biological activity. In contrast, IL-1α and IL-33 are synthesized as constitutively active cytokines, although their immunostimulatory activity can be further amplified by proteolytic processing (Afonina et al. 2011; Lefrancais et al. 2012).

Besides these agonistic molecules, several receptors, coreceptors, antagonists, and decoy receptors (e.g., IL-1R1, IL-18Rα, IL-18Rβ, IL-1Ra, IL-36Ra, IL-1R2, and IL-18BP) that either promote or dampen immune signaling by IL-1 family cytokines have been characterized (Table 1). Analogous to IL-1α and IL-1β that bind to IL-1R1, IL-1Ra functions as a receptor antagonist by binding with similar affinity to the receptor and blocking its signaling (Dripps et al. 1991; Schreuder et al. 1997). IL-36 binds to IL-36R, which akin to IL-1R1 uses IL-1RAcP as a coreceptor. IL-18 interacts with IL-18Rα and uses IL-18Rβ as a coreceptor. IL-18BP is a se-

creted protein that binds and neutralizes mature IL-18 in circulation (Novick et al. 1999). Its expression is up-regulated by a combination of proinflammatory cytokines, suggesting that it serves as a negative feedback loop to dampen IL-18 functions (Corbaz et al. 2002). Moreover, IL-1R2 was found to bind with high affinity to IL-1α and IL-1β, but because of the lack of a Toll/IL-1 receptor (TIR) domain, downstream signaling is blocked (McMahan et al. 1991). The IL-1 family thus not only contains potent proinflammatory cytokines, but also several checkpoints and mechanisms to dampen IL-1-mediated inflammation (Fig. 1).

Unlike IL-1α that is synthesized by epithelial cells of the gastrointestinal tract, lung, liver and kidney, endothelial cells, and keratinocytes, cells of the myeloid lineage are the main source of IL-1β. In contrast, IL-18 is constitutively expressed by most cell types. IL-1Ra is produced by cells that also express IL-1α and IL-1β, and the same holds true for IL-18 and IL-18BP (Dinarello 1996). The widespread expression profiles of these cytokines hints toward key roles in innate and adaptive immunity. In this review, we focus on members of the IL-1 cytokine family linked to inflammasome signaling. For further information on inflammasome-independent regulation of IL-1 family members, please see the review by Netea et al. (2015).

INFLAMMASOME-DEPENDENT PROCESSING OF IL-1 CYTOKINES

Unprocessed IL-1α is constitutively active but caspase-1-dependent proteolytic processing of

Cite this article as *Cold Spring Harb Perspect Biol* doi: 10.1101/cshperspect.a028563

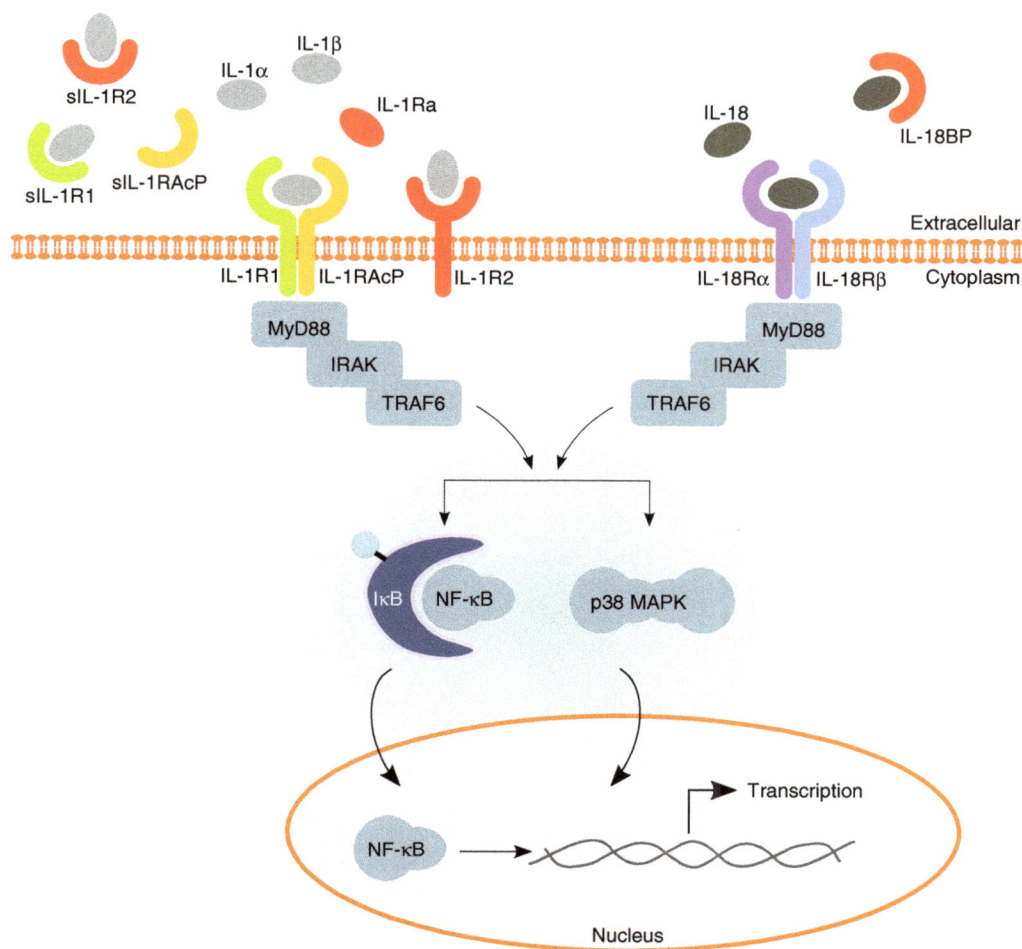

Figure 1. Interleukin (IL)-1α, IL-1β, and IL-18 receptor complexes and downstream signaling. Pro-IL-1α, mature IL-1α, and mature IL-1β can all bind to IL-1 receptor 1 (IL-1R1), which enables recruitment of the IL-1 receptor accessory protein (IL-1RAcP) coreceptor. Approximation of the intracellular Toll/IL-1 receptor (TIR) domains of the IL-1R complex results in recruitment of MyD88 followed by a cascade of downstream events that ultimately result in the activation of important signaling proteins, such as mitogen-activated protein kinases (MAPKs, e.g., p38), as well as transcription factors, including nuclear factor (NF)-κB, which control expression of a number of inflammatory genes. Signaling through the IL-1R complex is modulated by inhibitory actions of membrane-bound IL-1R2, soluble IL-1 receptor antagonist (IL-1Ra), and soluble forms of the receptors (sIL-1R1, sIL-1R2, and sIL-1RAcP). On binding of IL-18 to IL-18Rα, IL-18Rβ then is recruited to the complex. Similar to the IL-1R complex, approximation of the two receptors results in recruitment of MyD88 and a downstream cascade activating signaling proteins and transcription factors. Signaling through the IL-18R complex is modulated by inhibitory actions of IL-18 binding protein (IL-18BP).

IL-1β and IL-18 is required for gaining biological activity (Afonina et al. 2011). When produced locally in inflamed tissues, IL-1β mainly functions to stimulate leukocyte activation. Because systemic IL-1β is a prominent inducer of fever and acute phase proteins, which at high levels may even trigger systemic inflammatory response syndrome, its levels in circulation must be tightly regulated (Fig. 1). Indeed, IL-1β is hardly detected in serum, which is probably because of its short half-life, as well as its rapid neutralization by soluble IL-1R1 and IL-1R2 (Smith et al. 2003; Lachmann et al. 2009; Gabay et al. 2010). IL-18, on the other hand, is readily detectable in serum, and homeostatic levels appear to correlate with age (Kleiner et al. 2013). Similar to IL-1β, IL-18 is sequestered and neutralized in circulation, in this case by soluble IL-18BP (Dinarello et al. 2013).

Caspase-1, the founding member of the vertebrate cysteine aspartate-specific protease family, processes IL-1β into a 17 kDa mature fragment (Howard et al. 1991). Similarly, caspase-1-mediated proteolytic processing of IL-18 results in a mature product of 17.2 kDa that is released extracellularly (Ghayur et al. 1997; Gu et al. 1997). Not only are the cytokines IL-1β and IL-18 themselves expressed as inactive proforms, but caspase-1 is also expressed as a cytosolic inactive zymogen. It gains proteolytic activity through proximity-induced conformational changes imposed by its recruitment in inflammasome complexes. Under physiological conditions, this results in caspase-1 autocleavage and enhanced cleavage of IL-1β and IL-18, although genetic studies have shown that caspase-1 automaturation is dispensable for caspase-1 protease activity (Broz et al. 2010; Van Opdenbosch et al. 2014).

Inflammasomes are defined as caspase-1-activating complexes. They are assembled in response to infections, pathogen-associated molecular patterns (PAMPs), and cellular stress. Major outcomes of inflammasome activation are a lytic form of cell death called pyroptosis and the concomitant release of IL-1 cytokines and alarmins/danger-associated molecular patterns (DAMPs) such as high-mobility group box 1 protein (HMGB1). Several members of the nucleotide-binding domain and leucine-rich repeat (NLR)-containing receptor family, when responding to intracellular danger, initiate the formation of the inflammasome. Also, AIM2 (absent in melanoma 2), a member of the HIN200/AIM2-like receptor (ALR) family, is an inflammasome sensor (Fernandes-Alnemri et al. 2009; Rathinam et al. 2010; Sauer et al. 2010; Choubey 2012; Jin et al. 2012). More recently, a member of the tripartite motif

(TRIM) family, Pyrin/TRIM20, was shown to assemble a functional inflammasome (Xu et al. 2014). To date, five distinct inflammasomes, named after the sensor, have been genetically validated in mice studies (Fig. 2). Although in humans only one *NLRP1* gene exists, mice encode three paralogs: *Nlrp1a*, *Nlrp1b*, and *Nlrp1c* (Boyden and Dietrich 2006). Gain-of-function mutations in *Nlrp1a* were shown to cause leukopenia and anemia in mice as a result of unwarranted inflammasome activation in bone marrow cells (Masters et al. 2012). *Bacillus anthracis* lethal toxin is the only defined biochemical agent that activates the NLRP1b inflammasome (Boyden and Dietrich 2006). Contrastingly, the NLRP3 inflammasome responds to a large set of molecules and insults. Uniquely, the NLRP3 inflammasome requires a two-step mechanism for activation. First TLR-priming provides for nuclear factor (NF)-κB-mediated transcriptional up-regulation of NLRP3 itself and pro-IL-1β (Bauernfeind et al. 2009). This sets the scene for NLRP3 activation through incompletely understood mechanisms on subsequent exposure to pore-forming agents, crystals, β-amyloids, and many other products. Indeed, DAMPs like extracellular ATP and hyaluronic acid; medically relevant crystalline products such as alum, CPPD, MSU, silica, and asbestos; ionophores such as nigericin; and β-fibrils can all trigger assembly of the NLRP3 inflammasome (Mariathasan et al. 2006; Martinon et al. 2006; Halle et al. 2008; Hornung et al. 2008). Moreover, the major component of the outer membrane of Gram-negative bacteria, lipopolysaccharide (LPS), was shown to activate NLRP3 through a noncanonical pathway involving caspase-11. On detection of intracellular LPS, caspase-11 autonomously induces pyroptosis, and through the NLRP3 inflammasome triggers caspase-1-dependent IL-1β and IL-18 maturation (Kayagaki et al. 2011; Shi et al. 2014). Intracellular detection of bacterial flagellin or specific type III secretion systems (T3SS) of, for example, *Salmonella enterica* serovar Typhimurium results in activation of the NLRC4 inflammasome. Cytosolic recognition of these bacterial factors by members of the NLR family apoptosis inhibitory protein (NAIP) cluster within the NLR fam-

Cite this article as *Cold Spring Harb Perspect Biol* doi: 10.1101/cshperspect.a028563

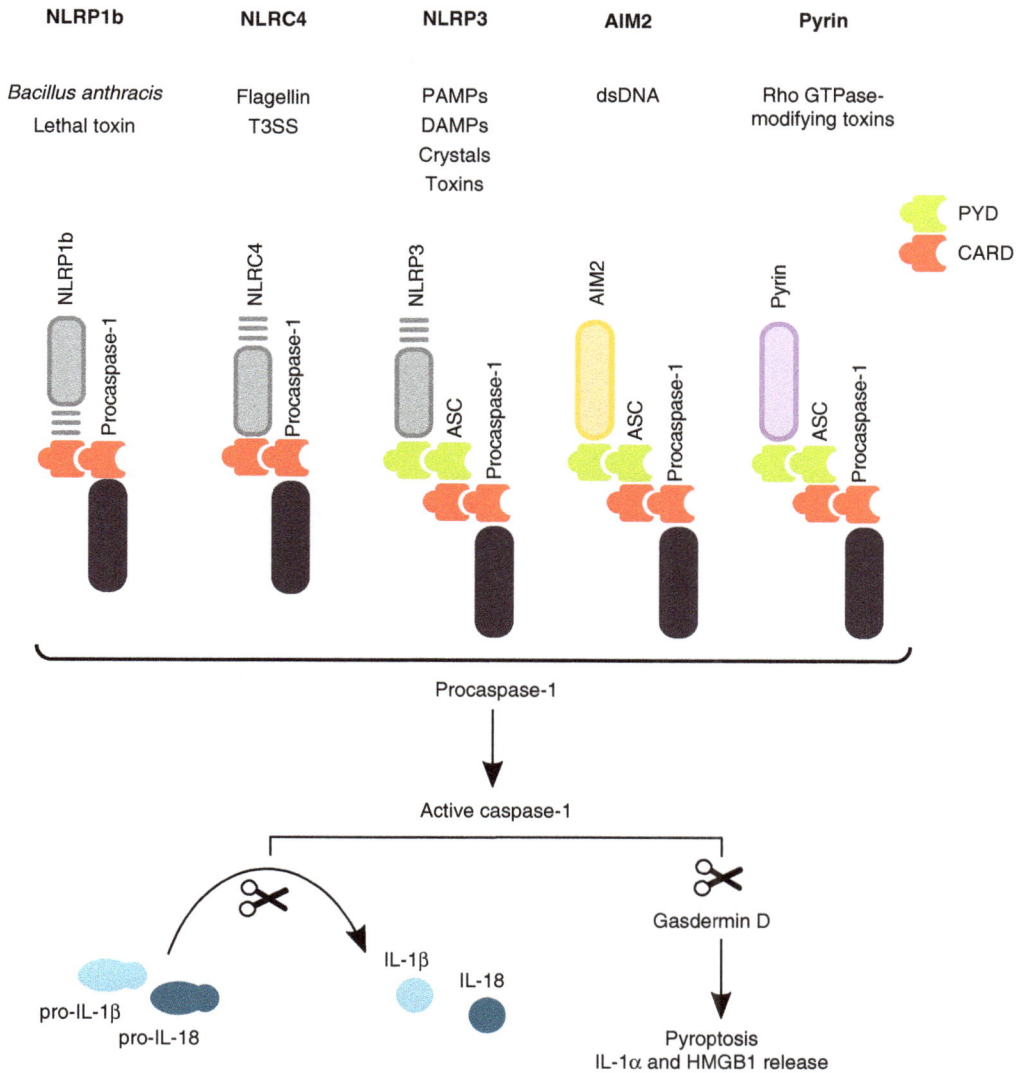

Figure 2. Schematic representation of inflammasome triggers and signaling events. Inflammasome-assembling proteins including the nucleotide-binding domain and leucine-rich repeat (NLR) containing family members NLRP1b, NLRC4, and NLRP3, as well as AIM2 and Pyrin detect, directly or through a secondary messenger, the presence of a pathogen and/or cellular damage. This results in their oligomerization and inflammasome assembly in which inactive procaspase-1 precursors are recruited. Although the PYD-based NLRP3, AIM2, and Pyrin platforms require the bipartite adaptor protein ASC to recruit caspase-1, the caspase recruitment domain (CARD)-based NLRP1b and NLRC4 inflammasomes recruit caspase-1 independently of ASC. Oligomerization of procaspase-1 leads to its autoactivation, and active caspase-1 catalyzes the maturation and secretion of the inflammatory cytokines interleukin (IL)-1β and IL-18. Additionally, active caspase-1 engages pyroptosis by cleaving its substrate gasdermin D. This lytic mode of programmed cell death coincides with the release of alarmins such as IL-1α and high-mobility group box 1 protein (HMGB1). T3SS, Type III secretion system; PAMPs, pathogen-associated molecular patterns; DAMPs, danger-associated molecular patterns; dsDNA, double-stranded DNA.

ily—along with Ser533 phosphorylation of NLRC4—mediates NLRC4 inflammasome activation (Mariathasan et al. 2004; Lightfield et al. 2011; Qu et al. 2012; Matusiak et al. 2015). On cytosolic recognition of viral (e.g., vaccinia virus), bacterial (e.g., *Francisella tularensis* and *Listeria monocytogenes*) or host-derived double-stranded DNA (dsDNA), the AIM2 inflammasome is assembled (Fernandes-Alnemri et al. 2010; Kim et al. 2010; Rathinam et al. 2010; Di Micco et al. 2016; Hu et al. 2016). Finally, it was recently established that RhoA GTPase-inactivating modifications induced by various bacterial toxins (e.g., *Clostridium difficile*) indirectly trigger activation of the Pyrin inflammasome (Xu et al. 2014).

Although a wide variety of PAMPs and DAMPs are able to trigger selective inflammasome assembly, the downstream effectors are largely equal in all inflammasomes (Fig. 2). The inflammasome adaptor apoptosis-associated speck-like protein containing a caspase recruitment domain, ASC, bridges interactions between inflammasome sensors and caspase-1 to promote caspase-1 oligomerization and the formation of a single micrometer-sized supramolecular fibril structure named the "ASC speck." ASC is essential for Pyrin and other PYD domain–containing sensors (NLRP3, AIM2) to recruit caspase-1, whereas the caspase recruitment domain-based sensors (NLRP1b and NLRC4) can also recruit caspase-1 directly as shown by the induction of caspase-1-mediated pyroptosis and cytokine secretion from macrophages of ASC-deficient mice (Broz et al. 2010; Van Opdenbosch et al. 2014).

INNATE AND ADAPTIVE IMMUNE STIMULATION BY INFLAMMASOME-DEPENDENT IL-1 FAMILY CYTOKINES

As discussed above, secreted IL-1α and IL-1β, which only share 26% homology at the protein level, are recruited to the same membrane-bound receptor (IL-1R), leading to similar immunological outcomes (March et al. 1985). An important difference, however, is that both full-length and mature IL-1α are able to bind IL-1R to induce signaling, while only mature IL-1β is able to do so (Mosley et al. 1987). IL-18 binds to IL-18Rα, which heterodimerizes with IL-18Rβ. Noteworthy, all IL-1 receptor family members contain a TIR domain responsible for recruitment of MyD88 and subsequent engagement of NF-κB and mitogen-activated protein kinase (MAPK)-dependent transcriptional reprogramming (Fig. 1) (Dinarello 2009). These receptors are differentially expressed on responder cells, enabling them to engage various facets of the immune system. Although IL-1α and IL-1β signal through the same receptors, these cytokines are biologically not fully redundant as studies using knockout mice showed IL-1α to be the predominant cytokine that promotes allergy responses and Th17 differentiation in the gut and the skin (Horai et al. 1998; Nakae et al. 2001; Chung et al. 2009). In contrast, a major role for IL-1β was attributed to induction of fever (Horai et al. 1998; Nakae et al. 2001; Chung et al. 2009). Apart from being a powerful endogenous pyrogen, IL-1β is also a potent recruiter and activator of neutrophils and macrophages. IL-1β is also well known to drive expression of adhesion molecules on immune cells to promote tissue infiltration of monocytes from circulation that instigate inflammatory responses (Wang et al. 1995). Additionally, lowered pain threshold, vasodilatation, and hypertension were reported to be regulated by IL-1 signaling. A role for IL-1 in the regulation of B-cell proliferation and antibody production was also suggested through its ability to stimulate production of IL-6 (Antoni et al. 1986). Originally, IL-18 was identified as interferon (IFN)-γ-inducing factor (IGIF) given its major role in the production of IFN-γ from T cells and natural killer (NK) cells. IL-18 was first seen as a Th1-promoting cytokine in the presence of IL-2 costimulation. IL-18 is now also known to promote severe airway inflammation through the production of Th2 cytokines by naïve or Th1 polarized T cells when combined with CD3 stimulation (Nakanishi et al. 2001; Hata et al. 2004). Taken together, these results show that inflammasome-dependent cytokines of the IL-1 family coordinate responses to PAMPs and DAMPs by impacting strongly on both innate and adaptive immune responses.

INFLAMMASOME-DEPENDENT IL-1 FAMILY CYTOKINES IN AUTOINFLAMMATORY DISEASES

The above illustrates the vital role of inflammasome-released cytokines and alarmins in sustaining homeostasis. However, when signaling is deregulated, it may become detrimental to the host, as shown by the extensive catalog of human diseases involving these cytokines and alarmins. Based on progressing insight in the genetic and functional etiology of noninfectious immune diseases, primary immunodeficiencies are increasingly regarded as a clinical continuum, ranging from monogenic autoinflammatory diseases that are mainly driven by aberrant innate immune signaling to equally rare monogenic autoimmune disorders that involve predominantly the adaptive immune response (McGonagle and McDermott 2006; Masters et al. 2009). The highly prevalent diseases that affect mankind are situated somewhere along this axis, involving both immune compartments to varying degrees. Intriguingly, inflammasome-released cytokines and alarmins appear to be implicated throughout the whole spectrum from autoinflammatory to autoimmune diseases, although inflammasome signaling has so far mainly been characterized in innate immune cells. This insight coincides with a steady increase in the number of clinical studies showing the efficacy of anti-IL-1 therapies in different diseases, although the underlying mechanisms and causal connections remain unclear in most cases. In the following sections, the contribution of inflammasome-released cytokines and alarmins in autoinflammatory diseases will be discussed. The interested reader is referred to other work for more detailed discussions on their roles in autoimmunity (Shaw et al. 2011; Yang and Chiang 2015).

As introduced before, autoinflammatory diseases are defined as illnesses caused by primary dysfunction of the innate immune system, and are thus frequently marked by the absence of pathogenic autoantibodies and autoreactive T cells (McGonagle and McDermott 2006; Masters et al. 2009). They are sometimes also referred to as "periodic fever syndromes" because many of these diseases feature recurrent fevers and episodes of systemic or organ-specific inflammation.

A clinical challenge is that efficient diagnosis is hampered by nonspecific symptoms that are shared by patients suffering from diseases with distinct etiologies. Moreover, patients suffering from autoinflammatory diseases with similar underlying mechanisms and that respond to particular therapies, may present with atypical or even distinctive symptoms. Thus, there is a growing need for functional assays and genetic markers to classify these diseases. Although this is work in progress that needs refinement whenever new information becomes available, classification based on the predominant proinflammatory cytokine(s) or inflammatory pathways involved may be sensible. As such, one may primarily distinguish diseases caused by excessive IL-1, type I IFN, tumor necrosis factor (TNF) production, and/or NF-κB signaling. Here, we focus on the IL-1-mediated autoinflammatory diseases (AIDs). The subset of IL-1-driven syndromes that are caused by mutations in core inflammasome proteins are further defined as "intrinsic inflammasomopathies" (Table 2). There are, however, also several diseases for which inflammasome involvement is hypothesized or in which mutations have been defined in proteins indirectly engaging inflammasome activation. They will be grouped as "extrinsic inflammasomopathies."

Intrinsic Inflammasomopathies

To date, autoinflammatory disorders caused by mutations in four out of the five well-established inflammasome sensors have been documented. Pathogenic NLRP1 mutations were shown to cause skin inflammatory and cancer susceptibility syndromes in the absence of recurring fever (Zhong et al. 2016) as well as an arthritis and dyskeratosis syndrome (Grandemange et al. 2016). Instead, mutations in NLRP3, NLRC4, and Pyrin are linked to systemic periodic fever syndromes. Caspase-1 variants also have been identified in patients suffering from recurrent systemic inflammation. Surprisingly, these *CASP1* variants with reduced or even abrogated

Table 2. Interleukin (IL)-1-mediated autoinflammatory diseases

Disease	Gene	Mode of inheritance	Treatments	Mouse model
Intrinsic inflammasomopathies				
CAPS	NLRP3	Autosomal dominant	Anti-IL-1	Nlrp3 knockin mice
FMF	MEFV	Autosomal recessive	Colchicine, anti-IL-1 in colchicine-resistent patients	Mefv knockin mice with mutant human B30.2 domain
NLRC4-AID	NLRC4	Autosomal dominant	Anti-IL-1, -IL-18, -IFN-γ (explorative, few patients)	–
DIRA	IL1RN	Autosomal recessive	Anti-IL-1	Il1rn-deficient mice
Extrinsic inflammasomopathies				
NLRP12AD	NLRP12	Autosomal dominant	Anti-IL-1 (temporary effect)	–
PAPA	PSTPIP1	Autosomal dominant	Anti-IL-1, -TNF	Pstpip1 knockin mice
TRAPS	TNFRSF1A	Autosomal dominant	Anti-IL-1, -TNF	Tnfrsf1a knockin mice
MKD/HIDS	MVK	Autosomal recessive	Anti-IL-1, -TNF	Heterozygous Mvk-deficient mice
Majeed syndrome	LPIN2	Autosomal recessive	Anti-IL-1	–

CAPS, Cryopyrin-associated periodic syndromes; FMF, familial Mediterranean fever; IFN, interferon; DIRA, deficiency of the IL-1 receptor antagonist; PAPA, pyogenic arthritis, pyoderma gangrenosum, and acne; TNF, tumor necrosis factor; TRAPS, TNF receptor-associated periodic syndrome; MKD, mevalonate kinase deficiency; HIDS, hyperimmunoglobulin D syndrome.

enzymatic activity nonetheless may create a proinflammatory environment that might be causal to the symptoms observed (Heymann et al. 2014). To our knowledge, however, no mutations in the inflammasome adaptor ASC have been reported to cause inflammatory disease.

Cryopyrin-Associated Periodic Syndromes (CAPS)

CAPS are a group of inherited inflammatory disorders encompassing three different phenotypes of increasing severity, namely, familial cold-induced autoinflammatory syndrome (FCAS), Muckle–Wells syndrome (MWS), and neonatal-onset multisystem inflammatory disease ([NOMID] also known as chronic infantile neurologic, cutaneous, articular [CINCA] syndrome). These disorders thus represent a spectrum of separate disease entities with overlapping clinical features of variable severity, even though sharing a common genetic cause. About 50 years after these syndromes were described, NLRP3 (CIAS1) was identified as the causal gene (Cuisset et al. 1999; Hoffman et al. 2001). More than 100 disease-causing variants in NLRP3 are

currently registered in the Infevers database (Milhavet et al. 2008). Most are missense mutations in exon 3, which encodes the central NACHT domain responsible for ATP binding and hydrolysis. CAPS disease is autosomal dominant with variable penetrance, and up to 70% of CAPS patients that appear NLRP3 mutation-negative by conventional Sanger DNA-sequencing methods may show somatic mosaicism for pathogenic NLRP3 mutations (Saito et al. 2008; Tanaka et al. 2011).

Mutant NLRP3 is thought to alter the balance for NLRP3 inflammasome nucleation, resulting in excessive inflammasome signaling in response to otherwise harmless environmental cues. This hypothesis is supported by the observation that cold exposure and low-dose LPS treatment of peripheral blood mononuclear cells (PBMCs) from CAPS patients trigger secretion of excessive IL-1β and IL-18 compared with levels of healthy controls (Janssen et al. 2004; Brydges et al. 2009). Furthermore, mouse strains harboring CAPS-associated mutations in Nlrp3 show elevated levels of IL-1β and IL-18 in serum and closely mimic human disease. Also, cultured cells from Nlrp3 mutant mice are hyperresponsive to inflammatory stimuli, similar to patients (Brydges et al. 2009). To assess the relative contribution of IL-1 and IL-18 signaling in CAPS, mutant mice were bred to mice deficient for the respective cytokine receptors. Notably, although both IL-1R1 and IL-18R were implicated, caspase-1 deletion appeared more effective in controlling ongoing inflammation. This suggests that release of pyroptosis-associated alarmins such as HMGB1 might also contribute to the disease. However, therapeutic neutralization of IL-1β is highly effective in controlling disease activity and leads to dramatic symptomatic improvement in the majority of CAPS patients despite the cytokine being largely undetectable in circulation of CAPS patients (Goldbach-Mansky et al. 2006; Gattorno et al. 2007; Brydges et al. 2013). Altogether, this suggests that excess IL-1β has a primary pathogenic role in CAPS disease, while additional inflammasome effector mechanisms may further support a vicious inflammatory cycle.

Familial Mediterranean Fever (FMF)

With an estimated 150,000 patients, FMF is considered the most common monogenic autoinflammatory disease worldwide. Most patients have autosomal recessive inheritance, although cases with apparent dominant disease also are relatively frequent. FMF usually has a childhood onset, and is characterized by recurrent attacks of fever associated with serositis. Its main long-term complication is amyloid A (AA) amyloidosis, a severe manifestation with poor prognosis. Since its suggested use in 1972, the microtubule polymerization inhibitor colchicine has become the gold standard for treatment in FMF, with an overall nonresponder rate of only 5%–10% (Goldfinger 1972; Zemer et al. 1986). Colchicine not only prevents FMF attacks, but also normalizes disease-associated complications (Zemer et al. 1986). FMF is particularly common around the Mediterranean basin, the Middle East, and the Caucasus, frequently affecting Jewish, Turkish, Armenian, Arab and Southern European, and North African populations. In these regions, the prevalence of FMF is between 1 in 500 and 1 in 1000.

Mutations in *MEFV* were shown to cause the disease in most FMF patients (Balow et al. 1997; French FMF Consortium 1997). To date, more than 310 MEFV sequence variants have been reported in the Infevers database (Milhavet et al. 2008). *MEFV* mutations are very common in Middle Eastern populations and the Mediterranean basin, with carrier rates reaching up to 1:5 individuals. This suggests an evolutionary benefit for heterozygous individuals living in these countries (Schaner et al. 2001; Fumagalli et al. 2009). Paradoxically, similar to apparently healthy individuals of the Mediterranean basin, only a single demonstrable *MEFV* mutation is found in as many as 20%–30% of patients diagnosed with FMF (Booty et al. 2009; Marek-Yagel et al. 2009; Ozen 2009). Moreover, a subset of 10%–20% of patients lacks *MEFV* mutations altogether (Ben-Zvi et al. 2015). Although the disease is less prevalent in Northern Europe (with estimated frequencies below 1:75,000), the disease has spread over the world with migrations of South European, North African, and Middle Eastern populations over the past centuries and in more recent times (Ozen and Bilginer 2014).

MEFV encodes Pyrin, a protein that is highly expressed in monocytes and neutrophils (Balow et al. 1997; French FMF Consortium 1997; Centola et al. 2000). Pyrin engages inflammasome activation in response to RhoA GTPase-inactivating bacterial toxins (Xu et al. 2014). Although this discovery sparked significant research in Pyrin regulation mechanisms and FMF etiology (Gao et al. 2016; Masters et al. 2016; Park et al. 2016; Van Gorp et al. 2016), the molecular mechanisms driving pathology in FMF are still poorly understood. As in most AID patients, IL-1β is barely detectable in the serum of FMF patients. IL-18, on the other hand, is elevated in serum of FMF patients, regardless of whether measured in patients experiencing an attack or when in remission (Haznedaroglu et al. 2005; Koga et al. 2016). These findings warrant further investigation of the potential role of IL-18 in FMF pathogenesis. Of note, mouse Pyrin lacks the carboxy-terminal B30.2 (PRY/SPRY) domain that is mutated in most FMF patients. Mice engineered to express Pyrin with a human B30.2 extension containing

FMF-associated mutations spontaneously develop IL-1-dependent inflammation (Chae et al. 2011). In agreement, disease symptoms in approximately 2/3 of FMF patients with inadequate response to colchicine were effectively treated with anti-IL-1 (Kuijk et al. 2007; Calligaris et al. 2008; Hashkes et al. 2012; van der Hilst et al. 2016).

Notably, despite the efficacious prophylactic use of colchicine in controlling FMF symptoms, FMF-associated mutations in the B30.2 domain were recently shown to dispense with the requirement for microtubules in Pyrin inflammasome activation (Van Gorp et al. 2016). Only FMF PBMCs activated the Pyrin inflammasome independently of microtubules because colchicine prevented Pyrin inflammasome activation in PBMCs of healthy individuals and patients suffering from other autoinflammatory diseases (CAPS and systemic juvenile idiopathic arthritis [sJIA]) (Van Gorp et al. 2016). The prophylactic efficacy of colchicine treatment is thus likely to be found downstream from Pyrin inflammasome activation. One likely explanation is that it may limit neutrophil recruitment into tissues by countering microtubule-dependent rolling, transmigration, and extravasation.

NLRC4-Associated Autoinflammatory Diseases (NLRC4-AIDs)

Several autoinflammatory patients were recently described with mutations in the inflammasome sensor NLRC4 (Canna et al. 2014; Kitamura et al. 2014; Romberg et al. 2014; Bracaglia et al. 2015; Kawasaki et al. 2016; Volker-Touw et al. 2016). NLRC4-AID may have a broad spectrum of clinical presentations. Patients have been reported with CAPS-like disease, or may suffer from episodic fever, enterocolitis, arthritis, cutaneous erythematous nodes, and/or urticarial rash. In some patients, NLRC4-AID may cause near-fatal autoinflammation associated with severe macrophage activation syndrome (MAS). The latter is a potentially lethal complication in rheumatic diseases associated with highly elevated ferritin and IL-18 levels in circulation. Although clinical presentation may vary considerably, all patients had an autosomal dominant inheritance of NLRC4 missense mutations (Canna et al. 2014; Kitamura et al. 2014; Romberg et al. 2014; Bracaglia et al. 2015; Volker-Touw et al. 2016). A single case with apparent de novo somatic mosaicism in NLRC4 has also been reported (Kawasaki et al. 2016). All disease-linked mutations are located within or close to the central NACHT domain. Mapping of these mutations on the crystal structure of murine NLRC4 (Hu et al. 2013) suggests that they may destabilize intramolecular interactions that keep NLRC4 in an autorepressed conformation (Canna et al. 2014). In agreement with a gain-of-function mode of operation, increased IL-18 and ferritin levels were detected in serum of several NLRC4-AID patients (Canna et al. 2014; Romberg et al. 2014; Bracaglia et al. 2015; Kawasaki et al. 2016; Volker-Touw et al. 2016). Unlike IL-18, IL-1β was barely detectable in circulation of these patients (Canna et al. 2014; Kawasaki et al. 2016; Volker-Touw et al. 2016).

Unlike in CAPS, clinical responses to anti-IL-1 therapy appear more disparate in NLRC4-AID. This is illustrated by clinical responses to anakinra—a recombinant IL-1Ra—in a large Dutch pedigree, which varied from resistance to complete remission (Volker-Touw et al. 2016). Moreover, anakinra treatment in another patient normalized some markers of systemic inflammation and enabled cessation of steroids, without affecting serum IL-18 levels (Canna et al. 2014). The above suggests that excessive IL-18 may contribute importantly to NLRC4-AID pathology. In agreement, an infant with NLRC4-AID and severe MAS that was refractory to anti-TNF and anti-IL-1β therapy was effectively treated with recombinant human IL-18BP, which binds and neutralizes bioactive IL-18 (Canna et al. 2016). IFN-γ, which is produced by T lymphocytes and NK cells in response to myeloid cell–derived IL-18 and IL-2, also may be a promising target for treatment of NLRC4-AID and associated MAS (Bracaglia et al. 2015). Experience from "emergency compassioned use" appears highly promising, and controlled studies therefore may expedite regulatory approval and adoption of these investigational therapeutic agents in NRC4-AID and other MAS-associated diseases.

Cite this article as *Cold Spring Harb Perspect Biol* doi: 10.1101/cshperspect.a028563

Deficiency of the IL-1 Receptor Antagonist

Deficiency of the IL-1 receptor antagonist (DIRA) is a very rare autosomal recessive disease caused by mutations in the *IL1RN* gene encoding IL-1Ra (Aksentijevich et al. 2009; Reddy et al. 2009). Pathogenic mutations result in defective IL-1Ra production, thus causing unopposed IL-1 signaling on engagement of the cognate receptor. IL-1R1 is ubiquitously expressed throughout tissues with particularly high expression levels noted in keratinocytes, possibly explaining the characteristic skin lesions and massive neutrophilic infiltration associated with the disease. DIRA is further associated with bone lesions and joint inflammation. If left untreated, DIRA presents a life-threatening condition resulting from the development of a systemic inflammatory response syndrome and death from multiorgan failure. Remarkably, however, recurrent fevers have not been reported to occur in these patients. Although lifelong treatment is required, patients respond very well to anti-IL-1 therapy.

Extrinsic Inflammasomopathies

Although intrinsic inflammasomopathies have molecular etiologies clearly tightening them to defined inflammasome signaling mechanisms as discussed above, extrinsic inflammasomopathies are disorders for which the potential link to aberrant inflammasome signaling still needs to be firmly established. As understanding of the roles of inflammasomes and IL-1 in these diseases progresses, some of these pathologies may require reclassification as intrinsic inflammasomopathies or inflammasome-independent syndromes in the future.

Majeed Syndrome

Majeed syndrome is a rare autosomal recessive disorder caused by mutations in *LPIN2*, a gene encoding a phosphatidic acid phosphatase (Ferguson et al. 2005). To date, 18 variants have been reported for Majeed syndrome in Infevers. Overlapping with the clinical presentation of DIRA patients, Majeed syndrome patients present with chronic recurrent multifocal osteomyelitis and neutrophilic skin lesions. Unlike DIRA, however, these symptoms are often accompanied by anemia and periodic fevers as part of the typical clinical picture of adult Majeed syndrome patients. Although the underlying molecular etiology of the disease is unclear, IL-1 likely is a key factor of pathogenesis in this syndrome as illustrated by anecdotal evidence showing a rapid and impressive clinical response to IL-1β-blocking therapies in two siblings with Majeed syndrome (Herlin et al. 2013).

NLRP12-Associated Autoinflammatory Disorder (NLRP12AD)

Missense mutations in *NLRP12* were reported to cause CAPS-like clinical manifestations (Jeru et al. 2008, 2011b; Borghini et al. 2011). To date, 35 *NLRP12* variants have been reported in the Infevers database. Akin to CAPS, NLRP12AD has an autosomal dominant inheritance, and PBMCs of NLRP12AD patients also were shown to have spontaneous and cold exposure–induced IL-1β release (Jeru et al. 2011a). Although the role of NLRP12 in inflammasome activation and secretion of IL-1β remains controversial, the clinical phenotype of NLRP12AD patients combined with cold-induced IL-1β release by ex vivo–stimulated PBMCs provided a rationale for exploring anti-IL-1 therapy in these patients. In contrast to CAPS patients in which anti-IL-1 is highly effective, anti-IL-1 initially led to marked but incomplete clinical improvement, and patients progressively relapsed after 3 months of treatment (Jeru et al. 2011a). Although larger studies are required, these findings suggest that NLRP12AD may differ mechanistically from CAPS. Further understanding of NLRP12's roles in the immune system and the underlying mechanisms may pave the way toward more effective therapeutic options for these patients.

Pyogenic Arthritis, Pyoderma Gangrenosum, and Acne Syndrome

Pyogenic arthritis, pyoderma gangrenosum, and acne (PAPA) syndrome is an autosomal dominant disease that is characterized by painful flares of recurrent arthritis with predominantly neutrophilic infiltration, and variable skin inflammation (Smith et al. 2010). In most patients, the disease arises from mutations in the proline–serine–threonine phosphatase-interacting protein 1 gene (*PSTPIP1*, also known as CD2-binding protein 1 [*CD2BP1*]) (Yeon et al. 2000; Wise et al. 2002). To date, 25 disease-associated variants of *PSTPIP1* have been reported in Infevers. PSTPIP1 is a cytoskeleton-associated protein with diverse binding partners, including the phosphatase rich in proline (P), glutamic acid (E), serine (S), and threonine (T) residues–type protein tyrosine phosphatase (PTP-PEST, also known as PTPN12), Wiskott–Aldrich syndrome protein (WASP), c-Abl kinase, Fas ligand, and the inflammasome protein Pyrin (Smith et al. 2010). Notably, PAPA-associated PSTPIP1 mutants show higher binding affinity for Pyrin compared with wild-type (WT) PSTPIP1 (Shoham et al. 2003), and disease-linked PSTPIP1 mutants were proposed to engage inflammasome activation through Pyrin (Yu et al. 2007; Waite et al. 2009). However, it is not clear why colchicine therapy, highly efficacious in FMF patients, does not alter disease progression in PAPA patients (Geusau et al. 2013). Nevertheless, anecdotal evidence suggests IL-1β blockade as a promising approach for controlling clinical manifestations in this ultrarare disease (Dierselhuis et al. 2005; Federici et al. 2013; Geusau et al. 2013), warranting further investigation in larger clinical trials.

Mevalonate Kinase Deficiency/Hyper-IgD Syndrome

In most patients, mevalonate kinase deficiency (MKD) is caused by homozygous loss-of-function mutations in the *MVK* gene encoding mevalonate kinase, although a subset of patients has no detectable MVK mutation, and others may display dominant autosomal disease associated with a single mutation. MVK is an enzyme of the cholesterol and isoprene biosynthesis pathway (van der Meer et al. 1984; Drenth et al. 1999; Houten et al. 1999). The disease was originally named hyperimmunoglobulin D syndrome (HIDS) because constitutively high levels

of immunoglobulin D (IgD) were observed in circulation of MKD patients, next to elevated levels of C-reactive protein, erythrocyte sedimentation rate, and other classical markers of ongoing inflammation. However, with the realization that not all MKD patients have increased serum IgD levels, that IgD levels do not appear to correlate with disease severity, and that elevated IgD levels are sometimes also detected in patients with other autoinflammatory diseases, the disease was renamed MKD to more accurately refer to the factor causing pathogenesis (Ammouri et al. 2007).

MKD is a very rare disease with currently 204 disease-associated *MVK* variants reported in the Infevers database. Most patients are of Caucasian background, with highest disease incidence in individuals of Dutch and French descent. Depending on the level of residual MVK enzymatic activity, the clinical spectrum ranges from mild to potentially lethal forms of mevalonic aciduria. In addition to recurrent fever, MKD patients may further display skin rash, arthralgia, myalgia, and arthritis. The threshold-dependent phenotype seen in patients is also reflected in mouse models, with MVK deficiency causing embryonic lethality, and haploinsufficient mice grossly phenocopying some serological parameters of the human disease (Hager et al. 2007).

Although the molecular etiology of MKD is still debated, it is clear that reduced MVK activity leads to build-up of mevalonic acid, and a shortage of cholesterol, vitamins, and other products of the isoprenoid biosynthesis pathway, which cause uncontrolled release of IL-1β through incompletely understood mechanisms. An attractive hypothesis suggests that defects in the mevalonate pathway may trigger unwarranted activation of the Pyrin inflammasome consequent to defective geranylgeranylation of RhoA GTPase (Mandey et al. 2006; Akula et al. 2016; Park et al. 2016). Surprisingly, however, MKD patients generally do not benefit from colchicine therapy, although on the other hand experience with IL-1 blockade has been impressive with ∼90% of patients having immediate and complete clinical responses (Ter Haar et al. 2013; Jesus and Goldbach-Mansky 2014).

TNF Receptor–Associated Periodic Syndrome (TRAPS)

TRAPS is an autosomal dominant disease caused by mutations in the *TNFRSF1A* gene coding for TNF receptor 1 (TNF-R1) (McDermott et al. 1999). More than 140 TRAPS-linked mutations in *TNFRSF1A* have been identified. TNF-R1 is a membrane-bound member of the death receptor family that initiates signaling on binding of its ligand TNF. Activation of TNF-R1 implies the formation of two spatially distinct TNF-R1 signaling complexes that have the capacity to signal either NF-κB activation (prosurvival) or programmed cell death. Most *TNFRSF1A* gene mutations that cause TRAPS result in accumulation of misfolded TNF-R1, and reduced expression of the mutated receptor at the plasma membrane. Starting from *TNFRSF1A* heterozygosity and the dominant inheritance of the disease, analysis of knockin mice with heterozygous mutations in TNF-R1 revealed that WT and mutant TNF-R1 act in concert from distinct cellular locations to potentiate inflammation in TRAPS (Simon et al. 2010). However, although mutations in TNF-R1 are the genetic basis of the disease, long-term use of TNF-inhibiting agents has been surprisingly ineffective in the clinic, with some biologicals even precipitating inflammatory attacks and worsening disease (Jacobelli et al. 2007; Nedjai et al. 2009; Quillinan et al. 2011; Bulua et al. 2012). Contrastingly, the use of both anakinra and the IL-1β-neutralizing monoclonal antibody canakinumab induced rapid control of symptoms in TRAPS patients with sustained clinical benefits during treatment (Simon et al. 2004; Gattorno et al. 2008, 2016; Brizi et al. 2012). Thus, although the pathogenesis of TRAPS is complex and still not completely understood, the success of anti-IL-1 therapies clearly indicates the importance of IL-1 signaling in TRAPS pathogenesis.

CONCLUDING REMARKS

As discussed above, autoinflammatory diseases may be caused by a myriad of genes regulating a diversity of innate immune and metabolic pathways. These defects converge on a select number of key inflammatory mediators, with inflammasome-produced IL-1β featuring prominently in this list. This is evidenced by the therapeutic efficacy of IL-1-blocking strategies in clinically diverse—though not all—autoinflammatory diseases. More recently, inflammasome-produced IL-18 and downstream IFN-γ production have emerged as potential therapeutic targets in the treatment of certain autoinflammatory syndromes and MAS. In addition, the existence of IL-1β/IL-18-independent inflammasome effector mechanisms, including pyroptosis and the associated release of alarmins such as HMGB1 and S100 proteins, highlight further avenues by which inflammasomes may contribute to autoinflammation. Undoubtedly, progressive understanding of the molecular underpinnings regulating inflammasome activation and production of the key cytokines IL-1β and IL-18 may provide much needed novel therapeutic targets for modulating inflammation and immunity. These endeavors should benefit patients suffering from autoinflammatory, as well as other immune-related diseases.

ACKNOWLEDGMENTS

We apologize to the authors whose work was omitted because of space limitations. We thank Dr. Lieselotte Vande Walle for help with artwork. N.V.O. is a postdoctoral fellow with the Fund for Scientific Research-Flanders. This work is supported by the European Research Council (Consolidator Grant No. 683144 and Proof-of-Concept Grant No. 727674) and the Baillet Latour Medical Research grant to M.L.

REFERENCES

Afonina IS, Tynan GA, Logue SE, Cullen SP, Bots M, Luthi AU, Reeves EP, McElvaney NG, Medema JP, Lavelle EC, et al. 2011. Granzyme B-dependent proteolysis acts as a switch to enhance the proinflammatory activity of IL-1α. *Mol Cell* **44:** 265–278.

Aksentijevich I, Masters SL, Ferguson PJ, Dancey P, Frenkel J, van Royen-Kerkhoff A, Laxer R, Tedgard U, Cowen EW, Pham TH, et al. 2009. An autoinflammatory disease with deficiency of the interleukin-1-receptor antagonist. *N Engl J Med* **360:** 2426–2437.

Akula MK, Shi M, Jiang Z, Foster CE, Miao D, Li AS, Zhang X, Gavin RM, Forde SD, Germain G, et al. 2016. Control of the innate immune response by the mevalonate pathway. *Nat Immunol* **17:** 922–929.

Ammouri W, Cuisset L, Rouaghe S, Rolland MO, Delpech M, Grateau G, Ravet N. 2007. Diagnostic value of serum immunoglobulinaemia D level in patients with a clinical suspicion of hyper IgD syndrome. *Rheumatology (Oxford)* **46:** 1597–1600.

Antoni G, Presentini R, Perin F, Tagliabue A, Ghiara P, Censini S, Volpini G, Villa L, Boraschi D. 1986. A short synthetic peptide fragment of human interleukin 1 with immunostimulatory but not inflammatory activity. *J Immunol* **137:** 3201–3204.

Balow JE Jr, Shelton DA, Orsborn A, Mangelsdorf M, Aksentijevich I, Blake T, Sood R, Gardner D, Liu R, Pras E, et al. 1997. A high-resolution genetic map of the familial Mediterranean fever candidate region allows identification of haplotype-sharing among ethnic groups. *Genomics* **44:** 280–291.

Bauernfeind FG, Horvath G, Stutz A, Alnemri ES, MacDonald K, Speert D, Fernandes-Alnemri T, Wu J, Monks BG, Fitzgerald KA, et al. 2009. Cutting edge: NF-κB activating pattern recognition and cytokine receptors license NLRP3 inflammasome activation by regulating *NLRP3* expression. *J Immunol* **183:** 787–791.

Ben-Zvi I, Herskovizh C, Kukuy O, Kassel Y, Grossman C, Livneh A. 2015. Familial Mediterranean fever without *MEFV* mutations: A case-control study. *Orphanet J Rare Dis* **10:** 34.

Booty MG, Chae JJ, Masters SL, Remmers EF, Barham B, Le JM, Barron KS, Holland SM, Kastner DL, Aksentijevich I. 2009. Familial Mediterranean fever with a single *MEFV* mutation: Where is the second hit? *Arthritis Rheum* **60:** 1851–1861.

Boraschi D, Lucchesi D, Hainzl S, Leitner M, Maier E, Mangelberger D, Oostingh GJ, Pfaller T, Pixner C, Posselt G, et al. 2011. IL-37: A new anti-inflammatory cytokine of the IL-1 family. *Eur Cytokine Netw* **22:** 127–147.

Borghini S, Tassi S, Chiesa S, Caroli F, Carta S, Caorsi R, Fiore M, Delfino L, Lasiglie D, Ferraris C, et al. 2011. Clinical presentation and pathogenesis of cold-induced autoinflammatory disease in a family with recurrence of an *NLRP12* mutation. *Arthritis Rheum* **63:** 830–839.

Boyden ED, Dietrich WF. 2006. *Nalp1b* controls mouse macrophage susceptibility to anthrax lethal toxin. *Nat Genet* **38:** 240–244.

Bracaglia C, Gatto A, Pardeo M, Lapeyre G, Ferlin W, Nelson R, de Min C, De Benedetti F. 2015. Anti interferon-γ (IFNγ) monoclonal antibody treatment in a patient carrying an *NLRC4* mutation and severe hemophagocytic lymphohistiocytosis. *Pediatr Rheumatol* **13:** 1–2.

Brizi MG, Galeazzi M, Lucherini OM, Cantarini L, Cimaz R. 2012. Successful treatment of tumor necrosis factor receptor-associated periodic syndrome with canakinumab. *Ann Intern Med* **156:** 907–908.

Broz P, von Moltke J, Jones JW, Vance RE, Monack DM. 2010. Differential requirement for caspase-1 autoproteolysis in pathogen-induced cell death and cytokine processing. *Cell Host Microbe* **8:** 471–483.

Brydges SD, Mueller JL, McGeough MD, Pena CA, Misaghi A, Gandhi C, Putnam CD, Boyle DL, Firestein GS, Horner AA, et al. 2009. Inflammasome-mediated disease animal models reveal roles for innate but not adaptive immunity. *Immunity* **30:** 875–887.

Brydges SD, Broderick L, McGeough MD, Pena CA, Mueller JL, Hoffman HM. 2013. Divergence of IL-1, IL-18, and cell death in NLRP3 inflammasomopathies. *J Clin Invest* **123:** 4695–4705.

Bulua AC, Mogul DB, Aksentijevich I, Singh H, He DY, Muenz LR, Ward MM, Yarboro CH, Kastner DL, Siegel RM, et al. 2012. Efficacy of etanercept in the tumor necrosis factor receptor-associated periodic syndrome: A prospective, open-label, dose-escalation study. *Arthritis Rheum* **64:** 908–913.

Calligaris L, Marchetti F, Tommasini A, Ventura A. 2008. The efficacy of anakinra in an adolescent with colchicine-resistant familial Mediterranean fever. *Eur J Pediatr* **167:** 695–696.

Canna SW, de Jesus AA, Gouni S, Brooks SR, Marrero B, Liu Y, DiMattia MA, Zaal KJ, Sanchez GA, Kim H, et al. 2014. An activating *NLRC4* inflammasome mutation causes autoinflammation with recurrent macrophage activation syndrome. *Nat Genet* **46:** 1140–1146.

Canna SW, Girard C, Malle L, de Jesus A, Romberg N, Kelsen J, Surrey LF, Russo P, Sleight A, Schiffrin E, et al. 2016. Life-threatening *NLRC4*-associated hyperinflammation successfully treated with IL-18 inhibition. *J Allergy Clin Immunol* **139:** 1698–1701.

Centola M, Wood G, Frucht DM, Galon J, Aringer M, Farrell C, Kingma DW, Horwitz ME, Mansfield E, Holland SM, et al. 2000. The gene for familial Mediterranean fever, *MEFV*, is expressed in early leukocyte development and is regulated in response to inflammatory mediators. *Blood* **95:** 3223–3231.

Chae JJ, Cho YH, Lee GS, Cheng J, Liu PP, Feigenbaum L, Katz SI, Kastner DL. 2011. Gain-of-function pyrin mutations induce NLRP3 protein-independent interleukin-1β activation and severe autoinflammation in mice. *Immunity* **34:** 755–768.

Choubey D. 2012. DNA-responsive inflammasomes and their regulators in autoimmunity. *Clin Immunol* **142:** 223–231.

Chung Y, Chang SH, Martinez GJ, Yang XO, Nurieva R, Kang HS, Ma L, Watowich SS, Jetten AM, Tian Q, et al. 2009. Critical regulation of early Th17 cell differentiation by interleukin-1 signaling. *Immunity* **30:** 576–587.

Corbaz A, ten Hove T, Herren S, Graber P, Schwartsburd B, Belzer I, Harrison J, Plitz T, Kosco-Vilbois MH, Kim SH, et al. 2002. IL-18-binding protein expression by endothelial cells and macrophages is up-regulated during active Crohn's disease. *J Immunol* **168:** 3608–3616.

Cuisset L, Drenth JP, Berthelot JM, Meyrier A, Vaudour G, Watts RA, Scott DG, Nicholls A, Pavek S, Vasseur C, et al. 1999. Genetic linkage of the Muckle–Wells syndrome to chromosome 1q44. *Am J Hum Genet* **65:** 1054–1059.

Dierselhuis MP, Frenkel J, Wulffraat NM, Boelens JJ. 2005. Anakinra for flares of pyogenic arthritis in PAPA syndrome. *Rheumatology (Oxford)* **44:** 406–408.

Di Micco A, Frera G, Lugrin J, Jamilloux Y, Hsu ET, Tardivel A, De Gassart A, Zaffalon L, Bujisic B, Siegert S, et al. 2016. AIM2 inflammasome is activated by pharmacological disruption of nuclear envelope integrity. *Proc Natl Acad Sci* **113:** E4671–E4680.

Dinarello CA. 1996. Biologic basis for interleukin-1 in disease. *Blood* **87:** 2095–2147.

Dinarello CA. 2009. Immunological and inflammatory functions of the interleukin-1 family. *Annu Rev Immunol* 27: 519–550.

Dinarello CA, Novick D, Kim S, Kaplanski G. 2013. Interleukin-18 and IL-18 binding protein. *Front Immunol* 4: 289.

Drenth JP, Cuisset L, Grateau G, Vasseur C, van de Velde-Visser SD, de Jong JG, Beckmann JS, van der Meer JW, Delpech M. 1999. Mutations in the gene encoding mevalonate kinase cause hyper-IgD and periodic fever syndrome. International Hyper-IgD Study Group. *Nat Genet* 22: 178–181.

Dripps DJ, Brandhuber BJ, Thompson RC, Eisenberg SP. 1991. Interleukin-1 (IL-1) receptor antagonist binds to the 80-kDa IL-1 receptor but does not initiate IL-1 signal transduction. *J Biol Chem* 266: 10331–10336.

Federici S, Martini A, Gattorno M. 2013. The central role of anti-IL-1 blockade in the treatment of monogenic and multi-factorial autoinflammatory diseases. *Front Immunol* 4: 351.

Ferguson PJ, Chen S, Tayeh MK, Ochoa L, Leal SM, Pelet A, Munnich A, Lyonnet S, Majeed HA, El-Shanti H. 2005. Homozygous mutations in *LPIN2* are responsible for the syndrome of chronic recurrent multifocal osteomyelitis and congenital dyserythropoietic anaemia (Majeed syndrome). *J Med Genet* 42: 551–557.

Fernandes-Alnemri T, Yu JW, Datta P, Wu J, Alnemri ES. 2009. AIM2 activates the inflammasome and cell death in response to cytoplasmic DNA. *Nature* 458: 509–513.

Fernandes-Alnemri T, Yu JW, Juliana C, Solorzano L, Kang S, Wu J, Datta P, McCormick M, Huang L, McDermott E, et al. 2010. The AIM2 inflammasome is critical for innate immunity to *Francisella tularensis*. *Nat Immunol* 11: 385–393.

French FMF Consortium. 1997. A candidate gene for familial Mediterranean fever. *Nat Genet* 17: 25–31.

Fumagalli M, Cagliani R, Pozzoli U, Riva S, Comi GP, Menozzi G, Bresolin N, Sironi M. 2009. A population genetics study of the familial Mediterranean fever gene: Evidence of balancing selection under an overdominance regime. *Genes Immun* 10: 678–686.

Gabay C, Lamacchia C, Palmer G. 2010. IL-1 pathways in inflammation and human diseases. *Nat Rev Rheumatol* 6: 232–241.

Gao W, Yang J, Liu W, Wang Y, Shao F. 2016. Site-specific phosphorylation and microtubule dynamics control Pyrin inflammasome activation. *Proc Natl Acad Sci* 113: E4857–4866.

Gattorno M, Tassi S, Carta S, Delfino L, Ferlito F, Pelagatti MA, D'Osualdo A, Buoncompagni A, Alpigiani MG, Alessio M, et al. 2007. Pattern of interleukin-1β secretion in response to lipopolysaccharide and ATP before and after interleukin-1 blockade in patients with *CIAS1* mutations. *Arthritis Rheum* 56: 3138–3148.

Gattorno M, Pelagatti MA, Meini A, Obici L, Barcellona R, Federici S, Buoncompagni A, Plebani A, Merlini G, Martini A. 2008. Persistent efficacy of anakinra in patients with tumor necrosis factor receptor-associated periodic syndrome. *Arthritis Rheum* 58: 1516–1520.

Gattorno M, Obici L, Cattalini M, Tormey V, Abrams K, Davis N, Speziale A, Bhansali SG, Martini A, Lachmann HJ. 2016. Canakinumab treatment for patients with active

recurrent or chronic TNF receptor-associated periodic syndrome (TRAPS): An open-label, phase II study. *Ann Rheum Dis* 76: 173–178.

Geusau A, Mothes-Luksch N, Nahavandi H, Pickl WF, Wise CA, Pourpak Z, Ponweiser E, Eckhart L, Sunder-Plassmann R. 2013. Identification of a homozygous *PSTPIP1* mutation in a patient with a PAPA-like syndrome responding to canakinumab treatment. *JAMA Dermatol* 149: 209–215.

Ghayur T, Banerjee S, Hugunin M, Butler D, Herzog L, Carter A, Quintal L, Sekut L, Talanian R, Paskind M, et al. 1997. Caspase-1 processes IFN-γ-inducing factor and regulates LPS-induced IFN-γ production. *Nature* 386: 619–623.

Goldbach-Mansky R, Dailey NJ, Canna SW, Gelabert A, Jones J, Rubin BI, Kim HJ, Brewer C, Zalewski C, Wiggs E, et al. 2006. Neonatal-onset multisystem inflammatory disease responsive to interleukin-1β inhibition. *N Engl J Med* 355: 581–592.

Goldfinger SE. 1972. Colchicine for familial Mediterranean fever. *N Engl J Med* 287: 1302.

Grandemange S, Sanchez E, Louis-Plence P, Tran Mau-Them F, Bessis D, Coubes C, Frouin E, Seyger M, Girard M, Puechberty J, et al. 2016. A new autoinflammatory and autoimmune syndrome associated with *NLRP1* mutations: NAIAD (NLRP1-associated autoinflammation with arthritis and dyskeratosis). *Ann Rheum Dis* 76: 1191–1198.

Gu Y, Kuida K, Tsutsui H, Ku G, Hsiao K, Fleming MA, Hayashi N, Higashino K, Okamura H, Nakanishi K, et al. 1997. Activation of interferon-γ inducing factor mediated by interleukin-1β converting enzyme. *Science* 275: 206–209.

Hager EJ, Tse HM, Piganelli JD, Gupta M, Baetscher M, Tse TE, Pappu AS, Steiner RD, Hoffmann GF, Gibson KM. 2007. Deletion of a single mevalonate kinase (Mvk) allele yields a murine model of hyper-IgD syndrome. *J Inherit Metab Dis* 30: 888–895.

Halle A, Hornung V, Petzold GC, Stewart CR, Monks BG, Reinheckel T, Fitzgerald KA, Latz E, Moore KJ, Golenbock DT. 2008. The NALP3 inflammasome is involved in the innate immune response to amyloid-β. *Nat Immunol* 9: 857–865.

Hashkes PJ, Spalding SJ, Giannini EH, Huang B, Johnson A, Park G, Barron KS, Weisman MH, Pashinian N, Reiff AO, et al. 2012. Rilonacept for colchicine-resistant or -intolerant familial Mediterranean fever: A randomized trial. *Ann Intern Med* 157: 533–541.

Hata H, Yoshimoto T, Hayashi N, Hada T, Nakanishi K. 2004. IL-18 together with anti-CD3 antibody induces human Th1 cells to produce Th1- and Th2-cytokines and IL-8. *Int Immunol* 16: 1733–1739.

Haznedaroglu S, Ozturk MA, Sancak B, Goker B, Onat AM, Bukan N, Ertenli I, Kiraz S, Calguneri M. 2005. Serum interleukin 17 and interleukin 18 levels in familial Mediterranean fever. *Clin Exp Rheumatol* 23: S77–S80.

Herlin T, Fiirgaard B, Bjerre M, Kerndrup G, Hasle H, Bing X, Ferguson PJ. 2013. Efficacy of anti-IL-1 treatment in Majeed syndrome. *Ann Rheum Dis* 72: 410–413.

Heymann MC, Winkler S, Luksch H, Flecks S, Franke M, Russ S, Ozen S, Yilmaz E, Klein C, Kallinich T, et al. 2014. Human procaspase-1 variants with decreased enzymatic activity are associated with febrile episodes and may contribute to inflammation via RIP2 and NF-κB signaling. *J Immunol* 192: 4379–4385.

Hoffman HM, Mueller JL, Broide DH, Wanderer AA, Kolodner RD. 2001. Mutation of a new gene encoding a putative pyrin-like protein causes familial cold autoinflammatory syndrome and Muckle–Wells syndrome. *Nat Genet* 29: 301–305.

Horai R, Asano M, Sudo K, Kanuka H, Suzuki M, Nishihara M, Takahashi M, Iwakura Y. 1998. Production of mice deficient in genes for interleukin (IL)-1α, IL-1β, IL-1α/β, and IL-1 receptor antagonist shows that IL-1β is crucial in turpentine-induced fever development and glucocorticoid secretion. *J Exp Med* 187: 1463–1475.

Hornung V, Bauernfeind F, Halle A, Samstad EO, Kono H, Rock KL, Fitzgerald KA, Latz E. 2008. Silica crystals and aluminum salts activate the NALP3 inflammasome through phagosomal destabilization. *Nat Immunol* 9: 847–856.

Houten SM, Kuis W, Duran M, de Koning TJ, van Royen-Kerkhof A, Romeijn GJ, Frenkel J, Dorland L, de Barse MM, Huijbers WA, et al. 1999. Mutations in *MVK*, encoding mevalonate kinase, cause hyperimmunoglobulinaemia D and periodic fever syndrome. *Nat Genet* 22: 175–177.

Howard AD, Kostura MJ, Thornberry N, Ding GJ, Limjuco G, Weidner J, Salley JP, Hogquist KA, Chaplin DD, Mumford RA, et al. 1991. IL-1-converting enzyme requires aspartic acid residues for processing of the IL-1β precursor at two distinct sites and does not cleave 31-kDa IL-1 α. *J Immunol* 147: 2964–2969.

Hu Z, Yan C, Liu P, Huang Z, Ma R, Zhang C, Wang R, Zhang Y, Martinon F, Miao D, et al. 2013. Crystal structure of NLRC4 reveals its autoinhibition mechanism. *Science* 341: 172–175.

Hu B, Jin C, Li HB, Tong J, Ouyang X, Cetinbas NM, Zhu S, Strowig T, Lam FC, Zhao C, et al. 2016. The DNA-sensing AIM2 inflammasome controls radiation-induced cell death and tissue injury. *Science* 354: 765–768.

Jacobelli S, Andre M, Alexandra JF, Dode C, Papo T. 2007. Failure of anti-TNF therapy in TNF receptor 1-associated periodic syndrome (TRAPS). *Rheumatology (Oxford)* 46: 1211–1212.

Janssen R, Verhard E, Lankester A, Ten Cate R, van Dissel JT. 2004. Enhanced interleukin-18 and interleukin-18 release in a patient with chronic infantile neurologic, cutaneous, articular syndrome. *Arthritis Rheum* 50: 3329–3333.

Jeru I, Duquesnoy P, Fernandes-Alnemri T, Cochet E, Yu JW, Lackmy-Port-Lis M, Grimprel E, Landman-Parker J, Hentgen V, Marlin S, et al. 2008. Mutations in *NALP12* cause hereditary periodic fever syndromes. *Proc Natl Acad Sci* 105: 1614–1619.

Jeru I, Hentgen V, Normand S, Duquesnoy P, Cochet E, Delwail A, Grateau G, Marlin S, Amselem S, Lecron JC. 2011a. Role of interleukin-1β in NLRP12-associated autoinflammatory disorders and resistance to anti-interleukin-1 therapy. *Arthritis Rheum* 63: 2142–2148.

Jeru I, Le Borgne G, Cochet E, Hayrapetyan H, Duquesnoy P, Grateau G, Morali A, Sarkisian T, Amselem S. 2011b. Identification and functional consequences of a recurrent *NLRP12* missense mutation in periodic fever syndromes. *Arthritis Rheum* 63: 1459–1464.

Jesus AA, Goldbach-Mansky R. 2014. IL-1 blockade in autoinflammatory syndromes. *Annu Rev Med* 65: 223–244.

Jin T, Perry A, Jiang J, Smith P, Curry JA, Unterholzner L, Jiang Z, Horvath G, Rathinam VA, Johnstone RW, et al. 2012. Structures of the HIN domain:DNA complexes reveal ligand binding and activation mechanisms of the AIM2 inflammasome and IFI16 receptor. *Immunity* 36: 561–571.

Kawasaki Y, Oda H, Ito J, Niwa A, Tanaka T, Hijikata A, Seki R, Nagahashi A, Osawa M, Asaka I, et al. 2016. Identification of a high-frequency somatic *NLRC4* mutation as a cause of autoinflammation by pluripotent cell-based phenotype dissection. *Arthritis Rheumatol* 69: 447–459.

Kayagaki N, Warming S, Lamkanfi M, Vande Walle L, Louie S, Dong J, Newton K, Qu Y, Liu J, Heldens S, et al. 2011. Non-canonical inflammasome activation targets caspase-11. *Nature* 479: 117–121.

Kim S, Bauernfeind F, Ablasser A, Hartmann G, Fitzgerald KA, Latz E, Hornung V. 2010. *Listeria monocytogenes* is sensed by the NLRP3 and AIM2 inflammasome. *Eur J Immunol* 40: 1545–1551.

Kitamura A, Sasaki Y, Abe T, Kano H, Yasutomo K. 2014. An inherited mutation in *NLRC4* causes autoinflammation in human and mice. *J Exp Med* 211: 2385–2396.

Kleiner G, Marcuzzi A, Zanin V, Monasta L, Zauli G. 2013. Cytokine levels in the serum of healthy subjects. *Mediators Inflamm* 2013: 434010.

Koga T, Migita K, Sato S, Umeda M, Nonaka F, Kawashiri SY, Iwamoto N, Ichinose K, Tamai M, Nakamura H, et al. 2016. Multiple serum cytokine profiling to identify combinational diagnostic biomarkers in attacks of familial Mediterranean fever. *Medicine (Baltimore)* 95: e3449.

Kuijk LM, Govers AM, Frenkel J, Hofhuis WJ. 2007. Effective treatment of a colchicine-resistant familial Mediterranean fever patient with anakinra. *Ann Rheum Dis* 66: 1545–1546.

Lachmann HJ, Lowe P, Felix SD, Rordorf C, Leslie K, Madhoo S, Wittkowski H, Bek S, Hartmann N, Bosset S, et al. 2009. In vivo regulation of interleukin 1β in patients with cryopyrin-associated periodic syndromes. *J Exp Med* 206: 1029–1036.

Lefrancais E, Roga S, Gautier V, Gonzalez-de-Peredo A, Monsarrat B, Girard JP, Cayrol C. 2012. IL-33 is processed into mature bioactive forms by neutrophil elastase and cathepsin G. *Proc Natl Acad Sci* 109: 1673–1678.

Lightfield KL, Persson J, Trinidad NJ, Brubaker SW, Kofoed EM, Sauer JD, Dunipace EA, Warren SE, Miao EA, Vance RE. 2011. Differential requirements for NAIP5 in activation of the NLRC4 inflammasome. *Infect Immun* 79: 1606–1614.

Lin H, Ho AS, Haley-Vicente D, Zhang J, Bernal-Fussell J, Pace AM, Hansen D, Schweighofer K, Mize NK, Ford JE. 2001. Cloning and characterization of IL-1HY2, a novel interleukin-1 family member. *J Biol Chem* 276: 20597–20602.

Mandey SH, Kuijk LM, Frenkel J, Waterham HR. 2006. A role for geranylgeranylation in interleukin-1β secretion. *Arthritis Rheum* 54: 3690–3695.

March CJ, Mosley B, Larsen A, Cerretti DP, Braedt G, Price V, Gillis S, Henney CS, Kronheim SR, Grabstein K, et al. 1985. Cloning, sequence and expression of two distinct human interleukin-1 complementary DNAs. *Nature* 315: 641–647.

Cite this article as *Cold Spring Harb Perspect Biol* doi: 10.1101/cshperspect.a028563

Marek-Yagel D, Berkun Y, Padeh S, Abu A, Reznik-Wolf H, Livneh A, Pras M, Pras E. 2009. Clinical disease among patients heterozygous for familial Mediterranean fever. *Arthritis Rheum* **60:** 1862–1866.

Mariathasan S, Newton K, Monack DM, Vucic D, French DM, Lee WP, Roose-Girma M, Erickson S, Dixit VM. 2004. Differential activation of the inflammasome by caspase-1 adaptors ASC and Ipaf. *Nature* **430:** 213–218.

Mariathasan S, Weiss DS, Newton K, McBride J, O'Rourke K, Roose-Girma M, Lee WP, Weinrauch Y, Monack DM, Dixit VM. 2006. Cryopyrin activates the inflammasome in response to toxins and ATP. *Nature* **440:** 228–232.

Martinon F, Petrilli V, Mayor A, Tardivel A, Tschopp J. 2006. Gout-associated uric acid crystals activate the NALP3 inflammasome. *Nature* **440:** 237–241.

Masters SL, Simon A, Aksentijevich I, Kastner DL. 2009. Horror autoinflammaticus: The molecular pathophysiology of autoinflammatory disease. *Annu Rev Immunol* **27:** 621–668.

Masters SL, Gerlic M, Metcalf D, Preston S, Pellegrini M, O'Donnell JA, McArthur K, Baldwin TM, Chevrier S, Nowell CJ, et al. 2012. NLRP1 inflammasome activation induces pyroptosis of hematopoietic progenitor cells. *Immunity* **37:** 1009–1023.

Masters SL, Lagou V, Jeru I, Baker PJ, Van Eyck L, Parry DA, Lawless D, De Nardo D, Garcia-Perez JE, Dagley LF, et al. 2016. Familial autoinflammation with neutrophilic dermatosis reveals a regulatory mechanism of pyrin activation. *Sci Transl Med* **8:** 332ra345.

Matusiak M, Van Opdenbosch N, Vande Walle L, Sirard JC, Kanneganti TD, Lamkanfi M. 2015. Flagellin-induced NLRC4 phosphorylation primes the inflammasome for activation by NAIP5. *Proc Natl Acad Sci* **112:** 1541–1546.

McDermott MF, Aksentijevich I, Galon J, McDermott EM, Ogunkolade BW, Centola M, Mansfield E, Gadina M, Karenko L, Pettersson T, et al. 1999. Germline mutations in the extracellular domains of the 55 kDa TNF receptor, TNFR1, define a family of dominantly inherited autoinflammatory syndromes. *Cell* **97:** 133–144.

McGonagle D, McDermott MF. 2006. A proposed classification of the immunological diseases. *PLoS Med* **3:** e297.

McMahan CJ, Slack JL, Mosley B, Cosman D, Lupton SD, Brunton LL, Grubin CE, Wignall JM, Jenkins NA, Brannan CI, et al. 1991. A novel IL-1 receptor, cloned from B cells by mammalian expression, is expressed in many cell types. *EMBO J* **10:** 2821–2832.

Milhavet F, Cuisset L, Hoffman HM, Slim R, El-Shanti H, Aksentijevich I, Lesage S, Waterham H, Wise C, Sarrauste de Menthiere C, et al. 2008. The Infevers autoinflammatory mutation online registry: Update with new genes and functions. *Hum Mutat* **29:** 803–808.

Mosley B, Urdal DL, Prickett KS, Larsen A, Cosman D, Conlon PJ, Gillis S, Dower SK. 1987. The interleukin-1 receptor binds the human interleukin-1α precursor but not the interleukin-1β precursor. *J Biol Chem* **262:** 2941–2944.

Nakae S, Naruse-Nakajima C, Sudo K, Horai R, Asano M, Iwakura Y. 2001. IL-1α, but not IL-1β, is required for contact-allergen-specific T cell activation during the sensitization phase in contact hypersensitivity. *Int Immunol* **13:** 1471–1478.

Nakanishi K, Yoshimoto T, Tsutsui H, Okamura H. 2001. Interleukin-18 regulates both Th1 and Th2 responses. *Annu Rev Immunol* **19:** 423–474.

Nedjai B, Hitman GA, Quillinan N, Coughlan RJ, Church L, McDermott MF, Turner MD. 2009. Proinflammatory action of the antiinflammatory drug infliximab in tumor necrosis factor receptor-associated periodic syndrome. *Arthritis Rheum* **60:** 619–625.

Netea MG, van de Veerdonk FL, van der Meer JW, Dinarello CA, Joosten LA. 2015. Inflammasome-independent regulation of IL-1-family cytokines. *Annu Rev Immunol* **33:** 49–77.

Novick D, Kim SH, Fantuzzi G, Reznikov LL, Dinarello CA, Rubinstein M. 1999. Interleukin-18 binding protein: A novel modulator of the Th1 cytokine response. *Immunity* **10:** 127–136.

Ozen S. 2009. Changing concepts in familial Mediterranean fever: Is it possible to have an autosomal-recessive disease with only one mutation? *Arthritis Rheum* **60:** 1575–1577.

Ozen S, Bilginer Y. 2014. A clinical guide to autoinflammatory diseases: Familial Mediterranean fever and next-of-kin. *Nat Rev Rheumatol* **10:** 135–147.

Palomo J, Dietrich D, Martin P, Palmer G, Gabay C. 2015. The interleukin (IL)-1 cytokine family—Balance between agonists and antagonists in inflammatory diseases. *Cytokine* **76:** 25–37.

Park YH, Wood G, Kastner DL, Chae JJ. 2016. Pyrin inflammasome activation and RhoA signaling in the autoinflammatory diseases FMF and HIDS. *Nat Immunol* **17:** 914–921.

Qu Y, Misaghi S, Izrael-Tomasevic A, Newton K, Gilmour LL, Lamkanfi M, Louie S, Kayagaki N, Liu J, Komuves L, et al. 2012. Phosphorylation of NLRC4 is critical for inflammasome activation. *Nature* **490:** 539–542.

Quillinan N, Mannion G, Mohammad A, Coughlan R, Dickie LJ, McDermott MF, McGonagle D. 2011. Failure of sustained response to etanercept and refractoriness to anakinra in patients with T50M TNF-receptor-associated periodic syndrome. *Ann Rheum Dis* **70:** 1692–1693.

Rathinam VA, Jiang Z, Waggoner SN, Sharma S, Cole LE, Waggoner L, Vanaja SK, Monks BG, Ganesan S, Latz E, et al. 2010. The AIM2 inflammasome is essential for host defense against cytosolic bacteria and DNA viruses. *Nat Immunol* **11:** 395–402.

Reddy S, Jia S, Geoffrey R, Lorier R, Suchi M, Broeckel U, Hessner MJ, Verbsky J. 2009. An autoinflammatory disease due to homozygous deletion of the IL1RN locus. *N Engl J Med* **360:** 2438–2444.

Romberg N, Al Moussawi K, Nelson-Williams C, Stiegler AL, Loring E, Choi M, Overton J, Meffre E, Khokha MK, Huttner AJ, et al. 2014. Mutation of *NLRC4* causes a syndrome of enterocolitis and autoinflammation. *Nat Genet* **46:** 1135–1139.

Saito M, Nishikomori R, Kambe N, Fujisawa A, Tanizaki H, Takeichi K, Imagawa T, Iehara T, Takada H, Matsubayashi T, et al. 2008. Disease-associated *CIAS1* mutations induce monocyte death, revealing low-level mosaicism in mutation-negative cryopyrin-associated periodic syndrome patients. *Blood* **111:** 2132–2141.

Sauer JD, Witte CE, Zemansky J, Hanson B, Lauer P, Portnoy DA. 2010. *Listeria monocytogenes* triggers AIM2-mediated pyroptosis upon infrequent bacteriolysis in the macrophage cytosol. *Cell Host Microbe* **7:** 412–419.

Schaner P, Richards N, Wadhwa A, Aksentijevich I, Kastner D, Tucker P, Gumucio D. 2001. Episodic evolution of pyrin in primates: Human mutations recapitulate ancestral amino acid states. *Nat Genet* **27:** 318–321.

Schreuder H, Tardif C, Trump-Kallmeyer S, Soffientini A, Sarubbi E, Akeson A, Bowlin T, Yanofsky S, Barrett RW. 1997. A new cytokine-receptor binding mode revealed by the crystal structure of the IL-1 receptor with an antagonist. *Nature* **386:** 194–200.

Sharma S, Kulk N, Nold MF, Graf R, Kim SH, Reinhardt D, Dinarello CA, Bufler P. 2008. The IL-1 family member 7b translocates to the nucleus and down-regulates proinflammatory cytokines. *J Immunol* **180:** 5477–5482.

Shaw PJ, McDermott MF, Kanneganti TD. 2011. Inflammasomes and autoimmunity. *Trends Mol Med* **17:** 57–64.

Shi J, Zhao Y, Wang Y, Gao W, Ding J, Li P, Hu L, Shao F. 2014. Inflammatory caspases are innate immune receptors for intracellular LPS. *Nature* **514:** 187–192.

Shoham NG, Centola M, Mansfield E, Hull KM, Wood G, Wise CA, Kastner DL. 2003. Pyrin binds the PSTPIP1/CD2BP1 protein, defining familial Mediterranean fever and PAPA syndrome as disorders in the same pathway. *Proc Natl Acad Sci* **100:** 13501–13506.

Simon A, Bodar EJ, van der Hilst JC, van der Meer JW, Fiselier TJ, Cuppen MP, Drenth JP. 2004. Beneficial response to interleukin 1 receptor antagonist in traps. *Am J Med* **117:** 208–210.

Simon A, Park H, Maddipati R, Lobito AA, Bulua AC, Jackson AJ, Chae JJ, Ettinger R, de Koning HD, Cruz AC, et al. 2010. Concerted action of wild-type and mutant TNF receptors enhances inflammation in TNF receptor 1-associated periodic fever syndrome. *Proc Natl Acad Sci* **107:** 9801–9806.

Smith DE, Hanna R, Della F, Moore H, Chen H, Farese AM, MacVittie TJ, Virca GD, Sims JE. 2003. The soluble form of IL-1 receptor accessory protein enhances the ability of soluble type II IL-1 receptor to inhibit IL-1 action. *Immunity* **18:** 87–96.

Smith EJ, Allantaz F, Bennett L, Zhang D, Gao X, Wood G, Kastner DL, Punaro M, Aksentijevich I, Pascual V, et al. 2010. Clinical, molecular, and genetic characteristics of PAPA Syndrome: A review. *Curr Genomics* **11:** 519–527.

Tanaka N, Izawa K, Saito MK, Sakuma M, Oshima K, Ohara O, Nishikomori R, Morimoto T, Kambe N, Goldbach-Mansky R, et al. 2011. High incidence of *NLRP3* somatic mosaicism in patients with chronic infantile neurologic, cutaneous, articular syndrome: Results of an International Multicenter Collaborative Study. *Arthritis Rheum* **63:** 3625–3632.

Taylor SL, Renshaw BR, Garka KE, Smith DE, Sims JE. 2002. Genomic organization of the interleukin-1 locus. *Genomics* **79:** 726–733.

Ter Haar N, Lachmann H, Ozen S, Woo P, Uziel Y, Modesto C, Kone-Paut I, Cantarini L, Insalaco A, Neven B, et al. 2013. Treatment of autoinflammatory diseases: Results from the Eurofever Registry and a literature review. *Ann Rheum Dis* **72:** 678–685.

van der Hilst J, Moutschen M, Messiaen PE, Lauwerys BR, Vanderschueren S. 2016. Efficacy of anti-IL-1 treatment in familial Mediterranean fever: A systematic review of the literature. *Biologics* **10:** 75–80.

van der Meer JW, Vossen JM, Radl J, van Nieuwkoop JA, Meyer CJ, Lobatto S, van Furth R. 1984. Hyperimmunoglobulinaemia D and periodic fever: A new syndrome. *Lancet* **1:** 1087–1090.

Van Gorp H, Saavedraa PHV, de Vasconcelosa NM, Van Opdenbosch N, Vande Walle L, Matusiak M, Prencipe G, Insalaco A, Van Hauwermeiren F, Demon D, et al. 2016. Familial Mediterranean fever mutations lift the obligatory requirement for microtubules in Pyrin inflammasome activation. *Proc Natl Acad Sci* **113:** 14384–14389.

Van Opdenbosch N, Gurung P, Vande Walle L, Fossoul A, Kanneganti TD, Lamkanfi M. 2014. Activation of the NLRP1b inflammasome independently of ASC-mediated caspase-1 autoproteolysis and speck formation. *Nat Commun* **5:** 3209.

Volker-Touw CM, de Koning HD, Giltay J, de Kovel C, van Kempen TS, Oberndorff K, Boes M, van Steensel MA, van Well GT, Blokx WA, et al. 2016. Erythematous nodes, urticarial rash and arthralgias in a large pedigree with *NLRC4*-related autoinflammatory disease, expansion of the phenotype. *Br J Dermatol* **176:** 244–248.

Waite AL, Schaner P, Richards N, Balci-Peynircioglu B, Masters SL, Brydges SD, Fox M, Hong A, Yilmaz E, Kastner DL, et al. 2009. Pyrin modulates the intracellular distribution of PSTPIP1. *PLoS ONE* **4:** e6147.

Wang X, Feuerstein GZ, Gu JL, Lysko PG, Yue TL. 1995. Interleukin-1β induces expression of adhesion molecules in human vascular smooth muscle cells and enhances adhesion of leukocytes to smooth muscle cells. *Atherosclerosis* **115:** 89–98.

Wise CA, Gillum JD, Seidman CE, Lindor NM, Veile R, Bashiardes S, Lovett M. 2002. Mutations in *CD2BP1* disrupt binding to PTP PEST and are responsible for PAPA syndrome, an autoinflammatory disorder. *Hum Mol Genet* **11:** 961–969.

Xu H, Yang J, Gao W, Li L, Li P, Zhang L, Gong YN, Peng X, Xi JJ, Chen S, et al. 2014. Innate immune sensing of bacterial modifications of Rho GTPases by the Pyrin inflammasome. *Nature* **513:** 237–241.

Yang CA, Chiang BL. 2015. Inflammasomes and human autoimmunity: A comprehensive review. *J Autoimmun* **61:** 1–8.

Yeon HB, Lindor NM, Seidman JG, Seidman CE. 2000. Pyogenic arthritis, pyoderma gangrenosum, and acne syndrome maps to chromosome 15q. *Am J Hum Genet* **66:** 1443–1448.

Yu JW, Fernandes-Alnemri T, Datta P, Wu J, Juliana C, Solorzano L, McCormick M, Zhang Z, Alnemri ES. 2007. Pyrin activates the ASC pyroptosome in response to engagement by autoinflammatory *PSTPIP1* mutants. *Mol Cell* **28:** 214–227.

Zemer D, Pras M, Sohar E, Modan M, Cabili S, Gafni J. 1986. Colchicine in the prevention and treatment of the amyloidosis of familial Mediterranean fever. *N Engl J Med* **314:** 1001–1005.

Zhong FL, Mamai O, Sborgi L, Boussofara L, Hopkins R, Robinson K, Szeverenyi I, Takeichi T, Balaji R, Lau A, et al. 2016. Germline *NLRP1* mutations cause skin inflammatory and cancer susceptibility syndromes via inflammasome activation. *Cell* **167:** 187–202.

Cytokines, Transcription Factors, and the Initiation of T-Cell Development

Hiroyuki Hosokawa and Ellen V. Rothenberg

Division of Biology and Biological Engineering, California Institute of Technology, Pasadena, California 91125

Correspondence: hiroyuki@caltech.edu; evroth@its.caltech.edu

Multipotent blood progenitor cells migrate into the thymus and initiate the T-cell differentiation program. T-cell progenitor cells gradually acquire T-cell characteristics while shedding their multipotentiality for alternative fates. This process is supported by extracellular signaling molecules, including Notch ligands and cytokines, provided by the thymic microenvironment. T-cell development is associated with dynamic change of gene regulatory networks of transcription factors, which interact with these environmental signals. Together with Notch or pre-T-cell-receptor (TCR) signaling, cytokines always control proliferation, survival, and differentiation of early T cells, but little is known regarding their cross talk with transcription factors. However, recent results suggest ways that cytokines expressed in distinct intrathymic niches can specifically modulate key transcription factors. This review discusses how stage-specific roles of cytokines and transcription factors can jointly guide development of early T cells.

The thymus is the organ specialized to make T cells. T cells originate from hematopoietic stem and precursor cells in the bone marrow or fetal liver, which migrate to the thymus and acquire T-cell identity. Relatively small numbers of T-cell progenitors migrate into the thymus per day, but they respond to the new environment by undergoing multiple rounds of proliferation while initiating the T-cell differentiation program (Rothenberg 2000; Petrie and Zuniga-Pflucker 2007; Rothenberg et al. 2008; Love and Bhandoola 2011; Naito et al. 2011; Thompson and Zúñiga-Pflücker 2011; Rothenberg 2014; Yui and Rothenberg 2014). They then undergo T-cell lineage commitment, begin T-cell receptor (TCR) rearrangements, and thus generate αβTCR- or γδTCR-expressing T cells. The αβ T cells further diverge into different sublineages, such as CD4 T cells, CD8 T cells, natural killer T (NKT) cells and regulatory T (Treg) cells, ultimately to act as a "conductor" of the immune system "orchestra."

Thymocytes are divided into multiple phenotypically distinct stages that are defined by the expression of CD4, CD8, and other markers (Hayday and Pennington 2007; Rothenberg et al. 2008; Yang et al. 2010; Naito et al. 2011; Yui and Rothenberg 2014). T-cell development is initiated from the subpopulation that lacks the expression of both CD4 and CD8, thus called double-negative (DN) cells, which then become $CD4^+ CD8^+$ double-positive (DP) and subsequently differentiate into mature CD4 or CD8 single-positive (SP) cells. The earliest T-cell precursors in the thymus, called early thymic progenitor (ETP) or Kit-high double-negative 1 ($KIT^{++} DN1; CD44^+ CD25^-$), still harbor the potential to gain access to non-T alternative fates. These cells start expressing T-cell markers in the next stage, DN2a (KIT^{++} CD44$^+$ CD25$^+$), but commitment to the T-cell lineage occurs only at the following stage, DN2b (Kit$^+$ CD44$^+$ CD25$^+$). Then in the DN3a (KIT$^-$ CD44$^-$ CD25$^+$) stage, *TCRβ* gene rearrangement begins. This process enables some cells to express either a pre-TCR (TCRβ with invariant pre-TCRα) or a γδTCR. Pre-TCR-mediated signal transduction triggers transition of DN3a cells through DN3b into DN4 (Kit$^-$ CD44$^-$ CD25$^-$), followed by progression to the DP stage. DP thymocytes undergo *TCRα* gene rearrangement and begin to express fully assembled αβTCR. Then, they are subjected to a selection process, which is known as positive selection, to identify cells that express TCR with potentially useful ligand specificities. Positively selected thymocytes are allowed to differentiate into either CD4 helper T cells or CD8 cytotoxic T cells, known as CD4/CD8-lineage choice.

The special feature of the thymic cortical environment is its dense presentation of Notch ligand, primarily Delta-like ligand 4 (DLL4) (Love and Bhandoola 2011). Very early in the ETP stage, T-cell precursors become not only influenced by Notch-DLL4 interaction but dependent on it for optimal growth and survival. NOTCH1 molecules on the surface of lymphoid precursors interact with DLL4 on thymic stromal cells, driving lymphoid precursors to initiate the T-cell-specific developmental program. Engagement of cell-surface NOTCH1 by environmental Notch ligands triggers the proteolytic release of intracellular NOTCH1, which travels to the nucleus to become a direct coactivator of DNA-bound recombining binding protein suppressor of hairless (RBPJ) and stimulates the expression of Notch target genes (Radtke et al. 2010). All of the events that establish the T-cell identity of precursors are driven directly or indirectly by Notch signaling (Schmitt and Zuniga-Pflucker 2002; Thompson and Zúñiga-Pflücker 2011).

THREE PHASES OF EARLY T-CELL DEVELOPMENT

Early T-cell precursor development can be divided usefully into three phases in which the first

two depend on Notch signaling and the third depends on signals from the pre-TCR. The first Notch-dependent phase involves the expansion of uncommitted T-cell precursors. The second Notch-dependent phase establishes the competence of the cells to express and depend on TCR complexes. The third phase, much less Notch-dependent, expands cells with well-assembled pre-TCR complexes and prepares them for full immunological repertoire selection. These stages of differentiation are shown in Figure 1.

Phase 1

Under the influence of Notch signaling, the ETP cells that are derived from KIT^{hi} interleukin-7 receptor (IL-7R)^{low} precursors develop into KIT^{hi} IL-7R^{hi} DN2a cells before undergoing T-cell commitment. Cytokines made by the thymic stroma act as growth factors for them and support extensive proliferative expansion of precommitment precursors (Love and Bhandoola 2011; Yui and Rothenberg 2014). Notch signaling not only turns on canonical Notch target genes such as *Hes1*, but also initiates the expression of the crucial T-cell specification factor coding genes *Gata3* and *Tcf7* (encoding TCF1 protein). Even before the cells are committed, ETP and DN2a cells begin to express some T-cell-specific genes. Together with Notch signaling, GATA3 and TCF1 antagonize the progenitor-specific factors in these pro-T cells and regulate specification and commitment of T cells (Rosenbauer et al. 2006; Hosoya et al. 2009; Germar et al. 2011; Weber et al. 2011; De Obaldia et al. 2013b; Garcia-Ojeda et al. 2013; Scripture-Adams et al. 2014).

Phase 2

Transcriptome analysis of developing T cells shows that the commitment-linked transition from phase 1 to phase 2 is marked by a large number of positive and negative gene regulation changes (Zhang et al. 2012). One element of these transcriptional changes is the activation of *Bcl11b* (Ikawa et al. 2010; Li et al. 2010a,b). BCL11B is a six zinc-finger transcriptional repressor that is turned on in the late DN2a stage by Notch signaling, TCF1, RUNX1, and

Cite this article as *Cold Spring Harb Perspect Biol* doi: 10.1101/cshperspect.a028621

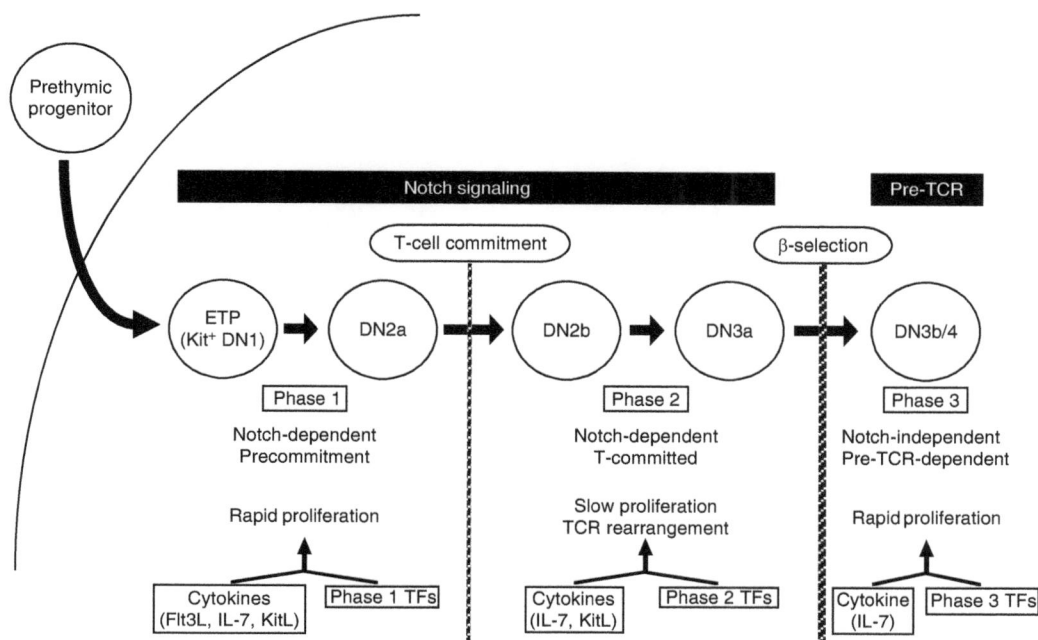

Figure 1. Roles of cytokines and transcription factors (TFs) in three phases of early T-cell development. Prethymic progenitor cells migrate into the thymus and begin T-cell differentiation program under the influence of Notch signaling. The earliest T-cell precursors in the thymus are called early thymic progenitor (ETP) or KIT-high double-negative 1 (KIT^{++} DN1; $CD44^+$ $CD25^-$) and they transit through DN2a, DN2b, DN3a, and DN3b/4 stages, followed by progression to DP stage. Early T-cell development can be divided into three phases based on the status of T-lineage commitment and Notch-dependency. The proliferation rate of the cells dynamically changes during the three phases and this reflects regulation by cytokine signaling and the phase-specific transcription factors. TCR, T-cell receptor; IL-7, interleukin-7.

GATA3 (Cismasiu et al. 2005; Tydell et al. 2007; Li et al. 2010b, 2013; Kueh et al. 2016). BCL11B is needed to finish the process of excluding cells from potential access to non-T-cell fates (Ikawa et al. 2010; Li et al. 2010a,b). It induces down-regulation of *Kit* expression, thus creating the DN2b phenotype, and direct binding of BCL11B at a possible *Kit* enhancer region may explain this ability to down-regulate *Kit* expression (H Hosokawa, HY Kueh, and EV Rothenberg, unpubl.). DN2b cells have a slower proliferation rate and begin to be desensitized to IL-7R signaling by a mechanism mediated by E proteins, and survival of DN2b cells becomes strictly Notch-dependent (Masuda et al. 2007; Wojciechowski et al. 2007; Yui et al. 2010). The subsequent DN3a stage is the peak period for expression of the Notch-dependent and E protein–dependent genes, such as recombina-

tion activating gene 1 (*Rag1*), *Rag2*, *Ptcra* (encoding pre-TCRα), and *Cd3e* (encoding CD3ε) (Takeuchi et al. 2001; Ikawa et al. 2006; Schwartz et al. 2006; Georgescu et al. 2008; Welinder et al. 2011). Therefore, DN3a cells are committed to T-cell fate even before expressing TCR on their surface, but are primed both for *TCRβ* (or *TCRγ* and *TCRδ*) gene rearrangement and for successful rearrangement to have an impact (Fig. 1).

Phase 3

The DN3a cells that have achieved successful V(D)J rearrangement for the *TCRβ* gene express pre-TCR and proceed to the DN3b stage. For DN2b and DN3a cells that rearrange both *TCRγ* and *TCRδ* productively, a separate pathway leads to development of several functionally distinct subsets of γδ T cells (Prinz et al.

2013; Vantourout and Hayday 2013). Although Notch signals are still required initially for passage through β-selection, cells that are able to receive signals through the pre-TCR transit from Notch-dependent to Notch-independent at this point. They strongly and rapidly turn off Notch target genes and IL-7R expression (Maillard et al. 2006; Taghon et al. 2006), becoming DN3b and then DN4 cells.

Signaling from pre-TCR triggers another shift in the gene regulatory network, transitioning into "phase 3." Despite the shutoff of IL-7R- and Notch-dependent growth-supporting systems, the DN4 cells enter a period of very rapid proliferation that is important for full phenotypic differentiation to the next stage (Kreslavsky et al. 2012). This is supported in part by chemokine signaling through CXCL12/ CXCR4 (Janas et al. 2010; Tussiwand et al. 2011). The proliferative burst also appears to be supported by very dynamic use of TCF1 and LEF1 in what may be a transient, self-limited canonical Wnt pathway response (Yu et al. 2010), although the details are still under debate. In the process, the cells begin to express not only the CD4 and CD8 coreceptors but also two new transcription factors, IKZF3 (Aiolos) and RORγt, while finally silencing the last of the DN-specific factors, like ERG and HES1.

In summary, early T-cell development can be divided into three phases based on the status of T-lineage commitment and Notch dependency: Notch-dependent precommitted (phase 1), Notch-dependent T-lineage committed (phase 2), and Notch-independent pre-TCR-dependent (phase 3) phases (Yui and Rothenberg 2014). Each of the phases is associated with distinct transcription factor ensembles that are supported by extracellular signaling provided from the microenvironment in the thymus. All three phases are essential for proper αβTCR$^+$ T-cell development.

ROLES OF ENVIRONMENTAL SIGNALS AND TRANSCRIPTION FACTORS IN THE INITIATION OF T-CELL PROGRAM—PHASE1

In phase 1, early T cells need to rapidly proliferate to maintain the pool size of pro-T cells,

because only small numbers of T-cell progenitors migrate into the thymus per day. The microenvironmental signals best established to support growth of initial thymic immigrants are Notch ligands and Kit ligand (also called stem-cell factor [SCF]). ETP cells have a NOTCH1$^+$ KIT^{hi} IL-7R^{low} phenotype initially and only later develop into NOTCH1$^+$ KIT^{hi} IL-7R^{hi} DN2a cells. Initially, the survival and proliferation of ETP cells mostly appear to be dependent on KIT/KIT ligand interaction (Waskow et al. 2002; Massa et al. 2006). Upon entry, thymic immigrants also express FLT3, but it seems to be down-regulated midway through the ETP stage, around the time when adult ETPs lose B-cell potential (Sambandam et al. 2005; Heinzel et al. 2007; Zhang et al. 2012; Mingueneau et al. 2013; Ramond et al. 2014). Although important prethymically, FLT3 does not appear to encounter its ligand in T-cell-promoting niches within the thymus. However, for cells escaping T-lineage commitment, FLT3 ligand may be important in much rarer, dendritic-cell-promoting intrathymic niches (Lyszkiewicz et al. 2015).

Notch1-DLL4 signaling is not only needed for viability and to induce genes encoding T-cell-specific regulatory factors, such as *Tcf7*, *Gata3*, and *Bcl11b*, but it is also required to antagonize alternative non-T-lineage development through a variety of pathways. A particularly sensitive target is the B-cell pathway (Mohtashami et al. 2010; Van de Walle et al. 2011), which may be blocked by Notch signaling through at least two distinct mechanisms (see below) (Hozumi et al. 2008). After this, ETP and DN2a cells still retain the potential to develop into dendritic cells, granulocytes, macrophages, and innate lymphocytes (ILCs), including NK cells and possibly mast cells (Shen et al. 2003; Schmitt et al. 2004; Balciunaite et al. 2005b; Taghon et al. 2007; Bell and Bhandoola 2008; Wada et al. 2008; Luc et al. 2012; Wong et al. 2012; De Obaldia et al. 2013a), but then all of these potentialities are also shut off at the transition from DN2a to DN2b under the continued influence of Notch signaling, as described below.

In view of their differential expression of KIT and IL-7R, it is notable that both ETP and

DN2a cells proliferate extensively before commitment, and both KIT ligand and IL-7 from the thymic microenvironment are crucial for this early expansion (Prockop and Petrie 2004; Massa et al. 2006; Calderon and Boehm 2012; Buono et al. 2016). This proliferation is driven first by a KITL-predominant microenvironment and then by an IL-7-predominant environment (Buono et al. 2016), matching the shifts in the ratio of KIT to IL-7R as the cells progress from ETP to DN2a cells. However, computational modeling suggests that the largest number of cell cycles may take place in the ETP stage (Manesso et al. 2013), before the cells activate strong expression of IL-7 receptors. This suggests that the earliest proliferative expansion could be a major function that depends not only on KIT signaling but also on another regulatory input in place of IL-7R, and one possibility is that this is supplied by the phase 1–specific transcription factors. Many of the transcription factors that are expressed only in phase 1 have roles in the proliferation, survival, and self-renewal of other hematopoietic cells and can be involved in the pathogenesis of leukemias (Yui and Rothenberg 2014).

Two phase 1–restricted transcription factor genes that have clear genetically defined roles in ETP-DN2-stage proliferative expansion are *Lyl1* (Zohren et al. 2012) and *Hhex* (Goodings et al. 2015; Jackson et al. 2015), whereas the factor encoded by a third gene, *Erg*, is specifically implicated in proliferation at the expense of differentiation (Knudsen et al. 2015). In addition, even PU.1 (encoded by *Spi1*, also known as *Sfpi1*), a phase 1–restricted transcription factor that seems most associated with alternative differentiation pathways, may also positively regulate growth of phase 1 pro-T cells, directly or indirectly. The expression of PU.1 continues in early DN cells from their prethymic precursors, persisting at high levels through multiple cell divisions in ETP and DN2a stages, but then decreasing sharply during T-cell lineage commitment (Yui et al. 2010). PU.1 is normally associated with the differentiation of dendritic cells (DCs), myeloid cells, and B cells, and its expression strongly correlates with the ability of uncommitted T-cell precursors to differentiate

to DC and myeloid cell fates when Notch signaling is withdrawn (Lefebvre et al. 2005; Franco et al. 2006; Laiosa et al. 2006; Carotta et al. 2010b; Del Real and Rothenberg 2013). However, in addition, it is important for the expression of FLT3 by prethymic precursors (Carotta et al. 2010a), and direct genomic analysis indicates that it regulates numerous G-protein-coupled receptors and signaling components as well as cytokine receptors in phase 1 pro-T cells (Champhekar et al. 2015). Any of these could help to explain how PU.1 also positively contributes to early T-cell development (Dakic et al. 2005; Champhekar et al. 2015).

GATA3 AND THE ESTABLISHMENT OF A T-CELL-SPECIFIC REGULATORY STATE

Despite the presence of non-T lineage factors, phase 1 is also the context in which the T-cell program is first triggered. Notch signaling in ETP cells initiates the expression of the crucial regulatory transcription factor genes *Gata3* and *Tcf7*, which then alter the developmental status of the cells (Hosoya et al. 2009; Germar et al. 2011). *Tcf7* is activated by Notch signaling directly, and its product, TCF1, primarily acts as a positive regulator of T-cell specification, collaborating with Notch to activate T-cell genes through a feedforward network circuit (De Obaldia and Bhandoola 2015). GATA3 and TCF1 (encoded by *Tcf7*) not only begin to antagonize the progenitor-specific factors, but they also collaborate with a different group of regulatory factors required for T-cell development that continue expression initiated from a prethymic stage, such as MYB, RUNX1, CBFβ, IKAROS, GFI1, and E2A. These are stably expressed as "legacy factors," but nevertheless function in developmentally modulated ways, as they are incorporated together with TCF1 and GATA3 into the network of factors regulating T-cell lineage commitment. Actually, some of these factors have also been identified as physically GATA3-interacting molecules in an early T-cell line (H Hosokawa and EV Rothenberg, unpubl.).

GATA3 has indispensable roles not only in early T-cell development but also throughout

T-cell development, with multiple functions in the thymus and periphery. It works differently in a broad variety of developmental stages, affecting early T-cell survival, growth, specification, and commitment and in priming DN2 cells for progression to β-selection (Pai et al. 2003; Hosoya et al. 2010; Garcia-Ojeda et al. 2013; Scripture-Adams et al. 2014; Tindemans et al. 2014). In one of the earliest steps in commitment, Notch signaling- and TCF1-induced *Gata3* expression blocks B-cell lineage developmental potential from ETP cells, very shortly after their immigration into the thymus (Heinzel et al. 2007; Hozumi et al. 2008; Garcia-Ojeda et al. 2013).

The levels of GATA3 gradually increase in DN2a-DN2b cells and contribute to both activation of *Bcl11b* and repression of *Sfpi1* (PU.1) expression to support T-cell specification and commitment (Taghon et al. 2007; Zhang et al. 2012; Li et al. 2013). Knockdown of *Gata3* expression in ETP and DN2 cells blocks appearance of the DN3 phenotype, with up-regulation of phase 1 genes, including *Sfpi1* and *Bcl11a*, and down-regulation of phase 2 genes, such as *Ets1*, *Zfpm1*, *Il9r*, *Il17rb*, and *Bcl11b* (Garcia-Ojeda et al. 2013; Scripture-Adams et al. 2014). Later, GATA3 is also required for full TCRβ locus activation needed to pass through β-selection, and for generation of CD4 SP T cells (Pai et al. 2003).

Therefore, GATA3 is a vitally important factor for early T-cell development, but at least in the murine system, it has a highly limited dose–response range in DN cells. This is an interesting contrast with its roles in mature Th2 cells, where overexpression is tolerated. Instead, increased concentrations of GATA3 are as toxic for mouse early T-cell precursors as loss of GATA3, and if the overexpressing cells can be kept alive, they can be driven into becoming mast cells (Taghon et al. 2007). Thus, T-cell identity is strongly dependent on GATA3 and also requires the preservation of strict limits to GATA3 expression.

While GATA3 at modest levels is nearly universal in T-lineage precursor cells from the ETP stage onward, its activity may be modulated in a stage- or signal-dependent way. Recently, several groups reported that posttranslational modifications of GATA3 regulate its functions (Cook and Miller 2010; Kitagawa et al. 2014; Hosokawa et al. 2015, 2016). Phosphorylation and acetylation regulate both stability and transcriptional activity of the related GATA1 factor (Hernandez-Hernandez et al. 2006). Thus, the functions of GATA3 could be strictly regulated by several distinct mechanisms in a stage-specific manner.

AVOIDING DIVERSION TO ALTERNATIVE FATES IN PHASE 1

Phase 1 cells, still being uncommitted, need continuous signaling from the microenvironment to stay in the T-cell pathway. Each kind of alternative pathway needs to be obstructed by a different mechanism. Within the thymus, Notch-DLL4 signaling rapidly induces antagonists of at least two alternative pathways, GATA3 to block B-cell development and HES1 to block myeloid development (De Obaldia et al. 2013b; Garcia-Ojeda et al. 2013; Scripture-Adams et al. 2014). Notch-induced TCF1 may also play a role in blocking access to DC development, at least (Kueh et al. 2016). Eventually, the cumulative action of the Notch-triggered regulatory cascade also induces BCL11B (Kueh et al. 2016) to block NK cell development (Li et al. 2010b). The up-regulation of these active antagonists of alternative fates can be viewed as "commitment by addition." The possibility of diversion to myeloid and dendritic cell fates is not eliminated at a cell-intrinsic level, however, until the cells transition to phase 2, through "commitment by subtraction," when the key enabling factor PU.1 is silenced. Until then, very selective expression of cytokines in the thymus microenvironment also provides a safety net. B cells can potentially grow in the IL-7 provided by cortical thymic epithelial cells (Calderon and Boehm 2012), if not aborted early by Notch-dependent mechanisms, but progenitors taking a pathway that might require myeloid growth factors cannot expand in the normal thymus where such factors are largely absent (Lyszkiewicz et al. 2015). In fact, these environmental restrictions are extremely effective, so that the majority of phase 1 cells appear not to give rise to any de-

Cite this article as *Cold Spring Harb Perspect Biol* doi: 10.1101/cshperspect.a028621

tectable myeloid progeny in vivo as long as their development in the thymus is undisturbed (Schlenner et al. 2010).

Other cytokines besides myeloid growth factors must also be kept away from developing pro-T cells. A fraction of phase 1 cells can respond to the addition of cytokines IL-33 and IL-7 in the presence of Notch signaling to become ILC2 (Wong et al. 2012). Signals from IL-33 can redirect the effects of transcription factors such as GATA3, BCL11B, and TCF1 to work in the ILC2 program, via induction of *Id2* and the gene that encodes orphan nuclear receptor RORα (Hoyler et al. 2012; Wong et al. 2012; Yang et al. 2013; Walker et al. 2015; Yu et al. 2015; Zhong and Zhu 2015). *Id2* expression is required to develop all types of ILCs, which do not use RAG-based gene rearrangement in their development program. ILCs keep the *Rag* genes off, at least in part because of neutralization of E protein activity by Id2 (Diefenbach et al. 2014; De Obaldia and Bhandoola 2015; Serafini et al. 2015). In contrast, high levels of BCL11B, exceeding those in any ILC subset, restrain *Id2* expression in cells taking the T-cell pathway (Longabaugh et al. 2017).

PROLIFERATION AND DIFFERENTIATION IN THE PHASE 1/PHASE 2 TRANSITION: ROLES OF IL-7/IL-7R SIGNALS

IL-7 plays a key role in the development of early T cells. Actually, mice lacking IL-7 or IL-7R show strongly impaired T-cell development (Peschon et al. 1994; DiSanto et al. 1995; von Freeden-Jeffry et al. 1995), and IL-7R-deficient T-cell precursors cannot fill available thymic niches (Prockop and Petrie 2004; Zietara et al. 2015). Defects in the IL-7R in human are associated with a severe combined immunodeficiency syndrome characterized by a complete lack of T cells (Noguchi et al. 1993; Puel et al. 1998). IL-7 signaling also has critical roles in both γδ T-cell development and in later CD8 lineage choice via induction of transcription factor Runx3 (Singer et al. 2008).

IL-7R is strongly up-regulated in DN2a, and the proliferative expansion of the cells at this stage becomes highly sensitive to the IL-7 level

(Wang et al. 2006). In addition, however, its signaling modulates the rate of developmental progression (Balciunaite et al. 2005a; Huang et al. 2005), and this implies an interaction with the T-cell specification gene-regulatory network. IL-7R signaling activates both signal transducers and activators of transcription (STAT)5 and phosphatidylinositol 3-kinase (PI3K) pathways. Both pathways are important: the PI3K pathway plays a crucial role in proliferation and survival of early T cells (Pallard et al. 1999) and STAT5A/5B are needed for full population expansion both before and after β-selection (Yao et al. 2006), as well as being indispensable for opening of the *Tcrg* loci for rearrangement to allow γδ T-cell development (Ye et al. 2001; Kang et al. 2004). Surprisingly, though, little is known about any other ways that STAT5A/5B interact with the genes in the T-cell specification gene network.

Two unresolved questions are: first, whether IL-7R contributes to the extensive proliferation that seems to occur in ETPs before IL-7R expression is detectable; and, second, why KIT and IL-7R signaling activate less rapid proliferation in DN2b cells than in DN2a cells despite similar expression of IL-7R. One report has suggested that KIT and IL-7R can interact physically and signal as a complex (Jahn et al. 2007). If responses to IL-7R are actually responses to an IL-7R/KIT complex, then a high level of KIT coexpression, only available in phase 1 cells, may be required to support full IL-7-mediated proliferation. Alternatively, IL-7R signaling may depend on phase 1 transcriptional regulatory factors to induce proliferation of early T cells.

Desensitization of IL-7R after the phase 1−phase 2 transition may be important in part because of cross talk of IL-7 signaling with transcription factors. There are positive and negative interactions of IL-7/IL-7R signals with key T-cell transcription factors. First, IL-7R-mediated PI3K/AKT signaling has been reported to induce phosphorylation of GATA3 in memory Th2 cells (Hosokawa et al. 2016), and AKT-mediated phosphorylation of GATA3 regulates organization of GATA3 complexes. Strict regulation of GATA3 expression and function is important for survival and proliferation of early T cells; thus, IL-7R-mediated phosphory-lation of GATA3 may have some role in stage-specific responses to IL-7. Second, IL-7/IL-7R signaling is known in some conditions to reduce expression of genes, including *Tcf7*, *Lef1*, and *Bcl11b*, encoding some phase 2 transcription factors that have important roles in the phase 2 regulatory network (Yu et al. 2004; Ikawa et al. 2010). This interaction may underlie the ability of high IL-7 to slow developmental progression to the DP stage, especially in cells developing from adult ETPs and bone marrow precursors (Balciunaite et al. 2005a; Huang et al. 2005).

Despite its relative desensitization during DN2b and DN3a stages, IL-7R activity remains a latent contributor to cell survival as long as *Il7r* is expressed. As cells enter β-selection, IL-7R signaling from existing cell-surface receptors still provides some support for proliferation and survival of DN3b and DN4 cells, even as transcription of *Il7r* itself is finally repressed. At this stage, IL-7/IL-7R signaling enhances survival and proliferative expansion via induction of prosurvival factor *Bcl2* and by sustaining expression of nutrient-transport protein-coding genes, such as *Cd98* and *Cd71* (Boudil et al. 2015). Therefore, sensitivity and reactivity to IL-7 are dramatically changed in three phases.

ROLES OF CYTOKINES AND TRANSCRIPTION FACTORS IN THE COMMITMENT OF T-CELL LINEAGE CELLS—PHASE 2

Right after *Bcl11b* expression is induced by the combination of Notch signaling, RUNX1, GATA3, and TCF1 cells transition to the DN2b stage (Yui et al. 2010; Li et al. 2013; Kueh et al. 2016). DN2b cells proliferate more slowly than DN2a, and their survival becomes strictly Notch-dependent (Yui et al. 2010). It is still not certain whether BCL11B directly represses most phase 1 regulatory genes or simply disables signaling that would otherwise maintain their expression (Longabaugh et al. 2017); but in either case, its action allows most phase 1–restricted genes to be silenced and enables the cells to enter a committed DN2b state (Ikawa et al. 2010; Li et al. 2010a,b).

As DN2a cells progress into the DN2b stage, they undergo dynamic shifts in the expression of other transcription factors as well. These are currently useful landmarks although their functional significance is still not fully understood. Some of the largest shifts are reciprocal regulation of members of the same factor family. For example, two ETS family transcription factor genes, *Ets1* and *Ets2*, are rapidly turned on as the ETS family gene *Sfpi1* is turned off; *Gfi1* is turned on as *Gfi1b* is turned off; *Bcl11b* is turned on as *Bcl11a* is turned off, and *Runx1* increases expression as *Runx2* and *Runx3* decrease (David-Fung et al. 2009). The silencing of *Sfpi1* expression, through a mechanism dependent on RUNX factors and possibly GATA3 (Taghon et al. 2007; Huang et al. 2008; Zarnegar et al. 2010; Scripture-Adams et al. 2014), is thought to be one of the molecular mechanisms of alternative lineage exclusion as noted above.

Several mechanisms can amplify Notch pathway signaling in DN3a cells. In addition to *Notch1*, *Notch3* is activated at DN3a, largely by a NOTCH1-derived signal, and this probably adds to the sensitivity for Notch pathway signaling. At the same time, the cells become desensitized to IL-7R-driven mitogenesis in a process that depends on the basic helix–loop–helix E protein transcription factors, E2A and HEB (Wojciechowski et al. 2007). Whereas E2A (encoded by *Tcf3*) is expressed stably, HEB levels (encoded by *Tcf12*) increase steadily through the DN stages so that an increasing amount of E2A/HEB heterodimer can be formed, which is important for β-selection (Barndt et al. 2000; Braunstein and Anderson 2012). The activity of E2A and HEB may be further boosted in DN3a by the silencing of a competitive binding partner, *Lyl1*, one of the phase 1 genes (Zhong et al. 2007; Yui et al. 2010). E proteins in T-cell precursors work in collaboration with Notch signals (Ikawa et al. 2006), not antagonistically as proposed in B cells (Nie et al. 2008), and Notch signaling levels can only be sustained in this context by strong E protein activity (Yashiro-Ohtani et al. 2009; Del Real and Rothenberg 2013). Expression of E protein–dependent and Notch-dependent genes, including *Ptcra*, *Rag1*, *Rag2*, terminal deoxynucleotidyl transferase (*Dntt*) and *Cd3e*, markedly increases as the proliferation slows and the cells proceed to the DN3a stage (Takeuchi et al. 2001; Ikawa et al. 2006; Schwartz et al. 2006; Georgescu et al. 2008; Welinder et al. 2011; Xu et al. 2013).

The E proteins also activate expression of growth-inhibitory factors, such as suppressor of cytokine signaling 1 (*Socs1*) and *Socs3*, which uncouple growth factor receptors like IL-7R from their signaling pathways. In parallel, they can directly inhibit cell-cycle activation genes, and induce cyclin-dependent kinase inhibitors (Schwartz et al. 2006). The cell-cycle arrest induced by E proteins is important to enable RAG1/RAG2-mediated recombination of the TCR genes to take place (Li et al. 1996) as well as for checkpoint control. Double knockout of the genes encoding E2A (*Tcf3*) and HEB (*Tcf12*) in DN3 stage not only prevents proliferative arrest but also enables DN3 cells to undergo reverse differentiation to return to a highly IL-7-responsive, DN2-cell-like phenotype (Wojciechowski et al. 2007). However, E protein effects on TCR expression go beyond their effects on the cell cycle. Singly, E2A (*Tcf3*)-deficient cells have a specific defect in *TCRβ* gene rearrangement (Agata et al. 2007) and are prone to leukemic transformation (Engel and Murre 2004).

To summarize, DN3a cells generated by phase 2 regulatory processes are committed to a T-cell fate that is αβ lineage biased and primed for RAG-mediated *TCRβ* gene rearrangement. To reach this stage, cells require Notch signaling, GATA3, TCF1, BCL11B, ETS1, RUNX1, and E2A (Oosterwegel et al. 1991; Schwartz et al. 2006; Li et al. 2010a,b; Germar et al. 2011; Zhang et al. 2012; Xu et al. 2013). The maximal expression of the *Cd3* gene cluster is regulated by E protein, GATA3, TCF1, and BCL11B, whereas the *Rag* genes are induced by E protein and GATA3. Together with Notch signaling, several transcription factors are involved in *TCRβ* gene rearrangement such as E2A, HEB, MYB, GATA3, and RUNX1. ETS1 collaborates with RUNX1 to activate the enhancer of the *TCRβ* gene (Gu et al. 2000). The expressions of *Hes1*, *Notch3*, and *Ptcra* are jointly regulated by E protein and Notch signaling. E proteins are required to promote and sustain NOTCH1 expression in DN3a stage, and Notch signaling can be antagonized by the E protein antagonist Id2. BCL11B at a high level keeps *Id2* expression silent and assists activation of *Dntt* and the *Cd3* gene cluster (Longabaugh et al. 2017) to support transition to the DN3a state. Thus, most of the T-cell genes are fully activated, and cell-cycle arrest is induced, by late phase 2 stage. At this stage, RAG1/RAG2-mediated programmed *TCRβ* (V(D)Jβ) gene rearrangement also takes place. In the mouse, only a minority of phase 2 cells rearrange the *Tcrg* and *Tcrd* genes instead. The cells expressing a functional pre-TCR on productive rearrangement of the *Tcrb* gene are able to pass through the β-selection checkpoint.

ROLES OF TRANSCRIPTION FACTORS IN PASSING THE β-SELECTION CHECKPOINT TO PHASE 3

Signaling through the newly expressed pre-TCR not only disrupts the quiescence of DN3a cells, but also rapidly shuts off the expression of Notch target genes in an IKAROS-dependent manner, and cells proceed to phase 3 (Chari and Winandy 2008; Kleinmann et al. 2008; Geimer Le Lay et al. 2014). The burst of proliferation that takes place on β-selection amplifies clones of cells with productive *Tcrb* rearrangements to maximize the chance for productive TCRαβ diversity. At the same time, it helps αβ T cells to undergo proliferation-dependent epigenetic changes and to dilute out previous stores of regulatory molecules that would otherwise interfere with their new regulatory state (Kreslavsky et al. 2012). Notch signal–dependent transcription becomes dispensable, and many of the transcription factors that participated in the phase 1 and phase 2 stages decrease or disappear (Tabrizifard et al. 2004; Yui and Rothenberg 2004). As cells become DP thymocytes, repressive histone marks accumulate at the promoters of phase 1 and Notch target genes (Zhang et al. 2012; Vigano et al. 2014), whereas other sites open (e.g., a set of DP-specific sites for ETS1) (Cauchy et al. 2015). The establishment of durable epigenetic changes at the T-cell and non-T-cell gene loci makes the program of T-cell gene expression sustained and irreversible (Zhang et al. 2012; Vigano et al. 2014).

As the cells proliferate and become DP, in response to a combination of TCF1 and β-catenin, with a required input from MYB, the cells distinctively activate expression of two new factors: RORγt (*Rorc*) and the Ikaros zinc-finger family member AIOLOS (*Ikzf3*). Although most universal T-cell properties have already been acquired by the cells, they need a special form of viability support to keep them alive in DP stage long enough to undergo several rounds of TCRα gene rearrangement. The combination of MYB, TCF1, high HEB/E2A, and RORγt helps to induce BCL-XL (*Bcl2l1*) expression in the future DP cells to sustain them through this process (Sun et al. 2000; D'Cruz et al. 2010; Yuan et al. 2010; Wang et al. 2011) and to arm them for positive selection. Thus, a new regulatory state prepares DP thymocytes for the complex positive and negative selection events that they must undergo.

REGULATING T-CELL PRECURSOR FLUX: GLIMPSES INTO THE EARLIEST PORT OF THYMOCYTE ENTRY

One domain in which microenvironmental cues may be most important for controlling T-cell generation remains relatively mysterious. This is the initial entry port for immigrants into the thymus. Early T-cell development is coordinated with migration through distinct thymic microenvironments (Fig. 2) (Petrie and Zuniga-Pflucker 2007; Love and Bhandoola 2011). In postnatal mammals, immigrant precursors initially enter the thymus through blood vessels near the corticomedullary junction (Fig. 2, right), drawn possibly by chemokine receptor signaling from CCR7 and CCR9 (Uehara et al. 2002; Schwarz et al. 2007; Krueger et al. 2010; Zlotoff et al. 2010). Although the numbers of cells involved are small, it appears that ETP cells undergo expansion with minimal differentiation in this corticomedullary junction region, and then, days later, differentiate into DN2a cells that can begin their migration from the

site of entry deep within the cortex to the outer rim of the cortex. β-selection occurs during the accumulation of the DN3 cells when they reach the extreme outer portion of the thymus. A directional reversal of migration back across the cortex toward the medulla occurs for the later stages of thymocyte development. Thus, for example, the different requirements for IL-7 may be accommodated in vivo by varying amounts of IL-7 in combination with other extracellular signals, including adhesive molecules, chemokines, and cytokines provided by different microenvironments (Zamisch et al. 2005; Alves et al. 2009; Griffith et al. 2009; Love and Bhandoola 2011).

This pattern is different in the early fetal thymus, where precursors enter directly into the thymic rudiment by migrating across the mesenchyme from the outside, before the organ is vascularized and before the capsule forms a barrier (Fig. 2) (Anderson et al. 2006). In the fetus, elegant mutant-rescue experiments indicate that the crucial genes that the epithelium must express to support T-cell differentiation are DLL4, KIT ligand (KITL), CXCL12, and CCL25 (i.e., ligands for NOTCH1, KIT, CXCR4, and CCR9, respectively) (Calderon and Boehm 2012). However, it is not clear that this fetal environment behaves equivalently to the postnatal one. First-wave fetal thymocytes differentiate faster than late-fetal or postnatal thymocytes (Watanabe et al. 1997; Ramond et al. 2014; Scripture-Adams et al. 2014), and they begin to carry out TCR gene rearrangement after fewer rounds of proliferation once they have entered the thymus (Lu et al. 2005). At least part of this difference is cell-intrinsic, because fetal thymocytes also differentiate faster when they are compared on the "level playing field" of the OP9-DLL1 artificial stroma coculture system (Schmitt and Zuniga-Pflucker 2002), and there are subtle but real differences in gene expression at corresponding developmental stages (David-Fung et al. 2006; Belyaev et al. 2012; Ramond et al. 2014; Scripture-Adams et al. 2014). First-wave fetal thymocytes also rearrange a particular subset of TCRγ genes that are almost never used after birth. However, in the adult thymus, the specific transition from

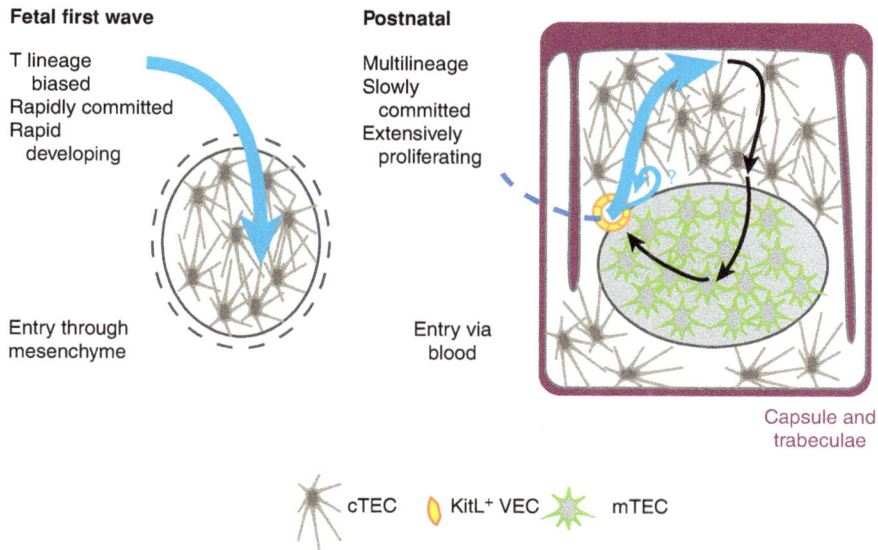

Figure 2. Fetal and postnatal pathways for progenitor entry into the thymus. Thymic epithelial function is established in embryonic life and maintained through postnatal life by genes controlled by the FOXN1 transcription factor, but the interactions between the thymic microenvironment and immigrating T-cell precursors are not the same in fetal and adult life. (*Left*) Schematic showing pathway of entry and cells involved in interactions of initial wave of T-cell precursors with the E12-14 fetal mouse thymic microenvironment. (*Right*) Same for pathways and interactions between postnatal T-cell precursors and the young adult (4–8 wk old) mouse thymus. Emphasis is on the cortex where progenitors undergo T-lineage commitment and the interactions controlling traffic into this domain. Medullary structure (pale green, *right* panel only) develops after the entry of the first-wave fetal thymic immigrants. Cyan arrows: major pathways of entering precursors, through T-cell lineage commitment. A thin cyan arrow indicates possible self-renewal pathway within entry compartment (*right*). Thin black arrows represent locations of T-cell receptor (TCR)-dependent phases of T-cell development in the postnatal thymus (*right*): β-selection, positive selection in the cortex leading to migration into the medulla, and medullary maturation events including selection and preparation for export. Locations of cortical thymic epithelial cells (cTEC), KITL$^+$ vascular endothelial cells (KITL$^+$VEC), medullary thymic epithelial cells (mTEC), and the connective tissue elements of the thymus (capsule and trabeculae, *right* panel only) are indicated. See text for details.

ETP to DN2 is disproportionately slower than in the fetus. In steady state, postnatal mouse ETPs take at least 7–10 days before they shift to the DN2a stage, although progression to subsequent stages may require only ~2 days per step (Porritt et al. 2003; Manesso et al. 2013). By contrast, the ETP to DN2a transition takes no more than 1–2 days in the fetus. This postnatal-specific extension of the ETP period is poorly reproduced in the OP9-DLL1 or OP9-DLL4 culture system, where most of the extended proliferation takes place in the DN2a/2b stages. In fact, one of the differences between "ETP" cells growing on OP9-DLL1 and those harvested ex vivo is the weaker expression of

FLT3, KIT, and several other stem or progenitor-cell regulatory genes in vitro that are normally ETP specific in vivo (Zhang et al. 2012; Mingueneau et al. 2013). What is clearly missing in the OP9-DLL1 culture is an equivalent of the entry compartment. Thus, it is exciting that there is evidence that particular niches in the adult thymic microenvironment not only limit the carrying capacity of the thymus (Zietara et al. 2015) but also preserve the immature ETP status.

Normally, all entering cells differentiate or die: self-renewing stem cells do not persist in the initial entry compartment. However, two groups showed recently that the niche itself can support extremely prolonged self-renewal in cases where there is no competition from new waves of immigrating precursors from the bone marrow (Martins et al. 2012; Peaudecerf et al. 2012). Thus, it is possible that the initial entry compartment, the ETP niche, intrinsically supports expansion and delays differentiation. If this is true, then it provides something distinctive other than a Notch ligand. The recent definition of markers that distinguish the cells forming these initial entry niches (Buono et al. 2016) can help greatly to characterize what these self-renewal cues may be.

Recently, cells expressing the membrane-bound form of Kit ligand (mKITL) were reported in a specific subdomain of the thymus, which may represent the entry port for thymus-seeding cells (Buono et al. 2016). Membrane-bound KITL is known to be more potent than secreted KITL, similar to other surface-tethered cytokines in the bone marrow microenvironment. Surprisingly, a subset of vascular endothelial cells (VECs) in the cortex specifically expresses both DLL4 Notch ligands and mKITL, and they act as a specific niche for ETP cells. VEC-specific deletion of the *KitL* gene results in a significant decrease of the number of ETP cells in the thymus. DN2a cells, on the other hand, are supported by IL-7-expressing cortical thymic epithelial cells (cTECs), which also express DLL4 but at lower levels than on the VECs, and

express a higher fraction of KITL in the less-potent soluble form (Buono et al. 2016). Both the KITL$^+$ VECs and the cTECs express one important lymphocyte-supporting chemokine, CXCL12 (ligand for the CXCR4 chemokine receptor), but the cTECs express much more CCL25 (ligand for the CCR9 chemokine receptor) (Buono et al. 2016), which is also important for T-cell development. These results suggest that early T cells migrate inside of the thymus through stage-specific niches that may provide them with differentially optimized regimes of extracellular signaling. Specific niches for DN2b, DN3, and DN4 have not been reported yet, but it is assumed that there are some niches to support their proliferation, survival, and construction of stage-specific gene regulatory networks.

Importantly, lacking the vascular endothelial entry compartment, the initial port of entry in the fetal thymus appears to be the immature cortical thymic epithelium itself. Thus, the initial compartment in which the ETP to DN2 transition occurs is different in the adult and fetal thymuses. One interesting possibility is that the lack of a VEC entry compartment itself promotes the rapid kinetics of fetal T-cell differentiation. This speculation remains to be tested, but one possibility is that signals provided by the adult entry niche actually retard dif-

Table 1. Expression of cytokine receptors in early T-cell precursors

	ETP	DN2a	DN2b	DN3a	DN3b	DN4
Flt3	High→low	Neg	Neg	Neg	Neg	Neg
Kit	High	High	Mid	Low	Neg	Neg
IL-7R	Low	High	High	High	High	Neg
IL-4R	High	High	High	Low	Low	Mid
IL-9R	Low (mid in fetal)	Low (mid in fetal)	Low (mid in fetal)	Low	Low	Mid
PDGFRβ	High	High	Mid	Mid	Low	Low
γ$_c$ (*Il2rg*)	High	High	High	High	High	High
IL-2Rα (CD25)	Low	High*	High*	High*	High→mid*	Neg
IL-2Rβ	Neg	Neg	Neg	Neg	Neg	Neg
PDGFRα	Neg	Neg	Neg	Neg	Neg	Neg

Data are based on RNA-seq (Zhang et al. 2012) and microarray (Mingueneau et al. 2013) measurements of messenger RNAs (mRNAs) encoding the indicated receptors. Data for early thymic progenitor (ETP) and DN2 subsets derived from fetal liver precursors were from (Zhang et al. 2012). The γc chain (CD132, encoded by *Il2rg*) is the common heterodimer partner of the interleukin-7 receptor (IL-7R)α, IL-4Rα, and IL-9Rα chains. Regarding IL-2Rα, asterisks indicate that these early T-lineage cells do not express functional IL-2 receptors even when they express high γc and IL-2Rα, because they do not express IL-2Rβ, which is a required functional component of both the IL-2R and the IL-15R. Neg: Negative (<1 fpkm).

ferentiation to enable more extensive proliferation, and that transition out of a potentially self-renewing early-ETP regulatory state depends on release from this compartment.

The nature of such signals is still far from clear. One approach is via analysis of receptors expressed by the ETPs themselves. Based on our RNA-seq data, we found that ETPs express *Il4ra*, *Pdgfrb*, and possibly *Il9r* (Table 1), receptors whose roles in early T-cell development have not been clarified yet (Zhang et al. 2012). Recent RNA-seq data suggest that the two niches for phase 1 cells, VEC and mTEC, differentially express Pdgf family ligands (Fig. 3). Whether or

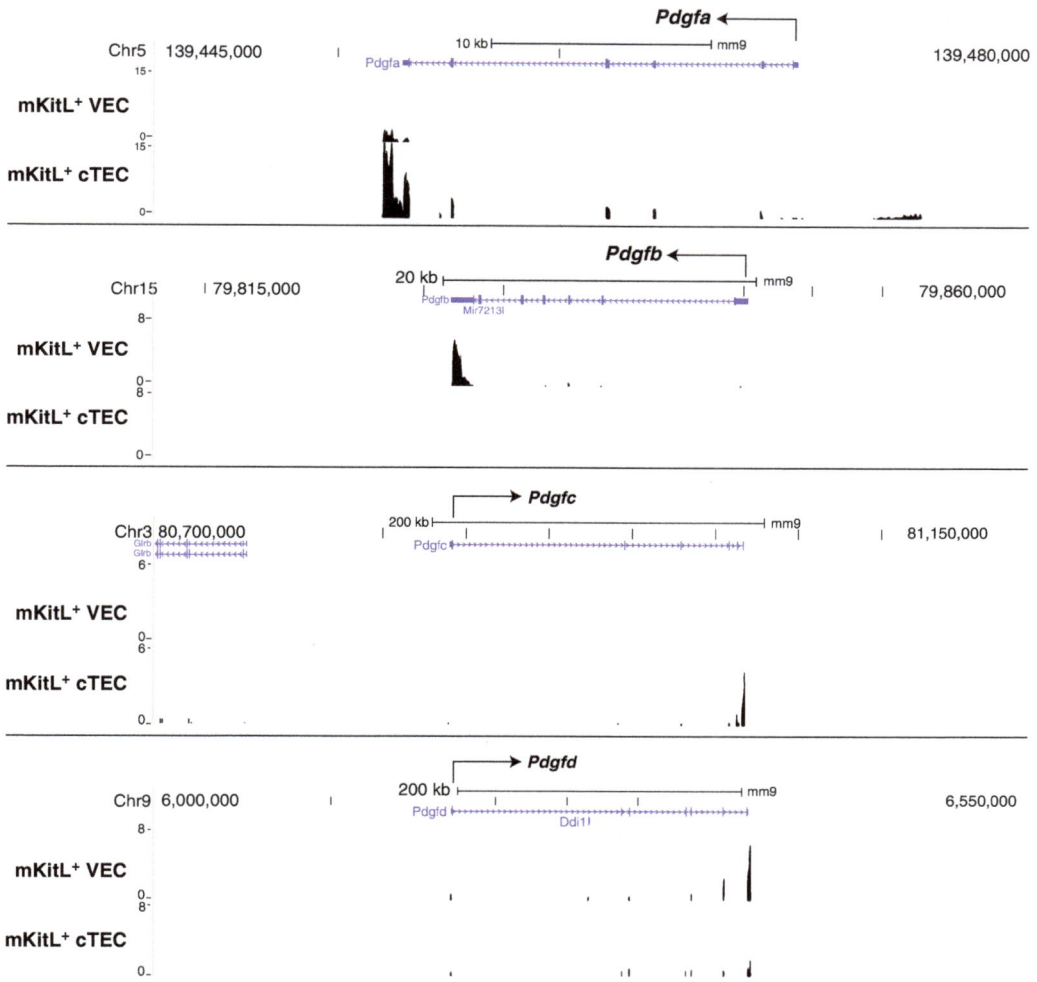

Figure 3. High specialization of different intrathymic phase 1 niches in postnatal mice: distinct gene expression patterns of Pdgf family ligands in niche cells for early entrants and differentiating phase 1 cells. RNA-sequencing tracks for the membrane-bound form of Kit ligand (mKITL)-expressing vascular endothelial cells (VECs) (early thymic progenitor [ETP] niche), and cortical thymic epithelial cells (cTECs) (DN2a niche) (Buono et al. 2016) were mapped against the murine (mm9) genome (National Center for Biotechnology Information [NCBI] build 37) on the University of California, Santa Cruz (UCSC) genome browser. Expression patterns of platelet-derived growth factor (PDGF) family ligands are shown. PDGF-B and -D RNAs were strongly detected in mKITL-expressing VEC cells, and PDGF-A and -C were preferentially expressed in mKITL-expressing cTEC.

not these particular ligand–receptor pairs control the starting gate for T-cell development, they offer fascinating hypotheses for exploration (H Hosokawa, M Yui, and EV Rothenberg, unpubl.).

CONCLUDING REMARKS

Early T-cell development is divided into three phases, based on Notch dependency and status of T-cell commitment, and all of the three phases are essential for proper T-cell development in the thymus. Inside of the thymus, pro-T cells undergo programmed migration and are supported by stage-specific microenvironments, which at a minimum provide Notch ligands and cytokine signaling. It is clear that stage-specific microenvironments are involved in the establishment of the stage-specific gene regulatory network. However, a great deal is yet to be explained and understood. Why is the effect of IL-7 on proliferative expansion of early T cells different in distinct stages? How are stage-specific transcription factors induced by extracellular signaling? And how are stage-specific roles of stably expressed transcription factors like GATA3 regulated by extracellular signaling? Are there any specific niches for DN2b, DN3, and DN4 cells as well as, possibly, for ETPs? What are the novel molecules provided by niches that support and regulate early T-cell development?

T-cell development requires a specific organ, the thymus, which provides well-regulated stage-specific microenvironments to T-cell progenitor cells. To understand T-cell development in perspective, it is important to address more deeply the molecular basis of cross talk between the extracellular signaling from the microenvironment and the cell-intrinsic gene regulatory networks in pro-T cells.

ACKNOWLEDGMENTS

The authors thank Jonas Ungerbäck and other members of the Rothenberg group for valuable discussions and advice and for sharing data before publication. Support was provided by the Manpei Suzuki Diabetes Foundation to H.H., the Albert Billings Ruddock Professorship to E.V.R., and grants from the United States Public Health Service (USPHS) to the Rothenberg Laboratory (R01 AI95943 and R01 HD76915).

REFERENCES

Agata Y, Tamaki N, Sakamoto S, Ikawa T, Masuda K, Kawamoto H, Murre C. 2007. Regulation of T cell receptor β gene rearrangements and allelic exclusion by the helix–loop–helix protein, E47. *Immunity* **27:** 871–884.

Alves NL, Richard-Le Goff O, Huntington ND, Sousa AP, Ribeiro VS, Bordack A, Vives FL, Peduto L, Chidgey A, Cumano A, et al. 2009. Characterization of the thymic IL-7 niche in vivo. *Proc Natl Acad Sci* **106:** 1512–1517.

Anderson G, Jenkinson WE, Jones T, Parnell SM, Kinsella FA, White AJ, Pongrac'z JE, Rossi SW, Jenkinson EJ. 2006. Establishment and functioning of intrathymic microenvironments. *Immunol Rev* **209:** 10–27.

Balciunaite G, Ceredig R, Fehling HJ, Zúñiga-Pflücker JC, Rolink AG. 2005a. The role of Notch and IL-7 signaling in early thymocyte proliferation and differentiation. *Eur J Immunol* **35:** 1292–1300.

Balciunaite G, Ceredig R, Rolink AG. 2005b. The earliest subpopulation of mouse thymocytes contains potent T, significant macrophage, and natural killer cell but no B-lymphocyte potential. *Blood* **105:** 1930–1936.

Barndt RJ, Dai M, Zhuang Y. 2000. Functions of E2A-HEB heterodimers in T-cell development revealed by a dominant negative mutation of HEB. *Mol Cell Biol* **20:** 6677–6685.

Bell JJ, Bhandoola A. 2008. The earliest thymic progenitors for T cells possess myeloid lineage potential. *Nature* **452:** 764–767.

Belyaev NN, Biro J, Athanasakis D, Fernandez-Reyes D, Potocnik AJ. 2012. Global transcriptional analysis of primitive thymocytes reveals accelerated dynamics of T cell specification in fetal stages. *Immunogenetics* **64:** 591–604.

Boudil A, Matei IR, Shih HY, Bogdanoski G, Yuan JS, Chang SG, Montpellier B, Kowalski PE, Voisin V, Bashir S, et al. 2015. IL-7 coordinates proliferation, differentiation and *Tcra* recombination during thymocyte β-selection. *Nat Immunol* **16:** 397–405.

Braunstein M, Anderson MK. 2012. HEB in the spotlight: Transcriptional regulation of T-cell specification, commitment, and developmental plasticity. *Clin Dev Immunol* **2012:** 678–705.

Buono M, Facchini R, Matsuoka S, Thongjuea S, Waithe D, Luis TC, Giustacchini A, Besmer P, Mead AJ, Jacobsen SE, et al. 2016. A dynamic niche provides Kit ligand in a stage-specific manner to the earliest thymocyte progenitors. *Nat Cell Biol* **18:** 157–167.

Calderon L, Boehm T. 2012. Synergistic, context-dependent, and hierarchical functions of epithelial components in thymic microenvironments. *Cell* **149:** 159–172.

Carotta S, Dakic A, D'Amico A, Pang SH, Greig KT, Nutt SL, Wu L. 2010a. The transcription factor PU.1 controls dendritic cell development and Flt3 cytokine receptor expression in a dose-dependent manner. *Immunity* **32:** 628–641.

Carotta S, Wu L, Nutt SL. 2010b. Surprising new roles for PU.1 in the adaptive immune response. *Immunol Rev* **238:** 63–75.

Cauchy P, Maqbool MA, Zacarias-Cabeza J, Vanhille L, Koch F, Fenouil R, Gut M, Gut I, Santana MA, Griffon A, et al. 2015. Dynamic recruitment of Ets1 to both nucleosome-occupied and -depleted enhancer regions mediates a transcriptional program switch during early T-cell differentiation. *Nucleic Acids Res* **44:** 3567–3585.

Champhekar A, Damle SS, Freedman G, Carotta S, Nutt SL, Rothenberg EV. 2015. Regulation of early T-lineage gene expression and developmental progression by the progenitor cell transcription factor PU.1. *Genes Dev* **29:** 832–848.

Chari S, Winandy S. 2008. Ikaros regulates Notch target gene expression in developing thymocytes. *J Immunol* **181:** 6265–6274.

Cismasiu VB, Adamo K, Gecewicz J, Duque J, Lin Q, Avram D. 2005. BCL11B functionally associates with the NuRD complex in T lymphocytes to repress targeted promoter. *Oncogene* **24:** 6753–6764.

Cook KD, Miller J. 2010. TCR-dependent translational control of GATA-3 enhances Th2 differentiation. *J Immunol* **185:** 3209–3216.

Dakic A, Metcalf D, Di Rago L, Mifsud S, Wu L, Nutt SL. 2005. PU.1 regulates the commitment of adult hematopoietic progenitors and restricts granulopoiesis. *J Exp Med* **201:** 1487–1502.

David-Fung ES, Yui MA, Morales M, Wang H, Taghon T, Diamond RA, Rothenberg EV. 2006. Progression of regulatory gene expression states in fetal and adult pro-T-cell development. *Immunol Rev* **209:** 212–236.

David-Fung ES, Butler R, Buzi G, Yui MA, Diamond RA, Anderson MK, Rowen L, Rothenberg EV. 2009. Transcription factor expression dynamics of early T-lymphocyte specification and commitment. *Dev Biol* **325:** 444–467.

D'Cruz LM, Knell J, Fujimoto JK, Goldrath AW. 2010. An essential role for the transcription factor HEB in thymocyte survival, *Tcra* rearrangement and the development of natural killer T cells. *Nat Immunol* **11:** 240–249.

Del Real MM, Rothenberg EV. 2013. Architecture of a lymphomyeloid developmental switch controlled by PU.1, Notch and Gata3. *Development* **140:** 1207–1219.

De Obaldia ME, Bhandoola A. 2015. Transcriptional regulation of innate and adaptive lymphocyte lineages. *Annu Rev Immunol* **33:** 607–642.

De Obaldia ME, Bell JJ, Bhandoola A. 2013a. Early T-cell progenitors are the major granulocyte precursors in the adult mouse thymus. *Blood* **121:** 64–71.

De Obaldia ME, Bell JJ, Wang X, Harly C, Yashiro-Ohtani Y, Delong JH, Zlotoff DA, Sultana DA, Pear WS, Bhandoola A. 2013b. T cell development requires constraint of the myeloid regulator C/EBP-α by the Notch target and transcriptional repressor Hes1. *Nat Immunol* **14:** 1277–1284.

Diefenbach A, Colonna M, Koyasu S. 2014. Development, differentiation, and diversity of innate lymphoid cells. *Immunity* **41:** 354–365.

DiSanto JP, Muller W, Guy-Grand D, Fischer A, Rajewsky K. 1995. Lymphoid development in mice with a targeted deletion of the interleukin 2 receptor γ chain. *Proc Natl Acad Sci* **92:** 377–381.

Engel I, Murre C. 2004. E2A proteins enforce a proliferation checkpoint in developing thymocytes. *EMBO J* **23:** 202–211.

Franco CB, Scripture-Adams DD, Proekt I, Taghon T, Weiss AH, Yui MA, Adams SL, Diamond RA, Rothenberg EV. 2006. Notch/Δ signaling constrains reengineering of pro-T cells by PU.1. *Proc Natl Acad Sci* **103:** 11993–11998.

Garcia-Ojeda ME, Klein Wolterink RG, Lemaitre F, Richard-Le Goff O, Hasan M, Hendriks RW, Cumano A, Di Santo JP. 2013. GATA-3 promotes T-cell specification by repressing B-cell potential in pro-T cells in mice. *Blood* **121:** 1749–1759.

Geimer Le Lay AS, Oravecz A, Mastio J, Jung C, Marchal P, Ebel C, Dembélé D, Jost B, Le Gras S, Thibault C, et al. 2014. The tumor suppressor Ikaros shapes the repertoire of Notch target genes in T cells. *Sci Signal* **7:** ra28.

Georgescu C, Longabaugh WJ, Scripture-Adams DD, David-Fung ES, Yui MA, Zarnegar MA, Bolouri H, Rothenberg EV. 2008. A gene regulatory network armature for T lymphocyte specification. *Proc Natl Acad Sci* **105:** 20100–20105.

Germar K, Dose M, Konstantinou T, Zhang J, Wang H, Lobry C, Arnett KL, Blacklow SC, Aifantis I, Aster JC, et al. 2011. T-cell factor 1 is a gatekeeper for T-cell specification in response to Notch signaling. *Proc Natl Acad Sci* **108:** 20060–20065.

Goodings C, Smith E, Mathias E, Elliott N, Cleveland SM, Tripathi RM, Layer JH, Chen X, Guo Y, Shyr Y, et al. 2015. Hhex is required at multiple stages of adult hematopoietic stem and progenitor cell differentiation. *Stem Cells* **33:** 2628–2641.

Griffith AV, Fallahi M, Nakase H, Gosink M, Young B, Petrie HT. 2009. Spatial mapping of thymic stromal microenvironments reveals unique features influencing T lymphoid differentiation. *Immunity* **31:** 999–1009.

Gu TL, Goetz TL, Graves BJ, Speck NA. 2000. Auto-inhibition and partner proteins, core-binding factor β (CBFβ) and Ets-1, modulate DNA binding by CBFα2 (AML1). *Mol Cell Biol* **20:** 91–103.

Hayday AC, Pennington DJ. 2007. Key factors in the organized chaos of early T cell development. *Nat Immunol* **8:** 137–144.

Heinzl K, Benz C, Martins VC, Haidl ID, Bleul CC. 2007. Bone marrow-derived hemopoietic precursors commit to the T cell lineage only after arrival in the thymic microenvironment. *J Immunol* **178:** 858–868.

Hernandez-Hernandez A, Ray P, Litos G, Ciro M, Ottolenghi S, Beug H, Boyes J. 2006. Acetylation and MAPK phosphorylation cooperate to regulate the degradation of active GATA-1. *EMBO J* **25:** 3264–3274.

Hosokawa H, Kato M, Tohyama H, Tamaki Y, Endo Y, Kimura MY, Tumes DJ, Motohashi S, Matsumoto M, Nakayama KI, et al. 2015. Methylation of Gata3 protein at Arg-261 regulates transactivation of the *Il5* gene in T helper 2 cells. *J Biol Chem* **290:** 13095–13103.

Hosokawa H, Tanaka T, Endo Y, Kato M, Shinoda K, Suzuki A, Motohashi S, Matsumoto M, Nakayama KI, Nakayama T. 2016. Akt1-mediated Gata3 phosphorylation controls the repression of IFNγ in memory-type Th2 cells. *Nat Commun* **7:** 11289.

Hosoya T, Kuroha T, Moriguchi T, Cummings D, Maillard I, Lim KC, Engel JD. 2009. GATA-3 is required for early T lineage progenitor development. *J Exp Med* **206:** 2987–3000.

Hosoya T, Maillard I, Engel JD. 2010. From the cradle to the grave: Activities of GATA-3 throughout T-cell development and differentiation. *Immunol Rev* **238:** 110–125.

Hoyler T, Klose CS, Souabni A, Turqueti-Neves A, Pfeifer D, Rawlins EL, Voehringer D, Busslinger M, Diefenbach A. 2012. The transcription factor GATA-3 controls cell fate and maintenance of type 2 innate lymphoid cells. *Immunity* **37:** 634–648.

Hozumi K, Negishi N, Tsuchiya I, Abe N, Hirano K, Suzuki D, Yamamoto M, Engel JD, Habu S. 2008. Notch signaling is necessary for GATA3 function in the initiation of T cell development. *Eur J Immunol* **38:** 977–985.

Huang J, Garrett KP, Pelayo R, Zúñiga-Pflücker JC, Petrie HT, Kincade PW. 2005. Propensity of adult lymphoid progenitors to progress to DN2/3 stage thymocytes with Notch receptor ligation. *J Immunol* **175:** 4858–4865.

Huang G, Zhang P, Hirai H, Elf S, Yan X, Chen Z, Koschmieder S, Okuno Y, Dayaram T, Growney JD, et al. 2008. PU.1 is a major downstream target of AML1 (RUNX1) in adult mouse hematopoiesis. *Nat Genet* **40:** 51–60.

Ikawa T, Kawamoto H, Goldrath AW, Murre C. 2006. E proteins and Notch signaling cooperate to promote T cell lineage specification and commitment. *J Exp Med* **203:** 1329–1342.

Ikawa T, Hirose S, Masuda K, Kakugawa K, Satoh R, Shibano-Satoh A, Kominami R, Katsura Y, Kawamoto H. 2010. An essential developmental checkpoint for production of the T cell lineage. *Science* **329:** 93–96.

Jackson JT, Nasa C, Shi W, Huntington ND, Bogue CW, Alexander WS, McCormack MP. 2015. A crucial role for the homeodomain transcription factor Hhex in lymphopoiesis. *Blood* **125:** 803–814.

Jahn T, Sindhu S, Gooch S, Seipel P, Lavori P, Leifheit E, Weinberg K. 2007. Direct interaction between Kit and the interleukin-7 receptor. *Blood* **110:** 1840–1847.

Janas ML, Varano G, Gudmundsson K, Noda M, Nagasawa T, Turner M. 2010. Thymic development beyond β-selection requires phosphatidylinositol 3-kinase activation by CXCR4. *J Exp Med* **207:** 247–261.

Kang J, DiBenedetto B, Narayan K, Zhao H, Der SD, Chambers CA. 2004. STAT5 is required for thymopoiesis in a development stage-specific manner. *J Immunol* **173:** 2307–2314.

Kitagawa K, Shibata K, Matsumoto A, Matsumoto M, Ohhata T, Nakayama KI, Niida H, Kitagawa M. 2014. Fbw7 targets GATA3 through cyclin-dependent kinase 2-dependent proteolysis and contributes to regulation of T-cell development. *Mol Cell Biol* **34:** 2732–2744.

Kleinmann E, Geimer Le Lay AS, Sellars M, Kastner P, Chan S. 2008. Ikaros represses the transcriptional response to Notch signaling in T-cell development. *Mol Cell Biol* **28:** 7465–7475.

Knudsen KJ, Rehn M, Hasemann MS, Rapin N, Bagger FO, Ohlsson E, Willer A, Frank AK, Sondergaard E, Jendholm J, et al. 2015. ERG promotes the maintenance of hematopoietic stem cells by restricting their differentiation. *Genes Dev* **29:** 1915–1929.

Kreslavsky T, Gleimer M, Miyazaki M, Choi Y, Gagnon E, Murre C, Sicinski P, von Boehmer H. 2012. β-Selection-induced proliferation is required for αβ T cell differentiation. *Immunity* **37:** 840–853.

Krueger A, Willenzon S, Lyszkiewicz M, Kremmer E, Förster R. 2010. CC chemokine receptor 7 and 9 double-deficient hematopoietic progenitors are severely impaired in seeding the adult thymus. *Blood* **115:** 1906–1912.

Kueh HY, Yui MA, Ng KKH, Zhang JA, Pease SS, Damle SS, Freedman G, Siu S, Bernstein ID, Elowitz MB, et al. 2016. Asynchronous combinatorial action of four regulatory factors activates *Bcl11b* for T cell commitment. *Nat Immunol* **17:** 956–965.

Laiosa CV, Stadtfeld M, Xie H, de Andres-Aguayo L, Graf T. 2006. Reprogramming of committed T cell progenitors to macrophages and dendritic cells by C/EBPα and PU.1 transcription factors. *Immunity* **25:** 731–744.

Lefebvre JM, Haks MC, Carleton MO, Rhodes M, Sinnathamby G, Simon MC, Eisenlohr LC, Garrett-Sinha LA, Wiest DL. 2005. Enforced expression of Spi-B reverses T lineage commitment and blocks β-selection. *J Immunol* **174:** 6184–6194.

Li Z, Dordai DI, Lee J, Desiderio S. 1996. A conserved degradation signal regulates RAG-2 accumulation during cell division and links V(D)J recombination to the cell cycle. *Immunity* **5:** 575–589.

Li L, Leid M, Rothenberg EV. 2010a. An early T cell lineage commitment checkpoint dependent on the transcription factor Bcl11b. *Science* **329:** 89–93.

Li P, Burke S, Wang J, Chen X, Ortiz M, Lee SC, Lu D, Campos L, Goulding D, Ng BL, et al. 2010b. Reprogramming of T cells to natural killer-like cells upon Bcl11b deletion. *Science* **329:** 85–89.

Li L, Zhang JA, Dose M, Kueh HY, Mosadeghi R, Gounari F, Rothenberg EV. 2013. A far downstream enhancer for murine Bcl11b controls its T-cell specific expression. *Blood* **122:** 902–911.

Longabaugh WJR, Zeng W, Zhang JA, Hosokawa H, Jansen C, Li L, Romero-Wolf M, Liu P, Kueh HY, Mortazavi A, Rothenberg EV. 2017. Bcl11b and combinatorial resolution of cell fate in the T-cell gene regulatory network. *Proc Natl Acad Sci* **114:** 5800–5807.

Love PE, Bhandoola A. 2011. Signal integration and crosstalk during thymocyte migration and emigration. *Nat Rev Immunol* **11:** 469–477.

Lu M, Tayu R, Ikawa T, Masuda K, Matsumoto I, Mugishima H, Kawamoto H, Katsura Y. 2005. The earliest thymic progenitors in adults are restricted to T, NK, and dendritic cell lineage and have a potential to form more diverse TCRβ chains than fetal progenitors. *J Immunol* **175:** 5848–5856.

Luc S, Luis TC, Boukarabila H, Macaulay IC, Buza-Vidas N, Bouriez-Jones T, Lutteropp M, Woll PS, Loughran SJ, Mead AJ, et al. 2012. The earliest thymic T cell progenitors sustain B cell and myeloid lineage potential. *Nat Immunol* **13:** 412–419.

Lyszkiewicz M, Zietara N, Fohse L, Puchalka J, Diestelhorst J, Witzlau K, Prinz I, Schambach A, Krueger A. 2015. Limited niche availability suppresses murine intrathymic dendritic-cell development from noncommitted progenitors. *Blood* **125:** 457–464.

Maillard I, Tu L, Sambandam A, Yashiro-Ohtani Y, Millholland J, Keeshan K, Shestova O, Xu L, Bhandoola A, Pear WS. 2006. The requirement for Notch signaling at the β-selection checkpoint in vivo is absolute and independent of the pre-T cell receptor. *J Exp Med* **203**: 2239–2245.

Manesso E, Chickarmane V, Kueh HY, Rothenberg EV, Peterson C. 2013. Computational modelling of T-cell formation kinetics: Output regulated by initial proliferation-linked deferral of developmental competence. *J R Soc Interface* **10**: 20120774.

Martins VC, Ruggiero E, Schlenner SM, Madan V, Schmidt M, Fink PJ, von Kalle C, Rodewald HR. 2012. Thymus-autonomous T cell development in the absence of progenitor import. *J Exp Med* **209**: 1409–1417.

Massa S, Balciunaite G, Ceredig R, Rolink AG. 2006. Critical role for c-kit (CD117) in T cell lineage commitment and early thymocyte development in vitro. *Eur J Immunol* **36**: 526–532.

Masuda K, Kakugawa K, Nakayama T, Minato N, Katsura Y, Kawamoto H. 2007. T cell lineage determination precedes the initiation of TCR β gene rearrangement. *J Immunol* **179**: 3699–3706.

Mingueneau M, Kreslavsky T, Gray D, Heng T, Cruse R, Ericson J, Bendall S, Spitzer MH, Nolan GP, Kobayashi K, et al. 2013. The transcriptional landscape of αβ T cell differentiation. *Nat Immunol* **14**: 619–632.

Mohtashami M, Shah DK, Nakase H, Kianizad K, Petrie HT, Zúñiga-Pflücker JC. 2010. Direct comparison of Dll1- and Dll4-mediated Notch activation levels shows differential lymphomyeloid lineage commitment outcomes. *J Immunol* **185**: 867–876.

Naito T, Tanaka H, Naoe Y, Taniuchi I. 2011. Transcriptional control of T-cell development. *Int Immunol* **23**: 661–668.

Nie L, Perry SS, Zhao Y, Huang J, Kincade PW, Farrar MA, Sun XH. 2008. Regulation of lymphocyte development by cell-type-specific interpretation of Notch signals. *Mol Cell Biol* **28**: 2078–2090.

Noguchi M, Yi H, Rosenblatt HM, Filipovich AH, Adelstein S, Modi WS, McBride OW, Leonard WJ. 1993. Interleukin-2 receptor γ chain mutation results in X-linked severe combined immunodeficiency in humans. *Cell* **73**: 147–157.

Oosterwegel M, van de Wetering M, Dooijes D, Klomp L, Winoto A, Georgopoulos K, Meijlink F, Clevers H. 1991. Cloning of murine TCF-1, a T cell-specific transcription factor interacting with functional motifs in the CD3-ε and T cell receptor α enhancers. *J Exp Med* **173**: 1133–1142.

Pai SY, Truitt ML, Ting CN, Leiden JM, Glimcher LH, Ho IC. 2003. Critical roles for transcription factor GATA-3 in thymocyte development. *Immunity* **19**: 863–875.

Pallard C, Stegmann AP, van Kleffens T, Smart F, Venkitaraman A, Spits H. 1999. Distinct roles of the phosphatidylinositol 3-kinase and STAT5 pathways in IL-7-mediated development of human thymocyte precursors. *Immunity* **10**: 525–535.

Peaudecerf L, Lemos S, Galgano A, Krenn G, Vasseur F, Di Santo JP, Ezine S, Rocha B. 2012. Thymocytes may persist and differentiate without any input from bone marrow progenitors. *J Exp Med* **209**: 1401–1408.

Peschon JJ, Morrissey PJ, Grabstein KH, Ramsdell FJ, Maraskovsky E, Gliniak BC, Park LS, Ziegler SF, Williams DE, Ware CB, et al. 1994. Early lymphocyte expansion is severely impaired in interleukin 7 receptor-deficient mice. *J Exp Med* **180**: 1955–1960.

Petrie HT, Zúñiga-Pflücker JC. 2007. Zoned out: Functional mapping of stromal signaling microenvironments in the thymus. *Annu Rev Immunol* **25**: 649–679.

Porritt HE, Gordon K, Petrie HT. 2003. Kinetics of steady-state differentiation and mapping of intrathymic-signaling environments by stem cell transplantation in nonirradiated mice. *J Exp Med* **198**: 957–962.

Prinz I, Silva-Santos B, Pennington DJ. 2013. Functional development of γδ T cells. *Eur J Immunol* **43**: 1988–1994.

Prockop SE, Petrie HT. 2004. Regulation of thymus size by competition for stromal niches among early T cell progenitors. *J Immunol* **173**: 1604–1611.

Puel A, Ziegler SF, Buckley RH, Leonard WJ. 1998. Defective IL7R expression in $T^-B^+NK^+$ severe combined immunodeficiency. *Nat Genet* **20**: 394–397.

Radtke F, Fasnacht N, MacDonald HR. 2010. Notch signaling in the immune system. *Immunity* **32**: 14–27.

Ramond C, Berthault C, Burlen-Defranoux O, de Sousa AP, Guy-Grand D, Vieira P, Pereira P, Cumano A. 2014. Two waves of distinct hematopoietic progenitor cells colonize the fetal thymus. *Nat Immunol* **15**: 27–35.

Rosenbauer F, Owens BM, Yu L, Tumang JR, Steidl U, Kutok JL, Clayton LK, Wagner K, Scheller M, Iwasaki H, et al. 2006. Lymphoid cell growth and transformation are suppressed by a key regulatory element of the gene encoding PU.1. *Nat Genet* **38**: 27–37.

Rothenberg EV. 2000. Stepwise specification of lymphocyte developmental lineages. *Curr Opin Genet Dev* **10**: 370–379.

Rothenberg EV. 2014. Transcriptional control of early T and B cell developmental choices. *Annu Rev Immunol* **32**: 283–321.

Rothenberg EV, Moore JE, Yui MA. 2008. Launching the T-cell-lineage developmental programme. *Nat Rev Immunol* **8**: 9–21.

Sambandam A, Maillard I, Zediak VP, Xu L, Gerstein RM, Aster JC, Pear WS, Bhandoola A. 2005. Notch signaling controls the generation and differentiation of early T lineage progenitors. *Nat Immunol* **6**: 663–670.

Schlenner SM, Madan V, Busch K, Tietz A, Laufle C, Costa C, Blum C, Fehling HJ, Rodewald HR. 2010. Fate mapping reveals separate origins of T cells and myeloid lineages in the thymus. *Immunity* **32**: 426–436.

Schmitt TM, Zúñiga-Pflücker JC. 2002. Induction of T cell development from hematopoietic progenitor cells by Delta-like-1 in vitro. *Immunity* **17**: 749–756.

Schmitt TM, Ciofani M, Petrie HT, Zúñiga-Pflücker JC. 2004. Maintenance of T cell specification and differentiation requires recurrent Notch receptor-ligand interactions. *J Exp Med* **200**: 469–479.

Schwartz R, Engel I, Fallahi-Sichani M, Petrie HT, Murre C. 2006. Gene expression patterns define novel roles for E47 in cell cycle progression, cytokine-mediated signaling, and T lineage development. *Proc Natl Acad Sci* **103**: 9976–9981.

Schwarz BA, Sambandam A, Maillard I, Harman BC, Love PE, Bhandoola A. 2007. Selective thymus settling regulated by cytokine and chemokine receptors. *J Immunol* **178**: 2008–2017.

Scripture-Adams DD, Damle SS, Li L, Elihu KJ, Qin S, Arias AM, Butler RR III, Champhekar A, Zhang JA, Rothenberg EV. 2014. GATA-3 dose-dependent checkpoints in early T cell commitment. *J Immunol* **193**: 3470–3491.

Serafini N, Vosshenrich CA, Di Santo JP. 2015. Transcriptional regulation of innate lymphoid cell fate. *Nat Rev Immunol* **15**: 415–428.

Shen HQ, Lu M, Ikawa T, Masuda K, Ohmura K, Minato N, Katsura Y, Kawamoto H. 2003. T/NK bipotent progenitors in the thymus retain the potential to generate dendritic cells. *J Immunol* **171**: 3401–3406.

Singer A, Adoro S, Park JH. 2008. Lineage fate and intense debate: Myths, models and mechanisms of CD4- versus CD8-lineage choice. *Nat Rev Immunol* **8**: 788–801.

Sun Z, Unutmaz D, Zou YR, Sunshine MJ, Pierani A, Brenner-Morton S, Mebius RE, Littman DR. 2000. Requirement for RORγ in thymocyte survival and lymphoid organ development. *Science* **288**: 2369–2373.

Tabrizifard S, Olaru A, Plotkin J, Fallahi-Sichani M, Livak F, Petrie HT. 2004. Analysis of transcription factor expression during discrete stages of postnatal thymocyte differentiation. *J Immunol* **173**: 1094–1102.

Taghon T, Yui MA, Pant R, Diamond RA, Rothenberg EV. 2006. Developmental and molecular characterization of emerging β- and γδ-selected pre-T cells in the adult mouse thymus. *Immunity* **24**: 53–64.

Taghon T, Yui MA, Rothenberg EV. 2007. Mast cell lineage diversion of T lineage precursors by the essential T cell transcription factor GATA-3. *Nat Immunol* **8**: 845–855.

Takeuchi A, Yamasaki S, Takase K, Nakatsu F, Arase H, Onodera M, Saito T. 2001. E2A and HEB activate the pre-TCR α promoter during immature T cell development. *J Immunol* **167**: 2157–2163.

Thompson PK, Zúñiga-Pflücker JC. 2011. On becoming a T cell, a convergence of factors kick it up a Notch along the way. *Semin Immunol* **23**: 350–359.

Tindemans I, Serafini N, Di Santo JP, Hendriks RW. 2014. GATA-3 function in innate and adaptive immunity. *Immunity* **41**: 191–206.

Tussiwand R, Engdahl C, Gehre N, Bosco N, Ceredig R, Rolink AG. 2011. The preTCR-dependent DN3 to DP transition requires Notch signaling, is improved by CXCL12 signaling and is inhibited by IL-7 signaling. *Eur J Immunol* **41**: 3371–3380.

Tydell CC, David-Fung ES, Moore JE, Rowen L, Taghon T, Rothenberg EV. 2007. Molecular dissection of prethymic progenitor entry into the T lymphocyte developmental pathway. *J Immunol* **179**: 421–438.

Uehara S, Grinberg A, Farber JM, Love PE. 2002. A role for CCR9 in T lymphocyte development and migration. *J Immunol* **168**: 2811–2819.

Van de Walle I, De Smet G, Gartner M, De Smedt M, Waegemans E, Vandekerckhove B, Leclercq G, Plum J, Aster JC, Bernstein ID, et al. 2011. Jagged2 acts as a Delta-like Notch ligand during early hematopoietic cell fate decisions. *Blood* **117**: 4449–4459.

Vantourout P, Hayday A. 2013. Six-of-the-best: Unique contributions of γδ T cells to immunology. *Nat Rev Immunol* **13**: 88–100.

Vigano MA, Ivanek R, Balwierz P, Berninger P, van Nimwegen E, Karjalainen K, Rolink A. 2014. An epigenetic profile of early T-cell development from multipotent progenitors to committed T-cell descendants. *Eur J Immunol* **44**: 1181–1193.

von Freeden-Jeffry U, Vieira P, Lucian LA, McNeil T, Burdach SE, Murray R. 1995. Lymphopenia in interleukin (IL)-7 gene-deleted mice identifies IL-7 as a nonredundant cytokine. *J Exp Med* **181**: 1519–1526.

Wada H, Masuda K, Satoh R, Kakugawa K, Ikawa T, Katsura Y, Kawamoto H. 2008. Adult T-cell progenitors retain myeloid potential. *Nature* **452**: 768–772.

Walker JA, Oliphant CJ, Englezakis A, Yu Y, Clare S, Rodewald HR, Belz G, Liu P, Fallon PG, McKenzie ANJ. 2015. Bcl11b is essential for group 2 innate lymphoid cell development. *J Exp Med* **212**: 875–882.

Wang H, Pierce LJ, Spangrude GJ. 2006. Distinct roles of IL-7 and stem cell factor in the OP9-DL1 T-cell differentiation culture system. *Exp Hematol* **34**: 1730–1740.

Wang R, Xie H, Huang Z, Ma J, Fang X, Ding Y, Sun Z. 2011. Transcription factor network regulating CD4+CD8+ thymocyte survival. *Crit Rev Immunol* **31**: 447–458.

Waskow C, Paul S, Haller C, Gassmann M, Rodewald HR. 2002. Viable c-Kit^{W/W} mutants reveal pivotal role for c-kit in the maintenance of lymphopoiesis. *Immunity* **17**: 277–288.

Watanabe Y, Aiba Y, Katsura Y. 1997. T cell progenitors in the murine fetal liver: Differences from those in the adult bone marrow. *Cell Immunol* **177**: 18–25.

Weber BN, Chi AW, Chavez A, Yashiro-Ohtani Y, Yang Q, Shestova O, Bhandoola A. 2011. A critical role for TCF-1 in T-lineage specification and differentiation. *Nature* **476**: 63–68.

Welinder E, Mansson R, Mercer EM, Bryder D, Sigvardsson M, Murre C. 2011. The transcription factors E2A and HEB act in concert to induce the expression of FOXO1 in the common lymphoid progenitor. *Proc Natl Acad Sci* **108**: 17402–17407.

Wojciechowski J, Lai A, Kondo M, Zhuang Y. 2007. E2A and HEB are required to block thymocyte proliferation prior to pre-TCR expression. *J Immunol* **178**: 5717–5726.

Wong SH, Walker JA, Jolin HE, Drynan LF, Hams E, Camelo A, Barlow JL, Neill DR, Panova V, Koch U, et al. 2012. Transcription factor RORα is critical for nuocyte development. *Nat Immunol* **13**: 229–236.

Xu W, Carr T, Ramirez K, McGregor S, Sigvardsson M, Kee BL. 2013. E2A transcription factors limit expression of Gata3 to facilitate T lymphocyte lineage commitment. *Blood* **121**: 1534–1542.

Yang Q, Jeremiah Bell J, Bhandoola A. 2010. T-cell lineage determination. *Immunol Rev* **238**: 12–22.

Yang Q, Monticelli LA, Saenz SA, Chi AW, Sonnenberg GF, Tang J, De Obaldia ME, Bailis W, Bryson JL, Toscano K, et al. 2013. T cell factor 1 is required for group 2 innate lymphoid cell generation. *Immunity* **38**: 694–704.

Yao Z, Cui Y, Watford WT, Bream JH, Yamaoka K, Hissong BD, Li D, Durum SK, Jiang Q, Bhandoola A, et al. 2006. Stat5a/b are essential for normal lymphoid development and differentiation. *Proc Natl Acad Sci* **103**: 1000–1005.

Yashiro-Ohtani Y, He Y, Ohtani T, Jones ME, Shestova O, Xu L, Fang TC, Chiang MY, Intlekofer AM, Blacklow SC, et al. 2009. Pre-TCR signaling inactivates Notch1 transcription by antagonizing E2A. *Genes Dev* **23**: 1665–1676.

Cite this article as *Cold Spring Harb Perspect Biol* doi: 10.1101/cshperspect.a028621

Ye SK, Agata Y, Lee HC, Kurooka H, Kitamura T, Shimizu A, Honjo T, Ikuta K. 2001. The IL-7 receptor controls the accessibility of the TCRγ locus by Stat5 and histone acetylation. *Immunity* **15:** 813–823.

Yu Q, Erman B, Park JH, Feigenbaum L, Singer A. 2004. IL-7 receptor signals inhibit expression of transcription factors TCF-1, LEF-1, and RORγt: Impact on thymocyte development. *J Exp Med* **200:** 797–803.

Yu Q, Sharma A, Sen JM. 2010. TCF1 and β-catenin regulate T cell development and function. *Immunol Res* **47:** 45–55.

Yu Y, Wang C, Clare S, Wang J, Lee SC, Brandt C, Burke S, Lu L, He D, Jenkins NA, et al. 2015. The transcription factor Bcl11b is specifically expressed in group 2 innate lymphoid cells and is essential for their development. *J Exp Med* **212:** 865–874.

Yuan J, Crittenden RB, Bender TP. 2010. c-Myb promotes the survival of CD4$^+$CD8$^+$ double-positive thymocytes through upregulation of Bcl-xL. *J Immunol* **184:** 2793–2804.

Yui MA, Rothenberg EV. 2004. Deranged early T cell development in immunodeficient strains of nonobese diabetic mice. *J Immunol* **173:** 5381–5391.

Yui MA, Rothenberg EV. 2014. Developmental gene networks: A triathlon on the course to T cell identity. *Nat Rev Immunol* **14:** 529–545.

Zamisch M, Moore-Scott B, Su DM, Lucas PJ, Manley N, Richie ER. 2005. Ontogeny and regulation of IL-7-expressing thymic epithelial cells. *J Immunol* **174:** 60–67.

Zarnegar MA, Chen J, Rothenberg EV. 2010. Cell-type-specific activation and repression of PU.1 by a complex of discrete, functionally specialized *cis*-regulatory elements. *Mol Cell Biol* **30:** 4922–4939.

Zhang JA, Mortazavi A, Williams BA, Wold BJ, Rothenberg EV. 2012. Dynamic transformations of genome-wide epigenetic marking and transcriptional control establish T cell identity. *Cell* **149:** 467–482.

Zhong C, Zhu J. 2015. Bcl11b drives the birth of ILC2 innate lymphocytes. *J Exp Med* **212:** 828.

Zhong Y, Jiang L, Hiai H, Toyokuni S, Yamada Y. 2007. Overexpression of a transcription factor LYL1 induces T- and B-cell lymphoma in mice. *Oncogene* **26:** 6937–6947.

Zietara N, Lyszkiewicz M, Puchalka J, Witzlau K, Reinhardt A, Forster R, Pabst O, Prinz I, Krueger A. 2015. Multicongenic fate mapping quantification of dynamics of thymus colonization. *J Exp Med* **212:** 1589–1601.

Zlotoff DA, Sambandam A, Logan TD, Bell JJ, Schwarz BA, Bhandoola A. 2010. CCR7 and CCR9 together recruit hematopoietic progenitors to the adult thymus. *Blood* **115:** 1897–1905.

Zohren F, Souroullas GP, Luo M, Gerdemann U, Imperato MR, Wilson NK, Göttgens B, Lukov GL, Goodell MA. 2012. The transcription factor Lyl-1 regulates lymphoid specification and the maintenance of early T lineage progenitors. *Nat Immunol* **13:** 761–769.

Cytokines and Long Noncoding RNAs

Susan Carpenter[1] and Katherine A. Fitzgerald[2,3]

[1]Department of Molecular, Cell and Developmental Biology, University of California, Santa Cruz, California 95064

[2]Program in Innate Immunity, Division of Infectious Diseases, University of Massachusetts Medical School, Worcester, Massachusetts 01655

[3]Centre for Molecular Inflammation Research, Department of Cancer Research and Molecular Medicine, NTNU, 7491 Trondheim, Norway

Correspondence: sucarpen@ucsc.edu

Cytokines and long noncoding RNAs (lncRNAs) are intertwined in the regulatory circuit controlling immunity. lncRNA expression levels are altered following cytokine stimulation in a cell-type-specific fashion. lncRNAs, in turn, regulate the inducible expression of cytokines following immune activation. These studies position lncRNAs as important regulators of gene expression within the complex pathways of the immune system. Our understanding of the functions of lncRNAs is just beginning. Current methodologies for functionally understanding how these transcripts mediate their effects are unable to keep up with the speed of genomic outputs cataloging thousands of these novel genes. In this review, we highlight the interplay between cytokines and lncRNAs and speculate on the future utility of these genes as potential biomarkers and therapeutic targets for the treatment of inflammatory disorders.

Cytokines are a large group of small proteins (5 to 20 kDa) that are critical in controlling a plethora of biological and physiological processes, including metabolism, inflammation, and blood pressure, to name just a few. They provide the necessary balance to maintain homeostasis at the cellular, tissue, and organismal level. Alterations in the balance between pro- and anti-inflammatory cytokines can have devastating consequences, leading to a myriad of inflammatory diseases (Vilcek and Feldmann 2004; Lin and Karin 2007). Cytokine gene expression is very tightly regulated at both the transcriptional and posttranscriptional level.

Although the role of noncoding RNAs such as microRNAs (miRNAs) as regulators of cytokine gene expression is well appreciated, much less is known regarding the contribution of other classes of noncoding RNAs to the control of cytokines. Regulation of long noncoding RNAs (lncRNAs) and cytokine expression is emerging as a reciprocal feedforward/feedback relationship, which will be detailed in this review.

RNA sequencing (RNA-seq) technology has revolutionized the field of genomics and provides a high-throughput approach to define and measure all transcripts expressed in a given cell type. RNA-seq allows for the characterization of annotated transcripts in addition to the identification of novel transcripts. lncRNAs are transcripts that exceed 200 nucleotides in length, a size cutoff that distinguishes these transcripts from small RNAs such as transfer RNA (tRNA), small nucleolar (snoRNA), and miRNAs. To date, there are close to 16,000 lncRNAs that have been cataloged in the human genome (Gencode Version 24). This number is likely to increase as lncRNAs are found to be highly restricted in their expression to specific cell types, and therefore the more sequencing that is performed the more lncRNAs will likely be identified. A recent study from Iyer et al. (2015) identified >58,000 lncRNAs from >7000 human RNA-seq data sets. Based on these data, it appears that lncRNAs greatly outnumber protein-coding genes in the human genome. lncRNAs can arise from regions between protein-coding genes, and these transcripts are referred to as "intergenic" lncRNAs. There are also a large number of antisense transcripts, with recent studies showing that the majority of protein-coding genes contain an antisense transcript owing to the bidirectionality of RNA-polymerase II (Seila et al. 2008; Flynn et al. 2011). lncRNAs can also arise from intronic regions or enhancer regions of the genome (eRNAs) (Lam et al. 2014; Kim et al. 2015). lncRNAs are often named based on their proximity to the nearest protein-coding gene; however, this naming scheme only provides information on the location within the genome where the lncRNA resides, and does not necessarily provide information on the function of a given lncRNA. Until such time as a consensus on nomenclature is reached, there are no formal guidelines on naming lncRNAs. Owing to the large number of lncRNAs present in the genome, it is unlikely that they will follow the naming scheme of miRNAs. However, the current trend of naming based on position has its drawbacks as it suggests their functions are based on or connected to the neighboring gene, which is not the case for *trans*-acting lncRNAs. lncRNAs were initially thought to represent transcriptional noise arising from imprecise transcriptional initiation; however, there is more and more evidence that these transcripts can have important regulatory functions in a wide variety of biological processes.

One of the most intriguing and also daunting aspects of lncRNAs is their wide variety of potential functions. lncRNAs can function through interactions with proteins, DNA, or RNA (Rinn and Chang 2012). Although lncRNAs are still being described and characterized, some transcripts have been studied for decades. For example, X-inactive specific transcript (Xist) is a large 20-kb transcript that has been studied intensively for the past 20 years. Xist is involved in dosage compensation and X chromosome inactivation in females (Lessing et al. 2013). We continue to learn interesting biology about this critical lncRNA. McHugh et al. (2015) elegantly showed that Xist can directly bind to the protein SHARP (SMRT and HDAC Associated Repressor Protein, also called SPEN) to silence the X chromosome. This was the first demonstration of a protein that binds to Xist required for silencing. Although Xist was previously shown to bind to the polycomb repressor complex 2 (PRC2), knockdown studies of PRC2 complex components failed to interfere with Xist-mediated silencing. Instead, SHARP binds to the SMRT corepressor, which activates HDAC3 mediating silencing and exclusion of RNA pol II from the X chromosome (McHugh et al. 2015). There is an exciting emerging role for Xist in T and B cells. Wang et al. 2016b showed that many immune-related genes are bi-allelically expressed in female lymphocytes, and it appeared that Xist RNA was down-regulated, thus enabling increased expression of these genes. They conclude that in female lymphocytes the X inactivation center is predisposed to become partially reactivated, which could account for why females display enhanced immunity over men and also why females are more prone to autoimmunity (Wang et al. 2016b).

lncRNAs also function as posttranscriptional regulators of gene expression involved in messenger RNA (mRNA) decay (1/2-sbsRNAs), mRNA stabilization (BACE-AS), as well as splicing regulation (MALAT1) (Rashid et al. 2016). A number of lncRNAs have been shown to act as miRNA sponges, and there is a database (miRSponge), which is a manually curated database for all miRNA sponges and competitive endogenous RNAs (ceRNAs) that have experimental data to support their function (Wang et al. 2015a).

A significant challenge in studying lncRNAs is to prove that the transcript is indeed noncoding. A number of bioinformatics tools have been used to determine the possibility of a transcript containing an open reading frame that will produce a protein or peptide. Codon substitution frequency and its most recent version (PhyloCSF) is a comparative genomics method that can analyze nucleotide sequence alignment across multiple species to determine whether a transcript is coding or noncoding (Lin et al. 2011). The development of ribosomal footprinting has increased our ability to experimentally validate whether a transcript is indeed noncoding (Ingolia et al. 2009, 2011, 2012). Ribosome profiling involves the mapping of translating ribosomes across any transcript by deep sequencing of the mRNA footprints that are occupied by ribosomes and thus protected from nuclease digestion (Ingolia et al. 2009). Analysis of the ribosome-protected RNA fragments allows for a map of translation to be identified at a single nucleotide level. Ingolia et al. (2014) have shown that many noncoding regions, including 5′ UTRs and lncRNAs, are found to produce footprints, raising the possibility that some of these regions could be translated producing proteins or small peptides. Guttman et al. (2013) showed that ribosome occupancy of lncRNAs and 5′ UTRs does not necessarily lead to the designation of these transcripts as coding. They developed the ribosome release score (RRS) that allows for lncRNAs and 5′ UTRs to be classified into defined categories that are unique to coding genes (Guttman et al. 2013). The release score indicates the time a ribosome is released once it hits a stop codon. These data indicate that the majority of noncoding transcripts do not function through the production of a protein. A number of additional studies have examined lncRNAs binding to ribosomes in human cells and zebrafish. Although many lncRNAs associate with ribosomes, it is hypothesized that the vast majority

of these transcripts fail to be translated, or if they generate small peptides, these are rapidly degraded possibly through the nonsense-mediated decay (NMD) pathway (Bánfai et al. 2012; Chew et al. 2013; Carlevaro-Fita et al. 2016).

Recent studies from the Olsen laboratory at the University of Texas Southwestern have shown that at least two lncRNAs can produce micropeptides in the mammalian system (Anderson et al. 2015; Nelson et al. 2016). Myoregulin (MLN) is a peptide encoded by a skeletal muscle–specific lncRNA. Genetic deletion of MLN-enhanced calcium handling in skeletal muscle in mice and improved exercise performance (Anderson et al. 2015). Dwarf open reading frame (DWORF) is a muscle-specific 34-amino-acid micropeptide encoded within an lncRNA. DWORF functions to regulate SERCA activity and enhances muscle contractility (Nelson et al. 2016). Here, we provide an overview of the cross talk between cytokines and lncRNAs and how these RNAs and the cytokines they regulate can impact the development and activation status of immune cells. We will also discuss the emerging interest in lncRNAs in disease and their potential as therapeutic targets.

lncRNAs, CYTOKINES, AND IMMUNE CELL DIFFERENTIATION

Cell differentiation is a tightly controlled process regulated both transcriptionally and posttranscriptionally. Cell-type specificity is governed through tight control of transcription factors and cytokine production. Recent studies have highlighted the importance of lncRNAs in controlling hematopoiesis through their ability to bind other RNAs, DNA, or proteins and modulate the chromatin state, mRNA stability, protein expression, and signaling. Interestingly, lncRNAs appear to be expressed in a much more cell-type-specific manner than transcription factors and other protein-coding genes, making them attractive targets to understand various stages of cell differentiation (Guttman et al. 2010; Cabili et al. 2011; Washietl et al. 2014).

Hematopoiesis involves the formation of all blood cell components. Hematopoietic stem cells (HSCs) are self-renewing cells that reside within the medulla of the bone marrow. HSCs differentiate in specific restricted stages to give rise to three lineages, erythroid cells (red blood cells and platelets), lymphocytes (T, B, and natural killer [NK] cells), and myeloid cells (monocytes, neutrophils, basophils, and eosinophils (Orkin and Zon 2008). The role of transcription factors in determining cell fate during hematopoiesis has been studied intensely for a number of years. Mouse knockouts of transcription factors confirm their importance in lineage specification. In addition, mutations within some transcription factors are associated with malignancies such as leukemia (Orkin and Zon 2008). Protein-based assays, including flow cytometry and, more recently, CyTOF, are revealing new markers of each cell lineage (Bendall et al. 2011). The contribution of noncoding RNA to these processes is also now emerging.

Cytokines are essential regulators of hematopoiesis. Many cytokines can be used in culture to differentiate HSCs into specific cell types. Cytokines function by binding to specific receptors present on HSCs, thus controlling self-renewal, differentiation, mobility, and death. In this section, we cover the emerging roles for lncRNAs within hematopoiesis and place them in the context of cytokine signaling and cell-type specificity for immune-cell subtypes.

ERYTHROCYTES

Recent studies have provided a comprehensive overview of the erythrocyte transcriptome (Alvarez-Dominguez et al. 2014; Doss et al. 2015). To differentiate cells from peripheral blood into erythrocytes, cells must be treated with interleukin (IL)-3, human stem cell factor, and erythropoietin. Close to 100 lncRNAs were shown to be differentially expressed during erythrocyte development (Alvarez-Dominguez et al. 2014). The top candidate lncRNAs from this study, based on their expression pattern and the transcription factors that regulate them (Gata1, Tal1, and KLF1), were all shown to play a role in controlling erythrocyte development. One of these RNAs (artificial long noncoding RNA [alncRNA]-EC7), an enhancer RNA neighboring the protein-coding gene

BAND3, was found to function in *cis* to control BAND3 expression, which is critical for erythrocyte differentiation. lncRNA-erythroid prosurvival (EPS) was also identified as a highly induced lncRNA during the terminal differentiation of erythrocytes. Knockdown of lncRNA-EPS led to apoptosis, whereas overexpression of the transcript promoted erythrocyte survival. lncRNA-EPS expression repressed expression of proapoptotic genes (Hu et al. 2011). A long intervening noncoding RNA (lincRNA)-EPS-deficient mouse was recently generated by removal of the 4-kb region harboring the lincRNA-EPS locus and replacing it with a neomycin resistance cassette. Surprisingly, this mouse did not manifest any developmental defect, and the process of erythropoiesis was intact (Atianand et al. 2016). This observation indicates potential redundancy regarding the contribution of lincRNA-EPS to erythropoiesis, suggesting that more than one lincRNA may be involved. An intriguing role for lincRNA-EPS in controlling immune signaling has been identified in this animal model (Atianand et al. 2016), and will be discussed in detail later in this review.

SHORT-LIVED MYELOID CELLS

In the peripheral blood, 70% of the leukocytes include neutrophils, eosinophils, and classical monocytes that represent the short-lived myeloid cells. Kotzin et al. (2016) have identified an lncRNA they named "Morrbid" (myeloid RNA regulator of Bim-induced death) that functions to control the short-lived myeloid cell life span. Morrbid is highly conserved across species. The authors generated a Morrbid knockout mouse and observed a dramatic decrease in the short-lived myeloid cell population circulating in the blood. They showed that Morrbid is specifically expressed in these myeloid cells where its expression levels are dependent on the common cytokine receptor β-chain cytokines (IL-3, IL-5, and granulocyte macrophage colony-stimulating factor [GMCSF]). Morrbid localized to the nucleus and bound to chromatin. Mechanistically, Morrbid functions in *cis* to suppress the expression of its neighboring proapoptotic gene

Bcl2l11 (Bim) by recruiting the PRC2 complex to the promoter of *Bcl2l11* in short-lived myeloid cells.

MACROPHAGES

Monocytes circulate in the periphery and, when activated by specific cytokines, differentiate into monocyte-derived macrophages. In vitro, human monocytes can be differentiated into classically activated macrophages or M1 cells using GMCSF or stimulation with lipopolysaccharide/interferon γ (LPS/IFN-γ). M2 cells (or alternatively activated cells) are obtained following macrophage colony-stimulating factor (MCSF) stimulation or activation with IL-4. The history behind the classification of macrophages into M1 and M2 is somewhat controversial and is reviewed extensively by Martinez and Gordon (2014) and Murray et al. (2014). PU.1 is the master transcription factor associated with macrophage identity. Polarization of M1 and M2 cells relies on distinct additional transcription factors. The M1 cell phenotype requires IRF5 and signal transducers and activators of transcription (STAT)1, whereas M2 cells use STAT6, IRF4, and PPARG. Activated M1 cells, which produce tumor necrosis factor (TNF), IL-6, and IL-12, display increased antimicrobial activity and antigen presentation, whereas M2 cells are associated with the production of IL-10 and IL-4, T helper (Th) 2 cell activation, and defense against parasitic infections (Lawrence and Natoli 2011).

Long noncoding monocyte (lnc-MC) is the first lncRNA associated with macrophage differentiation. lnc-MC is transcriptionally regulated by PU.1. It appears to be involved in differentiation in THP1 and HL-60 cells. lnc-MC functions as a competitive endogenous RNA (ceRNA) that sponges up miR-199a-5p, which in turn targets activin A receptor type 1B (ACVR1B). This protein is a known regulator of monocyte to macrophage differentiation; therefore, the sponging effect of lnc-MC on miR199-5p ensures its expression and allows for the differentiation program to occur (Chen et al. 2015).

Huang et al. (2016) differentiated macrophages into M1 (IFN-γ+LPS) and M2 (IL-4)

and profiled the expressed lncRNAs using an lncRNA-specific microarray. They identified 9343 lncRNAs with twofold differential expression in M1 macrophages and 4592 in M2 macrophages when compared to monocyte-derived macrophages (Huang et al. 2016). They focused on one lncRNA (TCONS_00019715) that was higher in M1 compared to M2 macrophages. Knocking down this lncRNA in human THP1 cells resulted in an increase in M2 markers compared to M1 markers. Whether this lncRNA is directly involved in M1 polarization remains to be determined.

DENDRITIC CELLS

Dendritic cells (DCs) are the professional antigen-presenting cells of the immune system. There are a large number of DC cell subtypes, including conventional DCs, plasmacytoid DCs, and monocyte-derived DCs (Shortman and Liu 2002). In mice, DCs can be cultured from bone marrow using GMCSF. Although, in humans, monocytes from the periphery are differentiated into DCs using GMCSF and IL-4 to produce monocyte-derived dendritic cells (MoDCs) (Shortman and Liu 2002).

Long noncoding dendritic cell (lnc-DC) was identified in human conventional DCs among a collection of lncRNAs whose expression increased during the process of DC differentiation. Knocking down lnc-DC in human monocytes inhibited their ability to differentiate into MoDCs. A similar result was observed when lnc-DC was knocked down in murine bone marrow–derived DCs, which resulted in an impairment in their ability to activate T cells. lnc-DC was shown to interact with STAT3 in the cytoplasm preventing SHP1 from dephosphorylating STAT3, suggesting that the function of lnc-DC is to enhance the activation of STAT3 in the cytoplasm, enabling the STAT3 transcriptional program to occur (Wang et al. 2014). There have been some questions raised concerning the identification and characterization of the murine ortholog of lnc-DC. The gene at this locus was previously identified as *Wdnm1-like* and possesses protein-coding capacity and is expressed in adipocytes (Wu and

Smas 2008). These data suggest that a functional lncRNA is present only in humans, whereas other mammals encode the protein-coding transcript *Wdnm1-like* (Dijkstra and Ballingall 2014).

T CELLS

Naïve Th cells differentiate into effector cells through the coordinated expression of specific transcription factors and cytokines. The transcription factors orchestrate complex epigenetic alterations of gene loci to specifically activate or repress genes required for lineage specification. Naïve $CD4^+$ T cells differentiate into effector T subsets, including Th1, Th2, Th17, and regulatory T cells (Tregs), which express the cytokines IFN-γ (Th1); IL-4, IL-5, and IL-13 (Th2); IL-17 (Th17); or transforming growth factor β (TGF-β) and IL-10 (Treg cells). In contrast, naïve $CD8^+$ T cells differentiate into cytotoxic CD8 T cells following IL-2 stimulation, and they produce IFN-γ and TNF. This differentiation step is essential for the adaptive immune system to provide protection against infection. Table 1 outlines the cytokines involved in T-cell differentiation, cytokines produced by each T-cell subset, and the lncRNAs identified to function in each subset (also depicted in Fig. 1).

CD8 T Cells

RNA-seq analysis of various T-cell subsets have identified a large collection of lncRNAs that are both specific for each subset and also play critical roles in T-cell development and differentiation. The first comprehensive study on lncRNAs in $CD8^+$ T cells identified >1000 lncRNAs in human and murine $CD8^+$ T cells that displayed cell-type-specific expression, with many close to protein-coding genes with known functions in T-cell biology (Pang et al. 2009). Patients suffering from severe asthma have increased activation of circulating $CD8^+$ T cells compared to $CD4^+$ T cells. 167 lncRNAs show differential expression in activated $CD8^+$ T cells; however, their roles in regulating these responses were not investigated (Tsitsiou et al. 2012).

lncRNA-CD244 (lncRNA-AS-GSTT1(1-72)) was found to be expressed at high levels in $CD8^+$ T cells that are infected with *Mycobacterium tuberculosis*, where it functions to impair T-cell antigen receptor (TCR) signaling. Knockdown of lncRNA-CD244 resulted in elevated expression of IFN-γ and TNF-α. Expression of this lncRNA was induced following activation with CD244 (a known TCR inhibitor that is induced following tuberculosis [TB] infection) and the lncRNA, in turn, represses expression of IFN-γ and TNF-α through its interactions with enhancer of zeste homolog 2 (EZH2), a major component of the PRC2 complex (Wang et al. 2015b).

Th1 Cells

Th1 cells require the transcription factor T-bet for their differentiation and production of IFN-γ. A number of studies have cataloged levels of lncRNAs in different T-cell subsets. Xia et al. (2014) studied lncRNA expressed in $CD4^-CD8^-$ (DN) cells compared to $CD4^+CD8^+$ (DP), and $CD4^+CD8^-$ and activated $CD4^+CD8^-$ T cells. They identified a total of 788 lncRNAs between all of these cell types and 746 mRNAs (Xia et al. 2014). Hu et al. (2013) systemically studied lncRNA expression in a number of T-cell subsets. They identified 354 lncRNAs that were expressed in Th1 cells. When they compared lncRNAs expressed in Th1, Th2, naïve $CD4^+$, Th17, or inducible Treg (iTreg) cells, they found only 100 lncRNAs that were common to all cell types. Nearly 50% of lncRNAs were unique to a given T-cell subset. Comparing this to >78% of mRNAs that are commonly expressed across all T-cell subtypes confirmed that lncRNAs are expressed in a more cell-type-specific manner than protein-coding genes (Hu et al. 2013). More than 56% of lncRNAs expressed in Th1 cells are controlled by the transcription factor STAT4, whereas STAT6 controls Th2-expressing lncRNAs.

NeST

NeST (nettoie *Salmonella* pas Theiler's [cleanup *Salmonella* not Theiler's], also called lincR-ifng-3′AS), was originally described as TMEVG, identified in Th1 cells, and expressed in a

Table 1. Cytokines and long noncoding RNAs (lncRNAs) involved in immune cell development and function

Cell type	Transcription factors	Cytokines required for differentiation	Cytokines produced	lncRNAs	Cell functions	References
Macrophage (M1)	PU1	GMCSF, LPS+IFN-γ	TNF, iNOS	lnc-MC	Proinflammatory cytokine production	Chen et al. 2015
Macrophage (M2)	PU1	MCSF, IL-4	IL-4, IL-13, IL-10	None identified to date	Resolution of inflammation	
Dendritic cells	PU1	GMCSF; IL-4		lnc-DC	Professional antigen-presenting cells	Wang et al. 2014
CD8^{+} T Cells	T-bet, Eomes	IL-2	IFN-γ, TNF, IL-2	lncRNA-CD244,	Antiviral activity	Wang et al. 2015b
Th1 cells	STAT1, 4, T-bet	IFN-γ, IL-12	IFN-γ, IL-4, IL-5, IL-13	NeST	Target intracellular pathogens	Collier et al. 2012; Gomez et al. 2013
Th2 cells	GATA-3, STAT6	IL-4	IL-4, IL-13, IL-5	lincR-Ccr2-5'AS, GATA3-AS, Th2-LCR	Fight parasitic infections	Hu et al. 2013; Spurlock et al. 2015
Th17 cells	Rorγt, BATF	IL-17	IL-17, IL-22	RMRP	Target extracelluar bacteria and fungi	Huang et al. 2015
Tregs	FoxP3	TGF-β, IL-2	TGF-β, IL-10	DQ786243	Maintain self-tolerance and homeostasis	Qiao et al. 2013

TGF-β, Transforming growth factor β; GMCSF, granulocyte macrophage colony-stimulating factor; LPS, lipopolysaccharide; IFN-γ, interferon γ; TNF, tumor necrosis factor; iNOS, inducible nitric oxide synthase; lnc-MC, Long noncoding monocyte; MCSF, macrophage colony-stimulating factor; IL, interleukin; lnc-DC, long noncoding dendritic cell; lncRNA, long noncoding RNA; STAT1, signal transducers and activators of transcription 1; LCR, locus control region; BATF, basic leucine zipper ATF-like; Treg, regulatory T cell.

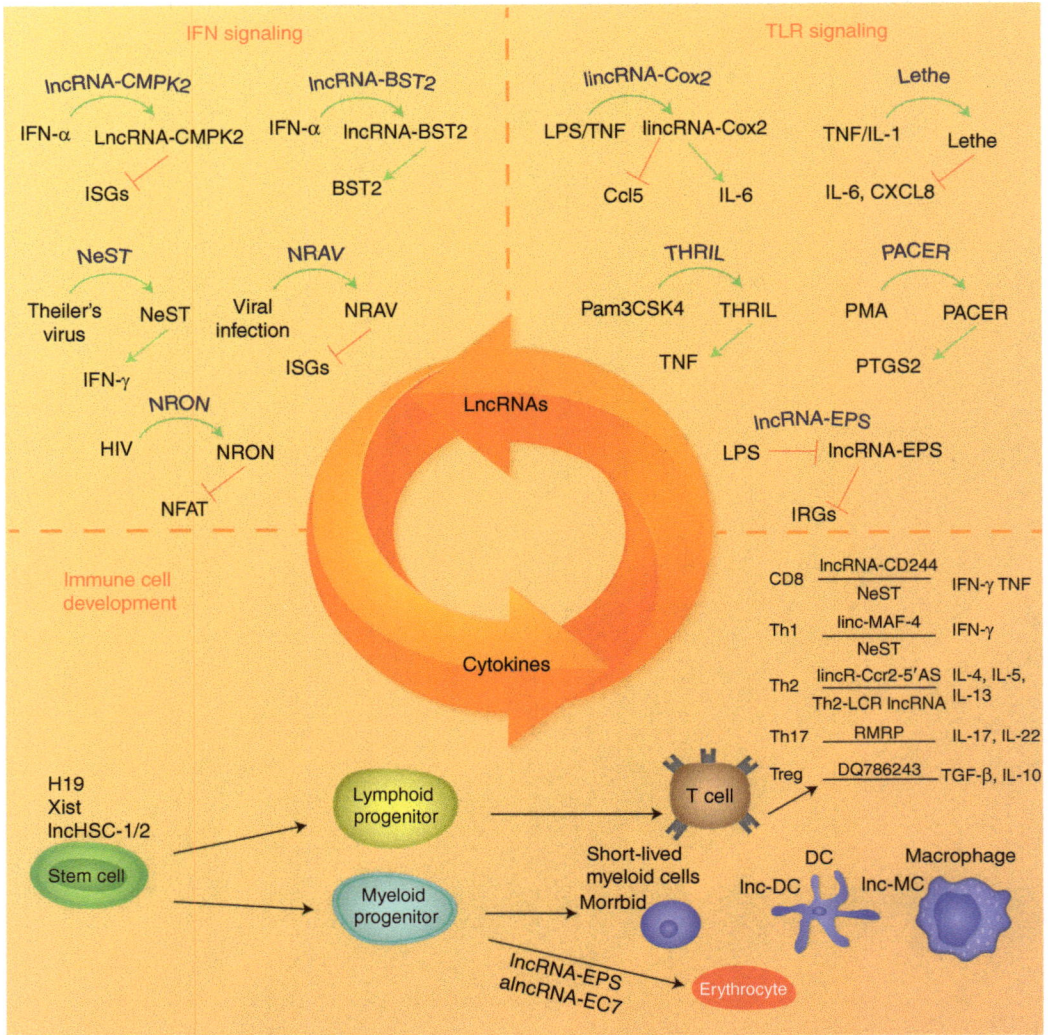

Figure 1. Cytokine/long noncoding RNA (lncRNA) interplay in controlling immune cell development and function. This figure shows lncRNAs that are controlling cytokine production and cytokines controlling lncRNA expression. (*A*) Interferon (IFN)/lncRNA regulatory loop, (*B*) lncRNAs induced by Toll-like receptor (TLR) signaling pathway to control cytokine production, and (*C*) those lncRNAs critical for immune cell development. TNF, Tumor necrosis factor; ISG, interferon-stimulated gene; IL, interleukin; NeST, nettoie *Salmonella* pas Theiler's; NRAV, negative regulator of antiviral response; THRIL, TNF-α and heterogeneous nuclear ribonucleoprotein L (hnRNPL)-related immunoregulatory lincRNA; PACER, p50-associated COX-2 extragenic RNA; NRON, noncoding repressor or nuclear factor of activated T (NFAT) cells; IRG, interferon regulated gene; Th, T helper cell; LCR, locus control region; EPS, erythroid prosurvival; HSC, hematopoietic stem cells; Xist, X-inactive specific transcript; alnc, artificial long noncoding; DC, dendritic cell; MC, monocyte.

manner dependent on T-bet (Hu et al. 2013). In more recent studies, NeST was identified as a gene associated with persistence of Theiler's virus in mice. In these studies, NeST was found to associate with the WDR subunit of mixed-lineage leukemia histone H3 Lys 4 (H3K4) methyltransferase to increase H3K4 methylation at the IFN-γ locus in CD8 T cells. Mice lacking NeST are capable of being persistently infected with Theiler's virus, while being resistant to lethal *Salmonella enterica* infection. These data were the first to identify roles for noncoding RNAs in host defense to infectious pathogens (Collier et al. 2012; Gomez et al. 2013). Recently, Li et al. (2016) have shown that expression of NeST is altered in patients with immune thrombocytopenia (IT). Peripheral blood mononuclear cells (PBMCs) from these patients showed decreased NeST expression, whereas IFN-γ was elevated. This was not the case for healthy controls in which NeST expression and IFN-γ expression are positively associated. They conclude that NeST and IFN-γ are in a regulatory loop in which NeST is initially needed for IFN-γ expression, and this increased expression subsequently inhibits NeST and IFN-γ expression levels in patients with IT (Li et al. 2016). NeST also shows elevated expression levels in patients suffering from Sjorgren's syndrome and ulcerative colitis (Padua et al. 2016; Wang et al. 2016a). The exact mechanism or role of NeST in the pathogenesis of these diseases remains to be determined.

Th2 Cells

The transcription factors STAT6 and GATA3 are critical for Th2 cell differentiation via their regulating the expression of many Th2-specific genes. LincR-Ccr2-5′AS is regulated by GATA3, with reduced expression in a GATA3 knockout Th2 cell. In a more global analysis, GATA3 regulated ∼28% of Th2-specific lncRNAs (Hu et al. 2013). LincR-Ccr2-5′AS can regulate migration of Th2 cells to the lung. Knocking down this lncRNA using short-hairpin (shRNA) did not inhibit expression of the Th2 cytokine IL-4 but did result in decreased expression of its

neighboring protein-coding genes, including *Ccr1-Ccr5*. There is an antisense lncRNA neighboring GATA3 (GATA3-AS) that is also Th2-cell restricted. The lncRNA Th2 locus control region (LCR) is also Th2-specific in expression and is localized at the 3′ end of the *Rad50* gene (Spurlock et al. 2015). This lncRNA is coexpressed with IL-4, IL-5, and IL-13, and knockdown of this lncRNA greatly impacted expression levels of these cytokines. The current mechanism for how lncRNA Th2-LCR controls these genes is through its recruitment of the WDR5-containing complexes that regulate H3K4me3 patterns at the affected target gene loci.

Th17

Th17 cells are specified by RORC (retinoic acid receptor-related orphan nuclear receptor γt) and BATF (basic leucine zipper ATF-like) transcription factors and they produce IL-17 and IL-22. Spurlock et al. (2015) identified 56 lncRNAs that are expressed in primary Th17 cells and 61 that are expressed in Th17 effector cells. These two cell states share a total of 31 lncRNAs. They found that lncRNAs are expressed in a highly lineage-specific manner with many Th17-specific lncRNAs encoded on chromosome 2, 16, and 17 (Spurlock et al. 2015).

Huang et al. (2015) have identified a role for lncRNA RMRP in controlling gene expression programs in Th17 cells. They identified DEAD-box helicase DD5 as the binding partner of the transcription factor RORγt that controls transcription of target genes in Th17 cells. The interaction between these two proteins is mediated by the lncRNA Rmrp. Rmrp is highly conserved, and mutations in Rmrp are linked to cartilage hair hyperplasia (CHH), a genetic disorder of bone growth that leads to short stature and other skeletal abnormalities, fine hair, and immune deficiencies. This is the first genetic disorder in which the causative mutation lies within an lncRNA. Mice carrying the point mutation present in CHH had reduced binding to DDX5, resulting in a decrease in RORC target genes (Huang et al. 2015). This discovery provides new insights in the regulation and complexity of Th17 cell functions.

Tregs

T regulatory cells act as suppressors of immune responses in which they function to down-regulate the actions of effector T cells. Natural Tregs originate in the thymus, whereas iTregs arise from naïve T cells in the periphery. Tregs require the transcription factor Foxp3 and the cytokine TGF-β for their differentiation. Tregs can produce TGF-β, IL-10, and IL-35, which all aid in suppressing immune responses. Hu et al. (2013) cataloged a total of 278 lncRNAs in iTregs; however, the functions for these genes in iTreg development or effector functions were not examined. DQ786243 is a lncRNA that was shown to be significantly overexpressed in patients with clinically activated Crohn's disease (CD). High expression levels of DQ786243 appear to affect expression of CREB and Foxp3 therefore impacting Treg functions. The results here are correlative, and the mechanism by which this lncRNA might impact gene expression within CD remains to be determined (Qiao et al. 2013).

CYTOKINE REGULATION OF lncRNAS

TNF-α Stimulation

Activation of mouse embryonic fibroblasts with TNF-α regulates the expression of 3596 protein-coding genes, 112 lncRNAs, and 54 pseudogene lncRNAs. Rapicavoli et al. (2013) identified *Lethe* (after the river of forgetfulness in Greek mythology) as a TNF-α-inducible and IL-1B-inducible lncRNA. *Lethe* is nuclear localized and binds to chromatin. *Lethe* acts as an inhibitor of nuclear factor κB (NF-κB)-mediated transcriptional activation by sequestering RelA (p65) and preventing its translocation into the nucleus (Rapicavoli et al. 2013). Increased NF-κB activity is associated with aging, and the authors showed that *Lethe* expression levels decrease dramatically with age because of increased NF-κB activity.

The Interferon/lncRNA Regulatory Loop

Activation of the interferon response (types I, II, and III) is critical for protection against infection by viruses and other pathogens. The type I IFNs bind to their receptor, IFN-α/BR, and trigger activation of the Janus kinase/signal transducers and activators of transcription (JAK-STAT) signaling pathway, leading to the production of interferon-stimulated genes (ISGs). IFN signaling can lead to the activation of a number of lncRNAs, which, in turn, act to control these signaling pathways. There were a number of high-throughput RNA-sequencing studies performed using a variety of viral infections, showing that 1000s of lncRNAs were regulated in virus-infected cells. Profiling of lncRNAs has been performed at a vast rate, yet the mechanistic understanding of these lncRNAs and their functions has lagged far behind. Most of these studies provide information on the differential regulation of lncRNAs, with no insight into the roles these lncRNAs might play in host defense. The database MONOCLdb contains annotated lncRNAs that are induced by IFN-α stimulation (Josset et al. 2014). The numbers of lncRNAs identified are outlined in Table 2. In this section, we will focus on IFN-regulated lncRNAs, with experimental data supporting biological activity in these pathways.

lncRNA-CMPK2

lncRNA-CMPK2 is a polyadenylated, spliced nuclear lncRNA that is induced by IFN-α in a number of cell lines in both humans and mice. lncRNA-CMPK2 has been shown to be elevated in the livers of patients with hepatitis C virus (HCV) infection (Kambara et al. 2014). lncRNA-CMPK2 can inhibit the IFN response as knockdown of this gene in hepatocytes resulted in a reduction in HCV replication and an up-regulation of ISGs. The mechanism by which this lncRNA mediates its inhibitory effects remains unclear.

lncRNA-BST2

Two groups identified the same IFN-α inducible lncRNA known as BST2 IFN-stimulated positive regulator (BISPR) (Barriocanal et al. 2015; Kambara et al. 2015). A bidirectional promoter of the protein-coding gene BST2

Cite this article as *Cold Spring Harb Perspect Biol* doi: 10.1101/cshperspect.a028589

Table 2. RNA-seq profiling of long noncoding RNAs (lncRNAs) following immune activation

Stimulant	Species	Cell/tissue type	lncRNAs (up-regulated)	lncRNAs (down-regulated)	References
TNF-α	Mouse	MEFs	166		Rapicavoli et al. 2013
Pam3CSK4	Mouse	BMDMs	62	62	Carpenter et al. 2013
Pam3CSK4	Human	THP1s (monocytic cell line)	156		Li et al. 2014
IFN-α	Human	Hepatocytes	∼120	∼100, exact number not reported	Kambara et al. 2014
IAV	Human	Alveolar cells	494	413	Ouyang et al. 2014
EV71	Human	Rhabdomyosarcoma cells	2990	1876	Yin et al. 2013
SARS-CoV	Mouse	Lung	1500		Peng et al. 2010

TNF-α, Tumor necrosis factor α; MEFs, murine embryo fibroblasts; BMDMs, bone-marrow-derived macrophages; IFN-α, interferon α; IAV, influenza A virus; SARS-CoV, severe acute respiratory syndrome coronavirus.

transcribes BISPR that then acts as a positive regulator for BST2 expression. Small interfering RNA (siRNA)-mediated knockdown of BISPR had a dramatic impact on BST2 protein expression levels.

NeST

As mentioned earlier, the lncRNA NeST (also known as TMEVPG1 and LincR-*Ifng*-3′AS′) is expressed in the Th1 subset of helper T cells and is critical for controlling infection with Theiler's virus. NeST controls the expression of IFN-γ through WDR5-dependent histone methylation at the *Ifng* locus in CD8$^+$ T cells.

NRAV

Negative regulator of antiviral response (NRAV) is a lncRNA that is dramatically down-regulated in cells following viral infection. It functions to suppress ISG expression through epigenetic regulation of these genes. Overexpressing human NRAV in mice results in enhanced influenza A virus (IAV) virulence (Ouyang et al. 2014).

NRON

The expression of lncRNA NRON (noncoding repressor or nuclear factor of activated T [NFAT] cells) is controlled by HIV. NRON is localized to the cytosol where it can bind to the transcription factor NFAT and repress its activity. Knockdown of NRON using siRNA resulted in enhanced HIV replication because of increased activity of NFAT (Imam et al. 2015).

lncRNA-CD244

lncRNA-CD244 (lncRNA-AS-GSTT1[1-72]) is found at high levels in CD8$^+$ T cells that are infected with *M. tuberculosis* where it inhibits T-cell signaling. Knocking down lncRNA-CD244 results in increased expression of IFN-γ and TNF-α. The authors find that this lncRNA expression is driven by CD244 signaling and functions to repress IFN-γ and TNF-α through its interactions with EZH2, a component of the PRC2 complex (Wang et al. 2015b).

lncRNAs THAT REGULATE CYTOKINE PRODUCTION

Innate immune signaling pathways are complex, consisting of feedforward and feedback mechanisms, which function in a temporally regulated manner to elicit effective host defenses and ensure that immune homeostasis is maintained. lncRNAs are now emerging as crucial players in controlling these inducible transcriptional networks. In this section, we will outline those lncRNAs with validated experimental data supporting their roles in controlling cytokine expression downstream from innate immune receptor activation.

lincRNA-Cox2

Studies in our laboratory identified lincRNA-Cox2 as a dynamically regulated lincRNA in murine macrophages stimulated with multiple Toll-like receptor (TLR) ligands, including lipopolysaccharide, which activates TLR4/MD2 and Pam3CSK4, a synthetic bacterial lipopeptide that acts as TLR1/2 ligand (Carpenter et al. 2013). Prior studies had identified this lincRNA in DCs exposed to lipopolysaccharide, which engages TLR4/MD2 (Guttman et al. 2009). lincRNA-Cox2 acts as both a repressor and an activator of distinct clusters of innate immune genes. lincRNA-Cox2 can form a complex with hnRNP-A/B and A2/B1 to mediate its repressive effects on interferon-stimulated genes. Knockdown of lincRNA-Cox2 also severely impaired the production of additional immune genes, including the proinflammatory cytokine IL-6; however, the precise molecular mechanism by which lincRNA-Cox2 mediates these latter effects in macrophages is still unclear.

Two recent studies have shed new light on the function of lincRNA-Cox2. Chen and colleagues have shown that TNF-α induced lincRNA-Cox2 functions to control IL-12B levels in intestinal epithelial cells. In a series of elegant biochemical studies, these authors showed that lincRNA-Cox2 facilitates the recruitment of Mi-2/nucleosome remodeling and deacetylate (Mi2/NuRD) complex to the IL-12B promoter region, resulting in inhibition of IL-12B expression (Tong et al. 2016). They also showed that lincRNA-Cox2 can promote immune gene activation through binding to the SWI/SNF chromatin remodeling complex. Both of these studies classify lincRNA-Cox2 as a primary response gene because they observed early induction (2 h) of this transcript in intestinal epithelial cells or RAWs (murine macrophage cell line). In our own studies, in macrophages, we found that lincRNA-Cox2 reached maximal expression following 5 h of stimulation with TLR2 ligands in primary murine macrophages. Indeed, we observe superinduction of lincRNA-Cox2 in murine macrophages treated with TLR2 ligands in the presence of cyclo-heximide (S Carpenter and KA Fitzgerald, unpubl.), indicating this gene may be a secondary response gene in murine macrophages. It is possible that lincRNA-Cox2 has varying expression profiles in different cell types. The exact mediators that collaborate with lincRNA-Cox2 to exert these regulatory functions remain to be further elucidated.

THRIL

TNF-α and heterogeneous nuclear ribonucleoprotein L (hnRNPL)-related immunoregulatory lincRNA (THRIL) was identified in the human monocyte cell line THP1 (Li et al. 2014). A total of 156 lncRNAs were found to be differentially regulated in THP1 cells following stimulation with the TLR1/2 ligand Pam3CSK4. THRIL is an antisense transcript that partially overlaps the 3′ UTR of the gene encoding BRI3-binding protein (Bri3bp). THRIL functions to regulate TNF-α expression through interactions with hnRNPL. RNA-seq analysis on THRIL knockdown cells showed that this lncRNA has a broad impact on expression of a large number of immune genes (454). Interestingly 98% of these genes were down-regulated when THRIL was knocked down, strongly suggesting that this lncRNA is critical for controlling basal expression of these genes (Li et al. 2014). The exact mechanism of action of this lincRNA and whether all of these effects are direct or indirectly regulated by THRIL remains to be clarified.

PACER

The p50-associated COX-2 extragenic RNA (PACER) is transcribed in the antisense direction directly upstream of the *PTGS2* (Cox2) transcription start site (Krawczyk and Emerson 2014). To date, there is no evidence that PACER is conserved across species. PACER functions to sequester p50, the repressive subunit of the NF-κB complex, preventing p50 from binding to the *PTGS2* promoter. The authors show that the presence of CTCF at the *PTGS2* promoter facilitates chromatin accessibility, resulting in transcription of PACER. PACER then functions to recruit p300 and RNA Pol II complexes, leading to the induction of Cox2. PACER's interaction with p50 permits access to the activating complex of NF-κB (p50-p65) at the *PTGS2* promoter following inflammatory stimulation (Krawczyk and Emerson 2014).

NEAT1

NEAT1 is a 4-kb unspliced nonpolyadenylated transcript critical for the formation of nuclear paraspeckles (Clemson et al. 2009). NEAT1 is evolutionarily conserved and the murine homolog displays two small regions of high sequence conservation (Clemson et al. 2009). NEAT1 is induced by influenza virus and herpes simplex virus in addition to PolyI:C (synthetic RNA ligand that activates TLR3). NEAT1 is critical for the expression of IL-8 through its interactions with the splicing protein splicing factor proline/glutamine-rich (SFPQ) (Imamura et al. 2014). Chromatin immunoprecipitation (ChIP) assays confirmed that at basal levels SFPQ binds to the promoter of the gene encoding IL-8 and inhibits transcription. NEAT1 can interact with SFPQ, resulting in the protein relocating to paraspeckles releasing its repression of IL-8.

Recent evidence suggests that NEAT1 expression is elevated in patients suffering from systemic lupus erythematosus (SLE), and the expression is predominantly present in monocytes. Knockdown of NEAT1 resulted in decreased expression of proinflammatory genes, including *IL6* and *CXCL10*, and the authors propose a mechanism by which NEAT1 affects the late-stage mitogen-activated protein kinases (MAPK) pathway following LPS stimulation. They conclude that NEAT1 could act as a regulatory contributor to the increased cytokine signature associated with the pathogenesis of SLE (Zhang et al. 2016).

lnc-IL-7R

lnc-IL-7R is an LPS-inducible transcript that overlaps the 3′ UTR of IL-7R, with both transcripts sharing the same polyA tail. Knockdown of lnc-IL-7R led to an increase in expression of a number of genes, including IL-7R, IL-8, VCAM-1, and E-selectin. This knockdown was accompanied by a decrease in the transcriptional histone repressor mark H3K27 trimethylation at the promoters of affected genes (Cui et al. 2014). The exact mechanism of how lnc-IL-7R regulates H3K27 at its target genes still requires mechanistic understanding.

IL-1β-RBT46

IL-1β-RBT46 originates from a region of bidirectional transcription upstream of the transcription start site of IL-1B. It is inducible following LPS stimulation or infection with *Listeria monocytogenes*. Knocking down this transcript with shRNA resulted in a reduction in IL-1B and CXCL8 expression at both the RNA and protein levels (Ilott et al. 2014).

NKILA

NF-κB-interacting lncRNA (NKILA) acts as a direct inhibitor of NF-κB signaling. NKILA binds to the NF-κB/IκB complex inhibiting phosphorylation and subsequent degradation of IκB, resulting in the p65 subunit of NF-κB remaining in the cytoplasm and therefore inhibiting signaling (Liu et al. 2015). A recent article has questioned some of these findings by Liu et al. and provided some additional information on this locus, showing that NKILA is transcribed in the opposite orientation to the protein-coding gene *PMEPA1* (Dijkstra and Alexander 2015).

AS-IL-1α

Studies in our own laboratory identified a natural antisense transcript that is transcribed at the IL-1A locus. The lncRNA contains partial sequence complementarity to IL-1A. AS-IL-1α is a nuclear localized transcript that is highly inducible following TLR stimulation or following *L. monocytogenes* infection in vivo (Chan et al. 2015). Knockdown studies have shown that AS-IL-1α is required for transcription of the *IL-1A* protein-coding gene.

lincRNA-EPS

The vast majority of studies of lncRNAs in immunity have focused on inducible lncRNAs and their role in controlling immune gene expression. However, it is clear from all of the RNA-seq data generated that there are just as many lncRNAs that are down-regulated following inflammatory stimulation of innate immune cells. One such example is lincRNA-EPS. As mentioned earlier, this lncRNA was first identified as an erythrocyte-expressed lncRNA crucial for during red blood cell development. A recent study expanded on this and showed that lincRNA-EPS has additional functions beyond those proposed for erythrocytes (Atianand et al. 2016). lincRNA-EPS is expressed in macrophages and DCs and is rapidly down-regulated following inflammatory stimulation. lincRNA-EPS-deficient mice display enhanced inflammation and succumb to endotoxin-mediated lethality faster than their corresponding littermate controls. lincRNA-EPS-deficient macrophages have elevated expression of ∼200 immune response genes. lincRNA-EPS associates with chromatin at regulatory regions of these genes where it controls nucleosome positioning to restrain transcription of these genes. lincRNA-EPS mediates some of these functions through its interaction with heterogeneous nuclear ribonucleoprotein L, which occurs via a *CANACA* motif, known to be enriched in

RNAs that bind hnRNPL located in its 3′ end (Atianand et al. 2016). The identification of lincRNA-EPS as a repressor of inflammatory responses further highlights the importance of lincRNAs in the immune system.

EMERGING ROLES FOR lncRNAs IN DISEASE

Remarkably, 93% of disease-associated SNPs lie in noncoding regions of the genome, some of which impact the expression of lncRNAs (Pennisi 2007; Kumar et al. 2013). Approximately 10% of autoimmune and immune-related disorder (AID) SNPs are present in lncRNA genes (Ricaño-Ponce and Wijmenga 2013). Gaining a better understanding of the molecular mechanisms of lncRNA could help define their roles in inflammatory and autoimmune disease pathologies. Many of the studies to date are descriptive and correlate expression of a particular lncRNA associated with specific diseases. Table 3 outlines lncRNAs that have been functionally associated with autoimmune disorders. Only one lncRNA with a known SNP has been shown to be the direct cause of a genetic disorder and that is lncRNA RMRP in CHH as mentioned previously. The majority of autoimmune conditions are not monogenic disorders, and therefore it is possible that many lncRNAs could impact disease pathogenesis in these conditions. Just this year, Lnc13 was identified and

Table 3. Disease-associated long coding RNAs (lncRNAs) and their functions

Disease condition	lncRNA	Function	References
Cartilage hair hypoplasia	RMRP	Controlling gene expression in Th17 cells	Huang et al. 2015
Celiac disease	Lnc13	Mediates repression of a subset of celiac-associated genes through its interaction with hnRNPD	Castellanos-Rubio et al. 2016
Kawasaki disease	THRIL	Regulates TNF-α through its interactions with hnRNPL in monocytes	Li et al. 2014
Psoriasis	PRINS	Down-regulates the antiapoptotic effects of G1P3 in keratinocytes	Sonkoly et al. 2005; Szegedi et al. 2010
Angelman syndrome	Ube3a-ATS	Silences paternal Ube3a through imprinting	Meng et al. 2015
Systemic lupus erythematosus	NEAT1	Regulates MAPK signaling	Zhang et al. 2016

TNF, Tumor necrosis factor; THRIL, TNF-α and heterogeneous nuclear ribonucleoprotein L (hnRNPL)-related immunoregulatory lincRNA; hnRNPL, heterogeneous nuclear ribonucleoprotein L; MAPK, mitogen-activated protein kinase.

Cite this article as *Cold Spring Harb Perspect Biol* doi: 10.1101/cshperspect.a028589

associated with susceptibility to celiac disease (Castellanos-Rubio et al. 2016). Similar to what has been reported for lincRNA-EPS, Lnc13 is down-regulated in myeloid cells exposed to inflammatory stimuli. This down-regulation, in turn, results in elevated expression levels of immune genes. Lnc13 partly overlaps with the protein-coding gene sIl18rap. The 5′ end of Lnc13 overlaps the 3′ end of IL-18rap. Like other lincRNAs, Lnc13 interacts with heterogeneous nuclear ribonucleoprotein D (hnRNPD), and this complex is involved in suppressing immune gene expression. Lnc13 is encoded within a celiac disease–associated haplotype. In patients suffering from celiac disease, the levels of Lnc13 are decreased in the small intestine, which could account for the increase in expression of inflammatory genes in these patients. In addition, the Lnc13 variant present in celiac disease patients interacts less efficiently with hnRNPD than its wild-type counterpart, which may account for its inability to mediate its repressor functions.

THERAPEUTIC TARGETING OF CYTOKINES AND lncRNAs

Cytokines are a major target for therapeutic intervention in a number of inflammatory conditions with targeted therapeutics for TNF-α, IFN, and IL-17 already in clinical use (Moreland et al. 1997; Lipsky et al. 2000; Plosker 2011). There is increasing interest in therapeutic targeting of noncoding RNAs. A number of companies are currently developing lncRNA therapeutic targeting methods, including RaNA Therapeutics, OPKU-CURNA, and IONIS Pharmaceuticals. All companies are taking similar approaches and targeting lncRNAs that typically mediate repression of genes through their interaction with repressor complexes such as the PRC2 complex to down-regulate expression of specific protein-coding genes. By interfering with the expression or function of these lncRNAs, the goal is to redirect expression of target protein-coding genes. RaNA Therapeutics just received a patent for Polycomb-associated noncoding RNAs, which function through the PRC2 complex (Lee et al. 2016).

All companies are making use of antisense technology to repress or interfere with lncRNAs and their target repressor complexes using antisense oligonucleotides. Specifically, they are developing antagoNATs or gapmers, which are single-stranded DNA oligos with locked nucleic acids (LNAs) that can act either as steric blockers (if LNAs are distributed throughout the oligo) or directly bind to lncRNAs and mediate their degradation through RNase H cleavage (if LNAs are placed at either end of the oligo). The LNAs help to protect the oligo against exonuclease cleavage.

Last year, IONIS Pharmaceuticals in collaboration with a laboratory at Baylor College of Medicine published a study describing the possible uses of lncRNA targeting in Angelman syndrome, a syndrome that is a monogenic disorder caused by maternal deficiency of the imprinted gene *UBE3A*, which is an E3 ubiquitin ligase. Patients suffering from Angelman syndrome possess one paternal copy of *UBE3A*; however, this is silenced by the lncRNA UBE3A-ATS (Meng et al. 2015). This disorder results in developmental delay, seizures, and ataxia and there are currently very few treatment options available. In this study, they made use of antisense oligonucleotides (ASOs) designed to target the lncRNA Ube3a-ATS in a mouse model of Angelman syndrome. Even partial restoration of the UBE3A protein resulted in an improvement in the cognitive defects of this condition. Like other oligo-based therapeutics (e.g., siRNA), one of the major obstacles is delivery and specificity of targeting the gene of interest in a specific location. The initial focus of many lncRNA therapeutics is on the central nervous system (CNS) diseases for which there are currently no treatment options. In the case of the Angelman syndrome study, they are delivering the oligo directly to the CNS with some success. Much work is needed to increase the efficiency of the delivery mechanisms and this will open a huge opportunity for RNA-based therapeutics.

CONCLUSIONS AND FUTURE PERSPECTIVES

lncRNAs are quickly emerging as key regulators of a wide variety of biological processes (Rinn and Chang 2012). Because lncRNAs display more cell-type specificity in expression patterns compared to protein-coding genes, it is estimated that the number of annotated lncRNAs will continue to expand as more sequencing is performed. As exciting as this field is, it is also a daunting challenge to begin to understand the biological importance of thousands of novel genes. Some of the challenges include understanding the exact annotations of full-length lncRNA transcripts. Chip-seq, Gro-seq, and RNA-seq technologies have greatly helped in the initial annotations. However, because lncRNAs are cell-type specific in expression patterns, we often need to perform these expensive and sometimes laborious experiments on each cell type of interest to fully understand the start sites of lncRNAs. This information becomes even more pertinent when using Cas9/CRISPR-based techniques such as CRISPRi and CRISPRa, which heavily rely on knowledge of the transcriptional start sites of target genes. Annotating lncRNAs are also made difficult owing to the levels of repeat elements present in their sequences (Kim et al. 2016). Novel long read technologies such as Pac-Bio and Oxford Nanopore sequencing will enable better definition of the full-length sequences and expressed isoforms of lncRNAs in our cells of interest. The more we understand about the primary sequences of lncRNAs, the easier it will be to determine their exact biological functions. Cas9/CRISPR techniques will allow for rapid high-throughput studies of lncRNAs in a variety of biological contexts.

An exciting avenue of lncRNA research relates to disease and possible use as therapeutic targets in the future. Because the majority of disease-associated SNPs lie within noncoding regions, it is critical to attempt to understand what role lncRNAs, enhancers, and enhancer RNAs play in disease pathogenesis. We outlined the current efforts to use lncRNAs for therapeutic manipulation; however, there is also growing interest in studying lncRNAs as biomarkers of disease. It is possible that lncRNA expression will be disease specific. LncRNAs have also been shown to be highly stable in bodily fluids and therefore offer an easy PCR-based target as a biomarker. To date, one lncRNA has been approved by the Food and Drug Administration (FDA) for use as a biomarker in prostate cancer, and this is the lncRNA PCA3 (de Kok et al. 2002). We believe this is just the beginning in terms of using these genes as biomarkers and possible therapeutic targets in the near future.

Once the genomic profiling phase of discovery is complete, we need to understand the specific functions for all these genes. Extensive functional studies are required to better understand the roles of lncRNA in controlling biological processes in addition to understanding their functions in disease settings. Future discoveries of lncRNAs will provide us with a greater understanding of the molecular mechanisms that govern the transcriptional regulation of genes in inflammatory and infectious diseases, thus providing a better platform from which to develop novel diagnostics and therapeutics.

REFERENCES

Alvarez-Dominguez JR, Hu W, Yuan B, Shi J, Park SS, Gromatzky AA, van Oudenaarden A, Lodish HF. 2014. Global discovery of erythroid long noncoding RNAs reveals novel regulators of red cell maturation. *Blood* **123:** 570–581.

Anderson DM, Anderson KM, Chang CL, Makarewich CA, Nelson BR, McAnally JR, Kasaragod P, Shelton JM, Liou J, Bassel-Duby R, et al. 2015. A micropeptide encoded by a putative long noncoding RNA regulates muscle performance. *Cell* **160:** 595–606.

Atianand MK, Hu W, Satpathy AT, Shen Y, Ricci EP, Alvarez-Dominguez JR, Bhatta A, Schattgen SA, McGowan JD, Blin J, et al. 2016. A long noncoding RNA lincRNA-EPS acts as a transcriptional brake to restrain inflammation. *Cell* **165:** 1672–1685.

Bánfai B, Jia H, Khatun J, Wood E, Risk B, Gundling WE, Kundaje A, Gunawardena HP, Yu Y, Xie L, et al. 2012. Long noncoding RNAs are rarely translated in two human cell lines. *Genome Res* **22:** 1646–1657.

Barriocanal M, Carnero E, Segura V. 2015. Long non-coding RNA BST2/BISPR is induced by IFN and regulates the expression of the antiviral factor tetherin. *Front Immunol* **5:** 1–13.

Bendall SC, Simonds EF, Qiu P, Amir E-AD, Krutzik PO, Finck R, Bruggner RV, Melamed R, Trejo A, Ornatsky OI, et al. 2011. Single-cell mass cytometry of differential immune and drug responses across a human hematopoietic continuum. *Science* **332:** 687–696.

Cabili MN, Trapnell C, Goff L, Koziol M, Tazon-Vega B, Regev A, Rinn JL. 2011. Integrative annotation of human large intergenic noncoding RNAs reveals global properties and specific subclasses. *Genes Dev* **25:** 1915–1927.

Cite this article as *Cold Spring Harb Perspect Biol* doi: 10.1101/cshperspect.a028589

Carlevaro-Fita J, Rahim A, Guigó R, Vardy LA, Johnson R. 2016. Cytoplasmic long noncoding RNAs are frequently bound to and degraded at ribosomes in human cells. *RNA* **22:** 867–882.

Carpenter S, Aiello D, Atianand MK, Ricci EP, Gandhi P, Hall LL, Byron M, Monks B, Henry-Bezy M, Lawrence JB, et al. 2013. A long noncoding RNA mediates both activation and repression of immune response genes. *Science* **341:** 789–792.

Castellanos-Rubio A, Fernandez-Jimenez N, Kratchmarov R, Luo X, Bhagat G, Green PHR, Schneider R, Kiledjian M, Bilbao JR, Ghosh S. 2016. A long noncoding RNA associated with susceptibility to celiac disease. *Science* **352:** 91–95.

Chan J, Atianand M, Jiang Z, Carpenter S, Aiello D, Elling R, Fitzgerald KA, Caffrey DR. 2015. Cutting edge: A natural antisense transcript, AS-IL1α, controls inducible transcription of the proinflammatory cytokine IL-1α. *J Immunol* **195:** 1359–1363.

Chen MT, Lin HS, Shen C, Ma YN, Wang F, Zhao HL, Yu J, Zhang JW. 2015. The PU.1-regulated long noncoding RNA lnc-MC controls human monocyte/macrophage differentiation through interaction with microRNA-199a-5p. *Mol Cell Biol* **35:** 3212–3224.

Chew GL, Pauli A, Rinn JL, Regev A, Schier AF, Valen E. 2013. Ribosome profiling reveals resemblance between long non-coding RNAs and 5′ leaders of coding RNAs. *Development* **140:** 2828–2834.

Clemson CM, Hutchinson JN, Sara SA, Ensminger AW, Fox AH, Chess A, Lawrence JB. 2009. An architectural role for a nuclear noncoding RNA: NEAT1 RNA is essential for the structure of paraspeckles. *Mol Cell* **33:** 717–726.

Collier SP, Collins PL, Williams CL, Boothby MR, Aune TM. 2012. Cutting edge: Influence of *Tmevpg1*, a long intergenic noncoding RNA, on the expression of *Ifng* by Th1 cells. *J Immunol* **189:** 2084–2088.

Cui H, Xie N, Tan Z, Banerjee S, Thannickal VJ, Abraham E, Liu G. 2014. The human long noncoding RNA lnc-IL7R regulates the inflammatory response. *Eur J Immunol* **44:** 2085–2095.

de Kok JB, Verhaegh GW, Roelofs RW, Hessels D, Kiemeney LA, Aalders TW, Swinkels DW, Schalken JA. 2002. *DD3^{PCA3}*, a very sensitive and specific marker to detect prostate tumors. *Cancer Res* **62:** 2695–2698.

Dijkstra JM, Alexander DB. 2015. The "NF-κB interacting long noncoding *RNA*" (*NKILA*) transcript is antisense to cancer-associated gene *PMEPA1*. *F1000Res* **4:** 96.

Dijkstra JM, Ballingall KT. 2014. Non-human *lnc-DC* orthologs encode Wdnm1-like protein. *F1000Res* **3:** 160.

Doss JF, Corcoran DL, Jima DD, Telen MJ, Dave SS, Chi JT. 2015. A comprehensive joint analysis of the long and short RNA transcriptomes of human erythrocytes. *BMC Genomics* **16:** 952.

Flynn RA, Almada AE, Zamudio JR, Sharp PA. 2011. Antisense RNA polymerase II divergent transcripts are P-TEFb dependent and substrates for the RNA exosome. *Proc Natl Acad Sci* **108:** 10460–10465.

Gomez JA, Wapinski OL, Yang YW, Bureau JF, Gopinath S, Monack DM, Chang HY, Brahic M, Kirkegaard K. 2013. The NeST long ncRNA controls microbial susceptibility and epigenetic activation of the interferon-γ locus. *Cell* **152:** 743–754.

Guttman M, Amit I, Garber M, French C, Lin MF, Feldser D, Huarte M, Zuk O, Carey BW, Cassady JP, et al. 2009. Chromatin signature reveals over a thousand highly conserved large non-coding RNAs in mammals. *Nature* **458:** 223–227.

Guttman M, Garber M, Levin JZ, Donaghey J, Robinson J, Adiconis X, Fan L, Koziol MJ, Gnirke A, Nusbaum C, et al. 2010. Ab initio reconstruction of cell type-specific transcriptomes in mouse reveals the conserved multi-exonic structure of lincRNAs. *Nat Biotechnol* **28:** 503–510.

Guttman M, Russell P, Ingolia NT, Weissman JS, Lander ES. 2013. Ribosome profiling provides evidence that large noncoding RNAs do not encode proteins. *Cell* **154:** 240–251.

Hu W, Yuan B, Flygare J, Lodish HF. 2011. Long noncoding RNA-mediated anti-apoptotic activity in murine erythroid terminal differentiation. *Genes Dev* **25:** 2573–2578.

Hu G, Tang Q, Sharma S, Yu F, Escobar TM, Muljo SA, Zhu J, Zhao K. 2013. Expression and regulation of intergenic long noncoding RNAs during T cell development and differentiation. *Nat Immunol* **14:** 1190–1198.

Huang W, Thomas B, Flynn RA, Gavzy SJ, Wu L, Kim SV, Hall JA, Miraldi ER, Ng CP, Rigo FW, et al. 2015. DDX5 and its associated lncRNA *Rmrp* modulate T$_H$17 cell effector functions. *Nature* **528:** 517–522.

Huang Z, Luo Q, Yao F, Qing C, Ye J, Deng Y, Li J. 2016. Identification of differentially expressed long non-coding RNAs in polarized macrophages. *Sci Rep* **6:** 19705.

Ilott NE, Heward JA, Roux B, Tsitsiou E, Fenwick PS, Lenzi L, Goodhead I, Hertz-Fowler C, Heger A, Hall N, et al. 2014. Long non-coding RNAs and enhancer RNAs regulate the lipopolysaccharide-induced inflammatory response in human monocytes. *Nat Commun* **5:** 3979.

Imam H, Shahr Bano A, Patel P, Holla P, Jameel S. 2015. The lncRNA NRON modulates HIV-1 replication in a NFAT-dependent manner and is differentially regulated by early and late viral proteins. *Sci Rep* **5:** 8639.

Imamura K, Imamachi N, Akizuki G, Kumakura M, Kawaguchi A, Nagata K, Kato A, Kawaguchi Y, Sato H, Yoneda M, et al. 2014. Long noncoding RNA NEAT1-dependent SFPQ relocation from promoter region to paraspeckle mediates IL8 expression upon immune stimuli. *Mol Cell* **53:** 393–406.

Ingolia NT, Ghaemmaghami S, Newman JRS, Weissman JS. 2009. Genome-wide analysis in vivo of translation with nucleotide resolution using ribosome profiling. *Science* **324:** 218–223.

Ingolia NT, Lareau LF, Weissman JS. 2011. Ribosome profiling of mouse embryonic stem cells reveals the complexity and dynamics of mammalian proteomes. *Cell* **147:** 789–802.

Ingolia NT, Brar GA, Rouskin S, McGeachy AM, Weissman JS. 2012. The ribosome profiling strategy for monitoring translation in vivo by deep sequencing of ribosome-protected mRNA fragments. *Nat Protoc* **7:** 1534–1550.

Ingolia NT, Brar GA, Stern-Ginossar N, Harris MS, Talhouarne GJS, Jackson SE, Wills MR, Weissman JS. 2014. Ribosome profiling reveals pervasive translation outside of annotated protein-coding genes. *Cell Rep* **8:** 1365–1379.

Iyer MK, Niknafs YS, Malik R, Singhal U, Sahu A, Hosono Y, Barrette TR, Prensner JR, Evans JR, Zhao S, et al. 2015. The landscape of long noncoding RNAs in the human transcriptome. *Nat Genet* **47**: 199–208.

Josset L, Tchitchek N, Gralinski LE, Ferris MT, Eisfeld AJ, Green RR, Thomas MJ, Tisoncik-Go J, Schroth GP, Kawaoka Y, et al. 2014. Annotation of long non-coding RNAs expressed in collaborative cross founder mice in response to respiratory virus infection reveals a new class of interferon-stimulated transcripts. *RNA Biol* **11**: 875–890.

Kambara H, Niazi F, Kostadinova L, Moonka DK, Siegel CT, Post AB, Carnero E, Barriocanal M, Anthony DD, Valadkhan S. 2014. Negative regulation of the interferon response by an interferon-induced long non-coding RNA. *Nucleic Acids Res* **42**: 10668–10680.

Kambara H, Gunawardane L, Zebrowski E, Valadkhan S. 2015. Regulation of interferon-stimulated gene BST2 by a lncRNA transcribed from a shared bidirectional promoter. *Front Immunol* **5**: 676.

Kim TK, Hemberg M, Gray JM. 2015. Enhancer RNAs: A class of long noncoding RNAs synthesized at enhancers. *Cold Spring Harb Perspect Biol* **7**: a018622.

Kim EZ, Wespiser AR, Caffrey DR. 2016. The domain structure and distribution of Alu elements in long noncoding RNAs and mRNAs. *RNA* **22**: 254–264.

Kotzin JJ, Spencer SP, McCright SJ, Kumar DBU, Collet MA, Mowel WK, Elliott EN, Uyar A, Makiya MA, Dunagin MC, et al. 2016. The long non-coding RNA *Morrbid* regulates *Bim* and short-lived myeloid cell lifespan. *Nature* **537**: 239–243.

Krawczyk M, Emerson BM. 2014. p50-associated COX-2 extragenic RNA (PACER) activates COX-2 gene expression by occluding repressive NF-κB complexes. *eLife* **3**: e01776.

Kumar V, Westra H-J, Karjalainen J, Zhernakova DV, Esko T, Hrdlickova B, Almeida R, Zhernakova A, Reinmaa E, Võsa U, et al. 2013. Human disease-associated genetic variation impacts large intergenic non-coding RNA expression. *PLoS Genet* **9**: e1003201.

Lam MTY, Li W, Rosenfeld MG, Glass CK. 2014. Enhancer RNAs and regulated transcriptional programs. *Trends Biochem Sci* **39**: 170–182.

Lawrence T, Natoli G. 2011. Transcriptional regulation of macrophage polarization: Enabling diversity with identity. *Nat Rev Immunol* **11**: 750–761.

Lee JT, Zhao J, Sarma K, Borowsky M, Ohsumi TK. 2016. Polycomb-associated non-coding RNAs. U.S. Patent 9,328,346 filed November 12, 2011, and issued May 3, 2016.

Lessing D, Anguera MC, Lee JT. 2013. X chromosome inactivation and epigenetic responses to cellular reprogramming. *Annu Rev Genomics Hum Genet* **14**: 85–110.

Li Z, Chao TC, Chang KY, Lin N, Patil VS, Shimizu C, Head SR, Burns JC, Rana TM. 2014. The long noncoding RNA *THRIL* regulates TNFα expression through its interaction with hnRNPL. *Proc Natl Acad Sci* **111**: 1002–1007.

Li H, Hao Y, Zhang D, Fu R, Liu W, Zhang X, Xue F, Yang R. 2016. Aberrant expression of long noncoding RNA TMEVPG1 in patients with primary immune thrombocytopenia. *Autoimmunity* **49**: 496–502.

Lin W-W, Karin M. 2007. A cytokine-mediated link between innate immunity, inflammation, and cancer. *J Clin Invest* **117**: 1175–1183.

Lin MF, Jungreis I, Kellis M. 2011. PhyloCSF: A comparative genomics method to distinguish protein coding and non-coding regions. *Bioinformatics* **27**: i275–i282.

Lipsky PE, van der Heijde DMFM, St Claire W, Furst DE, Breedveld FC, Kalden JR, Smolen JS, Weisman M, Emery P, Feldmann M, et al. 2000. Infliximab and methotrexate in the treatment of rheumatoid arthritis. Anti-tumor necrosis factor trial in rheumatoid arthritis with concomitant therapy study group. *N Engl J Med* **343**: 1594–1602.

Liu B, Sun L, Liu Q, Gong C, Yao Y, Lv X, Lin L, Yao H, Su F, Li D, et al. 2015. A cytoplasmic NF-κB interacting long noncoding RNA blocks IκB phosphorylation and suppresses breast cancer metastasis. *Cancer Cell* **27**: 370–381.

Martinez FO, Gordon S. 2014. The M1 and M2 paradigm of macrophage activation: Time for reassessment. *F1000Prime Rep* **6**: 13.

McHugh CA, Chen CK, Chow A, Surka CF, Tran C, McDonel P, Pandya-Jones A, Blanco M, Burghard C, Moradian A, et al. 2015. The *Xist* lncRNA interacts directly with SHARP to silence transcription through HDAC3. *Nature* **521**: 232–236.

Meng L, Ward AJ, Chun S, Bennett CF, Beaudet AL, Rigo F. 2015. Towards a therapy for Angelman syndrome by targeting a long non-coding RNA. *Nature* **518**: 409–412.

Moreland LW, Baumgartner SW, Schiff MH, Tindall EA, Fleischmann RM, Weaver AL, Ettlinger RE, Cohen S, Koopman WJ, Mohler K, et al. 1997. Treatment of rheumatoid arthritis with a recombinant human tumor necrosis factor receptor (p75)-Fc fusion protein. *N Engl J Med* **337**: 141–147.

Murray PJ, Allen JE, Biswas SK, Fisher EA, Gilroy DW, Goerdt S, Gordon S, Hamilton JA, Ivashkiv LB, Lawrence T, et al. 2014. Macrophage activation and polarization: Nomenclature and experimental guidelines. *Immunity* **41**: 14–20.

Nelson BR, Makarewich CA, Anderson DM, Winders BR, Troupes CD, Wu F, Reese AL, McAnally JR, Chen X, Kavalali ET, et al. 2016. A peptide encoded by a transcript annotated as long noncoding RNA enhances SERCA activity in muscle. *Science* **351**: 271–275.

Orkin SH, Zon LI. 2008. Hematopoiesis: An evolving paradigm for stem cell biology. *Cell* **132**: 631–644.

Ouyang J, Zhu X, Chen Y, Wei H, Chen Q, Chi X, Qi B, Zhang L, Zhao Y, Gao GF, et al. 2014. NRAV, a long noncoding RNA, modulates antiviral responses through suppression of interferon-stimulated gene transcription. *Cell Host Microbe* **16**: 616–626.

Padua DM, Mahurkar-Joshi S, Law IKM, Polytarchou C, Vu JP, Pisegna JR, Shih DQ, Iliopoulos D, Pothoulakis C. 2016. A long noncoding RNA signature for ulcerative colitis identifies IFNG-AS1 as an enhancer of inflammation. *Am J Physiol Gastrointest Liver Physiol* **311**: G446–G457.

Pang KC, Dinger ME, Mercer TR, Malquori L, Grimmond SM, Chen W, Mattick JS. 2009. Genome-wide identification of long noncoding RNAs in CD8$^+$ T cells. *J Immunol* **182**: 7738–7748.

Peng X, Gralinski L, Armour CD, Ferris MT, Thomas MJ, Proll S, Bradel-Tretheway BG, Korth MJ, Castle JC, Biery

MC, et al. 2010. Unique signatures of long noncoding RNA expression in response to virus infection and altered innate immune signaling. *MBio* **1:** e00206–10.

Pennisi E. 2007. Breakthrough of the year. Human genetic variation. *Science* **318:** 1842–1843.

Plosker GL. 2011. Interferon-β-1b: A review of its use in multiple sclerosis. *CNS Drugs* **25:** 67–88.

Qiao YQ, Huang ML, Xu AT, Zhao D, Ran ZH, Shen J. 2013. LncRNA DQ786243 affects Treg related CREB and Foxp3 expression in Crohn's disease. *J Biomed Sci* **20:** 87.

Rapicavoli NA, Qu K, Zhang J, Mikhail M, Laberge R-M, Chang HY. 2013. A mammalian pseudogene lncRNA at the interface of inflammation and anti-inflammatory therapeutics. *eLife* **2:** e00762.

Rashid F, Shah A, Shan G. 2016. Long non-coding RNAs in the cytoplasm. *Genomics Proteomics Bioinformatics* **14:** 73–80.

Ricaño-Ponce I, Wijmenga C. 2013. Mapping of immune-mediated disease genes. *Annu Rev Genomics Hum Genet* **14:** 325–353.

Rinn JL, Chang HY. 2012. Genome regulation by long noncoding RNAs. *Annu Rev Biochem* **81:** 145–166.

Seila AC, Calabrese JM, Levine SS, Yeo GW, Rahl PB, Flynn RA, Young RA, Sharp PA. 2008. Divergent transcription from active promoters. *Science* **322:** 1849–1851.

Shortman K, Liu YJ. 2002. Mouse and human dendritic cell subtypes. *Nat Rev Immunol* **2:** 151–161.

Sonkoly E, Bata-Csorgo Z, Pivarcsi A, Polyanka H, Kenderessy-Szabo A, Molnar G, Szentpali K, Bari L, Megyeri K, Mandi Y, et al. 2005. Identification and characterization of a novel, psoriasis susceptibility-related noncoding RNA gene, PRINS. *J Biol Chem* **280:** 24159–24167.

Spurlock CF, Tossberg JT, Guo Y, Collier SP, Crooke PS, Aune TM. 2015. Expression and functions of long noncoding RNAs during human T helper cell differentiation. *Nat Commun* **6:** 6932.

Szegedi K, Sonkoly E, Nagy N, Németh IB, Bata-Csorgo Z, Kemeny L, Dobozy A, Szell M. 2010. The anti-apoptotic protein G1P3 is overexpressed in psoriasis and regulated by the non-coding RNA, PRINS. *Exp Dermatol* **19:** 269–278.

Tong Q, Gong AY, Zhang XT, Lin C, Ma S, Chen J, Hu G, Chen XM. 2016. LincRNA-Cox2 modulates TNF-α-induced transcription of Il12b gene in intestinal epithelial cells through regulation of Mi-2/NuRD-mediated epigenetic histone modifications. *FASEB J* **30:** 1187–1197.

Tsitsiou E, Williams AE, Moschos SA, Patel K, Rossios C, Jiang X, Adams OD, Macedo P, Booton R, Gibeon D, et al.

2012. Transcriptome analysis shows activation of circulating CD8[+] T cells in patients with severe asthma. *J Allergy Clin Immunol* **129:** 95–103.

Vilcek J, Feldmann M. 2004. Historical review: Cytokines as therapeutics and targets of therapeutics. *Trends Pharmacol Sci* **25:** 201–209.

Wang P, Xue Y, Han Y, Lin L, Wu C, Xu S, Jiang Z, Xu J, Liu Q, Cao X. 2014. The STAT3-binding long noncoding RNA lnc-DC controls human dendritic cell differentiation. *Science* **344:** 310–313.

Wang P, Zhi H, Zhang Y, Liu Y, Zhang J, Gao Y, Guo M, Ning S, Li X. 2015a. miRSponge: A manually curated database for experimentally supported miRNA sponges and ceRNAs. *Database (Oxford)* **2015:** bav098.

Wang Y, Zhong H, Xie X, Chen CY, Huang D, Shen L, Zhang H, Chen ZW, Zeng G. 2015b. Long noncoding RNA derived from CD244 signaling epigenetically controls CD8[+] T-cell immune responses in tuberculosis infection. *Proc Natl Acad Sci* **112:** E3883–E3892.

Wang J, Peng H, Tian J, Ma J, Tang X, Rui K, Tian X, Wang Y, Chen J, Lu L, et al. 2016a. Upregulation of long noncoding RNA TMEVPG1 enhances T helper type 1 cell response in patients with Sjögren syndrome. *Immunol Res* **64:** 489–496.

Wang J, Syrett CM, Kramer MC, Basu A, Atchison ML, Anguera MC. 2016b. Unusual maintenance of X chromosome inactivation predisposes female lymphocytes for increased expression from the inactive X. *Proc Natl Acad Sci* **113:** E2029–E2038.

Washietl S, Kellis M, Garber M. 2014. Evolutionary dynamics and tissue specificity of human long noncoding RNAs in six mammals. *Genome Res* **24:** 616–628.

Wu Y, Smas CM. 2008. Wdnm1-like, a new adipokine with a role in MMP-2 activation. *Am J Physiol Endocrinol Metab* **295:** E205–E215.

Xia F, Dong F, Yang Y, Huang A, Chen S, Sun D, Xiong S, Zhang J. 2014. Dynamic transcription of long non-coding RNA genes during CD4[+] T cell development and activation. *PLoS ONE* **9:** e101588.

Yin Z, Guan D, Fan Q, Su J, Zheng W, Ma W, Ke C. 2013. lncRNA expression signatures in response to enterovirus 71 infection. *Biochem Biophys Res Commun* **430:** 629–633.

Zhang F, Wu L, Qian J, Qu B, Xia S, La T, Wu Y, Ma J, Zeng J, Guo Q, et al. 2016. Identification of the long noncoding RNA NEAT1 as a novel inflammatory regulator acting through MAPK pathway in human lupus. *J Autoimmun* **75:** 96–104.

Negative Regulation of Cytokine Signaling in Immunity

Akihiko Yoshimura, Minako Ito, Shunsuke Chikuma, Takashi Akanuma, and Hiroko Nakatsukasa

Department of Microbiology and Immunology, Keio University School of Medicine, Shinjuku-ku, Tokyo 160-8582, Japan

Correspondence: yoshimura@a6.keio.jp

Cytokines are key modulators of immunity. Most cytokines use the Janus kinase and signal transducers and activators of transcription (JAK-STAT) pathway to promote gene transcriptional regulation, but their signals must be attenuated by multiple mechanisms. These include the suppressors of cytokine signaling (SOCS) family of proteins, which represent a main negative regulation mechanism for the JAK-STAT pathway. Cytokine-inducible Src homology 2 (SH2)-containing protein (CIS), SOCS1, and SOCS3 proteins regulate cytokine signals that control the polarization of $CD4^+$ T cells and the maturation of $CD8^+$ T cells. SOCS proteins also regulate innate immune cells and are involved in tumorigenesis. This review summarizes recent progress on CIS, SOCS1, and SOCS3 in T cells and tumor immunity.

There are four types of the cytokine receptors: (1) receptors that activate nuclear factor (NF)-κB and mitogen-activated protein (MAP) kinases (mainly p38 and c-Jun amino-terminal kinase [JNK]), such as receptors for the tumor necrosis factor (TNF)-α family, the interleukin (IL)-1 family, including IL-1β, IL-18, and IL-33, and the IL-17 family; (2) receptors that activate the Janus kinase and signal transducers and activators of transcription (JAK-STAT) pathway—most cytokines belong to this family; (3) transforming growth factor (TGF)-β receptors carrying a serine/threonine kinase that activates Smad-family transcription factors; and (4) growth factor receptors in which cytoplasmic domain contains the tyrosine kinase domain. This latter family typically signals via the Ras extracellular signal-regulated kinase (ERK) pathway (see Fig. 1). Any receptor that activates intracellular signaling pathways has multiple negative feedback systems, which ensures transient activation of the pathway and downstream transcription factors. Typical negative regulators are shown in Figure 1. Lack of such negative regulators results in autoimmune diseases, autoinflammatory diseases, and some-times-fatal disorders, including cancer. Thus, negative feedback is essential for homeostasis.

Cytokine receptor signal regulators can be classified into three types: (1) proteins that physically suppress signal generation, (2) protein phosphatases, and (3) proteins recruiting degradation systems or processes such as proteasomes, autophagy, and endocytosis. All are multidomain proteins that bind to the receptors and/or signaling molecules through an Src ho-

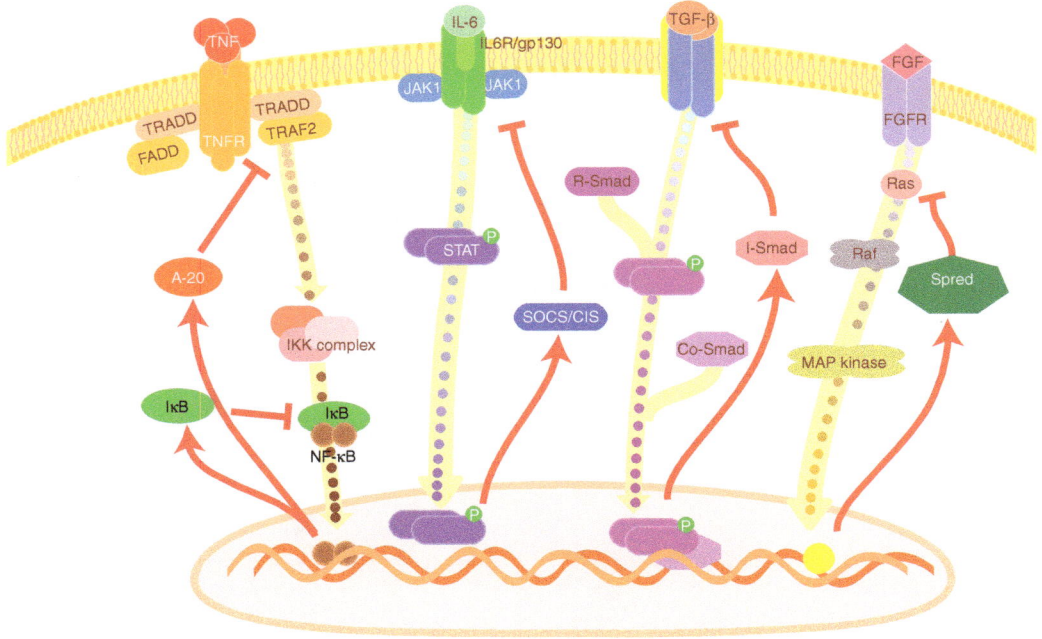

Figure 1. The cytokine signaling and their major negative regulators. There are four types of cytokine receptors: (1) receptors that activate nuclear factor (NF)-κB and mitogen-activated protein (MAP) kinases; (2) receptors that activate the Janus kinase and signal transducers and activators of transcription (JAK-STAT) pathway; (3) transforming growth factor (TGF)-β receptors; and (4) growth factor receptor family. Typical negative regulators are also shown. TNF, Tumor necrosis factor; TNFR, TNF receptor; FGF, fibroblast growth factor; FGFR, FGF receptor; IKK, IκB kinase; SOCS/CIS, suppressor of cytokine signaling/cytokine-inducible Src homology 2 (SH2)-containing protein; Spred, Sprouty-related protein with an EVH1 domain.

mology 2 (SH2) domain or other binding motifs and then suppress the signaling via other domains. For example, A20/TNFAIP3 is an important negative regulatory protein for the NF-κB pathway that interacts with NF-κB essential modulator (NEMO)/IκB kinase (IKK)γ and functions as a deubiquitinase (Shembade and Harhaj 2012). Sprouty-related protein with an EVH1 domain (Spred) family proteins suppress the Ras-ERK pathway by bridging the growth factor/cytokine receptors and NF-1, a Ras-GTPase-activating protein (Wakioka et al. 2001; Nonami et al. 2003; Yoshida et al. 2006; Hirata et al. 2016). Some negative-regulator proteins have two or more inhibitory domains; for example, suppressor of cytokine signaling (SOCS)1 and SOCS3 have an amino-terminal kinase-inhibitory region (KIR) that inhibits JAK tyrosine kinase activity and a carboxy-ter-

minal SOCS-box that recruits the ubiquitin-transferase complex. Because of space limitations in this review, we focus on proteins, especially SOCS proteins, which regulate signal transduction, but not on molecules interacting with the extracellular domain of the receptors or on transcription factors.

THE JAK-STAT PATHWAY

Cytokines play several essential roles in the development, differentiation, and function of myeloid and lymphoid cells. Some of them, including ILs, interferons (IFNs), and hematopoietic growth factors, activate the JAK-STAT pathway (O'Shea et al. 2002). In this pathway, cytokine binding results in receptor oligomerization, which initiates the activation of JAK kinases (JAK1, JAK2, JAK3, and TYK2). JAK3 is associated with IL-2 receptor γ (common cytokine receptor γ chain), and is activated by IL-2-related cytokines. The activated JAKs phosphorylate the receptor cytoplasmic domains, which creates docking sites for SH2-containing signaling proteins. The STAT proteins are the major substrates for JAKs. A large number of cytokines, growth factors, and hormonal factors activate the JAK-STAT pathway. For example, IFN-γ receptors activate JAK1 and JAK2, which then mainly phosphorylate and activate STAT1, whereas IL-6 binds to the IL-6 receptor α (IL-6Rα) chain and to gp130, both of which mainly activate JAK1 and STAT3 (Guschin et al. 1995). The anti-inflammatory cytokine IL-10 also activates STAT3 (Weber-Nordt et al. 1996). T helper (Th)1, Th2, and Th17 are induced by IL-12, IL-4, and IL-6/IL-23, and thus STAT4, STAT6, and STAT3 are essential

for Th1, Th2, and Th17 differentiation, respectively. IL-4 in combination with TGF-β has been shown to induce Th9 in vitro (Tamiya et al. 2013). IL-2/STAT5 is essential for regulatory T-cell (Treg) development, and IL-21/STAT3 is essential for follicular helper T (Tfh) cell differentiation (Vogelzang et al. 2008). IL-21 also regulates CD8[+] T cells (Gagnon et al. 2007) and Th17 cell differentiation (Bettelli et al. 2007).

THE CIS/SOCS FAMILY: MOLECULAR MECHANISMS

SOCS proteins and cytokine-inducible SH2-containing protein ([CIS] also known as CISH) molecules comprise a family of intracellular proteins (Fig. 2) (Yoshimura et al. 1995, 2007; Endo et al. 1997; Matsumoto et al. 1997; Tamiya

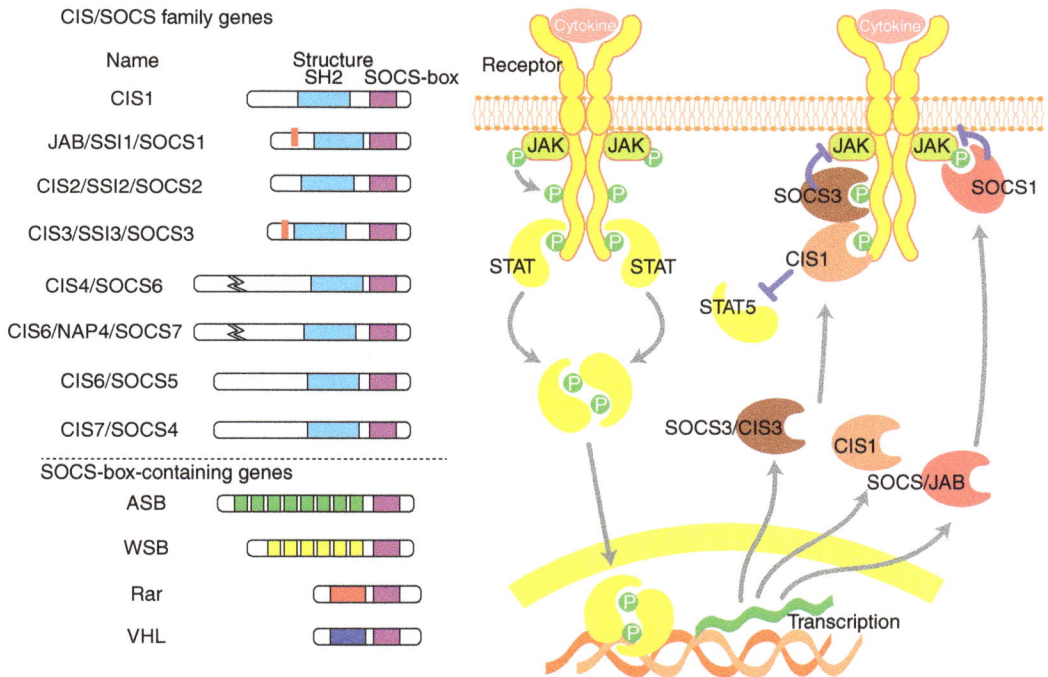

Figure 2. Structure and function of the suppressor of cytokine signaling (SOCS) family. The SOCS family consists of eight family members. All eight members share a central Src homology 2 (SH2) domain and a carboxy-terminal SOCS box. In addition, SOCS1 and SOCS3 possess a kinase inhibitory region (KIR) that inhibits Janus kinase (JAK) activity. (Right) The general mechanism of the action of cytokine-inducible SH2-containing protein (CIS), SOCS1, and SOCS3. STAT, Signal transducers and activators of transcription; JAB, Janus kinase-binding protein.

et al. 2011). There are eight CIS/SOCS family proteins: CIS, SOCS1, SOCS2, SOCS3, SOCS4, SOCS5, SOCS6, and SOCS7, each of which has a central SH2 domain, an amino-terminal domain of variable length and sequence, and a carboxy-terminal 40-amino-acid module known as the SOCS box (Fig. 2) (Hilton et al. 1998; Kamizono et al. 2001; Kamura et al. 2004). The SOCS box interacts with elongin B/C, Cullins, and the Really Interesting New Gene (RING)-finger domain-only protein RBX2, which recruits E2 ubiquitin transferase (Kamura et al. 2004). CIS/SOCS family proteins, as well as other SOCS-box-containing molecules, function as E3 ubiquitin ligases and mediate the degradation of proteins that are associated with these family members through their amino-terminal regions (Fig. 2).

The central SH2 domain determines the target of each SOCS and CIS protein. The SH2 domain of SOCS1 directly binds to the activation loop of JAKs (Yoshimura et al. 1995; Yasukawa et al. 1999). The SH2 domains of CIS, SOCS2, and SOCS3 bind to phosphorylated tyrosine residues of the activated cytokine receptors. SOCS3 binds to gp130-related cytokine receptors, including the phosphorylated tyrosine 757 (Tyr757) residue of gp130, the Tyr800 residue of IL-12 receptor β2, and the Tyr985 residue of the leptin receptor (Fig. 2). SOCS molecules bind to several tyrosine phosphorylated proteins and promote their degradation. SOCS1 binds to Mal, thereby negatively regulating Toll-like receptor (TLR) signaling (Mansell et al. 2006). SOCS3 has been shown to be an important regulator of insulin signaling, although binding to IRS1/2 (Shi et al. 2004; Torisu et al. 2007). SOCS3 also regulates chemokine signaling in B cells by interacting with focal adhesion kinase (FAK) (Le et al. 2007). SOCS2 binds active Pyk2 via pY402 and ubiquitinates it in natural killer (NK) cells (Lee et al. 2010). SOCS6 promotes p56Lck degradation via the proteasome in T cells (Choi et al. 2010).

In addition to general SOCS-box function in this family, both SOCS1 and SOCS3 have a unique amino-terminal motif that can inhibit JAK tyrosine kinase activity directly through their KIR. KIR has been proposed to function as a pseudosubstrate, and it is essential for the suppression of cytokine signals (Fig. 2) (Yasukawa et al. 1999; Kershaw et al. 2013). Recent study of the ternary cocrystal structure between mouse SOCS3, JAK2 kinase domain, and a fragment of gp130 supported this hypothesis (Fig. 3). The kinase-inhibitory region of SOCS3 occludes the substrate-binding groove on JAK2, and biochemical studies show that it blocks substrate association. SOCS3, and probably SOCS1, inhibits the catalytic activity of JAK1, JAK2, and TYK2, but not JAK3, because the SH2-KIR domain interacts with an evolutionarily conserved "GQM" sequence that is present in all vertebrate forms of JAK1, JAK2, and TYK2, but not JAK3, where it lines one edge of the substrate-binding groove (Babon et al. 2012). We have shown that a KIR-mutant SOCS1 functions as a dominant negative form not only for SOCS1 but also for SOCS3 (Hanada et al. 2001). Thus, the interaction between KIR and GQM motif is essential for the tight binding and for the inhibition of the tyrosine kinase activity. The SH2 domain of SOCS3 does not bind to JAKs with high affinity, but it is required for receptor binding to inhibit JAKs, whereas the SOCS1 SH2 domain has been shown to inhibit JAK kinase activity through direct binding of its SH2 domain to the activation loop of JAKs. Regulation of type I IFN signaling by SOCS1 was shown not to require any of the phosphorylation sites in the IFNAR1 receptor. Because receptors to which SOCS3 binds mostly activate STAT3, SOCS3 is an inhibitor that is relatively specific to STAT3. However, SOCS3 also inhibits STAT4, which is activated by IL-12 (Fig. 2, right) (Yamamoto et al. 2003).

A recent study suggested that alveolar macrophages secrete SOCS1 and SOCS3 in exosomes and microparticles, respectively, for uptake by alveolar epithelial cells and subsequent inhibition of STAT activation (Bourdonnay et al. 2015). Secretion and transcellular delivery of vesicular SOCS proteins was diminished by cigarette smoking, suggesting a novel mechanism of dysregulated inflammation by smoking.

Cite this article as *Cold Spring Harb Perspect Biol* doi: 10.1101/cshperspect.a028571

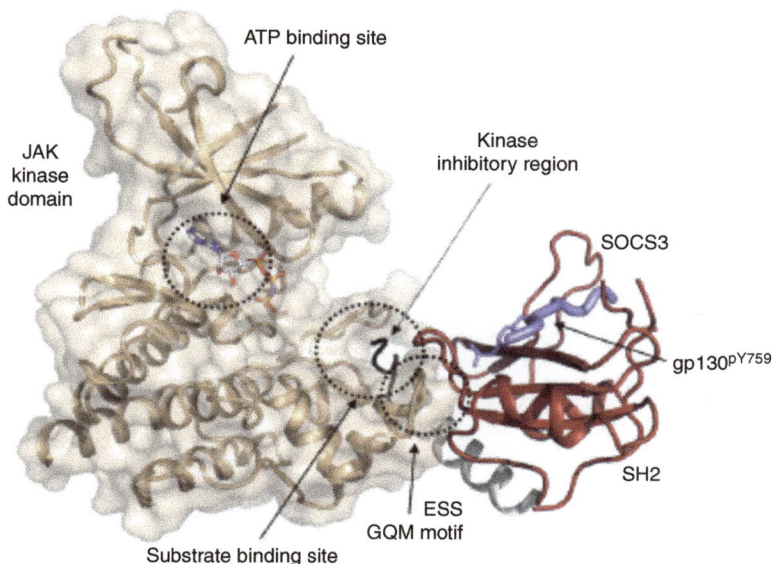

Figure 3. Structure of the complex of Janus kinase (JAK)2 and suppressor of cytokine signaling (SOCS)3, and gp130 phosphopeptide. SOCS3 binds the kinase domains of JAK1, JAK2, and TYK2 and inhibits its catalytic activity by blocking the substrate-binding site with its kinase inhibitory region. SOCS3 remains bound to gp130 while in complex with JAK (beige) and adenosine triphosphate (ATP) binding is unaffected. (Based on Kershaw et al. 2013, with permission, from the authors.)

CIS: INHIBITOR OF CYTOKINE AND T-CELL RECEPTOR (TCR) SIGNALING

CIS (also called CISH) was discovered as a rapid-inducible gene in response to various cytokines, including erythropoietin (EPO), IL-2, IL-3, and IL-5, which mostly activate STAT5 (Yoshimura et al. 1995; Matsumoto et al. 1997). CIS does not possess the KIR and cannot inhibit JAK tyrosine kinase activity directly. However, CIS binds to phosphorylated cytokine receptors, such as the EPO receptor, IL-2 receptor, murine IL-3 receptor β chain, prolactin receptor, and the growth hormone (GH) receptor, which mostly activate STAT5 (Yoshimura et al. 1995; Matsumoto et al. 1997; Aman et al. 1999; Ram and Waxman 2000; Endo et al. 2003). Thus, CIS is believed to suppress STAT5 by masking STAT5-binding phosphotyrosine motifs on the receptors, and also by inducing ubiquitin/proteasome-dependent degradation of the activated receptors (Fig. 2, right) (Okabe et al. 1999). In zebrafish, expression of cish.a

(there are two *cish* genes, *cish.a* and *cish.b* in zebrafish) was regulated by the JAK2-STAT5 pathway via conserved tetrameric STAT5-binding sites ($TTCN_3GAA....TTCN_3GAA$) in its promoter, and knockdown of *cish.a*, but not *cish.b*, resulted in enhanced embryonic erythropoiesis, myelopoiesis, and lymphopoiesis. This study showed conserved CIS functions for the control of hematopoiesis through STAT5 (Lewis et al. 2014).

However, recent studies using $Cish^{-/-}$ mice challenged this theory (Yang et al. 2013). *Cish*-deficient mice >10 months of age showed spontaneous allergic lung inflammation, including mucus impaction and more eosinophil infiltrates. T cells from $Cish^{-/-}$ mice expressed significantly higher IL-4 and IL-9 levels than wild-type (WT) cells, whereas IFN-γ and IL-17 levels were comparable. In vitro differentiation experiments indicated that *Cish* deficiency skewed T-cell differentiation toward Th2 and Th9 cells in the presence of IL-4. Higher amounts of STAT5 and STAT6 were recruited to the IL-4, IL-9, and GATA-3 promoters in $Cish^{-/-}$ T cells as compared with WT T cells. These data suggest that CIS is a negative feedback regulator of IL-4, although the precise biochemical mechanism remains to be clarified.

CIS has been shown to be involved in regulation of TCR signaling. CIS expression is induced by TCR stimulation in T cells, and T-cell activation from *Cish*-transgenic mice showed enhanced proliferative responses and prolonged survival following TCR stimulation (Li et al. 2000). Subsequent research showed that there was an interaction between CIS and activated protein kinase C (PKC) α, β, and θ, followed by promotion of the ERK pathway (Chen et al. 2003). Paradoxically, recent research indicates that genetic deletion of *Cish* in $CD8^{+}$ T cells enhances their expansion and function, resulting in pronounced and durable regression of established tumors (Palmer et al. 2015). In this study, CIS was shown to bind to the TCR intermediate PLC-γ1, targeting it for proteasomal degradation after TCR stimulation. The reason why *Cish* knockdown and overexpression show similar phenotypes in T cells has not been clarified, but similar paradoxical phenomena were observed in the regulation of GH signaling by SOCS2 (Greenhalgh et al. 2002). CIS and SOCS2 expression levels may be critical for regulation of signaling or they have different functions for various targets.

Another recent paper suggests that CIS is a critical negative regulator of IL-15 signaling in NK cells and that deletion of *Cish* enhances antitumor immunity (Delconte et al. 2016). CIS was rapidly induced in response to IL-15, and deletion of *Cish* rendered NK cells hypersensitive to IL-15, as evidenced by enhanced proliferation, survival, IFN-γ production, and cytotoxicity toward tumors. In this study, CIS has been shown to selectively interact with JAK1 and targeting JAK1 for proteasomal degradation. *Cish*-deficient mice were resistant to melanoma, prostate, and breast cancer metastasis in vivo, and this was intrinsic to NK cell activity. Although the mechanisms are different, these studies suggest possibilities for new cancer immunotherapies directed at blocking CIS function.

Human genetic linkage studies of *CISH* have shown an association between *CISH* genetic variants and susceptibility to bacteremia, malaria, and tuberculosis (Khor et al. 2010; Sun et al. 2014), as well as persistent hepatitis B virus (HBV) infection or clearance of HBV (Tong et al. 2012; Hu et al. 2014; Song et al. 2014). The molecular mechanism of *CISH*-related infection regulation remains to be clarified.

SOCS1 AND T CELLS

SOCS1 can inhibit almost all cytokines using JAKs because it binds to and directly inhibits JAKs. Thus, SOCS1 specificity is regulated by its induction. *Socs1* knockout (KO) mice and conditional KO (cKO) mice showed that SOCS1 plays an essential negative regulatory role in IFN-γ, IL-2, IL-4, and IL-7 functions (Yoshimura et al. 2012). Because *Socs1*-deficient mice showed an aberrant CD4/CD8 ratio, SOCS1 has been implicated in T-cell development in the thymus (Catlett and Hedrick 2005). SOCS1, SOCS3, and CIS were shown to be critical targets of ThPOK, which is essential for the $CD4^{+}$ lineage fate, thereby inhibiting $CD8^{+}$ lineage programming (Luckey et al. 2014). Transgenic SOCS1 expression in thymocytes, however, rescued defects in $CD4^{+}$ T-cell development in ThPOK-deficient mice (Luckey et al. 2014). SOCS1 and other SOCS molecules may be related to IL-7 sensitivity that determines the CD4/8 lineage decision.

SOCS1 plays critical roles in Th subset differentiation. $Socs1^{-/-}$ CD4-naïve T cells differentiated into Th1 cells, even under non-skewing conditions, whereas Th17 differentiation was strongly suppressed (Tanaka et al. 2008). Th17 suppression by *Socs1* deficiency is probably a result of hyperproduction and signal transduction of IFN-γ. STAT3 activation was reduced in *Socs1*-deficient T cells, mostly because of up-regulation of SOCS3 gene expression, which can account for reduced IL-6 responses and Th17 differentiation (Tanaka et al. 2008). In addition, $Socs1^{-/-}$ T cells were less responsive to TGF-β, although the mechanism has not been clarified (Tanaka et al. 2008).

Cite this article as *Cold Spring Harb Perspect Biol* doi: 10.1101/cshperspect.a028571

SOCS1 AND TREGs

SOCS1 also plays an important role in the regulation of Tregs. Several reports have suggested that Tregs lose the expression of their master transcription factor, Foxp3, under certain inflammatory conditions. Thus, inflammatory cytokine signaling, including IFN-γ and IL-6 signaling, play important roles in the pathogenic conversion of Tregs (Takahashi and Yoshimura 2014; Sekiya et al. 2016). SOCS1 has been reported to play an important role in Treg-cell integrity and function by protecting Tregs from excessive inflammatory cytokines (Takahashi et al. 2011). *Socs1* deficiency in Tregs did not affect in vitro suppression activity, but it impaired suppressive Treg function in vivo despite the increase in Tregs (Lu et al. 2009). *Socs1*-deficient Tregs lose Foxp3 expression and convert into Th1- or Th17-like cells, probably because of STAT1 and STAT3 hyperactivation. The increase in Treg number may be explained by hypersensitivity to IL-2/STAT5 signaling in *Socs1*-deficient Tregs. Recently, Ubc13 has been reported to be involved in suppressive activity by controlling Treg effector cytokine signaling molecules, including SOCS1 (Chang et al. 2012). Smad2/3-deficient Treg phenotypes were similar to those observed in *Socs1*-deficient Tregs (Takimoto et al. 2010). This may because hyperactivation of STAT1 inhibits the TGF-β/Smad pathway, which is important for Treg maintenance (Tanaka et al. 2008). Thus, interactive suppression of these molecules by STAT1 may be a mechanism of Foxp3 instability. Similarly, dysregulated STAT3 and STAT6 are thought to induce Treg instability (Feng et al. 2014). However, the role of SOCS3 in Tregs remains to be clarified. In humans, negative correlation between SOCS3 and Foxp3 levels was reported (Lan et al. 2013).

Socs1 is a target of miR-155 and miR-146a in Tregs (Lu et al. 2009, 2010, 2015). Lu et al. showed that, during thymic differentiation, up-regulation of Foxp3 is associated with high miR-155 expression, which in turn promotes the competitive fitness and proliferative potential of Treg cells by inducing *Socs1* down-regulation. miR-155 deficiency also attenuates liver ischemia-reperfusion injury through up-regulation of *Socs1*, which was associated with promotion of M2 macrophage polarization and suppression of Th17 differentiation (Tang et al. 2015). The importance of *Socs1* as the target gene for miR-155 was shown by disrupting the miR-155 binding site in the *Socs1* 3′ UTR in a murine germline, which shows that this axis is important for Treg and NK cell function (Lu et al. 2015). Conversely, miR-146a targets STAT1, thereby regulating SOCS1 expression (Lu et al. 2010).

SOCS1 AND ANTITUMOR IMMUNITY

Immune checkpoints are molecules in the immune system, which down-regulate the activation of T cells. They have garnered great interest in immunology and cancer science because immune-checkpoint molecules are involved in antitumor immunity and they are therapeutically important. The best known of these molecules are PD-1 and CTLA4, which negatively regulate TCR and costimulatory signals, respectively. Because SOCS1 is an important negative regulator of cytokine signaling, especially for IFN-γ and IL-12, which are essential cytokines for antitumor immunity, SOCS1 is now considered to be an immune checkpoint molecule for antitumor immunity. Previously, we and others showed that SOCS1-silenced dendritic cells (DCs) induce stronger antitumor immunity (Shen et al. 2004; Hanada et al. 2005; Chen et al. 2015). Myeloid-cell-specific *Socs1*-deficient mice were resistant to tumor growth in an IFN-γ-dependent manner (Hashimoto et al. 2009). In CD8[+] T cells, even though *Socs1* deficiency caused defective expansion following in vivo antigen stimulation (Ramanathan et al. 2010), *Socs1*-silenced CD8[+] T cells showed stronger antitumor activity (Dudda et al. 2013). Because SOCS1 is an important target of miR-155, miR-155 overexpression enhanced the antitumor response, and enforced *Socs1* expression in CD8[+] T cells phenocopied the miR-155 deficiency, whereas SOCS1 silenced augmented tumor destruction (Dudda et al. 2013). Similarly, miR-155 facilitates tumor growth modulation of myeloid-derived suppressive cells (MDSCs) through *Socs1* repression (Chen et al. 2015). These observations indicate that SOCS1 is a key regulator of antitumor immunity in both DCs and CD8$^+$ T cells. Thus, a SOSC1 inhibitor that suppresses SOCS–JAK interaction could be a potent enhancer of antitumor immunity (Ahmed et al. 2015; Chikuma et al. 2017).

SOCS1 AND TUMORIGENESIS

SOCS1 is a unique tumor-suppressor gene that regulates inflammation-related tumorigenesis (Hanada et al. 2006; Inagaki-Ohara et al. 2013). Silencing of *SOCS1* was frequently observed in hepatocellular carcinoma (HCC) (Yoshikawa et al. 2001) and also in premalignant hepatitis C virus (HCV)-infected patients (Yoshida et al. 2004). Subsequent studies showed increased susceptibility of *Socs1*$^{+/-}$ mice to the hepatocarcinogen diethylnitrosamine (DEN) (Yoshida et al. 2004). Liver injury is associated with hyperactivation of STAT1 and reduced activation of STAT3 (Ogata et al. 2006). Therefore, reduced expression of SOCS1 might enhance tissue injury and inflammation by hyperactivation of STAT1, promoting the turnover of epithelial cells and enhancing their susceptibility to oncogenesis. However, recent studies suggest a new role for SOCS1 in cancer. SOCS1 may regulate IFN-γ, IL-6, and hepatic growth factor signaling in the liver (Gui et al. 2015). In addition, SOCS1 has been shown to promote activation of the p53 tumor suppressor by a direct interaction (Mallette et al. 2010; Bouamar et al. 2015) and regulate p21$^{CIP1/WAF1}$ protein (p21) expression and stability (Yeganeh et al. 2016). SOCS1 interacts with p21 and promotes its ubiquitination and proteasomal degradation.

Decreased *SOCS1* gene expression could be a mechanism involved in promoter hypermethylation in human. *SOCS1* promoter hypermethylation is detected in various cancers. SOCS1 DNA hypermethylation is also frequently found in certain types of lymphomas and myelodysplastic syndrome (MDS). In these cases, the silencing of *SOCS1* leads to dysregulation of JAK-STAT signal transduction, and

therefore contributes to growth factor hypersensitivity. *SOCS1* gene loss-of-function mutations have been frequently observed in classical Hodgkin lymphoma (cHL) (Lennerz et al. 2015) and primary mediastinal and diffuse large B-cell lymphomas (DLBCL) (Schif et al. 2013). *SOCS1* deletion resulted in sustained JAK2-STAT5 activation, which may lead to dysregulated proliferation, whereas SOCS1 overexpression prevented tumor growth (Kamio et al. 2004; Tagami-Nagata et al. 2015).

SOCS1: GENOME-WIDE ASSOCIATION STUDIES

Recent genome-wide association studies (GWASs) revealed that *SOCS1* single-nucleotide polymorphisms (SNPs) are found in various diseases, including primary biliary cirrhosis (PBC) (Dong et al. 2015), multiple sclerosis (Disanto et al. 2014; de Lapuente et al. 2015; Leikfoss et al. 2015), leprosy (Liu et al. 2015), Crohn's disease (Ellinghaus et al. 2012), celiac disease (Dubois et al. 2010), and serum IgE levels (Mostecki et al. 2011). These data strongly suggest a role for SOCS1 in immune regulation and in human immunological disorders.

SOCS3: ESSENTIAL REGULATOR FOR STAT3-RELATED CYTOKINES

SOCS3 is highly specific for several key cytokines that are related to the gp130 family, because the SOCS3-SH2 domain has a high affinity for phosphorylated gp130. Conditional tissue deletion of SOCS3 showed a nonredundant ability to inhibit signaling from IL-6 and also from leukemia inhibitory factor (LIF), leptin, and granulocyte colony-stimulating factor (G-CSF) (Yoshimura et al. 2007). In macrophages, *Socs3* deficiency resulted in the conversion of the effects of IL-6 to those of IL-10, which is a potent inhibitor of macrophages and DCs (Yasukawa et al. 2003). This is probably because of sustained activation of STAT3 in the absence of *Socs3*. Macrophages expressing mutant gp130 that is unable to bind SOCS3 displayed similar sustained STAT3 activation and anti-inflammatory effects. However, mice lacking *Socs3* in the skin or mice carrying gp130 mutant develop exacerbated inflammation, chronic disease, and cancer (Inagaki-Ohara et al. 2013). Thus, the biological functions of the IL-6/STAT3 pathway are totally dependent on cell types.

ROLE OF SOCS3 IN HELPER T CELLS

SOCS3 in T cells regulates Th1/2/17 differentiation. SOCS3 expression in T cells is shown to positively correlate with the severity of human allergic diseases such as asthma and atopic dermatitis, because SOCS3 inhibits IL-12/Th1 development (Seki et al. 2003). SOCS3 also suppresses Th17 development because SOCS3 inhibits STAT3, which is essential for Th17 development (Chen et al. 2006; Tanaka et al. 2008). *Socs3* deficiency in T cells reduced atherosclerotic lesion development and vascular inflammation, which was dependent on IL-17, whereas SOCS3 overexpression in T cells reduces IL-17 and accelerates atherosclerosis (Taleb et al. 2009). SOCS3 overexpression by gene transfer could prevent the development of experimental arthritis and severe aortic aneurysm formation, which are highly dependent on Th17 (Shouda et al. 2001; Romain et al. 2013). However, *Socs3* deficiency in T cells showed different effects on Th1 and Th2 cells. T-cell-specific *Socs3*-cKO mice were resistant to Th1 and Th2 disease models. This is mostly because of higher IL-10 and TGF-β production from *Socs3*-deficient T cells (Kinjyo et al. 2006). *Socs3* deficiency in DCs also promotes induction of Foxp3$^+$ Tregs, which is dependent on higher production of TGF-β from *Socs3*$^{-/-}$ DCs (Kinjyo et al. 2006; Matsumura et al. 2007). A paradoxical effect of SOCS3 on T-cell regulation is mostly because STAT3 has a dual function—it promotes production of both inflammatory IL-17 and anti-inflammatory IL-10 and TGF-β.

SOCS3 AND CANCER

SOCS3 might also be involved in the development and progression of malignancies. SOCS3

expression levels were lower in HCV-positive tumors compared with nontumor regions (Ogata et al. 2006). Reduced SOCS3 messenger RNA (mRNA) and protein expression has been observed in various human cancers and is associated with constitutive STAT3 activation (Inagaki-Ohara et al. 2013). Recently, we reported that stomach tissue-specific deletion of *Socs3* resulted in gastric tumors, and this was dependent on leptin (Inagaki-Ohara et al. 2014). A *SOCS3* SNP was reported to be associated with human gastric cancer (Wang et al. 2016). Similarly, gp130 mutant mice carrying the Y757F mutant, which lost binding affinity to SOCS3, developed gastric tumors (Jenkins et al. 2005). In this case, IL-11 and TGF-β have been shown to play important roles (Judd et al. 2006). Loss of *Socs3* also promoted pancreatic cancer driven by the oncogenic Ras mutation (Lesina et al. 2011). SOCS3 overexpression suppressed growth of malignant fibrous histiocytoma cell lines by inhibiting STAT3 and IL-6 production. In addition, this study raised the possibility that small molecule inhibitors of JAK-STAT could be therapeutic for IL-6-producing tumors (Shouda et al. 2006). *SOCS3* mutation (or variant) in the SH2 domain was discovered in a patient with polycythemia vera (Suessmuth et al. 2009).

SOCS3 AND GWAS

GWAS studies have identified many SNPs of *SOCS3* underlying variations in plasma-lipid levels (Asselbergs et al. 2012), asthma (Hao et al. 2012), hypospadias (Karabulut et al. 2013), nonalcoholic fatty liver disease (Grigoryev et al. 2015), nonalcoholic steatohepatitis (Sharma et al. 2015), type 2 diabetes (Chambers et al. 2015), and body mass index (BMI) (Al Muftah et al. 2016). These studies suggest that SOCS3 in humans is important for adaptive immunity and for tissue injury and metabolism. SOCS3 has been shown to regulate IL-6 and tyrosine kinase receptors such as insulin and hepatocyte growth factor (HGF) (Sun et al. 2005; Tokumaru et al. 2005; Torisu et al. 2007).

CONCLUDING REMARKS

Over the past two decades, following the discovery of the SOCS family proteins, we have extended our understanding of the structure and function of these proteins. SOCS proteins act as simple negative-feedback regulators, and play a role in the fine-tuning of the immune response, inflammation, and metabolism, and also in cancer. Therapeutic trials using SOCS antisense oligonucleotide, short hairpin RNA (shRNA), and peptide mimetics are under investigation in animal models. Development of SOCS mimetics, based on structural analysis of the JAK–SOCS complex, is highly desirable.

ACKNOWLEDGMENTS

We thank Yuki Ushijima (Keio University) for manuscript preparation. This work is supported by special Grants-in-Aid from the Ministry of Education, Culture, Sports, Science and Technology of Japan (No. 25221305), Advanced Research & Development Programs for Medical Innovation (AMED-CREST), the Takeda Science Foundation, the Uehara Memorial Foundation, Mochida Memorial Foundation for Medical and Pharmaceutical Research, Kanae Foundation, and the SENSHIN Medical Research Foundation, Keio Gijuku Academic Developmental Funds. Disclosure: We have no conflict of interest.

REFERENCES

Ahmed CM, Larkin J IIII, Johnson HM. 2015. SOCS1 mimetics and antagonists: A complementary approach to positive and negative regulation of immune function. *Front Immunol* 6: 183.

Al Muftah WA, Al-Shafai M, Zaghlool SB, Visconti A, Tsai PC, Kumar P, Spector T, Bell J, Falchi M, Suhre K. 2016. Epigenetic associations of type 2 diabetes and BMI in an Arab population. *Clin Epigenetics* 8: 13.

Aman MJ, Migone TS, Sasaki A, Ascherman DP, Zhu M, Soldaini E, Imada K, Miyajima A, Yoshimura A, Leonard WJ. 1999. CIS associates with the interleukin-2 receptor β chain and inhibits interleukin-2-dependent signaling. *J Biol Chem* 274: 30266–30272.

Asselbergs FW, Guo YR, van Iperen EPA, Sivapalaratnam S, Tragante V, Lanktree MB, Lange LA, Almoguera B, Appelman YE, Barnard J, et al. 2012. Large-scale gene-centric meta-analysis across 32 studies identifies multiple lipid loci. *Am J Hum Genet* 91: 823–838.

Babon JJ, Kershaw NJ, Murphy JM, Varghese LN, Laktyushin A, Young SN, Lucet IS, Norton RS, Nicola NA. 2012. Suppression of cytokine signaling by SOCS3: Characterization of the mode of inhibition and the basis of its specificity. *Immunity* 36: 239–250.

Bettelli E, Korn T, Kuchroo VK. 2007. Th17: The third member of the effector T cell trilogy. *Curr Opin Immunol* 19: 652–657.

Bouamar H, Jiang DF, Wang L, Lin AP, Ortega M, Aguiar RCT. 2015. MicroRNA 155 control of p53 activity is context dependent and mediated by Aicda and Socs1. *Mol Cell Biol* 35: 1329–1340.

Bourdonnay E, Zaslona Z, Penke LRK, Speth JM, Schneider DJ, Przybranowski S, Swanson JA, Mancuso P, Freeman CM, Curtis JL, et al. 2015. Transcellular delivery of vesicular SOCS proteins from macrophages to epithelial cells blunts inflammatory signaling. *J Exp Med* 212: 729–742.

Catlett IM, Hedrick SM. 2005. Suppressor of cytokine signaling 1 is required for the differentiation of CD4$^+$ T cells. *Nat Immunol* 6: 715–721.

Chambers JC, Loh M, Lehne B, Drong A, Kriebel J, Motta V, Wahl S, Elliott HR, Rota F, Scott WR, et al. 2015. Epigenome-wide association of DNA methylation markers in peripheral blood from Indian Asians and Europeans with incident type 2 diabetes: A nested case-control study. *Lancet Diabetes Endo* 3: 526–534.

Chang JH, Xiao YC, Hu HB, Jin J, Yu JY, Zhou XF, Wu XF, Johnson HM, Akira SZ, Pasparakis M, et al. 2012. Ubc13 maintains the suppressive function of regulatory T cells and prevents their conversion into effector-like T cells. *Nat Immunol* 13: 481–U484.

Chen S, Anderson PO, Li L, Sjogren HO, Wang P, Li SL. 2003. Functional association of cytokine-induced SH2 protein and protein kinase C in activated T cells. *Int Immunol* 15: 403–409.

Chen Z, Laurence A, Kanno Y, Pacher-Zavisin M, Zhu BM, Tato C, Yoshimura A, Hennighausen L, O'Shea JJ. 2006. Selective regulatory function of Socs3 in the formation of IL-17-secreting T cells. *Proc Natl Acad Sci* 103: 8137–8142.

Chen SQ, Wang L, Fan J, Ye C, Dominguez D, Zhang Y, Curiel TJ, Fang DY, Kuzel TM, Zhang B. 2015. Host miR155 promotes tumor growth through a myeloid-derived suppressor cell-dependent mechanism. *Cancer Res* 75: 519–531.

Chikuma S, Kanamori M, Mise-Omata S, Yoshimura A. 2017. Suppressors of cytokine signaling: Potential immune checkpoint molecules for cancerimmunotherapy. *Cancer Sci* 108: 574–580.

Choi YB, Son M, Park M, Shin J, Yun Y. 2010. SOCS-6 negatively regulates T cell activation through targeting p56(lck) to proteasomal degradation. *J Biol Chem* 285: 7271–7280.

de Lapuente AL, Pinto-Medel MJ, Astobiza I, Alloza I, Comabella M, Malhotra S, Montalban X, Zettl UK, Rodriguez-Antiguedad A, Fernandez O, et al. 2015. Cell-specific effects in different immune subsets associated with SOCS1 genotypes in multiple sclerosis. *Mult Scler* 21: 1498–1512.

Delconte RB, Kolesnik TB, Dagley LF, Rautela J, Shi W, Putz EM, Stannard K, Zhang JG, Teh C, Firth M, et al. 2016.

CIS is a potent checkpoint in NK cell-mediated tumor immunity. *Nat Immunol* 17: 816.

Disanto G, Sandve GK, Ricigliano VAG, Pakpoor J, Berlanga-Taylor AJ, Handel AE, Kuhle J, Holden L, Watson CT, Giovannoni G, et al. 2014. DNase hypersensitive sites and association with multiple sclerosis. *Hum Mol Genet* 23: 942–948.

Dong M, Li JX, Tang RQ, Zhu P, Qiu F, Wang C, Qiu J, Wang L, Dai YP, Xu P, et al. 2015. Multiple genetic variants associated with primary biliary cirrhosis in a Han Chinese population. *Clin Rev Allerg Immunol* 48: 316–321.

Dubois PCA, Trynka G, Franke L, Hunt KA, Romanos J, Curtotti A, Zhernakova A, Heap GAR, Adany R, Aromaa A, et al. 2010. Multiple common variants for celiac disease influencing immune gene expression. *Nat Genet* 42: 295–302.

Dudda JC, Salaun B, Ji Y, Palmer DC, Monnot GC, Merck E, Boudousquie C, Utzschneider DT, Escobar TM, Perret R, et al. 2013. MicroRNA-155 is required for effector CD8[+] T cell responses to virus infection and cancer. *Immunity* 38: 742–753.

Ellinghaus D, Ellinghaus E, Nair RP, Stuart PE, Esko T, Metspalu A, Debrus S, Raelson JV, Tejasvi T, Belouchi M, et al. 2012. Combined analysis of genome-wide association studies for Crohn disease and psoriasis identifies seven shared susceptibility loci. *Am J Hum Genet* 90: 636–647.

Endo TA, Masuhara M, Yokouchi M, Suzuki R, Sakamoto H, Mitsui K, Matsumoto A, Tanimura S, Ohtsubo M, Misawa H, et al. 1997. A new protein containing an SH2 domain that inhibits JAK kinases. *Nature* 387: 921–924.

Endo T, Sasaki A, Minoguchi M, Joo A, Yoshimura A. 2003. CIS1 interacts with the Y532 of the prolactin receptor and suppresses prolactin-dependent STAT5 activation. *J Biochem* 133: 109–113.

Feng Y, Arvey A, Chinen T, van der Veeken J, Gasteiger G, Rudensky AY. 2014. Control of the inheritance of regulatory T cell identity by a *cis* element in the *Foxp3* locus. *Cell* 158: 749–763.

Gagnon J, Ramanathan S, Leblanc C, Ilangumaran S. 2007. Regulation of IL-21 signaling by suppressor of cytokine signaling-1 (SOCS1) in CD8[+] T lymphocytes. *Cell Signal* 19: 806–816.

Greenhalgh CJ, Metcalf D, Thaus AL, Corbin JE, Uren R, Morgan PO, Fabri LJ, Zhang JG, Martin HM, Willson TA, et al. 2002. Biological evidence that SOCS-2 can act either as an enhancer or suppressor of growth hormone signaling. *J Biol Chem* 277: 40181–40184.

Grigoryev DN, Cheranova DI, Chaudhary S, Heruth DP, Zhang LQ, Ye SQ. 2015. Identification of new biomarkers for acute respiratory distress syndrome by expression-based genome-wide association study. *BMC Pulm Med* 15: 95.

Gui Y, Yeganeh M, Donates YC, Tobelaim WS, Chababi W, Mayhue M, Yoshimura A, Ramanathan S, Saucier C, Ilangumaran S. 2015. Regulation of MET receptor tyrosine kinase signaling by suppressor of cytokine signaling 1 in hepatocellular carcinoma. *Oncogene* 34: 5718–5728.

Guschin D, Rogers N, Briscoe J, Witthuhn B, Watling D, Horn F, Pellegrini S, Yasukawa K, Heinrich P, Stark GR, et al. 1995. A major role for the protein tyrosine kinase JAK1 in the JAK/STAT signal transduction pathway in response to interleukin-6. *EMBO J* 14: 1421–1429.

Hanada T, Yoshida T, Kinjyo I, Minoguchi S, Yasukawa H, Kato S, Mimata H, Nomura Y, Seki Y, Kubo M, et al. 2001. A mutant form of JAB/SOCS1 augments the cytokine-induced JAK/STAT pathway by accelerating degradation of wild-type JAB/CIS family proteins through the SOCS-box. *J Biol Chem* 276: 40746–40754.

Hanada T, Tanaka K, Matsumura Y, Yamauchi M, Nishinakamura H, Aburatani H, Mashima R, Kubo M, Kobayashi T, Yoshimura A. 2005. Induction of hyper Th1 cell-type immune responses by dendritic cells lacking the suppressor of cytokine signaling-1 gene. *J Immunol* 174: 4325–4332.

Hanada T, Kobayashi T, Chinen T, Saeki K, Takaki H, Koga K, Minoda Y, Sanada T, Yoshioka T, Mimata H, et al. 2006. IFNγ-dependent, spontaneous development of colorectal carcinomas in SOCS1-deficient mice. *J Exp Med* 203: 1391–1397.

Hao K, Bosse Y, Nickle DC, Pare PD, Postma DS, Laviolette M, Sandford A, Hackett TL, Daley D, Hogg JC, et al. 2012. Lung eQTLs to help reveal the molecular underpinnings of asthma. *PloS Genet* 8: e1003029.

Hashimoto M, Ayada T, Kinjyo I, Hiwatashi K, Yoshida H, Okada Y, Kobayashi T, Yoshimura A. 2009. Silencing of SOCS1 in macrophages suppresses tumor development by enhancing antitumor inflammation. *Cancer Sci* 100: 730–736.

Hilton DJ, Richardson RT, Alexander WS, Viney EM, Willson TA, Sprigg NS, Starr R, Nicholson SE, Metcalf D, Nicola NA. 1998. Twenty proteins containing a C-terminal SOCS box form five structural classes. *Proc Natl Acad Sci* 95: 114–119.

Hirata Y, Brems H, Suzuki M, Kanamori M, Okada M, Morita R, Llano-Rivas I, Ose T, Messiaen L, Legius E, et al. 2016. Interaction between a domain of the negative regulator of the Ras-ERK pathway, SPRED1 protein, and the GTPase-activating protein-related domain of neurofibromin is implicated in Legius syndrome and neurofibromatosis type 1. *J Biol Chem* 291: 3124–3134.

Hu Z, Yang J, Wu Y, Xiong G, Wang Y, Yang J, Deng L. 2014. Polymorphisms in CISH gene are associated with persistent hepatitis B virus infection in Han Chinese population. *PloS ONE* 9: e100826.

Inagaki-Ohara K, Kondo T, Ito M, Yoshimura A. 2013. SOCS, inflammation, and cancer. *JAKSTAT* 2: e24053.

Inagaki-Ohara K, Mayuzumi H, Kato S, Minokoshi Y, Otsubo T, Kawamura YI, Dohi T, Matsuzaki G, Yoshimura A. 2014. Enhancement of leptin receptor signaling by SOCS3 deficiency induces development of gastric tumors in mice. *Oncogene* 33: 74–84.

Jenkins BJ, Grail D, Nheu T, Najdovska M, Wang B, Waring P, Inglese M, McLoughlin RM, Jones SA, Topley N, et al. 2005. Hyperactivation of Stat3 in gp130 mutant mice promotes gastric hyperproliferation and desensitizes TGF-β signaling. *Nat Med* 11: 845–852.

Judd LM, Bredin K, Kalantzis A, Jenkins BJ, Ernst M, Giraud AS. 2006. STAT3 activation regulates growth, inflammation, and vascularization in a mouse model of gastric tumorigenesis. *Gastroenterology* 131: 1073–1085.

Kamio M, Yoshida T, Ogata H, Douchi T, Nagata Y, Inoue M, Hasegawa M, Yonemitsu Y, Yoshimura A. 2004. SOC1 inhibits HPV-E7-mediated transformation by inducing degradation of E7 protein. *Oncogene* 23: 3107–3115.

Kamizono S, Hanada T, Yasukawa H, Minoguchi S, Kato R, Minoguchi M, Hattori K, Hatakeyama S, Yada M, Morita S, et al. 2001. The SOCS box of SOCS-1 accelerates ubiquitin-dependent proteolysis of TEL-JAK2. *J Biol Chem* **276**: 12530–12538.

Kamura T, Maenaka K, Kotoshiba S, Matsumoto M, Kohda D, Conaway RC, Conaway JW, Nakayama KI. 2004. VHL-box and SOCS-box domains determine binding specificity for Cul2-Rbx1 and Cul5-Rbx2 modules of ubiquitin ligases. *Genes Dev* **18**: 3055–3065.

Karabulut R, Turkyilmaz Z, Sonmez K, Kumas G, Ergun SG, Ergun MA, Basaklar AC. 2013. Twenty-four genes are upregulated in patients with hypospadias. *J Med Genet* **16**: 39–43.

Kershaw NJ, Murphy JM, Liau NP, Varghese LN, Laktyushin A, Whitlock EL, Lucet IS, Nicola NA, Babon JJ. 2013. SOCS3 binds specific receptor-JAK complexes to control cytokine signaling by direct kinase inhibition. *Nat Struct Mol Biol* **20**: 469–476.

Khor CC, Vannberg FO, Chapman SJ, Guo H, Wong SH, Walley AJ, Vukcevic D, Rautanen A, Mills TC, Chang KC, et al. 2010. CISH and susceptibility to infectious diseases. *N Engl J Med* **362**: 2092–2101.

Kinjyo I, Inoue H, Hamano S, Fukuyama S, Yoshimura T, Koga K, Takaki H, Himeno K, Takaesu G, Kobayashi T, et al. 2006. Loss of SOCS3 in T helper cells resulted in reduced immune responses and hyperproduction of interleukin 10 and transforming growth factor-β1. *J Exp Med* **203**: 1021–1031.

Lan F, Zhang N, Zhang J, Krysko O, Zhang QB, Xian JM, Derycke L, Qi YY, Li K, Liu SX, et al. 2013. Forkhead box protein 3 in human nasal polyp regulatory T cells is regulated by the protein suppressor of cytokine signaling 3. *J Allergy Clin Immun* **132**: 1314.

Le Y, Zhu BM, Harley B, Park SY, Kobayashi T, Manis JP, Luo HR, Yoshimura A, Hennighausen L, Silberstein LE. 2007. SOCS3 protein developmentally regulates the chemokine receptor CXCR4-FAK signaling pathway during B lymphopoiesis. *Immunity* **27**: 811–823.

Lee SH, Yun S, Piao ZH, Jeong M, Kim DO, Jung H, Lee J, Kim MJ, Kim MS, Chung JW, et al. 2010. Suppressor of cytokine signaling 2 regulates IL-15-primed human NK cell function via control of phosphorylated Pyk2. *J Immunol* **185**: 917–928.

Leikfoss IS, Keshari PK, Gustavsen MW, Bjolgerud A, Brorson IS, Celius EG, Spurkland A, Bos SD, Harbo HF, Berge T. 2015. Multiple sclerosis risk allele in CLEC16A acts as an expression quantitative trait locus for CLEC16A and SOCS1 in CD4+ T cells. *PloS ONE* **10**: e0132957.

Lennerz JK, Hoffmann K, Bubolz AM, Lessel D, Welke C, Ruther N, Viardot A, Moller P. 2015. Suppressor of cytokine signaling 1 gene mutation status as a prognostic biomarker in classical Hodgkin lymphoma. *Oncotarget* **6**: 29097–29110.

Lesina M, Kurkowski MU, Ludes K, Rose-John S, Treiber M, Kloppel G, Yoshimura A, Reindl W, Sipos B, Akira S, et al. 2011. Stat3/Socs3 activation by IL-6 transsignaling promotes progression of pancreatic intraepithelial neoplasia and development of pancreatic cancer. *Cancer Cell* **19**: 456–469.

Lewis RS, Noor SM, Fraser FW, Sertori R, Liongue C, Ward AC. 2014. Regulation of embryonic hematopoiesis by a cytokine-inducible SH2 domain homolog in zebrafish. *J Immunol* **92**: 5739–5748.

Li S, Chen S, Xu X, Sundstedt A, Paulsson KM, Anderson P, Karlsson S, Sjogren HO, Wang P. 2000. Cytokine-induced Src homology 2 protein (CIS) promotes T cell receptor-mediated proliferation and prolongs survival of activated T cells. *J Exp Med* **191**: 985–994.

Liu H, Irwanto A, Fu X, Yu GQ, Yu YX, Sun YH, Wang C, Wang ZZ, Okada Y, Low HQ, et al. 2015. Discovery of six new susceptibility loci and analysis of pleiotropic effects in leprosy. *Nat Genet* **47**: 267.

Lu LF, Thai TH, Calado DP, Chaudhry A, Kubo M, Tanaka K, Loeb GB, Lee H, Yoshimura A, Rajewsky K, et al. 2009. Foxp3-dependent microRNA155 confers competitive fitness to regulatory T cells by targeting SOCS1 protein. *Immunity* **30**: 80–91.

Lu LF, Boldin MP, Chaudhry A, Lin LL, Taganov KD, Hanada T, Yoshimura A, Baltimore D, Rudensky AY. 2010. Function of miR-146a in controlling Treg cell-mediated regulation of Th1 responses. *Cell* **142**: 914–929.

Lu LF, Gasteiger G, Yu IS, Chaudhry A, Hsin JP, Lu YH, Bos PD, Lin LL, Zawislak CL, Cho SL, et al. 2015. A single miRNA-mRNA interaction affects the immune response in a context- and cell-type-specific manner. *Immunity* **43**: 52–64.

Luckey MA, Kimura MY, Waickman AT, Feigenbaum L, Singer A, Park JH. 2014. The transcription factor ThPOK suppresses Runx3 and imposes CD4+ lineage fate by inducing the SOCS suppressors of cytokine signaling. *Nat Immunol* **15**: 638–645.

Mallette FA, Calabrese V, Ilangumaran S, Ferbeyre G. 2010. SOCS1, a novel interaction partner of p53 controlling oncogene-induced senescence. *Aging (Albany NY)* **2**: 445–452.

Mansell A, Smith R, Doyle SL, Gray P, Fenner JE, Crack PJ, Nicholson SE, Hilton DJ, O'Neill LA, Hertzog PJ. 2006. Suppressor of cytokine signaling 1 negatively regulates Toll-like receptor signaling by mediating Mal degradation. *Nat Immunol* **7**: 148–155.

Matsumoto A, Masuhara M, Mitsui K, Yokouchi M, Ohtsubo M, Misawa N, Miyajima A, Yoshimura A. 1997. CIS, a cytokine inducible SH2 protein, is a target of the JAK-STAT5 pathway and modulates STAT5 activation. *Blood* **89**: 3148–3154.

Matsumura Y, Kobayashi T, Ichiyama K, Yoshida R, Hashimoto M, Takimoto T, Tanaka K, Chinen T, Shichita T, Wyss-Coray T, et al. 2007. Selective expansion of Foxp3-positive regulatory T cells and immunosuppression by suppressors of cytokine signaling 3-deficient dendritic cells. *J Immunol* **179**: 2170–2179.

Mostecki J, Cassel SL, Klimecki WT, Stern DA, Knisz J, Iwashita S, Graves P, Miller RL, van Peer M, Halonen M, et al. 2011. A SOCS-1 promoter variant is associated with total serum IgE levels. *J Immunol* **187**: 2794–2802.

Nonami A, Kato R, Harada M, Yoshimura A. 2003. Spred-1 negatively regulates cytokine-induced MAP kinase activation in hematopoietic cells. *Blood* **102**: 833A–833A.

Ogata H, Kobayashi T, Chinen T, Takaki H, Sanada T, Minoda Y, Koga K, Takaesu G, Maehara Y, Iida M, et al. 2006. Deletion of the *SOCS3* gene in liver parenchymal cells promotes hepatitis-induced hepatocarcinogenesis. *Gastroenterology* **131**: 179–193.

Okabe S, Tauchi T, Morita H, Ohashi H, Yoshimura A, Ohyashiki K. 1999. Thrombopoietin induces an SH2-containing protein, CIS1, which binds to Mpl: Involvement of the ubiquitin proteosome pathway. *Exp Hematol* **27**: 1542–1547.

O'Shea JJ, Gadina M, Schreiber RD. 2002. Cytokine signaling in 2002: New surprises in the Jak/Stat pathway. *Cell* **109**: S121–S131.

Palmer DC, Guittard GC, Franco Z, Crompton JG, Eil RL, Patel SJ, Ji Y, Van Panhuys N, Klebanoff CA, Sukumar M, et al. 2015. Cish actively silences TCR signaling in CD8$^+$ T cells to maintain tumor tolerance. *J Exp Med* **212**: 2095–2113.

Ram PA, Waxman DJ. 2000. Role of the cytokine-inducible SH2 protein CIS in desensitization of STAT5b signaling by continuous growth hormone. *J Biol Chem* **275**: 39487–39496.

Ramanathan S, Dubois S, Gagnon J, Leblanc C, Mariathasan S, Ferbeyre G, Rottapel R, Ohashi PS, Ilangumaran S. 2010. Regulation of cytokine-driven functional differentiation of CD8 T cells by suppressor of cytokine signaling 1 controls autoimmunity and preserves their proliferative capacity toward foreign antigens. *J Immunol* **185**: 357–366.

Romain M, Taleb S, Dalloz M, Ponnuswamy P, Esposito B, Perez N, Wang Y, Yoshimura A, Tedgui A, Mallat Z. 2013. Overexpression of SOCS3 in T lymphocytes leads to impaired interleukin-17 production and severe aortic aneurysm formation in mice-brief report. *Arterioscler Thromb Vasc Biol* **33**: 581–584.

Schif B, Lennerz JK, Kohler CW, Bentink S, Kreuz M, Melzner I, Ritz O, Trumper L, Loeffler M, Spang R, et al. 2013. SOCS1 mutation subtypes predict divergent outcomes in diffuse large B-cell lymphoma (DLBCL) patients. *Oncotarget* **4**: 35–47.

Seki Y, Inoue H, Nagata N, Hayashi K, Fukuyama S, Matsumoto K, Komine O, Hamano S, Himeno K, Inagaki-Ohara K, et al. 2003. SOCS-3 regulates onset and maintenance of T$_H$2-mediated allergic responses. *Nat Med* **9**: 1047–1054.

Sekiya T, Nakatsukasa H, Lu Q, Yoshimura A. 2016. Roles of transcription factors and epigenetic modifications in differentiation and maintenance of regulatory T cells. *Microbes Infect* **18**: 378–386.

Sharma M, Mitnala S, Vishnubhotla RK, Mukherjee R, Reddy DN, Rao PN. 2015. The riddle of nonalcoholic fatty liver disease: Progression from nonalcoholic fatty liver to nonalcoholic steatohepatitis. *J Clin Exp Hepatol* **5**: 147–158.

Shembade N, Harhaj EW. 2012. Regulation of NF-κB signaling by the A20 deubiquitinase. *Cell Mol Immunol* **9**: 123–130.

Shen L, Evel-Kabler K, Strube R, Chen SY. 2004. Silencing of SOCS1 enhances antigen presentation by dendritic cells and antigen-specific anti-tumor immunity. *Nat Biotechnol* **22**: 1546–1553.

Shi H, Tzameli I, Bjorbaek C, Flier JS. 2004. Suppressor of cytokine signaling 3 is a physiological regulator of adipocyte insulin signaling. *J Biol Chem* **279**: 34733–34740.

Shouda T, Yoshida T, Hanada T, Wakioka T, Oishi M, Miyoshi K, Komiya S, Kosai K, Hanakawa Y, Hashimoto K, et al. 2001. Induction of the cytokine signal regulator SOCS3/CIS3 as a therapeutic strategy for treating inflammatory arthritis. *J Clin Invest* **108**: 1781–1788.

Shouda T, Hiraoka K, Komiya S, Hamada T, Zenmyo M, Iwasaki H, Isayama T, Fukushima N, Nagata K, Yoshimura A. 2006. Suppression of IL-6 production and proliferation by blocking STAT3 activation in malignant soft tissue tumor cells. *Cancer Lett* **231**: 176–184.

Song G, Rao H, Feng B, Wei L. 2014. Association between CISH polymorphisms and spontaneous clearance of hepatitis B virus in hepatitis B extracellular antigen-positive patients during immune active phase. *Chin Med J (Engl)* **127**: 1691–1695.

Suessmuth Y, Elliott J, Percy MJ, Inami M, Attal H, Harrison CN, Inokuchi K, McMullin MF, Johnston JA. 2009. A new polycythaemia vera-associated SOCS3 SH2 mutant (SOCS3^{F136L}) cannot regulate erythropoietin responses. *Brit J Haematol* **147**: 450–458.

Sun R, Jaruga B, Kulkarni S, Sun HY, Gao B. 2005. IL-6 modulates hepatocyte proliferation via induction of HGF/p21^{cip1}: Regulation by SOCS3. *Biochem Biophys Res Commun* **338**: 1943–1949.

Sun L, Jin YQ, Shen C, Qi H, Chu P, Yin QQ, Li JQ, Tian JL, Jiao WW, Xiao J, et al. 2014. Genetic contribution of CISH promoter polymorphisms to susceptibility to tuberculosis in Chinese children. *PloS ONE* **9**: e92020.

Tagami-Nagata N, Serada S, Fujimoto M, Tanemura A, Nakatsuka R, Ohkawara T, Murota H, Kishimoto T, Katayama I, Naka T. 2015. Suppressor of cytokine signaling-1 induces significant preclinical antitumor effect in malignant melanoma cells. *Exp Dermatol* **24**: 864–871.

Takahashi R, Yoshimura A. 2014. SOCS1 and regulation of regulatory T cells plasticity. *J Immunol Res* **2014**: 943149.

Takahashi R, Nishimoto S, Muto G, Sekiya T, Tamiya T, Kimura A, Morita R, Asakawa M, Chinen T, Yoshimura A. 2011. SOCS1 is essential for regulatory T cell functions by preventing loss of Foxp3 expression as well as IFN-γ and IL-17A production. *J Exp Med* **208**: 2055–2067.

Takimoto T, Wakabayashi Y, Sekiya T, Inoue N, Morita R, Ichiyama K, Takahashi R, Asakawa M, Muto G, Mori T, et al. 2010. Smad2 and Smad3 are redundantly essential for the TGF-β-mediated regulation of regulatory T plasticity and Th1 development. *J Immunol* **185**: 842–855.

Taleb S, Romain M, Ramkhelawon B, Uyttenhove C, Pasterkamp G, Herbin O, Esposito B, Perez N, Yasukawa H, Van Snick J, et al. 2009. Loss of SOCS3 expression in T cells reveals a regulatory role for interleukin-17 in atherosclerosis. *J Exp Med* **206**: 2067–2077.

Tamiya T, Kashiwagi I, Takahashi R, Yasukawa H, Yoshimura A. 2011. Suppressors of cytokine signaling (SOCS) proteins and JAK/STAT pathways: Regulation of T-cell inflammation by SOCS1 and SOCS3. *Arterioscler Thromb Vasc Biol* **31**: 980–985.

Tamiya T, Ichiyama K, Kotani H, Fukaya T, Sekiya T, Shichita T, Honma K, Yui K, Matsuyama T, Nakao T, et al. 2013. Smad2/3 and IRF4 play a cooperative role in IL-9-producing T cell induction. *J Immunol* **191**: 2360–2371.

Tanaka K, Ichiyama K, Hashimoto M, Yoshida H, Takimoto T, Takaesu G, Torisu T, Hanada T, Yasukawa H, Fukuyama S, et al. 2008. Loss of suppressor of cytokine signaling 1 in helper T cells leads to defective Th17 differentiation by enhancing antagonistic effects of IFN-γ on STAT3 and Smads. *J Immunol* **180**: 3746–3756.

Tang B, Wang ZR, Qi GY, Yuan SG, Yu SP, Li B, Wei YC, Huang Q, Zhai R, He SQ. 2015. MicroRNA-155 deficiency attenuates ischemia-reperfusion injury after liver transplantation in mice. *Transpl Int* **28**: 751–760.

Tokumaru S, Sayama K, Yamasaki K, Shirakata Y, Hanakawa Y, Yahata Y, Dai XJ, Tohyama M, Yang LJ, Yoshimura A, et al. 2005. SOCS3/CIS3 negative regulation of STAT3 in HGF-induced keratinocyte migration. *Biochem Biophys Res Commun* **327**: 100–105.

Tong HV, Toan NL, Song le H, Kremsner PG, Kun JF, Tp V. 2012. Association of CISH -292A/T genetic variant with hepatitis B virus infection. *Immunogenetics* **64**: 261–265.

Torisu T, Sato N, Yoshiga D, Kobayashi T, Yoshioka T, Mori H, Iida M, Yoshimura A. 2007. The dual function of hepatic SOCS3 in insulin resistance in vivo. *Genes Cells* **12**: 143–154.

Vogelzang A, McGuire HM, Yu D, Sprent J, Mackay CR, King C. 2008. A fundamental role for interleukin-21 in the generation of T follicular helper cells. *Immunity* **29**: 127–137.

Wakioka T, Sasaki A, Kato R, Shouda T, Matsumoto A, Miyoshi K, Tsuneoka M, Komiya S, Baron R, Yoshimura A. 2001. Spred is a Sprouty-related suppressor of Ras signalling. *Nature* **412**: 647–651.

Wang X, Li T, Li M, Cao N, Han J. 2016. The functional SOCS3 RS115785973 variant regulated by MiR-4308 promotes gastric cancer development in Chinese population. *Cell Physiol Biochem* **38**: 1796–1802.

Weber-Nordt RM, Riley JK, Greenlund AC, Moore KW, Darnell JE, Schreiber RD. 1996. Stat3 recruitment by two distinct ligand-induced, tyrosine-phosphorylated docking sites in the interleukin-10 receptor intracellular domain. *J Biol Chem* **271**: 27954–27961.

Yamamoto K, Yamaguchi M, Miyasaka N, Miura O. 2003. SOCS-3 inhibits IL-12-induced STAT4 activation by binding through its SH2 domain to the STAT4 docking site in the IL-12 receptor β2 subunit. *Biochem Biophys Res Commun* **310**: 1188–1193.

Yang XXO, Zhang HY, Kim BS, Niu XY, Peng J, Chen YH, Kerketta R, Lee YH, Chang SH, Corry DB, et al. 2013. The signaling suppressor CIS controls proallergic T cell development and allergic airway inflammation. *Nat Immunol* **14**: 732.

Yasukawa H, Misawa H, Sakamoto H, Masuhara M, Sasaki A, Wakioka T, Ohtsuka S, Imaizumi T, Matsuda T, Ihle JN, et al. 1999. The JAK-binding protein JAB inhibits Janus tyrosine kinase activity through binding in the activation loop. *EMBO J* **18**: 1309–1320.

Yasukawa H, Ohishi M, Mori H, Murakami M, Chinen T, Aki D, Hanada T, Takeda K, Akira S, Hoshijima M, et al. 2003. IL-6 induces an anti-inflammatory response in the absence of SOCS3 in macrophages. *Nat Immunol* **4**: 551–556.

Yeganeh M, Gui Y, Kandhi R, Bobbala D, Tobelaim WS, Saucier C, Yoshimura A, Ferbeyre G, Ramanathan S, Ilangumaran S. 2016. Suppressor of cytokine signaling 1-dependent regulation of the expression and oncogenic functions of p21 in the liver. *Oncogene* **35**: 4200–4211.

Yoshida T, Ogata H, Kamio M, Joo A, Shiraishi H, Tokunaga Y, Sata M, Nagai H, Yoshimura A. 2004. SOCS1 is a suppressor of liver fibrosis and hepatitis-induced carcinogenesis. *J Exp Med* **199**: 1701–1707.

Yoshida T, Hisamoto T, Akiba J, Koga H, Nakamura K, Tokunaga Y, Hanada S, Kumemura H, Maeyama M, Harada M, et al. 2006. Spreds, inhibitors of the Ras/ERK signal transduction, are dysregulated in human hepatocellular carcinoma and linked to the malignant phenotype of tumors. *Oncogene* **25**: 6056–6066.

Yoshikawa H, Matsubara K, Qian GS, Jackson P, Groopman JD, Manning JE, Harris CC, Herman JG. 2001. SOCS-1, a negative regulator of the JAK/STAT pathway, is silenced by methylation in human hepatocellular carcinoma and shows growth-suppression activity. *Nat Genet* **28**: 29–35.

Yoshimura A, Ohkubo T, Kiguchi T, Jenkins NA, Gilbert DJ, Copeland NG, Hara T, Miyajima A. 1995. A novel cytokine-inducible gene CIS encodes an SH2-containing protein that binds to tyrosine-phosphorylated interleukin 3 and erythropoietin receptors. *EMBO J* **14**: 2816–2826.

Yoshimura A, Naka T, Kubo M. 2007. SOCS proteins, cytokine signalling and immune regulation. *Nat Rev Immunol* **7**: 454–465.

Yoshimura A, Suzuki M, Sakaguchi R, Hanada T, Yasukawa H. 2012. SOCS, inflammation, and autoimmunity. *Front Immunol* **3**: 20.

T Helper Cell Differentiation, Heterogeneity, and Plasticity

Jinfang Zhu

Molecular and Cellular Immunoregulation Section, Laboratory of Immunology, National Institute of Allergy and Infectious Diseases, National Institutes of Health, Bethesda, Maryland 20892

Correspondence: jfzhu@niaid.nih.gov

Naïve CD4 T cells, on activation, differentiate into distinct T helper (Th) subsets that produce lineage-specific cytokines. By producing unique sets of cytokines, effector Th subsets play critical roles in orchestrating immune responses to a variety of infections and are involved in the pathogenesis of many inflammatory diseases including autoimmunity, allergy, and asthma. The differentiation of Th cells relies on the strength of T-cell receptor (TCR) signaling and signals triggered by polarizing cytokines that activate and/or up-regulate particular transcription factors. Several lineage-specific master transcription factors dictate Th cell fates and functions. Although these master regulators cross-regulate each other, their expression can be dynamic. Sometimes, they are even coexpressed, resulting in massive Th-cell heterogeneity and plasticity. Similar regulation mediated by these master regulators is also found in innate lymphoid cells (ILCs) that are innate counterparts of Th cells.

Cytokines are the central mediators of immune responses, and CD4 T helper (Th) cells are the professional cytokine-producing cells. By producing effector cytokines, Th cells play critical roles during adaptive immune responses to infections; distinct Th subsets are involved in protective immunity to different pathogens (Murphy and Reiner 2002; Zhu and Paul 2008; Zhu et al. 2010). There are different types of Th cells based on their cytokine profiles. Initially, type 1 T helper (Th1) and type 2 T helper (Th2) cell clones that preferentially produce interferon γ (IFN-γ) and interleukin (IL)-4, respectively, were reported (Mosmann et al. 1986; Mosmann and Coffman 1989; Paul and Seder 1994). A third major CD4 Th effector cell population Th17 that produces IL-17 was not discovered until decades later (Park et al. 2005; Acosta-Rodriguez et al. 2007; Weaver et al. 2007; Korn et al. 2009). All of the Th1, Th2, and Th17 cells are differentiated from naïve CD4 T cells when they are activated through T-cell receptor (TCR)-mediated signaling. Not only are the Th effector cells important for protective immunity, they can also induce inflammatory responses to self-antigens or to nonharmful allergens, resulting in autoimmunity or allergic diseases, respectively (Paul and Zhu 2010; Zhu et al. 2010). Interestingly, naïve CD4 T cells can also develop into regulatory T cells (Tregs), which are important for maintaining immune tolerance and for regulating the magnitude of immune responses (Shevach 2000; Chen et al. 2003; Sakaguchi 2004; Josefowicz et al. 2012; Abbas et al. 2013).

Th cells can produce the majority of the known cytokines. In addition to the signature effector cytokines, IFN-γ, IL-4, and IL-17A, Th cells may preferentially express many other important cytokines, such as lymphotoxin α for Th1; IL-5, IL-9, IL-13, and IL-24 for Th2; and IL-17F and IL-22 for Th17 cells. In addition, all of the Th subsets are capable of producing IL-2, IL-6, IL-10, IL-21, tumor necrosis factor α (TNF-α), and granulocyte macrophage colony-stimulating factor (GM-CSF). Furthermore, some regulatory functions of Tregs are mediated through production of anti-inflammatory cytokines such as transforming growth factor (TGF)-β, IL-10, and IL-35. Not only are Th cells professional cytokine producers, they can also respond to a variety of cytokines, including IL-1, IL-7, IL-12, IL-15, IL-18, IL-23, IL-27, IL-33, and type 1 IFNs, etc., that are produced by accessory cells. During differentiation, Th cells can also respond to their own cytokines, including IFN-γ and IL-4, resulting in powerful positive feedback or cross-inhibitory effects.

In this review, I will mainly focus on effector Th-cell differentiation, heterogeneity, and plasticity regulated by cytokines and transcription factors. Because innate lymphoid cells ([ILCs], an innate equivalent of Th cells) are also professional cytokine-producing cells (Artis and Spits 2015), the relationships between conventional Th cells and ILCs will also be discussed.

DISTINCT FUNCTIONS OF Th-CELL SUBSETS

Different Th-cell subsets have distinct immune functions in protective immunity. Th1 cells (Szabo et al. 2003) are mainly important for host defense against intracellular pathogens, including viruses, protozoa, and bacteria; they are also responsible for the development of certain forms of organ-specific autoimmunity. One of the major functions of Th1 cells is to activate macrophages through IFN-γ production.

Th2 cells are critical for mediating immune responses against extracellular parasites, including helminthes. These cells are also responsible for the pathogenesis of inflammatory asthmatic and allergic diseases. By producing IL-4, Th2 cells induce B-cell immunoglobulin (Ig) switching to IgG1 and IgE (Kopf et al. 1993); by producing IL-5, Th2 cells recruit eosinophils (Coffman et al. 1989); and by producing IL-13, Th2 cells can induce the movement of smooth muscle cells and mucus production by epithelial cells (Urban et al. 1998; Kuperman et al. 2002; Wynn 2003). IL-4 and IL-13 produced by Th2 cells can also induce alternatively activated macrophages (Gordon 2003).

Th17 cells are essential for orchestrating immune responses to extracellular bacteria and fungi. They are also responsible for different forms of autoimmunity, including psoriasis, multiple sclerosis, rheumatoid arthritis, and inflammatory bowel diseases (Ouyang et al. 2008; Korn et al. 2009). Th17 cells are also involved in severe asthma (McKinley et al. 2008). Th17 cells produce three major cytokines: IL-17A, IL-17F, and IL-22. IL-17A and IL-17F have redundant functions in diseases, but they may also have unique functions (Iwakura et al. 2011). The major function of IL-17A and IL-17F is to recruit and activate neutrophils. They can also stimulate different cell types to produce inflammatory cytokines, including IL-6. IL-22 is a critical cytokine for stimulating cells at mucosal barriers to produce antimicrobial peptides and proinflammatory cytokines and chemokines (Liang et al. 2006).

Treg cells include thymus-derived Treg (tTreg) cells and peripherally derived Treg (pTreg) cells (Shevach 2009; Sakaguchi et al. 2010; Josefowicz et al. 2012; Abbas et al. 2013). Together with the tTreg cells, pTreg cells play important roles in maintaining immune tolerance and regulating the magnitude of immune responses by controlling the differentiation and functions of T effector cells (Korn et al. 2009; Zhu et al. 2010; Crotty 2011; Bilate and Lafaille 2012).

CYTOKINE-MEDIATED SIGNAL TRANSDUCERS AND ACTIVATORS OF TRANSCRIPTION (STATs) ACTIVATION IS CRITICAL FOR Th-CELL-FATE DETERMINATION

Th-cell differentiation involves T-cell activation. Indeed, the strength of TCR signaling has a ma-

Cite this article as *Cold Spring Harb Perspect Biol* doi: 10.1101/cshperspect.a030338

jor impact on Th-cell-fate determination (Tao et al. 1997). Sometimes the TCR signaling strength may even have a dominant effect over adjuvants that usually create the cytokine environments that drive T-cell differentiation (Tubo et al. 2013; Nelson et al. 2014; van Panhuys et al. 2014). In particular, stimulation with low-dose peptide favors Th2-cell differentiation (Yamane et al. 2005), consistent with the finding that resting dendritic cells (DCs) promote Th2-cell differentiation (Stumbles et al. 1998; Everts et al. 2009; Steinfelder et al. 2009). TCR costimulation mediated by CD28, which is required for optimal T-cell activation, is also involved in Th-cell differentiation (Seder et al. 1994). In contrast, cytotoxic T-lymphocyte-associated protein 4 (CTLA-4) negatively regulates T-cell differentiation (Oosterwegel et al. 1999; Bour-Jordan et al. 2003).

It is well known that, besides TCR-mediated signaling, cytokine-mediated signals are critical for the differentiation of Th cells. For example, IL-12, mainly produced by antigen-presenting cells, including macrophages and DCs, induces Th1-cell differentiation (Hsieh et al. 1993). IL-12 activates a transcription factor, STAT4, in differentiating CD4 T cells (Hsieh et al. 1993; Kaplan et al. 1996b; Thierfelder et al. 1996). IFN-γ, which is produced by Th1 cells themselves, also promotes Th1-cell differentiation through activating STAT1 (Lighvani et al. 2001; Afkarian et al. 2002; Martin-Fontecha et al. 2004). Indeed, either IL-12 or IFN-γ is capable of inducing Th1-cell differentiation at least in vitro (Zhu et al. 2012). However, during *Toxoplasma gondii* infection, which elicits a robust Th1 response, IFN-γ signaling seems to be dispensable for generating IFN-γ-producing T cells, whereas IL-12 is essential.

IL-4 induces Th2-cell differentiation by activating STAT6 (Kaplan et al. 1996a; Shimoda et al. 1996; Takeda et al. 1996). A constitutively active STAT6 mutant is sufficient to replace IL-4 in inducing Th2-cell differentiation (Kurata et al. 1999; Zhu et al. 2001). However, under certain circumstances, Th2-cell differentiation in vivo particularly in response to parasite infection may occur in an IL-4/IL-4Rα/STAT6-independent manner (Finkelman et al. 2000;

Jankovic et al. 2000; Min et al. 2004; Voehringer et al. 2004; van Panhuys et al. 2008).

In addition to IL-4, IL-2 through activating STAT5 is critical for the differentiation of Th2 cells in vitro (Zhu et al. 2003; Cote-Sierra et al. 2004). IL-2-mediated STAT5 activation is detectable 24 h after T-cell activation with a low dose of TCR stimulation, which induces Th2 differentiation (Yamane et al. 2005). CD4 T-cell proliferation and survival only need low levels of STAT5 activation (Moriggl et al. 1999; Cote-Sierra et al. 2004); however, Th2-cell differentiation requires high levels of STAT5 activation (Kagami et al. 2000, 2001; Zhu et al. 2003; Cote-Sierra et al. 2004). Other cytokines such as IL-7 and thymic stromal lymphopoietin (TSLP) can also activate STAT5 in T cells. Indeed, TSLP expression may trigger the initiation of Th2-cell differentiation in vivo (Ito et al. 2005; Zhou et al. 2005; Liu 2006; Sokol et al. 2008). Although TSLP mainly acts on DCs (Ito et al. 2005; Liu et al. 2007), it may also directly stimulate naïve CD4 T cells to become Th2 cells (Al-Shami et al. 2005; Omori and Ziegler 2007; Rochman et al. 2007; He et al. 2008). IL-7 can promote Th2-cell differentiation in vitro; however, its physiological function during Th2 differentiation in vivo is still elusive. Activated STAT5 directly regulates IL-4 production by binding to the *Il4/Il13* locus at different regulatory regions (Zhu et al. 2003; Cote-Sierra et al. 2004; Liao et al. 2008). Furthermore, IL-2-mediated STAT5 activation induces IL-4Rα expression at the early stage of Th2-cell differentiation (Liao et al. 2008).

For Th17 and pTreg differentiation, TGF-β is involved (Chen et al. 2003; Korn et al. 2009). Together with IL-2-mediated STAT5 activation, TGF-β induces Treg differentiation; on the other hand, together with IL-6-mediated STAT3 activation, TGF-β induces Th17-cell differentiation (Bettelli et al. 2006; Veldhoen et al. 2006). IL-21 and IL-23 have a similar function as IL-6 in inducing STAT3 activation and, thus, Th17-cell differentiation. However, IL-6, IL-21, and IL-23 may be involved in the different stages of Th17-cell development and maintenance (Korn et al. 2009). IL-2 signaling is important for Treg generation. In the absence of IL-2, IL-2Rα, or IL-2Rβ, Treg-cell numbers are reduced and mice or humans bearing mutations in the genes encoding IL-2, IL-2Rα, or IL-2Rβ develop autoimmune disease (Fontenot et al. 2005; Caudy et al. 2007). In contrast, IL-2-mediated STAT5 activation suppresses Th17-cell differentiation (Laurence et al. 2007; Liao et al. 2011; Yang et al. 2011).

CYTOKINE-MEDIATED POSITIVE FEEDBACK DURING Th-CELL DIFFERENTIATION

The cytokine-mediated positive feedback mechanism is one of the basic principles of Th-cell differentiation. During Th1-cell differentiation, IFN-γ produced by developing Th1 cells may instruct IFN-γ nonproducers to produce IFN-γ. Similarly, IL-4 produced during Th2-cell differentiation may induce IL-4 expression by the previous IL-4 nonproducers. Furthermore, low amounts of IFN-γ or IL-4 derived from T cells may further induce these cells to produce high levels of IFN-γ or IL-4 in an autocrine manner. Therefore, Th1- and Th2-cell differentiation is enforced by the positive feedback loops. TGF-β produced by Treg cells may be important for the generation of pTreg cells (Andersson et al. 2008). Th17 cells are also capable of producing either TGF-β1 or TGF-β3, both of which may serve as positive feedback mechanisms for Th17-cell differentiation (Gutcher et al. 2011; Lee et al. 2012). Because TGF-β induces both Th17 and pTreg-cell differentiation and both cell types can produce TGF-β, how TGF-β exactly works in vivo may depend on other factors such as the presence of inflammatory cytokines and the stage of Th-cell differentiation.

MASTER TRANSCRIPTION FACTORS DICTATE T-CELL DIFFERENTIATION

Although networks of transcription factors regulate Th-cell differentiation (Zhu and Paul 2010b; Ciofani et al. 2012; Hu et al. 2013; Yosef et al. 2013), lineage-specific master transcription factors play the most critical roles during the differentiation process (Fig.1). T-bet is the master transcription factor for Th1-cell differentiation, and it directly regulates IFN-γ

production (Szabo et al. 2000, 2002). Because IFN-γ also induces T-bet expression, this explains the positive feedback for Th1-cell differentiation (Lighvani et al. 2001; Afkarian et al. 2002). IL-12 can also induce T-bet independent of IFN-γ signaling (Yang et al. 2007; Zhu et al. 2012). Thus, IL-12 and IFN-γ redundantly induce T-bet expression both in vitro and in vivo. In addition to IL-12 and IFN-γ, other cytokines such as IL-27 and type I IFNs are also capable of inducing T-bet, although their actual functions during Th-cell differentiation in vivo are still elusive (Zhu et al. 2012). T-bet is capable of inducing its own expression (Mullen et al. 2001); however, such autoregulation may not be required when either IL-12 or IFN-γ is present (Zhu et al. 2012). Nevertheless, T-bet collaborates with STAT4 to induce optimal IFN-γ production, and one of the main functions of T-bet in Th1-cell differentiation is to remodel the *Ifng* locus. Genome-wide study has shown that T-bet directly regulates a large number of Th1-specific genes (Zhu et al. 2012).

GATA3 is the master transcription factor for Th2-cell differentiation (Zhang et al. 1997; Zheng and Flavell 1997; Pai et al. 2004; Zhu et al. 2004). Retroviral expression of GATA3 is sufficient to induce endogenous GATA3 expression (Ouyang et al. 1998, 2000); however, GATA3 may not be required for its own expression when IL-4 is present (Wei et al. 2011). Unlike T-bet, which is not expressed in naïve T cells, GATA3 is expressed in naïve CD4$^+$ T cells at low levels, possibly because of its critical role during CD4 T-cell development in the thymus (Ho et al. 2009; Wei et al. 2011). STAT6 activation by IL-4 is one of the major inducers for GATA3 up-regulation. However, low-dose TCR stimulation is sufficient to up-regulate GATA3 expression independent of IL-4/STAT6 signaling (Yamane et al. 2005); this may offer a mechanism through which IL-4-independent Th2 differentiation may occur in vivo. Alternatively, the initiation of Th2-cell differentiation in vivo may occur with low amounts of GATA3 expression when STAT5 is highly activated (Zhu et al. 2003; Cote-Sierra et al. 2004; Rochman et al. 2009). Nevertheless, both IL-4-dependent and IL-4-independent Th2-cell differentiation re-

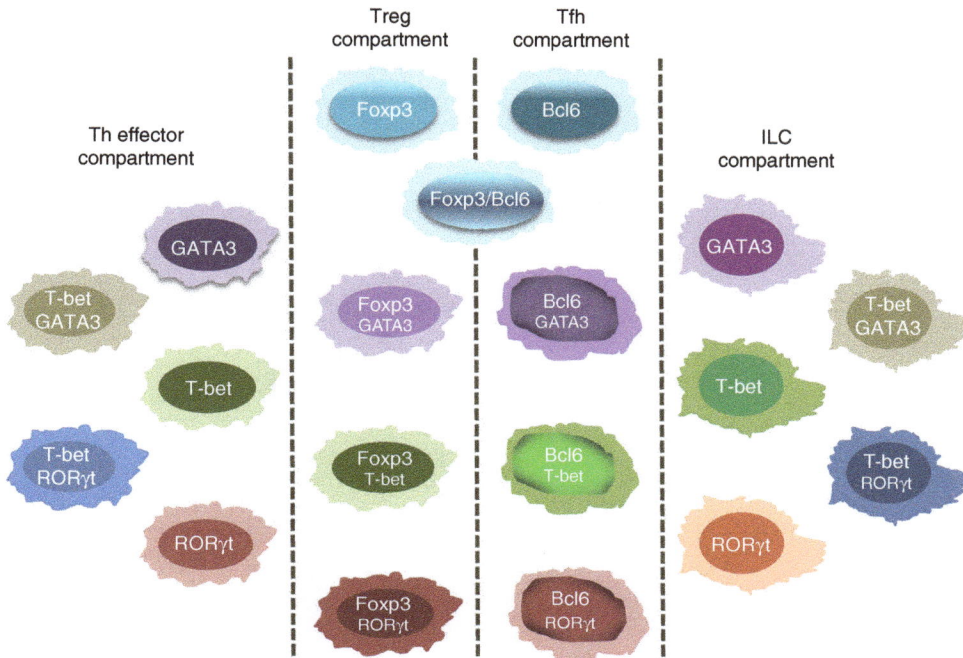

Figure 1. Combinatorial expression of master transcription factors determines the heterogeneity and functionality of effector T helper (Th) cells, follicular Th (Tfh) cells, regulatory T (Treg) cells, and innate lymphoid cell (ILC) subsets. T-bet, GATA3, and RORγt are the master transcription factors of distinct Th cell and ILC subsets. Within the Th effector compartment, T-bet/RORγt coexpressing cells have been found under inflammation conditions. These cells could be derived from either T-bet- or RORγt-single positive cells. During type 2 immune responses, GATA3/T-bet coexpressing cells are also found. Similarly, in the ILC compartment, NKp46[+] ILC3s coexpress T-bet and RORγt. Bcl6 is the master regulator for Tfh cells. Subsets of Tfh cells may either express low levels of effector master regulators, T-bet, GATA3, or RORγt, or have expressed one of these factors during their development. Interestingly, all of these master regulators, T-bet, GATA3, RORγt, and Bcl6, can be expressed by subsets of Foxp3-expressing Treg cells albeit at lower levels. Furthermore, although it is not indicated in the figure, all of these lymphocytes can express GATA3 but at various levels.

quire GATA3 (Zhu et al. 2004). Interestingly, during Th1 and/or Th17 differentiation, GATA3 expression is down-regulated (Zheng and Flavell 1997; Wei et al. 2011).

GATA3 may promote Th2-cell differentiation via different mechanisms (Yagi et al. 2011). Genome-wide profiling of GATA3 binding indicates that it directly binds to the *Il4/Il13* gene locus at various regions (Wei et al. 2011). Through its binding to the promoters of the *Il5* and the *Il13* genes, GATA3 induces *Il5* and *Il13* transcription (Yamashita et al. 2002; Tanaka et al. 2011). However, GATA3 mainly affects *Il4* expression through regulating epigenetic modifications at the Th2 cytokine gene locus.

Therefore, while GATA3 is required for the acquisition of IL-4-producing capacity by Th2 cells, in fully differentiated Th2 cells, GATA3 is no longer important for IL-4 production but it remains essential for the transcription of the *Il5* and *Il13* (Zhu et al. 2004). In addition, GATA3 directly regulates many other Th2-specific genes including *Il1rl1*, which encodes T1/ST2, the IL-33 receptor (Wei et al. 2011).

RORγt is the master transcription factor for Th17-cell differentiation (Ivanov et al. 2006). RORγt loss-of-function mutations in human patients results in susceptibility to both *Candida* and *Mycobacterium* infections (Okada et al. 2015). The induction of RORγt expression depends on STAT3 activation by IL-6, IL-21, and/ or IL-23 (Zhu and Paul 2008; Korn et al. 2009). RORγt directly regulates IL-17A and IL-17F expression. At the genome scale, RORγt binds to and regulates only a selected set of Th17-specific genes after BATF/IRF4 and STAT3 have initiated the Th17 differentiation program (Ciofani et al. 2012).

The master regulator for Treg generation and function is Foxp3 (Fontenot et al. 2003; Hori et al. 2003). Mutations in the human *FOXP3* gene are responsible for the immuno-dysregulation polyendocrinopathy enteropathy X-linked (IPEX) syndrome, and mutations in the *Foxp3* gene in mice, such as *Scurfy* mice, result in fatal autoimmune diseases. Genome-wide analysis indicates that Foxp3 directly binds to hundreds of genes, many of which are either positively or negatively regulated by Foxp3, especially in the thymus (Zheng et al. 2007). Some of Foxp3-regulated genes encode molecules that regulate gene expression and/or are involved in epigenetic modifications.

CROSS-REGULATION OF Th-CELL DIFFERENTIATION

During Th-cell differentiation toward a specific type, signals/pathways that induce such differentiation also repress the alternative lineage fates. At the transcriptional level, the transcription factors that are either activated or induced in one lineage often cross-regulate the expression and/or functions of the transcription factors that are involved in making other lineage decisions. For example, overexpression of T-bet suppresses *Gata3* transcription (Usui et al. 2006) and inhibits GATA3 function through direct protein–protein interaction (Hwang et al. 2005). Interestingly, T-bet and GATA3 binding sites colocalize at several critical Th1- or Th2-specific genes (Jenner et al. 2009; Kanhere et al. 2012; Zhu et al. 2012). It has also been shown that endogenous T-bet inhibits GATA3 function during Th1-cell differentiation, thus preventing the activation of a "default" program for Th2-cell differentiation (Zhu et al. 2012). On the other hand, during Th2-cell differentiation, GATA3 may suppress Th1-cell differenti-

ation by repressing STAT4 expression (Usui et al. 2003), and suppressing Runx3-mediation IFN-γ production (Yagi et al. 2010) as well as silencing the *Tbx21* locus (Wei et al. 2011). Cross-regulation has also been found between T-bet and RORγt (Lazarevic et al. 2011) and between Foxp3 and RORγt (Yang et al. 2008; Zhou et al. 2008). RORγt and Foxp3 may antagonize each other through protein–protein interaction (Yang et al. 2008; Zhang et al. 2008; Zhou et al. 2008).

NONCANONICAL Th SUBSETS THAT COEXPRESS MULTIPLE MASTER REGULATORS

Master regulators are usually expressed in a cell-type-specific manner (i.e., T-bet for Th1, GATA3 for Th2, RORγt for Th17, and Foxp3 for Treg cells) (Fig. 1). However, because both Th17- and Treg-cell differentiation require TGF-β, RORγt and Foxp3 are often coexpressed, presumably at early stages of Th17/Treg-cell differentiation. These cells may represent an intermediate stage before they eventually commit to either Th17 or Treg cells (Yang et al. 2008; Zhou et al. 2008; Komatsu et al. 2014). Interestingly, RORγt/Foxp3-coexpressing Treg cells, possibly representing a specialized Treg subset, are abundant in the large intestine (Ohnmacht et al. 2015; Sefik et al. 2015). Indeed, Treg-specific deletion of RORγt results in gut inflammation with an uncontrolled Th2 response (Ohnmacht et al. 2015).

In addition to RORγt, T-bet and GATA3 can be expressed by Treg subsets as well. Therefore, it has been suggested that Treg cells may hijack the master regulators of distinct T effector cells to specifically control a given type of immune response. According to this model, T-bet expression by Treg cells is critical for inhibiting Th1 responses (Koch et al. 2009). However, Treg-specific deletion of T-bet does not result in the development of Th1-related autoimmune diseases, indicating that T-bet is not required for Treg cells to control autoreactive Th1 cells at steady state (Yu et al. 2015). Because the functions of Treg cells during inflammation and/or immune responses may be different from their functions at steady state, it remains to be determined whether T-bet expression in Treg cells is important for their ability to limit Th1-related immune responses (Oldenhove et al. 2009; Yamaguchi et al. 2011).

GATA3 is expressed in all T cells, albeit at different levels. Interestingly, some Treg cells can express high levels of GATA3 (Wang et al. 2011; Wohlfert et al. 2011; Rudra et al. 2012; Yu et al. 2015), potentially explaining why a slight reduction in Foxp3 expression results in a Th2 phenotype in these "Treg cells" (Wan and Flavell 2007). In fact, even with normal levels of Foxp3 expression, GATA3 may induce the expression of many "Th2-specific" genes such as *Il1rl1* and *Ccr8* in Treg cells (Wei et al. 2011). One study has shown that GATA3 deletion in Treg cells results in spontaneous Th2-like autoimmunity (Wang et al. 2011), but other studies show that the mice with GATA3 deletion specifically in Treg cells do not develop Th2-specific systemic diseases (Wohlfert et al. 2011; Rudra et al. 2012; Yu et al. 2015).

Interestingly, T-bet- and GATA3-expressing Treg cells do not represent stable Treg subsets. The expression T-bet and GATA3 in Treg cells is dynamic (Yu et al. 2015). Although deletion of either T-bet or GATA3 in Treg cells does not affect the overall function of these cells, Treg cells lacking expression of both T-bet and GATA3 up-regulate RORγt expression and acquire IL-17-producing capacity. These Treg cells are unstable and many lose Foxp3 expression over time. Thus, dynamic expression of T-bet and GATA3 together with cross-regulation among T-bet, GATA3, RORγt, and Foxp3 are important for maintaining Treg functions (Yu et al. 2015).

In addition to the heterogeneity of Tregs, within T effector populations, multiple master regulators may also be expressed in the same cell. For example, T-bet/RORγt coexpressing cells are found in the gut and inflamed brain (Ivanov et al. 2006; Hirota et al. 2011). Many of these cells are capable of expressing both IFN-γ and IL-17 (Lee et al. 2009; Ghoreschi et al. 2010; Lexberg et al. 2010; Hirota et al. 2011). T-bet/RORγt (IFN-γ/IL-17) coexpressing T cells, which are specific for *Candida albi-*

cans antigens, are also found in human patients (Zielinski et al. 2012). Interestingly, these cells may represent an important population that is involved in immune responses to *Mycobacterium tuberculosis* (*Mtb*) infection in humans. RORγt-deficient patients fail to control *Mtb* infections (Okada et al. 2015). Besides T-bet/RORγt coexpressing cells, GATA3/T-bet coexpressing cells have been detected in helminth-infected animals (Peine et al. 2013) and GATA3/RORγt coexpressing cells are found in asthmatic mice and patients (Cosmi et al. 2010; Wang et al. 2010).

EPIGENETIC MODIFICATIONS AND CELL PLASTICITY

Different modifications at the histones binding to the genomic regions correlate with gene activation or silencing (Barski et al. 2007). Trimethylation at the lysine position 4 of histone 3 (H3K4me3), particularly in gene-promoter regions, is usually associated with active gene loci. On the other hand, H3K27me3 is generally a marker of repressed genes. Epigenetic status at the effector cytokine loci is usually reflected by H3K4me3 in cells that express such cytokines, or by H3K27me3 in cells where these cytokine genes are silenced (Wei et al. 2009). However, bivalent modifications with both H3K4me3 and H3K27me3, which are indicative for the genes poised for expression, are found in the promoters of the master regulator genes, such as *Tbx21*, *Gata3*, and *Rorc*, etc., even in the cells that do not express such transcription factors (Wei et al. 2009). Such bivalent modifications may allow the induction of a master regulator expression in cells of alternative lineage fate, resulting in the generation of master regulator coexpressing cells and cell plasticity. Therefore, a transient signal that alters the balance of key master regulators, which will cross-regulate each other when coexpressed, may ultimately result in a lineage switch.

Although there are some reports showing that Th1 and Th2 cells may alter their lineage fates (Hegazy et al. 2010), these cells are relatively stable (Zhu and Paul 2010a). Cell plasticity is more common for Th17 and Treg cells (Zhou et al. 2009a). Th17 cells may acquire IFN-γ-producing capacity in a T-bet-dependent manner (Mathur et al. 2006; Bending et al. 2009; Lee et al. 2009). Through a fate-mapping study, it has been shown that IFN-γ-producing cells found in the brain of autoimmune mice are largely derived from the cells that express IL-17 (Hirota et al. 2011). Th17 cells may also *trans*-differentiate into Treg cells when inflammation is resolved (Gagliani et al. 2015). Although several studies have shown that Treg cells are stable even under inflamed conditions (Rubtsov et al. 2010; Sakaguchi et al. 2013), other studies have shown the possible switch from Treg cells to Th1, Th2, or Th17 cells (Xu et al. 2007; Komatsu et al. 2009, 2014; Oldenhove et al. 2009; Zhou et al. 2009b; Noval Rivas et al. 2015).

RELATIONSHIP BETWEEN T EFFECTOR CELLS AND CYTOKINE-PRODUCING Tfh CELLS

A critical function of CD4 T cells during immune responses is to help B cells produce antibodies and Ig class switching (Fig. 1). It has been shown that CD4 Th cells that are found in the B-cell follicle, termed as Tfh cells, are critical for exerting such functions (Crotty 2011). Th-cell effects on B cells rely on cytokine production; although IL-4 is important for the Ig class switching to IgE and IgG1 (Kopf et al. 1993; King and Mohrs 2009; Reinhardt et al. 2009; Zaretsky et al. 2009), IFN-γ induces Ig switching to IgG2a/IgG2c. Even though IL-21 produced by Tfh cells is important for helping B cells, there are at least two different types of Tfh cells, with one type producing IFN-γ and the other IL-4. In fact, most of the IL-4-producing Th cells in vivo have a Tfh phenotype (King and Mohrs 2009). Different from conventional Th2 cells, however, IL-4-producing Tfh cells do not express IL-13 (Liang et al. 2012).

Tfh cells are considered as a fifth major Th-cell population that is different from conventional Th1, Th2, Th17, and Treg cells (Nurieva et al. 2008). Tfh-cell differentiation requires STAT3 activation, presumably by IL-21, but IL-2-mediated STAT5 activation suppresses Tfh-cell generation (Nurieva et al. 2008, 2012;

Johnston et al. 2012). The master transcription factor for Tfh cells is Bcl6 (Johnston et al. 2009; Nurieva et al. 2009). Interestingly, some Treg cells are also found in the B-cell follicle and these cells coexpress Foxp3 and Bcl6. These cells are named follicular regulatory T (Tfr) cells (Chung et al. 2011; Linterman et al. 2011). Tfr cells may play an important role in limiting the functions of Tfh cells in activating B cells.

Tfh cells do not express or express very low levels of T-bet and GATA3. However, IgE class switching completely depends on GATA3 expression, indicating that GATA3 is required for the development of IL-4-producing Tfh cells (Zhu et al. 2004). It is possible that low levels of GATA3 expression in Tfh cells are sufficient for these cells to produce IL-4 production (Yusuf et al. 2010; Liang et al. 2012). Alternatively, just like the dynamic expression of GATA3 found in Treg cells, as mentioned earlier (Yu et al. 2015), GATA3 may have been expressed at high levels at early stage of IL-4-producing Tfh-cell differentiation. Once the *Il4* locus is remodeled by GATA3, Tfh cells no longer require GATA3 for IL-4 production and thus do not need to express GATA3. This is consistent with the observation that in fully differentiated Th2 cells, *Gata3* deletion does not abolish IL-4 production (Zhu et al. 2004). Similarly, transient expression of T-bet during the early stage of Tfh-cell differentiation may allow these cells to acquire IFN-γ-producing capacity.

The developmental relationship between IFN-γ- or IL-4-producing Tfh cells and conventional Th1 or Th2 cells is still elusive. In vitro experiments suggest that conventional Th1 or Th2 cells can become IFN-γ- or IL-4-producing Tfh cells or differentiated Tfh cells can acquire IFN-γ- or IL-4-producing capacity (Lu et al. 2011). In vivo, activated IL-4-producing T cells may subsequently acquire a Tfh-cell phenotype through their interaction with B cells (Zaretsky et al. 2009). It is also possible that activated T cells may acquire IFN-γ- or IL-4-producing capacity and a Tfh phenotype simultaneously, and thus their fate to become either conventional Th1/Th2 effector cells or specific cytokine-producing Tfh cells may have been determined at very early stages of T-cell differentiation (Nakayamada et al. 2011; Johnston et al. 2012; Oestreich et al. 2012). It is interesting to point out that Tfh cells may be an important source for generating memory Th cells, which could subsequently give rise to conventional Th effector cells on reactivation (Luthje et al. 2012). In the future, genome-wide assessment of the transcriptomes and epigenomes of conventional Th cells, cytokine-producing Tfh-cell subsets and Th memory cells is necessary to further understand the relationship between these closely related cell types.

RELATIONSHIP BETWEEN Th CELLS AND INNATE LYMPHOID CELLS

During the past few years, ILCs have drawn much attention in the immunology field (Artis and Spits 2015). ILCs do not express antigen receptors, but they can respond to many inflammatory cytokines, such as IL-1, IL-12, IL-18, IL-23, IL-25, and IL-33, to produce their own cytokines, including IFN-γ, IL-5/13, and IL-17/22. ILCs express IL-7Rα and at least partially depend on IL-7 and/or TSLP for their development. Because of their distinct cytokine-producing capacity, ILCs are classified into group 1 ILCs (ILC1s) that produce IFN-γ, group 2 ILCs (ILC2s) that produce IL-5 and IL-13, and group 3 ILCs (ILC3s) that produce IL-17 and IL-22. Not only do ILC subsets produce similar cytokines to those produced by Th-cell subsets, but ILCs also use the same set of master regulators, namely, T-bet, GATA3, and RORγt for their development and function. For example, the Th2 master regulator, GATA3, is also critical for ILC2 development and the functional maintenance of these cells (Hoyler et al. 2012; Mjosberg et al. 2012; Furusawa et al. 2013; Klein Wolterink et al. 2013; Yang et al. 2013; Yagi et al. 2014). Although some researchers have classified natural killer (NK) cells into ILC1s based on their cytokine production, NK cells are in fact the innate counterparts of CD8 T cells in the adaptive system (Cortez and Colonna 2016; Spits et al. 2016). Both NK and CD8 T cells express transcription factor Eomes. Developmentally, all non-NK ILCs (or Th-like ILCs, or IL-7Rα-expressing ILCs) depend on GATA3,

whereas NK cells do not (Yagi et al. 2014); this mirrors the critical function of GATA3 in specifying CD4 but not CD8 lineage fate during T-cell development (Ho et al. 2009). In addition, GATA3 regulates IL-7Rα expression in both T cells and ILCs (Wang et al. 2013; Zhong et al. 2016).

Because of similarities between ILC and Th subsets (Yagi et al. 2014; Koues et al. 2016; Shih et al. 2016), a specific type of ILCs may participate in the same class of immune responses in a similar manner to corresponding distinct Th subsets. Therefore, ILC2s are important players during type 2 immune responses, including immunity against helminth infections (Fallon et al. 2006; Moro et al. 2010; Neill et al. 2010; Price et al. 2010) and during allergic lung and skin inflammation (Chang et al. 2011; Monticelli et al. 2011; Halim et al. 2012; Roediger et al. 2013; Kim et al. 2014). Similarly, ILC1s may participate in type 1 immune responses and ILC3s are important for controlling extracellular bacteria, which induce Th17 responses (Qiu et al. 2013; Klose et al. 2014; Sano et al. 2015). As mentioned above, ILCs mainly respond to cytokine stimulation. Interestingly, in fully differentiated Th cells, IL-1 family receptors IL-18R, IL-33R, and IL-1R, which are preferentially expressed by ILC1s, ILC2s, and ILC3s, respectively, are also selectively expressed by Th1, Th2, and Th17 cells, respectively (Guo et al. 2009, 2012). As a result, just like ILCs, Th cells can also respond to cytokine stimulation, which is independent of TCR stimulation, to produce effector cytokines in vivo (Guo et al. 2015).

One major difference between ILCs and Th cells is antigen specificity. Because ILCs do not recognize specific antigen and they are already developed even before possible microbial threats, these cells provide the first line of host defense to infections. Their unique position in tissue sites fits quite well with their functionality. Interestingly, the same type of ILCs and Th cells may cross talk to each other. For example, IL-13 produced by ILC2s on activation can induce the migration of DCs to the draining lymph nodes, and these migratory DCs preferentially induce Th2-cell differentiation (Halim et al. 2014). In addition, some ILC2s, by expressing major histocompatibility complex (MHC) class II, can activate T cells, and IL-2 produced by T cells may then act back onto ILC2s to influence their activation and cytokine production (Mirchandani et al. 2014; Oliphant et al. 2014). Cross talk between ILC3s and Th17 cells has also been reported (Sano et al. 2015). Nevertheless, the functions of ILCs and Th cells may be substantially redundant. Optimal activation of ILCs is sufficient to control infections in the absence of Th cells, as shown in several mouse models. Interestingly, however, humans without ILCs, which can result because of the failure of ILC reconstitution after bone marrow transplantation, are capable of controlling infections (Vely et al. 2016). However, the collaboration as well as labor division between ILCs and Th cells may allow the host to survive severe infections. Because of the activation of either ILCs or Th cells alone is sufficient to induce many inflammatory diseases, investigating the development and functions of both ILCs and Th subsets is clinically relevant.

CONCLUDING REMARKS

CD4 Th cells are professional cytokine-producing cells. To acquire a unique cytokine-producing profile, naïve CD4 T cells need to go through a differentiation process. During Th-cell differentiation, TCR and cytokine-mediated signaling pathways induce activation of STAT proteins followed by up-regulation of lineage-specific master transcription factors. Activated STAT proteins collaborate with lineage-specific master regulators such as T-bet, GATA3, and RORγt in epigenetically remodeling the respective cytokine loci and regulating cytokine gene expression. Although these master regulators are usually expressed in a cell-type-specific manner, they can often be coexpressed. Furthermore, effector cell–related transcription factors can be expressed by Foxp3-expressing Treg cells, resulting in massively heterogeneous Th effector and regulatory populations. Bivalent histone modifications at the gene loci of master regulators may explain the coexpression of multiple factors, and the coexpression, dynamic induction, and cross-regulation of these master regulators may determine the plasticity of Th cells. Some master regulator coexpressers are relatively stable, that is, T-bet/RORγt coexpressers. Importantly, these cells are found abundant in several inflammatory settings; they are considered as the most potent cells in inducing autoimmunity and they may be an important cell population to fight against *Mtb* infection in humans. How T-bet and RORγt, two mutually inhibitory transcription factors, may stably coexist in the same cell and what unique programs have been activated in these cells remain important questions. Investigation of Th- and ILC-cell heterogeneity and plasticity holds promise for finding specific treatments for a variety of human immunological diseases in the future. High-dimensional single-cell analyses, including single-cell RNAseq and CyTOF mass cytometry, may allow us to gain deeper insights into the immune responses mediated by Th cell and ILC subsets in autoimmunity, allergy, and infectious diseases.

ACKNOWLEDGMENTS

J.Z. is supported by the Division of Intramural Research (DIR), the National Institute of Allergy and Infectious Diseases (NIAID), and the National Institutes of Health (NIH).

REFERENCES

Abbas AK, Benoist C, Bluestone JA, Campbell DJ, Ghosh S, Hori S, Jiang S, Kuchroo VK, Mathis D, Roncarolo MG, et al. 2013. Regulatory T cells: Recommendations to simplify the nomenclature. *Nat Immunol* 14: 307–308.

Acosta-Rodriguez EV, Rivino L, Geginat J, Jarrossay D, Gattorno M, Lanzavecchia A, Sallusto F, Napolitani G. 2007. Surface phenotype and antigenic specificity of human interleukin 17-producing T helper memory cells. *Nat Immunol* 8: 639–646.

Afkarian M, Sedy JR, Yang J, Jacobson NG, Cereb N, Yang SY, Murphy TL, Murphy KM. 2002. T-bet is a STAT1-induced regulator of IL-12R expression in naive CD4+ T cells. *Nat Immunol* 3: 549–557.

Al-Shami A, Spolski R, Kelly J, Keane-Myers A, Leonard WJ. 2005. A role for TSLP in the development of inflammation in an asthma model. *J Exp Med* 202: 829–839.

Andersson J, Tran DQ, Pesu M, Davidson TS, Ramsey H, O'Shea JJ, Shevach EM. 2008. CD4+FoxP3+ regulatory T cells confer infectious tolerance in a TGF-β-dependent manner. *J Exp Med* 205: 1975–1981.

Cite this article as *Cold Spring Harb Perspect Biol* doi: 10.1101/cshperspect.a030338

Artis D, Spits H. 2015. The biology of innate lymphoid cells. *Nature* **517**: 293–301.

Barski A, Cuddapah S, Cui K, Roh TY, Schones DE, Wang Z, Wei G, Chepelev I, Zhao K. 2007. High-resolution profiling of histone methylations in the human genome. *Cell* **129**: 823–837.

Bending D, De la Pena H, Veldhoen M, Phillips JM, Uyttenhove C, Stockinger B, Cooke A. 2009. Highly purified Th17 cells from BDC2.5NOD mice convert into Th1-like cells in NOD/SCID recipient mice. *J Clin Invest* **119**: 565–572.

Bettelli E, Carrier YJ, Gao WD, Korn T, Strom TB, Oukka M, Weiner HL, Kuchroo VK. 2006. Reciprocal developmental pathways for the generation of pathogenic effector T_H17 and regulatory T cells. *Nature* **441**: 235–238.

Bilate AM, Lafaille JJ. 2012. Induced CD4$^+$Foxp3$^+$ regulatory T cells in immune tolerance. *Annu Rev Immunol* **30**: 733–758.

Bour-Jordan H, Grogan JL, Tang Q, Auger JA, Locksley RM, Bluestone JA. 2003. CTLA-4 regulates the requirement for cytokine-induced signals in T_H2 lineage commitment. *Nat Immunol* **4**: 182–188.

Caudy AA, Reddy ST, Chatila T, Atkinson JP, Verbsky JW. 2007. CD25 deficiency causes an immune dysregulation, polyendocrinopathy, enteropathy, X-linked-like syndrome, and defective IL-10 expression from CD4 lymphocytes. *J Allergy Clin Immunol* **119**: 482–487.

Chang YJ, Kim HY, Albacker LA, Baumgarth N, McKenzie AN, Smith DE, Dekruyff RH, Umetsu DT. 2011. Innate lymphoid cells mediate influenza-induced airway hyperreactivity independently of adaptive immunity. *Nat Immunol* **12**: 631–638.

Chen W, Jin W, Hardegen N, Lei KJ, Li L, Marinos N, McGrady G, Wahl SM. 2003. Conversion of peripheral CD4$^+$CD25$^-$ naive T cells to CD4$^+$CD25$^+$ regulatory T cells by TGF-β induction of transcription factor *Foxp3*. *J Exp Med* **198**: 1875–1886.

Chung Y, Tanaka S, Chu F, Nurieva RI, Martinez GJ, Rawal S, Wang YH, Lim H, Reynolds JM, Zhou XH, et al. 2011. Follicular regulatory T cells expressing Foxp3 and Bcl-6 suppress germinal center reactions. *Nat Med* **17**: 983–988.

Ciofani M, Madar A, Galan C, Sellars M, Mace K, Pauli F, Agarwal A, Huang W, Parkurst CN, Muratet M, et al. 2012. A validated regulatory network for Th17 cell specification. *Cell* **151**: 289–303.

Coffman RL, Seymour BW, Hudak S, Jackson J, Rennick D. 1989. Antibody to interleukin-5 inhibits helminth-induced eosinophilia in mice. *Science* **245**: 308–310.

Cortez VS, Colonna M. 2016. Diversity and function of group 1 innate lymphoid cells. *Immunol Lett* **179**: 19–24.

Cosmi L, Maggi L, Santarlasci V, Capone M, Cardilicchia E, Frosali F, Querci V, Angeli R, Matucci A, Fambrini M, et al. 2010. Identification of a novel subset of human circulating memory CD4$^+$ T cells that produce both IL-17A and IL-4. *J Allergy Clin Immunol* **125**: 222–230.e1–4.

Cote-Sierra J, Foucras G, Guo L, Chiodetti L, Young HA, Hu-Li J, Zhu J, Paul WE. 2004. Interleukin 2 plays a central role in Th2 differentiation. *Proc Natl Acad Sci* **101**: 3880–3885.

Crotty S. 2011. Follicular helper CD4 T cells (T_{FH}). *Annu Rev Immunol* **29**: 621–663.

Everts B, Perona-Wright G, Smits HH, Hokke CH, van der Ham AJ, Fitzsimmons CM, Doenhoff MJ, van der Bosch J, Mohrs K, Haas H, et al. 2009. Omega-1, a glycoprotein secreted by *Schistosoma mansoni* eggs, drives Th2 responses. *J Exp Med* **206**: 1673–1680.

Fallon PG, Ballantyne SJ, Mangan NE, Barlow JL, Dasvarma A, Hewett DR, McIlgorm A, Jolin HE, McKenzie AN. 2006. Identification of an interleukin (IL)-25-dependent cell population that provides IL-4, IL-5, and IL-13 at the onset of helminth expulsion. *J Exp Med* **203**: 1105–1116.

Finkelman FD, Morris SC, Orekhova T, Mori M, Donaldson D, Reiner SL, Reilly NL, Schopf L, Urban JF Jr. 2000. Stat6 regulation of in vivo IL-4 responses. *J Immunol* **164**: 2303–2310.

Fontenot JD, Gavin MA, Rudensky AY. 2003. Foxp3 programs the development and function of CD4$^+$CD25$^+$ regulatory T cells. *Nat Immunol* **4**: 330–336.

Fontenot JD, Rasmussen JP, Gavin MA, Rudensky AY. 2005. A function for interleukin 2 in Foxp3-expressing regulatory T cells. *Nat Immunol* **6**: 1142–1151.

Furusawa J, Moro K, Motomura Y, Okamoto K, Zhu J, Takayanagi H, Kubo M, Koyasu S. 2013. Critical role of p38 and GATA3 in natural helper cell function. *J Immunol* **191**: 1818–1826.

Gagliani N, Amezcua Vesely MC, Iseppon A, Brockmann L, Xu H, Palm NW, de Zoete MR, Licona-Limon P, Paiva RS, Ching T, et al. 2015. Th17 cells transdifferentiate into regulatory T cells during resolution of inflammation. *Nature* **523**: 221–225.

Ghoreschi K, Laurence A, Yang XP, Tato CM, McGeachy MJ, Konkel JE, Ramos HL, Wei L, Davidson TS, Bouladoux N, et al. 2010. Generation of pathogenic T_H17 cells in the absence of TGF-β signalling. *Nature* **467**: 967–971.

Gordon S. 2003. Alternative activation of macrophages. *Nat Rev* **3**: 23–35.

Guo L, Wei G, Zhu J, Liao W, Leonard WJ, Zhao K, Paul W. 2009. IL-1 family members and STAT activators induce cytokine production by Th2, Th17, and Th1 cells. *Proc Natl Acad Sci* **106**: 13463–13468.

Guo L, Junttila IS, Paul WE. 2012. Cytokine-induced cytokine production by conventional and innate lymphoid cells. *Trends Immunol* **33**: 598–606.

Guo L, Huang Y, Chen X, Hu-Li J, Urban JF Jr, Paul WE. 2015. Innate immunological function of T_H2 cells in vivo. *Nat Immunol* **16**: 1051–1059.

Gutcher I, Donkor MK, Ma Q, Rudensky AY, Flavell RA, Li MO. 2011. Autocrine transforming growth factor-β1 promotes in vivo Th17 cell differentiation. *Immunity* **34**: 396–408.

Halim TY, Krauss RH, Sun AC, Takei F. 2012. Lung natural helper cells are a critical source of Th2 cell-type cytokines in protease allergen-induced airway inflammation. *Immunity* **36**: 451–463.

Halim TY, Steer CA, Matha L, Gold MJ, Martinez-Gonzalez I, McNagny KM, McKenzie AN, Takei F. 2014. Group 2 innate lymphoid cells are critical for the initiation of adaptive T helper 2 cell-mediated allergic lung inflammation. *Immunity* **40**: 425–435.

He R, Oyoshi MK, Garibyan L, Kumar L, Ziegler SF, Geha RS. 2008. TSLP acts on infiltrating effector T cells to drive allergic skin inflammation. *Proc Natl Acad Sci* **105:** 11875–11880.

Hegazy AN, Peine M, Helmstetter C, Panse I, Frohlich A, Bergthaler A, Flatz L, Pinschewer DD, Radbruch A, Lohning M. 2010. Interferons direct Th2 cell reprogramming to generate a stable GATA-3⁺T-bet⁺ cell subset with combined Th2 and Th1 cell functions. *Immunity* **32:** 116–128.

Hirota K, Duarte JH, Veldhoen M, Hornsby E, Li Y, Cua DJ, Ahlfors H, Wilhelm C, Tolaini M, Menzel U, et al. 2011. Fate mapping of IL-17-producing T cells in inflammatory responses. *Nat Immunol* **12:** 255–263.

Ho IC, Tai TS, Pai SY. 2009. GATA3 and the T-cell lineage: Essential functions before and after T-helper-2-cell differentiation. *Nat Rev* **9:** 125–135.

Hori S, Nomura T, Sakaguchi S. 2003. Control of regulatory T cell development by the transcription factor *Foxp3*. *Science* **299:** 1057–1061.

Hoyler T, Klose CS, Souabni A, Turqueti-Neves A, Pfeifer D, Rawlins EL, Voehringer D, Busslinger M, Diefenbach A. 2012. The transcription factor GATA-3 controls cell fate and maintenance of type 2 innate lymphoid cells. *Immunity* **37:** 634–648.

Hsieh CS, Macatonia SE, Tripp CS, Wolf SF, O'Garra A, Murphy KM. 1993. Development of T$_H$1 CD4⁺ T cells through IL-12 produced by *Listeria*-induced macrophages. *Science* **260:** 547–549.

Hu G, Tang Q, Sharma S, Yu F, Escobar TM, Muljo SA, Zhu J, Zhao K. 2013. Expression and regulation of intergenic long noncoding RNAs during T cell development and differentiation. *Nat Immunol* **14:** 1190–1198.

Hwang ES, Szabo SJ, Schwartzberg PL, Glimcher LH. 2005. T helper cell fate specified by kinase-mediated interaction of T-bet with GATA-3. *Science* **307:** 430–433.

Ito T, Wang YH, Duramad O, Hori T, Delespesse GJ, Watanabe N, Qin FX, Yao Z, Cao W, Liu YJ. 2005. TSLP-activated dendritic cells induce an inflammatory T helper type 2 cell response through OX40 ligand. *J Exp Med* **202:** 1213–1223.

Ivanov II, McKenzie BS, Zhou L, Tadokoro CE, Lepelley A, Lafaille JJ, Cua DJ, Littman DR. 2006. The orphan nuclear receptor RORγt directs the differentiation program of proinflammatory IL-17⁺ T helper cells. *Cell* **126:** 1121–1133.

Iwakura Y, Ishigame H, Saijo S, Nakae S. 2011. Functional specialization of interleukin-17 family members. *Immunity* **34:** 149–162.

Jankovic D, Kullberg MC, Noben-Trauth N, Caspar P, Paul WE, Sher A. 2000. Single cell analysis reveals that IL-4 receptor/Stat6 signaling is not required for the in vivo or in vitro development of CD4⁺ lymphocytes with a Th2 cytokine profile. *J Immunol* **164:** 3047–3055.

Jenner RG, Townsend MJ, Jackson I, Sun K, Bouwman RD, Young RA, Glimcher LH, Lord GM. 2009. The transcription factors T-bet and GATA-3 control alternative pathways of T-cell differentiation through a shared set of target genes. *Proc Natl Acad Sci* **106:** 17876–17881.

Johnston RJ, Poholek AC, DiToro D, Yusuf I, Eto D, Barnett B, Dent AL, Craft J, Crotty S. 2009. Bcl6 and Blimp-1 are reciprocal and antagonistic regulators of T follicular helper cell differentiation. *Science* **325:** 1006–1010.

Johnston RJ, Choi YS, Diamond JA, Yang JA, Crotty S. 2012. STAT5 is a potent negative regulator of T$_{FH}$ cell differentiation. *J Exp Med* **209:** 243–250.

Josefowicz SZ, Lu LF, Rudensky AY. 2012. Regulatory T cells: Mechanisms of differentiation and function. *Annu Rev Immunol* **30:** 531–564.

Kagami S, Nakajima H, Kumano K, Suzuki K, Suto A, Imada K, Davey HW, Saito Y, Takatsu K, Leonard WJ, et al. 2000. Both Stat5a and Stat5b are required for antigen-induced eosinophil and T-cell recruitment into the tissue. *Blood* **95:** 1370–1377.

Kagami S, Nakajima H, Suto A, Hirose K, Suzuki K, Morita S, Kato I, Saito Y, Kitamura T, Iwamoto I. 2001. Stat5a regulates T helper cell differentiation by several distinct mechanisms. *Blood* **97:** 2358–2365.

Kanhere A, Hertweck A, Bhatia U, Gokmen MR, Perucha E, Jackson I, Lord GM, Jenner RG. 2012. T-bet and GATA3 orchestrate Th1 and Th2 differentiation through lineage-specific targeting of distal regulatory elements. *Nat Commun* **3:** 1268.

Kaplan MH, Schindler U, Smiley ST, Grusby MJ. 1996a. Stat6 is required for mediating responses to IL-4 and for development of Th2 cells. *Immunity* **4:** 313–319.

Kaplan MH, Sun YL, Hoey T, Grusby MJ. 1996b. Impaired IL-12 responses and enhanced development of Th2 cells in Stat4-deficient mice. *Nature* **382:** 174–177.

Kim BS, Wang K, Siracusa MC, Saenz SA, Brestoff JR, Monticelli LA, Noti M, Tait Wojno ED, Fung TC, Kubo M, et al. 2014. Basophils promote innate lymphoid cell responses in inflamed skin. *J Immunol* **193:** 3717–3725.

King IL, Mohrs M. 2009. IL-4-producing CD4⁺ T cells in reactive lymph nodes during helminth infection are T follicular helper cells. *J Exp Med* **206:** 1001–1007.

Klein Wolterink RG, Serafini N, van Nimwegen M, Vosshenrich CA, de Bruijn MJ, Fonseca Pereira D, Veiga Fernandes H, Hendriks RW, Di Santo JP. 2013. Essential, dose-dependent role for the transcription factor *Gata3* in the development of IL-5⁺ and IL-13⁺ type 2 innate lymphoid cells. *Proc Natl Acad Sci* **110:** 10240–10245.

Klose CS, Flach M, Mohle L, Rogell L, Hoyler T, Ebert K, Fabiunke C, Pfeifer D, Sexl V, Fonseca-Pereira D, et al. 2014. Differentiation of type 1 ILCs from a common progenitor to all helper-like innate lymphoid cell lineages. *Cell* **157:** 340–356.

Koch MA, Tucker-Heard G, Perdue NR, Killebrew JR, Urdahl KB, Campbell DJ. 2009. The transcription factor T-bet controls regulatory T cell homeostasis and function during type 1 inflammation. *Nat Immunol* **10:** 595–602.

Komatsu N, Mariotti-Ferrandiz ME, Wang Y, Malissen B, Waldmann H, Hori S. 2009. Heterogeneity of natural Foxp3⁺ T cells: A committed regulatory T-cell lineage and an uncommitted minor population retaining plasticity. *Proc Natl Acad Sci* **106:** 1903–1908.

Komatsu N, Okamoto K, Sawa S, Nakashima T, Oh-hora M, Kodama T, Tanaka S, Bluestone JA, Takayanagi H. 2014. Pathogenic conversion of Foxp3⁺ T cells into T$_H$17 cells in autoimmune arthritis. *Nat Med* **20:** 62–68.

Kopf M, Le Gros G, Bachmann M, Lamers MC, Bluethmann H, Kohler G. 1993. Disruption of the murine IL-4 gene blocks Th2 cytokine responses. *Nature* **362:** 245–248.

Korn T, Bettelli E, Oukka M, Kuchroo VK. 2009. IL-17 and Th17 cells. *Annu Rev Immunol* **27**: 485–517.

Koues OI, Collins PL, Cella M, Robinette ML, Porter SI, Pyfrom SC, Payton JE, Colonna M, Oltz EM. 2016. Distinct gene regulatory pathways for human innate versus adaptive lymphoid cells. *Cell* **165**: 1134–1146.

Kuperman DA, Huang X, Koth LL, Chang GH, Dolganov GM, Zhu Z, Elias JA, Sheppard D, Erle DJ. 2002. Direct effects of interleukin-13 on epithelial cells cause airway hyperreactivity and mucus overproduction in asthma. *Nat Med* **8**: 885–889.

Kurata H, Lee HJ, O'Garra A, Arai N. 1999. Ectopic expression of activated Stat6 induces the expression of Th2-specific cytokines and transcription factors in developing Th1 cells. *Immunity* **11**: 677–688.

Laurence A, Tato CM, Davidson TS, Kanno Y, Chen Z, Yao Z, Blank RB, Meylan F, Siegel R, Hennighausen L, et al. 2007. Interleukin-2 signaling via STAT5 constrains T helper 17 cell generation. *Immunity* **26**: 371–381.

Lazarevic V, Chen X, Shim JH, Hwang ES, Jang E, Bolm AN, Oukka M, Kuchroo VK, Glimcher LH. 2011. T-bet represses T$_H$17 differentiation by preventing Runx1-mediated activation of the gene encoding RORγt. *Nat Immunol* **12**: 96–104.

Lee YK, Turner H, Maynard CL, Oliver JR, Chen D, Elson CO, Weaver CT. 2009. Late developmental plasticity in the T helper 17 lineage. *Immunity* **30**: 92–107.

Lee Y, Awasthi A, Yosef N, Quintana FJ, Xiao S, Peters A, Wu C, Kleinewietfeld M, Kunder S, Hafler DA, et al. 2012. Induction and molecular signature of pathogenic T$_H$17 cells. *Nat Immunol* **13**: 991–999.

Lexberg MH, Taubner A, Albrecht I, Lepenies I, Richter A, Kamradt T, Radbruch A, Chang HD. 2010. IFN-γ and IL-12 synergize to convert in vivo generated Th17 into Th1/Th17 cells. *Eur J Immunol* **40**: 3017–3027.

Liang SC, Tan XY, Luxenberg DP, Karim R, Dunussi-Joannopoulos K, Collins M, Fouser LA. 2006. Interleukin (IL)-22 and IL-17 are coexpressed by Th17 cells and cooperatively enhance expression of antimicrobial peptides. *J Exp Med* **203**: 2271–2279.

Liang HE, Reinhardt RL, Bando JK, Sullivan BM, Ho IC, Locksley RM. 2012. Divergent expression patterns of IL-4 and IL-13 define unique functions in allergic immunity. *Nat Immunol* **13**: 58–66.

Liao W, Schones DE, Oh J, Cui Y, Cui K, Roh TY, Zhao K, Leonard WJ. 2008. Priming for T helper type 2 differentiation by interleukin 2-mediated induction of interleukin 4 receptor α-chain expression. *Nat Immunol* **9**: 1288–1296.

Liao W, Lin JX, Wang L, Li P, Leonard WJ. 2011. Modulation of cytokine receptors by IL-2 broadly regulates differentiation into helper T cell lineages. *Nat Immunol* **12**: 551–559.

Lighvani AA, Frucht DM, Jankovic D, Yamane H, Aliberti J, Hissong BD, Nguyen BV, Gadina M, Sher A, Paul WE, et al. 2001. T-bet is rapidly induced by interferon-γ in lymphoid and myeloid cells. *Proc Natl Acad Sci* **98**: 15137–15142.

Linterman MA, Pierson W, Lee SK, Kallies A, Kawamoto S, Rayner TF, Srivastava M, Divekar DP, Beaton L, Hogan JJ, et al. 2011. Foxp3$^+$ follicular regulatory T cells control the germinal center response. *Nat Med* **17**: 975–982.

Liu YJ. 2006. Thymic stromal lymphopoietin: Master switch for allergic inflammation. *J Exp Med* **203**: 269–273.

Liu YJ, Soumelis V, Watanabe N, Ito T, Wang YH, Malefyt Rde W, Omori M, Zhou B, Ziegler SF. 2007. TSLP: An epithelial cell cytokine that regulates T cell differentiation by conditioning dendritic cell maturation. *Annu Rev Immunol* **25**: 193–219.

Lu KT, Kanno Y, Cannons JL, Handon R, Bible P, Elkahloun AG, Anderson SM, Wei L, Sun H, O'Shea JJ, et al. 2011. Functional and epigenetic studies reveal multistep differentiation and plasticity of in vitro–generated and in vivo-derived follicular T helper cells. *Immunity* **35**: 622–632.

Luthje K, Kallies A, Shimohakamada Y, GT TB, Light A, Tarlinton DM, Nutt SL. 2012. The development and fate of follicular helper T cells defined by an IL-21 reporter mouse. *Nat Immunol* **13**: 491–498.

Martin-Fontecha A, Thomsen LL, Brett S, Gerard C, Lipp M, Lanzavecchia A, Sallusto F. 2004. Induced recruitment of NK cells to lymph nodes provides IFN-γ for T$_H$1 priming. *Nat Immunol* **5**: 1260–1265.

Mathur AN, Chang HC, Zisoulis DG, Kapur R, Belladonna ML, Kansas GS, Kaplan MH. 2006. T-bet is a critical determinant in the instability of the IL-17-secreting T-helper phenotype. *Blood* **108**: 1595–1601.

McKinley L, Alcorn JF, Peterson A, Dupont RB, Kapadia S, Logar A, Henry A, Irvin CG, Piganelli JD, Ray A, et al. 2008. T$_H$17 cells mediate steroid-resistant airway inflammation and airway hyperresponsiveness in mice. *J Immunol* **181**: 4089–4097.

Min B, Prout M, Hu-Li J, Zhu J, Jankovic D, Morgan ES, Urban JF Jr, Dvorak AM, Finkelman FD, LeGros G, et al. 2004. Basophils produce IL-4 and accumulate in tissues after infection with a Th2-inducing parasite. *J Exp Med* **200**: 507–517.

Mirchandani AS, Besnard AG, Yip E, Scott C, Bain CC, Cerovic V, Salmond RJ, Liew FY. 2014. Type 2 innate lymphoid cells drive CD4$^+$ Th2 cell responses. *J Immunol* **192**: 2442–2448.

Mjosberg J, Bernink J, Golebski K, Karrich JJ, Peters CP, Blom B, te Velde AA, Fokkens WJ, van Drunen CM, Spits H. 2012. The transcription factor GATA3 is essential for the function of human type 2 innate lymphoid cells. *Immunity* **37**: 649–659.

Monticelli LA, Sonnenberg GF, Abt MC, Alenghat T, Ziegler CG, Doering TA, Angelosanto JM, Laidlaw BJ, Yang CY, Sathaliyawala T, et al. 2011. Innate lymphoid cells promote lung-tissue homeostasis after infection with influenza virus. *Nat Immunol* **12**: 1045–1054.

Moriggl R, Topham DJ, Teglund S, Sexl V, McKay C, Wang D, Hoffmeyer A, van Deursen J, Sangster MY, Bunting KD, et al. 1999. Stat5 is required for IL-2-induced cell cycle progression of peripheral T cells. *Immunity* **10**: 249–259.

Moro K, Yamada T, Tanabe M, Takeuchi T, Ikawa T, Kawamoto H, Furusawa J, Ohtani M, Fujii H, Koyasu S. 2010. Innate production of T$_H$2 cytokines by adipose tissue-associated c-Kit$^+$Sca-1$^+$ lymphoid cells. *Nature* **463**: 540–544.

Mosmann TR, Coffman RL. 1989. T$_H$1 and T$_H$2 cells: Different patterns of lymphokine secretion lead to different functional properties. *Annu Rev Immunol* **7**: 145–173.

Mosmann TR, Cherwinski H, Bond MW, Giedlin MA, Coffman RL. 1986. Two types of murine helper T cell clone. I: Definition according to profiles of lymphokine activities and secreted proteins. *J Immunol* **136:** 2348–2357.

Mullen AC, High FA, Hutchins AS, Lee HW, Villarino AV, Livingston DM, Kung AL, Cereb N, Yao TP, Yang SY, et al. 2001. Role of T-bet in commitment of T_H1 cells before IL-12-dependent selection. *Science* **292:** 1907–1910.

Murphy KM, Reiner SL. 2002. The lineage decisions of helper T cells. *Nat Rev* **2:** 933–944.

Nakayamada S, Kanno Y, Takahashi H, Jankovic D, Lu KT, Johnson TA, Sun HW, Vahedi G, Hakim O, Handon R, et al. 2011. Early Th1 cell differentiation is marked by a Tfh cell–like transition. *Immunity* **35:** 919–931.

Neill DR, Wong SH, Bellosi A, Flynn RJ, Daly M, Langford TK, Bucks C, Kane CM, Fallon PG, Pannell R, et al. 2010. Nuocytes represent a new innate effector leukocyte that mediates type-2 immunity. *Nature* **464:** 1367–1370.

Nelson RW, Beisang D, Tubo NJ, Dileepan T, Wiesner DL, Nielsen K, Wuthrich M, Klein BS, Kotov DI, Spanier JA, et al. 2014. T cell receptor cross-reactivity between similar foreign and self-peptides influences naive cell population size and autoimmunity. *Immunity* **42:** 95–107.

Noval Rivas M, Burton OT, Wise P, Charbonnier LM, Georgiev P, Oettgen HC, Rachid R, Chatila TA. 2015. Regulatory T cell reprogramming toward a Th2-cell-like lineage impairs oral tolerance and promotes food allergy. *Immunity* **42:** 512–523.

Nurieva RI, Chung Y, Hwang D, Yang XO, Kang HS, Ma L, Wang YH, Watowich SS, Jetten AM, Tian Q, et al. 2008. Generation of T follicular helper cells is mediated by interleukin-21 but independent of T helper 1, 2, or 17 cell lineages. *Immunity* **29:** 138–149.

Nurieva RI, Chung Y, Martinez GJ, Yang XO, Tanaka S, Matskevitch TD, Wang YH, Dong C. 2009. Bcl6 mediates the development of T follicular helper cells. *Science* **325:** 1001–1005.

Nurieva RI, Podd A, Chen Y, Alekseev AM, Yu M, Qi X, Huang H, Wen R, Wang J, Li HS, et al. 2012. STAT5 protein negatively regulates T follicular helper (Tfh) cell generation and function. *J Biol Chem* **287:** 11234–11239.

Oestreich KJ, Mohn SE, Weinmann AS. 2012. Molecular mechanisms that control the expression and activity of Bcl-6 in T_H1 cells to regulate flexibility with a T_{FH}-like gene profile. *Nat Immunol* **13:** 405–411.

Ohnmacht C, Park JH, Cording S, Wing JB, Atarashi K, Obata Y, Gaboriau-Routhiau V, Marques R, Dulauroy S, Fedoseeva M, et al. 2015. Mucosal immunology. The microbiota regulates type 2 immunity through $ROR\gamma t^+$ T cells. *Science* **349:** 989–993.

Okada S, Markle JG, Deenick EK, Mele F, Averbuch D, Lagos M, Alzahrani M, Al-Muhsen S, Halwani R, Ma CS, et al. 2015. Immunodeficiencies. Impairment of immunity to *Candida* and *Mycobacterium* in humans with bi-allelic *RORC* mutations. *Science* **349:** 606–613.

Oldenhove G, Bouladoux N, Wohlfert EA, Hall JA, Chou D, Dos Santos L, O'Brien S, Blank R, Lamb E, Natarajan S, et al. 2009. Decrease of Foxp3$^+$ Treg cell number and acquisition of effector cell phenotype during lethal infection. *Immunity* **31:** 772–786.

Oliphant CJ, Hwang YY, Walker JA, Salimi M, Wong SH, Brewer JM, Englezakis A, Barlow JL, Hams E, Scanlon ST, et al. 2014. MHCII-mediated dialog between group 2 innate lymphoid cells and CD4$^+$ T cells potentiates type 2 immunity and promotes parasitic helminth expulsion. *Immunity* **41:** 283–295.

Omori M, Ziegler S. 2007. Induction of IL-4 expression in CD4$^+$ T cells by thymic stromal lymphopoietin. *J Immunol* **178:** 1396–1404.

Oosterwegel MA, Mandelbrot DA, Boyd SD, Lorsbach RB, Jarrett DY, Abbas AK, Sharpe AH. 1999. The role of CTLA-4 in regulating Th2 differentiation. *J Immunol* **163:** 2634–2639.

Ouyang W, Ranganath SH, Weindel K, Bhattacharya D, Murphy TL, Sha WC, Murphy KM. 1998. Inhibition of Th1 development mediated by GATA-3 through an IL-4-independent mechanism. *Immunity* **9:** 745–755.

Ouyang W, Lohning M, Gao Z, Assenmacher M, Ranganath S, Radbruch A, Murphy KM. 2000. Stat6-independent GATA-3 autoactivation directs IL-4-independent Th2 development and commitment. *Immunity* **12:** 27–37.

Ouyang W, Kolls JK, Zheng Y. 2008. The biological functions of T helper 17 cell effector cytokines in inflammation. *Immunity* **28:** 454–467.

Pai SY, Truitt ML, Ho IC. 2004. GATA-3 deficiency abrogates the development and maintenance of T helper type 2 cells. *Proc Natl Acad Sci* **101:** 1993–1998.

Park H, Li Z, Yang XO, Chang SH, Nurieva R, Wang YH, Wang Y, Hood L, Zhu Z, Tian Q, et al. 2005. A distinct lineage of CD4 T cells regulates tissue inflammation by producing interleukin 17. *Nat Immunol* **6:** 1133–1141.

Paul WE, Seder RA. 1994. Lymphocyte responses and cytokines. *Cell* **76:** 241–251.

Paul WE, Zhu J. 2010. How are T_H2-type immune responses initiated and amplified? *Nat Rev* **10:** 225–235.

Peine M, Rausch S, Helmstetter C, Frohlich A, Hegazy AN, Kuhl AA, Grevelding CG, Hofer T, Hartmann S, Lohning M. 2013. Stable T-bet$^+$GATA-3$^+$ Th1/Th2 hybrid cells arise in vivo, can develop directly from naive precursors, and limit immunopathologic inflammation. *PLoS Biol* **11:** e1001633.

Price AE, Liang HE, Sullivan BM, Reinhardt RL, Eisley CJ, Erle DJ, Locksley RM. 2010. Systemically dispersed innate IL-13-expressing cells in type 2 immunity. *Proc Natl Acad Sci* **107:** 11489–11494.

Qiu J, Guo X, Chen ZM, He L, Sonnenberg GF, Artis D, Fu YX, Zhou L. 2013. Group 3 innate lymphoid cells inhibit T-cell-mediated intestinal inflammation through aryl hydrocarbon receptor signaling and regulation of microflora. *Immunity* **39:** 386–399.

Reinhardt RL, Liang HE, Locksley RM. 2009. Cytokine-secreting follicular T cells shape the antibody repertoire. *Nat Immunol* **10:** 385–393.

Rochman I, Watanabe N, Arima K, Liu YJ, Leonard WJ. 2007. Cutting edge: Direct action of thymic stromal lymphopoietin on activated human CD4$^+$ T cells. *J Immunol* **178:** 6720–6724.

Rochman Y, Spolski R, Leonard WJ. 2009. New insights into the regulation of T cells by γ_c family cytokines. *Nat Rev* **9:** 480–490.

Roediger B, Kyle R, Yip KH, Sumaria N, Guy TV, Kim BS, Mitchell AJ, Tay SS, Jain R, Forbes-Blom E, et al. 2013. Cutaneous immunosurveillance and regulation of inflammation by group 2 innate lymphoid cells. *Nat Immunol* **14:** 564–573.

Rubtsov YP, Niec RE, Josefowicz S, Li L, Darce J, Mathis D, Benoist C, Rudensky AY. 2010. Stability of the regulatory T cell lineage in vivo. *Science* **329:** 1667–1671.

Rudra D, deRoos P, Chaudhry A, Niec RE, Arvey A, Samstein RM, Leslie C, Shaffer SA, Goodlett DR, Rudensky AY. 2012. Transcription factor Foxp3 and its protein partners form a complex regulatory network. *Nat Immunol* **13:** 1010–1019.

Sakaguchi S. 2004. Naturally arising CD4$^+$ regulatory t cells for immunologic self-tolerance and negative control of immune responses. *Annu Rev Immunol* **22:** 531–562.

Sakaguchi S, Miyara M, Costantino CM, Hafler DA. 2010. FOXP3$^+$ regulatory T cells in the human immune system. *Nat Rev* **10:** 490–500.

Sakaguchi S, Vignali DA, Rudensky AY, Niec RE, Waldmann H. 2013. The plasticity and stability of regulatory T cells. *Nat Rev* **13:** 461–467.

Sano T, Huang W, Hall JA, Yang Y, Chen A, Gavzy SJ, Lee JY, Ziel JW, Miraldi ER, Domingos AI, et al. 2015. An IL-23R/IL-22 circuit regulates epithelial serum amyloid A to promote local effector Th17 responses. *Cell* **163:** 381–393.

Seder RA, Germain RN, Linsley PS, Paul WE. 1994. CD28-mediated costimulation of interleukin 2 (IL-2) production plays a critical role in T cell priming for IL-4 and interferon γ production. *J Exp Med* **179:** 299–304.

Sefik E, Geva-Zatorsky N, Oh S, Konnikova L, Zemmour D, McGuire AM, Burzyn D, Ortiz-Lopez A, Lobera M, Yang J, et al. 2015. Mucosal immunology. Individual intestinal symbionts induce a distinct population of RORγ$^+$ regulatory T cells. *Science* **349:** 993–997.

Shevach EM. 2000. Regulatory T cells in autoimmmunity. *Annu Rev Immunol* **18:** 423–449.

Shevach EM. 2009. Mechanisms of Foxp3$^+$ T regulatory cell-mediated suppression. *Immunity* **30:** 636–645.

Shih HY, Sciume G, Mikami Y, Guo L, Sun HW, Brooks SR, Urban JF Jr, Davis FP, Kanno Y, O'Shea JJ. 2016. Developmental acquisition of regulomes underlies innate lymphoid cell functionality. *Cell* **165:** 1120–1133.

Shimoda K, van Deursen J, Sangster MY, Sarawar SR, Carson RT, Tripp RA, Chu C, Quelle FW, Nosaka T, Vignali DA, et al. 1996. Lack of IL-4-induced Th2 response and IgE class switching in mice with disrupted *Stat6* gene. *Nature* **380:** 630–633.

Sokol CL, Barton GM, Farr AG, Medzhitov R. 2008. A mechanism for the initiation of allergen-induced T helper type 2 responses. *Nat Immunol* **9:** 310–318.

Spits H, Bernink JH, Lanier L. 2016. NK cells and type 1 innate lymphoid cells: Partners in host defense. *Nat Immunol* **17:** 758–764.

Steinfelder S, Andersen JF, Cannons JL, Feng CG, Joshi M, Dwyer D, Caspar P, Schwartzberg PL, Sher A, Jankovic D. 2009. The major component in schistosome eggs responsible for conditioning dendritic cells for Th2 polarization is a T2 ribonuclease (omega-1). *J Exp Med* **206:** 1681–1690.

Stumbles PA, Thomas JA, Pimm CL, Lee PT, Venaille TJ, Proksch S, Holt PG. 1998. Resting respiratory tract dendritic cells preferentially stimulate T helper cell type 2 (Th2) responses and require obligatory cytokine signals for induction of Th1 immunity. *J Exp Med* **188:** 2019–2031.

Szabo SJ, Kim ST, Costa GL, Zhang XK, Fathman CG, Glimcher LH. 2000. A novel transcription factor, T-bet, directs Th1 lineage commitment. *Cell* **100:** 655–669.

Szabo SJ, Sullivan BM, Stemmann C, Satoskar AR, Sleckman BP, Glimcher LH. 2002. Distinct effects of T-bet in T_H1 lineage commitment and IFN-γ production in CD4 and CD8 T cells. *Science* **295:** 338–342.

Szabo SJ, Sullivan BM, Peng SL, Glimcher LH. 2003. Molecular mechanisms regulating Th1 immune responses. *Annu Rev Immunol* **21:** 713–758.

Takeda K, Tanaka T, Shi W, Matsumoto M, Minami M, Kashiwamura S, Nakanishi K, Yoshida N, Kishimoto T, Akira S. 1996. Essential role of Stat6 in IL-4 signalling. *Nature* **380:** 627–630.

Tanaka S, Motomura Y, Suzuki Y, Yagi R, Inoue H, Miyatake S, Kubo M. 2011. The enhancer HS2 critically regulates GATA-3-mediated Il4 transcription in T_H2 cells. *Nat Immunol* **12:** 77–85.

Tao X, Constant S, Jorritsma P, Bottomly K. 1997. Strength of TCR signal determines the costimulatory requirements for Th1 and Th2 CD4$^+$ T cell differentiation. *J Immunol* **159:** 5956–5963.

Thierfelder WE, van Deursen JM, Yamamoto K, Tripp RA, Sarawar SR, Carson RT, Sangster MY, Vignali DA, Doherty PC, Grosveld GC, et al. 1996. Requirement for Stat4 in interleukin-12-mediated responses of natural killer and T cells. *Nature* **382:** 171–174.

Tubo NJ, Pagan AJ, Taylor JJ, Nelson RW, Linehan JL, Ertelt JM, Huseby ES, Way SS, Jenkins MK. 2013. Single naive CD4$^+$ T cells from a diverse repertoire produce different effector cell types during infection. *Cell* **153:** 785–796.

Urban JF Jr, Noben-Trauth N, Donaldson DD, Madden KB, Morris SC, Collins M, Finkelman FD. 1998. IL-13, IL-4Rα, and Stat6 are required for the expulsion of the gastrointestinal nematode parasite *Nippostrongylus brasiliensis*. *Immunity* **8:** 255–264.

Usui T, Nishikomori R, Kitani A, Strober W. 2003. GATA-3 suppresses Th1 development by downregulation of Stat4 and not through effects on IL-12Rβ2 chain or T-bet. *Immunity* **18:** 415–428.

Usui T, Preiss JC, Kanno Y, Yao ZJ, Bream JH, O'Shea JJ, Strober W. 2006. T-bet regulates Th1 responses through essential effects on GATA-3 function rather than on *IFNG* gene acetylation and transcription. *J Exp Med* **203:** 755–766.

van Panhuys N, Tang SC, Prout M, Camberis M, Scarlett D, Roberts J, Hu-Li J, Paul WE, Le Gros G. 2008. In vivo studies fail to reveal a role for IL-4 or STAT6 signaling in Th2 lymphocyte differentiation. *Proc Natl Acad Sci* **105:** 12423–12428.

van Panhuys N, Klauschen F, Germain RN. 2014. T-cell-receptor-dependent signal intensity dominantly controls CD4$^+$ T cell polarization in vivo. *Immunity* **41:** 63–74.

Veldhoen M, Hocking RJ, Atkins CJ, Locksley RM, Stockinger B. 2006. TGFβ in the context of an inflammatory cytokine milieu supports de novo differentiation of IL-17-producing T cells. *Immunity* 24: 179–189.

Vely F, Barlogis V, Vallentin B, Neven B, Piperoglou C, Ebbo M, Perchet T, Petit M, Yessaad N, Touzot F, et al. 2016. Evidence of innate lymphoid cell redundancy in humans. *Nat Immunol* 17: 1291–1299.

Voehringer D, Shinkai K, Locksley RM. 2004. Type 2 immunity reflects orchestrated recruitment of cells committed to IL-4 production. *Immunity* 20: 267–277.

Wan YSY, Flavell RA. 2007. Regulatory T-cell functions are subverted and converted owing to attenuated Foxp3 expression. *Nature* 445: 766–770.

Wang YH, Voo KS, Liu B, Chen CY, Uygungil B, Spoede W, Bernstein JA, Huston DP, Liu YJ. 2010. A novel subset of CD4+ T_H2 memory/effector cells that produce inflammatory IL-17 cytokine and promote the exacerbation of chronic allergic asthma. *J Exp Med* 207: 2479–2491.

Wang Y, Su MA, Wan YY. 2011. An essential role of the transcription factor GATA-3 for the function of regulatory T cells. *Immunity* 35: 337–348.

Wang Y, Misumi I, Gu AD, Curtis TA, Su L, Whitmire JK, Wan YY. 2013. GATA-3 controls the maintenance and proliferation of T cells downstream of TCR and cytokine signaling. *Nat Immunol* 14: 714–722.

Weaver CT, Hatton RD, Mangan PR, Harrington LE. 2007. IL-17 family cytokines and the expanding diversity of effector T cell lineages. *Annu Rev Immunol* 25: 821–852.

Wei G, Wei L, Zhu J, Zang C, Hu-Li J, Yao Z, Cui K, Kanno Y, Roh TY, Watford WT, et al. 2009. Global mapping of H3K4me3 and H3K27me3 reveals specificity and plasticity in lineage fate determination of differentiating CD4+ T cells. *Immunity* 30: 155–167.

Wei G, Abraham BJ, Yagi R, Jothi R, Cui K, Sharma S, Narlikar L, Northrup DL, Tang Q, Paul WE, et al. 2011. Genome-wide analyses of transcription factor GATA3-mediated gene regulation in distinct T cell types. *Immunity* 35: 299–311.

Wohlfert EA, Grainger JR, Bouladoux N, Konkel JE, Oldenhove G, Ribeiro CH, Hall JA, Yagi R, Naik S, Bhairavabhotla R, et al. 2011. GATA3 controls Foxp3+ regulatory T cell fate during inflammation in mice. *J Clin Invest* 121: 4503–4515.

Wynn TA. 2003. IL-13 effector functions. *Annu Rev Immunol* 21: 425–456.

Xu L, Kitani A, Fuss I, Strober W. 2007. Cutting edge: Regulatory T cells induce CD4+CD25-Foxp3- T cells or are self-induced to become Th17 cells in the absence of exogenous TGF-β. *J Immunol* 178: 6725–6729.

Yagi R, Junttila IS, Wei G, Urban JF Jr, Zhao K, Paul WE, Zhu J. 2010. The transcription factor GATA3 actively represses RUNX3 protein-regulated production of interferon-γ. *Immunity* 32: 507–517.

Yagi R, Zhu J, Paul WE. 2011. An updated view on transcription factor GATA3-mediated regulation of Th1 and Th2 cell differentiation. *Int Immunol* 23: 415–420.

Yagi R, Zhong C, Northrup DL, Yu F, Bouladoux N, Spencer S, Hu G, Barron L, Sharma S, Nakayama T, et al. 2014. The transcription factor GATA3 is critical for the development of all IL-7Rα-expressing innate lymphoid cells. *Immunity* 40: 378–388.

Yamaguchi T, Wing JB, Sakaguchi S. 2011. Two modes of immune suppression by Foxp3+ regulatory T cells under inflammatory or non-inflammatory conditions. *Semin Immunol* 23: 424–430.

Yamane H, Zhu J, Paul WE. 2005. Independent roles for IL-2 and GATA-3 in stimulating naive CD4+ T cells to generate a Th2-inducing cytokine environment. *J Exp Med* 202: 793–804.

Yamashita M, Ukai-Tadenuma M, Kimura M, Omori M, Inami M, Taniguchi M, Nakayama T. 2002. Identification of a conserved GATA3 response element upstream proximal from the interleukin-13 gene locus. *J Biol Chem* 277: 42399–42408.

Yang Y, Ochando JC, Bromberg JS, Ding Y. 2007. Identification of a distant T-bet enhancer responsive to IL-12/Stat4 and IFNγ/Stat1 signals. *Blood* 110: 2494–2500.

Yang XO, Nurieva R, Martinez GJ, Kang HS, Chung Y, Pappu BP, Shah B, Chang SH, Schluns KS, Watowich SS, et al. 2008. Molecular antagonism and plasticity of regulatory and inflammatory T cell programs. *Immunity* 29: 44–56.

Yang XP, Ghoreschi K, Steward-Tharp SM, Rodriguez-Canales J, Zhu J, Grainger JR, Hirahara K, Sun HW, Wei L, Vahedi G, et al. 2011. Opposing regulation of the locus encoding IL-17 through direct, reciprocal actions of STAT3 and STAT5. *Nat Immunol* 12: 247–254.

Yang Q, Monticelli LA, Saenz SA, Chi AW, Sonnenberg GF, Tang J, De Obaldia ME, Bailis W, Bryson JL, Toscano K, et al. 2013. T cell factor 1 is required for group 2 innate lymphoid cell generation. *Immunity* 38: 694–704.

Yosef N, Shalek AK, Gaublomme JT, Jin H, Lee Y, Awasthi A, Wu C, Karwacz K, Xiao S, Jorgolli M, et al. 2013. Dynamic regulatory network controlling T_H17 cell differentiation. *Nature* 496: 461–468.

Yu F, Sharma S, Edwards J, Feigenbaum L, Zhu J. 2015. Dynamic expression of transcription factors T-bet and GATA-3 by regulatory T cells maintains immunotolerance. *Nat Immunol* 16: 197–206.

Yusuf I, Kageyama R, Monticelli L, Johnston RJ, Ditoro D, Hansen K, Barnett B, Crotty S. 2010. Germinal center T follicular helper cell IL-4 production is dependent on signaling lymphocytic activation molecule receptor (CD150). *J Immunol* 185: 190–202.

Zaretsky AG, Taylor JJ, King IL, Marshall FA, Mohrs M, Pearce EJ. 2009. T follicular helper cells differentiate from Th2 cells in response to helminth antigens. *J Exp Med* 206: 991–999.

Zhang DH, Cohn L, Ray P, Bottomly K, Ray A. 1997. Transcription factor GATA-3 is differentially expressed in murine Th1 and Th2 cells and controls Th2-specific expression of the interleukin-5 gene. *J Biol Chem* 272: 21597–21603.

Zhang F, Meng G, Strober W. 2008. Interactions among the transcription factors Runx1, RORγt and Foxp3 regulate the differentiation of interleukin 17-producing T cells. *Nat Immunol* 9: 1297–1306.

Zheng W, Flavell RA. 1997. The transcription factor GATA-3 is necessary and sufficient for Th2 cytokine gene expression in CD4 T cells. *Cell* 89: 587–596.

Zheng Y, Josefowicz SZ, Kas A, Chu TT, Gavin MA, Rudensky AY. 2007. Genome-wide analysis of Foxp3 target

Cite this article as *Cold Spring Harb Perspect Biol* doi: 10.1101/cshperspect.a030338

genes in developing and mature regulatory T cells. *Nature* **445**: 936–940.

Zhong C, Cui K, Wilhelm C, Hu G, Mao K, Belkaid Y, Zhao K, Zhu J. 2016. Group 3 innate lymphoid cells continuously require the transcription factor GATA-3 after commitment. *Nat Immunol* **17**: 169–178.

Zhou B, Comeau MR, De Smedt T, Liggitt HD, Dahl ME, Lewis DB, Gyarmati D, Aye T, Campbell DJ, Ziegler SF. 2005. Thymic stromal lymphopoietin as a key initiator of allergic airway inflammation in mice. *Nat Immunol* **6**: 1047–1053.

Zhou L, Lopes JE, Chong MM, Ivanov II, Min R, Victora GD, Shen Y, Du J, Rubtsov YP, Rudensky AY, et al. 2008. TGF-β-induced Foxp3 inhibits T_H17 cell differentiation by antagonizing RORγt function. *Nature* **453**: 236–240.

Zhou L, Chong MM, Littman DR. 2009a. Plasticity of CD4+ T cell lineage differentiation. *Immunity* **30**: 646–655.

Zhou X, Bailey-Bucktrout SL, Jeker LT, Penaranda C, Martinez-Llordella M, Ashby M, Nakayama M, Rosenthal W, Bluestone JA. 2009b. Instability of the transcription factor Foxp3 leads to the generation of pathogenic memory T cells in vivo. *Nat Immunol* **10**: 1000–1007.

Zhu J, Paul WE. 2008. CD4 T cells: Fates, functions, and faults. *Blood* **112**: 1557–1569.

Zhu J, Paul WE. 2010a. Heterogeneity and plasticity of T helper cells. *Cell research* **20**: 4–12.

Zhu J, Paul WE. 2010b. Peripheral CD4+ T-cell differentiation regulated by networks of cytokines and transcription factors. *Immunol Rev* **238**: 247–262.

Zhu J, Guo L, Watson CJ, Hu-Li J, Paul WE. 2001. Stat6 is necessary and sufficient for IL-4's role in Th2 differentiation and cell expansion. *J Immunol* **166**: 7276–7281.

Zhu J, Cote-Sierra J, Guo L, Paul WE. 2003. Stat5 activation plays a critical role in Th2 differentiation. *Immunity* **19**: 739–748.

Zhu J, Min B, Hu-Li J, Watson CJ, Grinberg A, Wang Q, Killeen N, Urban JF Jr, Guo L, Paul WE. 2004. Conditional deletion of Gata3 shows its essential function in T_H1-T_H2 responses. *Nat Immunol* **5**: 1157–1165.

Zhu J, Yamane H, Paul WE. 2010. Differentiation of effector CD4 T cell populations. *Annu Rev Immunol* **28**: 445–489.

Zhu J, Jankovic D, Oler AJ, Wei G, Sharma S, Hu G, Guo L, Yagi R, Yamane H, Punkosdy G, et al. 2012. The transcription factor T-bet is induced by multiple pathways and prevents an endogenous Th2 cell program during Th1 cell responses. *Immunity* **37**: 660–673.

Zielinski CE, Mele F, Aschenbrenner D, Jarrossay D, Ronchi F, Gattorno M, Monticelli S, Lanzavecchia A, Sallusto F. 2012. Pathogen-induced human T_H17 cells produce IFN-γ or IL-10 and are regulated by IL-1β. *Nature* **484**: 514–518.

Cytokine Signaling in the Development and Homeostasis of Regulatory T cells

Kevin H. Toomer[1] and Thomas R. Malek[1,2]

[1]Department of Microbiology and Immunology, Miller School of Medicine, University of Miami, Miami, Florida 33136

[2]Diabetes Research Institute, Miller School of Medicine, University of Miami, Miami, Florida 33136

Correspondence: tmalek@umail.miami.edu

Cytokine signaling is indispensable for regulatory T-cell (Treg) development in the thymus, and also influences the homeostasis, phenotypic diversity, and function of Tregs in the periphery. Because Tregs are required for establishment and maintenance of immunological self-tolerance, investigating the role of cytokines in Treg biology carries therapeutic potential in the context of autoimmune disease. This review discusses the potent and diverse influences of interleukin (IL)-2 signaling on the Treg compartment, an area of knowledge that has led to the use of low-dose IL-2 as a therapy to reregulate autoaggressive immune responses. Evidence suggesting Treg-specific impacts of the cytokines transforming growth factor β (TGF-β), IL-7, thymic stromal lymphopoietin (TSLP), IL-15, and IL-33 is also presented. Finally, we consider the technical challenges and knowledge limitations that must be overcome to bring other cytokine-based, Treg-targeted therapies into clinical use.

CD4$^+$Foxp3$^+$ regulatory T cells (Tregs) are indispensable for the maintenance of immunological self-tolerance because of their ability to suppress autoreactive T cells that escape negative selection in the thymus. Humans and mice lacking the signature Treg transcription factor Foxp3 succumb to fatal lymphoproliferative disease early in life. In keeping with this role, clinical and experimental data have implicated Treg dysfunction in the pathogenesis of many autoimmune diseases (Grant et al. 2015), which collectively account for an enormous burden of morbidity and mortality. Meanwhile, the nonspecific therapies used to control these conditions (e.g., corticosteroids) often have limited efficacy and severe adverse effects.

As with other T-cell lineages, many aspects of Treg biology are regulated by a vast and intricate network of cytokine signals. A detailed understanding of how these cytokines impact Treg differentiation, homeostasis, and suppressive function will facilitate new therapies that disrupt autoimmunity through precise targeting of its underlying molecular pathways. The recent use of low-dose interleukin (IL)-2 immunotherapy has already shown the promise of this cytokine-based approach. Low-dose IL-2, which selectively expands and activates the Treg compartment in vivo (Aoyama et al. 2012; Kosmaczewska 2014; Yu et al. 2015), has proven beneficial to patients with systemic lupus erythematosus (He et al. 2016), hepatitis C virus–induced vasculitis (Saadoun et al. 2011), alopecia areata

(Castela et al. 2014), and graft-versus-host disease (Koreth et al. 2011; Matsuoka et al. 2013). Early phase clinical trials have also been conducted in the setting of type 1 diabetes (Hartemann et al. 2013; Todd et al. 2016).

Numerous cytokines besides IL-2 are now known to affect Tregs, and behind each one lies a potential new tool for the immunotherapy arsenal. Nevertheless, preliminary efforts to manipulate these pathways have often been hobbled by poor efficacy or severe off-target effects, a challenge resulting from the highly pleiotropic nature of cytokine signals. Exploiting the full range of opportunity provided by cytokine-based therapies will require more advanced insight into the complexities of cytokine signaling as it relates to the Treg compartment. Accordingly, the purpose of this review is to describe how several extensively studied cytokines, IL-2, IL-7, thymic stromal lymphopoietin (TSLP), IL-15, transforming growth factor β (TGF-β), and IL-33, are known to influence Tregs.

INTERLEUKIN-2

Originally identified as a growth factor for T cells in vitro (Morgan et al. 1976), IL-2 is produced primarily by activated $CD4^+$ and $CD8^+$ T conventional (Tconv) cells (Malek 2008). The high-affinity IL-2 receptor (IL-2R) comprises three subunits, IL-2Rα (CD25), IL-2Rβ (CD122), and the common γ chain ($γ_c$ or CD132). Tregs and antigen-activated Tconv cells are the predominant populations that express all three simultaneously, positioning them to respond efficiently to IL-2 in vivo. Although Tregs do not produce IL-2, they express CD25 at uniquely high and consistent levels, whereas elevated CD25 expression in Tconv cells is generally a transient event during activation (Malek 2008). During signaling, IL-2 first binds to CD25, forming a dimer that recruits CD122 and $γ_c$ (Stauber et al. 2006). Following receptor engagement, the Janus kinase (JAK)1 and JAK3 associate with the cytoplasmic tails of IL-2Rβ and $γ_c$, leading to phosphorylation of the JAKs along with three key tyrosine residues in the tail of IL-2Rβ. These phosphotyrosines activate three major intracellular signaling pathways:

mitogen-activated protein kinase (MAPK) and phosphoinositide 3-kinase (PI3K), both mediated by the Shc adaptor protein, as well as the signal transducer and activator of transcription 5 (STAT5—composed of two similar proteins, STAT5A and STAT5B). These downstream signaling pathways govern survival, proliferation, and memory formation among antigen-activated Tconv lineages (Cheng et al. 2011), but are differentially regulated in the Treg compartment to support its suppressive function. Tregs are uniquely reliant on IL-2-dependent STAT5 activation, due in part to high levels of the lipid phosphatase and tensin homolog (PTEN), which acts to suppress IL-2-dependent PI3K and mechanistic target of rapamycin (mTOR) activity (Walsh et al. 2006; Huynh et al. 2015). IL-2 directly up-regulates Foxp3 and CD25 through STAT5 in a positive feedback mechanism to establish and maintain Treg transcriptional identity (Fontenot et al. 2005b; Burchill et al. 2007b; Feng et al. 2014). In conjunction with elevated CD25 expression, molecular adaptations in these pathways allow Treg development and homeostasis to be supported by a uniquely low threshold of IL-2 signaling, explaining the effectiveness of low-dose therapy (Yu et al. 2009). Tregs can also outcompete activated Tconv cells for available IL-2, and this cytokine deprivation has been highlighted as one mechanism of Treg-mediated immunosuppression (Pandiyan et al. 2007).

IL-2 is the predominant cytokine involved in regulating Treg development, stability, function, and peripheral homeostasis. Mouse models with abrogated expression of IL-2, CD25, or CD122 experience aggressive lymphoproliferation and fatal multiorgan autoimmunity (Sadlack et al. 1993, 1995; Suzuki et al. 1995; Willerford et al. 1995), and rare loss-of-function mutations in human CD25 produce a similar clinical syndrome in patients (Roifman 2000). This occurs because IL-2 signaling deficiency leads to a failure of Treg maturation and a consequent breakdown of self-tolerance (Fig. 1). Although $CD4^+Foxp3^+$ T cells are still present in IL-2/IL-2R-deficient mice, they show a ~50% numerical reduction, and Foxp3 expression on a per-cell basis is reduced twofold (Jo-

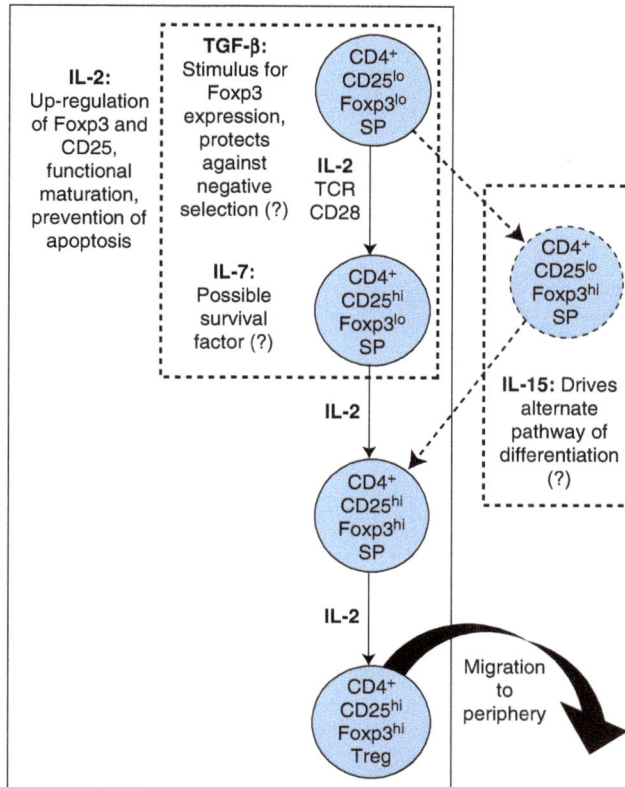

Figure 1. Cytokine contributions to regulatory T-cell (Treg) maturation in the thymus. Tregs develop from CD4 single-positive (SP) thymocyte precursors. Signaling through a high-affinity, self-reactive T-cell receptor (TCR) combined with CD28 costimulation is thought to trigger up-regulation of CD25. This potentiates the ability of Treg precursors to respond to interleukin (IL)-2, which prevents apoptosis, induces Foxp3 expression, and establishes Treg transcriptional identity and suppressive capacity. Boxes with dotted lines denote developmental stages where other cytokines may act. Dotted arrows denote a possible parallel pathway of Treg development. TGF-β, Transforming growth factor β.

sefowicz et al. 2012a). These Foxp3^{lo} cells have diminished expression of the suppressive mediators CTLA4, CD39, and CD73, suggesting that thymic Tregs generated in the setting of IL-2 deprivation lack adequate functional programming (Cheng et al. 2013). This idea is supported by adoptive transfer studies, as introducing a low number ($1-3 \times 10^5$) of CD4$^+$CD25$^+$ Tregs into neonatal $Il2rb^{-/-}$ mice provides full protection from autoimmunity, with treated animals retaining a donor-derived Treg compartment throughout life (Malek et al. 2002). Moreover, transgenic thymic rescue of IL-2Rβ signaling in these mice is sufficient to restore a normal Treg compartment and prevent autoim-

munity, further demonstrating a nonredundant requirement for the IL-2R in production of functional Tregs (Malek et al. 2000; Malek and Bayer 2004; Bayer et al. 2007).

Experiments using CD4 single-positive thymocytes confirm that IL-2 is an essential driver of Treg lineage commitment. Transgenic expression of a constitutively active *Stat5* transgene is sufficient to divert thymocyte precursors of naïve Tconv cells into the Treg lineage (Burchill et al. 2008). Mice deficient in the mediator TRAF3, which opposes IL-2R signaling, exhibit a two- to threefold increase in thymic Treg frequency (Yi et al. 2014). Active from the earliest stages of Treg differentiation, IL-2/IL-2R interaction is a critical inducer of Foxp3 expression in nascent Tregs (Bayer et al. 2007; Burchill et al. 2007a; Huehn et al. 2009), and also acts as a survival factor, up-regulating Bcl-2 to prevent their apoptosis (Tai et al. 2013). In vitro and in vivo experiments support a model whereby signals from a high-affinity, self-reactive T-cell receptor (TCR) and CD28 induce CD25 up-regulation in Treg precursors, leading to IL-2-driven induction of Foxp3 and resultant commitment to the Treg cell fate (Fig. 1) (Burchill et al. 2008; Lio and Hsieh 2008).

Although a majority of mature Tregs found in the periphery are of thymic origin, naïve Tconv cells can be instructed to differentiate into Tregs through a combination of molecular signals that includes IL-2. Treating cultured CD4$^+$CD25$^-$ T cells from humans or mice with the cytokine TGF-β (described below), in conjunction with TCR stimulation, can elicit Foxp3 expression and generate CD4$^+$CD25$^+$ "induced Tregs" (iTregs) with immunosuppressive function (Chen et al. 2003; Fantini et al. 2004). Although IL-2 on its own cannot convert naïve Tconv cells to a Treg cell fate (Xu et al. 2010), experiments using cytokine deprivation by antibody blockade and germline knockout (KO) show that IL-2 is essential for TGF-β-mediated induction of Foxp3, as well as subsequent iTreg clonal expansion and in vitro suppressor activity (Zheng et al. 2007). Adoptive transfer studies support the idea that iTreg persistence, transcriptional identity, and phenotypic stability are also IL-2-dependent in vivo (Shevach and Thornton 2014). iTregs are transcriptionally distinct from their thymic Treg counterparts on generation in vitro (Haribhai et al. 2009; Feuerer et al. 2010), and functional comparisons in some experimental models suggest that iTregs have lower phenotypic stability and suppressive capacity (Yadav et al. 2013; Shevach and Thornton 2014). These shortcomings have hindered the application of iTregs as a therapeutic tool, although cell-culture protocols for optimizing their potency and durability are a subject of continuing investigation (Kanamori et al. 2016; Schmidt et al. 2016).

Tconv precursors are also capable of differentiating into Tregs in vivo within peripheral tissue sites. Generation of these immuno-suppressive cells, termed "peripheral Tregs" (pTregs) to distinguish them from cell culture–derived iTregs, is favored by subimmunogenic doses of peptide antigen in conjunction with TGF-β and IL-2 (Povoleri et al. 2013). Mice that undergo thymectomy at 3 days of age to abolish thymic Treg maturation develop autoimmune disease, demonstrating that pTregs alone are insufficient to maintain self-tolerance (Sakaguchi et al. 1995; Asano et al. 1996). Nevertheless, pTregs possess functional specializations that make them uniquely suited to suppressing inflammation in certain peripheral sites, particularly mucosal surfaces (Bilate and Lafaille 2012; Josefowicz et al. 2012b; Yadav et al. 2013). pTregs are key contributors to immune homeostasis in the intestine, a tissue microenvironment that appears to be especially favorable for their differentiation (Tanoue et al. 2016). pTregs present at this interface between host and external environment are integral for maintaining tolerance to an enormous array of antigens derived from self, ingested material, and microbiota (Littman and Rudensky 2010; Harrison and Powrie 2013). Crucially, gut-localized CD4$^+$ T cells in humans and mice exhibit an unusual degree of phenotypic plasticity between the pTreg and proinflammatory Th17 lineages, with IL-2 serving as an important regulator of this balance (Fig. 2). In vivo, TGF-β may be involved in differentiation toward either the Treg or Th17 cell fate (Konkel and Chen 2011), and additional cytokine signals are required to tip the scales toward the transcriptional program of either Foxp3 or the Th17-defining transcription factor RORγt (Hoechst et al. 2011; Kleinewietfeld and Hafler 2013; Ueno et al. 2015). Exposing Tconv cells to TGF-β in an environment rich in proinflammatory signals (particularly IL-6) favors Th17 differentiation (Kimura and Kishimoto 2010; Fujimoto et al. 2011). Conversely, IL-2-dependent STAT5 signaling promotes TGF-β-mediated differentiation of pTregs in the intestine (Povoleri et al. 2013; Yadav et al. 2013), while impeding the generation of Th17 cells from naïve precursors (Laurence et al. 2007; Zheng et al. 2008). Since the shift of this dynamic equilibrium away from pTregs and

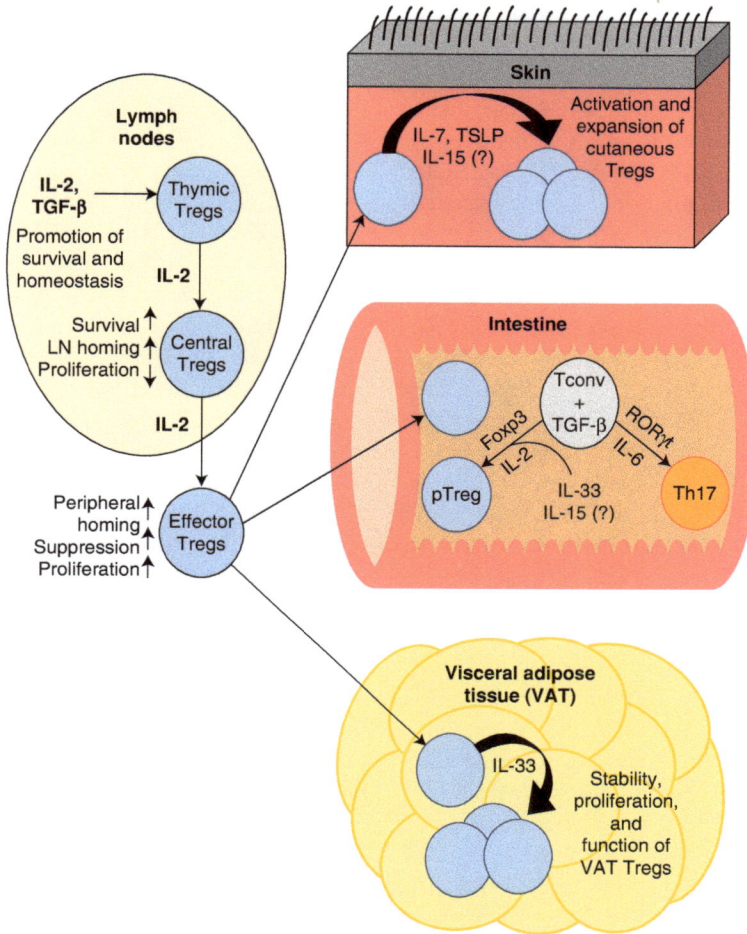

Figure 2. Effects of cytokine signaling on regulatory T cells (Tregs) in peripheral tissue sites. IL, Interleukin; TGF-β, transforming growth factor β; TSLP, thymic stromal lymphopoietin; LN, lymph node; Tconv, T conventional cell.

toward Th17 is essential to the pathogenesis of inflammatory bowel disease (Ueno et al. 2015), the role of IL-2 in this context has clear therapeutic implications.

Treg homeostasis does not require ongoing output from the thymus but is instead based on continuous self-renewal of thymic Tregs, which seed peripheral lymphoid tissues and undergo IL-2 driven expansion during early life (Bayer et al. 2005, 2007). There is substantial evidence that IL-2R signaling contributes to homeostatic maintenance of the peripheral Treg pool (Fig. 2), although the precise mechanisms have not been fully clarified. Many studies have positively

correlated CD25 with Foxp3 expression and phenotypic stability among Tregs in the periphery (Fontenot et al. 2005a; Barbi et al. 2014). IL-2 has also been found to participate in the epigenetic regulation that maintains Treg identity. Sustained heritability of Foxp3 expression depends on the *Foxp3* promoter as well as a *cis*-regulatory element within the *Foxp3* locus known as conserved noncoding sequence 2 (CNS2) or the Treg-specific demethylated region (TSDR). Both the *Foxp3* promoter and CNS2 contain STAT5-binding motifs, and CNS2 appears to function as an IL-2 sensor when this cytokine is present in limiting amounts. IL-2-dependent STAT5 signaling in Tregs enables CNS2 to interact with the *Foxp3* promoter, which prevents loss of Foxp3 expression during cell-cycle progression or exposure to proinflammatory cytokines (Zheng et al. 2010; Feng et al. 2014; Li et al. 2014). Cellular adoptive transfer and bone marrow chimera studies, using IL-2/IL-2R-deficient mouse models, show an IL-2-mediated contribution to the survival, engraftment, and proliferation of Tregs in lymphoid tissues (Bayer et al. 2005, 2007; Cheng et al. 2013). Notably, administering exogenous IL-2 to *Il2*$^{-/-}$ mice not only rescues the numerical deficit in Tregs, but also corrects many of their transcriptional abnormalities, including low expression of suppression mediators such as CTLA4, CD39, and CD73 (Fontenot et al. 2005b; Barron et al. 2010). One report identified the antiapoptotic protein Mcl-1, which is known to be regulated by IL-2 signaling, as an essential survival factor for Tregs in the periphery (Pierson et al. 2013). However, it should be noted that interpretation of findings from IL-2/IL-2R-deficient models is problematic because it is often unclear whether Treg defects are attributable to impaired survival in the periphery, defective maturation in the thymus, or a combination of both. One study attempted to isolate peripheral signaling effects using antibody blockade of IL-2 in thymectomized adult mice. This treatment led to a selective reduction in Treg numbers and induction of autoimmunity, establishing an unequivocal role for IL-2 in peripheral homeostasis (Setoguchi et al. 2005). Emerging genetic tools, which permit fate mapping of Foxp3$^+$ cells and inducible deletion of signaling components in mice, have provided new opportunities to clarify the nature and importance of the IL-2R for Tregs in the periphery, although a clear consensus has yet to emerge (Rubtsov et al. 2010; Chinen et al. 2016).

These ambiguities surrounding the contribution of peripheral IL-2 signaling may be partially explained by heterogeneity within the Treg compartment itself. Current evidence indicates that mature Tregs can be divided into phenotypically and functionally distinct subpopulations. "Central" Tregs (cTregs) express high levels of the lymphoid homing marker CD62L, and exhibit a phenotype geared toward long-term survival and diminished functional activation. CD62L^{lo} "effector" Tregs (eTregs), which originate from cTregs, show comparatively reduced survival, elevated expression of activation markers, and efficient trafficking to nonlymphoid tissues (Fig. 2) (Yuan et al. 2014). Preliminary evidence indicates that Treg subsets vary in their degree of dependence on IL-2R-mediated signals, although the exact nature of this variation is still uncertain. In a mouse model with attenuated IL-2Rβ signaling, which possessed a grossly normal Treg compartment, a highly activated subset of tissue-resident eTregs bearing the surface marker Klrg1 failed to develop (Cheng et al. 2012). However, the functionally immature Tregs from *Il2*$^{-/-}$ or *Il2ra*$^{-/-}$ mice are skewed toward highly activated phenotypes, a fact that may suggest that cTreg survival is more reliant on IL-2 than that of eTregs (Campbell and Koch 2011). Resolving these multiple sites of action will require an understanding of Treg heterogeneity on a scale more refined than the cTreg/eTreg dichotomy. Accordingly, the search for cell-surface molecules capable of delineating functionally specialized cTreg and eTreg subsets is ongoing (Toomer et al. 2016).

TRANSFORMING GROWTH FACTOR β

The TGF-β signaling pathway provides a versatile mechanism to coordinate essential cellular functions, including proliferation, differentiation, and morphogenesis (Massague 2012), and is key to shaping immune responses and carcinogenesis. The three mammalian isoforms (TGF-β1, TGF-β2, and TGF-β3) are constitutively expressed across many tissues, with cells of the immune system predominantly expressing TGF-β1 (Prud'homme 2007). Among the major downstream targets of TGF-β signal transduction are the Smad family of transcription factors, which interact with each other and additional cofactors to form a diverse array of DNA-binding complexes (Massague et al. 2005).

Consistent with this intrinsic complexity, TGF-β can exert a vast range of proinflammatory and anti-inflammatory effects on many types of immune cells (Prud'homme 2007; Li and Flavell 2008; Travis and Sheppard 2014). Yet remarkably, the *Tgfb1*$^{-/-}$ mouse was found to experience aggressive, multiorgan autoimmunity leading to premature death at 3–4 weeks of age (Kulkarni and Karlsson 1993), a phenotype strikingly similar to that of the *Foxp3* KO mouse. This observation suggested that TGF-β might exert its influence, in part, through the Treg compartment. In keeping with this idea, CD4$^+$CD25$^+$ T cells were found to express high levels of surface-bound TGF-β, and secreted this cytokine when activated (Nakamura et al. 2001; Green et al. 2003). Application of TGF-β-blocking antibodies or chemicals led to impairment of Treg-suppressive function (Aoki et al. 2005). Subsequently, a mouse model harboring a T-cell-specific deletion of TGF-β receptor II reproduced the lethal autoimmune phenotype, confirming that the central pathologic insult in this model is uncontrolled activation of self-reactive Tconv cells (Marie et al. 2006). This autoimmunity may be partially explained by the fact that TGF-β secreted from Tregs acts as a suppressive mediator in some models of inflammation (Schmidt et al. 2012). However, one study indicated that Tregs in the *Tgfb1*$^{-/-}$ mouse were diminished in number and had reduced expression of Foxp3. Following adoptive transfer into lymphopenic mice, both wild-type and *Tgfb1*$^{-/-}$ Tregs showed lower Foxp3 expression in hosts treated with TGF-β blocking antibody. Given that no phenotypic deficits were observed in CD4$^+$CD25$^+$ thymocytes, these findings suggest that TGF-β also makes a direct contribution to Treg maintenance and lineage commitment in the periphery (Fig. 2) (Marie et al. 2005). Notably, other studies do support a role for TGF-β in Treg thymic development. A T-cell-specific deletion, targeting the murine TGF-β receptor I, was found to temporarily block the appearance of CD4$^+$CD25$^+$Foxp3$^+$ thymocytes during the first week of postnatal life (Liu et al. 2008). Another study found evidence that TGF-β supports Treg development indirectly by curtailing negative selection of thymocyte precursors (Fig. 1) (Ouyang et al. 2010). The existence of new, more precise models of targeted genetic

deletion may help clarify the nature and extent of these TGF-β-mediated developmental effects (Chen and Konkel 2015).

As outlined previously, TGF-β is integral to the generation of Tregs from naïve Tconv precursors in vitro (iTregs) and in vivo (pTregs), an activity that has major physiological relevance within the intestinal microenvironment. The dual ability of this cytokine to provoke Th17 differentiation may be partly concentration-dependent, with larger amounts of TGF-β favoring Foxp3 induction (Omenetti and Pizarro 2015). Even at high levels, however, TGF-β alone cannot program the complete Treg transcriptional and functional profile. In addition to other cytokines (e.g., IL-2), the signals influencing Treg differentiation include retinoic acid, TCR affinity, dose and administration route of peptide antigen, and interactions with antigen-presenting cells (Povoleri et al. 2013; Yadav et al. 2013). As compared to thymic Tregs, iTregs boast a distinct pattern of epigenetic modifications at the *Foxp3* locus. In particular, the regulatory region known as conserved noncoding sequence 1 (CNS1) contains binding sites for the TGF-β downstream mediator Smad3, which acts cooperatively with other transcriptional regulators to induce Foxp3 expression (Tone et al. 2008). Deletion of CNS1 severely and selectively impairs iTreg differentiation, although it is dispensable for Foxp3 induction and Treg maturation in the thymus (Zheng et al. 2010; Josefowicz et al. 2012b).

INTERLEUKIN-7/THYMIC STROMAL LYMPHOPOIETIN

A critical factor for survival and maturation of T and B lymphocytes, IL-7 signals through a dimeric receptor consisting of the IL-7Rα chain (CD127) and the γ$_c$ (Fry and Mackall 2005; El Kassar and Gress 2010). IL-7Rα also forms one component of the dimeric receptor for the cytokine TSLP, a distant paralog of IL-7, which also exerts wide-ranging effects on the immune system (Ziegler et al. 2013). Unlike most cytokines, which are secreted by immune cells, IL-7 is produced mainly by nonlymphoid cells within lymphoid organs such as bone marrow stroma and thymic epithelium (Jiang et al. 2005). $Il7^{-/-}$ (von Freeden-Jeffry et al. 1995) and $Il7r^{-/-}$ (Peschon et al. 1994) mice exhibit lymphoid hypoplasia with deficiencies in T- and B-cell compartments, $\gamma\delta$ T-cell and natural killer T (NKT)-cell lineages (Maki et al. 1996; Boesteanu et al. 1997), and mutational loss of either IL-7Rα or the γ_c causes severe combined immunodeficiency (SCID) in man (Puel et al. 1998). IL-7R regulates survival and expansion of double-negative thymocytes, and is down-regulated during the double-positive stage (Alpdogan and van den Brink 2005; Fry and Mackall 2005; Xiong et al. 2013). Re-expressed in the single-positive stage, it is retained on most mature Tconv cells and serves as an important signal governing peripheral homeostasis (Alpdogan and van den Brink 2005; Bradley et al. 2005; Fry and Mackall 2005; Mackall et al. 2011).

Despite its general importance for T-cell maturation in the thymus, evidence suggests that IL-7 does not play a fundamental role in the differentiation of Tregs. In vitro, IL-7 triggers pSTAT5 phosphorylation (albeit to a lesser extent than IL-2) and supports maturation of $CD4^+CD25^+Foxp3^-$ thymic Treg precursors (Vang et al. 2008), but this contribution is less clear in vivo. Although a block in thymic Treg development has been reported in $Il7r^{-/-}$ mice, interpretation of this finding is complicated by generalized defects in T-cell production (Bayer et al. 2008; Mazzucchelli et al. 2008). $CD4^+Foxp3^+$ cells are virtually absent in mice doubly deficient in IL-2Rβ and IL-7Rα, indicating a developmental impairment more severe than that produced by IL-2/IL-2R deficiency alone (Bayer et al. 2008). Despite these findings, selective thymic reconstitution in $Il2rb/Il7r$ double KOs showed that restoration of IL-2Rβ signaling alone can rescue Treg production, completely compensating for a lack of IL-7Rα (Bayer et al. 2008). Thus, IL-7Rα is dispensable for the generation of a functionally mature Treg compartment. It is possible that IL-7 provides a contributory signal to support survival of thymic Tregs during the CD4 single-positive precursor stage, but discerning where and how this effect might operate will require further investigation (Fig. 1).

A number of contradictory results have been obtained regarding the effects of IL-7 on Tregs in the periphery. Mature Tregs have traditionally been considered to express uniformly low levels of CD127, which should render them largely insensitive to IL-7-mediated effects (Liu et al. 2006; Seddiki et al. 2006). In keeping with this idea, mouse studies have found IL-7 signaling to be dispensable for survival, proliferation, phenotypic stability, and suppressive capacity in the peripheral Treg compartment (Peffault de Latour et al. 2006; Mazzucchelli et al. 2008). In vivo experiments using IL-7-transgenic mice and exogenous cytokine administration show that Tregs in the periphery can proliferate in response to IL-7 when it is present at elevated levels (Simonetta et al. 2012). However, this pathway does not appear to play a major role in overall Treg homeostasis under normal physiological conditions. One study examining lymphopenia-driven proliferation of $CD4^+$ T cells found that Tconv expansion was closely tied to IL-7 levels, whereas expansion of Tregs was instead governed by IL-2 (Le Campion et al. 2012). There is some evidence to suggest an IL-7-mediated contribution to Treg suppression. One study evaluating a specialized population of immunosuppressive dendritic cells in the context of diabetes onset showed that dendritic cell (DC)-produced IL-7 acted as a survival factor for $CD4^+CD25^+$ Tregs in vitro, as well as in vivo following adoptive transfer into nonobese diabetic mice (Harnaha et al. 2006). A possible mechanism behind this effect is suggested by another study, which reported that IL-7 enabled Tregs in culture to maintain CD25 expression, potentiating their ability to respond efficiently to IL-2. Adoptive transfers into IL-7-deficient or IL-7-overexpressing mice showed that this effect was maintained in vivo as well (Simonetta et al. 2014).

It appears that IL-7 may exert a more pronounced regulatory effect on specialized subpopulations of Tregs than on the cell lineage as a whole. For example, this cytokine has been reported as a positive regulator of Treg survival and function in the setting of cutaneous immunosuppression (Fig. 2). Using a mouse model of inducible self-antigen expression in skin, one study reported that IL-7, rather than IL-2, was indispensable for the maintenance of highly activated cutaneous Foxp3$^+$ Tregs after self-antigen expression was turned off (Gratz et al. 2013). Another study examining Tregs in murine lymph nodes revealed that expression of the activation markers ICOS and CD103 defined a population of Tregs that express high levels of IL-7Rα (Simonetta et al. 2010). Tregs were found to up-regulate IL-7Rα during activation both in vitro and in vivo, suggesting an augmented responsiveness to IL-7-mediated signals (Simonetta et al. 2010). Both studies noted a CD127^{hi} phenotype in skin-resident Tregs characterized ex vivo, strengthening the idea that low expression of this receptor is not an inherent aspect of the Treg functional program (Simonetta et al. 2010; Gratz et al. 2013).

Emerging evidence suggests that IL-7Rα signaling through epithelium-derived TSLP is another important contributor to cutaneous Treg maintenance and function (Fig. 2). In one mouse model, cutaneous inflammation was initiated by a tamoxifen-inducible, tissue-specific deletion of the chromatin remodeler Mi-2β in the basal epidermis. This disruption of epigenetic regulation caused production of numerous inflammatory mediators and progression to a systemic autoimmune response, although the animals recovered completely (Kashiwagi et al. 2017). In this proinflammatory context, TSLP drove the proliferation of cutaneous Tregs as well as their differentiation into eTreg phenotypes, with corresponding up-regulation of transcripts related to activation, epithelial localization, and immunosuppression. Bone marrow chimera experiments showed that adoptively transferred Tregs with a deletion of *Crlf2* (which encodes the second "TSLPR" component of the heterodimeric TSLP receptor) were impaired in their ability to expand in skin following Mi-2β loss. Notably, in mice lacking TSLP-responsive Tregs, the progression to systemic autoimmunity went unchecked and resulted in a lethal phenotype (Kashiwagi et al. 2017). In another experiment, topical administration of a vitamin D3 analog to mice caused cutaneous Tregs to proliferate in a TSLP-dependent manner. This treatment reduced the incidence of autoimmune diabetes in nonobese diabetic mice, and ameliorated clinical symptoms in experimental autoimmune encephalomyelitis-induced mice (Leichner et al. 2017). These results suggest that IL-7Rα-mediated activation and expansion of cutaneous Tregs can influence systemic immune homeostasis, an intriguing possibility that may be exploited by future therapeutic approaches.

INTERLEUKIN-15

The IL-15 receptor (IL-15R) shares two of its three subunits in common with the IL-2R, comprising a high-affinity IL-15Rα chain, CD122, and the γ$_c$ (Olsen et al. 2007). However, this structural similarity belies a vast divergence of effects in vivo, as IL-15 is highly pleiotropic and influences a much broader array of immune cell types than IL-2. *Il15*$^{-/-}$ and *Il15ra*$^{-/-}$ mice exhibit numerical reductions in various innate and adaptive cell populations, including CD8$^+$ T cells, natural killer (NK) cells, NKT cells, and γδ intraepithelial lymphocytes. In addition to acting as a prominent lymphocyte survival factor (Lodolce et al. 2002; Alpdogan and van den Brink 2005), IL-15 provides a specific and potent stimulus for memory-phenotype (CD44^{hi}) CD8$^+$ T cells in vivo (Zhang et al. 1998), and appears to regulate homeostasis and acquisition of memory functions in naïve CD4$^+$ T cells (Chen et al. 2014). IL-15 signaling also promotes antigen presentation and production of type I cytokines (IL-12 and interferon γ [IFN-γ]) by dendritic cells and macrophages (Ohteki et al. 2001), and potentiates NK cell proliferation and cytotoxicity (Becknell and Caligiuri 2005).

As for IL-7, IL-15 signaling promotes STAT5 phosphorylation, Foxp3 expression, and maturation in CD4$^+$CD25^{hi} single-positive thymocytes cultured in vitro, albeit to a lesser extent than IL-2 (Lio and Hsieh 2008; Vang et al. 2008). Nevertheless, an IL-15-mediated contribution to thymic Treg development has been challenging to identify in vivo. The similarity in clinical phenotype between CD25-deficient and CD122-deficient mice shows that IL-15 cannot independently support development of a functional Treg compartment (Sakaguchi et al. 2008). IL-15-deficient mice have nearly normal numbers of Tregs, suggesting that this pathway is dispensable for their generation (Burchill et al. 2007b). Nevertheless, mice deficient in both IL-2 and IL-15 signaling pathways have a more profound deficit in Treg numbers than those lacking IL-2 alone (Burchill et al. 2007b; Sakaguchi et al. 2008). This finding suggests that IL-15 can contribute to Treg survival through IL-2Rβ/γc signaling when IL-2 is absent, although it evidently cannot rescue Treg function because the clinical course of lethal autoimmunity is unchanged. New mouse models for mapping the fates of thymocyte lineages in vivo suggest that IL-15 may make an ancillary contribution to Treg development. One such model featured deletion of the TCR signaling mediator Zap70, which abrogated thymic production of T cells. Reconstitution of T-cell development using a tetracycline-inducible Zap70 transgene allowed de novo–generated Tregs to be selectively analyzed by eliminating any mature Foxp3$^+$ cells of extra-thymic origin (Marshall et al. 2014). This approach corroborated the generally accepted model that Tregs originate from CD4$^+$CD25^{hi} single-positive thymocytes that up-regulate Foxp3 expression in response to IL-2. However, among CD4 single-positive cells, a second lineage of CD25$^-$Foxp3$^+$ precursors were found to be capable of developing efficiently into Tregs. Survival and maturation of these precursors was dependent on IL-15 produced by the thymic stroma, indicating that this cytokine may support an alternative pathway of Treg differentiation (Fig. 1) (Marshall et al. 2014).

Although the possible influence of IL-15 on mature Tregs in the periphery is only beginning to be explored, emerging evidence suggests that this cytokine may potentiate Treg-suppressive function in certain specialized tissue environments. Using an explant culture system, it was reported that resident Tregs in human skin can be selectively expanded by treatment with IL-15 (Fig. 2) (Clark and Kupper 2007). One recent experiment suggests that IL-15 may contribute to immune homeostasis in the gut by regulating the equilibrium between Treg and Th17 differ-

entiation. Purified Tregs that were adoptively transferred into an IL-15-deficient RAG^{KO} mouse model showed skewing toward a Th17 phenotype, with accelerated loss of Foxp3 expression and up-regulation of RORγt. This led to acquisition of Th17 effector functions by the transferred cells, fueling aggressive colitis in the recipient hosts (Fig. 2) (Tosiek et al. 2016). These findings highlight the possibility of targeting the IL-15 signaling axis therapeutically, to drive expansion or differentiation of tissue-resident Tregs.

INTERLEUKIN-33

IL-33 is constitutively expressed at high levels in epithelial cells, lymphoid cells, and stromal cells of both humans and mice, and is also up-regulated in response to inflammatory stimuli (Cayrol and Girard 2014). A member of the IL-1 family of cytokines, IL-33, signals by binding ST2, a member of the Toll-like/IL-1-receptor superfamily. The resulting dimer then forms a complex with the ubiquitously expressed receptor component IL-1R accessory protein, whose cytosolic domain activates MyD88 signaling to trigger production of inflammatory mediators (Miller 2011). Given its abundance in epithelium, IL-33 is believed to function as an "alarmin," rapidly activating the innate immune system when tissue is damaged by trauma or infectious insults (Chan et al. 2012). IL-33 expressed under homeostatic conditions is retained in the cell nucleus, which may serve to increase the propagation speed of this "danger" signal by allowing the cytokine to be released preformed from necrotic epithelial cells (Cayrol and Girard 2014). IL-33 is unique among IL-1 family cytokines in promoting Th2-polarized immune responses, suggesting a conserved evolutionary function in stimulating antiparasite immunity at mucosal surfaces (Miller 2011). Current evidence suggests that IL-33 is an important contributor to Th2-driven allergic diseases such as asthma (Borish and Steinke 2011; Cayrol and Girard 2014; Saluja et al. 2015) and atopic dermatitis (Cevikbas and Steinhoff 2012), as well as rheumatoid arthritis, psoriatic arthritis, and systemic lupus erythematosus

(Miller 2011; Duan et al. 2013; Macedo et al. 2016).

As discussed previously, Tregs are known to acquire functional adaptations and unique transcriptional profiles that support efficient suppressive activity in nonlymphoid tissue sites (Feuerer et al. 2010; Campbell and Koch 2011). IL-33 has recently been identified as a candidate cytokine that appears to support these tissue-specific adaptations. Despite its established role as a mediator of Th2-mediated inflammation at environmental interfaces, IL-33 seems to play a dichotomous role in inflammatory bowel disease. Although levels of IL-33 and ST2 are increased in the intestines and serum of inflammatory bowel disease (IBD) patients, experimental findings suggest that this signaling axis also supports epithelial repair and resolution of inflammatory injury (Pastorelli et al. 2011; Duan et al. 2012; Grobeta et al. 2012; Sedhom et al. 2013). The recently discovered capacity of IL-33 to promote Treg activity in the gut may clarify this apparent contradiction. ST2 is preferentially expressed on colonic Tregs from mice, and signaling through this receptor promotes survival, proliferation, and suppressive activity of these Tregs both in vitro and in vivo. In addition, IL-33 was found to enhance TGF-β-mediated conversion of naïve T cells into pTregs (Fig. 2) (Schiering et al. 2014).

A similar role for IL-33 has also been characterized in visceral adipose tissue (VAT). VAT-resident Tregs are a specialized population of considerable clinical importance, mediating the link between obesity-induced inflammation and its associated disease outcomes, particularly insulin resistance leading to type 2 diabetes (Feuerer et al. 2009). ST2 was found to be very highly expressed in murine Tregs from visceral adipose tissue compared to those from lymphoid organs. Moreover, this expression was specifically enriched among Tregs with a highly activated phenotype (Vasanthakumar et al. 2015). ST2-deficient mice exhibited clinical evidence of insulin resistance, drastically reduced Treg frequency in the VAT, and impaired maintenance of Treg functional and transcriptional identity. Notably, in vivo administration of IL-33 was sufficient to drive proliferation of VAT Tregs in normal mice, and this treatment also rescued the numerical deficit of VAT Tregs in obese mice with corresponding improvements in glucose tolerance (Fig. 2) (Vasanthakumar et al. 2015).

CONCLUDING REMARKS

As illustrated by the preceding examples, we have a reasonable understanding of how several cytokines shape the Treg compartment. IL-2 is a critical cytokine for Treg development while also importantly influencing Treg homeostasis, stability, and function. TGF-β provides critical signaling for the development of pTregs. Emerging data also support the notion that IL-7, TSLP, IL-15, and IL-33 help to shape the activity of Tregs in various nonlymphoid tissue sites. However, in some other instances, the search for causal links between a particular cytokine signal and certain aspects of Treg development and function has yielded contradictory results. In particular, the roles of IL-7, IL-15, and TGF-β in the development and peripheral lymphoid homeostasis of thymic Tregs are less clear. Factors that contribute to these inconsistencies include functional redundancy among cytokines and their receptor subunits, the capacity of many cytokines to exert pleiotropic effects, and experimental variables such as choice of genetic model, inflammatory milieu, or tissue site.

Therapeutic applications of these cytokines to boost Tregs and promote immune tolerance must take into consideration such redundancy and pleiotropy. The promising clinical results using low-dose IL-2 to enhance the Treg compartment are largely because of our current understanding that many critical aspects of Tregs are supported by low levels of IL-2R signaling, whereas T effector and memory cells require more extensive IL-2R signaling. As such, low-dose IL-2 decreases the pleotropic activity and minimizes nonspecific toxicity of IL-2. Nevertheless, many essential parameters such as dose–response relationships, optimal administration schedule, and duration of treatment remain undefined for low-dose IL-2 (Klatzmann and Abbas 2015).

Cite this article as *Cold Spring Harb Perspect Biol* doi: 10.1101/cshperspect.a028597

Therapies that exploit the Treg-promoting activities of IL-7, TSLP, IL-15, and IL-33 must also be devised to minimize their pleiotropic effects and render them selective toward Tregs. Unlike IL-2, we do not know whether signal thresholds for these cytokines might be favorably directed toward Tregs. Thus, a critical investigative frontier for this group of cytokines may be finding methods to deliver them with precision to minimize off-target effects. The evidence presented here suggests that even highly pleiotropic cytokines can exert defined and predictable influences on a Treg subpopulation specialized for a particular tissue compartment. Thus, rather than taking aim at the Treg pool in its entirety, it may be far more tractable to work toward precision delivery of an immunomodulatory agent to a peripheral anatomical site to engage one or more specialized tissue-resident Treg subsets with well-characterized signaling requirements. Accordingly, it appears that future progress in cytokine-based therapy will occur on a much finer scale.

ACKNOWLEDGMENTS

Our research is supported by grants from the National Institutes of Health (R01 DK093866, R01 AI055815, F31 AI124629), the American Diabetes Association (1-15-BS-125), and the Diabetes Institute Research Foundation. The authors declare no competing interests.

REFERENCES

Alpdogan O, van den Brink MR. 2005. IL-7 and IL-15: Therapeutic cytokines for immunodeficiency. *Trends Immunol* **26:** 56–64.

Aoki CA, Borchers AT, Li M, Flavell RA, Bowlus CL, Ansari AA, Gershwin ME. 2005. Transforming growth factor β (TGF-β) and autoimmunity. *Autoimmun Rev* **4:** 450–459.

Aoyama A, Klarin D, Yamada Y, Boskovic S, Nadazdin O, Kawai K, Schoenfeld D, Madsen JC, Cosimi AB, Benichou G, et al. 2012. Low-dose IL-2 for In vivo expansion of CD4$^+$ and CD8$^+$ regulatory T cells in nonhuman primates. *Am J Transplant* **12:** 2532–2537.

Asano M, Toda M, Sakaguchi N, Sakaguchi S. 1996. Autoimmune disease as a consequence of developmental abnormality of a T cell subpopulation. *J Exp Med* **184:** 387–396.

Barbi J, Pardoll D, Pan F. 2014. Treg functional stability and its responsiveness to the microenvironment. *Immunol Rev* **259:** 115–139.

Barron L, Dooms H, Hoyer KK, Kuswanto W, Hofmann J, O'Gorman WE, Abbas AK. 2010. Cutting edge: Mechanisms of IL-2-dependent maintenance of functional regulatory T cells. *J Immunol* **185:** 6426–6430.

Bayer AL, Yu A, Adeegbe D, Malek TR. 2005. Essential role for interleukin-2 for CD4$^+$CD25$^+$ T regulatory cell development during the neonatal period. *J Exp Med* **201:** 769–777.

Bayer AL, Yu A, Malek TR. 2007. Function of the IL-2R for thymic and peripheral CD4$^+$CD25$^+$ Foxp3$^+$ T regulatory cells. *J Immunol* **178:** 4062–4071.

Bayer AL, Lee JY, de la Barrera A, Surh CD, Malek TR. 2008. A function for IL-7R for CD4$^+$CD25$^+$Foxp3$^+$ T regulatory cells. *J Immunol* **181:** 225–234.

Becknell B, Caligiuri MA. 2005. Interleukin-2, interleukin-15, and their roles in human natural killer cells. *Adv Immunol* **86:** 209–239.

Bilate AM, Lafaille JJ. 2012. Induced CD4$^+$Foxp3$^+$ regulatory T cells in immune tolerance. *Annu Rev Immunol* **30:** 733–758.

Boesteanu A, Silva AD, Nakajima H, Leonard WJ, Peschon JJ, Joyce S. 1997. Distinct roles for signals relayed through the common cytokine receptor γ chain and interleukin 7 receptor α chain in natural T cell development. *J Exp Med* **186:** 331–336.

Borish L, Steinke JW. 2011. Interleukin-33 in asthma: How big of a role does it play? *Curr Allergy Asthma Rep* **11:** 7–11.

Bradley LM, Haynes L, Swain SL. 2005. IL-7: Maintaining T-cell memory and achieving homeostasis. *Trends Immunol* **26:** 172–176.

Burchill MA, Yang J, Vang KB, Farrar MA. 2007a. Interleukin-2 receptor signaling in regulatory T cell development and homeostasis. *Immunol Lett* **114:** 1–8.

Burchill MA, Yang J, Vogtenhuber C, Blazar BR, Farrar MA. 2007b. IL-2 receptor β-dependent STAT5 activation is required for the development of Foxp3$^+$ regulatory T cells. *J Immunol* **178:** 280–290.

Burchill MA, Yang J, Vang KB, Moon JJ, Chu HH, Lio CW, Vegoe AL, Hsieh CS, Jenkins MK, Farrar MA. 2008. Linked T cell receptor and cytokine signaling govern the development of the regulatory T cell repertoire. *Immunity* **28:** 112–121.

Campbell DJ, Koch MA. 2011. Phenotypic and functional specialization of FOXP3$^+$ regulatory T cells. *Nat Rev Immunol* **11:** 119–130.

Castela E, Le Duff F, Butori C, Ticchioni M, Hofman P, Bahadoran P, Lacour JP, Passeron T. 2014. Effects of low-dose recombinant interleukin 2 to promote T-regulatory cells in alopecia areata. *JAMA Dermatol* **150:** 748–751.

Cayrol C, Girard JP. 2014. IL-33: An alarmin cytokine with crucial roles in innate immunity, inflammation and allergy. *Curr Opin Immunol* **31:** 31–37.

Cevikbas F, Steinhoff M. 2012. IL-33: A novel danger signal system in atopic dermatitis. *J Invest Dermatol* **132:** 1326–1329.

Chan JK, Roth J, Oppenheim JJ, Tracey KJ, Vogl T, Feldmann M, Horwood N, Nanchahal J. 2012. Alarmins: Awaiting a clinical response. *J Clin Invest* **122:** 2711–2719.

Chen W, Konkel JE. 2015. Development of thymic Foxp3⁺ regulatory T cells: TGF-β matters. *Eur J Immunol* **45:** 958–965.

Chen W, Jin W, Hardegen N, Lei KJ, Li L, Marinos N, McGrady G, Wahl SM. 2003. Conversion of peripheral CD4⁺CD25⁻ naive T cells to CD4⁺CD25⁺ regulatory T cells by TGF-β induction of transcription factor Foxp3. *J Exp Med* **198:** 1875–1886.

Chen XL, Bobbala D, Cepero Donates Y, Mayhue M, Ilangumaran S, Ramanathan S. 2014. IL-15 trans-presentation regulates homeostasis of CD4⁺ T lymphocytes. *Cell Mol Immunol* **11:** 387–397.

Cheng G, Yu A, Malek TR. 2011. T-cell tolerance and the multi-functional role of IL-2R signaling in T-regulatory cells. *Immunol Rev* **241:** 63–76.

Cheng G, Yuan X, Tsai MS, Podack ER, Yu A, Malek TR. 2012. IL-2 receptor signaling is essential for the development of Klrg1⁺ terminally differentiated T regulatory cells. *J Immunol* **189:** 1780–1791.

Cheng G, Yu A, Dee MJ, Malek TR. 2013. IL-2R signaling is essential for functional maturation of regulatory T cells during thymic development. *J Immunol* **190:** 1567–1575.

Chinen T, Kannan AK, Levine AG, Fan X, Klein U, Zheng Y, Gasteiger G, Feng Y, Fontenot JD, Rudensky AY. 2016. An essential role for the IL-2 receptor in Treg cell function. *Nat Immunol* **17:** 1322–1333.

Clark RA, Kupper TS. 2007. IL-15 and dermal fibroblasts induce proliferation of natural regulatory T cells isolated from human skin. *Blood* **109:** 194–202.

Duan L, Chen J, Zhang H, Yang H, Zhu P, Xiong A, Xia Q, Zheng F, Tan Z, Gong F, et al. 2012. Interleukin-33 ameliorates experimental colitis through promoting Th2/Foxp3⁺ regulatory T-cell responses in mice. *Mol Med* **18:** 753–761.

Duan L, Chen J, Gong F, Shi G. 2013. The role of IL-33 in rheumatic diseases. *Clin Dev Immunol* **2013:** 924363.

El Kassar N, Gress RE. 2010. An overview of IL-7 biology and its use in immunotherapy. *J Immunotoxicol* **7:** 1–7.

Fantini MC, Becker C, Monteleone G, Pallone F, Galle PR, Neurath MF. 2004. Cutting edge: TGF-β induces a regulatory phenotype in CD4⁺CD25⁻ T cells through Foxp3 induction and down-regulation of Smad7. *J Immunol* **172:** 5149–5153.

Feng Y, Arvey A, Chinen T, van der Veeken J, Gasteiger G, Rudensky AY. 2014. Control of the inheritance of regulatory T cell identity by a *cis* element in the *Foxp3* locus. *Cell* **158:** 749–763.

Feuerer M, Herrero L, Cipolletta D, Naaz A, Wong J, Nayer A, Lee J, Goldfine AB, Benoist C, Shoelson S, et al. 2009. Lean, but not obese, fat is enriched for a unique population of regulatory T cells that affect metabolic parameters. *Nat Med* **15:** 930–939.

Feuerer M, Hill JA, Kretschmer K, von Boehmer H, Mathis D, Benoist C. 2010. Genomic definition of multiple ex vivo regulatory T cell subphenotypes. *Proc Natl Acad Sci* **107:** 5919–5924.

Fontenot JD, Dooley JL, Farr AG, Rudensky AY. 2005a. Developmental regulation of Foxp3 expression during ontogeny. *J Exp Med* **202:** 901–906.

Fontenot JD, Rasmussen JP, Gavin MA, Rudensky AY. 2005b. A function for interleukin 2 in Foxp3-expressing regulatory T cells. *Nat Immunol* **6:** 1142–1151.

Fry TJ, Mackall CL. 2005. The many faces of IL-7: From lymphopoiesis to peripheral T cell maintenance. *J Immunol* **174:** 6571–6576.

Fujimoto M, Nakano M, Terabe F, Kawahata H, Ohkawara T, Han Y, Ripley B, Serada S, Nishikawa T, Kimura A, et al. 2011. The influence of excessive IL-6 production in vivo on the development and function of Foxp3⁺ regulatory T cells. *J Immunol* **186:** 32–40.

Grant CR, Liberal R, Mieli-Vergani G, Vergani D, Longhi MS. 2015. Regulatory T-cells in autoimmune diseases: Challenges, controversies and—yet—unanswered questions. *Autoimmun Rev* **14:** 105–116.

Gratz IK, Truong HA, Yang SH, Maurano MM, Lee K, Abbas AK, Rosenblum MD. 2013. Cutting edge: Memory regulatory T cells require IL-7 and not IL-2 for their maintenance in peripheral tissues. *J Immunol* **190:** 4483–4487.

Green EA, Gorelik L, McGregor CM, Tran EH, Flavell RA. 2003. CD4⁺CD25⁺ T regulatory cells control anti-islet CD8⁺ T cells through TGF-β-TGF-β receptor interactions in type 1 diabetes. *Proc Natl Acad Sci* **100:** 10878–10883.

Grobeta P, Doser K, Falk W, Obermeier F, Hofmann C. 2012. IL-33 attenuates development and perpetuation of chronic intestinal inflammation. *Inflamm Bowel Dis* **18:** 1900–1909.

Haribhai D, Lin W, Edwards B, Ziegelbauer J, Salzman NH, Carlson MR, Li SH, Simpson PM, Chatila TA, Williams CB. 2009. A central role for induced regulatory T cells in tolerance induction in experimental colitis. *J Immunol* **182:** 3461–3468.

Harnaha J, Machen J, Wright M, Lakomy R, Styche A, Trucco M, Makaroun S, Giannoukakis N. 2006. Interleukin-7 is a survival factor for CD4⁺ CD25⁺ T-cells and is expressed by diabetes-suppressive dendritic cells. *Diabetes* **55:** 158–170.

Harrison OJ, Powrie FM. 2013. Regulatory T cells and immune tolerance in the intestine. *Cold Spring Harb Perspect Biol* **5:** a018341.

Hartemann A, Bensimon G, Payan CA, Jacqueminet S, Bourron O, Nicolas N, Fonfrede M, Rosenzwajg M, Bernard C, Klatzmann D. 2013. Low-dose interleukin 2 in patients with type 1 diabetes: A phase 1/2 randomised, double-blind, placebo-controlled trial. *Lancet Diabetes Endocrinol* **1:** 295–305.

He J, Zhang X, Wei Y, Sun X, Chen Y, Deng J, Jin Y, Gan Y, Hu X, Jia R, et al. 2016. Low-dose interleukin-2 treatment selectively modulates CD4⁺ T cell subsets in patients with systemic lupus erythematosus. *Nat Med* **22:** 991–993.

Hoechst B, Gamrekelashvili J, Manns MP, Greten TF, Korangy F. 2011. Plasticity of human Th17 cells and iTregs is orchestrated by different subsets of myeloid cells. *Blood* **117:** 6532–6541.

Huehn J, Polansky JK, Hamann A. 2009. Epigenetic control of FOXP3 expression: The key to a stable regulatory T-cell lineage? *Nat Rev Immunol* **9:** 83–89.

Huynh A, DuPage M, Priyadharshini B, Sage PT, Quiros J, Borges CM, Townamchai N, Gerriets VA, Rathmell JC, Sharpe AH, et al. 2015. Control of PI(3) kinase in Treg cells maintains homeostasis and lineage stability. *Nat Immunol* **16:** 188–196.

Jiang Q, Li WQ, Aiello FB, Mazzucchelli R, Asefa B, Khaled AR, Durum SK. 2005. Cell biology of IL-7, a key lympho-trophin. *Cytokine Growth Factor Rev* **16:** 513–533.

Josefowicz SZ, Lu LF, Rudensky AY. 2012a. Regulatory T cells: Mechanisms of differentiation and function. *Annu Rev Immunol* **30:** 531–564.

Josefowicz SZ, Niec RE, Kim HY, Treuting P, Chinen T, Zheng Y, Umetsu DT, Rudensky AY. 2012b. Extrathymi-cally generated regulatory T cells control mucosal T_H2 inflammation. *Nature* **482:** 395–399.

Kanamori M, Nakatsukasa H, Okada M, Lu Q, Yoshimura A. 2016. Induced regulatory T cells: Their development, stability, and applications. *Trends Immunol* **37:** 803–811.

Kashiwagi M, Hosoi J, Lai JF, Brissette J, Ziegler SF, Morgan BA, Georgopoulos K. 2017. Direct control of regulatory T cells by keratinocytes. *Nat Immunol* **18:** 334–343.

Kimura A, Kishimoto T. 2010. IL-6: Regulator of Treg/Th17 balance. *Eur J Immunol* **40:** 1830–1835.

Klatzmann D, Abbas AK. 2015. The promise of low-dose interleukin-2 therapy for autoimmune and inflammato-ry diseases. *Nat Rev Immunol* **15:** 283–294.

Kleinewietfeld M, Hafler DA. 2013. The plasticity of human Treg and Th17 cells and its role in autoimmunity. *Semin Immunol* **25:** 305–312.

Konkel JE, Chen W. 2011. Balancing acts: The role of TGF-β in the mucosal immune system. *Trends Mol Med* **17:** 668–676.

Koreth J, Matsuoka K, Kim HT, McDonough SM, Bindra B, Alyea EPIII, Armand P, Cutler C, Ho VT, Treister NS, et al. 2011. Interleukin-2 and regulatory T cells in graft-versus-host disease. *N Eng J Med* **365:** 2055–2066.

Kosmaczewska A. 2014. Low-dose interleukin-2 therapy: A driver of an imbalance between immune tolerance and autoimmunity. *Int J Mol Sci* **15:** 18574–18592.

Kulkarni AB, Karlsson S. 1993. Transforming growth factor-$β_1$ knockout mice. A mutation in one cytokine gene causes a dramatic inflammatory disease. *Am J Pathol* **143:** 3–9.

Laurence A, Tato CM, Davidson TS, Kanno Y, Chen Z, Yao Z, Blank RB, Meylan F, Siegel R, Hennighausen L, et al. 2007. Interleukin-2 signaling via STAT5 constrains T helper 17 cell generation. *Immunity* **26:** 371–381.

Le Campion A, Pommier A, Delpoux A, Stouvenel L, Auf-fray C, Martin B, Lucas B. 2012. IL-2 and IL-7 determine the homeostatic balance between the regulatory and con-ventional $CD4^+$ T cell compartments during peripheral T cell reconstitution. *J Immunol* **189:** 3339–3346.

Leichner TM, Satake A, Harrison VS, Tanaka Y, Archam-bault AS, Kim BS, Siracusa MC, Leonard WJ, Naji A, Wu GF, et al. 2017. Skin-derived TSLP systemically expands regulatory T cells. *J Autoimmun* doi: 10.1016/jjaut.2017.01.003.

Li MO, Flavell RA. 2008. TGF-β: A master of all T cell trades. *Cell* **134:** 392–404.

Li X, Liang Y, LeBlanc M, Benner C, Zheng Y. 2014. Function of a Foxp3 *cis*-element in protecting regulatory T cell identity. *Cell* **158:** 734–748.

Lio CW, Hsieh CS. 2008. A two-step process for thymic regulatory T cell development. *Immunity* **28:** 100–111.

Littman DR, Rudensky AY. 2010. Th17 and regulatory T cells in mediating and restraining inflammation. *Cell* **140:** 845–858.

Liu W, Putnam AL, Xu-Yu Z, Szot GL, Lee MR, Zhu S, Gottlieb PA, Kapranov P, Gingeras TR, Fazekas de St Groth B, et al. 2006. CD127 expression inversely corre-lates with FoxP3 and suppressive function of human $CD4^+$ T reg cells. *J Exp Med* **203:** 1701–1711.

Liu Y, Zhang P, Li J, Kulkarni AB, Perruche S, Chen W. 2008. A critical function for TGF-β signaling in the develop-ment of natural $CD4^+CD25^+Foxp3^+$ regulatory T cells. *Nat Immunol* **9:** 632–640.

Lodolce JP, Burkett PR, Koka RM, Boone DL, Ma A. 2002. Regulation of lymphoid homeostasis by interleukin-15. *Cytokine Growth Factor Rev* **13:** 429–439.

Macedo RB, Kakehasi AM, Melo de Andrade MV. 2016. IL33 in rheumatoid arthritis: Potential contribution to path-ogenesis. *Rev Bras Reumatol Engl Ed* **56:** 451–457.

Mackall CL, Fry TJ, Gress RE. 2011. Harnessing the biology of IL-7 for therapeutic application. *Nat Rev Immunol* **11:** 330–342.

Maki K, Sunaga S, Komagata Y, Kodaira Y, Mabuchi A, Kar-asuyama H, Yokomuro K, Miyazaki JI, Ikuta K. 1996. Interleukin 7 receptor-deficient mice lack γδ T cells. *Proc Natl Acad Sci* **93:** 7172–7177.

Malek TR. 2008. The biology of interleukin-2. *Annu Rev Immunol* **26:** 453–479.

Malek TR, Bayer AL. 2004. Tolerance, not immunity, cru-cially depends on IL-2. *Nat Rev Immunol* **4:** 665–674.

Malek TR, Porter BO, Codias EK, Scibelli P, Yu A. 2000. Normal lymphoid homeostasis and lack of lethal auto-immunity in mice containing mature T cells with severely impaired IL-2 receptors. *J Immunol* **164:** 2905–2914.

Malek TR, Yu A, Vincek V, Scibelli P, Kong L. 2002. CD4 regulatory T cells prevent lethal autoimmunity in IL-2Rβ-deficient mice. Implications for the nonredundant function of IL-2. *Immunity* **17:** 167–178.

Marie JC, Letterio JJ, Gavin M, Rudensky AY. 2005. TGF-β1 maintains suppressor function and Foxp3 expression in $CD4^+CD25^+$ regulatory T cells. *J Exp Med* **201:** 1061–1067.

Marie JC, Liggitt D, Rudensky AY. 2006. Cellular mecha-nisms of fatal early-onset autoimmunity in mice with the T cell-specific targeting of transforming growth fac-tor-β receptor. *Immunity* **25:** 441–454.

Marshall D, Sinclair C, Tung S, Seddon B. 2014. Differential requirement for IL-2 and IL-15 during bifurcated devel-opment of thymic regulatory T cells. *J Immunol* **193:** 5525–5533.

Massague J. 2012. TGFβ signalling in context. *Nat Rev Mol Cell Biol* **13:** 616–630.

Massague J, Seoane J, Wotton D. 2005. Smad transcription factors. *Genes Dev* **19:** 2783–2810.

Matsuoka K, Koreth J, Kim HT, Bascug G, McDonough S, Kawano Y, Murase K, Cutler C, Ho VT, Alyea EP, et al. 2013. Low-dose interleukin-2 therapy restores regulatory

T cell homeostasis in patients with chronic graft-versus-host disease. *Sci Transl Med* **5:** 179ra143.

Mazzucchelli R, Hixon JA, Spolski R, Chen X, Li WQ, Hall VL, Willette-Brown J, Hurwitz AA, Leonard WJ, Durum SK. 2008. Development of regulatory T cells requires IL-7Rα stimulation by IL-7 or TSLP. *Blood* **112:** 3283–3292.

Miller AM. 2011. Role of IL-33 in inflammation and disease. *J Inflamm (Lond)* **8:** 22.

Morgan DA, Ruscetti FW, Gallo R. 1976. Selective in vitro growth of T lymphocytes from normal human bone marrows. *Science* **193:** 1007–1008.

Nakamura K, Kitani A, Strober W. 2001. Cell contact-dependent immunosuppression by CD4+CD25+ regulatory T cells is mediated by cell surface-bound transforming growth factor β. *J Exp Med* **194:** 629–644.

Ohteki T, Suzue K, Maki C, Ota T, Koyasu S. 2001. Critical role of IL-15-IL-15R for antigen-presenting cell functions in the innate immune response. *Nat Immunol* **2:** 1138–1143.

Olsen SK, Ota N, Kishishita S, Kukimoto-Niino M, Murayama K, Uchiyama H, Toyama M, Terada T, Shirouzu M, Kanagawa O, et al. 2007. Crystal structure of the interleukin-15·interleukin-15 receptor α complex: Insights into *trans* and *cis* presentation. *J Biol Chem* **282:** 37191–37204.

Omenetti S, Pizarro TT. 2015. The Treg/Th17 axis: A dynamic balance regulated by the gut microbiome. *Front Immunol* **6:** 639.

Ouyang W, Beckett O, Ma Q, Li MO. 2010. Transforming growth factor-β signaling curbs thymic negative selection promoting regulatory T cell development. *Immunity* **32:** 642–653.

Pandiyan P, Zheng L, Ishihara S, Reed J, Lenardo MJ. 2007. CD4+CD25+Foxp3+ regulatory T cells induce cytokine deprivation-mediated apoptosis of effector CD4+ T cells. *Nat Immunol* **8:** 1353–1362.

Pastorelli L, De Salvo C, Cominelli MA, Vecchi M, Pizarro TT. 2011. Novel cytokine signaling pathways in inflammatory bowel disease: Insight into the dichotomous functions of IL-33 during chronic intestinal inflammation. *Therap Adv Gastroenterol* **4:** 311–323.

Peffault de Latour R, Dujardin HC, Mishellany F, Burlen-Defranoux O, Zuber J, Marques R, Di Santo J, Cumano A, Vieira P, Bandeira A. 2006. Ontogeny, function, and peripheral homeostasis of regulatory T cells in the absence of interleukin-7. *Blood* **108:** 2300–2306.

Peschon JJ, Morrissey PJ, Grabstein KH, Ramsdell FJ, Maraskovsky E, Gliniak BC, Park LS, Ziegler SF, Williams DE, Ware CB, et al. 1994. Early lymphocyte expansion is severely impaired in interleukin 7 receptor-deficient mice. *J Exp Med* **180:** 1955–1960.

Pierson W, Cauwe B, Policheni A, Schlenner SM, Franckaert D, Berges J, Humblet-Baron S, Schonefeldt S, Herold MJ, Hildeman D, et al. 2013. Antiapoptotic Mcl-1 is critical for the survival and niche-filling capacity of Foxp3+ regulatory T cells. *Nat Immunol* **14:** 959–965.

Povoleri GA, Scotta C, Nova-Lamperti EA, John S, Lombardi G, Afzali B. 2013. Thymic versus induced regulatory T cells—Who regulates the regulators? *Front Immunol* **4:** 169.

Prud'homme GJ. 2007. Pathobiology of transforming growth factor β in cancer, fibrosis and immunologic disease, and therapeutic considerations. *Lab Invest* **87:** 1077–1091.

Puel A, Ziegler SF, Buckley RH, Leonard WJ. 1998. Defective IL7R expression in T−B+NK+ severe combined immunodeficiency. *Nat Genet* **20:** 394–397.

Roifman CM. 2000. Human IL-2 receptor α chain deficiency. *Pediatr Res* **48:** 6–11.

Rubtsov YP, Niec RE, Josefowicz S, Li L, Darce J, Mathis D, Benoist C, Rudensky AY. 2010. Stability of the regulatory T cell lineage in vivo. *Science* **329:** 1667–1671.

Saadoun D, Rosenzwajg M, Joly F, Six A, Carrat F, Thibault V, Sene D, Cacoub P, Klatzmann D. 2011. Regulatory T-cell responses to low-dose interleukin-2 in HCV-induced vasculitis. *N Engl J Med* **365:** 2067–2077.

Sadlack B, Merz H, Schorle H, Schimpl A, Feller AC, Horak I. 1993. Ulcerative colitis-like disease in mice with a disrupted interleukin-2 gene. *Cell* **75:** 253–261.

Sadlack B, Lohler J, Schorle H, Klebb G, Haber H, Sickel E, Noelle RJ, Horak I. 1995. Generalized autoimmune disease in interleukin-2-deficient mice is triggered by an uncontrolled activation and proliferation of CD4+ T cells. *Eur J Immunol* **25:** 3053–3059.

Sakaguchi S, Sakaguchi N, Asano M, Itoh M, Toda M. 1995. Immunologic self-tolerance maintained by activated T cells expressing IL-2 receptor α-chains (CD25). Breakdown of a single mechanism of self-tolerance causes various autoimmune diseases. *J Immunol* **155:** 1151–1164.

Sakaguchi S, Yamaguchi T, Nomura T, Ono M. 2008. Regulatory T cells and immune tolerance. *Cell* **133:** 775–787.

Saluja R, Khan M, Church MK, Maurer M. 2015. The role of IL-33 and mast cells in allergy and inflammation. *Clin Transl Allergy* **5:** 33.

Schiering C, Krausgruber T, Chomka A, Frohlich A, Adelmann K, Wohlfert EA, Pott J, Griseri T, Bollrath J, Hegazy AN, et al. 2014. The alarmin IL-33 promotes regulatory T-cell function in the intestine. *Nature* **513:** 564–568.

Schmidt A, Oberle N, Krammer PH. 2012. Molecular mechanisms of Treg-mediated T cell suppression. *Front Immunol* **3:** 51.

Schmidt A, Eriksson M, Shang MM, Weyd H, Tegner J. 2016. Comparative analysis of protocols to induce human CD4+Foxp3+ regulatory T cells by combinations of IL-2, TGF-β, retinoic acid, rapamycin and butyrate. *PLoS ONE* **11:** e0148474.

Seddiki N, Santner-Nanan B, Martinson J, Zaunders J, Sasson S, Landay A, Solomon M, Selby W, Alexander SI, Nanan R, et al. 2006. Expression of interleukin (IL)-2 and IL-7 receptors discriminates between human regulatory and activated T cells. *J Exp Med* **203:** 1693–1700.

Sedhom MA, Pichery M, Murdoch JR, Foligne B, Ortega N, Normand S, Mertz K, Sanmugalingam D, Brault L, Grandjean T, et al. 2013. Neutralisation of the interleukin-33/ST2 pathway ameliorates experimental colitis through enhancement of mucosal healing in mice. *Gut* **62:** 1714–1723.

Setoguchi R, Hori S, Takahashi T, Sakaguchi S. 2005. Homeostatic maintenance of natural Foxp3+ CD25+ CD4+ regulatory T cells by interleukin (IL)-2 and induction of autoimmune disease by IL-2 neutralization. *J Exp Med* **201:** 723–735.

Shevach EM, Thornton AM. 2014. tTregs, pTregs, and iTregs: Similarities and differences. *Immunol Rev* **259:** 88–102.

Simonetta F, Chiali A, Cordier C, Urrutia A, Girault I, Bloquet S, Tanchot C, Bourgeois C. 2010. Increased CD127 expression on activated FOXP3+CD4+ regulatory T cells. *Eur J Immunol* **40:** 2528–2538.

Simonetta F, Gestermann N, Martinet KZ, Boniotto M, Tissieres P, Seddon B, Bourgeois C. 2012. Interleukin-7 influences FOXP3+CD4+ regulatory T cells peripheral homeostasis. *PLoS ONE* **7:** e36596.

Simonetta F, Gestermann N, Bloquet S, Bourgeois C. 2014. Interleukin-7 optimizes FOXP3+CD4+ regulatory T cells reactivity to interleukin-2 by modulating CD25 expression. *PLoS ONE* **9:** e113314.

Stauber DJ, Debler EW, Horton PA, Smith KA, Wilson IA. 2006. Crystal structure of the IL-2 signaling complex: Paradigm for a heterotrimeric cytokine receptor. *Proc Natl Acad Sci* **103:** 2788–2793.

Suzuki H, Kundig TM, Furlonger C, Wakeham A, Timms E, Matsuyama T, Schmits R, Simard JJ, Ohashi PS, Griesser H, et al. 1995. Deregulated T cell activation and autoimmunity in mice lacking interleukin-2 receptor β. *Science* **268:** 1472–1476.

Tai X, Erman B, Alag A, Mu J, Kimura M, Katz G, Guinter T, McCaughtry T, Etzensperger R, Feigenbaum L, et al. 2013. Foxp3 transcription factor is proapoptotic and lethal to developing regulatory T cells unless counterbalanced by cytokine survival signals. *Immunity* **38:** 1116–1128.

Tanoue T, Atarashi K, Honda K. 2016. Development and maintenance of intestinal regulatory T cells. *Nat Rev Immunol* **16:** 295–309.

Todd JA, Evangelou M, Cutler AJ, Pekalski ML, Walker NM, Stevens HE, Porter L, Smyth DJ, Rainbow DB, Ferreira RC, et al. 2016. Regulatory T cell responses in participants with Type 1 diabetes after a single dose of interleukin-2: A non-randomised, open label, adaptive dose-finding trial. *PLoS Med* **13:** e1002139.

Tone Y, Furuuchi K, Kojima Y, Tykocinski ML, Greene MI, Tone M. 2008. Smad3 and NFAT cooperate to induce Foxp3 expression through its enhancer. *Nat Immunol* **9:** 194–202.

Toomer KH, Yuan X, Yang J, Dee MJ, Yu A, Malek TR. 2016. Developmental progression and interrelationship of central and effector regulatory T cell subsets. *J Immunol* **196:** 3665–3676.

Tosiek MJ, Fiette L, El Daker S, Eberl G, Freitas AA. 2016. IL-15-dependent balance between Foxp3 and RORγt expression impacts inflammatory bowel disease. *Nat Commun* **7:** 10888.

Travis MA, Sheppard D. 2014. TGF-β activation and function in immunity. *Annu Rev Immunol* **32:** 51–82.

Ueno A, Ghosh A, Hung D, Li J, Jijon H. 2015. Th17 plasticity and its changes associated with inflammatory bowel disease. *World J Gastroenterol* **21:** 12283–12295.

Vang KB, Yang J, Mahmud SA, Burchill MA, Vegoe AL, Farrar MA. 2008. IL-2, -7, and -15, but not thymic stromal lymphopoietin, redundantly govern CD4+Foxp3+ regulatory T cell development. *J Immunol* **181:** 3285–3290.

Vasanthakumar A, Moro K, Xin A, Liao Y, Gloury R, Kawamoto S, Fagarasan S, Mielke LA, Afshar-Sterle S, Masters SL, et al. 2015. The transcriptional regulators IRF4, BATF and IL-33 orchestrate development and maintenance of adipose tissue-resident regulatory T cells. *Nat Immunol* **16:** 276–285.

von Freeden-Jeffry U, Vieira P, Lucian LA, McNeil T, Burdach SE, Murray R. 1995. Lymphopenia in interleukin (IL)-7 gene-deleted mice identifies IL-7 as a nonredundant cytokine. *J Exp Med* **181:** 1519–1526.

Walsh PT, Buckler JL, Zhang J, Gelman AE, Dalton NM, Taylor DK, Bensinger SJ, Hancock WW, Turka LA. 2006. PTEN inhibits IL-2 receptor-mediated expansion of CD4+ CD25+ Tregs. *J Clin Invest* **116:** 2521–2531.

Willerford DM, Chen J, Ferry JA, Davidson L, Ma A, Alt FW. 1995. Interleukin-2 receptor α chain regulates the size and content of the peripheral lymphoid compartment. *Immunity* **3:** 521–530.

Xiong J, Parker BL, Dalheimer SL, Yankee TM. 2013. Interleukin-7 supports survival of T-cell receptor-β-expressing CD4− CD8− double-negative thymocytes. *Immunology* **138:** 382–391.

Xu L, Kitani A, Strober W. 2010. Molecular mechanisms regulating TGF-β-induced Foxp3 expression. *Mucosal Immunol* **3:** 230–238.

Yadav M, Stephan S, Bluestone JA. 2013. Peripherally induced Tregs—Role in immune homeostasis and autoimmunity. *Front Immunol* **4:** 232.

Yi Z, Lin WW, Stunz LL, Bishop GA. 2014. The adaptor TRAF3 restrains the lineage determination of thymic regulatory T cells by modulating signaling via the receptor for IL-2. *Nat Immunol* **15:** 866–874.

Yu A, Zhu L, Altman NH, Malek TR. 2009. A low interleukin-2 receptor signaling threshold supports the development and homeostasis of T regulatory cells. *Immunity* **30:** 204–217.

Yu A, Snowhite I, Vendrame F, Rosenzwajg M, Klatzmann D, Pugliese A, Malek TR. 2015. Selective IL-2 responsiveness of regulatory T cells through multiple intrinsic mechanisms supports the development of low-dose IL-2 therapy in Type 1 diabetes. *Diabetes* **64:** 2172–2183.

Yuan X, Cheng G, Malek TR. 2014. The importance of regulatory T-cell heterogeneity in maintaining self-tolerance. *Immunol Rev* **259:** 103–114.

Zhang X, Sun S, Hwang I, Tough DF, Sprent J. 1998. Potent and selective stimulation of memory-phenotype CD8+ T cells in vivo by IL-15. *Immunity* **8:** 591–599.

Zheng SG, Wang J, Wang P, Gray JD, Horwitz DA. 2007. IL-2 is essential for TGF-β to convert naive CD4+CD25− cells to CD25+Foxp3+ regulatory T cells and for expansion of these cells. *J Immunol* **178:** 2018–2027.

Zheng SG, Wang J, Horwitz DA. 2008. Cutting edge: Foxp3+CD4+CD25+ regulatory T cells induced by IL-2 and TGF-β are resistant to Th17 conversion by IL-6. *J Immunol* **180:** 7112–7116.

Zheng Y, Josefowicz S, Chaudhry A, Peng XP, Forbush K, Rudensky AY. 2010. Role of conserved non-coding DNA elements in the Foxp3 gene in regulatory T-cell fate. *Nature* **463:** 808–812.

Ziegler SF, Roan F, Bell BD, Stoklasek TA, Kitajima M, Han H. 2013. The biology of thymic stromal lymphopoietin (TSLP). *Adv Pharmacol* **66:** 129–155.

Cytokine-Mediated Regulation of CD8 T-Cell Responses during Acute and Chronic Viral Infection

Masao Hashimoto, Se Jin Im, Koichi Araki, and Rafi Ahmed

Emory Vaccine Center and Department of Microbiology and Immunology, Emory University School of Medicine, Atlanta, Georgia 30322

Correspondence: rahmed@emory.edu

The common γ-chain cytokines, interleukin (IL)-2, IL-7, and IL-15, regulate critical aspects of antiviral CD8 T-cell responses. During acute infections, IL-2 controls expansion and differentiation of antiviral CD8 T cells, whereas IL-7 and IL-15 are key cytokines to maintain memory CD8 T cells long term in an antigen-independent manner. On the other hand, during chronic infections, in which T-cell exhaustion is established, precise roles of these cytokines in regulation of antiviral CD8 T-cell responses are not well defined. Nonetheless, administration of IL-2, IL-7, or IL-15 can increase function of exhausted CD8 T cells, and thus can be an attractive therapeutic approach. A new subset of stem-cell-like CD8 T cells, which provides a proliferative burst after programmed cell death (PD)-1 therapy, has been recently described during chronic viral infection. Further understanding of cytokine-mediated regulation of this CD8 T-cell subset will improve cytokine therapies to treat chronic infections and cancer in combination with immune checkpoint inhibitors.

CD8 T cells are critical components of adaptive immunity for controlling infections, and diverse sets of cytokines regulate antiviral CD8 T-cell responses (Rochman et al. 2009; Cox et al. 2013). In this review, we focus on the common γ-chain (γc) cytokines, interleukin (IL)-2, IL-7, and IL-15, which are well known to influence various key aspects of antiviral CD8 T-cell responses during acute and chronic infections. Potential therapeutic applications for modulating antiviral CD8 T-cell responses by targeting these cytokines are also discussed.

CD8 T-CELL RESPONSES DURING ACUTE VIRAL INFECTION

Upon infection, naïve CD8 T cells, whose T-cell receptors (TCRs) are specific to a given antigen, are stimulated by signals through TCRs (signal 1), costimulatory molecules (signal 2), and inflammatory cytokines (signal 3). Activated T cells undergo massive clonal expansion and differentiate into effector cells, which express cytotoxic molecules like perforin and granzyme B as well as produce antiviral cytokines such as interferon (IFN)-γ and tumor necrosis factor

(TNF)-α. Effector CD8 T cells show altered expression for chemokine and homing receptors, enabling their migration to peripheral tissues, where they function to lyse infected cells via cytotoxic molecules and inhibit viral replications by secreting effector cytokines, thereby contributing to viral clearance (Williams and Bevan 2007; Kaech and Cui 2012).

After eliminating the virus, most of the effector cells die by apoptosis, but some (5%–10%) survive, and memory cells are formed. The process from naïve to effector to memory CD8 T-cell differentiation includes the emergence of terminal effector (TE) and memory precursor (MP) cells. Both are highly functional effector cells, but MP cells preferentially survive during the contraction phase and differentiate into long-lived memory T cells. Once memory CD8 T cells are generated, they self-renew by homeostatic proliferation via signals from specific cytokines such as IL-7 and IL-15, and are maintained long term in the absence of antigen (>2 years in mice and >25 years in humans) (Lau et al. 1994; Homann et al. 2001; Hammarlund

et al. 2003; Fuertes Marraco et al. 2015). These memory CD8 T cells possess superior ability as compared with naïve CD8 T cells to exert rapid effector functions in response to previously encountered antigens (Barber et al. 2003; Byers et al. 2003; Wherry et al. 2003b).

IL-2

IL-2 is a key cytokine that regulates clonal expansion and effector/memory differentiation of antiviral CD8 T cells during acute infections. IL-2 is predominantly produced by CD4 T cells following antigen stimulation and, to a lesser extent, by other cell types such as CD8 T cells, natural killer (NK) cells, natural killer T (NKT) cells, activated dendritic cells (DCs), and mast cells. Three IL-2R subunits exist: IL-2Rα (CD25), IL-2Rβ (CD122, also known as IL-15Rβ), and IL-2Rγ (CD132, γc) (Fig. 1). Although CD25 is not expressed on resting CD8 T cells, its expression is induced after stimulation via the TCR or with IL-2. CD122 is constitutively expressed on broader cell types than CD25, including memory CD8 T

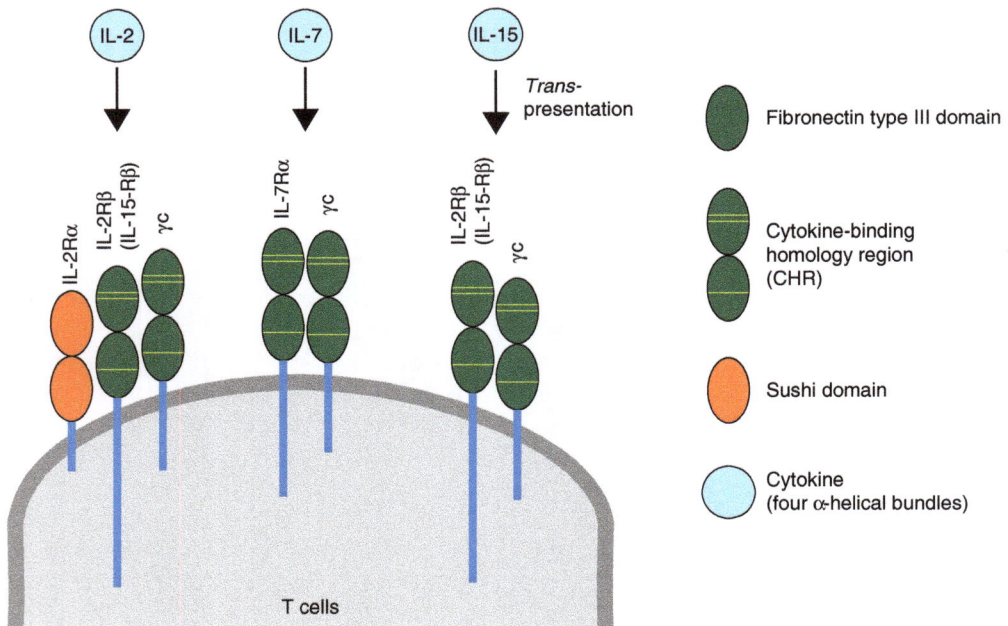

Figure 1. Cytokines (interleukin [IL]-2, IL-7, and IL-15) and their receptors. Each cytokine-binding homology region (CHR), composed of tandem fibronectin type III domains, contains three yellow lines that represent conserved disulfides and a WSXWS motif. γc, Common γ-chain.

Cite this article as *Cold Spring Harb Perspect Biol* doi: 10.1101/cshperspect.a028464

cells, NK cells, and NKT cells. Its expression is also induced on virtually all antigen-activated T cells. CD132 is much less strictly regulated than other IL-2R subunits and is constitutively expressed on virtually all hematopoietic cells and is shared by the receptors for IL-4, IL-7, IL-9, IL-15, and IL-21. IL-2 signals via two specific receptors formed by combinations of three IL-2R subunits: the intermediate-affinity ($K_d \sim 1$ nM) heterodimeric receptor, consisted of the IL-2Rβ and γc, and the high-affinity ($K_d \sim 10$ pM) heterotrimeric receptor, composed of the IL-2Rα, IL-2Rβ, and γc, and both of them signal through interaction of the intracellular domains of IL-2Rβ and γc with Janus kinase (JAK)1 and JAK3, respectively. Intermediate-affinity IL-2 receptors are expressed on resting T cells, and IL-2 binding to this receptor induces cell growth, cytolytic activity, and IL-2Rα transcription. As a result, high-affinity IL-2 receptors are formed in activated T cells, which further increase their responsiveness to IL-2 (Malek 2008; Boyman and Sprent 2012; Liao et al. 2013).

IL-2 mediates diverse pleiotropic functions, and numerous studies have shown that IL-2 signals influence antiviral CD8 T-cell responses in vivo. Initial studies using IL-2-deficient mice showed that antiviral CD8 T-cell expansion was slightly (about threefold) lower in IL-2-deficient mice than in wild-type (WT) mice after infection with lymphocytic choriomeningitis virus (LCMV), and the ability of antiviral CD8 T cells to produce IFN-γ was impaired in the absence of IL-2. Correspondingly, IL-2-deficient mice showed less efficient viral control than WT mice (Kundig et al. 1993; Cousens et al. 1995; Su et al. 1998). In addition, IL-2Rβ-deficient mice did not generate functional antiviral CD8 T-cell responses after LCMV infection (Suzuki et al. 1995). Overall, some caveats in these earlier studies are that (1) the concomitant autoimmunity associated with mice defective in IL-2 signaling may have bypassed normal physiological mechanisms (Malek 2008), and (2) competition between T cells for antigen and growth factors is not present in mice lacking IL-2 or its receptors (Bachmann et al. 2007). To avoid these issues, several approaches have been conducted to clarify the role of IL-2

on antiviral CD8 T-cell responses in vivo. Some studies used the adoptive transfer system, in which TCR-transgenic IL-2Rα-deficient and WT CD8 T cells were separately or cotransferred into recipient mice, followed by infection. In this setting, only IL-2Rα-deficient antigen-specific CD8 T cells failed to receive optimal IL-2 signals because of their inability to form heterotrimeric high-affinity IL-2 receptors during infections (D'Souza et al. 2002; D'Souza and Lefrancois 2003; Williams et al. 2006; Mitchell et al. 2010). These studies revealed that IL-2 signals were dispensable for priming of antigen-specific CD8 T cells, which was consistent with the idea that TCR ligation and costimulation, together with signals via inflammatory cytokines, induce their cellular divisions and expansions. Nevertheless, IL-2 signaling through the heterotrimeric IL-2 receptors is important for sustained expansion of antiviral CD8 T cells during LCMV or vesicular stomatitis virus (VSV) infection. Accordingly, IL-2Rα-deficient antiviral CD8 T cells expanded less efficiently (about twofold) than WT cells by the peak of the response after LCMV infection (Williams et al. 2006; Mitchell et al. 2010). Comparable results were derived from studies using bone marrow chimeric mice consisting of a mixture of cells from WT and IL-2Rα-deficient mice (Williams et al. 2006; Bachmann et al. 2007; Obar et al. 2010).

Other studies also examined the roles of IL-2 signals on antiviral CD8 T-cell responses by administering exogenous IL-2 in various settings. When IL-2 was given at optimal times during the expansion phase, the magnitude of antigen-specific CD8 T-cell responses were augmented after VSV infection or peptide-pulsed DC immunization (D'Souza and Lefrancois 2003; Kim et al. 2016). Likewise, when IL-2 was administered only during the contraction phase, it was beneficial most likely because of its promoting the proliferation and survival of antigen-specific CD8 T cells after viral and intracellular bacterial (*Listeria monocytogenes* [LM]) infections (Blattman et al. 2003; Rubinstein et al. 2008). IL-2 therapy also enhanced the proliferation of resting memory CD8 T cells in mice that had cleared LCMV infection. In contrast, however, giving IL-2 during the whole period of the expansion phase deteriorated antiviral CD8 T-cell responses during LCMV infection (Blattman et al. 2003).

Besides their role in primary T-cell responses, IL-2 signals are vital for potent secondary CD8 T-cell responses. Memory CD8 T cells deficient in IL-2Rα showed reduced ability to expand on rechallenge with viral and intracellular bacterial infections compared with WT counterparts (Williams et al. 2006; Bachmann et al. 2007; Mitchell et al. 2010). This poor expansion of IL-2Rα-deficient CD8 T cells was not a result of impaired cell division after reinfection, but rather because of their defective survival and accumulation during recall responses. Moreover, administration of IL-2/anti-IL-2 immunocomplexes during primary LCMV infection restored the recall responses of IL-2Rα-deficient memory CD8 T cells to LM infection as comparable to WT cells (Williams et al. 2006). These findings together show that IL-2 signals received during the primary responses are essential for programming of proliferation-potent memory CD8 T cells. Some rescue could also be seen if the IL-2/anti-IL-2 immunocomplexes were provided only during the secondary responses to LM infection (Williams et al. 2006). However, conflicting results were also obtained after LM infection in another study, in which IL-2Rα-deficient memory CD8 T cells generated by vaccinia virus (VV) or LM infection were able to mount a robust secondary response when rechallenged with LM (Obar et al. 2010). The reasons for these discrepancies among different studies are not known, and further studies are required.

IL-2 also functions as a critical differentiation factor dictating the fate of antigen-specific CD8 T cells. IL-2 signals during the expansion phase are influential for the formation of CD8 TE cells after viral and intracellular bacterial infections, which express effector molecules such as perforin, granzyme B, and IFN-γ, but produce low amounts of IL-2 (Williams et al. 2006; Bachmann et al. 2007; Kalia et al. 2010; Obar et al. 2010). IL-2Rα-deficient memory CD8 T cells also showed defects in their effector differentiation during secondary responses to VV or LM infection (Mitchell et al. 2010; Obar et al. 2010).

Prolonged IL-2 signaling during the expansion phase favors the generation of antiviral CD8 TE cells and there is a decrease in the number of CD8 MP cells during acute LCMV infection. This results in a reduction in number of long-lived memory CD8 T cells (Kalia et al. 2010). Although IL-2 production in the spleen was rapidly induced, and peaked at day 1 after LCMV infection, CD25 expression on antiviral CD8 T cells was uniformly induced around day 1–2 postinfection. CD25 expression was further augmented with increasing rounds of cell divisions, consistent with the fact that CD25 expression was rapidly up-regulated by TCR signals, and is maintained by positive feedback via enhanced IL-2 signals (Malek 2008). Of interest, heterogeneous populations of CD25^{lo} and CD25^{hi} cells were found among antiviral CD8 T cells around day 3–5 postinfection. Both cell populations showed similar direct ex vivo killing activity, indicating that they had differentiated into potent effector cells. However, the CD25^{hi} cell subset showed more pronounced down-regulation of CD62L, IL-7Rα (CD127), and CCR7 and also expressed modestly higher levels of granzyme B and perforin than the CD25^{lo} cell subset. After in vitro peptide stimulation, the CD25^{hi} cell subset produced substantially lower IL-2 than the CD25^{lo} cell subset. Moreover, when CD25^{lo} and CD25^{hi} CD8 T-cell subsets were transferred into infection-matched recipients, the CD25^{hi} cells proliferated rapidly, preferentially generated TE cells, and were more prone to die by apoptosis as compared with the CD25^{lo} cells. Conversely, CD25^{lo} cells preferentially up-regulated CD62L and CD127, and differentiated into long-lived functional memory CD8 T cells, which were capable of producing more IL-2 and showing enhanced recall responses than memory cells derived from the CD25^{hi} cell subset (Kalia et al. 2010). Similar results were also reported by in vitro studies, in which IL-2 signaling was essential for inducing the expression of cytotoxic molecules such as granzyme B and perforin. IL-2Rα-deficient effector CD8 T cells expressed fewer of those molecules, resulting in poor killing function (Pipkin et al. 2010).

Mechanistically, IL-2 signal-dependent regulation of TE versus MP fate during the primary

Cite this article as *Cold Spring Harb Perspect Biol* doi: 10.1101/cshperspect.a028464

response is partly mediated by several key transcriptional factors such as B-lymphocyte-induced maturation protein (Blimp)-1, which promotes the TE fate and represses IL-2 transcription in antiviral CD8 T cells. Blimp-1 is not expressed by naïve CD8 T cells, but up-regulated following the delivery of antigen- and IL-2-dependent signals (Gong and Malek 2007; Malek 2008; Kallies et al. 2009; Rutishauser et al. 2009; Kalia et al. 2010; Pipkin et al. 2010). CD8 TE cells expressed more Blimp-1 than MP cells during LCMV infection (Sarkar et al. 2008; Rutishauser et al. 2009). Furthermore, the CD25^{hi} CD8 T-cell subset, which preferentially differentiated into TE cells, expressed higher amounts of Blimp-1 than the CD25^{lo} subset early after LCMV infection. Blimp-1-deficient antiviral CD8 T cells showed impaired differentiation into TE cells, defective migration to inflammatory sites, and profound reduction of granzyme B expression, although they produced increased amounts of IL-2 (Kallies et al. 2009; Rutishauser et al. 2009; Kalia et al. 2010; Pipkin et al. 2010).

What is the critical source of IL-2 for antiviral CD8 T cells during acute infections? Although CD4 T cells are thought to be the major producers, activated antiviral CD8 T cells also express IL-2 early after LCMV or VSV infection (D'Souza and Lefrancois 2003, 2004; Sarkar et al. 2008). As antiviral CD8 T cells differentiate into effector cells, they lose the capacity to produce IL-2. However, during the memory phase, they gradually regain the ability to secrete IL-2 (Wherry et al. 2003a,b).

Depending on the situation, differential sources of IL-2 contribute to the formation of protective memory CD8 T cells. Although autocrine IL-2 produced by antiviral CD8 T cells themselves was not required for their initial cell-cycle initiation, its absence appeared to dampen their sustained proliferation and overall magnitude of the responses after VSV infection (D'Souza and Lefrancois 2003). In addition, autocrine IL-2 production, but not paracrine IL-2 derived from CD4 T cells or DCs, has been shown to be important for the generation of memory T cells with robust recall potential during VV infection (Feau et al. 2011). In contrast, IL-2-deficient CD8 T cells did not show any

impairment for the generation of primary and secondary responses after LCMV infection, suggesting that paracrine IL-2 was sufficient for the generation of potent memory CD8 T cells in this setting (Williams et al. 2006).

IL-7

IL-7 is produced mainly by stromal cells in lymphoid tissues. Besides its central role on T-cell development in the thymus and naïve T-cell homeostasis in the periphery, IL-7 regulates the formation of memory CD8 T cells primarily via promoting T-cell survival during acute infections (Link et al. 2007; Surh and Sprent 2008; Turley et al. 2010). The cellular receptors for IL-7 are formed by heterodimerization of IL-7Rα (CD127) and γc (Fig. 1). Although CD127 is highly expressed on naïve CD8 T cells, its expression is uniformly down-regulated after activation during the early phase of viral and intracellular bacterial infections. However, the small fraction (5%–10%) of effector CD8 T cells, referred to as MP cells, regain CD127 expression at the peak of the responses and differentiate into long-lived memory cells (Fig. 2) (Schluns et al. 2000; Kaech et al. 2003; Huster et al. 2004; Bachmann et al. 2005; Klonowski et al. 2006; Joshi et al. 2007; Sarkar et al. 2008).

Early in LCMV infection, when CD127 is uniformly down-regulated on antiviral CD8 T cells, a striking heterogeneity exists in terms of the expression of killer cell lectin-like receptor G1 (KLRG1). Although KLRG1^{hi} and KLRG1^{int} CD127^{lo} effector cells possess similar gene expression profiles and potent effector functions, they have distinct memory lineage fates. CD127^{lo}KLRG1^{hi} CD8 T cells are TE cells and the majority of them die by apoptosis after viral clearance, whereas MP cells are generated from some fractions among CD127^{lo}KLRG1^{int} CD8 T cells, which subsequently reexpress CD127, and preferentially differentiate into long-lived memory CD8 T cells (Joshi et al. 2007; Sarkar et al. 2008). However, these phenotypic discriminations are not exclusive criteria for the formation of MP cells, and not all CD127^{hi} effector CD8 T cells survive during the contraction phase. The generation of MP cells and their maturation into

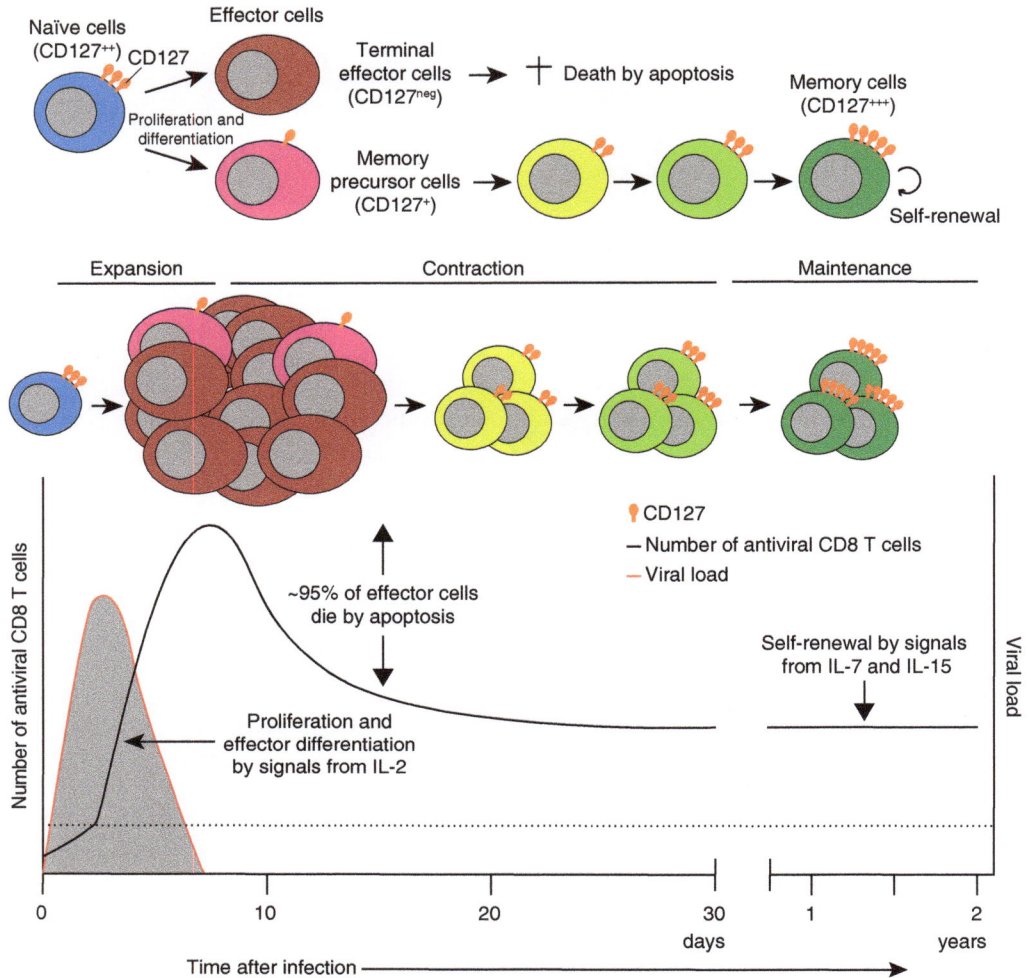

Figure 2. Cytokine-mediated control of antiviral CD8 T-cell responses and their differentiation from naïve to effector to memory cells during acute infections. The black solid line indicates the dynamics of the number of antiviral CD8 T cells. The red solid line indicates the dynamics of the viral load. The dotted line indicates the limit of detection. Levels of CD127 expression on antiviral CD8 T cells at different stages of T-cell differentiation are shown. IL, Interleukin.

long-lived memory CD8 T cells are also affected by the experimental models used, including the types of pathogen (tropism, dose, route, and inflammatory milieu) and precursor frequencies of antigen-specific CD8 T cells (Marzo et al. 2005; Obar et al. 2008; Kaech and Cui 2012).

CD127 is necessary but not sufficient for the formation of memory CD8 T cells. It has been speculated that CD8 MP cells have a survival advantage over CD8 TE cells during the contrac-

tion phase because of the increased IL-7 signaling. However, overexpression of CD127 on effector CD8 T cells did not prevent normal contraction of the response (Hand et al. 2007; Haring et al. 2008). Moreover, memory CD8 T cells can be generated in IL-7-deficient mice and CD127 expression was not regulated by IL-7 itself during VSV infection (Klonowski et al. 2006). In another study, a chimeric granulocyte macrophage colony-stimulating factor (GM-

CSF)/IL-7R complementary DNA (cDNA) was constructed by fusing the extracellular domain of GM-CSFRβ with the transmembrane and intracellular domains of IL-7Rα (Sun et al. 2006). When this was expressed on T cells by using retrovirus, GM-CSF triggered IL-7R signaling on T cells. Although IL-7 gene expression in the spleen was relatively constant, a burst of GM-CSF gene up-regulation was induced peaking around day 3 after LCMV infection. By using this model system, IL-7R signaling was enhanced on antiviral CD8 T cells during the expansion phase of LCMV infection, which allowed determining whether increased IL-7R signaling affected the formation of memory T cells. Unexpectedly, increased numbers of effector cells were generated; however, this did not lead to increased numbers of memory CD8 T cells (Sun et al. 2006). One possibility is that it might be the result of some redundant function of thymic stromal lymphopoietin (TSLP), a closely related cytokine that delivers signals through a receptor that includes CD127 and the TSLP receptor (Rochman et al. 2009). Although the role of TSLP in memory CD8 T-cell generation is largely unknown, TSLP can compensate for IL-7 deficiency in lymphopoiesis and enhance the survival and homeostatic proliferation of CD8 T cells (Chappaz et al. 2007).

Once memory CD8 T cells are formed, the expression of CD127 is necessary for their maintenance and survival (Fig. 2) (Schluns et al. 2000; Goldrath et al. 2002; Kaech et al. 2003; Huster et al. 2004; Bachmann et al. 2005, 2006). In a mouse model in which WT IL-7Rα was replaced with the Y449F mutant IL-7Rα (IL-7Rα^{449F}), which cannot transmit signals to activate signal transducers and activators of transcription 5 (STAT5), IL-7Rα^{449F} mutant naïve CD8 T cells showed equivalent expansions to WT counterparts after LM infection. They were able to differentiate into memory CD8 T cells and underwent a normal basal homeostatic proliferation in response to IL-15. However, these cells gradually disappeared, possibly because of insufficient survival signals from IL-7 (Osborne et al. 2007). Although prosurvival effects of IL-7 on lymphocytes have been well documented by inducing antiapoptotic mole-

cules such as B-cell leukemia/lymphoma (Bcl)-2 and myeloid cell leukemia (Mcl)-1 (Akashi et al. 1997; von Freeden-Jeffry et al. 1997; Opferman et al. 2003), defective maintenance of memory CD8 T cells found in IL-7Rα^{449F} mutant mice was independent of Bcl-2 (Osborne et al. 2007). These findings suggest that other IL-7-dependent mechanisms exist for the maintenance and survival of memory CD8 T cells.

A more recent study identified that IL-7 promoted memory CD8 T-cell survival by tailoring its metabolism via glycerol import through aquaporin 9 (AQP9) and triglyceride (TAG) synthesis and storage. IL-7 induced glycerol channel AQP9 expression in memory CD8 T cells, but not in naïve cells. AQP9 was preferentially expressed in memory, but not in naïve or effector CD8 T cells, and this selective induction of AQP9 and TAG synthesis in memory CD8 T cells might be another advantage for memory CD8 T cells to use glycerol and lipids more effectively over naïve cells. Importantly, AQP9, which imports glycerol, promotes TAG synthesis, and sustains ATP levels, was required for memory CD8 T-cell survival and self-renewal. AQP9 deficiency impaired glycerol import into memory CD8 T cells for esterification of fatty acid and synthesis and storage of TAG. These defects could be rescued by ectopic expression of TAG synthases, which restored lipid stores and memory T-cell survival (Cui et al. 2015). This study showed one IL-7-mediated mechanistic basis of how and where memory CD8 T cells obtained lipids to sustain fatty acid oxidation (FAO), which was known to be important for the maintenance of memory CD8 T cells (van der Windt et al. 2012; O'Sullivan et al. 2014).

In addition to IL-7-mediated survival of memory CD8 T cells, IL-7 can induce the proliferation of these cells when it is present in elevated levels, such as during lymphopenia or when exogenous IL-7 is administered (Schluns et al. 2000; Bradley et al. 2005; Nanjappa et al. 2008). During infections, the effects of exogenous IL-7 on antiviral CD8 T-cell responses depend on the timing of the therapy. IL-7 treatment during the expansion phase did not alter the numbers of effector and memory CD8 T cells after LCMV infection (Nanjappa et al.

2008). In contrast, IL-7 therapy restricted to the contraction phase resulted in an enhanced number of memory CD8 T cells after viral infection (LCMV or VV), DNA vaccination, and peptide-pulsed DC immunization (Nanjappa et al. 2008; Pellegrini et al. 2009). This enhanced generation of memory CD8 T cells was primarily associated with increased cellular proliferation, but not with the suppression of apoptosis (Nanjappa et al. 2008). Of importance, memory CD8 T cells generated by IL-7 therapy showed superior recall responses and improved viral or tumor control (Nanjappa et al. 2008; Pellegrini et al. 2009). Beneficial effects of IL-7 therapy at the contraction phase on enhanced memory CD8 T-cell formation was also found during secondary responses to LM infection (Nanjappa et al. 2008). Moreover, IL-7 therapy during the memory phase after LCMV infection induced increased numbers of antiviral CD8 T cells, but this effect was transient and not maintained long term (Nanjappa et al. 2008).

IL-15

IL-15 messenger RNA (mRNA) is expressed in a wide range of cell types: however, its protein expression is limited and mainly detected on monocytes/macrophages and DCs. IL-15 can bind to the IL-15Rα (CD215) with high affinity ($K_d > 10^{-11}$ M), which is not required for signaling. Rather, CD215 retains IL-15 on the cell surface, substantially increases the affinity of this cytokine for IL-2Rβ compared with its free form, and *trans*-presents it to neighboring cells that express IL-2Rβ and γc such as memory CD8 T cells (Fig. 1) (Waldmann 2006; Ring et al. 2012; Steel et al. 2012).

IL-15 seems to play minor roles for antiviral CD8 T cells during the expansion phase, and the magnitude of antiviral CD8 T-cell responses were only slightly reduced in IL-15- or IL-15Rα-deficient mice after infection (Becker et al. 2002; Schluns et al. 2002). However, IL-15 was critical for the survival of CD127^{lo} effector CD8 T-cell subsets after viral and intracellular bacterial infections (Yajima et al. 2006; Joshi et al. 2007; Rubinstein et al. 2008; Mitchell et al. 2010). The absence of IL-15 during the contrac-

tion phase (day 8–42), or throughout the course of infection (day 0–42), but not during the expansion phase only (day 0–8), diminished the persistence of KLRG1^{hi}CD127^{lo} effector CD8 T cells after LCMV infection (Mitchell et al. 2010). Accordingly, exogenous administration of IL-15 during the contraction phase promoted their survival after VSV or LM infection (Rubinstein et al. 2008).

Memory CD8 T cells formed in IL-15- or IL-15Rα-deficient mice after infection were phenotypically and functionally similar to those generated in IL-15-sufficient mice (Becker et al. 2002). However, IL-15 was critical for homeostatic proliferation of memory CD8 T cells generated after infection (Becker et al. 2002; Schluns et al. 2002). As a consequence, memory CD8 T cells gradually declined in number in the absence of IL-15 signals (Fig. 2) (Becker et al. 2002; Schluns et al. 2002; Tan et al. 2002). In contrast, administration of IL-15–IL-15Rα complexes could induce vigorous proliferation of memory CD8 T cells in vivo (Rubinstein et al. 2006; Stoklasek et al. 2006). Of note, although IL-15 is crucial for maintaining basal turnover of memory CD8 T cells, they can be maintained by IL-7 even in the absence of IL-15 under particular conditions such as lymphopenia, in which increased amounts of IL-7 are available in vivo because of the lack of competition with naïve T cells (Goldrath et al. 2002; Kieper et al. 2002; Tan et al. 2002).

Memory CD8 T cells induced after a primary acute infection expressed higher levels of surface protein and mRNA for IL-2Rβ than naïve cells. However, mRNA expression for IL-2Rβ was gradually decreased as memory CD8 T cells were boosted repeatedly by serial adoptive transfers of antigen-specific CD8 T cells into naïve mice and subsequent challenges (Wherry et al. 2003b; Wirth et al. 2010). Secondary memory CD8 T cells also showed decreased responsiveness to IL-15 in vitro compared with primary memory cells. Basal homeostatic proliferation was also progressively reduced with repetitive antigen challenges (Jabbari and Harty 2006; Sandau et al. 2010; Wirth et al. 2010). *Myc*, a downstream target of IL-15 signaling, was most strikingly down-regulated on memory CD8 T cells generated by multiple times of immunizations (Wirth et al. 2010). Accordingly, repeatedly stimulated memory CD8 T cells were thought to be less dependent on IL-15 signaling than primary memory cells. Nonetheless, secondary memory cells still depended on IL-15 for their survival and maintenance and their basal proliferation was impaired in IL-15-deficient mice (Sandau et al. 2010). In addition, transfer of secondary memory CD8 T cells into IL-15-deficient mice resulted in defective maintenance of these cells, which was linked to decreased expression of the antiapoptotic protein Bcl-2 (Sandau et al. 2010).

Although it remains to be fully resolved how IL-15 maintains memory CD8 T cells, recent studies highlighted the effect of IL-15 on metabolism of antigen-specific CD8 T cells (van der Windt et al. 2012; O'Sullivan et al. 2014). After LM infection, memory CD8 T cells had substantially more mitochondrial spare respiratory capacity (SRC) than naïve or effector CD8 T cells. SRC is an extra mitochondrial capacity available in cells to produce energy during an increase in energy demand, and thus thought to be associated with cellular survival and function. IL-15 enhanced SRC on memory CD8 T cells in vitro, which depends on mitochondrial FAO, by inducing expression of carnitine palmitoyl transferase 1a (CPT1a). CPT1a is a metabolic enzyme that controls the rate-limiting step to FAO, and SRC of memory CD8 T cells generated by coculture with IL-15 in vitro is impaired if CPT1a is suppressed by adding its inhibitor etomoxir. Consequently, survival of antigen-specific CD8 T cells during the contraction phase was improved when CPT1a was overexpressed by retroviral transduction of antigen-specific CD8 T cells during LM infection in vivo (van der Windt et al. 2012). A subsequent study showed that memory CD8 T cells used extracellular glucose, rather than extracellular fatty acids (FAs), to support FAO and oxidative phospholylation. To generate lipids, which were necessary for FAO, memory CD8 T cells relied on cell-intrinsic expression of the lysosomal acid lipase to mobilize FA for FAO (O'Sullivan et al. 2014). These results show that IL-15 signals promote CPT1a expression and FAO to support maintenance of antigen-specific memory CD8 T cells.

IL-15 also regulates recall responses of memory CD8 cells independent of antigen in several ways. First, by sensing inflammatory monocyte-derived bioactive IL-15, memory CD8 T cells generated after LM infection quickly exert effector functions by expressing IFN-γ and granzyme B when challenged with irrelevant pathogens such as murine cytomegalovirus (MCMV) or with LM lacking cognate antigen (Soudja et al. 2012). Second, the inflammatory milieu generated by infection with LCMV or pichinde virus (PV) drives memory CD8 T cells to enter the cell cycle. More specifically, IL-15 induced by type I IFN promotes memory CD8 T cell-cycle entry via activation of the mammalian target of rapamycin (mTOR) pathway, which prepares memory CD8 T cells for rapid division on subsequent antigen encounter (Richer et al. 2015). Third, IL-15 enhances memory T-cell trafficking to inflamed tissues by inducing core 2 O-glycans (ligands to P- and E-selectin) (Nolz and Harty 2014), which are expressed on activated endothelium and are important for leukocytes to extravasate into inflammatory sites (Ley et al. 2007). Naïve CD8 T cells did not express core 2 O-glycans, which are up-regulated on effector CD8 T cells after LCMV infection. However, their expression is transient and lost on the majority of memory cells. Following unrelated infectious or inflammatory challenges, however, memory CD8 T cells recruited by bystander inflammation synthesize core 2 O-glycans. Importantly, blocking the interaction of P- and E-selectin and those ligands by antibodies abrogates the nonspecific recruitment of memory CD8 T cells to inflamed sites. Stimulation with IL-15 increases the synthesis of those ligands on memory, but not on naïve CD8 T cells in vitro. Furthermore, IL-15-deficient mice expresses significantly lower levels of core 2 O-glycans than WT mice, and memory CD8 T cells in IL-15-deficient mice does not efficiently traffic to inflammatory tissues (Nolz and Harty 2014).

Recently, a new T-cell lineage, tissue resident memory T cells (T_{RM}), has been identified. In contrast to conventional memory T cells, which recirculate between blood and tissues, T_{RM} reside within tissues and may provide a frontline defense against infections reencountered (Schenkel and Masopust 2014). Besides its critical roles in regulating many aspects of circulating memory CD8 T cells, IL-15 is required to support the maintenance of CD8 T_{RM} after skin infection with herpes simplex virus type I (HSV-1) (Mackay et al. 2013). Conversely, memory CD8 T cells that express CD69, a canonical marker for T_{RM} in secondary lymphoid organs, are less dependent on IL-15 for their maintenance than circulating memory cells after LCMV infection (Schenkel et al. 2014). These results suggest that the regulation of T_{RM} may differ, depending on the context such as the anatomical location and the pathogen used.

CD8 T-CELL RESPONSES DURING CHRONIC VIRAL INFECTION

When host immune responses fail to control the virus and antigen persists, antiviral CD8 T cells differentiate into a state of "T-cell exhaustion," which is characterized by suboptimal effector functions and reduced proliferative potential. Exhausted CD8 T cells are distinct from naïve, effector, or memory CD8 T cells in terms of molecular and epigenetic signatures, including overexpression of several inhibitory receptors, altered cytokine signaling pathways, and dysregulated metabolism (Wherry et al. 2007; Wherry and Kurachi 2015; Pauken et al. 2016; Sen et al. 2016).

T-cell exhaustion is primarily a result of the persistence of high levels of antigens (Mueller and Ahmed 2009). Therefore, T-cell exhaustion occurs in various settings of antigen persistence, including human chronic infections and cancer, which is thought to be associated with unfavorable outcomes (Schietinger and Greenberg 2014; Wherry and Kurachi 2015). Accordingly, there is much interest in understanding the basic mechanisms of T-cell exhaustion and developing strategies for restoring function in exhausted T cells (Barber et al. 2006). Importantly, such attempts have already shown great promise in the clinic, as exemplified by programmed cell death (PD)-1-directed immunotherapy (Brahmer et al. 2012; Topalian et al. 2012). In addition to blocking the signals via inhibitory receptors, such as PD-1, targeting the actions of cytokines are at-

tractive approaches for improving antiviral immunity by modulating T-cell exhaustion (Blattman et al. 2003; Nanjappa et al. 2011; Pellegrini et al. 2011; West et al. 2013).

We have recently shown that exhausted CD8 T cells are comprised of two distinct PD-1-expressing T-cell subsets during chronic LCMV infection: stem-cell-like and terminally differentiated (exhausted) CD8 T-cell subsets (Fig. 3A) (Im et al. 2016). The stem-cell-like CD8 T cells, which are predominantly found in T-cell zones in lymphoid tissues, maintain their population by slow self-renewal, and differentiate into terminally differentiated CD8 T cells by antigen-driven proliferation. Terminally differentiated CD8 T cells, which have suboptimal proliferative capacity but express effector molecules such as granzyme B and perforin, migrate into the periphery, and are located at the major sites of infection in lymphoid and nonlymphoid tissues for controlling pathogens (Fig. 3A). These two CD8 T-cell subsets are also different from each other in terms of molecular signatures, including the expressions of cytokine receptors (e.g., CD127), costimulatory/inhibitory molecules, and transcription factors (Fig. 3A). Nonetheless, both T-cell subsets express PD-1, a major regulator of T-cell exhaustion, although the PD-1 expression level is lower on stem-cell-like CD8 T cells (PD-1$^+$) than terminally differentiated ones (PD-1^{++}). Importantly, the proliferative burst of exhausted CD8 T cells after PD-1 therapy is exclusively provided by stem-cell-like CD8 T cells (Fig. 3A) (Im et al. 2016). Similar observations have been made by others in chronic viral infections of mice, nonhuman primates, and humans (He et al. 2016; Leong et al. 2016; Utzschneider et al. 2016; Wu et al. 2016; Mylvaganam et al. 2017). Because these two CD8 T subsets in T-cell exhaustion have just been defined, it remains totally unknown how each cytokine regulates the biological attributes of these two distinct CD8 T-cell subsets, and future studies are of great importance (Fig. 3B).

IL-2

Similar to the setting of acute infections, IL-2 plays an important role in regulating antiviral

A

T-cell exhaustion

Stem-cell-like
PD-1+ CD8 T cells

Terminally differentiated
PD-1++ CD8 T cells

PD-1+

PD-1++

Costimulatory molecules
(CD28+, ICOS+)

Coinhibitory molecules
(Tim-3+, 2B4+)

Effector molecules
(perforin+, granzyme B+)

Transcription factors
(TCF-1+, Id3+, Bcl-6+)

Transcription factors
(Blimp-1+, Id2+)

Generation/ maintenance	Self-renewal	Generated after antigen-driven proliferation of stem-cell-like PD-1+ CD8 T cells
Proliferative capacity	Provides proliferative burst in response to PD-1 therapy	Minimal proliferation in response to PD-1 therapy
Molecular signature	CD8 MP cell-like CD4 Tfh cell-like Hematopoietic stem-cell-like Costimulatory molecules (+)	CD8 TE cell-like CD4 Th1 cell-like Coinhibitory molecules (+)
Location	Lymphoid tissues (T-cell zone)	Lymphoid and nonlymphoid tissues (major sites of infection)

B

Regulation by cytokines (?)

Stem-cell-like
PD-1+ CD8 T cells

Terminally differentiated
PD-1++ CD8 T cells

2. Proliferation/ differentiation

4. Effector function

1. Self-renewal

3. Migration

Lymphoid tissues
(T-cell zone)

Lymphoid and nonlymphoid tissues
(major sites of infection)

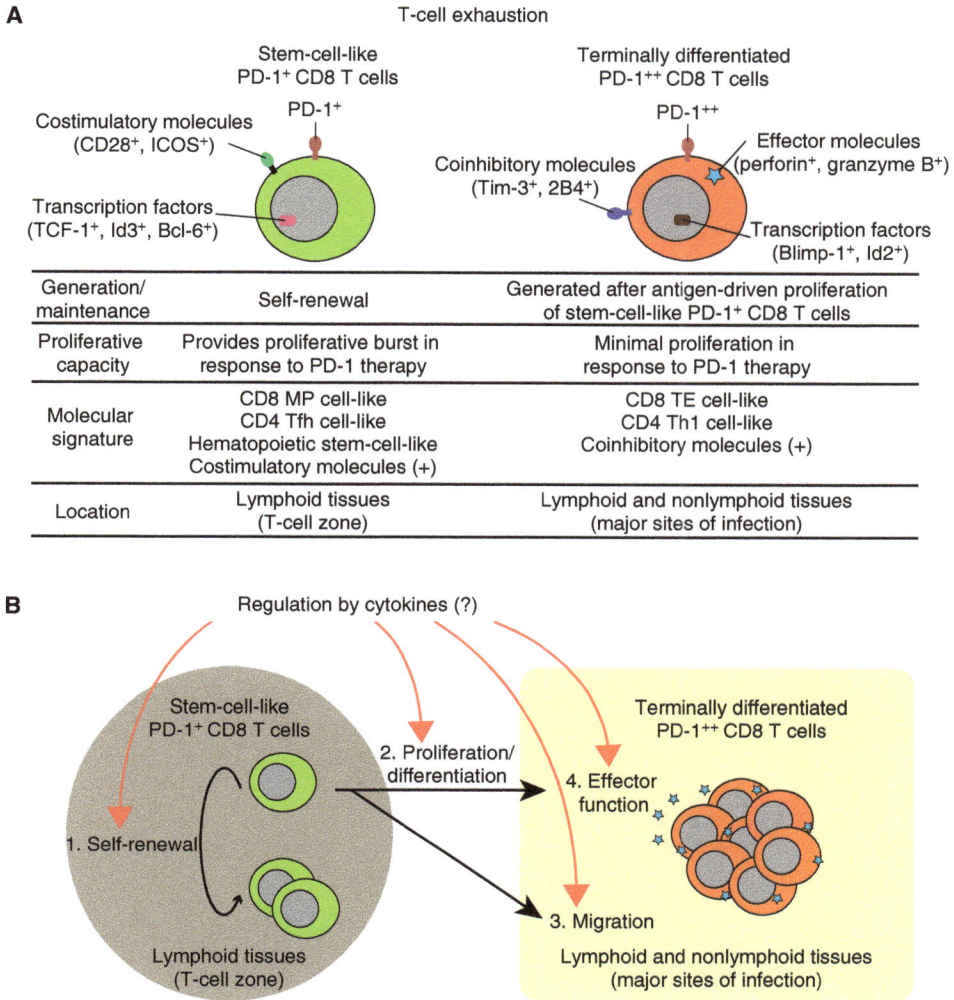

Figure 3. Composition of antiviral CD8 T cells during chronic infections and potential regulation by cytokines. (*A*) Two programmed cell death (PD)-1-expressing CD8 T-cell subsets in T-cell exhaustion and their biological characteristics/functions. Two (stem-cell-like and terminally differentiated) CD8 T-cell subsets are distinct from each other in terms of generation/maintenance, proliferative capacity, molecular signature, and location. (*B*) Potential regulation of antiviral CD8 T cells by cytokines during chronic infections. Cytokines may regulate two (stem-cell-like and terminally differentiated) CD8 T-cell subsets at four levels: (1) self-renewal of stem-cell-like CD8 T cells, (2) proliferation/differentiation of stem-cell-like CD8 T cells, (3) migration of terminally differentiated CD8 T cells into major sites of infection, and (4) effector function of terminally differentiated CD8 T cells. ICOS, Inducible T-cell costimulator; TCF-1, T-cell factor 1; Id3, inhibitor of DNA binding 3; Bcl-6, B-cell leukemia/lymphoma 6; Tim-3, T-cell immunoglobulin and mucin-domain containing-3; Blimp-1, B-lymphocyte-induced maturation protein; Id2, inhibitor of DNA binding 2; MP, memory precursor; TE, terminal effector; Tfh, T follicular helper; Th1, T helper type I.

CD8 T-cell responses during chronic infections. IL-2Rα-deficient antiviral CD8 T cells initially expanded but subsequently declined after LCMV or MCMV infection (Bachmann et al. 2007). From a therapeutic view, an initial study showed that IL-2 administration enhanced antiviral CD8 T-cell responses and accelerated viral clearance during LCMV infection (Blattman et al. 2003). IL-2 treatment is also beneficial for CD8 T-cell-mediated viral control during murine γ-herpesvirus 68 (MHV-68) infection (Molloy et al. 2009). Conversely, IL-2 therapy during human immunodeficiency virus (HIV) infection in humans increased CD4 T-cell counts of patients, but it did not lead to the reduction of the viral load (Kovacs et al. 1996). Similarly, IL-2 therapy combined with antiretroviral therapy (ART) increased CD4 T-cell counts of HIV-infected patients, yet had no clinical benefits compared with ART alone in two large randomized clinical trials (INSIGHT-ESPRIT Study Group et al. 2009). These studies together indicate that IL-2 has a potential to modulate immune responses to treat chronic infections; however, further improvements are required.

More recently, IL-2 administration with PD-1 blockade during chronic LCMV infection has been reported to induce striking synergistic effects for promoting expansion of antiviral CD8 T cells and reducing viral load (West et al. 2013). Remarkably, IL-2 treatment modulated the phenotype of exhausted CD8 T cells, decreasing the expression of inhibitory receptors, including PD-1 and increasing the expression of CD127, a critical molecule for T-cell survival. A whole picture of how IL-2 alone or IL-2 and PD-1 blockade modulates antiviral CD8 T cells during chronic infections still remains unclear and awaits further studies. Given that immune checkpoint inhibitors have great promise for treating various cancers, IL-2 therapy combined with immune checkpoint inhibitors may be an exciting approach to enhance T-cell immunity under conditions of chronic antigenic stimulations. As discussed above, a stem-cell-like CD8 T-cell subset has been defined among exhausted CD8 T cells during chronic LCMV infection. Given that a proliferative burst of exhausted CD8 T cells after PD-1 therapy comes exclusively from the stem-cell-like CD8 T-cell subset (Im et al. 2016; Utzschneider et al. 2016), it will be of great interest to evaluate how IL-2 therapy with or without immune checkpoint inhibitors acts on this T-cell subset (Fig. 3B).

IL-7 AND IL-15

One feature of exhausted CD8 T cells is their defective capacity to self-renew by homeostatic cytokines, such as IL-7 and IL-15, partly because they express lower levels of CD127 and CD122 than memory cells. Rather, maintenance and survival of exhausted CD8 T cells are critically dependent on their division in response to persistent antigens (Wherry et al. 2004; Shin et al. 2007). However, it should be noted that two (stem-cell-like and terminally differentiated) CD8 T-cell subsets exist in T-cell exhaustion, and future studies are necessary to reexamine how IL-7 and IL-15 influence these two subsets (Fig. 3B). Indeed, CD127 expression is higher in stem-cell-like CD8 T cells than in terminally differentiated ones, although it remains unknown whether the differential expression levels of CD127 between the two CD8 T-cell subsets impact on biological function in these cells.

Administration of IL-7 during chronic LCMV infection induced expansion of antiviral CD8 T cells. It also increased numbers of CD4, CD8, and B cells, and resulted in accelerated viral clearance in multiple organs (Nanjappa et al. 2011; Pellegrini et al. 2011). The outcome of IL-7 treatment in chronic LCMV infection depends on the treatment regimen. Administration of IL-7 during early contraction phase (day 8–15 postinfection) was less effective than treatment during the late contraction phase (day 16–25 postinfection) (Nanjappa et al. 2011). Extended duration of IL-7 treatment (day 8–30 postinfection) further augmented antiviral CD8 T-cell responses, which showed enhanced functionality, expressed increased levels of Bcl-2 and CD127, and reduced levels of inhibitory markers (Nanjappa et al. 2011; Pellegrini et al. 2011). Further studies are required to investigate how IL-7 therapy influences two (stem-cell-like and terminally differ-

Cite this article as *Cold Spring Harb Perspect Biol* doi: 10.1101/cshperspect.a028464

entiated) CD8 T-cell subsets in T-cell exhaustion (Fig. 3B).

In conjunction with the observations made in the chronic LCMV infection model, IL-7 treatment increased T-cell numbers during simian immunodeficiency virus (SIV) infection or HIV infection. However, it remains to be determined whether the IL-7-mediated increase in T-cell numbers leads to improved outcomes in human chronic infections (Fry et al. 2003; Nugeyre et al. 2003; Beq et al. 2006; Levy et al. 2009, 2012; Sereti et al. 2009; Vassena et al. 2012).

IL-15 treatment for 4 weeks during acute SIV infection induced a two- to threefold increase in numbers of SIV-specific CD8 T cells at week 2 and NK cells at week 1 postinfection. However, these effects did not contribute to improved viral control, and IL-15-treated animals had a 1-log and a 3-log higher viral load than untreated ones at week 6 and at week 20 postinfection, respectively (Mueller et al. 2008). In addition, IL-15 treatment for 4 weeks started at more than 9 months after SIV infection preferentially increased CD8 T cells with an effector memory phenotype; however, this did not result in improved viral control (Mueller et al. 2005). A more recent study showed that IL-15 treatment during ART delayed viral suppression and failed to enhance ART-induced immune reconstitution during chronic SIV infection (Lugli et al. 2011). Therefore, IL-15 treatment appears to be detrimental during SIV infection, and it might be possible that this is also the case in HIV infection.

CONCLUDING REMARKS

Our knowledge of how IL-2, IL-7, or IL-15 instructs various aspects of antiviral CD8 T-cell responses has shown great progress during the past few decades. Future mechanistic studies are needed to gain more insight into the role of these cytokines in antiviral immunity during acute and chronic infections. In particular, T-cell exhaustion has now been redefined as consisting of two distinct (stem-cell-like and terminally differentiated) CD8 T-cell subsets. Better understanding of the actions of each cytokine on these two CD8 T-cell subsets will contribute to the

development of novel cytokine-directed immunotherapies targeting chronic infections and cancer.

REFERENCES

Akashi K, Kondo M, von Freeden-Jeffry U, Murray R, Weissman IL. 1997. Bcl-2 rescues T lymphopoiesis in interleukin-7 receptor-deficient mice. *Cell* **89:** 1033–1041.

Bachmann MF, Wolint P, Schwarz K, Jager P, Oxenius A. 2005. Functional properties and lineage relationship of CD8⁺ T cell subsets identified by expression of IL-7 receptor and CD62L. *J Immunol* **175:** 4686–4696.

Bachmann MF, Beerli RR, Agnellini P, Wolint P, Schwarz K, Oxenius A. 2006. Long-lived memory CD8⁺ T cells are programmed by prolonged antigen exposure and low levels of cellular activation. *Eur J Immunol* **36:** 842–854.

Bachmann MF, Wolint P, Walton S, Schwarz K, Oxenius A. 2007. Differential role of IL-2R signaling for CD8⁺ T cell responses in acute and chronic viral infections. *Eur J Immunol* **37:** 1502–1512.

Barber DL, Wherry EJ, Ahmed R. 2003. Cutting edge: Rapid in vivo killing by memory CD8 T cells. *J Immunol* **171:** 27–31.

Barber DL, Wherry EJ, Masopust D, Zhu B, Allison JP, Sharpe AH, Freeman GJ, Ahmed R. 2006. Restoring function in exhausted CD8 T cells during chronic viral infection. *Nature* **439:** 682–687.

Becker TC, Wherry EJ, Boone D, Murali-Krishna K, Antia R, Ma A, Ahmed R. 2002. Interleukin 15 is required for proliferative renewal of virus-specific memory CD8 T cells. *J Exp Med* **195:** 1541–1548.

Beq S, Nugeyre MT, Fang RHT, Gautier D, Legrand R, Schmitt N, Estaquier J, Barre-Sinoussi F, Hurtrel B, Cheynier R, et al. 2006. IL-7 induces immunological improvement in SIV-infected rhesus macaques under antiviral therapy. *J Immunol* **176:** 914–922.

Blattman JN, Grayson JM, Wherry EJ, Kaech SM, Smith KA, Ahmed R. 2003. Therapeutic use of IL-2 to enhance antiviral T-cell responses in vivo. *Nat Med* **9:** 540–547.

Boyman O, Sprent J. 2012. The role of interleukin-2 during homeostasis and activation of the immune system. *Nat Rev Immunol* **12:** 180–190.

Bradley LM, Haynes L, Swain SL. 2005. IL-7: Maintaining T-cell memory and achieving homeostasis. *Trends Immunol* **26:** 172–176.

Brahmer JR, Tykodi SS, Chow LQ, Hwu WJ, Topalian SL, Hwu P, Drake CG, Camacho LH, Kauh J, Odunsi K, et al. 2012. Safety and activity of anti-PD-L1 antibody in patients with advanced cancer. *N Engl J Med* **366:** 2455–2465.

Byers AM, Kemball CC, Moser JM, Lukacher AE. 2003. Cutting edge: Rapid in vivo CTL activity by polyoma virus-specific effector and memory CD8⁺ T cells. *J Immunol* **171:** 17–21.

Chappaz S, Flueck L, Farr AG, Rolink AG, Finke D. 2007. Increased TSLP availability restores T- and B-cell compartments in adult IL-7 deficient mice. *Blood* **110:** 3862–3870.

Cousens LP, Orange JS, Biron CA. 1995. Endogenous IL-2 contributes to T cell expansion and IFN-γ production during lymphocytic choriomeningitis virus infection. *J Immunol* **155**: 5690–5699.

Cox MA, Kahan SM, Zajac AJ. 2013. Anti-viral CD8 T cells and the cytokines that they love. *Virology* **435**: 157–169.

Cui G, Staron MM, Gray SM, Ho PC, Amezquita RA, Wu J, Kaech SM. 2015. IL-7-induced glycerol transport and TAG synthesis promotes memory CD8$^+$ T cell longevity. *Cell* **161**: 750–761.

D'Souza WN, Lefrancois L. 2003. IL-2 is not required for the initiation of CD8 T cell cycling but sustains expansion. *J Immunol* **171**: 5727–5735.

D'Souza WN, Lefrancois L. 2004. Frontline: An in-depth evaluation of the production of IL-2 by antigen-specific CD8 T cells in vivo. *Eur J Immunol* **34**: 2977–2985.

D'Souza WN, Schluns KS, Masopust D, Lefrançois L. 2002. Essential role for IL-2 in the regulation of antiviral extralymphoid CD8 T cell responses. *J Immunol* **168**: 5566–5572.

Feau S, Arens R, Togher S, Schoenberger SP. 2011. Autocrine IL-2 is required for secondary population expansion of CD8$^+$ memory T cells. *Nat Immunol* **12**: 908–913.

Fry TJ, Moniuszko M, Creekmore S, Donohue SJ, Douek DC, Giardina S, Hecht TT, Hill BJ, Komschlies K, Tomaszewski J, et al. 2003. IL-7 therapy dramatically alters peripheral T-cell homeostasis in normal and SIV-infected nonhuman primates. *Blood* **101**: 2294–2299.

Fuertes Marraco SA, Soneson C, Cagnon L, Gannon PO, Allard M, Abed Maillard S, Montandon N, Rufer N, Waldvogel S, Delorenzi M, et al. 2015. Long-lasting stem cell-like memory CD8$^+$ T cells with a naïve-like profile upon yellow fever vaccination. *Sci Transl Med* **7**: 282ra248.

Goldrath AW, Sivakumar PV, Glaccum M, Kennedy MK, Bevan MJ, Benoist C, Mathis D, Butz EA. 2002. Cytokine requirements for acute and basal homeostatic proliferation of naïve and memory CD8$^+$ T cells. *J Exp Med* **195**: 1515–1522.

Gong D, Malek TR. 2007. Cytokine-dependent Blimp-1 expression in activated T cells inhibits IL-2 production. *J Immunol* **178**: 242–252.

Hammarlund E, Lewis MW, Hansen SG, Strelow LI, Nelson JA, Sexton GJ, Hanifin JM, Slifka MK. 2003. Duration of antiviral immunity after smallpox vaccination. *Nat Med* **9**: 1131–1137.

Hand TW, Morre M, Kaech SM. 2007. Expression of IL-7 receptor α is necessary but not sufficient for the formation of memory CD8 T cells during viral infection. *Proc Natl Acad Sci* **104**: 11730–11735.

Haring JS, Jing X, Bollenbacher-Reilley J, Xue HH, Leonard WJ, Harty JT. 2008. Constitutive expression of IL-7 receptor does not support increased expansion or prevent contraction of antigen-specific CD4 or CD8 T cells following *Listeria monocytogenes* infection. *J Immunol* **180**: 2855–2862.

He R, Hou S, Liu C, Zhang A, Bai Q, Han M, Yang Y, Wei G, Shen T, Yang X, et al. 2016. Follicular CXCR5- expressing CD8$^+$ T cells curtail chronic viral infection. *Nature* **537**: 412–428.

Homann D, Teyton L, Oldstone MB. 2001. Differential regulation of antiviral T-cell immunity results in stable CD8$^+$ but declining CD4$^+$ T-cell memory. *Nat Med* **7**: 913–919.

Huster KM, Busch V, Schiemann M, Linkemann K, Kerksiek KM, Wagner H, Busch DH. 2004. Selective expression of IL-7 receptor on memory T cells identifies early CD40L-dependent generation of distinct CD8$^+$ memory T cell subsets. *Proc Natl Acad Sci* **101**: 5610–5615.

Im SJ, Hashimoto M, Gerner MY, Lee J, Kissick HT, Burger MC, Shan Q, Hale JS, Lee J, Nasti TH, et al. 2016. Defining CD8$^+$ T cells that provide the proliferative burst after PD-1 therapy. *Nature* **537**: 417–421.

INSIGHT-ESPRIT Study Group; SILCAAT Scientific IN Committee; Abrams D, Levy Y, Losso MH, Babiker A, Collins G, Cooper DA, Darbyshire J, Emery S, et al. 2009. Interleukin-2 therapy in patients with HIV infection. *N Engl J Med* **361**: 1548–1559.

Jabbari A, Harty JT. 2006. Secondary memory CD8$^+$ T cells are more protective but slower to acquire a central-memory phenotype. *J Exp Med* **203**: 919–932.

Joshi NS, Cui W, Chandele A, Lee HK, Urso DR, Hagman J, Gapin L, Kaech SM. 2007. Inflammation directs memory precursor and short-lived effector CD8$^+$ T cell fates via the graded expression of T-bet transcription factor. *Immunity* **27**: 281–295.

Kaech SM, Cui W. 2012. Transcriptional control of effector and memory CD8$^+$ T cell differentiation. *Nat Rev Immunol* **12**: 749–761.

Kaech SM, Tan JT, Wherry EJ, Konieczny BT, Surh CD, Ahmed R. 2003. Selective expression of the interleukin 7 receptor identifies effector CD8 T cells that give rise to long-lived memory cells. *Nat Immunol* **4**: 1191–1198.

Kalia V, Sarkar S, Subramaniam S, Haining WN, Smith KA, Ahmed R. 2010. Prolonged interleukin-2Rα expression on virus-specific CD8$^+$ T cells favors terminal-effector differentiation in vivo. *Immunity* **32**: 91–103.

Kallies A, Xin A, Belz GT, Nutt SL. 2009. Blimp-1 transcription factor is required for the differentiation of effector CD8$^+$ T cells and memory responses. *Immunity* **31**: 283–295.

Kieper WC, Tan JT, Bondi-Boyd B, Gapin L, Sprent J, Ceredig R, Surh CD. 2002. Overexpression of interleukin (IL)-7 leads to IL-15–independent generation of memory phenotype CD8$^+$ T cells. *J Exp Med* **195**: 1533–1539.

Kim MT, Kurup SP, Starbeck-Miller GR, Harty JT. 2016. Manipulating memory CD8 T cell numbers by timed enhancement of IL-2 signals. *J Immunol* **197**: 1754–1761.

Klonowski KD, Williams KJ, Marzo AL, Lefrancois L. 2006. Cutting edge: IL-7-independent regulation of IL-7 receptor expression and memory CD8 T cell development. *J Immunol* **177**: 4247–4251.

Kovacs JA, Vogel S, Albert JM, Falloon J, Davey RT Jr, Walker RE, Polis MA, Spooner K, Metcalf JA, Baseler M, et al. 1996. Controlled trial of interleukin-2 infusions in patients infected with the human immunodeficiency virus. *N Engl J Med* **335**: 1350–1356.

Kundig TM, Schorle H, Bachmann MF, Hengartner H, Zinkernagel RM, Horak I. 1993. Immune responses in interleukin-2-deficient mice. *Science* **262**: 1059–1061.

Cite this article as *Cold Spring Harb Perspect Biol* doi: 10.1101/cshperspect.a028464

Lau LL, Jamieson BD, Somasundaram T, Ahmed R. 1994. Cytotoxic T-cell memory without antigen. *Nature* **369:** 648–652.

Leong YA, Chen Y, Ong HS, Wu D, Man K, Deleage C, Minnich M, Meckiff BJ, Wei Y, Hou Z, et al. 2016. CXCR5⁺ follicular cytotoxic T cells control viral infection in B cell follicles. *Nat Immunol* **17:** 1187–1196.

Levy Y, Lacabaratz C, Weiss L, Viard JP, Goujard C, Lelievre JD, Boue F, Molina JM, Rouzioux C, Avettand-Fenoel V, et al. 2009. Enhanced T cell recovery in HIV-1-infected adults through IL-7 treatment. *J Clin Invest* **119:** 997–1007.

Levy Y, Sereti I, Tambussi G, Routy JP, Lelievre JD, Delfraissy JF, Molina JM, Fischl M, Goujard C, Rodriguez B, et al. 2012. Effects of recombinant human interleukin 7 on T-cell recovery and thymic output in HIV-infected patients receiving antiretroviral therapy: Results of a phase I/IIa randomized, placebo-controlled, multicenter study. *Clin Infect Dis* **55:** 291–300.

Ley K, Laudanna C, Cybulsky MI, Nourshargh S. 2007. Getting to the site of inflammation: The leukocyte adhesion cascade updated. *Nat Rev Immunol* **7:** 678–689.

Liao W, Lin JX, Leonard WJ. 2013. Interleukin-2 at the crossroads of effector responses, tolerance, and immunotherapy. *Immunity* **38:** 13–25.

Link A, Vogt TK, Favre S, Britschgi MR, Acha-Orbea H, Hinz B, Cyster JG, Luther SA. 2007. Fibroblastic reticular cells in lymph nodes regulate the homeostasis of naïve T cells. *Nat Immunol* **8:** 1255–1265.

Lugli E, Mueller YM, Lewis MG, Villinger F, Katsikis PD, Roederer M. 2011. IL-15 delays suppression and fails to promote immune reconstitution in virally suppressed chronically SIV-infected macaques. *Blood* **118:** 2520–2529.

Mackay LK, Rahimpour A, Ma JZ, Collins N, Stock AT, Hafon ML, Vega-Ramos J, Lauzurica P, Mueller SN, Stefanovic T, et al. 2013. The developmental pathway for CD103⁺CD8⁺ tissue-resident memory T cells of skin. *Nat Immunol* **14:** 1294–1301.

Malek TR. 2008. The biology of interleukin-2. *Annu Rev Immunol* **26:** 453–479.

Marzo AL, Klonowski KD, Le Bon A, Borrow P, Tough DF, Lefrancois L. 2005. Initial T cell frequency dictates memory CD8⁺ T cell lineage commitment. *Nat Immunol* **6:** 793–799.

Mitchell DM, Ravkov EV, Williams MA. 2010. Distinct roles for IL-2 and IL-15 in the differentiation and survival of CD8⁺ effector and memory T cells. *J Immunol* **184:** 6719–6730.

Molloy MJ, Zhang W, Usherwood EJ. 2009. Cutting edge: IL-2 immune complexes as a therapy for persistent virus infection. *J Immunol* **182:** 4512–4515.

Mueller SN, Ahmed R. 2009. High antigen levels are the cause of T cell exhaustion during chronic viral infection. *Proc Natl Acad Sci* **106:** 8623–8628.

Mueller YM, Petrovas C, Bojczuk PM, Dimitriou ID, Beer B, Silvera P, Villinger F, Cairns JS, Gracely EJ, Lewis MG, et al. 2005. Interleukin-15 increases effector memory CD8⁺ T cells and NK Cells in simian immunodeficiency virus-infected macaques. *J Virol* **79:** 4877–4885.

Mueller YM, Do DH, Altork SR, Artlett CM, Gracely EJ, Katsetos CD, Legido A, Villinger F, Altman JD, Brown CR, et al. 2008. IL-15 treatment during acute simian immunodeficiency virus (SIV) infection increases viral set point and accelerates disease progression despite the induction of stronger SIV-specific CD8⁺ T cell responses. *J Immunol* **180:** 350–360.

Mylvaganam GH, Rios D, Abdelaal HM, Iyer S, Tharp G, Mavinger M, Hicks S, Chahroudi A, Ahmed R, Bosinger SE, et al. 2017. Dynamics of SIV-specific CXCR5⁺ CD8 T cells during chronic SIV infection. *Proc Natl Acad Sci* **114:** 1976–1981.

Nanjappa SG, Walent JH, Morre M, Suresh M. 2008. Effects of IL-7 on memory CD8 T cell homeostasis are influenced by the timing of therapy in mice. *J Clin Invest* **118:** 1027–1039.

Nanjappa SG, Kim EH, Suresh M. 2011. Immunotherapeutic effects of IL-7 during a chronic viral infection in mice. *Blood* **117:** 5123–5132.

Nolz JC, Harty JT. 2014. IL-15 regulates memory CD8⁺ T cell O-glycan synthesis and affects trafficking. *J Clin Invest* **124:** 1013–1026.

Nugeyre MT, Monceaux V, Beq S, Cumont MC, Fang RHT, Chene L, Morre M, Barre-Sinoussi F, Hurtrel B, Israel N. 2003. IL-7 stimulates T cell renewal without increasing viral replication in simian immunodeficiency virus-infected macaques. *J Immunol* **171:** 4447–4453.

Obar JJ, Khanna KM, Lefrancois L. 2008. Endogenous naïve CD8⁺ T cell precursor frequency regulates primary and memory responses to infection. *Immunity* **28:** 859–869.

Obar JJ, Molloy MJ, Jellison ER, Stoklasek TA, Zhang W, Usherwood EJ, Lefrancois L. 2010. CD4⁺ T cell regulation of CD25 expression controls development of short-lived effector CD8⁺ T cells in primary and secondary responses. *Proc Natl Acad Sci* **107:** 193–198.

Opferman JT, Letai A, Beard C, Sorcinelli MD, Ong CC, Korsmeyer SJ. 2003. Development and maintenance of B and T lymphocytes requires antiapoptotic MCL-1. *Nature* **426:** 671–676.

Osborne LC, Dhanji S, Snow JW, Priatel JJ, Ma MC, Miners MJ, Teh HS, Goldsmith MA, Abraham N. 2007. Impaired CD8 T cell memory and CD4 T cell primary responses in IL-7Rα mutant mice. *J Exp Med* **204:** 619–631.

O'Sullivan D, van der Windt GJ, Huang SC, Curtis JD, Chang CH, Buck MD, Qiu J, Smith AM, Lam WY, DiPlato LM, et al. 2014. Memory CD8⁺ T cells use cell-intrinsic lipolysis to support the metabolic programming necessary for development. *Immunity* **41:** 75–88.

Pauken KE, Sammons MA, Odorizzi PM, Manne S, Godec J, Khan O, Drake AM, Chen Z, Sen DR, Kurachi M, et al. 2016. Epigenetic stability of exhausted T cells limits durability of reinvigoration by PD-1 blockade. *Science* **354:** 1160–1165.

Pellegrini M, Calzascia T, Elford AR, Shahinian A, Lin AE, Dissanayake D, Dhanji S, Nguyen LT, Gronski MA, Morre M, et al. 2009. Adjuvant IL-7 antagonizes multiple cellular and molecular inhibitory networks to enhance immunotherapies. *Nat Med* **15:** 528–536.

Pellegrini M, Calzascia T, Toe JG, Preston SP, Lin AE, Elford AR, Shahinian A, Lang PA, Lang KS, Morre M, et al. 2011. IL-7 engages multiple mechanisms to overcome chronic

M. Hashimoto et al.

viral infection and limit organ pathology. *Cell* **144**: 601–613.

Pipkin ME, Sacks JA, Cruz-Guilloty F, Lichtenheld MG, Bevan MJ, Rao A. 2010. Interleukin-2 and inflammation induce distinct transcriptional programs that promote the differentiation of effector cytolytic T cells. *Immunity* **32**: 79–90.

Richer MJ, Pewe LL, Hancox LS, Hartwig SM, Varga SM, Harty JT. 2015. Inflammatory IL-15 is required for optimal memory T cell responses. *J Clin Invest* **125**: 3477–3490.

Ring AM, Lin JX, Feng D, Mitra S, Rickert M, Bowman GR, Pande VS, Li P, Moraga I, Spolski R, et al. 2012. Mechanistic and structural insight into the functional dichotomy between IL-2 and IL-15. *Nat Immunol* **13**: 1187–1195.

Rochman Y, Spolski R, Leonard WJ. 2009. New insights into the regulation of T cells by γc family cytokines. *Nat Rev Immunol* **9**: 480–490.

Rubinstein MP, Kovar M, Purton JF, Cho JH, Boyman O, Surh CD, Sprent J. 2006. Converting IL-15 to a superagonist by binding to soluble IL-15Rα. *Proc Natl Acad Sci* **103**: 9166–9171.

Rubinstein MP, Lind NA, Purton JF, Filippou P, Best JA, McGhee PA, Surh CD, Goldrath AW. 2008. IL-7 and IL-15 differentially regulate CD8+ T-cell subsets during contraction of the immune response. *Blood* **112**: 3704–3712.

Rutishauser RL, Martins GA, Kalachikov S, Chandele A, Parish IA, Meffre E, Jacob J, Calame K, Kaech SM. 2009. Transcriptional repressor Blimp-1 promotes CD8+ T cell terminal differentiation and represses the acquisition of central memory T cell properties. *Immunity* **31**: 296–308.

Sandau MM, Kohlmeier JE, Woodland DL, Jameson SC. 2010. IL-15 regulates both quantitative and qualitative features of the memory CD8 T cell pool. *J Immunol* **184**: 35–44.

Sarkar S, Kalia V, Haining WN, Konieczny BT, Subramaniam S, Ahmed R. 2008. Functional and genomic profiling of effector CD8 T cell subsets with distinct memory fates. *J Exp Med* **205**: 625–640.

Schenkel JM, Masopust D. 2014. Tissue-resident memory T cells. *Immunity* **41**: 886–897.

Schenkel JM, Fraser KA, Masopust D. 2014. Cutting edge: Resident memory CD8 T cells occupy frontline niches in secondary lymphoid organs. *J Immunol* **192**: 2961–2964.

Schietinger A, Greenberg PD. 2014. Tolerance and exhaustion: Defining mechanisms of T cell dysfunction. *Trends Immunol* **35**: 51–60.

Schluns KS, Kieper WC, Jameson SC, Lefrancois L. 2000. Interleukin-7 mediates the homeostasis of naïve and memory CD8 T cells in vivo. *Nat Immunol* **1**: 426–432.

Schluns KS, Williams K, Ma A, Zheng XX, Lefrancois L. 2002. Cutting edge: Requirement for IL-15 in the generation of primary and memory antigen-specific CD8 T cells. *J Immunol* **168**: 4827–4831.

Sen DR, Kaminski J, Barnitz RA, Kurachi M, Gerdemann U, Yates KB, Tsao HW, Godec J, LaFleur MW, Brown FD, et al. 2016. The epigenetic landscape of T cell exhaustion. *Science* **354**: 1165–1169.

Sereti I, Dunham RM, Spritzler J, Aga E, Proschan MA, Medvik K, Battaglia CA, Landay AL, Pahwa S, Fischl MA, et al. 2009. IL-7 administration drives T cell-cycle entry and expansion in HIV-1 infection. *Blood* **113**: 6304–6314.

Shin H, Blackburn SD, Blattman JN, Wherry EJ. 2007. Viral antigen and extensive division maintain virus-specific CD8 T cells during chronic infection. *J Exp Med* **204**: 941–949.

Soudja SM, Ruiz AL, Marie JC, Lauvau G. 2012. Inflammatory monocytes activate memory CD8+ T and innate NK lymphocytes independent of cognate antigen during microbial pathogen invasion. *Immunity* **37**: 549–562.

Steel JC, Waldmann TA, Morris JC. 2012. Interleukin-15 biology and its therapeutic implications in cancer. *Trends Pharmacol Sci* **33**: 35–41.

Stoklasek TA, Schluns KS, Lefrancois L. 2006. Combined IL-15/IL-15R immunotherapy maximizes IL-15 activity in vivo. *J Immunol* **177**: 6072–6080.

Su HC, Cousens LP, Fast LD, Slifka MK, Bungiro RD, Ahmed R, Biron CA. 1998. CD4+ and CD8+ T cell interactions in IFN-γ and IL-4 responses to viral infections: Requirements for IL-2. *J Immunol* **160**: 5007–5017.

Sun JC, Lehar SM, Bevan MJ. 2006. Augmented IL-7 signaling during viral infection drives greater expansion of effector T cells but does not enhance memory. *J Immunol* **177**: 4458–4463.

Surh CD, Sprent J. 2008. Homeostasis of naïve and memory T cells. *Immunity* **29**: 848–862.

Suzuki H, Kundig T, Furlonger C, Wakeham A, Timms E, Matsuyama T, Schmits R, Simard J, Ohashi P, Griesser H, et al. 1995. Deregulated T cell activation and autoimmunity in mice lacking interleukin-2 receptor β. *Science* **268**: 1472–1476.

Tan JT, Ernst B, Kieper WC, LeRoy E, Sprent J, Surh CD. 2002. Interleukin (IL)-15 and IL-7 jointly regulate homeostatic proliferation of memory phenotype CD8+ cells but are not required for memory phenotype CD4+ cells. *J Exp Med* **195**: 1523–1532.

Topalian SL, Hodi FS, Brahmer JR, Gettinger SN, Smith DC, McDermott DF, Powderly JD, Carvajal RD, Sosman JA, Atkins MB, et al. 2012. Safety, activity, and immune correlates of anti-PD-1 antibody in cancer. *N Engl J Med* **366**: 2443–2454.

Turley SJ, Fletcher AL, Elpek KG. 2010. The stromal and haematopoietic antigen-presenting cells that reside in secondary lymphoid organs. *Nat Rev Immunol* **10**: 813–825.

Utzschneider DT, Charmoy M, Chennupati V, Pousse L, Ferreira DP, Calderon-Copete S, Danilo M, Alfei F, Hofmann M, Wieland D, et al. 2016. T cell factor 1-expressing memory-like CD8+ T cells sustain the immune response to chronic viral infections. *Immunity* **45**: 415–427.

van der Windt GJ, Everts B, Chang CH, Curtis JD, Freitas TC, Amiel E, Pearce EJ, Pearce EL. 2012. Mitochondrial respiratory capacity is a critical regulator of CD8+ T cell memory development. *Immunity* **36**: 68–78.

Vassena L, Miao H, Cimbro R, Malnati MS, Cassina G, Proschan MA, Hirsch VM, Lafont BA, Morre M, Fauci AS, et al. 2012. Treatment with IL-7 prevents the decline of circulating CD4+ T cells during the acute phase of SIV infection in rhesus macaques. *PLoS Pathog* **8**: e1002636.

von Freeden-Jeffry U, Solvason N, Howard M, Murray R. 1997. The earliest T lineage–committed cells depend on IL-7 for Bcl-2 expression and normal cell cycle progression. *Immunity* **7**: 147–154.

Waldmann TA. 2006. The biology of interleukin-2 and interleukin-15: Implications for cancer therapy and vaccine design. *Nat Rev Immunol* **6**: 595–601.

West EE, Jin HT, Rasheed AU, Penaloza-Macmaster P, Ha SJ, Tan WG, Youngblood B, Freeman GJ, Smith KA, Ahmed R. 2013. PD-L1 blockade synergizes with IL-2 therapy in reinvigorating exhausted T cells. *J Clin Invest* **123**: 2604–2615.

Wherry EJ, Kurachi M. 2015. Molecular and cellular insights into T cell exhaustion. *Nat Rev Immunol* **15**: 486–499.

Wherry EJ, Blattman JN, Murali-Krishna K, van der Most R, Ahmed R. 2003a. Viral persistence alters CD8 T-cell immunodominance and tissue distribution and results in distinct stages of functional impairment. *J Virol* **77**: 4911–4927.

Wherry EJ, Teichgraber V, Becker TC, Masopust D, Kaech SM, Antia R, von Andrian UH, Ahmed R. 2003b. Lineage relationship and protective immunity of memory CD8 T cell subsets. *Nat Immunol* **4**: 225–234.

Wherry EJ, Barber DL, Kaech SM, Blattman JN, Ahmed R. 2004. Antigen-independent memory CD8 T cells do not develop during chronic viral infection. *Proc Natl Acad Sci* **101**: 16004–16009.

Wherry EJ, Ha SJ, Kaech SM, Haining WN, Sarkar S, Kalia V, Subramaniam S, Blattman JN, Barber DL, Ahmed R. 2007. Molecular signature of CD8$^+$ T cell exhaustion during chronic viral infection. *Immunity* **27**: 670–684.

Williams MA, Bevan MJ. 2007. Effector and memory CTL differentiation. *Annu Rev Immunol* **25**: 171–192.

Williams MA, Tyznik AJ, Bevan MJ. 2006. Interleukin-2 signals during priming are required for secondary expansion of CD8$^+$ memory T cells. *Nature* **441**: 890–893.

Wirth TC, Xue HH, Rai D, Sabel JT, Bair T, Harty JT, Badovinac VP. 2010. Repetitive antigen stimulation induces stepwise transcriptome diversification but preserves a core signature of memory CD8$^+$ T cell differentiation. *Immunity* **33**: 128–140.

Wu T, Ji Y, Moseman EA, Xu HC, Manglani M, Kirby M, Anderson SM, Handon R, Kenyon E, Elkahloun A, et al. 2016. The TCF1-Bcl6 axis counteracts type I interferon to repress exhaustion and maintain T cell stemness. *Sci Immunol* **1**: eaai8593.

Yajima T, Yoshihara K, Nakazato K, Kumabe S, Koyasu S, Sad S, Shen H, Kuwano H, Yoshikai Y. 2006. IL-15 regulates CD8$^+$ T cell contraction during primary infection. *J Immunol* **176**: 507–515.

Innate Lymphoid Cells (ILCs): Cytokine Hubs Regulating Immunity and Tissue Homeostasis

Maho Nagasawa, Hergen Spits, and Xavier Romero Ros

Department of Experimental Immunology, Academic Medical Center at the University of Amsterdam, 1105 BA Amsterdam, Netherlands

Correspondence: hergen.spits@amc.uva.nl

Innate lymphoid cells (ILCs) have emerged as an expanding family of effector cells particularly enriched in the mucosal barriers. ILCs are promptly activated by stress signals and multiple epithelial- and myeloid-cell-derived cytokines. In response, ILCs rapidly secrete effector cytokines, which allow them to survey and maintain the mucosal integrity. Uncontrolled action of ILCs might contribute to tissue damage, chronic inflammation, metabolic diseases, autoimmunity, and cancer. Here we discuss the recent advances in our understanding of the cytokine network that modulate ILC immune responses: stimulating cytokines, signature cytokines secreted by ILC subsets, autocrine cytokines, and cytokines that induce cell plasticity.

Innate lymphoid cells (ILCs) are innate lymphocytes that play important roles in immune defense against microbes, regulation of adaptive immunity, tissue remodeling, and repair and homeostasis of hematopoietic and nonhematopoietic cell types. ILCs are present in all tissues, but they are particularly enriched in mucosal surfaces. Unlike adaptive lymphocytes, ILCs do not possess rearranged antigen-specific cell receptors (T-cell receptor [TCR] or B-cell receptor [BCR]), but they mirror T helper (Th) cell diversity regarding the secretion of signature cytokines and key transcription factors that regulate their differentiation and functions (Spits et al. 2013). ILCs develop from the common lymphoid progenitor (CLP) early in life and seed various tissues to become tissue-resident lymphocytes (Diefenbach et al. 2014;

Klose et al. 2014; Gasteiger et al. 2015). ILCs cross talk with the resident tissue by sensing the cytokines present in their microenvironments and subsequently secreting a plethora of cytokines that regulate innate immunity and homeostasis of hematopoietic and nonhematopoietic cells in the tissues (Artis and Spits 2015). ILC dysregulation contributes to several pathological conditions, such as inflammatory bowel disease (IBD), chronic obstructive pulmonary disease (COPD), asthma, psoriasis, and atopic dermatitis (Artis and Spits 2015; Eberl et al. 2015). In this review, we will focus on the noncytotoxic ILC subsets, which are also termed as helper ILCs. We discuss the cytokines that influence ILC biology and the role of effector cytokines produced by ILCs in health and disease.

ILC FAMILY AND SUBSETS

All ILCs express the common cytokine receptor γ chain (γc) together with interleukin (IL)-7 receptor (R)α (IL-7Rα, also called CD127), whereas killer ILCs, natural killer (NK) cells, and intraepithelial (ie) ILC1s, lack the expression of IL-7R but instead express IL-2Rβ (also known as CD122) (Artis and Spits 2015; Eberl et al. 2015). ILCs are divided into three groups based on the expression of specific transcription factors and cell-surface molecules as well as their ability to secrete key cytokines (Table 1). Group 1 ILCs comprising NK cells and ILC1s produce interferon (IFN)-γ in response to IL-12 and are dependent on T-bet; group 2 ILCs (ILC2s) preferential-

ly produce type 2 cytokines (IL-5, IL-4, IL-9, and IL-13) in response to IL-33, IL-25, and thymic stromal lymphopoietin (TSLP) and rely on GATA3 as their key transactional factor; group 3 ILCs include ILC3s and lymphoid tissue inducer (LTi) cells endowed with the ability to secrete IL-17 and IL-22 in response to IL-1β and IL-23, and are functionally dependent on the transcription factor RAR-related orphan receptor γt (RORγt).

CYTOKINES REQUIRED FOR ILC DEVELOPMENT AND MAINTENANCE

ILC ontogeny has been intensively studied particularly in mouse (Fig. 1). ILCs differentiate

Table 1. Mouse and human innate lymphoid cell (ILC) phenotypes

| Group | Progenitor | Mouse | | Human | | |
		Cell-surface molecules	TFs	Cell-surface molecules	TFs	Cytokines
1	cNK	CD122, CD49b, NK1.1, NKG2A, NKp46, IL-12RB1, CD25, KLRG1	EOMES T-bet	CD122, NKG2A, NKp46, NKp44$^{+/-}$, IL-12RB1, CD25, KLRG1	EOMES T-bet	IFN-γ, TNF
	ieILC1	CD122, CD90, NK1.1, NKG2A, NKp46, IL-12RB1	EOMES T-bet	CD122, NKG2A, NKp46, NKp44, IL-12RB1	EOMES T-bet	IFN-γ
	ILC1	CD127, CD122, CD90, CD49a, NK1.1, NKp46, IL-12RB1	T-bet	CD127, CD161, IL-12RB1, KLRG1, ICOS, CD4$^{+/-}$	T-bet	IFN-γ, TNF
2	ILC2	CD127, CD90, CRTH2, CD117$^{+/-}$, CD25, ST2, TSLPR, IL-17RB, KLRG1, ICOS, MHCII	GATA3	CD127, CD161, CRTH2, CD117$^{+/-}$, CD25, ST2, TSLPR, IL17RB, KLRG1, ICOS, CCR6, MHCII	GATA3	IL-4, IL-5, IL-9, IL-13
3	ILC3 NCR$^+$	CD127, CD90, CD117, NKp46, CD25, ICOS, IL-23R, L-1R1	RORγt T-bet	CD127, CD161, CD117, NKp46, NKp44, IL-12RB1, CD25, ICOS, CCR6, IL-23R, IL-1R1, MHCII	RORγt	IL-22
	ILC3 NCR$^-$	CD127, CD90, CD25, ICOS, IL-23R, IL-1R1, MHCII	RORγt T-bet	CD127, CD161, CD117, IL-12RB1, CD25, ICOS, CCR6, IL-23R, IL-1R1	RORγt	IL-17A
	LTi	CD127, CD90, CD117, CD25, CCR6, IL-23R, IL-1R1, MCHII, CD4$^{+/-}$, NRP1	RORγt	CD127, CD161, CD117, CD25, CCR6, IL-23R, IL-1R1, NRP1	RORγt	IL-17A, IL-22

ICOS, Inducible T-cell costimulator; KLRG1, killer cell lectin-like receptor subfamily G member 1; NCR, natural cytotoxicity receptor; GATA3, GATA-binding factor 3; EOMES, eomesodermin; TSLPR, thymic stromal lymphopoietin receptor; TFs; transcription factors; IFN, interferon; TNF, tumor necrosis factor; IL, interleukin; RORγt, RAR-related orphan receptor γt.

 Cite this article as *Cold Spring Harb Perspect Biol* doi: 10.1101/cshperspect.a030304

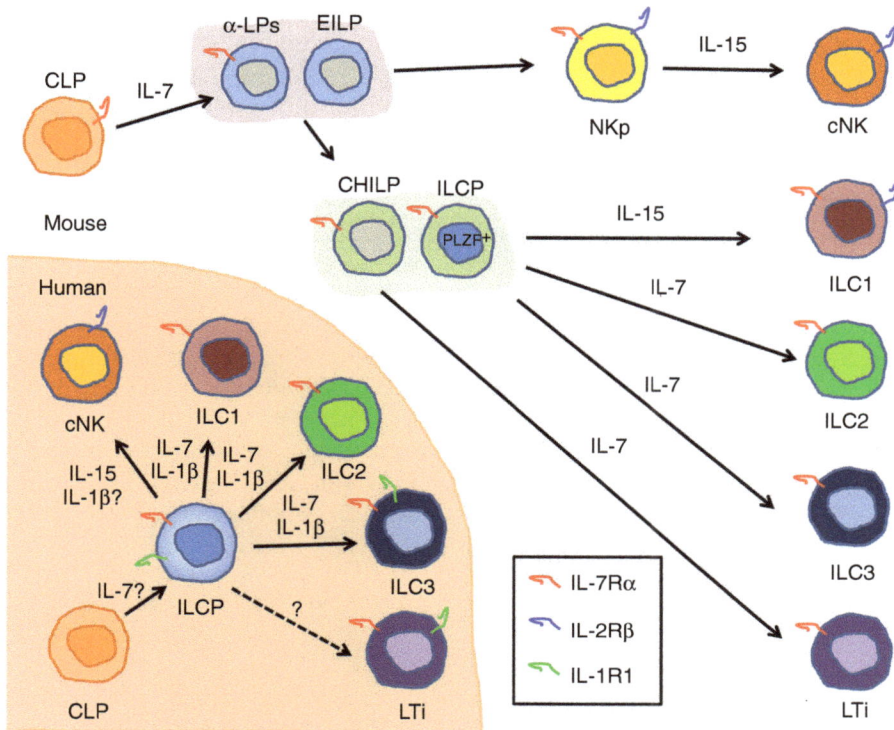

Figure 1. Cytokines involved in mouse and human innate lymphoid cells (ILCs) development. All ILCs are derived from a common lymphoid progenitor (CLP). In the mouse, interleukin (IL)-7 is essential for the development of ILC2, ILC3, and lymphoid tissue inducer (LTi) cells, and IL-15 is required for development of natural killer (NK) cells. IL-15 is important for ILC1 development but some ILC1 can develop independent of IL-15. For humans, IL-7 and IL-15 also play an important role in ILC and NK-cell development, and some studies indicate the involvement of IL-1β as, thus far, identified ILC progenitors (ILCPs) all express its receptor. α-LP, α-Lymphoid progenitor; EILP, early innate lymphoid progenitor; CHILP, common helper-like ILC progenitor; PLZF, promyelocytic leukemia zinc-finger protein; NKp, NK progenitor; cNK, conventional NK.

from the CLP, which resides in the bone marrow (Possot et al. 2011; Hoyler et al. 2012). CLPs further develop into progenitors expressing the α4β7 integrin (called α lymphoid progenitors [α-LPs]) and the early innate lymphoid progenitors (EILPs) that are able to produce NK cells and all helper ILCs but have lost the potential to differentiate into B and T cells (Yu et al. 2014; Yang et al. 2015). α-LPs and EILPs develop into the common helper-like ILC progenitors (CHILPs) able to differentiate into all helper ILC subsets, including LTi cells, but not into NK cells (Klose et al. 2014). The CHILP population contains a subset of cells expressing pro-myelocytic leukemia zinc-finger protein (PLZF),

so-called ILC progenitors (ILCPs), which have lost the potential to develop into LTi cells but retain the ability to generate helper ILCs (Constantinides et al. 2014). ILC development is dependent on the expression of the common cytokine receptor γc, which is shared by the IL-2, IL-4, IL-7, IL-9, IL-15, and IL-21 receptors (Rochman et al. 2009). IL-7 and IL-15 are the γc family cytokines that are important for ILC development. NK cells express IL-2/IL-15Rβ chain (IL-2Rβ or CD122) and IL-15Rα but not IL-7Rα chain (or CD127), and their development and/or maintenance is dependent on IL-15 (Kennedy et al. 2000), whereas all helper ILCs express IL-7Rα and particularly ILC2s and ILC3s require IL-7 for their development and/or maintenance (Moro et al. 2010; Neill et al. 2010; Vonarbourg et al. 2010; Hoyler et al. 2012). Recently, however, it was reported that functional ILC2s and ILC3s are present in the lamina propria but not in other organs of $Il7ra^{-/-}$ mice, indicating that the requirement for IL-7R is incomplete. IL-15 could sustain these IL-7R-independent ILC2s and ILC3s (Robinette et al. 2017). IL-7Rα and IL-2Rβ are expressed on ILC1s, but IL-7 signaling is not required for their development and maintenance because mice lacking IL-7Rα displayed normal ILC1 development. In contrast, ILC1 cell numbers were significantly reduced in IL-15-deficient mice but not absent, suggesting that ILC1 development partially relies on additional factor(s) besides IL-15 (Klose et al. 2014).

ILC development depends on the transcriptional regulator, inhibitor of DNA-binding 2 (Id2), as Id2 deficiency leads to developmental block of all ILCs (Yokota et al. 1999; Boos et al. 2007; Moro et al. 2010). Id2 expression in CHILPs is controlled by the basic leucine zipper transcription factor NFIL3, promoting CHILP commitment toward ILCp and facilitating the development of ILC1s, ILC2s, and ILC3s (Yu et al. 2014; Xu et al. 2015). NFIL3 expression is controlled by IL-7/signal transducers and activators of transcription (STAT)5 signaling. Indeed, IL-7 and IL-15 are differentially required by ILC subsets, governing their differentiation and/or survival. Thus, CHILPs as well as ILCPs are characterized by their expression of Id2 and give rise to various ILC subsets, whereas NK cells develop from Id2-negative CLPs and require IL-15 signaling (Diefenbach et al. 2014; Huntington 2014; Klose et al. 2014).

The intermediate cellular stages of human ILC development are less clear than those in mice (Fig. 1). An IL-1R1$^+$CD34$^+$RORγt$^+$ hematopoietic progenitor cell (HPC) was found in tonsil and intestinal lamina propria that gives rise to ILC3s (Montaldo et al. 2014), but another study showed that a similar population of Lin$^-$CD34$^+$CD45RA$^+$CD117$^+$IL-1R1$^+$ in secondary lymphoid tissues could develop into all ILC subsets, including NK cells (Scoville et al. 2016). Most recently, Lin$^-$CD127$^+$CD117$^+$IL-1R1$^+$ ILC precursors were found in the peripheral tissues and circulation, which could generate all ILC subsets (Lim et al. 2017). These ILC precursors may be the innate version of naïve T cells that can develop into all Th cell subsets under the influence of cytokines. Also, in humans, ILC development depends on γc because patients who are deficient in this receptor (i.e., patients with X-linked severe combined immunodeficiency [XSCID]) lack helper ILC and NK cells (Vely et al. 2016). In one study, it was shown that *IL7R*-deficient SCID patients lack ILC3s (Vonarbourg et al. 2010), indicating that, like in mice, human ILC3s require IL-7 for development. It is to be expected that patients with XSCID also have diminished ILC1s and ILC2s but this has yet to be reported.

CYTOKINES: THE INS AND OUTS OF ILCs

Group 1 ILCs

Group 1 ILCs include NK cells and ILC1s (Table 1). These cell types are distinct because their developmental trajectories are different (Constantinides et al. 2014; Klose et al. 2014) Whereas NK cells are dedicated cytotoxic cells expressing cytotoxic molecules like perforin and granzymes, ILC1s are in general noncytotoxic, although ILC1 subsets with weak cytotoxic activity have been described. This might be comparable with Th1 cells that under certain conditions mediate cytotoxic activity as well. In naïve mice, ILC1 express CD127, CD49a, and the transcription factor T-bet but not the related factor Eomes, whereas NK cells lack CD127 and express Eomes and CD49b. However, as reviewed (Spits et al. 2016), it is difficult to unambiguously distinguish ILC1s from NK cells, particularly in mice with ongoing immune responses.

ILC1s produce IFN-γ and tumor necrosis factor (TNF) (Fig. 2) (Vonarbourg et al. 2010; Bernink et al. 2013; Fuchs et al. 2013; Daussy et al. 2014; Klose et al. 2014). Like NK cells, ILC1s are activated by IL-12, IL-18, and IL-15 (Satoh-Takayama et al. 2009; Fuchs et al. 2013), IL-12 and IL-18 are produced by monocytes and dendritic cells (DCs) activated during infections (Kang et al. 2008), whereas IL-15 is expressed

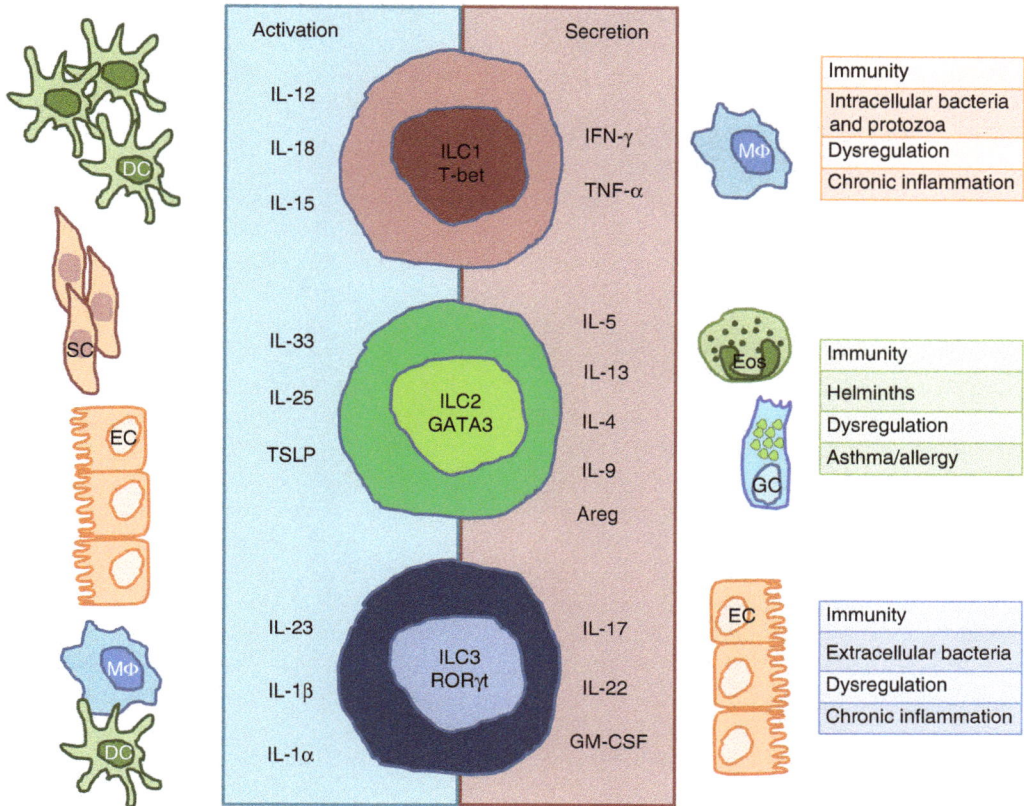

Figure 2. Innate lymphoid cell (ILC) subsets and cytokines. ILC1s depend on transcription factor T-bet for their function and produce interferon (IFN)-γ and tumor necrosis factor (TNF)-α in response to interleukin (IL)-12, IL-15, and IL-18 derived from dendritic cells (DCs). ILC2s require GATA3 and are endowed with the ability to secrete IL-4, IL-5, IL-9, IL-13, and amphiregulin (Areg) in response to IL-25, IL-33, and thymic stromal lymphopoietin (TSLP) derived from epithelial cells and tissue-resident immune cells. ILC3s depend on the transcription factor RAR-related orphan receptor γt (RORγt) and secrete IL-17, IL-22, and granulocyte macrophage colony-stimulating factor (GM-CSF) in response to IL-23, IL-1β, and IL-1α derived from macrophages and DCs. ILC immune responses are shaped to cope with different types of pathogens, although their dysregulation might lead to chronic inflammation. SC, Stromal cell; EC, epithelial cell; Eos, eosinophils; GC, goblet cell.

not only by activated monocytes and macrophages, but also by a wide variety of tissues, including placenta, skeletal muscle, kidney, epithelial, and fibroblast cell lines, which describe its pleiotropic function (Grabstein et al. 1994). IFN-γ plays a crucial role in innate and adaptive immunity against viral and intracellular bacterial infections by inhibiting viral replication directly and promoting macrophage activation, which increases phagocytosis and directly or indirectly up-regulates both major histocompatibility complex (MHC) class I and class II an-

tigen presentation (Schoenborn and Wilson 2007). ILC1s accumulate in a variety of inflammatory diseases, including Crohn's disease, and the resulting increased levels of IFN-γ production may contribute to chronical intestinal inflammation such as IBD (Bernink et al. 2013; Fuchs et al. 2013).

Group 2 ILCs

Both in human and mouse, ILC2s are characterized by their high expression of transcription factor GATA3, which is crucial for their ability to produce the Th2 type of cytokines IL-4, IL-5, and IL-13 (Hoyler et al. 2012; Mjosberg et al. 2012). Mouse ILC2s can be identified by the combined expression of IL-7Rα, CD25, CD90, SCA-1, ICOS, and KLRG1, but they lack the lineage markers CD3, B220, CD11b, Ter119, and Gr-1 (Moro et al. 2010; Neill et al. 2010; Price et al. 2010), and NK1.1 and NKp46, which are expressed on part of the ILC3s and ILC1s. In humans, ILC2s uniformly express IL-7Rα, CD161, CRTH2, and CD25 (Table 1) (Mjosberg et al. 2011, 2012; Monticelli et al. 2011).

ILC2s produce IL-5 and IL-13 in response to IL-33, IL-25, or both IL-33 and IL-25 (Moro et al. 2010; Neill et al. 2010; Price et al. 2010), and TSLP enhances production of those cytokines (Fig. 2) (Kim et al. 2013). ILC2-activated cytokines are produced by many cell types. IL-33 is constitutively expressed in airway epithelial cells and its expression can be induced in keratinocytes, endothelial cells, fibroblast, smooth muscle cells, macrophages, and DCs on stimulation (Schmitz et al. 2005; Carriere et al. 2007). TSLP is primarily expressed by epithelial cells and keratinocytes in the skin, gut, and lungs (Reche et al. 2001), and involved in the regulation of type 2 cytokine-driven inflammatory processes occurring at the barrier surface (West et al. 2012; Ziegler 2012). IL-25 is produced by activated Th2 cells, macrophages, mast cells, eosinophils, basophils, lung, intestinal, and skin epithelial cells, and endothelial cell fibroblasts (Fort et al. 2001; Zaph et al. 2008; Valizadeh et al. 2015). More recently, specialized cell types within the epithelial layer of the intestine, called Tuft cells, were found to produce IL-25 (von Moltke et al. 2016).

Recently, the existence of a second ILC2 subtype was reported, which was called inflammatory ILC2 (iILC2). iILC2s appear in vivo on injection into mice of IL-25 and were distinct from IL-33-responsive natural ILC2 (nILC2) (Huang et al. 2015). Like nILC2s, iILC2s contribute to expulsion of the parasite *Nippostrongylus brasiliensis* by producing IL-13 but they also partially protect mice against the fungus *Candida albicans* by producing IL-17. Because iILC2s can develop into nILC2s in vivo and in vitro, it was

speculated that iILC2s are a transient precursor that arises on inflammation. However, it is as-yet unknown from which precursor iILC2s develop.

IL-5, produced by ILC2s, plays a crucial role in the development, activation, and survival of eosinophils (Lopez et al. 1988; Yamaguchi et al. 1988; Dent et al. 1990; Kopf et al. 1996). Eosinophils are implicated in the pathogenesis of variable inflammatory processes, including helminth, bacterial, and viral infections, tissue injury, tumor immunity, and allergic diseases (Hogan et al. 2008). IL-13 has both pro- and anti-inflammatory effects, depending on the target cells (Wynn 2003), and its role in mucosal immunology is particularly well appreciated. For instance, IL-13 induces hyperplasia and mucus production by goblet cells and production of the eosinophils chemoattractant eotaxin by epithelial cells (Grunig et al. 1998; Zhu et al. 1999; Pope et al. 2001; Mishra and Rothenberg 2003). Consistent with these reports, ILC2s derived IL-13 induce mucus production, helminth expulsion, eotaxin production, eosinophil recruitment, airway hyperreactivity, and pulmonary fibrosis (Schmitz et al. 2005; Bhandoola and Sambandam 2006; Kang et al. 2008; Rochman et al. 2009; Diefenbach et al. 2014; Roediger et al. 2015).

Activation of ILC2s is regulated by cytokines produced by ILC2s, including IL-4 and IL-9. Although ILC2s are able to produce IL-4 in vitro (Moro et al. 2010; Barlow et al. 2012; Mjosberg et al. 2012; Salimi et al. 2013), there is little evidence that IL-4 is secreted in vivo during steady-state or helminth infection (Price et al. 2010; Liang et al. 2011; Roediger et al. 2015), suggesting that production of IL-4, IL-5, and IL-13 are regulated differentially. This notion is consistent with the observations that Leukotriene D4 promotes calcium signaling, NFAT activation, and IL-4 secretion (Doherty et al. 2013), and that prostaglandin D2 (PGD2), which interacts with CRTH2 and triggers mobilization and enhancement of NFAT, induced a massive production of IL-4 (Xue et al. 2014). These findings indicate that IL-4 requires an additional NFAT-activating signal to be efficiently produced. IL-4 is an autocrine factor not only produced by ILC2s but also able to coactivate these cells. IL-4 acts in synergy with IL-33 to induce ILC2 proliferation and production of IL-5 and IL-13, although IL-4 by itself was insufficient to induce production of IL-5 and IL-13. Basophils and eosinophils cross talk with ILC2s by producing IL-4, which enhances the activities of ILC2s. Recently, we showed the existence of a positive feedback loop in the interaction of ILC2s and eosinophils; IL-4 produced by eosinophils coactivates ILC2s to produce IL-5, which activates eosinophils (Bal et al. 2016). This positive feedback loop may play a role in the inflammatory airway disease chronic rhinosinusitis (CRS). In the inflamed nasal polyps of CRS patients, eosinophils colocalize with ILC2 (Bal et al. 2016).

ILC2s are predominant producers of IL-9 (Wilhelm et al. 2011; Licona-Limon et al. 2013). IL-9 fate reporter mice infected with *N. brasiliensis* displayed a higher IL-9 expression in lung ILC2s when compared to T cells. Interestingly, IL-9R, a member of the γc receptor family, is expressed by ILC2s at much higher levels than by T cells (Price et al. 2010). Turner et al. (2013) showed that *Il9r*-deficient mice infected with *N. brasiliensis* showed a reduced number of ILC2s and impaired production of IL-13, IL-5, and amphiregulin, resulting in a reduction of eosinophil recruitment, delay of worm expulsion, and diminished tissue repair. In line with this study, it has been shown that IL-9 acts in an autocrine fashion on ILC2s, enhancing the production of IL-5 and IL-13 in papain-induced airway inflammation (Wilhelm et al. 2011). Furthermore, the IL-9 signal was crucial for the survival of activated ILC2s, which was mediated by the antiapoptotic protein BCL2 (Turner et al. 2013). Interestingly, IL-9 production seems to be differently regulated than IL-5 and IL-13. In lung ILC2s stimulated with IL-33, IL-9 was up-regulated in T-bet-deficient mice without affecting IL-5 and IL-13, suggesting that T-bet mainly suppresses IL-9 production (Matsuki et al. 2017). Furthermore, the expression of IL-9 in ILC2 seems to be transient (Wilhelm et al. 2011; Licona-Limon et al. 2013), indicative of a tighter regulated expression. Multiple other molecules, including IRF4, transforming growth factor (TGF)-β, Itk, or STAT6

regulate IL-9 expression in T cells (Jabeen and Kaplan 2012; Yao et al. 2013). Of these, IRF4 modulates IL-9 secretion by ILC2s (Mohapatra et al. 2016). However, there are also differences between factors that trigger IL-9 secretion by T cells and ILCs. For example, IL-25 stimulates IL-9 production by T cells (Angkasekwinai et al. 2010) but has no effect on IL-9 production by ILC2s (Wilhelm et al. 2011).

IL-9 produced by mucosal ILC2s enhances mucus production in goblet cells and activates mast cells in inflammatory lung diseases (Jabeen and Kaplan 2012; Ealey et al. 2017). Consistent with this, blocking IL-9 in chronic models of lung inflammation inhibits mastocytosis and airway remodeling (Kearley et al. 2011). Similarly, in cystic fibrosis, a disease characterized by high levels of IL-9, ILC2s play a key role in promoting lung inflammation (Moretti et al. 2017). All of these studies highlight the relevance of IL-9 as autocrine cytokine acting as a promotor of ILC2 function involved in pathogens removal, tissue repair in the recovery phase of lung inflammation, as well as their key role in chronic lung inflammatory diseases (Fig. 3).

Cytokines can also negatively regulate cytokine secretion by ILC2. Type I and II IFNs have been shown to inhibit the secretion of signature cytokines by ILC2s in a manner dependent on the ISGF3 complex (STAT1, STAT2, and IRF9) for type I IFNs or STAT1 activation for type II IFN-γ (Molofsky et al. 2015; Duerr et al. 2016; Moro et al. 2016). In addition, IL-27, a member of the IL-12/23 family of cytokines, can inhibit ILC2 cytokine secretion and ILC2-mediated pathology via STAT1 (Duerr et al. 2016; Moro et al. 2016).

Group 3 ILCs

Group 3 ILCs are dependent on transcription factor RAR-RORγt and comprise a heterogeneous population of ILC3s and LTi cells (Cella et al. 2009; Cupedo et al. 2009; Artis and Spits 2015). In mouse, three subsets of ILC3s can be distinguished based on the cell-surface expression of chemokine receptors NKp46 and CCR6. CCR6[+] ILC3s comprise CD4[+] and CD4[−] LTi cells (Mebius et al. 1997; Eberl et al. 2004;

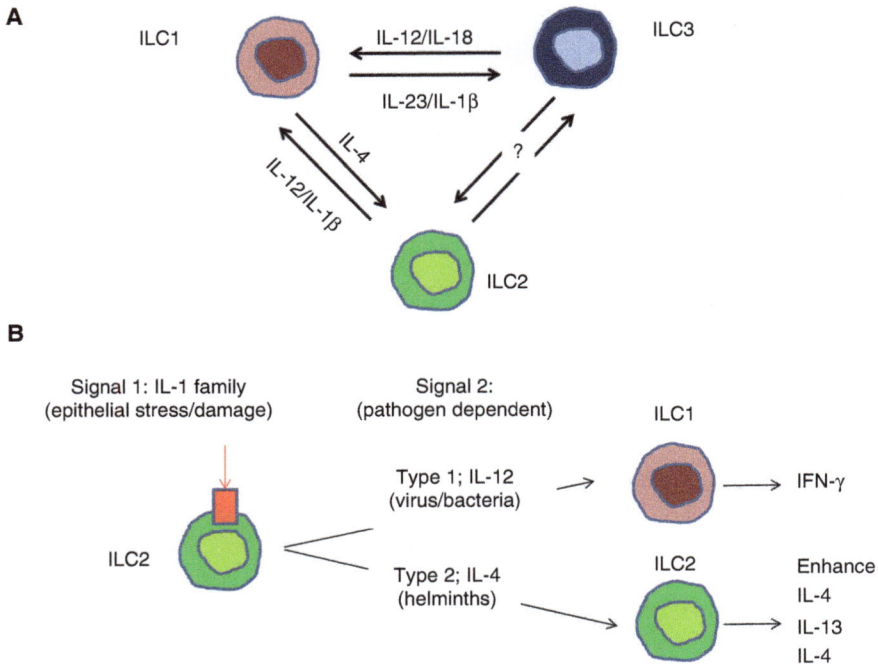

Figure 3. Innate lymphoid cell (ILC) plasticity is governed by the cytokine milieu. ILCs can switch between fully polarized subsets to rapidly adapt to changes occurring in their environment. (*A*) Interleukin (IL)-12 and IL-18 drives the transdifferentiation of ILC3s into ILC1s. This process is reversible because ILC1s convert to ILC3s in the presence of IL-23 and IL-1β. ILC2s undergo cell plasticity in the presence of IL-1β and IL-12 to convert into interferon (IFN)-γ-secreting ILC1s. IL-4 can revert this transdifferentiation process and convert ILC1s into ILC2s. (*B*) Double hit model of ILC2s plasticity. ILC2s require of two types of signals to undergo transdifferentiation. Signal 1, secreted by epithelial cells on stress or damage conditions, is induced by IL-1 family members and trigger modifications in the chromatin architecture. DNA reprogramming will be orchestrated by the signal 2 or "driving cytokines." IL-12 drives ILC2s plasticity toward ILC1s to better cope with intracellular bacteria or virus, whereas IL-4 enhances type 2 immune responses against helminths.

Sawa et al. 2010; van de Pavert et al. 2014), and CCR6$^-$ ILC3s can be divided into two subsets based on the presence or absence of the natural cytotoxicity receptor (NCR) NKp46 (Table 1) (Artis and Spits 2015). Human ILC3s express CD117 (also known as c-Kit) and two subsets can be defined by their expression of NKp44 (Hoorweg et al. 2012; Hazenberg and Spits 2014). Almost all human ILC subsets express CCR6 (Mjosberg et al. 2011; Hazenberg and Spits 2014; Shikhagaie et al. 2017); therefore, CCR6 expression does not define human LTi cells. However, neuropilin 1 (NRP1, also known as CD304 or BDCA4) was exclusively found on ILC3s isolated from lymphoid tissues but not from peripheral blood or skin, and NRP1$^+$

ILC3 had LTi activity in vitro (Shikhagaie et al. 2017). *Nrp1* transcripts were found on mouse LTi cells isolated from adult intestine (Robinette et al. 2015). Together, these findings indicate that NRP1 is a conserved marker for LTi cells.

Whereas IL-2 and IL-7 maintain cell survival and proliferation of ILC3 subsets, ILC3s are activated by IL-1α, IL-1β, and IL-23 to produce IL-22, IL-17A, IL-17F, and granulocyte macrophage colony-stimulating factor (GM-CSF) (Fig. 2). The combination of IL-1β and IL-23 are particularly efficient in inducing the cytokine production by ILC3s (Zheng et al. 2008; Cella et al. 2009, 2010; Takatori et al. 2009; Hughes et al. 2010; Kim et al. 2014; Longman et al. 2014; Hernandez et al. 2015). IL-1β

Cite this article as *Cold Spring Harb Perspect Biol* doi: 10.1101/cshperspect.a030304

and IL-23 are produced and secreted by DCs and macrophages in response to exogenous or endogenous signals (Lopez-Castejon and Brough 2011; Teng et al. 2015). In humans, IL-17 is predominantly produced by NKp44$^-$ILC3s, whereas IL-22 is produced by NKp44$^+$ILC3s (Cupedo et al. 2009; Hoorweg et al. 2012). IL-17 and IL-22 are known to have distinct roles in inducing an innate response in epithelial cells, as IL-17 is responsible for inflammatory tissue responses and involved in several autoimmune diseases, whereas IL-22 is involved in tissue protection and repair (Ouyang and Valdez 2008; Eyerich et al. 2010). However, there are also reports suggesting that IL-22 can be involved in inflammatory diseases such as inflammatory bowel disease and psoriasis (Wolk et al. 2010; Maloy and Powrie 2011).

CYTOKINES THAT MODULATE ILC PLASTICITY

Th cell subsets often display plasticity, for example under certain conditions a Th17 cell can adopt features of Th1 cells (O'Shea and Paul 2010). Although ILCs were initially classified under stable phenotypes (ILC1/2 and ILC3s), it is now clear that they also show considerable plasticity (Fig. 3). Each ILC subset senses inductive cytokines (e.g., IL-25/TSLP/IL-33 activate ILC2s) that trigger their core transcriptional program, resulting in the secretion of signature cytokines (e.g., IL-5 secreted by ILC2s) tailored to combat specific microbial pathogens. However, ILCs are particularly sensitive for cell reprograming by certain cytokines ("switching cytokines"), which enable the cell to acquire other functions to better cope with changes in their microenvironment that require a different ILC response (Fig. 3A).

Early studies of Colonna and Cella provided the first indication of ILC plasticity, showing that IL-22-secreting-NCR$^+$ILC3 activated with IL-1β and IL-23 induce the production of IL-22 or IFN-γ, depending on whether these cells are cultured in IL-7 or IL-2 plus IL-7, respectively (Cella et al. 2010). Moreover, NCR$^+$ILC3s cultured with IL-2 plus IL-7 became responsive to IL-12, which might further promote IFN-γ

secretion and immunopathology. In addition, human IL-22-producing ILC3s can produce IL-2, IL-5, and IL-13 and IL-22 after activation of TLR2 ligands (Crellin et al. 2010). Together, these studies with human ILCs suggested that differential activation signals can change the cytokine-production pattern of ILC3s. Using mouse models, Vonarbourg et al. (2010) showed that IL-7 stabilize RORγt expression by IL-22-producing NCR$^+$ILC3s, whereas IL-12 and IL-15 accelerate RORγt loss accompanied with an acquisition of T-bet and the capacity to produce IFN-γ. IL-12 drives NCR$^+$ILC3 into IFN-γ-producing-ILC1s by down-regulating RORγt and up-regulating T-bet, a process that is enhanced in the presence of IL-18 and/or IL-1β (Vonarbourg et al. 2010). Human ILC3s were shown to convert into ILC1s under the influence of IL-1β and IL-12. Transdifferentiation of ILC3s into ILC1s is reversible because ILC1s convert to ILC3s in the presence of IL-23 and IL-1β, a process accelerated by retinoic acid (Bernink et al. 2013, 2015). IL-12-producing CD14$^+$ intestinal DCs switch ILC3s into IFN-γ-producing ILC1s, whereas conversely, retinoic acid and IL-23-producing CD14$^-$ DCs induce the differentiation of ILC1s into ILC3s. ILC3 into ILC1 conversion probably occurs in inflammatory conditions in vivo, because an increased frequency of inflammatory ILC1s inversely correlates with the frequency of ILC3s in inflamed intestinal resection specimens from Crohn's disease patients (Bernink et al. 2013). A study using immunodeficient BALB/c mice with a human immune system confirmed that an intestinal inflammation resulted in a shift of ILC3s into ILC1s. Overall, these data support the idea that bidirectional ILC3, ILC1 plasticity is involved in the regulation of innate immunity in the gut mucosa.

ILC2s also show plasticity. IL-1β plays a key role in ILC2 transdifferentiation. IL-1β is a potent activator of ILC2s (Bal et al. 2016; Lim et al. 2016; Ohne et al. 2016; Silver et al. 2016), enhancing the expression of the receptors for the epithelial cytokines IL-33, IL-25, and TSLP and triggering IL-5 and IL-13 secretion. IL-1β also induces transcriptome changes, resulting in an induction of transcription of TBX1 and IL12RB2 in ILC2s, setting the stage for conversion into IFN-γ-secreting ILC1s in response to IL-12. IL-12RB1 is critical for the transdifferentiation because ILC2s from patients with biallelic mutations in the *IL12RB1* gene did not acquire the capacity to produce IFN-γ and lacked ILC1s (Lim et al. 2016). The combination of IL-1β and IL-12 favor epigenetic modifications (increase of H3K9ac and a decrease of H3K27me3 at the *IL12RB2* locus) that enhance IL12Rβ2 expression in ILC2s and perpetuate the acquisition of the ILC1 phenotype (Ohne et al. 2016). Interestingly, whereas IL-1β increases accessibility at the *TBX21* and *IFNG* loci, the *IL5* and *IL13* loci remain accessible as well, which suggested that ILC2-derived ILC1s could differentiate back to ILC2s. Indeed, when ILC2-derived ILC1s were cultured with IL-4, these cells revert to ILC2s (Ohne et al. 2016). ILC2 plasticity might occur in the inflamed lungs of COPD patients that show an increase of ILC1 subsets, whereas the sizes of the ILC3 and ILC2 pools decrease (Bal et al. 2016). The proportion of ILC1s also increases in the peripheral blood of COPD patients, and the degree of this increase is correlated with disease severity (Silver et al. 2016).

The high degree of ILC plasticity raises the question as to the physiological relevance of this phenomenon. As ILCs are mostly tissue resident with a limited influx of cells from the periphery, plasticity enables ILC to quickly adapt to changing environments. For example, IL-22-producing ILC3s representing most ILCs in a noninflamed gut can rapidly change into INF-γ-producing ILC1s on inflammation to help neutralize pathogenic bacteria such as *Salmonella enterica* (Klose et al. 2013). After resolution of the infection, these ILC1s may revert to IL-22-producing ILC3, restoring the homeostatic phenotype. ILC plasticity might be induced by IL-1β or equivalent factors (signal 1), whereas cytokines such as IL-12 determine the final outcome (signal 2) (Fig. 3B). This way, ILCs effectively tailor their immune response against pathogens.

THE ROLE OF ILC-PRODUCED CYTOKINES IN TUMOR IMMUNITY

Resident ILCs are among the first line of defense against pathogens and likewise ILCs might also establish the first cytokine dialog with the nascent tumor in the mucosa, promoting or inhibiting tumor growth. The tumor microenvironment may trigger tailored immune responses by ILC1s, ILC2s, and ILC3s. Along these lines, a novel ILC1-like cell expressing NK1.1, CD49a, CD103, Granzyme, and TNF-related apoptosis-inducing ligand (TRAIL) with strong antitumor potential has been identified in mice. These cells may play a key role in immune surveillance of tumors (Dadi et al. 2016).

Colorectal cancer (CRC) tissues show high levels of IL-17A, IL-17F, and IL-22 as well as the Th17-polarizing cytokines IL-1β, IL-6, IL-21, and TGF-β (West et al. 2015). The abundance of several of these cytokines correlate with the disease stage (West et al. 2015). Collectively, these cytokines promote tumor growth and proliferation, resistance to apoptosis, angiogenesis, gene instability, invasiveness, and metastasis. Interestingly, ILC3s, the major ILC subset in the gut, has been shown to promote tumor growth in a mouse model of bacteria-induced CRC in an IL-22-dependent manner (Kirchberger et al. 2013). Depletion of ILC3s protect mice from IBD-associated CRC. Blocking IL-17 did not did not prevent tumor formation, whereas blocking IL-22 decreases tumor burden. Although IL-22 transgenic mice do not display a clear increase of spontaneous tumors (Park et al. 2011; Sabat et al. 2014), formation a carcinogen-induced hepatocellular carcinoma was enhanced in IL-22 transgenic mice, whereas tumor formation was reduced in IL-22-deficient mice (Sabat et al. 2014). In line with these studies, deficiencies in IL22BP, a natural neutralizing molecule of IL-22, increase the tumor burden in IBD-associated CRC (Huber et al. 2012). These observations suggest that IL-22 acts to accelerate the development of ongoing inflammation-induced tumors.

IL-12 that mediated the repression of subcutaneous melanoma (transplantation of B16 melanoma cell line into C57BL/6 mice) was independent of adaptive immunity, NK T cells, or NK cells. Instead, the crucial antitumor activity was mediated by NKp46+ILC3, which differentiated into IFN-γ-secreting ILC1 apparently induced by IL-12. The mechanism of tumor rejection is unclear. Tumor rejection did not involve IFN-γ or other analyzed ILC-secreting cytokines (Eisenring et al. 2010).

In a mouse model of melanoma, which the tumor cells (B16F10) metastasize into the lung, the numbers of innate IL-5-producing cells were increased in response to tumor invasion. Furthermore, eosinophils, which possess antitumor activities, were located around the melanoma clusters set in the capillary vessels of alveolar walls. The key effect of IL-5 was revealed in analyses of the antitumor response in IL-5- and IL-5R-deficient mice or following treatment with blocking anti-IL-5 antibodies. In the absence of IL-5, increased tumor metastasis and reduced numbers of eosinophils in the lung were observed, suggesting that in this model eosinophil recruitment by ILC2s is critical to suppress tumor metastasis (Ikutani et al. 2012).

FUTURE PERSPECTIVES

Over the past decade, ILCs have emerged as important regulators of tissue homeostasis and innate immune responses. ILCs form a first line of defense against infectious pathogens, providing protection before the adaptive immune system is mobilized. ILCs may also amplify ongoing immune responses. As discussed here, cytokines play critical roles in mediating the functional activities of ILCs, serving as activators as well as effector molecules. ILC responses induced by cytokines need to be tightly regulated to prevent an uncontrolled cytokine storm. Although certain cytokines have been identified that inhibit ILC activation, the mechanisms of ILC control are incompletely understood. Some cytokines by themselves do not activate ILCs but either amplify or inhibit secretion of signature cytokines; the mechanisms for this are yet to be elucidated. ILCs also display plasticity, changing completely their cytokine-production profile on cues provided by their microenvironment. Plasticity has been observed in various inflammatory and metabolic diseases but it is yet unclear whether ILC plasticity is involved in causing a disease or amplifying the pathology or is merely a consequence of the disease process. Understanding the contribution of ILC plasticity to

diseases requires relevant mouse models in which plasticity can be manipulated by genetic modification specifically in ILCs. This is quite a challenge because it is difficult to selectively target ILCs in the presence of an intact adaptive immune system. Such models may help us to investigate whether small molecular compounds or biologicals that modify the functions of ILCs are efficacious as therapeutics.

ACKNOWLEDGMENTS

H.S. is supported by an advanced grant of the European Research Council (ERC) (No. 341038).

REFERENCES

Angkasekwinai P, Chang SH, Thapa M, Watarai H, Dong C. 2010. Regulation of IL-9 expression by IL-25 signaling. *Nat Immunol* **11:** 250–256.

Artis D, Spits H. 2015. The biology of innate lymphoid cells. *Nature* **517:** 293–301.

Bal SM, Bernink JH, Nagasawa M, Groot J, Shikhagaie MM, Golebski K, van Drunen CM, Lutter R, Jonkers RE, Hombrink P, et al. 2016. IL-1β, IL-4 and IL-12 control the fate of group 2 innate lymphoid cells in human airway inflammation in the lungs. *Nat Immunol* **17:** 636–645.

Barlow JL, Bellosi A, Hardman CS, Drynan LF, Wong SH, Cruickshank JP, McKenzie AN. 2012. Innate IL-13-producing nuocytes arise during allergic lung inflammation and contribute to airways hyperreactivity. *J Allergy Clin Immunol* **129:** 191–198, e191–e194.

Bernink JH, Peters CP, Munneke M, te Velde AA, Meijer SL, Weijer K, Hreggvidsdottir HS, Heinsbroek SE, Legrand N, Buskens CJ, et al. 2013. Human type 1 innate lymphoid cells accumulate in inflamed mucosal tissues. *Nat Immunol* **14:** 221–229.

Bernink JH, Krabbendam L, Germar K, de Jong E, Gronke K, Kofoed-Nielsen M, Munneke JM, Hazenberg MD, Villaudy J, Buskens CJ, et al. 2015. Interleukin-12 and -23 control plasticity of CD127+ group 1 and group 3 innate lymphoid cells in the intestinal lamina propria. *Immunity* **43:** 146–160.

Bhandoola A, Sambandam A. 2006. From stem cell to T cell: One route or many? *Nat Rev Immunol* **6:** 117–126.

Boos MD, Yokota Y, Eberl G, Kee BL. 2007. Mature natural killer cell and lymphoid tissue-inducing cell development requires Id2-mediated suppression of E protein activity. *J Exp Med* **204:** 1119–1130.

Carriere V, Roussel L, Ortega N, Lacorre DA, Americh L, Aguilar L, Bouche G, Girard JP. 2007. IL-33, the IL-1-like cytokine ligand for ST2 receptor, is a chromatin-associated nuclear factor in vivo. *Proc Natl Acad Sci* **104:** 282–287.

Cella M, Fuchs A, Vermi W, Facchetti F, Otero K, Lennerz JK, Doherty JM, Mills JC, Colonna M. 2009. A human natural killer cell subset provides an innate source of IL-22 for mucosal immunity. *Nature* **457**: 722–725.

Cella M, Otero K, Colonna M. 2010. Expansion of human NK-22 cells with IL-7, IL-2, and IL-1β reveals intrinsic functional plasticity. *Proc Natl Acad Sci* **107**: 10961–10966.

Constantinides MG, McDonald BD, Verhoef PA, Bendelac A. 2014. A committed precursor to innate lymphoid cells. *Nature* **508**: 397–401.

Crellin NK, Trifari S, Kaplan CD, Satoh-Takayama N, Di Santo JP, Spits H. 2010. Regulation of cytokine secretion in human CD127⁺ LTi-like innate lymphoid cells by Toll-like receptor 2. *Immunity* **33**: 752–764.

Cupedo T, Crellin NK, Papazian N, Rombouts EJ, Weijer K, Grogan JL, Fibbe WE, Cornelissen JJ, Spits H. 2009. Human fetal lymphoid tissue-inducer cells are interleukin 17-producing precursors to RORC⁺ CD127⁺ natural killer-like cells. *Nat Immunol* **10**: 66–74.

Dadi S, Chhangawala S, Whitlock BM, Franklin RA, Luo CT, Oh SA, Toure A, Pritykin Y, Huse M, Leslie CS, et al. 2016. Cancer immunosurveillance by tissue-resident innate lymphoid cells and innate-like T cells. *Cell* **164**: 365–377.

Daussy C, Faure F, Mayol K, Viel S, Gasteiger G, Charrier E, Bienvenu J, Henry T, Debien E, Hasan UA, et al. 2014. T-bet and Eomes instruct the development of two distinct natural killer cell lineages in the liver and in the bone marrow. *J Exp Med* **211**: 563–577.

Dent LA, Strath M, Mellor AL, Sanderson CJ. 1990. Eosinophilia in transgenic mice expressing interleukin 5. *J Exp Med* **172**: 1425–1431.

Diefenbach A, Colonna M, Koyasu S. 2014. Development, differentiation, and diversity of innate lymphoid cells. *Immunity* **41**: 354–365.

Doherty TA, Khorram N, Lund S, Mehta AK, Croft M, Broide DH. 2013. Lung type 2 innate lymphoid cells express cysteinyl leukotriene receptor 1, which regulates T$_H$2 cytokine production. *J Allergy Clin Immunol* **132**: 205–213.

Duerr CU, McCarthy CD, Mindt BC, Rubio M, Meli AP, Pothlichet J, Eva MM, Gauchat JF, Qureshi ST, Mazer BD, et al. 2016. Type I interferon restricts type 2 immunopathology through the regulation of group 2 innate lymphoid cells. *Nat Immunol* **17**: 65–75.

Ealey KN, Moro K, Koyasu S. 2017. Are ILC2s Jekyll and Hyde in airway inflammation? *Immunol Rev* **278**: 207–218.

Eberl G, Marmon S, Sunshine MJ, Rennert PD, Choi Y, Littman DR. 2004. An essential function for the nuclear receptor RORγt in the generation of fetal lymphoid tissue inducer cells. *Nat Immunol* **5**: 64–73.

Eberl G, Colonna M, Di Santo JP, McKenzie AN. 2015. Innate lymphoid cells. Innate lymphoid cells: A new paradigm in immunology. *Science* **348**: aaa6566.

Eisenring M, vom Berg J, Kristiansen G, Saller E, Becher B. 2010. IL-12 initiates tumor rejection via lymphoid tissue-inducer cells bearing the natural cytotoxicity receptor NKp46. *Nat Immunol* **11**: 1030–1038.

Eyerich S, Eyerich K, Cavani A, Schmidt-Weber C. 2010. IL-17 and IL-22: Siblings, not twins. *Trends Immunol* **31**: 354–361.

Fort MM, Cheung J, Yen D, Li J, Zurawski SM, Lo S, Menon S, Clifford T, Hunte B, Lesley R, et al. 2001. IL-25 induces IL-4, IL-5, and IL-13 and Th2-associated pathologies in vivo. *Immunity* **15**: 985–995.

Fuchs A, Vermi W, Lee JG, Lonardi S, Gilfillan S, Newberry RD, Cella M, Colonna M. 2013. Intraepithelial type 1 innate lymphoid cells are a unique subset of IL-12- and IL-15-responsive IFN-γ-producing cells. *Immunity* **38**: 769–781.

Gasteiger G, Fan X, Dikiy S, Lee SY, Rudensky AY. 2015. Tissue residency of innate lymphoid cells in lymphoid and non-lymphoid organs. *Science* **350**: 981–985.

Grabstein KH, Eisenman J, Shanebeck K, Rauch C, Srinivasan S, Fung V, Beers C, Richardson J, Schoenborn MA, Ahdieh M, et al. 1994. Cloning of a T cell growth factor that interacts with the β chain of the interleukin-2 receptor. *Science* **264**: 965–968.

Grunig G, Warnock M, Wakil AE, Venkayya R, Brombacher F, Rennick DM, Sheppard D, Mohrs M, Donaldson DD, Locksley RM, et al. 1998. Requirement for IL-13 independently of IL-4 in experimental asthma. *Science* **282**: 2261–2263.

Hazenberg MD, Spits H. 2014. Human innate lymphoid cells. *Blood* **124**: 700–709.

Hernandez PP, Mahlakoiv T, Yang I, Schwierzeck V, Nguyen N, Guendel F, Gronke K, Ryffel B, Holscher C, Dumoutier L, et al. 2015. Interferon-λ and interleukin 22 act synergistically for the induction of interferon-stimulated genes and control of rotavirus infection. *Nat Immunol* **16**: 698–707.

Hogan SP, Rosenberg HF, Moqbel R, Phipps S, Foster PS, Lacy P, Kay AB, Rothenberg ME. 2008. Eosinophils: Biological properties and role in health and disease. *Clin Exp Allergy* **38**: 709–750.

Hoorweg K, Peters CP, Cornelissen F, Aparicio-Domingo P, Papazian N, Kazemier G, Mjosberg JM, Spits H, Cupedo T. 2012. Functional differences between human NKp44⁻ and NKp44⁺ RORC⁺ innate lymphoid cells. *Front Immunol* **3**: 72.

Hoyler T, Klose CS, Souabni A, Turqueti-Neves A, Pfeifer D, Rawlins EL, Voehringer D, Busslinger M, Diefenbach A. 2012. The transcription factor GATA-3 controls cell fate and maintenance of type 2 innate lymphoid cells. *Immunity* **37**: 634–648.

Huang Y, Guo L, Qiu J, Chen X, Hu-Li J, Siebenlist U, Williamson PR, Urban JF Jr, Paul WE. 2015. IL-25-responsive, lineage-negative KLRG1ʰⁱ cells are multipotential "inflammatory" type 2 innate lymphoid cells. *Nat Immunol* **16**: 161–169.

Huber S, Gagliani N, Zenewicz LA, Huber FJ, Bosurgi L, Hu B, Hedl M, Zhang W, O'Connor W Jr, Murphy AJ, et al. 2012. IL-22BP is regulated by the inflammasome and modulates tumorigenesis in the intestine. *Nature* **491**: 259–263.

Hughes T, Becknell B, Freud AG, McClory S, Briercheck E, Yu J, Mao C, Giovenzana C, Nuovo G, Wei L, et al. 2010. Interleukin-1β selectively expands and sustains interleukin-22⁺ immature human natural killer cells in secondary lymphoid tissue. *Immunity* **32**: 803–814.

Huntington ND. 2014. The unconventional expression of IL-15 and its role in NK cell homeostasis. *Immunol Cell Biol* **92**: 210–213.

Cite this article as *Cold Spring Harb Perspect Biol* doi: 10.1101/cshperspect.a030304

Ikutani M, Yanagibashi T, Ogasawara M, Tsuneyama K, Yamamoto S, Hattori Y, Kouro T, Itakura A, Nagai Y, Takaki S, et al. 2012. Identification of innate IL-5-producing cells and their role in lung eosinophil regulation and antitumor immunity. *J Immunol* **188:** 703–713.

Jabeen R, Kaplan MH. 2012. The symphony of the ninth: The development and function of Th9 cells. *Curr Opin Immunol* **24:** 303–307.

Kang MJ, Lee CG, Lee JY, Dela Cruz CS, Chen ZJ, Enelow R, Elias JA. 2008. Cigarette smoke selectively enhances viral PAMP- and virus-induced pulmonary innate immune and remodeling responses in mice. *J Clin Invest* **118:** 2771–2784.

Kearley J, Erjefalt JS, Andersson C, Benjamin E, Jones CP, Robichaud A, Pegorier S, Brewah Y, Burwell TJ, Bjermer L, et al. 2011. IL-9 governs allergen-induced mast cell numbers in the lung and chronic remodeling of the airways. *Am J Respir Crit Care Med* **183:** 865–875.

Kennedy MK, Glaccum M, Brown SN, Butz EA, Viney JL, Embers M, Matsuki N, Charrier K, Sedger L, Willis CR, et al. 2000. Reversible defects in natural killer and memory CD8 T cell lineages in interleukin 15-deficient mice. *J Exp Med* **191:** 771–780.

Kim BS, Siracusa MC, Saenz SA, Noti M, Monticelli LA, Sonnenberg GF, Hepworth MR, Van Voorhees AS, Comeau MR, Artis D. 2013. TSLP elicits IL-33-independent innate lymphoid cell responses to promote skin inflammation. *Sci Transl Med* **5:** 170ra116.

Kim HY, Lee HJ, Chang YJ, Pichavant M, Shore SA, Fitzgerald KA, Iwakura Y, Israel E, Bolger K, Faul J, et al. 2014. Interleukin-17-producing innate lymphoid cells and the NLRP3 inflammasome facilitate obesity-associated airway hyperreactivity. *Nat Med* **20:** 54–61.

Kirchberger S, Royston DJ, Boulard O, Thornton E, Franchini F, Szabady RL, Harrison O, Powrie F. 2013. Innate lymphoid cells sustain colon cancer through production of interleukin-22 in a mouse model. *J Exp Med* **210:** 917–931.

Klose CS, Kiss EA, Schwierzeck V, Ebert K, Hoyler T, d'Hargues Y, Goppert N, Croxford AL, Waisman A, Tanriver Y, et al. 2013. A T-bet gradient controls the fate and function of CCR6⁻RORγt⁺ innate lymphoid cells. *Nature* **494:** 261–265.

Klose CS, Flach M, Mohle L, Rogell L, Hoyler T, Ebert K, Fabiunke C, Pfeifer D, Sexl V, Fonseca-Pereira D, et al. 2014. Differentiation of type 1 ILCs from a common progenitor to all helper-like innate lymphoid cell lineages. *Cell* **157:** 340–356.

Kopf M, Brombacher F, Hodgkin PD, Ramsay AJ, Milbourne EA, Dai WJ, Ovington KS, Behm CA, Kohler G, Young IG, et al. 1996. IL-5-deficient mice have a developmental defect in CD5⁺ B-1 cells and lack eosinophilia but have normal antibody and cytotoxic T cell responses. *Immunity* **4:** 15–24.

Liang HE, Reinhardt RL, Bando JK, Sullivan BM, Ho IC, Locksley RM. 2011. Divergent expression patterns of IL-4 and IL-13 define unique functions in allergic immunity. *Nat Immunol* **13:** 58–66.

Licona-Limon P, Henao-Mejia J, Temann AU, Gagliani N, Licona-Limon I, Ishigame H, Hao L, Herbert DR, Flavell RA. 2013. Th9 cells drive host immunity against gastrointestinal worm infection. *Immunity* **39:** 744–757.

Lim AI, Menegatti S, Bustamante J, Le Bourhis L, Allez M, Rogge L, Casanova JL, Yssel H, Di Santo JP. 2016. IL-12 drives functional plasticity of human group 2 innate lymphoid cells. *J Exp Med* **213:** 569–583.

Lim AI, Li Y, Lopez-Lastra S, Stadhouders R, Paul F, Casrouge A, Serafini N, Puel A, Bustamante J, Surace L, et al. 2017. Systemic human ILC precursors provide a substrate for tissue ILC differentiation. *Cell* **168:** 1086–1100, e1010.

Longman RS, Diehl GE, Victorio DA, Huh JR, Galan C, Miraldi ER, Swaminath A, Bonneau R, Scherl EJ, Littman DR. 2014. CX₃CR1⁺ mononuclear phagocytes support colitis-associated innate lymphoid cell production of IL-22. *J Exp Med* **211:** 1571–1583.

Lopez AF, Sanderson CJ, Gamble JR, Campbell HD, Young IG, Vadas MD. 1988. Recombinant human interleukin 5 is a selective activator of human eosinophil function. *J Exp Med* **167:** 219–224.

Lopez-Castejon G, Brough D. 2011. Understanding the mechanism of IL-1β secretion. *Cytokine Growth Factor Rev* **22:** 189–195.

Maloy KJ, Powrie F. 2011. Intestinal homeostasis and its breakdown in inflammatory bowel disease. *Nature* **474:** 298–306.

Matsuki A, Takatori H, Makita S, Yokota M, Tamachi T, Suto A, Suzuki K, Hirose K, Nakajima H. 2017. T-bet inhibits innate lymphoid cell-mediated eosinophilic airway inflammation by suppressing IL-9 production. *J Allergy Clin Immunol* **139:** 1355–1367.

Mebius RE, Rennert P, Weissman IL. 1997. Developing lymph nodes collect CD4⁺CD3⁻ LTβ⁺ cells that can differentiate to APC, NK cells, and follicular cells but not T or B cells. *Immunity* **7:** 493–504.

Mishra A, Rothenberg ME. 2003. Intratracheal IL-13 induces eosinophilic esophagitis by an IL-5, eotaxin-1, and STAT6-dependent mechanism. *Gastroenterology* **125:** 1419–1427.

Mjosberg JM, Trifari S, Crellin NK, Peters CP, van Drunen CM, Piet B, Fokkens WJ, Cupedo T, Spits H. 2011. Human IL-25- and IL-33-responsive type 2 innate lymphoid cells are defined by expression of CRTH2 and CD161. *Nat Immunol* **12:** 1055–1062.

Mjosberg J, Bernink J, Golebski K, Karrich JJ, Peters CP, Blom B, te Velde AA, Fokkens WJ, van Drunen CM, Spits H. 2012. The transcription factor GATA3 is essential for the function of human type 2 innate lymphoid cells. *Immunity* **37:** 649–659.

Mohapatra A, Van Dyken SJ, Schneider C, Nussbaum JC, Liang HE, Locksley RM. 2016. Group 2 innate lymphoid cells utilize the IRF4-IL-9 module to coordinate epithelial cell maintenance of lung homeostasis. *Mucosal Immunol* **9:** 275–286.

Molofsky AB, Van Gool F, Liang HE, Van Dyken SJ, Nussbaum JC, Lee J, Bluestone JA, Locksley RM. 2015. Interleukin-33 and interferon-γ counter-regulate group 2 innate lymphoid cell activation during immune perturbation. *Immunity* **43:** 161–174.

Montaldo E, Teixeira-Alves LG, Glatzer T, Durek P, Stervbo U, Hamann W, Babic M, Paclik D, Stolzel K, Grone J, et al. 2014. Human RORγt⁺CD34⁺ cells are lineage-specified progenitors of group 3 RORγt⁺ innate lymphoid cells. *Immunity* **41:** 988–1000.

Monticelli LA, Sonnenberg GF, Abt MC, Alenghat T, Ziegler CG, Doering TA, Angelosanto JM, Laidlaw BJ, Yang CY, Sathaliyawala T, et al. 2011. Innate lymphoid cells promote lung-tissue homeostasis after infection with influenza virus. *Nat Immunol* **12:** 1045–1054.

Moretti S, Renga G, Oikonomou V, Galosi C, Pariano M, Iannitti RG, Borghi M, Puccetti M, De Zuani M, Pucillo CE, et al. 2017. A mast cell-ILC2-Th9 pathway promotes lung inflammation in cystic fibrosis. *Nat Commun* **8:** 14017.

Moro K, Yamada T, Tanabe M, Takeuchi T, Ikawa T, Kawamoto H, Furusawa J, Ohtani M, Fujii H, Koyasu S. 2010. Innate production of T_H2 cytokines by adipose tissue-associated c-Kit⁺Sca-1⁺ lymphoid cells. *Nature* **463:** 540–544.

Moro K, Kabata H, Tanabe M, Koga S, Takeno N, Mochizuki M, Fukunaga K, Asano K, Betsuyaku T, Koyasu S. 2016. Interferon and IL-27 antagonize the function of group 2 innate lymphoid cells and type 2 innate immune responses. *Nat Immunol* **17:** 76–86.

Neill DR, Wong SH, Bellosi A, Flynn RJ, Daly M, Langford TK, Bucks C, Kane CM, Fallon PG, Pannell R, et al. 2010. Nuocytes represent a new innate effector leukocyte that mediates type-2 immunity. *Nature* **464:** 1367–1370.

Ohne Y, Silver JS, Thompson-Snipes L, Collet MA, Blanck JP, Cantarel BL, Copenhaver AM, Humbles AA, Liu YJ. 2016. IL-1 is a critical regulator of group 2 innate lymphoid cell function and plasticity. *Nat Immunol* **17:** 646–655.

O'Shea JJ, Paul WE. 2010. Mechanisms underlying lineage commitment and plasticity of helper CD4⁺ T cells. *Science* **327:** 1098–1102.

Ouyang W, Valdez P. 2008. IL-22 in mucosal immunity. *Mucosal Immunol* **1:** 335–338.

Park O, Wang H, Weng H, Feigenbaum L, Li H, Yin S, Ki SH, Yoo SH, Dooley S, Wang FS, et al. 2011. In vivo consequences of liver-specific interleukin-22 expression in mice: Implications for human liver disease progression. *Hepatology* **54:** 252–261.

Pope SM, Brandt EB, Mishra A, Hogan SP, Zimmermann N, Matthaei KI, Foster PS, Rothenberg ME. 2001. IL-13 induces eosinophil recruitment into the lung by an IL-5- and eotaxin-dependent mechanism. *J Allergy Clin Immunol* **108:** 594–601.

Possot C, Schmutz S, Chea S, Boucontet L, Louise A, Cumano A, Golub R. 2011. Notch signaling is necessary for adult, but not fetal, development of RORγt⁺ innate lymphoid cells. *Nat Immunol* **12:** 949–958.

Price AE, Liang HE, Sullivan BM, Reinhardt RL, Eisley CJ, Erle DJ, Locksley RM. 2010. Systemically dispersed innate IL-13-expressing cells in type 2 immunity. *Proc Natl Acad Sci* **107:** 11489–11494.

Reche PA, Soumelis V, Gorman DM, Clifford T, Liu M, Travis M, Zurawski SM, Johnston J, Liu YJ, Spits H, et al. 2001. Human thymic stromal lymphopoietin preferentially stimulates myeloid cells. *J Immunol* **167:** 336–343.

Robinette ML, Fuchs A, Cortez VS, Lee JS, Wang Y, Durum SK, Gilfillan S, Colonna M, et al. 2015. Transcriptional programs define molecular characteristics of innate lymphoid cell classes and subsets. *Nat Immunol* **16:** 306–317.

Robinette ML, Bando JK, Song W, Ulland TK, Gilfillan S, Colonna M. 2017. IL-15 sustains IL-7R-independent ILC2 and ILC3 development. *Nat Commun* **8:** 14601.

Rochman Y, Spolski R, Leonard WJ. 2009. New insights into the regulation of T cells by γc family cytokines. *Nat Rev Immunol* **9:** 480–490.

Roediger B, Kyle R, Tay SS, Mitchell AJ, Bolton HA, Guy TV, Tan SY, Forbes-Blom E, Tong PL, Koller Y, et al. 2015. IL-2 is a critical regulator of group 2 innate lymphoid cell function during pulmonary inflammation. *J Allergy Clin Immunol* **136:** 1653–1663, e1651–e1657.

Sabat R, Ouyang W, Wolk K. 2014. Therapeutic opportunities of the IL-22-IL-22R1 system. *Nat Rev Drug Discov* **13:** 21–38.

Salimi M, Barlow JL, Saunders SP, Xue L, Gutowska-Owsiak D, Wang X, Huang LC, Johnson D, Scanlon ST, McKenzie AN, et al. 2013. A role for IL-25 and IL-33-driven type-2 innate lymphoid cells in atopic dermatitis. *J Exp Med* **210:** 2939–2950.

Satoh-Takayama N, Dumoutier L, Lesjean-Pottier S, Ribeiro VS, Mandelboim O, Renauld JC, Vosshenrich CA, Di Santo JP. 2009. The natural cytotoxicity receptor NKp46 is dispensable for IL-22-mediated innate intestinal immune defense against *Citrobacter rodentium*. *J Immunol* **183:** 6579–6587.

Sawa S, Cherrier M, Lochner M, Satoh-Takayama N, Fehling HJ, Langa F, Di Santo JP, Eberl G. 2010. Lineage relationship analysis of RORγt⁺ innate lymphoid cells. *Science* **330:** 665–669.

Schmitz J, Owyang A, Oldham E, Song Y, Murphy E, McClanahan TK, Zurawski G, Moshrefi M, Qin J, Li X, et al. 2005. IL-33, an interleukin-1-like cytokine that signals via the IL-1 receptor-related protein ST2 and induces T helper type 2-associated cytokines. *Immunity* **23:** 479–490.

Schoenborn JR, Wilson CB. 2007. Regulation of interferon-γ during innate and adaptive immune responses. *Adv Immunol* **96:** 41–101.

Scoville SD, Mundy-Bosse BL, Zhang MH, Chen L, Zhang X, Keller KA, Hughes T, Chen L, Cheng S, Bergin SM, et al. 2016. A progenitor cell expressing transcription factor RORγt generates all human innate lymphoid cell subsets. *Immunity* **44:** 1140–1150.

Shikhagaie MM, Bjorklund AK, Mjosberg J, Erjefalt JS, Cornelissen AS, Ros XR, Bal SM, Koning JJ, Mebius RE, Mori M, et al. 2017. Neuropilin-1 is expressed on lymphoid tissue residing LTi-like group 3 innate lymphoid cells and associated with ectopic lymphoid aggregates. *Cell Rep* **18:** 1761–1773.

Silver JS, Kearley J, Copenhaver AM, Sanden C, Mori M, Yu L, Pritchard GH, Berlin AA, Hunter CA, Bowler R, et al. 2016. Inflammatory triggers associated with exacerbations of COPD orchestrate plasticity of group 2 innate lymphoid cells in the lungs. *Nat Immunol* **17:** 626–635.

Spits H, Artis D, Colonna M, Diefenbach A, Di Santo JP, Eberl G, Koyasu S, Locksley RM, McKenzie AN, Mebius RE, et al. 2013. Innate lymphoid cells—A proposal for uniform nomenclature. *Nat Rev Immunol* **13:** 145–149.

Spits H, Bernink JH, Lanier L. 2016. NK cells and type 1 innate lymphoid cells: Partners in host defense. *Nat Immunol* **17:** 758–764.

Takatori H, Kanno Y, Watford WT, Tato CM, Weiss G, Ivanov II, Littman DR, O'Shea JJ. 2009. Lymphoid tissue inducer-like cells are an innate source of IL-17 and IL-22. *J Exp Med* **206:** 35–41.

Teng MW, Bowman EP, McElwee JJ, Smyth MJ, Casanova JL, Cooper AM, Cua DJ. 2015. IL-12 and IL-23 cytokines: From discovery to targeted therapies for immune-mediated inflammatory diseases. *Nat Med* **21:** 719–729.

Turner JE, Morrison PJ, Wilhelm C, Wilson M, Ahlfors H, Renauld JC, Panzer U, Helmby H, Stockinger B. 2013. IL-9-mediated survival of type 2 innate lymphoid cells promotes damage control in helminth-induced lung inflammation. *J Exp Med* **210:** 2951–2965.

Valizadeh A, Khosravi A, Zadeh LJ, Parizad EG. 2015. Role of IL-25 in Immunity. *J Clin Diagn Res* **9:** OE01–OE04.

van de Pavert SA, Ferreira M, Domingues RG, Ribeiro H, Molenaar R, Moreira-Santos L, Almeida FF, Ibiza S, Barbosa I, Goverse G, et al. 2014. Maternal retinoids control type 3 innate lymphoid cells and set the offspring immunity. *Nature* **508:** 123–127.

Vely F, Barlogis V, Vallentin B, Neven B, Piperoglou C, Ebbo M, Perchet T, Petit M, Yessaad N, Touzot F, et al. 2016. Evidence of innate lymphoid cell redundancy in humans. *Nat Immunol* **17:** 1291–1299.

Vonarbourg C, Mortha A, Bui VL, Hernandez PP, Kiss EA, Hoyler T, Flach M, Bengsch B, Thimme R, Holscher C, et al. 2010. Regulated expression of nuclear receptor RORγt confers distinct functional fates to NK cell receptor-expressing RORγt⁺ innate lymphocytes. *Immunity* **33:** 736–751.

von Moltke J, Ji M, Liang HE, Locksley RM. 2016. Tuft-cell-derived IL-25 regulates an intestinal ILC2-epithelial response circuit. *Nature* **529:** 221–225.

West EE, Kashyap M, Leonard WJ. 2012. TSLP: A Key regulator of asthma pathogenesis. *Drug Discov Today Dis Mech* doi: 10.1016/j.ddmec.2012.09.003.

West NR, McCuaig S, Franchini F, Powrie F. 2015. Emerging cytokine networks in colorectal cancer. *Nat Rev Immunol* **15:** 615–629.

Wilhelm C, Hirota K, Stieglitz B, Van Snick J, Tolaini M, Lahl K, Sparwasser T, Helmby H, Stockinger B. 2011. An IL-9 fate reporter demonstrates the induction of an innate IL-9 response in lung inflammation. *Nat Immunol* **12:** 1071–1077.

Wolk K, Witte E, Witte K, Warszawska K, Sabat R. 2010. Biology of interleukin-22. *Semin Immunopathol* **32:** 17–31.

Wynn TA. 2003. IL-13 effector functions. *Annu Rev Immunol* **21:** 425–456.

Xu W, Domingues RG, Fonseca-Pereira D, Ferreira M, Ribeiro H, Lopez-Lastra S, Motomura Y, Moreira-Santos L,

Bihl F, Braud V, et al. 2015. NFIL3 orchestrates the emergence of common helper innate lymphoid cell precursors. *Cell Rep* **10:** 2043–2054.

Xue L, Salimi M, Panse I, Mjosberg JM, McKenzie AN, Spits H, Klenerman P, Ogg G. 2014. Prostaglandin D2 activates group 2 innate lymphoid cells through chemoattractant receptor-homologous molecule expressed on T$_H$2 cells. *J Allergy Clin Immunol* **133:** 1184–1194.

Yamaguchi Y, Hayashi Y, Sugama Y, Miura Y, Kasahara T, Kitamura S, Torisu M, Mita S, Tominaga A, Takatsu K. 1988. Highly purified murine interleukin 5 (IL-5) stimulates eosinophil function and prolongs in vitro survival. IL-5 as an eosinophil chemotactic factor. *J Exp Med* **167:** 1737–1742.

Yang Q, Li F, Harly C, Xing S, Ye L, Xia X, Wang H, Wang X, Yu S, Zhou X, et al. 2015. TCF-1 upregulation identifies early innate lymphoid progenitors in the bone marrow. *Nat Immunol* **16:** 1044–1050.

Yao W, Zhang Y, Jabeen R, Nguyen ET, Wilkes DS, Tepper RS, Kaplan MH, Zhou B. 2013. Interleukin-9 is required for allergic airway inflammation mediated by the cytokine TSLP. *Immunity* **38:** 360–372.

Yokota YM, Mori A, Sugawara S, Adachi S, Nishikawa S, Gruss P. 1999. Development of peripheral lymphoid organs and natural killer cells depends on the helix–loop–helix inhibitor Id2. *Nature* **397:** 702–706.

Yu X, Wang Y, Deng M, Li Y, Ruhn KA, Zhang CC, Hooper LV. 2014. The basic leucine zipper transcription factor NFIL3 directs the development of a common innate lymphoid cell precursor. *eLife* **3:** e04406.

Zaph C, Du Y, Saenz SA, Nair MG, Perrigoue JG, Taylor BC, Troy AE, Kobuley DE, Kastelein RA, Cua DJ, et al. 2008. Commensal-dependent expression of IL-25 regulates the IL-23-IL-17 axis in the intestine. *J Exp Med* **205:** 2191–2198.

Zheng Y, Valdez PA, Danilenko DM, Hu Y, Sa SM, Gong Q, Abbas AR, Modrusan Z, Ghilardi N, de Sauvage FJ, et al. 2008. Interleukin-22 mediates early host defense against attaching and effacing bacterial pathogens. *Nat Med* **14:** 282–289.

Zhu Z, Homer RJ, Wang Z, Chen Q, Geba GP, Wang J, Zhang Y, Elias JA. 1999. Pulmonary expression of interleukin-13 causes inflammation, mucus hypersecretion, subepithelial fibrosis, physiologic abnormalities, and eotaxin production. *J Clin Invest* **103:** 779–788.

Ziegler SF. 2012. Thymic stromal lymphopoietin and allergic disease. *J Allergy Clin Immunol* **130:** 845–852.

Development, Diversity, and Function of Dendritic Cells in Mouse and Human

David A. Anderson III,[1] Kenneth M. Murphy,[1,2] and Carlos G. Briseño[1]

[1]Department of Pathology and Immunology, Washington University in St. Louis, School of Medicine, St. Louis, Missouri 63110

[2]Howard Hughes Medical Institute, Washington University in St. Louis, School of Medicine, St. Louis, Missouri 63110

Correspondence: kmurphy@pathology.wustl.edu

The study of murine dendritic cell (DC) development has been integral to the identification of specialized DC subsets that have unique requirements for their form and function. Advances in the field have also provided a framework for the identification of human DC counterparts, which appear to have conserved mechanisms of development and function. Multiple transcription factors are expressed in unique combinations that direct the development of classical DCs (cDCs), which include two major subsets known as cDC1s and cDC2s, and plasmacytoid DCs (pDCs). pDCs are potent producers of type I interferons and thus these cells are implicated in immune responses that depend on this cytokine. Mouse models deficient in the cDC1 lineage have revealed their importance in directing immune responses to intracellular bacteria, viruses, and cancer through the cross-presentation of cell-associated antigen. Models of transcription factor deficiency have been used to identify subsets of cDC2 that are required for T helper (Th)2 and Th17 responses to certain pathogens; however, no single factor is known to be absolutely required for the development of the complete cDC2 lineage. In this review, we will discuss the current state of knowledge of mouse and human DC development and function and highlight areas in the field that remain unresolved.

DEVELOPMENT AND FUNCTION OF MURINE AND HUMAN DENDRITIC CELL SUBSETS

Classical dendritic cells (cDCs) and plasmacytoid DCs (pDCs) make up the two major subsets of DCs that exist in mice and humans. Among cDCs in mice, two major lineages have been identified and are referred to as cDC1s and cDC2s (Guilliams et al. 2014). cDC1s express high levels of IRF8 and are dependent on *Irf8* (Schiavoni et al. 2002; Aliberti et al. 2003), *Batf3*

(Hildner et al. 2008; Edelson et al. 2010), *Id2* (Hacker et al. 2003; Kusunoki et al. 2003), *Nfil3* (Kashiwada et al. 2011), and *Bcl6* for their development (Ohtsuka et al. 2011; Watchmaker et al. 2014). cDC2s express IRF4 and also IRF8 but at levels lower than cDC1 cells, and can be subdivided into at least two functionally distinct subsets that either require the transcription factors Notch2 or KLF4 (Satpathy et al. 2013; Schlitzer et al. 2013; Tussiwand et al. 2015). pDCs also express high levels of IRF8, similar to levels expressed by cDC1s, but depend on the transcription factor E2-2 for their development (Cisse et al. 2008). cDCs and pDCs develop from a common progenitor in the bone marrow (BM), known as the macrophage DC progenitor (MDP), which has both DC and macrophage potential (Fogg et al. 2006; Auffray et al. 2009). Restriction to the DC lineage occurs downstream of the MDP at a stage defined as the common DC progenitor (CDP) (Naik et al. 2007; Onai et al. 2013), which can give rise to both pDCs and cDCs. Cells in the gate that defined the CDP were characterized by expression of intermediate levels of c-Kit and by expression of both Flt3 ($CD135^+$) and M-CSFR ($CD115^+$), differing from the MDP that expresses c-Kit at high levels. The CDP is negative for expression of CD11c and MHC class II molecules. Subsequent studies identified populations within BM that appeared to be a common progenitor of cDCs, termed pre-cDCs, that were first identified in the spleen (Naik et al. 2006) and later independently identified in the BM (Liu et al. 2009). A common marker of both pre-cDCs was the expression of CD11c, and in the BM these cells were defined as expressing Flt3.

Identification of Distinct Committed Progenitors of cDC1 and cDC2 in Murine Bone Marrow

The identification of progenitors with potential for only one type of cDC initially relied on the use of a reporter for the gene *Zbtb46*, which had been previously identified as a marker for cDCs (Meredith et al. 2012; Satpathy et al. 2012a). One study found that the expression of the $Zbtb46^{gfp}$ reporter allele by immature cells in the BM was associated with commitment of these cells to the cDC lineage and the exclusion of pDC potential (Satpathy et al. 2012a). However, that study did not examine the various subpopulations of cells expressing *Zbtb46*. Subsequently, it was recognized that *Zbtb46* was expressed heterogeneously in BM cells, by populations of cells that expressed intermediate levels of c-Kit, similar to expression levels in the CDP, but also by cells that lacked c-Kit expression. The majority of the c-Kit^{int} population expressing $Zbtb46^{gfp}$ expressed Flt3 ($CD135^+$) but did not express M-

CSFR ($CD115^-$). However, it was discovered that these two populations represented a divergence in the potential for cDC subsets, with the c-Kit^{int} population being committed to the cDC1 lineage and the c-Kit$^{-/lo}$ population being committed to the cDC2 lineage. These populations were referred to as pre-cDC1 and pre-cDC2 cells, respectively (Grajales-Reyes et al. 2015). An intriguing finding of this study was that the pre-cDC1 cell could develop even in BM of $Batf3^{-/-}$ mice, indicating that specification of the pre-cDC1 did not require BATF3. It appeared that the action of BATF3 was rather late in the developmental process and acted to maintain the expression of the *Irf8* gene by interaction with IRF8 at an enhancer site. In a contemporaneous study, single-cell RNA-Seq on pre-DCs, which were defined as Lin$^-$CD11c$^+$MHCII$^-$ CD135$^+$CD172a$^-$, revealed heterogeneity in expression of SiglecH and Ly6C that could be used to identify pre-cDC1 and pre-cDC2 progenitors. Pre-cDC1 cells were identified as SiglecH$^-$ Ly6C$^-$ and pre-cDC2s were SiglecH$^-$Ly6C$^+$ (Table 1) (Schlitzer et al. 2015). The SiglecH$^+$ fraction of the pre-DC was found to give rise to all DC subsets, including pDCs and cDCs, independent of Ly6C expression. The expression of Ly6C indicated the potential for cDC2, but the levels of c-Kit or M-CSRF were unspecified. It is not clear at this time whether the populations described in these two studies represent the same stages of cDC1 and cDC2 specification, and this will await additional analysis. These two studies also differ in the interpretation of whether a clonally identifiable pre-cDC exists. Schlitzer et al. (2015) suggested the existence of a true pre-cDC stage lacking pDC potential but retaining the ability to generate both types of cDCs. Grajales-Reyes et al. (2015) showed that cKit^{int} $Zbtb46^{gfp+}$ cells that arise from the CDP but lack functional *Batf3* can divert into the cDC2, but not pDC lineage. Thus, conceivably, a natural rate of failure in commitment specification of pre-cDC1 could explain some cDC2 development, but no homogeneous population lacking pDC potential and retaining both cDC1 and cDC2 potential has emerged. Notably, the previously defined pre-cDC in the BM retained substantial pDC potential (Liu et al. 2009) and

Cite this article as *Cold Spring Harb Perspect Biol* doi: 10.1101/cshperspect.a028613

Table 1. Markers of lineage committed murine pre-DCs

Surface marker	Pre-cDCl (Grajales-Reyes et al. 2015)	Pre-cDCl (Schlitzer et al. 2015)	Pre-cDC2 (Grajales-Reyes et al. 2015)	Pre-cDC2 (Schlitzer et al. 2015)
MHCII	int	−	−	−
Zbtb46	+	?	+	?
CDllc	+	+	+	+
CD24	+	+	−	−
CD115	−	?	+	?
CD117	int	?	−	?
CD135	+	+	+	+
CD172a	int	−	int	−
Ly6C	?	−	?	+
SiglecH	−	−	?	−

Summary of surface marker expression used by two independent groups to identify murine preclassical dendritic cells (cDCs) committed to cDCl or cDC2 lineages. Question marks indicate that the expression for a marker was not reported for or used to define the population in the study.

was heterogeneous for c-Kit and MHCII expression. Previously defined pre-cDCs and recently defined pre-cDC1 and pre-cDC2 emerge from the BM, can be identified in the blood, and seed peripheral tissues where they are thought to proliferate, although the extent of local proliferation has been difficult to quantify in vivo (Liu and Nussenzweig 2010; Grajales-Reyes et al. 2015). There are a number of remaining unanswered questions regarding the development of these progenitors, particularly with respect to the mechanisms of transcriptional control. Below we will discuss the extension of these findings to human DCs, but first we describe several additional aspects of development and function in murine DCs.

cDC1 Development and Function

Single deficiencies in *Irf8*, *Batf3*, *Nfil3*, *Id2*, and *Bcl6* are all associated with the loss of cDC1s in lymphoid and nonlymphoid tissues (Schiavoni et al. 2002; Aliberti et al. 2003; Hacker et al. 2003; Kusunoki et al. 2003; Hildner et al. 2008; Edelson et al. 2010; Kashiwada et al. 2011; Ohtsuka et al. 2011; Watchmaker et al. 2014). The mechanisms underlying the requirement for these factors are still a matter of active investigation. Recently, BATF3 was found to function in cDC1 development by acting in the maintenance of the high levels of IRF8 that are already expressed

in the CDP. This function of BATF3 is exerted through its interaction with IRF8 at a specific enhancer site in the *Irf8* gene locus that mediates transcriptional autoactivation. This enhancer appears to become activated during specification of the pre-cDC1 cell upon the induction of *Batf3* expression at the final stages of cDC1 specification in the CDP. Deficiency in *Batf3* did not cause a loss of the identifiable pre-cDC1 cells in the BM, indicating that BATF3 is not required for the initial specification process per se but results in the subsequent decay of IRF8 protein levels and diversion of the specified pre-cDC1 cells into the cDC2 lineage (Grajales-Reyes et al. 2015). This result explains the loss cDC1s in all lymphoid and nonlymphoid tissues in *Batf3*$^{-/-}$ mice under homeostatic conditions. This also explains why cDC1s in *Batf3*$^{-/-}$ mice can be restored under other conditions, such as during infection or treatment with interleukin (IL)-12 through the induction of *Batf* that can compensate for *Batf3* deficiency and rescue cDC1 development (Tussiwand et al. 2012). In *Itgax*-Cre *Irf8*$^{fl/fl}$ mice, in which *Irf8* is deleted downstream of the CDP, cDC1s exhibit reduced survival in peripheral tissues, demonstrating that *Irf8* is also required during the terminal stages of cDC1 development (Sichien et al. 2016).

With the knowledge that cDC1 development is *Irf8*-dependent, observations that either deficiency in *Irf8* or DC-dependent IL-12 production results in susceptibility to *Toxoplasma gondii* shed light on the specialized functions of cDC1 cells (Scharton-Kersten et al. 1997; Liu et al. 2006). It was first demonstrated that these defects were intrinsic to cDC1s through the generation of *Batf3*$^{-/-}$:*Il12p40*$^{-/-}$ mixed BM chimeras, in which IL-12 production is only deficient in cDC1s (Mashayekhi et al. 2011). It was also revealed by using *Batf3*$^{-/-}$ mice that cDC1s are required for cross-presentation of exogenous antigen to CD8 T cells, which in turn is required for antiviral and antitumor responses (Hildner et al. 2008). While normally conferring protection to pathogens, cDC1s can also confer susceptibility in certain cases, such as in the setting of blood-borne infection by *Listeria monocytogenes*. In this particular setting, cDC1s act as the primary cellular target of infection that leads to the spread of the pathogen into the lymphoid areas of the spleen, where massive cellular loss occurs as a result (Edelson et al. 2011a). Surprisingly, mice lacking BATF3, and thus lacking cDC1s, were remarkably resistant to intravenous infection by *L. monocytogenes* (Edelson et al. 2011a). Although the adaptive function of cDC1s is largely considered to be restricted to immune responses mediated by CD8 T cells and T helper (Th)1 cells, *Batf3*$^{-/-}$ mice have enhanced Th2 responses to helminth infection, a phenomenon attributed to loss of constitutive expression of IL-12 by DC1s (Everts et al. 2016).

cDC2 Development and Function

Although several transcription factors are implicated in regulating cDC2 development, there is, to our knowledge, no mutant mouse model in which cDC2 development is selectively ablated. This is in contrast to several single transcription factors whose deletion can ablate cDC1 development. RelB was the first transcription factor to be implicated in the development of cDC2s (Burkly et al. 1995; Weih et al. 1995). Germline deletion of *Relb* in mice causes a multifaceted phenotype that includes splenomegaly, extramedullary hematopoiesis, multiorgan inflammation, myeloid hyperplasia, and disturbed development of thymic and splenic cDCs (Burkly et al. 1995; Weih et al. 1995). Initial studies performed to determine the cell-intrinsic requirements for RelB in DC development claimed that cDCs did not develop in wild-type (WT) chimeras reconstituted with *Relb*$^{-/-}$ BM; however, little to no evidence was ever provided in these studies to support such a statement (Burkly et al. 1995; DeKoning et al. 1997; Gerloni et al. 1998a,b). Subsequent independent analysis showed that thymic CD8α$^+$ cDC1s develop normally in *Relb*$^{-/-}$ WT chimeras, but suggested that there was a cell-intrinsic requirement for RelB in the development of CD8α$^-$Dec205$^-$ cDC2s (Wu et al. 1998). Another report confirmed reduced cDC numbers in *Relb*$^{-/-}$ mice but did not establish a cell-intrinsic requirement for their development or function (Kobayashi et al. 2003). Very recently, our analysis concluded that the majority of cDCs show no cell-intrinsic requirement for RelB for their development (Briseno et al. 2017) with one exception. There was a cell-intrinsic requirement for RelB only in the development of the CD4$^+$Esam$^+$ cDC2 subset of the spleen (Briseno et al. 2017), a subset that is also dependent on Notch2 signaling (Satpathy et al. 2013) and lymphotoxin β (LT-β) receptor signaling (Kabashima et al. 2005). This subset of cDC2 cells appears to represent a terminal maturational stage of cDC2 cells (Satpathy et al. 2013) that develops in response to Notch ligands expressed in specific lymphoid tissue niches (Fasnacht et al. 2014). This similarity between the phenotypes caused by deficiency in *Relb* and LT-β receptor might suggest that RelB could act downstream of this receptor in the final maturation of cDC2 progenitors in such lymphoid niches. However, the majority of cDCs in lymph nodes and peripheral organs showed no cell-autonomous requirement for RelB in their development. There was, however, a role for RebB in nonhematopoietic tissues that regulated the myeloid compartment. Specifically, *Relb*$^{-/-}$ recipient chimeras reconstituted with WT BM showed a similarly abnormal myelopoiesis to that observed in germline *Relb*$^{-/-}$ mice, indicating that the initially reported abnormality of myeloid cells in *Relb*$^{-/-}$ mice was a result of loss of

RelB in nonhematopoietic, radio-resistant cells. Although $Relb^{-/-}$ cDCs are able to activate T cells against cell-associated antigens (Briseno et al. 2017), a role for RelB in other cDC functions has not been excluded, although these are unknown at present. A novel floxed allele for conditional deletion of *Relb* should facilitate such studies (De Silva et al. 2016).

The functional specialization of cDC2 subsets has been revealed using models of transcription factor deletion. *Itgax*-Cre-mediated deletion of *Notch2* and the signaling partner, *Rbpj*, results in the loss of cDC2s in the spleen that are CD11b⁺ESAM⁺, and cDC2s in the intestinal lamina propia and mesenteric lymph nodes that are CD103⁺CD11b⁺ (Satpathy et al. 2013). Loss of this subset is associated with susceptibility to *Citrobacter* infection. Mortality at the early time point of day 10 after infection in $Notch2^{fl/fl}$ *Itgax*-Cre mice suggested a role for the *Notch2*-dependent cDC2 subset in innate defense, in addition to its expected role in adaptive immunity. Using mixed chimeras of *Itgax*-Cre $Notch2^{fl/fl}$ and $Il23a^{-/-}$ BM, it was found that IL-23 production by the *Notch2*-dependent cDC2 subset is required during *Citrobacter* infection (Satpathy et al. 2013). IL-23 is known to activate ILC3 cells for production of IL-22, a cytokine that is required to maintain the barrier function of intestinal epithelial cells (Zheng et al. 2008; Sonnenberg and Artis 2015). Reduced numbers of Th17 cells has also been observed in models where the development or function of cDC2s is impaired. *Itgax*-Cre $Irf4^{fl/fl}$ mice show a defect in the production of Th17-polarizing cytokines on immunization and reduced Th17 populations at homeostasis (Persson et al. 2013; Schlitzer et al. 2013). A specific deficiency in transforming growth factor β (TGF-β) or IL-6 in CD11c-expressing cells is also sufficient to reduce Th17 polarization following infection with *Streptococcus pyogenes* (Persson et al. 2013; Schlitzer et al. 2013; Linehan et al. 2015). A specific requirement for cytokine production by cDC2s, rather than other CD11c-expressing subsets, in Th17 polarization has not been established.

At steady state, expression of *Klf4* is required for the development or function of a subset of migratory cDC2 cells that are CD11b⁻CD24⁻ in the skin-draining lymph node and CD24⁺ CD172a⁺Mgl2⁺ in the lung. A loss of these cells in *Itgax*-Cre $Klf4^{fl/fl}$ mice correlates with enhanced susceptibility to helminth infection and enhanced lung inflammation during house dust mite challenge (Tussiwand et al. 2015). These results are consistent with previous studies that used an *Mgl2*-DTR and *Itgax*-Cre-mediated deletion of *Irf4* that attributed reduced Th2 responses to *Irf4*-expressing cDC2s (Gao et al. 2013; Kumamoto et al. 2013). Still, it is not clear that the missing population is directly responsible for inducing Th2 responses, and the mechanism underlying these phenomena are unknown. Recent studies suggest that ILC2s promote the migration of cDCs to draining lymph nodes, and that cDC2s express chemokines that attract memory Th2 cells on rechallenge (Halim et al. 2014, 2016). Further, normal Th2 responses are driven by cytokines, including IL-13, that are produced by ILC2 cells in response to IL-25 produced in tissues, for example, by epithelial tuft cells in response to certain stimuli (Van Dyken et al. 2016; von Moltke et al. 2016). Thus, it is conceivable that Th2 responses may rely on T cells that reach the tissues in a sufficiently nonpolarized state to respond to this "tissue checkpoint" (Van Dyken et al. 2016). Because BATF3- and Notch2-dependent DCs have been associated with Th1 and Th17 responses, perhaps KLF4-dependent DCs simply provide for activation of T cells without strong polarization, allowing for flexible T-cell responses in tissues.

pDC Development and Function

E2-2, encoded by *Tcf4*, is a member of the E family of basic helix–loop–helix transcription factors (Kee 2009). In both mice and humans, E2-2 is required for the specification of CDPs to pDCs (Cisse et al. 2008). Induced deletion of E2-2 in mature pDCs results in the acquisition of cDC-like properties, such as dendritic morphology, MHCII and CD8α expression, and the ability to induce proliferation of allogeneic CD4⁺ T cells (Ghosh et al. 2010). Deletion of E2-2 in pDCs also induces the expression of ID2.

MTG16, a transcriptional cofactor of the ETO protein family, represses the expression of ID2 in pre-DCs and mature pDCs (Ghoshi et al. 2014). The proteins encoded by *Tcf4* are expressed as multiple isoforms (Corneliussen et al. 1991), TCF4s (short) and TCF4L (long) (Sepp et al. 2011). TCF4L contains activation domain 1 (AD1), which can interact with both p300 and the corepressor RUNX1T1 (Zhang et al. 2004). Within the immune system, TCF4s is expressed in many cells, including cDCs, B cells, and pDCs; however, TCF4L expression is restricted to pDCs (Grajkowska et al. 2017). Loss of TCF4L caused a reduction of pDCs in the BM and spleen similar to that observed in $Mtg16^{-/-}$ mice. The induction of *Tcf4* expression in pDCs is regulated by a proximal pDC-specific 3′ enhancer that requires TCF4 to maintain a positive feedback loop. TCF4L induction occurs at the CDP stage of development, but the stage at which it is required for development remains unclear. This would be aided by the identification of the clonogenic pDC progenitor; however, so far, there has only been identification of populations of BM cells that show relative enrichment for pDCs, and no population that is clonogencially restricted to the pDC lineage has been reported to date (Schlitzer et al. 2011). Similarly, the transcriptional basis for pDC specification and commitment awaits identification.

One mechanism proposed for pDC specification is the expression of ID2, which is required for cDC1 development. Recently, we and others identified *Zeb2*, a Zinc-finger homeodomain transcription factor (Vandewalle et al. 2009) to be required for pDC development (Scott et al. 2016; Wu et al. 2016b). Germline deletion of *Zeb2* causes embryonic lethality in mice as a result of its action during the epithelial–mesenchymal transition (Higashi et al. 2002; Van de Putte et al. 2003), which involves the repression of E-cadherin (Comijn et al. 2001; Vandewalle et al. 2005). In the nervous system, ZEB2 regulates myelination by modulating the actions of Smad proteins, which are activated members of the TGF-β superfamily known as bone morphogenic proteins (Weng et al. 2012). In oligodendrocyte precursors, *Zeb2* expression is low and activated Smads bind P300, a coactivator histone

acetyltransferase, inducing expression of negative regulatory genes such as *Id2* and *Hes1*. However, in differentiating oligodendrocytes, OLIG1 and OLIG2 induce the expression of ZEB2, which binds to and represses Smad-P300 complexes thus blocking *Id2* and *Hes1* expression (Weng et al. 2012). Within DC development, ZEB2 appears to act as a negative regulator of ID2. We found that deletion of *Zeb2* in DCs using *Itgax*-Cre caused slightly higher expression of *Id2* in cDC2s compared with *Id2* expression in WT cDC2 cells. Overexpression of *Zeb2* in BM cultures stimulated with FLT3L caused strongly increased pDC development while restricting the frequency of cDC1 cells (Wu et al. 2016b). The role of ZEB2 in cDC2s is still unclear. If specification to the pDC and cDC lineages is dependent *Zeb2* and *Id2*, respectively, it is unclear how cDC2s develop in $Id2^{-/-}$ mice (Hacker et al. 2003; Kusunoki et al. 2003). In summary, it is unclear currently whether ID2 acts simply to exclude pDC potential from cells arising from the CDP population, for example, by preventing runaway E2-2 expression (Grajkowska et al. 2017) or, alternatively, whether it acts to support cDC1 development in some way. In either case, the actual mechanism has not been identified.

Other factors have been implicated in pDC development. Deletion of *Runx2*, a Runt family transcription factor that is required for osteoblast development (Komori et al. 1997; Otto et al. 1997), causes reduced expression of CCR5 on pDCs, thus impairing their egress from the BM to the periphery (Sawai et al. 2013). Previously, it was thought that deletion of *Irf8* prevented pDC development (Schiavoni et al. 2002). However, a recent study showed that pDCs develop in *Itgax*-Cre $Irf8^{fl/fl}$ mice but exhibit an abnormal phenotype and altered transcriptional profile (Sichien et al. 2016). This result does not rule out a requirement for *Irf8* in the development of pDCs prior to the expression of CD11c; however, given the altered phenotype of pDCs in *Itgax*-Cre $Irf8^{fl/fl}$ mice, it is conceivable that pDCs may still be present in $Irf8^{-/-}$ mice. A reevaluation of the dependence on *Irf8* for pDC development may thus be warranted.

Cite this article as *Cold Spring Harb Perspect Biol* doi: 10.1101/cshperspect.a028613

Human DCs

Recent efforts to identify human counterparts of murine DCs suggest that their development is conserved across species (Dutertre et al. 2014). The current understanding of the cellular stages of DC development in mice, particularly progenitors developing in the BM, has been reviewed recently (Murphy et al. 2016). Identification of the human counterparts has been challenging because of relative limitations in access to samples such as BM compared with mice. Human cDC1s are identified by the expression of CD141, Clec9a, and XCR1 (Bachem et al. 2010; Crozat et al. 2010; Poulin et al. 2010). Like murine cDC1s, these cells express IRF8, produce IL-12, and have superior capacity to cross-present (Haniffa et al. 2012). Human cDC2s can be identified by the expression of CD1c and BDCA1, and, like their mouse counterparts, express IRF4, produce IL-23, and induce differentiation of Th17 cells in response to *Aspergillus fumigatus* (Schlitzer et al. 2013). Recently, cDC2 cells in peripheral blood were segregated into two distinct groups based on CD5 expression (Yin et al. 2017). $CD5^{high}$ cDC2s expressed high levels of IRF4 and were potent inducers of T-cell activation. The ontogeny of $CD5^{lo}$ cells, however, appears unclear. $CD5^{lo}$ DCs express high levels of *MafB*, which in mice is highly expressed in monocytes and macrophages but not cDCs (Satpathy et al. 2012b; Wu et al. 2016a). A large cohort of human lymphoid tissue samples was used to confirm the broad tissue distribution of cDC1s and cDC2s and the conservation of cDC migratory phenotypes between mice and humans based on the expression of CCR7 and higher MHCII (Granot et al. 2017).

With the application of single-cell RNA-Seq (scRNA-Seq), several studies have identified human DC progenitors in BM and blood. Three independent studies described human pre-cDC progenitors with cDC1 and cDC2 potential (Breton et al. 2015a,b; See et al. 2017; Villani et al. 2017), using different surface markers (Table 2). However, whether these populations are related has not been tested. A side-by-side comparison of pre-cDCs identified by See et al.

Table 2. Markers of humans pre-cDCs in peripheral blood

Surface marker	Pre-cDC (Breton et al. 2015a,b)	Pre-cDC (Villani et al. 2017)	Pre-cDC (See et al. 2017)
HLA-DR	+	+	+
CDllc	lo		lo
CD14	−	−	−
CD34	−	int	−
CD45RA	+	+	+
CD100	?	+	?
CD115	−	−	?
CD116	+		+
CD117	+	+	−
CD123	int	−	+
CD135	+	−	+

Summary of surface marker expression used by three independent groups to identify distinct populations of pre-classical dendritic cells (cDCs) in peripheral blood from humans. Question marks indicate that the marker's expression was not reported in the study.

and Breton et al. showed that the former was more abundant, had higher expression of CD303, and lower expression of CD117 (See et al. 2017). However, it has been suggested that pre-cDCs are heterogeneous and composed of cells already specified to each subset of the cDC lineage, similar to that observed in mice (Breton et al. 2016; Grajales-Reyes et al. 2015). Along these lines, progenitors with predominantly cDC1 or cDC2 potential have also been identified (Table 3). The first set of committed pre-cDC progenitors were distinguished based on surface expression of CD172a, cDC1 progenitors being $CD172a^-$ and cDC2 progenitors $CD172a^+$ (Breton et al. 2016). See et al. (2017) independently defined two populations of pre-cDCs that are $CD33^+CD45RA^+CD123^{lo}$. The first is $CADM1^+$ and gives rise to cDC1s, and the second is $CD1c^+$ and gives rise to cDC2s (See et al. 2017). As defined by See et al. (2017), circulating pre-cDC1 and pre-cDC2 cells were morphologically similar to mature cDC1 and cDC2, secreted cytokines after TLR activation, and induced T-cell proliferation during allogeneic responses in vitro.

A novel DC population in circulation identified by scRNA-Seq has recently been proposed and is referred to as the AS-DC based on the

Table 3. Markers of cDC1 and cDC2 committed human pre-cDCs

Surface marker	Pre-cDC1 (Breton et al. 2016)	Pre-cDC1 (See et al. 2017)	Pre-cDC2 (Breton et al. 2016)	Pre-cDC2 (See et al. 2017)
HLA-DR	+	+	+	+
Cadm1	?	+/int	?	−
CD1c	−	−	−	+
CD14	−	−	−	−
CD33	?	+	?	+
CD34	−	−	−	−
CD45RA	+	+	+	+
CD115	−	?	?	?
CD116	+	?	+	?
CD117	+/int	?	+/int	?
CD123	−	+/−	−	−
CD135	+	+	+	+
CD172a	−	?	+	?

Summary of flow cytometry analysis of surface marker expression used to identify classical dendritic cell (cDC)1 and cDC2 committed pre-cDCs in humans. Question marks indicate the expression for a marker that was not reported for or used to define the population in the study.

expression of AXL and SIGLEC6 (Villani et al. 2017). The gene signatures observed for this population clustered between pDCs and cDC2s. The authors of the study suggest this population does not represent an intermediate stage of pre-cDCs because DC progenitors do not induce the expression of AXL or SIGLEC6 in culture (Villani et al. 2017). However, the pre-cDC reported by See et al. (2017) expresses the genes that encode these markers. Interrogation of the molecular mechanisms that control human cDC development is limited but gene expression analysis of the progenitors identified to date suggest conservation between mouse and human. For example, specification of cDC1 and cDC2 progenitors is associated with the differential expression of known regulators of mouse DC development, including BATF3, ID2, TCF4, IRF4, ZEB2, and IRF8 (Breton et al. 2016; See et al. 2017).

Human pDCs produce high levels of type I interferons (IFNs) during responses to viral infection (Cella et al. 1999; Siegal et al. 1999). They can also activate $CD4^+$ and $CD8^+$ T cells in response to influenza virus (Fonteneau et al. 2003). However, heterogeneity within the bulk pDC population was later recognized using CD2 as a marker to distinguish two distinct pDC populations (Matsui et al. 2009). $CD2^+$ pDCs secrete high levels of IL-12p40, induce surface expression of CD80, and induce proliferation of naïve allogeneic $CD4^+$ T cells. These cells more closely resemble cDCs than pDCs. This $CD2^+$ population was further refined using CD5 and CD81 to identify the pDC population capable of secreting IL-12 and activating $CD4^+$ T cells (Zhang et al. 2017a). A separate study showed that CD56 expression identified a myeloid DC population within the $CD2^+$ pDC gate. This novel population did not produce IFN-α, and instead secreted IL-12 and activated T cells. Transcriptomic analysis showed that $CD2^+$ $CD56^+$ pDCs were more closely related to cDCs than to pDCs (Yu et al. 2015). Further, transcriptomic analysis of $CD56^+$ pDCs showed they were closely related to blastic plasmacytoid DC neoplasms (BPDCNs). These observations were further confirmed by Villani et al. (2017), in which the AS-DC population shared some transcriptomic characteristics with pDCs. Functionally, they were potent inducers of T-cell proliferation and secreted high levels of IL-12. These multiple lines of evidence suggest that the cDC-like function attributed to a subset of human pDCs is the result of analysis of heterogeneous populations composed of cDCs and pDCs in early studies of human pDC function.

CONTEMPORARY ANALYSIS OF PARADIGMS IN cDC DEVELOPMENT AND FUNCTION

Identification of cDCs in Vivo

Shared surface marker expression among cells of the myeloid lineage has complicated the discrimination of DC subsets from other myeloid lineages. Recent analyses have proposed a simplified set of markers to discriminate DC subsets across tissues by defining cDCs as Lin⁻CD11c⁺ MHCII⁺CD26⁺CD64⁻, among which cDC1s and cDC2s can be identified as XCR1⁺ and CD172a⁺, respectively (Guilliams et al. 2016). Surface-marker-independent methods of discriminating lineages have been helpful in resolving the origin of myeloid cells in vivo. Expression of the transcription factor, ZBTB46, is restricted to the cDC lineage and can be used to identify cDCs and their progenitors in lymphoid and nonlymphoid tissues (Satpathy et al. 2012a; Grajales-Reyes et al. 2015). Alternatively, the transcription factor MafB is expressed by cells of the monocyte and macrophage lineage (Aziz et al. 2009; Gautier et al. 2012). A novel lineage-tracing reagent, MafB-mCherry-Cre mice, marks cells that express MafB (mCherry⁺) or have expressed MafB during development (YFP⁺ when crossed to R26-stop-YFP mice), and can thus be used in combination with Zbtb46 to discriminate between macrophage and DC lineages in vivo (Wu et al. 2016a). Interestingly, it was found that, among the tissues examined, Langerhans cells (LCs) in skin-draining lymph nodes were the only lineage to express Zbtb46 and also be marked by Mafb-driven lineage tracing.

Mo-DCs and GM-DCs

Numerous studies have suggested that under inflammatory conditions monocytes have the potential to differentiate into cDCs. From in vitro studies, it is known that monocytes from mice or humans cultured with granulocyte macrophage colony-stimulating factor (GM-CSF) and IL-4, referred to as Mo-DCs, acquire characteristics of cDCs (Caux et al. 1992; Inaba et al. 1992, 1993; Romani et al. 1994; Sallusto and Lanzavecchia 1994). Similarities include the expression of canonical surface markers, such as CD11c and MHCII (Leon et al. 2004), and DC-specific transcription factors, such as Zbtb46 and Mycl1 (Satpathy et al. 2012a; Wumesh et al. 2014). Upon treatment of GM-CSF, monocytes rapidly induce the expression of IRF4 (Lehtonen et al. 2005), which is required for their differentiation into cDC-like cells that express Zbtb46 and MHCII (Briseno et al. 2016). Mo-DCs have the ability to cross-prime CD8 T cells to cell-associated antigens in vitro; however, cross-presentation by the subset specialized for this activity in vivo, cDC1s, is Irf4-independent (Vander et al. 2014; Briseno et al. 2016). GM-CSF-derived DCs (GM-DCs) have been used extensively in studies surveying DC function. Many of the known actors involved in cross-presentation were first identified in GM-DCs and are reviewed here (Theisen and Murphy 2017). However, it was recently reported that BM cultures stimulated with GM-CSF produce heterogeneous populations and has thus casted doubt over physiological relevance GM-DCs to in vivo cDC subsets (Helft et al. 2015). To obtain populations of pDCs and cDCs that more closely resemble in vivo counterparts, an alternative in vitro culture system that uses whole BM or purified progenitors in Flt3L was developed (Naik et al. 2005). We recently identified Rab43 to be involved in the cross-presentation of cell-associated antigen by cDC1s but not by GM-DCs (Kretzer et al. 2016). Therefore, it may be necessary to evaluate the function of molecules previously reported in GM-DCs to regulate vesicular trafficking and cross-presentation, including but not limited to RAC2 (Savina et al. 2009), RAB11A (Nair-Gupta et al. 2014), RAB3B (Zou et al. 2009), and SEC22B (Cebrian 2011).

The precise role of GM-DCs in promoting CD8⁺ T-cell responses via cross-presentation is unclear because Zbtb46-expressing Mo-DCs have yet to be distinguished in vivo from bona fide CD11b⁺ DCs, and thus a model to selectively deplete them is unavailable. Recent work that replicated in vivo models of putative Mo-DC differentiation by house dust mite challenge did not identify Zbtb46-expressing cells that were marked by *MafB*-driven lineage tracing (Wu et al. 2016a). Therefore, the developmental origins of the Mo-DCs in vivo remain elusive. Notwithstanding, GM-DCs have been demonstrated to be a viable option in the generation of tumor vaccines (Linette and Carreno 2013). Human Mo-DCs generated with GM-CSF and IL-4 have been used as the basis for therapeutic cancer vaccines (Palucka and Bancereau 2013; Carreno et al. 2015). Vaccines based on human Mo-DCs pulsed with tumor-specific peptides can initiate CD8 T-cell responses and induce clinical responses in melanoma, renal cell carcinoma, and malignant glioma (Nestle et al. 1998; Holtl et al. 1999; Thurner et al. 1999; Timmerman et al. 2002). Human Mo-DCs generated ex vivo can also elicit broad CD8$^+$ T-cell responses against tumor antigens, and to a class of subdominant neoantigens in patients with melanoma (Carreno et al. 2015).

cDC Maturation

An additional area of study to emerge from the use of GM-DCs is DC maturation. The term "mature" was first used to describe the adherent fraction of DCs isolated from the spleen of mice (Steinman and Cohn 1973). The process of "maturation" was later described as the acquisition of T-cell stimulatory capacity of LCs and DCs after isolation and ex vivo culture (Schuler et al. 1985; Witmer-Pack et al. 1987; Heufler et al. 1988). The capacity to stimulate T cells was correlated with the induction of costimulatory molecules, such as CD80 and CD86 (Inaba et al. 1994), and chemokine receptors, such as CCR7 (Sallusto et al. 1998; Sozzani et al. 1998). Immature DCs in vitro most closely resemble resident DCs in vivo and are identified as CD11c^{hi} MHCII$^+$. Mature DCs in vitro most closely resemble migratory DCs in vivo and are identified as CD11c$^+$MHCII^{hi}. Consistent with this correlation, migratory DCs in vivo express elevated levels of the canonical maturation markers CCR7, CD80, CD86, and CD40. CCR7 is required for the migration of cDCs to draining lymphoid organs (Forster et al. 1999; Ohl et al. 2004), CD80 and CD86 are required for stimulation of naïve T cells (Steinman et al. 2003), and

CD40 is required to receive CD4 T-cell help (Bennett et al. 1998; Schoenberger et al. 1998). Therefore, the study of maturation in vitro has led to important in vivo discoveries regarding fundamental DC biology. However, recent analysis of maturation in vivo calls for a revision of previously established paradigms regarding functional differences between immature and mature DCs.

Low expression of costimulatory molecules on immature DCs formed the basis for a hypothesis that immature DCs are specialized at tolerance induction and have thus been referred to as tolerogenic (Morelli and Thomson 2007). Recent in vivo evidence is contrary to this distinction. It was recently shown that mature cDC1s are the sole population capable of cross-presenting thymic epithelial-cell-derived self-antigens, and that BATF3-dependent cDC1s are required to induce a subset of *Aire*-dependent natural regulatory T (Treg) cells (Perry et al. 2014; Ardouin et al. 2016). Peripheral Treg induction in the small intestine lamina propria is also induced by cDC1s that have taken up host-derived antigen and migrate to draining lymph nodes. These cells express higher levels of *Ccr7* and undergo transcriptional reprogramming associated with maturation (Cummings et al. 2016). This is consistent with results from the examination of draining lymph nodes of the oral mucosa, where migratory cDC1s are most efficient at inducing oral tolerance to dietary antigens (Esterhazy et al. 2016). Contrary to a role for cDC1s in the induction of tolerance, cDC1s can also be essential for the initiation of auto-immunity in mice with genetic backgrounds predisposed to the development of diabetes (Ferris et al. 2014).

Given that functional distinctions between immature and mature DCs may not be consistent with phenomena that occur in vivo, the process of maturation may be better conceptualized as a stage of cDC development. cDC progenitors are known to egress from the BM specified to either the cDC1 or cDC2 lineage (Grajales-Reyes et al. 2015). In peripheral tissues, immature cDCs exhibit phenotypes associated with cell division and proliferation (Liu et al. 2007). Work to identify the factors that regulate DC proliferation remains an active area of research. In secondary lymphoid organs, deficiency in the DC-specific transcription factor, L-Myc, results in a reduction in the number of cDC1s, reduction in DNA replication associated with cell division, reduction in priming of antigen-specific CD8 T cells, and enhanced resistance to infection with *L. monocytogenes* (Wumesh et al. 2014). The growth factor, Flt3L, has been shown to expand BM progenitors of cDCs and increase the population size of cDCs in lymphoid organs (Waskow et al. 2008). Although a requirement for GM-CSF in cDC1 development is debated (Edelson et al. 2011b; Greter et al. 2012), treatment with GM-CSF in vivo and in vitro in combination with Flt3L is sufficient to expand populations of cDCs (Daro et al. 2000; Mayer et al. 2014).

The extent to which local growth factor concentrations and milieus influence proliferation of cDCs at steady state and during inflammation in peripheral tissues is not well understood, and may be made redundant by constant recruitment of progenitors from the BM (Liu and Nussenzweig 2010). Whole transcriptome analysis of mature CCR7$^+$ and immature CCR7$^-$ cDC1s sorted from the thymus and spleen at steady state revealed broad transcriptional reprogramming that includes differential expression of genes associated with exit from the cell cycle (Ardouin et al. 2016). Similar transcriptional changes have been detected in cDCs from a variety of tissues when resident and migratory counterparts are compared (Manh et al. 2013). Identification of cells that have undergone cell-cycle exit or entered a state of quiescence is commonly used to uncover stages at which terminal differentiation occurs during the ontogeny of cellular lineages (Massague 2004; Buttitta and Edgar 2007; Coller 2011). Therefore, cell-cycle exit on cDC maturation suggests that this process represents a terminal stage in the developmental program of cDCs. The factors necessary to induce maturation in vivo remain largely unknown, and much of the work conducted to date has focused on the cell-extrinsic influence of host- and commensal-derived stimuli.

Both host- and microbiota-derived factors have been reported to be sufficient to induce DC maturation. In a model of vaccination using mice deficient in IFNAR, it was shown that type I IFN in response to poly-IC acted directly on DCs to induce maturation of splenic DCs and induce Th1 immunity to a model antigen of HIV (Longhi et al. 2009). In models of viral infection and tumor rejection, the action of type I IFN on cDCs was required for optimal CD8 T-cell priming and Th1-cell polarization (Brewitz et al. 2017; Diamond et al. 2011; Fuertes et al. 2011). Recent whole transcriptome analysis has demonstrated that transcriptional reprogramming is conserved between maturation at homeostasis or under inflammatory conditions of poly-IC injection or viral infection, and occurs independently of IFNAR signaling (Ardouin et al. 2016). Although signaling through IFNAR may be sufficient to induce maturation in the spleen (Longhi et al. 2009), where at least 90% of DCs exhibit an immature phenotype (Ardouin et al. 2016), such signals are not necessary to execute the transcriptional program that occurs during this process.

Signaling cascades initiated by engagement of the receptors for IL-1β, tumor necrosis factor α (TNF-α), CD40L, and LT converge on activation of canonical and noncanonical nuclear factor (NF)-κB (Jost and Ruland 2007). As discussed above, it was recently shown that a requirement of RelB in the development of cDCs is largely cell-extrinsic, with the exception of a splenic cDC2 subset that is also Notch2- and LTβR-dependent (Kabashima et al. 2005; Satpathy et al. 2013; Briseno et al. 2017). However, this does not rule out a DC-intrinsic role for RelB or the remaining NF-κB family members in cDC maturation. In vivo analysis of p50-deficient mice showed no effect on the expression of CD80 or CD86; however, they have a defect in Th2 cell differentiation during helminth infection, which is now known to be regulated by the KLF4-dependent cDC2 subset (Artis et al. 2005; Tussiwand et al. 2015). An independent study also showed that cRel, p50, and RelA are dispensable for development of cDCs and the expression of CD80 and CD86. However, the deletion of p50 and RelA together led to a significant loss of CD11c$^+$ cells in the spleen (Ouaaz et al. 2002). Therefore, NF-κB family members may have compensatory roles in cDC development and function. Defining such combinatorial complexity in NF-κB activity has been difficult to define in vivo, because up to 15 dimer combinations are possible with 13 reported to date (Smale 2012; Zhang et al. 2017b). Modules of genes that are known NF-κB targets are differentially expressed during maturation (Manh et al. 2013; Ardouin et al. 2016); however, definition of the cell-intrinsic requirements for NF-κB in cDC function remains a hurdle to overcome in this field.

Homeostatic interactions between commensal microbiota and the host-immune system have been linked to various immunological disorders in patients with mutations of pattern-recognition receptors (PRRs) (Hooper et al. 2012). Direct signals from microbiota acting on DCs at steady state and during infection have been suggested to regulate DC maturation (Steinman et al. 2003). Although there is mounting evidence for the regulation of immune homeostasis through interactions between the host and commensal microbiota (Belkaid and Hand 2014), evidence for the regulation of DC maturation is limited. DC-specific deletion of *Traf6*, which encodes a signaling adaptor downstream of various PRRs, results in defective inflammatory cytokine production on stimulation with CpG and LPS. These mice also develop spontaneous inflammation of the small intestine that is associated with aberrant Th2 cell priming, which can be rescued by treatment with antibiotics (Han et al. 2013). *Traf6*$^{-/-}$ mice have defective induction of maturation markers on DCs in vivo when treated with LPS or CD40L and, therefore, it was proposed to be necessary for DC maturation (Kobayashi et al. 2003). However, in single and double knockout mice of *MyD88* and *Ticam1*, no effect on the development or maturation of DCs was observed (Wilson et al. 2008). Such conflicting results are difficult to interpret in light of complex cross talk that may establish redundancy in signaling pathways downstream of PRRs (Lee and Kim 2007). Therefore, ablation of commensal microbiota is widely used as a strategy to probe the impact of steady-state microbial signals on immune homeostasis. To that end, WT specific pathogen-free (SPF) and germ-free mice showed no significant differences in the core transcriptional program associated with maturation in vivo for thymic cDC1s (Ardouin et al. 2016). In addition, the core maturation programs that occur at homeostasis or under inflammation induced by poly-IC or STAg overlap broadly (Ardouin et al. 2016). As opposed to a model that focuses on cell-extrinsic stimuli, representation of maturation as a cell-intrinsic developmental program can provide an alternative framework to discover novel mechanisms that regulate DC development and function.

CONCLUSION

In summary, analysis of the molecular events that underlie distinct forms of DCs in the mouse have advanced over the past 8 years, with the identification of several transcription factors required for some, but not all, DC subsets. Notably, while several factors appear to be required for cDC1 and pDC development, there has been no single factor whose ablation selectively prevents cDC2 development. It is true that Notch2 is required for the normal functioning of cDC2 in response to certain pathogens and that KLF4 is required for cDC2 support of Th2 type responses, but, in each case, cDC2 cells develop in the absence of these factors. It is not clear that these results mean that cDC2 development is a "default" pathway, because it may turn out that a mechanism will be found that is necessary for development of all forms of cDC2. Progress arising from analysis of the murine system has also provided a basis for analysis of human DCs, which now are recognized as being structurally similar to their murine counterparts, at least in certain fundamental ways. These studies in both systems promise to provide a basis for future rational therapeutic interventions to complement the current progress in immunotherapy.

REFERENCES

Aliberti J, Schulz O, Pennington DJ, Tsujimura H, Sousa RE, Ozato K, Sher A. 2003. Essential role for ICSBP in the in vivo development of murine CD8α$^+$ dendritic cells. *Blood* **101:** 305–310.

Ardouin L, Luche H, Chelbi R, Carpentier S, Shawket A, Montanana AF, Santa MC, Grenot P, Alexandre Y, Gregoire C, et al. 2016. Broad and largely concordant molecular changes characterize tolerogenic and immunogenic dendritic cell maturation in thymus and periphery. *Immunity* **45:** 305–318.

Artis D, Kane CM, Fiore J, Zaph C, Shapira S, Joyce K, Macdonald A, Hunter C, Scott P, Pearce EJ. 2005. Dendritic cell-intrinsic expression of NF-κB1 is required to promote optimal Th2 cell differentiation. *J Immunol* **174:** 7154–7159.

Auffray C, Fogg DK, Narni-Mancinelli E, Senechal B, Trouillet C, Saederup N, Leemput J, Bigot K, Campisi L, Abitbol M, et al. 2009. CX3CR1⁺ CD115⁺ CD135⁺ common macrophage/DC precursors and the role of CX3CR1 in their response to inflammation. *J Exp Med* **206:** 595–606.

Aziz A, Soucie E, Sarrazin S, Sieweke MH. 2009. MafB/c-Maf deficiency enables self-renewal of differentiated functional macrophages. *Science* **326:** 867–871.

Bachem A, Guttler S, Hartung E, Ebstein F, Schaefer M, Tannert A, Salama A, Movassaghi K, Opitz C, Mages HW, et al. 2010. Superior antigen cross-presentation and XCR1 expression define human CD11c⁺CD141⁺ cells as homologues of mouse CD8⁺ dendritic cells. *J Exp Med* **207:** 1273–1281.

Belkaid Y, Hand TW. 2014. Role of the microbiota in immunity and inflammation. *Cell* **157:** 121–141.

Bennett SR, Carbone FR, Karamalis F, Flavell RA, Miller JF, Heath WR. 1998. Help for cytotoxic-T-cell responses is mediated by CD40 signalling. *Nature* **393:** 478–480.

Breton G, Lee J, Liu K, Nussenzweig MC. 2015a. Defining human dendritic cell progenitors by multiparametric flow cytometry. *Nat Protoc* **10:** 1407–1422.

Breton G, Lee J, Zhou YJ, Schreiber JJ, Keler T, Puhr S, Anandasabapathy N, Schlesinger S, Caskey M, Liu K, et al. 2015b. Circulating precursors of human CD1c⁺ and CD141⁺ dendritic cells. *J Exp Med* **212:** 401–413.

Breton G, Zheng S, Valieris R, da Silva IT, Satija R, Nussenzweig MC. 2016. Human dendritic cells (DCs) are derived from distinct circulating precursors that are precommitted to become CD1c⁺ and CD141⁺ DCs. *J Exp Med* **213:** 2861–2870.

Brewitz A, Eickhoff S, Dahling S, Quast T, Bedoui S, Kroczek RA, Kurts C, Garbi N, Barchet W, Iannacone M, et al. 2017. CD8⁺ T cells orchestrate pDC-XCR1⁺ dendritic cell spatial and functional cooperativity to optimize priming. *Immunity* **46:** 205–219.

Briseno CG, Haldar M, Kretzer NM, Wu X, Theisen DJ, Wumesh KC, Durai V, Grajales-Reyes GE, Iwata A, Bagadia P, et al. 2016. Distinct transcriptional programs control cross-priming in classical and monocyte-derived dendritic cells. *Cell Rep* **15:** 2462–2474.

Briseno CG, Gargaro M, Durai V, Davidson JT, Theisen DJ, Anderson DA III, Novack DV, Murphy TL, Murphy KM. 2017. Deficiency of transcription factor RelB perturbs myeloid and DC development by hematopoietic-extrinsic mechanisms. *Proc Natl Acad Sci* **114:** 3957–3962.

Burkly L, Hession C, Ogata L, Reilly C, Marconi LA, Olson D, Tizard R, Cate R, Lo D. 1995. Expression of relB is required for the development of thymic medulla and dendritic cells. *Nature* **373:** 531–536.

Buttitta LA, Edgar BA. 2007. Mechanisms controlling cell cycle exit upon terminal differentiation. *Curr Opin Cell Biol* **19:** 697–704.

Carreno BM, Magrini V, Becker-Hapak M, Kaabinejadian S, Hundal J, Petti AA, Ly A, Lie WR, Hildebrand WH, Mardis ER, et al. 2015. Cancer immunotherapy. A dendritic cell vaccine increases the breadth and diversity of melanoma neoantigen-specific T cells. *Science* **348:** 803–808.

Caux C, Dezutter-Dambuyant C, Schmitt D, Banchereau J. 1992. GM-CSF and TNF-α cooperate in the generation of dendritic Langerhans cells. *Nature* **360:** 258–261.

Cella M, Jarrossay D, Facchetti F, Alebardi O, Nakajima H, Lanzavecchia A, Colonna M. 1999. Plasmacytoid monocytes migrate to inflamed lymph nodes and produce large amounts of type I interferon [see comments]. *Nat Med* **5:** 919–923.

Cisse B, Caton ML, Lehner M, Maeda T, Scheu S, Locksley R, Holmberg D, Zweier C, den Hollander NS, Kant SG, et al. 2008. Transcription factor E2-2 is an essential and specific regulator of plasmacytoid dendritic cell development 1. *Cell* **135:** 37–48.

Coller HA. 2011. Cell biology. The essence of quiescence. *Science* **334:** 1074–1075.

Comijn J, Berx G, Vermassen P, Verschueren K, van Grunsven L, Bruyneel E, Mareel M, Huylebroeck D, Van Roy F. 2001. The two-handed E box binding zinc finger protein SIP1 downregulates E-cadherin and induces invasion. *Mol Cell* **7:** 1267–1278.

Corneliussen B, Thornell A, Hallberg B, Grundstrom T. 1991. Helix–loop–helix transcriptional activators bind to a sequence in glucocorticoid response elements of retrovirus enhancers. *J Virol* **65:** 6084–6093.

Crozat K, Guiton R, Contreras V, Feuillet V, Dutertre CA, Ventre E, Vu Manh TP, Baranek T, Storset AK, Marvel J, et al. 2010. The XC chemokine receptor 1 is a conserved selective marker of mammalian cells homologous to mouse CD8α⁺ dendritic cells. *J Exp Med* **207:** 1283–1292.

Cummings RJ, Barbet G, Bongers G, Hartmann BM, Gettler K, Muniz L, Furtado GC, Cho J, Lira SA, Blander JM. 2016. Different tissue phagocytes sample apoptotic cells to direct distinct homeostasis programs. *Nature* **539:** 565–569.

Daro E, Pulendran B, Brasel K, Teepe M, Pettit D, Lynch DH, Vremec D, Robb L, Shortman K, McKenna HJ, et al. 2000. Polyethylene glycol-modified GM-CSF expands CD11b^high^CD11c^high^ but notCD11b^low^CD11c^high^ murine dendritic cells in vivo: A comparative analysis with Flt3 ligand. *J Immunol* **165:** 49–58.

DeKoning J, DiMolfetto L, Reilly C, Wei Q, Havran WL, Lo D. 1997. Thymic cortical epithelium is sufficient for the development of mature T cells in relB-deficient mice. *J Immunol* **158:** 2558–2566.

De Silva NS, Silva K, Anderson MM, Bhagat G, Klein U. 2016. Impairment of mature B cell maintenance upon combined deletion of the alternative NF-κB transcription factors RELB and NF-κB2 in B Cells. *J Immunol* **196:** 2591–2601.

Diamond MS, Kinder M, Matsushita H, Mashayekhi M, Dunn GP, Archambault JM, Lee H, Arthur CD, White JM, Kalinke U, et al. 2011. Type I interferon is selectively required by dendritic cells for immune rejection of tumors. *J Exp Med* **208:** 1989–2003.

Dutertre CA, Wang LF, Ginhoux F. 2014. Aligning bona fide dendritic cell populations across species. *Cell Immunol* **291:** 3–10.

Edelson BT, Wumesh KC, Juang R, Kohyama M, Benoit LA, Klekotka PA, Moon C, Albring JC, Ise W, Michael DG, et al. 2010. Peripheral CD103[+] dendritic cells form a unified subset developmentally related to CD8α[+] conventional dendritic cells. *J Exp Med* **207:** 823–836.

Edelson BT, Bradstreet TR, Hildner K, Carrero JA, Frederick KE, Wumesh KC, Belizaire R, Aoshi T, Schreiber RD, Miller MJ, et al. 2011a. CD8a[+] dendritic cells are an obligate cellular entry point for productive infection by *Listeria monocytogenes*. *Immunity* **35:** 236–248.

Edelson BT, Bradstreet TR, Wumesh KC, Hildner K, Herzog JW, Sim J, Russell JH, Murphy TL, Unanue ER, Murphy KM. 2011b. Batf3-dependent CD11b[low/-] peripheral dendritic cells are GM-CSF-independent and are not required for Th cell priming after subcutaneous immunization. *PLoS ONE* **6:** e25660.

Esterhazy D, Loschko J, London M, Jove V, Oliveira TY, Mucida D. 2016. Classical dendritic cells are required for dietary antigen-mediated induction of peripheral Treg cells and tolerance. *Nat Immunol* **17:** 545–555.

Everts B, Tussiwand R, Dreesen L, Fairfax KC, Huang SC, Smith AM, O'Neill CM, Lam WY, Edelson BT, Urban JF Jr, et al. 2016. Migratory CD103[+] dendritic cells suppress helminth-driven type 2 immunity through constitutive expression of IL-12. *J Exp Med* **213:** 35–51.

Fasnacht N, Huang HY, Koch U, Favre S, Auderset F, Chai Q, Onder L, Kallert S, Pinschewer DD, MacDonald HR, et al. 2014. Specific fibroblastic niches in secondary lymphoid organs orchestrate distinct Notch-regulated immune responses. *J Exp Med* **211:** 2265–2279.

Ferris ST, Carrero JA, Mohan JF, Calderon B, Murphy KM, Unanue ER. 2014. A minor subset of Batf3-dependent antigen-presenting cells in islets of Langerhans is essential for the development of autoimmune diabetes. *Immunity* **41:** 657–669.

Fogg DK, Sibon C, Miled C, Jung S, Aucouturier P, Littman DR, Cumano A, Geissmann F. 2006. A clonogenic bone marrow progenitor specific for macrophages and dendritic cells. *Science* **311:** 83–87.

Fontenau JF, Gilliet M, Larsson M, Dasilva I, Munz C, Liu YJ, Bhardwaj N. 2003. Activation of influenza virus-specific CD4[+] and CD8[+] T cells: A new role for plasmacytoid dendritic cells in adaptive immunity. *Blood* **101:** 3520–3526.

Forster R, Schubel A, Breitfeld D, Kremmer E, Renner-Muller I, Wolf E, Lipp M. 1999. CCR7 coordinates the primary immune response by establishing functional microenvironments in secondary lymphoid organs. *Cell* **99:** 23–33.

Fuertes MB, Kacha AK, Kline J, Woo SR, Kranz DM, Murphy KM, Gajewski TF. 2011. Host type I IFN signals are required for antitumor CD8[+] T cell responses through CD8α[+] dendritic cells. *J Exp Med* **208:** 2005–2016.

Gao Y, Nish SA, Jiang R, Hou L, Licona-Limon P, Weinstein JS, Zhao H, Medzhitov R. 2013. Control of T helper 2 responses by transcription factor IRF4-dependent dendritic cells. *Immunity* **39:** 722–732.

Gautier EL, Shay T, Miller J, Greter M, Jakubzick C, Ivanov S, Helft J, Chow A, Elpek KG, Gordonov S, et al. 2012. Gene-

expression profiles and transcriptional regulatory pathways that underlie the identity and diversity of mouse tissue macrophages. *Nat Immunol* **13:** 1118–1128.

Gerloni M, Lo D, Ballou WT, Zanetti M. 1998a. Immunological memory after somatic transgene immunization is positively affected by priming with GM-CSF and does not require bone marrow-derived dendritic cells. *Eur J Immunol* **28:** 1832–1838.

Gerloni M, Lo D, Zanetti M. 1998b. DNA immunization in relB-deficient mice discloses a role for dendritic cells in IgM→IgG1 switch in vivo. *Eur J Immunol* **28:** 516–524.

Ghosh HS, Cisse B, Bunin A, Lewis KL, Reizis B. 2010. Continuous expression of the transcription factor e2-2 maintains the cell fate of mature plasmacytoid dendritic cells. *Immunity* **33:** 905–916.

Ghosh HS, Ceribelli M, Matos I, Lazarovici A, Bussemaker HJ, Lasorella A, Hiebert SW, Liu K, Staudt LM, Reizis B. 2014. ETO family protein Mtg16 regulates the balance of dendritic cell subsets by repressing Id2. *J Exp Med* **211:** 1623–1635.

Grajales-Reyes GE, Iwata A, Albring J, Wu X, Tussiwand R, Wumesh KC, Kretzer NM, Briseno CG, Durai V, Bagadia P, et al. 2015. Batf3 maintains autoactivation of Irf8 for commitment of a CD8α[+] conventional DC clonogenic progenitor. *Nat Immunol* **16:** 708–717.

Grajkowska LT, Ceribelli M, Lau CM, Warren ME, Tiniakou I, Nakandakari HS, Bunin A, Haecker H, Mirny LA, Staudt LM, et al. 2017. Isoform-specific expression and feedback regulation of E protein TCF4 control dendritic cell lineage specification. *Immunity* **46:** 65–77.

Granot T, Senda T, Carpenter DJ, Matsuoka N, Weiner J, Gordon CL, Miron M, Kumar BV, Griesemer A, Ho SH, et al. 2017. Dendritic cells display subset and tissue-specific maturation dynamics over human life. *Immunity* **46:** 504–515.

Greter M, Helft J, Chow A, Hashimoto D, Mortha A, Agudo-Cantero J, Bogunovic M, Gautier EL, Miller J, Leboeuf M, et al. 2012. GM-CSF controls nonlymphoid tissue dendritic cell homeostasis but is dispensable for the differentiation of inflammatory dendritic cells. *Immunity* **36:** 1031–1046.

Guilliams M, Ginhoux F, Jakubzick C, Naik SH, Onai N, Schraml BU, Segura E, Tussiwand R, Yona S. 2014. Dendritic cells, monocytes and macrophages: A unified nomenclature based on ontogeny. *Nat Rev Immunol* **14:** 571–578.

Guilliams M, Dutertre CA, Scott CL, McGovern N, Sichien D, Chakarov S, Van Gassen S, Chen J, Poidinger M, De Prijck S, et al. 2016. Unsupervised high-dimensional analysis aligns dendritic cells across tissues and species. *Immunity* **45:** 669–684.

Hacker C, Kirsch RD, Ju XS, Hieronymus T, Gust TC, Kuhl C, Jorgas T, Kurz SM, Rose-John S, Yokota Y, et al. 2003. Transcriptional profiling identifies Id2 function in dendritic cell development. *Nat Immunol* **4:** 380–386.

Halim TY, Steer CA, Matha L, Gold MJ, Martinez-Gonzalez I, McNagny KM, McKenzie AN, Takei F. 2014. Group 2 innate lymphoid cells are critical for the initiation of adaptive T helper 2 cell-mediated allergic lung inflammation. *Immunity* **40:** 425–435.

Halim TY, Hwang YY, Scanlon ST, Zaghouani H, Garbi N, Fallon PG, McKenzie AN. 2016. Group 2 innate lymphoid cells license dendritic cells to potentiate memory T_H2 cell responses. *Nat Immunol* **17**: 57–64.

Han D, Walsh MC, Cejas PJ, Dang NN, Kim YF, Kim J, Charrier-Hisamuddin L, Chau L, Zhang Q, Bittinger K, et al. 2013. Dendritic cell expression of the signaling molecule TRAF6 is critical for gut microbiota-dependent immune tolerance. *Immunity* **38**: 1211–1222.

Haniffa M, Shin A, Bigley V, McGovern N, Teo P, See P, Wasan PS, Wang XN, Malinarich F, Malleret B, et al. 2012. Human tissues contain CD141[hi] cross-presenting dendritic cells with functional homology to mouse CD103[+] nonlymphoid dendritic cells. *Immunity* **37**: 60–73.

Helft J, Bottcher J, Chakravarty P, Zelenay S, Huotari J, Schraml BU, Goubau D, Sousa RE. 2015. GM-CSF mouse bone marrow cultures comprise a heterogeneous population of CD11c[+]MHCII[+] macrophages and dendritic cells. *Immunity* **42**: 1197–1211.

Heufler C, Koch F, Schuler G. 1988. Granulocyte/macrophage colony-stimulating factor and interleukin 1 mediate the maturation of murine epidermal Langerhans cells into potent immunostimulatory dendritic cells. *J Exp Med* **167**: 700–705.

Higashi Y, Maruhashi M, Nelles L, Van de PT, Verschueren K, Miyoshi T, Yoshimoto A, Kondoh H, Huylebroeck D. 2002. Generation of the floxed allele of the SIP1 (Smad-interacting protein 1) gene for Cre-mediated conditional knockout in the mouse. *Genesis* **32**: 82–84.

Hildner K, Edelson BT, Purtha WE, Diamond M, Matsushita H, Kohyama M, Calderon B, Schraml BU, Unanue ER, Diamond MS, et al. 2008. Batf3 deficiency reveals a critical role for CD8α[+] dendritic cells in cytotoxic T cell immunity. *Science* **322**: 1097–1100.

Holtl L, Rieser C, Papesh C, Ramoner R, Herold M, Klocker H, Radmayr C, Stenzl A, Bartsch G, Thurnher M. 1999. Cellular and humoral immune responses in patients with metastatic renal cell carcinoma after vaccination with antigen pulsed dendritic cells. *J Urol* **161**: 777–782.

Hooper LV, Littman DR, Macpherson AJ. 2012. Interactions between the microbiota and the immune system. *Science* **336**: 1268–1273.

Inaba K, Inaba M, Romani N, Aya H, Deguchi M, Ikehara S, Muramatsu S, Steinman RM. 1992. Generation of large numbers of dendritic cells from mouse bone marrow cultures supplemented with granulocyte/macrophage colony-stimulating factor. *J Exp Med* **176**: 1693–1702.

Inaba K, Inaba M, Deguchi M, Hagi K, Yasumizu R, Ikehara S, Muramatsu S, Steinman RM. 1993. Granulocytes, macrophages, and dendritic cells arise from a common major histocompatibility complex class II-negative progenitor in mouse bone marrow. *Proc Natl Acad Sci* **90**: 3038–3042.

Inaba K, Witmer-Pack M, Inaba M, Hathcock KS, Sakuta H, Azuma M, Yagita H, Okumura K, Linsley PS, Ikehara S, et al. 1994. The tissue distribution of the B7-2 costimulator in mice: Abundant expression on dendritic cells in situ and during maturation in vitro. *J Exp Med* **180**: 1849–1860.

Jost PJ, Ruland J. 2007. Aberrant NF-κB signaling in lymphoma: Mechanisms, consequences, and therapeutic implications. *Blood* **109**: 2700–2707.

Kabashima K, Banks TA, Ansel KM, Lu TT, Ware CF, Cyster JG. 2005. Intrinsic lymphotoxin-β receptor requirement for homeostasis of lymphoid tissue dendritic cells. *Immunity* **22**: 439–450.

Kashiwada M, Pham NL, Pewe LL, Harty JT, Rothman PB. 2011. NFIL3/E4BP4 is a key transcription factor for CD8α[+] dendritic cell development. *Blood* **117**: 6193–6197.

Kee BL. 2009. E and ID proteins branch out. *Nat Rev Immunol* **9**: 175–184.

Kobayashi T, Walsh PT, Walsh MC, Speirs KM, Chiffoleau E, King CG, Hancock WW, Caamano JH, Hunter CA, Scott P, et al. 2003. TRAF6 is a critical factor for dendritic cell maturation and development. *Immunity* **19**: 353–363.

Komori T, Yagi H, Nomura S, Yamaguchi A, Sasaki K, Deguchi K, Shimizu Y, Bronson RT, Gao YH, Inada M, et al. 1997. Targeted disruption of Cbfa1 results in a complete lack of bone formation owing to maturational arrest of osteoblasts. *Cell* **89**: 755–764.

Kretzer NM, Theisen DJ, Tussiwand R, Briseno CG, Grajales-Reyes GE, Wu X, Durai V, Albring J, Bagadia P, Murphy TL, et al. 2016. RAB43 facilitates cross-presentation of cell-associated antigens by CD8α[+] dendritic cells. *J Exp Med* **213**: 2871–2883.

Kumamoto Y, Linehan M, Weinstein JS, Laidlaw BJ, Craft JE, Iwasaki A. 2013. CD301b[+] dermal dendritic cells drive T helper 2 cell-mediated immunity. *Immunity* **39**: 733–743.

Kusunoki T, Sugai M, Katakai T, Omatsu Y, Iyoda T, Inaba K, Nakahata T, Shimizu A, Yokota Y. 2003. T_H2 dominance and defective development of a CD8[+] dendritic cell subset in Id2-deficient mice. *J Allergy Clin Immunol* **111**: 136–142.

Lee MS, Kim YJ. 2007. Signaling pathways downstream of pattern-recognition receptors and their cross talk. *Annu Rev Biochem* **76**: 447–480.

Lehtonen A, Veckman V, Nikula T, Lahesmaa R, Kinnunen L, Matikainen S, Julkunen I. 2005. Differential expression of IFN regulatory factor 4 gene in human monocyte-derived dendritic cells and macrophages. *J Immunol* **175**: 6570–6579.

Leon B, Martínez del Hoyo G, Parrillas V, Vargas HH, Sanchez-Mateos P, Longo N, Lopez-Bravo M, Ardavin C. 2004. Dendritic cell differentiation potential of mouse monocytes: Monocytes represent immediate precursors of CD8[−] and CD8[+] splenic dendritic cells. *Blood* **103**: 2668–2676.

Linehan JL, Dileepan T, Kashem SW, Kaplan DH, Cleary P, Jenkins MK. 2015. Generation of Th17 cells in response to intranasal infection requires TGF-β1 from dendritic cells and IL-6 from CD301b[+] dendritic cells. *Proc Natl Acad Sci* **112**: 12782–12787.

Linette GP, Carreno BM. 2013. Dendritic cell-based vaccines: Shining the spotlight on signal 3. *Oncoimmunology* **2**: e26512.

Liu K, Nussenzweig MC. 2010. Origin and development of dendritic cells. *Immunol Rev* **234**: 45–54.

Liu CH, Fan YT, Dias A, Esper L, Corn RA, Bafica A, Machado FS, Aliberti J. 2006. Cutting edge: Dendritic cells are essential for in vivo IL-12 production and development of resistance against *Toxoplasma gondii* infection in mice 1. *J Immunol* **177**: 31–35.

Liu K, Waskow C, Liu X, Yao K, Hoh J, Nussenzweig M. 2007. Origin of dendritic cells in peripheral lymphoid organs of mice. *Nat Immunol* **8:** 578–583.

Liu K, Victora GD, Schwickert TA, Guermonprez P, Meredith MM, Yao K, Chu FF, Randolph GJ, Rudensky AY, Nussenzweig M. 2009. In vivo analysis of dendritic cell development and homeostasis. *Science* **324:** 392–397.

Longhi MP, Trumpfheller C, Idoyaga J, Caskey M, Matos I, Kluger C, Salazar AM, Colonna M, Steinman RM. 2009. Dendritic cells require a systemic type I interferon response to mature and induce CD4$^+$ Th1 immunity with poly IC as adjuvant. *J Exp Med* **206:** 1589–1602.

Manh TP, Alexandre Y, Baranek T, Crozat K, Dalod M. 2013. Plasmacytoid, conventional, and monocyte-derived dendritic cells undergo a profound and convergent genetic reprogramming during their maturation. *Eur J Immunol* **43:** 1706–1715.

Mashayekhi M, Sandau MM, Dunay IR, Frickel EM, Khan A, Goldszmid RS, Sher A, Ploegh HL, Murphy TL, Sibley LD, et al. 2011. CD8a$^+$ dendritic cells are the critical source of interleukin-12 that controls acute infection by *Toxoplasma gondii* tachyzoites. *Immunity* **35:** 249–259.

Massague J. 2004. G$_1$ cell-cycle control and cancer. *Nature* **432:** 298–306.

Matsui T, Connolly JE, Michnevitz M, Chaussabel D, Yu CI, Glaser C, Tindle S, Pypaert M, Freitas H, Piqueras B, et al. 2009. CD2 distinguishes two subsets of human plasmacytoid dendritic cells with distinct phenotype and functions. *J Immunol* **182:** 6815–6823.

Mayer CT, Ghorbani P, Nandan A, Dudek M, Arnold-Schrauf C, Hesse C, Berod L, Stuve P, Puttur F, Merad M, et al. 2014. Selective and efficient generation of functional Batf3-dependent CD103$^+$ dendritic cells from mouse bone marrow. *Blood* **124:** 3081–3091.

Meredith MM, Liu K, Darrasse-Jeze G, Kamphorst AO, Schreiber HA, Guermonprez P, Idoyaga J, Cheong C, Yao KH, Niec RE, et al. 2012. Expression of the zinc finger transcription factor zDC (Zbtb46, Btbd4) defines the classical dendritic cell lineage. *J Exp Med* **209:** 1153–1165.

Morelli AE, Thomson AW. 2007. Tolerogenic dendritic cells and the quest for transplant tolerance. *Nat Rev Immunol* **7:** 610–621.

Murphy TL, Grajales-Reyes GE, Wu X, Tussiwand R, Briseno CG, Iwata A, Kretzer NM, Durai V, Murphy KM. 2016. Transcriptional control of dendritic cell development. *Annu Rev Immunol* **34:** 93–119.

Naik SH, Proietto AI, Wilson NS, Dakic A, Schnorrer P, Fuchsberger M, Lahoud MH, O'Keeffe M, Shao QX, Chen WF, et al. 2005. Cutting edge: Generation of splenic CD8$^+$ and CD8$^-$ dendritic cell equivalents in Fms-like tyrosine kinase 3 ligand bone marrow cultures. *J Immunol* **174:** 6592–6597.

Naik SH, Metcalf D, van Nieuwenhuijze A, Wicks I, Wu L, O'Keeffe M, Shortman K. 2006. Intrasplenic steady-state dendritic cell precursors that are distinct from monocytes. *Nat Immunol* **7:** 663–671.

Naik SH, Sathe P, Park HY, Metcalf D, Proietto AI, Dakic A, Carotta S, O'Keeffe M, Bahlo M, Papenfuss A, et al. 2007. Development of plasmacytoid and conventional dendritic cell subtypes from single precursor cells derived in vitro and in vivo. *Nat Immunol* **8:** 1217–1226.

Nair-Gupta P, Baccarini A, Tung N, Seyffer F, Florey O, Huang Y, Banerjee M, Overholtzer M, Roche PA, Tampe R, et al. 2014. TLR signals induce phagosomal MHC-I delivery from the endosomal recycling compartment to allow cross-presentation. *Cell* **158:** 506–521.

Nestle FO, Alijagic S, Gilliet M, Sun Y, Grabbe S, Dummer R, Burg G, Schadendorf D. 1998. Vaccination of melanoma patients with peptide- or tumor lysate-pulsed dendritic cells. *Nat Med* **4:** 328–332.

Ohl L, Mohaupt M, Czeloth N, Hintzen G, Kiafard Z, Zwirner J, Blankenstein T, Henning G, Forster R. 2004. CCR7 governs skin dendritic cell migration under inflammatory and steady-state conditions. *Immunity* **21:** 279–288.

Ohtsuka H, Sakamoto A, Pan J, Inage S, Horigome S, Ichii H, Arima M, Hatano M, Okada S, Tokuhisa T. 2011. Bcl6 is required for the development of mouse CD4$^+$ and CD8a$^+$ dendritic cells. *J Immunol* **186:** 255–263.

Onai N, Kurabayashi K, Hosoi-Amaike M, Toyama-Sorimachi N, Matsushima K, Inaba K, Ohteki T. 2013. A clonogenic progenitor with prominent plasmacytoid dendritic cell developmental potential. *Immunity* **38:** 943–957.

Otto F, Thornell AP, Crompton T, Denzel A, Gilmour KC, Rosewell IR, Stamp GW, Beddington RS, Mundlos M, Olsen BR, et al. 1997. Cbfa1, a candidate gene for cleidocranial dysplasia syndrome, is essential for osteoblast differentiation and bone development. *Cell* **89:** 765–771.

Ouaaz F, Arron J, Zheng Y, Choi Y, Beg AA. 2002. Dendritic cell development and survival require distinct NF-κB subunits. *Immunity* **16:** 257–270.

Palucka K, Banchereau J. 2013. Dendritic-cell-based therapeutic cancer vaccines. *Immunity* **39:** 38–48.

Perry JS, Lio CW, Kau AL, Nutsch K, Yang Z, Gordon JI, Murphy KM, Hsieh CS. 2014. Distinct contributions of Aire and antigen-presenting-cell subsets to the generation of self-tolerance in the thymus. *Immunity* **41:** 414–426.

Persson EK, Uronen-Hansson H, Semmrich M, Rivollier A, Hagerbrand K, Marsal J, Gudjonsson S, Hakansson U, Reizis B, Kotarsky K, et al. 2013. IRF4 transcription-factor-dependent CD103$^+$CD11b$^+$ dendritic cells drive mucosal T helper 17 cell differentiation. *Immunity* **38:** 958–969.

Poulin LF, Salio M, Griessinger E, Anjos-Afonso F, Craciun L, Chen JL, Keller AM, Joffre O, Zelenay S, Nye E, et al. 2010. Characterization of human DNGR-1$^+$ BDCA3$^+$ leukocytes as putative equivalents of mouse CD8a$^+$ dendritic cells. *J Exp Med* **207:** 1261–1271.

Romani N, Gruner S, Brang D, Kampgen E, Lenz A, Trockenbacher B, Konwalinka G, Fritsch O, Steinman RM, Schuler G. 1994. Proliferating dendritic cell progenitors in human blood. *J Exp Med* **180:** 83–93.

Sallusto F, Lanzavecchia A. 1994. Efficient presentation of soluble antigen by cultured human dendritic cells is maintained by granulocyte/macrophage colony-stimulating factor plus interleukin 4 and downregulated by tumor necrosis factor α. *J Exp Med* **179:** 1109–1118.

Sallusto F, Schaerli P, Loetscher P, Schaniel C, Lenig D, Mackay CR, Qin S, Lanzavecchia A. 1998. Rapid and coordinated switch in chemokine receptor expression during dendritic cell maturation. *Eur J Immunol* **28:** 2760–2769.

Satpathy AT, Wumesh KC, Albring JC, Edelson BT, Kretzer NM, Bhattacharya D, Murphy TL, Murphy KM. 2012a. Zbtb46 expression distinguishes classical dendritic cells and their committed progenitors from other immune lineages. *J Exp Med* **209:** 1135–1152.

Satpathy AT, Wu X, Albring JC, Murphy KM. 2012b. Re(de)fining the dendritic cell lineage. *Nat Immunol* **13:** 1145–1154.

Satpathy AT, Briseno CG, Lee JS, Ng D, Manieri NA, Wumesh KC, Wu X, Thomas SR, Lee WL, Turkoz M, et al. 2013. Notch2-dependent classical dendritic cells orchestrate intestinal immunity to attaching-and-effacing bacterial pathogens. *Nat Immunol* **14:** 937–948.

Savina A, Peres A, Cebrian I, Carmo N, Moita C, Hacohen N, Moita LF, Amigorena S. 2009. The small GTPase Rac2 controls phagosomal alkalinization and antigen crosspresentation selectively in CD8$^+$ dendritic cells. *Immunity* **30:** 544–555.

Sawai CM, Sisirak V, Ghosh HS, Hou EZ, Ceribelli M, Staudt LM, Reizis B. 2013. Transcription factor Runx2 controls the development and migration of plasmacytoid dendritic cells. *J Exp Med* **210:** 2151–2159.

Scharton-Kersten T, Contursi C, Masumi A, Sher A, Ozato K. 1997. Interferon consensus sequence binding protein-deficient mice display impaired resistance to intracellular infection due to a primary defect in interleukin 12 p40 induction. *J Exp Med* **186:** 1523–1534.

Schiavoni G, Mattei F, Sestili P, Borghi P, Venditti M, Morse HC III, Belardelli F, Gabriele L. 2002. ICSBP is essential for the development of mouse type I interferon-producing cells and for the generation and activation of CD8α$^+$ dendritic cells. *J Exp Med* **196:** 1415–1425.

Schlitzer A, Loschko J, Mair K, Vogelmann R, Henkel L, Einwachter H, Schiemann M, Niess JH, Reindl W, Krug A. 2011. Identification of CCR9$^-$ murine plasmacytoid DC precursors with plasticity to differentiate into conventional DCs. *Blood* **117:** 6562–6570.

Schlitzer A, McGovern N, Teo P, Zelante T, Atarashi K, Low D, Ho AW, See P, Shin A, Wasan PS, et al. 2013. IRF4 transcription factor-dependent CD11b$^+$ dendritic cells in human and mouse control mucosal IL-17 cytokine responses. *Immunity* **38:** 970–983.

Schlitzer A, Sivakamasundari V, Chen J, Sumatoh HR, Schreuder J, Lum J, Malleret B, Zhang S, Larbi A, Zolezzi F, et al. 2015. Identification of cDC1- and cDC2-committed DC progenitors reveals early lineage priming at the common DC progenitor stage in the bone marrow. *Nat Immunol* **16:** 718–728.

Schoenberger SP, Toes RE, van der Voort EI, Offringa R, Melief CJ. 1998. T-cell help for cytotoxic T lymphocytes is mediated by CD40–CD40L interactions. *Nature* **393:** 480–483.

Schuler G, Romani N, Steinman RM. 1985. A comparison of murine epidermal Langerhans cells with spleen dendritic cells. *J Invest Dermatol* **85:** 99s–106s.

Scott CL, Soen B, Martens L, Skrypek N, Saelens W, Taminau J, Blancke G, Van Isterdael G, Huylebroeck D, Haigh J, et al. 2016. The transcription factor Zeb2 regulates development of conventional and plasmacytoid DCs by repressing Id2. *J Exp Med* **213:** 897–911.

See P, Dutertre CA, Chen J, Gunther P, McGovern N, Irac SE, Gunawan M, Beyer M, Handler K, Duan K, et al. 2017. Mapping the human DC lineage through the integration of high-dimensional techniques. *Science* doi: 10.1126/science.aag3009.

Sepp M, Kannike K, Eesmaa A, Urb M, Timmusk T. 2011. Functional diversity of human basic helix–loop–helix transcription factor TCF4 isoforms generated by alternative 5′ exon usage and splicing. *PLoS ONE* **6:** e22138.

Sichien D, Scott CL, Martens L, Vanderkerken M, Van Gassen S, Plantinga M, Joeris T, De Prijck S, Vanhoutte L, Vanheerswynghels M, et al. 2016. IRF8 transcription factor controls survival and function of terminally differentiated conventional and plasmacytoid dendritic cells, respectively. *Immunity* **45:** 626–640.

Siegal FP, Kadowaki N, Shodell M, Fitzgerald-Bocarsly PA, Shah K, Ho S, Antonenko S, Liu YJ. 1999. The nature of the principal type 1 interferon-producing cells in human blood. *Science* **284:** 1835–1837.

Smale ST. 2012. Dimer-specific regulatory mechanisms within the NF-κB family of transcription factors. *Immunol Rev* **246:** 193–204.

Sonnenberg GF, Artis D. 2015. Innate lymphoid cells in the initiation, regulation and resolution of inflammation. *Nat Med* **21:** 698–708.

Sozzani S, Allavena P, D'Amico G, Luini W, Bianchi G, Kataura M, Imai T, Yoshie O, Bonecchi R, Mantovani A. 1998. Differential regulation of chemokine receptors during dendritic cell maturation: A model for their trafficking properties. *J Immunol* **161:** 1083–1086.

Steinman RM, Cohn ZA. 1973. Identification of a novel cell type in peripheral lymphoid organs of mice. I: Morphology, quantitation, tissue distribution. *J Exp Med* **137:** 1142–1162.

Steinman RM, Hawiger D, Nussenzweig MC. 2003. Tolerogenic dendritic cells. *Annu Rev Immunol* **21:** 685–711.

Theisen D, Murphy K. 2017. The role of cDC1s in vivo: CD8 T cell priming through cross-presentation. *F1000Res* **6:** 98.

Thurner B, Haendle I, Roder C, Dieckmann D, Keikavoussi P, Jonuleit H, Bender A, Maczek C, Schreiner D, von den DP, et al. 1999. Vaccination with mage-3A1 peptide-pulsed mature, monocyte-derived dendritic cells expands specific cytotoxic T cells and induces regression of some metastases in advanced stage IV melanoma. *J Exp Med* **190:** 1669–1678.

Timmerman JM, Czerwinski DK, Davis TA, Hsu FJ, Benike C, Hao ZM, Taidi B, Rajapaksa R, Caspar CB, Okada CY, et al. 2002. Idiotype-pulsed dendritic cell vaccination for B-cell lymphoma: Clinical and immune responses in 35 patients. *Blood* **99:** 1517–1526.

Tussiwand R, Lee WL, Murphy TL, Mashayekhi M, Wumesh KC, Albring JC, Satpathy AT, Rotondo JA, Edelson BT, Kretzer NM, et al. 2012. Compensatory dendritic cell development mediated by BATF–IRF interactions. *Nature* **490:** 502–507.

Tussiwand R, Everts B, Grajales-Reyes GE, Kretzer NM, Iwata A, Bagaitkar J, Wu X, Wong R, Anderson DA, Murphy TL, et al. 2015. Klf4 expression in conventional dendritic cells is required for T helper 2 cell responses. *Immunity* **42:** 916–928.

Van de Putte T, Maruhashi M, Francis A, Nelles L, Kondoh H, Huylebroeck D, Higashi Y. 2003. Mice lacking ZFHX1B, the gene that codes for Smad-interacting protein-1, reveal a role for multiple neural crest cell defects in the etiology of Hirschsprung disease-mental retardation syndrome. *Am J Hum Genet* **72:** 465–470.

Vander LB, Khan AA, Hackney JA, Agrawal A, Lesch J, Zhou M, Lee WP, Park S, Xu M, DeVoss J, et al. 2014. Transcriptional programming of dendritic cells for enhanced MHC class II antigen presentation. *Nat Immunol* **15:** 161–167.

Vandewalle C, Comijn J, De Craene B, Vermassen P, Bruyneel E, Andersen H, Tulchinsky E, Van Roy F, Berx G. 2005. SIP1/ZEB2 induces EMT by repressing genes of different epithelial cell–cell junctions. *Nucleic Acids Res* **33:** 6566–6578.

Vandewalle C, Van Roy F, Berx G. 2009. The role of the ZEB family of transcription factors in development and disease. *Cell Mol Life Sci* **66:** 773–787.

Van Dyken SJ, Nussbaum JC, Lee J, Molofsky AB, Liang HE, Pollack JL, Gate RE, Haliburton GE, Ye CJ, Marson A, et al. 2016. A tissue checkpoint regulates type 2 immunity. *Nat Immunol* **17:** 1381–1387.

Villani AC, Satija R, Reynolds G, Sarkizova S, Shekhar K, Fletcher J, Griesbeck M, Butler A, Zheng S, Lazo S, et al. 2017. Single-cell RNA-seq reveals new types of human blood dendritic cells, monocytes, and progenitors. *Science* doi: 10.1126/science.aah4573.

von Moltke J, Ji M, Liang HE, Locksley RM. 2016. Tuft-cell-derived IL-25 regulates an intestinal ILC2-epithelial response circuit. *Nature* **529:** 221–225.

Waskow C, Liu K, Darrasse-Jeze G, Guermonprez P, Ginhoux F, Merad M, Shengelia T, Yao K, Nussenzweig M. 2008. The receptor tyrosine kinase Flt3 is required for dendritic cell development in peripheral lymphoid tissues. *Nat Immunol* **9:** 676–683.

Watchmaker PB, Lahl K, Lee M, Baumjohann D, Morton J, Kim SJ, Zeng R, Dent A, Ansel KM, Diamond B, et al. 2014. Comparative transcriptional and functional profiling defines conserved programs of intestinal DC differentiation in humans and mice. *Nat Immunol* **15:** 98–108.

Weih F, Carrasco D, Durham SK, Barton DS, Rizzo CA, Ryseck RP, Lira SA, Bravo R. 1995. Multiorgan inflammation and hematopoietic abnormalities in mice with a targeted disruption of RelB, a member of the NF-κ B/Rel family. *Cell* **80:** 331–340.

Weng Q, Chen Y, Wang H, Xu X, Yang B, He Q, Shou W, Chen Y, Higashi Y, van dB V, et al. 2012. Dual-mode modulation of Smad signaling by Smad-interacting protein Sip1 is required for myelination in the central nervous system. *Neuron* **73:** 713–728.

Wilson NS, Young LJ, Kupresanin F, Naik SH, Vremec D, Heath WR, Akira S, Shortman K, Boyle J, Maraskovsky E, et al. 2008. Normal proportion and expression of maturation markers in migratory dendritic cells in the absence of germs or Toll-like receptor signaling. *Immunol Cell Biol* **86:** 200–205.

Witmer-Pack MD, Olivier W, Valinsky J, Schuler G, Steinman RM. 1987. Granulocyte/macrophage colony-stimulating factor is essential for the viability and function of cultured murine epidermal Langerhans cells. *J Exp Med* **166:** 1484–1498.

Wu L, D'Amico A, Winkel KW, Suter M, Lo D, Shortman K. 1998. RelB is essential for the development of myeloid-related CD8α-dendritic cells but not of lymphoid-related CD8α$^+$ dendritic cells. *Immunity* **9:** 839–847.

Wu X, Briseno CG, Durai V, Albring JC, Haldar M, Bagadia P, Kim KW, Randolph GJ, Murphy TL, Murphy KM. 2016a. Mafb lineage tracing to distinguish macrophages from other immune lineages reveals dual identity of Langerhans cells. *J Exp Med* **213:** 2553–2565.

Wu X, Briseno CG, Grajales-Reyes GE, Haldar M, Iwata A, Kretzer NM, Wumesh KC, Tussiwand R, Higashi Y, Murphy TL, et al. 2016b. Transcription factor Zeb2 regulates commitment to plasmacytoid dendritic cell and monocyte fate. *Proc Natl Acad Sci* **113:** 14775–14780.

Wumesh KC, Satpathy AT, Rapaport AS, Briseno CG, Wu X, Albring JC, Russler-Germain EV, Kretzer NM, Durai V, Persaud SP, et al. 2014. L-Myc expression by dendritic cells is required for optimal T-cell priming. *Nature* **507:** 243–247.

Yin X, Yu H, Jin X, Li J, Guo H, Shi Q, Yin Z, Xu Y, Wang X, Liu R, et al. 2017. Human blood CD1c$^+$ dendritic cells encompass CD5^{high} and CD5^{low} subsets that differ significantly in phenotype, gene expression, and functions. *J Immunol* **198:** 1553–1564.

Yu H, Zhang P, Yin X, Yin Z, Shi Q, Cui Y, Liu G, Wang S, Piccaluga PP, Jiang T, Zhang L. 2015. Human BDCA2$^+$CD123$^+$CD56$^+$ dendritic cells (DCs) related to blastic plasmacytoid dendritic cell neoplasm represent a unique myeloid DC subset. *Protein Cell* **6:** 297–306.

Zhang J, Kalkum M, Yamamura S, Chait BT, Roeder RG. 2004. E protein silencing by the leukemogenic AML1-ETO fusion protein. *Science* **305:** 1286–1289.

Zhang H, Gregorio JD, Iwahori T, Zhang X, Choi O, Tolentino LL, Prestwood T, Carmi Y, Engleman EG. 2017a. A distinct subset of plasmacytoid dendritic cells induces activation and differentiation of B and T lymphocytes. *Proc Natl Acad Sci* **114:** 1988–1993.

Zhang Q, Lenardo MJ, Baltimore D. 2017b. 30 Years of NF-κB: A blossoming of relevance to human pathobiology. *Cell* **168:** 37–57.

Zheng Y, Valdez PA, Danilenko DM, Hu Y, Sa SM, Gong Q, Abbas AR, Modrusan Z, Ghilardi N, de Sauvage FJ, et al. 2008. Interleukin-22 mediates early host defense against attaching and effacing bacterial pathogens. *Nat Med* **14:** 282–289.

Zou L, Zhou J, Zhang J, Li J, Liu N, Chai L, Li N, Liu T, Li L, Xie Z, et al. 2009. The GTPase Rab3b/3c-positive recycling vesicles are involved in cross-presentation in dendritic cells. *Proc Natl Acad Sci* **106:** 15801–15806.

Cite this article as *Cold Spring Harb Perspect Biol* doi: 10.1101/cshperspect.a028613

Cancer Inflammation and Cytokines

Maria Rosaria Galdiero,[1] Gianni Marone,[1,2] and Alberto Mantovani[3,4]

[1]Department of Translational Medical Sciences (DiSMeT) and Center for Basic and Clinical Immunology Research (CISI), University of Naples Federico II, 80131 Naples, Italy

[2]Institute of Experimental Endocrinology and Oncology "Gaetano Salvatore" (IEOS), National Research Council (CNR), 80131 Naples, Italy

[3]Istituto di Ricovero e Cura a Carattere Scientifo (IRCCS), Istituto Clinico Humanitas, Rozzano, Milan, Italy

[4]Humanitas University, Rozzano, Milan, Italy

Correspondence: alberto.mantovani@humanitasresearch.it

Chronic inflammation is a well-recognized tumor-enabling capability, which allows nascent tumors to escape immunosurveillance. A number of soluble and cellular inflammatory mediators take part in the various phases of cancer initiation and progression, giving rise to a fatal conspiracy, which is difficult to efficiently overcome. Tumor-associated macrophages (TAMs) are pivotal players of the tumor microenvironment and, because of their characteristic plasticity, can acquire a number of distinct phenotypes and contribute in different ways to the various phases of cancerogenesis. Tumor-associated neutrophils (TANs) are also emerging as important components of the tumor microenvironment, given their unexpected heterogeneity and plasticity. TAMs and TANs are both integrated in cancer-related inflammation and an ever better understanding of their functions can be useful to tailor the use of anticancer therapeutic approaches and patient follow-up.

Following the revision of the paradigm proposed by Hanahan and Weinberg (2000), it is now well recognized that chronic inflammation represents an enabling characteristic of cancer. Indeed, even if the presence of an immune infiltrate in and around the tumors was already known for a long time (Dvorak 1986), it was largely attributed to an effort of the immune system to combat tumors. In contrast, experimental evidence proved that cancer-related inflammation (CRI) had the unexpected effect of promoting tumorigenesis and progression, favoring nascent neoplasias to acquire all the hallmark capabilities of cancer, including the evasion from immunosurveillance. This revision has drastically changed the theoretical and therapeutic approach to cancer, expanding the focus from the tumor cell to the tumor microenvironment (TME).

Tumor-associated macrophages (TAMs) are a key component of the TME and are important mediators of the link between inflammation and cancer. These cells are present in different amounts and phenotypes in almost all tumor types and usually represent the main conductors of CRI. Indeed, they are characterized by plasticity, allowing them to acquire distinct phenotypes in response to different signals from the microenvironment.

In addition to macrophages, there is now evidence that neutrophils also can play several roles in the various phases of cancer development (Galdiero et al. 2013a). Indeed, contrary to what it has always been thought, they represent an unexpectedly heterogeneous population, with a spectrum of roles in CRI (Granot and Jablonska 2015).

In this review, we will recapitulate the main biological aspects of TAMs and tumor-associated neutrophils (TANs) and their roles in cancer initiation and progression. We will evaluate their role(s) as prognostic and predictive biomarkers in human cancers and we will explore the functions of these tumor-infiltrating immune cells as means or targets of old and new anticancer therapeutic approaches.

INFLAMMATION AND CANCER: A FATAL CONSPIRACY

Inflammation is an ancestral physiological response, working as a defense mechanism to combat pathogens, contain damage, and promote tissue repair. In acute inflammation, this reaction is self-limiting and sufficient to reestablish homeostasis. The resolution of inflammation is an active process, which includes cellular determinants and molecules that are locally active mediators, namely resolvins and protectins (Serhan 2010). When the inflammatory response is turned off, tissue remodeling is optimized to restore the local physiological conditions. In some circumstances, this mechanism is deranged and gives rise to chronic inflammation. This is the case of a nascent tumor, which prevents the resolution process, given the production of inflammatory molecules and recruitment of inflammatory cells, which persistently subverts the local tissue homeostasis (Dvorak 1986).

Chronic inflammation is now a well-recognized tumor-enabling capability, which can promote cancer development (Balkwill and Mantovani 2001; Hanahan and Weinberg 2011). About 20% of cancers are induced by chronic inflammation. Soluble and cellular in-

flammatory mediators are responsible for tumor initiation and progression (e.g., stomach, colon, skin, liver, breast, lung, and head/neck) (Al Murri et al. 2006; Bornstein et al. 2009; Barash et al. 2010; Grivennikov et al. 2010; Watanabe et al. 2012; Liang et al. 2013; Alam et al. 2016; Lund et al. 2016).

Tumor-related inflammatory responses vary depending on the context but, in general, tend to promote tumor progression (Mantovani et al. 2008; Galdiero et al. 2013b; Varricchi et al. 2017). Tumors can induce inflammatory reactions through several mechanisms. First, tumor and stromal cells release chemotactic factors that recruit macrophages and neutrophils (Bonavita et al. 2015). Moreover, the tumor can physically damage the normal tissue and release damage-associated molecular patterns, which activate granulocytes. These recruited cells release inflammatory molecules, amplifying the response. In addition, acidification of the TME has been associated with certain key features of cancer aggressiveness, including invasion, evasion from the immune system, increased angiogenesis, and resistance to therapy (Granja et al. 2017). Indeed, uncontrolled growth requires adaptations in energy metabolism to fuel cell proliferation. Thus, cancer growth leads to the production of high amounts of lactic acid, which is responsible for the acidification of the microenvironment. In contrast to normal mammalian cells, cancer cells present increased glycolysis independently of the oxygen levels ("aerobic glycolysis" or "Warburg effect"). As a consequence, high amounts of protons are generated and, to cope with this, cancer cells export protons to the microenvironment, allowing them to survive in the hostile environment that they have created (Granja et al. 2017). In addition, growing tumors increase oxygen consumption as a result of their increased metabolism (Stylianopoulos et al. 2012). The resulting hypoxia induces the production of cytokines and angiogenic growth factors, which give rise to neo-angiogenesis and lymphangiogenesis and recruit macrophages. These inflammatory processes persist as long as the tumor grows, thus giving rise to a fatal conspiracy increasingly difficult to overcome.

ROLES OF TAMs IN TUMOR GROWTH AND PROGRESSION

Macrophages are the most represented leukocytes in the TME (Mantovani et al. 2002). Classically viewed as terminally differentiated cells, they were thought to derive from circulating monocytes and to differentiate at sites of inflammation under the influence of growth factors, such as macrophage colony-stimulating factor (M-CSF) or granulocyte macrophage colony-stimulating factor (GM-CSF) (Allavena et al. 2008). However, several investigations have described, at least in mice, a self-renewing population of macrophages, derived from embryonic precursors that spread to tissues before birth and can locally proliferate and differentiate independently on bone marrow–derived monocytes (Davies et al. 2011; Jenkins et al. 2011; Robbins et al. 2013; Ginhoux et al. 2016). A few studies in atherosclerosis and cancer indicate that macrophage proliferation also exists in humans; however, their contribution to cancer development is still unclear (Bottazzi et al. 1990; Lutgens et al. 1999; Campbell et al. 2011).

During the last decades, increasing evidence has highlighted the multifunctional properties of macrophages, which are now considered highly plastic cells, which can modify their phenotype in response to microenvironmental signals, with classical M1 and alternative M2 polarization states as the reference paradigm (Galdiero et al. 2013b; Bonavita et al. 2015).

Chemotactic molecules involved in monocyte recruitment at the tumor site include CCL2 and CCL5, vascular endothelial growth factors (VEGFs), and M-CSF. Besides their chemotactic functions, these factors contribute to macrophage polarization toward specific phenotypes (Kitamura et al. 2015). In a transgenic mouse model in which CCL2 was overexpressed specifically in mammary epithelial cells, there was increased macrophage infiltration, increased expression of extracellular matrix (ECM) remodeling genes, such as matrix metalloproteases (MMPs) and lipoxygenase (LOX), and increased stromal density. In addition, CCL2 transgenic mice displayed an increased susceptibility to 7,12-dimethylbenz(a)anthracene (DMBA)-in-duced carcinogenesis, thus suggesting that CCL2 overexpression increases mammary stromal density and breast cancer risk (Sun et al. 2017). Also, M-CSF is a classical monocyte chemoattractant, which also favors macrophage survival and skewing toward a tumor-promoting "M2-like" phenotype (Pyonteck et al. 2013).

Tumor-infiltrating T and B cells as well as stromal cells can release factors activating classic M1 macrophages, able to recognize and eliminate nascent tumor cells in line with the "elimination phase" of the immunoediting (Dunn et al. 2002). However, if this process is not successful, tumors can evolve and, along with tumor progression, macrophage can divert through an M2/M2-like phenotype, which sustains many aspects of tumor growth and progression in line with the "escape phase" of the immunoediting (Dunn et al. 2002). This phenomenon can be driven directly by tumor cells or indirectly by "already corrupted" immune cells releasing M2-skewing molecules, such as interleukin (IL)-4, IL-13, immunocomplexes, transforming growth factor (TGF)-β, or M-CSF. Recently, in a murine model of breast cancer, overexpression of IL-23p19 was associated with increased tumor growth, pulmonary metastasis, and reduced survival. IL-23p19 overexpressing tumors displayed increased expression of MMP-9, CD31, and ki67, thus suggesting a higher ECM remodeling and proliferative activity. Moreover, tumors displayed decreased percentages of $CD4^+$ and $CD8^+$ T cells, as well as increased infiltration of M2-like macrophages expressing VEGF and TGF-β and neutrophils expressing IL-10 and VEGF. These findings suggested that IL-23 promoted infiltration of M2-like macrophages and neutrophils endowed with immunosuppressive capacity (Nie et al. 2017). M2 macrophages are classically characterized by a high production of chemokines, including CCL17, CCL22, or CCL24, involved in the recruitment of T helper (Th)2 cells, regulatory T cells (Tregs), eosinophils, and basophils, as well as a high production of IL-10. M2 macrophages produce low levels of IL-12 and are mainly involved in immunoregulatory networks, regulating tissue remodeling, and angiogenesis (Mantovani et al. 2013).

TAMs can acquire a wide range of activation states, depending on the tumor-related cellular and molecular network. Thus, the pathways of TAM activation vary among the various tumor types and, in some circumstances, within the same tumor (Ruffell et al. 2012). For example, in distinct tumor areas the variable access to oxygen is responsible for various levels of activation of metabolic pathways involved in tuning macrophage phenotypes (Movahedi et al. 2010; Henze and Mazzone 2016).

Despite the fine modulation of macrophage activation states in distinct tumors, M2-like polarization usually represents a common determinant. Indeed, TAMs display a number of M2-resembling functions, which ultimately are beneficial to cancer progression. Indeed, TAMs promote tumor cell growth through the production of growth factors such as epidermal growth factor (EGF), which induces breast cancer cell proliferation (Qian and Pollard 2010). In addition, TAMs produce high levels of reactive oxygen and nitrogen species, which contribute to DNA damage and genetic instability of cancer cells (Bonavita et al. 2015). Moreover, TAMs promote tumor-invasive behavior and metastatic progression. Indeed, they release proteolytic enzymes, such as MMPs, involved in ECM digestion and remodeling thus favoring tumor cell invasion (Allavena and Mantovani 2012). In addition to tissue remodeling, TAMs also promote angiogenesis and lymphangiogenesis, producing angiogenic/lymphangiogenic factors such as VEGF-A, VEGF-C, TGF-β, as well as proangiogenic chemokines such as CCL2 and CXCL8 (Hotchkiss et al. 2003; Murdoch et al. 2008; Granata et al. 2010; Schmidt and Carmeliet 2010). Tumor-associated hypoxia induces a proangiogenic program in TAMs, through the up-regulation of hypoxia-inducible factor (HIF)-1α, as well as through the production of adenosine, which in turn promotes the release of proangiogenic and lymphangiogenic factors by human macrophages (Granata et al. 2010). Finally, TAMs promote tumor progression by suppressing antitumor immunity. Indeed, TAMs produce immunosuppressive molecules (e.g., TGF-β, IL-10, indoleamine 2,3-dioxygenase [IDO], and arginase-1), which

suppress adaptive T-cell immune responses and favor Treg recruitment and functions (Ruffell et al. 2012; Noy and Pollard 2014). In a mouse model of colitis-associated cancer (CAC), macrophages produced IL-17, which increased survival and immunosuppressive activity of granulocytic myeloid-derived suppressor cells (G-MDSCs) thus fostering tumor progression (Zhang et al. 2016). TAMs also express programmed cell death protein 1 (PD-1) ligands PD-L1 and PD-L2, which bind on T cells and activate the inhibitory PD-1 immune checkpoint in T cells (Kryczek et al. 2006; Wang et al. 2011). Moreover, TAMs could also express B7-H4 and VISTA, which likely exert similar functions (Deng et al. 2016; Wang et al. 2016b).

ROLES OF NEUTROPHILS IN TUMOR GROWTH AND PROGRESSION

Experimental models and epidemiological studies have shed new light on neutrophil roles in modulating tumor behavior. Indeed, TANs are pivotal players in CRI and can exert antitumoral or protumoral functions. Moreover, they are endowed with unsuspected plasticity (Fridlender et al. 2009; Mantovani 2009; Granot and Jablonska 2015). In murine models of cancer, neutrophils were driven by TGF-β to acquire a protumoral phenotype (Fridlender et al. 2009). Indeed, TGF-β inhibition led to the tumor infiltration of neutrophils with increased cytotoxicity against tumor cells, high expression of tumor necrosis factor α (TNF-α), CCL3, and intercellular adhesion molecule 1 (ICAM-1), and low levels of arginase-1. TGF-β inhibition also promoted a T-cell antitumor response, which involved neutrophils as effector cells (Fridlender et al. 2009). In this seminal paper, neutrophils were proposed to polarize in two distinct activation states: an antitumor N1 or a protumor N2 phenotype in response to signals derived from TME. In an in vivo model of melanoma and fibrosarcoma, mice lacking interferon (IFN)-β showed an infiltration of proangiogenic neutrophils, characterized by a high expression of CXCR4, VEGF-A, and MMP-9 (Jablonska et al. 2010). These results suggested a pivotal role for type I IFNs in polarizing neutrophils toward an N1 antitumor phenotype (Granot and Jablonska 2015).

Within the TME, CXC chemokines produced by tumor and stromal cells and associated with cancer growth and progression also retain neutrophil-recruiting functions (Keeley et al. 2010; Lazennec and Richmond 2010; Mantovani et al. 2011). For instance, murine models showed a central role for CXCR2 in promoting lung and pancreatic cancers (Keane et al. 2004; Ijichi et al. 2011). Indeed, inflammation-induced and spontaneous carcinogenesis were suppressed following CXCR2 abrogation or neutrophil depletion in mice (Jamieson et al. 2012). Moreover, CXCL17 promoted cancer growth together with the increased infiltration of a myeloid subset of $CD11b^+Gr1^+F4/80^-$ cells in a murine model of graft tumor (Matsui et al. 2012). In a conditional genetic murine model of lung cancer driven by K-ras activation and p53 inactivation, TAM and TAN precursors accumulated in the spleen and relocated from the spleen to the tumor, suggesting a role for the spleen as reservoir for tumor-promoting myeloid cells (Cortez-Retamozo et al. 2012). In humans, head and neck squamous cell carcinoma (HNSCC) cell lines produced CXCL8 and macrophage-inhibiting factor (MIF), which recruited neutrophils through the engagement of CXCR2 (Dumitru et al. 2011; Trellakis et al. 2011b). Hepatocellular carcinoma cells recruited neutrophils through the production of CXCL8 (Kuang et al. 2011).

Neutrophils play important roles in tumor initiation. Indeed, neutrophil-derived oxygen and nitrogen derivatives are responsible for DNA point mutations and promoted genetic instability (Gungor et al. 2010). Moreover, the MPO-derived hypochlorous acid HOCl activated MMPs and inactivated the tissue inhibitor of proteases (TIMP-1), thus promoting ECM remodeling as well as invasive and metastatic behavior of cancer cells (De Larco et al. 2004).

Granule proteins were also involved in tumor progression. For instance, neutrophil elastase (NE) taken up by lung cancer cells degraded the phosphatidylinositol-4,5-bisphosphate 3-kinase (PI3K) inhibitor, insulin receptor substrate 1 (IRS-1). This event unleashed PI3K activation and platelet-derived growth factor receptor (PDGFR) signaling, thus favoring tumor cell proliferation (Houghton et al. 2010). NE was also involved in neutrophil-related epithelial-to-mesenchymal transition (EMT) (Grosse-Steffen et al. 2012). On the contrary, NE taken up by breast cancer cells cleaved cyclin E, which was then presented in a truncated form in HLA-I context and efficiently activated a cytotoxic T lymphocytes–mediated antitumor response (Mittendorf et al. 2012). More recently, NE uptake increased the responsiveness of breast cancer cells to adaptive immunity by up-regulation of HLA class I (Chawla et al. 2016). Neutrophils released the cytokine oncostatin M, which up-regulated VEGF production in breast cancer cells, promoting tumor cell detachment and invasiveness (Queen et al. 2005). In bronchoalveolar carcinoma patients, hepatocyte growth factor (HGF) in broncholavage fluid correlated with neutrophil infiltration and was associated with poor prognosis (Wislez et al. 2003; Imai et al. 2005). In HNSCC patients, neutrophil infiltration correlated with the expression of Cortactin and with poor clinical outcome (Dumitru et al. 2013).

On the contrary, neutrophils also release TRAIL, which retains important antitumoral activities (Cassatella 2006; Hewish et al. 2010). Indeed, in bladder cancer, *Mycobacterium bovis* bacillus Calmette–Guerin (BCG) induced the release of TRAIL from neutrophils, which accounted for the anticancer effects of BCG (Kemp et al. 2005). Moreover, in chronic myeloid leukemia (CML) patients, IFN-α stimulation induced the release of TRAIL from neutrophils, which favored apoptosis of leukemic cells (Tecchio et al. 2004; Tanaka et al. 2007).

In surgically resected lung cancer patients, TANs produced the proinflammatory molecules CCL2, CCL3, CXCL8, and IL-6, stimulated T-cell proliferation, and IFN-γ release, mainly in a contact-dependent manner (Eruslanov et al. 2014). Neutrophils up-regulated the expression of costimulatory molecules (e.g., CD86 and OX40L), amplifying a positive feedback loop, which suggested an antitumor role for TANs in early stages human lung cancers (Eruslanov et al. 2014).

Neutrophils can also play a dual role in modulating metastatic behavior of cancer cells and angiogenesis. Melanoma-derived CXCL8 up-regulated β2-integrin expression on neutrophils, which interacted with ICAM-1 expressed by melanoma cells, allowing neutrophils to carry tumor cells to metastatic sites (Huh et al. 2010). Neutrophil extracellular traps (NETs) also captured circulating tumor cells and favored their engraftment to distant organ sites (Cools-Lartigue et al. 2013). In contrast, in a murine model of transplanted breast cancer, under the influence of granulocyte colony stimulating factor (G-CSF) and tumor-derived CCL2, neutrophils inhibited breast metastasis in the premetastatic lung in an H_2O_2-dependent manner (Granot et al. 2011).

Neutrophils are a major source of VEGF-A, which is also responsible for the angiogenic activity exerted by CXCL1 in vivo (Scapini et al. 2004). Neutrophils express high levels of MMP-9, which releases the active form of VEGF-A from the ECM (Nozawa et al. 2006; Kuang et al. 2011; Dumitru et al. 2012). Interestingly, neutrophils release MMP-9 in a TIMP-free manner, which further enhanced the proangiogenic and proinvasive activity of MMP-9 (Ardi et al. 2007). Unexpectedly, intratumoral delivery of MMP-9 decreased tumor growth and angiogenesis in a murine model of breast cancer, suggesting that MMP-9 also retains antiangiogenic functions (Leifler et al. 2013). In a tumor xenograft murine model, under the influence of G-CSF, neutrophils released the proangiogenic molecule Bv8, and its neutralization significantly impaired angiogenesis and tumor growth (Shojaei et al. 2007). Interestingly, tumors resistant to anti-VEGF therapy showed high infiltration of neutrophils and drug resistance was associated with G-CSF-induced Bv8 neutrophil expression (Shojaei et al. 2008, 2009). In contrast, neutrophils also express a number of antiangiogenic molecules. For example, NE cleaved VEGF and fibroblast growth factor 2 (FGF-2), giving rise to the angiostatin-like fragments from plasminogen, which suppressed VEGF- and FGF-2-induced angiogenesis (Scapini et al. 2004; Ai et al. 2007).

Neutrophil plasticity and heterogeneity have been highlighted by several recent observations in mice and in cancer patients. Indeed, circulat-

ing neutrophils are usually purified on a discontinuous density gradient (Ficoll). Following this separation, neutrophils are found in the high-density (HD) granulocytic fraction, whereas peripheral blood mononuclear cells (PBMCs) segregate in the low-density (LD) mononuclear fraction (Boyum 1968). However, an increasing number of studies shows that in chronic inflammatory conditions such as HIV, autoimmunity, and cancer, neutrophils could also be found in the LD fraction (Schmielau and Finn 2001; Rodriguez et al. 2009; Denny et al. 2010; Cloke et al. 2012). Moreover, the percentage of low-density neutrophils (LDNs) increases with cancer progression, and these cells retain T-cell-suppressive properties and include both mature and immature granulocytes (Mishalian et al. 2017). Immature granulocytes found in LD fraction have always been considered as G-MDSCs. MDSCs are a heterogeneous subset of myeloid cells, expanding in peripheral blood and spleen of tumor-bearing mice and cancer patients, and characterized by the capacity to suppress T-cell activation and proliferation (Gabrilovich and Nagaraj 2009; Peranzoni et al. 2010). Because G-MDSCs and neutrophils are both of myeloid origin, have similar morphological aspects and surface markers, as well as tumor-promoting properties, there is no clear consensus on the differences between these populations of cells. A transcriptomic analysis of peripheral neutrophils, TANs, and G-MDSCs in tumor-bearing mice found that TANs and G-MDSCs are distinct populations and that naïve neutrophils and G-MDSCs are more closely related to each other than to TANs (Fridlender et al. 2012). An interesting study performed on tumor-bearing mice as well as on breast and lung cancer patients showed that circulating neutrophils in cancer consist of two distinct subsets: mature segmented high-density neutrophils (HDNs) and LDNs. Within LDNs, two further subsets could be distinguished: a mature segmented one and a banded immature one, namely, G-MDSC. Both in tumor-bearing mice and cancer patients, the LDN fraction increased along with tumor progression (Sagiv et al. 2015). Although HDNs displayed antitumor functions, LDNs showed reduced chemotaxis, phagocytosis, oxidative burst, no significant cytotoxic activity against tumor cells, and significantly impaired T-cell activity and proliferation. These cancer-promoting activities were shared by both mature and immature (G-MDSCs) LDNs. Moreover, in this study, beyond this heterogeneity, the authors proposed an important plasticity, showing that HDNs can progress through the LDN transition under the influence of TGF-β and acquire T-cell-suppressive properties, thus suggesting that part of the LDN fraction is a subset of highly activated mature neutrophils but with reduced inflammatory properties. They also proposed that LDNs can switch to HDNs, but to a lesser extent than the opposite transition (Sagiv et al. 2015). These observations suggest that neutrophils are not terminally differentiated as previously thought. Indeed, they highlight the potential heterogeneity and plasticity of circulating neutrophils in cancer development and call for a rigorous reassessment of neutrophil characterization in cancer patients.

A schematic view of the roles of TAMs and TANs in CRI is summarized in Figure 1.

TAMs AND TANs AS PROGNOSTIC/PREDICTIVE BIOMARKERS IN CANCER PATIENTS

High TAM infiltration was associated with poor clinical outcome in a variety of human cancers (Bingle et al. 2002; Qian and Pollard 2010; Zhang et al. 2012). In breast cancer patients, a high macrophage infiltration was associated with high tumor grade and poor outcome (Campbell et al. 2011). Similarly, in bladder cancer patients, high TAM density correlated with advanced disease stage and poor survival (Hanada et al. 2000). In contrast, a positive correlation was found between TAM infiltration and patient survival in high-grade osteosarcoma patients (Buddingh et al. 2011) and TAMs positively correlated with tumor cell apoptosis and CD8$^+$ infiltration in gastric cancer (Ohno et al. 2003). Some apparently controversial results can be explained considering that macrophages within a tumor are not homogeneous, and the TAM phenotype can vary within the same tumor. Moreover, there is a huge variability related on the techniques used to identify TAMs in tissues (CD68$^+$, CD203$^+$, CD206$^+$, stabilin1$^+$ cells, etc.), which may contribute to the variability of the results among different studies.

Interestingly, in follicular lymphoma patients, CD68$^+$CD163$^+$ TAM infiltration was associated with an adverse outcome in patients treated with first-line systemic treatment, including rituximab, cyclophosphamide, vincristine, and prednisone, but with favorable outcome in patients treated with rituximab, cyclophosphamide, doxorubicin, vincristine, and prednisone. These results suggest that CD163$^+$ macrophage density predicts the outcome in follicular lymphoma, but their prognostic impact is highly dependent on treatment received. This interesting study highlights the potential role of TAMs as a predictive marker of chemotherapy response (Kridel et al. 2015). Indeed, the vast majority of cancer patients are treated with old and new generation cytoreductive drugs or radiotherapy. This aspect is often neglected in epidemiological studies. Thus, the prognostic significance of TAM infiltration loses its value if it is not related to the administered therapy.

With regard to solid tumors, data on the predictive role of TAMs are missing. Indeed, most studies do not mention the therapeutic regimen administered to the patients, nor take account of this parameter in the statistical evaluation. The only study investigating the role of TAMs as predictors of chemotherapy response has been performed on pancreas cancer patients. This study revealed that TAM density correlated with a worse prognosis and increased distant metastasis only in patients who did not receive chemotherapy; indeed, gemcitabine administration restrained TAM protumour prognostic significance (Di Caro et al. 2016). Thus, TAMs retain important predictive significance in the response to chemotherapy in cancer patients and can be an additional tool to stratify patients for chemotherapy after surgery.

The relationship between TAN infiltration and prognosis in human cancer has been previously discussed (Donskov 2013). Neutrophil in-

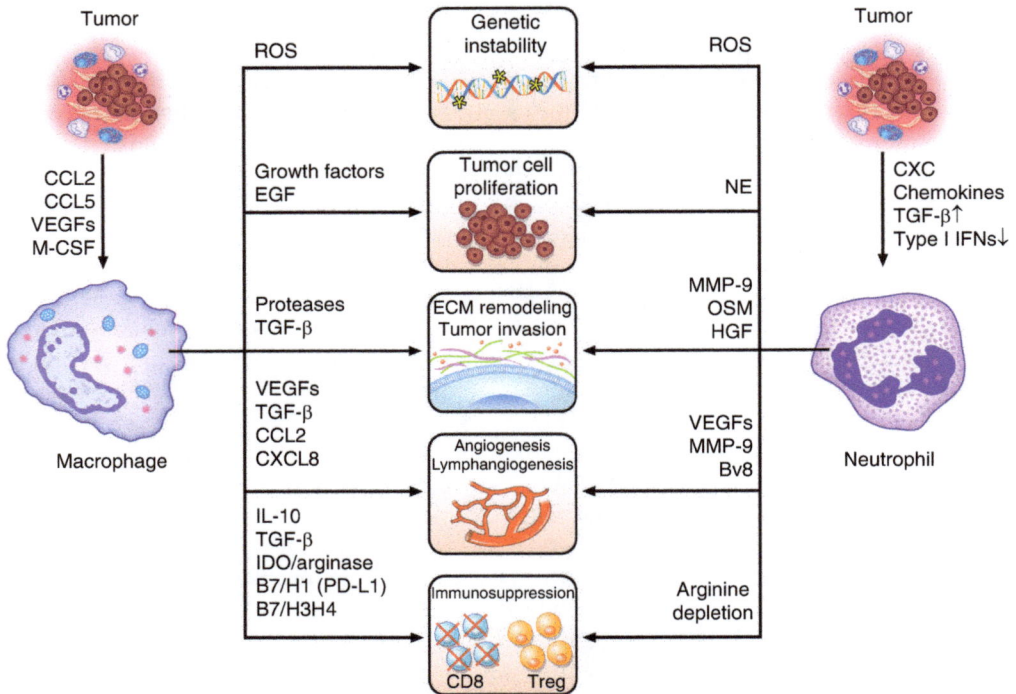

Figure 1. Tumor-associated macrophages (TAMs) and tumor-associated neutrophils (TANs) as key regulators of the tumor-related inflammation. Neoplastic and stromal cells recruit macrophages and neutrophils, favoring their polarization toward a protumor phenotype. In turn, TAMs and TANs induce genetic instability (through the release of reactive oxygen species [ROS]), favor tumor growth (through the production of growth factors and neutrophil elastase [NE]), promote the remodeling of the extracellular matrix (ECM) and tumor cell invasive capabilities (through the release of proteases, transforming growth factor β [TGF-β], hepatocyte growth factor [HGF], and oncostatin M [OSM]), support angiogenesis and lymphangiogenesis (through the release of vascular endothelial growth factors [VEGFs], matrix metalloprotease [MMP]-9, and Bv8), and suppress antitumoral adaptive immunity (through arginine depletion and expression of suppressive soluble and membrane molecules, such as interleukin (IL)-10 and programmed cell death protein 1 ligand (PD-L1). See text for details. IDO, Indoleamine 2,3-dioxygenase; M-CSF, macrophage colony-stimulating factor; EGF, epidermal growth factor; Treg, regulatory T cell; IFN, interferon.

filtration within human cancers has been correlated with poor clinical outcome in patients with metastatic and localized clear cell carcinomas, bronchioloalveolar carcinoma, liver cancer, colorectal carcinoma (CRC), and HNSCC (Wislez et al. 2003; Jensen et al. 2009; Kuang et al. 2011; Trellakis et al. 2011a; Rao et al. 2012). In addition, high neutrophil infiltration has been associated with high tumor grade in human gliomas and with aggressive pancreatic tumors (Fossati et al. 1999; Reid et al. 2011). In contrast, TANs have been associated with better prognosis in gastric cancer (Caruso et al. 2002) and CRC (Droeser et al. 2013; Galdiero et al.

2016). As discussed for TAMs, these apparently controversial results may depend on the type/subtype of tumors and on the techniques used to identify neutrophils within the tumors (e.g., hematoxylin–eosin stain versus immunohistochemistry) (Donskov 2013).

As for TAMs, studies evaluating the predictive value of TANs in human settings are missing. The only published comparison of the association of TANs with outcome in patients who received chemotherapy after tumor resection versus those who did not receive chemotherapy was performed on patients with CRC. In stage III patients, TAN infiltration was associated with better response to 5-fluorouracil (5FU)-based chemotherapy but with poor prognosis in patients treated with surgery alone (Galdiero et al. 2016). These results suggest a dual clinical significance of TANs, depending on the administration of chemotherapy, and make necessary a rigorous evaluation of the role of TAN density as a predictive factor for response to therapy in human cancer (Galdiero et al. 2016).

Several studies have evaluated the prognostic and predictive value of neutrophil-to-lymphocyte ratio (NLR) in peripheral blood of cancer patients. NLR is commonly used as a measure of systemic inflammation, and it has been shown to predict patient clinical outcome in a number of human cancers, such as rectal (Shen et al. 2017), esophageal (Nakamura et al. 2017), prostate (Gokce et al. 2016), pancreatic (Kadokura et al. 2016), and breast cancer (Ethier et al. 2017). Overall, a high NLR score was associated with worse survival and retained a more consistent prognostic value among patients with an advanced disease stage, who are also more likely to receive chemotherapy treatments or who are not operable (Guthrie et al. 2013). The advantage of this score is that it can be easily evaluated, but its prognostic power remains controversial. Indeed, NLR is not a specific biomarker because it might be confounded by other comorbidities (Di Caro et al. 2014). Moreover, it is important to remember that, because of the well-known heterogeneity of neutrophil subsets, circulating neutrophils may not faithfully mirror the tumor-related ones. Thus, further studies aimed at investigating circulating neutrophil-related markers that more likely reflect the TME are needed to identify more specific diagnostic biomarkers of tumor detection.

ROLES OF TAMs AND TANs IN ANTICANCER THERAPEUTIC RESPONSES

Given the protumor functions of TAMs, a number of therapeutic strategies have been evaluated based on their targeting. These approaches are designed to limit TAM recruitment, inhibit their protumor functions, and reeducate them toward an antitumor phenotype.

CCL2 inhibition reduced tumor growth and metastasis in experimental models of prostate, breast, lung, liver cancer, or melanoma. In combination with chemotherapy, anti-CCL2 antibodies improved the therapeutic efficacy of the drugs (Loberg et al. 2007; Lu and Kang 2009; Fridlender et al. 2011; Moisan et al. 2014). Anti-CCL2 antibodies have entered phase I and II clinical trials in patients with solid tumors, but showed controversial results (Pienta et al. 2013; Sandhu et al. 2013; Brana et al. 2015). M-CSF (CSF-1) is the main growth and differentiation factor for monocytes and macrophages and is expressed by several tumors. A number of small molecules and antibodies directed against CSF-1 receptor (CSF-1R) have been evaluated in preclinical settings and clinical trials (Manthey et al. 2009; Ries et al. 2014). The humanized antibody emactuzumab showed efficacy in patients with various malignancies and was promising in patients with diffuse-type tenosynovial giant-cell tumor (Ries et al. 2014). Pexidartinib, a small CSF-1R inhibitor did not show efficacy in glioblastoma patients (Butowski et al. 2016). However, when anti-CSF-1R drugs were combined with traditional anticancer therapy, the results were enhanced. For example, in a transgenic model of gemcitabine-resistant pancreatic tumor, the anti-CSF-1R inhibitor GW2580 enhanced the efficacy of gemcitabine through the elimination of TAMs, which were responsible for drug resistance (Weizman et al. 2014). In a transgenic model of breast cancer, inhibition of CSF-1/CSF-1R axis enhanced the therapeutic effect of paclitaxel, inhibited metastatic spreading, and increased intratumoral T-cell infiltration (DeNardo et al. 2011). Thus, targeting TAMs appears to be a promising complementary strategy to enhancing the therapeutic power of conventional anticancer therapies.

Trabectedin is an European Medicines Evaluation Agency (EMEA)-approved natural product with antitumor activity (Germano et al. 2010). Indeed, trabectedin activated the TRAIL-dependent apoptotic pathway selectively in monocytes, because of their low expression of TRAIL decoy receptors (Liguori et al. 2016). In murine models and human sarcoma patients, trabectedin treatment resulted in a reduction of TAM infiltration and angiogenesis (Germano et al. 2010), thus suggesting a promising role in TAM-targeted antitumor therapies.

TAM reeducation toward an antitumor phenotype represents a desirable goal. In this line, IFN-γ administration has been proposed in ovarian cancer patients (Colombo et al. 1992). This treatment led to the systemic antitumor cytotoxic activation and clinical response, but the real efficacy of IFN-γ immunotherapy is still poorly understood. In a murine model of pancreatic cancer, the fully human CD40 agonist antibody CP-870,893, in combination with gemcitabine chemotherapy, induced a switch in TAM phenotype from a tumor-promoting to an antitumor profile, with enhanced antigen-presenting activities that impaired tumor growth. In a phase II clinical trial in patients with advanced pancreatic cancer, 19% of patients had partial responses and 52% had a period of disease stabilization (Beatty et al. 2011).

Myeloid cells can also influence the effectiveness of chemotherapeutic drugs. Indeed, it is now well known that chemotherapeutic drugs exert their effects not only by acting on the tumor cell itself, but also on tumor-related immune cells. Actually, some chemotherapeutic drugs, such as doxorubicin, determine an "immunogenic cell death." Tumor cell death induces the expression of "danger signals" (i.e., calreticulin, adenosine triphosphate [ATP], high-mobility group box 1 [HMGB-1]), which recruit and activate myeloid dendritic-cell-like cells. These cells are particularly efficient in capturing and presenting tumor cell antigens and give rise to an effective antitumor immune response (Galluzzi et al. 2012; Ma et al. 2013).

TAMs can also limit the effectiveness of chemotherapeutic drugs, such as paclitaxel and doxorubicin. For example, in murine models of breast and lung cancer, following chemotherapy M2-like macrophages accumulated in perivascular areas of tumors and favored tumor neo-angiogenesis in a CXCL12/CXCR4-dependant manner (Hughes et al. 2015). Anti-VEGF therapies are also associated with the accumulation of myeloid cells in perivascular areas as a consequence of the local hypoxia induced by the antiangiogenic therapy. In some circumstances, these cells activate an alternative program and produce proangiogenic molecules such as Bv8, which overcome the antiangiogenic effect of the drug (Murdoch et al. 2008; Shojaei et al. 2009). Thus, targeting TAMs can be an effective therapeutic strategy that is complementary to current chemotherapeutic and antiangiogenic therapies and can efficiently improve their effectiveness.

Immunotherapy using checkpoint inhibitors is an established part of the therapeutic strategies for an increasing number of cancers (Sharma and Allison 2015). Macrophages can express PD-L1 and PD-L2, which can be upregulated under the influence of proinflammatory stimuli and hypoxia (Noman et al. 2014). The predictive power of these molecules on TAMs needs to be carefully evaluated. To what extent the expression of PD-L1 on macrophages can contribute to the therapeutic efficacy of immune checkpoint inhibitors is not yet understood.

The evaluation of TANs as therapeutic targets is still limited because a role of these cells in cancer development is a recent concept. Considering the tumor-promoting functions of TANs, targeting these cells could be desirable. However, their depletion could lead to deleterious "side effects." Indeed, neutrophils play a pivotal role in host defense against infections and their depletion could give rise to immunosuppression. TAN neutralization could be obtained by inhibiting their recruitment or their effector molecules. In a murine model of fibrosarcoma and prostate cancer, TAN recruitment inhibition through CXCL8/IL-8 blockage significantly reduced angiogenesis and tumor growth (Bekes et al. 2011). In addition, in murine inflammation-driven and spontaneous carcinogenesis, CXCR2 deletion and/or inhibition blocked tumor development (Jamieson et al. 2012). Repertaxin, a small molecule inhibitor of CXCR1 and CXCR2, selectively targeted human breast cancer stem cells and inhibited tumor growth in xenograft murine models (Ginestier et al. 2010). More recently, the combination of repertaxin and 5-FU was shown to increase gastric cancer cell apoptosis and inhibited cellular proliferation, migration, and invasion (Wang et al. 2016a). Clinical trials are currently investigating the role of repertaxin in breast cancer patients, alone or in combination with chemotherapeutic drugs (paclitaxel) (www .clinicaltrials.gov).

The inhibitor of NE sivelastat efficiently suppressed breast cancer cell proliferation and enhanced the antitumor effect of trastuzumab, through restoring the expression of Her2/Neu (Nawa et al. 2012). Genetic deficiency and chemical inhibition of NE significantly reduced the incidence of ultraviolet-B-induced tumors in mice (Starcher et al. 1996). The NE inhibitor ONO-5046 inhibited both primary and metastatic growth of non-small-cell lung cancer (NSCLC) in severe combined immunodeficiency (SCID) mice (Inada et al. 1998). NE inhibitors are currently undergoing clinical trials for treatment of cystic fibrosis and respiratory diseases (www.clinicaltrials.gov), and these results could also be useful for cancer research.

CONCLUDING REMARKS

There is compelling evidence that cellular and humoral components of the TME have a large impact on cancer initiation and progression and on the resilience of most tumors in the face of therapy. Macrophages and neutrophils are both integrated within CRI and can take part in the various phases of tumor initiation and progression. Cancer cells as well as TAMs and neutrophils can release a plethora of protumorigenic and proangiogenic cytokines/chemokines. Targeting these mediators as well as blocking protumor functions could be useful for inhibiting tumor growth. On the other hand, fostering anticancer immune responses by blocking immunosuppressive molecules (e.g., TGF-β, IL-10, CTLA-4, PD-1, PD-L1), expressed either by cancer cells or by tumor-infiltrating immune cells, appears a promising therapeutic strategy in different tumors.

In conclusion, a deeper insight into the molecular mechanisms regulating the link between tumor-infiltrating immune cells and cancer cells could lead to the finding of new prognostic/predictive biomarkers, as well as a wider view of cancer immunotherapy, in an even more personalized therapeutic approach.

ACKNOWLEDGMENTS

This work is supported in part by grants from the Regione Campania CISI Laboratory Project, the CRèME Project, and the TIMING Project. We thank Fabrizio Fiorbianco for the figure. A.M. is supported by grants from the Associazione Italiana per la Ricerca sul Cancro (AIRC), an IG grant, and a 5X1000 grant, and by the Italian Ministry of Health.

REFERENCES

Ai S, Cheng XW, Inoue A, Nakamura K, Okumura K, Iguchi A, Murohara T, Kuzuya M. 2007. Angiogenic activity of bFGF and VEGF suppressed by proteolytic cleavage by neutrophil elastase. *Biochem Biophys Res Commun* **364:** 395–401.

Alam M, Khan M, Veledar E, Pongprutthipan M, Flores A, Dubina M, Nodzenski M, Yoo SS. 2016. Correlation of inflammation in frozen sections with site of nonmelanoma skin cancer. *JAMA Dermatol* **152:** 173–176.

Allavena P, Mantovani A. 2012. Immunology in the clinic review series; focus on cancer: Tumour-associated macrophages: Undisputed stars of the inflammatory tumour microenvironment. *Clin Exp Immunol* **167:** 195–205.

Allavena P, Sica A, Garlanda C, Mantovani A. 2008. The Yin–Yang of tumor-associated macrophages in neoplastic progression and immune surveillance. *Immunol Rev* **222:** 155–161.

Al Murri AM, Bartlett JM, Canney PA, Doughty JC, Wilson C, McMillan DC. 2006. Evaluation of an inflammation-based prognostic score (GPS) in patients with metastatic breast cancer. *Br J Cancer* **94:** 227–230.

Ardi VC, Kupriyanova TA, Deryugina EI, Quigley JP. 2007. Human neutrophils uniquely release TIMP-free MMP-9 to provide a potent catalytic stimulator of angiogenesis. *Proc Natl Acad Sci* **104:** 20262–20267.

Balkwill F, Mantovani A. 2001. Inflammation and cancer: Back to Virchow? *Lancet* **357:** 539–545.

Barash H, E RG, Edrei Y, Ella E, Israel A, Cohen I, Corchia N, Ben-Moshe T, Pappo O, Pikarsky E, et al. 2010. Accelerated carcinogenesis following liver regeneration is associated with chronic inflammation-induced double-strand DNA breaks. *Proc Natl Acad Sci* **107:** 2207–2212.

Beatty GL, Chiorean EG, Fishman MP, Saboury B, Teitelbaum UR, Sun W, Huhn RD, Song W, Li D, Sharp LL, et al. 2011. CD40 agonists alter tumor stroma and show efficacy against pancreatic carcinoma in mice and humans. *Science* **331:** 1612–1616.

Bekes EM, Schweighofer B, Kupriyanova TA, Zajac E, Ardi VC, Quigley JP, Deryugina EI. 2011. Tumor-recruited neutrophils and neutrophil TIMP-free MMP-9 regulate coordinately the levels of tumor angiogenesis and efficiency of malignant cell intravasation. *Am J Pathol* **179:** 1455–1470.

Bingle L, Brown NJ, Lewis CE. 2002. The role of tumour-associated macrophages in tumour progression: Implications for new anticancer therapies. *J Pathol* **196:** 254–265.

Bonavita E, Galdiero MR, Jaillon S, Mantovani A. 2015. Phagocytes as corrupted policemen in cancer-related inflammation. *Adv Cancer Res* **128:** 141–171.

Bornstein S, White R, Malkoski S, Oka M, Han G, Cleaver T, Reh D, Andersen P, Gross N, Olson S, et al. 2009. Smad4 loss in mice causes spontaneous head and neck cancer with increased genomic instability and inflammation. *J Clin Invest* **119:** 3408–3419.

Bottazzi B, Erba E, Nobili N, Fazioli F, Rambaldi A, Mantovani A. 1990. A paracrine circuit in the regulation of the proliferation of macrophages infiltrating murine sarcomas. *J Immunol* **144:** 2409–2412.

Boyum A. 1968. Isolation of mononuclear cells and granulocytes from human blood. Isolation of monuclear cells by one centrifugation, and of granulocytes by combining centrifugation and sedimentation at 1 g. *Scand J Clin Lab Invest Suppl* **97:** 77–89.

Brana I, Calles A, LoRusso PM, Yee LK, Puchalski TA, Seetharam S, Zhong B, de Boer CJ, Tabernero J, Calvo E. 2015. Carlumab, an anti-C-C chemokine ligand 2 monoclonal antibody, in combination with four chemotherapy regimens for the treatment of patients with solid tumors: An open-label, multicenter phase 1b study. *Target Oncol* **10:** 111–123.

Buddingh EP, Kuijjer ML, Duim RA, Burger H, Agelopoulos K, Myklebost O, Serra M, Mertens F, Hogendoorn PC, Lankester AC, et al. 2011. Tumor-infiltrating macrophages are associated with metastasis suppression in high-grade osteosarcoma: A rationale for treatment with macrophage activating agents. *Clin Cancer Res* **17:** 2110–2119.

Butowski N, Colman H, De Groot JF, Omuro AM, Nayak L, Wen PY, Cloughesy TF, Marimuthu A, Haidar S, Perry A, et al. 2016. Orally administered colony stimulating factor 1 receptor inhibitor PLX3397 in recurrent glioblastoma: An Ivy Foundation Early Phase Clinical Trials Consortium phase II study. *Neuro Oncol* **18:** 557–564.

Campbell MJ, Tonlaar NY, Garwood ER, Huo D, Moore DH, Khramtsov AI, Au A, Baehner F, Chen Y, Malaka DO, et al. 2011. Proliferating macrophages associated with high grade, hormone receptor negative breast cancer and poor clinical outcome. *Breast Cancer Res Treat* **128:** 703–711.

Caruso RA, Bellocco R, Pagano M, Bertoli G, Rigoli L, Inferrera C. 2002. Prognostic value of intratumoral neutrophils in advanced gastric carcinoma in a high-risk area in northern Italy. *Mod Pathol* **15:** 831–837.

Cassatella MA. 2006. On the production of TNF-related apoptosis-inducing ligand (TRAIL/Apo-2L) by human neutrophils. *J Leukoc Biol* **79:** 1140–1149.

Chawla A, Alatrash G, Philips AV, Qiao N, Sukhumalchandra P, Kerros C, Diaconu I, Gall V, Neal S, Peters HL, et al. 2016. Neutrophil elastase enhances antigen presentation by upregulating human leukocyte antigen class I expression on tumor cells. *Cancer Immunol Immunother* **65:** 741–751.

Cloke T, Munder M, Taylor G, Muller I, Kropf P. 2012. Characterization of a novel population of low-density granulocytes associated with disease severity in HIV-1 infection. *PLoS ONE* **7:** e48939.

Colombo N, Peccatori F, Paganin C, Bini S, Brandely M, Mangioni C, Mantovani A, Allavena P. 1992. Anti-tumor and immunomodulatory activity of intraperitoneal IFN-γ in ovarian carcinoma patients with minimal residual tumor after chemotherapy. *Int J Cancer* **51:** 42–46.

Cools-Lartigue J, Spicer J, McDonald B, Gowing S, Chow S, Giannias B, Bourdeau F, Kubes P, Ferri L. 2013. Neutrophil extracellular traps sequester circulating tumor cells and promote metastasis. *J Clin Invest* doi: 10.1172/JCI67484.

Cortez-Retamozo V, Etzrodt M, Newton A, Rauch PJ, Chudnovskiy A, Berger C, Ryan RJ, Iwamoto Y, Marinelli B, Gorbatov R, et al. 2012. Origins of tumor-associated macrophages and neutrophils. *Proc Natl Acad Sci* **109:** 2491–2496.

Davies LC, Rosas M, Smith PJ, Fraser DJ, Jones SA, Taylor PR. 2011. A quantifiable proliferative burst of tissue macrophages restores homeostatic macrophage populations after acute inflammation. *Eur J Immunol* **41:** 2155–2164.

De Larco JE, Wuertz BR, Furcht LT. 2004. The potential role of neutrophils in promoting the metastatic phenotype of tumors releasing interleukin-8. *Clin Cancer Res* **10:** 4895–4900.

DeNardo DG, Brennan DJ, Rexhepaj E, Ruffell B, Shiao SL, Madden SF, Gallagher WM, Wadhwani N, Keil SD, Junaid SA, et al. 2011. Leukocyte complexity predicts breast cancer survival and functionally regulates response to chemotherapy. *Cancer Discov* **1:** 54–67.

Deng J, Le Mercier I, Kuta A, Noelle RJ. 2016. A New VISTA on combination therapy for negative checkpoint regulator blockade. *J Immunother Cancer* **4:** 86.

Denny MF, Yalavarthi S, Zhao W, Thacker SG, Anderson M, Sandy AR, McCune WJ, Kaplan MJ. 2010. A distinct subset of proinflammatory neutrophils isolated from patients with systemic lupus erythematosus induces vascular damage and synthesizes type I IFNs. *J Immunol* **184:** 3284–3297.

Di Caro G, Marchesi F, Galdiero MR, Grizzi F. 2014. Immune mediators as potential diagnostic tools for colorectal cancer: From experimental rationale to early clinical evidence. *Expert Rev Mol Diagn* **14:** 387–399.

Di Caro G, Cortese N, Castino GF, Grizzi F, Gavazzi F, Ridolfi C, Capretti G, Mineri R, Todoric J, Zerbi A, et al. 2016. Dual prognostic significance of tumour-associated macrophages in human pancreatic adenocarcinoma treated or untreated with chemotherapy. *Gut* **65:** 1710–1720.

Donskov F. 2013. Immunomonitoring and prognostic relevance of neutrophils in clinical trials. *Semin Cancer Biol* **23:** 200–207.

Droeser RA, Hirt C, Eppenberger-Castori S, Zlobec I, Viehl CT, Frey DM, Nebiker CA, Rosso R, Zuber M, Amicarella F, et al. 2013. High myeloperoxidase positive cell infiltration in colorectal cancer is an independent favorable prognostic factor. *PLoS ONE* **8:** e64814.

Dumitru CA, Gholaman H, Trellakis S, Bruderek K, Dominas N, Gu X, Bankfalvi A, Whiteside TL, Lang S, Brandau S. 2011. Tumor-derived macrophage migration inhibitory factor modulates the biology of head and neck cancer cells via neutrophil activation. *Int J Cancer* **129:** 859–869.

Dumitru CA, Fechner MK, Hoffmann TK, Lang S, Brandau S. 2012. A novel p38–MAPK signaling axis modulates neutrophil biology in head and neck cancer. *J Leukoc Biol* **91:** 591–598.

Dumitru CA, Bankfalvi A, Gu X, Eberhardt WE, Zeidler R, Lang S, Brandau S. 2013. Neutrophils activate tumoral CORTACTIN to enhance progression of orohypopharynx carcinoma. *Front Immunol* **4:** 33.

Dunn GP, Bruce AT, Ikeda H, Old LJ, Schreiber RD. 2002. Cancer immunoediting: From immunosurveillance to tumor escape. *Nat Immunol* **3:** 991–998.

Dvorak HF. 1986. Tumors: Wounds that do not heal. Similarities between tumor stroma generation and wound healing. *N Engl J Med* **315:** 1650–1659.

Eruslanov EB, Bhojnagarwala PS, Quatromoni JG, Stephen TL, Ranganathan A, Deshpande C, Akimova T, Vachani A, Litzky L, Hancock WW, et al. 2014. Tumor-associated neutrophils stimulate T cell responses in early-stage human lung cancer. *J Clin Invest* **124:** 5466–5480.

Ethier JL, Desautels D, Templeton A, Shah PS, Amir E. 2017. Prognostic role of neutrophil-to-lymphocyte ratio in breast cancer: A systematic review and meta-analysis. *Breast Cancer Res* **19:** 2.

Fossati G, Ricevuti G, Edwards SW, Walker C, Dalton A, Rossi ML. 1999. Neutrophil infiltration into human gliomas. *Acta Neuropathol* **98:** 349–354.

Fridlender ZG, Kapoor V, Buchlis G, Cheng G, Sun J, Wang LC, Singhal S, Snyder LA, Albelda SM. 2011. Monocyte chemoattractant protein-1 blockade inhibits lung cancer tumor growth by altering macrophage phenotype and activating CD8$^+$ cells. *Am J Respir Cell Mol Biol* **44:** 230–237.

Fridlender ZG, Sun J, Mishalian I, Singhal S, Cheng G, Kapoor V, Horng W, Fridlender G, Bayuh R, Worthen GS, et al. 2012. Transcriptomic analysis comparing tumor-associated neutrophils with granulocytic myeloid-derived suppressor cells and normal neutrophils. *PloS ONE* **7:** e31524.

Gabrilovich DI, Nagaraj S. 2009. Myeloid-derived suppressor cells as regulators of the immune system. *Nat Rev Immunol* **9:** 162–174.

Galdiero MR, Bonavita E, Barajon I, Garlanda C, Mantovani A, Jaillon S. 2013a. Tumor associated macrophages and neutrophils in cancer. *Immunobiology* **218:** 1402–1410.

Galdiero MR, Garlanda C, Jaillon S, Marone G, Mantovani A. 2013b. Tumor associated macrophages and neutrophils in tumor progression. *J Cell Physio* **228:** 1404–1412.

Galdiero MR, Bianchi P, Grizzi F, Di Caro G, Basso G, Ponzetta A, Bonavita E, Barbagallo M, Tartari S, Polentarutti N, et al. 2016. Occurrence and significance of tumor-associated neutrophils in patients with colorectal cancer. *Int J Cancer* **139:** 446–456.

Galluzzi L, Senovilla L, Zitvogel L, Kroemer G. 2012. The secret ally: Immunostimulation by anticancer drugs. *Nat Rev Drug Discov* **11:** 215–233.

Germano G, Frapolli R, Simone M, Tavecchio M, Erba E, Pesce S, Pasqualini F, Grosso F, Sanfilippo R, Casali PG, et al. 2010. Antitumor and anti-inflammatory effects of trabectedin on human myxoid liposarcoma cells. *Cancer Res* **70:** 2235–2244.

Ginestier C, Liu S, Diebel ME, Korkaya H, Luo M, Brown M, Wicinski J, Cabaud O, Charafe-Jauffret E, Birnbaum D, et al. 2010. CXCR1 blockade selectively targets human breast cancer stem cells in vitro and in xenografts. *J Clin Invest* **120:** 485–497.

Ginhoux F, Schultze JL, Murray PJ, Ochando J, Biswas SK. 2016. New insights into the multidimensional concept of macrophage ontogeny, activation and function. *Nat Immunol* **17:** 34–40.

Gokce MI, Tangal S, Hamidi N, Suer E, Ibis MA, Beduk Y. 2016. Role of neutrophil-to-lymphocyte ratio in prediction of Gleason score upgrading and disease upstaging in low-risk prostate cancer patients eligible for active surveillance. *Can Urol Assoc J* **10:** E383–E387.

Granata F, Frattini A, Loffredo S, Staiano RI, Petraroli A, Ribatti D, Oslund R, Gelb MH, Lambeau G, Marone G, et al. 2010. Production of vascular endothelial growth factors from human lung macrophages induced by group IIA and group X secreted phospholipases A2. *J Immunol* **184:** 5232–5241.

Granja S, Tavares-Valente D, Queiros O, Baltazar F. 2017. Value of pH regulators in the diagnosis, prognosis and treatment of cancer. *Semin Cancer Biol* doi: 10.1016/j.semcancer.2016.12.003.

Granot Z, Jablonska J. 2015. Distinct functions of neutrophil in cancer and its regulation. *Mediators Inflamm* **2015:** 701067.

Granot Z, Henke E, Comen EA, King TA, Norton L, Benezra R. 2011. Tumor entrained neutrophils inhibit seeding in the premetastatic lung. *Cancer Cell* **20:** 300–314.

Grivennikov SI, Greten FR, Karin M. 2010. Immunity, inflammation, and cancer. *Cell* **140:** 883–899.

Grosse-Steffen T, Giese T, Giese N, Longerich T, Schirmacher P, Hansch GM, Gaida MM. 2012. Epithelial-to-mesenchymal transition in pancreatic ductal adenocarcinoma and pancreatic tumor cell lines: The role of neutrophils and neutrophil-derived elastase. *Clin Dev Immunol* **2012:** 720768.

Gungor N, Knaapen AM, Munnia A, Peluso M, Haenen GR, Chiu RK, Godschalk RW, van Schooten FJ. 2010. Genotoxic effects of neutrophils and hypochlorous acid. *Mutagenesis* **25:** 149–154.

Guthrie GJ, Charles KA, Roxburgh CS, Horgan PG, McMillan DC, Clarke SJ. 2013. The systemic inflammation-based neutrophil-lymphocyte ratio: Experience in patients with cancer. *Crit Rev Oncol Hematol* **88:** 218–230.

Hanada T, Nakagawa M, Emoto A, Nomura T, Nasu N, Nomura Y. 2000. Prognostic value of tumor-associated macrophage count in human bladder cancer. *Int J Urol* **7:** 263–269.

Hanahan D, Weinberg RA. 2000. The hallmarks of cancer. *Cell* **100:** 57–70.

Hanahan D, Weinberg RA. 2011. Hallmarks of cancer: The next generation. *Cell* **144:** 646–674.

Henze AT, Mazzone M. 2016. The impact of hypoxia on tumor-associated macrophages. *J Clin Invest* **126:** 3672–3679.

Hewish M, Lord CJ, Martin SA, Cunningham D, Ashworth A. 2010. Mismatch repair deficient colorectal cancer in the era of personalized treatment. *Nat Rev Clin Oncol* **7:** 197–208.

Hotchkiss KA, Ashton AW, Klein RS, Lenzi ML, Zhu GH, Schwartz EL. 2003. Mechanisms by which tumor cells and monocytes expressing the angiogenic factor thymidine phosphorylase mediate human endothelial cell migration. *Cancer Res* **63:** 527–533.

Houghton AM, Rzymkiewicz DM, Ji H, Gregory AD, Egea EE, Metz HE, Stolz DB, Land SR, Marconcini LA, Kliment CR, et al. 2010. Neutrophil elastase-mediated degradation of IRS-1 accelerates lung tumor growth. *Nat Med* **16:** 219–223.

Hughes R, Qian BZ, Rowan C, Muthana M, Keklikoglou I, Olson OC, Tazzyman S, Danson S, Addison C, Clemons M, et al. 2015. Perivascular M2 macrophages stimulate tumor relapse after chemotherapy. *Cancer Res* **75:** 3479–3491.

Huh SJ, Liang S, Sharma A, Dong C, Robertson GP. 2010. Transiently entrapped circulating tumor cells interact with neutrophils to facilitate lung metastasis development. *Cancer Res* **70:** 6071–6082.

Ijichi H, Chytil A, Gorska AE, Aakre ME, Bierie B, Tada M, Mohri D, Miyabayashi K, Asaoka Y, Maeda S, et al. 2011. Inhibiting Cxcr2 disrupts tumor-stromal interactions and improves survival in a mouse model of pancreatic ductal adenocarcinoma. *J Clin Invest* **121:** 4106–4117.

Imai Y, Kubota Y, Yamamoto S, Tsuji K, Shimatani M, Shibatani N, Takamido S, Matsushita M, Okazaki K. 2005. Neutrophils enhance invasion activity of human cholangiocellular carcinoma and hepatocellular carcinoma cells: An in vitro study. *J Gastroenterol Hepatol* **20:** 287–293.

Inada M, Yamashita J, Nakano S, Ogawa M. 1998. Complete inhibition of spontaneous pulmonary metastasis of human lung carcinoma cell line EBC-1 by a neutrophil elastase inhibitor (ONO-5046.Na). *Anticancer Res* **18:** 885–890.

Jablonska J, Leschner S, Westphal K, Lienenklaus S, Weiss S. 2010. Neutrophils responsive to endogenous IFN-β regulate tumor angiogenesis and growth in a mouse tumor model. *J Clin Invest* **120:** 151–164.

Jamieson T, Clarke M, Steele CW, Samuel MS, Neumann J, Jung A, Huels D, Olson MF, Das S, Nibbs RJ, et al. 2012. Inhibition of CXCR2 profoundly suppresses inflammation-driven and spontaneous tumorigenesis. *J Clin Invest* **122:** 3127–3144.

Jenkins SJ, Ruckerl D, Cook PC, Jones LH, Finkelman FD, van Rooijen N, MacDonald AS, Allen JE. 2011. Local macrophage proliferation, rather than recruitment from the blood, is a signature of TH2 inflammation. *Science* **332:** 1284–1288.

Jensen HK, Donskov F, Marcussen N, Nordsmark M, Lundbeck F, von der Maase H. 2009. Presence of intratumoral neutrophils is an independent prognostic factor in localized renal cell carcinoma. *J Clin Oncol* **27:** 4709–4717.

Kadokura M, Ishida Y, Tatsumi A, Takahashi E, Shindo H, Amemiya F, Takano S, Fukasawa M, Sato T, Enomoto N. 2016. Performance status and neutrophil-lymphocyte ratio are important prognostic factors in elderly patients

with unresectable pancreatic cancer. *J Gastrointest Oncol* **7:** 982–988.

Keane MP, Belperio JA, Xue YY, Burdick MD, Strieter RM. 2004. Depletion of CXCR2 inhibits tumor growth and angiogenesis in a murine model of lung cancer. *J Immunol* **172:** 2853–2860.

Keeley EC, Mehrad B, Strieter RM. 2010. CXC chemokines in cancer angiogenesis and metastases. *Adv Cancer Res* **106:** 91–111.

Kemp TJ, Ludwig AT, Earel JK, Moore JM, Vanoosten RL, Moses B, Leidal K, Nauseef WM, Griffith TS. 2005. Neutrophil stimulation with *Mycobacterium bovis* bacillus Calmette–Guerin (BCG) results in the release of functional soluble TRAIL/Apo-2L. *Blood* **106:** 3474–3482.

Kitamura T, Qian BZ, Soong D, Cassetta L, Noy R, Sugano G, Kato Y, Li J, Pollard JW. 2015. CCL2-induced chemokine cascade promotes breast cancer metastasis by enhancing retention of metastasis-associated macrophages. *J Exp Med* **212:** 1043–1059.

Kridel R, Xerri L, Gelas-Dore B, Tan K, Feugier P, Vawda A, Canioni D, Farinha P, Boussetta S, Moccia AA, et al. 2015. The prognostic impact of CD163-positive macrophages in follicular lymphoma: A study from the BC Cancer Agency and the Lymphoma Study Association. *Clin Cancer Res* **21:** 3428–3435.

Kryczek I, Wei S, Zou L, Zhu G, Mottram P, Xu H, Chen L, Zou W. 2006. Cutting edge: Induction of B7-H4 on APCs through IL-10: Novel suppressive mode for regulatory T cells. *J Immunol* **177:** 40–44.

Kuang DM, Zhao Q, Wu Y, Peng C, Wang J, Xu Z, Yin XY, Zheng L. 2011. Peritumoral neutrophils link inflammatory response to disease progression by fostering angiogenesis in hepatocellular carcinoma. *J Hepatol* **54:** 948–955.

Lazennec G, Richmond A. 2010. Chemokines and chemokine receptors: New insights into cancer-related inflammation. *Trends Mol Med* **16:** 133–144.

Leifler KS, Svensson S, Abrahamsson A, Bendrik C, Robertson J, Gauldie J, Olsson AK, Dabrosin C. 2013. Inflammation induced by MMP-9 enhances tumor regression of experimental breast cancer. *J Immunol* **190:** 4420–4430.

Liang J, Nagahashi M, Kim EY, Harikumar KB, Yamada A, Huang WC, Hait NC, Allegood JC, Price MM, Avni D, et al. 2013. Sphingosine-1-phosphate links persistent STAT3 activation, chronic intestinal inflammation, and development of colitis-associated cancer. *Cancer Cell* **23:** 107–120.

Liguori M, Buracchi C, Pasqualini F, Bergomas F, Pesce S, Sironi M, Grizzi F, Mantovani A, Belgiovine C, Allavena P. 2016. Functional TRAIL receptors in monocytes and tumor-associated macrophages: A possible targeting pathway in the tumor microenvironment. *Oncotarget* **7:** 41662–41676.

Loberg RD, Ying C, Craig M, Day LL, Sargent E, Neeley C, Wojno K, Snyder LA, Yan L, Pienta KJ. 2007. Targeting CCL2 with systemic delivery of neutralizing antibodies induces prostate cancer tumor regression in vivo. *Cancer Res* **67:** 9417–9424.

Lu X, Kang Y. 2009. Chemokine (C-C motif) ligand 2 engages CCR2+ stromal cells of monocytic origin to promote breast cancer metastasis to lung and bone. *J Biol Chem* **284:** 29087–29096.

Lund AW, Medler TR, Leachman SA, Coussens LM. 2016. Lymphatic vessels, inflammation, and immunity in skin cancer. *Cancer Discov* **6:** 22–35.

Lutgens E, de Muinck ED, Kitslaar PJ, Tordoir JH, Wellens HJ, Daemen MJ. 1999. Biphasic pattern of cell turnover characterizes the progression from fatty streaks to ruptured human atherosclerotic plaques. *Cardiovasc Res* **41:** 473–479.

Ma Y, Adjemian S, Mattarollo SR, Yamazaki T, Aymeric L, Yang H, Portela Catani JP, Hannani D, Duret H, Steegh K, et al. 2013. Anticancer chemotherapy-induced intratumoral recruitment and differentiation of antigen-presenting cells. *Immunity* **38:** 729–741.

Manthey CL, Johnson DL, Illig CR, Tuman RW, Zhou Z, Baker JF, Chaikin MA, Donatelli RR, Franks CF, Zeng L, et al. 2009. JNJ-28312141, a novel orally active colony-stimulating factor-1 receptor/FMS-related receptor tyrosine kinase-3 receptor tyrosine kinase inhibitor with potential utility in solid tumors, bone metastases, and acute myeloid leukemia. *Mol Cancer Ther* **8:** 3151–3161.

Mantovani A. 2009. The yin-yang of tumor-associated neutrophils. *Cancer Cell* **16:** 173–174.

Mantovani A, Sozzani S, Locati M, Allavena P, Sica A. 2002. Macrophage polarization: Tumor-associated macrophages as a paradigm for polarized M2 mononuclear phagocytes. *Trends Immunol* **23:** 549–555.

Mantovani A, Allavena P, Sica A, Balkwill F. 2008. Cancer-related inflammation. *Nature* **454:** 436–444.

Mantovani A, Cassatella MA, Costantini C, Jaillon S. 2011. Neutrophils in the activation and regulation of innate and adaptive immunity. *Nat Rev Immunol* **11:** 519–531.

Mantovani A, Biswas SK, Galdiero MR, Sica A, Locati M. 2013. Macrophage plasticity and polarization in tissue repair and remodelling. *J Pathol* **229:** 176–185.

Matsui A, Yokoo H, Negishi Y, Endo-Takahashi Y, Chun NA, Kadouchi I, Suzuki R, Maruyama K, Aramaki Y, Semba K, et al. 2012. CXCL17 expression by tumor cells recruits CD11b$^+$Gr1 high F4/80$^-$ cells and promotes tumor progression. *PloS ONE* **7:** e44080.

Mishalian I, Granot Z, Fridlender ZG. 2017. The diversity of circulating neutrophils in cancer. *Immunobiology* **222:** 82–88.

Mittendorf EA, Alatrash G, Qiao N, Wu Y, Sukhumalchandra P, St John LS, Philips AV, Xiao H, Zhang M, Ruisaard K, et al. 2012. Breast cancer cell uptake of the inflammatory mediator neutrophil elastase triggers an anticancer adaptive immune response. *Cancer Res* **72:** 3153–3162.

Moisan F, Francisco EB, Brozovic A, Duran GE, Wang YC, Chaturvedi S, Seetharam S, Snyder LA, Doshi P, Sikic BI. 2014. Enhancement of paclitaxel and carboplatin therapies by CCL2 blockade in ovarian cancers. *Mol Oncol* **8:** 1231–1239.

Movahedi K, Laoui D, Gysemans C, Baeten M, Stange G, Van den Bossche J, Mack M, Pipeleers D, In't Veld P, De Baetselier P, et al. 2010. Different tumor microenvironments contain functionally distinct subsets of macrophages derived from Ly6C^{high} monocytes. *Cancer Res* **70:** 5728–5739.

Murdoch C, Muthana M, Coffelt SB, Lewis CE. 2008. The role of myeloid cells in the promotion of tumour angiogenesis. *Nat Rev Cancer* **8:** 618–631.

Nakamura K, Yoshida N, Baba Y, Kosumi K, Uchihara T, Kiyozumi Y, Ohuchi M, Ishimoto T, Iwatsuki M, Sakamoto Y, et al. 2017. Elevated preoperative neutrophil-to-lymphocytes ratio predicts poor prognosis after esophagectomy in T1 esophageal cancer. *Int J Clin Oncol* doi: 10.1007/s10147-017-100-5.

Nawa M, Osada S, Morimitsu K, Nonaka K, Futamura M, Kawaguchi Y, Yoshida K. 2012. Growth effect of neutrophil elastase on breast cancer: Favorable action of sivelestat and application to anti-HER2 therapy. *Anticancer Res* **32:** 13–19.

Nie W, Yu T, Sang Y, Gao X. 2017. Tumor-promoting effect of IL-23 in mammary cancer mediated by infiltration of M2 macrophages and neutrophils in tumor microenvironment. *Biochem Biophys Res Commun* **482:** 1400–1406.

Noman MZ, Desantis G, Janji B, Hasmim M, Karray S, Dessen P, Bronte V, Chouaib S. 2014. PD-L1 is a novel direct target of HIF-1α, and its blockade under hypoxia enhanced MDSC-mediated T cell activation. *J Exp Med* **211:** 781–790.

Noy R, Pollard JW. 2014. Tumor-associated macrophages: From mechanisms to therapy. *Immunity* **41:** 49–61.

Nozawa H, Chiu C, Hanahan D. 2006. Infiltrating neutrophils mediate the initial angiogenic switch in a mouse model of multistage carcinogenesis. *Proc Natl Acad Sci* **103:** 12493–12498.

Ohno S, Inagawa H, Dhar DK, Fujii T, Ueda S, Tachibana M, Suzuki N, Inoue M, Soma G, Nagasue N. 2003. The degree of macrophage infiltration into the cancer cell nest is a significant predictor of survival in gastric cancer patients. *Anticancer Res* **23:** 5015–5022.

Peranzoni E, Zilio S, Marigo I, Dolcetti L, Zanovello P, Mandruzzato S, Bronte V. 2010. Myeloid-derived suppressor cell heterogeneity and subset definition. *Curr Opin Immunol* **22:** 238–244.

Pienta KJ, Machiels JP, Schrijvers D, Alekseev B, Shkolnik M, Crabb SJ, Li S, Seetharam S, Puchalski TA, Takimoto C, et al. 2013. Phase 2 study of carlumab (CNTO 888), a human monoclonal antibody against CC-chemokine ligand 2 (CCL2), in metastatic castration-resistant prostate cancer. *Invest New Drugs* **31:** 760–768.

Pyonteck SM, Akkari L, Schuhmacher AJ, Bowman RL, Sevenich L, Quail DF, Olson OC, Quick ML, Huse JT, Teijeiro V, et al. 2013. CSF-1R inhibition alters macrophage polarization and blocks glioma progression. *Nat Med* **19:** 1264–1272.

Qian BZ, Pollard JW. 2010. Macrophage diversity enhances tumor progression and metastasis. *Cell* **141:** 39–51.

Queen MM, Ryan RE, Holzer RG, Keller-Peck CR, Jorcyk CL. 2005. Breast cancer cells stimulate neutrophils to produce oncostatin M: Potential implications for tumor progression. *Cancer Res* **65:** 8896–8904.

Rao HL, Chen JW, Li M, Xiao YB, Fu J, Zeng YX, Cai MY, Xie D. 2012. Increased intratumoral neutrophil in colorectal carcinomas correlates closely with malignant phenotype and predicts patients' adverse prognosis. *PloS ONE* **7:** e30806.

Reid MD, Basturk O, Thirabanjasak D, Hruban RH, Klimstra DS, Bagci P, Altinel D, Adsay V. 2011. Tumor-infiltrating neutrophils in pancreatic neoplasia. *Mod Pathol* **24**: 1612–1619.

Ries CH, Cannarile MA, Hoves S, Benz J, Wartha K, Runza V, Rey-Giraud F, Pradel LP, Feuerhake F, Klaman I, et al. 2014. Targeting tumor-associated macrophages with anti-CSF-1R antibody reveals a strategy for cancer therapy. *Cancer Cell* **25**: 846–859.

Robbins CS, Hilgendorf I, Weber GF, Theurl I, Iwamoto Y, Figueiredo JL, Gorbatov R, Sukhova GK, Gerhardt LM, Smyth D, et al. 2013. Local proliferation dominates lesional macrophage accumulation in atherosclerosis. *Nat Med* **19**: 1166–1172.

Rodriguez PC, Ernstoff MS, Hernandez C, Atkins M, Zabaleta J, Sierra R, Ochoa AC. 2009. Arginase I-producing myeloid-derived suppressor cells in renal cell carcinoma are a subpopulation of activated granulocytes. *Cancer Res* **69**: 1553–1560.

Ruffell B, Affara NI, Coussens LM. 2012. Differential macrophage programming in the tumor microenvironment. *Trends Immunol* **33**: 119–126.

Sagiv JY, Michaeli J, Assi S, Mishalian I, Kisos H, Levy L, Damti P, Lumbroso D, Polyansky L, Sionov RV, et al. 2015. Phenotypic diversity and plasticity in circulating neutrophil subpopulations in cancer. *Cell Rep* **10**: 562–573.

Sandhu SK, Papadopoulos K, Fong PC, Patnaik A, Messiou C, Olmos D, Wang G, Tromp BJ, Puchalski TA, Balkwill F, et al. 2013. A first-in-human, first-in-class, phase I study of carlumab (CNTO 888), a human monoclonal antibody against CC-chemokine ligand 2 in patients with solid tumors. *Cancer Chemother Pharmacol* **71**: 1041–1050.

Scapini P, Morini M, Tecchio C, Minghelli S, Di Carlo E, Tanghetti E, Albini A, Lowell C, Berton G, Noonan DM, et al. 2004. CXCL1/macrophage inflammatory protein-2-induced angiogenesis in vivo is mediated by neutrophil-derived vascular endothelial growth factor-A. *J Immunol* **172**: 5034–5040.

Schmidt T, Carmeliet P. 2010. Blood-vessel formation: Bridges that guide and unite. *Nature* **465**: 697–699.

Schmielau J, Finn OJ. 2001. Activated granulocytes and granulocyte-derived hydrogen peroxide are the underlying mechanism of suppression of T-cell function in advanced cancer patients. *Cancer Res* **61**: 4756–4760.

Serhan CN. 2010. Novel lipid mediators and resolution mechanisms in acute inflammation: To resolve or not? *Am J Pathol* **177**: 1576–1591.

Sharma P, Allison JP. 2015. The future of immune checkpoint therapy. *Science* **348**: 56–61.

Shen J, Zhu Y, Wu W, Zhang L, Ju H, Fan Y, Zhu Y, Luo J, Liu P, Zhou N, et al. 2017. Prognostic role of neutrophil-to-lymphocyte ratio in locally advanced rectal cancer treated with neoadjuvant chemoradiotherapy. *Med Sci Monit* **23**: 315–324.

Shojaei F, Wu X, Zhong C, Yu L, Liang XH, Yao J, Blanchard D, Bais C, Peale FV, van Bruggen N, et al. 2007. Bv8 regulates myeloid-cell-dependent tumour angiogenesis. *Nature* **450**: 825–831.

Shojaei F, Singh M, Thompson JD, Ferrara N. 2008. Role of Bv8 in neutrophil-dependent angiogenesis in a transgenic model of cancer progression. *Proc Natl Acad Sci* **105**: 2640–2645.

Shojaei F, Wu X, Qu X, Kowanetz M, Yu L, Tan M, Meng YG, Ferrara N. 2009. G-CSF-initiated myeloid cell mobilization and angiogenesis mediate tumor refractoriness to anti-VEGF therapy in mouse models. *Proc Natl Acad Sci* **106**: 6742–6747.

Starcher B, O'Neal P, Granstein RD, Beissert S. 1996. Inhibition of neutrophil elastase suppresses the development of skin tumors in hairless mice. *J Invest Dermatol* **107**: 159–163.

Stylianopoulos T, Martin JD, Chauhan VP, Jain SR, Diop-Frimpong B, Bardeesy N, Smith BL, Ferrone CR, Hornicek FJ, Boucher Y, et al. 2012. Causes, consequences, and remedies for growth-induced solid stress in murine and human tumors. *Proc Natl Acad Sci* **109**: 15101–15108.

Sun X, Glynn DJ, Hodson LJ, Huo C, Britt K, Thompson EW, Woolford L, Evdokiou A, Pollard JW, Robertson SA, et al. 2017. CCL2-driven inflammation increases mammary gland stromal density and cancer susceptibility in a transgenic mouse model. *Breast Cancer Res* **19**: 4.

Tanaka H, Ito T, Kyo T, Kimura A. 2007. Treatment with IFNα in vivo up-regulates serum-soluble TNF-related apoptosis inducing ligand (sTRAIL) levels and TRAIL mRNA expressions in neutrophils in chronic myelogenous leukemia patients. *Eur J Haematol* **78**: 389–398.

Tecchio C, Huber V, Scapini P, Calzetti F, Margotto D, Todeschini G, Pilla L, Martinelli G, Pizzolo G, Rivoltini L, et al. 2004. IFNα-stimulated neutrophils and monocytes release a soluble form of TNF-related apoptosis-inducing ligand (TRAIL/Apo-2 ligand) displaying apoptotic activity on leukemic cells. *Blood* **103**: 3837–3844.

Trellakis S, Bruderek K, Dumitru CA, Gholaman H, Gu X, Bankfalvi A, Scherag A, Hutte J, Dominas N, Lehnerdt GF, et al. 2011a. Polymorphonuclear granulocytes in human head and neck cancer: Enhanced inflammatory activity, modulation by cancer cells and expansion in advanced disease. *Int J Cancer* **129**: 2183–2193.

Trellakis S, Farjah H, Bruderek K, Dumitru CA, Hoffmann TK, Lang S, Brandau S. 2011b. Peripheral blood neutrophil granulocytes from patients with head and neck squamous cell carcinoma functionally differ from their counterparts in healthy donors. *Int J Immunopathol Pharmacol* **24**: 683–693.

Varricchi G, Galdiero MR, Marone G, Granata F, Borriello F, Marone G. 2017. Controversial role of mast cells in skin cancers. *Exp Dermatol* **26**: 11–17.

Wang L, Rubinstein R, Lines JL, Wasiuk A, Ahonen C, Guo Y, Lu LF, Gondek D, Wang Y, Fava RA, et al. 2011. VISTA, a novel mouse Ig superfamily ligand that negatively regulates T cell responses. *J Exp Med* **208**: 577–592.

Wang J, Hu W, Wang K, Yu J, Luo B, Luo G, Wang W, Wang H, Li J, Wen J. 2016a. Repertaxin, an inhibitor of the chemokine receptors CXCR1 and CXCR2, inhibits malignant behavior of human gastric cancer MKN45 cells in vitro and in vivo and enhances efficacy of 5-fluorouracil. *Int J Oncol* **48**: 1341–1352.

Wang L, Heng X, Lu Y, Cai Z, Yi Q, Che F. 2016b. Could B7-H4 serve as a target to activate anti-cancer immunity? *Int Immunopharmacol* **38**: 97–103.

Watanabe M, Kato J, Inoue I, Yoshimura N, Yoshida T, Mukoubayashi C, Deguchi H, Enomoto S, Ueda K, Maekita T, et al. 2012. Development of gastric cancer in nonatrophic stomach with highly active inflammation identified by serum levels of pepsinogen and *Helicobacter pylori* antibody together with endoscopic rugal hyperplastic gastritis. *Int J Cancer* **131:** 2632–2642.

Weizman N, Krelin Y, Shabtay-Orbach A, Amit M, Binenbaum Y, Wong RJ, Gil Z. 2014. Macrophages mediate gemcitabine resistance of pancreatic adenocarcinoma by upregulating cytidine deaminase. *Oncogene* **33:** 3812–3819.

Wislez M, Rabbe N, Marchal J, Milleron B, Crestani B, Mayaud C, Antoine M, Soler P, Cadranel J. 2003. Hepa-tocyte growth factor production by neutrophils infiltrating bronchioloalveolar subtype pulmonary adenocarcinoma: Role in tumor progression and death. *Cancer Res* **63:** 1405–1412.

Zhang QW, Liu L, Gong CY, Shi HS, Zeng YH, Wang XZ, Zhao YW, Wei YQ. 2012. Prognostic significance of tumor-associated macrophages in solid tumor: A meta-analysis of the literature. *PloS ONE* **7:** e50946.

Zhang Y, Wang J, Wang W, Tian J, Yin K, Tang X, Ma J, Xu H, Wang S. 2016. IL-17A produced by peritoneal macrophages promote the accumulation and function of granulocytic myeloid-derived suppressor cells in the development of colitis-associated cancer. *Tumour Biol* **37:** 15883–15891.

Cytokines in Cancer Immunotherapy

Thomas A. Waldmann

Lymphoid Malignancies Branch, Center for Cancer Research, National Cancer Institute, National Institutes of Health, Clinical Center, Bethesda, Maryland 20892-1374

Correspondence: tawald@helix.nih.gov

Cytokines that control the immune response were shown to have efficacy in preclinical murine cancer models. Interferon (IFN)-α is approved for treatment of hairy cell leukemia, and interleukin (IL)-2 for the treatment of advanced melanoma and metastatic renal cancer. In addition, IL-12, IL-15, IL-21, and granulocyte macrophage colony-stimulating factor (GM-CSF) have been evaluated in clinical trials. However, the cytokines as monotherapy have not fulfilled their early promise because cytokines administered parenterally do not achieve sufficient concentrations in the tumor, are often associated with severe toxicities, and induce humoral or cellular checkpoints. To circumvent these impediments, cytokines are being investigated clinically in combination therapy with checkpoint inhibitors, anticancer monoclonal antibodies to increase the antibody-dependent cellular cytotoxicity (ADCC) of these antibodies, antibody cytokine fusion proteins, and anti-CD40 to facilitate tumor-specific immune responses.

Cytokines are major regulators of the innate and adaptive immune systems that allow cells of the immune systems to communicate over short distances in paracrine and autocrine fashion. They control proliferation, differentiation, effector functions, and survival of leukocytes. In light of the ability of the immune system to recognize and destroy cancer cells, cytokines have been explored in the treatment of cancer (Goldstein and Laszlo 1988; Dranoff 2004; Lee and Margolin 2011; Nicholas and Lesinski 2011; Ardolino et al. 2015). In recent years, a number of cytokines, including interleukin (IL)-2, IL-12, IL-15, IL-21, granulocyte macrophage colony-stimulating factor (GM-CSF), and interferon (IFN)-α have been shown to have efficacy in preclinical murine cancer models (Fig. 1). This preclinical work has supported the evaluation of these cytokines in clinical trials. IFN-α was the first cytokine approved for the treatment of a human cancer (hairy cell leukemia [HCL]) in 1986. IL-2 was approved for the treatment of metastatic renal cell cancer in 1992 and advanced melanoma in 1998. Nevertheless, cytokines as monotherapy have not fulfilled the initial excitement they induced.

Soluble cytokines normally act over short distances in a paracrine or autocrine fashion (Rochman et al. 2009). Therefore, when cytokines are administered parenterally, large quantities must be administered or there is a failure to achieve effective concentrations of cytokine within the tumor. These large quantities are often associated with severe toxicities, especially

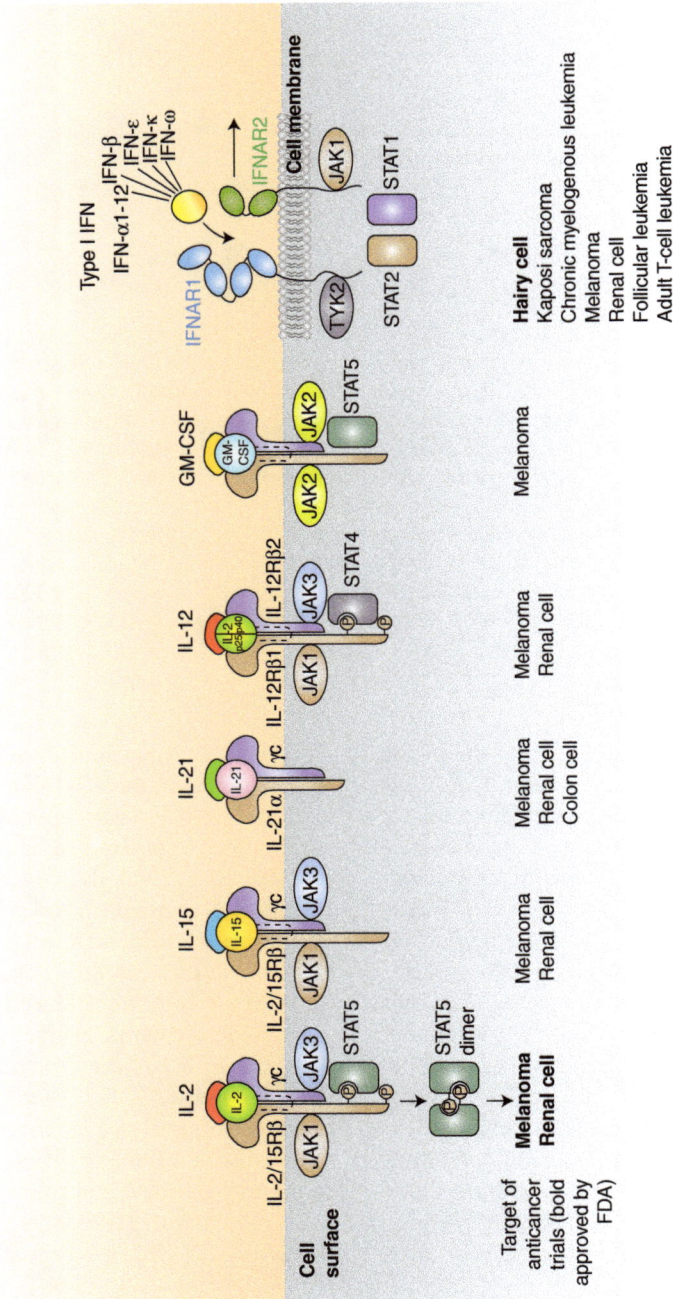

Figure 1. Interleukin (IL)-2, IL-15, IL-21, granulocyte macrophage colony-stimulating factor (GM-CSF), and type 1 interferons (IFNs) have been evaluated in clinical trials for the immunotherapy of cancer. Targets of anticancer trials are shown at the *bottom* with those approved by the U.S. Food and Drug Administration (FDA) in bold. IFNAR, Interferon-α/β receptor; JAK, Janus kinase; STAT, signal transducers and activators of transcription; TYK, tyrosine kinase.

flu-like symptoms that include fever, malaise, hypotension, fatigue, nausea, anorexia, neutropenia, and neuropsychiatric symptoms. In the case of IL-2, a major additional impediment is the induction of the capillary leak syndrome. An additional impediment is that, for a number of cytokines, the positive actions are paralleled by the cytokine induction of immunological checkpoints that include the secretion of inhibitory factors such as IL-10 and transforming growth factor (TGF)-β, the expression of inhibitors such as triosephosphate isomerase (TIM) on the cell surface, the induction of regulatory cells, including regulatory T (Treg) cells, and myeloid suppressor cells, as well as one of the intracellular suppressors of cytokine signaling (SOCS) proteins that terminate signaling (Alexander et al. 2004). Furthermore, although cytokines such as IL-15 dramatically augment the number and activation of natural killer (NK) cells, such cells are inhibited by the interaction of self-class I-A,B recognition with killer-cell immunoglobulin (Ig)-like receptors (KIRs) and of major histocompatibility complex (MHC) self-class I-E with NKG2A.

These impediments are being addressed with new strategies to augment their modest success. Because monotherapy may not be optimal, combination therapies are being evaluated with cytokines to achieve meaningful tumor responses. These approaches include cytokine engineering to augment cytokine activity, perilesional injection of cytokines, antibody–cytokine fusion proteins, infusion of cytokines along with antibodies to checkpoint proteins, the use of cytokines with anticancer vaccines, and cytokines associated with anticancer monoclonal antibodies to increase the antibody-dependent cellular cytotoxicity (ADCC) of these antibodies, thereby augmenting their efficacy (Becker et al. 1996; Carter 2001; Boyman et al. 2006; Schrama et al. 2006; Boder 2012; Levin et al. 2012; Young et al. 2014; Spangler et al. 2015). The goal of this paper is to review the cytokines involved in cancer immunotherapy and to discuss their biology and clinical application. The paper will also emphasize combinations of biological agents, novel delivery mechanisms, and directions for future investigation.

INTERLEUKIN-2

IL-2 is a 15.5-kDa globular glycoprotein of 133 amino acids (Waldmann 1986; Taniguchi and Minami 1993; Liao et al. 2011a; Boyman and Sprent 2012). It consists of four antiparallel amphipathic α helices. The three subunits of the IL-2 receptor include the γ chain (γc) (CD132) shared with IL-4, IL-7, IL-9, IL-15, and IL-21, IL-2Rβ (CD122) shared with IL-15, and the (CD25) IL-2Rα chain (Fig. 2). Three different IL-2R complexes exist and consist of combinations of the three receptor subunits (Waldmann 1991; Noguchi et al. 1993; Taniguchi and Minami 1993; Damjanovich et al. 1997; Rochman et al. 2009; Liao et al. 2011a). IL-2Rα alone binds IL-2 with low-affinity $K_d 10^{-8}$ M and does not transduce a signal. The heterodimeric IL-2Rβγ that binds IL-2 with intermediate affinity $K_d 10^9$ M can transduce an intracellular signal, and the heterotrimeric IL-2Rα, β, and γ binds IL-2 with high affinity. The IL-2 receptor signals through the Janus kinase (JAK)1, JAK3, and signal transducers and activators of transcription (STAT)5 molecules (Wang et al. 2005). IL-2 has paradoxical functions. It both acts as a T-cell growth factor during the initiation of the immune response, and has a crucial role in terminating T-cell responses for the maintenance of self-tolerance. IL-2 promotes the clonal expansion of antigen-activated CT8 T cells, and is a growth factor for CD4$^+$ T cells as well as NK cells (Boyman and Sprent 2012). IL-2 facilitates the production of Ig synthesis by B cells that have been stimulated with anti-Ig M (IgM) or by CD40 ligation. As noted above, IL-2 plays a crucial role in the negative regulation of T-cell responses. Although IL-2 signals are not essential for Treg-cell development in the thymus, they are critical for maintenance of Tregs in the periphery (Sakaguchi et al. 1995). IL-2 and all three IL-2 receptor chains (α, β, and γ) are required for high-affinity IL-2 binding and for Foxp3 (forkhead boxp3) transcription. In addition, IL-2 plays a pivotal role in Fas-mediated activation-induced cell death (AICD) of CD4 T cells (Leonardo 1996). In the phenomenon of AICD, receptor-mediated stimulation of CD4 T cells by antigen at high concentrations induces

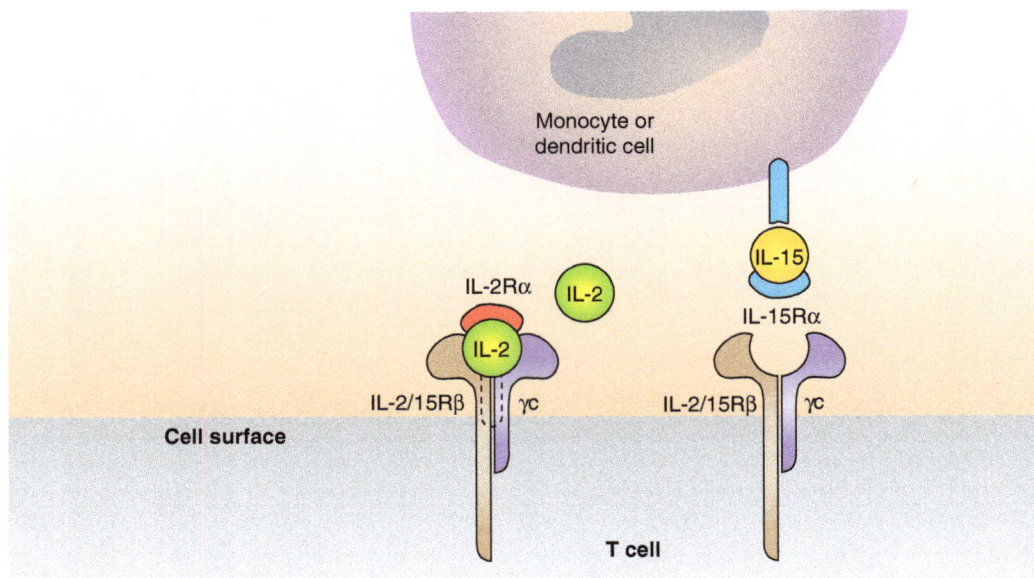

Figure 2. Model of interaction of interleukin (IL)-2 and IL-15 with their receptors. IL-2 is predominantly a secreted cytokine that binds to preformed high-affinity heterotrimeric receptors. In contrast, IL-15 is a membrane-associated molecule that signals as part of an immunological synapse between antigen-presenting cells and natural killer (NK) cells, $\gamma\Delta$, and CD8 T cells. IL-15Rα on the surface of activated monocytes or dendritic cells (DCs) presents IL-15 in *trans* to cells that express IL-2/IL-15Rβ and γ chain (γc), thereby allowing signaling through these complexes. (Based on figures in Waldmann 2006.)

the expression of IL-2 and the IL-2 receptor that, in turn, interact to yield T-cell activation and cycling. Antigen stimulation of the cycling T cells at this stage through the T-cell antigen receptor increases the transcription and surface expression of the death effector molecule Fas ligand (FasL) and suppresses the inhibition of Fas signaling FLIP (FLICE inhibitory protein) leading to T-cell death.

The conclusions concerning the negative regulatory role of IL-2 derived from ex vivo functional studies were supported by analysis of mice with disruptive cytokine–cytokine receptor genes. IL-2$^{-/-}$ and IL-2R$\alpha^{-/-}$-deficient mice develop massive enlargement of peripheral lymphoid organs associated with polyclonal T- and B-cell expansion owing to the impairment of AICD and Tregs (Sadlack et al. 1994). IL-2Rα-deficient mice develop autoimmune diseases such as hemolytic anemia and inflammatory bowel disease.

THE RESULTS OF TREATMENT OF PATIENTS WITH METASTATIC RENAL CELL CARCINOMA AND MALIGNANT MELANOMA WHO RECEIVED HIGH-DOSE RECOMBINANT IL-2 THERAPY

Rosenberg and coworkers use high doses of IL-2 (720,000 IU/kg/q every 8 h for up to 14 doses over 5 days) (Rosenberg et al. 1989). Such high-dose IL-2 therapy resulted in clinical responses in metastatic renal carcinoma (Rosenberg et al. 1989; Fyfe et al. 1995; Atkins et al. 1999). Based on this clinical study, the overall objective response was 14%, with 5% complete responses and 9% partial responses. The responses observed were of long-term duration with complete remissions (CRs) in patients who were still in response many months after therapy.

High-dose IL-2 therapy activates not only the IL-2-specific high-affinity α, β, and γ receptors but also activates cells expressing intermediate affinity IL-2Rβ and γc (e.g., NK cells, T cells, monocytes/macrophages, and B cells). This global activation of IL-2Rβ and γc results in secondary proinflammatory cytokine release, and therefore is presumed responsible for the severe life-threatening side effects associated with high-dose IL-2, which include hypotension and the capillary leak syndrome (Dutcher et al. 2014). At least grade 3 toxicities in at least 20% of patients included nausea and vomiting, mental status changes, and oliguria. Given the long duration of the IL-2 therapy responses, high-dose IL-2 therapy represents a meaningful addition to the therapeutic armamentarium for patients in intensive care hospital settings.

LOW-DOSE IL-2 THERAPY TARGETING THE HIGH-AFFINITY IL-2Rα, β, γ RECEPTOR-BEARING CELLS

Trials were designed by Soiffer et al. (1992, 1996) and extended by Caligiuri et al. (1993) and Fehniger and coworkers (2000) based on the known affinity of the IL-2Rα, β, γ receptor. One basis for this IL-2 therapy was derived from studies showing that the CD56^{bright} NK cells expressed the high-affinity IL-2 receptor. Low-dose IL-2 therapy provided via continuous intravenous (i.v.) infusion over the course of 90 days was associated with a significant expansion of CD56^{+} CD3^{-} NK cells (450%–900% increase) (Caligiuri et al. 1993). Although such prolonged low-dose IL-2 therapy was successful in expanding the number of NK cells, these NK cells were not activated but required higher amounts of IL-2 in vivo to kill tumor cells. Therefore, low-dose IL-2 to expand NK cells has been combined with intermediate pulses of IL-2 to provide activation of the expanded NK cell pool.

A number of approaches have been used to augment the efficacy of cytokines such as IL-2. Structure-based cytokine engineering has opened new opportunities for cytokines as drugs with the focus on the immunotherapeutic cytokines, IL-2, IFN, and IL-4 (Wang et al. 2005). Using in vitro evolution, the Garcia Laboratory (Levin et al. 2012) eliminated the functional requirement of IL-2 for IL-2Rα (CD25)

expression by engineering an IL-2 "superkine" with increased binding affinity for IL-2Rβ (Fig. 3). The evolved mutations were principally in the core of the cytokine that stabilized IL-2, reducing the flexibility of a helix in the IL-2Rβ-binding site into an optimized receptor-binding confirmation resembling that when bound to CD25. The mutations in the IL-2 superkine recapitulated the functional role of CD25 by eliciting potent phosphorylation of STAT5. When compared to IL-2, the superkine induced augmented expression of cytotoxic T cells, leading to improved antitumor responses in murine models with proportionately less expansion of Treg cells and reduced pulmonary edema (Levin et al. 2012; Spangler et al. 2015).

An alternative approach to augment IL-2 action was provided by Boyman et al. (2006). They induced selective stimulation of T cells with anti-IL-2 immune complexes (Fig. 3). Certain monoclonal antibodies directed toward IL-2 inhibited its action. However, other antibodies coupled with IL-2 caused massive (>100-fold) expansion of CD8 cells in vivo, whereas others selectively stimulated CD4^{+} Tregs. These actions were a result, in part, of increasing the survival T1/2 of IL-2. Thus, different cytokine–antibody complexes selectively boost or inhibit the immune response.

Yet another approach to augment the action and to increase the specificity of activity-involved antibody–cytokine fusion proteins (Becker et al. 1996; Reisfeld and Gillies 1996; Lode and Reisfeld 2000; Penichet and Morrison 2001; Jin et al. 2008). Becker et al. (1996) used an antibody–IL-2 fusion protein to eradicate established tumors by augmenting activated host-immune cells, particularly CD8^{+} T cells. Young et al. (2014) also used antibody–cytokine fusion proteins containing IL-2, IL-12, IL-21, tumor necrosis factor (TNF)-α, and IFN-α, -β, and -γ to guide cytokines specifically to tumor sites where they stimulated an antitumor response while avoiding the systemic toxicities of free cytokine therapy. These antitumor cytokine fusion proteins have shown antitumor activity in preclinical and early-phase clinical studies.

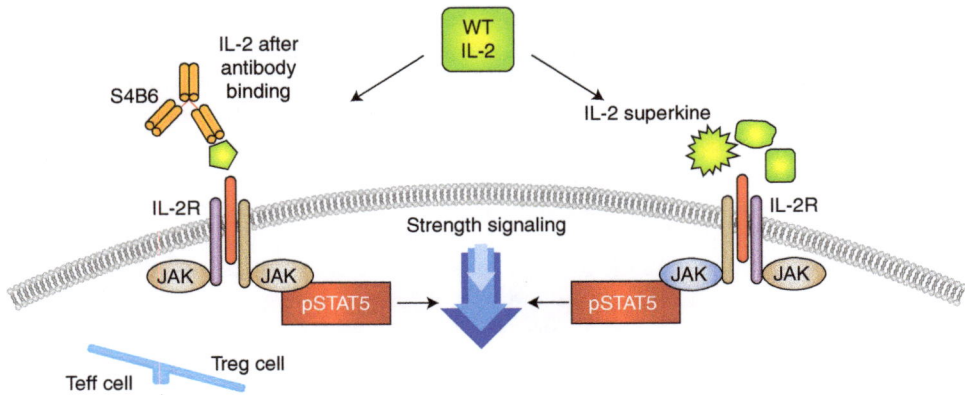

Figure 3. Different approaches to change interleukin (IL)-2 conformation to favor effector over negative regulatory function. One strategy for selectively modulating the effects of IL-2 is the development of cytokine-directed antibodies that bias activity toward specific T-cell subsets. The anti-IL-2 antibody S4B6 blocked the IL-2:IL-2Rα interaction while also conformationally stabilizing the IL-2:IL-2Rβ interaction, thus stimulating all IL-2-responsive immune cells, particularly IL-2Rβ^{hi} effector cells favored over IL-2Rα expressing regulatory T (Treg) cells. The *right* side shows a different mutational approach to generate a distinct IL-2R superkine variance, which uses mutations to stabilize IL-2, including a flexible helix in the IL-2Rβ-binding site, into an optimized receptor-binding confirmation resembling that when bound to CD25 (IL-2Rα) (Levin et al. 2012). The evolved mutations in super-2 recapitulated the functional role of CD25 by eliciting potential phosphorylation of signal transducers and activators of transcription 5 (STAT5) and vigorous proliferation of T cells irrespective of CD25 expression. Compared to IL-2, super-IL-2 induced superior expansion of cytotoxic T cells, leading to improved antitumor responses in vivo and eliciting proportionally less expansion of Treg cells and reduced pulmonary edema. WT, Wild type.

INTERFERONS

IFNs were first identified in 1957 by Alick Isaacs and Jean Lindenmann (Isaacs and Lindenmann 2015). The IFNs are classified by their ability to bind to specific receptors for type I (IFN-α and -β), type II (IFN-γ), and more recently described type III IFNs.

Type I IFNs

Type I IFNs comprise a family of cytokines that are synthesized by a variety of cells in response to viral infection, immune stimulation, and certain chemical inducers. IFN-α and -β are encoded on chromosome 9 (Goldstein and Laszlo 1988). Twenty IFNs have been identified in humans. Most of the subtypes belong to the IFN-α group. There are two subtypes of β (I and II) and only one type II IFN-γ. The type I IFNs signal through a common pair of receptors, IFN-α receptors IFNAR1 and IFNAR2 (Constantinescu et al. 1994; Muller et al. 1994a,b). The IFN receptors principally signal through receptor-associated Tyk2 and JAK1 to initiate the multiple STAT1 and STAT2 phosphorylation cascades. Hundreds of genes associated with antiviral and antiproliferative functions are induced in response to different IFNs. Type I IFNs induce the expression of MHC class I molecules on tumor cells, mediate the maturation of a subset of dendritic cells (DCs), have antiangiogenic properties, activate B and T cells, increase cytotoxic cell numbers, and induce apoptosis of tumor cells.

Clinical Application of IFN-α

IFN-α was approved for the treatment of some hematological malignancies, AIDS-related Kaposi sarcoma, HCL, chronic myelogenous leukemia (CML), and adjuvant therapy for patients with high-risk stage II and stage III melanoma, and has been used in the treatment of renal cell cancer (Gutterman et al. 1980; Kirkwood and Ernstoff 1984; Chronic Myeloid Leukemia

Cite this article as *Cold Spring Harb Perspect Biol* doi: 10.1101/cshperspect.a028472

Trialists' Collaborative Group 1997; Windbichler et al. 2000; Tagliaferri et al. 2005; Lange et al. 2011). However, in many cases, IFN has been replaced or relegated to therapy in second-line treatment by the development of novel reagents or combinations (Goldman et al. 2003).

IFN-α treatment resulted in substantial and sustained improvement in granulocyte, platelet counts, and hemoglobin levels in 77% of treated patients with HCL and at least some improvement in minor responses occurred in 90% (Ratain et al. 1985). The percentage of patients with HCL who required blood cell or platelet transfusions decreased significantly during treatment. The median time to relapse estimated using the Kaplan–Meier method was earlier in an observation group compared to an IFN-α-treated group. Although IFN-α is approved for the treatment of patients with HCL, it has been relegated to second-line treatment because nucleoside analog drugs have replaced it as primary therapy.

AIDS-Related Kaposi Sarcoma

IFN-α was evaluated in clinical trials in 144 patients with AIDS-related Kaposi sarcoma. A dose of 30 million units/M^2 was administered subcutaneously three times a week. 44% of asymptomatic patients responded versus 7% of symptomatic patients. The median time to response was approximately 2 mo and 1 mo, respectively, for asymptomatic and symptomatic patients. The median duration of response was approximately 3 mo and 1 mo, respectively, for the two groups of patients.

Follicular Lymphoma

In a randomized control trial 130 patients received CHVP (cyclophosphamide, doxorubicin, teniposide, and prednisone) chemotherapy and 135 patients received CHVP therapy plus IFN-α at 5 million units/M^2 subcutaneously three times a week for the duration of 18 months. The group receiving a combination of IFN-α therapy and CHVP had a significantly longer progression-free survival than the CHVP-alone group of 2.9 years versus 1.5 years ($p < 001$).

Malignant Melanoma

One hundred and forty-three patients received IFN-α at 30 million IU/M^2 five times a week for 4 weeks, followed by 10 million IU/M^2 subcutaneously three times a week for 48 weeks (Kirkwood et al. 2001). IFN-α therapy was begun ≤ 56 days after surgical resection (Kirkwood et al. 1996). The remaining 137 patients were observed. IFN-α therapy produced a significant increase in the relapse-free and overall survival. Median time to relapse for the IFN-α-treated patients versus observation patients was 1.72 years versus 0.98 years ($p < 0.01$). In an analysis of multiple studies, there was a significant improvement in disease-free survival in 10 of 17 comparisons and overall survival in four of 14 comparisons. In a meta-analysis of seven randomized trials, IFN-α produced a statistically significantly better survival than those involving either hydroxyurea or busulfan. IFN-α produced variable results in patients with metastatic renal cell carcinoma (Amato 1999; Gollob et al. 2000; Flanigan et al. 2001). However, sunitinib produced longer progression-free survival, and response rates were higher in patients receiving sunitinib versus those receiving IFN-α. In addition to its use in malignancy, IFN-α was shown to be of value in condylomata acuminata.

Bazarbachi et al. (2010) reported that azidothymidine (AZT) and IFN-α plus or minus arsenic therapy was associated with a response in the majority of patients with human T-cell lymphotropic virus 1 (HTLV-1)-associated adult T-cell leukemia/lymphoma (ATLL). In a meta-analysis, 100% of patients with smoldering and chronic forms of this leukemia survived 5 years, whereas 82% of patients with acute leukemia survived 5 years when treated primarily with this combination. However, for patients with lymphomatous ATLL, those who received prior intensive chemotherapy and manifested mutations of p53 or high levels of IRF4 in the malignant cells did not respond to AZT IFN-γ therapy.

Attempts have been made to enhance IFN antiviral activity using DNA shuffling. For example, DNA sequences from all IFN-α subtypes were combined to generate a shuffled library that was screened for function based on antiproliferative and antiviral activity (Stemmer 1994; Brideau-Andersen et al. 2007; Spangler et al. 2015). This screening identified two shuffled proteins, B9X25 and B9X14, with a 20- to 70-fold improvement in antiviral potencies compared to IFN-α2. The shuffled proteins had 9- to 100-fold increases in IFNAR complex affinity, respectively, compared to wild-type IFN-α2. Unfortunately, this and other shuffled IFNs did not advance into the clinic because of immunogenicity resulting from the numerous mutations found in the shuffled IFN products, which generated many new potential T-cell epitopes.

Toxicity

There was considerable toxicity and morbidity associated with IFN use (Quesada et al. 1986; Zaidi and Merlino 2011). Acute symptoms with IFN administration involved flu-like constitutional symptoms, including fever, fatigue, headaches, gastrointestinal symptoms, and myalgias. IFN-α also produced increases in hepatic enzymes, particularly during high-dose i.v. administration. Thrombocytopenia, leukopenia, and neutropenia were common and can require a dose reduction. Quite serious were neuropsychiatric issues, including depression (45%), confusion (10%), and mania (less than 1%). IFN-α-induced depression was highly significant and in some cases suicides were reported. Diffuse electroencephalographic slowing has been shown. Thus, patients treated with high-dose IFN should be treated with antidepressants.

INTERLEUKIN-12

IL-12 is a four-bundle α-helix heterodimeric cytokine encoded by two separate genes: *IL-12A* (p35) and *IL-12B* (p40) (Trinchieri 1995, 2003). The active heterodimer referred to as p70 is formed following protein synthesis. The functional high-affinity IL-12R is comprised of the units IL-12Rβ1 and IL-12Rβ2. IL-12 p40 binds primarily to the IL-12Rβ1 subunit, whereas IL-12 p35 interacts with IL-12Rβ2 representing the signaling transducing component. IL-12 activates both innate (NK) and adaptive cytotoxic (T-lymphocyte) immunity and inhibits angio-genesis. IL-12 is involved in the differentiation of naïve T helper (Th)0 cells into Th1 cells and stimulates the production of IFN-γ from plasmocytoid DCs and T cells. IL-12 augments the activity of cytotoxic T cells and enhances B-cell survival. IL-12 induces the production of the chemokine-inducible protein-10 (IP-10 or CXCL10), which mediates its antiangiogenesis effect (Chan et al. 1992; Lee and Margolin 2011). Because of its ability to induce innate and adaptive immune responses and its antiangiogenic activity, there has been considerable interest in evaluating IL-12 as a potential anticancer drug (Lasek et al. 2014). Limitations in its use are its induction of negative immunoregulatory IL-10 and TIM3.

IL-12 showed efficacy in a series of preclinical studies in experimental models in mice. A series of clinical trials were initiated with IL-12 (Gollob et al. 2000). In a phase I trial, the maximum tolerated dose (MTD) was 500 ng/kg/d. In a phase II trial, severe side effects of treatment developed in 12 of 17 enrolled patients leading to the death of two patients (Lasek et al. 2014). This resulted in the immediate halting of all trials with IL-12 by the U.S. Food and Drug Administration (FDA). An explanation for the different tolerability in phase I versus the phase II trial was a change in the dosing schedule. In the phase I trial, a single dose of IL-12 was administered before the multidose regimen. This priming dose was found to be critical for protection from the severe toxicity. After several months of suspension, clinical trials were resumed in several centers. In small nonrandomized trials, IL-12 showed modest efficacy in cutaneous T-cell lymphoma (CTCL), Hodgkin and non-Hodgkin lymphoma, and Kaposi sarcoma. More recently, alternative strategies have been attempted, including intraperitoneal administration of an IL-12 plasmid, tumor injections of an IL-12-expressing adenovirus vector, the combination of IL-12 with a vaccine containing tumor cells fused with DCs, adoptive immunotherapy with IL-12-engineered lymphoid cells, as well as IL-12 in conjunction with an anticancer monoclonal antibody to increase its ADCC (Lasek et al. 2014). In some trials, IL-12 was used with anticytotoxic T-lymphocyte antigen 4 (CTLA-4) and anti-PD1 checkpoint inhibitors. The value of IL-12 in the treatment of cancer will await the results of these studies.

INTERLEUKIN-21

IL-21 is composed of four α-helical bundles. IL-21 is a member of the common γ IL-2, IL-4, IL-7, IL-9, IL-15, IL-21 family of cytokines and uses the common γc as well as its cytokine-specific IL-21Rα chain (Spolski and Leonard 2008). IL-21 regulates both innate and adaptive immune responses. IL-21 has a major role in B-cell differentiation into plasma cells and in the development of T-follicular helper (Tfh) cells. It induces a program of CD8$^+$ T cells that leads to enhanced survival, antiviral activity, and antitumor activity (Ma et al. 2003; Wang et al. 2003; Zeng et al. 2005; Cappuccio et al. 2006; Skak et al. 2008; Spolski and Leonard 2014). IL-21 has a major role in the development of Tfh cells and can promote the development of Th17 cells.

In light of its ability to enhance cytotoxic activity of CD8$^+$ T cells and NK cells, IL-21 has been evaluated in the treatment of cancer (Skak et al. 2008). In studies in mice, IL-21 inhibited the growth of melanomas and fibrosarcomas (Wang et al. 2003). IL-21 has been evaluated in phase I/II clinical trials as a single agent for melanoma, renal cell cancer, and metastatic colorectal cancer (Bhatia et al. 2014). IL-21 was combined with cetuximab (Erbitux), an antibody targeting epidermal growth factor receptor (EGFR) in targeting ADCC against tumors (Steele et al. 2012). In phase I trials, the combination of IL-21 and cetuximab against stage IV colorectal cancer stable disease was achieved in 60% of patients, but the clinical trial was terminated when IL-21 was shown to have a role in chronic inflammatory bowel disease and that IL-21 appears to have a major role in promoting the inflammation-induced development of colon cancer (Steele et al. 2012).

IL-21R has been shown to be expressed in diverse hematopoietic malignancies. In chronic lymphocytic leukemia (CLL), follicular lymphoma, diffuse large B-cell lymphoma (DLBCL), and mantle cell lymphoma, the apoptotic effects

of IL-21 make it a candidate for use as a single agent, as well as in combination with tumor-specific antibodies. IL-21 has been used in combination with the anti-CD20 rituximab in phase I clinical trials with clinical responses seen in eight of 19 patients (Timmerman et al. 2012). In contrast, the growth-promoting effects of IL-21 in multiple myeloma and Hodgkin lymphoma suggest that blocking the IL-21 signaling pathway using IL-21R-specific antibodies and an IL-21R-Fc fusion protein or via JAK inhibitors might be of therapeutic value in these hematopoietic malignancies (Spolski and Leonard 2014).

GRANULOCYTE MACROPHAGE COLONY-STIMULATING FACTOR

GM-CSF is a 23-kDa glycoprotein that has a four-α-helical bundle structure that binds to a heterodimeric receptor composed of subunits belonging to the type 1 cytokine receptor family. GM-CSF is the product of activated T lymphocytes, fibroblasts, endothelial cells, macrophages, and stromal cells. GM-CSF stimulates the survival of hematopoietic colony-forming cells of neutrophil, eosinophil, macrophage, megakaryocyte, and erythroid linages. GM-CSF stimulates antigen presentation to the immune system. It does this by its direct effects on DCs and macrophage production, with induction of the expression of the class II MHC and Fc receptors on macrophages and DCs. GM-CSF stimulates the capacity of neutrophils, monocytes, and macrophages to mediate ADCC. GM-CSF has been evaluated in the immunotherapy of melanoma (Mach and Dranoff 2000; Li et al. 2006; Kaufman et al. 2014). In a seminal study, a panel of recombinant retroviral vectors expressing various cytokines, costimulatory molecules, or adhesion molecules was used to infect murine B16 melanoma cells. Infected cells were irradiated and injected subcutaneously into immunocompetent hosts, followed by a subsequent challenge with wild-type B16 cells (Dranoff et al. 1993; Mach and Dranoff 2000). The GM-CSF-secreting tumor vaccines conveyed 90% protection, whereas in vaccines expressing IL-2, IFN-γ failed to mediate antitumor protection. Analysis of the vaccination site revealed an influx of dividing monocytes and granulocytes, as well as an increase in lymphocytes in tumor-draining lymph nodes, suggesting direct augmentation of antigen presentation and T-cell priming against the tumor. In light of the evidence of antitumor activity in preclinical models of melanoma, GM-CSF has been evaluated in completely resected stage III/IV melanoma patients; however, data from these studies have not shown consistent efficacy. GM-CSF was evaluated as intratumoral monotherapy with only modest effects—with only one partial response in two small studies but with reduced lesion size in six of seven patients and a reduction in cutaneous metastases in five of seven patients in another study. Given the evidence from preclinical studies, GM-CSF has been evaluated as an adjunct to cancer vaccines in a number of clinical studies (Dranoff 2002; Gupta and Emens 2010). In contrast to the data from murine studies, the adjuvant effect of GM-CSF in human trials was inconsistent. Two randomized perspective trials suggested that the addition of GM-CSF to melanoma vaccines did not improve cellular immune responses and indeed may have had negative effects. The inconsistent effects may be caused by contrasting effects of GM-CSF, inducing DC maturation on the one hand and induction of myeloid suppressor cells on the other (Kaufman et al. 2014).

INTERLEUKIN-15

IL-15 is a 14- to 15-kDa member of the four-α-helix bundle family of cytokines. IL-15 expression is controlled at the levels of transcription, translation, and intracellular trafficking. In particular, IL-15 is posttranslationally regulated by multiple controlling elements that impede translation, including 13 upstream AUGs of the 5′-UTR, two unusual signal peptides, and the carboxyl terminus of the mature protein (Bamford et al. 1998). The IL-15 receptor includes the γc subunit shared with IL-2, IL-4, IL-7, IL-15, and IL-21, and IL-2Rβ shared with IL-2 as well as an IL-15-specific subunit IL-15Rα. This receptor signals through the JAK1/3 and STAT5/STAT3 system.

IL-15 and IL-2 have several similar functions as a consequence of their sharing of receptor components IL-2/IL-15Rβ and γc and their use of common JAK1/3 STAT3/5 signaling molecules (Fehniger and Caligiuri 2001; Waldmann et al. 2001; Fehniger et al. 2002; Waldmann 2006, 2015; Steel et al. 2012). These functions include stimulating the proliferation of activated T cells and their differentiation into defined effector T-cell subsets following antigen-mediated activation. Furthermore, the two cytokines facilitate the production of CTLs and Ig synthesis by B cells that have been stimulated with IgM and specific antibodies or by agonistic anti-CD40. The two cytokines also stimulate the generation and proliferation of NK cells (Carson et al. 1994, 1997). In addition to these similarities, there are distinctions between the functions of IL-2 and IL-15 in the adaptive immune response. IL-2 acts as a T-cell growth factor during initiation of a murine response but it also has a crucial role in the termination of T-cell immune responses by AICD and by the action of Tregs (Sakaguchi et al. 1995; Leonardo 1996). In contrast with IL-2, IL-15 has no major net effect on the maintenance of the fitness of Foxp3 expressing T regs. IL-2 and IL-15 also have distinct roles in AICD. IL-2 is a critical determinant in the choice between proliferation and death (Leonardo 1996). In contrast, IL-15 is an antiapoptotic factor in several systems, in particular, in IL-15 transgenic mice, IL-2-induced AICD is inhibited (Marks-Konczalik et al. 2000). Furthermore, IL-15 promotes the maintenance of CD8$^+$ CD44^{hi} T cells memory phenotype cells.

These observations from ex vivo functional studies were supported by the analysis of mice with disrupted cytokine and cytokine-receptor genes. As noted above, IL-2-, IL-2Rα-, and IL-2/IL-15Rβ-deficient mice developed a marked enlargement of peripheral lymphoid organs that was associated with polyclonal expansions of T- and B-cell populations (Sadlack et al. 1994). A dysregulated proliferation reflects the impairment of Treg fitness in AICD. In contrast, mice that were deficient in IL-15 or its private receptor, IL-15Rα, did not develop lymphoid enlargement, increased serum Ig concentrations, or autoimmune disease (Kennedy et al.

2000). Instead, such mice had a marked reduction in the number of thymic and peripheral NK cells, NK T (NKT) cells, γ/δ T cells, and intestinal intraepithelial lymphocytes.

A most critical factor in the functional differences between IL-2 and IL-15 involves the fact that IL-2 is predominantly a secreted molecule, which in its soluble form or when linked to an extracellular matrix binds to preformed high-affinity heterotrimeric receptors at the surface of activated cells. In contrast, IL-15 is only secreted along with IL-15Rα in small quantities and is membrane-bound (Fig. 2) (Dubois et al. 2002). IL-15 induces signaling in the context of cell–cell contact at an immunological synapse. Stimulation of monocytes or DCs with IFN together with activation of NF-κB through ligation of CD40 or TLR4 induces the coordinate simultaneous expression of IL-15 and IL-15Rα. The IL-15 and IL-15Rα expressed by monocytes and DCs then become associated on the cell surface with IL-15 presented by IL-15Rα in *trans* to cells that express IL-2/IL-15Rβ and γc but not IL-15Rα (Dubois et al. 2002). Such targets of IL-15/IL-15Rα *trans*-presentation include NK cells and CD8 memory T cells. The distinctions between IL-15 and IL-2, including the fact that IL-15 does not yield a net stimulation of Tregs, AICD, or extensive capillary leak syndrome, suggests that IL-15 may be superior to IL-2 in the treatment of malignancy.

IL-15 AS AN IMMUNOTHERAPEUTIC AGENT: IL-15 IN PRECLINICAL IMMUNOTHERAPY MODELS

IL-15 proved of value in the therapy of neoplasia in a number of murine models (Munger et al. 1995; Evans et al. 1997; Fehniger and Caligiuri 2001; Fehniger et al. 2002; Klebanoff et al. 2004; Roychowdhury et al. 2004; Kobayashi et al. 2005; Waldmann 2006; Dubois et al. 2008; Bessard et al. 2009; Zhang et al. 2009; Steel et al. 2012; Yu et al. 2012; Zhang et al. 2012). In particular, whereas following i.v. administration of MC38 syngeneic colon carcinoma cells wild-type mice died of pulmonary metastases within 6 wk, IL-15 transgenic mice survived for more than 8 mo following infusions of the tumor cells.

Furthermore, Klebanoff and coworkers (2004) showed that IL-15 enhanced the in vivo activity of tumor-related CD8$^+$ T cells in the T-cell receptor transgenic mouse (pmel-1) whose CD8$^+$ T cells recognized an epitope derived from the melanoma antigen gp100. In our studies, IL-15 was shown to prolong the survival of mice with established DC26 and MC38 colon cancers and with the TRAMP-C2 prostatic cancer (Zhang et al. 2009, 2012). On the basis of animal and laboratory trials, great interest was generated among leading immunotherapeutic experts participating in the National Cancer Institute (NCI) Immunotherapy Agent Workshop 2007 that ranked IL-15 as the most promising unavailable immunotherapeutic agent to be brought to therapeutic trials.

TOXICITY, PHARMACOKINETICS, IMMUNOGENICITY, AND IMPACT ON ELEMENTS OF THE NORMAL IMMUNE SYSTEM OF RECOMBINANT IL-15 IN RHESUS MACAQUES

The safety of IL-15 was evaluated in rhesus macaques (Mueller et al. 2005; Berger et al. 2009; Lugli et al. 2010; Waldmann et al. 2011). Human IL-15 was administered to rhesus macaques by subcutaneous (s.c.) injection for up to 14 days. Daily administration of IL-15 for 14 days caused reversible severe neutropenia, anemia, weight loss, generalized skin rash, and a massive expansion of T cells. Recombinant human (rh)IL-15 (produced in *Escherichia coli*) produced by the Biopharmaceutical Development Program, NCI was administered at a dosing schedule of 12 daily i.v. bolus infusions at doses of 10, 20, and 50 mcg/kg/d to rhesus macaques (Lugli et al. 2010; Waldmann et al. 2011). The only biological meaningful laboratory abnormality was a grade 3/4 transient neutropenia. This neutropenia was shown to be secondary to a redistribution of neutrophils in that bone marrow examination showed increased marrow cellularity, including cells of the neutrophil series. A 12-d bolus of i.v. administration of 20 mcg/kg/d of IL-15 to rhesus macaques was associated with a four- to eightfold increase in the number of circulating NK, stem, central, and effector memory CD8 T cells. Subsequently, alternative routes of administration were evaluated, including continuous i.v. (c.i.v.) infusion and s.c. administration of IL-15 (Sneller et al. 2011). The administration of IL-15 by c.i.v. at 20 mcg/kg/d for 10 days led to an approximately 10-fold increase in the number of circulating NK cells and a massive 80- to 100-fold increase in the number of circulating effector memory CD8 T cells. s.c. infusions at 20 mcg/kg/d for 10 days led to a more modest 10-fold increase in the number of circulating effector memory CD8 T cells. No vascular leak syndrome, hemodynamic instability, or renal failure was observed in these studies.

CLINICAL TRIALS USING IL-15 IN THE TREATMENT OF CANCER

Five clinical trials were initiated using *E. coli* rhIL-15 in the treatment of cancer. The primary goal of our published trial: "A phase I study of recombinant IL-15 in adults with refractory metastatic malignant melanoma and metastatic renal cancer" was to determine the safety, adverse event profile, dose-limiting toxicity (DLT), and MTD of IL-15 administered as a daily bolus infusion for 12 days (Conlon et al. 2015). The initial patient receiving 3 µg/kg/d developed grade 3 hypotension and another patient developed grade 3 thrombocytopenia. Therefore, the protocol was amended to add lower doses of 1.0 and 0.3 µg/kg/d. Two of the four patients given the 1.0 µg/kg/d had persistent grade 3 alanine aminotransferase and aspartate aminotransferase elevations that were dose limiting. All nine patients with IL-15 administered at 0.3 µg/kg/d received all 12 doses without DLT. Thus, the MTD of IL-15 was determined to be 0.3 µg/kg/d. Posttreatment in patients given 3 µg/kg/d doses of IL-15 resulted in fever and rigors beginning 2.5 to 4 h after the start of IL-15 infusions and blood pressure drops to a nadir of ~20 mm/Hg below pretreatment levels 5 to 9 h after the infusion. These changes were concurrent with a maximum of 50-fold elevations of circulating IL-6 and IFN-γ concentrations. Flow cytometry of peripheral blood lymphocytes revealed a dramatic efflux of NK and effector memory T cells

from the circulating blood within minutes of IL-15 administration followed by influx and hyperproliferation, leading to a 10-fold expansion of NK and γ/δ T cells that ultimately return to baseline. Furthermore, there were significant increases in the number of CD8 memory phenotype T cells. In this first-in-human phase I trial there were no responses, with stable disease as the best response. However, five patients manifested a decrease between 10% and 30% of their marker lesions and two patients had clearing of lung lesions. Subsequently, alternative dosing strategies were used, including c.i.v. and s.c. infusions, so that a lower C_{max} would be achieved in trials that have been completed (Waldmann 2015).

When rhIL-15 was administered subcutaneously Monday–Friday for 2 weeks for a given dose, the C_{max} was approximately 10-fold less than that observed with bolus infusion. In general, the administration was well tolerated with the MTD of 3.0 mcg/kg/d, which was 10-fold greater than that observed with bolus infusions (unpubl.). Among 22 patients, there was one serious adverse event at 2.0 mcg/kg/d: grade 2 pancreatitis in a patient with metastatic melanoma 3 days after completing cycle one. The mean fold increase in the circulating NK cells was 11.8-fold in the cohort receiving a 3.0 mcg/kg dose. In contrast, the mean fold increase in the circulating CD8 T cells was relatively modest with 3.2-fold for the 3.0 mcg/kg dose cohort.

In patients receiving rhIL-15 by continuous i.v. infusion for 10 days, the MTD was 2.0 mcg/kg (K Conlon and TA Waldmann, unpubl.). The patients manifested a marked reduction in numbers of circulating NK cell numbers early during the infusion followed by a return to normal and 5- to 10-fold increased levels. Interestingly, in addition, they showed a phenomenon characterized by a major burst in the number of NK cells that occurred 1 to 2 days following cessation of the 10-day i.v. infusion with a 30-fold increase in the circulating total NK numbers and more than a 350-fold increase in the number of CD56[bright] NK cells. In a correlative study, the latter cells were shown to be effective as cytotoxic agents in diverse assays, including ADCC (S Dubois and TA Waldmann, unpubl.). These latter results support the use of continuous i.v. infusion of IL-15 with anticancer monoclonal antibodies to augment their ADCC and efficacy as anticancer agents (see below).

IL-15 has been engineered to exert both increased agonistic and inhibitory effects. Mortier and colleagues (2006) designed a truncated version of IL-15Rα extracellular domain fused to IL-15 that activated IL-15Rα-deficient cells by stabilizing the signaling-complex formation analogous to the effects of IL-2 superkine on cells that lack IL-2Rα. The IL-15 fusion generated by Mortier and colleagues enhanced the proliferative and antiapoptotic effects of IL-15. Zhu et al. (2009) generated IL-15 agonists by increasing the affinity of the cytokine for IL-2Rβ, thereby impacting interface stability. Another focus of IL-15 engineering has been to produce antagonists to counteract its immunostimulatory activity. Pettit and colleagues (1997) identified the Q108 residue of IL-15 critical for γc interaction, so that when this residue is deleted, one thereby abrogates cytokine-mediated proliferation. Thus, engineered examples of IL-2 and IL-15 could have therapeutic utility in many aspects of immune regulation.

IL-15/IL-15Rα

Although IL-15 may show efficacy in the treatment of patients with metastatic malignancy, it is not optimal when used in monotherapy for cancer. A particular challenge is that there is only low-level expression of IL-15Rα on resting DCs (Chen et al. 2012a). Physiologically, IL-15 is produced as a heterodimer in association with IL-15Rα following stimulation of antigen-presenting cells with IFN, agonistic anti-CD40, or Toll-like receptor signaling (Schluns et al. 2004; Mortier et al. 2008). In mice, it is only the heterodimer that is stably produced and transported to the surface of the cell. On cleavage from the cell surface, minor quantities of IL-15Rα/IL-15 were associated in the serum as the sole form of circulating IL-15. To address the issue of deficient IL-15Rα, both IL-15/IL-15Rα and IL-5Rα IgFc had been produced and entered into clinical trials that evaluated patients with metastatic malignancy (Tinhofer et al.

2000; Chertova et al. 2013; Pharmaceutical Business Review 2013).

It is clear from the results of clinical trials with IL-15 or IL-15/IL-15Rα that to make a major impact for cancer therapy, IL-15 must be administered in combination therapy where it optimizes the action of agents that already have an action, albeit inadequate, in the treatment of cancer.

AGENTS TO RELIEVE CHECKPOINTS ON THE IMMUNE SYSTEM TO OPTIMIZE IL-15

IL-15 is associated with the expression of immunological checkpoints, including IL-10, TIGIT, and the expression of PD1 on CD8 T cells (Yu et al. 2010). In addition, IL-15 was shown to be critical in the maintenance of CD122[+] CD8[+] Tregs. To address the issue of induced checkpoints, IL-15 was administered with agents to remove such checkpoints with antibodies directed toward CTLA-4 and programmed death ligand-1 (PD-L1) (Yu et al. 2010, 2012). In the CT26, MC38 colon carcinoma, and TRAMP-C2 prostatic cancer syngeneic tumor models, IL-15 alone provided only modest antitumor activity. The addition of either anti-CTLA-4 or anti-PD-L1 alone in association with IL-15 did not augment the action of IL-15. However, tumor-bearing mice receiving IL-15 in combination with both anticheckpoint antibodies together manifested a marked prolongation of survival.

COMBINATION THERAPY USING IL-15 WITH ANTICANCER MONOCLONAL ANTIBODIES TO AUGMENT THEIR ANTIBODY-DEPENDENT CELLULAR CYTOTOXICITY

The predominant approaches involving cytokines when used alone are based on the hypothesis that the host is making an immune response albeit inadequate to their tumor, and this action can be augmented by the administration of the cytokine. However, these cytokines could also be used in drug combinations where an additional coadministered drug provides specificity directed toward the tumor. In particular, IL-15 could be used with anticancer vaccines, cellular therapy, or with tumor-directed monoclonal antibodies (Sondel and Hank 1997; Moga et al. 2008; Roberti et al. 2011). Given the capacity of IL-15 to dramatically increase the number and activation state of NK cells and monocytes, a very attractive antitumor combination strategy would be to use the optimal IL-15 agent in conjunction with an antitumor monoclonal antibody to augment its ADCC. Vincent and colleagues (2014) reported highly potent anti-CD20-RIL-I immunocytokine targeting of established human B-cell lymphomas in severe combined immune deficiency (SCID) mice. Clinical trials have been initiated to evaluate this strategy. Rituximab in association with IL-15 is being used in the treatment of relapsed and refractory follicular lymphoma. Furthermore, we have initiated a clinical trial using alemtuzumab (CAMPATH-1/anti-CD52) along with rhIL-15 in the treatment of patients with HTLV-1-associated adult T-cell leukemia (ATL), a disorder in which alemtuzumab alone has provided responses.

IL-15 PLUS AGONISTIC ANTI-CD40 ANTIBODY

It has been shown that the administration of γ cytokines leads to the induction of intracellular checkpoints, including cytokine-inducible SH2-containing protein (CIS) and SOCS-3 (Sckisel et al. 2015). This expression of SOCS3 in turn leads to inadequate CD4 help and the induction of "helpless" CD8 T cells. In a series of studies, it was shown that an agonistic anti-CD40 or CD40 ligand can substitute for inadequate CD4 cells, thereby leading to the generation of antigen-specific CD8 cytotoxic T cells (Bennett et al. 1998; Ridge et al. 1998; Schoenberger et al. 1998; Sckisel et al. 2015). In the murine syngeneic TRAMP-C2 tumor model, we showed that either IL-15 alone or with an agonistic anti-CD40 antibody (FGK 4.5) prolonged the survival of the TRAMP-C2 tumor-bearing mice (Zhang et al. 2009, 2012). Moreover, we showed that the combination of IL-15 with anti-CD40 produced markedly additive effects that were curative in the majority of mice when compared

to either agent alone. This combination appeared to circumvent the problem of "helpless" CD8 T cells, whereas administration of IL-15 or anti-CD40 alone did not augment the number of tumor-specific tetramer-positive CD8 T cells in the TRAMP-C2 model system; administration of the combination of IL-15 plus the agonistic anti-CD40 antibody was associated with a meaningful increase in the number of TRAMP-C2 tumor-specific SPAS-1/SNC9-H8 tetramer-positive CD8 T cells. It is hoped that with the diverse approaches discussed IL-15 will take a place in the combination treatment of cancer (Zhang et al. 2009, 2012).

CYTOKINE DISORDERS ASSOCIATED WITH MALIGNANCY

The studies that have just been discussed focus on the use of cytokines to increase patient immune responses to their tumor. Other efforts have the opposite goal of diminishing cytokine action in situations in which disorders of their expression play a pathogenic role in the malignancy. The γ cytokines and their signaling pathway appear frequently disordered in lymphoid malignancy. In the case of T-cell malignancy, abnormal activation of the γ-cytokine receptor JAK/STAT system in the presence and absence of mutations of these proteins was shown to be pervasive in a proportion of patients with diverse T-cell malignancies as assessed by pSTAT3 and pSTAT5 phosphorylation and nuclear translocation (Chen et al. 2012b; Kucuk et al. 2015; TA Waldmann, unpubl). Activating mutations predominantly of the SH2 domains of STAT3 and STAT5 and the pseudokinase domains of JAK1 and JAK3 were described in some but not all cases with activation of the signaling pathway. However, activating JAK and STAT mutations were not sufficient to initiate leukemic cell proliferation but rather only augmented signals from upstream in a cytokine–cytokine receptor pathway. Even with mutations of JAK and STAT, activation required the full pathway, including the cytokine and cytokine receptor acting as a scaffold and docking site for the required downstream JAK/STAT elements.

The fundamental insight that disorders of the γc/JAK/STAT system are pervasive in T-cell malignancy suggests novel therapeutic approaches discussed below. In addition to activating mutations of the receptor and JAK/STAT system, certain T-cell malignancies were shown to have an increased expression of select γc cytokines that led to augmented STAT/JAK signaling. Such increases in γ cytokine production were observed in HTLV-1-associated ATL (Tendler et al. 1990; Migone et al. 1995; Chen et al. 2008; Ju et al. 2011). ATL is an aggressive T-cell lymphoproliferative disorder characterized by the presence of malignant CD4, CD25-expressing cells in the peripheral blood and in lymphoid and other tissues. Epidemiological studies showed clear association of disease with the presence of the retrovirus HTLV-1. The retrovirus HTLV-1 encodes a 40-kDa protein Tax that transactivates two autocrine (IL-2/IL-2Rα, IL-15/IL-15Rα) and one paracrine (IL-9) γ cytokine. Associated with these two autocrine and paracrine pathways, the ex vivo leukemic cells proliferate spontaneously. These cytokine–cytokine receptor loops lead to activation of the JAK1, JAK3, and STAT5 signaling pathway, and the associated ex vivo spontaneous proliferation that was inhibited by tofacitinib.

IL-15 has been reported to play a role in CTCL (Dobbeling et al. 1998; Leroy et al. 2001). IL-15 expression was shown in CTCL tumor tissues, and IL-15 was stimulatory for CTCL cells in vitro. Analysis ex vivo of Sézary cells and CTCL cell lines by reverse transcription polymerase chain reaction (RT-PCR) indicated that these cells expressed IL-15 in messenger RNA (mRNA). In CTCL cells, there was down-regulation of ZEB1, a candidate tumor suppressor that normally inhibits IL-2 and IL-15 expression (Nakahata et al. 2010). Furthermore, there was a report of hypermethylation of the ZEB1-binding site in the promoter of IL-15 (Mishra et al. 2015). Thus, IL-15 appears to be a growth and viability factor for CTCL cell lines, and may play an important role in CTCL biology.

Gene-expression analysis of human angioimmunoblastic T-cell lymphoma (AITL), a malignancy of Tfh cells, showed that essentially all cases expressed elevated levels of transcripts for IL-21, IL-21R, and a series of genes associated with Tfh cell development and function (Jain et al. 2015).

THE CYTOKINE–CYTOKINE RECEPTOR AND JAK/STAT SYSTEM AS TARGETS FOR TREATMENT OF MALIGNANCY

As just noted, disorders of the γ cytokine, JAK/STAT signaling pathway are pervasive in T-cell malignancy, suggesting new molecular targets and novel therapeutic opportunities that may revolutionize the treatment of these tumors, which are usually associated with a very poor prognosis. Such therapeutic approaches include targeting the γc cytokines directly, agents that block cytokine receptor interactions, and JAK kinase inhibitors (Fig. 4). Such diverse approaches include the use of neutralizing antibodies to γc-receptor-dependent cytokines (IL-2, IL-15, and IL-21 antibodies), blocking antibodies to their receptors, small molecule inhibitors interdicting cytokine–cytokine receptor interaction, and JAK kinase inhibitors (Ferrari-Lacraz et al. 2001, 2004). Antibodies directed to IL-15 have been evaluated in rheumatoid arthritis and may be of value in the treatment of CTCL in which disorders of IL-15 have been identified. The specific α chains of the γc cytokines and IL-2/IL-15Rβ shared by IL-2 and IL-15 have been shown to be valuable targets in the treatment of select T-cell malignancies (Vincenti et al. 1998; Waldmann et al. 2001, 2007; Berkowitz et al. 2014). Monoclonal antibodies directed toward the specific IL-2 receptor subunit IL-2Rα have been approved by the U.S. FDA for use in the prevention of organ transplant rejection and in the treatment of autoimmunity. Furthermore, anti-IL-2Rα (daclizumab) anti-Tac (Zenapax/Zinbryta) has been shown to be of value in the treatment of patients with smoldering and chronic ATL (Berkowitz et al. 2014). The rationale for this target was that IL-2Rα was not expressed by most normal resting cells with the exception of Tregs but was expressed by various T-cell leukemias, including ATL. An antibody to IL-2/IL-15Rβ, Hu-Mikβ1 has been used to inhibit IL-15 action in patients with large granular lymphocytic leukemia (LGL) (Waldmann et al. 2013).

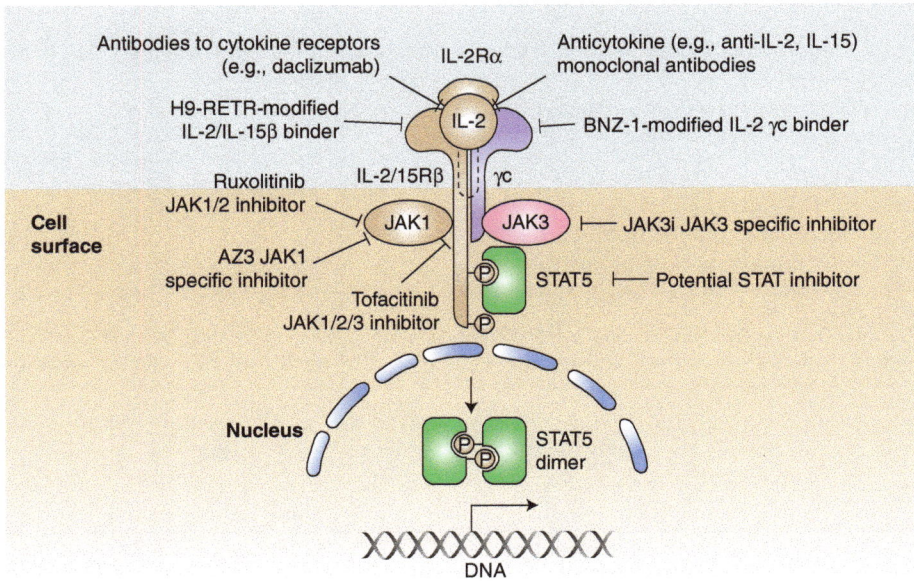

Figure 4. A series of approaches has been developed to block γ chain (γc) cytokine, Janus kinase/signal transducers and activators of transcription (JAK/STAT) signaling in T-cell malignancies and autoimmune disorders. Such therapeutic approaches include targeting the γc cytokines directly, agents that block cytokine receptor interactions, and JAK kinase inhibitors. H9-RETR, the RETR mutant of H9 super-IL-2. (This figure was previously published as Figure 4 in Waldmann et al. 2017. It is now reproduced with permission.)

Recently, two modifications of IL-2 have been generated that block binding of normal IL-2 and IL-15 to IL-2/IL-15Rβ and γc, thereby simultaneously inhibiting the actions of both IL-2 and IL-15 (Fig. 5). BNZ-1 binds only to the γc but not IL-2/IL-15Rβ (Nata et al. 2015). In parallel, the RETR mutant of H9 super-IL-2 (denoted H9-RETR) binds tightly to IL-2/IL-15Rβ but not γc (Mitra et al. 2015). Both agents prevent the heterodimerization of IL-2Rβ with γc that is required for signaling. These agents are being evaluated in murine xenograft models of ATL.

As noted above, a subset of patients with virtually all forms of T-cell malignancies require the activation of JAKs for malignant T-cell proliferation, thus providing the scientific basis for the use of JAK inhibitors in the treatment of patients with T-cell malignancy (Koskela et al. 2012; Crescenzo et al. 2015). The JAK1/2 inhibitor ruxolitinib and the JAK1/2/3 inhibitor tofacitinib diminished the proliferation of malignant T-cell lines and of ex vivo malignant leukemic cells supporting the use of such inhib-

itors (Leonard and O'Shea 1998; Ju et al. 2011; O'Shea et al. 2013). A clinical trial using the JAK1/2 inhibitor ruxolitinib is underway in patients with smoldering and chronic ATL wherein HTLV-1 Tax transactivates IL-2, IL-15, and IL-9 systems (Chen et al. 2008; K Conlon and TA Waldmann, unpubl.). With rare exceptions, T-cell malignancies are associated with activation of γc cytokine, JAK1/3, STAT3/5 signaling but not activation of JAK2. However, both tofacitinib and ruxolitinib inhibit JAK2 in addition to the desired JAK1 and JAK3. This off-target JAK2 inhibition interferes with the signaling mediated by thrombopoietin, erythropoietin IL-3, IL-5, and GM-CSF, leading to thrombocytopenia, anemia, and neutropenia, which in turn are associated with an increased incidence of infections, including herpes zoster and tuberculosis. In light of the toxicities caused by JAK2 inhibition, agents with greater JAK specificity are being developed. Filgotinib, which preferentially inhibits JAK1 is in the clinic. However, JAK1 activates and inhibits cytokines, including

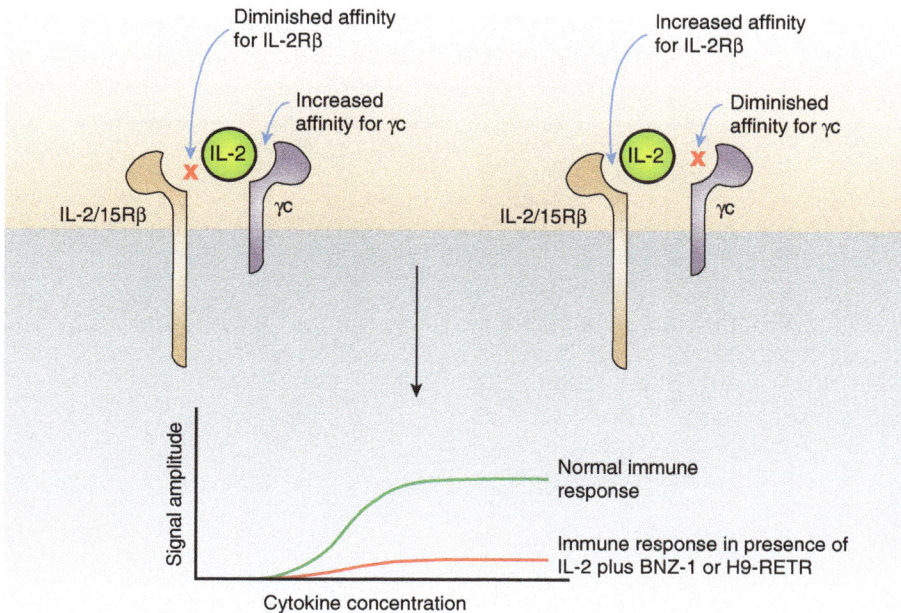

Figure 5. Modified interleukin (IL)-2 to simultaneously inhibit IL-2 and IL-15 action. Two modifications of IL-2 have been generated that block binding of normal IL-2 and IL-15 to IL-2/IL-15Rβ and γ chain (γc), thereby simultaneously inhibiting the actions of both IL-2 and IL-15. BNZ-1 (*left*) binds only to the γc but not IL-2/IL-15Rβ (Nata et al. 2015). In parallel, the RETR mutant of H9 superkine IL-2 (H9-RETR, *right*) binds tightly to IL-2/IL-15Rβ but not γc (Mitra et al. 2015). Both agents prevent the heterodimerization of IL-2Rβ with γc, which is required for signaling.

IFNs, in addition to γc cytokines. JAK3 is an alternative attractive target because its expression is restricted to cells of the hematopoietic lineage and is exclusively associated with the common γ chain. Mutations of JAK3 in humans or mice resulted in abnormalities restricted to SCID (Noguchi et al. 1993; Macchi et al. 1995; Russell et al. 1995; Liao et al. 2011b).

Nevertheless, Haan and coworkers (2011) have reported that JAK1 has a dominant role over JAK3 in signal transduction through γc-containing cytokine receptors and conclude that a selective adenosine triphosphate (ATP) competitive JAK3 kinase inhibitor would not be effective as a therapeutic agent. In contrast, Smith et al. (2016) showed that a selective covalent inhibitor of JAK3 (JAK3i) blocked IL-2-stimulated proliferation with great selectivity. This observation reflected the temporal disassociation of

IL-2 signaling, whereas JAK3i revealed a basic biphasic role for JAK3 catalytic activity in CD4[+] T cells. The JAK3i blocked a second wave of IL-2-mediated signaling. Thus, the results of Smith et al. contradict the conclusions of Haan et al. and reveal the preferential requirement for JAK3 kinase activity in a second wave of IL-2-mediated signaling, indicating that it would be a very attractive target in the treatment of T-cell malignancy.

CONCLUSIONS AND FUTURE DIRECTIONS

Cytokines play pivotal roles in the actions of the innate and adaptive immune system that normally are directed toward eliminating invading pathogens and the prevention of the development of neoplasia. Cytokines acting in autocrine and paracrine fashion regulate the immune system and play important roles in the control of this system. Currently, IFN-α, IFN-β, IL-2, and GM-GSF are approved for various clinical indications, including anticancer treatment, and others, such as IL-12, IL-15, and IL-21, are undergoing clinical evaluation. Despite efforts to develop systemic monotherapy with cytokines for cancer, this approach has several limitations that must be overcome for cytokines to play a dominant role in the immunotherapy of cancer. There is a failure to achieve adequate concentrations of cytokines at the immune cells in the tumor bed. To address this limitation, antibody cytokine fusion proteins have been generated to deliver the cytokine primarily to the tumor bed. Also, the cytokines often elicit immune checkpoints; for example, IL-2 stimulates the survival of Tregs. One strategy that has been used is genetic engineering to modify the IL-2 molecule so that it binds tightly to the β and γ chain and does not require the presence of the α chain that is expressed by Tregs. An alternative strategy might be to use IL-2 with an antibody to CCR4 (e.g., mogamulizumab) that is expressed predominantly by Tregs among normal cells. A limitation of GM-CSF is the elicitation of granulocyte-suppressive cells. Most of the cytokines, including γ cytokines, rapidly elicit the intracellular expression of one of the eight SOCS family of proteins that dephosphorylate JAKs and STATs (Sckisel et al. 2015). This yields inadequate helper CD4 cells yielding "helpless" antigen-nonspecific CD8 T cells. In the case of IL-15, this is being addressed by the administration of IL-15 with anti-CD40 that eliminates the requirement for CD4 helper cells (Sckisel et al. 2015). Cytokines such as IL-2 and especially IL-15 dramatically augment the number and state of activation of NK cells. However, the action of these cells is aborted by the interaction of an inhibitory KIR and NKG2A receptor with self-class 1 MHC and A or B and E, respectively, which is expressed on the tumor cells. To address this impediment, cytokines, including IL-15, are being coadministered with antitumor monoclonal antibodies to augment the NK-mediated ADCC of the anticancer monoclonal antibodies.

It is hoped that with novel combination approaches, cytokines will ultimately play a major role in cancer immunotherapy.

ACKNOWLEDGMENTS

This work is supported by the Intramural Research Program of the National Cancer Institute, Center for Cancer Research, National Institutes of Health.

REFERENCES

Alexander WS, Hilton DJ. 2004. The role of suppressors of cytokine signaling (SOCS) proteins in regulation of the immune response. *Annu Rev Immunol* **22:** 503–529.

Amato R. 1999. Modest effect of interferon α on metastatic renal-cell carcinoma. *Lancet* **353:** 6–7.

Ardolino M, Hsu J, Raulet DH. 2015. Cytokine treatment in cancer immunotherapy. *Oncotarget* **6:** 19346–19347.

Atkins MB, Lotze MT, Dutcher JP, Fisher RI, Weiss G, Margolin K, Abrams J, Sznol M, Parkinson D, Hawkins M, et al. 1999. High-dose recombinant interleukin-2 therapy for patients with metastatic melanoma: Analysis of 270 patients treated between 1985 and 1993. *J Clin Oncol* **17:** 2105–2116.

Bamford RN, DeFilippis AP, Azimi N, Kurys G, Waldmann TA. 1998. The 5′ untranslated region, signal peptide and the coding sequence of the carboxyl terminus of IL-15 participate in its multifaceted translation control. *J Immunol* **160:** 4418–4426.

Bazarbachi A, Plumelle Y, Ramos JC, Tortevoye P, Otrock Z, Taylor G, Gessain A, Harrington W, Panelatti G, Hermine O. 2010. Meta-analysis on the use of zidovudine and interferon-α in adult T-cell leukemia/lymphoma showing improved survival in the leukemic subtypes. *J Clin Oncol* **28:** 4177–4183.

Becker JC, Varki N, Gillies SD, Furukawa K, Reisfeld RA. 1996. An antibody-interleukin 2 fusion protein overcomes tumor heterogeneity by induction of a cellular immune response. *Proc Natl Acad Sci* **93:** 7826–7831.

Bennett SRM, Carbone FR, Karamalis F, Flavell RA, Miller JFAP, Heath WR. 1998. Help for cytotoxic-T-cell responses is mediated by CD40 signalling. *Nature* **393:** 478–480.

Berger C, Berger M, Hackman RC, Gough M, Elliott C, Jensen MC, Riddell SR. 2009. Safety and immunologic effects of IL-15 administration in nonhuman primates. *Blood* **114:** 2417–2426.

Berkowitz JL, Janik JE, Stewart DM, Jaffe ES, Stetler-Stevenson M, Shih JH, Fleisher TA, Turner M, Urquhart NE, Wharfe GH, et al. 2014. Safety, efficacy, and pharmacokinetics/pharmacodynamics of daclizumab (anti-CD25) in patients with adult T-cell leukemia/lymphoma. *Clin Immunol* **155:** 176–187.

Bessard A, Sole V, Bouchaud G, Quemener A, Jacques Y. 2009. High antitumor activity of RLI, an interleukin-15 (IL-15)-IL-15 receptor α fusion protein, in metastatic melanoma and colorectal cancer. *Mol Cancer Ther* **8:** 2736–2745.

Bhatia S, Curti B, Ernstoff MS, Gordon M, Heath EI, Miller WH Jr, Puzanov I, Quinn DI, Flaig TW, VanVeldhuizen P, et al. 2014. Recombinant interleukin-21 plus sorafenib for metastatic renal cell carcinoma: A phase 1/2 study. *J Immunother Cancer* doi: 10.1186/2051-1426-2-2.

Boder ET. 2012. Protein engineering: Tighter ties that bind. *Nature* **484:** 463–464.

Boyman O, Sprent J. 2012. The role of interleukin-2 during homeostasis and activation of the immune system. *Nat Rev Immunol* **12:** 180–190.

Boyman O, Kovar M, Rubinstein MP, Surh CD, Sprent J. 2006. Selective stimulation of T cell subsets with antibody-cytokine immune complexes. *Science* **311:** 1924–1927.

Brideau-Andersen AD, Huang XJ, Sun SCC, Chen TT, Stark D, Sas IJ, Zadik L, Dawes GN, Guptill DR, McCord R, et al. 2007. Directed evolution of gene-shuffled IFN-α molecules with activity profiles tailored for treatment of chronic viral diseases. *Proc Natl Acad Sci* **104:** 8269–8274.

Caligiuri MA, Murray C, Robertson MJ, Wang E, Cochran K, Cameron C, Schow P, Ross ME, Klumpp TR, Soiffer RJ, et al. 1993. Selective modulation of human natural killer cells in vivo after prolonged infusion of low dose recombinant interleukin 2. *J Clin Invest* **91:** 123–132.

Cappuccio A, Elishmereni M, Agur Z. 2006. Cancer immunotherapy by interleukin-21: Potential treatment strategies evaluated in a mathematical model. *Cancer Res* **66:** 7293–7300.

Carson WE, Giri JG, Lindemann MJ, Linett ML, Ahdieh M, Paxton R, Anderson D, Eisenmann J, Grabstein K, Caligiuri MA. 1994. Interleukin (IL)-15 is a novel cytokine that activates human natural-killer-cells via components of the IL-2 receptor. *J Exp Med* **180:** 1395–1403.

Carson WE, Fehniger TA, Haldar S, Eckhert K, Lindemann MJ, Lai CF, Croce CM, Baumann H, Caligiuri MA. 1997. Potential role for interleukin-15 in the regulation of human natural killer cell survival. *J Clin Invest* **99:** 937–943.

Carter P. 2001. Improving the efficacy of antibody-based cancer therapies. *Nat Rev Cancer* **1:** 118–129.

Chan SH, Kobayahi M, Santoli D, Perussia B, Trinchieri G. 1992. Mechanisms of IFN-γ induction by natural killer cell stimulatory factor (NKSF/IL-12). Role of transcription and mRNA stability in the synergistic interaction between NKSF and IL-2. *J Immunol* **148:** 92–98.

Chen J, Petrus M, Bryant BR, Nguyen VP, Stamer M, Goldman CK, Bamford R, Morris JC, Janik JE, Waldmann TA. 2008. Induction of the IL-9 gene by HTLV-1 Tax stimulates the spontaneous proliferation of primary adult T-cell leukemia cells by a paracrine mechanism. *Blood* **111:** 5163–5172.

Chen J, Petrus M, Bamford R, Shih JH, Morris JC, Janik JE, Waldmann TA. 2012a. Increased serum soluble IL-15R α levels in T-cell large granular lymphocyte leukemia. *Blood* **119:** 137–143.

Chen E, Staudt LM, Green AR. 2012b. Janus kinase deregulation in leukemia and lymphoma. *Immunity* **36:** 529–541.

Chertova E, Bergamaschi C, Chertov O, Sowder R, Bear J, Roser JD, Beach RK, Lifson JD, Felber BK, Pavlakis GN. 2013. Characterization and favorable in vivo properties of heterodimeric soluble IL-15·IL-15Rα cytokine compared to IL-15 monomer. *J Biol Chem* **288:** 18093–18103.

Chronic Myeloid Leukemia Trialists' Collaborative Group. 1997. Interferon α versus chemotherapy for chronic myeloid leukemia: A meta-analysis of seven randomized trials. *J Natl Cancer Inst* **89:** 1616–1620.

Conlon KC, Lugli E, Welles HC, Rosenberg SA, Fojo AT, Morris JC, Fleisher TA, Dubois SP, Perera LP, Stewart DM, et al. 2015. Redistribution, hyperproliferation, activation of natural killer cells and CD8 T cells, and cytokine production during first-in-human clinical trial of recombinant human interleukin-15 in patients with cancer. *J Clin Oncol* **33:** 74–82.

Constantinescu SN, Croze E, Wang C, Murti A, Basu L, Mullersman JE, Pfeffer LM. 1994. Role of interferon α/β receptor chain 1 in the structure and transmembrane signaling of the interferon α/β receptor complex. *Proc Natl Acad Sci* **91:** 9602–9606.

Crescenzo R, Abate F, Lasorsa E, Tabbo F, Gaudiano M, Chiesa N, Di Giacomo F, Spaccarotella E, Barbarossa L, Ercole E, et al. 2015. Convergent mutations and kinase fusions lead to oncogenic STAT3 activation in anaplastic large cell lymphoma. *Cancer Cell* **27:** 516–532.

Damjanovich S, Bene L, Matko J, Alileche A, Goldman CK, Sharrow S, Waldmann TA. 1997. Preassembaly of interleukin 2 (IL-2) receptor subunits on resting Kit 225 K6 T cells and their modulation by IL-2, IL-7, and IL-15: A fluorescence resonance energy transfer study. *Proc Natl Acad Sci* **94:** 13134–13139.

Dobbeling U, Dummer R, Laine E, Potoczna N, Gin JZ, Burg G. 1998. Interleukin-15 is an autocrine/paracrine viability factor for cutaneous T-cell lymphoma cells. *Blood* **92:** 252–258.

Dranoff G. 2002. GM-CSF-based cancer vaccines. *Immunol Rev* **188:** 147–154.

Dranoff G. 2004. Cytokines in cancer pathogenesis and cancer therapy. *Nat Rev Cancer* **4:** 11–22.

Dranoff G, Jaffee E, Lazenby A, Golumbek P, Levitsky H, Brose K, Jackson V, Hamada H, Pardoll D, Mulligan RC. 1993. Vaccination with irradiated tumor cells engineered to secrete murine granulocyte-macrophage colony-stimulating factor stimulates potent, specific, and long-lasting anti-tumor immunity. *Proc Natl Acad Sci* **90:** 3539–3543.

Dubois S, Mariner J, Waldmann TA, Tagaya Y. 2002. IL-15Rα recycles and presents IL-15 in *trans* to neighboring cells. *Immunity* **17:** 537–547.

Dubois S, Patel HJ, Zhang M, Waldmann TA, Muller JR. 2008. Preassociation of IL-15 with IL-15Rα-IgG1-Fc enhances its activity on proliferation of NK and CD8+/CD44high T cells and its antitumor action. *J Immunol* **180:** 2099–2106.

Dutcher JP, Schwartzentruber DJ, Kaufman HL, Agarwala SS, Tarhini AA, Lowder JN, Atkins MB. 2014. High dose interleukin-2 (Aldesleukin): Expert consensus on best management practices-2014. *J ImmunoTherapy Cancer* **2:** 26.

Evans R, Fuller JA, Christianson G, Krupke DM, Trout AB. 1997. IL-15 mediates anti-tumor effects after cyclophosphamide injection of tumor-bearing mice and enhances adoptive immunotherapy: The potential role of NK cell subpopulations. *Cell Immunol* **179:** 66–73.

Fehniger TA, Caligiuri MA. 2001. Interleukin-15: Biology and relevance to human disease. *Blood* **97:** 14–32.

Fehniger TA, Bluman EM, Porter MM, Mrózek E, Cooper MA, VanDeusen JB, Frankel SR, Stock W, Caligiuri MA. 2000. Potential mechanisms of human natural killer cell expansion in vivo during low-dose IL-2 therapy. *J Clin Invest* **106:** 117–124.

Fehniger TA, Cooper MA, Caligiuri MA. 2002. Interleukin-2 and interleukin-15: Immunotherapy for cancer. *Cytokine Growth Factor Rev* **13:** 169–183.

Ferrari-Lacraz S, Zheng XX, Kim YS, Li YS, Maslinski W, Li XC, Strom TB. 2001. An antagonist IL-15/Fc protein prevents costimulation blockade-resistant rejection. *J Immunol* **167:** 3478–3485.

Ferrari-Lacraz S, Zanelli E, Neuberg M, Donskoy E, Kim YS, Zheng XX, Hancock WW, Maslinski W, Li XC, Strom TB, et al. 2004. Targeting IL-15 receptor-bearing cells with an antagonist mutant IL-15/Fc protein prevents disease development and progression in murine collagen-induced arthritis. *J Immunol* **173:** 5818–5826.

Flanigan RC, Salmon SE, Blumenstein BA, Bearman SI, Roy V, McGrath PC, Caton JR Jr, Nikhil M, Crawford ED. 2001. Nephrectomy followed by interferon α-2b compared with interferon α-2b alone for metastatic renal-cell cancer. *N Eng J Med* **345:** 1655–1659.

Fyfe G, Fisher RI, Rosenberg SA, Sznol M, Parkinson DR, Louie AC. 1995. Results of treatment of 255 patients with metastatic renal cell carcinoma who received high-dose recombinant interleukin-2 therapy. *J Clin Oncol* **13:** 688–696.

Goldman JM, Marin D, Olavarria E, Apperley JF. 2003. Clinical decisions for chronic myeloid leukemia in the imatinib era. *Semin Hemtol* **40:** 98–103.

Goldstein D, Laszio J. 1988. The role of interferon in cancer therapy: A current perspective. *CA Cancer J Clin* **38:** 258–277.

Gollob JA, Mier JW, Veenstra K, McDermott DF, Clancy D, Clancy M, Atkins MB. 2000. Phase I trial of twice-weekly intravenous interleukin 12 in patients with metastatic renal cell cancer or malignant melanoma: Ability to maintain IFN-γ induction is associated with clinical response. *Clin Cancer Res* **6:** 1678–1692.

Gupta R, Emens LA. 2010. GM-CSM-secreting vaccines for solid tumors: Moving forward. *Discov Med* **10:** 52–60.

Gutterman JU, Blumenschein GR, Alexanian R, Yap HY, Budzdar AU, Cabanillas F, Hortobagyi GN, Hersh EM, Rasmussen SL, Harmon M, et al. 1980. Leukocyte interferon-induced tumor regression in human metastatic breast cancer, multiple myeloma and malignant lymphoma. *Ann Intern Med* **93:** 399–406.

Haan C, Rolvering C, Raulf F, Kapp M, Druckes P, Thoma G, Behrmann I, Zerwes HG. 2011. Jak1 has a dominant role over Jak3 in signal transduction through γc-containing cytokine receptors. *Chem Biol* **18:** 314–323.

Isaacs A, Lindenmann J. 2015. Virus interference. I: The interferon. *J Immunol* **195:** 1911–1920.

Jain S, Chen J, Nicolae A, Wang H, Shin DM, Adkins EB, Sproule TJ, Leeth CM, Sakai T, Kovalchuk AL, et al. 2015. IL-21 driven neoplasms in SJL mice mimic some key features of human angioimmunoblastic T-cell lymphoma. *Am J Pathol* **185:** 3102–3114.

Jin GH, Hirano T, Murakami M. 2008. Combination treatment with IL-2 and anti-IL-2 mAbs reduces tumor metastasis via NK cell activation. *Int Immunol* **20:** 783–789.

Ju W, Zhang ML, Jiang JK, Thomas CJ, Oh U, Bryant BR, Chen J, Sato N, Tagaya Y, Morris JC, et al. 2011. CP-690,550, a therapeutic agent, inhibits cytokine-mediated Jak3 activation and proliferation of T cells from patients with ATL and HAM/TSP. *Blood* **117:** 1938–1946.

Kaufman HL, Ruby CE, Hughes T, Slingluff CL Jr. 2014. Current status of ganulocyte-macrophage colony-stimulating factor in the immunotherapy of melanoma. *J Immunother Cancer* doi: 10.1186/2051-1426-2-11.

Kennedy MK, Glaccum M, Brown SN, Butz EA, Viney JL, Embers M, Matsuki N, Charrier K, Sedger L, Willis CR, et al. 2000. Reversible defects in natural killer and memory CD8 T cell lineages in interleukin 15-deficient mice. *J Exp Med* **191:** 771–780.

Kirkwood JM, Ernstoff MS. 1984. Interferons in the treatment of human cancer. *J Clin Oncol* **2:** 336–352.

Kirkwood JM, Strawderman MH, Ernstoff MS, Smith TJ, Bordern EC, Blum RH. 1996. Interferon α-2b adjuvant therapy of high-risk resected cutaneous melanoma: The Eastern Cooperative Oncology Group Trial EST 1684. *J Clin Oncol* **14:** 7–17.

Kirkwood JM, Ibrahim JG, Sosman JA, Sondak VK, Agarwala SS, Ernstoff MS, Rao U. 2001. High-dose interferon α-2b significantly prolongs relapse-free and overall survival compared with the GM2-KLH/QS-21 vaccine in patients with resected stage IIB-III melanoma: Results of intergroup trial E1694/S9512/C509801. *J Clin Oncol* **19:** 2370–2380.

Klebanoff CA, Finkelstein SE, Surman DR, Lichtman MK, Gattinoni L, Theoret MR, Grewal N, Spiess PJ, Antony PA, Palmer DC, et al. 2004. Il-15 enhances the in vivo antitumor activity of tumor-reactive CD8+ T cells. *Proc Natl Acad Sci* **101:** 1969–1974.

Kobayashi H, Dubois S, Sato N, Sabzevari H, Sakai Y, Waldmann TA, Tagaya Y. 2005. Role of *trans*-cellular IL-15 presentation in the activation of NK cell-mediated killing, which leads to enhanced tumor immunosurveillance. *Blood* **105:** 721–727.

Koskela HLM, Eldfors S, Ellonen P, van Adrichem AJ, Kuusanmaki H, Andersson EI, Lagstrom S, Clemente MJ, Olson T, Jalkanen SE, et al. 2012. Somatic STAT3 mutations in large granular lymphocytic leukemia. *N Engl J Med* **366:** 1905–1913.

Kucuk C, Jiang B, Hu XZ, Zhang WY, Chan JKC, Xiao WM, Lack N, Alkan C, Williams JC, Avery KN, et al. 2015. Activating mutations of STAT5B and STAT3 in lymphomas derived from γδ-T or NK cells. *Nat Commun* doi: 10.1038/ncomms7025.

Lange F, Rateitschak K, Fitzner B, Pöhland R, Wolkenhauer O, Jaster R. 2011. Studies on mechanisms of interferon-γ action in pancreatic cancer using a data-driven and model-based approach. *Mol Cancer* doi: 10.1186/1476-4598-10-13.

Lasek W, Zagozdzon R, Jakobisiak M. 2014. Interleukin 12: Still a promising candidate for tumor immunotherapy? *Cancer Immunol Immunother* **63:** 419–435.

Lee S, Margolin K. 2011. Cytokines in cancer immunotherapy. *Cancers* **3:** 3856–3893.

Leonard WJ, O'Shea JJ. 1998. Jaks and STATs: Biological implications. *Annu Rev Immunol* **16:** 293–322.

Leonardo MJ. 1996. Fas and the art of lymphocyte maintenance. *J Exp Med* **183:** 721–724.

Leroy S, Dubois S, Tenaud I, Chebassier N, Godard A, Jacques Y, Dreno B. 2001. Interleukin-15 expressions in cutaneous T-cell lymphoma (mycosis fungoides and Sézary syndrome). *Br J Dematol* **144:** 1016–1023.

Levin AM, Bates DL, Ring AM, Krieg C, Lin JT, Su L, Moraga I, Raeber ME, Bowman GR, Novick P, et al. 2012. Exploiting a natural conformational switch to engineer an interleukin-2 "superkine." *Nature* **484:** 529–533.

Li B, Lalani AS, Harding TC, Luan B, Koprivnikar K, Tu GH, Prell R, VanRoey MJ, Simons AD, Jooss K. 2006. Vascular endothelial growth factor blockade reduces intratumoral regulatory T cells and enhances the efficacy of a GM-CSF-secreting cancer immunotherapy. *Clin Cancer Res* **12:** 6808–6816.

Liao W, Lin JX, Leonard WJ. 2011a. IL-2 family cytokines: New insights into the complex roles of IL-2 as a broad regulator of T helper cell differentiation. *Curr Opin Immunol* **23:** 598–604.

Liao W, Lin JX, Wang L, Li P, Leonard WJ. 2011b. Modulation of cytokine receptors by IL-2 broadly regulates differentiation into helper T cell lineages. *Nat Immunol* **12:** 551–559.

Lode HN, Reisfield RA. 2000. Targeted cytokines for cancer immunotherapy. *Immunol Res* **21:** 279–288.

Lugli E, Goldman CK, Perera LP, Smedley J, Pung R, Yovandich JL, Jason L, Creekmore SP, Waldmann TA, Roederer M. 2010. Transient and persistent effects of IL-15 on lymphocyte homeostasis in nonhuman primates. *Blood* **116:** 3238–3248.

Ma HL, Whitters MJ, Konz RF, Senices M, Young DA, Grusby MJ, Collins M, Dunussi-Joannopoulos K. 2003. IL-21 activates both innate and adaptive immunity to generate potent antitumor responses that require perforin but are independent of IFN-γ. *J Immunol* **171:** 608–615.

Macchi P, Villa A, Giliani S, Sacco MG, Frattini A, Porta F, Ugazio AG, Johnston JA, Candotti F, O'Shea JJ, et al. 1995. Mutations of *JAK-3* gene in patients with autosomal severe combined immune-deficiency (SCID). *Nature* **377:** 6544–6568.

Mach N, Dranoff G. 2000. Cytokine-secreting tumor cell vaccines. *Curr Opin Immunol* **12:** 571–575.

Marks-Konczalik J, Dubois S, Losi JM, Sabzevari H, Yamada N, Feigenbaum L, Waldmann TA, Tagaya Y. 2000. IL-2-induced activation-induced cell death is inhibited in IL-15 transgenic mice. *Proc Natl Acad Sci* **97:** 11445–11450.

Migone JX, Cereseto A, Mulloy JC, O'Shea JJ, Franchini G, Leonard WJ. 1995. Constitutively activated Jak-STAT pathways in T-cells transformed with HTLV-1. *Science* **269:** 79–81.

Mishra A, Kwiatkowski S, Sullivan L, Grinshpun L, Russo G, Porcu P, Caligiuri M. 2015. Epigenetic disruption of ZEB1 binding causes constructive activation of IL-15 in cutaneous T-cell lymphoma. *Blood* **126:** 2.

Mitra S, Ring AM, Amarnath S, Spangler JB, Li P, Ju W, Fischer S, Oh J, Spolski R, Weiskopf K, et al. 2015. Interleukin-2 activity can be fine tuned with engineered receptor signaling clamps. *Immunity* **42:** 826–838.

Moga E, Alvareza E, Canto E, Vidal S, Rodriguez-Sanchez JL, Sierra J, Briones J. 2008. NK cells stimulated with IL-15 or CpG ODN enhance rituximab-dependent cellular cytotoxicity against B-cell lymphoma. *Exp Hematol* **36:** 69–77.

Mortier E, Quemener A, Vusio P, Lorenzen I, Boublik Y, Grotzinger J, Plet A, Jacques Y. 2006. Soluble interleukin-15 receptor α (IL-15Rα)-sushi as a selective and potent agonist of IL-15 action through IL-15R β/γ—Hyperagonist IL-15·IL-15Rα fusion proteins. *J Biol Chem* **281:** 1612–1619.

Mortier E, Woo T, Advincula R, Gozalo S, Ma A. 2008. IL-15Rα chaperones IL-15 to stable dendritic cell membrane complexes that activate NK cells via *trans* presentation. *J Exp Med* **205:** 1213–1225.

Mueller YM, Petrovas C, Bojczuk PM, Dimitriou LD, Beer B, Silvera P, Villinger F, Cairns JS, Gracely EJ, Lewis MG, et al. 2005. Interleukin-15 increases effector memory CD8[+] T cells and NK cells in simian immunodeficiency virus-infected macaques. *J Virol* **79:** 4877–4885.

Muller M, Ibelgaufts H, Kerr IM. 1994a. Interferon response pathways—A paradigm for cytokine signaling? *J Viral Hepat* **1:** 87–103.

Muller U, Steinhoff U, Reis LFL, Hemmi S, Pavlovic J, Zinkernagel RM, Aguet M. 1994b. Functional-role of type-I and type-II interferons in antiviral defense. *Science* **264:** 1918–1921.

Munger W, Dejoy SQ, Jeyaseelan R, Torley LW, Grabstein KH, Eisenmann J, Paxton R, Cox T, Wick MM, Kerwar SS. 1995. Studies evaluating the antitumor-activity and toxicity of interleukin-15, a new T-cell growth-factor—Comparison with interleukin-2. *Cell Immunol* **165:** 289–293.

Nakahata S, Yamazaki S, Nakauchi H, Morishita K. 2010. Downregulation of ZEB1 and overexpression of Smad7 contribute to resistance to TGF-β-mediated growth suppression in adult T-cell leukemia/lymphoma. *Oncogene* **29:** 4157–4169.

Nata T, Basheer A, Cocchi F, van Besien R, Massoud R, Jacobson S, Azimi N, Tagaya Yl. 2015. Targeting the binding interface on a shared receptor subunit of a cytokine family enables the inhibition of multiple member cytokines with selectable target spectrum. *J Biol Chem* **290:** 22338–22351.

Nicholas C, Lesinski GB. 2011. Immunomodulatory cytokines as therapeutic agents for melanoma. *Immunotherapy* **3:** 673–690.

Noguchi M, Yi HF, Rosenblatt HM, Flipovich AH, Adelstein S, Modi WS, Mcbride OW, Leonard WJ. 1993. Interleukin-2 receptor γ chain mutation results in X-linked severe combined immunodeficiency in humans. *Cell* **73:** 147–157.

O'Shea JJ, Holland SM, Staudt LM. 2013. Review article mechanism of disease. JAKS and STATs in immunity, immunodeficiency, and cancer. *N Engl J Med* **368:** 161–170.

Penichet ML, Morrison SL. 2001. Antibody-cytokine fusion proteins for the therapy of cancer. *J Immunol Methods* **248:** 91–101.

Pettit DK, Bonnert TP, Eisenman J, Srinivasan S, Paxton R, Beers C, Lynch D, Miller B, Yost J, Grabstein KH, et al. 1997. Structure-function studies of interleukin 15 using site-specific mutagenesis, polyethylene glycol conjugation, and homology modeling. *J Biol Chem* **272:** 2312–2318.

Pharmaceutical Business Review. 2013. Altor launches clinical trial of IL-15 superagonist protein complex against metastatic melanoma. clinicaltrials.pharmaceutical-business-review.com/news/altor-launches-clinical-trial-of-il-15-superagonist-protein-complex-against-metastatic-melanoma-190813.

Quesada JR, Talpaz M, Rios A, Kurzrock R, Gutterman JU. 1986. Clinical toxicity of interferons in cancer patients: A review. *J Clin Oncol* **4**: 234–243.

Ratain MJ, Golomb HM, Vardiman JW, Vokes EE, Jacobs RH, Daly K. 1985. Treatment of hairy-cell leukemia with recombinant α-2 interferon. *Blood* **65**: 644–648.

Reisfeld RA, Gillies SD. 1996. Antibody-interleukin 2 fusion proteins: A new approach to cancer therapy. *J Clin Lab Anal* **10**: 160–166.

Ridge JP, Di Rosa F, Matzinger P. 1998. A conditioned dendritic cell can be a temporal bridge between a CD4+ T-helper and a T-killer cell. *Nature* **393**: 474–478.

Roberti MP, Barrio MM, Bravo Al, Rocca YS, Arriaga JM, Bianchini M, Mordoh J, Levy EM. 2011. IL-15 and IL-2 increase cetuximab-mediated cellular cytotoxicity against triple negative breast cancer cell lines expressing EGFR. *Breast Cancer Res Treat* **130**: 465–575.

Rochman R, Leonard WJ. 2009. New insights into the regulation of T cells by γc family cytokines. *Nat Rev Immunol* **9**: 480–490.

Rosenberg SA, Lotze MT, Yang JC, Aebersold PM, Linehan WM, Seipp CA, White DE. 1989. Experience with the use of high-dose interleukin-2 in the treatment of 652 cancer-patients. *Ann Surg* **210**: 474–485.

Roychowdhury S, May KF Jr, Tzou KS, Lin T, Bhatt D, Freud AG, Guimond M, Ferketich AK, Liu Y, Caligiuri MA. 2004. Failed adoptive immunotherapy with tumor-specific T cells: Reversal with low-dose interleukin 15 but not low-dose interleukin 2. *Cancer Res* **64**: 8062–8067.

Russell SM, Tayebi N, Nakajima H, Riedy MC, Roberts JL Aman MJ, Migone TS, Noguchi M, Markert ML, Buckley RH, et al. 1995. Mutation of JAK3 in a patient with SCID: essential role of JAK3 in lymphoid development. *Science* **270**: 797–800.

Sadlack B, Kuhn R, Schorle H, Rajewsky K, Muller W, Horak I. 1994. Development and proliferation of lymphocytes in mice deficient for both interleukin-2 and interleukin-4. *Eur J Immunol* **24**: 281–284.

Sakaguchi S, Sakaguchi N, Asano M, Itoh M, Toda M. 1995. Immunological self-tolerance maintained by activated T-cells expressing Il-2 receptor α-chains (CD25)—Breakdown of a single mechanism of self-tolerance causes various autoimmune diseases. *J Immunol* **155**: 1151–1164.

Schluns KS, Klonowski KD, Lefrancois L. 2004. Transregulation of memory CD8 T-cell proliferation by IL-15Rα+ bone marrow-derived cells. *Blood* **103**: 988–994.

Schoenberger SP, Toes REM, van der Voort EIH, Offringa R, Melief CJM. 1998. T-cell help for cytotoxic T lymphocytes is mediated by CD40-CD40L interactions. *Nature* **393**: 480–483.

Schrama D, Reisfeld RA, Becker JC. 2006. Antibody targeted drugs as cancer therapeutics. *Nat Rev Drug Discov* **5**: 147–159.

Sckisel GD, Bouchlaka MN, Monjazeb AM, Crittenden M, Curti BD, Wilkins DEC, Alderson KA, Sungur CM, Ames E, Mirsoian A, et al. 2015. Out-of-sequence signal 3 paralyzes primary CD4+ T-cell-dependent immunity. *Immunity* **43**: 240–250.

Skak K, Kragh M, Hausman D, Smyth MJ, Sivakumar PV. 2008. Interleukin 21: Combination strategies for cancer therapy. *Nat Rev Drug Discov* **7**: 231–240.

Smith GA, Uchida K, Weiss A, Taunton J. 2016. Essential biphasic role for JAK3 catalytic activity in IL-2 receptor signaling. *Nat Chem Biol* **12**: 373.

Sneller MC, Kopp WC, Engelke KJ, Yovandich JL, Creekmore SP, Waldmann TA, Lane HC. 2011. IL-15 administered by continuous infusion to rhesus macaques induces massive expansion of CD8+ T effector memory population in peripheral blood. *Blood* **118**: 6845–6848.

Soiffer RJ, Murray C, Cochran K, Cameron C, Wang E, Schow PW, Daley JF, Ritz J. 1992. Clinical and immunological effects of prolonged infusion of low-dose recombinant interleukin-2 after autologous and T-cell depleted allogeneic bone-marrow transplantation. *Blood* **79**: 517–526.

Soiffer RJ, Murray C, Shapiro C, Collins H, Chartier S, Lazo S, Ritz J. 1996. Expansion and manipulation of natural killer cells in patients with metastatic cancer by low-dose continuous infusion and intermittent bolus administration of interleukin 2. *Clin Cancer Res* **2**: 493–499.

Sondel PM, Hank JA. 1997. Combination therapy with interleukin-2 and antitumor monoclonal antibodies. *Cancer J Sci Am* **3**: S121–S127.

Spangler JB, Moraga I, Mendoza JL, Garcia KC. 2015. Insights into cytokine-receptor interactions from cytokine engineering. *Annu Rev Immunol* **33**: 139–167.

Spolski R, Leonard WJ. 2008. Interleukin-21: Basic biology and implications for cancer and autoimmunity. *Annu Rev Immunol* **26**: 57–79.

Spolski R, Leonard WJ. 2014. Interleukin-21: A double-edged sword with therapeutic potential. *Nat Rev Drug Discov* **13**: 379–395.

Steel JC, Waldmann TA, Morris JC. 2012. Interleukin-15 biology and its therapeutic implications in cancer. *Trends Pharmacol Sci* **33**: 35–41.

Steele N, Anthony A, Saunders M, Esmarck B, Ehmrooth E, Kristjansen PEG, Nihlen A, Hansen LT, Cassidy J. 2012. A phase 1 trial of recombinant human IL-21 in combination with cetuximab in patients with metastatic colorectal cancer. *Br J Cancer* **106**: 793–798.

Stemmer WPC. 1994. Rapid evolution of a protein in-vitro by DNA shuffling. *Nature* **370**: 389–391.

Tagliaferri P, Caraglia M, Budillon A, Marra M, Vitale G, Viscomi C, Masciari S, Tassone P, Abbruzzese A, Venuta S. 2005. New pharmacokinetic and pharmacodynamic tools for interferon-α (IFN-α) treatment of human cancer. *Cancer Immunol Immunother* **54**: 1–10.

Taniguchi T, Minami Y. 1993. The IL-2/IL-2 receptor system—A current overview. *Cell* **73**: 5–8.

Tendler CL, Greenberg SJ, Blattner WA, Manns A, Murphy E, Fleisher T, Hanchard B, Morgan O, Burton JD, Nelson DL, et al. 1990. Transactivation of interleukin-2 and its receptor induces immune activation in human T-cell lymphotropic virus type-I associated myelopathy—Pathogenic implications and a rationale for immunotherapy. *Proc Natl Acad Sci* **87**: 5218–5222.

Cite this article as *Cold Spring Harb Perspect Biol* doi: 10.1101/cshperspect.a028472

Timmerman JM, Byrd JC, Andorsky DJ, Yamada RE, Kramer J, Muthusamy N, Hunder N, Pagel JM. 2012. A phase 1 dose-finding trial of recombinant interleukin-21 and rituximab in relapsed and refractory low grade B-cell lymphoproliferative disorders. *Clin Cancer Res* **18:** 5752–5760.

Tinhofer I, Marschitz I, Henn T, Egle A, Greil R. 2000. Expression of functional interleukin-15 receptor and autocrine production of interleukin-15 as mechanisms of tumor propagation in multiple myeloma. *Blood* **95:** 610–618.

Trinchieri G. 1995. Interleukin-12: A proinflammatory cytokine with immunoregulatory functions that bridge innate resistance and antigen-specific adaptive immunity. *Annu Rev Immunol* **13:** 251–276.

Trinchieri G. 2003. Interleukin-12 and the regulation of innate resistance and adaptive immunity. *Nat Rev Immunol* **3:** 133–146.

Vincent M, Teppaz G, Lajoie L, Sole V, Bessard A, Maillasson M, Loisel S, Bechard D, Clemenceau B, Thibault G, et al. 2014. Highly potent anti-CD20-RLI immunocytokine targeting established human B lymphoma in SCID mouse. *MABS* **6:** 1026–1037.

Vincenti F, Kirkman R, Light S, Bumgardner G, Pescovitz M, Halloran P, Neylan J, Wilkinson A, Ekberg H, Gastron R, et al. 1998. Interleukin-2-receptor blockade with daclizumab to prevent acute rejection in renal transplantation. *N Engl J Med* **338:** 161–165.

Waldmann TA. 1986. The structure, function, and expression of interleukin-2 receptors on normal and malignant lymphocytes. *Science* **232:** 727–732.

Waldmann TA. 1991. The interleukin-2 receptor. *J Biol Chem* **266:** 2681–2684.

Waldmann TA. 2006. The biology of interleukin-2 and interleukin-15: Implications for cancer therapy and vaccine design. *Nat Rev Immunol* **6:** 595–601.

Waldmann TA. 2007. Anti-Tac (daclizumab, Zenapax) in the treatment of leukemia, autoimmune diseases, and in the prevention of allograft rejection: A 25-year personal odyssey. *J Clin Immunol* **27:** 1–18.

Waldmann TA. 2015. The shared and contrasting roles of IL-2 and IL-15 in the life and death of normal and neoplastic lymphocytes: Implications for cancer therapy. *Cancer Immunol Res* **3:** 219–227.

Waldmann TA, Chen J. 2017. Disorders of the JAK/STAT pathway in T-cell lymphoma, pathogenesis: Implications for immunotherapy. *Annu Rev Immunol* **35:** 533–550.

Waldmann TA, Dubois S, Tagaya Y. 2001. Contrasting roles of IL-2 and IL-15 in the life and death of lymphocytes: Implications for immunotherapy. *Immunity* **14:** 105–110.

Waldmann TA, Lugli E, Roederer M, Perera LP, Smedley JV, Macallister RP, Goldman CK, Bryant BR, Decker JM, Fleisher TA, et al. 2011. Safety (toxicity), pharmacokinetics, immunogenicity, and impact on elements of the normal immune system of recombinant human IL-15 in rhesus macaques. *Blood* **117:** 4787–4795.

Waldmann TA, Conlon KC, Stewart DM, Worthy TA, Janik JE, Fleisher TA, Albert PS, Figg WD, Spencer SD, Raffeld M, et al. 2013. Phase 1 trial of IL-15 *trans* presentation blockade using humanized Mik-B-1 mAb in patients with T-cell large granular lymphocytic leukemia. *Blood* **121:** 476–484.

Wang G, Tschoi M, Spolski R, Lou YY, Ozaki K, Feng CG Kim G, Leonard WJ, Hwu P. 2003. In vivo antitumor activity of interleukin 21 mediated by natural killer cells. *Cancer Res* **63:** 9016–9022.

Wang XQ, Rickert M, Garcia KC. 2005. Structure of the quaternary complex of interleukin-2 with its α, β, and γ_c receptors. *Science* **310:** 1159–1163.

Windbichler GH, Hausmaninger H, Stummvoll W, Graf AH, Kainz C, Lahodny J, Denison U, Müller-Holzner E, Marth C. 2000. Interferon-γ in the first-line therapy of ovarian cancer: A randomized phase III trial. *Br J Cancer* **82:** 1138–1144.

Young PA, Morrison SL, Timmerman JM. 2014. Antibody-cytokine fusion proteins for treatment of cancer: Engineering cytokines for improved efficacy and safety. *Semin Oncol* **41:** 623–636.

Yu P, Steel JC, Zhang ML, Morris JC, Waldmann TA. 2010. Simultaneous blockade of multiple immune system inhibitory checkpoints enhances antitumor activity mediated by interleukin-15 in a murine metastatic colon carcinoma model. *Clin Cancer Res* **16:** 6019–6028.

Yu P, Steel JC, Zhang ML, Morris JC, Waitz R, Fasso M, Allison JP, Waldmann TA. 2012. Simultaneous inhibition of two regulatory T-cell subsets enhanced interleukin-15 efficacy in a prostate tumor model. *Proc Natl Acad Sci* **109:** 6187–6192.

Zaidi MR, Merlino G. 2011. The two faces of interferon-γ in cancer. *Clin Cancer Res* **17:** 6118–6124.

Zeng R, Spolski R, Finkelstein SE, Oh SK, Kovanen PE, Hinrichs CS, Pise-Masison CA, Radonovich MF, Brady JN, Restifo NP, et al. 2005. Synergy of IL-21 and IL-15 in regulating CD8$^+$ T cell expansion and function. *J Exp Med* **201:** 139–148.

Zhang ML, Yao ZS, Dubois S, Ju W, Muller JR, Waldmann TA. 2009. Interleukin-15 combined with an anti-CD40 antibody provides enhanced therapeutic efficacy for murine models of colon cancer. *Proc Natl Acad Sci* **106:** 7513–7518.

Zhang ML, Ju W, Yao ZS, Yu P, Wei BR, Simpson RM, Waitz R, Fasso M, Allison JP, Waldmann TA. 2012. Augmented IL-15Rα expression by CD40 activation is critical in synergistic CD8 T cell-mediated antitumor activity of anti-CD40 antibody with IL-15 in TRAMP-C2 tumors in mice. *J Immunol* **188:** 6156–6164.

Zhu XY, Marcus WD, Xu WX, Lee HI, Han KP, Egan JO, Yovandich JL, Rhode PR, Wong HC. 2009. Novel human interleukin-15 agonists. *J Immunol* **183:** 3598–3607.

The Future of the Cytokine Discipline

Joost J. Oppenheim

Cellular Immunology, Cancer and Inflammation Program, National Cancer Institute, National Institutes of Health, Frederick, Maryland 21702

Correspondence: oppenhej@mail.nih.gov

The study of cytokines has evolved from the detection of functional activities present in tissue culture supernatants to the characterization of the three-dimensional molecular structures of the cytokines and their receptors. Investigators studying cytokines need to have specialized expertise in using cytokine assays, assessing their receptor interactions, signal transduction, gene activation, and biological effects, and in the therapeutic utilization of agonists and antagonists. Cytokinology can therefore be considered a discipline. In this article, I have considered studies leading to the identification of novel cytokines, potential producers of cytokine mimics such as viruses and the microbiome, and the complex interactions of the cytokine network with our vital functions. Our ever-increasing success in using cytokines and, in particular, cytokine inhibitors therapeutically suggest that cytokinology will eventually become an independent discipline.

I have been invited to contribute an article on the future of cytokine studies presumably because I have spent more time than most in trying to advance our research into the future. My previous paper on the past, present, and future of cytokines did not cause any embarrassment, probably because it was read by few and ignored by all (Oppenheim 2001). Nevertheless, my crystal ball is cloudy and I have always continued to be surprised by developments that have circumvented, sideswiped, or completely ignored my predictions. The past, however, provides prologue to the future. Although I am much better at reviewing the past than in predicting the future, I will try to predict some of the miracles of tomorrow based on what has happened yesterday and today.

DEFINITION

We should begin by defining the scope of cytokine studies. The definition of cytokines can be broadly based on their function as intercellular signals (Oppenheim 2001). They can be either cell-surface associated or secreted proteins that interact with specific cell-surface receptors resulting in the mobilization and or modulation of target cells. Although I dislike use of the term "modulation," it is appropriate because cytokines can either have stimulating or suppressive effects on host cells. Although intercellular signals by molecular entities such as lipids, carbohydrates, DNA, or RNA are generally not considered cytokines, they can function as such when complexed by cytokines. Those intercellular signals that are produced by specialized cells in tissues or glands and are persistently released into the bloodstream are considered endocrine hormones and growth factors that act at a distance from their site of origin. They are the subject of endocrinologists. In contrast, cytokines usually, but not always, act locally over a short range in tissues either in a paracrine or even in an autocrine fashion. When cytokines are produced in abundance because of overwhelming infections, autoimmune reactions, or injuries, they can spill over into the serum and have global effects, such as fever and behavioral changes, and can have deleterious effects, such as "cytokine storm, cytokine release syndrome or macrophage activation syndrome" (Grom et al. 2016). This potentially lethal syndrome can be treated with anti-inflammatory reagents such as steroids, anti-interleukin (IL)-6, or anakinra, an antagonist of IL-1 (Schulert and Grom 2015).

HISTORICAL BACKGROUND

Over the past 70+ years, the field of cytokine research has developed in a miraculous manner beginning with the initial observation in 1944 that "pyrexins" in inflammatory exudates caused the systemic response of fever (Menkin 1944). This led in 1953 to the identification of bacterial-derived endotoxin as the inducer of an endogenous pyrogen present in extracts of leukocytes (Bennett and Beeson 1953). At the same time, intercellular nerve growth factors were discovered (Levi-Montalcini and Hamburger 1953). This was followed in 1957 by the discovery of nonantibody cell-derived antiviral factors called interferons (IFNs) (Isaacs and Lindenmann 1957). The development of improved in vitro tissue culture techniques led immunologists to discover "blastogenic factors" in 1965, which activated lymphocytes to proliferate (Kasakura and Lowenstein 1965). This was soon followed by the detection in tissue culture supernatants of macrophage inhibitory factor in 1966 (Bloom and Bennett 1966; David 1966) and a cytotoxic factor called lymphotoxin in 1967 (Ruddle and Waksman 1967). By 1978, about 100 different cytokine activities produced

by a wide variety of cell types had been described (Waksman 1978). These biological activities were biochemically uncharacterized and ungraciously termed "lymphodrek" by yours truly (Oppenheim and Gery 1993). The cytokine field was so chaotic and deficient in molecular characterization that peer review committees stopped further funding of cytokine studies for over a decade. Fortuitously, miraculous progress by molecular biologists and biochemists rescued the cytokine field and made it possible to obtain pure recombinant preparations of cytokines by the 1980s. This led to an explosive expansion in the identification and molecular characterization of the hundreds of cytokines known to be involved in regulating our development, host defense, and repair processes. In addition, the structural elucidation of cytokine receptors led to the classification of cytokines into families based on the identities of their receptors (Oppenheim and Feldmann 2001).

NOVEL CYTOKINES

Of course it is easy to predict that even more cytokines will be discovered, especially because gene chip assays are revealing the novel targets and links of our ever-expanding cytokine network. One example of this is the identification of alarmins, a subset of danger-associated molecular patterns (DAMPs), as initiators of host defense. The alarmins are constitutively available endogenous molecules released from the granules, cytosol, or nucleus of cells by cell injury or other stimuli. In the extracellular milieu, they function as intercellular signals that mobilize cells by interacting with chemotactic receptors such as G-protein-coupled receptors (GiPCRs) and, in addition, activate cells by interacting with receptors such as Toll-like receptors (TLRs) (Nie et al. 2016). It has been very surprising that chromatin-binding proteins such as HMGB1, HMGN1, IL-33, and IL-1α also have extracellular functions and are potent in inducing the production of proinflammatory cytokines. Other alarmins are antimicrobial peptides (AMPs), such as defensins and cathelicidins, which also have immune-enhancing host-defense functions. It is likely that more AMPs and chromatin binders will be identified to have alarmin functions. In addition, a number of thoroughly studied immunoenhancing moieties such as S100 A8/9, some heat shock proteins, and complement components also behave as alarmins and can be considered as such.

There are other intercellular signals that behave more like cytokines than hormones, which should be considered as important topics for studies by cytokinologists such as the neuropeptides. The fact that some cytokines have dramatic effects on peripheral and central nervous systems functions such as inducing fever, fatigue, and malaise undoubtedly involves the induction of a number of neuropeptide signals. Conversely, it is also quite reasonable to predict that neuropeptides can induce the production of some cytokines (Karin and Clevers 2016).

CYTOKINE MIMICS

I must also consider the possible effects of the microbiome on the cytokine network. It is clear that persistent microbial stimulation serves to optimize our immune responses and this undoubtedly results in the induction of cytokines. There is also considerable evidence that viruses produce many cytokine mimics to their own advantage. Whether bacteria also do this has not been as clearly documented, but one can predict that evolution has created some microorganisms that directly respond to cytokines or have endogenous genes that encode for cytokine mimics/antagonists (Elenkov 2000).

CYTOKINES ASSAYS

One of the major issues about cytokines is that no one can define what is a "normal" range for serum cytokines. While pharmaceutical companies may have such data on thousands of individuals, it may be impossible to define what "normal" represents because it is very likely that variables such as age, race, sex, and time of day of serum sampling may all impact the results. Nevertheless, threshold levels of cytokines will be identified that, when crossed, serve as a marker for disease. Thus, one can predict

that more sensitive cytokine assays will eventually be included in normal blood tests.

BIOLOGICAL EFFECTS OF CYTOKINES

Despite heroic attempts to utilize computers to circumvent the need to sacrifice animals to solve biological problems, it is clear that cytokinology is an experimental science. Hypotheses need to be challenged and tested in the laboratory. Nevertheless, systems biology can play an important role in helping us understand the potentially voluminous results. A cytokine is a signal that never occurs alone, it is part of a complex milieu. Depending on the mix, we may repel an invader or self-destruct. The integration of cytokine signals requires a systems approach that cytokinologists must be equipped to study. Our current experimental tools can generate such an overwhelming amount of data that adequate analysis requires the application of systems biology.

There have been recent reports of patients having antibodies to cytokines, such as IFN-γ or IL-6. If these antibodies are neutralizing, it could potentially impact the overall host immune response and might certainly contribute to the patients' resistance to infection (Kampmann et al. 2005). In fact, those patients with anti-IFN-γ antibodies had a major immunodeficiency condition. Thus, new assays will be needed to assess the presence of these endogenous antibodies and whether they neutralize the cytokine activity. Means of overcoming these autocytokine reactions need to be developed.

CYTOKINE SIGNAL TRANSDUCTION

An incredible amount of effort and expertise has been devoted to identifying the intracellular transduction pathways resulting in gene activation initiated by cytokine–receptor interactions. Although many cytokines share these signaling pathways, the fact that they interact with unique receptors on unique subsets of cells accounts for the specificity of cytokine functions. Consequently, targeting shared signal transducers for therapeutic purposes has been difficult because of the undesirably broad inhibitory effects. Despite this, there has been some notable commercial success in developing inhibitors of some of the Janus kinase (JAK) signal transducers as anti-inflammatory therapeutics (O'Shea et al. 2015). It is a safe prediction that more targetable signal transducers and effective inhibitors will be identified. These can be antagonistic antibodies or inhibitors linked to aptamers targeted to cell-surface proteins (i.e., aptamers are small RNA or DNA molecules that bind specifically to target proteins). Aptamers might also be used to inhibit a cytokine activity by specifically binding and neutralizing their targets. Studies of the inhibitory effects or microRNAs (miRNAs) that bind cytokine messenger RNAs (mRNAs) may also yield a better means of inhibiting cytokine production with the goal of reducing inflammation.

CYTOKINE MUTATIONS

A number of primary genetic cytokine deficiencies have also been identified, such as γ_c and IL-7 receptor deficiencies as causes of severe combined immunodeficiency (SCID). cKit gain-of-function mutations and IFN-γ receptor mutations have a deleterious phenotype; CXCR4 deficiency has a phenotype, etc. With the development of better and easier methods for genetic engineering, perhaps manipulation of genes by CRISPR on other techniques will lead to therapeutic benefits in the future by the correction of mutations in cytokines or their receptors or even elimination of these genes in appropriate target cells.

THERAPEUTIC CYTOKINE AGONISTS

The course of cytokine research has led to some other surprising results. For example, many attempts have been made to utilize cytokines therapeutically with limited success. A number of proinflammatory cytokines such as tumor necrosis factor (TNF) and IL-1 have been thoroughly tested for their clinical antitumor effects with disastrous undesirable side effects such as fever and shaking chills, hypotension, and shock. Only the therapeutic use of IFNs to treat a number of diseases such as multiple sclerosis, virally induced chronic hepatitis, and malignant melanoma and leukemias has met with a modi-

cum of success (Antonelli et al. 2015). This is despite the fact IFN has a number of undesirable and unpleasant side effects such as arthralgias, depressions, fatigue, muscle aches, and anorexia. These undesirable consequences of IFN's clinical use have been elucidated by an experimental attempt to treat adenovirus-induced upper respiratory infections (URIs) by administration of IFN-α in nose drops. This study actually showed that the IFNs administered without the adenovirus caused the nasal congestion and stuffiness, sneezing, and other symptoms of the URIs. These observations taught us that the toxic side effects of cytokines such as TNF, IFN, IL-12, and IL-1, rather than being a result of the infectious agent, were responsible for our disease symptoms. Attempts to use antagonists of these cytokines to reduce these symptoms backfired because the consequent reduction in host defense enabled the infectious agents to flourish and cause more severe damage. Fortunately, IFN is at present often being replaced by more effective therapies of liver disease and cancers.

It's tough to make predictions, especially about the future.—Yogi Berra

Research often results in unanticipated effects. For example, the CXCR4 blocker AMD3100 was developed as a human immunodeficiency virus (HIV) drug, but is now actually used to mobilize hematopoietic stem cells (HSCs) from bone marrow. Cytokines have also met with modest therapeutic success in the treatment of various cancers as immunoenhancing stimulants. Cytokines such as IL-2, IL-7, and IL-15 are used to grow and maintain lymphoid cells in vitro. In addition, cytokines and growth factors have been used to grow and differentiate stem cells and develop organoids. Perhaps in the future we will use cytokines to culture entire tissues and organs.

Many of the proinflammatory cytokines were evaluated for their adjuvant capacity, and granulocyte macrophage colony-stimulating factor (GM-CSF) was found to be the most potent in attracting dendritic cells to inflammatory tumor sites and thus in enhancing antitumor immunity (Jinushi et al. 2008). More recently, some alarmins that activate various TLRs have been shown to be even more potent in augmenting adaptive antitumor immunity. Consequently, the role of these cytokine-like agonists may prove of therapeutic benefit, especially when utilized in an adjuvant setting. They can expand preexisting antitumor immunity, which has been shown to exist by the use of checkpoint inhibitors (Davar and Kirkwood 2016).

Another concern requiring a comment is that, although the pro- and anti-inflammatory effects of cytokines have been the predominant focus of study, the role of the anti-inflammatory cytokines such as transforming growth factor β (TGF-β), IL-10, epidermal growth factor (EGF), vascular endothelial growth factor (VEGF), etc., in healing repair and fibrotic scarring, probably deserves more attention. After all, the aftermath of inflammation is the restoration of homeostasis. Future studies of those growth factors and cytokines that are engaged in this task may lead to therapeutic benefits.

THERAPEUTIC CYTOKINE ANTAGONISTS

A major surprise to cytokine experts was the serendipitous discovery that antagonistic antibodies to some of the proinflammatory cytokines could provide major benefits in reducing the inflammation associated with a number of autoimmune and autoinflammatory diseases. This came about based on futile attempts to use antagonistic anti-TNF to treat septic shock by Jan Vilcek at New York University in collaboration with investigators at a small company, Centocor (Vilcek 2016). In desperation, the antibodies were then provided to investigators as possible treatments of autoimmune diseases such as rheumatoid arthritis and Crohn's disease, resulting in amazing success (Feldmann and Maini 2010). Although this has not cured these diseases, it has provided considerable symptom relief to millions of patients. Similarly, neutralizing antibodies to IL-6, IL-12, IL-23, and IL-1 have benefited patients with selected autoimmune and autoinflammatory diseases. The therapeutic applications of cytokine antagonists have proven useful in diverse disciplines such as rheumatology, oncology, and genetic defects. Although the pharmaceutical industry dislikes developing antibodies or receptors that need to

be injected, the many attempts to develop peptide or protein antagonists that can be administered in a pill form have failed. However, based on the obvious advantages of this means of administering drugs, I hope that more and better oral cytokine antagonists will be developed.

Another promising application of cytokine antagonists is in the realm of checkpoint inhibitors. A number of cytokines such as IL-10, IL-27, and TGF-β have regulatory immunosuppressive effects. Although they are presumably intended to maintain homeostasis and to prevent autoimmune reactions, unfortunately, they also block antitumor responses. Attempts made in various animal tumor models to utilize antibodies or other antagonists of these cytokines have had modest antitumor effects. More research needs to be done to establish their utility. In particular, they will probably contribute to antitumor therapy in conjunction with cocktails of tumor-enhancing reagents and more conventional antitumor therapeutics such as surgery, radiation, and chemotherapy. The same probably holds for checkpoint inhibitors such as anti-PD1, anti-CTLA4, and others that have impressive antitumor benefits as stand-alone therapeutics in a subset of patients, but are more effective when given in a cocktail with other therapeutics. Of course, one can argue that these cell markers are not really cytokines, but since they are shed and many cytokines have membrane-associated forms, I do not see any need to disown them.

This also raises the need for novel and rapid methods to identify more soluble cytokine receptors. Such receptors by binding cytokines could act as antagonists and decrease the availability of cytokines in the serum or alternatively might even enhance cytokine biological activity when complexed to their soluble receptors as has been reported for IL-6 and IL-15. It is also likely that "superkines" will be genetically engineered that have specific affinity for cytokine receptor chains, thus increasing their biological activity and specificity, as has recently been shown for IL-2 and its interaction with the IL-2 receptor β chain. Similarly, I predict that new hybrid chemokines will be generated that attract and focus on specific immune cell subsets on a targeted tissue (e.g., CD4$^+$ and CD8$^+$ T cells or T regulatory cells [Tregs]) and thus reduce systemic toxicity that could occur upon intravenous infusion of cytokines/chemokines.

CONCLUSIONS

The current fragmentation of the field of cytokines into subgroups that focus on studies of chemotaxis, TNF, and hematopoietic factors will probably be a temporary phase as indicated by the fusion of the IFN and cytokine societies. I foresee cytokinology becoming a more coherent field of study. Based on the ever-burgeoning therapeutic successes of cytokine agonists and antagonists, will cytokinology develop into an autonomous discipline such as endocrinology? Many diverse human conditions have been identified, resulting in primary or secondary cytokine deficiencies and cytokine overproduction. In agreement with the consequences of under- and overproduction of endocrine hormones, abnormalities in the production and effects of cytokines are of concern to a variety of medical subspecialties. In the case for immunology, a subspecialty for allergy and immunology has already been established. Thus, cytokinology may be considered a subsection of immunology or, like endocrinology, may eventually become entirely independent.

Currently, the clinical use of cytokines is largely based on direct infusion of the cytokines or anticytokine antibodies into the patients. In some cases, such as IL-12, this has resulted in serious toxicity or even death. However, in the future, there will be more targeted cytokine therapy such as cytokines in skin creams to promote tissue healing, cytokines directed toward specific organs or immune cell subsets to suppress an immune response in that organ, or cytokines directly targeted to tumors, an approach that has been under study for quite some time without much clinical success as yet.

I probably speak for my many cytokinologist colleagues in expressing our gratitude for having had the opportunity to contribute to the dynamic field of cytokine research because of the support of the National Institutes of Health and other funding agencies. As a result, cytokinology has developed into an immunological discipline.

Fortunately, immunologists from an initial restricted focus on the role of antibody specificity in adaptive immunity have been sufficiently flexible to expand their horizons to encompass the role of innate immunity and nonspecific cytokines. By analogy, I optimistically expect students of cytokines to expand their horizons and include studies of all intercellular host defense signals in their discipline. The cytokine network connects and interacts with lipid moieties: nucleotides, growth factors, neuropeptides, alarmins, hormones, and microbial reagents. Although as investigators we have to focus on our own research topic, we must be open to consideration of how cytokines interdigitate with the many other systems engaged in maintaining host integrity. Furthermore, as indicated in this article, we will never solve all of the problems, but we have to keep in mind the clinical utility of focusing our studies on the identification of the therapeutic benefit of cytokine agonists and antagonists with the goal of helping patients.

ACKNOWLEDGMENTS

I am grateful for the constructive critical discussion and modifications of this review by Drs. Howard Young and Scott Durum, and for the superb editorial assistance of Ms. Sharon Livingstone.

REFERENCES

Antonelli G, Scagnolari C, Moschella F, Proietti E. 2015. Twenty-five years of type I interferon-based treatment: A critical analysis of its therapeutic use. *Cytokine Growth Factor Rev* **26**: 121–131.

Bennett IL Jr, Beeson PB. 1953. Studies on the pathogenesis of fever. II: Characterization of fever-producing substances from polymorphonuclear leukocytes and from the fluid of sterile exudates. *J Exp Med* **98**: 493–508.

Bloom BR, Bennett B. 1966. Mechanism of a reaction in vitro associated with delayed-type hypersensitivity. *Science* **153**: 80–82.

Davar D, Kirkwood JM. 2016. Adjuvant therapy of melanoma. *Cancer Treatment Res* **167**: 181–208.

David JR. 1966. Delayed hypersensitivity in vitro: Its mediation by cell-free substances formed by lymphoid cell-antigen interaction. *Proc Natl Acad Sci* **56**: 72–77.

Elenkov I. 2000. Neuroendocrine effects on immune system. In *Endotext* (ed. De Groot LJ, et al.). MDText.com, South Dartmouth, MA.

Cite this article as *Cold Spring Harb Perspect Biol* doi: 10.1101/cshperspect.a028498

Feldmann M, Maini RN. 2010. Anti-TNF therapy, from rationale to standard of care: What lessons has it taught us? *J Immunol* **185**: 791–794.

Grom AA, Horne A, De Benedetti F. 2016. Macrophage activation syndrome in the era of biologic therapy. *Nat Rev Rheumatol* **12**: 259–268.

Isaacs A, Lindenmann J. 1957. Virus interference. I: The interferon. *Proc R Soc Ser B Biol Sci* **147**: 258–267.

Jinushi M, Hodi FS, Dranoff G. 2008. Enhancing the clinical activity of granulocyte-macrophage colony-stimulating factor-secreting tumor cell vaccines. *Immunol Rev* **222**: 287–298.

Kampmann B, Hemingway C, Stephens A, Davidson R, Goodsall A, Anderson S, Nicol M, Scholvinck E, Relman D, Waddell S, et al. 2005. Acquired predisposition to mycobacterial disease due to autoantibodies to IFN-γ. *J Clin Invest* **115**: 2480–2488.

Karin M, Clevers H. 2016. Reparative inflammation takes charge of tissue regeneration. *Nature* **529**: 307–315.

Kasakura S, Lowenstein L. 1965. A factor stimulating DNA synthesis derived from the medium of leukocyte cultures. *Nature* **208**: 794–795.

Levi-Montalcini R, Hamburger V. 1953. A diffusible agent of mouse sarcoma, producing hyperplasia of sympathetic-ganglia and hyperneurotization of viscera in the chick embryo. *J Exp Zool* **123**: 233–287.

Menkin V. 1944. Chemical basis of fever. *Science* **100**: 337–338.

Nie Y, Yang D, Oppenheim JJ. 2016. Alarmins and antitumor immunity. *Clin Ther* **38**: 1042–1053.

Oppenheim JJ. 2001. Cytokines: Past, present, and future. *Int J Hematol* **74**: 3–8.

Oppenheim JJ, Feldmann M. 2001. Introduction to the role of cytokines and cytokine production by cells in inflammation and immunity. In *Cytokine reference*, pp. 3–20. Academic, Orlando, Fl.

Oppenheim JJ, Gery I. 1993. From lymphodrek to interleukin 1 (IL-1). *Immunol Today* **14**: 232–234.

O'Shea JJ, Schwartz DM, Villarino AV, Gadina M, McInnes IB, Laurence A. 2015. The JAK-STAT pathway: Impact on human disease and therapeutic intervention. *Annu Rev Med* **66**: 311–328.

Ruddle NH, Waksman BH. 1967. Cytotoxic effect of lymphocyte–antigen interaction in delayed hypersensitivity. *Science* **157**: 1060–1062.

Schulert GS, Grom AA. 2015. Pathogenesis of macrophage activation syndrome and potential for cytokine-directed therapies. *Annu Rev Med* **66**: 145–159.

Vilcek J. 2016. *Love and science*, pp. 1–264. Seven Stories, New York.

Waksman BH. 1978. Modulation of immunity by soluble mediators. *Pharmacol Ther* **2**: 623–672.

Index

www.ingramcontent.com/pod-product-compliance
Lightning Source LLC
Chambersburg PA
CBHW051659210326
41598CB00004B/4